Interactive Activities

Fun and educational tools await you at *Essentials* Online Learning Center. Each chapter offers quizzes, flashcards and other engaging activities designed to reinforce learning.

You can view animations of key biological processes online, and test yourself with a follow-up quiz. Spanish-speaking students can now view a Spanish version of the animation online.

Virtual Labs help you to apply the scientific method by allowing you to repeat an experiment done by a modern researcher, and study the original papers they wrote regarding their work.

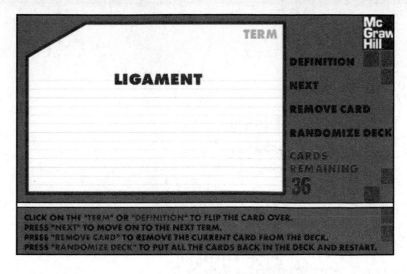

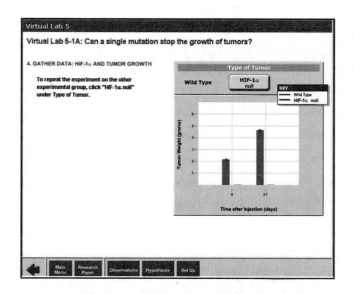

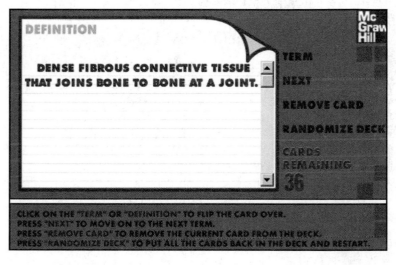

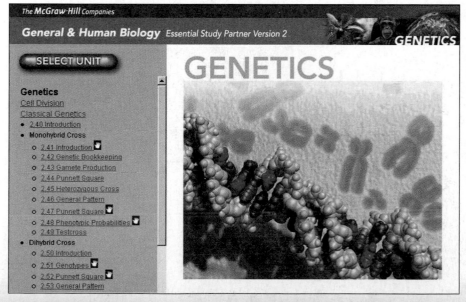

Essentials Online Learning Center is your portal to exclusive interactive study tools like McGraw-Hill's Essential Study Partner. The Essential Study Partner offers art activities, animations, and quizzes that are embedded within tutorials designed to walk you through key concepts in biology.

ESSENTIALS OF

The Living World

George B. Johnson

Washington University
St. Louis, Missouri

Boston   Burr Ridge, IL   Dubuque, IA   Madison, WI   New York   San Francisco   St. Louis
Bangkok   Bogotá   Caracas   Kuala Lumpur   Lisbon   London   Madrid   Mexico City
Milan   Montreal   New Delhi   Santiago   Seoul   Singapore   Sydney   Taipei   Toronto

Higher Education

ESSENTIALS OF THE LIVING WORLD

Published by McGraw-Hill, a business unit of The McGraw-Hill Companies, Inc., 1221 Avenue of the Americas, New York, NY 10020. Copyright © 2006 by The McGraw-Hill Companies, Inc. All rights reserved. No part of this publication may be reproduced or distributed in any form or by any means, or stored in a database or retrieval system, without the prior written consent of The McGraw-Hill Companies, Inc., including, but not limited to, in any network or other electronic storage or transmission, or broadcast for distance learning.

Some ancillaries, including electronic and print components, may not be available to customers outside the United States.

✪ This book is printed on recycled, acid-free paper containing 10% postconsumer waste.

3 4 5 6 7 8 9 0 DOW/DOW 0 9 8 7 6

ISBN-13: 978-0-07-305238-0
ISBN-10: 0-07-305238-8

Editorial Director: *Kent A. Peterson*
Sponsoring Editor: *Thomas C. Lyon*
Senior Developmental Editor: *Anne L. Winch*
Director of Development: *Kristine Tibbetts*
Marketing Manager: *Tamara Maury*
Project Manager: *Jodi Rhomberg*
Production Supervisor: *Kara Kudronowicz*
Senior Media Project Manager: *Jodi K. Banowetz*
Lead Media Technology Producer: *John J. Theobald*
Senior Coordinator of Freelance Design: *Michelle D. Whitaker*
Cover/Interior Designer: *Christopher Reese*
Senior Photo Research Coordinator: *Lori Hancock*
Photo Research: *Don Murie/Meyers Photo Art*
Supplement Producer: *Brenda A. Ernzen*
Compositor: *Carlisle Communications, Ltd.*
Typeface: *10.5/12 Times Roman*
Printer: *R. R. Donnelley Willard, OH*

(USE) Cover Images: Chimpanzee: ©*Eye Wire, Animal Life;* Hawk: ©*PJules Frazier/Getty Images;* Human chromosome: ©*Adrian T. Sumner/Getty Images;* Snake, Snail, & Lizard: ©*Photodisc 0S 50* Bird of paradise X-ray: ©*Nicholas Veasey/Getty Images;* Hepatitis A Virus: ©*BSIP/Photo Researcher, Inc.;* Nautilus shell: © *Stephen Johnson/Getty Images;* Sunset Moth: © *Davies & Starr/Getty Images*

The credits section for this book begins on page C-1 and is considered an extension of the copyright page.

Library of Congress Cataloging-in-Publication Data

Johnson, George B. (George Brooks), 1942–
  Essentials of the living world / George B. Johnson. — 1st ed.
    p. cm.
  Includes bibliographical references and index.
  ISBN 0–07–305238–8 (alk. paper)
  1. Biology—Textbooks. I. Title.

QH308.2.J6199   2006
570—dc22   20040592626
                               CIP

www.mhhe.com

Brief Contents

Contents

Boxed Readings

Preface

I have been teaching biology to college freshmen in Washington University classrooms for over thirty years, and my writing of *Essentials of The Living World* is a product of that long educational journey. I first put finger to keyboard in 1995, several years after having been assigned the teaching of nonmajors biology in my department. My initial experience teaching biology to nonmajors had been discouraging. While my students were bright and very interested in biology, they were put off by the flood of information, and particularly by the mass of unfamiliar terminology. When you don't know what the words mean, it's easy to slip into thinking that the subject matter is difficult, when actually the ideas are simple, easy to grasp, and fun to consider.

How I Came to Write This Text

The available textbooks weren't much help. They had lots of pretty pictures, but were dense in content and terminology. In large measure, this reflected the fact that these texts had been cobbled together from longer majors texts—chopping out material to shorten the book had produced choppy chapters that were difficult for my students to follow, and way too hard. A wall of terminology and detail stood between my students and the ideas that form the core of biology.

So I decided to write my own book. I had already written a successful majors text with my friend Peter Raven, but in writing a nonmajors biology text I promised myself I would not repeat the fundamental mistake I had seen in other texts. Rather than prepare another "cut down majors text," I set out to write a whole new book from scratch, based on my experience in the classroom and aimed squarely at nonmajors students.

Writing *The Living World* was one of the most enjoyable experiences of my life. Organizing lectures for the classroom for thirty years had taught me that biology is at its core a set of ideas, and if students can master these ideas, the rest comes easy to them. So I set out to write a text that focused on concepts rather than terminology and information, a book that would be easy for students to learn from. Sorting out how best to teach each key idea was for me both challenging and an enormous amount of fun.

It's been very gratifying to see *The Living World* become a popular textbook for nonmajor's students, and to receive positive feedback from instructors and students who have used the text in their course. In recent years, the feedback has included an increasing number of instructors asking for a shorter, less expensive text; one that places less emphasis on botany and diversity. This *Essentials* version of *The Living World* is a response to that request. By removing some chapters, and combining and shortening many others, I have arrived at a lean but effective text, seven chapters shorter, that retains all the essential elements of *The Living World*.

What Sets This Text Apart

The following characteristics help distinguish *Essentials* from other textbooks written for nonmajors.

Writing Style

1. *Using Analogies.* I tried to write *Essentials* in an informal, friendly way, to engage as well as to teach. My principal tools to counteract the tendency of new terminology to intimidate wary students were analogies, relating the matter at hand to things we all know. As science, analogies are not exact, trading precision for clarity, but the classroom has taught me that if I do my job right, the key idea is not compromised by the analogy I use to explain it, but rather revealed.

2. *Focusing on Key Processes.* However clearly it is written, there is no way a text can avoid the fact that some processes like photosynthesis and the Krebs cycle are complex. To aid in a student's learning of complex ideas, I have prepared special "This is how it works" Process Boxes for some four dozen important processes that students encounter in introductory biology. Each of these Process Boxes walks the student through a complex process, one step at a time, so that the central idea is not lost in the details.

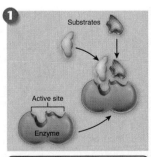
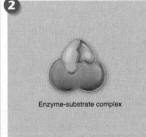
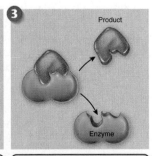

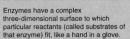

1 Enzymes have a complex three-dimensional surface to which particular reactants (called substrates of that enzyme) fit, like a hand in a glove.

2 An enzyme and its substrate(s) bind tightly together, forming an enzyme-substrate complex. The binding brings key atoms near each other and stresses key covalent bonds.

3 As a result, a chemical reaction occurs within the active site, forming the product. The product then diffuses away, freeing the enzyme to work again.

3. *Trimming Away Detail.* A third barrier stands between students and biology, and that is the mass of information typically presented in an introductory biology text. To make the ideas of biology more accessible to students, I attempted to address ideas and concepts rather than detailed information, trying to teach *how* things work and *why* things happen the way they do rather than merely naming parts or giving information.

4. *Creating Educated Citizens.* In writing *The Living World,* I endeavored in the first edition, and in each subsequent edition, to relate what the student is learning to the biology each student ought to know to live as an informed citizen in the twenty-first century.

I have continued this in the *Essentials* version, including an entire chapter devoted to "The New Biology," with discussions of genomics, gene technology and the impact of genetic engineering on agriculture and medicine, reproductive cloning, stem cells and gene therapy.

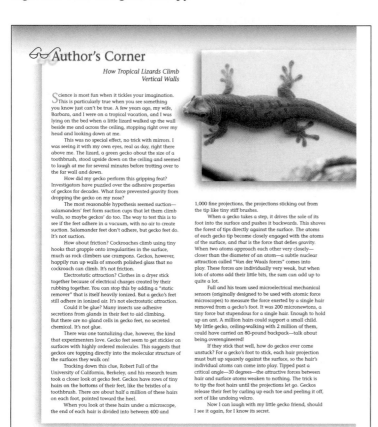

Author's Corner

How Tropical Lizards Climb Vertical Walls

Science is most fun when it tickles your imagination. This is particularly true when you see something you know just can't be true. A few years ago, my wife, Barbara, and I were on a tropical vacation, and I was lying on the bed when a little lizard walked up the wall beside me and across the ceiling, stopping right over my head and looking down at me.

This was no special effect, no trick with mirrors. I was seeing it with my own eyes, real as day, right there above me. The lizard, a green gecko about the size of a toothbrush, stood upside down on the ceiling and seemed to laugh at me for several minutes before trotting over to the far wall and down.

How did my gecko perform this gripping feat? Investigators have puzzled over the adhesive properties of geckos for decades. What force prevented gravity from dropping the gecko on my nose?

The most reasonable hypothesis seemed suction—salamanders' feet form suction cups that let them climb walls, so maybe geckos' do too. The way to test this is to see if the feet adhere in a vacuum, with no air to create suction. Salamander feet don't adhere, but gecko feet do. It's not suction.

How about friction? Cockroaches climb using tiny hooks that grapple onto irregularities in the surface, much as rock climbers use crampons. Geckos, however, happily run up walls of smooth polished glass that no cockroach can climb. It's not friction.

Electrostatic attraction? Clothes in a dryer stick together because of electrical charges created by their rubbing together. You can stop this by adding a "static remover" that is itself heavily ionized. But a gecko's feet still adhere in ionized air. It's not electrostatic attraction.

Could it be glue? Many insects use adhesive secretions from glands in their feet to aid climbing. But there are no gland cells in gecko feet, no secreted chemical. It's not glue.

There was one tantalizing clue, however, the kind that experimenters love. Gecko feet seem to get stickier on surfaces with highly ordered molecules. This suggests that geckos are tapping directly into the molecular structure of the surfaces they walk on!

Tracking down this clue, Robert Full of the University of California, Berkeley, and his research team took a closer look at gecko feet. Geckos have rows of tiny hairs on the bottoms of their feet, like the bristles of a toothbrush. There are about half a million of these hairs on each foot, pointed toward the heel.

When you look at these hairs under a microscope, the end of each hair is divided into between 400 and 1,000 fine projections, the projections sticking out from the tip like tiny stiff brushes.

When a gecko takes a step, it drives the sole of its foot into the surface and pushes it backwards. This shoves the forest of tips directly against the surface. The atoms of each gecko tip become closely engaged with the atoms of the surface, and *that* is the force that defies gravity. When two atoms approach each other very closely—closer than the diameter of an atom—a subtle nuclear attraction called "Van der Waals forces" comes into play. These forces are individually very weak, but when lots of atoms add their little bits, the sum can add up to quite a lot.

Full and his team used microelectrical mechanical sensors (originally designed to be used with atomic force microscopes) to measure the force exerted by a single hair removed from a gecko's foot. It was 200 micronewtons, a tiny force but stupendous for a single hair. Enough to hold up an ant. A million hairs could support a small child. My little gecko, ceiling-walking with 2 million of them, could have carried an 80-pound backpack—talk about being overengineered!

If they stick that well, how do geckos ever come unstuck? For a gecko's foot to stick, each hair projection must butt up squarely against the surface, so the hair's individual atoms can come into play. Tipped past a critical angle—30 degrees—the attractive forces between hair and surface atoms weaken to nothing. The trick is to tip the foot hairs until the projections let go. Geckos release their feet by curling up each toe and peeling it off, sort of like undoing velcro.

Now I can laugh with my little gecko friend, should I see it again, for I know its secret.

Throughout the book, I have written new full-page boxed readings to make connections to the everyday world: *A Closer Look* essays examine important new advances; *Science In Action* essays focus on how scientific analysis is carried out; and *Author's Corner* essays take a more personal view of how science relates to our everyday lives.

Current issues are of great interest to many students of introductory biology. In a new end-of-chapter element called *Exploring Current Issues* I provide these students with ways to explore the issues raised in the chapter in more depth. These include references to articles from scientific publications on topics of interest, links to articles written by me which expand on these topics, and videos of lectures on these topics presented by me to my students at Washington University in my course "Biology and Society."

Chapter Organization

I have made an effort to organize this text according to what was most successful in my own classroom. These decisions have created important differences between this text and its competitors.

1. *Centered Around the Learning Module.* I wrote *Essentials* to fit one and two-page spreads, so that each learning module begins with a new heading on the upper left-hand page and ends with a summary statement at the bottom of the right-hand page. I believe this format makes it easier for students to understand how the chapter content fits together conceptually, and the feedback I have received from our many reviewers seems to support that belief. This method of content organization also allows instructors to customize the text to their courses, as they can clearly point out which modules will be included in lecture.

2. *Clearer Teaching of Photosynthesis.* I have deliberately combined photosynthesis and cellular respiration into a single chapter in *Essentials,* not because metabolism is unimportant, but because teaching this difficult material in the classroom has taught me that students more easily grasp the complex metabolic activities of organisms when they explore photosynthesis and cellular respiration together, the many similarities of the two processes revealing their underlying unity.

3. *Beginning with Evolution and Ecology.* It is no accident that *Essentials* begins with a chapter on evolution and ecology. These ideas, central to biology, provide the student a framework within which to explore the world of the cell and gene function which occupy the initial third of the text. Students learn about cells and genes much more readily when they are presented in an evolutionary context, as biology rather than as molecular machinery.

4. *Presenting biological diversity as a story.* In traditional texts, evolution and diversity are taught as separate subjects. In *Essentials,* Evolution and Diversity are no longer treated as separate sections of the text. I have chosen instead to combine these areas into one continuous narrative, presenting biological diversity as an evolutionary journey. It is a lot more fun to teach this way, and students learn a great deal more, too.

Art

The first introduction of the student to idealized representations of plant and animal cells is particularly important in a nonmajors' text, because the beginning student must make a mental reference to that cell throughout his or her biology course, as important cellular processes and the cell's landscape are referenced again and again. For that reason I have put a great deal of thought into those representations, to remove as many areas for student misconception and confusion as possible. The resulting illustrations, seen on pages 78 and 79, are both instructive and attractive and are examples of the attention given to the art in this text.

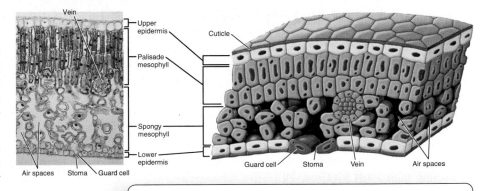

Combination Figures

These figures combine a photo or micrograph with a line drawing, to make the connection between conceptual art and what the student may encounter in a lab (see Figure 26.18 A leaf in cross section, page 617).

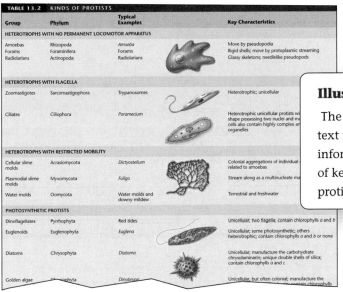

Illustrated Tables

The inclusion of figures within many of the tables in the text now makes it easier for students to understand the information at a glance, and helps remind the student of key structures or processes (see table 13.2 Kinds of protists, page 302).

Process Boxes

This "how it works" feature provides a step-by-step description that walks the student through a compact summary of an important concept (see figure 24.3 How steroid hormones work, page 570).

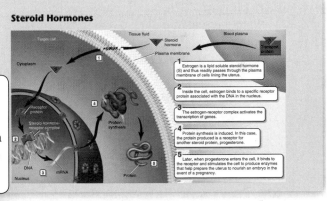

Biochemistry Pathway Icons

Found in the discussions of respiration and photosynthesis, these icons help students follow complex metabolic processes by highlighting the step currently under discussion (see figure 6.10 Chemiosmosis in a chloroplast, page 121).

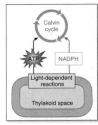
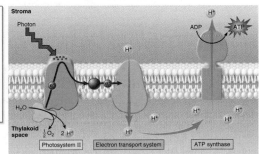

Teaching and Learning Supplements

McGraw-Hill offers various tools and technology products to support *Essentials of The Living World*. Students can order supplemental study materials by contacting their local bookstore or by calling 800-262-4729. Instructors can obtain teaching aids by calling the Customer Service Department at 800-338-3987, visiting our website at www.mhhe.com/biology, or contacting their local McGraw-Hill sales representative.

For The Instructor:

Digital Content Manager CD-ROM

This multimedia collection of visual resources allows instructors to utilize artwork from the text in multiple formats to create customized classroom presentations, visually based tests and quizzes, dynamic course website content, or attractive printed support material. The digital assets on this cross-platform CD-ROM include:

Art Library Color-enhanced, digital files of all illustrations in the book, plus the same art saved in unlabeled and gray scale versions, can be readily incorporated into lecture presentations, exams, or custom-made classroom materials. Upsized labels make the images appropriate for use in large lecture halls.

TextEdit Art Library Selected line art is placed into a PowerPoint presentation that allows the user to revise and/or move or delete labels as desired for creation of customized presentations and/or for testing purposes.

Active Art Library Active Art consists of art files that have been converted to a format that allows the artwork to be edited inside of PowerPoint. Each piece can be broken down to its core elements, grouped or ungrouped, and edited to create customized illustrations.

Animations Library Full color presentations of key biological processes have been brought to life via animation. These animations offer flexibility for instructors and were designed to be used in lecture. Instructors can pause, rewind, fast forward, and turn the audio off or on to create dynamic lecture presentations. The animations are now also available with Spanish narration and text.

PowerPoint Lecture Outlines These ready-made presentations combine art and lecture notes for each of the 27 chapters of the book. The presentations can be used as they are, or can be customized to reflect your preferred lecture topics and organization.

PowerPoint Outlines The art photos and tables for each chapter are inserted into blank PowerPoint presentations to which you can add your own notes.

Photo Library Like the Art Library, digital files of all photographs from the book are available.

Table Library Every table that appears in the book is provided in electronic form.

Additional Photo Library Over 700 photos, not found in *Essentials*, are available for use in creating lecture presentations.

TextEdit Art Library with modified labels

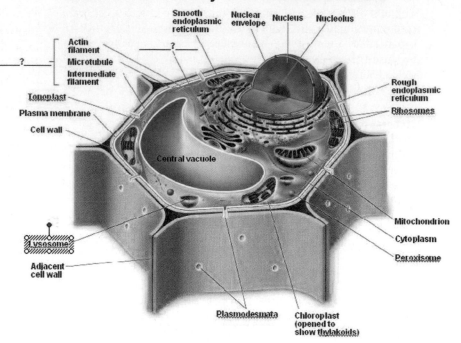

Art Library with upsized labels

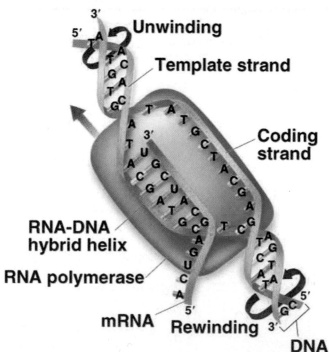

Biology Digitized Videos DVD

McGraw-Hill is pleased to offer adopting instructors a new presentation tool—digitized biology video clips on DVD! Licensed from some of the highest-quality science video producers in the world, these brief segments range from 15 seconds to two minutes in length and cover all areas of general biology from cells to ecosystems. Engaging and informative, McGraw-Hill's digitized biology videos will help capture students' interest while illustrating key biological concepts and processes such as how cilia and flagella work and how some plants have evolved into carnivores.

Alex the African Grey Parrot correctly answers complex questions in a video segment that vividly introduces the subject of animal cognition and behavior.

Instructor's Testing and Resource CD-ROM

The cross-platform CD-ROM contains the Instructor's Manual and Test Item File, both available in Word and PDF formats. The manual contains chapter outlines, learning objectives, key terms, lecture suggestions, additional critical thinking questions, and sources for visual resources. The Test Bank offers questions that can be used for homework assignments or the preparation of exams. The computerized test bank allows the user to quickly create customized exams. Instructors can search questions by topic, format, or difficulty level; edit existing questions or add new ones; and scramble questions and answer keys for multiple versions of the same test.

Transparencies

A set of 700 transparency overheads includes every piece of line art and table in *The Living World*. The images are printed with better visibility and contrast than ever before, and labels are large and bold for clear projection.

Online Learning Center
www.mhhe.com/tlwessentials

Instructor's resources at this site include access to online laboratories, case studies, newsfeeds, and the Course Integration Guide as well as Word and PDF versions of the Instructor's Manual. An Active Art demo teaches you how to create your own art for presentation from the Active Art pieces found on the Digital Content Manager CD-ROM. Adopters of *Essentials* also have access to PageOut, McGraw-Hill's exclusive tool for creating a course website. PageOut requires no knowledge of coding and the website you create is hosted by McGraw-Hill.

Course Delivery Systems

With help from our partners, WebCT, Blackboard, TopClass, eCollege, and other course management systems, instructors can take complete control over their course content. These course cartridges also provide online testing and powerful student tracking features. The *Essentials of The Living World* Online Learning Center is available within all of these platforms.

For the Student:

Online Learning Center
www.mhhe.com/tlwessentials

The site includes quizzes for each chapter, animations, interactive activities, and answers to questions from the text. Turn to the inside cover of the text to learn more about the exciting features provided for students through the enhanced *Essentials of The Living World* Online Learning Center.

Student Study Guide

This student resource contains activities and questions to help reinforce chapter concepts. The guide provides students with tips and strategies for mastering the chapter content, concept outlines, key terms, and sample quizzes.

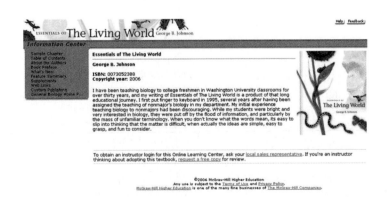

Acknowledgements

Every author labors on the shoulders of many others. When I first set out to write a textbook, I had no idea how much work remained to be done after I had finished writing. An army of editors, spelling and grammar checkers, photo researchers, and artists assembles the final manuscript, and another even larger army transforms this manuscript into a bound book. I cannot begin to thank them all. Tom Lyon and Anne Winch were my editorial team, the people I worked with every day; their boss, Kent Peterson, put out any fires I created out of excess enthusiasm or sheer pig-headedness. Jodi Rhomberg spearheaded the production team, balancing many balls in the air. The new art program was done by Imagineering. They did a superb job, despite my occasionally breathing fire down their necks. When the manuscript was ready and the art finished, the book went to Carlisle Communications, where another team headed by Cindy Sweeney went to work to prepare the book for printing, with Jodi Gaherty expertly composing the pages into learning modules. My own longtime developmental editors and right arms Megan Jackman and Liz Sievers have again played an invaluable role in overseeing every detail of a complex revision. Their intelligence and perseverance continue to play a major role in the quality of this book. I would also like to thank JodyLee Estrada Duek of Pima Community College, who provided end-of-chapter questions, and Jennifer Warner of the University of North Carolina-Charlotte, who contributed the Additional Resources for the *Exploring Current Issues* feature. Last but not least, I would like to extend a special thanks to Michael Lange, the editor-in-chief at McGraw-Hill, for his continued strong support of this project.

Reviewers of *The Living World, Fourth Edition*

Over twenty years of authoring have taught me the great value of reviewers in improving my texts. My colleagues around the country have provided numerous suggestions on how to improve this fourth edition. Many teachers and students using the previous edition have also suggested ways to improve it. Even teachers who chose not to adopt the previous edition often drew my attention to something they did not like. All of these instructors and students have much to teach me, pointing out ways to improve presentation, clarify explanations, and add or expand on important topics. The instructors listed below provided detailed comments. I have tried to listen carefully to all of you. Every one of you has my heartfelt thanks.

Christa Behrendt-Adam
Missouri Western State College

D. Daryl Adams
Minnesota State University

Sylvester Allred
Northern Arizona University

Norris Armstrong
University of Georgia

Amir M. Assadi-Rad
San Joaquin Delta College

Bert Atsma
Union County College

D. S. K. Ballal
Tennessee Technological University

David Bass
University of Central Oklahoma

James Enderby Bidlack
University of Central Oklahoma

Charles L. Biles
East Central University

Michael J. Bodri
Northwestern State University of Louisiana

Richard Boutwell
Missouri Western State College

Robert Boyd
Auburn University

Marguerite Brickman
University of Georgia

Katherine Buhrer
Tidewater Community College

Sharon K. Bullock
Virginia Commonwealth University

David Byres
Florida Community College Jacksonville

Jane E. Caldwell
West Virginia University

Beth Campbell
Itawamba Community College

Ruth Chesnut
Eastern Illinois University

Barry Chess
Pasadena City College

Guided Tour

Instructive Art Program

The core of every biology textbook is its art program, and George Johnson has worked hard to create a dynamic program of full-color illustrations and photographs that support and further clarify the text explanations. Brilliantly rendered and meticulously reviewed for accuracy and consistency, the carefully conceived illustrations and accompanying photos provide concrete, visual reinforcement of the topics discussed throughout the text.

Process Boxes

Process Boxes break down complex processes into a series of smaller steps, allowing you to track the key occurrences and learn them as you go.

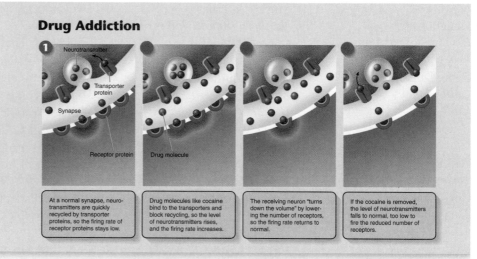

Figure 23.10   How drug addiction works.

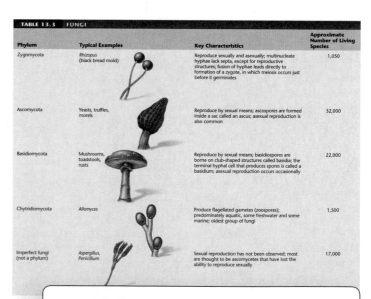

TABLE 13.3	FUNGI		
Phylum	**Typical Examples**	**Key Characteristics**	**Approximate Number of Living Species**
Zygomycota	*Rhizopus* (black bread mold)	Reproduce sexually and asexually; multinucleate hyphae lack septa, except for reproductive structures; fusion of hyphae leads directly to formation of a zygote, in which meiosis occurs just before it germinates	1,050
Ascomycota	Yeasts, truffles, morels	Reproduce by sexual means; ascospores are formed inside a sac called an ascus; asexual reproduction is also common	32,000
Basidiomycota	Mushrooms, toadstools, rusts	Reproduce by sexual means; basidiospores are borne on club-shaped structures called basidia; the terminal hyphal cell that produces spores is called a basidium; asexual reproduction occurs occasionally	22,000
Chytridiomycota	*Allomyces*	Produce flagellated gametes (zoospores); predominately aquatic, some freshwater and some marine; oldest group of fungi	1,500
Imperfect fungi (not a phylum)	*Aspergillus*, *Penicillium*	Sexual reproduction has not been observed; most are thought to be ascomycetes that have lost the ability to reproduce sexually	17,000

TABLE 14.1	PLANT PHYLA		
Phylum	**Typical Examples**	**Key Characteristics**	**Approximate Number of Living Species**
NONVASCULAR PLANTS			
Hepaticophyta (liverworts)	*Marchantia*	Without true vascular tissues; lack true roots and leaves; live in moist habitats and obtain water and nutrients by osmosis and diffusion; require water for fertilization; gametophyte is dominant structure in the life cycle; the three phyla were once grouped together	15,600
Anthocerophyta (hornworts)	*Anthoceros*		
Bryophyta (mosses)	*Polytrichum, Sphagnum* (hairy cap and peat moss)		
SEEDLESS VASCULAR PLANTS			
Lycophyta (lycopods)	*Lycopodium* (club mosses)	Seedless vascular plants similar in appearance to mosses but diploid; require water for fertilization; sporophyte is dominant structure in life cycle; found in moist woodland habitats	1,150
Pterophyta (ferns)	*Azolla, Sphaeropteris* (water and tree ferns) *Equisetum* (horsetails) *Psilotum* (whisk ferns)	Seedless vascular plants; require water for fertilization; sporophytes diverse in form and dominate the life cycle	11,000

Illustrated Tables

The inclusion of figures within many of the tables in the text now make it easier for you to understand the table information at a glance, and helps remind you of key structures or processes.

Combination Figures

Line drawings are often combined with photographs to facilitate visualization of structures.

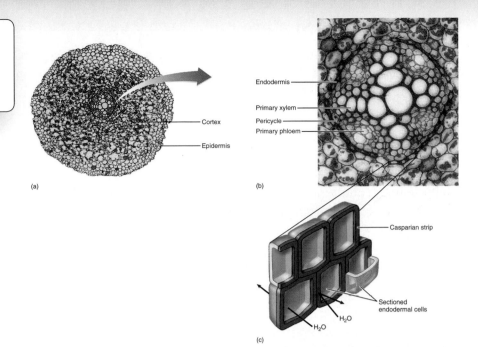

(a)

(b)

- Endodermis
- Primary xylem
- Pericycle
- Primary phloem
- Casparian strip
- Sectioned endodermal cells

H_2O H_2O H_2O

(c)

Multi-Level Perspective

Illustrations depicting complex structures or processes combine macroscopic and microscopic views to help you see the relationship between increasingly detailed images.

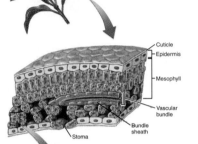

- Cuticle
- Epidermis
- Mesophyll
- Vascular bundle
- Bundle sheath
- Stoma

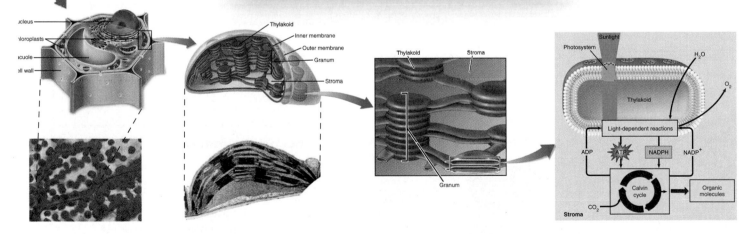

Nucleus
Chloroplasts
Vacuole
Cell wall

- Thylakoid
- Inner membrane
- Outer membrane
- Granum
- Stroma

Thylakoid   Stroma

Granum

Sunlight
Photosystem
H_2O
O_2
Thylakoid
Light-dependent reactions
ADP   ATP   NADPH   NADP$^+$
Calvin cycle
Organic molecules
CO_2
Stroma

Biochemistry Pathway Icons

These icons are paired with more detailed illustrations to assist you in keeping the big picture in mind when learning complex metabolic processes. The icon highlights which step the main illustration represents, and where that step occurs in the complete process.

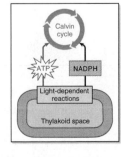

Calvin cycle
ATP   NADPH
Light-dependent reactions
Thylakoid space

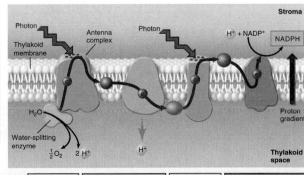

Stroma
Photon   Antenna complex   Photon H^+ + NADP$^+$ NADPH
Thylakoid membrane
H_2O Proton gradient
Water-splitting enzyme $\frac{1}{2} O_2$ 2 H^+ H^+ Thylakoid space

| Photosystem II | Electron transport system | Photosystem I | Electron transport system |

The Learning System

This text is designed to help you learn in a systematic fashion. Simple facts are the building blocks for developing explanations of more complex concepts. The text discussion is presented within a supporting framework of learning aids that help organize studying, reinforce learning, and promote problem-solving skills.

12
Exploring Biological Diversity

Chapter-at-a-Glance

The Classification of Organisms
12.1 The Invention of the Linnaean System
Biologists name organisms using a two-word "binomial" system.

12.2 Species Names
Every kind of organism is assigned a unique Latin name.

12.3 Higher Categories
The higher groups into which an organism is placed reveal a great deal about the organism.

12.4 What Is a Species?
Species are groups of similar organisms. Animal species tend not to interbreed with individuals of other species, while plants often do so.

Inferring Phylogeny
12.5 How to Build a Family Tree
Traditional and cladistic interpretations of an organism's evolutionary history differ in the emphasis they place on key traits.

In 1799, the skin of a most unusual animal was sent to England by Captain John Hunter, governor of the British penal colony in New South Wales (Australia). Covered in soft fur, it was less than two feet long. As it had mammary glands with which to suckle its y[...] a mammal, but in other ways it seemed very [...] Males have internal testes, and females have [...] and reproductive tract opening called a cloa[...] reptiles do, and like reptilian eggs, the yolk [...] egg does not divide. It thus seemed a confu[...] mammalian and reptilian traits. Adding to t[...] its appearance: It has a tail not unlike that o[...] not unlike that of a duck, and webbed feet! [...] had mixed together body parts at random—

Chapter-at-a-Glance

Each chapter begins with an outline that gives you an overview of the content contained within that chapter. Reviewing the outline before reading the chapter will help focus your attention on the major concepts you should take away from the chapter.

12.2 Species Names

A group of organisms at a particular level in a classification system is called a **taxon** (plural, **taxa**), and the branch of biology that identifies and names such groups of organisms is called **taxonomy.** Taxonomists are in a real sense detectives, biologists who must use clues of appearance and behavior to [...] organisms.

[...] among taxonomists through[...]sms can have the same name. [...]ored, a language spoken by no [...]e names. Because the scientific [...]me anywhere in the world, this [...]d precise way of communicat[...] particular biologist is Chinese, [...]his is a great improvement over [...]ich often vary from one place to [...]ure 12.2, corn in Europe refers

to the plant Americans call wheat; a bear is a large placental omnivore in the United States but a koala (a vegetarian marsupial) in Australia; and a robin is a very different bird in Europe and North America.

By convention, the first word of the binomial name is the genus to which the organism belongs. This word is always capitalized. The second word, called the *epithet,* refers to the particular species and is not capitalized. The two words together are called the **scientific name,** or species name, and are written in italics. The system of naming animals, plants, and other organisms established by Linnaeus has served the science of biology well for nearly 250 years.

> **12.2** By convention, the first part of a binomial species name identifies the genus to which the species belongs, and the second part distinguishes that particular species from other species in the genus.

Numbered Headings

The numbered headings employed in the modules form the backbone of the Chapter-at-a-Glance outline. This consistency makes it easier to identify the key concepts for each chapter, and to then manage the supporting details for each concept.

Section Summaries

Each module ends with a summary intended to reinforce the key concepts from that section. Reviewing the summary after reading the section will indicate whether you learned the main ideas presented in the module.

(a)

(b)

(c)

Figure 12.2 Common names make poor labels.
The common names corn (a), bear (b), and robin (c) bring clear images to our minds (photos on *top*), but the images would be very different to someone living in Europe or Australia (photos on *bottom*). There, the same common names are used to label very different species.

Modular Format

Each page or two-page spread in *Essentials of The Living World* is organized as an independent module, with its own numbered heading at the top of the left-hand page, and a highlighted summary at the bottom of the right-hand page. This system organizes the information in the chapter within a clear conceptual framework, which in turn helps you learn and retain the material.

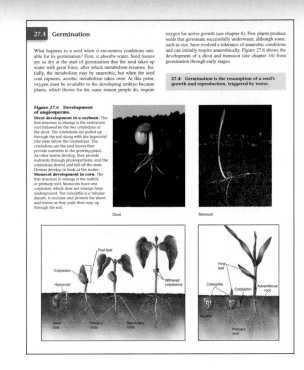

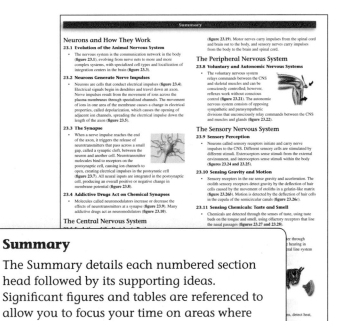

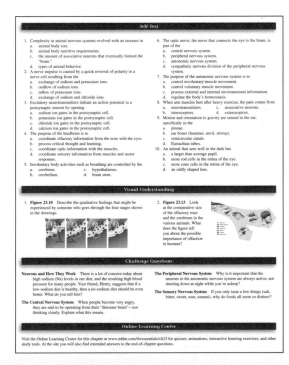

Summary

The Summary details each numbered section head followed by its supporting ideas. Significant figures and tables are referenced to allow you to focus your time on areas where you need additional study.

End-of-chapter Questions

Each chapter concludes with a set of questions designed to test your knowledge of the content, including multiple choice questions, illustration-based questions, and application questions. Answers to the multiple choice questions are in a section at the end of the book, while extended answers for all of the questions are found on *Essentials of The Living World* Online Learning Center at www.mhhe.com/tlwessentials. At the site you can take an interactive version of the end-of-chapter quiz that provides you with instructional feedback.

1

The Science of Biology

These Antarctic penguins share many properties with you and all living things. Their bodies are made up of cells, just as yours is. They grow by eating, as you do, although their diet is limited to fishes they catch in the cold Antarctic waters. The sky above them shields them from the sun's harmful UV radiation, just as the sky above you shields you. Not in the summer, however. In the Antarctic summer an "ozone hole" appears, depleting the ozone above these penguins and exposing them to the danger of UV radiation. Scientists are analyzing this situation by a process of observation and experiment, rejecting ideas that do not match their data, and they are learning what is going on. The study of biology is a matter of observing carefully, and asking the right questions.

1.1 The Diversity of Life

In its broadest sense, biology is the study of living things—the science of life. The living world teems with a breathtaking variety of creatures—whales, butterflies, mushrooms, and mosquitoes—all of which can be categorized into six groups, or **kingdoms,** of organisms (figure 1.1).

Biologists study the diversity of life in many different ways. They live with gorillas, collect fossils, and listen to whales. They isolate bacteria, grow mushrooms, and examine the structure of fruit flies. They read the messages encoded in the long molecules of heredity and count how many times a hummingbird's wings beat each second. In the midst of all this diversity, it is easy to lose sight of the key lesson of biology, which is that all living things have much in common.

1.1 The living world is very diverse, but all living things share many key properties.

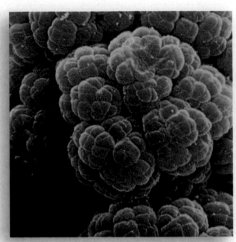

Archaea. This kingdom of prokaryotes (the simplest of cells that do not have nuclei) includes this methanogen, which manufactures methane as a result of its metabolic activity.

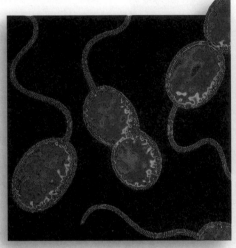

Bacteria. This group is the second of the two prokaryotic kingdoms. Shown here are purple sulfur bacteria, which are able to convert light energy into chemical energy.

Protista. Most of the unicellular eukaryotes (those whose cells contain a nucleus) are grouped into this kingdom, and so are the algae pictured here.

Fungi. This kingdom contains nonphotosynthetic multicellular organisms that digest their food externally, such as these mushrooms.

Plantae. This kingdom contains photosynthetic multicellular organisms that are terrestrial, such as the flowering plant pictured here.

Animalia. Organisms in this kingdom are nonphotosynthetic multicellular organisms that digest their food internally, such as this ram.

Figure 1.1 The six kingdoms of life.
Biologists categorize all living things into six major categories called *kingdoms*. Each kingdom is profoundly different from the others.

Properties of Life

Biology is the study of life—but what does it mean to be alive? What are the properties that define a living organism? This is not as simple a question as it seems because some of the most obvious properties of living organisms are also properties of many nonliving things—for example, *complexity* (a computer is complex), *movement* (clouds move in the sky), and *response to stimulation* (a soap bubble pops if you touch it). To appreciate why these three properties, so common among living things, do not help us to define life, imagine a mushroom standing next to a television: the television seems more complex than the mushroom, the picture on the television screen is moving while the mushroom just stands there, and the television responds to a remote control device while the mushroom continues to just stand there—yet it is the mushroom that is alive.

All living things share five more basic properties, passed down over millions of years from the first organisms to evolve on earth: *cellular organization, metabolism, homeostasis, growth and reproduction,* and *heredity.*

1. **Cellular organization.** All living things are composed of one or more cells. A cell is a tiny compartment with a thin covering called a *membrane.* Some cells have simple interiors, while others are complexly organized, but all are able to grow and reproduce. Many organisms possess only a single cell (figure 1.2); your body contains about 100 trillion—that's how many centimeters long a string would be wrapped around the world 1,600 times!

2. **Metabolism.** All living things use energy. Moving, growing, thinking—everything you do requires energy. Where does all this energy come from? It is captured from sunlight by plants and algae. To get the energy that powers our lives, we extract it from plants or from plant-eating animals. The transfer of energy from one form to another in cells is an example of *metabolism* (figure 1.3). All organisms require energy to grow,

Figure 1.3   Metabolism.
This kingfisher obtains the energy it needs to move, grow, and carry out its body processes by eating fish. It metabolizes this food using chemical processes that occur within cells.

and all organisms transfer this energy from one place to another within cells using special energy-carrying molecules called ATP molecules.

3. **Homeostasis.** All living things maintain stable internal conditions so that their complex processes can be better coordinated. While the environment often varies a lot, organisms act to keep their interior conditions relatively constant; a process called *homeostasis.* Your body acts to maintain an internal temperature of 37°C (98.5°F), however hot or cold the weather might be.

4. **Growth and reproduction.** All living things grow and reproduce. Bacteria increase in size and simply split in two, as often as every 15 minutes, while more complex organisms grow by increasing the number of cells and reproduce sexually (some, like the bristlecone pine of California, have reproduced after 4,600 years).

5. **Heredity.** All organisms possess a genetic system that is based on the replication and duplication of a long molecule called *DNA (deoxyribonucleic acid).* The information that determines what an individual organism will be like is contained in a code that is dictated by the order of the subunits making up the DNA molecule, just as the order of letters on this page determines the sense of what you are reading. Each set of instructions within the DNA is called a *gene.* Together, the genes determine what the organism will be like. Because DNA is faithfully copied from one generation to the next, any change in a gene is also preserved and passed on to future generations. The transmission of characteristics from parent to offspring is a process called *heredity.*

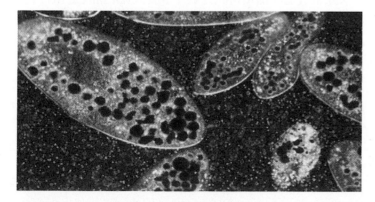

Figure 1.2   Cellular organization.
These paramecia are complex single-celled protists that have just ingested several yeast cells. Like these paramecia, many organisms consist of just a single cell, while others are composed of trillions of cells.

1.2   All living things possess cells that carry out metabolism, maintain stable internal conditions, reproduce themselves, and use DNA to transmit hereditary information to offspring.

The Organization of Life

The organisms of the living world function and interact with each other at many levels (figure 1.4).

A Hierarchy of Increasing Complexity

A key factor in organizing these interactions is the degree of complexity.

Cellular Level.   There is a hierarchy of increasing complexity within cells.

1. **Molecules.**   Atoms, the fundamental elements of matter, are joined together into complex clusters called **molecules.** DNA, which stores the hereditary information in all living organisms, is a complex biological molecule, called macromolecules.
2. **Organelles.**   Complex biological molecules are assembled into tiny compartments within cells called **organelles,** within which cellular activities are organized. The nucleus is an organelle within which the cell's DNA is stored.
3. **Cells.**   Organelles and other elements are assembled in the membrane-bounded units we call **cells.** Cells are the smallest level of organization that can be considered alive.

All living organisms are composed of cells. The organism is the basic unit of life. Many organisms are composed of single cells. Bacteria are single cells, for example. All animals and plants, as well as most fungi and algae, are multicellular—composed of more than one cell.

Organismal Level.   At the organismal level, cells are organized into three levels of complexity.

1. **Tissues.**   The most basic level is that of **tissues,** which are groups of similar cells that act as a functional unit. Nerve tissue is one kind of tissue, composed of cells called neurons that are specialized to carry electrical signals from one place to another in the body.
2. **Organs.**   Tissues, in turn, are grouped into **organs,** which are body structures composed of several different tissues grouped together in a structural and functional unit. Your brain is an organ composed of nerve cells and a variety of connective tissues that form protective coverings and distribute blood.
3. **Organ Systems.**   At the third level of organization, organs are grouped into **organ systems.** The nervous system, for example, consists of sensory organs, the brain and spinal cord, and neurons that convey signals to and from them.

Populational Level.   Organisms are organized into several higher hierarchical levels within the living world.

1. **Population.**   The most basic of these is the **population,** which is a group of organisms of the same species living in the same place. A flock of geese living together on a pond is a population.
2. **Species.**   All the populations of a particular kind of organism together form a **species,** its members similar in appearance and able to interbreed. All Canada geese, whether found in Canada, Minnesota, or Missouri, are basically the same, members of the species *Branta canadensis.*
3. **Community.**   At a higher level of biological organization, a **community** consists of all the populations of different species living together in one place. Geese, for example, may share their pond with ducks, fish, grasses, and many kinds of insects. All interact in a single pond community.
4. **Ecosystem.**   At the highest tier of biological organization, a biological community and the soil and water within which it lives together constitute an ecological system, or **ecosystem.**

Emergent Properties

At each higher level in the living hierarchy, novel properties emerge, properties that were not present at the simpler level of organization. These **emergent properties** result from the way in which components interact, and often cannot be guessed just by looking at the parts themselves. You have the same array of cell types as a giraffe, for example. Examining a collection of your cells gives little clue of what you are like as an animal.

The emergent properties of life are not magical or supernatural. They are the natural consequence of the hierarchy or structural organization which is the hallmark of life. Water and ice are both made of H_2O molecules, but one is liquid and the other solid because the H_2O molecules in ice are more organized.

Functional properties emerge from more complex organization. Metabolism is an emergent property of life. The chemical reactions within a cell arise from interactions between molecules that are orchestrated by the orderly environment of the cell's interior. Consciousness is an emergent property of the brain that results from the interactions of many neurons in different parts of the brain.

1.3   Cells, multicellular organisms, and ecological systems each are organized in a hierarchy of increased complexity. Life's hierarchical organization is responsible for the emergent properties that characterize so many aspects of the living world.

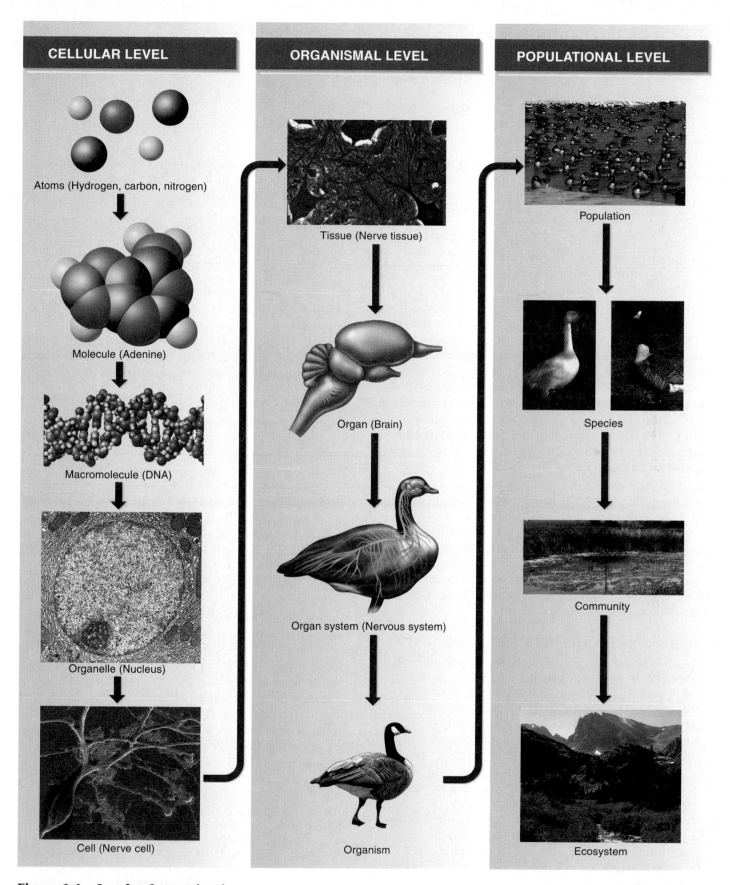

CELLULAR LEVEL

Atoms (Hydrogen, carbon, nitrogen)

Molecule (Adenine)

Macromolecule (DNA)

Organelle (Nucleus)

Cell (Nerve cell)

ORGANISMAL LEVEL

Tissue (Nerve tissue)

Organ (Brain)

Organ system (Nervous system)

Organism

POPULATIONAL LEVEL

Population

Species

Community

Ecosystem

Figure 1.4   Levels of organization.

A traditional and very useful way to sort through the many ways in which the organisms of the living world interact is to organize them in terms of levels of organization, proceeding from the very small and simple to the very large and complex. Here we examine organization within the cellular, organismal, and populational levels.

1.4 Biological Themes

Just as every house is organized into thematic areas such as bedroom, kitchen, and bathroom, so the living world is organized by major *themes,* such as how energy flows within the living world from one part to another. As you study biology in this text, five general themes will emerge repeatedly, themes that serve to both unify and explain biology as a science (table 1.1):

1. evolution;
2. the flow of energy;
3. cooperation;
4. structure determines function;
5. homeostasis.

Evolution

Evolution is genetic change in a species over time. Charles Darwin was an English naturalist who, in 1859, proposed the idea that this change is a result of a process called **natural selection.** Simply stated, those organisms whose characteristics make them better able to survive the challenges of their environment live to reproduce, passing their favorable characteristics on to their offspring. Darwin was thoroughly familiar with variation in domesticated animals (in addition to many nondomesticated organisms), and he knew that varieties of pigeons could be selected by breeders to exhibit exaggerated characteristics, a process called **artificial selection.** We now know that the characteristics selected are passed on through generations because DNA is transmitted from parent to offspring. Darwin visualized how selection in nature could be similar to that which had produced the different varieties of pigeons. Thus, the many forms of life we see about us on earth today, and the way we ourselves are constructed and function, reflect a long history of natural selection. Evolution will be explored in more detail in chapters 2 and 11.

The Flow of Energy

All organisms require energy to carry out the activities of living—to build bodies and do work and think thoughts. All of the energy used by most organisms comes from the sun and is gradually used up as it passes in one direction through ecosystems. The simplest way to understand the flow of energy through the living world is to look at who uses it. The first stage of energy's journey is its capture by green plants, algae, and some bacteria in photosynthesis. Plants then serve as a source of life-driving energy for animals that eat them. Other animals may then eat the plant eaters. At each stage, some energy is used, some is transferred, and much is lost. The flow of energy is a key factor in shaping ecosystems, affecting how many and what kinds of animals live in a community.

Cooperation

Cooperation between different kinds of organisms has played a critical role in the evolution of life on earth. For example, organisms of two different species that live in direct contact form a type of relationship called **symbiosis.** Animal cells possess organelles that are the descendants of symbiotic bacteria, and symbiotic fungi helped plants first invade land from the sea. The coevolution of flowering plants and insects, where changes in flowers influenced insect evolution and in turn, changes in insects influenced flower evolution, has been responsible for much of life's great diversity.

Structure Determines Function

One of the most obvious lessons of biology is that biological structures are very well suited to their functions. You will see this at every level of organization: within cells, the shape of the proteins called enzymes that cells use to carry out chemical reactions are precisely suited to match the chemicals the enzymes must manipulate. Within the many kinds of organisms in the living world, body structures seem carefully designed to carry out their functions—the long tongue with which a moth sucks nectar from a deep flower is one example. The superb fit of structure to function in the living world is no accident. Life has existed on earth for over 2 billion years, a long time for evolution to favor changes that better suit organisms to meet the challenges of living. It should come as no surprise to you that after all this honing and adjustment, biological structures carry out their functions well.

Homeostasis

The high degree of specialization we see among complex organisms is only possible because these organisms act to maintain a relatively stable internal environment, a process called homeostasis. Without this constancy, many of the complex interactions that need to take place within organisms would be impossible, just as a city cannot function without rules and order. Maintaining homeostasis in a body as complex as yours requires a great deal of signaling back-and-forth between cells.

As already stated, you will encounter these biological themes repeatedly in this text. But just as a budding architect must learn more than the parts of buildings, so your study of biology should teach you more than a list of themes, concepts, and parts of organisms. Biology is a dynamic science that will affect your life in many ways, and that lesson is one of the most important you will learn. It is also an awful lot of fun.

1.4 **The five general themes of biology are (1) evolution, (2) the flow of energy, (3) cooperation, (4) structure determines function, and (5) homeostasis.**

TABLE 1.1   BIOLOGICAL THEMES

Cooperation   Latin American ants live within the hollow thorns of certain species of acacia trees. The nectar at the bases of the leaves and at the tips of the leaflets provide food. The ants supply the trees with organic nutrients and protection.

Evolution   Charles Darwin's studies of artificial selection in pigeons provided key evidence that selection could produce the sorts of changes predicted by his theory of evolution. The differences that have been obtained by artificial selection of the wild European rock pigeon (*top*) and such domestic races as the red fantail (*middle*) and the fairy swallow (*bottom*), with its fantastic tufts of feathers around its feet, are indeed so great that the birds probably would, if wild, be classified in different major groups.

The Flow of Energy   Energy passes from plants to plant-eating animals to animal-eating animals, such as this eagle.

Homeostasis   Homeostasis often involves water balance. All complex organisms need water—some, like this hippo, luxuriate in it. Others, like the kangaroo rat that lives in arid conditions where water is scarce, obtain water from food and never drink.

Structure Determines Function   With its long tongue, the hummingbird clear-wing moth is able to reach the nectar deep within these flowers.

1.5 How Scientists Think

Deductive Reasoning

Science is a process of investigation, using observation, experimentation, and reasoning. Not all investigations are scientific. For example, when you want to know how to get to Chicago from St. Louis, you do not conduct a scientific investigation—instead, you look at a map to determine a route. In other investigations, you make individual decisions by applying a "guide" of accepted general principles. This is called **deductive reasoning.** Deductive reasoning, using general principles to explain specific observations, is the reasoning of mathematics, philosophy, politics, and ethics; deductive reasoning is also the way a computer works. All of us rely on deductive reasoning to make everyday decisions. We use general principles as the basis for examining and evaluating these decisions (figure 1.5).

Inductive Reasoning

Where do general principles come from? Religious and ethical principles often have a religious foundation; political principles reflect social systems. Some general principles, however, are not derived from religion or politics but from observation of the physical world around us. If you drop an apple, it will fall, whether or not you wish it to and despite any laws you may pass forbidding it to do so. Science is devoted to discovering the general principles that govern the operation of the physical world.

How do scientists discover such general principles? Scientists are, above all, observers: they look at the world to understand how it works. It is from observations that scientists determine the principles that govern our physical world.

This way of discovering general principles by careful examination of specific cases is called **inductive reasoning** (figure 1.5). Inductive reasoning first became popular about 400 years ago, when Isaac Newton, Francis Bacon, and others began to conduct experiments and from the results infer general principles about how the world operates. The experiments were sometimes quite simple. Newton's consisted simply of releasing an apple from his hand and watching it fall to the ground. This simple observation is the stuff of science. From a host of particular observations, each no more complicated than the falling of an apple, Newton inferred a general principle—that all objects fall toward the center of the earth. This principle was a possible explanation, or **hypothesis,** about how the world works. Like Newton, scientists today formulate hypotheses, and observations are the materials on which they build them.

> **1.5** Science uses inductive reasoning to infer general principles from detailed observation.

INDUCTIVE REASONING

Observations of Specific Events

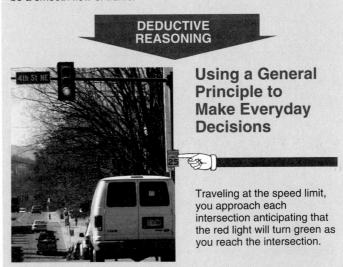

Driving down the street at the speed limit, you observe that the red traffic light turns green just as you approach the intersection.

Maintaining the same speed, you observe the same event at the next several intersections: the traffic lights turn green just as you approach the intersections. When you speed up, however, the light doesn't change until after you reach the intersection.

INDUCTIVE REASONING

Formation of a General Principle

You conclude that the traffic lights along this street are "timed" to change in the time it takes your car, traveling at the speed limit, to traverse the distance between them.

DEDUCTIVE REASONING

An Accepted General Principle

When traffic lights along city streets are "timed" to change at the time interval it takes traffic to pass between them, the result will be a smooth flow of traffic.

DEDUCTIVE REASONING

Using a General Principle to Make Everyday Decisions

Traveling at the speed limit, you approach each intersection anticipating that the red light will turn green as you reach the intersection.

Figure 1.5 Inductive and deductive reasoning.
An inference is a conclusion drawn from specific observations. A deduction is a conclusion drawn from general principles. In this hypothetical example, a driver who is not aware of the general control and programming of traffic signals can use inductive reasoning to determine that the traffic signals are timed as the driver encounters similar timing of signals at several intersections. In contrast, a driver who assumes that the traffic signals are timed can use deductive reasoning to expect that the traffic lights will change predictably at intersections.

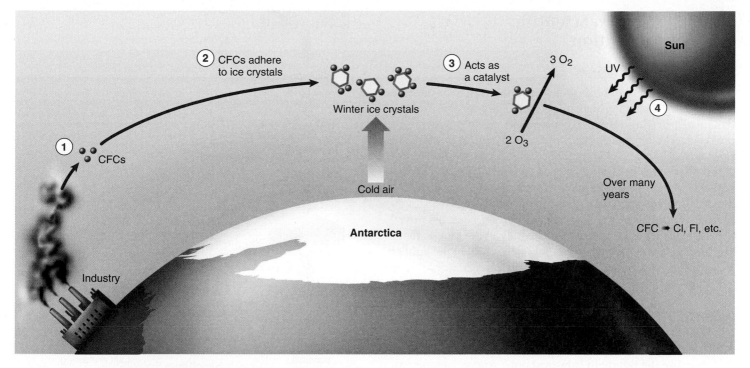

Figure 1.6   How CFCs attack and destroy ozone.
CFCs are stable chemicals that accumulate in the atmosphere as a by-product of industrial society (*1*). In the intense cold of the Antarctic, these CFCs adhere to tiny ice crystals in the upper atmosphere (*2*), where they catalytically destroy ozone (*3*). As a result, more harmful UV radiation reaches the earth's surface (*4*).

1.6   Science in Action: A Case Study

In 1985 Joseph Farman, a British earth scientist working in Antarctica, made an unexpected discovery. Scanning the Antarctic sky, he found far less ozone (O_3, a form of oxygen gas) than should be there—a 30% drop from a reading recorded five years earlier in the Antarctic!

At first it was argued that this thinning of the ozone (soon dubbed the "ozone hole") was an as-yet-unexplained weather phenomenon. Evidence soon mounted, however, implicating synthetic chemicals as the culprit. Detailed analysis of chemicals in the Antarctic atmosphere revealed a surprisingly high concentration of chlorine, a chemical known to destroy ozone. The source of the chlorine was a class of chemicals called **chlorofluorocarbons (CFCs).** CFCs have been manufactured in large amounts since they were invented in the 1920s, largely for use as coolants in air conditioners, propellants in aerosols, and foaming agents in making Styrofoam. CFCs were widely regarded as harmless because they are chemically unreactive under normal conditions. But in the atmosphere over Antarctica, CFCs condense onto tiny ice crystals; warmed by the sun in the spring, they attack and destroy ozone without being used up (figure 1.6).

The thinning of the ozone layer in the upper atmosphere 25 to 40 kilometers above the surface of the earth is a serious matter. The ozone layer protects life from the harmful ultraviolet (UV) rays from the sun that bombard the earth continuously. Like invisible sunglasses, the ozone layer filters out these dangerous rays. When UV rays damage the DNA in skin cells, it can lead to skin cancer. It is estimated that every 1% drop in the atmospheric ozone concentration leads to a 6% increase in skin cancers.

The world currently produces less than 200,000 tons of CFCs annually, down from 1986 levels of 1.1 million tons. As scientific observations have become widely known, governments have rushed to correct the situation. By 1990, worldwide agreements to phase out production of CFCs by the end of the century had been signed. Production of CFCs declined by 86% in the following ten years.

Nonetheless, most of the CFCs manufactured since they were invented are still in use in air conditioners and aerosols and have not yet reached the atmosphere. As these CFCs move slowly upward through the atmosphere, the problem can be expected to continue. Ozone depletion is still producing major ozone holes over the Antarctic.

But the worldwide reduction in CFC production is having a major impact. The period of maximum ozone depletion will peak in the next few years, and researchers' models predict that after that the situation should gradually improve, and that the ozone later will recover by the middle of the 21st century. Clearly, global environmental problems can be solved by concerted action.

> **1.6   Industrially produced CFCs catalytically destroy ozone in the upper atmosphere.**

1.7 Stages of a Scientific Investigation

How Science Is Done

How do scientists establish which general principles are true from among the many that might be? They do this by systematically testing alternative proposals. If these proposals prove inconsistent with experimental observations, they are rejected as untrue. After making careful observations concerning a particular area of science, scientists construct a hypothesis, which is a suggested explanation that accounts for those observations. A hypothesis is a proposition that might be true. Those hypotheses that have not yet been disproved are retained. They are useful because they fit the known facts, but they are always subject to future rejection if, in the light of new information, they are found to be incorrect.

We call the test of a hypothesis an experiment. Suppose that a room appears dark to you. To understand why it appears dark, you propose several hypotheses. The first might be, "The room appears dark because the light switch is turned off." An alternative hypothesis might be, "The room appears dark because the lightbulb is burned out." And yet another alternative hypothesis might be, "I am going blind." To evaluate these hypotheses, you would conduct an experiment designed to eliminate one or more of the hypotheses. For example, you might reverse the position of the light switch. If you do so and the light does not come on, you have disproved the first hypothesis. Something other than the setting of the light switch must be the reason for the darkness. Note that a test such as this does not prove that any of the other hypotheses are true; it merely demonstrates that one of them is not. A successful experiment is one in which one or more of the alternative hypotheses is demonstrated to be inconsistent with the results and is thus rejected.

As you proceed through this text, you will encounter a great deal of information, often accompanied by explanations. These explanations are hypotheses that have withstood the test of experiment. Many will continue to do so; others will be revised as new observations are made. Biology, like all science, is in a constant state of change, with new ideas appearing and replacing old ones.

The Scientific Process

Joseph Farman, who first reported the ozone hole, is a practicing scientist, and what he was doing in Antarctica was science. Science is a particular way of investigating the world, of forming general rules about why things happen by observing particular situations. A scientist like Farman is an observer, someone who looks at the world in order to understand how it works.

Scientific investigations can be said to have six stages: (1) observing what is going on, (2) forming a set of hypotheses, (3) making predictions, (4) testing them, (5) carrying out controls, and (6) forming conclusions after eliminating one or more of the hypotheses (figure 1.7).

1. **Observation.** The key to any successful scientific investigation is careful **observation.** Farman and other

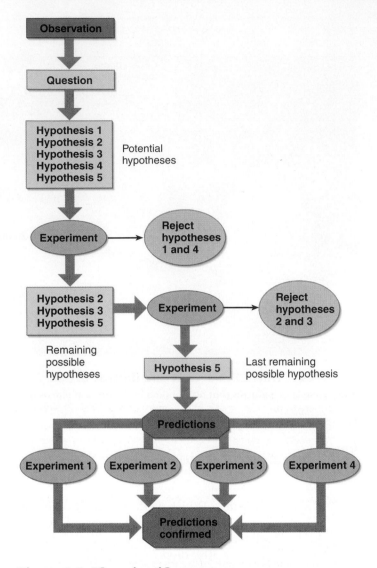

Figure 1.7 The scientific process.
This diagram illustrates the stages of a scientific investigation. First, observations are made. Then a number of potential explanations (hypotheses) are suggested in response to an observation. Experiments are conducted to eliminate any hypotheses. Next, predictions are made based on the remaining hypotheses, and further experiments (including control experiments) are carried out in an attempt to eliminate one or more of the hypotheses. Finally, any hypothesis that is not eliminated is retained. If it is validated by numerous experiments and stands the test of time, a hypothesis may eventually become a theory.

scientists had studied the skies over the Antarctic for many years, noting a thousand details about temperature, light, and levels of chemicals. Had these scientists not kept careful records of what they observed, Farman might not have noticed that ozone levels were dropping.

2. **Hypothesis.** When the unexpected drop in ozone was reported, environmental scientists made a guess about what was destroying the ozone—that perhaps the culprit was CFCs. We call such a guess a hypothesis. A hypothesis is a guess that might be true. What the scientists guessed was that chlorine from CFCs

Figure 1.8   The ozone hole.
The swirling colors represent different concentrations of ozone over the South Pole as viewed from a satellite on September 15, 2001. As you can easily see, there is an "ozone hole" (the *purple* areas) over Antarctica covering an area about the size of the United States.

released into the atmosphere was reacting chemically with ozone over the Antarctic, converting ozone (O_3) into oxygen gas (O_2) and in the process removing the ozone shield from our earth's atmosphere. Often, scientists will form **alternative hypotheses** if they have more than one guess about what they observe. In this case, there were several other hypotheses advanced to explain the ozone hole (figure 1.8). One suggestion explained it as the result of convection. A hypothesis was proposed that the seeming depletion of ozone was in fact a normal consequence of the spinning of the earth; the ozone spun away from the polar regions much as water spins away from the center as a clothes washer moves through its spin cycle. Another hypothesis was that the ozone hole was a transient phenomenon, due perhaps to sunspots, and would soon disappear.

3. **Predictions.** If the CFC hypothesis is correct, then several consequences can reasonably be expected. We call these expected consequences **predictions.** A prediction is what you expect to happen if a hypothesis is true. The CFC hypothesis predicts that if CFCs are responsible for producing the ozone hole, then it should be possible to detect CFCs in the upper Antarctic atmosphere as well as the chlorine released from CFCs that attack the ozone.

4. **Testing.** Scientists set out to test the CFC hypothesis by attempting to verify some of its predictions. We

call the test of a hypothesis an **experiment.** To test the hypothesis, atmospheric samples were collected from the stratosphere over 6 miles up by a high-altitude balloon. Analysis of the samples revealed CFCs, as predicted. Were the CFCs interacting with the ozone? The samples contained free chlorine and fluorine, confirming the breakdown of CFC molecules. The results of the experiment thus support the hypothesis.

5. **Controls.** Events in the upper atmosphere can be influenced by many factors. We call each factor that might influence a process a **variable.** To evaluate alternative hypotheses about one variable, all the other variables must be kept constant so that we do not get misled or confused by these other influences. This is done by carrying out two experiments in parallel: in the first experimental test, we alter one variable in a known way to test a particular hypothesis; in the second, called a **control experiment,** we do *not* alter that variable. In all other respects, the two experiments are the same. To further test the CFC hypothesis, scientists carried out control experiments in which the key variable was the amount of CFCs in the atmosphere. Working in laboratories, scientists reconstructed the atmospheric conditions, solar bombardment, and extreme temperatures found in the sky far above the Antarctic. If the ozone levels fell without addition of CFCs to the chamber, then CFCs could not be what was attacking the ozone, and the CFC hypothesis must be wrong. Carefully monitoring the chamber, however, scientists detected no drop in ozone levels in the absence of CFCs. The result of the control was thus consistent with the predictions of the hypothesis.

6. **Conclusion.** A hypothesis that has been tested and not rejected is tentatively accepted. The hypothesis that CFCs released into the atmosphere are destroying the earth's protective ozone shield is now supported by a great deal of experimental evidence and is widely accepted. While other factors have also been implicated in ozone depletion, destruction by CFCs is clearly the dominant phenomenon. A collection of related hypotheses that have been tested many times and not rejected is called a **theory.** A theory indicates a higher degree of certainty; however, in science, nothing is "certain." The theory of the ozone shield—that ozone in the upper atmosphere shields the earth's surface from harmful UV rays by absorbing them—is supported by a wealth of observation and experimentation and is widely accepted. The explanation for the destruction of this shield is still at the hypothesis stage.

1.7   Science progresses by systematically eliminating potential hypotheses that are not consistent with observation.

Author's Corner

Where Are All My Socks Going?

All my life, for as far back as I can remember, I have been losing socks. Not pairs of socks, mind you, but single socks. I first became aware of this peculiar phenomenon when as a young man I went away to college. When Thanksgiving rolled around that first year, I brought an enormous duffle bag of laundry home. My mother, instead of braining me, dumped the lot into the washer and dryer, and so discovered what I had not noticed—that few of my socks matched anymore.

That was forty years ago, but it might as well have been yesterday. All my life, I have continued to lose socks. This last Christmas I threw out a sock drawer full of socks that didn't match, and took advantage of sales to buy a dozen pairs of brand-new ones. Last week, when I did a body count, three of the new pairs had lost a sock!

Enough. I have set out to solve the mystery of the missing socks. How? The way Sherlock Holmes would have, scientifically. Holmes worked by eliminating those possibilities that he found not to be true. A scientist calls possibilities "hypotheses" and, like Sherlock, rejects those that do not fit the facts. Sherlock tells us that when only one possibility remains unrejected, then—however unlikely—it must be true.

Hypothesis 1: It's the socks. I have four pairs of socks bought as Christmas gifts but forgotten until recently. Deep in my sock drawer, they have remained undisturbed for five months. If socks disappear because of some intrinsic property (say the manufacturer has somehow designed them to disappear to generate new sales), then I could expect at least one of these undisturbed ones to have left the scene by now. However, when I looked, all four pairs were complete. Undisturbed socks don't disappear. Thus I reject the hypothesis that the problem is caused by the socks themselves.

Hypothesis 2: The "little people" take them. In folklore, elves and leprechauns are said to inhabit Irish houses and borrow items like socks. I am Irish. Perhaps my house is infested with wee "little people." So I called my roommate of forty years ago, now living in California. A Frenchman, there is not an Irish drop of blood in his body, and no self-respecting leprechaun would be caught dead in his house. Does he too lose socks? Yep. Thus I reject the hypothesis that the loss of socks is caused by Irish "little people."

Hypothesis 3: Transformation, a fanciful suggestion by science fiction writer Avram Davidson in his 1958 story "Or All the Seas with Oysters" that I cannot get out of the quirky corner of my mind. I discard the socks I have worn each evening in a laundry basket in my closet. Over many years, I have noticed a tendency for socks I have placed in the closet to disappear. Over that same long period, as my socks are disappearing, there is something in my closet that seems to multiply—COAT HANGERS! Socks are larval coat hangers! To test this outlandish hypothesis, I had only to move the laundry basket out of the clos-

et. Several months later, I was still losing socks, so this hypothesis is rejected.

Hypothesis 4: Static cling. The missing single socks may have been hiding within the sleeves of sweat shirts or jackets, inside trouser legs, or curled up within seldom-worn garments. Rubbing around in the dryer, socks can garner quite a bit of static electricity, easily enough to cause them to cling to other garments. Socks adhering to the outside of a shirt or pant leg are soon dislodged, but ones that find themselves within a sleeve, leg, or fold may simply stay there, not "lost" so much as misplaced. However, after a diligent search, I did not run across any previously lost socks hiding in the sleeves of my winter garments or other seldom-worn items, so I reject this hypothesis.

Hypothesis 5: I lose my socks going to or from the laundry. Perhaps in handling the socks from laundry basket to the washer/dryer and back to my sock drawer, a sock is occasionally lost. To test this hypothesis, I have pawed through the laundry coming into the washer. No single socks. Perhaps the socks are lost after doing the laundry, during folding or transport from laundry to sock drawer. If so, there should be no single socks coming out of the dryer. But there are! The singletons are first detected among the dry laundry, before folding. Thus I eliminate the hypothesis that the problem arises from mishandling the laundry.

Hypothesis 6: I lose them during washing. It seems the problem is in the laundry room. Perhaps the washing machine is somehow "eating" my socks. I looked in the washing machine to see if a sock could get trapped inside, or chewed up by the machine, but I can see no possibility. The clothes slosh around in a closed metal container with water passing in and out through little holes no wider than a pencil. No sock could slip through such a hole. There is a thin gap between the rotating cylinder and the top of the washer through which an errant sock might escape, but my socks are too bulky for this route. So I eliminate the hypothesis that the washing machine is the culprit.

Hypothesis 7: I lose them during drying. Perhaps somewhere in the drying process socks are being lost. I stuck my head in our clothes dryer to see if I could see any socks, and I couldn't. However, as I look, I can see a place a sock could go—behind the drying wheel! A clothes dryer is basically a great big turning cylinder with dry air blowing through the middle. The edges of the turning cylinder don't push hard against the side of the machine. Just maybe, every once in a while, a sock might get pulled through, sucked into the back of the machine.

To test this hypothesis, I should take the back of the dryer off and look inside to see if it is stuffed with my missing socks. My wife, knowing my mechanical abilities, is not in favor of this test. Thus, until our dryer dies and I can take it apart, I shall not be able to reject hypothesis 7. Lacking any other likely hypothesis, I take Sherlock Holmes' advice and tentatively conclude that the dryer is the culprit.

A theory is a unifying explanation for a broad range of observations. Thus we speak of the theory of gravity, the theory of evolution, and the theory of the atom. Theories are the solid ground of science, that of which we are the most certain. There is no absolute truth in science, however, only varying degrees of uncertainty. The possibility always remains that future evidence will cause a theory to be revised. A scientist's acceptance of a theory is always provisional. For example, in another scientist's experiment, evidence that is inconsistent with a theory may be revealed. As information is shared throughout the scientific community, previous hypotheses and theories may be modified, and scientists may formulate new ideas.

Very active areas of science are often alive with controversy, as scientists grope with new and challenging ideas. This uncertainty is not a sign of poor science but rather of the push and pull that is the heart of the scientific process. The hypothesis that the world's climate is growing warmer due to humanity's excessive production of carbon dioxide (CO_2), for example, has been quite controversial, although the weight of evidence has increasingly supported the hypothesis.

The word theory is thus used very differently by scientists than by the general public. To a scientist, a theory represents that of which he or she is most certain; to the general public, the word theory implies a *lack* of knowledge or a guess. How often have you heard someone say, "It's only a theory!"? As you can imagine, confusion often results. In this text the word theory will always be used in its scientific sense, in reference to a generally accepted scientific principle.

The Scientific "Method"

It was once fashionable to claim that scientific progress is the result of applying a series of steps called the **scientific method;** that is, a series of logical "either/or" predictions tested by experiments to reject one alternative. The assumption was that trial-and-error testing would inevitably lead one through the maze of uncertainty that always slows scientific progress. If this were indeed true, a computer would make a good scientist—but science is not done this way! If you ask successful scientists like Farman how they do their work, you will discover that without exception they design their experiments with a pretty fair idea of how they will come out. Environmental scientists understood the chemistry of chlorine and ozone when they formulated the CFC hypothesis, and they could imagine how the chlorine in CFCs would attack ozone molecules. A hypothesis that a successful scientist tests is not just any hypothesis. Rather, it is a "hunch" or educated guess in which the scientist integrates all that he or she knows. The scientist also allows his or her imagination full play, in an attempt to get a sense of what *might* be true. It is because insight

Figure 1.9   Nobel Prize winner.
Sherwood Roland, along with Mario Molena, won the 1998 Nobel Prize for discovering how CFCs act to catalytically break down atmospheric ozone in the stratosphere, the chemistry responsible for the "ozone hole" over the Antarctic.

and imagination play such a large role in scientific progress that some scientists are so much better at science than others (figure 1.9)—just as Beethoven and Mozart stand out among composers.

The Limitations of Science

Scientific study is limited to organisms and processes that we are able to observe and measure. Supernatural and religious phenomena are beyond the realm of scientific analysis because they cannot be scientifically studied, analyzed, or explained. Supernatural explanations can be used to explain any result, and cannot be disproven by experiment or observation. Scientists in their work are limited to objective interpretations of observable phenomena.

It is also important to recognize that there are practical limits to what science can accomplish. While scientific study has revolutionized our world, it cannot be relied upon to solve all problems. For example, we cannot pollute the environment and squander its resources today, in the blind hope that somehow science will make it all right sometime in the future. Nor can science restore an extinct species. Science identifies solutions to problems when solutions exist, but it cannot invent solutions when they don't.

1.8   A scientist does not follow a fixed method to form hypotheses but relies also on judgment and intuition.

1.9 Four Theories Unify Biology as a Science

The Cell Theory: Organization of Life

As was stated at the beginning of this chapter, all organisms are composed of cells, life's basic units (figure 1.10). Cells were discovered by Robert Hooke in England in 1665. Hooke was using one of the first microscopes, one that magnified 30 times. Looking through a thin slice of cork, he observed many tiny chambers, which reminded him of monks' cells in a monastery. Not long after that, the Dutch scientist Anton van Leeuwenhoek used microscopes capable of magnifying 300 times, and discovered an amazing world of single-celled life in a drop of pond water. However, it took almost two centuries before biologists fully understood the significance of cells. In 1839, the German biologists Matthias Schleiden and Theodor Schwann, summarizing a large number of observations by themselves and others, concluded that all living organisms consist of cells. Their conclusion forms the basis of what has come to be known as the **cell theory.** Later, biologists added the idea that all cells come from other cells. The cell theory, one of the basic ideas in biology, is the foundation for understanding the reproduction and growth of all organisms. The nature of cells and how they function is discussed in detail in chapter 4.

The Gene Theory: Molecular Basis of Inheritance

Even the simplest cell is incredibly complex, more intricate than a computer. The information that specifies what a cell is like—its detailed plan—is encoded in a long cablelike mol-

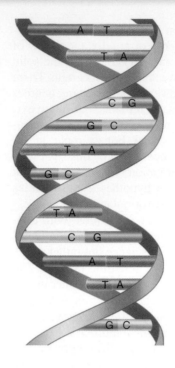

Figure 1.11 Genes are made of DNA.
Winding around each other like the rails of a spiral staircase, the two strands of a DNA molecule make a double helix. Because of its size and shape, the nucleotide represented by the letter A can only pair with the nucleotide represented by the letter T, and likewise G can only pair with C.

ecule called **DNA** (deoxyribonucleic acid). Researchers Jim Watson and Francis Crick discovered in 1953 that each DNA molecule is formed from two long chains of building blocks, called nucleotides, wound around each other (figure 1.11). The two chains face each other, like two lines of people holding hands. The chains contain information in the same way this sentence does—as a sequence of letters. There are four different nucleotides in DNA, and the sequence in which they occur encodes the information. Specific sequences of several hundred to many thousand nucleotides make up a *gene,* a discrete unit of information. A gene might encode a particular protein, or a different kind of unique molecule called RNA, or a gene might act to regulate other genes. The **gene theory** states that the proteins and RNA molecules encoded by an organism's genes determine what it will be like (figure 1.12). The entire set of DNA instructions that specifies a cell is called its **genome.** The sequence of the human genome, 3 *billion* nucleotides long, was decoded in 2001, a triumph of scientific investigation. How genes function is the subject of chapter 9. In chapters 10 and 11 we explore how detailed knowledge of genes is revolutionizing biology, and having an impact on the lives of all of us.

The Theory of Heredity: Unity of Life

The storage of hereditary information in genes composed of DNA is common to all living things. The **theory of heredity** first advanced by Gregor Mendel in 1865 states that the genes of an organism are inherited as discrete units. A triumph of experimental science developed long before genes and DNA were understood, Mendel's theory of heredity is the subject of chapter 8. Soon after Mendel's theory gave rise

Figure 1.10 Life in a drop of pond water.
All organisms are composed of cells. Some organisms, including these protists, are single-celled, while others, such as plants, worms, and mushrooms, consist of many cells.

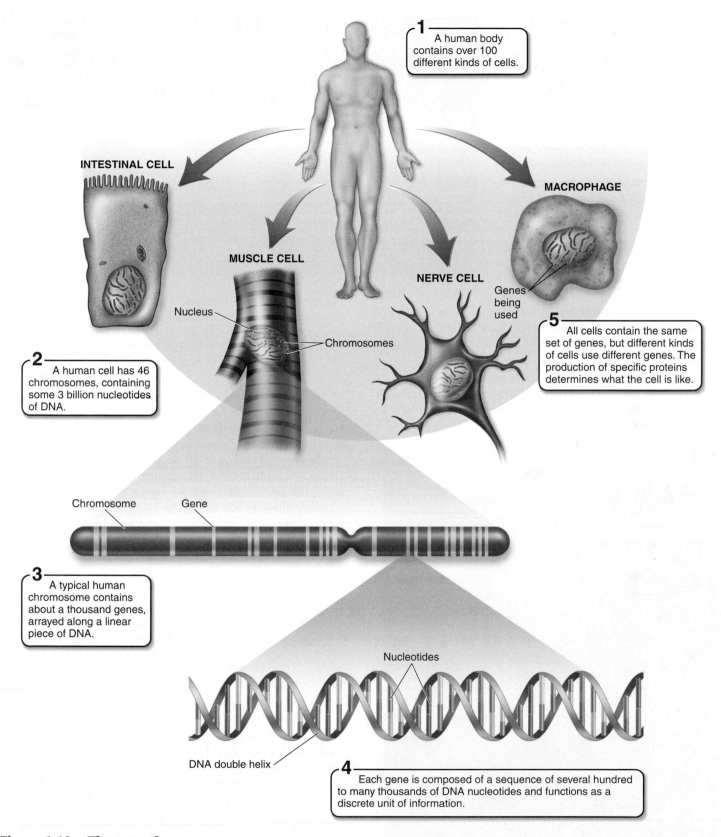

1 A human body contains over 100 different kinds of cells.

INTESTINAL CELL

MACROPHAGE

MUSCLE CELL

NERVE CELL

Genes being used

Nucleus

Chromosomes

2 A human cell has 46 chromosomes, containing some 3 billion nucleotides of DNA.

5 All cells contain the same set of genes, but different kinds of cells use different genes. The production of specific proteins determines what the cell is like.

Chromosome   Gene

3 A typical human chromosome contains about a thousand genes, arrayed along a linear piece of DNA.

Nucleotides

DNA double helix

4 Each gene is composed of a sequence of several hundred to many thousands of DNA nucleotides and functions as a discrete unit of information.

Figure 1.12   The gene theory.

The gene theory states that what an organism is like is determined in large measure by its genes. Here you see how the many kinds of cells in the body of each of us are each determined by which genes are used in making each particular kind of cell.

to the field of genetics, other biologists proposed what has come to be called the **chromosomal theory of inheritance,** which in its simplest form states that the genes of Mendel's theory are physically located on chromosomes, and that it is because chromosomes are parceled out in a regular manner during reproduction that Mendel's regular patterns of inheritance are seen. In modern terms, the two theories state that genes are a component of a cell's chromosomes (figure 1.13), and that the regular duplication of these chromosomes in meiosis is responsible for the pattern of inheritance we call Mendelian segregation. Sometimes a character is conserved essentially unchanged in a long line of descent, reflecting a fundamental role in the biology of the organism, one not easily changed once adopted.

The Theory of Evolution: Diversity of Life

The unity of life, which we see in the retention of certain key characteristics among many related life-forms, contrasts with the incredible diversity of living things that have evolved to

Figure 1.14   Prosimians.
Humans and apes are primates. Among the earliest primates to evolve were prosimians, small nocturnal (night-active) insect eaters. These lemurs, prosimians native to Madagascar, show the features characteristic of all primates: grasping fingers and toes and binocular vision.

fill the varied environments of earth. The **theory of evolution,** advanced by Charles Darwin in 1859, attributes the diversity of the living world to natural selection. Those organisms best able to respond to the challenges of living will leave more offspring, he argued, and thus become more common. It is because the world offers diverse opportunities that it contains so many different life-forms. An essential component of Darwin's theory is, in his own words, that evolution is a process of "descent with modification," that all living organisms are related to one another in a common tree of descent, a family tree of life. For example, the first primate was a small arboreal mammal that lived some 65 million years ago. About 40 million years ago, the primates split into two groups: one group, prosimians (figure 1.14), has changed little since then, while the other, monkeys, apes, and our family line, has continued to evolve. Today scientists can decipher each of all the thousands of genes (the genome) of an organism. By comparing genomes of different organisms, researchers can literally reconstruct the tree of life (figure 1.15).

Biologists divide the living world's astonishing diversity of organisms into three great groups, called *domains:* Bacteria, Archaea, and Eukarya. Bacteria and Archaea each consist of one kingdom of prokaryotes (single-celled organisms with little internal structure). Within the Eukarya, or eukaryotes (organisms composed of a complexly organized cell or multiple complex cells), are four more kingdoms (see figure 1.1 and figure 1.16).

The simplest and most ancient of the eukaryotes is the Kingdom Protista, mostly composed of tiny unicellular organisms. The protists are the most diverse of the four eukaryotic kingdoms, and gave rise to the other three. Kingdom Plantae, the land plants, arose from a kind of photosynthetic protist called a green algae. Kingdom Fungi, the mushrooms and

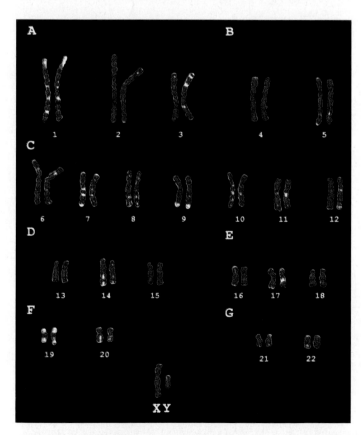

Figure 1.13   Human chromosomes.
The chromosomal theory of inheritance states that genes are located on chromosomes. This human karyotype (an ordering of chromosomes) shows banding patterns on chromosomes that represent clusters of genes.

Figure 1.15   The tree of life.

Biologists who compare genomes have reached a broad consensus about the major branches of the tree of life. The tree illustrated here was proposed in 2003. It shows that crocodiles seem to be more closely related to birds than to other reptiles. Fungi are thought to share a common ancestor with animals because both have a single whip-like flagellum on their reproductive cells. Thus, although they first appeared before plants, fungi seem to be more closely related to animals than plants.

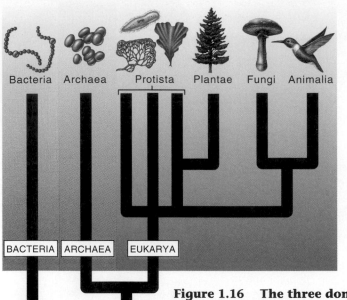

Bacteria Archaea Protista Plantae Fungi Animalia

BACTERIA ARCHAEA EUKARYA

yeasts, arose from a protist that has not yet been clearly identified. Kingdom Animalia, of which we are a member, is thought to have arisen from a kind of protist called a choanoflagellate. The simplest animals, sponges, have special flagellated cells called choanocytes that line the body interior. By waving the choanocyte flagella, the sponge draws water into its interior through pores, filtering out food particles from the water as it passes. Each choanocyte cell of a sponge closely resembles a choanoflagellate protist, its presumed ancestor. The kingdoms of life are discussed in detail in chapter 12.

> **1.9**   The theories uniting biology state that cellular organisms store hereditary information in DNA. Sometimes DNA alterations occur, which when preserved result in evolutionary change. Today's biological diversity is the product of a long evolutionary journey.

Figure 1.16   The three domains of life.
Biologists categorize all living things into three overarching groups called domains: Bacteria, Archaea, and Eukarya. Domain Bacteria contains the kingdom Bacteria, and domain Archaea contains the kingdom Archaea. Domain Eukarya is composed of four more kingdoms: Protista, Plantae, Fungi, and Animalia.

Exploring Current Issues

Additional Resources

Go to your campus library to find the following articles, which further develop some of the concepts found in this chapter.

Guttman, B. S. (2004). The real method of scientific discovery: scientists don't sit around in their labs trying to establish generalizations. Instead they engage in mystery-solving essentially like that of detective work, and it often involves a creative, imaginative leap. *Skeptical Inquirer,* 28(1), 45–48.

Karoly, D. J. (2003). Ozone and climate change. *Science,* 302(5643), 236–238.

Kerr, R. A. (2003). Life's diversity may truly have leaped since the dinosaurs. *Science,* 300(5622), 1067–1069.

Shermer, M. (2001). Baloney detection. (Distinguishing between pseudoscience and science.) *Scientific American,* 285(5), 36.

Simpson, S. (2002). A push from above: the ozone hole may be stirring up Antarctica's climate. *Scientific American,* 287(2), 18–20.

Biology and Society Lecture: Introduction to AIDS

In this lecture Professor Johnson explores the AIDS epidemic. The year 2001 marked the twentieth anniversary of this epidemic, which has claimed the lives of over 24 million people worldwide. AIDS is caused by a virus called HIV, which attacks and destroys the immune system. It is transmitted by sex, needles, and anything else that transfers white blood cells. AIDS is fatal—there is no cure.

Biology and Society Lecture: The Challenge of Curing AIDS

In this lecture Professor Johnson overviews the challenges of developing an AIDS vaccine that targets the HIV virus and the often discouraging results. He also discusses new, more promising approaches of various treatments that involve training the body's own immune defenses to combat the HIV virus.

Find these lectures, delivered by the author to his class at Washington University, online at www.mhhe.com/tlwessentials/exp1.

Summary

Biology and the Living World

1.1 The Diversity of Life

- Biology is the study of life. Although all living organisms share common characteristics, they are also diverse and are therefore categorized into six groups called kingdoms.

- The six kingdoms are Archaea, Bacteria, Protista, Fungi, Plantae, and Animalia **(figure 1.1).**

1.2 Properties of Life

- All living organisms share five basic properties: cellular organization, all living organisms are composed of cells; metabolism, all living organisms use energy; homeostasis, all living organisms maintain stable internal conditions; growth and reproduction, all living organisms grow and reproduce; heredity, all

living organisms possess genetic information that determines how each organism looks and functions, and this information is passed on to future generations.

1.3 The Organization of Life

- Living organisms exhibit increasing levels of complexity within their cells, within their bodies, and within ecosystems (**figure 1.4**).

- Novel properties that appear in each level of the hierarchy of living organisms are called emergent properties. These properties are the natural consequences of ever more complex structural organization.

1.4 Biological Themes

- Five themes emerge from the study of biology: evolution, the flow of energy, cooperation, structure determines function, and homeostasis. These themes are used to examine the similarities and differences among organisms (**table 1.1**).

The Scientific Process

1.5 How Scientists Think

- Scientists use reasoning when examining the world. Deductive reasoning is the process of using general principles to explain individual observations. Inductive reasoning is the process of using specific observations to formulate general principles (**figure 1.5**).

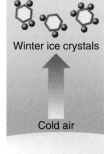

Winter ice crystals

Cold air

Antarctica

1.6 Science in Action: A Case Study

- The scientific investigation of the "ozone hole" revealed that industrially produced CFCs were responsible for the thinning of the ozone layer in the earth's atmosphere (**figure 1.6**).

1.7 Stages of a Scientific Investigation

- Scientific investigations involve using observations to formulate hypotheses. Hypotheses are possible explanations of these observations that can be used in forming predictions that can be tested experimentally. Some hypotheses are rejected based on experimentation, while others are tentatively accepted.

- Scientific investigations use a series of six stages, called the scientific process, to study a scientific question. These stages are

observations, forming hypotheses, making predictions, testing, establishing controls, and drawing conclusions (**figure 1.7**).

1.8 Theory and Certainty

- Hypotheses that hold up to testing over time are sometimes combined into statements called theories. Theories carry a higher degree of certainty, although no theory in science is absolute.

- The process of science was once viewed as a series of "either/or" predictions that were tested experimentally. This process, referred to as the scientific "method," did not take into account the importance of insight and imagination that are necessary to good scientific investigations.

Core Ideas of Biology

1.9 Four Theories Unify Biology as a Science

- There are four unifying themes in the study of biology: cell theory, gene theory, the theory of heredity, and the theory of evolution.

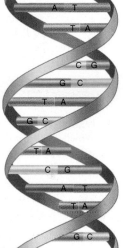

- The cell theory states that all living organisms are composed of cells, which grow and reproduce to form other cells (**figure 1.10**).

- The gene theory states that long molecules in the cell, called DNA, encode instructions for producing cellular components. These instructions, organized into discrete units called genes, determine how an organism looks and functions (**figure 1.12**).

- The theory of heredity states that the genes of an organism are passed as discrete units from parent to offspring through a process of heredity (**figure 1.13**).

- The theory of evolution states that modifications in genes that are passed from parent to offspring result in changes in future generations. These changes lead to greater diversity among organisms over time, ultimately leading to the formation of new groups of organisms (**figures 1.15 and 1.16**).

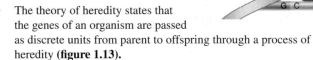

Self-Test

1. Biologists categorize all living things based on related characteristics into large groups, called
   a. kingdoms.　c. courses.
   b. planets.　d. territories.
2. Living things can be distinguished from nonliving things because they have
   a. complexity.　c. cellular organization.
   b. movement.　d. response to a stimulus.
3. Living things are organized. Choose the answer that illustrates this complexity, and is arranged from smallest to largest.
   a. cell, atom, molecule, tissue, organelle, organ, organ system, organism, population, species, community, ecosystem
   b. atom, molecule, organelle, cell, tissue, organ, organ system, organism, population, species, community, ecosystem
   c. atom, molecule, organelle, cell, tissue, organ, organ system, organism, community, population, species, ecosystem

   d. atom, molecule, cell wall, cell, organ, organelle, organism, species, population, community, ecosystem
4. At each higher level in the hierarchy of living things, properties occur that were not present at the simpler levels. These properties are referred to as
   a. novelistic properties.　c. incremental properties.
   b. complex properties.　d. emergent properties.
5. The five general biological themes include
   a. evolution, energy flow, competition, structure determines function, and homeostasis.
   b. evolution, energy flow, cooperation, structure determines function, and homeostasis.
   c. evolution, growth, competition, structure determines function, and homeostasis.
   d. evolution, growth, cooperation, structure determines function, and homeostasis.

6. When you are trying to understand something new, you begin by observation, and then put the observations together in a logical fashion to form a general principle. This method is called
   a. inductive reasoning.   c. theory production.
   b. rule enhancement.   d. deductive reasoning.

7. When trying to figure out explanations for observations, you usually construct a series of possible hypotheses. Then you make predictions of what will happen if each hypothesis is true, and
   a. test each hypothesis, using appropriate controls, to determine the truth.
   b. test each hypothesis, using appropriate controls, to rule out as many as possible.
   c. use logic to determine which hypothesis is most likely true.
   d. use logic to determine which hypotheses are most likely false.

8. Which of the following statements is correct regarding a hypothesis?
   a. After sufficient testing, you can conclude that it is true.
   b. After sufficient testing, you can conclude that it is probably true.
   c. After sufficient testing, you can accept it as probable, being aware that it may be revised or rejected in the future.
   d. You never have any degree of certainty that it is true; there are too many variables.

9. Cell theory states that
   a. all organisms have cell walls and all cell walls come from other cells.
   b. all cellular organisms undergo sexual reproduction.
   c. all living organisms use cells for energy, either their own or they ingest cells of other organisms.
   d. all living organisms consist of cells, and all cells come from other cells.

10. The gene theory states that all the information that specifies what a cell is and what it does
    a. is passed down, unchanged, from mother to offspring.
    b. is passed down, unchanged, from parents to offspring.
    c. is contained in a long molecule called DNA.
    d. is contained in the body's nucleus.

Visual Understanding

1. **Figure 1.4**   Refer to figure 1.4, a copy of which is shown here. Which level(s) will be affected by cancer? Why? Which levels will be affected by air and water pollution? Why?

2. **Figure 1.5**   You notice that on cloudy days people often carry umbrellas, folded or in a case. You also note that when umbrellas are open there are many car accidents. You conclude that open umbrellas cause car accidents. Referring back to figure 1.5, explain the type of reasoning used to reach this conclusion, and why it can sometimes be a problem.

Challenge Questions

Biology and the Living World   You are the biologist in a group of scientists who have traveled to a distant star system and landed on a planet. You see an astounding array of shapes and forms. You have three days to take samples of living things before returning to earth. How do you decide what is alive?

The Scientific Process   St. John's wort is an herb that has been used for hundreds of years as a remedy for mild depression. How might a modern-day scientist research its effectiveness?

Core Ideas of Biology   Explain how the four unifying themes of biology encompass all of the levels of complexity, from cellular through organismal and population levels.

Online Learning Center

Visit the Online Learning Center for this chapter at www.mhhe.com/tlwessentials/ch1 for quizzes, animations, interactive learning exercises, and other study tools. At the site you will also find extended answers to the end-of-chapter questions.

2
Evolution and Ecology

These four finches live on the Galápagos Islands, a cluster of volcanic islands far out to sea off the coast of South America. All descendants of a single ancestral migrant, blown to the islands from the mainland long ago, the Galápagos finches gave Darwin valuable clues about how natural selection shapes the evolution of species. The two upper finches are ground finches, their different bills adapting them to eat different-sized seeds. On the lower left is a woodpecker finch, a kind of tree finch that carries around a cactus spine, which it uses to probe for insects in deep crevices. On the lower right is a warbler finch that like its namesake eats crawling insects. Each of these species utilizes food resources differently. Evolution and ecology tell us different but related things about the diversity of the living world: ecological interactions generate the selective pressures that shape the evolution of groups like Darwin's finches.

The great diversity of life on earth—ranging from bacteria to elephants and roses—is the result of a long process of **evolution,** the change that occurs in organisms' characteristics through time. In 1859, the English naturalist Charles Darwin (1809–82; figure 2.1) first suggested an explanation for why evolution occurs, a process he called **natural selection.** Biologists soon became convinced Darwin was right and now consider evolution one of the central concepts of the science of biology. A second key concept, and one that closely relates to evolution, is that of **ecology,** how organisms live in their environment. It has been said that evolution is the consequence of ecology over time. Ecology is of increasing concern to all of us, as a growing human population places ever-greater stress on our planet. In this chapter, we introduce these two key related concepts, evolution and ecology, to provide a foundation as you begin to explore the living world. Both are revisited in more detail later.

2.1 Darwin's Voyage on HMS *Beagle*

The theory of evolution proposes that a population can change over time, sometimes forming a new species (a group of organisms that possess similar characteristics that allow them to successfully interbreed). This famous theory provides a good example of how a scientist develops a hypothesis—in this case, a hypothesis of how evolution occurs and how, after much testing, it is eventually accepted as a theory.

Charles Robert Darwin was an English naturalist who, after 30 years of study and observation, wrote one of the most famous and influential books of all time. This book, *On the Origin of Species by Means of Natural Selection, or The Preservation of Favoured Races in the Struggle for Life,* created a sensation when it was published, and the ideas Darwin expressed in it have played a central role in the development of human thought ever since.

In Darwin's time, most people believed that the various kinds of organisms and their individual structures resulted from direct actions of the Creator. Species were thought to be specially created and unchangeable over the course of time. In contrast to these views, a number of earlier philosophers had presented the view that living things must have changed during the history of life on earth. Darwin proposed a concept he called natural selection as a coherent, logical explanation for this process. Darwin's book, as its title indicates, presented a conclusion that differed sharply from conventional wisdom. Although his theory did not challenge the existence of a Divine Creator, Darwin argued that this Creator did not simply create things and then leave them forever unchanged. Instead, Darwin's God expressed Himself through the operation of natural laws that produced change over time—evolution.

Figure 2.1   The theory of evolution by natural selection was proposed by Charles Darwin.
This rediscovered photograph appears to be the last ever taken of the great biologist. It was taken in 1881, the year before Darwin died.

The story of Darwin and his theory begins in 1831, when he was 22 years old. A small British naval vessel, HMS *Beagle* (figure 2.2) was about to set sail on a five-year navigational mapping expedition around the coasts of South America (figure 2.3). The young (26-year-old) captain of HMS *Beagle,* unable by British naval tradition to have social contact with his crew, and anticipating a voyage that would last many years, wanted a gentleman companion, someone to talk to. Indeed, the *Beagle*'s previous skipper had broken down and shot himself to death after three solitary years away from home.

On the recommendation of one of his professors at Cambridge University, Darwin, son of a wealthy doctor and very much a gentleman, was selected to serve as the captain's companion, primarily to share his table at mealtime during every shipboard dinner of the long voyage. Darwin paid his own expenses, and even brought along a manservant.

Darwin took on the role of ship's naturalist (the official naturalist, a man named Robert McKormick, left the ship be-

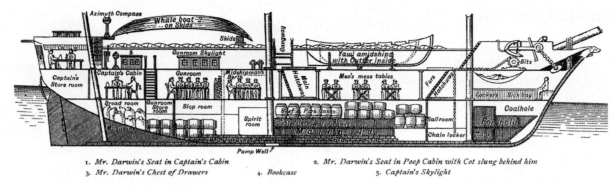

1. *Mr. Darwin's Seat in Captain's Cabin*   2. *Mr. Darwin's Seat in Poop Cabin with Cot slung behind him*
3. *Mr. Darwin's Chest of Drawers*   4. *Bookcase*   5. *Captain's Skylight*

Figure 2.2   Cross section of HMS *Beagle.*

HMS *Beagle,* a 10-gun brig of 242 tons, only 90 feet in length, had a crew of 74 people! After he first saw the ship, Darwin wrote to his college professor Henslow: "The absolute want of room is an evil that nothing can surmount."

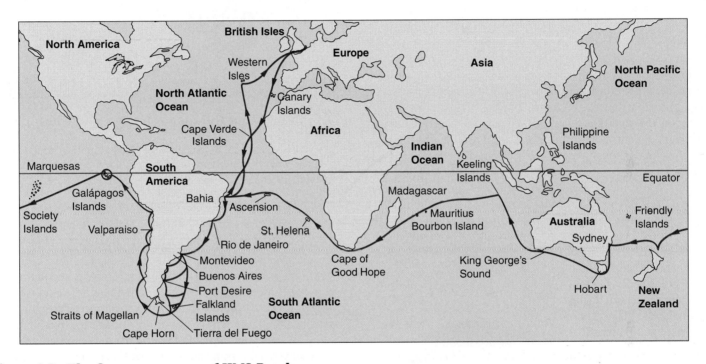

Figure 2.3   The five-year voyage of HMS *Beagle.*

Although the ship sailed around the world, most of the time was spent exploring the coasts and coastal islands of South America, such as the Galápagos Islands. Darwin's studies of the animals of these islands played a key role in the eventual development of his theory of evolution by means of natural selection.

fore the first year was out). During this long voyage, Darwin had the chance to study a wide variety of plants and animals on continents and islands and in distant seas. He was able to explore the biological richness of the tropical forests, examine the extraordinary fossils of huge extinct mammals in Patagonia at the southern tip of South America, and observe the remarkable series of related but distinct forms of life on the **Galápagos Islands.** Such an opportunity clearly played an important role in the development of his thoughts about the nature of life on earth.

When Darwin returned from the voyage at the age of 27, he began a long period of study and contemplation. During the next 10 years, he published important books on several different subjects, including the formation of oceanic islands

from coral reefs and the geology of South America. He also devoted eight years of study to barnacles, a group of small marine animals with shells that inhabit rocks and pilings, eventually writing a four-volume work on their classification and natural history. In 1842, Darwin and his family moved out of London to a country home at Down, in the county of Kent. In these pleasant surroundings, Darwin lived, studied, and wrote for the next 40 years.

2.1   Darwin was the first to propose natural selection as the mechanism of evolution that produced the diversity of life on earth.

2.2 Darwin's Evidence

One of the obstacles that had blocked the acceptance of any theory of evolution in Darwin's day was the incorrect notion, widely believed at that time, that the earth was only a few thousand years old. The discovery of thick layers of rocks, evidences of extensive and prolonged erosion, and the increasing numbers of diverse and unfamiliar fossils discovered during Darwin's time made this assertion seem less and less likely. The great geologist Charles Lyell (1797–1875), whose *Principles of Geology* (1830) Darwin read eagerly as he sailed on HMS *Beagle,* outlined for the first time the story of an ancient world of plants and animals in flux. In this world, species were constantly becoming extinct while others were emerging. It was this world that Darwin sought to explain.

What Darwin Saw

When HMS *Beagle* set sail, Darwin was fully convinced that species were immutable. Indeed, it was not until two or three years after his return that he began to seriously consider the possibility that they could change. Nevertheless, during his five years on the ship, Darwin observed a number of phenomena that were of central importance to him in reaching his ultimate conclusion. For example, in the rich fossil beds of southern South America, he observed fossils of extinct armadillos similar in form to the armadillos that still lived in the same area (figure 2.4). Why would similar living and fossil organisms be in the same area unless the earlier form had given rise to the other? Later, Darwin's observations would be strengthened by the discovery of other examples of fossils that show intermediate characteristics, pointing to successive change.

Repeatedly, Darwin saw that the characteristics of similar species varied somewhat from place to place. These geographical patterns suggested to him

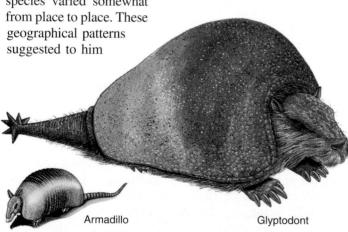

Armadillo   Glyptodont

Figure 2.4   Fossil evidence of evolution.

The now-extinct glyptodont was a large 2,000-kilogram South American armadillo (about the size of a small car), much larger than the modern armadillo, which weighs an average of about 4.5 kilograms and is about the size of a house cat. The similarity of fossils such as the glyptodonts to living organisms found in the same regions suggested to Darwin that evolution had taken place.

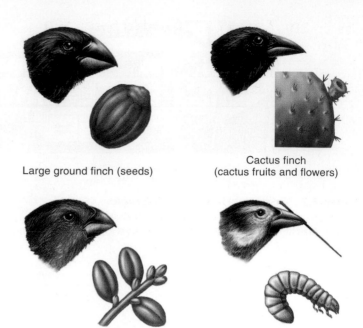

Large ground finch (seeds)

Cactus finch (cactus fruits and flowers)

Vegetarian finch (buds)

Woodpecker finch (insects)

Figure 2.5   Four Galápagos finches and what they eat.

Darwin observed 14 different species of finches on the Galápagos Islands, differing mainly in their beaks and feeding habits. These four finches eat very different food items, and Darwin surmised that the very different shapes of their bills represented evolutionary adaptations improving their ability to do so.

that organismal lineages change gradually as individuals move into new habitats. On the Galápagos Islands, 900 kilometers (540 miles) off the coast of Ecuador, Darwin encountered a variety of different finches on the islands. The 14 species, although related, differed slightly in appearance, particularly in their beaks (figure 2.5). Darwin felt it most reasonable to assume all these birds had descended from a common ancestor blown by winds from the South American mainland several million years ago. Eating different foods on different islands, the species had changed in different ways as the generations descended from the common ancestor—"descent with modification," Darwin called it—evolution.

In a more general sense, Darwin was struck by the fact that the plants and animals on these relatively young volcanic islands resembled those on the nearby coast of South America. If each one of these plants and animals had been created independently and simply placed on the Galápagos Islands, why didn't they resemble the plants and animals of islands with similar climates, such as those off the coast of Africa, for example? Why did they resemble those of the adjacent South American coast instead?

> **2.2   The fossils and patterns of life that Darwin observed on the voyage of HMS *Beagle* eventually convinced him that evolution had taken place.**

The Theory of Natural Selection

It is one thing to observe the results of evolution but quite another to understand how it happens. Darwin's great achievement lies in his formulation of the hypothesis that evolution occurs because of natural selection.

Darwin and Malthus

Of key importance to the development of Darwin's insight was his study of Thomas Malthus's *Essay on the Principle of Population* (1798). In his book, Malthus pointed out that populations of plants and animals (including human beings) tend to increase geometrically, while the ability of humans to increase their food supply increases only arithmetically. A geometric progression is one in which the elements increase by a constant factor; for example, in the progression 2, 6, 18, 54, . . . each number is three times the preceding one. An arithmetic progression, in contrast, is one in which the elements increase by a constant difference; in the progression 2, 4, 6, 8, . . . each number is two greater than the preceding one (figure 2.6).

Because populations increase geometrically, virtually any kind of animal or plant, if it could reproduce unchecked, would cover the entire surface of the world within a surprisingly short time. Instead, populations of species remain fairly constant year after year, because death limits population numbers. Malthus's conclusion provided the key ingredient that

was necessary for Darwin to develop the hypothesis that evolution occurs by natural selection.

Natural Selection

Sparked by Malthus's ideas, Darwin saw that although every organism has the potential to produce more offspring than can survive, only a limited number actually do survive and produce further offspring. Combining this observation with what he had seen on the voyage of HMS *Beagle,* as well as with his own experiences in breeding domestic animals, Darwin made an important association (figure 2.7): those individuals that possess physical, behavioral, or other attributes that help them live in their environment are more likely to survive than those that do not have these characteristics. By surviving, they gain the opportunity to pass on their favorable characteristics to their offspring. As the frequency of these characteristics increases in the population, the nature of the population as a whole will gradually change. Darwin called this process selection. The driving force he identified has often been referred to as survival of the fittest. However, this is not to say the biggest or the strongest always survive. These characteristics may be favorable in one environment but less favorable in another. The organisms that are "best suited" to their particular environment, and therefore produce more offspring than others in the population, are the "fittest."

Darwin was thoroughly familiar with variation in domesticated animals and began *On the Origin of Species* with a detailed discussion of pigeon breeding. He knew that breeders selected certain varieties of pigeons and other animals, such as dogs, to produce certain characteristics, a process

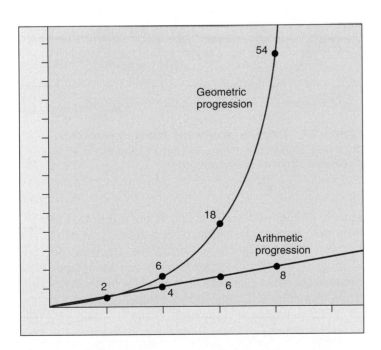

Figure 2.6 Geometric and arithmetic progressions.
An arithmetic progression increases by a constant difference (for example, units of 1 or 2 or 3), while a geometric progression increases by a constant factor (for example, by 2 or by 3 or by 4). Malthus contended that the human growth curve was geometric, but the human food production curve was only arithmetic. Can you see the problems this difference would cause?

"Can we doubt . . . that individuals having any advantage, however slight, over others, would have the best chance of surviving and procreating their kind? On the other hand, we may feel sure that any variation in the least degree injurious would be rigidly destroyed. This preservation of favorable variations, I call Natural Selection."

Figure 2.7 An excerpt from Charles Darwin's *On the Origin of Species.*

Darwin called **artificial selection.** Once this had been done, the animals would breed true for the characteristics that had been selected. Darwin had also observed that the differences purposely developed between domesticated races or breeds were often greater than those that separated wild species. Domestic pigeon breeds, for example, show much greater variety than all of the hundreds of wild species of pigeons found throughout the world. Such relationships suggested to Darwin that evolutionary change could occur in nature too. Surely if pigeon breeders could foster such variation by "artificial selection," nature through environmental pressures could do the same, playing the breeder's role in selecting the next generation—a process Darwin called **natural selection.**

Darwin's theory provides a simple and direct explanation of biological diversity, or why animals are different in different places: because habitats differ in their requirements and opportunities, the organisms with characteristics favored locally by natural selection will tend to vary in different places.

Darwin Drafts His Argument

Darwin drafted the overall argument for evolution by natural selection in a preliminary manuscript in 1842. After showing the manuscript to a few of his closest scientific friends, however, Darwin put it in a drawer and for 16 years turned to other research. No one knows for sure why Darwin did not publish his initial manuscript—it is very thorough and outlines his ideas in detail. Some historians have suggested that Darwin was wary of igniting public, and even private, criticism of his evolutionary ideas—there could have been little doubt in his mind that his theory of evolution by natural selection would spark controversy. Others have proposed that Darwin was simply refining his theory, although there is little evidence he altered his initial manuscript in all that time.

Wallace Has the Same Idea

The stimulus that finally brought Darwin's theory into print was an essay he received in 1858. A young English naturalist named Alfred Russel Wallace (1823–1913) sent the essay to Darwin from Malaysia; it concisely set forth the theory of evolution by means of natural selection, a theory Wallace had developed independently of Darwin. Like Darwin, Wallace had been greatly influenced by Malthus's 1798 book. Colleagues of Wallace, knowing of Darwin's work, encouraged him to communicate with Darwin. After receiving Wallace's essay, Darwin arranged for a joint presentation of their ideas at a seminar in London. Darwin then completed his own book, expanding the 1842 manuscript that he had written so long ago, and submitted it for publication.

Publication of Darwin's Theory

Darwin's book appeared in November 1859 and caused an immediate sensation. Many people were deeply disturbed by the suggestion that human beings were descended from the same ancestor as apes (figure 2.8). Although people had long accepted that humans closely resembled apes in many characteristics, the possibility that there might be a direct evolutionary relationship was unacceptable to many. Darwin did not

Figure 2.8   Darwin greets his monkey ancestor.
In his time, Darwin was often portrayed unsympathetically, as in this drawing from an 1874 publication.

actually discuss this idea in his book, but it followed directly from the principles he outlined. In a subsequent book, *The Descent of Man*, Darwin presented the argument directly, building a powerful case that humans and living apes have common ancestors. Darwin's arguments for the theory of evolution by natural selection were so compelling, however, that his views were almost completely accepted within the intellectual community of Great Britain after the 1860s.

> **2.3   The fact that populations do not really expand geometrically implies that nature acts to limit population numbers. The traits of organisms that survive to produce more offspring will be more common in future generations—a process Darwin called natural selection.**

2.4 The Beaks of Darwin's Finches

Darwin's Galápagos finches played a key role in his argument for evolution by natural selection. He collected 31 specimens of finch from three islands when he visited the Galápagos Islands in 1835. Darwin, not an expert on birds, had trouble identifying the specimens. He believed by examining their bills that his collection contained wrens, "gross-beaks," and blackbirds. You can see Darwin's sketches of four of these birds in figure 2.9.

The Importance of the Beak

Upon Darwin's return to England, ornithologist John Gould examined the finches. Gould recognized that Darwin's collection was in fact a closely related group of distinct species, all similar to one another except for their bills. In all, 14 species are now recognized, 13 from the Galápagos and one from far-distant Cocos Island. The two ground finches with the larger bills in figure 2.10 feed on seeds that they crush in their beaks, whereas those with narrower bills eat insects. Other species include fruit and bud eaters, and species that feed on cactus fruits and the insects they attract; some populations of the sharp-beaked ground finch even include "vampires" that creep up on seabirds and uses their sharp beaks to drink their

Figure 2.9 Darwin's own sketches of Galápagos finches.

From Darwin's *Journal of Researches:* (1) large ground finch, *Geospiza magnirostris;* (2) medium ground finch, *Geospiza fortis;* (3) small tree finch, *Camarhynchus parvulus;* (4) warbler finch, *Certhidea olivacea.*

blood. Perhaps most remarkable are the tool users, woodpecker finches that pick up a twig, cactus spine, or leaf stalk, trim it into shape with their bills, and then poke it into dead branches to pry out grubs.

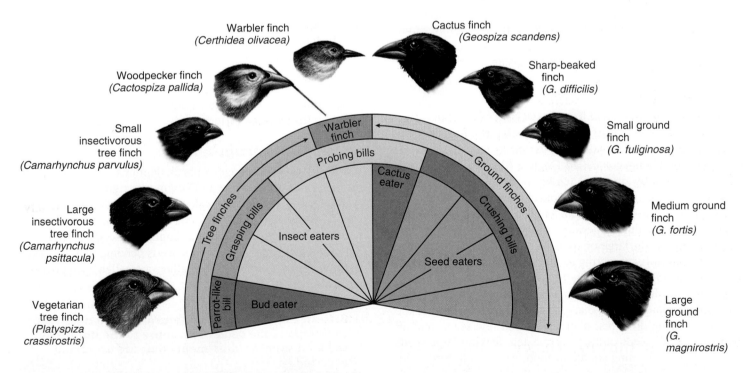

Figure 2.10 A diversity of finches on a single island.

Ten species of Darwin's finches from Isla Santa Cruz, one of the Galápagos Islands. The ten species show differences in bills and feeding habits. These differences presumably arose when the finches arrived and encountered habitats lacking small birds. Scientists concluded that all of these birds derived from a single common ancestor.

The correspondence between the beaks of the 14 finch species and their food source immediately suggested to Darwin that evolution had shaped them:

"Seeing this gradation and diversity of structure in one small, intimately related group of birds, one might really fancy that from an original paucity of birds in this archipelago, one species has been taken and modified for different ends."

Was Darwin Wrong?

If Darwin's suggestion that the beak of an ancestral finch had been "modified for different ends" is correct, then it ought to be possible to see the different species of finches acting out their evolutionary roles, each using its bill to acquire its particular food specialty. The four species that crush seeds within their bills, for example, should feed on different seeds, with those with stouter beaks specializing on harder-to-crush seeds.

Many biologists visited the Galápagos after Darwin, but it was 100 years before any tried this key test of his hypothesis. When the great naturalist David Lack finally set out to do this in 1938, observing the birds closely for a full five months, his observations seemed to contradict Darwin's proposal! Lack often observed many different species of finch feeding together on the same seeds. His data indicated that the stout-beaked species and the slender-beaked species were feeding on the very same array of seeds.

We now know that it was Lack's misfortune to study the birds during a wet year, when food was plentiful. The size of the finch's beak is of little importance in such flush times; slender and stout beaks work equally well to gather the abundant tender small seeds. Later work revealed a very different picture during dry years, when few seeds are available.

A Closer Look

Starting in 1973, Peter and Rosemary Grant of Princeton University and generations of their students have studied the medium ground finch, *Geospiza fortis,* on a tiny island in the center of the Galápagos called Daphne Major. These finches feed preferentially on small tender seeds, abundantly available in wet years. The birds resort to larger, drier seeds that are harder to crush when small seeds are hard to find. Such lean times come during periods of dry weather, when plants produce few seeds, large or small.

By carefully measuring the beak shape of many birds every year, the Grants were able to assemble for the first time a detailed portrait of evolution in action. The Grants found that beak depth changed from one year to the next in a predictable fashion. During droughts, plants produced few seeds, and all available small seeds quickly were eaten, leaving large seeds as the major remaining source of food. As a result, birds with large beaks survived better, because they were better able to

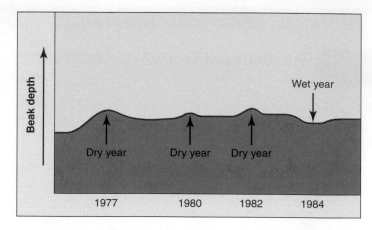

Figure 2.11 Evidence that natural selection alters beak size in *Geospiza fortis*.
In dry years, when only large, tough seeds were available, the mean beak size increased. In wet years, when many small seeds were available, smaller beaks became more common.

break open these large seeds. Consequently, the average beak depth of birds in the population increased the next year because this next generation included offspring of the large-beaked birds that survived. The average beak size decreased again when wet seasons returned because the larger beak size was no longer more favorable when seeds were plentiful and so smaller-beaked birds survived to reproduce (figure 2.11).

Could these changes in beak dimension reflect the action of natural selection? An alternative possibility might be that the changes in beak depth do not reflect changes in gene frequencies but rather are simply a response to diet, with poorly fed birds having stouter beaks. To rule out this possibility, the Grants measured the relation of parent bill size to offspring bill size, examining many broods over several years. The depth of the bill was passed down faithfully from one generation to the next, suggesting the differences in bill size indeed reflected gene differences.

Support for Darwin

If the year-to-year changes in beak depth can be predicted by the pattern of dry years, then Darwin was right after all—natural selection influences beak size based on available food supply. Birds with stout beaks have an advantage during dry periods, for they can break the large, dry seeds that are the only food available. When small seeds become plentiful once again with the return of wet weather, a smaller beak proves a more efficient tool for harvesting smaller seeds.

2.4 In Darwin's finches, natural selection adjusts the shape of the beak in response to the nature of the food supply, adjustments that are occurring even today.

⚜ Science in Action

Evolution Repeats Itself in Caribbean Lizards

Darwin would have been puzzled at the average American's reluctance to accept his theory of evolution. The evidence supporting Darwin's theory is clear, and every year more supporting evidence accumulates.

There is a sticky point in Darwin's argument, however. If evolution is indeed guided by natural selection, as Darwin claims, then two environments that are similar should select in the same way—similar habitats should select for the same sorts of critters, all else being equal.

Is Darwin right? Do two communities of animals living in similar habitats evolve to be the same? Does evolution repeat itself at the community level?

This is not an easy question to answer, simply because it's difficult to find an array of similar but independent habitats to compare.

But not impossible. A team of researchers led by Washington University biology professor Jonathan Losos has spent the last several years studying lizards of the genus *Anolis* (commonly called "anoles"), which live on large Caribbean islands. He has focused on Puerto Rico, Cuba, Haiti, and Jamaica. All four islands are inhabited by a diverse array of anole lizards (there are 57 species on Cuba alone), and all four islands have quite similar habitats and vegetation.

Unlike rats and cockroaches, which are generalists and much the same wherever you find them, anole lizards are specialists. In Puerto Rico, for example, one slender anole species with a long tail lives only in the grass. On narrow twigs at the base of trees you find a different species, also slender, but with stubby legs. On the higher branches of the tree a third species is found, of stocky build and long legs. High up in the leafy canopy of the tree lives a fourth giant green species.

Do the four Caribbean islands have similar lizard communities? Yes. If you go to Cuba, to Haiti, or to Jamaica, you can find on each island a species that looks nearly identical to each of the specialists on Puerto Rico, living in the same type of habitat and behaving in much the same manner.

Does this striking similarity of anole communities on the four islands indicate that the "Darwin experiment" has given the same result four times running?

The striking similarity of anole communities living on the four islands might be explained two different ways:

Hypothesis A   Lizards migrated between the islands. A specialist anole like the one that lives in grass may have evolved only once, but then travelled to the other islands, perhaps on floating driftwood. If this is true, the similarity of communities is not the result of evolution repeating itself, but just a matter of specialists finding their way to the habitats they prefer.

Hypothesis B   Lizards evolved in parallel on the four islands. The anole communities on the four islands may have evolved their similarity independently, evolution taking the same course again and again.

Working with Allan Larson of Washington University and Todd Jackman (now at Villanova University), Losos was able to choose between these two hypotheses by looking at the DNA of the lizards. The team compared several genes from more than 50 anole species. Points of similarity allowed them to construct a "phylogenetic tree," a family tree that showed who was related to who.

If hypothesis A is correct, then all the leaf specialists should be closely related to one another, whatever island they live on. The same would be expected for the four twig species, and also for the branch and canopy species.

On the other hand, if hypothesis B is correct, then a leaf specialist on one island should be more closely related to the other lizards on the same island, regardless of their specialty, than to a leaf specialist on another island.

Has evolution repeated itself? Yes. The DNA data are clear-cut: specialist species on one island are not closely related to the same specialists elsewhere, and are closely related to other anoles inhabiting the same island. Hypothesis B is correct. The four lizard communities evolved independently to be similar to one another.

The Losos research team has gone on to examine the functional consequences of Caribbean anole specializations, to see if natural selection can reasonably explain how each species has evolved. Why do some anole species have long legs, for example, while others have short stubby ones? These studies, involving both field and laboratory experiments, are science at its very best, insightful and fun. The rich picture of lizard evolution that is emerging would have delighted Darwin.

2.5 How Natural Selection Produces Diversity

Darwin believed that each Galápagos finch species had adapted to the particular foods and other conditions on the particular island it inhabited. Because the islands presented different opportunities, a cluster of species resulted. Presumably, the ancestor of Darwin's finches reached these islands before other land birds, so that when it arrived, all of the niches where birds occur on the mainland were unoccupied. A *niche* is what a biologist calls the way a species makes a living—the biological (that is, other organisms) and physical (climate, food, shelter, etc.) conditions with which an organism interacts as it attempts to survive and reproduce. As the new arrivals to the Galápagos moved into vacant niches and adopted new lifestyles, they were subjected to diverse sets of selective pressures. Under these circumstances, the ancestral finches rapidly split into a series of populations, some of which evolved into separate species.

The phenomenon by which a cluster of species change, as they occupy a series of different habitats within a region, is called *adaptive radiation.* Adaptive radiation occurred among the 14 species of Darwin's finches on the Galápagos Islands and Cocos Island (figure 2.12). Such species clusters are often particularly impressive on island groups, in series of lakes, or in other sharply discontinuous habitats.

The descendants of the original finches that reached the Galápagos Islands now occupy many different kinds of habitats on the islands. The 14 species that inhabit the Galápagos Islands and Cocos Island occupy four types of niches:

1. **Ground finches.** There are six species of *Geospiza* ground finches. Most of the ground finches feed on seeds. The size of their bills is related to the size of the seeds they eat. Some of the ground finches feed primarily on cactus flowers and fruits and have longer, larger, more pointed bills.
2. **Tree finches.** There are five species of insect-eating tree finches. Four species have bills that are suitable for feeding on insects. The woodpecker finch has a chisel-like beak. This unique bird carries around a twig or a cactus spine, which it uses to probe for insects in deep crevices.
3. **Vegetarian finch.** The very heavy bill of this bud-eating bird is used to wrench buds from branches.
4. **Warbler finches.** These unusual birds play the same ecological role in the Galápagos woods that warblers play on the mainland, searching continually over the leaves and branches for insects. They have a slender, warblerlike beak.

2.5 Darwin's finches, all derived from one similar mainland species, have radiated widely on the Galápagos Islands, filling unoccupied niches in a variety of ways.

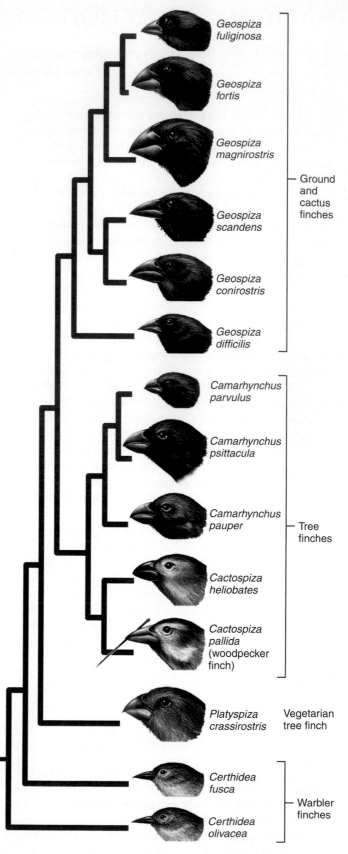

Figure 2.12 An evolutionary tree of Darwin's finches.

This family tree was constructed by comparing DNA of the 14 species. Their position at the base of the finch tree suggests that warbler finches were among the first adaptive types to evolve in the Galápagos.

2.6 What Is Ecology?

Darwin's finches teach us an important lesson about evolution, which is that understanding how natural selection occurs is really a matter of understanding how species adapt to particular niches. As we saw on the preceding pages, the diversity of finches Darwin found on the Galápagos Islands arose as a consequence of the availability of a variety of niches on the islands, each fostering the evolution of a new finch species. In a very real sense, the nature of the habitats the finches invaded, and the ways in which different populations of finches came to utilize these habitats, determined the course of the evolutionary radiation that followed. Biologists who study the nature of niches, and how the species that occupy them interact, are called ecologists.

The word **ecology** was coined in 1866 by the great German biologist Ernst Haeckel to describe the study of how organisms interact with their environment. It comes from the Greek words *oikos* (house, place where one lives) and *logos* (study of). Our study of ecology, then, is a study of the house in which we live. Do not forget this simple analogy built into the word ecology—most of our environmental problems could be avoided if we treated the world in which we live the same way we treat our own homes. Would you pollute your own house?

Levels of Ecological Organization

Ecologists consider groups of organisms at five progressively more encompassing levels of organization.

1. **Populations.** Individuals of the same species that live together are members of a population. They potentially interbreed with one another, share the same habitat, and use the same pool of resources the habitat provides.

2. **Communities.** Populations of different species that live together in the same place are called communities. Different species typically use different resources within the habitat they share.

3. **Ecosystems.** A community and the nonliving factors with which it interacts is called an **ecosystem** (figure 2.13). An ecosystem is affected by the flow of energy, ultimately derived from the sun, and the cycling of the essential elements on which the lives of its constituent organisms depend.

4. **Biomes.** Biomes are major terrestrial assemblages of plants, animals, and microorganisms that occur over wide geographical areas that have distinct physical characteristics. Examples include deserts, tropical forests, and grasslands. Similar groupings occur in marine and freshwater habitats.

5. **The biosphere.** All the world's biomes, along with its marine and freshwater assemblages,

Figure 2.13   A Galápagos Island ecosystem.
The Galápagos Islands are named after the giant tortoise, seen here climbing through the scrub vegetation of Isabela, the largest of the islands. Although this scene may at first glance seem barren, the island ecosystem is rich with plant and animal species. Sharing common resources, the giant tortoises, Darwin's finches, and other species possess many adaptations that promote their mutual survival.

together constitute an interactive system we call the biosphere. Changes in one biome can have profound consequences for others.

Some ecologists, called *population ecologists,* focus on a particular species and how its populations grow. Other ecologists, called *community ecologists,* study how the different species living in a place interact with one another. Still other ecologists, called *systems ecologists,* are interested in how biological communities interact with their physical environment.

> **2.6   Ecology is the study of how the organisms that live in a place interact with each other and with their physical habitat.**

A Closer Look at Ecosystems

In a sense, evolutionary biologists and ecologists study the same things from different perspectives. An evolutionary biologist focuses on the changes that occur in a species, while an ecologist focuses on how that species is interacting with other species and its physical environment. The one occurs as a result of the other.

Just as the population is the fundamental unit of evolution, so ecological systems, or ecosystems, are the fundamental units of ecology. An **ecosystem** is basically a biological community and the physical environment in which it lives. It is the most complex biological system that a biologist can study.

Energy Flows Through Ecosystems

All of the organisms within a community require energy to grow, reproduce, and carry out all of the many other activities of living. Almost all of this energy comes, ultimately, from the sun, captured by photosynthesis. The plants, algae, and microbes that capture it are in turn consumed by plant-eating animals called herbivores, and some of the energy captured from sunlight is passed on to the herbivore. The herbivore may then be eaten by a meat-eating animal called a carnivore, which captures some of the energy of the herbivore. Energy thus flows through the ecosystem, from plant to herbivore to carnivore, a system called a **food chain.**

Unfortunately, much of the energy is used up or lost as heat at each step of this food chain. Thus, because only so much energy arrives from the sun, food chains can be only so long—typically three or four steps. Can you see why there is no lion-eating top carnivore on the African savanna? It would need to consume so many lions to sustain itself, that the available lions would be quickly depleted.

Materials Cycle Within Ecosystems

The raw materials that make up organisms—the carbon, nitrogen, phosphorus, and other atoms—are not used up when the organisms die. Instead, as the organisms decompose, the materials of their bodies pass back into the ecosystem, where they can be used to make other organisms. The materials thus cycle between organisms and the physical environment.

Major Ecosystems

While many aspects of the environment act to limit the distribution of particular species, the two most important are rainfall and temperature. Particular organisms are adapted to particular combinations of rainfall and temperature, and every place with that combination of rainfall and temperature will tend to have organisms with similar adaptations living there. On land, these physical conditions with their similar sets of plants and animals are referred to as **biomes.** The United States contains a variety of quite different biomes (figure 2.14). The biomes mentioned here will be discussed in detail in chapter 16.

2.7   An ecosystem is a dynamic ecological system that consumes energy and cycles materials.

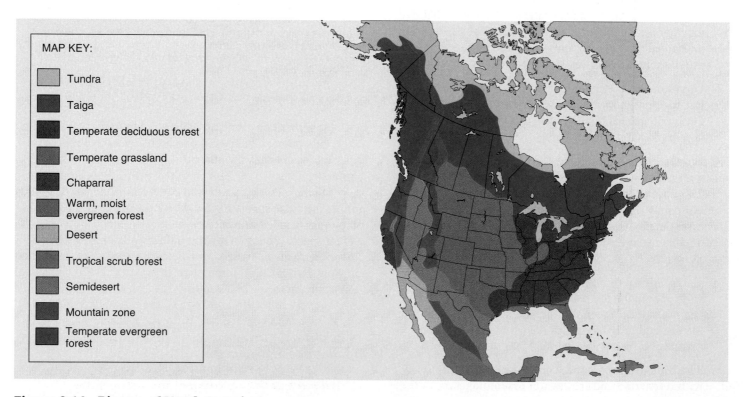

MAP KEY:

Tundra

Taiga

Temperate deciduous forest

Temperate grassland

Chaparral

Warm, moist evergreen forest

Desert

Tropical scrub forest

Semidesert

Mountain zone

Temperate evergreen forest

Figure 2.14   Biomes of North America.

Figure 2.15   Resource partitioning among lizard species.

Species of *Anolis* lizards in the Caribbean partition their tree habitats in a variety of ways. Some species of anoles occupy the canopy of trees (*a*), others use twigs on the periphery (*b*), and still others are found at the base of the trunk (*c*). In addition, some use grassy areas in the open (*d*). This same pattern of resource partitioning has evolved independently on different Caribbean islands.

2.8 How Species Evolve to Occupy Different Niches Within an Ecosystem

Each organism in an ecosystem confronts the challenge of survival in a different way. As we discussed on page 30, the **niche** an organism occupies is the sum total of all the ways it uses the resources of its environment. The niche of a species may be thought of as its biological role in the community. A niche may be described in terms of space utilization, food consumption, temperature range, appropriate conditions for mating, requirements for moisture, and other factors. *Niche* is not synonymous with **habitat,** the place where an organism lives. *Habitat* is a place, *niche* a pattern of living.

Resource Partitioning

Competition is the struggle of two organisms to use the same resource when there is not enough of the resource to satisfy both. When two species compete for the same resource, the species that utilizes the resource more efficiently will eventually outcompete the other in that location and drive it to extinction there. Ecologists call this *the principle of competitive exclusion:* no two species with the same niche can coexist. Persistent competition between two species is rare in nature. Either one species drives the other to extinction, or natural selection favors changes that reduce the competition between them. In **resource partitioning,** species that live in the same geographical area avoid competition by living in different portions of the habitat or by using different food or other resources (figure 2.15).

The changes that evolve in two species to reduce niche overlap—that is, to lessen the degree to which they compete for the same resources—are called **character displacements.** Character displacement can be seen clearly among Darwin's finches. The two Galápagos finches in figure 2.16 have bills of similar size when each is living on an island where the other

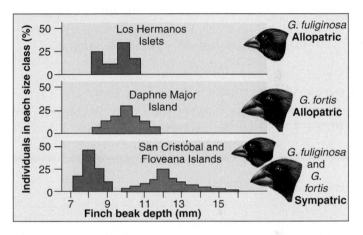

Figure 2.16   Character displacement in Darwin's finches.

These two species of Galápagos finches (genus *Geospiza*) have bills of similar sizes when living apart (allopatric) but different sizes when living together (sympatric).

does not occur. On islands where they are found living together, the two species have evolved beaks of different sizes, one adapted to larger seeds, the other to smaller ones. In essence, the two finches have subdivided the food niche, creating two new smaller niches. By partitioning the available food resources, the two species have avoided direct competition with each other, and so are able to live together in the same habitat.

The 14 species of Darwin's finches seen in figure 2.12 differ largely in bill size and shape, suggesting that partitioning of food resources has been a principal factor in the evolution of these species.

2.8   Species that live together partition available resources, reducing competition between them.

Patterns of Population Growth

No ecosystem has unlimited resources. As a general rule, natural populations of a species will grow in numbers until all the resources available to their niche are being used. Before that limit is reached, population size can increase rapidly. The rate at which a population will increase when there are no limits on its rate of growth is called the **innate capacity for increase,** or **biotic potential.** This theoretical rate is almost impossible to calculate, however, because there are usually limits to growth. What biologists in fact calculate is the **realized rate of population increase** (abbreviated r). This parameter is defined as the number of individuals added to the population minus the number lost from it. The number added to it equals the birthrate plus the number of **immigrants** (new individuals entering and residing with the population), while the number lost from it equals the death rate plus the number of **emigrants** (individuals leaving the population). Thus:

$$r = (\textbf{birth} + \textbf{immigration}) - (\textbf{death} + \textbf{emigration})$$

Exponential Growth

A population's innate capacity for growth is constant, determined largely by the organism's physiology. Its actual

growth, on the other hand, is not a constant, because r depends on both the birthrate and the death rate, and both of these factors may change as the population increases in size. Thus to get the population growth rate, r must be adjusted for population size:

$$\textbf{population growth rate} = rN$$

where r is the realized rate of population increase and N is the number of individuals in the population. In general, the number of individuals grows rapidly at first, as the number of individuals is increasing exponentially (figure 2.17). This type of growth is called **exponential growth.** As a population increases and begins to exhaust its resources, the death rate rises, and the rate of increase slows. Eventually, just as many individuals are dying as are being born. The early rapid phase of population growth lasts only for a short period, usually when an organism reaches a new habitat where resources are abundant. Examples of this pattern include algae colonizing a newly formed pond and the first terrestrial organisms arriving on a recently formed island.

Carrying Capacity

No matter how rapidly new populations grow, they eventually reach an environmental limit imposed by shortages of some important factor, such as space, light, water, or nutrients. A

Time (hours)	Number of bacteria
10.0	1,048,576
9.5	524,288
9.0	262,144
8.5	131,072
8.0	65,536
7.5	32,768
7.0	16,384
6.5	8,192
6.0	4,096
5.5	2,048
5.0	1,024
4.5	512
4.0	256
3.5	128
3.0	64
2.5	32
2.0	16
1.5	8
1.0	4
0.5	2
0.0	1

Figure 2.17   Exponential growth in a population of bacteria.

In just 10 hours, this population grew from one individual to over 1 million!

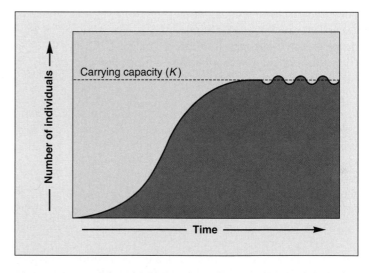

Figure 2.18   The sigmoid growth curve.

The sigmoid growth curve begins with a period of exponential growth like that shown in figure 2.17. When the population approaches its environmental limits, growth slows and finally stabilizes, fluctuating around the carrying capacity (K) of the environment.

population ultimately stabilizes at a certain size, called the **carrying capacity** of the particular place where the population lives. The carrying capacity, symbolized by K, is the number of individuals that can be supported at that place indefinitely. As carrying capacity is approached, the population's rate of growth slows greatly. The growth of a specific population, which is always limited by one or more factors in the environment, can be approximated by the following *logistic growth equation:*

$$\text{population growth rate } = rN \left(\frac{K - N}{K} \right)$$

In other words, the growth of the population under consideration equals the ideal rate of increase (r multiplied by N, the number of individuals present at any one time), adjusted for the amount of resources still available. The adjustment is made by multiplying rN by the fraction of K still unused (K minus N, divided by K). As N increases, the fraction by which r is multiplied becomes smaller and smaller, and the rate of increase of the population declines. Graphically, this relationship is the S-shaped **sigmoid growth curve,** characteristic of biological populations (figure 2.18).

Figure 2.20 An animal with a *K*-selected life history.
Elephants exhibit mainly *K*-selected adaptations, including only a few offspring per brood (typically one), prolonged parental care extending many years, slow growth and maturation, and life spans as long as a hundred years.

Life History Strategies

Most populations exhibit characteristics that lie on a continuous spectrum between traits that favor exponential growth by r and traits that keep population size under K. The particular set of adaptations that adjusts an organism's reproductive rate to its environment is called its **life history strategy** (see chapter 17).

Life history strategies will sometimes favor rapid growth, particularly those of organisms occupying rapidly changing environments. Most mosquitos, for example, reproduce in temporary puddles and exhibit *r*-selected life histories (figure 2.19). Many weeds exhibit *r*-selected life histories, rapidly invading disturbed vegetation or lawns; indeed, r life history strategies are often called "weedy growth."

Other life history strategies favoring slow growth are often exhibited by organisms competing for limited resources. Elephants reproduce in a crowded African savanna community and exhibit *K*-selected life histories (figure 2.20).

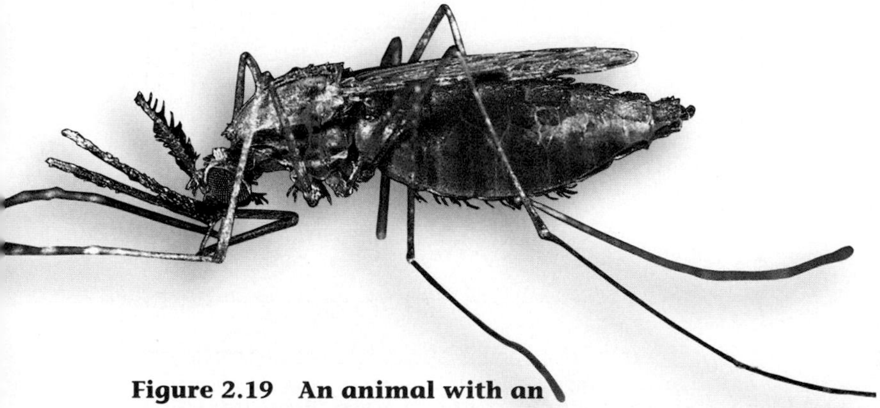

Figure 2.19 An animal with an *r*-selected life history.
Mosquitos exhibit mainly *r*-selected adaptations, including hundreds of offspring per brood, no investment in parental care (eggs are laid on water and abandoned), rapid growth (young mature within hours), and a short life span numbered in weeks.

2.9 Population growth is limited by the ability of the environment to support the population. Organisms in transient environments are often adapted to reproduce rapidly, while those in stable environments tend to reproduce more slowly.

Human Populations

Humans exhibit many *K*-selected life history traits, including small brood size, late reproduction, and a high degree of parental care. These life history traits evolved during the early history of human ancestors, when the limited resources available from the environment controlled population size. Throughout most of human history, our populations have been regulated by food availability, disease, and predators. While unusual disturbances, including floods, plagues, and droughts, no doubt affected the pattern of human population growth, the overall size of the human population grew only slowly during our early history. Two thousand years ago, perhaps 130 million people populated the earth. It took a thousand years for that number to double, and it was 1650 before it had doubled again, to about 500 million. For over 16 centuries, the human population was characterized by very slow growth. In this respect, human populations resembled many other species with predominantly *K*-selected life history adaptations.

The Advent of Exponential Growth

Starting in the early 1700s, changes in technology have given humans more control over their food supply, enabled them to develop superior weapons to ward off predators, and led to the development of cures for many diseases. At the same time, improvements in shelter and storage capabilities have made humans less vulnerable to climatic uncertainties. These changes allowed humans to expand the carrying capacity of the habitats in which they lived and thus to escape the confines of logistic growth and reenter the exponential phase of the sigmoidal growth curve.

Responding to the lack of environmental constraint, the human population has grown explosively over the last 300 years. While the birthrate is currently about 22 per 1,000 per year, the death rate has fallen dramatically, from 29 per 1,000 per year to its present level of about 9 per 1,000 per year. This difference between birth and death rates (13 per 1,000) means that the population is growing at the rate of 1.3% per year.

A 1.3% annual growth rate may not seem large, but it has produced a current human population of over 6.3 billion people (figure 2.21)! At this growth rate, 80 million people are added to the world population annually, and the human population will double in 53 years. As we will discuss in chapter 18, both the current human population level and the projected growth rate have potential consequences for our future that are extremely grave.

Population Pyramids

While the human population as a whole continues to grow rapidly, this growth is not occurring uniformly over the planet (table 2.1). Some countries, like Mexico, are currently growing rapidly; its birthrate greatly exceeds its death rate (figure 2.22). Other countries are growing much more slowly. The rate

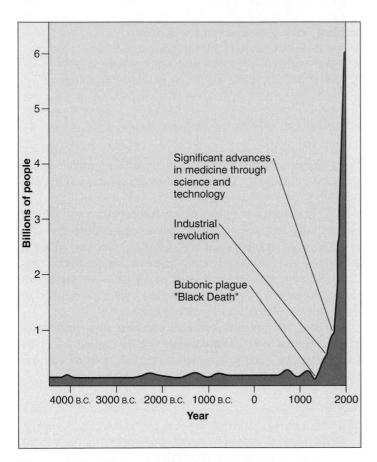

Figure 2.21 History of human population size.
Temporary increases in death rate, even severe ones like the Black Death of the 1400s, have little lasting impact. Explosive growth began with the Industrial Revolution in the 1700s, which produced a significant long-term lowering of the death rate. The current population is over 6 billion, and at the current rate will double in 53 years.

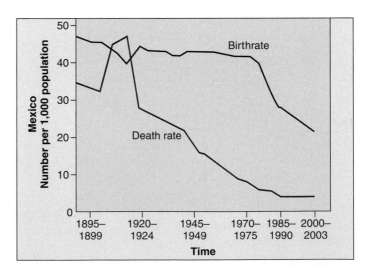

Figure 2.22 Why Mexico's population is growing.
The death rate (*red line*) in Mexico has been falling, while the birthrate (*blue line*) remained fairly steady until 1970. The difference between birth and death rates has fueled a high growth rate. Efforts begun in 1970 to reduce the birthrate have been quite successful. Although the growth rate remains rapid, it is expected to begin leveling off in the near future as the birthrate continues to drop.

TABLE 2.1	A COMPARISON OF 2003 POPULATION DATA IN DEVELOPED AND DEVELOPING COUNTRIES		
	United States (highly developed)	Brazil (moderately developed)	Ethiopia (developing)
Fertility rate	2.1	2.0	5.5
Doubling time at current rate (yr)	115	57.5	35
Infant mortality rate (infant deaths/1,000 births)	6.8	32	103
Life expectancy (yr)	77	71	41
Per capita income (U.S. dollar equivalent)	$36,300	$7,600	$700

Source: Population Reference Bureau

at which a population can be expected to grow in the future can be assessed graphically by means of a population pyramid—a bar graph displaying the numbers of people in each age category. Males are conventionally shown to the left of the vertical age axis and females to the right. In most human population pyramids, the number of older females is disproportionately large compared with the number of older males, because females in most regions have a longer life expectancy than males.

Viewing such a pyramid, one can predict demographic trends in births and deaths. In general, rectangular "pyramids" are characteristic of countries whose populations are stable; their numbers are neither growing nor shrinking. A triangular pyramid is characteristic of a country that will exhibit rapid future growth, as most of its population has not yet entered the child-bearing years. Inverted triangles are characteristic of populations that are shrinking.

Examples of population pyramids for the United States and Kenya are shown in figure 2.23. In the somewhat more rectangular population pyramid for the United States in 2000, the cohort (group of individuals) 35 to 54 years old represents the "baby boom." When the media refers to the "graying of America," they are referring to the aging of this disproportionately large cohort that will impact the health-care system and other age-related systems in the future. The very triangular pyramid of Kenya, by contrast, predicts explosive future growth. The population of Kenya is predicted to double in less than 20 years.

The Level of Consumption in the Developed World Is Also a Problem

The world population is expected to stabilize sometime in this century at about 10 billion. We in the developed countries of the world need to pay more attention to lessening the impact each of us makes, because, even though the vast majority of the world's population is in developing countries, the vast majority of resource consumption occurs in the developed world. Indeed, the wealthiest 20% of the world's population accounts for 86% of the world's consumption of resources and produces 53% of the world's carbon dioxide emissions, whereas the poorest 20% of the world is responsible for only 1.3% of consumption and 3% of CO_2 emissions.

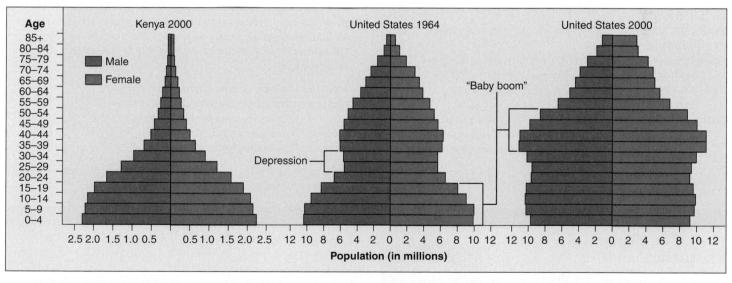

Figure 2.23 Population pyramids from 2000.

Population pyramids are graphed according to a population's age distribution. Kenya's pyramid has a broad base because of the great number of individuals below child-bearing age. When all of the young people begin to bear children, the population will experience rapid growth. The 2000 U.S. pyramid demonstrates a larger number of individuals in the "baby boom" cohort—the pyramid bulges because of an increase in births between 1945 and 1964, as shown at the base of the 1964 pyramid. The 25 to 34 cohort in the 1964 pyramid represents people born during the Depression and is smaller in size than the cohorts in the preceding and following years.

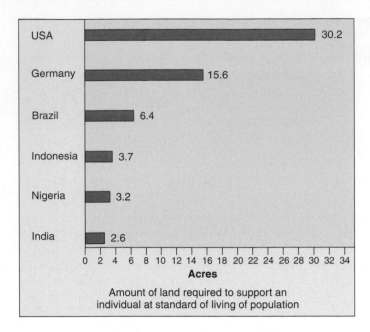

Figure 2.24 **Ecological footprint of individuals in different countries.**

The ecological footprint calculates how much land is required to support a person through his or her life, including the acreage used for production of food, forest products, and housing, in addition to the forest required to absorb the carbon dioxide produced by the combustion of fossil fuels.

One way of quantifying this disparity is by calculating what has been termed the **ecological footprint,** which is the amount of productive land required to support an individual at the standard of living of a particular population through the course of his or her life. As figure 2.24 illustrates, the ecological footprint of an individual in the United States is more than 10 times greater than that of someone in India. Based on these measurements, researchers have calculated that resource use by humans is now 1/3 greater than the amount that nature can sustainably replace; if all humans lived at the standard of living in the developed world, two additional planet earths would be needed.

2.10 The human population is growing at a rate of 1.3% annually, and at that rate will double in 53 years.

Exploring Current Issues

Additional Resources

Go to your campus library to find the following articles, which further develop some of the concepts found in this chapter.

Clark, A.G., *et al.* (2003) Inferring nonneutral evolution from human-chimp-mouse orthologous gene trios. *Science,* 302(5652), 1960–1964.

Cohen, J.E. (2003) Human population: the next half century. *Science,* 302(5648), 1172–1176.

Grant, P.R. and B.R. Grant. (2002). Unpredictable evolution in a 30-year study of Darwin's finches. *Science,* 296(5568), 707–712.

Ruse, M. (2004) Natural selection vs. intelligent design. *USA Today* (Magazine), 132(2704), 32–35.

Thorne, A.G. and M.H. Wolpoff. (2003) The multiregional evolution of humans: both fossil and genetic evidence argues that ancient ancestors of various human groups lived where they are found today. *Scientific American,* June 15, 46–53.

In the News: Darwin's Critics

While Darwin's theory of evolution by natural selection is almost universally accepted by scientists, it has proven quite a bit more controversial among the general public. Objections to the teaching of evolution in American public schools have focused on a family of arguments with which every student of biology should become familiar. Many of the criticisms of evolution were raised in the years soon after Darwin's publication of *On the Origin of Species.* More recent criticisms doubt that evolution could produce the complexity we see at the cellular and molecular level.

Article 1. 140 Years without Darwin are enough.
Article 2. Answering evolution's critics.
Article 3. Keeping Darwin out of schoolrooms.
Article 4. Darwinism at the cellular level.
Article 5. Darwin and charter schools.

Find these articles, written by the author, online at www.mhhe.com/tlwessentials/exp2.

Summary

Evolution

2.1 Darwin's Voyage on the HMS *Beagle*

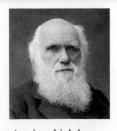

- Darwin's voyage on the HMS *Beagle* allowed him the opportunity to study a large variety of animal and plant species across the globe. Years after his voyage, he published the book *On the Origin of Species,* in which he

proposed natural selection as the mechanism that underlies the process of evolution (**figure 2.3**).

- The theory of evolution through natural selection is overwhelmingly accepted by scientists but is viewed as controversial by some in the general public.

2.2 Darwin's Evidence

- Darwin observed fossils in South America of extinct species that resembled living species. On the Galápagos Islands, Darwin observed finches that differed slightly in appearance between islands but resembled finches found on the South American mainland. These observations laid the groundwork for his proposal of evolution through natural selection (**figures 2.4 and 2.5**).

2.3 The Theory of Natural Selection

- Key to Darwin's hypothesis of evolution by natural selection was the observation by Malthus that the food supply limits population growth. A population, be it plants, animals, etc., grows only as large as that which can live off of the available food (**figure 2.6**).

- Using Malthus's observations and his own observations, Darwin proposed that individuals that are better suited to their environments survive to produce offspring, gaining the opportunity to pass their characteristics on to future generations. Therefore, future generations may be different from ancestral populations, a process known as evolution.

- Spurred by a similar proposal put forth years later by Alfred Wallace, Darwin published his book in 1859.

Darwin's Finches: Evolution in Action

2.4 The Beaks of Darwin's Finches

- By observing the different sizes and shapes of beaks in the closely related finches of the Galápagos Islands and correlating the beaks with the types of food consumed by the different birds, Darwin concluded that the birds' beaks were modified from an ancestral species based on the food available, each suited to its food supply (**figure 2.10**).

- Research has since supported Darwin's hypothesis that natural selection influences beak size in island finches based on the available food supply (**figure 2.11**).

2.5 How Natural Selection Produces Diversity

- The 14 species of finches found on the islands off the coast of South America evolved from a mainland species that adapted to different niches, a process called adaptive radiation (**figure 2.12**).

Ecology

2.6 What Is Ecology?

- Ecology is the study of how organisms interact with each other

and with their physical environment. There are five levels of ecological organization: populations, communities, ecosystems, biomes, and the biosphere.

2.7 A Closer Look at Ecosystems

- An ecosystem is a physical environment that contains a community of various organisms. These organisms interact with each other and with their physical environment, extracting energy and raw materials from it. Similar ecosystems found throughout the world are called biomes (**figure 2.14**).

2.8 How Species Evolve to Occupy Different Niches Within an Ecosystem

- Resource partitioning is the process whereby two competitive species coexist by utilizing different portions of the habitat or different resources such as food. The varying beak sizes seen in Darwin's finches suggest that the bird species evolved in part due to resource partitioning (**figures 2.15 and 2.16**).

Populations and How They Grow

2.9 Patterns of Population Growth

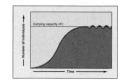

- Populations grow exponentially until limited by environmental conditions. A population will stabilize at a size, called its carrying capacity, that can be supported by its environment (**figure 2.18**).

- Organisms exposed to an ever-changing environment tend to reproduce and grow rapidly, exhibiting an *r*-selected life history. Organisms exposed to more stable environmental conditions tend to reproduce and grow more slowly, exhibiting a *k*-selected life history.

2.10 Human Populations

- Technology has allowed the human population to grow exponentially for the last 300 years to a current population of over 6 billion (**figure 2.21**).

- Human populations grow at different rates, with underdeveloped countries' populations growing more rapidly than developed countries' populations. However, it takes more resources to support populations in developed countries versus those in underdeveloped countries (**figures 2.22 and 2.23**).

Self-Test

1. The theory of evolution states that
   a. once a species is formed, it remains stable over time, and populations do not change.
   b. populations always change over time, forming new species.
   c. populations can change over time, sometimes forming new species.
   d. populations that change over time usually go extinct, such as the dinosaurs.

2. A key observation made by Darwin was
   a. a species always looked almost exactly the same wherever he saw it; traits were preserved.
   b. the characteristics of a species varied in different places; there were geographic patterns.

   c. the characteristics of a species varied in different places; it was unpredictable.
   d. everywhere he went there were wildly different and unrelated organisms.

3. Darwin was greatly influenced by Thomas Malthus, who pointed out
   a. populations increase geometrically.
   b. populations increase arithmetically.
   c. populations are capable of geometric increase, yet remain at constant levels.
   d. the food supply usually increases faster than the population that depends on it.

4. Darwin proposed that individuals with traits that help them live in their immediate environment are more likely to survive and reproduce than individuals without those traits. He called this
   a. natural selection.
   b. the principle of population growth.
   c. the theory of evolution.
   d. Malthusian growth.
5. A great deal of research has been done on Darwin's finches over the last 70 years. The research
   a. seems to often contradict Darwin's original ideas.
   b. seems to agree with Darwin's original ideas.
   c. does not show any clear patterns that support or refute Darwin's original ideas.
   d. sometimes supports his ideas and sometimes does not.
6. The way a species makes its living—that is, the biological and physical conditions in which it exists—is called its
   a. population.
   b. territory.
   c. community.
   d. niche.
7. In the levels of ecological organization, the lowest level, composed of individuals of a single species who live near each other, share the same resources, and can potentially interbreed is called a

   a. population.   c. ecosystem.
   b. community.   d. biome.
8. Within an ecosystem
   a. materials flow through once and are lost, while energy cycles and recycles.
   b. materials and energy both flow through once and are lost.
   c. materials and energy both cycle and recycle.
   d. energy flows through once and is lost, while materials cycle and recycle.
9. A major principle of ecology states that no two species can occupy exactly the same niche. One will utilize resources more efficiently than the other and will drive the second species to extinction. This is the principle of
   a. competition.   c. community organization.
   b. natural selection.   d. competitive exclusion.
10. Biotic potential is
   a. the ability of a species to compete with other species within a particular habitat.
   b. the rate at which an individual of a species can grow and reproduce.
   c. the rate at which the population can grow and reproduce if there are no limits.
   d. the rate at which the population can grow and reproduce under natural limits.

Visual Understanding

1. **Figure 2.15** Using Darwin's reasoning, explain how these four species of lizards, all closely related, came to be separate species on a Caribbean island.

2. **Figure 2.23** In the 1960s and 1970s the United States needed huge numbers of new schools, homes, and then jobs for the children of the "baby boom" that occurred after World War II. Looking at where the baby-boom bulge is by the year 2000, what goods and services are they going to need in the 2010s and 2020s?

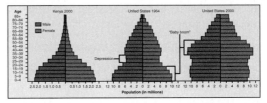

Challenge Questions

Evolution   Many evolutionary pathways seem to have moved from larger to smaller organisms over time, such as the glyptodont and the armadillo, the mammoth and the elephant. Yet other organisms, such as the tiny eohippus and the horse, have increased in size. Explain how this can occur.

Darwin's Finches: Evolution in Action   Your friend Gustavo tells you that he has heard about the work of the famous scientist, David Lack, and that it proves Darwin was wrong about evolution and modification over time. What do you say?

Ecology   *Chthamalus* is a small barnacle that lives on rocky shores. It can survive on rocks that are underwater, or on rocks that

are exposed during low tide. *Balanus* is a larger barnacle that can outcompete *Chthamalus* if they both want the same spot. *Balanus* can only live on rocks that are seldom exposed to air. Explain what happens when they are both present.

Populations and How They Grow   People in the United States require an average of over 30 acres of land to support each individual, while in Indonesia, Nigeria, and India each person needs only about 2.5 to 4 acres. The figure for the United States is also about double that of some other highly industrialized countries, such as Germany. Explain the differences, and suggest ways to decrease the U.S. figures.

Online Learning Center

Visit the Online Learning Center for this chapter at www.mhhe.com/tlwessentials/ch2 for quizzes, animations, interactive learning exercises, and other study tools. At the site you will also find extended answers to the end-of-chapter questions.

3

The Chemistry of Life

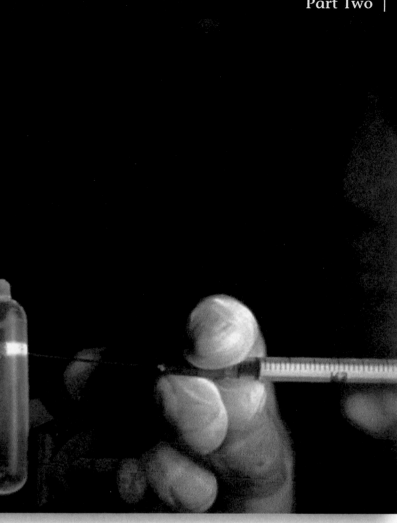

Using a syringe, this researcher is gently removing a glowing band of DNA, to be used in an experiment studying heredity. DNA, the carrier of an organism's genes, is one of several kinds of very large molecules found in all organisms. A molecule is a collection of tiny atoms linked together. Incredibly, every atom in your body was at one time part of a star. On earth, atoms are the basic chemical elements. Only a few are found in any significant numbers in living things. An essential atom for life is carbon, which can assemble into DNA and other very large molecules. Interacting with water, these long carbon chains twist about each other, or fold up into compact masses. Much of the chemistry that goes on in organisms, determining what each individual is like, depends on the actions of large folded molecules called proteins. By promoting particular chemical reactions, proteins trigger the production of structural materials like carbohydrates and energy storage molecules like lipids. Because DNA encodes the information needed to assemble each protein present in an organism, it is the library of life.

3.1 Atoms

Organisms are chemical machines, and to understand them we must learn a little chemistry. Any substance in the universe that has mass and occupies space is defined as **matter.** All matter is composed of extremely small particles called **atoms.** An atom is the smallest particle into which a substance can be divided and still retain its chemical properties. The terms *mass* and *weight* are often used interchangeably, but they have slightly different meanings. Mass refers to the amount of a substance, whereas weight refers to the force gravity exerts on a substance. Hence, an object has the same mass whether it is on the earth or the moon, but its weight will be greater on the earth, because the earth's gravitational force is greater than the moon's. For example, an astronaut weighing 180 pounds on earth will weigh about 30 pounds on the moon. He didn't

lose any significant mass during his flight to the moon, there is just less gravitational pull on his mass.

Every atom has the same basic structure: a core nucleus of protons and neutrons surrounded by a cloud of electrons (figure 3.1). At the center of every atom is a small, very dense nucleus formed of two types of subatomic particles, **protons** and **neutrons.** Whizzing around the core is an orbiting cloud of a third kind of subatomic particle, the **electron.** Neutrons have no electrical charge, whereas protons have a positive charge and electrons a negative one. In a neutral atom (that is, one that carries no electrical charge), there is an orbiting electron for every proton in the nucleus. The electron's negative charge balances the proton's positive charge.

The number of protons in the nucleus of an atom is called the **atomic number.** For example, the atomic number of carbon is 6 because it has six protons. Neutrons are similar to protons in mass, and the number of protons and neutrons in the nucleus of an atom is called the **atomic mass.** A carbon atom that has six protons and six neutrons has an atomic mass of 12. Although precise measurements of atomic mass are often presented in tables, it is also common to round the atomic mass to an integer value. The mass of atoms and subatomic particles is measured in units called *daltons.* A proton's mass is approximately 1 dalton (actually 1.007 daltons), as is a neutron's mass (1.009 daltons). In contrast, an electron's mass is only 1/1,840 of a dalton, so its contribution to the overall mass of an atom is negligible. Atoms with the same atomic number (that is, the same number of protons) have the same chemical properties and are said to belong to the same **element.** Formally speaking, an element is any substance that cannot be broken down into any other substance by ordinary chemical means. The atomic numbers and mass numbers of some of the most common elements on earth are shown in table 3.1.

Electrons Determine What Atoms Are Like

Electrons have very little mass (only 1/1,840 the mass of a proton). Of all the mass contributing to your weight, the portion that is contributed by electrons is less than the mass of your eyelashes. And yet electrons determine the chemical behavior of atoms because they are the parts of atoms that come close enough to each other in nature to interact. Almost all the volume of an atom is empty space. Protons and neutrons lie at the core of this space, while orbiting electrons are very far from the nucleus. If the nucleus of an atom were the size of an apple, the orbit of the nearest electron would be more than a mile out!

Electrons Carry Energy

Because electrons are negatively charged, they are attracted to the positively charged nucleus, but they also repel the negative charges of each other. It takes work to keep them in orbit, just as it takes work to hold an apple in your hand when gravity is pulling the apple down toward the ground. The apple in

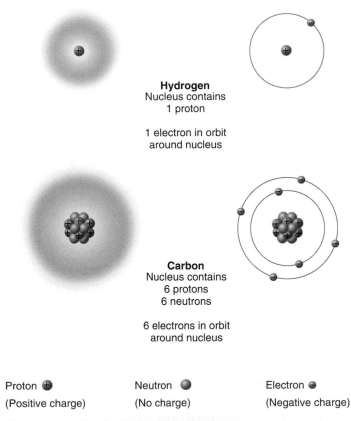

Hydrogen
Nucleus contains
1 proton

1 electron in orbit
around nucleus

Carbon
Nucleus contains
6 protons
6 neutrons

6 electrons in orbit
around nucleus

Proton ⊕   Neutron ●   Electron ⊖
(Positive charge)   (No charge)   (Negative charge)

Figure 3.1 Basic structure of atoms.
All atoms have a nucleus consisting of protons and neutrons, except hydrogen, the smallest atom, which has only one proton and no neutrons in its nucleus. Carbon, for example, has six protons and six neutrons in its nucleus. Electrons spin around the nucleus in orbitals a far distance away from the nucleus. The electrons determine how atoms react with each other.

TABLE 3.1	ELEMENTS COMMON IN LIVING ORGANISMS		
Element	**Symbol**	**Atomic Number**	**Atomic Mass**
Hydrogen	H	1	1.008
Carbon	C	6	12.011
Nitrogen	N	7	14.007
Oxygen	O	8	15.999
Sodium	Na	11	22.989
Magnesium	Mg	12	24.305
Phosphorus	P	15	30.974
Sulfur	S	16	32.064
Chlorine	Cl	17	35.453
Potassium	K	19	39.098
Calcium	Ca	20	40.080
Iron	Fe	26	55.847
Iodine	I	53	126.904

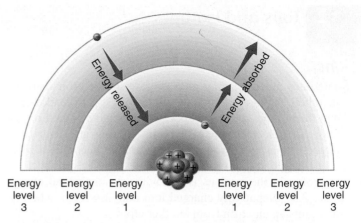

Figure 3.2 The electrons of atoms possess potential energy.

Electrons circulate rapidly around the nucleus in paths called orbitals. Energy level 1 is the lowest potential energy level because it is closest to the nucleus. When an electron absorbs energy, it moves from level 1 to the next higher energy level (level 2). When an electron loses energy, it falls to a lower energy level closer to the nucleus.

your hand is said to possess **energy,** the ability to do work, because of its position—if you were to release it, the apple would fall. Similarly, electrons have energy of position, called potential energy (figure 3.2). It takes work to oppose the attraction of the nucleus, so moving the electron farther out to a more distant orbit requires an input of energy and results in an electron with greater potential energy. Moving an electron in toward the nucleus has the opposite effect; energy is released, and the electron has less potential energy. Consider again an apple held in your hand. If you carry the apple up to a second-story window, it has a greater potential energy when you drop it, compared to when it is dropped at ground level. Similarly, if you lower the apple until it is six inches from the ground, it has less potential energy. Cells use the potential energy of atoms to drive chemical reactions, as we will discuss in chapter 5.

While the energy levels of an atom are often visualized as well-defined circular orbits around a central nucleus, such a simple picture is not realistic. These energy levels, called *electron shells,* often consist of complex three-dimensional shapes, and the exact location of an individual electron at any given time is impossible to specify. However, some locations are more probable than others, and it is often possible to say where an electron is *most likely* to be located. The volume of space around a nucleus where an electron is most likely to be found is called the **orbital** of that electron.

Each electron shell has a specific number of orbitals, and each orbital can hold up to two electrons. The first shell in any atom contains one orbital. Helium, for example, has one electron shell with one orbital that corresponds to the lowest energy level. The orbital contains two electrons (figure 3.3*a*). In atoms with more than one electron shell, the second shell contains four orbitals and holds up to eight electrons. Nitrogen has two electron shells; the first one is completely filled with two electrons, but the four orbitals in the second electron

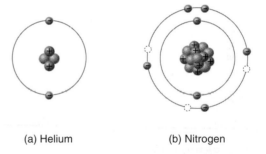

(a) Helium   (b) Nitrogen

Figure 3.3 Electrons in electron shells.

(*a*) An atom of helium has two protons, two neutrons, and two electrons. The electrons fill the one orbital in its one electron shell. (*b*) An atom of nitrogen has seven protons, seven neutrons, and seven electrons. Two electrons fill the orbital in the innermost electron shell, and five electrons occupy orbitals in the second electron shell. The orbitals in the second electron shell can hold up to eight electrons; therefore there are three vacancies in the outer electron shell of a nitrogen atom.

shell are not filled because nitrogen's second shell contains only five electrons (figure 3.3*b*). In atoms with more than two electron shells, subsequent shells also contain up to four orbitals and a maximum of eight electrons. Atoms with incomplete electron orbitals tend to be more reactive because they lose, gain, or share electrons in order to fill their outermost electron shell. This losing, gaining, or sharing of electrons is the basis for chemical reactions in which chemical bonds form between atoms. Chemical bonds will be discussed later in this chapter.

3.1 Atoms, the smallest particles into which a substance can be divided, are composed of electrons orbiting a nucleus composed of protons and neutrons. Electrons determine the chemical behavior of atoms.

3.2 Ions and Isotopes

Ions

Sometimes an atom may gain or lose an electron from its outer shell. Atoms in which the number of electrons does not equal the number of protons because they have gained or lost one or more electrons are called **ions.** All ions are electrically charged. An atom of sodium, for example, becomes a positively charged sodium ion when it loses an electron, because one proton in the nucleus is left with an unbalanced charge (figure 3.4). Negatively charged ions can also form, when an atom gains an electron from another atom.

Isotopes

The number of neutrons in an atom of a particular element can vary without changing the chemical properties of the element.

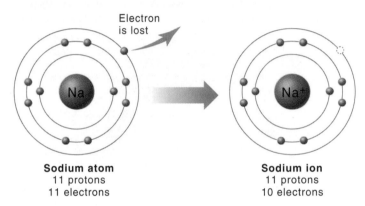

Sodium atom
11 protons
11 electrons

Sodium ion
11 protons
10 electrons

Figure 3.4 Making a sodium ion.

An electrically neutral sodium atom has 11 protons and 11 electrons. Sodium ions bear one positive charge when they ionize and lose one electron. Sodium ions have 11 protons and only 10 electrons.

Atoms that have the same number of protons but different numbers of neutrons are called **isotopes.** Isotopes of an atom have the same atomic number but differ in their atomic mass. Most elements in nature exist as mixtures of different isotopes. For example, there are three isotopes of the element carbon, all of which possess six protons (figure 3.5). The most common isotope of carbon (99% of all carbon) has six neutrons. Because its total mass is 12 (six protons plus six neutrons), it is referred to as carbon-12. The isotope carbon-14 is rare (1 in 1 trillion atoms of carbon) and unstable, such that its nucleus tends to break up into particles with lower atomic numbers, a process called **radioactive decay.** Radioactive isotopes are used in medicine and in dating fossils.

Medical Uses of Radioactive Isotopes

When most people hear the word "radioactive" they picture atomic bombs exploding into mushroom clouds and the devastation that results. While it is true that the radiation emitted from radioactive isotopes can damage cells of the human body, it is also true that isotopes can be used in many medical procedures. Short-lived isotopes, those that decay fairly rapidly and produce harmless products, are commonly used as tracers in the body. A **tracer** is a radioactive substance that is taken up and used by the body. Emissions from the radioactive isotope tracer are detected using special laboratory equipment, and can reveal key diagnostic information about the functioning of the body. For example, a "picture" can be taken of the thyroid gland using a radioactive iodine tracer. The thyroid uses iodine in the production of thyroid hormones, and takes up the radioactive tracer as if it were normal iodine. An overactive thyroid gland may be a symptom of cancer, and can easily be detected using radioactive tracers. There are many other uses of radioactive isotopes in medicine both in detection and treatment of disorders.

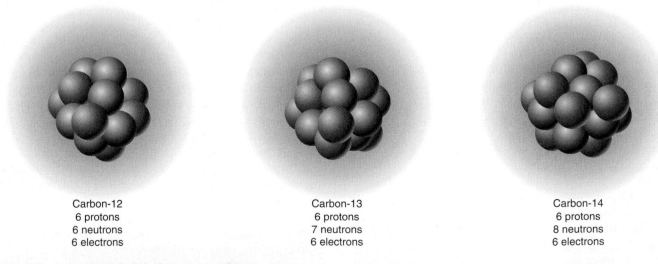

Carbon-12
6 protons
6 neutrons
6 electrons

Carbon-13
6 protons
7 neutrons
6 electrons

Carbon-14
6 protons
8 neutrons
6 electrons

Figure 3.5 Isotopes of the element carbon.

The three most abundant isotopes of carbon are carbon-12, carbon-13, and carbon-14. The yellow "clouds" in the diagrams represent the orbiting electrons, whose numbers are the same for all three isotopes. Protons are shown in purple, and neutrons are shown in pink.

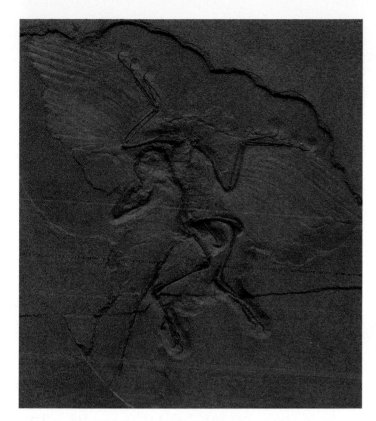

Figure 3.6 Fossil of a transitional bird, *Archaeopteryx*.
This well-preserved fossil of *Archaeopteryx*, a fossil link between reptiles and birds, is about 150 million years old. It was discovered within two years of the publication of *On the Origin of Species*.

Dating Fossils

Fossils are created when the remains, footprints, or other traces of organisms become buried in sand or sediment. Over time, the calcium in bone and other hard tissues becomes mineralized as the sediment is converted to rock. A fossil is any record of prehistoric life—generally taken to mean older than 10,000 years (figure 3.6). By dating the rocks in which fossils occur, biologists can get a very good idea of how old the fossils are. Rocks are usually dated by measuring the degree of radioactive decay of certain radioactive atoms among rock-forming minerals. A radioactive atom is one whose nucleus contains so many neutrons and protons that it is unstable and eventually flies apart, creating more stable atoms of another element. Because the rate of decay of a radioactive element (the percent of atoms that undergo decay in a minute) is constant, scientists can use the amount of radioactive decay to date fossils. The older the fossil, the greater the fraction of its radioactive atoms that have decayed.

A widely employed method of dating fossils less than 50,000 years old is the carbon-14 (^{14}C) **radioisotopic dating** method (figure 3.7). Most carbon atoms have an atomic mass of 12 (^{12}C). However, a tiny but fixed proportion of the carbon atoms in the atmosphere consists of carbon atoms with an atomic mass of 14 (^{14}C). This isotope of carbon is created by the bombardment of nitrogen-14 atoms with cosmic rays. This proportion of ^{14}C is captured by plants in photosynthesis,

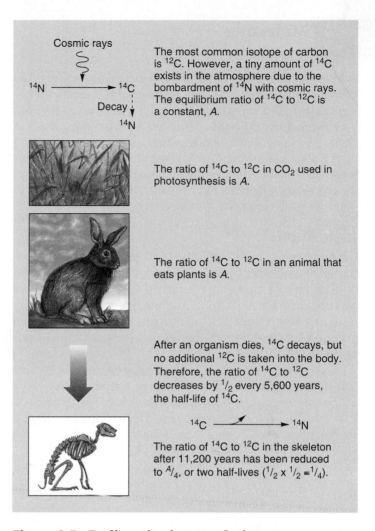

Figure 3.7 Radioactive isotope dating.
This diagram illustrates radioactive dating using carbon-14, a short-lived isotope.

and is the proportion present in the carbon molecules of animal bodies, all of which come ultimately from plants. After a plant or animal dies, its ^{14}C gradually decays over time back to nitrogen-14 (^{14}N). It takes 5,600 years for half of the ^{14}C present in a sample to be converted to ^{14}N by this process. This length of time is called the **half-life** of the isotope. Because the half-life of an isotope is a constant that never changes, the extent of radioactive decay allows you to date a sample. Thus a sample that had a quarter of its original proportion of ^{14}C remaining would be approximately 11,200 years old (two half-lives).

For fossils older than 50,000 years, there is too little ^{14}C remaining to measure precisely, and scientists instead examine the decay of potassium-40 (^{40}K) into argon-40 (^{40}Ar), which has a half-life of 1.3 billion years.

3.2 Isotopes of an element differ in the number of neutrons they contain, but all have the same chemical properties.

3.3 Molecules

A **molecule** is a group of atoms held together by energy. The energy acts as "glue," ensuring that the various atoms stick to one another. The energy or force holding two atoms together is called a **chemical bond.** There are three principal kinds of chemical bonds: ionic bonds, where the force is generated by the attraction of oppositely charged ions; covalent bonds, where the force results from the sharing of electrons; and hydrogen bonds, where the force is generated by the attraction of opposite partial electrical charges.

Ionic Bonds

Chemical bonds called **ionic bonds** form when atoms are attracted to each other by opposite electrical charges. Just as the positive pole of a magnet is attracted to the negative pole of another, so an atom can form a strong link with another atom if they have opposite electrical charges. Because an atom with an electrical charge is an ion, these bonds are called ionic bonds.

Everyday table salt is built of ionic bonds. The sodium and chlorine atoms of table salt are ions, sodium having given up the sole electron in its outermost shell (the shell underneath has eight) and chlorine having gained an electron to complete its outermost shell (figure 3.8). As a result of this electron hopping, sodium atoms in table salt are positive sodium ions and chlorine atoms are negative chloride ions. Because each ion is attracted electrically to all surrounding ions of opposite charge, this causes the formation of an elaborate matrix of sodium and chloride ionic bonds—a crystal. That is why table salt is composed of tiny crystals and is not a powder.

The two key properties of ionic bonds that make them form crystals are that they are strong (although not as strong as covalent bonds) and that they are *not* directional. A charged atom is attracted to the electrical field contributed by all nearby atoms of opposite charge. Ionic bonds do not play an important part in most biological molecules because of this lack of directionality. Complex, stable shapes require the more specific associations made possible by directional bonds.

Covalent Bonds

Strong chemical bonds called **covalent bonds** form between two atoms when they share electrons. Most of the atoms in your body are linked to other atoms by covalent bonds. Why do atoms in molecules share electrons? All atoms seek to fill up their outermost shell of orbiting electrons, which in all atoms (except tiny hydrogen and helium) takes eight electrons. For example, an atom with six outer shell electrons seeks to share them with an atom that has two outer shell electrons or with two atoms that have single outer shell electrons. If only one pair of electrons is shared between two atoms, the

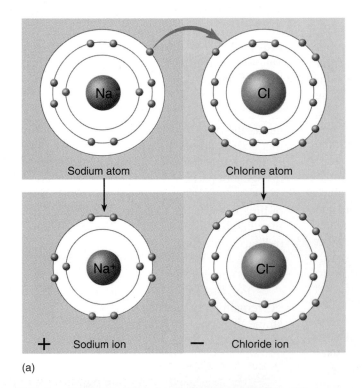

(a)

NaCl crystal

(b)

Figure 3.8 The formation of the ionic bond in table salt.

(a) When a sodium atom donates an electron to a chlorine atom, the sodium atom, lacking that electron, becomes a positively charged sodium ion. The chlorine atom, having gained an extra electron, becomes a negatively charged chloride ion. (b) Sodium chloride forms a highly regular lattice of alternating sodium ions and chloride ions. You are familiar with these crystals as everyday table salt.

=✹=Science in Action

The Case of the Dying Racehorse Foals

Chemistry often plays a role in fiction, particularly detective stories involving deadly poisons like cyanide. In real life, it is sometimes difficult to tell the difference between scientific investigation and good detective work. In the spring of the year 2000, scientists were called in to solve a mystery that Sherlock Holmes would have enjoyed.

Our mystery starts in April in Kentucky, the Bluegrass State. At this time of year, the lush pastures of Kentucky's many horse farms become filled with thoroughbred mares and their leggy foals. Kentucky is the acknowledged center of thoroughbred horse breeding in the United States, a $1.2 billion industry. Normally about 10,000 thoroughbred foals are born in Kentucky each year.

In April of 2000, in horse farms all over Kentucky, newly pregnant mares began to abort unborn fetuses in the first trimester of pregnancy. Foals also began to die unexpectedly soon after birth, some surviving only a few days. A certain number of stillborn or weak foals is to be expected, but not this wave of death.

There were few clues as to the cause of the problem. It seemed to affect all horse breeds and bloodlines, and so probably wasn't due to bad genes. A similar problem arose in 1980–81, when many unexplained horse deaths occurred. That mystery has never been solved.

Veterinarians dubbed the illness causing hundreds of Kentucky mares to deliver stillborn or weak foals "mare reproductive loss syndrome."

How widespread was this syndrome? An alarming 497 of the year's foals had died by May. The number of lost fetuses was unknown, but much larger—roughly one out of 20 foals due to be born that spring had aborted or been stillborn. During May, the University of Kentucky Diagnostic Laboratory was receiving 20+ dead foals/fetuses a day. Financial losses were estimated to be at least $225 million. By the end of the summer the wave of foal death subsided.

Breeders, after eliminating heredity and veterinary procedures as possible causes, focused on pastures. The weather that spring in Kentucky had been abnormal, with warm weather followed by a hard freeze and unusually dry conditions. This suggested a working hypothesis: perhaps the unusual weather had encouraged mold to grow on the grass the horses had been eating. Mold can produce powerful mycotoxins, poisons that could easily harm the reproduction of animals eating the toxins.

Testing this hypothesis, University of Kentucky Agriculture Department investigators looked for mold and mycotoxins on the grass of pastures where foals were dying. They didn't find any. The mycotoxin hypothesis was wrong.

The investigators did find something else, however—something unexpected. Cyanide.

Cyanide is a deadly poison to all vertebrates. In high concentrations, cyanide binds to a protein called cytochrome oxidase that carries out a key step in respiration. Like throwing a monkey wrench into the gears of an engine, this causes respiration to screech to a halt, no longer able to manufacture the energy-rich ATP molecules that fuel the body's activities. Within minutes, the brain dies.

The low cyanide concentrations found in Kentucky grass were not enough to kill an adult horse, but were high enough to kill fetuses and weaken foals.

How did cyanide get onto Kentucky's pastures? Sherlock Holmes would have loved the answer: caterpillar feces! Here is the train of events that transpired, as the University of Kentucky scientists pieced it together.

Black cherry trees are common all over Kentucky, and their leaves naturally contain traces of cyanide. Analyzing the leaves, the scientists discovered that the warm dry weather followed by a frost had caused the cyanide levels in the tree leaves to be much higher than usual.

Eastern tent caterpillars eat the leaves of black cherry trees. They are ferocious eaters, and can defoliate a tree in days, eating every leaf. Agronomists had reported a heavy infestation of Eastern tent caterpillars in Kentucky that spring. Laden with cyanide from the leaves they had eaten, the caterpillars left cyanide-laced feces on the surrounding grass. Eating the grass poisoned the mares, killing their foals.

The caterpillar infestation now is over, and so is the problem. In future, breeders will know what to look for. This case is solved.

covalent bond is called a *single covalent bond*. If two pairs of electrons are shared, it is called a *double covalent bond;* and if three pairs of electrons are shared it is called a *triple covalent bond*. Water (H_2O), for example, is a molecule in which oxygen (six outer electrons) forms single covalent bonds with two hydrogens (one outer electron each) (figure 3.9a). Because the carbon atom has four electrons in its outermost shell, carbon can form as many as four covalent bonds in its attempt to fully populate its outermost shell of electrons. Because there are many ways four covalent bonds can form, carbon atoms participate in many different kinds of molecules.

The two key properties of covalent bonds that make them ideal for their molecule-building role in living systems are that (1) they are strong, involving the sharing of lots of energy, and (2) they are very directional—bonds form between two specific atoms, rather than a generalized attraction of one atom for its neighbors.

Hydrogen Bonds

Weak chemical bonds of a very special sort called **hydrogen bonds** play a key role in biology. To understand them we need to look at covalent bonds again briefly. When a covalent bond forms between two atoms, one nucleus may be much better at attracting the shared electrons than the other—in water, for example, the shared electrons are much more strongly attracted to the oxygen atom than to the hydrogen atoms. When this happens, shared electrons spend more time in the vicinity of the more strongly attracting atom, which as a result becomes somewhat negative in charge; they spend less time in the vicinity of the less strongly attracting atom or atoms, and these become somewhat positive in charge (figure 3.9b). The charges are not full electrical charges like ions possess but rather tiny *partial charges,* signified by the Greek letter delta (δ). What you end up with is a sort of molecular magnet, with positive and negative ends, or "poles." Molecules like this are said to be *polar*. A hydrogen bond occurs when the positive end of one **polar molecule** is attracted to the negative end of another, like two magnets drawn to each other (figure 3.10).

Two key properties of hydrogen bonds cause them to play an important role in biological molecules, the molecules found in living organisms. First, they are weak and so are not effective over long distances like more powerful covalent and ionic bonds. Second, as a result of their weakness, hydrogen bonds are highly directional—polar molecules must be very close for the weak attraction to be effective. Hydrogen bonds are too weak to actually form stable molecules by themselves. Instead, they act like Velcro, forming a tight bond by the additive effects of *many* weak interactions. Hydrogen bonds stabilize the shapes of many important biological molecules by causing certain parts of a molecule to be attracted to other parts. Later in this chapter, we will discuss the role of hydrogen bonding in maintaining the structures of large biological molecules such as proteins and DNA.

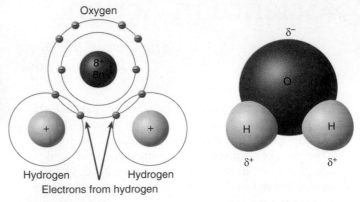

(a) Electron shells in a water molecule

(b) Distribution of partial charges in a water molecule

Figure 3.9 Water molecules contain two covalent bonds.

Each water molecule is composed of one oxygen atom and two hydrogen atoms. (a) The oxygen atom shares one electron with each participating hydrogen atom. (b) A three-dimensional representation of a water molecule showing the partial charges of the molecule.

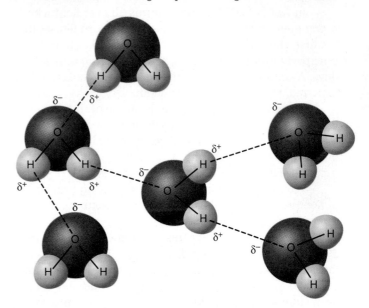

Figure 3.10 Hydrogen bonding in water molecules.

Hydrogen bonds, indicated by the dashed lines, hold water molecules together. The partial positive charge of the hydrogen atoms of one molecule is attracted to the partial negative charge of the oxygen atom of another molecule. This attraction of partial charges holds water molecules together.

3.3 Molecules are atoms linked together by chemical bonds. Ionic bonds form when electrically charged ions attract each other. Most biological molecules are held together by covalent bonds, which are strong linkages created by the sharing of electrons. Hydrogen bonds are important but weaker bonds.

Author's Corner

How Tropical Lizards Climb Vertical Walls

Science is most fun when it tickles your imagination. This is particularly true when you see something you know just can't be true. A few years ago, my wife, Barbara, and I were on a tropical vacation, and I was lying on the bed when a little lizard walked up the wall beside me and across the ceiling, stopping right over my head and looking down at me.

This was no special effect, no trick with mirrors. I was seeing it with my own eyes, real as day, right there above me. The lizard, a green gecko about the size of a toothbrush, stood upside down on the ceiling and seemed to laugh at me for several minutes before trotting over to the far wall and down.

How did my gecko perform this gripping feat? Investigators have puzzled over the adhesive properties of geckos for decades. What force prevented gravity from dropping the gecko on my nose?

The most reasonable hypothesis seemed suction—salamanders' feet form suction cups that let them climb walls, so maybe geckos' do too. The way to test this is to see if the feet adhere in a vacuum, with no air to create suction. Salamander feet don't adhere, but gecko feet do. It's not suction.

How about friction? Cockroaches climb using tiny hooks that grapple onto irregularities in the surface, much as rock climbers use crampons. Geckos, however, happily run up walls of smooth polished glass that no cockroach can climb. It's not friction.

Electrostatic attraction? Clothes in a dryer stick together because of electrical charges created by their rubbing together. You can stop this by adding a "static remover" that is itself heavily ionized. But a gecko's feet still adhere in ionized air. It's not electrostatic attraction.

Could it be glue? Many insects use adhesive secretions from glands in their feet to aid climbing. But there are no gland cells in gecko feet, no secreted chemical. It's not glue.

There was one tantalizing clue, however, the kind that experimenters love. Gecko feet seem to get stickier on surfaces with highly ordered molecules. This suggests that geckos are tapping directly into the molecular structure of the surfaces they walk on!

Tracking down this clue, Robert Full of the University of California, Berkeley, and his research team took a closer look at gecko feet. Geckos have rows of tiny hairs on the bottoms of their feet, like the bristles of a toothbrush. There are about half a million of these hairs on each foot, pointed toward the heel.

When you look at these hairs under a microscope, the end of each hair is divided into between 400 and 1,000 fine projections, the projections sticking out from the tip like tiny stiff brushes.

When a gecko takes a step, it drives the sole of its foot into the surface and pushes it backwards. This shoves the forest of tips directly against the surface. The atoms of each gecko tip become closely engaged with the atoms of the surface, and *that* is the force that defies gravity. When two atoms approach each other very closely—closer than the diameter of an atom—a subtle nuclear attraction called "Van der Waals forces" comes into play. These forces are individually very weak, but when lots of atoms add their little bits, the sum can add up to quite a lot.

Full and his team used microelectrical mechanical sensors (originally designed to be used with atomic force microscopes) to measure the force exerted by a single hair removed from a gecko's foot. It was 200 micronewtons, a tiny force but stupendous for a single hair. Enough to hold up an ant. A million hairs could support a small child. My little gecko, ceiling-walking with 2 million of them, could have carried an 80-pound backpack—talk about being overengineered!

If they stick that well, how do geckos ever come unstuck? For a gecko's foot to stick, each hair projection must butt up squarely against the surface, so the hair's individual atoms can come into play. Tipped past a critical angle—30 degrees—the attractive forces between hair and surface atoms weaken to nothing. The trick is to tip the foot hairs until the projections let go. Geckos release their feet by curling up each toe and peeling it off, sort of like undoing velcro.

Now I can laugh with my little gecko friend, should I see it again, for I know its secret.

3.4 Hydrogen Bonds Give Water Unique Properties

Three-fourths of the earth's surface is covered by liquid water. About two-thirds of your body is water, and you cannot exist long without it. All other organisms also require water. It is no accident that tropical rain forests are bursting with life, whereas dry deserts seem almost lifeless except after rain. The chemistry of life, then, is water chemistry.

Water has a simple atomic structure, an oxygen atom linked to two hydrogen atoms by single covalent bonds. The chemical formula for water is thus H_2O. It is because the oxygen atom attracts the shared electrons more strongly than the hydrogen atoms that water is a *polar molecule* and so can form *hydrogen bonds*. Water's ability to form hydrogen bonds is responsible for much of the organization of living chemistry, from membrane structure to how proteins fold.

The weak hydrogen bonds that form between a hydrogen atom of one water molecule and the oxygen atom of another produce a lattice of hydrogen bonds within liquid water. Each of these bonds is individually very weak and short-lived—a single bond lasts only 1/100,000,000,000 of a second. However, like the grains of sand on a beach, the cumulative effect of large numbers of these bonds is enormous and is responsible for many of the important physical properties of water (table 3.2).

Heat Storage

The temperature of any substance is a measure of how rapidly its individual molecules are moving. Because of the many hydrogen bonds that water molecules form with one another, a large input of thermal energy is required to disrupt the organization of liquid water and raise its temperature. Because of this, water heats up more slowly than almost any other compound and holds its temperature longer. That is a major reason why your body is able to maintain a relatively constant internal temperature.

Ice Formation

If the temperature is low enough, very few hydrogen bonds break in water. Instead, the lattice of these bonds assumes a crystal-like structure, forming a solid we call ice (figure 3.11). Interestingly, ice is less dense than water—that is why icebergs and ice cubes float. Why is ice less dense? Because the hydrogen bonds space the water molecules apart, preventing them from approaching each other more closely.

High Heat of Vaporization

If the temperature is high enough, many hydrogen bonds break in water, with the result that the liquid is changed into vapor. A considerable amount of heat energy is required to do this—every gram of water that evaporates from your skin removes 2,452 joules of heat from your body, which is equal to

TABLE 3.2	THE PROPERTIES OF WATER
Property	**Explanation**
Heat storage	Hydrogen bonds require considerable heat before they break, minimizing temperature changes.
Ice formation	Water molecules in an ice crystal are spaced relatively far apart because of hydrogen bonding.
High heat of vaporization	Many hydrogen bonds must be broken for water to evaporate.
Cohesion	Hydrogen bonds hold molecules of water together.
High polarity	Water molecules are attracted to ions and polar compounds.

Water molecules   Hydrogen bonds

Figure 3.11   Ice formation.
When water cools below 0°C, it forms a regular crystal structure that floats. The individual water molecules are spaced apart and held in position by hydrogen bonds.

the energy released by lowering the temperature of 586 grams of water 1°C. That is why sweating cools you off.

Cohesion

Because water molecules are very polar, they are attracted to other polar molecules—hydrogen bonds bind polar molecules to each other. When the other polar molecule is another water molecule, the attraction is called **cohesion.** The surface tension of water is created by cohesion (figure 3.12). When the other polar molecule is a different substance, the attraction is called **adhesion.** Capillary action—such as water moving up

(a)

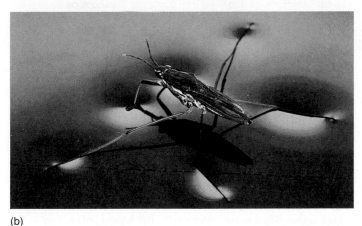

(b)

Figure 3.12   Cohesion.

(a) Cohesion allows water molecules to stick together and form droplets. (b) Surface tension is a property derived from cohesion—that is, water has a "strong" surface due to the force of its hydrogen bonds. Some insects, such as this water strider, literally walk on water.

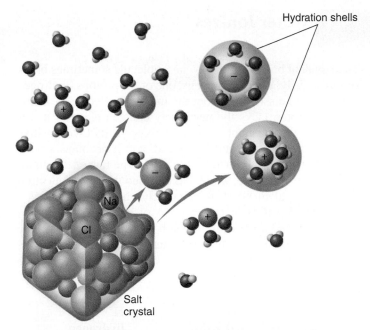

Figure 3.13   How salt dissolves in water.

Salt is soluble in water because the partial charges on water molecules are attracted to the charged sodium and chloride ions. The water molecules surround the ions, forming what are called hydration shells. When all of the ions have been separated from the crystal, the salt is said to be dissolved.

into a paper towel—is created by adhesion. Water clings to any substance, such as paper fibers, with which it can form hydrogen bonds. Adhesion is why things get "wet" when they are dipped in water and why waxy substances do not—they are composed of nonpolar molecules.

High Polarity

Water molecules in solution always tend to form the maximum number of hydrogen bonds possible. Polar molecules form hydrogen bonds and are attracted to water molecules. Polar molecules are called **hydrophilic** (from the Greek *hydros*, water, and *philic*, loving), or water-loving, molecules. Water molecules gather closely around any molecule that exhibits an electrical charge, whether a full charge (ion) or partial charge (polar molecule). When a salt crystal dissolves

in water, what really happens is that individual ions break off from the crystal and become surrounded by water molecules attracted to the negative charge of the chloride ions and the positive charge of the sodium ions. Water molecules orient around each ion like a swarm of bees attracted to honey, and this shell of water molecules prevents the ions from reassociating with the crystal. Similar shells of water form around all polar molecules, and polar molecules that dissolve in water in this way are said to be **soluble** in water (figure 3.13).

Nonpolar molecules like oil do not form hydrogen bonds and are not water-soluble. When nonpolar molecules are placed in water, the water molecules shy away, instead forming hydrogen bonds with other water molecules. The nonpolar molecules are forced into association with one another, crowded together to minimize their disruption of the hydrogen bonding of water. It seems almost as if the nonpolar compounds shrink from contact with water, and for this reason they are called **hydrophobic** (from the Greek *hydros*, water, and *phobos*, fearing). Many biological structures are shaped by such hydrophobic forces, as will be discussed later in this chapter.

> **3.4** Water molecules form a network of hydrogen bonds in liquid and dissolve other polar molecules. Many of the key properties of water arise because it takes considerable energy to break liquid water's many hydrogen bonds.

3.5 Water Ionizes

The covalent bonds within a water molecule sometimes break spontaneously. When it happens, one of the protons (hydrogen atom nuclei) dissociates from the molecule. Because the dissociated proton lacks the negatively charged electron it was sharing in the covalent bond with oxygen, its own positive charge is no longer counterbalanced, and it becomes a positively charged ion, **hydrogen ion** (H^+). The rest of the dissociated water molecule, which has retained the shared electron from the covalent bond, is negatively charged and forms a **hydroxide ion** (OH^-). This process of spontaneous ion formation is called **ionization.** It can be represented by a simple chemical equation, in which the chemical formulas for water and the two ions are written down, with an arrow showing the direction of the dissociation:

$$\begin{array}{ccccc} \text{H}_2\text{O} & \rightleftarrows & \text{OH}^- & + & \text{H}^+ \\ \text{water} & & \text{hydroxide} & & \text{hydrogen} \\ & & \text{ion} & & \text{ion} \end{array}$$

Because covalent bonds are strong, spontaneous ionization is not common. In a liter of pure water at 25°C, only roughly 1 molecule out of each 550 million is ionized at any instant in time, corresponding to 1/10,000,000 (that is, 10^{-7}) of a mole of hydrogen ions. (A mole is a measurement of weight. One mole of any object is the weight of 6.022×10^{23} units of that object). The concentration of H^+ in water can be written more easily by simply counting the number of decimal places after the digit "1" in the denominator:

$$[\text{H}^+] = \frac{1}{10,000,000}$$

pH

A more convenient way to express the hydrogen ion concentration of a solution is to use the **pH scale** (figure 3.14). This scale defines pH as the negative logarithm of the hydrogen ion concentration in the solution:

$$\text{pH} = -\log\ [\text{H}^+]$$

Since the logarithm of the hydrogen ion concentration is simply the exponent of the molar concentration of H^+, the pH equals the exponent times -1. Thus, pure water, with an $[H^+]$ of 10^{-7} mole/liter, has a pH of 7. Recall that for every hydrogen ion formed when water dissociates, a hydroxide ion is also formed, meaning that the dissociation of water produces H^+ and OH^- in equal amounts. Therefore, a pH value of 7 indicates neutrality—a balance between H^+ and OH^-—on the pH scale.

Note that the pH scale is *logarithmic,* which means that a difference of 1 on the pH scale represents a 10-fold change in hydrogen ion concentration. This means that a solution

Figure 3.14   The pH scale.
A fluid is assigned a value according to the number of hydrogen ions present in a liter of that fluid. The scale is logarithmic, so that a change of only 1 means a 10-fold change in the concentration of hydrogen ions; thus lemon juice is 100 times more acidic than tomatoes, and seawater is 10 times more basic than pure water.

with a pH of 4 has *10 times* the concentration of H^+ present in one with a pH of 5.

Acids.   Any substance that dissociates in water to increase the concentration of H^+ is called an acid. Acidic solutions have pH values below 7. The stronger an acid is, the more H^+ it produces and the lower its pH. For example, hydrochloric acid (HCl), which is abundant in your stomach, ionizes completely in water. This means that a dilution of 10^{-1} mole/liter of HCl will dissociate to form 10^{-1} mole/liter of H^+, giving the solution a pH of 1. The pH of champagne, which bubbles because of the carbonic acid dissolved in it, is about 3.

Bases.   A substance that combines with H^+ when dissolved in water is called a base. By combining with H^+, a base lowers the H^+ concentration in the solution. Basic (or alkaline) solutions, therefore, have pH values above 7. Very strong bases, such as sodium hydroxide (NaOH), have pH values of 12 or more.

Buffers

The pH inside almost all living cells, and in the fluid surrounding cells in multicellular organisms, is fairly close to 7. The many proteins that govern metabolism are all extremely sensitive to pH, and slight alterations in pH change their shape and so disrupt their activities. For this reason it is important that a cell maintain a constant pH level. The pH of your blood, for example, is 7.4, and you would survive only a few minutes if it were to fall to 7.0 or rise to 7.8.

Yet the chemical reactions of life constantly produce acids and bases within cells. Furthermore, many animals eat substances that are acidic or basic; Coca-Cola, for example, is acidic, and egg white is basic. What keeps an organism's pH constant? Cells contain chemical substances called buffers that minimize changes in concentrations of H^+ and OH^- (figure 3.15).

A **buffer** is a substance that can take up or release hydrogen ions into solution as the hydrogen ion concentration of a solution changes. Hydrogen ions are donated to the solution when their concentration falls and taken from the solution when their concentration rises. What sort of substance will act in this way? Within organisms, most buffers consist of pairs of substances, one an acid and the other a base. The key buffer in human blood is an acid-base pair consisting of *carbonic acid* (acid) and *bicarbonate* (base). These two substances interact in a pair of reversible reactions. First, carbon dioxide (CO_2) and H_2O join to form carbonic acid (H_2CO_3), which in a second reaction dissociates to yield bicarbonate ion (HCO_3^-) and H^+ (figure 3.16). If some acid or other substance adds H^+ to the blood, the HCO_3^- acts as a base and removes the excess H^+ by forming H_2CO_3. Similarly, if a basic substance removes H^+ from the blood, H_2CO_3 dissociates, releasing more H^+ into the blood. The forward and reverse reactions that interconvert H_2CO_3 and HCO_3^- thus stabilize the blood's pH.

The interaction of carbon dioxide and water has the important consequence that significant amounts of carbon enter into water solution from air in the form of carbonic acid. Biologists believe that life first evolved in the early oceans, which were rich in carbon because of this reaction.

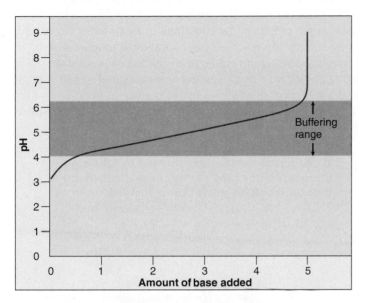

Figure 3.15 Buffers minimize changes in pH.
Adding a base to a solution neutralizes some of the acid present and so raises the pH. Thus, as the curve moves to the right, reflecting more and more base, it also rises to higher pH values. What a buffer does is to make the curve rise or fall very slowly over a portion of the pH scale, called the "buffering range" of that buffer.

> **3.5** A tiny fraction of water molecules spontaneously ionize at any moment, forming H^+ and OH^-. The pH of a solution is the negative logarithm of the H^+ concentration in the solution. Thus, low pH values indicate high H^+ concentrations (acidic solutions), and high pH values indicate low H^+ concentrations (basic solutions).

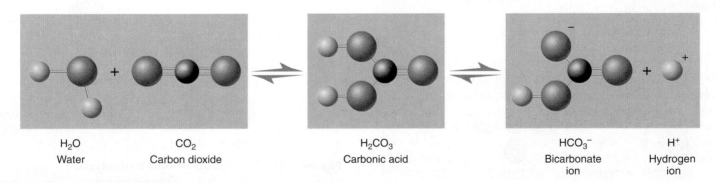

Figure 3.16 The buffering action of carbonic acid and bicarbonate.
Carbon dioxide and water combine chemically to form carbonic acid (H_2CO_3), which dissociates in water, freeing H^+. These reversible reactions occur in your blood. Now if some acid or other substance adds H^+ to your blood, the bicarbonate ion acts as a base and removes the excess H^+ from the solution, forming H_2CO_3. Similarly, if some process removes H^+ from your blood, the carbonic acid dissociates, releasing more hydrogen ions into the solution.

3.6 Forming Macromolecules

An **organic molecule** is a molecule formed by living organisms that consists of a carbon-based core with special groups attached. These groups of atoms have special chemical properties and are referred to as *functional groups*. Functional groups tend to act as units during chemical reactions and to confer specific chemical properties on the molecules that possess them. Five principal functional groups are listed in figure 3.17.

The bodies of organisms contain thousands of different kinds of organic molecules, but much of the body is made of just four kinds: *proteins, nucleic acids, carbohydrates,* and *lipids.* Called **macromolecules** because they can be very large, these four are the building materials of cells, the "bricks and mortar" that make up the bodies of cells and the machinery that runs within them.

The body's macromolecules are assembled by sticking smaller bits together, much as a train is built by linking railcars together. A molecule built up of long chains of similar subunits is called a **polymer** (table 3.3).

Making (and Breaking) Macromolecules

The four different kinds of macromolecules (proteins, nucleic acids, carbohydrates, and lipids) all put their subunits together in the same way: a covalent bond is formed between two subunits in which a hydroxyl group (OH) is removed from one subunit and a hydrogen (H) is removed from the other. This process is called **dehydration synthesis** because, in effect, the removal of the OH and H groups constitutes removal of a molecule of water—the word *dehydration* means "taking away water" (figure 3.18a). This process requires the help of a special class of proteins called **enzymes** to facilitate the positioning of the molecules so that the correct chemical bonds are stressed and broken. The process of tearing down a molecule, such as the protein or fat contained in food that is consumed, is essentially the reverse of dehydration synthesis: instead of removing a water molecule, one is added. When a water molecule comes in, a hydrogen becomes attached to one subunit and a hydroxyl to another, and the covalent bond is broken. The breaking up of a polymer in this way is called **hydrolysis** (figure 3.18b).

> **3.6** Macromolecules are formed by linking subunits together into long chains, removing a water molecule as each link is formed.

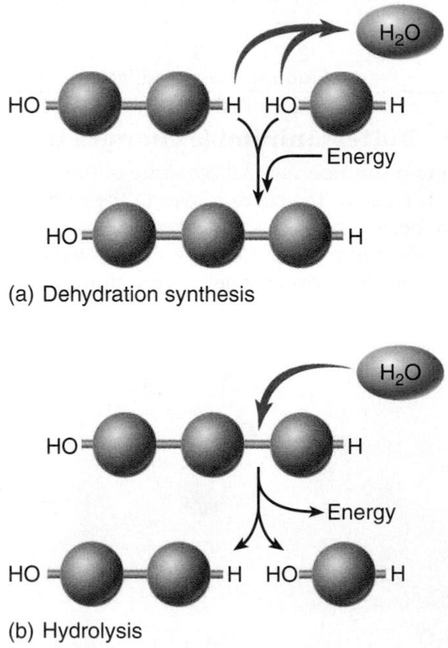

Group	Structural Formula	Ball-and-Stick Model	Found In
Hydroxyl	—OH		Carbohydrates
Carbonyl	$\begin{array}{c} -C- \\ \parallel \\ O \end{array}$		Lipids
Carboxyl	$-C\overset{\displaystyle O}{\underset{\displaystyle OH}{\big<}}$		Proteins
Amino	$-N\overset{\displaystyle H}{\underset{\displaystyle H}{\big<}}$		Proteins
Phosphate	$\begin{array}{c} O^- \\ \mid \\ -O-P-O^- \\ \parallel \\ O \end{array}$		DNA, ATP

Figure 3.17   Five principal functional groups.
Most chemical reactions that occur within organisms involve transferring a functional group from one molecule to another or breaking a carbon-carbon bond.

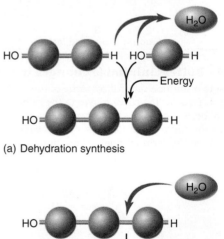

(a) Dehydration synthesis

(b) Hydrolysis

Figure 3.18   Dehydration and hydrolysis.
(a) Biological molecules are formed by linking subunits. The covalent bond between subunits is formed in dehydration synthesis, a process during which a water molecule is eliminated. (b) Breaking such a bond requires the addition of a water molecule, a reaction called hydrolysis.

TABLE 3.3	MACROMOLECULES	
Monomer	**Polymer**	**Cellular structure**
Amino Acid	Polypeptide	Intermediate filament
Nucleotide	DNA strand	Chromosome
Monosaccharide	Starch	Starch grains in a chloroplast
Fatty acid	Fat molecule	Adipose cells with fat droplets

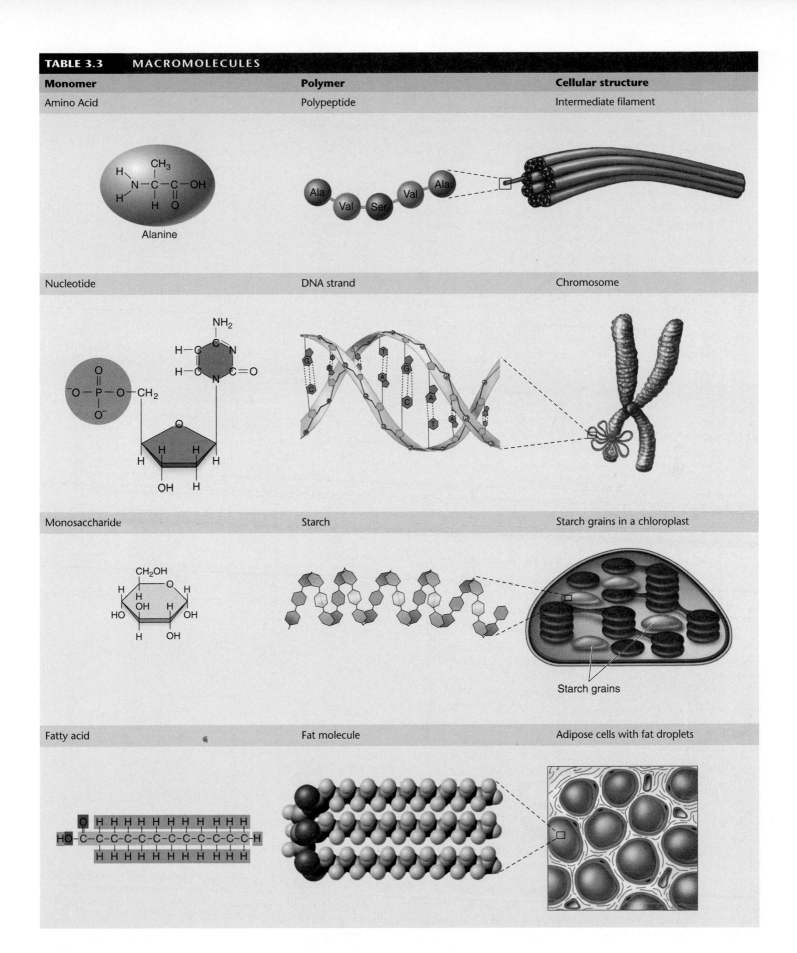

Alanine

Ala Val Ser Val Ala

Starch grains

3.7 Proteins

Complex macromolecules called **proteins** are a major group of biological macromolecules within the bodies of organisms. Perhaps the most important proteins are *enzymes,* which have the key role in cells of lowering the energy required to initiate particular chemical reactions. Other proteins play structural roles. Cartilage, bones, and tendons all contain a structural protein called collagen. Keratin, another structural protein, forms the horns of a rhinoceros and the feathers of a bird. Still other proteins act as chemical messengers within the brain and throughout the body (table 3.4).

Despite their diverse functions, all proteins have the same basic structure: a long polymer chain made of subunits called amino acids. **Amino acids** are small molecules with a simple basic structure: a central carbon atom to which an amino group ($-NH_2$), a carboxyl group ($-COOH$), a hydrogen atom (H), and a functional group, designated "R," are bonded. There are 20 common kinds of amino acids that differ from one another by the identity of their functional R group.

There are four general groups of amino acids (figure 3.19). Six of the amino acid functional groups are nonpolar, differing chiefly in size—the most bulky contain ring structures, and amino acids containing them are called aromatic. Another six are polar but uncharged, and these differ from one another in the strength of their polarity. Five more are polar and are capable of ionizing to a charged form. The remaining three possess special chemical groups that are important in forming links between protein chains or in forming kinks in their shapes.

An individual protein is made by linking specific amino acids together in a particular order, just as a word is made by linking a specific sequence of letters of the alphabet together in a particular order. The covalent bond linking two amino acids together is called a **peptide bond** (figure 3.20), and long chains of amino acids linked by peptide bonds are called **polypeptides.** The hemoglobin proteins in your red blood cells are each composed of four polypeptide chains. The hemoglobin functions as a carrier of oxygen from your lungs to the cells of your body.

Protein Structure

Some proteins form long, thin fibers, whereas others are globular, with the long strands of polypeptides coiled up and folded back on themselves or intertwined with other polypeptides. The shape of a protein is very important because it determines the protein's function. If we picture a polypeptide as a long strand similar to a reed, a protein might be the basket woven from it. Importantly, the sequence of amino acids in a polypeptide determines the protein's structure. There are four general levels of protein structure, primary, secondary, tertiary, and quaternary; all are ultimately determined by the sequence of amino acids.

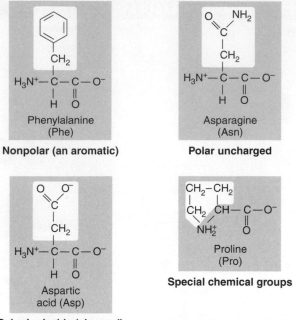

Figure 3.19   Examples of amino acids.
There are four general groups of amino acids that differ in their functional groups (highlighted in *white*).

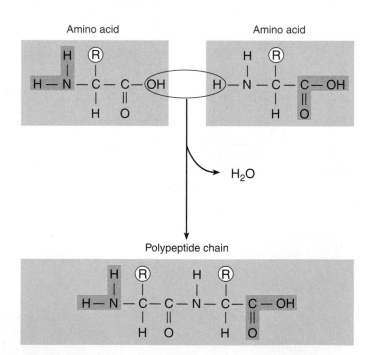

Figure 3.20   The formation of a peptide bond.
Every amino acid has the same basic structure, with an amino group ($-NH_2$) at one end and a carboxyl group ($-COOH$) at the other. The only variable is the functional, or "R," group. Amino acids are linked by dehydration synthesis to form peptide bonds. Chains of amino acids linked in this way are called polypeptides and are the basic structural components of proteins.

Function	Class of Protein	Examples	Examples of Use
TABLE 3.4	THE MANY FUNCTIONS OF PROTEINS		
Enzyme catalysis	Enzymes	Hydrolytic enzymes	Cleave macromolecules
		Proteases	Break down proteins
		Polymerases	Produce nucleic acids
		Kinases	Phosphorylate sugars and proteins
Defense	Immunoglobulins	Antibodies	Mark foreign proteins for elimination
	Toxins	Snake venom	Block nerve function
	Cell surface antigens	MHC proteins	"Self" recognition
Transport	Circulating transporters	Hemoglobin	Carries O_2 and CO_2 in blood
		Myoglobin	Carries O_2 and CO_2 in muscle
		Cytochromes	Electron transport
	Membrane transporters	Sodium-potassium pump	Excitable membranes
		Proton pump	Chemiosmosis
		Glucose transporter	Transport sugar into cells
Support	Fibers	Collagen	Forms cartilage
		Keratin	Forms hair, nails
		Fibrin	Forms blood clots
Motion	Muscle	Actin	Contraction of muscle fibers
		Myosin	Contraction of muscle fibers
Regulation	Osmotic proteins	Serum albumin	Maintains osmotic concentration of blood
	Gene regulators	*lac* repressor	Regulates transcription
	Hormones	Insulin	Controls blood glucose levels
		Vasopressin	Increases water retention by kidneys
		Oxytocin	Regulates uterine contractions and milk let down
Storage	Ion binding	Ferritin	Stores iron, especially in spleen
		Casein	Stores ions in milk
		Calmodulin	Binds calcium ions

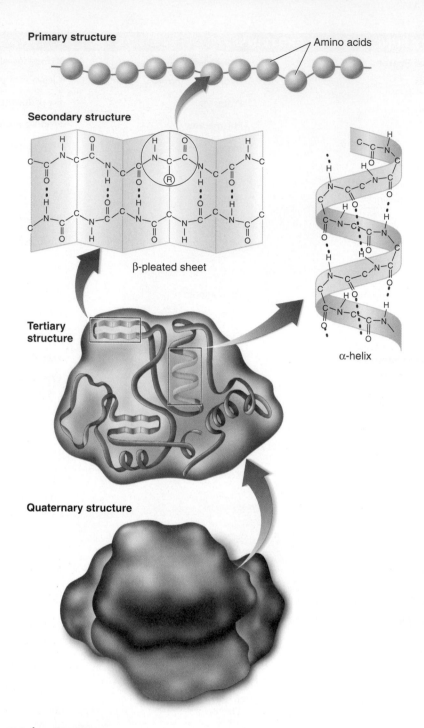

Figure 3.21 Levels of protein structure.

The *primary structure* of a protein is its sequence of amino acids. Twisting or pleating of the chain of amino acids, called *secondary structure,* is due to the formation of localized hydrogen bonds (*red*) within the chain. More complex folding of the chain is referred to as *tertiary structure.* Two or more protein chains associated together form a *quaternary structure.*

The sequence of amino acids of a polypeptide chain is termed the polypeptide's *primary structure* (figure 3.21). The amino acids are linked together by peptide bonds. Because some of the amino acids are nonpolar and others are polar, a protein chain folds up in solution as the nonpolar regions are forced together. To understand this, recall the polar properties of water. Water is a polar molecule that is attracted to and forms hydrogen bonds with other polar molecules but repels nonpolar molecules. This polar attraction and repulsion will push nonpolar amino acid functional groups away from the watery environment, leaving the polar amino acid functional groups to interact with water molecules and each other. Hydrogen bonds forming between amino acids stabilize the folding of the polypeptide. This initial folding is called the *secondary structure* of a protein. Hydrogen bonding within this secondary structure can fold the polypeptide into coils and sheets.

Very
mati
store
long
nucl
part:
orga
tion
by t
AR]

perr
hum
peat
ine

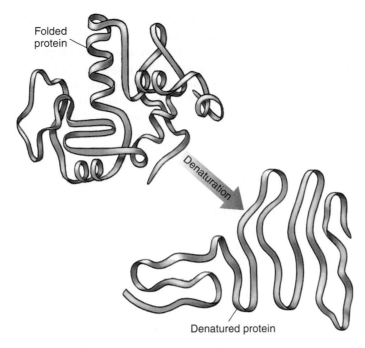

Folded
protein

Denaturation

Denatured protein

Figure 3.22   Protein denaturation.
Changes in a protein's environment, such as variations in
temperature or pH, can cause a protein to unfold and lose its shape
in a process called denaturation. In this denatured state, proteins
are biologically inactive.

The final three-dimensional shape, or *tertiary structure*, of the
protein, folded and twisted in the case of a globular molecule,
is determined by exactly where in a protein chain the nonpolar amino acids occur. Again, the repulsion of the nonpolar
amino acids by water will force these amino acids toward the
interior of the globular protein, leaving the polar amino acids
exposed to the outside. When a protein is composed of more
than one polypeptide chain, the spatial arrangement of the
several component chains is called the *quaternary structure*
of the protein.

The polar nature of the watery environment in the cell
influences how the polypeptide folds into the functional protein. If the polar nature of the protein's environment changes by either increasing temperature or lowering pH, both of
which alter hydrogen bonding, the protein may unfold. When
this happens the protein is said to be **denatured** (figure 3.22).
When the polar nature of the solvent is reestablished, some
proteins may spontaneously refold.

Many structural proteins form long cables that have
architectural roles in cells, providing strength and determining shape (figure 3.23). The globular proteins called enzymes
have three-dimensional shapes with grooves or depressions
that precisely fit a particular sugar or other chemical (figure
3.24); once in the groove, the chemical is encouraged to undergo a reaction—often, one of its chemical bonds is stressed
as the chemical is bent by the enzyme, like a foot in a flexing
shoe. This process of enhancing chemical reactions is called
catalysis, and proteins are the catalytic agents of cells, determining what chemical processes take place and where and
when.

H—

H—

(b)

Fi
(a)
ph
ac
(b)
DI
us

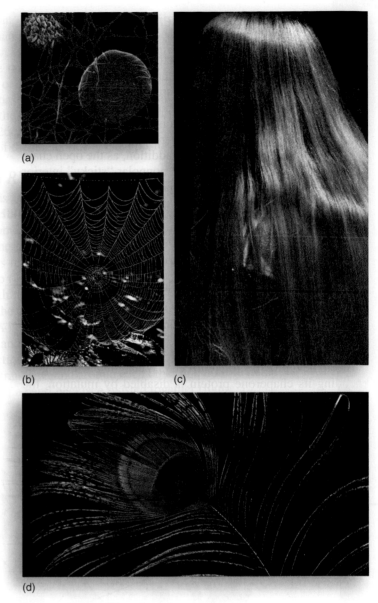

(a)

(b)   (c)

(d)

Figure 3.23   Structural proteins.
This class of proteins has a wide variety of functions. (*a*) Fibrin: traps
red blood cells in forming a blood clot; (*b*) silk: forms a spider's web;
(*c*) keratin: in human hair; (*d*) keratin: in a peacock's feather.

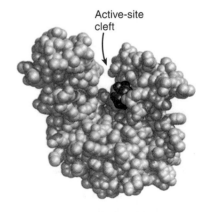

Active-site
cleft

Figure 3.24   Globular protein: an enzyme.
This enzyme (*blue*) has a deep groove that binds a specific chemical
(*red*) at a site on the enzyme called the active site. The bound
molecule then undergoes a chemical reaction catalyzed by the
enzyme.

How (
no aci
that n(
intera(
protei
tein cl
ror w(
folds :
portio
terme
enviro
protei

Cha

How
usual
from
teins (
mal c
that h
enco(
bacte
teins.
right

kinds
heat :
expo:

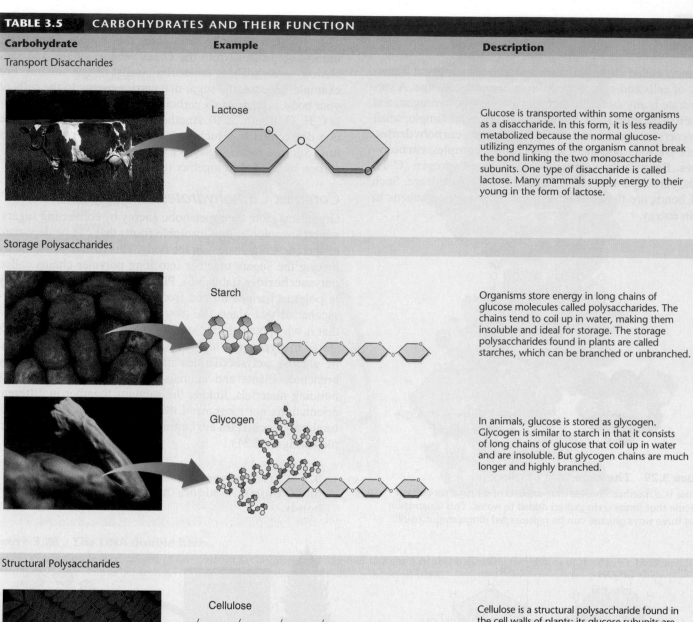

TABLE 3.5 CARBOHYDRATES AND THEIR FUNCTION

Carbohydrate	Example	Description
Transport Disaccharides		
	Lactose	Glucose is transported within some organisms as a disaccharide. In this form, it is less readily metabolized because the normal glucose-utilizing enzymes of the organism cannot break the bond linking the two monosaccharide subunits. One type of disaccharide is called lactose. Many mammals supply energy to their young in the form of lactose.
Storage Polysaccharides		
	Starch	Organisms store energy in long chains of glucose molecules called polysaccharides. The chains tend to coil up in water, making them insoluble and ideal for storage. The storage polysaccharides found in plants are called starches, which can be branched or unbranched.
	Glycogen	In animals, glucose is stored as glycogen. Glycogen is similar to starch in that it consists of long chains of glucose that coil up in water and are insoluble. But glycogen chains are much longer and highly branched.
Structural Polysaccharides		
	Cellulose	Cellulose is a structural polysaccharide found in the cell walls of plants; its glucose subunits are joined in a way that cannot be broken down readily. Cleavage of the links between the glucose subunits in cellulose requires an enzyme most organisms lack. Some animals, such as cows, are able to digest cellulose by means of bacteria and protists they harbor in their digestive tract which provide the necessary enzymes.
	Chitin	Chitin is a type of structural polysaccharide found in the external skeletons of many invertebrates, including insects and crustaceans, and in the cell walls of fungi. Chitin is a modified form of cellulose with a nitrogen group added to the glucose units. When cross-linked by proteins, it forms a tough, resistant surface material.

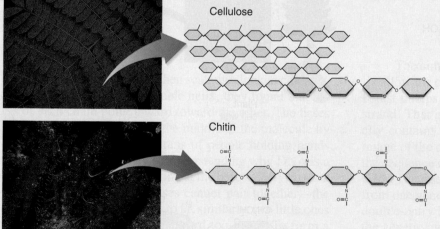

Figu

This
prote
aggr(
unfo

3.10 Lipids

For long-term storage, organisms usually convert glucose into fats, another kind of storage molecule that contains more energy-rich C–H bonds than carbohydrates. Fats and all other biological molecules that are not soluble in water but soluble in oil are called **lipids**. Lipids are insoluble in water not because they are long chains like starches but rather because they are nonpolar. In water, fat molecules cluster together because they cannot form hydrogen bonds with water molecules. This is why oil forms into a layer on top of water when the two substances are mixed. This is also what drives the formation of membranes that surround cells (figure 3.32).

Fats

Fat molecules are composed of two subunits: fatty acids and glycerol. A **fatty acid** is a long chain of carbon and hydrogen atoms (called a hydrocarbon) ending in a carboxyl (—COOH) group. The three carbons of glycerol form the backbone to which three fatty acids are attached in the dehydration reaction that forms the fat molecule. Because there are three fatty acids, the resulting fat molecule is called a *triacylglycerol,* or **triglyceride** (figure 3.33*a*).

Fatty acids with all internal carbon atoms forming covalent bonds with two hydrogen atoms contain the maximum number of hydrogen atoms. Fats composed of these fatty acids are said to be **saturated** (figure 3.33*b*). On the other hand, fats composed of fatty acids that have double bonds between one or more pairs of carbon atoms contain fewer than the maximum

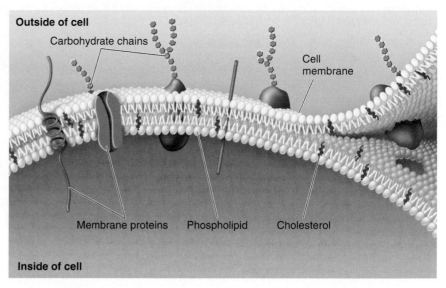

Outside of cell
Carbohydrate chains
Cell membrane
Membrane proteins   Phospholipid   Cholesterol
Inside of cell

Figure 3.32   Lipids are a key component of biological membranes.

Lipids are one of the most common molecules in the human body, because the membranes of all the body's ten trillion cells are composed largely of lipids called phospholipids. Membranes also contain cholesterol, another type of lipid.

number of hydrogen atoms and are called **unsaturated** (figure 3.33*c*). Many plant fatty acids are unsaturated oils. Animal fats, in contrast, are often saturated and occur as hard fats.

Other Types of Lipids

Your body also contains other types of lipids that play many roles in cells in addition to energy storage. The membranes of cells are made of a modified fat called a **phospholipid.** Phospholipids have a polar group at one end and two long tails that

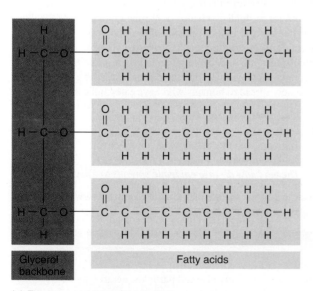

Glycerol backbone

Fatty acids

(a) Fat molecule (triacylglycerol)

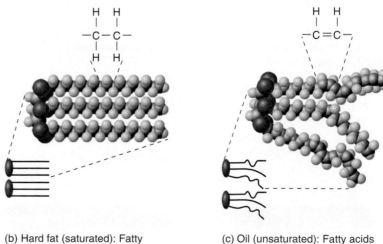

(b) Hard fat (saturated): Fatty acids with single bonds between all carbon pairs

(c) Oil (unsaturated): Fatty acids that contain double bonds between one or more pairs of carbon atoms

Figure 3.33   Saturated and unsaturated fats.

(*a*) Fat molecules are triacylglycerols, each containing a three-carbon glycerol attached to three fatty acid tails. (*b*) Most animal fats are "saturated" (every carbon atom carries the maximum load of hydrogens). Their fatty acid chains fit closely together, and these triacylglycerols form immobile arrays called hard fats. (*c*) Most plant fats are unsaturated, which prevents close association between triacylglycerols and produces oils.

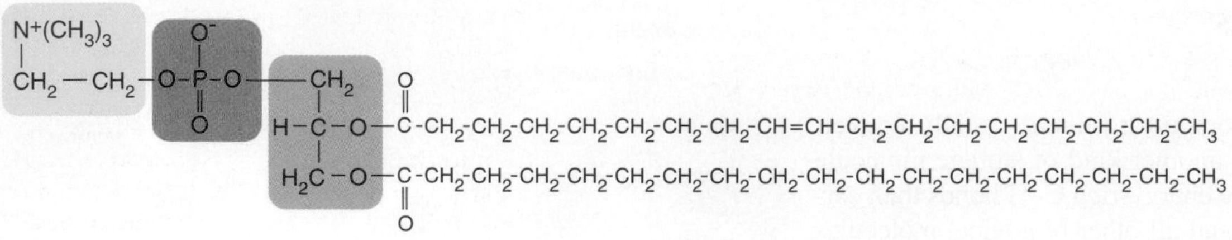

(a) Phospholipid

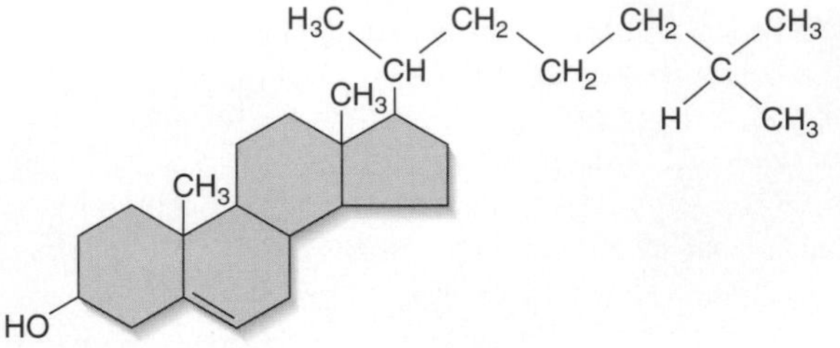

(b) Steroid (cholesterol)

Figure 3.34   Types of lipid molecules.
(*a*) A phospholipid is similar to a fat molecule with the exception that one of the fatty acid tails is replaced with a polar group linked to the glycerol by a phosphate group. (*b*) Steroids like cholesterol are lipids with complex ring structures.

are strongly nonpolar (figure 3.34*a*). In water, the nonpolar ends of phospholipids group together, forming two layers of molecules with the nonpolar tails pointed inside—a lipid bilayer (see figure 3.32 and chapter 4). Membranes also contain a quite different kind of lipid called a **steroid,** composed of four carbon rings (figure 3.34*b*). Most animal cell membranes contain the steroid cholesterol. Excess saturated fat intake can cause plugs of cholesterol to form in the blood vessels, which may lead to blockage, high blood pressure, stroke, or

heart attack. Male and female sex hormones are also steroids. Other important biological lipids include rubber, waxes, and pigments, such as the chlorophyll that makes plants green and the retinal that your eyes use to detect light.

> **3.10** Lipids are not water-soluble. Fats contain chains of fatty acid subunits and can store energy.

Exploring Current Issues

Additional Resources

Go to your campus library or look online to find the following articles, which further develop some of the concepts found in this chapter.

Centers for Disease Control and Prevention. (2004). Frequently asked questions about food irradiation. *Retrieved from* www.cdc.gov/ncidod/dbmd/diseaseinfo/foodirradiation.htm.

Jaffe, S. (2003). Biologically derived hydrogen: future fuel? Researchers turn to *Thermotoga neapolitana* and *Chlamydomonas reinhardtii* to fill gas tanks and heat homes. *The Scientist,* 17(8), 28.

Stephan, M.A. (2004). The floodgates open for aquaporins. *The Scientist,* 18(1), 26.

U.S. Food and Drug Administration. (2003). Revealing trans fats. *FDA Consumer Magazine,* 37(5).

Willet, W.C. and M.J. Stampfer. (2003). Rebuilding the food pyramid. *Scientific American,* 288(1), 64.

In the News: Mad Cow Disease

We are accustomed to thinking of proteins as enzymes and structural molecules. It has come as something of a shock to discover that a protein can be responsible for infectious disease. Called "prions," these very stable proteins are misfolded versions of normal brain proteins that have the unfortunate ability to induce others of their kind to similarly misfold. A chain reaction of misfolding results, eventually destroying the brain. An epidemic spread among cows by feeding them protein supplements prepared from infected animals, and the prions responsible for mad cow disease have now spread to humans who have eaten infected cows.

Article 1.   *Mad cows and prions.*
Article 2.   *Do prions threaten our blood supply?*
Article 3.   *The growing epidemic of mad cow disease.*
Article 4.   *Mad deer disease.*
Article 5.   *The mad cow disease epidemic spreads to Europe.*

Find these articles, written by the author, online at www.mhhe.com/tlwessentials/exp3.

Summary

Some Simple Chemistry

3.1 Atoms

- An atom is the smallest particle that retains the chemical properties of its substance. Atoms contain a core nucleus of protons and neutrons with electrons orbiting around the nucleus (**figure 3.1**).

- Protons are positively charged particles, neutron particles carry no charge, and electrons are negatively charged particles. Electrons also carry energy and determine the chemical behavior of the atom (**figure 3.2**).

- Most electron shells hold up to eight electrons and atoms will undergo chemical reactions in order to fill the outer electron shell, either by gaining, losing, or sharing electrons (**figure 3.3**).

3.2 Ions and Isotopes

- Ions are atoms that have gained one or more electrons (negative ions) or lost one or more electrons (positive ions) (**figure 3.4**).

- Isotopes are atoms that have the same number of protons but differing numbers of neutrons. Some are unstable and have applications in medicine and dating fossils (**figures 3.5 and 3.7**).

3.3 Molecules

- Molecules form when atoms are held together with energy. The force holding atoms together is called a chemical bond. There are three main types of chemical bonds.

- Ionic bonds form when ions of opposite charge are attracted to each other (**figure 3.8**). Covalent bonds form when two atoms share electrons (**figure 3.9**). Hydrogen bonds form between polar molecules. Polar molecules are held together by covalent bonds in which the shared electrons are unevenly distributed around their nuclei, giving them a slightly positive end and a slightly negative end. Hydrogen bonds form when the positive end of one molecule is attracted to the negative end of another (**figure 3.10**).

Water: Cradle of Life

3.4 Hydrogen Bonds Give Water Unique Properties

- Water molecules are polar molecules that form hydrogen bonds with each other and with other polar molecules. Many of the physical properties of water, such as heat storage, ice formation, high heat of vaporization, cohesion, and being a highly polar solvent, are attributed to this hydrogen bonding (**figures 3.11–3.13**).

3.5 Water Ionizes

- Water molecules dissociate forming negatively charged hydroxide ions and positively charged hydrogen ions. This property of water is significant because the concentration of hydrogen ions in a solution determines its pH.

- A solution with a higher hydrogen ion concentration is an acid with specific chemical properties and a solution with a lower hydrogen ion concentration is a base with different chemical properties (**figure 3.14**). Buffers control changes in pH.

Macromolecules

3.6 Forming Macromolecules

- Large molecules called macromolecules are formed by dehydration reactions in which molecular subunits called monomers are linked together through covalent bonding that produces a water molecule as a product of the reaction (**figure 3.18**).

- Amino acids are the subunits that link together to form proteins. Nucleotide subunits link together to form nucleic acids. Monosaccharide subunits link together to form carbohydrates. Fatty acids link together to form lipids (**table 3.3**).

3.7 Proteins

- Proteins are macromolecules that carry out many functions in the cell (**table 3.4**). They are produced by the linking together of amino acids with covalent bonds referred to as peptide bonds. The sequence of amino acids is the primary structure of the protein. The chain of amino acids then folds into its final shape, exhibiting different levels of complexity, from secondary to tertiary to quaternary structures (**figure 3.21**).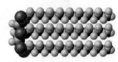

- Chaperone proteins aid in the folding of the amino acid chain into the correct shape. If misfolded, the protein will not function properly (**figure 3.25**).

3.8 Nucleic Acids

- Nucleic acids, such as DNA and RNA, are long chains of nucleotides. They function as information storage molecules in the cell, storing the information needed to build proteins. The sequence of A, T, G, and C nucleotides in DNA determines the sequence of amino acids in a protein (**figures 3.27 and 3.28**).

3.9 Carbohydrates

- Carbohydrates, also referred to as sugars, are macromolecules that serve two primary functions in the cell: structural framework and energy storage. Carbohydrates such as cellulose and chitin provide structural integrity to cells and bodies of organisms. Carbohydrates such as starch and glycogen provide a means of storing energy in the cell. Starch and glycogen are broken down in the cells when energy is needed (**table 3.5**).

3.10 Lipids

- Lipids are large nonpolar molecules that are insoluble in water. They make up the lipid bilayer of the plasma membrane of cells and function in long-term energy storage (**figures 3.32 and 3.33**).

Self-Test

1. The smallest particle into which a substance can be divided and still retain all of its chemical properties is
   a. matter.
   c. a molecule.
   b. an atom.
   d. mass.
2. An atom that has gained or lost one or more electrons is
   a. an isotope.
   c. an ion.
   b. a neutron.
   d. radioactive.
3. Atoms are held together by a force called a bond. The three types of bonds discussed in chapter 3 are
   a. positive, negative, and neutral.
   b. ionic, doric, and corinthian.
   c. magnetic, electric, and radioactive.
   d. ionic, covalent, and hydrogen.

4. Water has some very unusual properties. These properties occur because of the
   a.   hydrogen bonds between the individual water molecules.
   b.   covalent bonds within each individual water molecule.
   c.   hydrogen bonds within each individual water molecule.
   d.   ionic bonds between the individual water molecules.
5. Water sometimes ionizes, a single molecule breaking apart into an H^+ ion and an OH^- ion. Other materials may dissociate in water, resulting in an increase of either (1) H^+ ions or (2) OH^- ions in the solution. We call the results
   a.   (1) acids and (2) bases.
   b.   (1) bases and (2) acids.
   c.   (1) neutral solutions and (2) neutronic solutions.
   d.   (1) hydrogen solutions and (2) hydroxide solutions.
6. The four kinds of organic macromolecules are
   a.   hydroxyls, carboxyls, aminos, and phosphates.
   b.   proteins, carbohydrates, lipids, and nucleic acids.
   c.   ATP, ADP, DNA, and RNA.
   d.   carbon, hydrogen, oxygen, and nitrogen.

7. Your body is filled with many types of proteins. Each type has a distinctive sequence of amino acids that determines both its specialized _____ and its unique _____.
   a.   number, weight.   c.   structure, function.
   b.   length, mass.   d.   charge, pH.
8. Nucleic acids
   a.   are the energy source for our bodies.
   b.   act on other molecules, breaking them apart or building new ones, to help us function.
   c.   are only found in a few, specialized locations within the body.
   d.   are information storage devices found in every cell in the body.
9. Carbohydrates are used for
   a.   structure and for energy.   c.   fat storage and for hair.
   b.   information storage.   d.   hormones and for enzymes.
10. Lipids are used for
   a.   structure and for energy.
   b.   information storage.
   c.   energy storage and for hormones.
   d.   enzymes and for hormones.

Visual Understanding

1. **Figure 3.7**   Based on the figure, explain why it is difficult to use carbon-14 dating on things that are older than about 50,000 years.

2. **Figure 3.12**   How can these things—drops of water clinging to a web and insects walking on water—occur?

Challenge Questions

Some Simple Chemistry   What happens if an atom of an element gains or loses electrons? What happens if an atom of an element gains or loses neutrons? What happens if an atom of an element gains or loses protons?

Water: Cradle of Life   You are on a 10-day backpacking trip with a small group of friends. Yvonne has washed out a set of water bottles with bleach. Before she can rinse them, Carlos, hot and thirsty, picks one up and drinks it, drops it, and begins to choke. What is the problem, and what can you do for him?

Macromolecules   Gloria, who is six years old, has become very ill with the flu. She has been vomiting, has diarrhea, and is running a very high fever. Explain the two major biochemical problems that Gloria faces.

Online Learning Center

Visit the Online Learning Center for this chapter at www.mhhe.com/tlwessentials/ch3 for quizzes, animations, interactive learning exercises, and other study tools. At the site you will also find extended answers to the end-of-chapter questions.

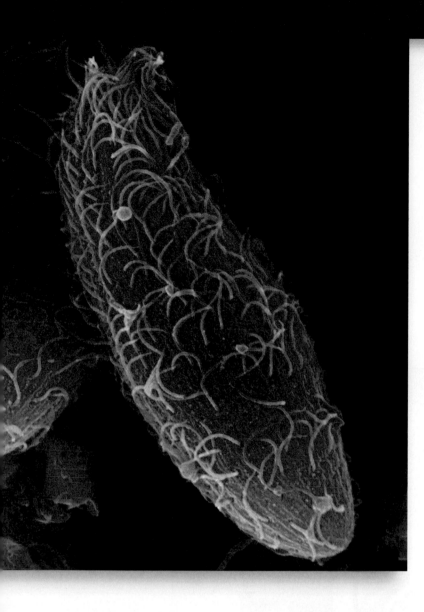

4

Cells

This alarming looking creature is the single-celled organism *Dileptus,* magnified a thousand times. Too small to see with the unaided eye, *Dileptus* is one of hundreds of inhabitants of a drop of pond water. Everything that a living organism does to survive and prosper, *Dileptus* must do with only the equipment this tiny cell provides. Just as you move about using legs to walk, so *Dileptus* uses the hairlike projections (called cilia) that cover its surface to propel itself through the water. Just as your brain is the control center of your body, so the compartment called the nucleus, deep within the interior of *Dileptus,* controls the many activities of this complex and very active cell. *Dileptus* has no mouth, but it takes in food particles and other molecules through its surface. This versatile protist is capable of leading a complex life because its interior is subdivided into compartments, in each of which it carries out different activities. Functional specialization is the hallmark of this cell's interior, a powerful approach to cellular organization which is shared by all eukaryotes.

4.1 Cells

Hold your finger up and look at it closely. What do you see? Skin. It looks solid and smooth, creased with lines and flexible to the touch. But if you were able to remove a bit and examine it under a microscope (figure 4.1), it would look very different—a sheet of tiny, irregularly shaped bodies crammed together like tiles on a floor. What you would see are epithelial cells; in fact, every tissue of your body is made of cells, as are the bodies of all organisms. Some organisms are composed of a single cell, and some are composed of many cells. All cells, however, are very small. In this chapter we look more closely

at cells and learn something of their internal structure and how they communicate with their environment.

The Cell Theory

Because cells are so small, no one observed them until microscopes were invented in the mid-seventeenth century. Robert Hooke first described cells in 1665, when he used a microscope he had built to examine a thin slice of nonliving plant tissue called cork. Hooke observed a honeycomb of tiny, empty (because the cells were dead) compartments. He called the compartments in the cork *cellulae* (Latin, small rooms), and the term has come down to us as **cells.** For another century and a half, however, biologists failed to rec-

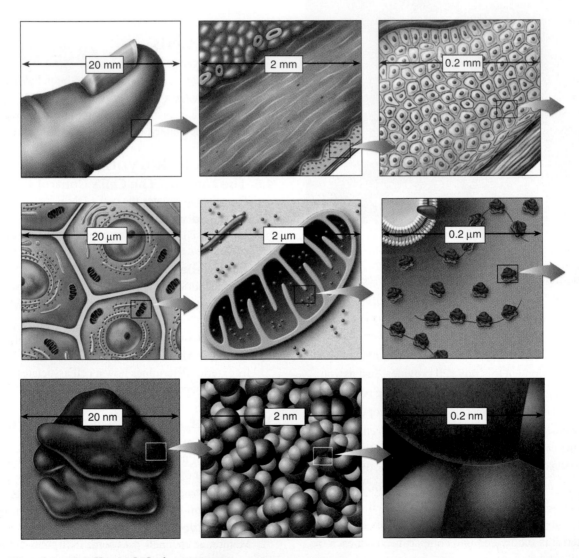

Figure 4.1 The size of cells and their contents.

This diagram shows the size of human skin cells, organelles, and molecules. In general, the diameter of a human skin cell is 20 micrometers (μm), of a mitochondrion is 2 μm, of a ribosome is 20 nanometers (nm), of a protein molecule is 2 nm, and of an atom is 0.2 nm.

ognize the importance of cells. In 1838, botanist Matthias Schleiden made a careful study of plant tissues and developed the first statement of the cell theory. He stated that all plants "are aggregates of fully individualized, independent, separate beings, namely the cells themselves." In 1839, Theodor Schwann reported that all animal tissues also consist of individual cells.

The idea that all organisms are composed of cells is called the **cell theory.** In its modern form, the cell theory includes three principles:

1. All organisms are composed of one or more cells, within which the processes of life occur.
2. Cells are the smallest living things. Nothing smaller than a cell is considered alive.
3. Cells arise only by division of a previously existing cell. Although life likely evolved spontaneously in the environment of the early earth, biologists have concluded that no additional cells are originating spontaneously at present. Rather, life on earth represents a continuous line of descent from those early cells.

Most Cells Are Very Small

Cells are not all the same size. Individual cells of the marine alga *Acetabularia,* for example, are up to 5 centimeters long—as long as your little finger. In contrast, the cells of your body are typically from 5 to 20 micrometers in diameter, too small to see with the naked eye. It would take anywhere from 100 to 400 human cells to span the diameter of the head of a pin. The cells of bacteria are even smaller than yours, only a few micrometers thick.

Why Aren't Cells Larger?

Why are most cells so tiny? Most cells are small because larger cells do not function as efficiently. In the center of every cell is a command center that must issue orders to all parts of the cell, directing the synthesis of certain enzymes, the entry of ions and molecules from the exterior, and the assembly of new cell parts. These orders must pass from the core to all parts of the cell, and it takes them a very long time to reach the periphery of a large cell. For this reason, an organism made up of relatively small cells is at an advantage over one composed of larger cells.

Another reason cells are not larger is the advantage of having a greater **surface-to-volume ratio.** As cell size increases, volume grows much more rapidly than surface area (figure 4.2). For a round cell, surface area increases as the square of the diameter, whereas volume increases as the cube. Thus a cell with 10 times greater diameter would have 100 (10^2) times the surface area but 1,000 (10^3) times the volume. A cell's surface provides the interior's only opportunity to interact with the environment with substances passing into and out of the cell across its surface, and large cells have far less surface for each unit of volume than do small ones.

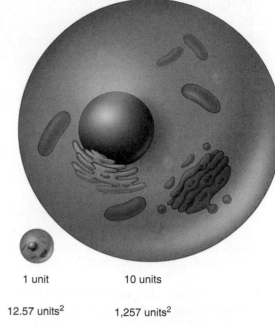

Cell radius (r)	1 unit	10 units
Surface area ($4\pi r^2$)	12.57 units^2	1,257 units^2
Volume ($\frac{4}{3}\pi r^3$)	4.189 units^3	4,189 units^3

Figure 4.2 Surface-to-volume ratio.
As a cell gets larger, its volume increases at a faster rate than its surface area. If the cell radius increases by 10 times, the surface area increases by 100 times, but the volume increases by 1,000 times. A cell's surface area must be large enough to meet the needs of its volume.

Some larger cells, however, function quite efficiently in part because they have structural features that increase surface area. Cells in the nervous system, for example, called neurons, are long slender cells, some extending more than a meter in length. These cells efficiently interact with their environment because although they are long, they are thin, some less than 1 micrometer in diameter, and so their interior regions are not far from the surface at any given point.

Another structural feature that increases the surface area of a cell are small "fingerlike" projections called microvilli. The cells that line the small intestines of the human digestive system are covered with microvilli that dramatically increase the surface area of the cells.

An Overview of Cell Structure

All cells are surrounded by a delicate membrane that controls the permeability of the cell to water and dissolved substances. A semifluid matrix called **cytoplasm** fills the interior of the cell. It used to be thought that the cytoplasm was uniform, like Jell-O, but we now know that it is highly organized. Your cells, for example, have an internal framework that both gives the cell its shape and positions components and materials within its interior. In the following sections, we explore the membranes that encase all living cells and then examine in detail their interiors.

Visualizing Cells

How many cells are big enough to see with the unaided eye? Other than egg cells, not many are (figure 4.3). Most are less than 50 micrometers in diameter, far smaller than the period at the end of this sentence.

The Resolution Problem. How do we study cells if they are too small to see? The key is to understand why we can't see them. The reason we can't see such small objects is the limited resolution of the human eye. **Resolution** is defined as the minimum distance two points can be apart and still be distinguished as two separated points. When two objects are closer together than about 100 micrometers, the light reflected from each strikes the same "detector" cell at the rear of the eye. Only when the objects are farther than 100 micrometers apart will the light from each strike different cells, allowing your eye to resolve them as two objects rather than one.

Microscopes. One way to increase resolution is to increase magnification, so that small objects appear larger. Robert Hooke and Antony van Leeuwenhoek used glass lenses to magnify small cells and cause them to appear larger than the 100-micrometer limit imposed by the human eye. The glass lens adds additional focusing power. Because the glass lens makes the object appear closer, the image on the back of the eye is bigger than it would be without the lens.

Modern light microscopes use two magnifying lenses (and a variety of correcting lenses) to achieve very high magnification and clarity (table 4.1). The first lens focuses the image of the object on the second lens, which magnifies it again and focuses it on the back of the eye. Microscopes that magnify in stages using several lenses are called **compound microscopes.** They can resolve structures that are separated by more than 200 nanometers (nm).

Increasing Resolution. Light microscopes, even compound ones, are not powerful enough to resolve many structures within cells. For example, a membrane is only 5 nanometers thick. Why not just add another magnifying stage to the microscope and so increase its resolving power? Because when two objects are closer than a few hundred nanometers, the light beams reflecting from the two images start to overlap. The only way two light beams can get closer together and still be resolved is if their wavelengths are shorter.

One way to avoid overlap is by using a beam of electrons rather than a beam of light. Electrons have a much shorter wavelength, and a microscope employing electron beams has 1,000 times the resolving power of a light microscope. **Transmission electron microscopes,** so called because the electrons used to visualize the specimens are transmitted through the material, are capable of resolving objects only 0.2 nanometer apart—just twice the diameter of a hydrogen atom!

A second kind of electron microscope, the **scanning electron microscope,** beams the electrons onto the surface of the specimen. The electrons reflected back from the surface of the specimen, together with other electrons that the specimen itself emits as a result of the bombardment, are amplified and transmitted to a screen, where the image can be viewed and photographed. Scanning electron microscopy yields striking three-dimensional images and has improved our understanding of many biological and physical phenomena (see table 4.1).

Visualizing Cell Structure by Staining Specific Molecules. A powerful tool for the analysis of cell structure has been the

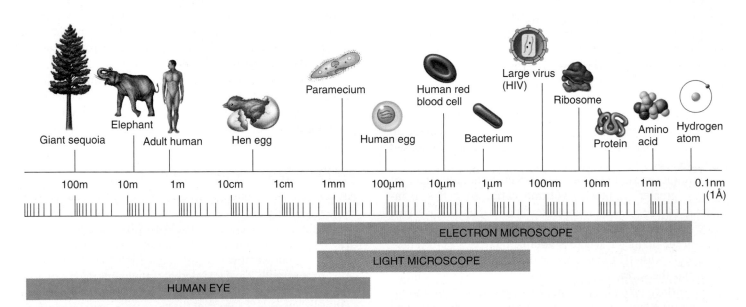

Figure 4.3   A scale of visibility.

Most cells are microscopic in size, although vertebrate eggs are typically large enough to be seen with the unaided eye. Prokaryotic cells are generally 1 to 2 micrometers (μm) across.

TABLE 4.1   TYPES OF MICROSCOPES

Light Microscopes

28.36 μm

Bright-field microscope: Light is simply transmitted through a specimen in culture, giving little contrast. Staining specimens improves contrast but requires that cells be fixed (not alive), which can cause distortion or alteration of components.

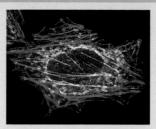

Fluorescence microscope: A set of filters transmits only light that is emitted by fluorescently stained molecules or tissues.

67.74 μm

Dark-field microscope: Light is directed at an angle toward the specimen; a condenser lens transmits only light reflected off the specimen. The field is dark, and the specimen is light against this dark background.

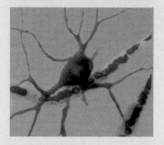

Confocal microscope: Light from a laser is focused to a point and scanned across the specimen in two directions. Clear images of one plane of the specimen are produced, while other planes of the specimen are excluded and do not blur the image. Fluorescent dyes and false coloring enhances the image.

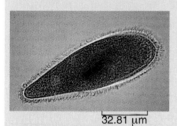

32.81 μm

Phase-contrast microscope: Components of the microscope bring light waves out of phase, which produces differences in contrast and brightness when the light waves recombine.

Electron Microscopy

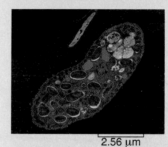

2.56 μm

Transmission electron microscope: A beam of electrons is passed through the specimen. Electrons that pass through are used to form an image. Areas of the specimen that scatter electrons appear dark. False coloring enhances the image.

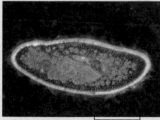

26.6 μm

Differential-interference-contrast microscope: Out-of-phase light waves to produce differences in contrast are combined with two beams of light travelling close together, which create even more contrast, especially at the edges of structures.

6.76 μm

Scanning electron microscope: An electron beam is scanned across the surface of the specimen, and electrons are knocked off the surface. Thus, the surface topography of the specimen determines the contrast and the content of the image. False coloring enhances the image.

---

use of stains that bind to specific molecular targets. This approach has been used in the analysis of tissue samples, or histology, for many years and has been improved dramatically with the use of antibiotics that bind to very specific molecular structures. This process, called immunocytochemistry, uses antibodies generated in animals such as rabbits or mice. When these animals are injected with specific proteins, they will produce antibodies that specifically bind to the injected protein, which can be purified from their blood. These purified antibodies can then be chemically bonded to enzymes, stains, or fluorescent molecules that glow when exposed to specific wavelengths of light. When cells are washed in a solution con-

taining the antibodies, they bind to cellular structures that contain the target molecule and can be seen with light microscopy. This approach has been used extensively in the analysis of cell structure and function.

4.1   All living things are composed of one or more cells, each a small volume of cytoplasm surrounded by a cell membrane. Most cells and their components are so small they can only be viewed using microscopes.

4.2 The Plasma Membrane

Encasing all living cells is a delicate sheet of molecules called the **plasma membrane.** It would take more than 10,000 of these sheets, which are about 7 nanometers thick, piled on top of one another to equal the thickness of this sheet of paper. However, the sheets are not simple in structure, like a soap bubble's skin. Rather, they are made up of a diverse collection of proteins floating within a lipid framework like small boats bobbing on the surface of a pond. Regardless of the kind of cell they enclose, all plasma membranes have the same basic structure of proteins embedded in a sheet of lipids, called the **fluid mosaic model.**

The lipid layer that forms the foundation of a plasma membrane is composed of modified fat molecules called **phospholipids.** One end of a phospholipid molecule has a phosphate chemical group attached to it, making it extremely polar (and thus water-soluble), whereas the other end is composed of two long fatty acid chains that are strongly nonpolar (and thus water-insoluble) (figure 4.4).

Imagine what happens when a collection of phospholipid molecules is placed in water. A structure called a **lipid bilayer** forms spontaneously (figure 4.5). How can this happen? The long nonpolar tails of the phospholipid molecules are pushed away by the water molecules that surround them, shouldered aside as the water molecules seek partners that can form hydrogen bonds. After much shoving and jostling, every phospholipid molecule ends up with its polar head facing water and its nonpolar tail facing away from water. The phospholipid molecules form a *double* layer. Because there are two layers with the tails facing each other, no tails are ever in contact with water.

Because the interior of a lipid bilayer is completely nonpolar, it repels any water-soluble molecules that attempt to pass through it, just as a layer of oil stops the passage of a drop of water (that's why ducks do not get wet).

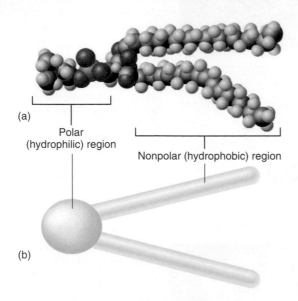

Figure 4.4 Phospholipid structure.

One end of a phospholipid molecule is polar and the other is nonpolar. (*a*) The molecular structure is shown by colored spheres representing individual atoms (*black* for carbon, *blue* for hydrogen, *red* for oxygen, and *yellow* for phosphorus). (*b*) The phospholipid is often depicted diagrammatically as a ball with two tails.

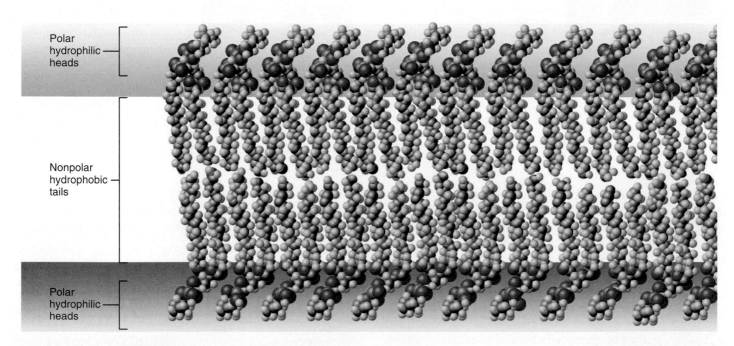

Figure 4.5 The lipid bilayer.

The basic structure of every plasma membrane is a double layer of lipids. This diagram illustrates how phospholipids aggregate to form a bilayer with a nonpolar interior.

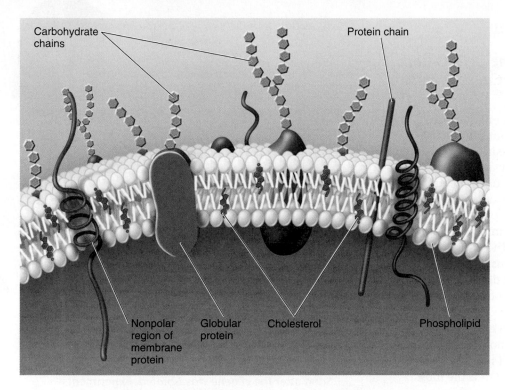

Figure 4.6 Proteins are embedded within the lipid bilayer.
A variety of proteins protrude through the lipid bilayer of animal cells. Membrane proteins function as channels, receptors, and cell surface markers. Carbohydrate chains are often bound to these proteins and to phospholipids in the membrane itself as well. These chains serve as distinctive identification tags, unique to particular types of cells.

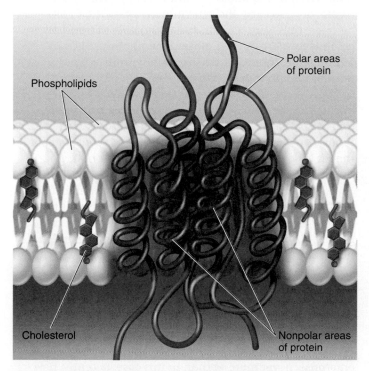

Figure 4.7 Nonpolar regions lock proteins into membranes.
A spiral helix of nonpolar amino acids (*red*) extends across the nonpolar lipid interior, while polar (*purple*) portions of the protein protrude out from the bilayer. The protein cannot move in or out because such a movement would drag polar segments of the protein into the nonpolar interior of the membrane.

It was once thought that plasma membranes were uniform seas of phospholipid, but we now know that they contain zones called **lipid rafts** that are heavily enriched in cholesterol and therefore more tightly packed than the surrounding membrane. Lipid rafts are involved in how cells move and communicate with each other. Interestingly, recent work indicates that they are also the sites where deadly Ebola virus enters human cells.

Proteins Within the Membrane

The second major component of every biological membrane is a collection of **membrane proteins** that float within the lipid bilayer (figure 4.6). In plasma membranes, these proteins provide channels through which molecules and information pass. Some membrane proteins are fixed into position but others are not fixed; instead, they move freely. Some membranes are crowded with proteins, packed tightly side by side. In other membranes the proteins are sparsely distributed.

Some membrane proteins project up from the surface of the plasma membrane like buoys, often with carbohydrate chains or lipids attached to their tips like flags. These **cell surface proteins** act as markers to identify particular types of cells or as beacons to bind specific hormones or proteins to the cell. The CD4 protein by which AIDS viruses dock onto human white blood cells is such a beacon.

Other proteins extend all the way across the bilayer, providing channels across which polar ions and molecules can pass into and out of the cell. How do these **transmembrane proteins** manage to span the membrane, rather than just floating on the surface in the way that a drop of water floats on oil? The part of the protein that actually traverses the lipid bilayer is a specially constructed spiral helix of nonpolar amino acids (figure 4.7). Water responds to nonpolar amino acids much as it does to nonpolar lipid chains, and the helical spiral is held within the lipid interior of the bilayer, anchored there by the strong tendency of water to avoid contact with these nonpolar amino acids. Many transmembrane proteins lock this arrangement in place by positioning amino acids with electrical charges (which are very polar) at the two ends of the helical region.

4.2 All cells are encased within a delicate lipid bilayer sheet, the plasma membrane, within which are embedded a variety of proteins that act as markers or channels through the membrane.

A Closer Look

Membrane Defects Can Cause Disease

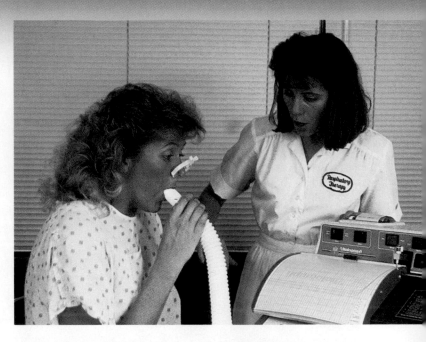

The year 1993 marked an important milestone in the treatment of human disease. That year the first attempt was made to cure **cystic fibrosis** (**CF**), a deadly genetic disorder, by transferring healthy genes into sick individuals. Cystic fibrosis is a fatal disease in which the body cells of affected individuals secrete a thick mucus that clogs the airways of the lungs. The cystic fibrosis patient in the photograph is breathing into a Vitalograph, a device that measures lung function. These same secretions block the ducts of the pancreas and liver so that the few patients who do not die of lung disease die of liver failure. Cystic fibrosis is usually thought of as a children's disease because until recently few affected individuals lived long enough to become adults. Even today half die before their mid-twenties. There is no known cure.

Cystic fibrosis results from a defect in a single gene that is passed down from parent to child. It is the most common fatal genetic disease of Caucasians. One in 20 individuals possesses at least one copy of the defective gene. Most of these individuals are not afflicted with the disease; only those children who inherit a copy of the defective gene from each parent succumb to cystic fibrosis—about 1 in 2,500 infants.

Cystic fibrosis has proven difficult to study. Many organs are affected, and until recently it was impossible to identify the nature of the defective gene responsible for the disease. In 1985 the first clear clue was obtained. An investigator, Paul Quinton, seized on a commonly observed characteristic of cystic fibrosis patients, that their sweat is abnormally salty, and performed the following experiment. He isolated a sweat duct from a small piece of skin and placed it in a solution of salt (NaCl) that was three times as concentrated as the NaCl inside the duct. He then monitored the movement of ions. Diffusion tends to drive both the sodium (Na^+) and the chloride (Cl^-) ions into the duct because of the higher outer ion concentrations. In skin isolated from normal individuals, Na^+ and Cl^- both entered the duct, as expected. In skin isolated from cystic fibrosis individuals, however, only Na^+ entered the duct—no Cl^- entered. For the first time, the molecular nature of cystic fibrosis became clear. Water accompanies chloride, and was not entering the ducts because chloride was not, creating thick mucus. Cystic fibrosis is a defect in a plasma membrane protein called CFTR (cystic fibrosis transmembrane conductance regulator) that normally regulates passage of Cl^- into and out of the body's cells.

The defective *cf* gene was isolated in 1987, and its position on a particular human chromosome (chromosome 7) was pinpointed in 1989. Interestingly, many cystic fibrosis patients produce a CFTR protein with a normal amino acid sequence. The *cf* mutation in these cases appears to interfere with how the CFTR protein folds, preventing it from folding into a functional shape.

Soon after the *cf* gene was isolated, experiments were begun to see if it would be possible to cure cystic fibrosis by gene therapy—that is, by transferring healthy *cf* genes into the cells with defective ones. In 1990 a working *cf* gene was successfully transferred into human lung cells growing in tissue culture, using adenovirus, a cold virus, to carry the gene into the cells. The CFTR-defective cells were "cured," becoming able to transport chloride ions across their plasma membranes. Then in 1991 a team of researchers successfully transferred a normal human *cf* gene into the lung cells of a living animal—a rat. The *cf* gene was first inserted into the adenovirus genome because adenovirus is a cold virus and easily infects lung cells. The treated virus was then inhaled by the rat. Carried piggyback, the *cf* gene entered the rat lung cells and began producing the normal human CFTR protein within these cells!

These results were very encouraging, and at first the future for all cystic fibrosis patients seemed bright. Clinical tests using adenovirus to introduce healthy *cf* genes into cystic fibrosis patients were begun with much fanfare in 1993.

They were not successful. There were insurmountable problems with the adenovirus being used to transport the *cf* gene into cystic fibrosis patients. The difficult and frustrating challenge that cystic fibrosis researchers had faced was not over. Research into clinical problems is often a time-consuming and frustrating enterprise, never more so than in this case. Recently, new ways of introducing the healthy *cf* gene have been tried with better results. The long, slow journey toward a cure has taught us not to leap to the assumption that a cure is now at hand, but the steady persistence of researchers has taken us a long way, and again the future for cystic fibrosis patients seems bright.

4.3 Prokaryotic Cells

There are two major kinds of cells: prokaryotes and eukaryotes. **Prokaryotes** have a relatively uniform cytoplasm that is not subdivided by interior membranes into separate compartments. They do not, for example, have special membrane-bounded compartments, called *organelles,* or a *nucleus* (a membrane-bounded compartment that holds the hereditary information). All bacteria and archaea are prokaryotes; all other organisms are eukaryotes.

Prokaryotes are the simplest cellular organisms. Over 5,000 species are recognized, but doubtless many times that number actually exist and have not yet been described. Although these species are diverse in form, their organization is fundamentally similar (figure 4.8): small cells typically about 1 to 10 micrometers thick; enclosed like all cells by a plasma membrane, but with no distinct interior compartments. Outside of almost all bacteria is a *cell wall,* a framework of carbohydrates cross-linked into a rigid structure. In some bacteria another layer called the *capsule* encloses the cell wall. Bacterial cells assume many shapes or can adhere in chains and masses (figure 4.9), but individual cells are separate from one another.

If you were able to magnify your vision and peer into a prokaryotic cell, you would be struck by its simple organization. The entire interior of the cell, the cytoplasm, is one unit, with no internal support structure (the rigid wall supports the cell's shape) and no internal compartments bounded by membranes. Scattered throughout the cytoplasm of prokaryotic cells are small structures called *ribosomes,* the sites where proteins are made. Ribosomes are not considered organelles because they lack a membrane boundary. The DNA is found in a region of the cytoplasm called the *nucleoid region.* Although the DNA is localized in this region of the cytoplasm, it is not considered a nucleus because the nucleoid region and its associated DNA are not enclosed within an internal membrane.

Some prokaryotes use a *flagellum* (plural, *flagella*) to move. Flagella are long, threadlike structures projecting from the surface of a cell. They are used in locomotion and feeding. Prokaryotic flagella are protein fibers that extend out from the cell. There may be one or more per cell, or none, depending on the species. Bacteria can swim at speeds up to 20 cell diameters per second by rotating their flagella like screws.

Pili (singular, **pilus**) are other hairlike structures that occur on the cells of some prokaryotes. They are shorter than prokaryotic flagella, up to several micrometers long, and about 7.5 to 10 nanometers thick. Pili help the prokaryotic cells attach to appropriate substrates and exchange genetic information.

> **4.3** Prokaryotic cells lack a nucleus and do not have an extensive system of interior membranes.

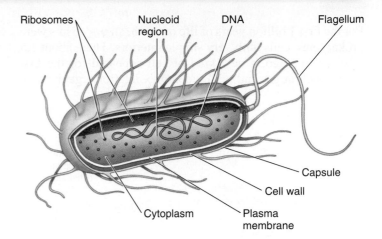

Ribosomes Nucleoid region DNA Flagellum

Capsule

Cell wall

Cytoplasm Plasma membrane

Figure 4.8 Organization of a prokaryotic cell.
Prokaryotic cells lack internal compartments. Not all prokaryotic cells have a flagellum or a capsule like the one illustrated here, but all do have a nucleoid region, ribosomes, a plasma membrane, cytoplasm, and a cell wall.

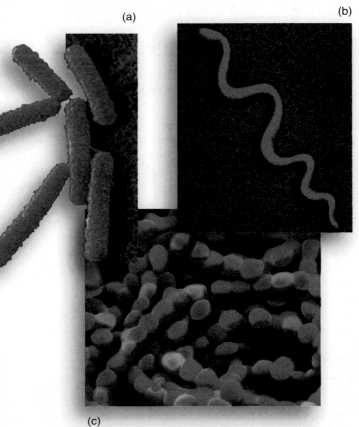

(a) (b)

(c)

Figure 4.9 Bacterial cells have different shapes.
(*a*) *Bacillus* is a rod-shaped bacterium. (*b*) *Treponema* is a coil-shaped bacterium; rotation of internal filaments produces a cork screw movement. (*c*) *Streptococcus* is a more or less spherical bacterium in which the individuals adhere in chains.

4.4 Eukaryotic Cells

For the first 1 billion years of life on earth, all organisms were prokaryotes, cells with very simple interiors. Then, about 1.5 billion years ago, a new kind of cell appeared for the first time, much larger and with a complex interior organization.

All cells alive today except bacteria and archaea are of this new kind. Unlike prokaryotes, these big cells have many membrane-bounded interior compartments and a variety of **organelles** (specialized structures within which particular cell processes occur). One of the organelles is very visible when these cells are examined with a microscope, filling the center of the cell like the pit of a peach. Seeing it, the English

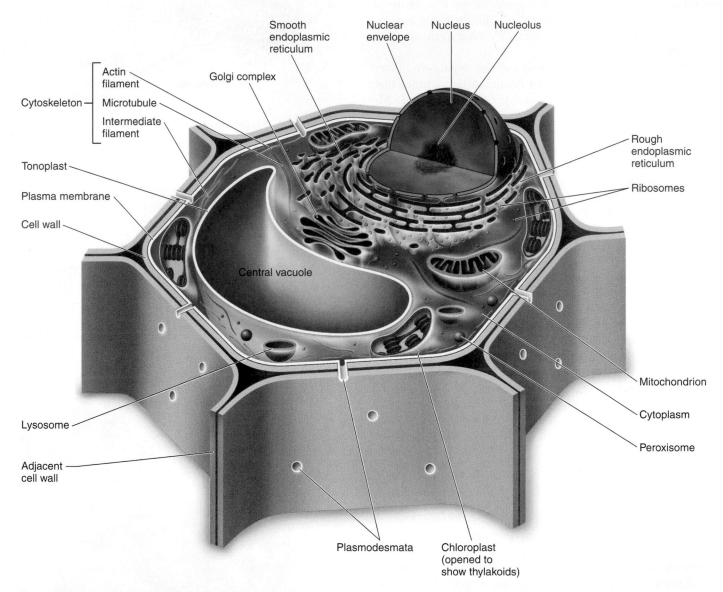

Figure 4.10   Structure of a plant cell.
Most mature plant cells contain large central vacuoles that occupy a major portion of the internal volume of the cell and organelles called chloroplasts, within which photosynthesis takes place. The cells of plants, fungi, and some protists have cell walls, although the composition of the walls varies among the groups. Plant cells have cytoplasmic connections through openings in the cell wall called plasmodesmata. Flagella occur in sperm of a few plant species, but are otherwise absent in plant and fungal cells. Centrioles are also absent in plant and fungal cells.

botanist Robert Brown in 1831 called it the *nucleus* (plural, *nuclei*), from the Latin word for "kernel." Inside the nucleus, the DNA is wound tightly around proteins and packaged into compact units called *chromosomes.* All cells with nuclei are called **eukaryotes** (from the Greek words *eu,* true, and *karyon,* nut), whereas bacteria and archaea are called *prokaryotes* ("before the nut").

Both plant cells (figure 4.10) and animal cells (figure 4.11) are eukaryotic. The hallmark of the eukaryotic cell is compartmentalization, achieved by an extensive *endomembrane system* that weaves through the cell interior, creating organelles and a variety of *vesicles* (small membrane-bounded sacs that store and transport materials). These many closed-off compartments allow different processes to proceed simultaneously without interferring with one another, just as rooms do in a house. All eukaryotic cells are supported within by an internal protein scaffold, the *cytoskeleton.* The cells of plants and fungi have strong exterior *cell walls* composed of cellulose or chitin fibers, while the cells of animals lack cell walls. We now journey into the interior of a typical eukaryotic cell and explore its complex organization.

4.4 Eukaryotic cells have a system of interior membranes and membrane-bounded organelles that subdivide the interior into functional compartments.

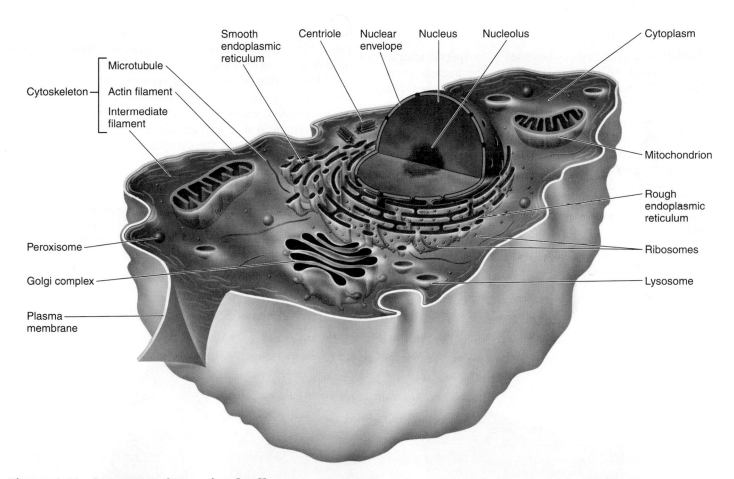

Figure 4.11   Structure of an animal cell.
In this generalized diagram of an animal cell, the plasma membrane encases the cell, which contains the cytoskeleton and various cell organelles and interior structures suspended in a semifluid matrix called the cytoplasm. Some kinds of animal cells possess fingerlike projections called microvilli. Other types of eukaryotic cells, for example many protist cells, may possess flagella, which aid in movement, or cilia, which can have many different functions.

4.5 The Nucleus: The Cell's Control Center

The many parts of the cell (table 4.2) are remarkably similar from plants to animals, or from paramecia to primates. Organelles look similar and carry out similar functions in all organisms. It is possible that these shared properties were derived from common ancestral cells several billion years ago.

If you were to journey far into the interior of one of your cells, you would eventually reach the center of the cell. There you would find, cradled within a network of fine filaments like a ball in a basket, the **nucleus** (figure 4.12). The nucleus is the command and control center of the cell, directing all of its activities. It is also the genetic library where the hereditary information is stored.

Nuclear Membrane

The surface of the nucleus is bounded by a special kind of membrane called the **nuclear envelope.** The nuclear envelope is actually *two* membranes, one outside the other, like a sweater over a shirt. Scattered over the surface of this envelope are selective openings called **nuclear pores.** Nuclear pores form when the two membrane layers of the nuclear envelope pinch together. A nuclear pore is not an empty opening like the hole in a doughnut; rather, it has many proteins embedded within it that permit proteins and RNA to pass into and out of the nucleus.

Chromosomes

In both prokaryotes and eukaryotes, all hereditary information specifying cell structure and function is encoded in DNA.

However, unlike prokaryotic DNA, the DNA of eukaryotes is divided into several segments and associated with protein, forming **chromosomes.** The proteins in the chromosome permit the DNA to wind tightly and condense during cell division. Under a light microscope, these condensed chromosomes are readily seen in dividing cells as densely staining rods. After cell division, eukaryotic chromosomes uncoil and fully extend into threadlike strands called **chromatin** that can no longer be distinguished individually with a light microscope within the nucleoplasm.

Nucleolus

To make its many proteins, the cell employs a special structure called a **ribosome,** which reads the RNA copy of a gene and uses that information to direct the construction of a protein. Ribosomes are made up of several special forms of RNA called ribosomal RNA, or rRNA, bound up within a complex of several dozen different proteins.

One region of the nucleus appears darker than the rest; this darker region is called the **nucleolus.** There a cluster of several hundred genes encode rRNA where the ribosome subunits assemble. These subunits leave the nucleus through the nuclear pores and enter the cytoplasm, where final assembly of ribosomes takes place.

> **4.5** The nucleus is the command center of the cell, issuing instructions that control cell activities. It also stores the cell's hereditary information.

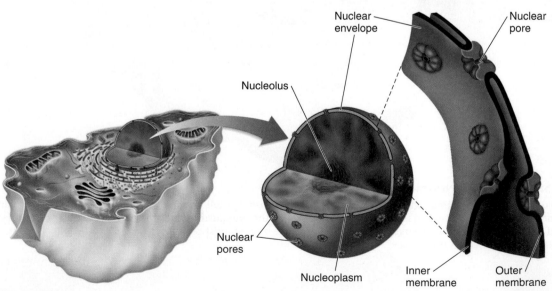

Figure 4.12 The nucleus.
The nucleus is composed of a double membrane, called a nuclear envelope, enclosing a fluid-filled interior containing the chromosomes. In cross section, the individual nuclear pores are seen to extend through the two membrane layers of the envelope; the dark material within the pore is protein, which acts to control access through the pore.

TABLE 4.2 EUKARYOTIC CELL STRUCTURES AND THEIR FUNCTIONS

Structure		Description	Function
Structural Elements			
Cell wall		Outer layer of cellulose or chitin; or absent	Protection; support
Cytoskeleton		Network of protein filaments	Structural support; cell movement
Flagella and cilia		Cellular extensions with 9 + 2 arrangement of pairs of microtubules	Motility or moving fluids over surfaces
Plasma Membrane and Endomembrane System			
Plasma membrane		Lipid bilayer in which proteins are embedded	Regulates what passes into and out of cell; cell-to-cell recognition
Endoplasmic reticulum		Network of internal membranes	Forms compartments and vesicles; participates in protein and lipid synthesis
Nucleus		Structure (usually spherical) surrounded by double membrane that contains chromosomes	Control center of cell; directs protein synthesis and cell reproduction
Golgi complex		Stacks of flattened vesicles	Packages proteins for export from the cell; forms secretory vesicles
Lysosomes		Vesicles derived from Golgi complex that contain hydrolytic digestive enzymes	Digest worn-out organelles and cell debris; play role in cell death
Peroxisomes		Vesicles formed from the ER containing oxidative and other enzymes	Isolate particular chemical activities from rest of cell
Energy-Producing Organelles			
Mitochondria		Bacteria-like elements with double membrane	Sites of oxidative metabolism; provides ATP for cellular energy
Chloroplast		Bacteria-like organelle found in plants and algae; complex inner membrane consists of stacked vesicles	Site of photosynthesis
Elements of Gene Expression			
Chromosomes		Long threads of DNA that form a complex with protein	Contain hereditary information
Nucleolus		Site of genes for rRNA synthesis	Assembles ribosomes
Ribosomes		Small, complex assemblies of protein and RNA, often bound to endoplasmic reticulum	Sites of protein synthesis

The Endomembrane System

Surrounding the nucleus within the interior of the eukaryotic cell is a tightly packed mass of membranes. They fill the cell, dividing it into compartments, channeling the transport of molecules through the interior of the cell and providing the surfaces on which enzymes act. The system of internal compartments created by these membranes in eukaryotic cells constitutes the most fundamental distinction between the cells of eukaryotes and prokaryotes.

Endoplasmic Reticulum: The Transportation System

The extensive system of internal membranes is called the **endoplasmic reticulum,** often abbreviated **ER** (figure 4.13). The term *endoplasmic* means "within the cytoplasm," and the term *reticulum* is a Latin word meaning "little net." The ER, weaving in sheets through the interior of the cell, creates a series of channels and interconnections, and it also isolates some spaces as membrane-enclosed sacs called **vesicles.**

The surface of the ER is the place where the cell makes proteins intended for export (such as enzymes secreted from the cell surface). The surface of those regions of the ER devoted to the synthesis of such transported proteins is heavily studded with ribosomes and appears pebbly, like the surface of sandpaper, when seen through an electron microscope. For this reason, these regions are called **rough ER.** Regions in which ER-bounded ribosomes are relatively scarce are correspondingly called **smooth ER.** The surface of the smooth ER is embedded with enzymes that aid in the manufacture of carbohydrates and lipids.

The Golgi Complex: The Delivery System

As new molecules are made on the surface of the ER, they are passed from the ER to flattened stacks of membranes called **Golgi bodies.** These structures are named for Camillo Golgi, the nineteenth-century Italian physician who first called attention to them. The number of Golgi bodies a cell contains ranges from 1 or a few in protists, to 20 or more in animal cells and several hundred in plant cells. Golgi bodies function in the collection, packaging, and distribution of molecules manufactured in the cell. Scattered through the cytoplasm, Golgi bodies are collectively referred to as the **Golgi complex** (figure 4.14).

The proteins and lipids that are manufactured on the ER membranes are transported through the channels of the ER, or as vesicles budded off from it, into the Golgi bodies. Within the Golgi bodies, many of these molecules become tagged with carbohydrates. The molecules collect at the ends of the membranous folds of the Golgi bodies; these folds are given the special name *cisternae* (Latin, collecting vessels). Vesicles

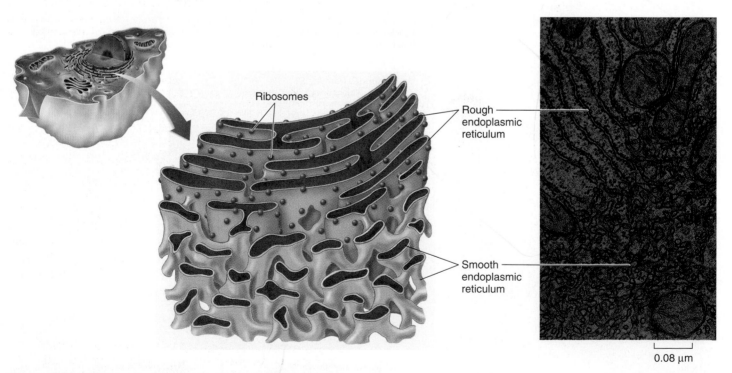

Ribosomes

Rough endoplasmic reticulum

Smooth endoplasmic reticulum

0.08 μm

Figure 4.13   The endoplasmic reticulum (ER).
The endoplasmic reticulum provides the cell with an extensive system of internal membranes for the synthesis and transport of materials. Ribosomes are associated with only one side of the rough ER; the other side is the boundary of a separate compartment within the cell into which the ribosomes extrude newly made proteins destined for secretion. Smooth endoplasmic reticulum has few to no bound ribosomes.

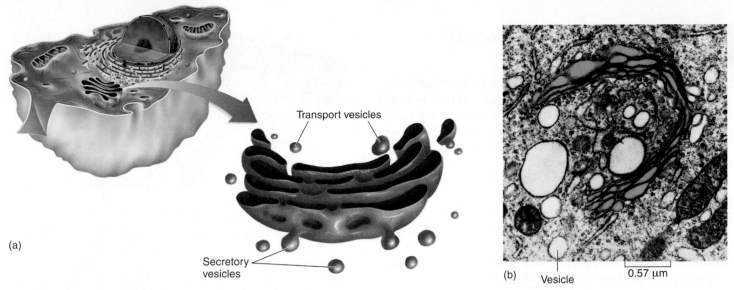

(a)

(b) Vesicle

0.57 μm

Transport vesicles

Secretory vesicles

Figure 4.14   Golgi complex.
This vesicle-forming system, called the Golgi complex after its discoverer, is an integral part of the cell's internal membrane system. The Golgi complex processes and packages materials for transport to another region within the cell and/or for export from the cell. It receives material for processing in transport vesicles on one side and sends the material packaged in secretory vesicles off the other side. (a) Diagram of a Golgi complex. (b) Micrograph of a Golgi complex showing vesicles.

that pinch off from the cisternae carry the molecules to the different compartments of the cell and to the inner surface of the plasma membrane (figure 4.15), where molecules to be secreted are released to the outside.

Lysosomes: Recycling Centers

Other organelles called **lysosomes** arise from the Golgi complex and contain a concentrated mix of the powerful

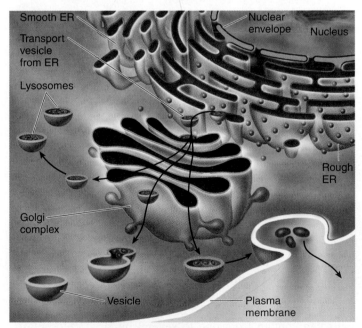

Smooth ER
Transport vesicle from ER
Lysosomes
Golgi complex
Vesicle
Nuclear envelope
Nucleus
Rough ER
Plasma membrane

Figure 4.15   How the endomembrane system works.
A highly efficient highway system within the cell, the endomembrane system transports material from the ER to the Golgi and from there to other destinations.

enzymes that break down macromolecules. Lysosomes are also the recycling centers of the cell, digesting worn-out cell components to make way for newly formed ones while recycling the proteins and other materials of the old parts. Large organelles called mitochondria are replaced in some human tissues every 10 days, with lysosomes digesting the old ones as the new ones are produced. In addition to breaking down organelles and other structures within cells, lysosomes also eliminate particles (including other cells) that the cell has engulfed.

Peroxisomes: Chemical Specialty Shops

The interior of the eukaryotic cell contains a variety of membrane-bounded spherical organelles derived from the ER that carry out particular chemical functions. Almost all eukaryotic cells, for example, contain **peroxisomes,** vesicles that contain two sets of enzymes and that are about the same size as lysosomes. One set found in plant seeds converts fats to carbohydrates, and the other set found in all eukaryotes detoxifies various potentially harmful molecules—strong oxidants—that form in cells. They do this by using molecular oxygen to remove hydrogen atoms from specific molecules. These chemical reactions would be very destructive to the cell if not confined to the peroxisomes.

4.6   An extensive system of interior membranes organizes the interior of the cell into functional compartments that manufacture and deliver proteins and carry out a variety of specialized chemical processes.

Organelles That Contain DNA

Eukaryotic cells contain several kinds of complex, cell-like organelles that contain their own DNA and appear to have been derived from ancient bacteria assimilated by ancestral eukaryotes in the distant past. The two principal kinds are mitochondria (which occur in the cells of all but a very few eukaryotes) and chloroplasts (which do not occur in animal cells—they occur only in algae and plants).

Mitochondria: Powerhouses of the Cell

Eukaryotic organisms extract energy from organic molecules ("food") in a complex series of chemical reactions called **oxidative metabolism,** which takes place only in their mitochondria. **Mitochondria** (singular, **mitochondrion**) are sausage-shaped organelles about the size of a bacterial cell (figure 4.16). Mitochondria are bounded by two membranes. The outer membrane is smooth and apparently derives from the plasma membrane of the host cell that first took up the bacterium long ago. The inner membrane, apparently the plasma membrane of the bacterium that gave rise to the mitochondrion, is bent into numerous folds called **cristae** (singular, **crista**) that resemble the folded plasma membranes in various groups of bacteria. The cristae partition the mitochondrion into two compartments, an inner **matrix** and an outer compartment. As you will learn in chapter 6, this architecture is critical to successfully carrying out oxidative metabolism.

During the 1.5 billion years in which mitochondria have existed in eukaryotic cells, most of their genes have been transferred to the chromosomes of the host cells. But mitochondria still have some of their original genes, contained in a circular, closed, naked molecule of DNA (called mitochondrial DNA, or mtDNA) that closely resembles the circular DNA molecule of a bacterium. On this mtDNA are several genes that produce some of the proteins essential for oxidative metabolism. In both mitochondria and bacteria, the circular DNA molecule is replicated during the process of division. When a mitochondrion divides, it copies its DNA located in the matrix and splits into two by simple fission, dividing much as bacteria do.

Chloroplasts: Energy-Capturing Centers

All photosynthesis in plants and algae takes place within another bacteria-like organelle, the **chloroplast** (figure 4.17). There is strong evidence that chloroplasts, like mitochondria, were derived by symbiosis from bacteria. A chloroplast is bounded, like a mitochondrion, by two membranes, the inner derived from the original bacterium and the outer resembling the host cell's ER. Chloroplasts are larger than mitochondria, and their inner membranes have a more complex organization. The inner membranes are fused to form stacks of closed vesicles called **thylakoids.** The light-powered reactions of photosynthesis take place within the thylakoids. The thyla-

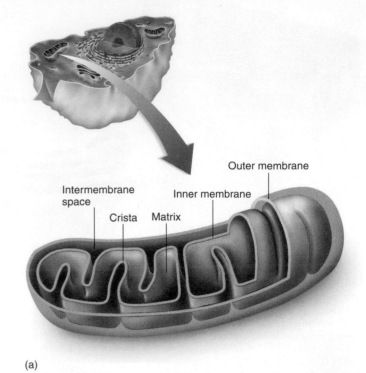

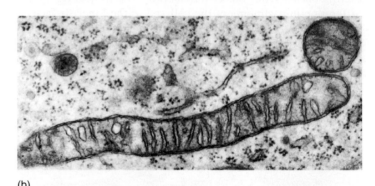

(a)

(b)

Figure 4.16   Mitochondria.
The mitochondria of a cell are sausage-shaped organelles within which oxidative metabolism takes place, and energy is extracted from food using oxygen. (*a*) A mitochondrion has a double membrane. The inner membrane is shaped into folds called cristae. The space within the cristae is called the matrix. The cristae greatly increase the surface area for oxidative metabolism. (*b*) Micrograph of two mitochondria, one in cross section, the other cut lengthwise.

koids are stacked on top of one another to form a column called a **granum** (plural, **grana**). The interior of a chloroplast is bathed with a semiliquid substance called the **stroma.**

Like mitochondria, chloroplasts have a circular DNA molecule. On this DNA are located many of the genes coding for the proteins necessary to carry out photosynthesis. Plant cells can contain from one to several hundred chloroplasts, depending on the species. Neither mitochondria nor chloroplasts can be grown in a cell-free culture; they are totally dependent on the cells within which they occur.

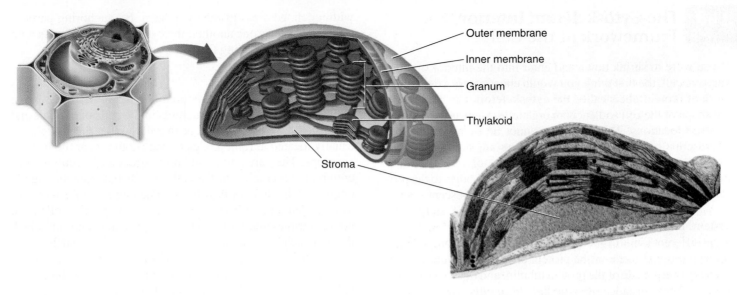

Figure 4.17   A chloroplast.
Bacteria-like organelles called chloroplasts are the sites of photosynthesis in photosynthetic eukaryotes. Like mitochondria, they have a complex system of internal membranes on which chemical reactions take place. The inner membrane of a chloroplast is fused to form stacks of closed vesicles called thylakoids. Photosynthesis occurs within these thylakoids. Thylakoids are stacked one on top of the other in columns called grana. The interior of the chloroplast is bathed in a semiliquid substance called the stroma.

Endosymbiosis

Symbiosis is a close relationship between organisms of different species that live together. The theory of **endosymbiosis** proposes that some of today's eukaryotic organelles evolved by a symbiosis in which one cell of a prokaryotic species was engulfed by and lived inside the cell of another species of prokaryote that was a precursor to eukaryotes (figure 4.18). According to the endosymbiont theory, the engulfed prokaryotes provided their hosts with certain advantages associated with their special metabolic abilities. Two key eukaryotic organelles just described are believed to be the descendants of these endosymbiotic prokaryotes: mitochondria, which are thought to have originated as bacteria capable of carrying out oxidative metabolism; and chloroplasts, which apparently arose from photosynthetic bacteria.

The endosymbiont theory is supported by a wealth of evidence. Both mitochondria and chloroplasts are surrounded by two membranes; the inner membrane probably evolved from the plasma membrane of the engulfed bacterium, while the outer membrane is probably derived from the plasma membrane or endoplasmic reticulum of the host cell. Mitochondria are about the same size as most bacteria, and the cristae formed by their inner membranes resemble the folded membranes in various groups of bacteria. Mitochondrial ribosomes are also similar to bacterial ribosomes in size and structure. Both mitochondria and chloroplasts contain circular molecules of DNA similar to those in bacteria. Finally, mitochondria divide by simple fission, splitting in two just as bacterial cells do, and they apparently replicate and partition their DNA in much the same way as bacteria.

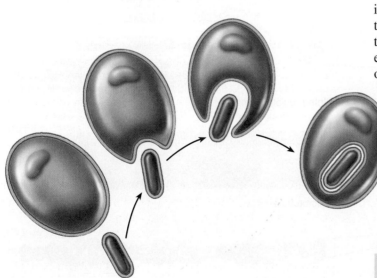

Figure 4.18   Endosymbiosis.
This figure shows how a double membrane may have been created during the symbiotic origin of mitochondria or chloroplasts.

4.7   Eukaryotic cells contain several complex organelles that have their own DNA and are thought to have arisen by endosymbiosis from ancient bacteria.

4.8 The Cytoskeleton: Interior Framework of the Cell

If you were to shrink down and enter into the interior of a eukaryotic cell, the first thing you would encounter is a dense network of protein fibers called the **cytoskeleton,** which supports the shape of the cell and anchors organelles such as the nucleus to fixed locations. This network cannot be seen with a light microscope because the individual fibers are single chains of protein, much too fine for microscopes to resolve. To "see" the cytoskeleton, scientists attach fluorescent antibodies to the protein fibers and then photograph them under fluorescent light.

The protein fibers of the cytoskeleton are a dynamic system, constantly being formed and disassembled. There are three different kinds of protein fibers (figure 4.19): long, slender **microfilaments** made of the protein actin, hollow tubes called **microtubules** made of the protein tubulin, and thick ropes of intertwined protein called **intermediate filaments.** The microfilaments, microtubules, and intermediate filaments are anchored to membrane proteins embedded within the plasma membrane.

The cytoskeleton plays a major role in determining the shape of animal cells, which lack rigid cell walls. Because filaments can form and dissolve readily, the shape of an animal cell can change rapidly. If you examine the surface of an animal cell with a microscope, you will often find it alive with motion, projections shooting out from the surface and then retracting, only to shoot out elsewhere moments later.

The cytoskeleton is not only responsible for the cell's shape, but it also provides a scaffold both for ribosomes to carry out protein synthesis and for enzymes to be localized within defined areas of the cytoplasm. By anchoring particular enzymes near one another, the cytoplasm participates with organelles in organizing the cell's activities.

Centrioles

Complex structures called **centrioles** (figure 4.20) assemble microtubules from tubulin subunits in the cells of animals and most protists. Centrioles occur in pairs within the cytoplasm, usually located at right angles to one another near the nuclear envelope. They are among the most structurally complex microtubular assemblies of the cell. In cells that contain flagella or cilia, each cilium or flagellum is anchored by a form of centriole called a basal body. Most animal and protist cells have both centrioles and basal bodies; higher plants and fungi lack them, instead organizing microtubules without such structures. Although they lack a membrane, centrioles resemble spirochete bacteria in many other respects. Some biologists believe that centrioles, like mitochondria and chloroplasts, originated as symbiotic bacteria.

Cell Movement

Essentially, all cell motion is tied to the movement of actin microfilaments, microtubules, or both. Intermediate filaments act as intracellular tendons, preventing excessive stretching of cells, and actin microfilaments play a major role in determining the shape of cells. Because actin microfilaments can form and dissolve so readily, they enable some cells to change shape quickly. If you look at the surfaces of such cells under a microscope, you will find them moving and changing shape.

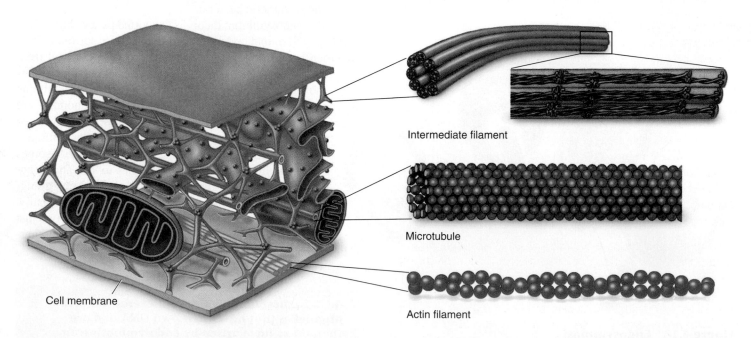

Intermediate filament

Microtubule

Actin filament

Figure 4.19 The three protein fibers of the cytoskeleton.
Organelles such as mitochondria are anchored to fixed locations in the cytoplasm by the cytoskeleton, a network of protein fibers. *Intermediate filaments* are composed of overlapping proteins that allows for a ropelike structure that provides tremendous mechanical strength to the cell. *Microtubules* are composed of tubulin protein subunits arranged side by side to form a tube. Microtubules are comparatively stiff cytoskeletal elements that function in intracellular transport and stabilization of cell structure. *Actin microfilaments* are made of two strands of the fibrous protein actin twisted together and usually occur in bundles. Actin microfilaments are responsible for cell movement.

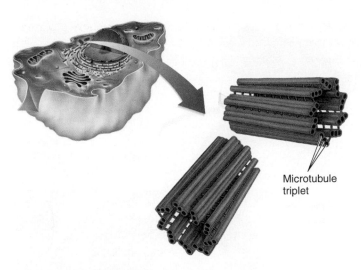

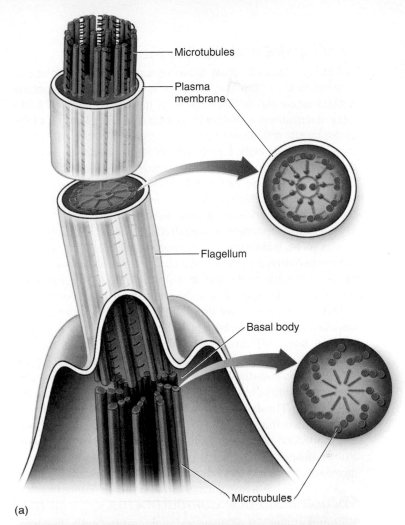

Microtubules

Plasma membrane

Flagellum

Basal body

Microtubules

(a)

Figure 4.20 Centrioles.

Centrioles anchor and assemble microtubules. In their anchoring capacity, they can be seen as the basal bodies of eukaryotic flagella. In their organizing capacity, they function during cell division to indicate the plane along which the cell separates. Centrioles usually occur in pairs and are located in the cell in characteristic planes. One centriole typically lies parallel to the cell surface, while another lies perpendicular to the surface. Centrioles are composed of nine triplets of microtubules.

Microtubule triplet

Some Cells Crawl. It is the arrangement of actin microfilaments within the cell cytoplasm that allows cells to "crawl," literally! Crawling is a significant cellular phenomenon, essential to inflammation, clotting, wound healing, and the spread of cancer. White blood cells in particular exhibit this ability. Produced in the bone marrow, these cells are released into the circulatory system and then eventually crawl out of capillaries and into the tissues to destroy potential pathogens. The crawling mechanism is an exquisite example of cellular coordination.

Actin microfilaments play a role in other types of cell movement. For example, during animal cell reproduction (see chapter 7), chromosomes move to opposite sides of a dividing cell because they are attached to shortening microtubules. The cell then pinches in two when a belt of actin microfilaments contracts like a purse string. Muscle cells also use actin microfilaments to contract their cytoskeletons. The fluttering of an eyelash, the flight of an eagle, and the awkward crawling of a baby all depend on these cytoskeletal movements within muscle cells.

Swimming with Flagella and Cilia. Flagella (singular, flagellum) are fine, long, threadlike organelles protruding from the cell surface. Each flagellum arises from a structure called a **basal body** and consists of a circle of nine microtubule pairs surrounding two central ones (figure 4.21). This **9 + 2 arrangement** is a fundamental feature of eukaryotes and apparently evolved early in their history. Even in cells that lack flagella, derived structures with the same 9 + 2 arrangement often occur, like in the sensory hairs of the human ear. If flagella are numerous and organized in dense rows, they are called **cilia.** Cilia do not differ from flagella in their structure, but cilia are usually short. In humans, we find a single long fla-

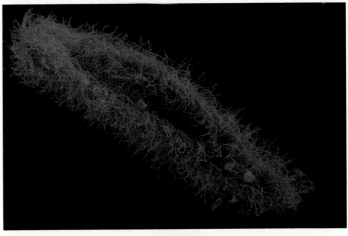

(b)

Figure 4.21 Flagella and cilia.

(a) A eukaryotic flagellum springs directly from a basal body and is composed of a ring of nine pairs of microtubules with two microtubules in its core. (b) The surface of this paramecium is covered with a dense forest of cilia.

gellum on each sperm cell that propels the cell in a swimming motion, and dense mats of cilia project from cells that line our breathing tube, the trachea, to move mucus and dust particles out of the respiratory tract into the throat (where we can expel these unneeded contaminants by spitting or swallowing).

Moving Materials Within the Cell

All eukaryotic cells must move materials from one place to another in the cytoplasm. Most cells use the endomembrane system as an intracellular highway; the Golgi complex packages materials into vesicles that come from the channels of the endoplasmic reticulum to the far reaches of the cell. However, this highway is only effective over short distances. When a cell has to transport materials through long extensions like the axon of a nerve cell, the ER highways are too slow. For these situations, eukaryotic cells have developed high-speed locomotives that run along microtubular tracks.

Four components are required: (1) a vesicle or organelle that is to be transported, (2) a motor molecule that provides the energy-driven motion, (3) a connector molecule that connects the vesicle to the motor molecule, and (4) microtubules on which the vesicle will ride like a train on a rail. As nature's tiniest motors, these motor proteins literally pull the transport vesicles along the microtubular tracks. The motor protein **kinesin** uses ATP to power its movement toward the cell periphery, dragging the vesicle with it as it travels along the microtubule. Another motor protein, **dynein** (figure 4.22), directs movement in the opposite direction, inward toward the cell's center. The destination of a particular transport vesicle and its contents is thus determined by the nature of the linking protein embedded within the vesicle's membrane. Like possessing a ticket to one of two destinations, if a vesicle links to kinesin, it moves outward; if it links to dynein, it moves inward.

Vacuoles: Storage Compartments

Within the interiors of plant and many protist cells, the cytoskeleton positions not only organelles, but also storage com-

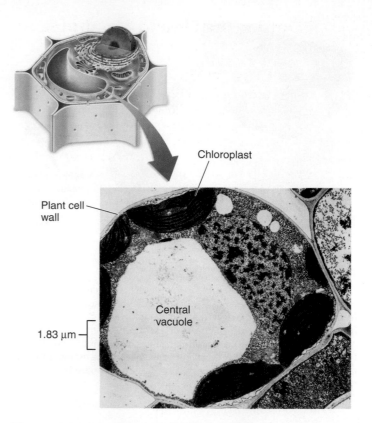

Figure 4.23   A plant central vacuole.

A plant's central vacuole stores dissolved substances and can increase in size to increase the surface area of a plant cell.

partments that are not membrane-bounded, called **vacuoles.** The center of a plant cell usually contains a large, apparently empty space, called the *central vacuole* (figure 4.23). This vacuole is not really empty; it contains large amounts of water and other materials, such as sugars, ions, and pigments. The central vacuole functions as a storage center for these important substances and also helps to increase the surface-to-volume ratio of the plant cell outside the vacuole by applying pressure to the plasma membrane. The plasma membrane expands outward under this pressure, thereby increasing its surface area.

In protists like *Paramecium,* cells contain a **contractile vacuole** near the cell surface that accumulates excess water. This vacuole is bounded by actin microfilaments and has a small pore that opens onto the outside of the cell. By rhythmic ATP-powered contractions, it pumps accumulated water out through the pore.

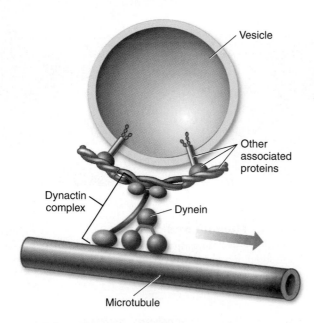

Figure 4.22   Molecular motors.

Vesicles that are transported within cells are attached with connector molecules, such as the dynactin complex shown here, to motor molecules, like dynein, which move along microtubules.

> **4.8**   The cytoskeleton is a latticework of protein fibers that determines a cell's shape and anchors organelles to particular locations within the cytoplasm. Cells can move by changing their shape.

4.9 Outside the Plasma Membrane

Cell Walls Offer Protection and Support

Plants, fungi, and many protists cells share a characteristic with bacteria that is not shared with animal cells—that is, they have **cell walls,** which protect and support their cells. Eukaryotic cell walls are chemically and structurally different from bacterial cell walls. In plants, cell walls are composed of fibers of the polysaccharide cellulose, while in fungi they are composed of chitin. The **primary walls** of plant cells are laid down when the cell is still growing, and between the walls of adjacent cells is a sticky substance called the **middle lamella,** which glues the cells together (figure 4.24). Some plant cells produce strong **secondary walls,** which are deposited inside the primary walls of fully expanded cells.

An Extracellular Matrix Surrounds Animal Cells

As we discussed, many types of eukaryotic cells possess a cell wall exterior to the plasma membrane that protects the cell, maintains its shape, and prevents excessive water uptake. Ani- mal cells are the great exception, lacking the cell walls that encase plants, fungi, and most protists. Instead, animal cells secrete an elaborate mixture of **glycoproteins** (proteins with short chains of sugars attached to them) into the space around them, forming the **extracellular matrix (ECM)** (figure 4.25).

The fibrous protein collagen, the same protein in finger- nails and hair, is abundant in the ECM. Strong fibers of colla- gen and another fibrous protein, elastin, are embedded within a complex web of other glycoproteins called proteoglycans, which form a protective layer over the cell surface.

The ECM is attached to the plasma membrane by a third kind of glycoprotein, **fibronectin.** Fibronectin molecules bind not only to ECM glycoproteins but also to proteins called **integrins,** which are an integral part of the plasma mem- brane. Integrins extend into the cytoplasm, where they are attached to the microfilaments of the cytoskeleton. Linking ECM and cytoskeleton, integrins allow the ECM to influence cell behavior in important ways, altering gene expression and cell migration patterns by a combination of mechanical and chemical signalling pathways. In this way, the ECM can help coordinate the behavior of all the cells in a particular tissue.

> **4.9** Plant and protist cells encase themselves within a strong cell wall. In animal cells, which lack a cell wall, the cytoskeleton is linked by integrin proteins to a web of glycoproteins called the extracellular matrix.

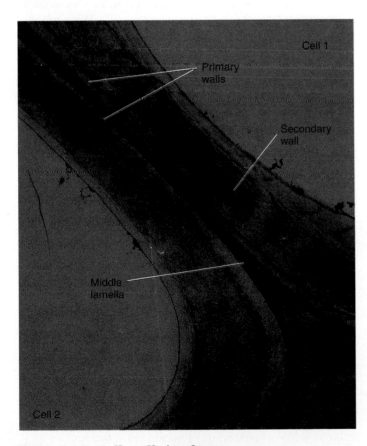

Figure 4.24 Cell walls in plants.
Plant cell walls are thick, strong, and rigid. Primary cell walls are laid down when the cell is young. Thicker secondary cell walls may be added later when the cell is fully grown. The middle lamella lies between the walls of adjacent cells and glues the cells together.

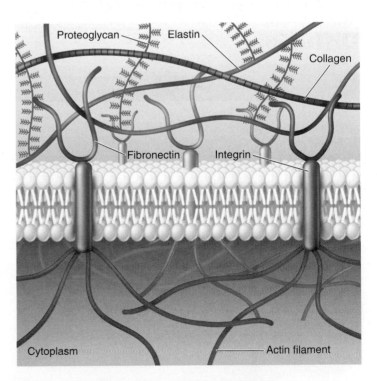

Figure 4.25 The extracellular matrix.
Animal cells are surrounded by an extracellular matrix composed of various glycoproteins that give the cells support, strength, and resilience.

4.10 Diffusion and Osmosis

For cells to survive, food particles, water, and other materials must pass into the cell, and waste materials must be eliminated. All of this moving back and forth across the cell's plasma membrane occurs in one of three ways: (1) water and other substances diffuse through the membrane; (2) food particles and sometimes liquids are engulfed by the membrane folding around them; or (3) proteins in the membrane act as doors that admit certain molecules only. First we will examine diffusion.

Diffusion

How a molecule moves—just where it goes—is totally random, like shaking marbles in a cup, so if two kinds of molecules are added together, they soon mix. The random motion of molecules always tends to produce uniform mixtures when a substance moves from regions where its concentration is high to regions where its concentration is lower (that is, *down* the **concentration gradient**). How does a molecule "know" in what direction to move? It doesn't—molecules don't "know" anything. A molecule is equally likely to move in any direction and is constantly changing course in random

ways. There are simply more molecules able to move from where they are common than from where they are scarce. This mixing process is called **diffusion** (figure 4.26). Diffusion is the net movement of molecules down a concentration gradient toward regions of lower concentration (that is, where there are relatively fewer of them) as a result of random motion. Eventually, the substance will achieve a state of **equilibrium,** where there is no net movement toward any particular direction. The individual molecules of the substance are still in motion, but there is no overall directionality of the motion.

Osmosis

Diffusion allows molecules like oxygen, carbon dioxide, and nonpolar lipids to cross the plasma membrane. The movement of water molecules is not blocked because there are many small channels, called aquaporins, that pass water freely through the membrane.

As in diffusion, water passes into and out of a cell down its concentration gradient, a process called **osmosis.** However, the movement of water into and out of a cell is dependent upon the concentration of other substances in solution. To understand how water moves into and out of a cell, let's focus on the water molecules already present inside a cell. What are they

Diffusion

1 Lump of sugar — A lump of sugar is dropped into a beaker of water.

2 Sugar molecule — Sugar molecules begin to break off from the lump.

3 More and more sugar molecules move away and randomly bounce around.

4 Eventually, all of the sugar molecules become evenly distributed throughout the water.

Figure 4.26   How diffusion works.
Diffusion is the mixing process that spreads molecules through the cell interior. To see how diffusion works, visualize a simple experiment in which a lump of sugar is dropped into a beaker of water.

Osmosis

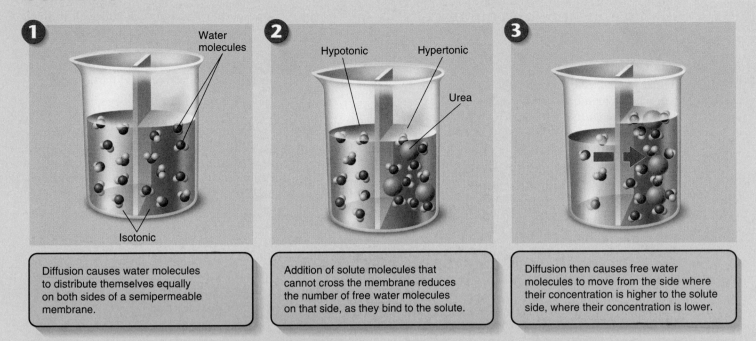

1 Diffusion causes water molecules to distribute themselves equally on both sides of a semipermeable membrane.

2 Addition of solute molecules that cannot cross the membrane reduces the number of free water molecules on that side, as they bind to the solute.

3 Diffusion then causes free water molecules to move from the side where their concentration is higher to the solute side, where their concentration is lower.

Figure 4.27 How osmosis works.
Osmosis is the net movement of water across a membrane toward the side with less "free" water. To visualize this, imagine adding polar urea molecules to one side of a vessel of water divided in the middle by a semipermeable membrane (lets water pass but not larger molecules). When such a polar solute is added, the water molecules that gather around each urea molecule are no longer free to diffuse across the membrane—in effect, the polar solute has reduced the number of free water molecules. Because the side of the membrane with less solute (the left) has more unbound water molecules than the side on the right with more solute, water moves by diffusion from the left to the right.

doing? Many of them are interacting with the sugars, proteins, and other polar molecules inside. Remember, water is very polar itself and readily interacts with other polar molecules. These social water molecules are not randomly moving about as they were outside; instead, they remain clustered around the polar molecules they are interacting with. As a result, while water molecules keep coming into the cell by random motion, they don't randomly come out again. Because more water molecules come in than go out, there is a net movement of water into the cell (figure 4.27).

The concentration of *all* molecules dissolved in a solution (called **solutes**) is called the osmotic concentration of the solution. If two solutions have unequal osmotic concentrations, the solution with the higher solute concentration is said to be **hypertonic** (Greek *hyper,* more than), and the solution with the lower one is **hypotonic** (Greek *hypo,* less than). If the osmotic concentrations of the two solutions are equal, the solutions are **isotonic** (Greek *iso,* the same).

Movement of water into a cell by osmosis creates pressure, called **osmotic pressure,** which can cause a cell to swell and burst. Most cells cannot withstand osmotic pressure unless their plasma membranes are braced to resist the swelling. If placed in pure water, they soon burst like over-inflated balloons (figure 4.28). That is why the cells of so many kinds of organisms have cell walls to stiffen their exteriors. In animals, the fluids bathing the cells have as many polar molecules dissolved in them as the cells do, so the problem doesn't arise.

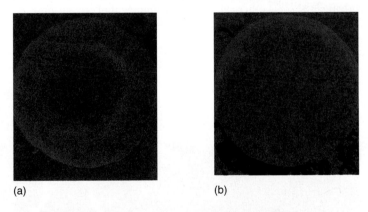

(a) (b)

Figure 4.28 Osmotic pressure in a red blood cell.
(a) Normally, a red blood cell has a flattened, pillowlike appearance, the concentrations of solutes inside the cell and in the surrounding fluid are the same, and there is no net movement of water across the membrane. (b) If placed in pure water, a red blood cell swells and will ultimately burst because the concentration of solutes is higher inside the cell, and there is a net movement of water into the cell.

4.10 Random movements of molecules cause them to mix uniformly in solution, a process called diffusion. Water associated with polar solutes is not free to diffuse, and there is a net movement of water across a membrane toward the side with less "free" water, a process called osmosis.

Bulk Passage into and out of Cells

Endocytosis and Exocytosis

The cells of many eukaryotes take in food and liquids by extending their plasma membranes outward toward food particles. The membrane engulfs the particle and forms a vesicle—a membrane-bordered sac—around it. This process is called **endocytosis** (figure 4.29).

The reverse of endocytosis is **exocytosis,** the discharge of material from vesicles at the cell surface (figure 4.30). In plant cells, exocytosis is an important means of exporting the materials needed to construct the cell wall through the plasma membrane. Among protists, the discharge of a contractile vacuole is a form of exocytosis. In animal cells, exocytosis provides a mechanism for secreting many hormones, neurotransmitters, digestive enzymes, and other substances.

Phagocytosis and Pinocytosis

If the material the cell takes in is particulate (made up of discrete particles), such as an organism or some other fragment of organic matter (figure 4.29a), the process is called **phagocytosis** (Greek *phagein,* to eat, and *cytos,* cell). If the material

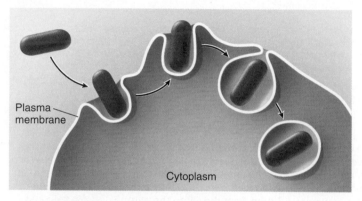

(a) Phagocytosis

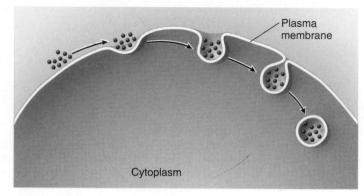

(b) Pinocytosis

Figure 4.29 Endocytosis.
Endocytosis is the process of engulfing material by folding the plasma membrane around it, forming a vesicle. (*a*) When the material is an organism or some other relatively large fragment of organic matter, the process is called phagocytosis. (*b*) When the material is a liquid, the process is called pinocytosis.

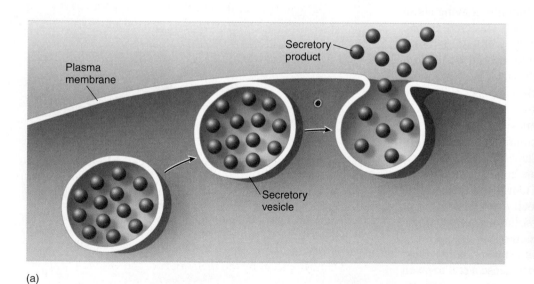

(a)

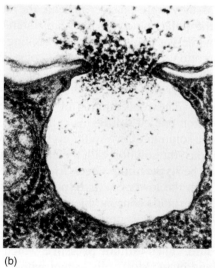

(b)

Figure 4.30 Exocytosis.
Exocytosis is the discharge of material from vesicles at the cell surface. (*a*) Proteins and other molecules are secreted from cells in small pockets called secretory vesicles, whose membranes fuse with the plasma membrane, thereby allowing the secretory vesicles to release their contents to the cell surface. (*b*) In the photomicrograph, you can see exocytosis taking place explosively.

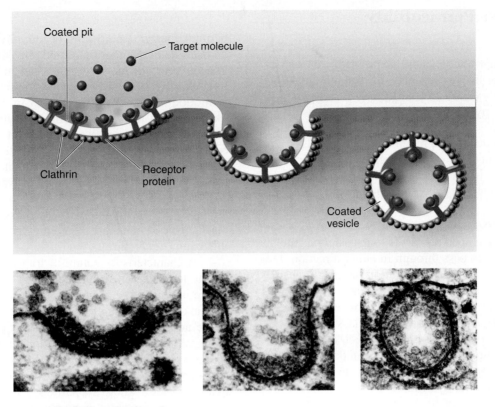

Figure 4.31 Receptor-mediated endocytosis.
Cells that undergo receptor-mediated endocytosis have pits coated with the protein clathrin that initiate endocytosis when target molecules bind to receptor proteins in the plasma membrane. In the photomicrographs, a coated pit appears in the plasma membrane of a developing egg cell, covered with a layer of proteins (80,000×). When an appropriate collection of molecules gathers in the coated pit, the pit deepens, and eventually seals off to form a vesicle.

the cell takes in is liquid (figure 4.29b), it is called **pinocytosis** (Greek *pinein,* to drink). Pinocytosis is common among animal cells. Mammalian egg cells, for example, "nurse" from surrounding cells; the nearby cells secrete nutrients that the maturing egg cell takes up by pinocytosis. Virtually all eukaryotic cells constantly carry out these kinds of endocytosis, trapping particles and extracellular fluid in vesicles and ingesting them. Endocytosis rates vary from one cell type to another. They can be surprisingly high: some types of white blood cells ingest 25% of their cell volume each hour!

Receptor-Mediated Endocytosis

Specific molecules are often transported into eukaryotic cells through **receptor-mediated endocytosis.** Molecules to be transported first bind to specific receptors in the plasma membrane. The transport process is specific to only molecules that have a shape that fits snugly into the receptor. The plasma membrane of a particular kind of cell contains a characteristic battery of receptor types, each for a different kind of molecule.

The portion of the receptor molecule inside the membrane is trapped in an indented pit coated with the protein clathrin. The pits act like molecular mousetraps, closing over to form an internal vesicle when the right molecule enters the

pit (figure 4.31). The trigger that releases the trap is the binding of the properly fitted target molecule to a receptor embedded in the membrane of the pit. When binding occurs, the cell reacts by initiating endocytosis. The process is highly specific and very fast.

One type of molecule that is taken up by receptor-mediated endocytosis is called low density lipoprotein (LDL). The LDL molecules bring cholesterol into the cell where it can be incorporated into membranes. Cholesterol plays a key role in determining the stiffness of the body's membranes. In the human genetic disease called hypercholesterolemia, the receptors lack tails and so are never caught in the clathrin-coated pits and, thus, are never taken up by the cells. The cholesterol stays in the bloodstream of affected individuals, coating their arteries and leading to heart attacks.

It is important to understand that receptor-mediated endocytosis in itself does not bring substances directly into the cytoplasm of a cell. The material taken in is still separated from the cytoplasm by the membrane of the vesicle.

4.11 The plasma membrane can engulf materials by endocytosis, folding the membrane around the material to encase it within a vesicle.

4.12 | Selective Permeability

From the point of view of efficiency, the problem with endocytosis is that it is expensive to carry out—the cell must make and move a lot of membrane. Also, endocytosis is not picky—in pinocytosis particularly, engulfing liquid does not allow the cell to choose which molecules come in. Cells solve this problem by using proteins in the plasma membrane as channels to pass molecules into and out of the cell. Because each kind of channel passes only a certain kind of molecule, the cell can control what enters and leaves, an ability called **selective permeability.**

Selective Diffusion

Some channels act like open doors. As long as a molecule fits the channel, it is free to pass through in either direction. Diffusion tends to equalize the concentration of such molecules on both sides of the membrane, with the molecules moving toward the side where they are scarcest. This mechanism of transport is called **selective diffusion.** One class of selectively open channels consists of ion channels, which are pores that span the membrane. Ions that fit the pore can diffuse through it in either direction. Such ion channels play an essential role in signaling by the nervous system.

Facilitated Diffusion

Most diffusion occurs through use of a special carrier protein. This protein binds only certain kinds of molecules, such as a particular sugar, amino acid, or ion, physically binding them on one side of the membrane and releasing them on the other. The direction of the molecule's net movement depends on its concentration gradient across the membrane. If the concentration is greater in the cytoplasm, the molecule is more likely to bind to the carrier on the cytoplasmic side of the membrane and be released on the extracellular side. If the concentration of the molecule is greater outside in the fluid surrounding the cell, the net movement will be from outside to inside. Thus the net movement always occurs from high concentration to low, just as it does in simple diffusion, but the process is facilitated by the carriers. For this reason, this mechanism of transport is given a special name, **facilitated diffusion** (figure 4.32).

A characteristic feature of transport by carrier proteins is that its rate can be saturated. If the concentration of a substance is progressively increased, the rate of transport of the substance increases up to a certain point and then levels off. There are a limited number of carrier proteins in the membrane, and when the concentration of the transported substance is raised high enough, all the carriers will be in use. The transport system is then said to be "saturated." When an investigator wishes to know if a particular substance is being transported across a membrane by a carrier system, or is diffusing across, he or she conducts experiments to see if the transport system can be saturated. If it can be saturated, it is carrier-mediated; if it cannot be saturated, it is not.

Facilitated Diffusion

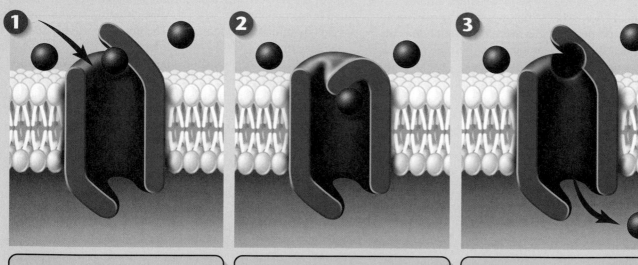

1 Particular molecules can bind to special protein carriers in the plasma membrane.

2 The protein carrier helps (facilitates) the diffusion process and does not require energy.

3 The molecule is released on the far side of the membrane. Protein carriers transport only certain molecules across the membrane but will take them in either direction down their concentration gradients.

Figure 4.32 How facilitated diffusion works.

Active Transport

Other channels through the plasma membrane are closed doors. These channels open only when energy is provided. They are designed to enable the cell to maintain high or low concentrations of certain molecules, much more or less than exists outside the cell. If the doors were open, the molecules would simply flood in or out by facilitated diffusion. Instead, like motor-driven turnstiles, the channels operate only when energy is provided, and they move a certain substance only in one direction (up its concentration gradient). The operation of these one-way, energy-requiring channels results in **active transport,** the movement of molecules across a membrane to a region of higher concentration by the expenditure of energy.

You might think that the plasma membrane possesses all sorts of active transport channels for the transport of important sugars, amino acids, and other molecules, but in fact, almost all of the active transport in cells is carried out by only two kinds of channels, the sodium-potassium pump and the proton pump.

The Sodium-Potassium Pump. The first of these, the **sodium-potassium (Na$^+$-K$^+$) pump,** expends metabolic energy to actively pump sodium ions (Na$^+$) in one direction, out of cells, and potassium ions (K$^+$) in one direction, into cells (figure 4.33). More than one-third of all the energy expended by your body's cells is spent driving Na$^+$-K$^+$ pump channels. This energy is derived from *adenosine triphosphate (ATP),* a

molecule we will learn about in chapter 5. Each channel can move over 300 sodium ions per second when working full tilt. As a result of all this pumping, there are far fewer sodium ions in the cell. This concentration gradient, paid for by the expenditure of considerable metabolic energy in the form of ATP molecules, is exploited by your cells in many ways. Two of the most important are (1) the conduction of signals along nerve cells (discussed in detail in chapter 23) and (2) the pulling into the cell of valuable molecules such as sugars and amino acids *against* their concentration gradient!

We will focus for a moment on this second process. The plasma membranes of many cells are studded with facilitated diffusion channels, which offer a path for sodium ions that have been pumped out by the Na$^+$-K$^+$ pump to diffuse back in. There is a catch, however; these channels require that the sodium ions have a partner in order to pass through—like a dancing party where only couples are admitted through the door. These special channels won't let sodium ions across unless another molecule tags along, crossing hand in hand with the sodium ion. In some cases the partner molecule is a sugar, in others an amino acid or other molecule. Because so many sodium ions are trying to get back in, this diffusion pressure drags in the partner molecules as well, even if they are already in high concentration within the cell. In this way, sugars and other actively transported molecules enter the cell—via special **coupled channels** (figure 4.34). Their movement is in fact a form of facilitated diffusion driven by the active transport of sodium ions.

Sodium-Potassium Pump

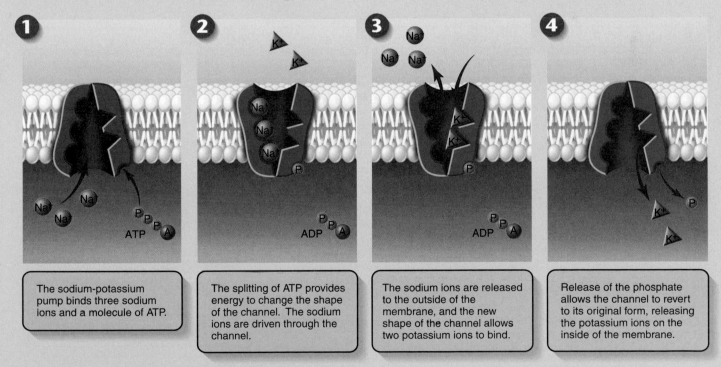

1 The sodium-potassium pump binds three sodium ions and a molecule of ATP.

2 The splitting of ATP provides energy to change the shape of the channel. The sodium ions are driven through the channel.

3 The sodium ions are released to the outside of the membrane, and the new shape of the channel allows two potassium ions to bind.

4 Release of the phosphate allows the channel to revert to its original form, releasing the potassium ions on the inside of the membrane.

Figure 4.33 How the sodium-potassium pump works.

The Proton Pump. The second major active transport channel is the **proton pump,** a complex channel that expends metabolic energy to pump protons across membranes (figure 4.35). Just as in the sodium-potassium pump, this creates a diffusion pressure that tends to drive protons back across again, but in this case the only channels open to them are not coupled channels but rather channels that make ATP, the energy currency of the cell. This pump is the key to cell metabolism, which is the way cells convert photosynthetic energy or chemical energy from food into ATP. Its activity is referred to as **chemiosmosis.** We discuss chemiosmosis at greater length in chapter 6. Table 4.3 summarizes the mechanisms for transport across cell membranes that we have discussed.

How Cells Get Information

A cell's ability to respond appropriately to changes in its environment is a key element in its ability to survive. Cells have evolved a variety of ways of sensing things about them. Almost all cells sense their environment primarily by detecting chemical or electrical signals. To do this, cells employ a battery of special proteins called cell surface proteins embedded within the plasma membrane. The many proteins that protrude from the surface of a cell are the cell's only contact with the outside world—its only avenue of communication with its environment.

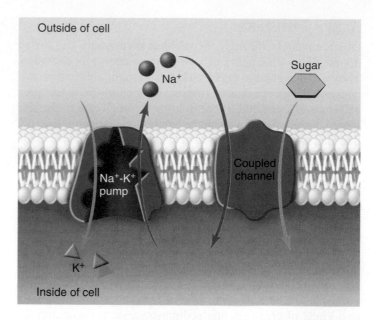

Figure 4.34 A coupled channel.
The active transport of a sugar molecule into a cell typically takes place in two stages—facilitated diffusion of the sugar coupled to active transport of sodium ions. The sodium-potassium pump keeps the Na⁺ concentration higher outside the cell than inside. For Na⁺ to diffuse back in through the coupled channel requires the simultaneous transport of a sugar molecule as well. Because the concentration gradient for Na⁺ is steeper than the opposing gradient for sugar, Na⁺ and sugar move into the cell.

Proton Pump

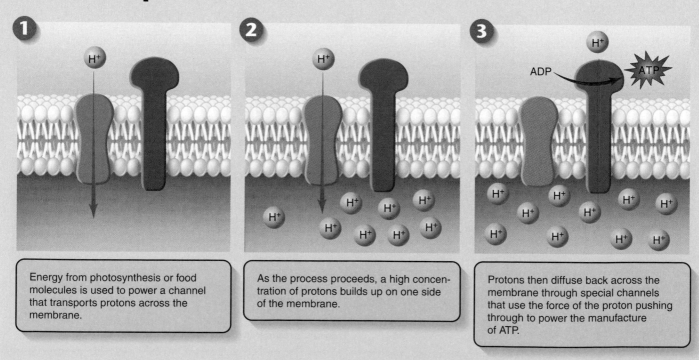

1. Energy from photosynthesis or food molecules is used to power a channel that transports protons across the membrane.

2. As the process proceeds, a high concentration of protons builds up on one side of the membrane.

3. Protons then diffuse back across the membrane through special channels that use the force of the proton pushing through to power the manufacture of ATP.

Figure 4.35 How the proton pump works.

TABLE 4.3 MECHANISMS FOR TRANSPORT ACROSS CELL MEMBRANES

Process	Passage Through Membrane	How It Works	Example
PASSIVE PROCESSES			
Diffusion			
Direct		Random molecular motion produces net migration of molecules toward region of lower concentration.	Movement of oxygen into cells
Protein channel		Polar molecules pass through a protein channel.	Movement of ions in or out of cell
Facilitated Diffusion			
Protein carrier		Molecule binds to carrier protein in membrane and is transported across; net movement is toward region of lower concentration.	Movement of glucose into cells
Osmosis			
Aquaporins		Diffusion of water across differentially permeable membrane.	Movement of water into cells placed in a hypotonic solution
ACTIVE PROCESSES			
Endocytosis			
Membrane vesicle			
Phagocytosis		Particle is engulfed by membrane, which folds around it and forms a vesicle.	Ingestion of bacteria by white blood cells
Pinocytosis		Fluid droplets are engulfed by membrane, which forms vesicles around them.	"Nursing" of human egg cells
Receptor-mediated endocytosis		Endocytosis is triggered by a specific receptor.	Cholesterol uptake
Exocytosis Membrane vesicle		Vesicles fuse with plasma membrane and eject contents.	Secretion of mucus
Active Transport			
Protein carrier			
Na^+-K^+ pump		Carrier expends energy to export a substance across a membrane against its concentration gradient.	Na^+ and K^+ against their concentration gradients
Coupled transport		Molecules are transported across a membrane against their concentration gradients by the cotransport of another substance down its concentration gradient.	Coupled uptake of glucose into cells against its concentration gradient
Proton pump		Protons are pumped across membranes against their concentration gradient. The proton gradient then drives the formation of ATP through another channel.	Proton pump in chemiosmosis

Sensing Chemical Information. Cells sense chemical information by means of cell surface proteins called **receptor proteins** projecting from their plasma membranes. These proteins bind a particular kind of molecule, but they do not provide a channel for the molecule to enter the cell. What the receptors do transmit into the cell is information. Often the information is about the presence of other cells in the vicinity—sensing cellular identity is the basis of the immune system, which defends your body from infection. In other instances the information may be a chemical signal sent from other cells.

Your body uses chemical signals called **hormones,** which provide a good example of how receptor proteins work. The end of a receptor protein exposed to the cell surface has a shape that fits to a specific hormone molecule, like insulin. Most of your cells have only a few insulin receptors, but your liver cells possess as many as 100,000 each! When an insulin molecule encounters an insulin receptor on the surface of a liver cell, the insulin molecule binds to the receptor. This binding produces a change in the shape of the other end of the receptor protein (the end protruding into the interior of the cell), just as stamping hard on your foot causes your mouth to open. This change in receptor shape at the interior end initiates a change in cell activity—in this case, the end protruding into the cytoplasm begins to add phosphate groups to proteins, and so it activates a variety of cell processes involved with regulating glucose levels in the blood.

Sensing Voltage. Many cells can sense electrical as well as chemical information. Embedded within their plasma membranes are special channels for sodium or other ions, channels that are usually closed. Unlike the sodium-potassium pump, these closed channels do not open in response to chemical energy. What does open these channels is voltage. Like little magnets, they flip open or shut in response to electrical signals.

It is not difficult to understand how **voltage-sensitive channels** work. The center of the protein that provides the channel through the membrane is occupied by a voltage-sensitive "door"—a portion of the protein containing charged amino acids. When a voltage charge in the vicinity changes, the door flips up out of the way and the channel is open to the passage of sodium or other ions. Voltage-sensitive channels play many important roles within excitable cells of muscle tissue and the nervous system (see chapter 23).

Sensing Information Within the Cell. In eukaryotic cells, which have many compartments, it is very important that the different parts of the cell be able to sense what is going on elsewhere in the cytoplasm. This sort of communication is provided by the diffusion of molecules within the cytoplasm. Some of the chemical signals are molecules used in metabolism; others are within-cell hormones; and still others are ions, particularly those that indicate cell pH.

4.12 Cells are selectively permeable, admitting only certain molecules. Facilitated diffusion is selective transport across a membrane in the direction of lower concentration. Active transport is energy-driven transport across a membrane toward a region of higher concentration. Cells obtain information about their surroundings from a battery of proteins protruding from the cell membrane.

Exploring Current Issues

Additional Resources

Go to your campus library or look online to find the following articles, which further develop some of the concepts found in this chapter.

Dartmouth Medical School. (2002). Drink at least eight glasses of water a day—really? *Dartmouth Medical School News. Retrieved from www.dartmouth.edu/dms/news/2002_h2/08aug2002_water.shtml.*

Nadis, S. (2003). The cells that rule the seas. *Scientific American,* 289(6), 52.

Pray, L. (2003). Microbial multicellularity. *The Scientist,* 17(23), 20.

Spinney, L. (2004). The gene chronicles: your DNA doesn't just reveal where you came from, it also tells stories about the way your ancestors lived. *New Scientist,* 181(2433), 40.

Westphal, S. P. (2003). Rewrite the textbooks. *New Scientist,* 178(2401), 22.

Biology and Society Lecture: Cystic Fibrosis— A Membrane Disorder

Cystic fibrosis is the most common fatal gene disorder of Caucasians. The body cells of affected individuals secrete a thick mucus that clogs the airways of the lungs and blocks the ducts of the pancreas and liver. Most patients do not survive past their mid-twenties. There is no known cure.

Cystic fibrosis results from a defect in a gene encoding a plasma membrane protein called CFTR (*cystic fibrosis transmembrane conductance regulator*). This protein channel regulates the passage of chloride ions into and out of the body's cells. A defective CFTR channel leads to a buildup of chloride ions within lung, pancreas, and liver cells, causing water to move into the cells by osmosis. Removing water from the surrounding mucus causes it to thicken, clogging the passageways. Attempts are under way to cure cystic fibrosis with gene therapy, using a virus to ferry healthy CFTR genes into patients lacking them.

Find this lecture, delivered by the author to his class at Washington University, online at www.mhhe.com/tlwessentials/exp4.

Summary

The World of Cells

4.1 Cells

- Cells are the smallest living structure. They consist of the cytoplasm enclosed in a plasma membrane. Organisms may be composed of a single cell or multiple cells.

- Materials pass into and out of cells across the plasma membrane. A smaller cell has a larger surface-to-volume ratio, which increases the area through which materials may pass (**figure 4.2**). Because cells are so small, a microscope is needed to view and study them (**table 4.1**).

4.2 The Plasma Membrane

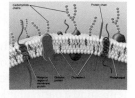

- The plasma membrane that encloses all cells consists of a double layer of lipids, called the lipid bilayer, in which proteins are embedded. The structure of the plasma membrane is called the fluid mosaic model (**figure 4.6**).

- The lipid bilayer is made up of special lipid molecules called phospholipids (**figure 4.4**), which have a polar (water-soluble) end and a nonpolar (water-insoluble) end. The bilayer forms because the nonpolar ends move away from the watery surroundings, forming the two layers (**figure 4.5**). Membrane proteins are either attached to the surface of the cell or are embedded within the membrane (**figure 4.7**).

Kinds of Cells

4.3 Prokaryotic Cells

- Prokaryotic cells are simple cellular organisms that lack nuclei or other internal organelles and are usually encased in a rigid cell wall (**figures 4.8 and 4.9**).

4.4 Eukaryotic Cells

- Eukaryotic cells are larger and more structurally complex compared with prokaryotic cells. They contain nuclei and have internal membrane systems (**figures 4.10 and 4.11**).

Tour of a Eukaryotic Cell

4.5 The Nucleus: The Cell's Control Center

- The nucleus is the command and control center of the cell. It contains the cell's DNA, which encodes the hereditary information that runs the cell (**figure 4.12**).

4.6 The Endomembrane System

- The endomembrane system is a collection of interior membranes that organize and divide the cell's interior into functional areas. The endoplasmic reticulum is a transport system that modifies and moves proteins and other molecules produced in the ER to the Golgi complex (**figure 4.13**). The Golgi complex is a delivery system that carries molecules to the surface of the cell where they are released to the outside (**figure 4.14 and 4.15**).

4.7 Organelles That Contain DNA

- Mitochondria and chloroplasts are cell-like organelles that appear to be ancient bacteria that formed endosymbiotic relationships with early eukaryotic cells (**figure 4.18**). The mitochondrion is called the powerhouse of the cell because it is the site of oxidative metabolism, an energy-extracting process (**figure 4.16**). Chloroplasts are the site of photosynthesis and are present in plants and algal cells (**figure 4.17**).

4.8 The Cytoskeleton: Interior Framework of the Cell

- The interior of the cell contains a network of protein fibers, called the cytoskeleton, that supports the shape of the cell and anchors organelles in place (**figure 4.19**).

- Cells are dynamic structures. Centrioles, microtubules, and molecular motors move materials around inside the cell. Cilia and flagella propel the cell through its environment (**figures 4.20–4.22**).

4.9 Outside the Plasma Membrane

- The cells of plants, fungi, and many protists have cell walls that serve a similar function as prokaryotic cell walls, but are composed of different molecules (**figure 4.24**). Animal cells lack cell walls but contain an outer layer of glycoproteins, called the extracellular matrix (**figure 4.25**).

Transport Across Plasma Membranes

4.10 Diffusion and Osmosis

- Materials pass into and out of the cell passively through diffusion and osmosis. Diffusion is the movement of molecules from an area of high concentration to an area of low concentration (**figure 4.26**). Molecules pass into and out of cells down their concentration gradients. Osmosis is the movement of water into and out of the cell, driven by differing concentrations of solute. Water molecules move to areas of higher solute concentrations (**figure 4.27**).

4.11 Bulk Passage into and out of Cells

- Larger structures or larger quantities of material move into and out of the cell through endocytosis and exocytosis, respectively (**figures 4.29 and 4.30**). Receptor-mediated endocytosis is a selective transport process, bringing in only those substances that are able to bind to specific receptors (**figure 4.31**).

4.12 Selective Permeability

- Selective transport of materials across the membrane is accomplished by facilitated diffusion and active transport.

- Facilitated diffusion is driven by the concentration gradient, transporting substances down their concentration gradient, but substances must bind to a membrane transporter, called a carrier, in order to pass across the membrane (**figure 4.32**).

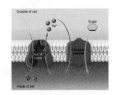

- Active transport involves the input of energy to transport substances against (or up) their concentration gradients. Examples include the sodium-potassium pump (**figure 4.33**), coupled channels (**figure 4.34**), and the proton pump (**figure 4.35**).

- Cells gain information from their environment through specialized membrane protein receptors and channels.

---

1. Cell theory includes the principle that
   a. cells are the smallest living things. Nothing smaller than a cell is considered alive.
   b. all cells are surrounded by cell walls that protect them.
   c. all organisms are made up of many cells arranged in specialized, functional groups.
   d. all cells are made of smaller subunits called organelles. Nothing smaller than an organelle is considered alive.

2. The plasma membrane is
   a. a carbohydrate layer that surrounds groups of cells, called tissues, to protect them.
   b. a double lipid layer with proteins inserted in it, which surrounds every cell individually.
   c. a thin sheet of structural proteins that lines the inside of some body cavities.
   d. composed of blood plasma that has solidified into a protective barrier.

3. Organisms that have cells with a relatively uniform cytoplasm and no organelles are called _____, and organisms whose cells have organelles and a nucleus are called _____.
   a. cellulose, nuclear
   b. flagellated, streptococcal
   c. eukaryotes, prokaryotes
   d. prokaryotes, eukaryotes

4. Within the nucleus of a cell you can find
   a. a nucleolus.
   b. many ribosomes.
   c. a cytoskeleton.
   d. all of these.

5. The endomembrane system within a cell includes the
   a. cytoskeleton and the ribosomes.
   b. prokaryotes and the eukaryotes.
   c. endoplasmic reticulum and the Golgi complex.
   d. mitochondria and the chloroplasts.

6. Until fairly recently, it was thought that only the nucleus of each cell contained DNA. We now know that DNA is also carried in the
   a. cytoskeleton and the ribosomes.
   b. prokaryotes and the eukaryotes.
   c. endoplasmic reticulum and the Golgi bodies.
   d. mitochondria and the chloroplasts.

7. Which of the following statements is true?
   a. All cells have a cell wall for protection and structure.
   b. Eukaryotic cells in plants and fungi, and all prokaryotes, have a cell wall.
   c. There is a second membrane composed of structural carbohydrates surrounding all cells.
   d. Prokaryotes and all cells of eukaryotic animals have a cell wall.

8. If you put a drop of food coloring into a glass of water, the drop of color will
   a. fall to the bottom of the glass and sit there unless you stir the water; this is because of hydrogen bonds.
   b. float on the top of the water, like oil, unless you stir the water; this is because of surface tension.
   c. instantly disperse throughout the water; this is because of osmosis.
   d. slowly disperse throughout the water; this is because of diffusion.

9. When large molecules such as food particles need to get into a cell, they cannot easily pass through the plasma membrane, and so they move across the membrane through the processes of
   a. diffusion and osmosis.
   b. endocytosis and phagocytosis.
   c. exocytosis and pinocytosis.
   d. permeability and reception.

10. Active transport of certain molecules involves
    a. diffusion and osmosis.
    b. endocytosis and phagocytosis.
    c. energy and specialized pumps or channels.
    d. permeability and reception.

Visual Understanding

1. **Figure 4.3** The first microscope was used in about 1590. Electron microscopes came into common use about 70 years ago. Just over 100 years ago most physicians did not wash up between patients, even when someone had just died, or was very sick. Explain why it took so long to convince doctors to wash their hands.

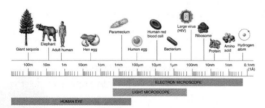

2. **Figure 4.6** A hormone molecule is a messenger, circulating through the bloodstream. It needs to find only a particular subset of cells to deliver its message. How does the hormone find the correct cells?

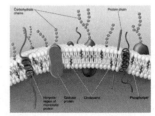

Challenge Questions

World of Cells You are designing a new single-celled organism. Discuss the problems of size, getting molecules such as nutrients and wastes in and out, temperature, and energy.

Kinds of Cells Antibiotics are medicines that target bacterial infections in vertebrates. How can an antibiotic kill all the bacterial cells and not harm vertebrate cells; what part of the bacterial cell must antibiotics be targeting and why?

Tour of a Eukaryotic Cell Compare the cellular organelles and other structures to the parts of a city—for example, the nucleus is city hall and the DNA is all the city's laws and instructions.

Transport Across Cell Membranes Compare the mechanisms required for a cell to obtain all the different kinds of molecules that it needs.

Online Learning Center

Visit the Online Learning Center for this chapter at www.mhhe.com/tlwessentials/ch4 for quizzes, animations, interactive learning exercises, and other study tools. At the site you will also find extended answers to the end-of-chapter questions.

5
Energy and Life

A ll of life is driven by energy. It took energy for these mice to climb up the wheat stalks. Their mousely activities while perched on the stalks—looking for danger, generating body heat, wiggling their whiskers—take energy. The energy comes from the wheat kernels and other foods these mice eat. By breaking the chemical bonds of carbohydrates and other molecules in the wheat kernels, and transferring the energy of these bonds to those of a "molecular currency" called ATP, the mice are able to capture the chemical energy in their food and put it to work. The cells of the mice perform this feat with the aid of enzymes, which are proteins with highly specialized shapes. Each enzyme has a shape with a surface cavity called an active site, into which some chemical in the cell fits precisely, like a foot fits into a shoe of the proper size. When the chemical nudges in, the enzyme responds by bending, stressing particular covalent bonds in the chemical and so triggering a specific chemical reaction. The chemistry of life is enzyme chemistry.

5.1 The Flow of Energy in Living Things

Studying the chemistry of the cell may seem to you an uninteresting way to study biology. Indeed, few subjects put off biology students more quickly than chemistry. The prospect of encountering chemical equations seems at the same time both difficult and boring. Here we will try to keep the pain to a minimum, to avoid equations where possible, and to stress ideas rather than formulas. However, there is no avoiding some study of cell chemistry, for the same reason that a successful race car driver must learn how the engine of a car works. We are chemical machines, powered by chemical energy, and if we are to understand ourselves, we must "look under the hood" at the chemical machinery of our cells and see how it operates.

As described in chapter 3, **energy** is defined as the ability to do work. It can be considered to exist in two states: kinetic energy and potential energy. Kinetic energy is the energy of motion. Objects that are not in the process of moving but have the capacity to do so are said to possess potential energy, or stored energy. Hence, a boulder perched on a hilltop has potential energy; as it begins to roll downhill, some of its potential energy is converted into kinetic energy (figure 5.1). All of the work carried out by living organisms involves the transformation of potential energy to kinetic energy.

Energy exists in many forms: mechanical energy, heat, sound, electric current, light, or radioactive radiation. Because it can exist in so many forms, there are many ways to measure energy. The most convenient is in terms of heat, because all other forms of energy can be converted into heat. In fact, the study of energy is called **thermodynamics,** meaning heat changes.

Energy flows into the biological world from the sun, which shines a constant beam of light on the earth. It is estimated that the sun provides the earth with more than 13×10^{23} calories per year, or 40 million billion calories per second! Plants, algae, and certain kinds of bacteria capture a fraction of this energy through photosynthesis. In photosynthesis, energy garnered from sunlight is used to combine small molecules (water and carbon dioxide) into more complex molecules (sugars). These complex sugar molecules have potential energy due to the arrangement of their atoms. This potential energy, in the form of chemical energy, does the work in cells. Recall from chapter 3 that an atom consists of a central nucleus surrounded by one or more orbiting electrons, and a covalent bond forms when two atomic nuclei share electrons. Breaking such a bond requires energy to pull the nuclei apart. Indeed, the strength of a covalent bond is measured by the amount of energy required to break it. For example, it takes 98.8 kcal to break 1 mole (6.023×10^{23}) of carbon–hydrogen (C–H) bonds.

All the chemical activities within cells can be viewed as a series of chemical reactions between molecules. A **chemical reaction** is the making or breaking of chemical bonds—gluing atoms together to form new molecules or tearing molecules apart and sometimes sticking the pieces onto other molecules.

> **5.1** Energy is the capacity to do work, either actively (kinetic energy) or stored for later use (potential energy).

(a) Potential energy

(b) Kinetic energy

Figure 5.1 Potential and kinetic energy.
Objects that have the capacity to move but are not moving have potential energy, while objects that are in motion have kinetic energy. (*a*) The energy required to move the ball up the hill is stored as potential energy. (*b*) This stored energy is released as kinetic energy as the ball rolls down the hill.

5.2 The Laws of Thermodynamics

Running, thinking, singing, reading these words—all activities of living organisms involve changes in energy. A set of universal laws we call the laws of thermodynamics govern all energy changes in the universe, from nuclear reactions to the buzzing of a bee.

The First Law of Thermodynamics

The first of these universal laws, the **first law of thermodynamics,** concerns the amount of energy in the universe. It states that energy can change from one state to another (from potential to kinetic, for example) but it can never be destroyed, nor can new energy be made. The total amount of energy in the universe remains constant.

A lion eating a giraffe is in the process of acquiring energy. Rather than creating new energy or capturing the energy in sunlight, the lion is merely transferring some of the potential energy stored in the giraffe's tissues to its own body (just as the giraffe obtained the potential energy stored in the plants it ate while it was alive). Within any living organism, this chemical potential energy can be shifted to other molecules and stored in chemical bonds, or it can be converted into other states, such as kinetic energy, or into other forms such as light, or electrical energy. During each conversion, some of the energy dissipates into the environment as **heat energy,** a measure of the random motions of molecules (and, hence, a measure of one form of kinetic energy). Energy continuously flows through the biological world in one direction, with new energy from the sun constantly entering the system to replace the energy dissipated as heat.

Heat can be harnessed to do work only when there is a heat gradient—that is, a temperature difference between two areas; this is how a steam engine functions. A boiler heats up water to create steam. The steam is pumped into the cylinder of a steam engine where it moves a piston. The piston does the work of the steam engine, such as moving a lever that turns a

Disorder happens "spontaneously"

Organization requires energy

Figure 5.3   Entropy in action.

As time elapses, a child's room becomes more disorganized. It takes energy to clean it up.

wheel (figure 5.2). Cells are too small to maintain significant internal temperature differences, so heat energy is incapable of doing the work of cells. Thus, although the total amount of energy in the universe remains constant, the energy available to do useful work in a cell decreases, as progressively more of it dissipates as heat.

The Second Law of Thermodynamics

The **second law of thermodynamics** concerns this transformation of potential energy into heat, or random molecular motion. It states that the disorder in a closed system like the universe is continuously increasing. Put simply, disorder is more likely than order. For example, it is much more likely that a column of bricks will tumble over than that a pile of bricks will arrange themselves spontaneously to form a column. In general, energy transformations proceed spontaneously to convert matter from a more ordered, less stable form, to a less ordered, more stable form (figure 5.3).

Entropy

Entropy is a measure of the degree of disorder of a system, so the second law of thermodynamics can also be stated simply as "entropy increases." When the universe formed 10 to 20 billion years ago, it held all the potential energy it will ever have. It has become progressively more disordered ever since, with every energy exchange increasing the amount of entropy in the universe.

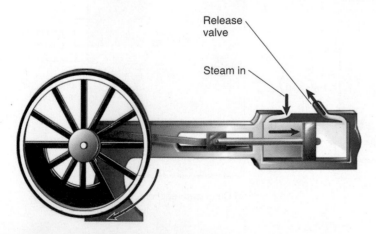

Release valve

Steam in

Figure 5.2   A steam engine.

In a steam engine, heat is used to produce steam. The heat gradient created by the steam is then used to do work. In this simple engine, the expanding steam pushes against a piston, turning a wheel and so rotating the engine's axle.

5.2   The first law of thermodynamics states that energy cannot be created or destroyed; it can only undergo conversion from one form to another. The second law of thermodynamics states that disorder (entropy) in the universe is increasing. Life converts energy from the sun to other forms of energy that drive life processes; the energy is never lost, but as it is used, more and more of it is converted to heat, the energy of random molecular motion.

5.3 Chemical Reactions

In a chemical reaction, the molecules that you start with are called **reactants,** or sometimes **substrates,** whereas the molecules that you end up with, after the reaction is over, are called the **products** of the reaction. Not all chemical reactions are equally likely to occur. Just as a boulder is more likely to roll downhill than uphill, so a reaction is more likely to occur if it releases energy than if it needs to have energy supplied. Reactions that release energy, ones in which the products contain less energy than the reactants, tend to occur spontaneously and are called **exergonic** (figure 5.4b). By contrast, reactions in which the products contain more energy than the reactants are called **endergonic** (figure 5.4a).

Activation Energy

If all chemical reactions that release energy tend to occur spontaneously, it is fair to ask, "Why haven't all exergonic reactions occurred already?" Clearly they have not. If you ignite gasoline, it burns with a release of energy. So why doesn't all the gasoline in all the automobiles in the world just burn up right now? It doesn't because the burning of gasoline, and almost all other chemical reactions, requires an input of energy to get it started—a kick in the pants such as a match or spark plug. Even if the product contains or stores less energy than the reactants, it is first necessary to break existing bonds in the reactants, and this takes energy. The extra energy required to

destabilize existing chemical bonds and so initiate a chemical reaction is called **activation energy** (figure 5.4b). Thus, to roll a boulder downhill, you must first nudge it out of the hole it sits in. Activation energy is simply a chemical nudge.

Catalysis

Just as all the gasoline in the world doesn't burn up right now because the combustion reaction has a sizable activation energy, so all the exergonic reactions in your cells don't spontaneously happen either. Each reaction waits for something to nudge it along, to supply it with sufficient activation energy to get going. One way to make an exergonic reaction more likely to happen is to lower the necessary activation energy. Like digging away the ground below your boulder, lowering activation energy reduces the nudge needed to get things started. The process of lowering the activation energy of a reaction is called **catalysis.** Catalysis cannot make an endergonic reaction occur spontaneously—you cannot avoid the need to supply energy—but it can make a reaction, endergonic or exergonic, proceed much faster, because almost all the time required for chemical reactions is tied up in overcoming the activation energy barrier (figure 5.4c).

> **5.3 Chemical reactions occur when the covalent bonds linking atoms together are formed or broken. It takes energy to initiate chemical reactions.**

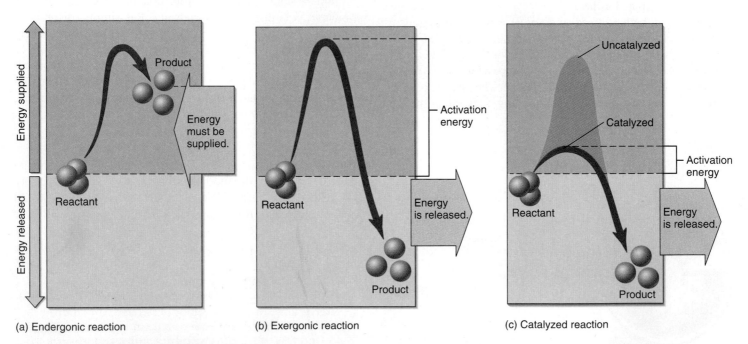

(a) Endergonic reaction   (b) Exergonic reaction   (c) Catalyzed reaction

Figure 5.4 Chemical reactions and catalysis.

(a) The products of endergonic reactions contain more energy than the reactants. (b) The products of exergonic reactions contain less energy than the reactants, but exergonic reactions do not necessarily proceed rapidly because it takes energy to get them going. The "hill" in this energy diagram represents energy that must be supplied to destabilize existing chemical bonds. (c) Catalyzed reactions occur faster because the amount of activation energy required to initiate the reaction—the height of the energy hill that must be overcome—is lowered.

5.4 How Enzymes Work

Proteins called **enzymes** are the catalysts used by cells to touch off particular chemical reactions. By controlling which enzymes are present, and when they are active, cells are able to control what happens within themselves, just as a conductor controls the music an orchestra produces by dictating which instruments play when.

An enzyme works by binding to a specific molecule and stressing the bonds of that molecule in such a way as to make a particular reaction more likely. The key to this activity is the shape of the enzyme (figure 5.5). An enzyme is specific for a particular reactant because the enzyme surface provides a mold that very closely fits the shape of the desired reactant. Other molecules that fit less perfectly simply don't adhere to the enzyme's surface. The site on the enzyme surface where the reactant fits is called the **active site.** The site on the reactant that binds to an enzyme is called the **binding site.** Proteins are not rigid. The binding of the reactant induces the enzyme to change its shape slightly, leading to an induced fit between the enzyme and its reactant, like a glove molding to a hand.

An enzyme lowers the activation energy of a particular reaction. It may encourage the breaking of a particular chemical bond in the reactant or substrate, weakening the bond by drawing away some of its electrons. Alternatively, an enzyme may encourage the formation of a link between two reactants

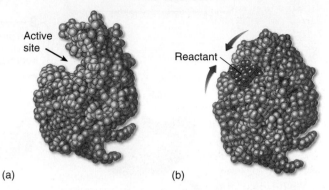

(a)   (b)

Figure 5.5 Enzyme shape determines its activity.
(a) A groove runs through the lysozyme enzyme (*blue* in this diagram) that fits the shape of the reactant (in this case, a chain of sugars). (b) When such a chain of sugars, indicated in *yellow*, slides into the groove, it induces the protein to change its shape slightly and embrace the substrate more intimately. This induced fit causes a chemical bond between two sugar molecules within the chain to break.

by holding them near each other (figure 5.6). Regardless of the type of reaction, the enzyme is not affected by the chemical reaction and is available to be used again (figure 5.7).

5.4 Enzymes are proteins that catalyze chemical reactions within cells.

How Enzymes Work

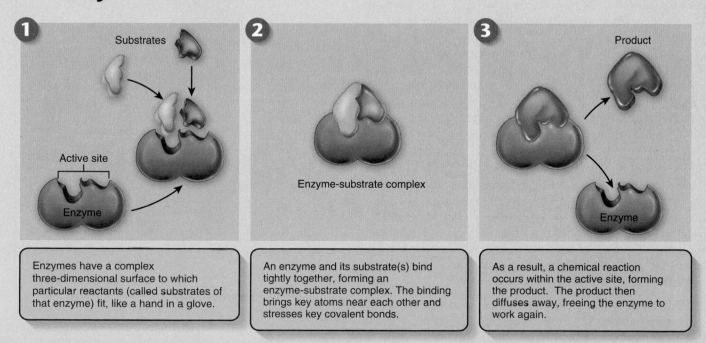

1 Enzymes have a complex three-dimensional surface to which particular reactants (called substrates of that enzyme) fit, like a hand in a glove.

2 An enzyme and its substrate(s) bind tightly together, forming an enzyme-substrate complex. The binding brings key atoms near each other and stresses key covalent bonds.

3 As a result, a chemical reaction occurs within the active site, forming the product. The product then diffuses away, freeing the enzyme to work again.

Figure 5.6 How enzymes work.

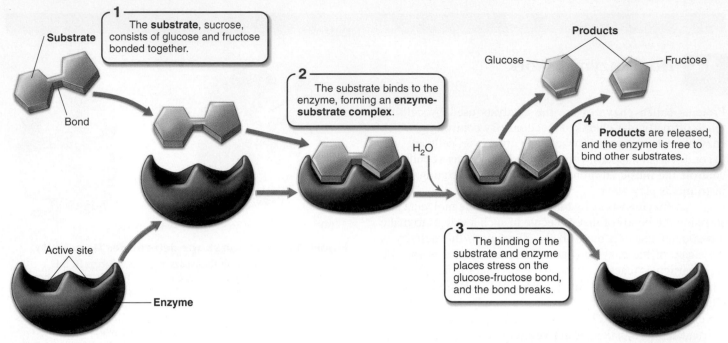

1 The **substrate**, sucrose, consists of glucose and fructose bonded together.

Substrate

Bond

2 The substrate binds to the enzyme, forming an **enzyme-substrate complex**.

H_2O

Products

Glucose

Fructose

4 **Products** are released, and the enzyme is free to bind other substrates.

3 The binding of the substrate and enzyme places stress on the glucose-fructose bond, and the bond breaks.

Active site

Enzyme

Figure 5.7   The catalytic cycle of an enzyme.

Enzymes increase the speed with which chemical reactions occur, but they are not altered themselves as they do so. In the reaction illustrated here, the enzyme sucrase is splitting the sugar sucrose (present in most candy) into two simpler sugars: glucose and fructose. (1) First, the sucrose substrate binds to the active site of the enzyme, fitting into a depression in the enzyme surface. (2) The binding of sucrose to the active site forms an enzyme-substrate complex and induces the sucrase molecule to alter its shape, fitting more tightly around the sucrose. (3) Amino acid residues in the active site, now in close proximity to the bond between the glucose and fructose components of sucrose, break the bond. (4) The enzyme releases the resulting glucose and fructose fragments, the products of the reaction, and is then ready to bind another molecule of sucrose and run through the catalytic cycle once again.

5.5   Factors Affecting Enzyme Activity

Temperature

Temperature and pH can have a major influence on the action of enzymes (figure 5.8). Enzyme activity is affected by any change in condition that alters the enzyme's three-dimensional shape. When the temperature increases, the bonds that determine enzyme shape are too weak to hold the enzyme's peptide chains in the proper position and the enzyme denatures. This is why an extremely high fever in humans can be fatal. As a result, enzymes function best within an optimum temperature range. This range is relatively narrow for most human enzymes.

pH

In addition, most enzymes also function within an optimal pH range, because the structural bonds of enzymes are also sensitive to hydrogen ion (H^+) concentration. Most human enzymes, such as the protein-degrading enzyme trypsin, work best within the range of pH 6 to 8. However, some enzymes, such as the digestive enzyme pepsin, are able to function in very acidic environments such as the stomach, but can't function at higher pHs such as those found in blood.

> **5.5   Enzymes are sensitive to temperature and pH, because both of these variables influence protein shape.**

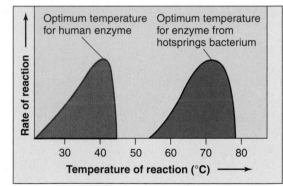

(a)

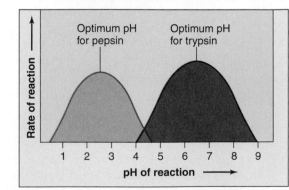

(b)

Figure 5.8   Enzymes are sensitive to their environment.

The activity of an enzyme is influenced by both (a) temperature and (b) pH. Most human enzymes, such as the protein-degrading enzyme trypsin, work best at temperatures of about 40°C and within a pH range of 6 to 8.

Allosteric Enzyme Regulation

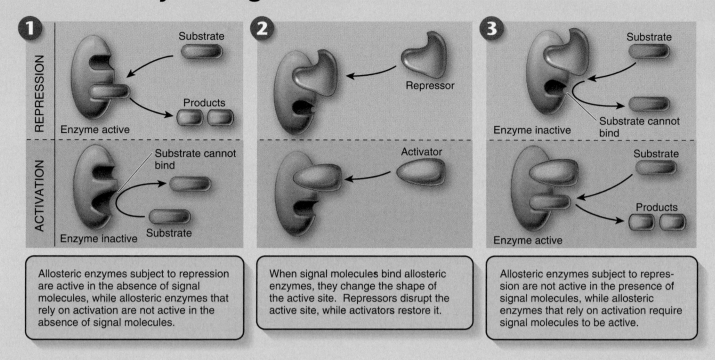

1

REPRESSION

Substrate

Products

Enzyme active

ACTIVATION

Substrate cannot bind

Enzyme inactive   Substrate

Allosteric enzymes subject to repression are active in the absence of signal molecules, while allosteric enzymes that rely on activation are not active in the absence of signal molecules.

2

Repressor

Activator

When signal molecules bind allosteric enzymes, they change the shape of the active site. Repressors disrupt the active site, while activators restore it.

3

Substrate

Substrate cannot bind

Enzyme inactive

Substrate

Products

Enzyme active

Allosteric enzymes subject to repression are not active in the presence of signal molecules, while allosteric enzymes that rely on activation require signal molecules to be active.

Figure 5.9   How cells control enzymes.
Many enzymes are allosteric—that is, they can change their shape. Allosteric enzymes have two binding sites, one for the substrate and another for a signal molecule. When a signal molecule binds to its site, the shape of the enzyme changes. This change disrupts some active sites (repression) and restores others to a proper configuration (activation).

5.6   How Cells Regulate Enzymes

Because an enzyme must have a precise shape to work correctly, it is possible for the cell to control when an enzyme is active by altering its shape. Many enzymes have shapes that can be altered by the binding to their surfaces of "signal" molecules. Such enzymes are called *allosteric* (Latin, other shape). When a signal molecule binds to an enzyme, the shape of the enzyme may be altered such that the reactants no longer fit into the active site. In this case, the signal molecule acts to inhibit enzyme activity and is called a **repressor.** In other cases, the enzyme may not be able to bind the reactants *unless* the signal molecule is bound to the enzyme. Thus, the signal molecule functions as an **activator** (figure 5.9). The site where the signal molecule binds to the enzyme surface is called the **allosteric site.**

Enzymes are often regulated by a mechanism called *feedback inhibition,* where the product of the reaction acts as the repressor. Enzyme inhibition can occur in two ways: *competitive inhibitors* compete with the substrate for the same active site; *noncompetitive inhibitors* bind to the enzyme at the allosteric site, changing the shape of the enzyme and making it unable to bind to the substrate (figure 5.10).

Many drugs and antibiotics work by inhibiting enzymes. Statin drugs like Lipitor lower cholesterol by inhibiting a key enzyme cells use to make cholesterol. The antibiotic penicillin inhibits an enzyme bacteria use in making cell walls. As humans lack this enzyme, we are not harmed by the drug.

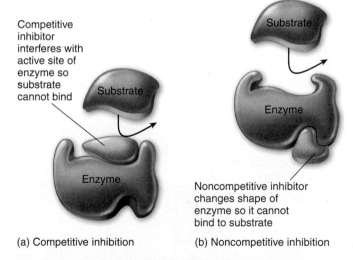

Competitive inhibitor interferes with active site of enzyme so substrate cannot bind

Substrate

Enzyme

Substrate

Enzyme

Noncompetitive inhibitor changes shape of enzyme so it cannot bind to substrate

(a) Competitive inhibition   (b) Noncompetitive inhibition

Figure 5.10   How enzymes can be inhibited.
(a) In competitive inhibition, the inhibitor interferes with the active site of the enzyme. (b) In noncompetitive inhibition, the inhibitor binds to the enzyme at a place away from the active site, effecting a conformational change in the enzyme so that it can no longer bind to its substrate. In feedback inhibition, the inhibitor molecule is the product of the reaction.

5.6   An enzyme's activity can be affected by signal molecules that bind to them, changing their shape.

5.7 ATP: The Energy Currency of the Cell

Cells use energy to do all those things that require work, but how does the cell use energy from the sun or the potential energy stored in molecules to power its activities? The sun's radiant energy and the energy stored in molecules are energy sources, but like money that is invested in stocks and bonds or real estate, these energy sources cannot be used directly to run a cell, any more than money invested in stocks can be used to buy a candy bar at the store. To be useful, the energy from the sun or food molecules must first be converted to a source of energy that a cell can use, like someone converting stocks and bonds to ready cash. The "cash" molecule in the body is **adenosine triphosphate (ATP).** ATP is the energy currency of the cell.

Structure of the ATP Molecule

Each ATP molecule is composed of three parts: (1) a sugar that serves as the backbone to which the other two parts are attached; (2) adenine, one of the four nucleotide bases in DNA and RNA; and (3) a chain of three phosphates (figure 5.11).

Because phosphates have negative electrical charges, it takes considerable chemical energy to hold the line of three phosphates next to one another at the end of ATP. Like a coiled spring, the phosphates are poised to push apart. It is for this reason that the chemical bonds linking the phosphates are such chemically reactive bonds.

When the endmost phosphate is broken off an ATP molecule, a sizable packet of energy is released. The reaction converts ATP to adenosine diphosphate, ADP. The second phosphate group can also be removed, yielding additional energy and leaving adenosine monophosphate (AMP). Most energy exchanges in cells involve cleavage of only the outermost bond, converting ATP into ADP and P_i, inorganic phosphate (figure 5.12).

$$ATP \rightarrow ADP + P_i + energy$$

Exergonic reactions require activation energy and endergonic reactions require the input of even more energy, and so these reactions in the cell are usually coupled with the breaking of the phosphate bond in ATP. Because almost all endergonic reactions in cells require less energy than is released by this reaction, ATP is able to power many of the cell's activities, producing heat as a by-product. Table 5.1 summarizes cellular activities powered by the breakdown of ATP.

Cells use two different but complementary processes to convert energy from the sun and potential energy found in food molecules into ATP, which are the subjects of chapter 6, *"How Cells Acquire Energy."* Plant cells, as well as algae and some bacteria, use the process of **photosynthesis** to convert the energy from the sun into potential energy stored in molecules of ATP and into sugar molecules that are built by the cell using

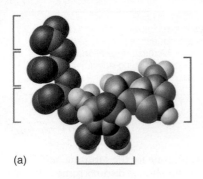

(a)

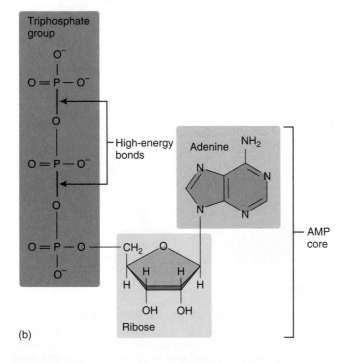

(b)

Figure 5.11 The parts of an ATP molecule.
The model (*a*) and structural diagram (*b*) both show that ATP consists of three phosphate groups attached to a ribose (five-carbon sugar) molecule. The ribose molecule is also attached to an adenine molecule (also one of the nucleotides of DNA and RNA). When the outermost phosphate group is split off from the ATP molecule, a great deal of energy is released. The resulting molecule is ADP, or adenosine diphosphate. The second phosphate group can also be removed, yielding additional energy and leaving AMP (adenosine monophosphate).

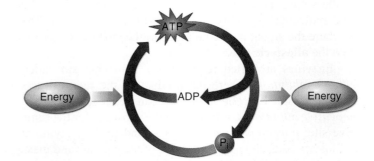

Figure 5.12 The ATP-ADP cycle.
In mitochondria and chloroplasts, chemical or photosynthetic energy is harnessed to form ATP from ADP and inorganic phosphate. When ATP is used to drive the living activities of cells, the molecule is cleaved back to ADP and inorganic phosphate, which are then available to form new ATP molecules.

TABLE 5.1   HOW CELLS USE ATP ENERGY TO POWER CELLULAR WORK

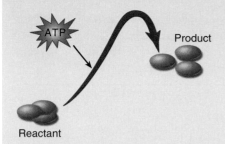

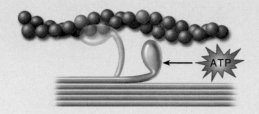

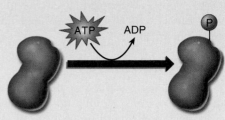

Biosynthesis

Cells use the energy released from the exergonic hydrolysis of ATP to drive endergonic reactions like those of protein synthesis, an approach called energy coupling.

Contraction

In muscle cells, filaments of protein repeatedly slide past each other to achieve contraction of the cell. An input of ATP is required for the filaments to reset and slide again.

Chemical Activation

Proteins can become activated when a high-energy phosphate from ATP attaches to the protein, activating it. Other types of molecules can also become phosphorylated by transfer of a phosphate from ATP.

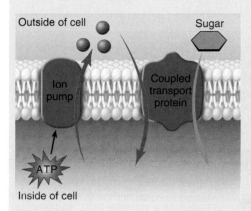

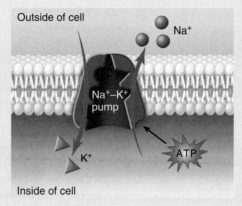

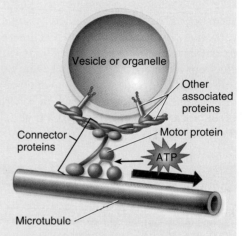

Importing Metabolites

Metabolite molecules such as amino acids and sugars can be transported into cells against their concentration gradients by coupling the intake of the metabolite to the inward movement of an ion moving down its concentration gradient, this ion gradient being established using ATP.

Active Transport: Na$^+$– K$^+$ Pump

Most animal cells maintain a low internal concentration of Na$^+$ relative to their surroundings, and a high internal concentration of K$^+$. This is achieved using a protein called the sodium-potassium pump, which actively pumps Na$^+$ out of the cell and K$^+$ in, using energy from ATP.

Cytoplasmic Transport

Within a cell's cytoplasm, vesicles or organelles can be dragged along microtubular tracks using molecular motor proteins, which are attached to the vesicle or organelle with connector proteins. The motor proteins use ATP to power their movement.

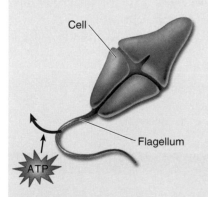

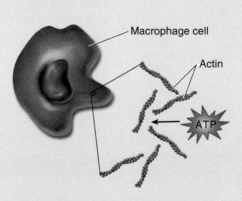

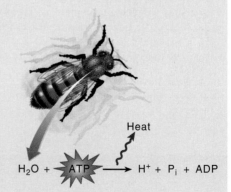

Flagellar Movements

Microtubules within flagella slide past each other to produce flagellar movements. ATP powers the sliding of the microtubules.

Cell Crawling

Actin filaments in a cell's cytoskeleton continually assemble and disassemble to achieve changes in cell shape and to allow cells to crawl over substrates or engulf materials. The dynamic character of actin is controlled by ATP molecules bound to actin filaments.

Heat Production

The hydrolysis of the ATP molecule releases heat. Reactions that hydrolyze ATP often take place in mitochondria or in contracting muscle cells and may be coupled to other reactions. The heat generated by these reactions can be used to maintain an organism's temperature.

ATP. All cells use the process of **cellular respiration,** which occurs in the mitochondrion, to break down the sugar molecule glucose. The potential energy of glucose is converted into molecules of ATP used by the cell as an energy source to power chemical reactions.

Oxidation-Reduction

In many of the reactions of photosynthesis and cellular respiration, electrons actually pass from one atom or molecule to another. When an atom or molecule loses an electron, it is said to be **oxidized,** and the process by which this occurs is called oxidation. The name reflects the fact that in biological systems, oxygen, which attracts electrons strongly, is the most common electron acceptor. Conversely, when an atom or molecule gains an electron, it is said to be **reduced,** and the process is called reduction. Oxidation and reduction always take place together, because every electron that is lost by an atom through oxidation is gained by some other atom through reduction. Therefore, chemical reactions of this sort are called oxidation-reduction (redox) reactions (figure 5.13). Because electrons maintain their potential energy of position, energy is transferred from one molecule to another through the transfer of electrons via redox reactions. In general, the reduced form of an organic molecule thus has a higher level of energy than the oxidized form. Oxidation-reduction reactions play a key

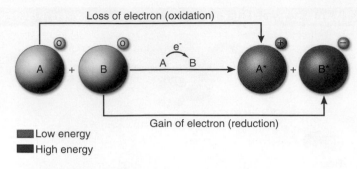

Loss of electron (oxidation)

Gain of electron (reduction)

■ Low energy
■ High energy

Figure 5.13 Redox reactions.

Oxidation is the loss of an electron; reduction is the gain of an electron. Here the charges of molecules A and B are shown in small circles to the upper right of each molecule. Molecule A loses energy as it loses an electron, while molecule B gains energy as it gains an electron.

role in the flow of energy through biological systems because the electrons that pass from one atom to another carry energy with them.

> **5.7   Cells store chemical energy in ATP and use ATP to drive chemical reactions. Energy is often transferred with electrons. Oxidation is the loss of an electron; reduction is the gain of one.**

Exploring Current Issues

Additional Resources

Go to your campus library or look online to find the following articles, which further develop some of the concepts found in this chapter.

Halasz, S. and J. DeBrosse. (2002). Supplements pose danger to teens. *Retrieved from http://www1.ncaa.org/membership/ed_outreach/ nutrition-performance/about/articles/supplement.html.*

Lucentini, J. (2003). Enzymatic alter-egos unmasked. *The Scientist,* 17(19), 23.

Minkel, J.R. (2003). The meaning of life: are we nothing more than energy-shredding machines? *New Scientist,* 176(2363), 30.

Seppa, N. (2003). Enzyme may reveal cancer susceptibility. *Science News,* 164(11), 164.

Wilson, J.F. (2002). Enzyme role found for aging gene. *The Scientist,* 16(22), 36.

In the News: Aging

Given enough food to live on, and protection from infectious disease, humans live quite a long time, often for 80 years or more. But they do eventually die. Is this merely a matter of our bodies wearing out, or is our eventual death somehow programmed into the human blueprint? Theories abound. Many involve the progressive accumulation of damage to DNA, as genes that prolong life often affect DNA repair processes. Other theories involve the progressive loss of telomeric DNA from the ends of chromosomes with successive cell divisions. Still other theories focus on caloric restriction, arguing for prolonging life by reducing the efficiency with which energy is gleaned from food.

Article 1. *Aging may be the body's way of preventing the development of cancer.*

Article 2. *Unraveling the mystery of aging.*

Article 3. *A gene mutation called "I'm not dead yet" may hold the secret of longer life.*

Find these articles, written by the author, online at www.mhhe.com/ tlwessentials/exp5.

Summary

Cells and Energy

5.1 The Flow of Energy in Living Things

- Energy is the ability to do work. Energy exists in two states: kinetic energy and potential energy.

- Kinetic energy is the energy of motion. Potential energy is stored energy which

exists in objects that aren't in motion but have the capacity to move (**figure 5.1**). All of the work carried out by living things involves the transformation of potential energy into kinetic energy.

- Energy flows from the sun to the earth where it is trapped by photosynthetic organisms and stored in carbohydrates as potential energy. This energy is transferred during chemical reactions.

5.2 The Laws of Thermodynamics

- The laws of thermodynamics describe changes in energy in our universe. The first law of thermodynamics explains that energy can never be created or destroyed, only changed from one state to another. The total amount of energy in the universe remains constant.

- The second law of thermodynamics explains that the conversion of potential energy into random molecular motion, or disorder, is constantly increasing. Entropy, which is a measure of disorder in a system, is constantly increasing such that disorder is more likely than order. Energy must be used to maintain order (**figure 5.3**).

Cell Chemistry

5.3 Chemical Reactions

- Chemical reactions involve the breaking or formation of covalent bonds. The starting molecules are called the reactants and the molecules produced by the reaction are called the products.

- Some chemical reactions are more likely to occur than others. Chemical reactions that release more energy than the energy put into them are called exergonic reactions (**figure 5.4b**). Chemical reactions in which the products contain more potential energy than the reactants are called endergonic reactions (**figure 5.4a**).

- All chemical reactions require an input of energy, even exergonic reactions that occur spontaneously. This energy required to start a reaction is called activation energy. A chemical reaction proceeds faster when its activation energy is lowered, and this occurs through a process called catalysis (**figure 5.4c**).

Enzymes

5.4 How Enzymes Work

- Enzymes are cellular proteins that lower the activation energy of chemical reactions in the cell. Enzymes are catalysts.

- An enzyme binds the reactants, also called substrate, of a reaction to its active site, increasing the likelihood that chemical bonds will break or form—therefore the reaction will require less energy. The enzyme lowers the activation energy of the reaction (**figures 5.5 and 5.6**). The enzyme is not affected by the reaction and can be used over and over again (**figure 5.7**).

5.5 Factors Affecting Enzyme Activity

- In the cell, chemical reactions are regulated by controlling which enzymes are present in the cell. Other factors can also affect enzyme function such as temperature and pH, and so most enzymes have an optimal temperature and pH range (**figure 5.8**).

- Temperatures above the optimal range can disrupt the bonds that hold the enzyme in its proper shape, decreasing its ability to catalyze a chemical reaction. The bonds that hold the enzyme's shape are also affected by hydrogen ion concentrations, and so increasing or decreasing the pH (hydrogen ion concentrations) can disrupt the enzyme's function.

5.6 How Cells Regulate Enzymes

- An enzyme can be deactivated or activated in the cell as a means of regulation by temporarily altering the enzyme's shape. An enzyme can be deactivated by when a molecule, called a repressor, binds to the enzyme, altering the shape of the active site so that it can not bind the substrate. Some enzymes need to be activated, or turned on, in order to bind to their substrate. A molecule called an activator binds to the enzyme, changing the shape of the active site so that it is able to bind the substrate (**figures 5.9 and 5.10**).

- Enzymes are often regulated by feedback inhibition, a process whereby the product of the reaction functions as a repressor. When enough of the product has been produced, it binds to the enzyme and stops the chemical reaction.

How Cells Use Energy

5.7 ATP: The Energy Currency of the Cell

- Cells require energy to do work. This cellular energy is stored in molecules of ATP (**table 5.1**). ATP contains a sugar, an adenine, and a chain of three phosphates. The three phosphates are held together with high-energy bonds that are chemically reactive (**figure 5.11**). When the endmost phosphate bond breaks, considerable energy is released. A cell uses this energy to drive other reactions in the cell (**figure 5.12**).

- Coupled reactions, called oxidation-reduction or redox reactions, involve the transfer of electrons from one atom or molecule to another. The atom or molecule that loses an electron is said to be oxidized and loses energy. The atom or molecule that gains the electron is said to be reduced and gains energy (**figure 5.13**).

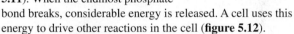

Self-Test

1. The ability to do work is the definition for
   a. thermodynamics.   c. energy.
   b. radiation.   d. entropy.
2. The first law of thermodynamics
   a. says that energy recycles constantly, as organisms use and reuse it.
   b. is concerned with how heat energy can be used to do work.
   c. is a formula for measuring entropy.
   d. says that energy can change forms, but cannot be made or destroyed.

3. The second law of thermodynamics
   a. says that energy recycles constantly, as organisms use and reuse it.
   b. says that entropy, or disorder, continually increases in a closed system.
   c. is a formula for measuring entropy.
   d. says that energy can change forms, but cannot be made nor destroyed.

4. Chemical reactions that occur spontaneously are called
   a. exergonic and release energy.
   b. exergonic and require energy.
   c. endergonic and release energy.
   d. endergonic and require energy.
5. The catalysts that help an organism carry out needed chemical reactions are called
   a. hormones.   c. reactants.
   b. enzymes.   d. substrates.
6. Factors that affect the activity of an enzyme molecule include
   a. peptides and energy.
   b. substrates and reactants.
   c. temperature and pH.
   d. entropy and thermodynamics.
7. In order for an enzyme to work properly
   a. it must have a particular shape.
   b. the temperature must be within certain limits.
   c. the pH must be within certain limits.
   d. All of these are true.

8. One way a cell can stop an enzyme from working when the cell does not need more of the product is that the cell
   a. changes its internal temperature to denature the enzyme.
   b. changes its internal pH to inhibit enzyme activity.
   c. changes its shape to inhibit enzyme activity.
   d. produces repressor molecules that alter the enzyme shape.
9. In competitive inhibition
   a. an enzyme molecule has to compete with other enzyme molecules for the necessary substrate.
   b. an enzyme molecule has to compete with other enzyme molecules for the necessary energy.
   c. an inhibitor molecule competes with the substrate for the same binding site on the enzyme.
   d. two different substrates compete for the same binding site on the enzyme.
10. The source of immediate, or "ready cash," energy in the body is
    a. ATP molecules.   c. enzyme molecules.
    b. DNA molecules.   d. product molecules.

Visual Understanding

1. **Figure 5.8** You eat a hamburger. Salivary amylase begins to digest the carbohydrates in the bun while you are still chewing. Pepsin works in your stomach to digest the protein, and trypsin is active in your small intestine to break the bonds between specific amino acids. How does the optimum pH for pepsin and trypsin reflect this chain of events?

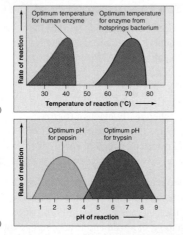

2. **Table 5.1** ATP comes primarily from the breakdown of glucose. If your blood glucose level drops, what sorts of problems can that cause?

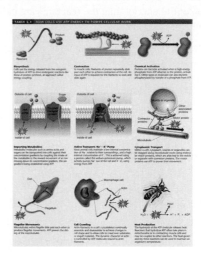

Challenge Questions

Cells and Energy Living organisms on the surface of the planet and the top layer of the ocean get their energy from the photosynthetic plants and bacteria that capture the sun's energy and transform it to molecules that other organisms can use. At the bottom of some deep ocean trenches there are entire colonies of many types of organisms. What do they use for energy?

Cell Chemistry Why do you need enzymes to help your cells do their jobs?

Enzymes What sorts of things can keep an enzyme from doing its job?

How Cells Use Energy After a rain your new iron fenceposts are rusting. Explain how this is an example of a redox reaction, just like the ones that occur in your body. Now can you think of a reason why blood coming from the lungs is bright red?

Online Learning Center

Visit the Online Learning Center for this chapter at www.mhhe.com/tlwessentials/ch5 for quizzes, animations, interactive learning exercises, and other study tools. At the site you will also find extended answers to the end-of-chapter questions.

6

How Cells Acquire Energy

In this forest glade you can literally see the pulse of life flowing through the organisms of an ecosystem. Sunlight beams down, a stream of energy in the form of packets of light called photons. Everywhere the light falls there are plants: trees and shrubs and flowers and grasses, all with green leaves intercepting the energy as it rains down. In the cells of each leaf are organelles called chloroplasts that contain light-gathering pigments in their membranes. These pigments, notably the pigment chlorophyll, which makes leaves green, absorb photons of light and use the energy to strip electrons from water molecules. The chloroplasts use these electrons to reduce CO_2—that is, to add hydrogens (a hydrogen atom, you will recall, is just a proton with an associated electron)—and so make organic molecules. In the roots and other tissues of the plants, the opposite process is taking place. Organic molecules are being broken down in the process of cellular respiration to provide energy to power growth and cellular activities. These reactions take place largely in another kind of organelle called a mitochondrion. Together, chloroplasts and mitochondria carry out a cycle of energy driven by the power of sunlight.

6.1   An Overview of Photosynthesis

Life is powered by sunshine. All of the energy used by almost all living cells comes ultimately from the sun, captured by plants, algae, and some bacteria through the process of **photosynthesis.** Thus, life is only possible because our earth is awash in energy streaming inward from the sun. Each day, the radiant energy that reaches the earth is equal to that of about 1 million Hiroshima-sized atomic bombs. About 1% of it is captured by photosynthesis and provides the energy that drives us all. The other 99%, which is not absorbed by the photosynthetic machinery in the cells, is reflected back into space, warms the land and water, or is used in the water cycle (evaporation and precipitation).

Photosynthesis occurs within the plasma membrane in bacteria, in the cells of algae, and within leaves in plants. Recall from chapter 4 that the cells of plant leaves contain organelles called chloroplasts that actually carry out photosynthesis (figure 6.1). No other structure in a plant other than chloroplasts is able to carry out photosynthesis. Photosynthesis takes place in three stages: (1) capturing energy from sunlight; (2) using the energy to make ATP and another key molecule, NADPH; and (3) using the ATP and NADPH to power the synthesis of carbohydrates from CO_2 in the air.

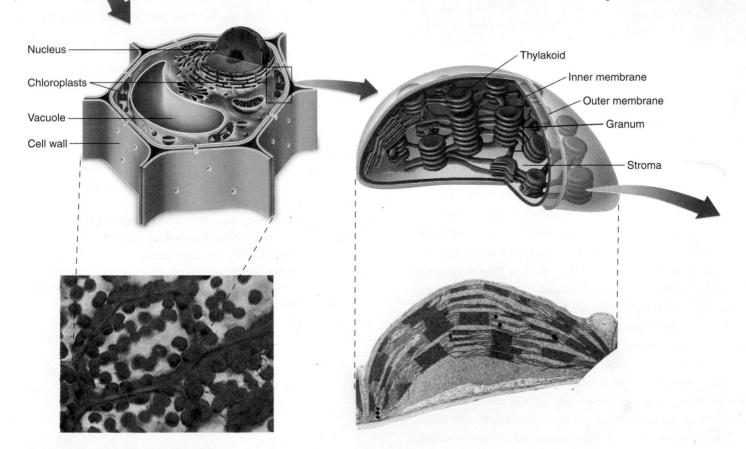

Figure 6.1   Journey into a leaf.

Plant cells within leaves contain chloroplasts, in which thylakoid membranes are stacked. Within the thylakoid, chlorophyll pigments grouped in photosystems drive the reactions of photosynthesis.

The first two stages take place only in the presence of light and are commonly called the **light-dependent reactions.** The third stage, the formation of organic molecules from atmospheric CO_2, is called the **Calvin cycle.** This stage is also referred to as the **light-independent reactions** because it doesn't require light directly. However, these reactions do depend indirectly on the light-dependent ones, as they require the ATP and NADPH produced by the light-dependent reactions; perhaps for this reason, several of the enzymes involved in the Calvin cycle are activated by light-dependent reactions.

The overall process of photosynthesis may be summarized by the following simple equation:

$$6\,CO_2 + 12\,H_2O + light \rightarrow C_6H_{12}O_6 + 6\,H_2O + 6\,O_2$$

carbon   water   energy   glucose   water   oxygen dioxide

Inside the Chloroplast

The chloroplast is the site of all three stages of photosynthesis. The internal membranes of chloroplasts are organized into flattened sacs called *thylakoids,* and often numerous thylakoids are stacked on top of one another in columns called *grana.* Surrounding the thylakoid membrane system is a semiliquid substance called *stroma.* In the membranes of thylakoids, chlorophyll pigments are grouped together in a network called a **photosystem** (see figure 6.7).

The photosystem is the starting point of photosynthesis, acting as an antenna to capture photons (units of light energy).

A lattice of proteins anchors pigment molecules in precise positions on the antenna. A pigment molecule is a molecule that absorbs light. The primary pigment molecule in most photosystems is **chlorophyll.** The photosystem lattice positions the chlorophyll molecules around one another so that when a photon of light strikes any chlorophyll molecule in the photosystem, the excitation passes from one chlorophyll to another. This is not a chemical reaction, in which an electron physically passes between atoms. Rather, it is energy that passes between chlorophylls. A crude analogy to this form of energy transfer is the initial "break" in a game of pool. If the cue ball squarely hits the point of the triangular array of 15 pool balls, the two balls at the far corners of the triangle fly off, and none of the central balls move at all. The kinetic energy is transferred through the central balls to the most distant ones.

Eventually the energy arrives at a key chlorophyll molecule that is touching a membrane-bound protein. An excited electron is then transferred from that key chlorophyll to an acceptor molecule in the membrane, which passes it in turn to a series of other proteins that put the energy to work making ATP and NADPH in the light-dependent reactions and building organic molecules in the Calvin cycle. We will discuss these stages of photosynthesis in the following sections.

> **6.1** **Photosynthesis uses energy from sunlight to power the synthesis of organic molecules from CO_2 in the air. In plants, photosynthesis takes place in specialized compartments within chloroplasts.**

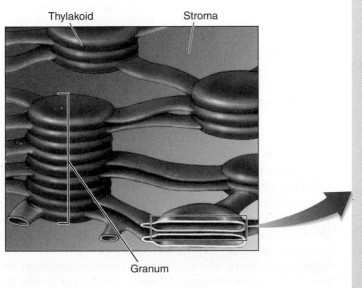

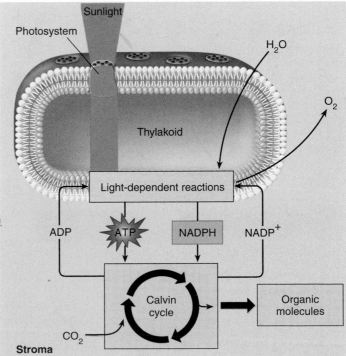

6.2 How Plants Capture Energy from Sunlight

Where is the energy in light? What is there about sunlight that a plant can use to create chemical bonds? The revolution in physics in the twentieth century taught us that light actually consists of tiny packets of energy called **photons.** When light shines on your hand, your skin is being bombarded by a stream of these photons smashing onto its surface.

Sunlight contains photons of many energy levels, only some of which we "see." Some of the photons in sunlight carry a great deal of energy—for example, X rays and ultraviolet light. Others such as radio waves carry very little. High-energy photons have shorter wavelengths than low-energy photons. Our eyes perceive photons carrying intermediate amounts of energy as **visible light** (figure 6.2), because the retinal pigment molecules in our eyes, which are different from chlorophyll, absorb only those photons of intermediate wavelength. Plants are even more picky, absorbing mainly blue and red light and reflecting back what is left of the visible light, which is why they appear green (figure 6.3).

Visible light represents only a small portion of the range of photon energies in sunlight. We call the full range of these photons the **electromagnetic spectrum** (see figure 6.2). The

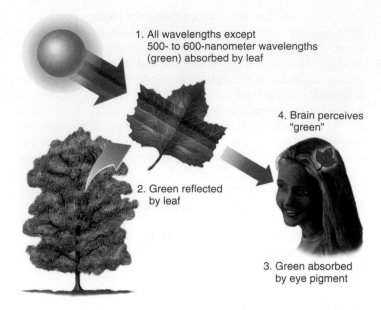

Figure 6.3   Why are plants green?

A leaf containing chlorophyll absorbs a broad range of photons—all the colors in the spectrum except for the photons around 500 to 600 nanometers. The leaf reflects these colors. These reflected wavelengths are absorbed by the visual pigments in our eyes, and our brains perceive the reflected wavelengths as "green."

highest-energy photons, which have the shortest wavelengths, are gamma rays with wavelengths of less than 1 nanometer; the lowest-energy photons, with wavelengths of thousands of meters, are radio waves.

How can a leaf or a human eye choose which photons to absorb? The answer to this important question has to do with the nature of atoms. Remember that electrons spin in particular orbits around the atomic nucleus, at different energy levels. Atoms absorb light by boosting electrons to higher energy levels, using the energy in the photon to power the move. Boosting the electron requires just the right amount of energy, no more and no less, just as when climbing a ladder you must raise your foot just so far to climb a rung. A particular kind of atom absorbs only certain photons of light, those with the appropriate amount of energy.

Pigments

As mentioned earlier, molecules that absorb light energy are called **pigments.** When we speak of visible light, we refer to those wavelengths that the pigment within human eyes, called *retinal,* can absorb—roughly from 380 nanometers (violet) to 750 nanometers (red). Other animals use different pigments for vision and thus "see" a different portion of the electromagnetic spectrum. For example, the pigment in insect eyes absorbs at shorter wavelengths than retinal. That is why bees can see ultraviolet light, which we cannot see, but are blind to red light, which we can see.

As noted, the main pigment in plants that absorbs light is chlorophyll. Its two forms, chlorophyll *a* and chlorophyll *b,* are similar in structure, but slight differences in

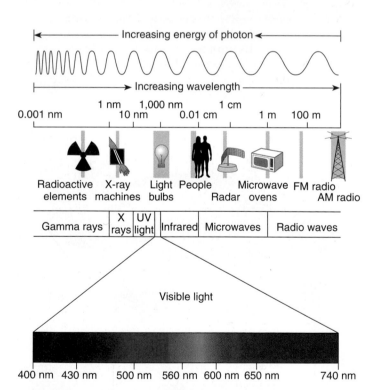

Figure 6.2   Photons of different energy: the electromagnetic spectrum.

Light is composed of packets of energy called photons. Some of the photons in light carry more energy than others. Light, a form of electromagnetic energy, is conveniently thought of as a wave. The shorter the wavelength of light, the greater the energy of its photons. Visible light represents only a small part of the electromagnetic spectrum, that with wavelengths between about 400 and 740 nanometers.

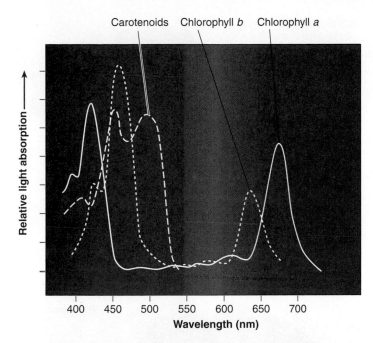

Carotenoids   Chlorophyll *b*   Chlorophyll *a*

Figure 6.4   Absorption spectra of chlorophylls and carotenoids.

The peaks represent wavelengths of sunlight strongly absorbed by the two common forms of photosynthetic pigment, chlorophyll *a* and chlorophyll *b,* and by accessory pigments called carotenoids. Chlorophylls absorb predominantly violet-blue and red light, in two narrow bands of the spectrum, while they reflect the green light in the middle of the spectrum. Carotenoids absorb mostly blue and green light and reflect orange and yellow light.

their chemical "side groups" produce slight differences in their absorption spectra. An absorption spectrum is a graph indicating how effectively a pigment absorbs different wavelengths of visible light (figure 6.4). While chlorophyll absorbs fewer kinds of photons than our visual pigment retinal, it is much more efficient at capturing them. Chlorophyll molecules capture photons with a metal ion (magnesium) that lies at the center of a complex carbon ring. Photons excite electrons of the magnesium ion, which are then channeled away by the carbon atoms.

While chlorophyll is the primary pigment involved in photosynthesis, plants contain other pigments called *accessory pigments.* These pigments absorb light of wavelengths not captured by chlorophyll. **Carotenoids** are a group of pigments that capture violet to blue-green light, some wavelengths not efficiently absorbed by chlorophyll (see figure 6.4).

Accessory pigments give color to flowers, fruits, and vegetables but are also present in leaves, their presence usually masked by chlorophyll. During the warm months, plants are actively producing food through photosynthesis, their cells filled with chlorophyll-containing chloroplasts. In the fall, the days become shorter and cooler and, for many species, leaves stop their food-making process. The chlorophyll molecules break down and are not replaced. When this happens, the colors reflected by the accessory pigments become visible. The leaves turn colors of yellow, orange, and red (figure 6.5).

6.2   Plants use pigments like chlorophyll to capture photons of blue and red light, reflecting photons of green wavelengths.

Oak leaf in summer

Oak leaf in autumn

Figure 6.5   Fall colors are produced by pigments such as carotenoids.

During the spring and summer, chlorophyll masks the presence in leaves of other pigments called carotenoids. Cool temperatures in the fall cause leaves of deciduous trees to cease manufacturing chlorophyll. With chlorophyll no longer present to reflect green light, the orange and yellow light reflected by carotenoids gives bright colors to the autumn leaves.

6.3 Organizing Pigments into Photosystems

The light-dependent reactions of photosynthesis occur on membranes. In photosynthetic bacteria, the proteins involved in the light-dependent reactions are embedded within the plasma membrane. In algae, intracellular membranes contain the proteins that drive the light-dependent reactions. In plants, photosynthesis occurs in specialized organelles called chloroplasts and the proteins involved in the light-dependent reactions are embedded in the thylakoid membranes inside the chloroplasts. The chlorophyll and accessory pigment molecules are embedded in a matrix of proteins within the thylakoid membrane (figure 6.6). This protein-pigment complex makes up the photosystem. The light-dependent reactions take place in five stages, which will be discussed in detail later:

1. **Capturing light.** A photon of light of the appropriate wavelength is captured by a pigment molecule and the excitation energy is passed from one chlorophyll molecule to another.

2. **Exciting an electron.** The excitation energy is funneled to a key chlorophyll *a* molecule called the **reaction center.** The excitation energy causes the transfer of an excited electron from the reaction center to another molecule that is an electron acceptor. The reaction center replaces this "lost" electron with an electron from the breakdown of a water molecule. Oxygen is produced as a by-product of this reaction.

3. **Electron transport.** The excited electron is then shuttled along a series of electron-carrier molecules embedded in the membrane. This is called the **electron transport system.** As the electron passes along the electron transport system, the energy from the electron is "siphoned" out in small amounts. This energy is used to pump hydrogen ions, called protons, across the membrane, building up a high concentration of protons on one side of the membrane.

4. **Making ATP.** The high concentration of protons can be used as an energy source to make ATP. Protons are only able to move back across the membrane via special channels, the protons flooding through them, like water through a dam. The kinetic energy that is released by the movement of protons is transferred to potential energy in the building of ATP molecules from ADP. This process, called **chemiosmosis,** makes the ATP that will be used in the Calvin cycle to make carbohydrates.

5. **Making NADPH.** The electron leaves the electron transport system and enters another photosystem where it is "reenergized" by the absorption of another

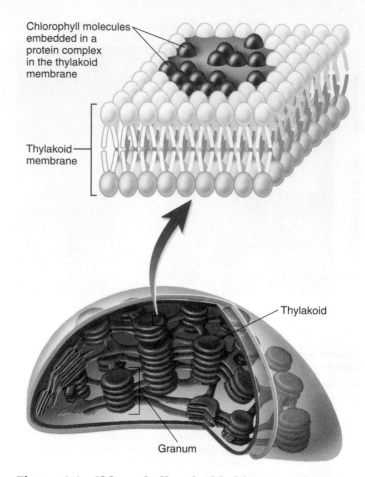

photon of light. This energized electron enters another electron transport system, where it is again shuttled along a series of electron-carrier molecules. The result of this electron transport system is not the synthesis of ATP, but rather the formation of NADPH. The electron is transferred to a molecule, NADP$^+$ and a hydrogen ion that forms NADPH. This molecule is important in the synthesis of carbohydrates in the Calvin cycle.

Figure 6.6   Chlorophyll embedded in a membrane.
Chlorophyll molecules are embedded in a network of proteins that hold the pigment molecules in place. The proteins are embedded within the membranes of thylakoids.

Architecture of a Photosystem

In all but the most primitive bacteria, light is captured by photosystems. Like a magnifying glass focusing light on a precise point, a photosystem channels the excitation energy gathered by any one of its pigment molecules to a specific chlorophyll *a* molecule, the reaction center chlorophyll (figure 6.7). This molecule then passes the energy, in the form of an excited electron, out of the photosystem to drive the synthesis of ATP and organic molecules.

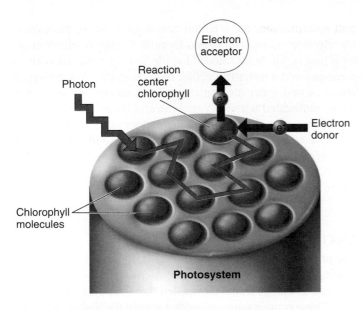

Figure 6.7 How a photosystem works.

When light of the proper wavelength strikes any pigment molecule within a photosystem, the light is absorbed and its excitation energy is then transferred from one molecule to another within the cluster of pigment molecules until it encounters the reaction center, which exports the energy as high-energy electrons to an acceptor molecule.

Plants Use Two Photosystems

Green plants and algae use two photosystems, photosystems I and II (figure 6.8). The first photosystem produces the ATP energy needed to build sugar molecules. The light energy that it captures is used in the stages 1 and 2 discussed on page 118 to transfer the energy of a photon of light to an excited electron. The energy of this electron is then used by the electron transport system in stage 3 to produce ATP in stage 4.

The second photosystem powers the production of the hydrogen atoms needed to build sugars and other organic molecules from CO_2 (which has no hydrogen atoms). The second photosystem is used in stage 5 to energize an electron that drives the production of NADPH from $NADP^+$.

The photosystems are not numbered in the order in which they are used. Photosystem II actually acts first in the series, and photosystem I acts second. The confusion arises because the photosystems were named in the order in which they were discovered, and photosystem I was discovered before photosystem II.

> **6.3** Photon energy is captured by pigments that employ it to excite electrons that are channeled away to do the chemical work of producing ATP and NADPH.

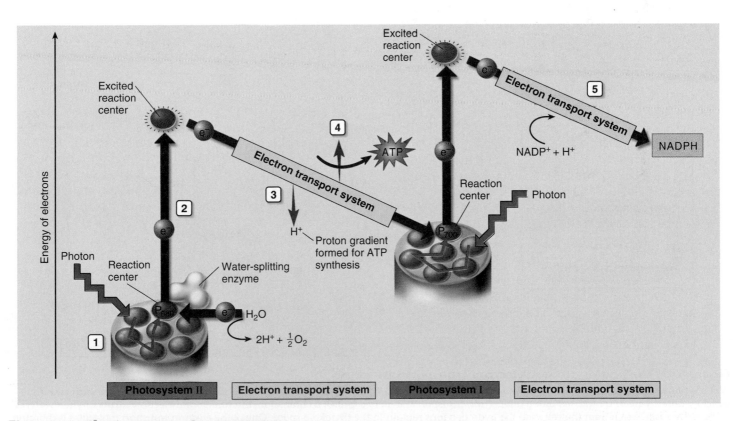

Figure 6.8 Plants use two photosystems.

In stage 1, a photon excites pigment molecules in photosystem II. In stage 2, a high-energy electron from photosystem II is transferred to the electron transport system. In stage 3, the excited electron is used to pump a proton across the membrane. In stage 4, the concentration gradient of protons is used to produce a molecule of ATP. In stage 5, the ejected electron then passes to photosystem I, which uses it, with a photon of light energy, to drive the formation of NADPH.

6.4 How Photosystems Convert Light to Chemical Energy

Plants use the two photosystems discussed in series, first one and then the other, to produce both ATP and NADPH. This two-stage process is called **noncyclic photophosphorylation,** because the path of the electrons is not a circle—the electrons ejected from the photosystems do not return to them, but rather end up in NADPH. The photosystems are replenished instead with electrons obtained by splitting water. As described earlier, photosystem II acts first. High-energy electrons generated by photosystem II are used to synthesize ATP and then passed to photosystem I to drive the production of NADPH.

Photosystem II

The reaction center of photosystem II consists of more than 10 transmembrane protein subunits. The *antenna complex,* which is the portion of the photosystem that contains all the pigment molecules, consists of some 250 molecules of chlorophyll *a* and accessory pigments bound to several protein chains. The antenna complex captures energy from a photon and funnels it to a reaction center. The reaction center gives up an excited electron to a primary electron acceptor in the electron trans-

port system. After the reaction center gives up an electron to the electron transport system, there is an empty electron orbital that needs to be filled. This electron is replaced with an electron from a water molecule. In photosystem II the oxygen atoms of two water molecules bind to a cluster of manganese atoms embedded within an enzyme and bound to the reaction center. In a way that is poorly understood, this enzyme splits water, removing electrons one at a time to fill the holes left in the reaction center by departure of light-energized electrons. As soon as four electrons have been removed from the two water molecules, O_2 is released.

Electron Transport System

The primary electron acceptor for the light-energized electrons leaving photosystem II passes the excited electron to a series of electron-carrier molecules called the electron transport system. These proteins are embedded within the thylakoid membrane; one of them is a "proton pump" protein (discussed in chapter 4). That is, the energy of the electron is used by this protein to pump a proton into the thylakoid space (the area inside the thylakoid). A nearby protein in the membrane then carries the now energy-depleted electron on to photosystem I (figure 6.9).

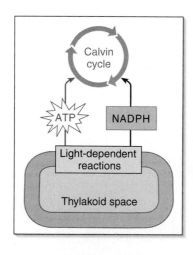

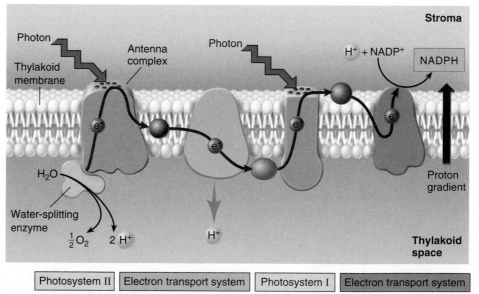

Figure 6.9 The photosynthetic electron transport system.
When a photon of light strikes a pigment molecule in photosystem II, it excites an electron. This electron is coupled to a proton stripped from water by an enzyme and is passed along an electron transport system (*red arrow*), a chain of membrane-bound electron carriers. When water is split, oxygen is released from the cell, and the hydrogen ions remain in the thylakoid space. One of the electron transport molecules is a proton pump, where the energy supplied by the photon is used to transport a proton across the membrane into the thylakoid. The concentration of hydrogen ions within the thylakoid thus increases further. When photosystem I absorbs another photon of light, its pigment passes a second high-energy electron to another electron transport system, which generates NADPH.

Making ATP: Chemiosmosis

Each thylakoid is a closed compartment into which protons are pumped from the stroma by the electron transport system. The thylakoid membrane is impermeable to protons, so protons build up inside the thylakoid space, creating a very large concentration gradient. As you may recall from chapter 4, substances move from areas of high concentration to areas of lower concentration in a process called diffusion. Here protons diffuse back out of the thylakoid space toward lower concentration through special channel proteins called *ATP synthases*. These channels protrude like knobs on the external surface of the thylakoid membrane. As protons pass out of the thylakoid through the ATP synthase channels, ADP is phosphorylated to ATP and released into the stroma, the fluid matrix inside the chloroplast (figure 6.10). The stroma contains the enzymes that catalyze the reactions of the light-independent reactions.

Photosystem I

Photosystem I accepts an electron from the electron transport system (see figure 6.9). The reaction center of photosystem I is a membrane complex consisting of at least 13 protein subunits. Energy is fed to it by an antenna complex consisting of 130 chlorophyll *a* and accessory pigment molecules. The electron arriving from the electron transport system has by no means lost all of its light-excited energy; almost half remains. Thus, the absorption of another photon of light energy by photosystem I boosts the electron leaving its reaction center to a very high energy level.

Making NADPH

Like photosystem II, photosystem I passes electrons to an electron transport system. When two of these electrons reach the end of this electron transport system, they are then donated to a molecule of $NADP^+$ to form NADPH. This reaction involves an $NADP^+$, two electrons, and a proton (see figure 6.9). Because the reaction occurs on the stromal side of the membrane and involves the uptake of a proton in forming NADPH, it contributes further to the proton gradient established during photosynthetic electron transport.

Products of the Light-Dependent Reactions

The light-dependent reactions can be seen more as a stepping stone, rather than an end point of photosynthesis. All of the products of the light-dependent reactions are either waste products, such as oxygen, or are ultimately used elsewhere in the cell. The ATP and NADPH produced in the light-dependent reactions end up being passed on to the Calvin cycle in the stroma of the chloroplast. There, ATP is used to power chemical reactions that build carbohydrates. NADPH is used in the Calvin cycle as a source of "reducing power," the hydrogens used in building carbohydrates. The next section discusses the Calvin cycle of photosynthesis.

6.4 The light-dependent reactions of photosynthesis produce the ATP and NADPH needed to build organic molecules, and release O_2 as a by-product of stripping hydrogen atoms and their associated electrons from water molecules.

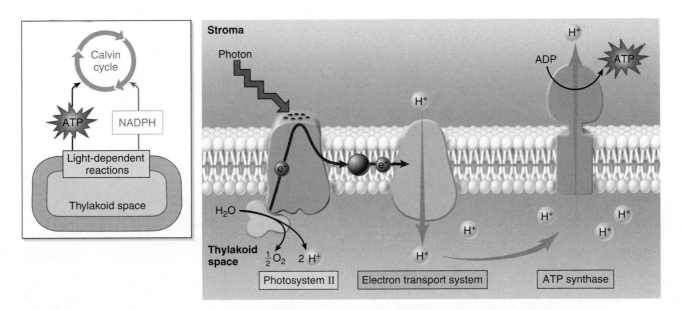

Figure 6.10 Chemiosmosis in a chloroplast.
One of the proteins of the photosynthetic electron transport system pumps protons into the interior of the thylakoid. ATP is produced on the outside surface of the membrane (stroma side), as protons diffuse back out of the thylakoid through ATP synthase channels.

6.5 | Building New Molecules

The Calvin Cycle

Stated very simply, photosynthesis is a way of making organic molecules from carbon dioxide (CO_2). To build organic molecules, cells use raw materials provided by the light-dependent reactions:

1. **Energy.** ATP (provided by photosystem II) drives the endergonic reactions.
2. **Reducing power.** NADPH (provided by photosystem I) provides a source of hydrogens and the energetic electrons needed to bind them to carbon atoms.

The actual assembly of new molecules employs a complex battery of enzymes in what is called the **Calvin cycle** (figure 6.11), or **C_3 photosynthesis.** A carbon atom from a carbon dioxide molecule is first added to a five-carbon sugar, producing two three-carbon sugars. This process is called "fixing carbon" because it attaches a carbon atom that was in a gas to an organic molecule (figure 6.12). Then, in a long series of reactions, the carbons are shuffled about. Eventually some of the resulting molecules are channeled off to make sugars, while others are used to re-form the original five-carbon sugar, which is then available to restart the cycle. The cycle has to "turn" six times in order to form a new glucose molecule, because each turn of the cycle adds only one carbon atom from CO_2, and glucose is a six-carbon sugar.

The Calvin Cycle

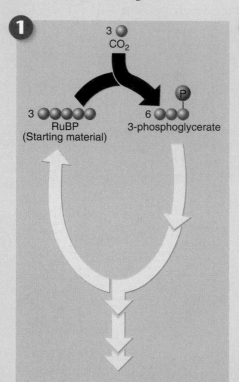

The Calvin cycle begins when a carbon atom from a CO_2 molecule is added to a five-carbon molecule (the starting material). The resulting six-carbon molecule is unstable and immediately splits into three-carbon molecules. (Three "turns" of the cycle are indicated here with three molecules of CO_2 entering the cycle)

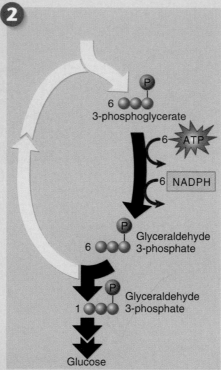

Then, through a series of reactions, energy from ATP and hydrogens from NADPH (the products of the light reactions) are added to the three-carbon molecules. The now-reduced three-carbon molecules either combine to make glucose or are used to make other molecules.

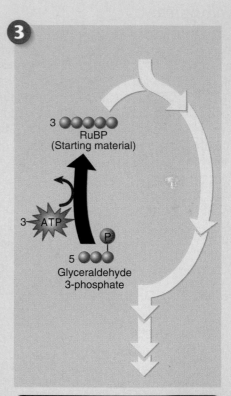

Most of the reduced three-carbon molecules are used to regenerate the five-carbon starting material, thus completing the cycle.

Figure 6.11 How the Calvin cycle works.
The Calvin cycle takes place in the stroma of the chloroplasts. The NADPH and the ATP that were generated by the light-dependent reactions are used in the Calvin cycle to build carbohydrate molecules. The number of carbon atoms at each stage is indicated by the number of balls. It takes six turns of the cycle to make one six-carbon molecule of glucose.

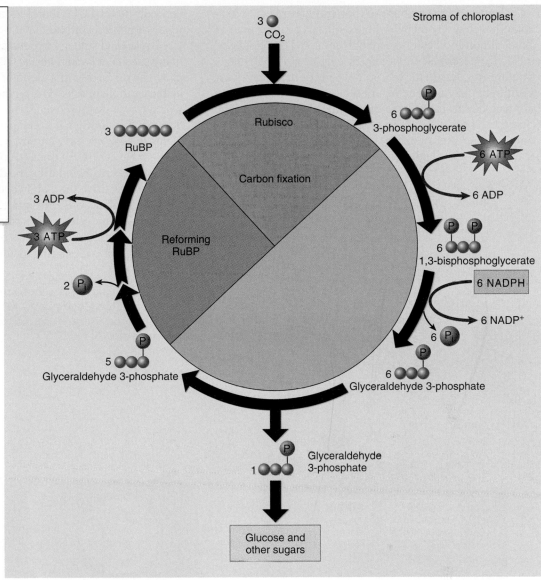

Figure 6.12 The Calvin cycle.

For every three molecules of CO_2 that enter the cycle, one molecule of the three-carbon compound glyceraldehyde 3-phosphate (G3P) is produced. Notice that the process requires energy stored in ATP and NADPH, which are generated by the light-dependent reactions. This process occurs in the stroma of the chloroplast. The large 16-subunit enzyme that catalyzes the reaction, RuBP carboxylase, or **rubisco,** is the most abundant protein in chloroplasts and is thought to be the most abundant protein on earth.

C_4 Photosynthesis

Many plants have trouble carrying out C_3 photosynthesis when the weather is hot. As temperature increases, plants partially close their leaf openings, called **stomata** (singular, stoma), to conserve water. As a result, CO_2 and O_2 are not able to enter and exit the leaves through these openings. The concentration of CO_2 in the leaves falls, while the concentration of O_2 in the leaves rises. Under these conditions rubisco, the enzyme that carries out the first step of the Calvin cycle, engages in **photorespiration,** where the enzyme incorporates O_2, not CO_2, into the cycle and when this occurs, CO_2 is ultimately released as a by-product (figure 6.13). Photorespiration thus short-circuits the successful performance of the Calvin cycle.

Some plants are able to adapt to climates with higher temperatures by performing **C_4 photosynthesis.** In this process, plants such as sugarcane, corn, and many grasses are able to fix carbon using different types of cells and chemical reactions within their leaves and thereby avoiding a reduction in photosynthesis due to higher temperatures.

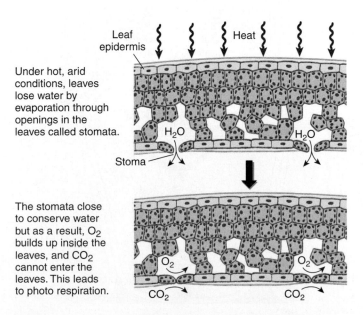

Under hot, arid conditions, leaves lose water by evaporation through openings in the leaves called stomata.

The stomata close to conserve water but as a result, O_2 builds up inside the leaves, and CO_2 cannot enter the leaves. This leads to photo respiration.

Figure 6.13 Photorespiration in hot weather.

C_4 plants fix CO_2 first as the four-carbon molecule oxaloacetate (hence the name, C_4 photosynthesis), rather than as the three-carbon molecule phosphoglycerate of C_3 photosynthesis, such as described previously (see figure 6.12). C_4 plants carry out this process in the mesophyll cells of their leaves using a different enzyme. The oxaloacetate is then converted to malate, which is transferred to the bundle-sheath cells of the leaf and there broken down to regenerate CO_2, which enters the Calvin cycle (figure 6.14a). The bundle-sheath cells are impermeable to CO_2 and therefore hold it within them. The concentration of CO_2 increases and thus decreases the occurrence of photorespiration.

A second strategy to decrease photorespiration is used by many succulent (water-storing) plants such as cacti and pineapples. The mode of initial carbon fixation is called **crassulacean acid metabolism (CAM)** after the plant family Crassulaceae in which it was first discovered. In these plants, the stomata open during the night when it's cooler and close during the day. CAM plants initially fix CO_2 into organic compounds at night, using the C_4 pathway. These organic compounds accumulate at night and are broken down during the day, releasing CO_2. These high levels of CO_2 drive the Calvin cycle and decrease photorespiration. CAM plants differ from C_4 plants in that the C_4 pathway and the Calvin cycle occur in the same cell, a mesophyll cell, but they occur at different times of the day, the C_4 cycle at night and the Calvin cycle during the day (figure 6.14b).

> **6.5** In a series of reactions that do not directly require light, cells use ATP and NADPH provided by photosystems II and I to assemble new organic molecules.

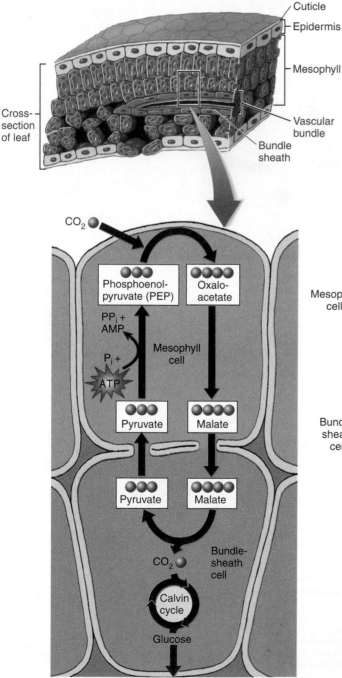

(a) C_4 pathway

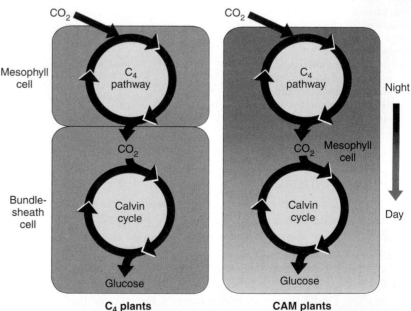

(b) C_4 versus CAM pathways

Figure 6.14 Carbon fixation in C_4 and CAM plants.
(a) This process is called the C_4 pathway because the first molecule formed in the pathway contains four carbons. (b) A comparison of C_4 and CAM plants. Both C_4 and CAM plants utilize the C_4 and C_3 pathways. In C_4 plants, the pathways are separated spatially; the C_4 pathway takes place in the mesophyll cells and the C_3 pathway (the Calvin cycle) in the bundle-sheath cells. In CAM plants, the two pathways occur in mesophyll cells but are separated temporally; the C_4 pathway is utilized at night and the C_3 pathway during the day.

A Closer Look

The Energy Cycle

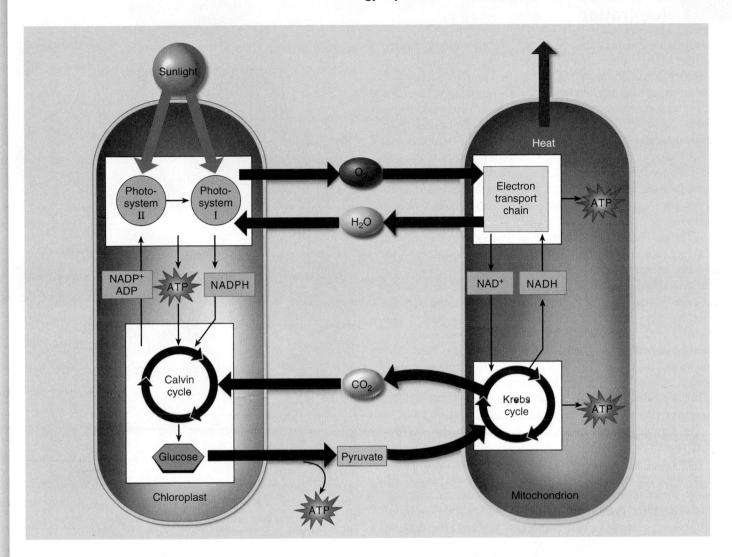

The energy-capturing metabolisms of the chloroplasts studied in this section of the chapter and the mitochondria studied in the chapter's next section are intimately related. Photosynthesis carried out by chloroplasts uses the products of cellular respiration as starting substrates, and cellular respiration carried out by mitochondria uses the products of photosynthesis as its starting substrates. Together, photosynthesis and cellular respiration form an energy cycle, as diagrammed here.

The evolutionary history of the energy cycle can be seen in its elements. The Calvin cycle of photosynthesis uses part of the glycolytic pathway of cellular respiration, run in reverse, to produce glucose. The principal proteins involved in electron transport in chloroplasts are related to those in mitochondria, and in many cases are actually the same. Biologists believe the glycolysis stage of cellular respiration, which takes place in the cytoplasm, converting glucose to pyruvate and requiring no oxygen, to be the most ancient process in the energy cycle. The chloroplast's oxygen-generating photosynthesis is thought to have evolved next, followed later by the oxidative reactions of the mitochondria's cellular respiration.

Photosynthesis is but one aspect of plant biology, although it is an important one. In later chapters we will examine plants in more detail. We have treated photosynthesis here, in a section devoted to cell biology, because photosynthesis arose long before plants did, and all organisms depend directly or indirectly on photosynthesis for the energy that powers their lives.

6.6 An Overview of Cellular Respiration

In both plants and animals, and in fact in almost all organisms, the energy for living is obtained by breaking down the organic molecules originally produced in plants. The ATP energy and reducing power invested in building the organic molecules are retrieved by stripping away the energetic electrons and using them to make ATP. When electrons are stripped away from chemical bonds, the food molecules are being oxidized (remember, oxidation is the loss of electrons). The oxidation of foodstuffs to obtain energy is called **cellular respiration.** Do not confuse the term cellular respiration with the breathing of oxygen gas that your lungs carry out, which is called simply respiration.

The cells of plants fuel their activities with sugars and other fuel molecules, just as yours do; only the chloroplasts carry out photosynthesis. No light shines on roots below the ground, and yet root cells are just as alive as the cells in the stem and leaves. Why would plant cells manufacture organic molecules only to turn around and break them down? The organic molecules produced by plants serve two functions, to build plant tissue and to store energy for future need. The plant breaks down "storage molecules" during cellular respiration. Nonphotosynthetic organisms eat plants, extracting energy from plant tissue and plant storage molecules in cellular respiration. Other animals eat these animals (figure 6.15).

In aerobic respiration, which requires oxygen, ATP is formed as electrons are harvested, transferred along an electron transport chain (similar to the electron transport system in photosynthesis), and eventually donated to oxygen gas. Eukaryotes produce the majority of their ATP from glucose in this way. Chemically, there is little difference between this oxidation of carbohydrates in a cell and the burning of wood in a fireplace. In both instances, the reactants are carbohydrates and oxygen, and the products are carbon dioxide, water, and energy:

$$C_6H_{12}O_6 \;+\; 6\,O_2 \rightarrow 6\,CO_2 \;+\; 6\,H_2O \;+\; \text{energy}$$
$$\text{(heat or ATP)}$$

Cellular respiration is carried out in two stages. The first stage uses coupled reactions to make ATP. This stage, **glycolysis,** takes place in the cell's cytoplasm and does not require oxygen. This ancient energy-extracting process is thought to have evolved over 2 billion years ago, when there was no oxygen in the earth's atmosphere. The second stage, which includes the Krebs cycle, harvests electrons from chemical bonds and uses their energy to power the production of ATP.

Figure 6.15 Lion at lunch.
Energy that this lion extracts from its meal of giraffe will be used to power its roar, fuel its running, and build a bigger lion.

This stage, **oxidation,** takes place only within mitochondria. It is far more powerful than glycolysis at recovering energy from food molecules and is where the bulk of the energy used by eukaryotic cells is extracted (figure 6.16).

> **6.6 Cellular respiration is the dismantling of food molecules to obtain energy. In aerobic respiration, the cell harvests energy from glucose molecules in two stages, glycolysis and oxidation. Oxygen is the final electron acceptor.**

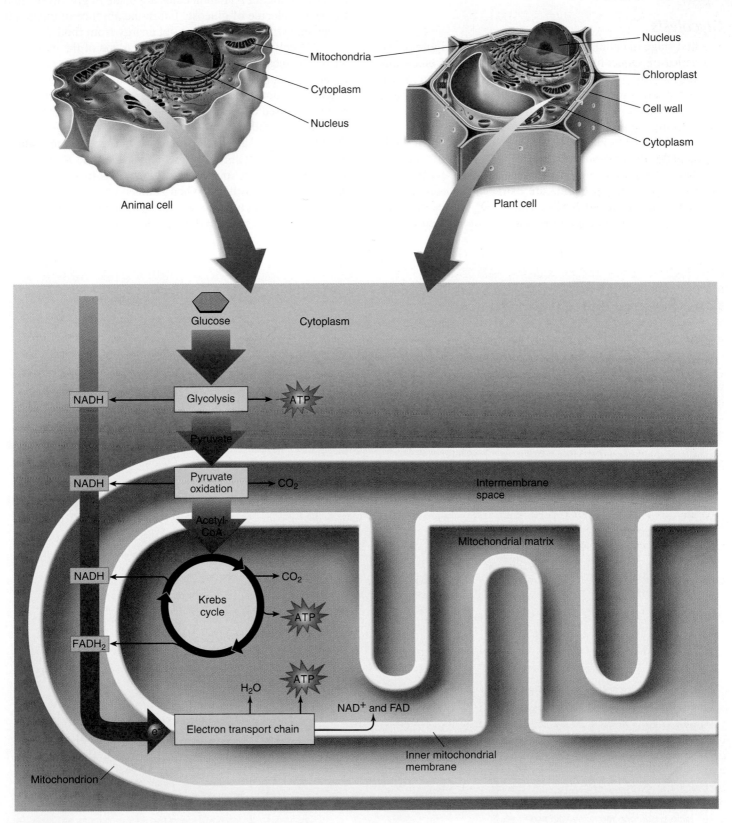

Figure 6.16 An overview of aerobic respiration.

6.7 Using Coupled Reactions to Make ATP

Glycolysis

The first stage in cellular respiration, glycolysis (figure 6.17), is a series of sequential biochemical reactions, a *biochemical pathway*. In 10 enzyme-catalyzed reactions (figure 6.18), the six-carbon sugar glucose is cleaved into two three-carbon molecules called pyruvate. In each of two coupled reactions the breaking of a chemical bond contributes enough energy to enable the formation of an ATP molecule from ADP. This transfer of a phosphate group from a substrate to ADP is called **substrate-level phosphorylation.** In the process, electrons and hydrogen atoms are extracted and donated to a carrier molecule called NAD$^+$. The NAD$^+$ carries the electrons as NADH to join the other electrons extracted during oxidative respiration, discussed in the following section. Only a small number of ATP molecules are made in glycolysis, two for each molecule of glucose, but in the absence of oxygen it is the only way organisms can get energy from food.

Glycolysis is thought to have been one of the earliest of all biochemical processes to evolve. Every living creature is capable of carrying out glycolysis. Glycolysis, although inefficient, was not discarded during the course of evolution but rather was used as the starting point for the further extraction of energy by oxidation. Nature did not, so to speak, go back to the drawing board and design metabolism from scratch. Rather, new reactions, which make up what is called the *Krebs cycle,* were added onto the old, just as renovations to a house build upon what is already there. When a homeowner wants to expand the living space in a house, a room is added on to the existing house; the house isn't torn down, redesigned, and rebuilt.

Overview of Glycolysis

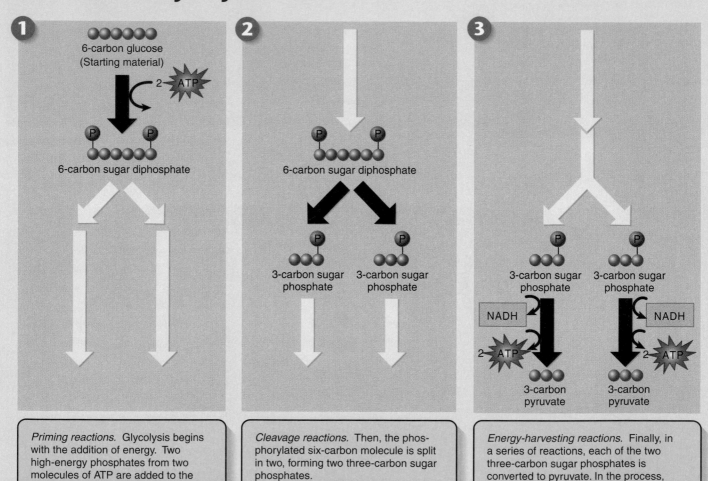

1 6-carbon glucose (Starting material)

2 — ATP

6-carbon sugar diphosphate

Priming reactions. Glycolysis begins with the addition of energy. Two high-energy phosphates from two molecules of ATP are added to the six-carbon molecule glucose, producing a six-carbon molecule with two phosphates.

2 6-carbon sugar diphosphate

3-carbon sugar phosphate   3-carbon sugar phosphate

Cleavage reactions. Then, the phosphorylated six-carbon molecule is split in two, forming two three-carbon sugar phosphates.

3 3-carbon sugar phosphate   3-carbon sugar phosphate

NADH   NADH

2 — ATP   2 — ATP

3-carbon pyruvate   3-carbon pyruvate

Energy-harvesting reactions. Finally, in a series of reactions, each of the two three-carbon sugar phosphates is converted to pyruvate. In the process, an energy-rich hydrogen is harvested as NADH, and two ATP molecules are formed for each pyruvate.

Figure 6.17 How glycolysis works.

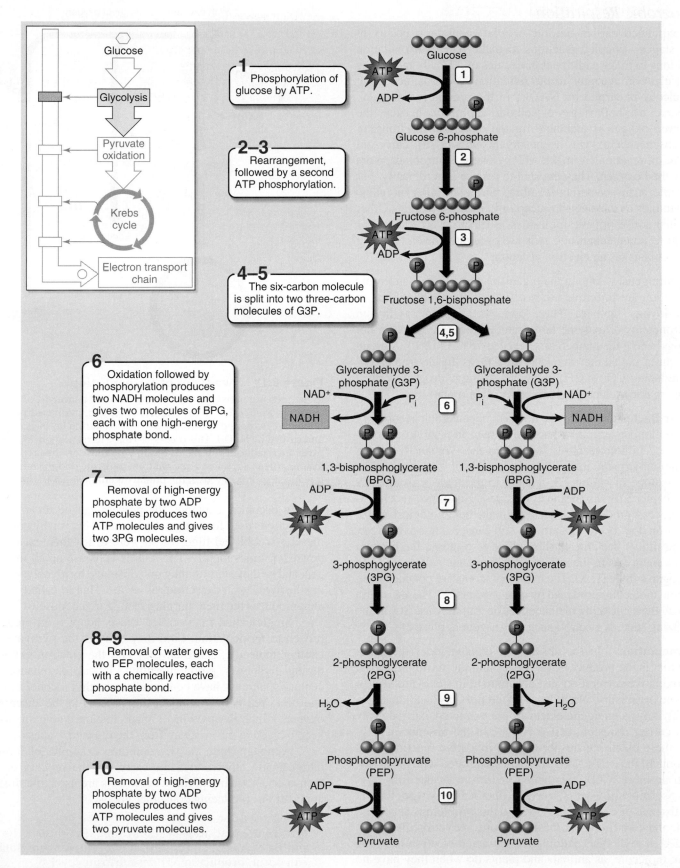

Figure 6.18   Glycolysis.

The process of glycolysis involves ten enzyme-catalyzed reactions.

Anaerobic Respiration

As explained earlier, aerobic cellular respiration occurs in two stages—coupled reactions to make ATP, and oxidation reactions to make additional molecules of ATP. Oxygen is the final electron acceptor in the oxidation reactions, accepting the electrons carried by NADH. In the presence of oxygen, cells can use both stages of cellular respiration, because the required oxygen is available for the oxidation reactions. In the absence of oxygen, some organisms can still carry out oxidation reactions to make ATP by using electron acceptors other than oxygen. They are said to respire anaerobically. For example, many bacteria use sulfur, nitrate, or other inorganic compounds as the electron acceptor in place of oxygen. Other organisms use organic molecules as electron acceptors. For example, some eukaryotic cells use pyruvate, the end product of glycolysis, as an electron acceptor.

Methanogens.   Among the organisms that practice anaerobic respiration are primitive archaea such as the thermophilic, or "heat-loving," archaea. These prokaryotic organisms live in environments of extreme temperature; some are found in hot springs where the temperatures exceed 78°C (172°F). A group of archaea, called methanogens, use CO_2 as the electron acceptor, reducing CO_2 to CH_4 (methane), using hydrogens derived from organic molecules produced by other organism.

Sulfur Bacteria.   Evidence of a second anaerobic respiratory process among primitive bacteria is seen in a group of rocks about 2.7 billion years old, known as the Woman River iron formation. Organic material in these rocks is enriched for the light isotope of sulfur, ^{32}S, relative to the heavier isotope ^{34}S (see chapter 3 for discussion of isotopes). No known geochemical process produces such enrichment, but biological sulfur reduction does occur in a process still carried out today by certain primitive bacteria. In this sulfate respiration, the bacteria derive energy from the reduction of inorganic sulfate (SO_4) to hydrogen sulfide (H_2S). The hydrogen atoms are obtained from organic molecules produced by other organisms. These bacteria thus do the same thing methanogens do, but they use SO_4 as the oxidizing (that, is electron-accepting) agent in place of CO_2.

Fermentation.   In the absence of oxygen, does the pyruvate that is the product of glycolysis and the starting material for oxidative respiration just accumulate in the cytoplasm of aerobic organisms? No. It has a different fate. Recall that during glycolysis energetic electrons are extracted, carried away by a carrier molecule called NAD^+. In the absence of oxygen, these electrons are not used in oxidative reactions, and so soon all the cell's NAD^+ becomes saturated with electrons. With no more NAD^+ available to carry away electrons, glycolysis cannot proceed. Clearly, to obtain energy from food in the absence of oxygen, a solution to this problem is needed. A home must be found for these electrons, recycling the NAD^+ needed in glycolysis. Adding the extracted electron to an organic molecule, as animals and plants do when they have no oxygen to take it, is called **fermentation.**

Two types of fermentation are common among eukaryotes (figure 6.19). Animals such as ourselves simply add the extracted electrons to pyruvate, forming lactate. Later, when

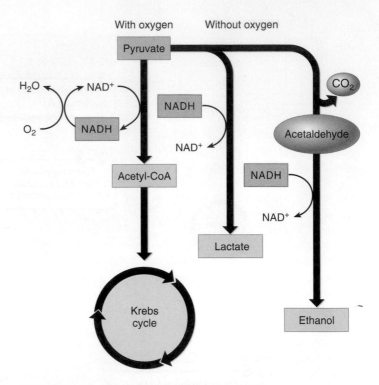

Figure 6.19   Two types of fermentation.
In the presence of oxygen, pyruvate is oxidized to acetyl-CoA and enters the Krebs cycle. In the absence of oxygen, pyruvate instead is reduced, accepting the electrons extracted during glycolysis and carried by NADH. This process is called fermentation. When pyruvate is reduced directly, as it is in your muscles, the product is lactate; when CO_2 is first removed from pyruvate and the remainder is reduced, as it is in yeasts, the product is ethanol.

oxygen becomes available, the process can be reversed and the electrons used for energy production. This is why your arm muscles would feel tired if you were to lift this text up and down 100 times rapidly. The muscle cells use up all the oxygen, and so they start running on ATP made by glycolysis, storing the pyruvate and electrons as lactate. This so-called oxygen debt produces the tired, burning feeling in the muscle.

Single-celled fungi called yeasts adopt a different approach to fermentation. First they convert the pyruvate into another molecule, and then they add the electron extracted during glycolysis, producing ethyl alcohol, or ethanol. For centuries, humans have consumed such ethyl alcohol in wine and beer. Yeasts only conduct this process in the absence of oxygen, which is why wine is made in closed containers—to keep oxygen in the air away from the crushed grapes.

Although these are two common examples of fermentation, many other organisms carry out the process of fermentation and many of these organisms are used in industrial applications, producing consumer goods.

6.7   In the first stage of respiration, glycolysis, cells shuffle chemical bonds so that two coupled reactions can occur, producing ATP by substrate-level phosphorylation. When electrons that are also a product of glycolysis are not donated to oxygen, they are added to inorganic or organic molecules.

Author's Corner

Fad Diets and Impossible Dreams

In my mind I will always weigh 165 pounds, as I did the day I married. The bathroom scale tells a different story, somehow finding another 30 pounds. I did not ask for that weight, do not want it, and am constantly looking for a way to get rid of it. I have not found this to be a lonely search—it seems like everyone I know past the flush of youth is trying to lose weight, too. And, like many, I have been seduced by fad diets, investing hope only to harvest frustration. The much discussed Atkins' diet was the fad diet I tried. As a scientist I should have known better, but so many people seemed to use it—*Dr. Atkins' Diet Revolution* is one of the 10 best-selling books in history, and was (and is) prominently displayed in every bookstore I enter. The reason this diet doesn't deliver on its promise of pain-free weight loss is well understood by science, but not by the general public. Only hope and hype make it a perpetual best seller.

The secret of the Atkins' diet, stated simply, is to avoid carbohydrates. Atkins' basic proposition is that your body, if it does not detect blood glucose (from metabolizing carbohydrates), will think it is starving and start to burn body fat, even if there is lots of fat already circulating in your bloodstream. You may eat all the fat and protein you want, all the steak and eggs and butter and cheese, and you will still burn fat and lose weight—just don't eat any carbohydrates, any bread or pasta or potatoes or fruit or candy. Despite the title of Atkins' book, this diet is hardly revolutionary. A basic low-carbohydrate diet was first promoted over a century ago in the 1860s by William Banting, an English casket maker, in his best-selling book *Letter on Corpulence.* Books promoting low-carbohydrate diets have continued to be best sellers ever since. I even found one on my mother's bookshelf, in the guise of Dr. Herman Taller's 1961 *Calories Don't Count.*

When I tried the Atkins' diet I lost 10 pounds in three weeks. In three months it was all back, and then some. So what happened? Where did the pounds go, and why did they come back? The temporary weight loss turns out to have a simple explanation: because carbohydrates act as water sponges in your body, forcing your body to become depleted of carbohydrates causes your body to lose water. The 10 pounds I lost on this diet was not fat weight but water, quickly regained with the first starchy foods I ate.

The Atkins' diet is the sort of diet the American Heart Association tells us to avoid (all those saturated fats and cholesterol), and it is difficult to stay on. If you do hang in there, you will lose weight, simply because you eat less. Other popular diets these days, *The Zone* diet of Dr. Barry Sears and *The South Beach Diet* of Dr. Arthur Agatston, are also low-carbohydrate diets, although not as extreme as the Atkins' diet. Like the Atkins' diet, they work not for the bizarre reasons claimed by their promoters, but simply because they are low-calorie diets.

In teaching my students at Washington University about fad diets, I tell them there are two basic laws that no diet can successfully violate:

1. **All calories are equal.**
2. **(calories in) − (calories out) = fat.**

The fundamental fallacy of the Atkins' diet, the Zone diet, the South Beach diet, and indeed of all fad diets, is the idea that somehow carbohydrate calories are different from fat and protein calories. This is scientific foolishness. Every calorie you eat contributes equally to your eventual weight, whether it comes from carbohydrate, fat, or protein.

To the extent these diets work at all, they do so because they obey the second law. By reducing calories in, they reduce fat. If that were all there was to it, I'd go out and buy Sears' book. Unfortunately, losing weight isn't that simple, as anyone who has seriously tried already knows. The problem is that your body will not cooperate. If you try to lose weight by exercising and eating less, your body will attempt to compensate by metabolizing more efficiently. It has a fixed weight, what obesity researchers call a "set point," a weight to which it will keep trying to return. A few years ago, a group of researchers at Rockefeller University in New York, in a landmark study, found that if you lose weight, your metabolism slows down and becomes more efficient, burning fewer calories to do the same work—your body will do everything it can to gain the weight back! Similarly, if you gain weight, your metabolism speeds up. In this way your body uses its own natural weight control system to keep your weight at its set point. No wonder it's so hard to lose weight!

Clearly our bodies don't keep us at one weight all our adult lives. It turns out your body adjusts its fat thermostat—its set point—depending on your age, food intake and amount of physical activity. Adjustments are slow, however, and it seems to be a great deal easier to move the body's set point up than to move it down. Apparently higher levels of fat reduce the body's sensitivity to the leptin hormone that governs how efficiently we burn fat. That is why you can gain weight, despite your set point resisting the gain—your body still issues leptin alarm calls to speed metabolism, but your brain doesn't respond with as much sensitivity as it used to. Thus the fatter you get, the less effective your weight control system becomes.

This doesn't mean that we should give up and learn to love our fat. Rather, now that we are beginning to understand the biology of weight gain, we must accept the hard fact that we cannot beat the requirements of the two diet laws. The real trick is not to give up. Eat less and exercise more, and keep at it. In one year, or two, or three, your body will readjust its set point to reflect the new reality you have imposed by constant struggle. There simply isn't any easy way to lose weight.

6.8 Harvesting Electrons from Chemical Bonds

The first step of oxidative respiration in the mitochondrion is the oxidation of the three-carbon molecule called pyruvate, which is the end product of glycolysis. The cell harvests pyruvate's considerable energy in two steps: first, by oxidizing pyruvate to form acetyl-CoA, and then by oxidizing acetyl-CoA in the Krebs cycle.

Producing Acetyl-CoA

Pyruvate is oxidized in a single reaction that cleaves off one of pyruvate's three carbons. This carbon then departs as CO_2 (figure 6.20). This reaction produces a two-carbon fragment called an acetyl group, as well as a hydrogen and electrons, which reduces NAD^+ to NADH (figure 6.21). Pyruvate dehydrogenase, the complex of enzymes that removes CO_2 from pyruvate, is one of the largest enzymes known. It contains 60 subunits! In the course of the reaction, the acetyl group removed from pyruvate combines with a cofactor called coenzyme A (CoA), forming a compound known as **acetyl-CoA.** Acetyl-CoA is also generated by the breakdown of proteins and lipids (see figure 6.20). If the cell has plentiful supplies of ATP, acetyl-CoA is funneled into fat synthesis, with its energetic electrons preserved for later needs. If the cell needs ATP now, the fragment is directed into ATP production through the Krebs cycle.

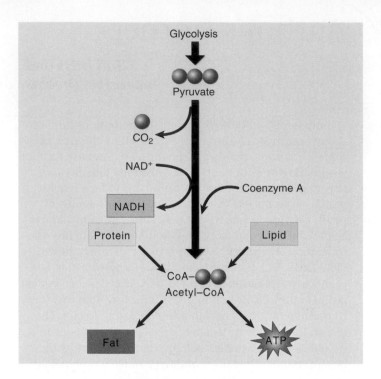

Figure 6.20 Producing acetyl-CoA.
Pyruvate, the three-carbon product of glycolysis, is oxidized to the two-carbon molecule acetyl-CoA, in the process losing one carbon atom as CO_2 and an electron (donated to NAD^+ to form NADH). Almost all the molecules you use as foodstuffs are converted to acetyl-CoA; the acetyl-CoA is then channeled into fat synthesis or into ATP production, depending on your body's needs.

Transferring Hydrogen Atoms

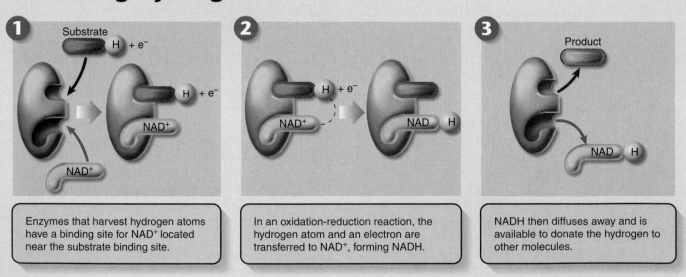

1 Enzymes that harvest hydrogen atoms have a binding site for NAD^+ located near the substrate binding site.

2 In an oxidation-reduction reaction, the hydrogen atom and an electron are transferred to NAD^+, forming NADH.

3 NADH then diffuses away and is available to donate the hydrogen to other molecules.

Figure 6.21 How NAD⁺ works.
Cells use NAD^+ to carry hydrogen atoms and energetic electrons from one molecule to another. Molecules that gain energetic electrons are ~~s~~id to be reduced, while ones that lose energetic electrons are said to be oxidized. NAD^+ oxidizes energy-rich molecules by acquiring their ~~h~~ydrogens (in the figure, this proceeds 1 → 2 → 3) and then reduces other molecules by giving the hydrogens to them (in the figure, this ~~proceeds~~ 3 → 2 → 1).

A Closer Look

Metabolic Efficiency and the Length of Food Chains

In the earth's ecosystems, the organisms that carry out photosynthesis are often consumed as food by other organisms. We call these "organism-eaters" *heterotrophs*. Humans are heterotrophs, as no human photosynthesizes.

It is thought that the first heterotrophs were ancient bacteria living in a world where photosynthesis had not yet introduced much oxygen into the oceans or atmosphere. The only mechanism they possessed to harvest chemical energy from their food was glycolysis. Neither oxygen-generating photosynthesis nor the oxidative stage of cellular respiration had evolved yet. It has been estimated that a heterotroph limited to glycolysis, as these ancient bacteria were, captures only 3.5% of the energy in the food it consumes. Hence, if such a heterotroph preserves 3.5% of the energy in the photosynthesizers it consumes, then any other heterotrophs that consume the first heterotroph will capture through glycolysis 3.5% of the energy in it, or 0.12% of the energy available in the original photosynthetic organisms. A very large base of photosynthesizers would thus be needed to support a small number of heterotrophs.

When organisms became able to extract energy from organic molecules by oxidative cellular respiration, this constraint became far less severe, because the efficiency of oxidative respiration is estimated to be about 32%. This increased efficiency results in the transmission of much more energy from one trophic level to another than does glycolysis. (A trophic level is a step in the movement of energy through an ecosystem.) The efficiency of oxidative cellular respiration has made possible the evolution of food chains, in which photosynthesizers are consumed by heterotrophs, which are consumed by other heterotrophs, and so on.

Even with this very efficient oxidative metabolism, approximately two-thirds of the available energy is lost at each trophic level, and that puts a limit on how long a food chain can be. Most food chains, like the East African grassland ecosystem illustrated here, involve only three or rarely four trophic levels. Too much energy is lost at each transfer to allow chains to be much longer than that. For example, it would be impossible for a large human population to subsist by eating lions captured from the grasslands of East Africa; the amount of grass available there would not support enough zebras and other herbivores to maintain the number of lions needed to feed the human population. Thus, the ecological complexity of our world is fixed in a fundamental way by the chemistry of oxidative cellular respiration.

Photosynthesizers. The grass under this yellow fever tree grows actively during the hot, rainy season, capturing the energy of the sun and storing it in molecules of glucose, which are then converted into starch and stored in the grass.

Herbivores. This zebra consumes the grass and transfers some of its stored energy into its own body.

Carnivores. The lion feeds on zebras and other animals, capturing part of their stored energy and storing it in its own body.

Scavengers. This hyena and the vultures occupy the same stage in the food chain as the lion. They are also consuming the body of the dead zebra, which has been abandoned by the lion.

Refuse utilizers. These butterflies, mostly *Precis octavia,* are feeding on the material left in the hyena's dung after the food the hyena consumed had passed through its digestive tract.

A food chain in the savannas, or open grasslands, of East Africa.
At each of these levels in the food chain, only about a third or less of the energy present is used by the recipient.

Overview of the Krebs Cycle

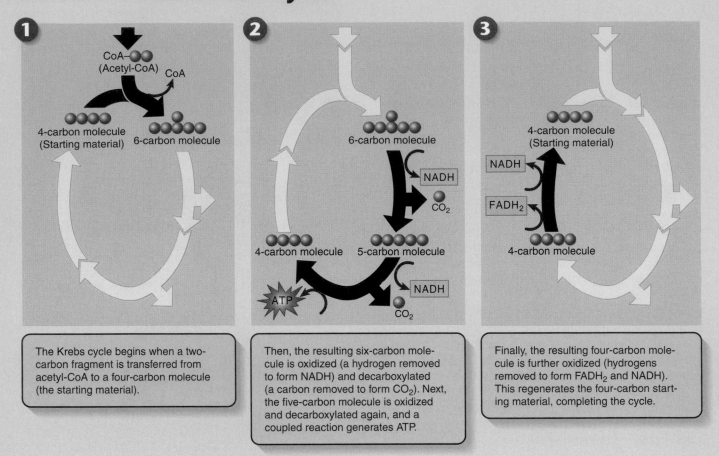

Figure 6.22 **How the Krebs cycle works.**

The Krebs Cycle

The next stage in oxidative respiration is called the **Krebs cycle,** named after the man who discovered it. The Krebs cycle (not to be confused with the Calvin cycle in photosynthesis) takes place within the mitochondrion and occurs in three stages (figure 6.22):

Stage 1. Acetyl-CoA joins the cycle, binding to a four-carbon molecule and producing a six-carbon molecule.

Stage 2. Two carbons are removed as CO_2, their electrons donated to NAD^+, and a four-carbon molecule is left. A molecule of ATP is also produced.

Stage 3. More electrons are extracted forming NADH and $FADH_2$, and the four-carbon starting material is regenerated.

The cycle starts when the two-carbon acetyl-CoA fragment produced from pyruvate is stuck onto a four-carbon sugar called oxaloacetate. Then, in rapid-fire order, a series of eight additional reactions occur. When it is all over, two carbon atoms have been expelled as CO_2, one ATP molecule has been made in a coupled reaction, eight more energetic electrons have been harvested and taken away as NADH or on [ot]her carriers, such as $FADH_2$, which serves the same function

as NADH, and we are left with the same four-carbon sugar we started with. The process of reactions is a cycle—that is, a circle of reactions (figure 6.23). In each turn of the cycle, a new acetyl group replaces the two CO_2 molecules lost, and more electrons are extracted. Note that a single glucose molecule produces *two* turns of the cycle, one for each of the two pyruvate molecules generated by glycolysis.

In the process of cellular respiration, glucose is entirely consumed. The six-carbon glucose molecule is first cleaved into a pair of three-carbon pyruvate molecules during glycolysis. One of the carbons of each pyruvate is then lost as CO_2 in the conversion of pyruvate to acetyl-CoA, and the other two carbons are lost as CO_2 during the oxidations of the Krebs cycle. All that is left to mark the passing of the glucose molecule into six CO_2 molecules is its energy, preserved in four ATP molecules and electrons carried by 10 NADH and two $FADH_2$ carriers.

6.8 **The end product of glycolysis, pyruvate, is oxidized to the two-carbon acetyl-CoA, yielding a pair of electrons plus CO_2. Acetyl-CoA then enters the Krebs cycle, yielding ATP, many energized electrons, and two CO_2 molecules.**

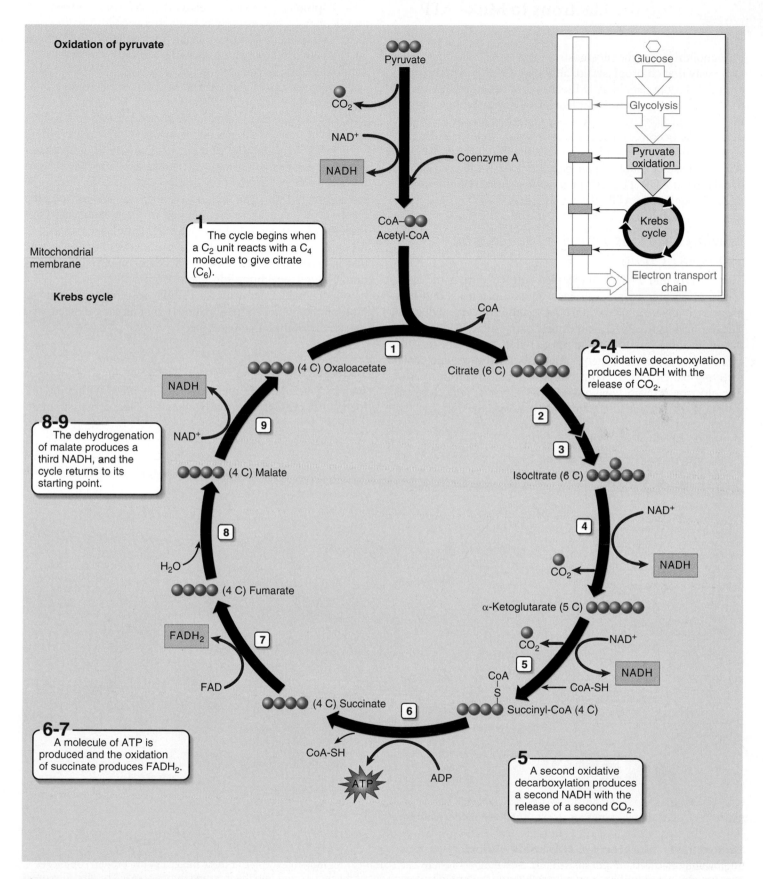

Oxidation of pyruvate

Pyruvate

CO_2

NAD^+

NADH

Coenzyme A

CoA—Acetyl-CoA

Mitochondrial membrane

Krebs cycle

1 The cycle begins when a C_2 unit reacts with a C_4 molecule to give citrate (C_6).

Glucose

Glycolysis

Pyruvate oxidation

Krebs cycle

Electron transport chain

1

CoA

2-4 Oxidative decarboxylation produces NADH with the release of CO_2.

(4 C) Oxaloacetate

NADH

Citrate (6 C)

2

9

NAD^+

3

8-9 The dehydrogenation of malate produces a third NADH, and the cycle returns to its starting point.

(4 C) Malate

Isocltrate (6 C)

4

NAD^+

8

CO_2

NADH

H_2O

α-Ketoglutarate (5 C)

(4 C) Fumarate

CO_2

NAD^+

FADH₂

7

5

NADH

FAD

CoA—S

CoA-SH

6-7 A molecule of ATP is produced and the oxidation of succinate produces $FADH_2$.

(4 C) Succinate

6

Succinyl-CoA (4 C)

CoA-SH

ATP

ADP

5 A second oxidative decarboxylation produces a second NADH with the release of a second CO_2.

Figure 6.23   The Krebs cycle.

This series of nine enzyme-catalyzed reactions takes place within the mitochondrion.

6.9 Using the Electrons to Make ATP

Mitochondria use chemiosmosis to make ATP in much the same way that chloroplasts do, although the proton pumps are oriented the opposite way. Mitochondria use energetic electrons extracted from food molecules to power proton pumps that shove protons across the inner mitochondrial membrane. As protons become far more scarce inside than outside, the concentration gradient drives protons back in through special ATP synthase channels. Their passage powers the production of ATP from ADP. The ATP then passes out of the mitochondrion through open ATP-passing channels.

Moving Electrons Through the Electron Transport Chain

The NADH and $FADH_2$ molecules formed during the first stages of aerobic respiration each contain electrons and hydrogens that were gained when NAD^+ and FAD were reduced. The NADH and $FADH_2$ molecules carry their electrons to the inner mitochondrial membrane, where they transfer the electrons to a series of membrane-associated molecules collectively called the **electron transport chain** (figure 6.24). The electron transport chain works much as does the electron transport system you encountered in photosynthesis.

A protein complex receives the electrons and using a mobile carrier, passes the electrons to a second protein complex. This protein complex, along with others in the chain, operates as a proton pump, driving a proton out across the membrane into the intermembrane space.

The electron is then carried by another carrier to a third protein complex. This complex uses four such electrons to reduce two oxygen atoms, which then combine with four hydrogen ions to form two molecules of water.

It is the availability of a plentiful supply of electron acceptor molecules (often oxygen) that makes oxidative respiration possible. The electron transport chain used in aerobic respiration is similar to, and may well have evolved from, the electron transport system employed in photosynthesis. Photosynthesis is thought to have preceded cellular respiration in the evolution of biochemical pathways, generating the oxygen that is necessary as the electron acceptor in cellular respiration. Natural selection didn't start from scratch and design a new biochemical pathway for cellular respiration; instead, it built on the photosynthetic pathway that already existed, and uses many of the same reactions.

Producing ATP: Chemiosmosis

In eukaryotes, aerobic metabolism takes place within the mitochondria present in virtually all cells. The internal compart-

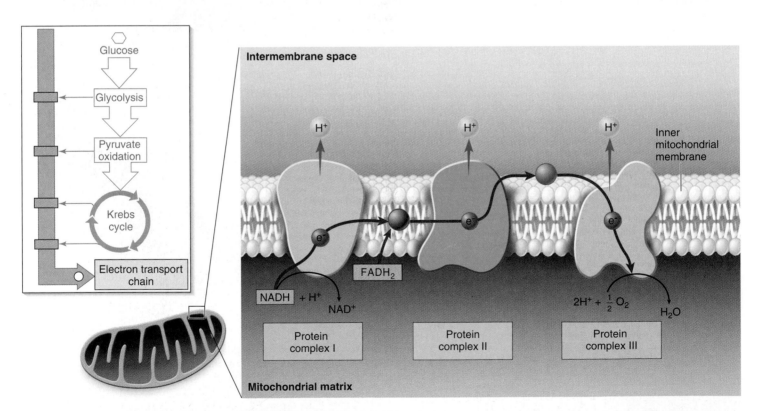

Figure 6.24 The electron transport chain.
High-energy electrons are transported (*red arrows*) along a chain of electron-carrier molecules. Three of these molecules are protein complexes that use portions of the electrons' energy to pump protons (*blue arrows*) out of the matrix and into the intermembrane space. The electrons are finally donated to oxygen to form water.

ment, or **matrix,** of a mitochondrion contains the enzymes that carry out the reactions of the Krebs cycle. As described earlier, the electrons harvested by oxidative respiration are passed along the electron transport chain and the energy they release transports protons out of the matrix and into the outer compartment, sometimes called the **intermembrane space.** Proton pumps in the inner mitochondrial membrane accomplish the transport. The electrons contributed by NADH activate three of these proton pumps, and those contributed by FADH$_2$ activate two. As the proton concentration in the outer compartment rises above that in the matrix, the concentration gradient induces the protons to reenter the matrix by diffusion through special proton channels called ATP synthases, embedded in the inner mitochondrial membrane. When the protons pass through, these channels synthesize ATP from ADP + P$_i$ within the matrix. The ATP is then transported by facilitated diffusion out of the mitochondrion and into the cell's cytoplasm. It is because the chemical formation of ATP is driven by a diffusion force similar to osmosis that this process is referred to as **chemiosmosis** (figure 6.25).

Thus, the electron transport chain uses electrons harvested in aerobic respiration to pump a large number of protons across the inner mitochondrial membrane. Their subsequent reentry into the mitochondrial matrix drives the synthesis of ATP by chemiosmosis. Figure 6.26 summarizes the overall process.

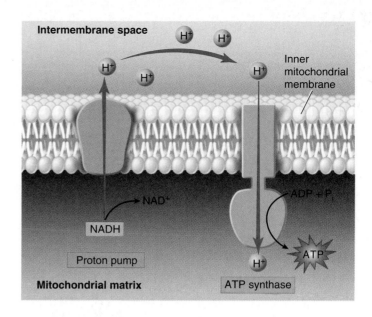

Figure 6.25 Chemiosmosis.

NADH transports high-energy electrons harvested from macromolecules to "proton pumps" that use the energy to pump protons out of the mitochondrial matrix. As a result, the concentration of protons outside the inner mitochondrial membrane rises, inducing protons to diffuse back into the matrix. Many of the protons pass through ATP synthase channels that couple the reentry of protons to the production of ATP.

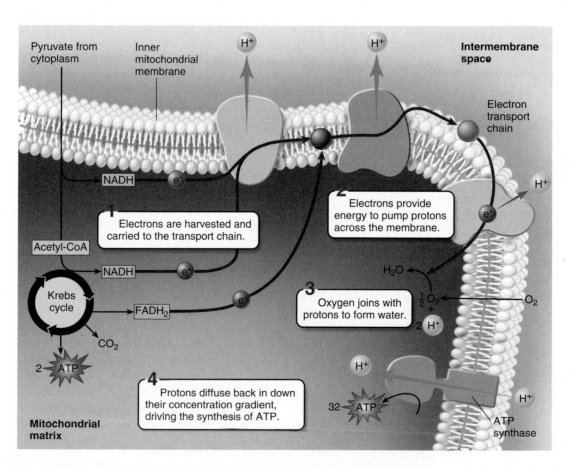

Figure 6.26 An overview of the electron transport chain and chemiosmosis.

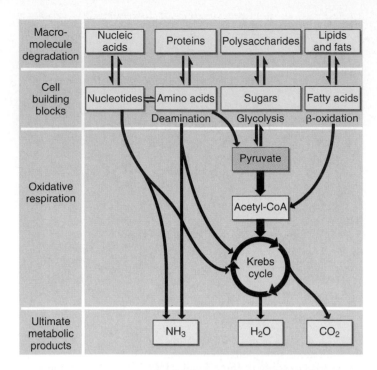

Macro-molecule degradation	Nucleic acids	Proteins	Polysaccharides	Lipids and fats
Cell building blocks	Nucleotides ⇄	Amino acids	Sugars	Fatty acids
		Deamination	Glycolysis	β-oxidation

Pyruvate

Acetyl-CoA

Krebs cycle

| Ultimate metabolic products | | NH_3 | H_2O | CO_2 |

Figure 6.27   How cells obtain energy from foods.
Most organisms extract energy from organic molecules by oxidizing them. The first stage of this process, breaking down macromolecules into their subunits, yields little energy. The second stage, cellular respiration, extracts energy, primarily in the form of high-energy electrons. The subunit of many carbohydrates, glucose, readily enters glycolysis and passes through the biochemical pathways of oxidative respiration. However, the subunits of other macromolecules must be converted into products that can enter the biochemical pathways found in oxidative respiration.

Other Sources of Energy

We have considered in detail the fate of a molecule of glucose, a simple sugar, in cellular respiration. But how much of what you eat is sugar? As a more realistic example of the food you eat, consider the fate of a fast food hamburger. The hamburger you eat is composed of carbohydrates, fats, protein, and many other molecules. This diverse collection of complex molecules is broken down by the process of digestion in your stomach and intestines into simpler molecules. Macromolecules are broken down by digestion into their subunits (building blocks), carbohydrates into glucose, fats into fatty acids, and proteins into amino acids. These breakdown reactions produce little or no energy themselves, but prepare the way for cellular respiration—that is, glycolysis and oxidative metabolism.

We have seen what happens to the glucose. What happens to the amino acids and fatty acids? These subunits undergo chemical modifications that convert them into products that feed into cellular respiration. For example, proteins are first broken down into their individual amino acids. A series of reactions removes the nitrogen side groups and converts the rest of the amino acid into a molecule that takes part in the Krebs cycle. Thus the proteins and fats in the hamburger also become important sources of energy (figure 6.27).

6.9   The electrons harvested by oxidizing food molecules are used to power proton pumps that chemiosmotically drive the production of ATP.

Exploring Current Issues

Additional Resources

Go to your campus library or look online to find the following articles, which further develop some of the concepts found in this chapter.

Carpenter, B. (2004). Fighting for a forgotten forest. *U.S. News & World Report,* 136(5), 58.

Demmig-Adams, B. and W. W. Adams III. (2002). Antioxidants in photosynthesis and human nutrition. *Science,* 298(5601), 2149.

Hunter, P. (2004). A new resolution for photosystem II. A 3.5 Å structure reveals fine features of the enzyme's catalytic core. *The Scientist,* 18(3), 25.

U.S. Food and Drug Administration. (2004). Questions and answers about the FDA's actions on dietary supplements containing ephedrine alkaloids. *Retrieved from www.fda.gov/oc/initiatives/ephedra/february2004/qa_020604.html.*

Weart, S. (2003). The carbon dioxide greenhouse effect. *Retrieved from www.aip.org/history/climate/co2.htm.*

Biology and Society Lecture: Aging—Does Metabolism Limit Life Span?

All the activities of life—growth, communication, reproduction—require energy. It thus should come as no surprise that researchers now suggest aging is related to changes in the way we process metabolic energy. All humans die. After puberty, the rate of death increases exponentially with age. A variety of hypotheses have been advanced to explain why. Many focus on damage and wear-and-tear on the cells. Others point to cell division; every time a cell divides, it loses material from the tips of its chromosome—eventually the chromosome can no longer divide. But what might be the most interesting hypothesis is that of genetic control. Single gene mutations can double the life span of fruit flies. When researchers isolated the gene involved, it proved to encode a protein involved in moving products of food metabolism across membranes to where the food's processing takes place. Surveys of very-long-lived humans also point to a single gene, whose function is eagerly sought.

Find this lecture delivered by the author to his class at Washington University online at www.mhhe.com/tlwessentials/exp6.

Photosynthesis

6.1 An Overview of Photosynthesis

- Photosynthesis is the process whereby energy from the sun is captured and used to build carbohydrates from CO_2 gas.

- Photosynthesis is a series of chemical reactions that occurs in two steps: the light-dependent reactions that produce ATP and NADPH occur on the thylakoid membranes of chloroplasts in plants and the light-independent reactions, or the Calvin cycle, that synthesize carbohydrates occur in the stroma (**figure 6.1**).

6.2 How Plants Capture Energy from Sunlight

- Pigments are molecules that capture light energy. Energy present in visible light is captured by chlorophyll and other accessory pigments present in chloroplasts (**figures 6.2 and 6.4**).

6.3 Organizing Pigments into Photosystems

- The protein and pigment molecules involved in photosynthesis are embedded in membranes: the plasma membrane in certain bacteria, internal membranes in algae, and thylakoid membranes of chloroplasts in plants. These embedded proteins and pigments make up a photosystem (**figure 6.6**).

- Light energy is captured by the photosystem and used to excite an electron that is passed to an electron transport system where it is used to generate ATP and NADPH, both which are used in the Calvin cycle. Plants utilize two photosystems that occur in series. Photosystem II leads to the formation of ATP and photosystem I leads to the formation of NADPH (**figures 6.7 and 6.8**).

6.4 How Photosystems Convert Light to Chemical Energy

- In plants, the antenna complex of photosystem II first captures energy from the sun. The energy is transferred to a chlorophyll molecule called the reaction center, which gives up an excited electron to a series of proteins within the membrane called the electron transport system. The lost electron is replenished with an electron from a water molecule.

- The excited electron is passed from one protein to another in the electron transport system, where energy from the electron is used to operate a proton pump that pumps hydrogen ions across the membrane against a concentration gradient (**figure 6.9**).

- The electron is then transferred to a second photosystem, photosystem I, where it gets an energy boost from the capture of another photon of light. This reenergized electron is passed along another series of proteins to an ultimate electron acceptor, $NADP^+$. $NADP^+ + e-$ and a H^+ produces NADPH, which is shuttled to the Calvin cycle (**figure 6.9**).

- The hydrogen ion concentration gradient is used as a source of energy to generate molecules of ATP. This energy is used to drive H^+ back across the membrane through specialized channel proteins called ATP synthases, which catalyze the formation of ATP (**figure 6.10**).

6.5 Building New Molecules

- The products of the light-dependent reactions, ATP and NADPH, are shuttled to the stroma where they are used in the Calvin cycle.

- The Calvin cycle is a series of enzymes that use the energy from ATP and electrons and hydrogens from NADPH to build carbohydrates using CO_2 from the air (**figures 6.11 and 6.12**).

- In hot dry weather the levels of O_2 increase in the leaves and CO_2 levels drop. Under these conditions, the Calvin cycle, also called C_3 photosynthesis, is disrupted because O_2 rather than CO_2 enters the Calvin cycle in a process called photorespiration (**figure 6.13**). C_4 and CAM plants reduce the effects of photorespiration by modifying the carbon-fixation step (**figure 6.14**).

Cellular Respiration

6.6 An Overview of Cellular Respiration

- Cellular respiration is the process of harvesting energy from glucose molecules and storing it as cellular energy in ATP. Cellular respiration is carried out in two stages: glycolysis occurring in the cytoplasm and oxidation occurring in the mitochondria (**figure 6.16**).

6.7 Using Coupled Reactions to Make ATP

- Glycolysis is a series of 10 chemical reactions in which glucose is broken down into two three-carbon pyruvate molecules. Two exergonic reactions are coupled with a reaction that leads to the formation of ATP. This is called substrate-level phosphorylation (**figures 6.17 and 6.18**).

- Glycolysis does not require oxygen and is therefore referred to as anaerobic respiration. There are other forms of anaerobic respiration, including fermentation (**figure 6.19**).

6.8 Harvesting Electrons from Chemical Bonds

- The two molecules of pyruvate formed in glycolysis are passed into the mitochondrion where they are converted into two molecules of acetyl-coenzyme A (**figure 6.20**). Some of the energy in pyruvate is transferred to NAD^+ to produce NADH, which will be used in a later step (**figure 6.21**).

- Acetyl-CoA enters a series of chemical reactions called the Krebs cycle, where one molecule of ATP is produced and energy is harvested in the form of electrons that are transferred to molecules of NAD^+ and FAD to produce NADH and $FADH_2$, respectively (**figures 6.22 and 6.23**).

6.9 Using the Electrons to Make ATP

- The energy stored in NADH and $FADH_2$ is harvested by the electron transport chain to make ATP. NADH and $FADH_2$ are transported to the inner mitochondrial membrane where they transfer electrons to the electron transport chain. The electrons are passed along the electron transport chain, which drives proton pumps that pump H^+ across the inner membrane, creating a H^+ concentration gradient (**figure 6.24**).

- As in photosynthesis, ATP is formed when H^+ passes back across the membrane through ATP synthase channels. Thus, the energy harvested from glucose is stored in ATP (**figures 6.25 and 6.26**).

- Other food sources are also used in oxidative respiration. Macromolecules are broken down into intermediate products that enter cellular respiration in different reaction steps (**figure 6.27**).

1. The energy that is used by almost all living things on our planet comes from the sun. It is captured by plants, algae, and some bacteria through the process of
   a. thylakoid.
   c. photosynthesis.
   b. chloroplasts.
   d. the Calvin cycle.
2. Plants capture sunlight
   a. with netlike structures in their leaves that catch and hold photons.
   b. with molecules called pigments that absorb photons and use their energy.
   c. with sticky molecules that absorb photons and transfer their energy.
   d. by angling their leaves slightly to the sun's rays for more absorbance of photons.
3. Once a plant has initially captured the energy of a photon
   a. a series of reactions occurs in chloroplast membranes of the cell.
   b. the energy is transferred through several steps into a molecule of ATP.
   c. several pigments, including two kinds of chlorophyll, may be involved.
   d. All of these are true.
4. Plants use two photosystems to produce ATP and NADPH. The electrons used in these photosystems
   a. recycle through the system constantly, with energy added from the photons.
   b. recycle through the system several times, using energy added from photons, and then are lost due to entropy.
   c. only go through the system once; they are obtained by splitting a water molecule.
   d. only go through the system once; they are obtained from the photon.
5. The purpose of photosynthesis is to
   a. obtain the energy to make organic molecules from carbon dioxide.

   b. obtain the energy to make organic molecules from starches.
   c. make energy.
   d. All of these are true.
6. Many plants cannot carry out the typical C_3 photosynthesis in hot weather, so some plants
   a. use the ATP cycle.
   b. use C_4 photosynthesis or CAM.
   c. wait until evening to carry out photosynthesis.
   d. All of these are true for different plants.
7. In animals, the energy for life is obtained by cellular respiration. This involves
   a. breaking down the organic molecules that were consumed.
   b. capturing photons from plants.
   c. utilizing ATP that was produced by plants.
   d. breaking down CO_2 that was produced by plants.
8. Every living creature on this planet is capable of carrying out the rather inefficient biochemical process of glycolysis, which
   a. makes glucose using the energy from ATP.
   b. makes ATP by splitting glucose and capturing the energy.
   c. phosphorylates ATP to make ADP using the energy from photons.
   d. makes glucose using oxygen and carbon dioxide and water.
9. After glycolysis the pyruvate molecules go to the
   a. nucleus of the cell and provide energy.
   b. membranes of the cell and are broken down in the presence of CO_2 to make more ATP.
   c. mitochondria of the cell and are broken down in the presence of O_2 to make more ATP.
   d. Golgi bodies and are packaged and stored until needed.
10. The vast majority of the ATP molecules produced within a cell are produced
   a. during photosynthesis.
   b. during glycolysis.
   c. during the Krebs cycle.
   d. from the electron transport chain.

Visual Understanding

1. **Figure 6.4** Why do most leaves have more than one type of pigment?

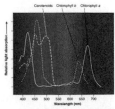

2. **Figure 6.27** Your friend Yevgeny wants to go on a low-carbohydrate diet so that he can lose some of the "baby fat" he's still carrying. He asks your advice; what do you tell him?

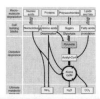

Challenge Questions

Photosynthesis If you were going to design a plant that would survive in the deserts of Arizona and New Mexico, how would you balance its need for CO_2 with its need to avoid water loss in the hot summer temperatures?

Cellular Respiration If cellular respiration were the stock market (you're investing ATPs and getting ATP dividends), where would you get the most return on your investment: glycolysis, the Krebs cycle, or the electron transport chain? Explain your answer.

Online Learning Center

Visit the Online Learning Center for this chapter at www.mhhe.com/tlwessentials/ch6 for quizzes, animations, interactive learning exercises, and other study tools. At the site you will also find extended answers to the end-of-chapter questions.

7

How Cells Divide

T hese cells are dividing; several stages in the cell division process are shown. You can see the chromosomes being drawn to opposite ends of the cells by microtubules too tiny to be visible to our eyes. Some human cells divide frequently, particularly those subjected to a lot of wear and tear. The epithelial cells of your skin divide so often that your skin replaces itself every two weeks. The lining of your stomach is replaced every few days! Nerve cells, on the other hand, can live for 100 years without dividing. Cells use a battery of genes to regulate when and how frequently they divide. If some of these genes become disabled, a cell may begin to divide ceaselessly, a condition we call cancer. Exposure to DNA-damaging chemicals such as those in cigarette smoke greatly increase the chance this sort of event will occur in the tissues exposed to the smoke, which is why smokers will more likely get lung cancer than colon cancer. Most of the cells of the body divide by a process called mitosis, but a more complex division called meiosis takes place in germ-line cells to form haploid gametes, eggs and sperm. Meiosis is a more intricate process than mitosis, but utilizes the same cell machinery working in much the same way.

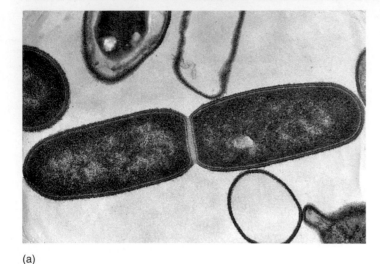

(a)

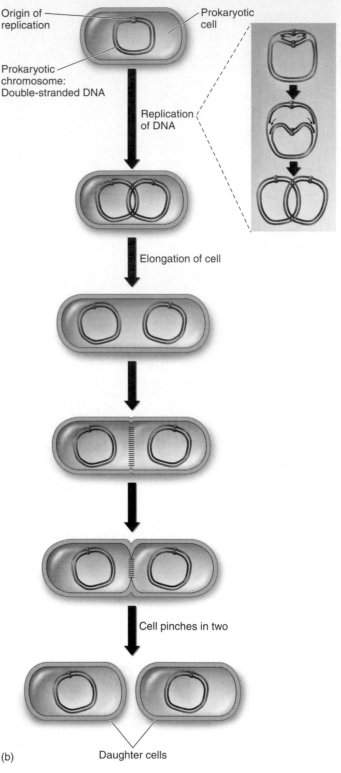

Origin of replication

Prokaryotic cell

Prokaryotic chromosome: Double-stranded DNA

Replication of DNA

Elongation of cell

Cell pinches in two

(b)

Daughter cells

7.1 Prokaryotes Have a Simple Cell Cycle

All species reproduce, passing their hereditary information on to their offspring. In this chapter, we begin our consideration of heredity with a look at how cells reproduce. Cell division in prokaryotes takes place in two stages, which together make up a **simple cell cycle.** First the DNA is copied, and then the cell splits by a process called **binary fission.**

In prokaryotes, the hereditary information—that is, the genes that specify the prokaryote—is encoded in a single circle of DNA. Before the cell itself divides, the DNA circle makes a copy of itself. Starting at one point, the origin of replication, the double helix of DNA begins to unzip, exposing the two strands (figure 7.1). A new double helix is then formed on each naked strand by placing on each exposed nucleotide its complementary nucleotide (that is, A with T, G with C, as discussed in chapter 3). DNA replication is discussed in more detail in chapter 9. When the unzipping has gone all the way around the circle, the cell possesses two copies of its hereditary information.

When the DNA has been copied, the cell grows, resulting in elongation. The newly replicated DNA molecules are partitioned toward each end of the cell. This partitioning process involves DNA sequences near the origin of replication, and results in these sequences being attached to the membrane. When the cell reaches an appropriate size, the prokaryotic cell begins to split into two equal halves. New plasma membrane and cell wall are added at a point between where the two DNA copies are partitioned. As the growing plasma membrane pushes inward, the cell is constricted in two, eventually forming two **daughter cells.** Each contains one of the circles of DNA and is a complete living cell in its own right.

7.1 Prokaryotes divide by binary fission after the DNA has replicated.

Figure 7.1 Cell division in prokaryotes.

(a) Prokaryotes divide by a process of binary fission. Here, a cell has divided in two and is about to be pinched apart by the growing plasma membrane. (b) Before the cell splits, the circular DNA molecule of a prokaryote initiates replication at a single site, called the origin of replication, moving out in both directions. When the two moving replication points meet on the far side of the molecule, its replication is complete. The cell then undergoes binary fission, where the cell divides into two daughter cells.

7.2 Eukaryotes Have a Complex Cell Cycle

The evolution of the eukaryotes introduced several additional factors into the process of cell division. Eukaryotic cells are much larger than prokaryotic cells, and they contain much more DNA. Eukaryotic DNA is contained in a number of linear chromosomes, whose organization is much more complex than that of the single, circular DNA molecules in prokaryotes. A **chromosome** is a single, long DNA molecule packaged with proteins into a compact shape. In chromosomes, DNA forms a complex with packaging proteins called histones and is wound into tightly condensed coils.

Cell division in eukaryotes is more complex than in prokaryotes, both because eukaryotes contain far more DNA and because it is packaged differently. The cells of eukaryotic organisms either undergo mitosis or meiosis to divide up the DNA. **Mitosis** is the mechanism of cell division that occurs in an organism's nonreproductive cells, or **somatic cells.** A second process, called **meiosis,** divides the DNA in cells that participate in sexual reproduction, or **germ cells.** Meiosis results in the production of gametes, such as sperm and eggs, and is discussed later in this chapter.

The events that prepare the eukaryotic cell for division and the division process itself constitute a **complex cell cycle** (figure 7.2):

G₁ phase. This "first gap" phase is the cell's primary growth phase. For most organisms, this phase occupies the major portion of the cell's life span.

S phase. In this "synthesis" phase, the DNA replicates, producing two copies of each chromosome.

G₂ phase. Cell division preparation continues with the replication of mitochondria, chromosome condensation, and the synthesis of microtubules. G_1, S, and G_2 are collectively referred to as **interphase,** or the period between cell divisions.

M phase. In mitosis, a microtubular apparatus binds to the chromosomes and moves them apart.

C phase. In cytokinesis, the cytoplasm divides, creating two daughter cells.

> **7.2** Eukaryotic cells divide by separating duplicate copies of their chromosomes into daughter cells.

The Cell Cycle

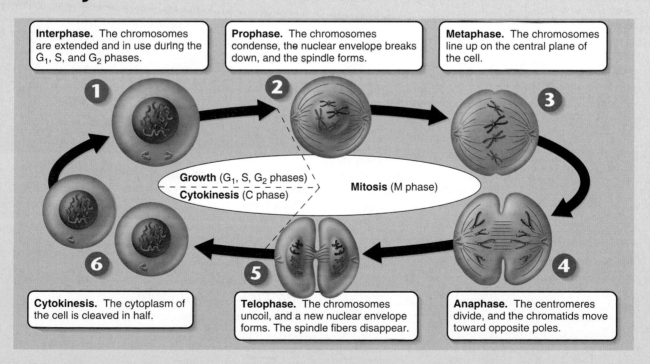

Interphase. The chromosomes are extended and in use during the G_1, S, and G_2 phases.

Prophase. The chromosomes condense, the nuclear envelope breaks down, and the spindle forms.

Metaphase. The chromosomes line up on the central plane of the cell.

Growth (G_1, S, G_2 phases)
Cytokinesis (C phase)

Mitosis (M phase)

Cytokinesis. The cytoplasm of the cell is cleaved in half.

Telophase. The chromosomes uncoil, and a new nuclear envelope forms. The spindle fibers disappear.

Anaphase. The centromeres divide, and the chromatids move toward opposite poles.

Figure 7.2 How the cell cycle works.
The G_1, S, and G_2 phases occur during interphase, while the cell is growing and preparing to divide. Then, during mitosis, the cell's nucleus divides. Finally, the cell's cytoplasm divides (cytokinesis) with the formation of two daughter cells. Human cells growing in culture typically have a 22-hour cell cycle. Most cell types take about 80 minutes in this 22 hours to complete cell division: prophase—23 minutes, metaphase—29 minutes, anaphase—10 minutes, telophase—14 minutes, and cytokinesis—4 minutes. The proportion of the cell cycle spent in any one phase of mitosis varies considerably in different tissues.

7.3 Chromosomes

Discovery of Chromosomes

Chromosomes were first observed by the German embryologist Walther Fleming in 1882, while he was examining the rapidly dividing cells of salamander larvae. When Fleming looked at the cells through what would now be a rather primitive light microscope, he saw minute threads within their nuclei that appeared to be dividing lengthwise. Fleming called their division *mitosis,* based on the Greek word *mitos,* meaning "thread."

Chromosome Number

Since their initial discovery, chromosomes have been found in the cells of all eukaryotes examined. Their number may vary enormously from one species to another. A few kinds of organisms—such as the Australian ant *Myrmecia* spp.; the plant *Haplopappus gracilis,* a relative of the sunflower that grows in North American deserts; and the fungus *Penicillium*—have only 1 pair of chromosomes, while some ferns have more than 500 pairs. Most eukaryotes have between 10 and 50 chromosomes in their body cells.

Homologous Chromosomes

Chromosomes exist in somatic cells as pairs, called **homologous chromosomes,** or **homologues.** Homologues carry information about the same traits at the same locations on each chromosome but the information can vary between homologues, which will be discussed in chapter 8. Cells that have two of each type of chromosome are called **diploid cells.** One chromosome of each pair is inherited from the mother and the other from the father. Before cell division, each homologous chromosome replicates, resulting in two identical copies, called **sister chromatids** (figure 7.3), that remain joined together at a special linkage site called the **centromere.** For example, human body cells have a total of 46 chromosomes, 23 pairs of homologous chromosomes. In their duplicated state, before mitosis, there are still only 23 pairs of chromosomes, but each chromosome has duplicated and each consists of two sister chromatids, for a total of 92 chromatids.

The Human Karyotype

The 46 human chromosomes can be paired as homologues by comparing size, shape, location of centromeres, and so on (figure 7.4). Each chromosome contains thousands of genes that play important roles in determining how a person's body develops and functions. For this reason, possession of all the chromosomes is essential to survival. Humans missing even one chromosome, a condition called monosomy, do not usually survive embryonic development. Nor does the human embryo develop properly with an extra copy of any one chromosome, a condition called trisomy. For all but a few of the smallest chromosomes, trisomy is fatal; even in those cases, serious problems result.

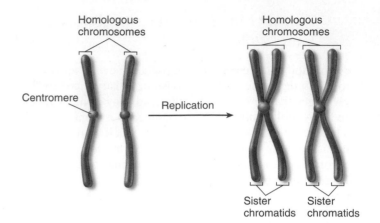

Figure 7.3   The difference between homologous chromosomes and sister chromatids.

Homologous chromosomes are a pair of the same chromosome— say, chromosome number 16. Sister chromatids are the two replicas of a single chromosome held together by the centromeres after DNA replication. A duplicated chromosome looks somewhat like an X.

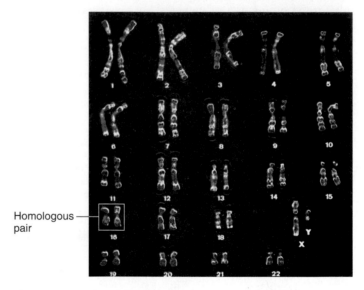

Figure 7.4   The 46 chromosomes of a human.

In this presentation, photographs of the individual chromosomes of a human male have been cut out and paired with their homologues, creating an organized display called a *karyotype.* The chromosomes are in a duplicated state, and the sister chromatids can actually be seen in many of the homologous pairs.

Chromosome Structure

Chromosomes are composed of **chromatin,** a complex of DNA and protein; most are about 40% DNA and 60% protein. A significant amount of RNA is also associated with chromosomes because chromosomes are the sites of RNA synthesis. The DNA of a chromosome is one very long, double-stranded fiber that extends unbroken through the entire length of the chromosome. A typical human chromosome contains about 140 million (1.4×10^8) nucleotides in its DNA. Furthermore, if the strand of DNA from a single chromosome were laid out

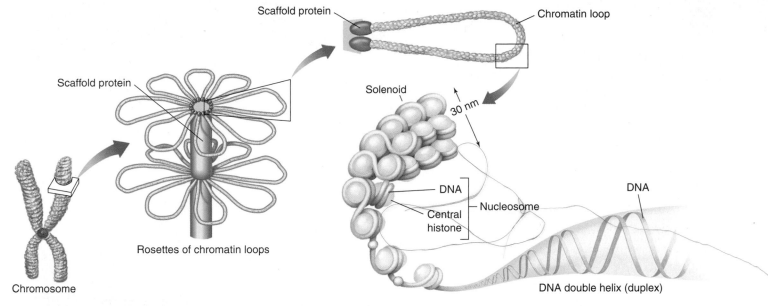

Figure 7.5 levels are labeled:
Scaffold protein — Chromatin loop — Solenoid — 30 nm — DNA — Nucleosome — Central histone — DNA — DNA double helix (duplex) — Scaffold protein — Rosettes of chromatin loops — Chromosome

Figure 7.5 Levels of eukaryotic chromosomal organization.
Compact, rod-shaped chromosomes are in fact highly wound-up molecules of DNA. The arrangement illustrated here is one of many possibilities.

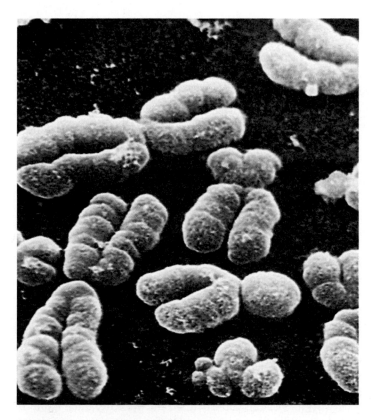

Figure 7.6 Human chromosomes.
The photograph (×950) shows human chromosomes as they appear immediately before nuclear division. Each DNA strand has already replicated, forming identical copies held together by the centromere.

in a straight line, it would be about 5 centimeters (2 inches) long. The amount of information in one human chromosome would fill about 2,000 printed books of 1,000 pages each! Fitting such a strand into a nucleus is like cramming a string the length of a football field into a baseball—and that's only 1 of 46 chromosomes! In the cell, however, the DNA is coiled, allowing it to fit into a much smaller space than would otherwise be possible.

Chromosome Coiling

The DNA of eukaryotes is divided into several chromosomes. Because the phosphate groups of DNA molecules have negative charges, it is impossible to just tightly wind up DNA because all the negative charges would simply repel one another. Instead, the DNA helix is wrapped around proteins with positive charges called **histones,** the positive (histone) and negative (DNA) charges counteract each other so that the complex has no net charge. Every 200 nucleotides, the DNA duplex is coiled around a core of eight histone proteins, forming a complex known as a **nucleosome.** The nucleosomes are further coiled into a solenoid. This solenoid is then organized into looped domains. The final organization of the chromosome is not known, but it appears to involve further radial looping into rosettes around a preexisting scaffolding of protein (figure 7.5). This complex of DNA and histone proteins, coiled tightly, forms a compact chromosome (figure 7.6).

> **7.3** All eukaryotic cells store their hereditary information in chromosomes, but different kinds of organisms use very different numbers of chromosomes to store this information.

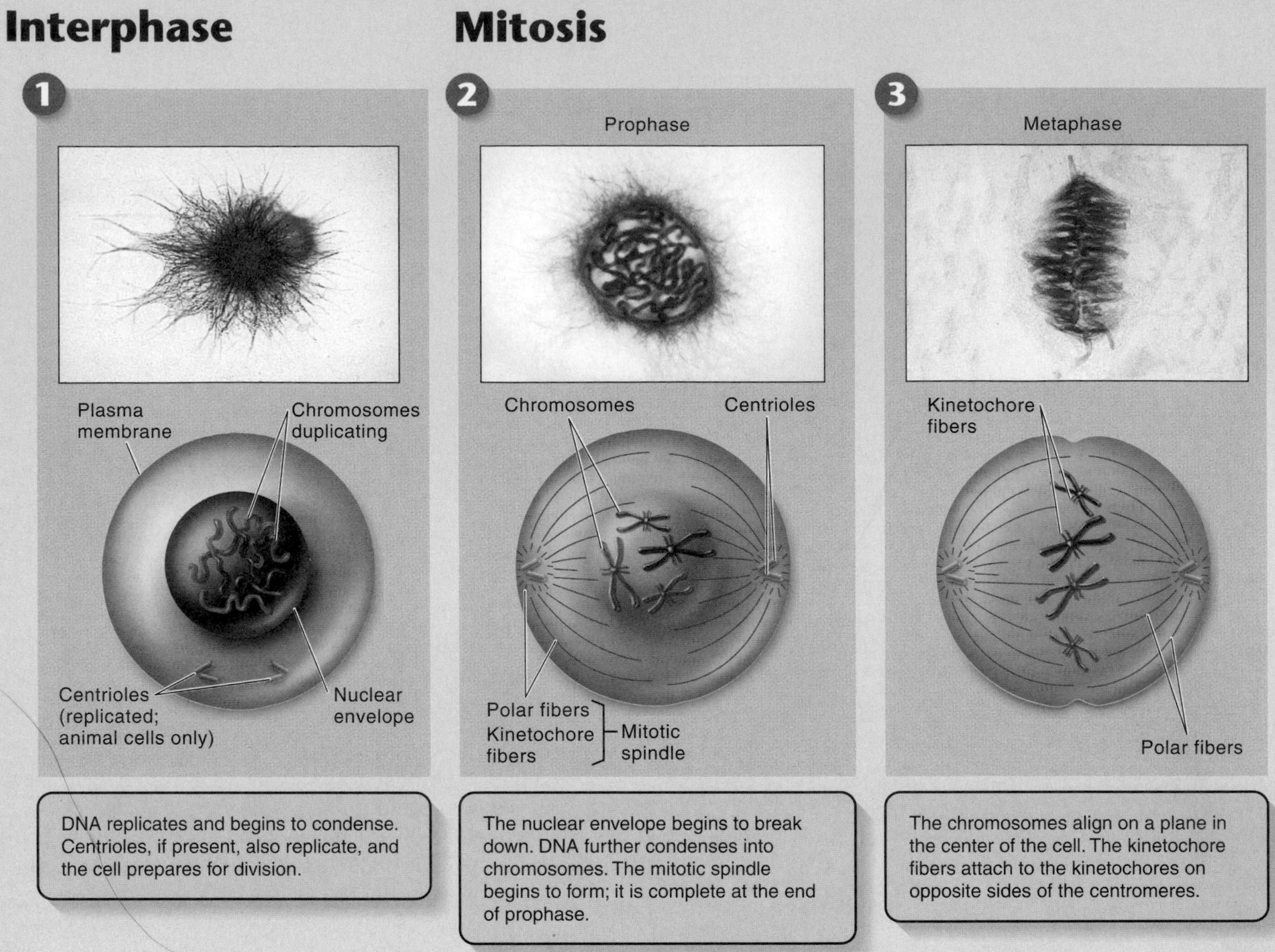

Interphase

Mitosis

1

Plasma membrane

Chromosomes duplicating

Centrioles (replicated; animal cells only)

Nuclear envelope

DNA replicates and begins to condense. Centrioles, if present, also replicate, and the cell prepares for division.

2 Prophase

Chromosomes

Centrioles

Polar fibers

Kinetochore fibers — Mitotic spindle

The nuclear envelope begins to break down. DNA further condenses into chromosomes. The mitotic spindle begins to form; it is complete at the end of prophase.

3 Metaphase

Kinetochore fibers

Polar fibers

The chromosomes align on a plane in the center of the cell. The kinetochore fibers attach to the kinetochores on opposite sides of the centromeres.

Figure 7.7 How cell division works.
Cell division in eukaryotes begins in interphase, carries through the four stages of mitosis, and ends with cytokinesis. The chromosomes, stained *blue* in the photos, are being drawn to the poles of the dividing cell by microtubules, stained *red* in the photos.

7.4 Cell Division

Interphase

When cell division begins, in the interphase portion of the cell cycle, chromosomes replicate and then begin to wind up tightly, a process called **condensation.** Daughter chromosomes are held together by a complex of proteins called cohesin.

Mitosis

Interphase is not a phase of mitosis, but it sets the stage for cell division. It is followed by nuclear division, called *mitosis.* Although the process of mitosis is continuous, with the stages flowing smoothly one into another, for ease of study, mitosis is traditionally subdivided into four stages: prophase, metaphase, anaphase, and telophase (figure 7.7).

Prophase: Mitosis Begins.   In **prophase,** the individual condensed chromosomes first become visible with a light microscope. As the replicated chromosomes condense, the nucleolus disappears and the cell dismantles the nuclear membrane and begins to assemble the apparatus it will use to pull the replicated daughter chromosomes to opposite ends ("poles") of the cell. In the center of an animal cell, the pair of centrioles starts to separate; the two centrioles move apart toward opposite poles of the cell, forming between them as they move apart a network of protein cables called the **spindle.** Each cable is called a spindle fiber and is made of microtubules, which are long, hollow tubes of protein. Plant cells lack centrioles and instead brace the ends of the spindle with a support structure called an aster.

As condensation of the chromosomes continues, a second group of microtubules extends out from the centromere of each chromosome from a disk of protein called a **kinetochore.**

Cytokinesis

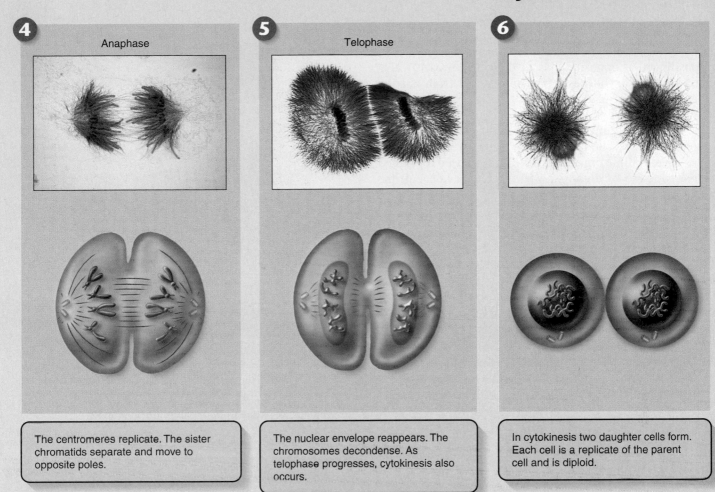

4 Anaphase

The centromeres replicate. The sister chromatids separate and move to opposite poles.

5 Telophase

The nuclear envelope reappears. The chromosomes decondense. As telophase progresses, cytokinesis also occurs.

6

In cytokinesis two daughter cells form. Each cell is a replicate of the parent cell and is diploid.

The two sets of microtubules extend out from opposite sides of the kinetochore toward opposite poles of the cell. Each set of microtubules continues to grow longer until it makes contact with the pole toward which it is growing. When the process is complete, one daughter chromosome of each pair is attached by microtubules to one pole and the other daughter to the other pole. Also in this stage, the nuclear envelope begins to break down and the mitotic spindle that will be used to separate the sister chromatids is assembled. In animal cells and those of most protists, the spindle fibers are associated with the centrioles. After the centrioles are replicated, the pairs separate and move to opposite poles of the cell.

Metaphase: Alignment of the Chromosomes.   The second phase of mitosis, **metaphase,** begins when the chromosomes, each consisting of a pair of chromatids, align in the center of the cell along an imaginary plane that divides the cell in half, referred to as the equatorial plane. Microtubules attached to the kinetochores of the centromeres are fully extended back toward the opposite poles of the cell.

Anaphase: Separation of the Chromatids.   In **anaphase,** enzymes cleave the cohesin link holding sister chromatids together, the kinetochores split, and the daughter chromosomes are freed from each other. Cell division is now simply a matter of reeling in the microtubules, dragging to the poles the daughter chromosomes like fish on the end of a line—at the poles, the ends of the microtubules are dismantled, one bit after another, making the tubes shorter and shorter and so drawing the chromosome attached to the far end closer and closer to the pole. They move rapidly toward opposite poles of the cell. When they finally arrive, each pole has one complete set of chromosomes. This is the shortest stage of mitosis.

Telophase: Re-formation of the Nuclei.   The only tasks that remain in **telophase** are the dismantling of the stage and the removal of the props. The mitotic spindle is disassembled, and a nuclear envelope forms around each set of chromosomes while they begin to uncoil and the nucleolus reappears.

Cytokinesis

At the end of telophase, mitosis is complete. The cell has divided its replicated chromosomes into two nuclei, which are positioned at opposite ends of the cell. Mitosis is also referred to as **karyokinesis.** You may recall from chapter 4 that the nucleus is also referred to as *karyon* (Latin for "kernel"); therefore, karyokinesis is the division of the nucleus. Following mitosis, **cytokinesis,** the division of the cytoplasm, occurs, and the cell is cleaved into roughly equal halves. Cytoplasmic organelles have already been replicated and reassorted to the areas that will separate and become the daughter cells.

In animal cells, which lack cell walls, cytokinesis is achieved by pinching the cell in two with a contracting belt of actin filaments. As contraction proceeds, a **cleavage furrow** becomes evident around the cell's circumference, where the cytoplasm is being progressively pinched inward by the decreasing diameter of the actin belt (figure 7.8*a*). The cleavage furrow deepens until the cell is literally pinched in two.

Plant cells have rigid walls that are far too strong to be deformed by actin filament contraction. A different approach to cytokinesis has therefore evolved in plants. Plant cells assemble membrane components in their interior, at right angles to the mitotic spindle. This expanding partition, called a **cell plate,** grows outward until it reaches the interior surface of the plasma membrane and fuses with it, at which point it has effectively divided the cell in two (figure 7.8*b*). Cellulose, the major component of cell walls, is then laid down on the new membranes, creating two new cells.

Cell Death

Despite the ability to divide, no cell lives forever. The ravages of living slowly tear away at a cell's machinery. To some degree damaged parts can be replaced, but no replacement process is perfect. And sometimes the environment intervenes. If food supplies are cut off, for example, animal cells cannot obtain the energy necessary to maintain their lysosome membranes. The cells die, digested from within by their own enzymes.

During fetal development, many cells are programmed to die. In human embryos, hands and feet appear first as "paddles," but the skin cells between bones die on schedule to form the separated toes and fingers (figure 7.9). In ducks, this cell death is not part of the developmental program, which is why ducks have webbed feet and you don't.

Human cells appear to be programmed to undergo only so many cell divisions and then die, following a plan written into the genes. In tissue culture, cell lines divide about 50 times, and then the entire population of cells dies off. Even if some of the cells are frozen for years, when they are thawed they simply resume where they left off and die on schedule. Only cancer cells appear to thwart these instructions, dividing endlessly. All other cells in your body contain a hidden clock that keeps time by counting cell divisions, and when the alarm goes off the cells die.

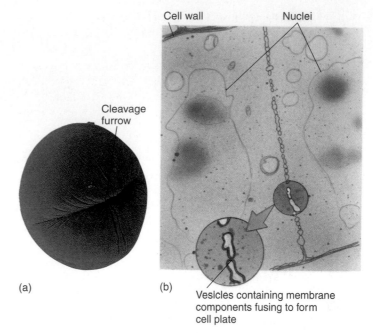

(a) (b)

Figure 7.8   Cytokinesis.
The division of cytoplasm that occurs after mitosis is called cytokinesis and cleaves the cell into roughly equal halves. (*a*) In an animal cell, such as this sea urchin egg, a cleavage furrow forms around the dividing cell. (*b*) In this dividing plant cell, a cell plate is forming between the two newly forming daughter cells.

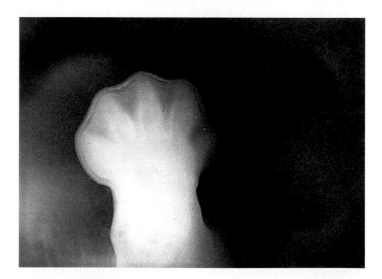

Figure 7.9   Programmed cell death.
In the human embryo, programmed cell death results in the formation of fingers and toes from paddlelike hands and feet.

7.4   The eukaryotic cell cycle starts in interphase with the condensation of replicated chromosomes; in mitosis, these chromosomes are drawn by microtubules to opposite ends of the cell; in cytokinesis, the cell is split into two daughter cells.

7.5 Controlling the Cell Cycle

The events of the cell cycle are coordinated in much the same way in all eukaryotes. The control system human cells use first evolved among the protists over a billion years ago; today, it operates in essentially the same way in fungi as it does in humans.

The goal of controlling any cyclic process is to adjust the duration of the cycle to allow sufficient time for all events to occur. In principle, a variety of methods can achieve this goal. For example, an internal clock can be employed to allow adequate time for each phase of the cycle to be completed. This is how many organisms control their daily activity cycles. The disadvantage of using such a clock to control the cell cycle is that it is not very flexible. One way to achieve a more flexible and sensitive regulation of a cycle is simply to let the completion of each phase of the cycle trigger the beginning of the next phase, as a runner passing a baton starts the next leg in a relay race. Until recently, biologists thought this type of mechanism controlled the cell division cycle. However, we now know that eukaryotic cells employ a separate, centralized controller to regulate the process: at critical points in the cell cycle, further progress depends upon a central set of "go/no-go" switches that are regulated by feedback from the cell.

This mechanism is the same one engineers use to control many processes. For example, the furnace that heats a home in the winter typically goes through a daily heating cycle. When the daily cycle reaches the morning "turn on" checkpoint, sensors report whether the house temperature is below the set point (for example, 70°F). If it is, the thermostat triggers the furnace, which warms the house. If the house is already at least that warm, the thermostat does not start the furnace. Similarly, the cell cycle has key checkpoints where feedback signals from the cell about its size and the condition of its chromosomes can either trigger subsequent phases of the cycle or delay them to allow more time for the current phase to be completed.

Three principal checkpoints control the cell cycle in eukaryotes (figure 7.10):

1. **Cell growth is assessed at the G_1 checkpoint.** Located near the end of G_1, just before entry into S phase, this checkpoint makes the key decision of whether the cell should divide, delay division, or enter a resting stage (figure 7.11). In yeasts, where researchers first studied this checkpoint, it is called START. If conditions are favorable for division, the cell begins to copy its DNA, initiating S phase. The G_1 checkpoint is where the more complex eukaryotes typically arrest the cell cycle if environmental conditions make cell division impossible or if the cell passes into an extended resting period called G_0.

2. **DNA replication is assessed at the G_2 checkpoint.** The second checkpoint, which occurs at the end of G_2, triggers the start of M phase. If this checkpoint is

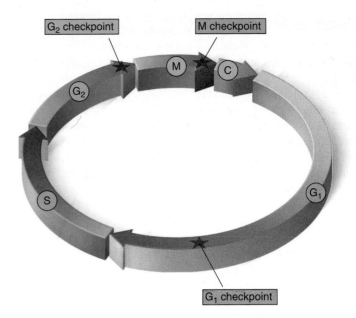

Figure 7.10 Control of the cell cycle.
Cells use a centralized control system to check whether proper conditions have been achieved before passing three key checkpoints in the cell cycle.

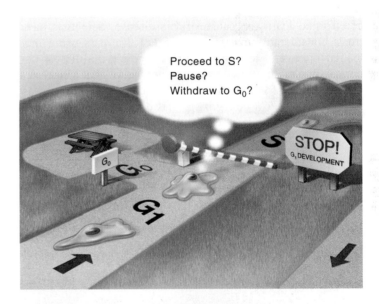

Figure 7.11 The G_1 checkpoint.
Feedback from the cell determines whether the cell cycle will proceed to the S phase, pause, or withdraw into G_0 for an extended rest period.

passed, the cell initiates the many molecular processes that signal the beginning of mitosis.

3. **Mitosis is assessed at the M checkpoint.** Occurring at metaphase, the third checkpoint triggers the exit from mitosis and cytokinesis and the beginning of G_1.

> **7.5 The complex cell cycle of eukaryotes is controlled by feedback at three checkpoints.**

7.6 What Is Cancer?

Cancer is a growth disorder of cells. It starts when an apparently normal cell begins to grow in an uncontrolled way, spreading out to other parts of the body (figure 7.12). The result is a cluster of cells, called a **tumor**, that constantly expands in size. Benign tumors are completely enclosed by normal tissue and are said to be encapsulated. These tumors do not spread to other parts of the body and are therefore noninvasive. Malignant tumors are invasive and not encapsulated. Because they are not enclosed by normal tissue, cells are able to break away from the tumor and spread to other areas of the body. Cells that leave the tumor and spread throughout the body, forming new tumors at distant sites, are called **metastases** (figure 7.13). Cancer is perhaps the most devastating and deadly disease. Of the children born in 1985, one-third will contract cancer at some time during their lives; one-fourth of the male children and one-third of the female children will someday die of cancer. Most of us have had family or friends affected by the disease. In 2004, 563,700 Americans died of cancer.

Not surprisingly, researchers are expending a great deal of effort to learn the cause of this disease. Scientists have made considerable progress in the last 30 years using molecular biological techniques, and the rough outlines of understanding are now emerging. We now know that cancer is a gene disorder of somatic tissue, in which damaged genes fail to properly control cell growth and division. The cell division cycle is regulated by a sophisticated group of proteins.

Cancer results from the damage of these genes encoding these proteins. Damage to DNA, such as damage to these genes, is called **mutation.**

Most cancers are the direct result of mutations in growth-regulating genes. There are two general classes of genes that are usually involved in cancer: proto-oncogenes and tumor-suppressor genes. Genes known as **proto-oncogenes** encode proteins that stimulate cell division. Mutations to these genes can cause cells to divide excessively. Mutated proto-oncogenes become cancer-causing genes called **oncogenes.**

The second class of cancer-causing genes are called **tumor-suppressor genes.** Cell division is normally turned off in healthy cells by proteins encoded by tumor-suppressor genes. Mutations to these genes essentially "release the brakes," allowing the cell containing the mutated gene to divide uncontrolled.

Cancer can be caused by chemicals or environmental factors such as UV rays that damage DNA, or in some instances by viruses that circumvent the cell's normal growth and division controls. Whatever the immediate cause, however, all cancers are characterized by unrestrained cell growth and division. The cell cycle never stops in a cancerous line of cells. Cancer cells are virtually immortal—until the body in which they reside dies.

7.6   Cancer is unrestrained cell growth and division caused by damage to genes regulating the cell division cycle.

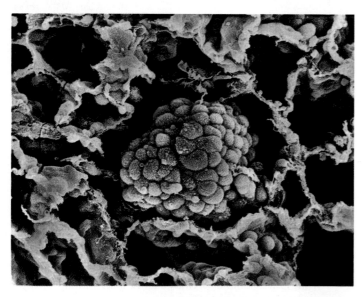

Figure 7.12   Lung cancer cells (×300).
These cells are from a tumor located in the alveolus (air sac) of a human lung.

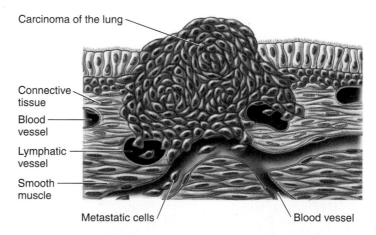

Carcinoma of the lung

Connective tissue

Blood vessel

Lymphatic vessel

Smooth muscle

Metastatic cells

Blood vessel

Figure 7.13   Portrait of a cancer.
This ball of cells is a carcinoma (cancer tumor) developing from epithelial cells that line the interior surface of a human lung. As the mass of cells grows, it invades surrounding tissues, eventually penetrating lymphatic and blood vessels, both of which are plentiful within the lung. These vessels carry metastatic cancer cells throughout the body, where they lodge and grow, forming new masses of cancerous tissue.

7.7 Cancer and Control of the Cell Cycle

As already mentioned, cancer results from damaged genes failing to control cell division. Researchers have identified several of these genes but one particular gene seems to be a key regulator of the cell cycle. Working independently, scientists researching such diverse fields as genetics, molecular biology, cell biology, and cancer have repeatedly identified what has proven to be the same gene! Officially dubbed *p53* (researchers italicize the gene symbol to differentiate it from the protein) and popularly referred to as the "guardian angel gene," this gene plays a key role in the G_1 checkpoint of cell division. The gene's product, the p53 protein, monitors the integrity of DNA, checking that it has been successfully replicated and is undamaged. If the p53 protein detects damaged DNA, it halts cell division and stimulates the activity of special enzymes to repair the damage. Once the DNA has been repaired, p53 allows cell division to continue. In cases where the DNA cannot be repaired, p53 then directs the cell to kill itself, activating an apoptosis (cell suicide) program.

By halting division in damaged cells, *p53* prevents the development of many mutated cells, and it is therefore considered a tumor-suppressor gene (even though its activities are not limited to cancer prevention). Scientists have found that *p53* is entirely absent or damaged beyond use in the majority of cancerous cells they have examined! It is precisely because *p53* is nonfunctional that these cancer cells are able to repeatedly undergo cell division without being halted at the G_1 checkpoint (figure 7.14). To test this, scientists administered healthy p53 protein to rapidly dividing cancer cells in a petri dish: the cells soon ceased dividing and died. Scientists have further reported that cigarette smoke causes mutations in the *p53* gene. This study reinforced the strong link between smoking and cancer and is described in chapter 20.

7.7 Mutations disabling key elements of the G_1 checkpoint are associated with many cancers.

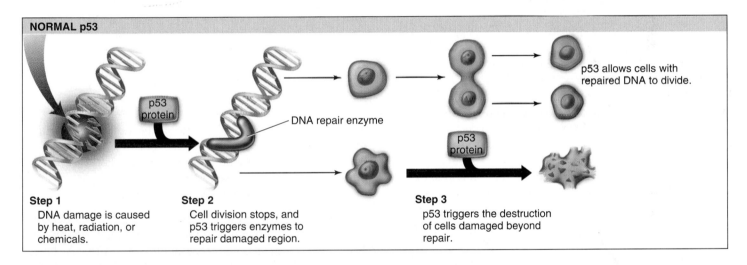

NORMAL p53

Step 1
DNA damage is caused by heat, radiation, or chemicals.

Step 2
Cell division stops, and p53 triggers enzymes to repair damaged region.

DNA repair enzyme

Step 3
p53 triggers the destruction of cells damaged beyond repair.

p53 allows cells with repaired DNA to divide.

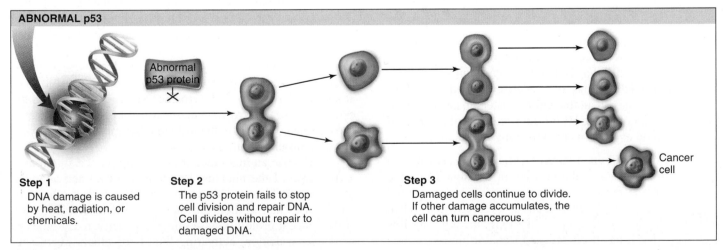

ABNORMAL p53

Abnormal p53 protein

Step 1
DNA damage is caused by heat, radiation, or chemicals.

Step 2
The p53 protein fails to stop cell division and repair DNA. Cell divides without repair to damaged DNA.

Step 3
Damaged cells continue to divide. If other damage accumulates, the cell can turn cancerous.

Cancer cell

Figure 7.14 Cell division and p53 protein.
Normal p53 protein monitors DNA, destroying cells with irreparable damage to their DNA. Abnormal p53 protein fails to stop cell division and repair DNA. As damaged cells proliferate, cancer develops.

7.8 Curing Cancer

Potential cancer therapies are being developed on many fronts (figure 7.15). Some act to prevent the start of cancer within cells. Others act outside cancer cells, preventing tumors from growing and spreading.

Preventing the Start of Cancer

Many promising cancer therapies act within potential cancer cells, focusing on different stages of the cell's "Shall I divide?" decision-making process.

1. Receiving the Signal to Divide. The first step in the decision process is receiving a "divide" signal, usually a small protein called a growth factor released from a neighboring cell. The growth factor is received by a protein receptor on the cell surface. Like banging on a door, its arrival signals that it's time to divide. Mutations that increase the number of receptors on the cell surface amplify the division signal and so lead to cancer. Over 20% of breast cancer tumors prove to overproduce a protein called HER2 associated with the receptor for epidermal growth factor (EGF).

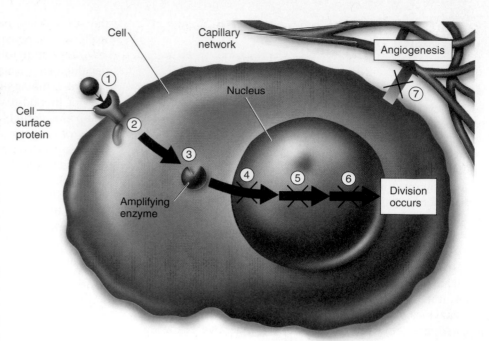

Figure 7.15 New molecular therapies for cancer target seven different stages in the cancer process.
(*1*) On the cell surface, a growth factor signals the cell to divide. (*2*) Just inside the cell, a protein relay switch passes on the divide signal. (*3*) In the cytoplasm, enzymes amplify the signal. In the nucleus, (*4*) a "brake" preventing DNA replication is released, (*5*) proteins check that the replicated DNA is not damaged, and (*6*) other proteins rebuild chromosome tips so DNA can replicate. (*7*) The new tumor promotes angiogenesis, the formation of new blood vessels that promote growth.

Therapies directed at this stage of the decision process utilize the human immune system to attack cancer cells. Special protein molecules called *monoclonal antibodies,* created by genetic engineering, are the therapeutic agents. These monoclonal antibodies are designed to seek out and stick to HER2. Like waving a red flag, the presence of the monoclonal antibody calls down attack by the immune system on the HER2 cell. Because breast cancer cells overproduce HER2, they are killed preferentially. The biotechnology research company Genentech's recently approved monoclonal antibody, called herceptin, has given promising results in clinical tests.

Up to 70% of colon, prostate, lung, and head/neck cancers have excess copies of a related receptor, epidermal growth factor 1 (HER1). The monoclonal antibody C225, directed against HER1, has succeeded in shrinking 22% of advanced, previously incurable colon cancers in early clinical trials. Apparently blocking HER1 interferes with the ability of tumor cells to recover from chemotherapy or radiation.

2. Passing the Signal via a Relay Switch. The second step in the decision process is the passage of the signal into the cell's interior, the cytoplasm. This is carried out in normal cells by a protein called Ras that acts as a relay switch. When growth factor binds to a receptor like EGF, the adjacent Ras protein acts like it has been "goosed," contorting into a new shape. This new shape is chemically active, and initiates a chain of reactions that passes the "divide" signal inward toward the nucleus. Mutated forms of the Ras protein behave like a relay switch stuck in the "ON" position, continually instructing the cell to divide when it should not. Thirty percent of all cancers have a mutant form of Ras. So far, no effective therapies have been developed targeting this step.

3. Amplifying the Signal. The third step in the decision process is the amplification of the signal within the cytoplasm. Just as a TV signal needs to be amplified in order to be received at a distance, so a "divide" signal must be amplified if it is to reach the nucleus at the interior of the cell, a very long journey at a molecular scale. To get a signal all the way into the nucleus, the cell employs a sort of pony express. The "ponies" in this case are enzymes called tyrosine kinases. These enzymes add phosphate groups to proteins, but only at a particular amino acid, tyrosine. No other enzymes in the cell do this, so the tyrosine kinases form an elite core of signal carriers not confused by the myriad other molecular activities going on around them.

Cells use an ingenious trick to amplify the signal as it moves toward the nucleus. Ras, when "ON," activates the initial protein kinase. This protein kinase activates other protein kinases that in their turn activate still others. The trick is that once a protein kinase enzyme is activated, it goes to work like a demon, activating hoards of others every second! And each and every one it activates behaves the same way too, activating still more, in a cascade of ever-widening effect. At each stage of the relay, the signal is amplified a thousandfold.

Mutations stimulating any of the protein kinases can dangerously increase the already amplified signal and lead to cancer. Some 15 of the cell's 32 internal tyrosine kinases have been implicated in cancer. Five percent of all cancers, for example, have a mutant hyperactive form of the protein kinase Src. The trouble begins when a mutation causes one of the tyrosine kinases to become locked into the "ON" position, sort of like a stuck doorbell that keeps ringing and ringing.

To cure the cancer, you have to find a way to shut the bell off. Each of the signal carriers presents a different problem, as you must quiet it without knocking out all the other signal pathways the cell needs. The cancer therapy drug Gleevec, a monoclonal antibody, has just the right shape to fit into a groove on the surface of the tyrosine kinase called "abl." Mutations locking abl "ON" are responsible for chronic myelogenous leukemia, a lethal form of white blood cell cancer. Gleevec totally disables abl. In clinical trials, blood counts revert to normal in more than 90% of cases.

4. Releasing the Brake. The fourth step in the decision process is the removal of the "brake" the cell uses to restrain cell division. In healthy cells this brake, a tumor-suppressor protein called Rb, blocks the activity of a protein called E2F. When free, E2F enables the cell to copy its DNA. Normal cell division is triggered to begin when Rb is inhibited, unleashing E2F. Mutations that destroy Rb release E2F from its control completely, leading to ceaseless cell division. Forty percent of all cancers have a defective form of Rb.

Therapies directed at this stage of the decision process are only now being attempted. They focus on drugs able to inhibit E2F, which should halt the growth of tumors arising from inactive Rb. Experiments in mice in which the E2F genes have been destroyed provide a model system to study such drugs, which are being actively investigated.

5. Checking That Everything Is Ready. The fifth step in the decision process is the mechanism used by the cell to ensure that its DNA is undamaged and ready to divide. This job is carried out in healthy cells by the tumor-suppressor protein p53, which inspects the integrity of the DNA. When it detects damaged or foreign DNA, p53 stops cell division and activates the cell's DNA repair systems. If the damage doesn't get repaired in a reasonable time, p53 pulls the plug, triggering events that kill the cell. In this way, mutations such as those that cause cancer are either repaired or the cells containing them eliminated. If p53 is itself destroyed by mutation, future damage accumulates unrepaired. Among this damage are mutations that lead to cancer. Fifty percent of all cancers have a disabled p53. Fully 70% to 80% of lung cancers have a mutant inactive p53—the chemical benzo[a]pyrene in cigarette smoke is a potent mutagen of p53.

A promising new therapy using adenovirus (responsible for mild colds) is being targeted at cancers with a mutant p53. To grow in a host cell, adenovirus must use the product of its gene *E1B* to block the host cell's p53, thereby enabling replication of the adenovirus DNA. This means that while mutant adenovirus without *E1B* cannot grow in healthy cells, the mutants should be able to grow in, and destroy, cancer cells with defective p53. When human colon and lung cancer cells are introduced into mice lacking an immune system and allowed to produce substantial tumors, 60% of the tumors simply disappear when treated with *E1B*-deficient adenovirus, and do not reappear later. Initial clinical trials have been less encouraging, as many people possess antibodies to adenovirus, so the adenovirus is destroyed by their immune systems before it can invade the cancerous cells.

6. Stepping on the Gas. Cell division starts with replication of the DNA. In healthy cells, another tumor suppressor "keeps the gas tank nearly empty" for the DNA replication process by inhibiting production of an enzyme called telomerase. Without this enzyme, a cell's chromosomes lose material from their tips, called telomeres. Every time a chromosome is copied, more tip material is lost. After some 30 divisions, so much is lost that copying is no longer possible. Cells in the tissues of an adult human have typically undergone 25 or more divisions. Cancer can't get very far with only the five remaining cell divisions, so inhibiting telomerase is a very effective natural break on the cancer process. It is thought that almost all cancers involve a mutation that destroys the telomerase inhibitor, releasing this break and making cancer possible. It should be possible to block cancer by reapplying this inhibition. Cancer therapies that inhibit telomerase are just beginning clinical trials.

Preventing the Spread of Cancer

7. Stopping Tumor Growth. Once a cell begins cancerous growth, it forms an expanding tumor. As the tumor grows ever-larger, it requires an increasing supply of food and nutrients, obtained from the body's blood supply. To facilitate this necessary grocery shopping, tumors leak out substances into the surrounding tissues that encourage the formation of small blood vessels, a process called angiogenesis. Chemicals that inhibit this process are called *angiogenesis inhibitors.* Two such natural angiogenesis inhibitors, angiostatin and endostatin, caused tumors to regress to microscopic size in mice, but initial human trials were disappointing.

Laboratory drugs are more promising. A monoclonal antibody drug called Avastin, targeted against a blood vessel growth promoting substance called vascular endothelial growth factor (VEGF), destroys the ability of VEGF to carry out its blood-vessel-forming job. Given to hundreds of advanced colon cancer patients in 2003 as part of a large clinical trial, Avastin improved colon cancer patients' chance of survival by 50% over chemotherapy.

7.8 Understanding of how mutations produce cancer has progressed to the point where promising potential therapies are being clinically tested.

7.9 Discovery of Meiosis

Only a few years after Walther Fleming's discovery of chromosomes in 1882, Belgian cytologist Pierre-Joseph van Beneden was surprised to find different numbers of chromosomes in different types of cells in the roundworm *Ascaris*. Specifically, he observed that the **gametes** (eggs and sperm) each contained two chromosomes, whereas the *somatic* (nonreproductive) cells of embryos and mature individuals each contained four.

Fertilization

From his observations, van Beneden proposed in 1887 that an egg and a sperm, each containing half the complement of chromosomes found in other cells, fuse to produce a single cell called a **zygote.** The zygote, like all of the somatic cells ultimately derived from it, contains two copies of each chromosome. The fusion of gametes to form a new cell is called **fertilization** or **syngamy.**

Meiosis

It was clear even to early investigators that gamete formation must involve some mechanism that reduces the number of chromosomes to half the number found in other cells. If it did not, the chromosome number would double with each fertilization, and after only a few generations, the number of chromosomes in each cell would become impossibly large. For example, in just 10 generations, the 46 chromosomes present in human cells would increase to over 47,000 (46×2^{10}) chromosomes.

The number of chromosomes does not explode in this way because of a special reduction division that occurs during gamete formation, producing cells with half the normal number of chromosomes. The subsequent fusion of two of these cells ensures a consistent chromosome number from one generation to the next. This reduction division process, known as *meiosis,* is the subject of this section.

The Sexual Life Cycle

Meiosis and fertilization together constitute a cycle of reproduction. Two sets of chromosomes are present in the somatic cells of adult individuals, making them **diploid** cells (Greek, *di,* two), but only one set is present in the gametes, which are thus **haploid** (Greek, *haploos,* one). Reproduction that involves this alternation of meiosis and fertilization is called **sexual reproduction** (figure 7.16). Some organisms however, reproduce by mitotic division and don't involve the fusion of gametes. Reproduction in these organisms is referred to as **asexual reproduction.** The dividing prokaryote in figure 7.1*a* is undergoing asexual reproduction. Some organisms are able to reproduce both asexually and sexually (figure 7.17).

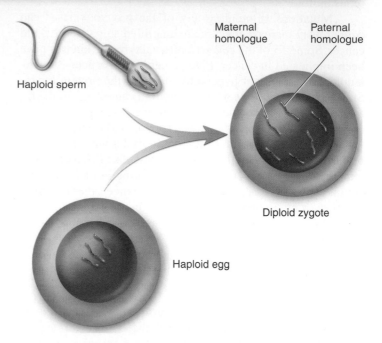

Figure 7.16 Diploid cells carry chromosomes from two parents.
A diploid cell contains two versions of each chromosome, a maternal homologue contributed by the haploid egg of the mother, and a paternal homologue contributed by the haploid sperm of the father.

Figure 7.17 Sexual and asexual reproduction.
Not all organisms reproduce exclusively asexually or sexually; some do both. The strawberry reproduces both asexually (runners) and sexually (flowers).

7.9 Meiosis is a process of cell division in which the number of chromosomes in certain cells is halved during gamete formation.

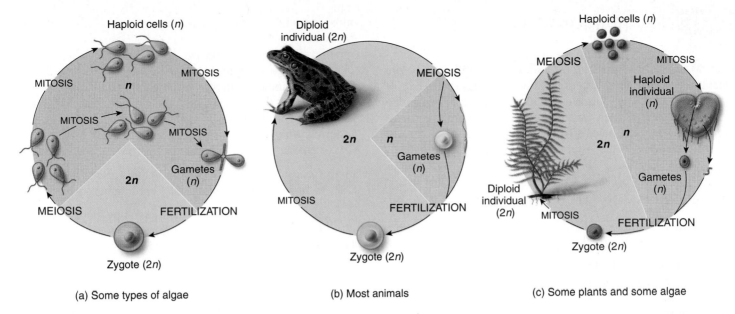

Figure 7.18 Three types of sexual life cycles.
In sexual reproduction, haploid cells or organisms alternate with diploid cells or organisms.

(a) Some types of algae

(b) Most animals

(c) Some plants and some algae

7.10 The Sexual Life Cycle

Somatic Tissues

The life cycles of all sexually reproducing organisms follow the same basic pattern of alternation between the diploid and haploid chromosome numbers (figure 7.18). After fertilization in most animals, the resulting zygote begins to divide by mitosis. This single diploid cell eventually gives rise to all of the cells in the adult. These cells are called **somatic** cells, from the Latin word for "body." Except when rare accidents occur, or in special variation-creating situations such as occur in the immune system, every one of the adult's somatic cells is genetically identical to the zygote's.

In unicellular eukaryotic organisms, including most protists, individual cells function as gametes, fusing with other gamete cells. The zygote may undergo mitosis, or it may divide immediately by meiosis to give rise to haploid individuals. In plants, the haploid cells that meiosis produces divide by mitosis, forming a multicellular haploid phase. Cells of this haploid phase eventually differentiate into eggs or sperm.

Germ-Line Tissues

In animals, the cells that will eventually undergo meiosis to produce gametes are set aside from somatic cells early in the course of development. These cells are often referred to as **germ-line** cells. Both the somatic cells and the gamete-producing germ-line cells are diploid, but while somatic cells undergo mitosis to form genetically identical, diploid daughter cells, germ-line cells undergo meiosis, producing haploid gametes (figure 7.19).

> **7.10 In the sexual life cycle, there is an alternation of diploid and haploid generations.**

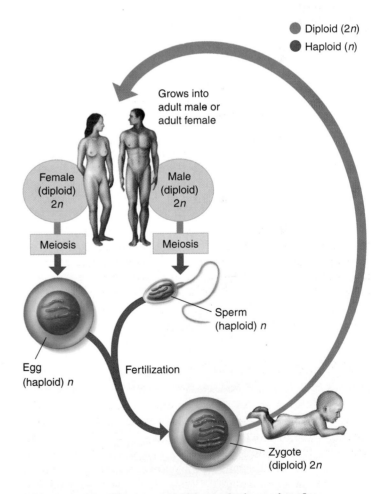

Figure 7.19 The sexual life cycle in animals.
In animals, the completion of meiosis is followed soon by fertilization. Thus, the vast majority of the life cycle is spent in the diploid stage. In this text, n stands for haploid and $2n$ stands for diploid. Germ-line cells are set aside early in development and undergo meiosis to form haploid gametes (eggs or sperm). The rest of the body cells are called somatic cells.

The Stages of Meiosis

Just as in mitosis, the chromosomes have replicated before meiosis begins, during a period called interphase. The first of the two divisions of meiosis, called **meiosis I,** serves to separate the two versions of each chromosome (the homologus chromosomes); the second, **meiosis II,** serves to separate the two replicas of each version (the sister chromatids). Thus when meiosis is complete, what started out as one diploid cell ends up as four haploid cells. Because there was one replication of DNA but *two* cell divisions, the process halves the number of chromosomes.

Meiosis I

Meiosis I is traditionally divided into four stages:

1. **Prophase I.** The two versions of each chromosome pair up and exchange segments.
2. **Metaphase I.** The chromosomes align on a central plane.
3. **Anaphase I.** One version of each chromosome moves to a pole of the cell, and the other version moves to the opposite pole.
4. **Telophase I.** Individual chromosomes gather together at each of the two poles.

In *prophase I,* individual chromosomes first become visible, as viewed with a light microscope, as their DNA coils more and more tightly. Because the chromosomes (DNA) have replicated before the onset of meiosis, each of the threadlike chromosomes actually consists of two sister chromatids associated along their lengths (a process called *sister chromatid cohesion*) and joined at their centromeres. The two homologous chromosomes thus line up side by side, and **crossing over** is initiated, in which DNA is exchanged between the two nonsister chromatids of homologous chromosomes (figure 7.20). Two elements hold the homologous chromosomes together:

1. Cohesion between sister chromatids and
2. Crossovers between nonsister chromatids (homologues).

Late in prophase, the nuclear envelope disperses.

In *metaphase I,* the spindle apparatus forms, but because homologues are held close together by crossovers, spindle fibers can attach to only the outward-facing kinetochore of each centromere. For each pair of homologues, the orientation on the spindle axis is random; which homologue is oriented toward which pole is a matter of chance. Like shuffling a deck of cards, many combinations are possible—in fact, 2 raised to a power equal to the number of chromosome pairs. In a hypothetical cell that has three chromosome pairs, there are eight possible orientations (2^3). Each orientation results in gametes with different combinations of parental chromosomes. This process is called **independent assortment** (figure 7.21).

In *anaphase I,* the spindle attachment is complete, and homologues are pulled apart and move toward opposite poles. Sister chromatids are not separated at this stage. Because the orientation along the spindle equator is random, the chromosome that a pole receives from each pair of homologues is also random with respect to all chromosome pairs. At the end of anaphase I, each pole has half as many chromosomes as were present in the cell when meiosis began. Remember that the chromosomes replicated and thus contained two sister chromatids before the start of meiosis. The function of the meiotic stages up to this point has not been to reduce the number of chromosomes but to allow for the exchange of genetic material in crossing over.

In *telophase I,* the chromosomes gather at their respective poles to form two chromosome clusters. After an interval of variable length, meiosis II occurs, in which the number of chromosomes is reduced (figure 7.22).

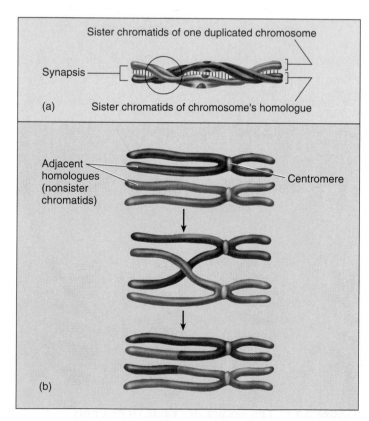

Figure 7.20 Crossing over.
In crossing over, the two homologues of each chromosome exchange portions. (*a*) The *circle* highlights the complex events that occur during close pairing of the two homologues, a process called synapsis. (*b*) During the crossing over process, nonsister chromatids that are next to each other exchange chromosome arms.

Figure 7.21 Independent assortment.
Independent assortment occurs because the orientation of chromosomes (*blue* in this micrograph) on the metaphase plate is random. Each of the many possible orientations results in gametes with different combinations of parental chromosomes. The metaphase I cell shown above is from an Oregon newt (*Taricha granulosa*).

Meiosis

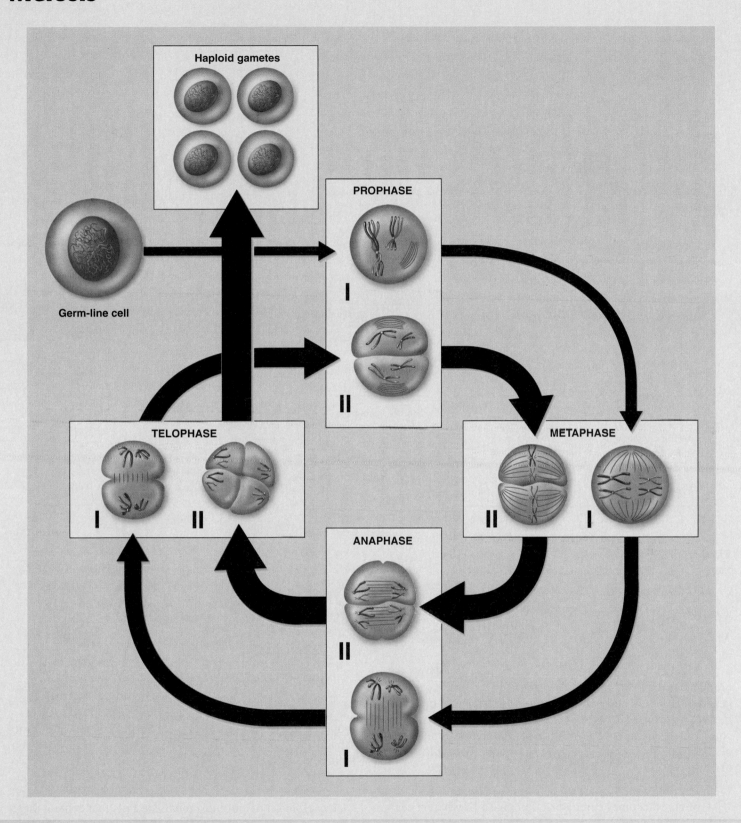

Figure 7.22 How meiosis works.

Meiosis consists of two rounds of cell division similar to mitosis, and produces four haploid cells. Meiosis I is represented in the outer circle and meiosis II is represented in the inner circle.

Meiosis I

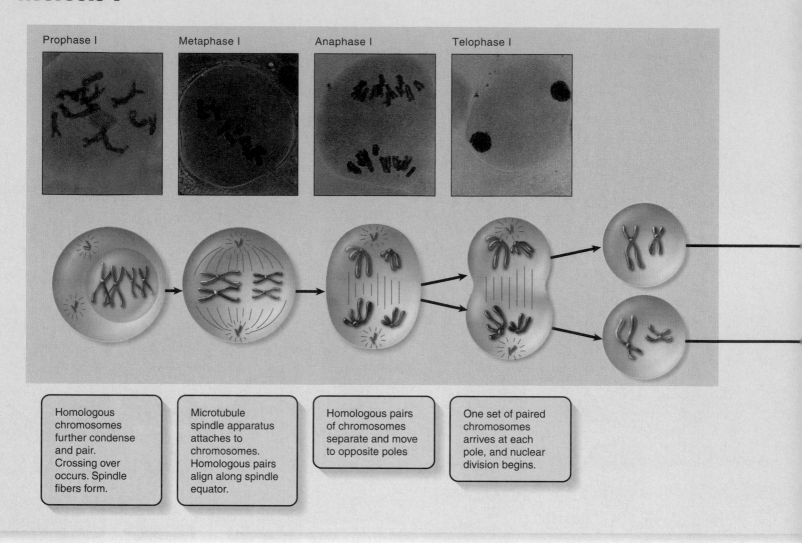

Prophase I	Metaphase I	Anaphase I	Telophase I

Homologous chromosomes further condense and pair. Crossing over occurs. Spindle fibers form.

Microtubule spindle apparatus attaches to chromosomes. Homologous pairs align along spindle equator.

Homologous pairs of chromosomes separate and move to opposite poles

One set of paired chromosomes arrives at each pole, and nuclear division begins.

Figure 7.23 Meiosis.
Meiosis results in four haploid daughter cells.

Meiosis II

After a brief interphase, in which no DNA synthesis occurs, the second meiotic division begins.

Meiosis II is simply a mitotic division involving the products of meiosis I, except that the sister chromatids are not genetically identical, as they are in mitosis, because of crossing over. At the end of anaphase I, each pole has a haploid complement of chromosomes, each of which is still composed of two sister chromatids attached at the centromere. Like meiosis I, meiosis II is divided into four stages:

1. **Prophase II.** At the two poles of the cell, the clusters of chromosomes enter a brief prophase II, where a new spindle forms.
2. **Metaphase II.** In metaphase II, spindle fibers bind to both sides of the centromeres.

3. **Anaphase II.** The spindle fibers contract, splitting the centromeres and moving the sister chromatids to opposite poles.
4. **Telophase II.** Finally, the nuclear envelope re-forms around the four sets of daughter chromosomes.

The main outcome of the four stages of meiosis II—prophase II, metaphase II, anaphase II, and telophase II—is to separate the sister chromatids. The final result of this division is four cells containing haploid sets of chromosomes. No two are alike, because of the crossing over in prophase I. The nuclei are then reorganized, and nuclear envelopes form around each haploid set of chromosomes. The cells that contain these haploid nuclei may develop directly into gametes, as they do in animals. Alternatively, they may themselves divide mitotically, as they do in plants, fungi, and many protists, eventually

Meiosis II

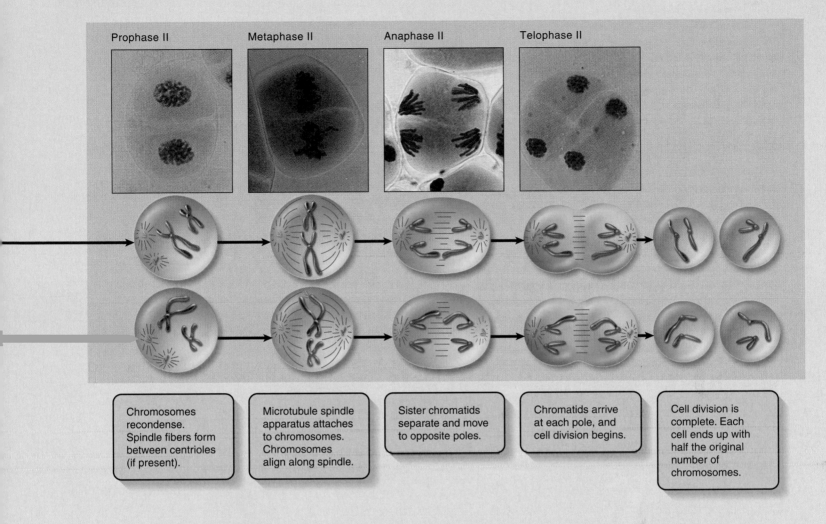

Prophase II — Chromosomes recondense. Spindle fibers form between centrioles (if present).

Metaphase II — Microtubule spindle apparatus attaches to chromosomes. Chromosomes align along spindle.

Anaphase II — Sister chromatids separate and move to opposite poles.

Telophase II — Chromatids arrive at each pole, and cell division begins.

Cell division is complete. Each cell ends up with half the original number of chromosomes.

producing greater numbers of gametes or, as in the case of some plants and insects, adult individuals of varying ploidy. Figure 7.23 summarizes the various stages of meiosis.

The Important Role of Crossing Over

If you think about it, the key to meiosis is that the replicas of each chromosome are not separated from each other in the first division. Why not? What prevents microtubules from attaching to them and pulling them to opposite poles of the cell, just as eventually happens later in the second meiotic division? The answer is the crossing over that occurred early in the first division. By exchanging segments, the two versions of each chromosome are tied together by strands of DNA, like two people sharing a belt. It is because microtubules can gain access to only one side of each replicated chromosome that they cannot pull the two replicas apart! Imagine two people dancing closely—you can tie a rope to the back of each person's belt, but you cannot tie a second rope to

their belt buckles because the two dancers are facing each other and are very close. In just the same way, microtubules cannot attach to the inside replica of chromosomes because crossing over holds chromosome replicas together like dancing partners.

7.11 During meiosis I, homologous chromosomes move toward opposite poles in anaphase I, and individual chromosomes cluster at the two poles in telophase I. At the end of meiosis II, each of the four haploid cells contains one copy of every chromosome in the set, rather than two. Because of crossing over, no two cells are the same. These haploid cells may develop directly into gametes, as in animals, or they may divide by mitosis, as in plants, fungi, and many protists.

7.12 Comparing Meiosis and Mitosis

The mechanism of meiosis varies in important details in different organisms. This is particularly true of chromosomal separation mechanisms, which differ substantially in protists and fungi from the process in plants and animals that we describe here. Although meiosis and mitosis have much in common, meiosis has two unique features: synapsis and reduction division.

Synapsis

The first of these two features happens early during the first nuclear division. Following chromosome replication, homologous chromosomes or homologues *pair all along their lengths*, sister chromatids held together by special proteins called *cohesin proteins*. While homologues are thus physically joined, *genetic exchange occurs at one or more points between them* (figure 7.24a). The process of forming these complexes of homologous chromosomes is called *synapsis*, and the exchange process between paired homologues is called *crossing over*. Chromosomes are then drawn together along the equatorial plane of the dividing cell; subsequently, homologues are pulled by microtubules toward opposite poles of the cell. When this process is complete, the cluster of chromosomes at each pole contains one of the two homologues of each chromosome. Each pole is haploid, containing half the number of chromosomes present in the original diploid cell. Sister chromatids do not separate from each other in the first nuclear division, so each homologue is still composed of two chromatids.

Reduction Division

The second unique feature of meiosis is that *the chromosome homologues do not replicate between the two nuclear divisions*, so that chromosome assortment in the second division separates sister chromatids of each chromosome into different daughter cells (figure 7.24b).

In most respects, the second meiotic division is identical to a normal mitotic division. However, because of the crossing over that occurred during the first division, the sister chromatids in meiosis II are not identical to each other.

Meiosis is a continuous process, but it is most easily studied when we divide it into arbitrary stages, just as we did for mitosis. Like mitosis, the two stages of meiosis, meiosis I and II, are subdivided into prophase, metaphase, anaphase, and telophase. In meiosis, however, prophase I is more complex than in mitosis. Figure 7.25 compares and contrasts meiosis and mitosis.

> **7.12** In meiosis, homologous chromosomes become intimately associated and do not replicate between the two nuclear divisions.

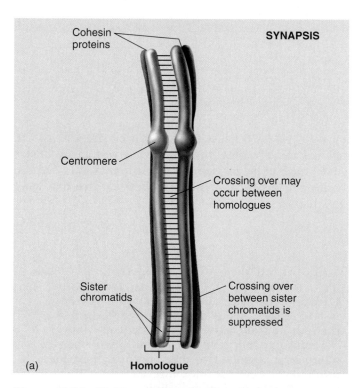

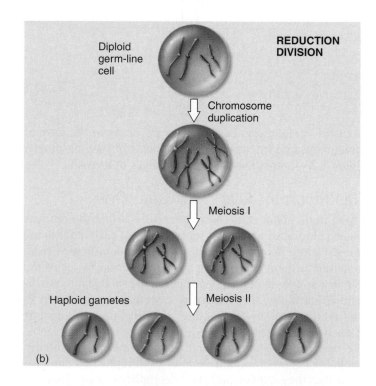

Figure 7.24 Unique features of meiosis.

(*a*) Synapsis draws homologous chromosomes together, all along their lengths, creating a situation where two homologues can physically exchange portions of arms, a process called crossing over. (*b*) Reduction division, omitting a chromosome duplication before meiosis II, produces haploid gametes, thus ensuring that the chromosome number remains the same as the parents, following fertilization.

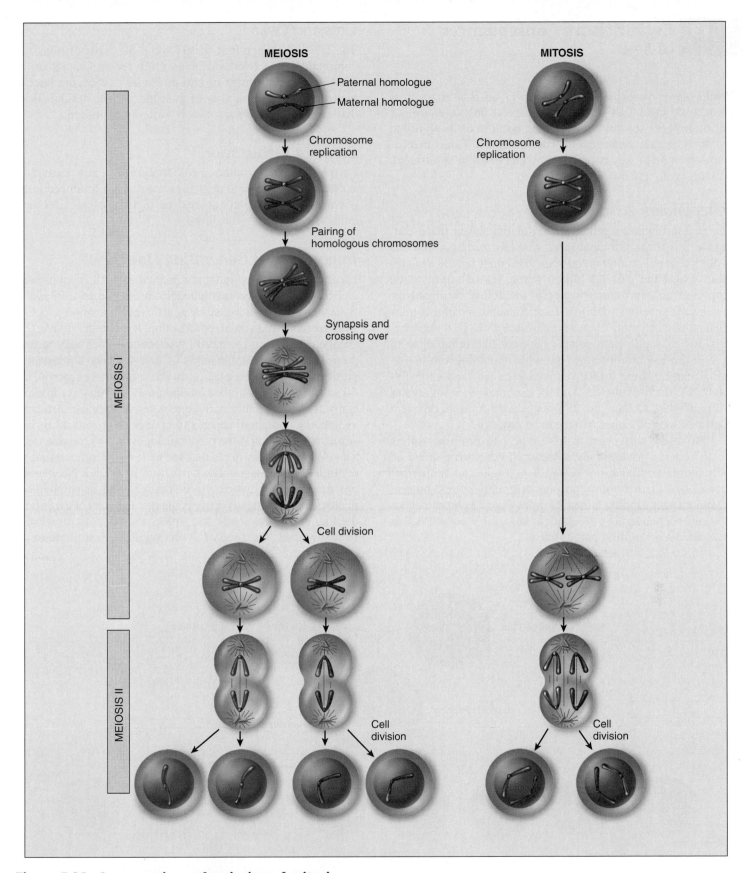

Figure 7.25 A comparison of meiosis and mitosis.

Meiosis involves two nuclear divisions with no DNA replication between them. It thus produces four daughter cells, each with half the original number of chromosomes. Crossing over occurs in prophase I of meiosis. Mitosis involves a single nuclear division after DNA replication. It thus produces two daughter cells, each containing the original number of chromosomes.

7.13 Evolutionary Consequences of Sex

While our knowledge of how sex evolved is sketchy, it is abundantly clear that sexual reproduction has an enormous impact on how species evolve today, because of its ability to rapidly generate new genetic combinations. Three mechanisms each make key contributions: independent assortment, crossing over, and random fertilization.

Independent Assortment

The reassortment of genetic material that takes place during meiosis is the principal factor that has made possible the evolution of eukaryotic organisms, in all their bewildering diversity, over the past 1.5 billion years. Sexual reproduction represents an enormous advance in the ability of organisms to generate genetic variability. To understand, recall that most organisms have more than one chromosome. In human beings, for example, each gamete receives one homologue of each of the 23 chromosomes, but which homologue of a particular chromosome it receives is determined randomly (figure 7.26). Each of the 23 pairs of chromosomes segregates independently, so there are 2^{23} (more than 8 million) different possible kinds of gametes that can be produced.

To make this point to his class, one professor offers an "A" course grade to any student who can write down all the possible combinations of heads and tails (an "either/or" choice, like that of a chromosome migrating to one pole or the other) with flipping a coin 23 times (like 23 chromosomes moving independently). No student has ever won an "A," as there are over 8 million possibilities.

Crossing Over

The DNA exchange that occurs when the arms of nonsister chromatids cross over adds even more recombination to the independent assortment of chromosomes that occurs later in meiosis. Thus, the number of possible genetic combinations that can occur among gametes is virtually unlimited.

Random Fertilization

Furthermore, because the zygote that forms a new individual is created by the fusion of two gametes, each produced independently, fertilization squares the number of possible outcomes ($2^{23} \times 2^{23} = 70$ trillion).

Importance of Generating Diversity

Paradoxically, the evolutionary process is both revolutionary and conservative. It is revolutionary in that the pace of evolutionary change is quickened by genetic recombination, much of which results from sexual reproduction. It is conservative in that change is not always favored by selection, which may instead preserve existing combinations of genes. These conservative pressures appear to be greatest in some asexually reproducing organisms that do not move around freely and that live in especially demanding habitats. In vertebrates, on the other hand, the evolutionary premium appears to have been on versatility, and sexual reproduction is the predominant mode of reproduction.

Whatever the forces that led to sexual reproduction, its evolutionary consequences have been profound. No genetic process generates diversity more quickly; and, as you will see in chapter 8, genetic diversity is the raw material of evolution, the fuel that drives it and determines its potential directions. In many cases, the pace of evolution appears to increase as

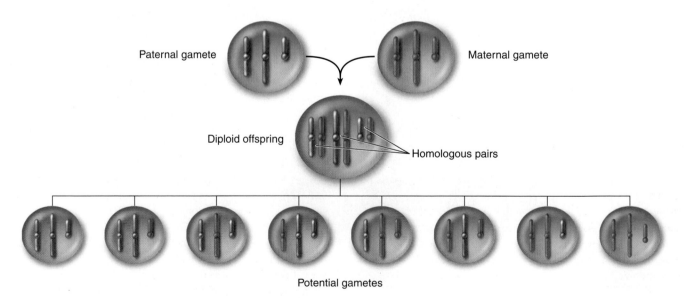

Figure 7.26 Independent assortment increases genetic variability.
Independent assortment contributes new gene combinations to the next generation because the orientation of chromosomes on the metaphase plate is random. In the cell shown here with three chromosome pairs, there are eight different gametes that can result, each with different combinations of parental chromosomes.

A Closer Look

Why Sex?

Not all reproduction is sexual. In **asexual reproduction**, an individual inherits all of its chromosomes from a single parent and is, therefore, genetically identical to its parent. Prokaryotic cells reproduce asexually, undergoing binary fission to produce two daughter cells containing the same genetic information. Most protists reproduce asexually except under conditions of stress; then they switch to sexual reproduction. Among plants and fungi, asexual reproduction is common. In animals, asexual reproduction often involves the budding off of a localized mass of cells, which grows by mitosis to form a new individual.

Even when meiosis and the production of gametes occur, there may still be reproduction without sex. The development of an adult from an unfertilized egg, called **parthenogenesis**, is a common form of reproduction in arthropods. Among bees, for example, fertilized eggs develop into diploid females, but unfertilized eggs develop into haploid males. Parthenogenesis even occurs among the vertebrates. Some lizards, fishes, and amphibians are capable of reproducing in this way; their unfertilized eggs undergo a mitotic nuclear division without cell cleavage to produce a diploid cell, which then develops into an adult.

If reproduction can occur without sex, why does sex occur at all? This question has generated considerable discussion, particularly among evolutionary biologists. Sex is of great evolutionary advantage for populations or species, which benefit from the variability generated in meiosis by random orientation of chromosomes and by crossing over. However, evolution occurs because of changes at the level of *individual* survival and reproduction, rather than at the population level, and no obvious advantage accrues to the progeny of an individual that engages in sexual reproduction. In fact, recombination is a destructive as well as a constructive process in evolution. The segregation of chromosomes during meiosis tends to disrupt advantageous combinations of genes more often than it creates new, better adapted combinations; as a result, some of the diverse progeny produced by sexual reproduction will not be as well adapted as their parents were. In fact, the more complex the adaptation of an individual organism, the less likely that recombination will improve it, and the more likely that recombination will disrupt it. It is, therefore, a puzzle to know what a well-adapted individual gains from participating in sexual reproduction, as *all* of its progeny could maintain its successful gene combinations if that individual simply reproduced asexually.

The Red Queen Hypothesis. One evolutionary advantage of sex may be that it allows populations to "store" forms of a trait that are currently bad but have promise for reuse at some time in the future. Because populations are constrained by a changing physical and biological environment, selection is constantly acting against such traits. But in sexual species, selection can never get rid of those variants sheltered by more dominant forms of the trait. The evolution of most sexual species, most of the time, thus manages to keep pace with ever-changing physical and biological constraints. This "treadmill evolution" is sometimes called the "Red Queen hypothesis," after the Queen of Hearts in Lewis Carroll's Through the Looking Glass, who tells Alice, "Now, here, you see, it takes all the running you can do, to keep in the same place."

The DNA Repair Hypothesis. Several geneticists have suggested that sex occurs because only a diploid cell can effectively repair certain kinds of chromosome damage, particularly double-strand breaks in DNA. Both radiation and chemical events within cells can induce such breaks. As organisms became larger and longer-lived, it must have become increasingly important for them to be able to repair such damage. The synaptonemal complex, which in early stages of meiosis precisely aligns pairs of homologous chromosomes, may well have evolved originally as a mechanism for repairing double-strand damage to DNA, using the undamaged homologous chromosome as a template to repair the damaged chromosome. A transient diploid phase would have provided an opportunity for such repair. In yeast, mutations that inactivate the repair system for double-strand breaks of the chromosomes also prevent crossing over, suggesting a common mechanism for both synapsis and repair processes.

Muller's Ratchet. The geneticist Herman Muller pointed out in 1965 that asexual populations incorporate a kind of mutational ratchet mechanism—once harmful mutations arise, asexual populations have no way of eliminating them, and they accumulate over time, like turning a ratchet. Sexual populations, on the other hand, can employ recombination to generate individuals carrying fewer mutations, which selection can then favor. Sex may just be a way to keep the mutational load down.

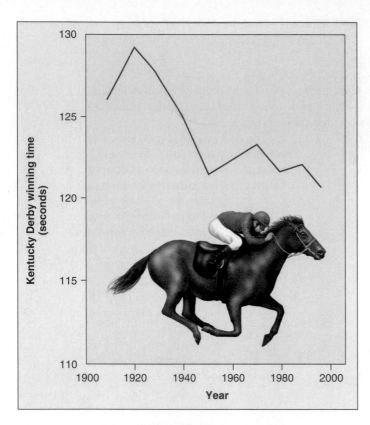

Figure 7.27 Selection for increased speed in racehorses is no longer effective.
Kentucky Derby winning times have not improved significantly since 1950.

the level of genetic diversity increases. Programs for selecting larger stature in domesticated animals such as cattle and sheep, for example, proceed rapidly when new genetic combinations arise by crossing different varieties.

Breeding programs can only accomplish so much, however. There are limits to what can be accomplished, because genes often affect more than one aspect of an individual, setting limits on how much a character can be altered. For example, selecting for large clutch size (more eggs) in barnyard chickens eventually leads to eggs with thinner shells that break more easily. For this reason we do not have chickens that lay twice as many eggs as the best layers do now, or gigantic cattle that yield twice as much meat, or corn with an ear at the base of every leaf instead of just at the base of a few leaves.

Breeding programs are also limited by the amount of genetic variability available. Over 80% of the gene pool of thoroughbred horses racing today goes back to 31 known ancestors from the late eighteenth century. Despite intense selection of thoroughbreds by horse breeders for speed, their performance times have not improved for the last 50 years (figure 7.27). The years of intense selection have removed genetic variation from the population at a rate greater than it could be replenished by mutation. Now no genetic variation remains, and rapid evolutionary change is not possible.

> **7.13 Sexual reproduction increases genetic variability through independent assortment in metaphase I of meiosis, crossing over in prophase I of meiosis, and random fertilization.**

Exploring Current Issues

Additional Resources

Go to your campus library or look online to find the following articles, which further develop some of the concepts found in this chapter.

Bunk, S. (2003). Cancer's other conduit. *The Scientist,* 17(17), 33.

Gibbs, W. W. (2003). Untangling the roots of cancer. *Scientific American,* 289(1), 56.

Hunt, P. A. and T. J. Hassold. (2002). Sex matters in meiosis. *Science,* 296(5576), 2181.

Lewis, R. (2003). Breast cancer: the big picture emerges. *The Scientist,* 17(3), 24.

National Institutes of Health. (2003). Cancers. *Retrieved from www.health.nih.gov/search.asp/4.*

Biology and Society Lecture: **Progress in Curing Cancer**

Cancer is a cell growth defect, a breakdown in the cell's control of proliferation. Progress in curing cancer is focused on the elements of this breakdown. Two sorts of damage to genes must occur to produce cancer. A crude analogy can be made to the accelerators and brakes in a car. Chemical signals called growth factors pass a "divide" signal to genes in the nucleus—these signals are the accelerators. Any change to this information pathway will promote cancer. Cells also have three safeguards against uncontrolled cell division—these are the brakes. All of these safeguards are controlled by genes, and the inactivation of these genes promotes cancer. To cure cancer, the accelerators and the brakes must be made to function properly.

Find this lecture, delivered by the author to his class at Washington University, online at www.mhhe.com/tlwessentials/exp7.

Summary

Cell Division

7.1 Prokaryotes Have a Simple Cell Cycle

* Prokaryotic cells divide in a two-step process: DNA replication and binary fission, producing two daughter cells that are genetically identical to the parent cell (**figure 7.1**).

7.2 Eukaryotes Have a Complex Cell Cycle

* Eukaryotes have a more complex cell cycle. The replicated DNA is packaged into chromosomes that are distributed into daughter cells (**figure 7.2**).

7.3 Chromosomes

- Eukaryotic DNA is organized into chromosomes, and two chromosomes that carry copies of the same genes are called homologous chromosomes (**figures 7.3 and 7.4**).

- DNA in the nucleus of eukaryotic cells is packaged into compact structures called chromosomes. The DNA wraps around histone proteins, and the DNA-histone complex is tightly coiled, forming a compact chromosome (**figures 7.5 and 7.6**).

7.4 Cell Division

- When a eukaryotic cell begins to divide, it enters a stage in the cell cycle called interphase. During interphase, the DNA replicates and condenses into chromosomes. The chromosomes stay attached to their replicated partners and are called sister chromatids.

- Interphase is followed by mitosis, which consists of four phases: prophase, metaphase, anaphase, and telophase (**figure 7.7**).

- Prophase signals the beginning of mitosis. The chromosomes condense, the nucleolus and nuclear membrane disappear, and centrioles when present begin forming the spindle. Microtubules also form from the kinetochores, which are located at the centromere area of the chromosome.

- Metaphase involves the alignment of sister chromatids along the equatorial plane. Microtubules connect the kinetochores of sister chromatids to each of the poles.

- During anaphase, the microtubules shorten, pulling the sister chromatids apart and toward the pole, such that each pole receives one copy of each chromosome.

- Telophase signals the completion of nuclear division. The microtubule spindle is dismantled, the nuclear membrane and nucleolus re-forms.

- Following mitosis, the cell separates into two daughter cells in a process called cytokinesis (**figure 7.8**).

7.5 Controlling the Cell Cycle

- The cell cycle is controlled at three checkpoints (**figure 7.10**). At the G_1 checkpoint, before entering the S phase, the cell either initiates division or enters a period of rest called G_0 (**figure 7.11**).

- If cell division is initiated, DNA replication begins and is checked at the G_2 checkpoint. Mitosis is either initiated or delayed depending on the successful replication of DNA. The third checkpoint is toward the end of mitosis, before cytokinesis.

Cancer and the Cell Cycle

7.6 What Is Cancer?

- Cancer is a growth disorder of cells, where there is a loss of control over cell division. Mutations in proto-oncogenes and tumor-suppressor genes lead to uncontrolled growth—cancer (**figure 7.13**).

7.7 Cancer and Control of the Cell Cycle

- Mutations of the *p53* gene are found in many types of cancer cells. This gene plays a key role in the G_1 checkpoint, checking the accuracy of DNA replication (**figure 7.14**).

7.8 Curing Cancer

- Various aspects of cancer cells are targeted for possible cures. One approach is to stop the cancer before it starts, which includes turning off the signal to divide, or to stop the uncontrolled division of cells. Another approach focuses on preventing cancer cells from spreading to other parts of the body (**figure 7.15**).

Meiosis

7.9 Discovery of Meiosis

- In sexually reproducing organisms, the number of chromosomes in gametes must be halved to maintain the correct number of chromosomes in offspring (**figure 7.16**).

7.10 The Sexual Life Cycle

- Germ-line cells of an organism produce gametes that contain half the number of chromosomes found in its somatic cells (**figure 7.19**).

7.11 The Stages of Meiosis

- Meiosis involves two nuclear divisions, meiosis I and meiosis II, each containing a prophase, metaphase, anaphase and telophase.

- Prophase I is distinguished by the exchange of genetic material between homologous chromosomes, a process called crossing over (**figure 7.20**).

- During metaphase I, homologous chromosomes align along the equatorial plane. The homologous chromosomes separate during anaphase I, which differs from mitosis where sister chromatids separate in anaphase (**figure 7.22**).

- Meiosis II involves the separation of sister chromatids but because homologous pairs were separated during meiosis I, each daughter cell has only one-half the number of chromosomes (**figure 7.23**).

7.12 Comparing Meiosis and Mitosis

- Two processes that distinguish meiosis from mitosis are, first, synapsis and crossing over, which results in daughter cells that are not genetically identical, and, second, reduction division, the halving of the number of chromosomes (**figures 7.24 and 7.25**).

7.13 Evolutionary Consequences of Sex

- Sexual reproduction results in the introduction of genetic variation in future generations through independent assortment, crossing over, and random fertilization (**figure 7.26**).

Self-Test

---

1. Prokaryotes reproduce new cells by
   a. copying DNA then undergoing binary fission.
   b. sexual reproduction involving meiosis.
   c. copying DNA then undergoing mitosis.
   d. copying DNA then undergoing the M phase.

2. In eukaryotes, the genetic material is found in chromosomes
   a. and the more complex the organism, the more pairs of chromosomes it has.
   b. and a few organisms have only one chromosome.

c. and most eukaryotes have between 10 and 50 pairs of chromosomes.

d. and most eukaryotes have between 2 and 10 pairs of chromosomes.

3. In mitosis, when the duplicated chromosomes line up in the center of the cell, that stage is called
   a. prophase.
   b. metaphase.
   c. anaphase.
   d. telophase.

4. When cell division becomes unregulated, and a cluster of cells begins to grow without regard for the normal controls, that is / they are
   a. a tumor.
   b. a cancer.
   c. metastases
   d. oncogenes.

5. An egg and a sperm unite to form a new organism. To prevent the new organism from having twice as many chromosomes as its parents
   a. half of the chromosomes in the new organism quickly die off, leaving the correct number.
   b. half of the chromosomes from the egg, and half from the sperm, are ejected from the new cell.
   c. the large egg contains all the chromosomes, the tiny sperm only contributes some DNA.
   d. the germ cells went through meiosis; the egg and sperm only have half the parental chromosomes.

6. In organisms that have sexual life cycles, there is a time when there are
   a. 1*n* gametes (haploid), then next there are 2*n* zygotes (diploid).
   b. 2*n* gametes (haploid), then next there are 1*n* zygotes (diploid).
   c. 2*n* gametes (diploid), then next there are 1*n* zygotes (haploid).
   d. 1*n* gametes (diploid), then next there are 2*n* zygotes (haploid).

7. The purpose of meiosis I is to
   a. duplicate all chromosomes.
   b. randomly separate the homologous pairs, called independent assortment.
   c. separate the duplicated sister chromatids.
   d. divide the original material into four complete haploid cells.

8. The purpose of meiosis II is to
   a. duplicate all chromosomes.
   b. randomly separate the homologous pairs, called independent assortment.
   c. separate the duplicated sister chromatids.
   d. divide the original material into four complete haploid cells.

9. The purpose of mitosis is to _____, while the purpose of meiosis is to _____.
   a. make new cells, and only germ-line cells do it; make eggs or sperm, and all body cells do it
   b. make eggs or sperm, and only germ-line cells do it; make new cells, and all body cells do it
   c. make eggs or sperm, and only somatic cells do it; make new cells, and all body cells do it
   d. make new cells, and all body cells do it; make eggs or sperm, and only germ-line cells do it

10. A major consequence of sex and meiosis is that each species
   a. remains pretty much the same because the chromosomes are carefully duplicated and passed to the next generation.
   b. has a lot of genetic reassortment due to processes in meiosis II.
   c. has a lot of genetic reassortment due to processes in meiosis I.
   d. has a lot of genetic reassortment due to processes in telophase II.

Visual Understanding

1. **Figure 7.4** This shows a set of human chromosomes. At what stage of the cell cycle are such photos taken, and why?

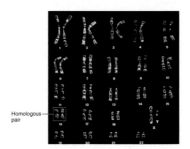

Homologous pair

2. **Figure 7.20** How is it that, in meiosis, you can end up with four "daughter cells" that are all genetically different from one another?

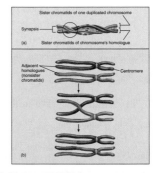

Challenge Questions

Cell Division Why does the DNA need to change periodically from a long, double-helix chromatin molecule into a tightly wound-up chromosome? What does it do at each stage that it cannot do at the other?

Cancer and the Cell Cycle Despite all we know about cancer today, some types of cancers are still increasing in frequency. Lung cancer in women is one of those. What reason(s) might there be for this increasing problem?

Meiosis You only need one set of instructions for your body to do all the jobs it needs to carry out. So why aren't organisms simply haploid all their lives?

Online Learning Center

Visit the Online Learning Center for this chapter at www.mhhe.com/tlwessentials/ch7 for quizzes, animations, interactive learning exercises, and other study tools. At the site you will also find extended answers to the end-of-chapter questions.

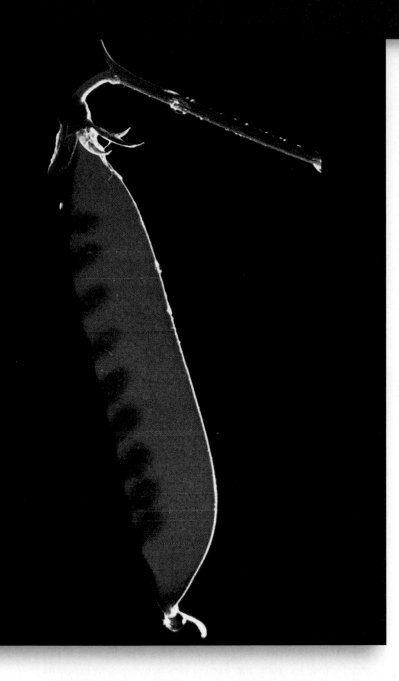

8

Foundations of Genetics

In this pea pod, you can see the shadowy outlines of seeds that will form part of the next generation of this pea plant. While the seeds appear similar to one another, the plants they produce may differ in significant ways. This is because the process of meiosis that produced the seeds contributes chromosomes from both parents, in effect "shuffling the deck of cards" so that a progeny plant will have some characteristics from one parent and some from the other. Just over a hundred years ago, Gregor Mendel first described this process, before anyone knew what genes, or chromosomes, were. We now understand the process of heredity in considerable detail, and can begin to devise ways of treating some of the disorders that arise in people when particular genes are damaged in germ-line tissue.

8.1 Mendel and the Garden Pea

When you were born, many things about you resembled your mother or father. This tendency for traits to be passed from parent to offspring is called **heredity. Traits** are alternative forms of a **character,** or heritable feature. How does heredity happen? Before DNA and chromosomes were discovered, this puzzle was one of the greatest mysteries of science. The key to understanding the puzzle of heredity was found in the garden of an Austrian monastery over a century ago by a monk named Gregor Mendel (figure 8.1). Mendel used the scientific process described in chapter 1 as a powerful way of analyzing the problem. Crossing pea plants with one another, Mendel made observations that allowed him to form a simple but powerful hypothesis that accurately predicted patterns of heredity—that is, how many offspring would be like one parent and how many like the other. When Mendel's rules, introduced in chapter 1 as the theory of heredity, became widely known, investigators all over the world set out to discover the physical mechanism responsible for them. They learned that hereditary traits are instructions carefully laid out in the DNA a child receives from each parent. Mendel's solution to the puzzle of heredity was the first step on this journey of understanding and one of the greatest intellectual accomplishments in the history of science.

Early Ideas About Heredity

Mendel was not the first person to try to understand heredity by crossing pea plants. Over 200 years earlier British farmers had performed similar crosses and obtained results similar to Mendel's. They observed that in crosses between two types— tall and short plants, say—one type would disappear in one generation, only to reappear in the next. In the 1790s, for example, the British farmer T. A. Knight crossed a variety of the garden pea that had purple flowers with one that had white flowers. All the offspring of the cross had purple flowers. If two of these offspring were crossed, however, some of *their* offspring were purple and some were white. Knight noted that the purple had a "stronger tendency" to appear than white, but he did not count the numbers of each kind of offspring.

Mendel's Experiments

Gregor Mendel was born in 1822 to peasant parents and was educated in a monastery. He became a monk and was sent to the University of Vienna to study science and mathematics. Although he aspired to become a scientist and teacher, he failed his university exams for a teaching certificate and returned to the monastery, where he spent the rest of his life, eventually becoming abbot. Upon his return, Mendel joined an informal neighborhood science club, a group of local farmers and others interested in science. Under the patronage of a local nobleman, each member set out to undertake scientific investigations, which were then discussed at meetings

Figure 8.1 Gregor Mendel.
The key to understanding the puzzle of heredity was solved by Mendel by cultivating pea plants in the garden of his monastery in Brunn, Austria.

and published in the club's own journal. Mendel undertook to repeat the classic series of crosses with pea plants done by Knight and others, but this time he intended to count the numbers of each kind of offspring in the hope that the numbers would give some hint of what was going on. Quantitative approaches to science—measuring and counting—were just becoming fashionable in Europe.

Mendel's Experimental System: The Garden Pea

Mendel chose to study the garden pea because several of its characteristics made it easy to work with:

1. Many varieties were available. Mendel selected seven pairs of lines that differed in easily distinguished traits (including the white versus purple flowers that Knight had studied 60 years earlier).
2. Mendel knew from the work of Knight and others that he could expect the infrequent version of a character to disappear in one generation and reappear in the next. He knew, in other words, that he would have something to count.
3. Pea plants are small, easy to grow, produce large numbers of offspring, and mature quickly.
4. The reproductive organs of peas are enclosed within their flowers (figure 8.2), and, left alone, the flowers do not open. They simply fertilize themselves with their own pollen (male gametes). To carry out a cross, Mendel had only to pry the petals apart, reach in with

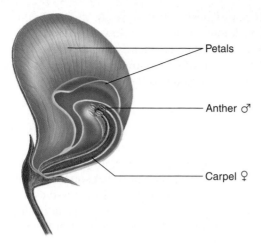

Figure 8.2 The garden pea.
Because it is easy to cultivate and because there are many distinctive varieties, the garden pea, *Pisum sativum,* was a popular choice as an experimental subject in investigations of heredity for as long as a century before Mendel's studies. Also, in a pea flower, the petals enclose the male (anther) and female (carpel) parts, (shown in this cross sectional view), ensuring that self-fertilization will take place unless the flower is disturbed.

a scissors, and snip off the male organs (anthers); he could then dust the female organs with pollen from another plant to make the cross.

Mendel's Experimental Design

Mendel's experimental design was the same as Knight's, only Mendel counted his plants. The crosses were carried out in three steps:

1. Mendel began by letting each variety self-fertilize for several generations. This ensured that each variety was **true-breeding** and contained no other varieties. The white flower variety, for example, produced only white flowers and no purple ones in each generation. Mendel called these lines the **P generation** (P for parental).
2. Mendel then conducted his experiment: he crossed two pea varieties exhibiting alternative traits, such as white versus purple flowers (figure 8.3). The offspring that resulted he called the F_1 **generation** (F_1 for "first filial" generation, from the Latin word for "son" or "daughter").
3. Finally, Mendel allowed the plants produced in the crosses of step 2 to self-fertilize, and he counted the numbers of each kind of offspring that resulted in this F_2 ("second filial") **generation.**

8.1 Mendel studied heredity by crossing true-breeding garden peas that differed in easily scored alternative traits and then allowing the offspring to self-fertilize.

Mendel's Experimental Design

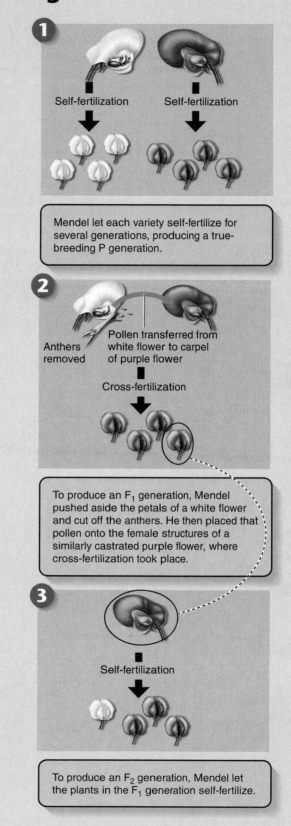

1 Mendel let each variety self-fertilize for several generations, producing a true-breeding P generation.

2 Anthers removed. Pollen transferred from white flower to carpel of purple flower. Cross-fertilization.

To produce an F_1 generation, Mendel pushed aside the petals of a white flower and cut off the anthers. He then placed that pollen onto the female structures of a similarly castrated purple flower, where cross-fertilization took place.

3 Self-fertilization

To produce an F_2 generation, Mendel let the plants in the F_1 generation self-fertilize.

Figure 8.3 How Mendel conducted his experiments.

8.2 What Mendel Observed

Mendel experimented with a variety of traits in the garden pea and repeatedly made similar observations. In all, Mendel examined seven pairs of contrasting traits. For each pair of contrasting varieties that Mendel crossed (table 8.1), he observed the same result. We will examine in detail Mendel's crosses with flower color.

The F_1 Generation

In the case of flower color, when Mendel crossed purple and white flowers, all the F_1 generation plants he observed were purple; he did not see the contrasting trait, white flowers. Mendel called the trait expressed in the F_1 plants **dominant** and the trait not expressed **recessive.** In this case, purple flower color was dominant and white flower color recessive. Mendel studied several other characters in addition to flower color, and for every pair of contrasting traits Mendel examined, one proved to be dominant and the other recessive.

The F_2 Generation

After allowing individual F_1 plants to mature and self-fertilize, Mendel collected and planted the seeds from each plant to see what the offspring in the F_2 generation would look like. Mendel found (as Knight had earlier) that some F_2 plants exhibited white flowers, the recessive trait. The recessive trait had disappeared in the F_1 generation, only to reappear in the F_2 generation. It must somehow have been present in the F_1 individuals but unexpressed!

TABLE 8.1 SEVEN CHARACTERS MENDEL STUDIED IN HIS EXPERIMENTS

Character			F_2 Generation	
Dominant Form	×	Recessive Form	Dominant:Recessive	Ratio
Purple flowers	×	White flowers	705:224	3.15:1 (3/4:1/4)
Yellow seeds	×	Green seeds	6022:2001	3.01:1 (3/4:1/4)
Round seeds	×	Wrinkled seeds	5474:1850	2.96:1 (3/4:1/4)
Green pods	×	Yellow pods	428:152	2.82:1 (3/4:1/4)
Inflated pods	×	Constricted pods	882:299	2.95:1 (3/4:1/4)
Axial flowers	×	Terminal flowers	651:207	3.14:1 (3/4:1/4)
Tall plants	×	Dwarf plants	787:277	2.84:1 (3/4:1/4)

Figure 8.4 Round versus wrinkled seeds.
One of the differences among varieties of pea plants that Mendel studied was the shape of the seed. In some varieties the seeds were round, whereas in others they were wrinkled.

At this stage Mendel instituted his radical change in experimental design. He *counted* the number of each type among the F_2 offspring. He believed the proportions of the F_2 types would provide some clue about the mechanism of heredity. In the cross between the purple-flowered F_1 plants, he counted a total of 929 F_2 individuals (see table 8.1). Of these, 705 (75.9%) had purple flowers and 224 (24.1%) had white flowers. Approximately one-fourth of the F_2 individuals exhibited the recessive form of the trait. Mendel carried out similar experiments with other traits, such as round versus wrinkled seeds (figure 8.4) and obtained the same result: three-fourths of the F_2 individuals exhibited the dominant form of the character, and one-fourth displayed the recessive form. In other words, the dominant:recessive ratio among the F_2 plants was always close to 3:1.

A Disguised 1:2:1 Ratio

Mendel let the F_2 plants self-fertilize for another generation and found that the one-fourth that were recessive were true-breeding—future generations showed nothing but the recessive trait. Thus, the white F_2 individuals described previously showed only white flowers in the F_3 generation. Among the three-fourths of the F_2 plants that had shown the dominant trait, only one-third of the individuals were true-breeding. The others showed both traits in the F_3 generation—and when Mendel counted their numbers, he found the ratio of dominant to recessive to again be 3:1! From these results Mendel concluded that the 3:1 ratio he had observed in the F_2 generation was in fact a disguised 1:2:1 ratio (figure 8.5):

1   **2**   **1**

true-breeding : not-true-breeding : true-breeding

dominant   **dominant**   **recessive**

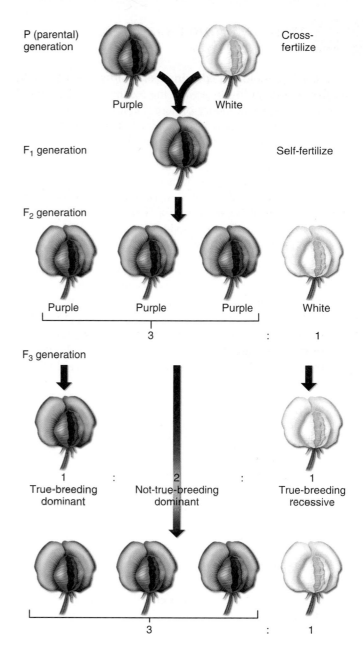

Figure 8.5 The F_2 generation is a disguised 1:2:1 ratio.
By allowing the F_2 generation to self-fertilize, Mendel found from the offspring (F_3) that the ratio of F_2 plants was one true-breeding dominant, two not-true-breeding dominant, and one true-breeding recessive.

8.2 When Mendel crossed two contrasting varieties and counted the offspring in the subsequent generations, he observed that all of the offspring in the first generation exhibited one (dominant) trait, and none exhibited the other (recessive) trait. In the following generation, 25% were true-breeding for the dominant trait, 50% were not-true-breeding and appeared dominant, and 25% were true-breeding for the recessive trait.

8.3 Mendel Proposes a Theory

To explain his results, Mendel proposed a simple set of hypotheses that would faithfully predict the results he had observed. Now called Mendel's theory of heredity, it has become one of the most famous theories in the history of science. Mendel's theory is composed of five simple hypotheses:

Hypothesis 1: *Parents do not transmit traits directly to their offspring.* Rather, they transmit information about the traits, what Mendel called merkmal (the German word for "factor"). These factors act later, in the offspring, to produce the trait. In modern terminology, we call Mendel's factors **genes.**

Hypothesis 2: *Each parent contains two copies of the factor governing each trait.* The two copies may or may not be the same. If the two copies of the factor are the same (both encoding purple or both white, for example) the individual is said to be **homozygous.** If the two copies of the factor are different (one encoding purple, the other white, for example), the individual is said to be **heterozygous.**

Hypothesis 3: *Alternative forms of a factor lead to alternative traits.* Alternative forms of a factor are called **alleles.** Mendel used lowercase letters to represent recessive alleles and uppercase letters to represent dominant ones. Thus, in the case of purple flowers, the dominant purple flower allele is represented as *P* and the recessive white flower allele is represented as *p*. In modern terms, we call the appearance of an individual, such as possessing white flowers, its **phenotype.** Appearance is determined by which alleles of the flower-color gene the plant receives from its parents, and we call those particular alleles its **genotype.** Thus a pea plant might have the phenotype "white flower" and the genotype *pp*.

Hypothesis 4: *The two alleles that an individual possesses do not affect each other,* any more than two letters in a mailbox alter each other's contents. Each allele is passed on unchanged when the individual matures and produces its own gametes (egg and sperm). At the time, Mendel did not know that his factors were carried from parent to offspring on chromosomes (figure 8.6).

Hypothesis 5: *The presence of an allele does not ensure that a trait will be expressed in the individual that carries it.* In heterozygous individuals, only the dominant allele achieves expression; the recessive allele is present but unexpressed.

These five hypotheses, taken together, constitute Mendel's model of the hereditary process. Many traits in humans exhibit dominant or recessive inheritance similar to the traits Mendel studied in peas (table 8.2).

Analyzing Mendel's Results

To analyze Mendel's results, it is important to remember that each trait is determined by the inheritance of alleles from the parents, one allele from the mother and the other from the father. These alleles, present on chromosomes, are distributed to gametes during meiosis. Each gamete receives one copy of each chromosome, and therefore one of the alleles.

Consider again Mendel's cross of purple-flowered with white-flowered plants. Like Mendel, we will assign the symbol *P* to the dominant allele, associated with the production of purple flowers, and the symbol *p* to the recessive allele,

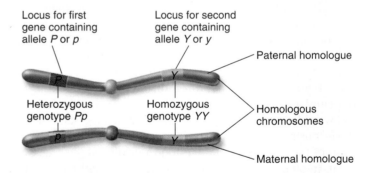

Figure 8.6   Genes on homologous chromosomes.
Mendel did not know that the factors (genes) he was studying are carried on chromosomes, homologous chromosomes carrying the same array of genes. Today biologists call alternative forms of a gene "alleles" and the location of a gene on the chromosome is called its "locus."

TABLE 8.2	SOME DOMINANT AND RECESSIVE TRAITS IN HUMANS		
Recessive Traits	**Phenotypes**	**Dominant Traits**	**Phenotypes**
Common baldness	M-shaped hairline receding with age	Mid-digital hair	Presence of hair on middle segment of fingers
Albinism	Lack of melanin pigmentation	Brachydactyly	Short fingers
Alkaptonuria	Inability to metabolize homogenistic acid	Phenylthiocarbamide (PTC) sensitivity	Ability to taste PTC as bitter
Red-green color blindness	Inability to distinguish red and green wavelengths of light	Camptodactyly	Inability to straighten the little finger
		Polydactyly	Extra fingers and toes

associated with the production of white flowers. By convention, genetic traits are usually assigned a letter symbol referring to their more common forms, in this case "*P*" for purple flower color. The dominant allele is written in uppercase, as *P;* the recessive allele (white flower color) is assigned the same symbol in lowercase, *p.*

In this system, the genotype of an individual true-breeding for the recessive white-flowered trait would be designated *pp*. In such an individual, both copies of the allele specify the white-flowered phenotype. Similarly, the genotype of a true-breeding purple-flowered individual would be designated *PP*, and a heterozygote would be designated *Pp* (dominant allele first). Using these conventions, and denoting a cross between two strains with ×, we can symbolize Mendel's original cross as *pp* × *PP*.

Punnett Squares

The possible results from a cross between a true-breeding, white-flowered plant (*pp*) and a true-breeding, purple-flowered plant (*PP*) can be visualized with a **Punnett square.** In a Punnett square, the possible gametes of one individual are listed along the horizontal side of the square, while the possible gametes of the other individual are listed along the vertical side. The genotypes of potential offspring are represented by the cells within the square (figure 8.7).

The frequency that these genotypes occur in the offspring is usually expressed by a **probability.** For example, in a cross between a homozygous white-flowered plant (*pp*) and a homozygous purple-flowered plant (*PP*), *Pp* is the only possible genotype for all individuals in the F_1 generation (figure 8.8). Because *P* is dominant to *p*, all individuals in the F_1 generation have purple flowers. When individuals from the F_1 generation are crossed, the probability of obtaining a homozygous dominant (*PP*) individual in the F_2 is 25% be-cause one-fourth of the possible genotypes are *PP*. Similarly, the probability of an individual in the F_2 generation being homozygous recessive (*pp*) is 25%. Because the heterozygous genotype (*Pp*) occurs in half of the cells within the square, the probability of obtaining a heterozygous (*Pp*) individual in the F_2 is 50%.

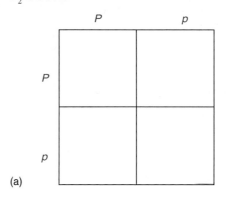

(a)

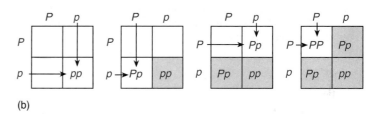

(b)

Figure 8.7   A Punnett square.

A Punnett square analysis is an easy way to determine all the possible genotypes of a particular cross. (*a*) The possible gametes for one parent are placed along the top of the square, and the possible gametes for the other parent are placed along the side of the square. (*b*) The Punnett square can be used like a mathematical table to visualize the genotypes of all potential offspring.

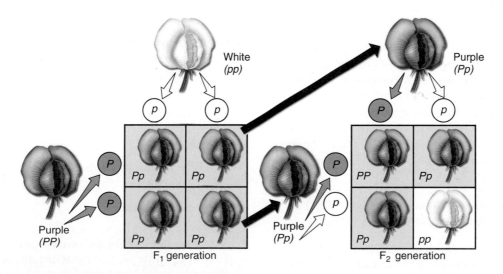

Figure 8.8   How Mendel analyzed flower color.

The only possible offspring of the first cross are *Pp* heterozygotes, purple in color. These individuals are known as the F_1 generation. When two heterozygous F_1 individuals cross, three kinds of offspring are possible: *PP* homozygotes (purple flowers); *Pp* heterozygotes (also purple flowers), which may form two ways; and *pp* homozygotes (white flowers). Among these individuals, known as the F_2 generation, the ratio of dominant phenotype to recessive phenotype is 3:1.

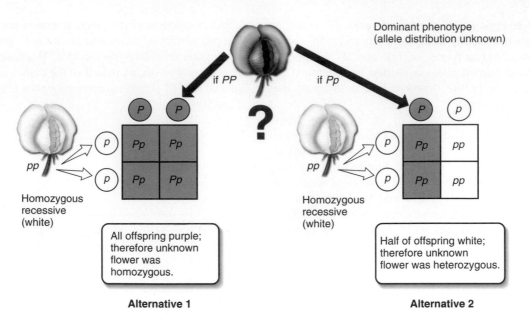

Dominant phenotype
(allele distribution unknown)

if *PP*

if *Pp*

pp

Homozygous
recessive
(white)

	P	P
p	Pp	Pp
p	Pp	Pp

?

pp

Homozygous
recessive
(white)

	P	p
p	Pp	pp
p	Pp	pp

All offspring purple;
therefore unknown
flower was
homozygous.

Half of offspring white;
therefore unknown
flower was heterozygous.

Alternative 1

Alternative 2

Figure 8.9 How Mendel used the testcross to detect heterozygotes.

To determine whether an individual exhibiting a dominant phenotype, such as purple flowers, was homozygous (*PP*) or heterozygous (*Pp*) for the dominant allele, Mendel devised the testcross. He crossed the individual in question with a known homozygous recessive (*pp*)—in this case, a plant with white flowers.

The Testcross

How did Mendel know which of the purple-flowered individuals in the F_2 generation (or the P generation) were homozygous (*PP*) and which were heterozygous (*Pp*)? It is not possible to tell simply by looking at them. For this reason, Mendel devised a simple and powerful procedure called the **testcross** to determine an individual's actual genetic composition. Consider a purple-flowered plant. It is impossible to tell whether such a plant is homozygous or heterozygous simply by looking at its phenotype. To learn its genotype, you must cross it with some other plant. What kind of cross would provide the answer? If you cross it with a homozygous dominant individual, all of the progeny will show the dominant phenotype whether the test plant is homozygous or heterozygous. It is also difficult (but not impossible) to distinguish between the two possible test plant genotypes by crossing with a heterozygous individual. However, if you cross the test plant with a homozygous recessive individual, the two possible test plant genotypes will give totally different results (figure 8.9):

Alternative 1: unknown individual homozygous (*PP*).
    PP × *pp:* all offspring have purple flowers (*Pp*).

Alternative 2: unknown individual heterozygous (*Pp*).
    Pp × *pp:* one-half of offspring have white flowers (*pp*) and one-half have purple flowers (*Pp*).

To perform his testcross, Mendel crossed heterozygous F_1 individuals back to the parent homozygous for the recessive trait. He predicted that the dominant and recessive traits would appear in a 1:1 ratio, and that is what he observed.

For each pair of alleles he investigated, Mendel observed phenotypic F_2 ratios of 3:1 (see table 8.1) and testcross ratios very close to 1:1, just as his model predicted.

Testcrosses can also be used to determine the genotype of an individual when two genes are involved. Mendel carried out many two-gene crosses, some of which we will soon discuss. He often used testcrosses to verify the genotypes of particular dominant-appearing F_2 individuals. Thus an F_2 individual showing both dominant traits (*A_ B_*) might have any of the following genotypes: *AABB, AaBB, AABb,* or *AaBb*. By crossing dominant-appearing F_2 individuals with homozygous recessive individuals (that is, *A_ B_* × *aabb*), Mendel was able to determine if either or both of the traits bred true among the progeny and so determine the genotype of the F_2 parent.

AABB	trait A breeds true	trait B breeds true
AaBB		trait B breeds true
AABb	trait A breeds true	
AaBb		

8.3 The genes that an individual has are referred to as its genotype; the outward appearance of the individual is referred to as its phenotype. The phenotype is determined by the alleles inherited from the parents. Analyses using Punnett squares determine all possible genotypes of a particular cross.

8.4 Mendel's Laws

Mendel's First Law: Segregation

Mendel's model brilliantly predicts the results of his crosses, accounting in a neat and satisfying way for the ratios he observed. Similar patterns of heredity have since been observed in countless other organisms. Traits exhibiting this pattern of heredity are called *Mendelian traits*. Because of its overwhelming importance, Mendel's theory is often referred to as Mendel's first law, or the **law of segregation.** In modern terms, Mendel's first law states that *only one allele specifying an alternative trait can be carried in a particular gamete, and gametes combine randomly in forming offspring.*

Mendel's Second Law: Independent Assortment

Mendel went on to ask if the inheritance of one factor, such as flower color, influences the inheritance of other factors, such as plant height. To investigate this question, he first established a series of true-breeding lines of peas that differed from one another with respect to two of the seven pairs of characteristics he had studied. Second, he crossed contrasting pairs of true-breeding lines. For example, homozygous individuals with round, yellow seeds crossed with individuals that are homozygous for wrinkled, green seeds produce offspring that have round, yellow seeds and are heterozygous for both of these traits. Such F_1 individuals are said to be **dihybrid** (figure 8.10).

Mendel then allowed the dihybrid individuals to self-fertilize. If the segregation of alleles affecting seed shape and seed color were independent, the probability that a particular pair of seed-shape alleles would occur together with a particular pair of seed-color alleles would simply be a product of the two individual probabilities that each pair would occur separately. For example, the probability of an individual with wrinkled, green seeds appearing in the F_2 generation would be equal to the probability of an individual with wrinkled seeds (1 in 4) multiplied by the probability of an individual with green seeds (1 in 4), or 1 in 16.

In his dihybrid crosses, Mendel found that the frequency of phenotypes in the F_2 offspring closely matched the 9:3:3:1 ratio predicted by a Punnett square analysis. He concluded that for the pairs of traits he studied, the inheritance of one trait does not influence the inheritance of the other trait, a result often referred to as Mendel's second law, or the **law of independent assortment.** We now know that this result is only valid for genes not located near one another on the same chromosome. Thus in modern terms Mendel's second law is often stated as follows: *Genes located on different chromosomes are inherited independently of one another.*

Mendel's paper describing his results was published in the journal of his local scientific society in 1866. Unfortunately, his paper failed to arouse much interest, and his work was forgotten. Sixteen years after his death, in 1900, several investigators independently rediscovered Mendel's pioneering paper. They came across it while they were searching the literature in preparation for publishing their own findings, which were similar to those Mendel had quietly presented more than three decades earlier.

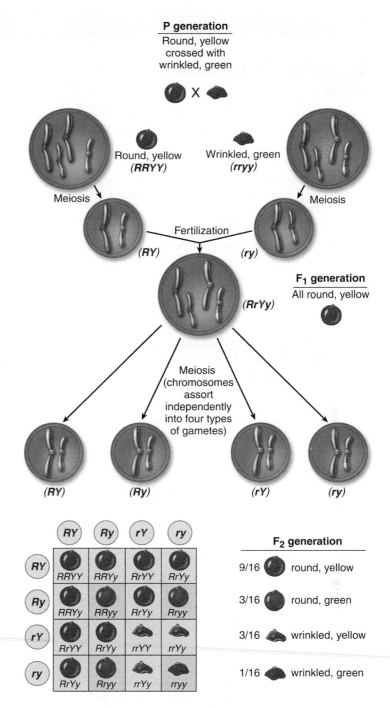

Figure 8.10 Analysis of a dihybrid cross.
This dihybrid cross shows round (*R*) versus wrinkled (*r*) seeds and yellow (*Y*) versus green (*y*) seeds. The ratio of the four possible combinations of phenotypes in the F_2 generation is predicted to be 9:3:3:1, the ratio that Mendel found.

8.4 Mendel's theories of segregation and independent assortment are so well supported by experimental results that they are considered "laws."

8.5 How Genes Influence Traits

It is useful, before considering Mendelian genetics further, to gain a brief overview of how genes work (figure 8.11). With this in mind, we will now sketch, in broad strokes, a picture of how a Mendelian trait is influenced by a particular gene, how a gene can be altered by mutation, and the potential long-term evolutionary consequences of such an alteration.

From DNA to Protein

Each cell of an individual contains a set of DNA molecules, called its genome, that determines what the individual will be like. As you learned in chapter 3, DNA molecules are composed of two strands twisted about each other, each the mirror image of the other. Each strand is a long chain of nucleotide subunits linked together like so many pearls in a necklace. There are four kinds of nucleotides, and like an alphabet with four letters the order of nucleotides determines the message encoded in the DNA of a gene.

The human genome contains about 25,000 genes. The DNA of the human genome is parcelled out into 23 pairs of chromosomes, each chromosome containing from 1,000 to 2,000 different genes.

Individual genes are "read" from the chromosomal DNA by enzymes that create an RNA strand of the same sequence (except U is substituted for T). This RNA transcript leaves the cell nucleus and acts as a work order for protein production in other parts of the cell. But, in eukaryotic cells, the RNA transcript has more information than is needed, so it is first "edited" to remove unnecessary bits before it leaves the nucleus. For example, the initial RNA gene transcript encoding the beta-subunit of the protein hemoglobin is 1,660 nucleotides long; after "editing", the resulting "messenger" RNA is 1,000 nucleotides long.

After an RNA transcript is edited, it leaves the nucleus as messenger RNA (mRNA) and is delivered to ribosomes in the cytoplasm. Each ribosome is a tiny protein-assembly plant, and uses the sequence of the messenger RNA to determine the amino acid sequence of a particular polypeptide. In the case of beta-hemoglobin, the messenger RNA encodes a polypeptide strand of 146 amino acids.

How Proteins Determine the Phenotype

As we have seen in chapter 3, amino acid strands spontaneously fold in water into complex three-dimensional shapes. The beta-hemoglobin polypeptide folds into a compact mass that associates with three others to form an active hemoglobin protein molecule. In red blood cells, each hemoglobin molecule binds oxygen (a process described fully in chapter 20) in the oxygen-rich environment of the lungs, and releases oxygen in the oxygen-poor environment of active tissues.

The oxygen-binding efficiency of the hemoglobin proteins in a person's bloodstream has a great deal to do with how well the body functions, particularly under conditions of strenuous physical activity, when delivery of oxygen to the body's muscles is the chief factor limiting the activity.

As a general rule, genes influence the phenotype by specifying the kind of proteins present in the body, which determines in large measure how that body functions.

How Mutation Alters Phenotype

A change in the identity of a single nucleotide within a gene, called a mutation, can have a profound effect if the change alters the identity of the amino acid encoded there. When a mutation of this sort occurs, the new version of the protein may fold differently, altering or destroying its function. For example, how well the hemoglobin protein performs its oxygen-binding duties depends a great deal on the precise shape that the protein assumes when it folds. A change in the identity of a single amino acid can have a drastic impact on that final shape. In particular, a change in the sixth amino acid of beta-hemoglobin from glutamic acid to valine causes the hemoglobin molecules to aggregate into stiff rods that deform blood cells into a sickle shape and can no longer carry oxygen efficiently. The resulting sickle-cell anemia can be fatal.

Natural Selection for Alternative Phenotypes Leads to Evolution

Because random mutations occur in all genes occasionally, populations usually contain several versions of a gene, usually all but one of them rare. Sometimes the environment changes in such a way that one of the rare versions functions better under the new conditions. When that happens, natural selection will favor the rare allele, which will then become more common. The sickle-cell version of the beta hemoglobin gene, rare throughout most of the world, is common in Central Africa because heterozygous individuals obtain enough functional hemoglobin from their one normal allele to get along, but are resistant to malaria, a deadly disease common there, due to their other sickle-cell allele.

Over time, natural selection will act to preserve more and more of the mutations that occur within species. As a general rule, the longer two species have been separated evolutionarily, the more differences will accumulate between them in their genes. Thus, a chimpanzee has beta-hemoglobin molecules with the same amino acid sequence as humans, whereas a Rhesus monkey (also a primate) has 8 differences, a mouse (also a mammal) has 27 differences, and a chicken (also a vertebrate) has 45 differences.

8.5 Genes determine phenotypes by specifying the amino acid sequences, and thus the functional shapes, of the proteins that carry out cell activities. Mutations, by altering protein sequence, can change a protein's function and thus alter the phenotype in evolutionarily significant ways.

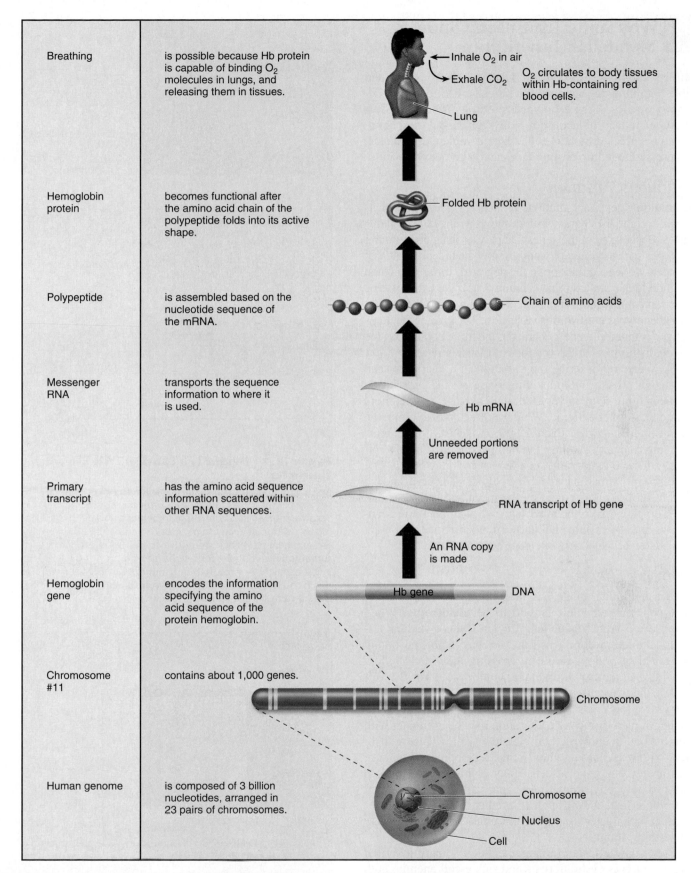

Breathing	is possible because Hb protein is capable of binding O_2 molecules in lungs, and releasing them in tissues.	Inhale O_2 in air / Exhale CO_2 — O_2 circulates to body tissues within Hb-containing red blood cells. / Lung
Hemoglobin protein	becomes functional after the amino acid chain of the polypeptide folds into its active shape.	Folded Hb protein
Polypeptide	is assembled based on the nucleotide sequence of the mRNA.	Chain of amino acids
Messenger RNA	transports the sequence information to where it is used.	Hb mRNA
		Unneeded portions are removed
Primary transcript	has the amino acid sequence information scattered within other RNA sequences.	RNA transcript of Hb gene
		An RNA copy is made
Hemoglobin gene	encodes the information specifying the amino acid sequence of the protein hemoglobin.	Hb gene DNA
Chromosome #11	contains about 1,000 genes.	Chromosome
Human genome	is composed of 3 billion nucleotides, arranged in 23 pairs of chromosomes.	Chromosome / Nucleus / Cell

Figure 8.11 The journey from DNA to phenotype.

What an organism is like is determined in large measure by its genes. Here you see how one gene of the 25,000 in the human genome plays a key role in enabling you to breathe. The many steps on the journey from gene to trait are the subject of chapter 9.

8.6 Why Some Traits Don't Show Mendelian Inheritance

Scientists attempting to confirm Mendel's theory often had trouble obtaining the same simple ratios he had reported. Often the expression of the genotype is not straightforward. Most phenotypes reflect the action of many genes, and the phenotype can be affected by alleles that lack complete dominance, are expressed together, or influence each other's expression.

Continuous Variation

When multiple genes act jointly to influence a character such as height or weight, the character often shows a range of small differences. Because all of the genes that play a role in determining phenotypes such as height or weight segregate independently of each other, we see a gradation in the degree of difference when many individuals are examined (figure 8.12). We call this type of inheritance **polygenic** (many genes) and we call this gradation in phenotypes **continuous variation.** The greater the number of genes that influence a character, the more continuous the expected distribution of the versions of that character.

How can we describe the variation in a character such as the height of the individuals in figure 8.12*a*? Individuals range from quite short to very tall, with average heights more common than either extreme. What we often do is to group the variation into categories. Each height, in inches, is a separate phenotypic category. Plotting the numbers in each height category produces a histogram, such as that in figure 8.12*b*. The histogram approximates an idealized bell-shaped curve, and the variation can be characterized by the mean and spread of that curve. Compare this to the inheritance of plant height in Mendel's peas; they were either tall or dwarf, no intermediate height plants existed because only one gene controlled that trait.

Pleiotropic Effects

Often, an individual allele has more than one effect on the phenotype. Such an allele is said to be **pleiotropic.** When the pioneering French geneticist Lucien Cuenot studied yellow fur in mice, a dominant trait, he was unable to obtain a true-breeding yellow strain by crossing individual yellow mice with one another. Individuals homozygous for the yellow allele died, because the yellow allele was pleiotropic: one effect was yellow color, but another was a lethal developmental defect. A pleiotropic gene alteration may be dominant with respect to one phenotypic consequence (yellow fur) and recessive with respect to another (lethal developmental defect). In pleiotropy, one gene affects many characters, in marked contrast to polygeny, where many genes affect one character. Pleiotropic effects are difficult to predict, because the genes that affect a character often perform other functions we may know nothing about.

Pleiotropic effects are characteristic of many inherited disorders, such as cystic fibrosis and sickle-cell anemia, both discussed later in this chapter. In these disorders, multiple symptoms can be traced back to a single gene defect. In cystic fibrosis (figure 8.13), patients exhibit clogged blood vessels,

(a)

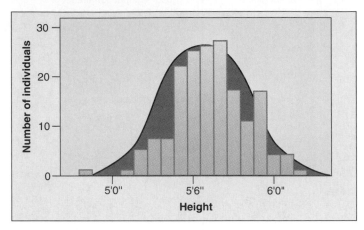

(b)

Figure 8.12 Height is a continuously varying character.

(*a*) This photograph shows the variation in height among students of the 1914 class of the Connecticut Agricultural College. Because many genes contribute to height and tend to segregate independently of each other, there are many possible combinations of those genes. (*b*) The cumulative contribution of different combinations of alleles for height forms a continuous spectrum of possible heights—a random distribution, in which the extremes are much rarer than the intermediate values. This is quite different from the 3:1 ratio seen in Mendel's F_2 peas.

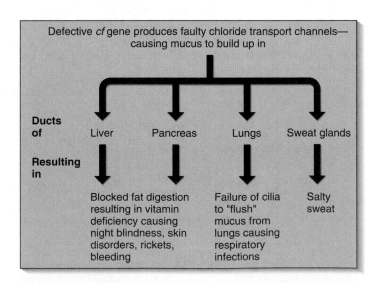

Figure 8.13 Pleiotropic effects of the cystic fibrosis gene, *cf*.

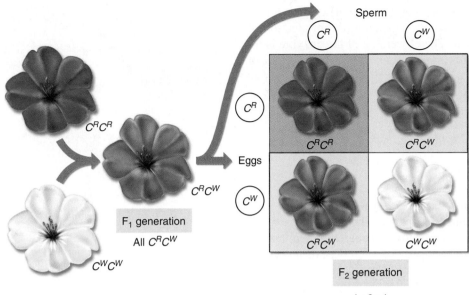

Sperm

C^R C^W

Eggs

C^R

C^W

C^RC^R C^RC^W

C^RC^W C^WC^W

F₁ generation

All C^RC^W

C^RC^R

C^WC^W

F₂ generation

1 : 2 : 1

C^RC^R:C^RC^W:C^WC^W

Figure 8.14 Incomplete dominance.
In a cross between a red-flowered Japanese four o'clock, genotype C^RC^R, and a white-flowered one (C^WC^W), neither allele is dominant. The heterozygous progeny have pink flowers and the genotype C^RC^W. If two of these heterozygotes are crossed, the phenotypes of their progeny occur in a ratio of 1:2:1 (red:pink:white).

overly sticky mucus, salty sweat, liver and pancreas failure, and a battery of other symptoms. All are pleiotropic effects of a single defect, a mutation in a gene that encodes a chloride ion transmembrane channel. In sickle-cell anemia, a defect in the oxygen-carrying hemoglobin molecule causes anemia, heart failure, increased susceptibility to pneumonia, kidney failure, enlargement of the spleen, and many other symptoms. It is usually difficult to deduce the nature of the primary defect from the range of its pleiotropic effects.

Incomplete Dominance

Not all alternative alleles are fully dominant or fully recessive in heterozygotes. Some pairs of alleles exhibit **incomplete dominance** and produce a heterozygous phenotype that is intermediate between those of the parents. For example, the cross of red- and white-flowered Japanese four o'clocks described in figure 8.14 produced red-, pink-, and white-flowered F₂ plants in a 1:2:1 ratio—heterozygotes are intermediate in color.

Environmental Effects

The degree to which many alleles are expressed depends on the environment. Some alleles are heat-sensitive, for example. Traits influenced by such alleles are more sensitive to temperature or light than are the products of other alleles. The arctic foxes in figure 8.15, for example, make fur pigment only when the weather is warm. Can you see why this trait would be an advantage for the fox? Imagine a fox that didn't possess this trait and was white all year round. It would be very visible to predators in the summer, standing out against its darker surroundings. Similarly, the *ch* allele in Himalayan

(a)

(b)

Figure 8.15 Environmental effects on an allele.
(*a*) An arctic fox in winter has a coat that is almost white, so it is difficult to see the fox against a snowy background. (*b*) In summer, the same fox's fur darkens to a reddish brown, so that it resembles the color of the surrounding tundra. Heat-sensitive alleles control this color change.

rabbits and Siamese cats encodes a heat-sensitive version of tyrosinase, one of the enzymes mediating the production of melanin, a dark pigment. The ch version of the enzyme is inactivated at temperatures above about 33°C. At the surface of the main body and head, the temperature is above 33°C and the tyrosinase enzyme is inactive, while it is more active at body extremities such as the tips of the ears and tail, where the temperature is below 33°C. The dark melanin pigment this enzyme produces causes the ears, snout, feet, and tail to be black.

Epistasis

In some situations, two or more genes interact with each other, such that one gene contributes to or masks the expression of the other gene. This becomes apparent when analyzing dihybrid crosses involving these traits. Recall that when individuals heterozygous for two different genes mate (a dihybrid cross), offspring may display the dominant phenotype for both genes, either one of the genes, or for neither gene. Sometimes, however, an investigator cannot find four phenotype classes because two or more of the genotypes express the same phenotypes.

As was stated earlier, few phenotypes are the result of the action of one gene. Most traits reflect the action of many genes, some that act sequentially or jointly. **Epistasis** is an interaction between the products of two genes in which one of the genes modifies the phenotypic expression produced by the other. For example, some commercial varieties of corn, *Zea mays,* exhibit a purple pigment called anthocyanin in their seed coats, while others do not. In 1918, geneticist R. A. Emerson crossed two true-breeding corn varieties, neither exhibiting anthocyanin pigment. Surprisingly, all of the F_1 plants produced purple seeds.

When two of these pigment-producing F_1 plants were crossed to produce an F_2 generation, 56% were pigment producers and 44% were not. What was happening? Emerson correctly deduced that two genes were involved in producing pigment, and that the second cross had thus been a dihybrid cross like those performed by Mendel. Mendel had predicted 16 equally possible ways gametes could combine with each other, resulting in genotypes with a phenotypic ratio of 9:3:3:1 (9 + 3 + 3 + 1 = 16). How many of these were in each of the two types Emerson obtained? He multiplied the fraction that were pigment producers (0.56) by 16 to obtain 9 and multiplied the fraction that were not (0.44) by 16 to obtain 7. Thus, Emerson had a **modified ratio** of 9:7 instead of the usual 9:3:3:1 ratio (figure 8.16).

Why Was Emerson's Ratio Modified? It turns out that either one of the two genes that contribute to kernel color can block the expression of the other. One of the genes (*B*) produces an enzyme that permits colored pigment to be produced only if a dominant allele (*BB* or *Bb*) is present. The other gene (*A*) produces an enzyme that in its dominant form (*AA* or *Aa*) allows the pigment to be deposited on the seed coat color.

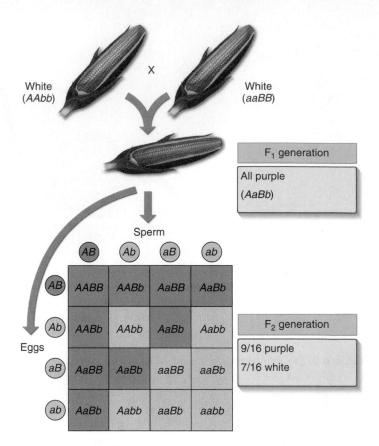

Figure 8.16 How epistasis affects kernel color.
The purple pigment found in some varieties of corn is the result of two genes. Unless a dominant allele is present at each of the two loci, no pigment is expressed.

Thus, an individual with two recessive alleles for gene *A* (no pigment deposition) will have white seed coats even though it is able to manufacture the pigment because it possesses dominant alleles for gene *B* (purple pigment production). Similarly, an individual with dominant alleles for gene *A* (pigment can be deposited) will also have white seed coats if it has only recessive alleles for gene *B* (pigment production) and cannot manufacture the pigment.

To produce and deposit pigment, a plant must possess at least one functional copy of each enzyme gene. Of the 16 genotypes predicted by random assortment, 9 contain at least one dominant allele of both genes; they produce purple progeny. The remaining 7 genotypes lack dominant alleles at either or both loci (3 + 3 + 1 = 7) and so are phenotypically the same (nonpigmented), giving the phenotypic ratio of 9:7 that Emerson observed.

Other Examples of Epistasis.   In many animals, coat color is the result of epistatic interactions among genes. Coat color in Labrador retrievers, a breed of dog, is due primarily to the interaction of two genes. The *E* gene determines if dark pigment will be deposited in the fur or not. If a dog has the

A Closer Look

Does Environment Affect I.Q.?

The influence of environment on the expression of genetic traits is especially hard to study when a number of different genes affect the trait. Nowhere has this difficulty led to more controversy than in studies of I.Q. scores. I.Q. is a controversial measure of general intelligence based on a written test that many feel to be biased toward white middle-class America. However well or poorly I.Q. scores measure intelligence, a person's I.Q. score has been believed for some time to be determined largely by his or her genes.

How did science come to that conclusion? Scientists measure the degree to which genes influence a multigene trait by using an off-putting statistical measure called the *variance*. Variance is defined as the square of the standard deviation (a measure of the degree-of-scatter of a group of numbers around their mean value), and has the very desirable property of being additive—that is, the total variance is equal to the sum of the variances of the factors influencing it.

The degree of gene influence, or *heritability,* is defined as the fraction of the total variance that is genetic:

$$H = \text{Variance (genes)/Variance (total)}$$

What factors contribute to the total Variance? There are three. The first factor is variation at the gene level, some gene combinations leading to higher I.Q. scores than others. This is the term "Variance (genes)" in the simple equation above. The second factor is variation at the environmental level, some environments leading to higher I.Q. scores than others. This term is expressed as "Variance (environment)." The third factor is what a statistician calls the covariance, the degree to which the value of one factor influences the value of others. In this instance, covariance expresses the degree to which environment affects genes. The covariance term is expressed "Variance (coVar)."

Now the heritability of I.Q. can be expressed in terms that can be measured:

$$H = \frac{\text{Variance (genes)}}{\text{Variance (genes)} + \text{Variance (environment)} + \text{Variance (coVar)}}$$

The environmental contributions to variance in I.Q. can be measured by comparing the I.Q. scores of identical twins reared together with those reared apart (any differences should reflect environmental influences). The genetic contributions can be measured by comparing identical twins reared together (which are 100% genetically identical) with fraternal twins reared together (which are 50% genetically identical). Any differences should reflect genes, as twins share identical prenatal conditions in the womb and are raised in virtually identical environmental circumstances, so when traits are more commonly shared between identical twins than fraternal twins, the difference is likely genetic.

When these sorts of "twin studies" have been done in the past, researchers have uniformly reported that I.Q. is highly heritable, with values of H typically reported as being around 0.7 (a very high value). While it didn't seem significant at the time, almost all the twins available for study over the years have come from middle-class or wealthy families.

The study of I.Q. has proven controversial, because I.Q. scores are often different when social and racial groups are compared. What is one to make of the observation that I.Q. scores of poor children measure lower as a group than do scores of children of middle-class and wealthy families? This difference has led to the controversial suggestion by some that the poor are genetically inferior.

What should we make of such a harsh conclusion? To make a judgment, we need to focus for a moment on the fact that these measures of the heritability of I.Q. have all made a critical assumption, one to which population geneticists, who specialize in these sorts of things, object strongly. The assumption is that environment does not affect gene expression—that is, that the covariance term in the preceding equation is zero.

Recent studies have allowed a direct assessment of this assumption. Importantly, it proves to be flat wrong.

In November of 2003, researchers reported an analysis of twin data from a study carried out in the late 1960s. The National Collaborative Prenatal Project, funded by the National Institutes of Health, enrolled nearly 50,000 pregnant women, most of them black and quite poor, in several major U.S. cities. Researchers collected abundant data, and gave the children I.Q. tests seven years later. Although not designed to study twins, this study was so big that many twins were born, 623 births. Seven years later, 320 of these pairs were located and given I.Q. tests. This thus constitutes a huge "twin study," the first ever conducted of I.Q. among the poor.

When the data were analyzed, the results were unlike any ever reported. The heritability of I.Q. was different in different environments! Most notably, the influence of genes on I.Q. was far less in conditions of poverty, where environmental limitations seem to block the expression of genetic potential. Specifically, for families of high socioeconomic status, H= 0.72, much as reported in previous studies, but for families raised in poverty, H = 0.10, a very low value, indicating genes were making little contribution to observed I.Q. scores. The lower a child's socioeconomic status, the less impact genes had on I.Q.

These data say that the genetic contributions to I.Q. don't mean much in an impoverished environment. Clearly, improvements in the growing and learning environments of poor children can be expected to have a major impact on their I.Q. scores. Additionally, these data argue that the controversial differences reported in mean I.Q scores between racial groups may well reflect no more than poverty, and are no more inevitable.

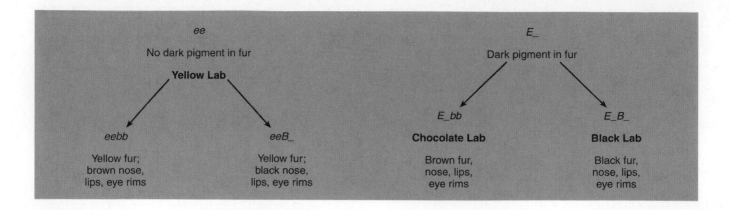

Figure 8.17 The effect of epistatic interactions on coat color in dogs.
The coat color seen in Labrador retrievers is an example of the interaction of two genes, each with two alleles. The *E* gene determines if the pigment will be deposited in the fur, and the *B* gene determines how dark the pigment will be.

genotype *ee,* no pigment will be deposited in the fur, and it will be yellow. If a dog has the genotype *EE* or *Ee* (*E_*), pigment will be deposited in the fur.

A second gene, the *B* gene, determines how dark the pigment will be. Dogs with the genotype *E_bb* will have brown fur and are called chocolate labs. Dogs with the genotype *E_B_* will have black fur. But, even in yellow dogs, the *B* gene does have some effect. Yellow dogs with the genotype *eebb* will have brown pigment on their nose, lips, and eye rims, while yellow dogs with the genotype *eeB_* will have black pigment in these areas. The interaction among these alleles is illustrated in figure 8.17. The genes for coat color in this breed have been found, and a genetic test is available to determine the coat color in a litter of puppies.

Codominance

A gene may have more than two alleles in a population, and in fact most genes possess several different alleles. Often, there isn't a dominant allele; instead, the effects of each allele are expressed. The result is an organism that expresses more than one phenotype. In these cases, the alleles are considered **codominant.**

Codominance is seen in the color patterning of some animals. For example, the "roan" pattern is a coloring pattern exhibited in some varieties of horses and cattle. A roan animal expresses both white and colored hairs on at least part of its body. This intermingling of the different colored hairs creates either an overall lighter color, or patches of lighter and darker colors (figure 8.18). The roan pattern results from a heterozygous genotype, such as produced by mating of a homozygous white and homozygous colored. Could the intermediate color be the result of incomplete dominance? No. The heterozygous that receives a white allele and a colored allele does not have individual hairs that are a mix of the two colors; rather, both alleles are being expressed, with the result that the animal has some hairs that are white and some that are colored.

A human gene that exhibits more than one dominant allele is the gene that determines ABO blood type. This gene encodes an enzyme that adds sugar molecules to lipids on the surface of red blood cells. These sugars act as recognition markers for cells in the immune system and are called cell surface antigens. The gene that encodes the enzyme, designated *I,* has three common alleles: I^B, whose product adds the sugar

Figure 8.18 Codominance in color patterning.
This roan horse (named "Rivers Top Gun" by its owners) is heterozygous for coat color. The offspring of a cross between a white homozygote and a black homozygote, it expresses both phenotypes. Some of the hairs on its body are white and some are black. This codominance creates patches of black and white, or it can create an overall coloring that is a lighter shade of black. But in the latter case, the effect is due to the intermingling of white and black hairs and not from a mixing of the two colors to produce grey hairs as would be seen in incomplete dominance.

galactose; I^A, whose product adds galactosamine; and i, which codes for a protein that does not add a sugar.

Different combinations of the three I gene alleles occur in different individuals because each person possesses two copies of the chromosome bearing the I gene and may be homozygous for any allele or heterozygous for any two. An individual heterozygous for the I^A and I^B alleles produces both forms of the enzyme and adds both galactose and galactosamine to the surfaces of red blood cells. Because both alleles are expressed simultaneously in heterozygotes, the I^A and I^B alleles are codominant. Both I^A and I^B are dominant over the i allele because both I^A or I^B alleles lead to sugar addition and the i allele does not. The different combinations of the three alleles produce four different phenotypes (figure 8.19):

1. Type A individuals add only galactosamine. They are either $I^A I^A$ homozygotes or $I^A i$ heterozygotes.
2. Type B individuals add only galactose. They are either $I^B I^B$ homozygotes or $I^B i$ heterozygotes.
3. Type AB individuals add both sugars and are $I^A I^B$ heterozygotes.
4. Type O individuals add neither sugar and are ii homozygotes.

These four different cell surface phenotypes are called the **ABO blood groups** or, less commonly, the Landsteiner

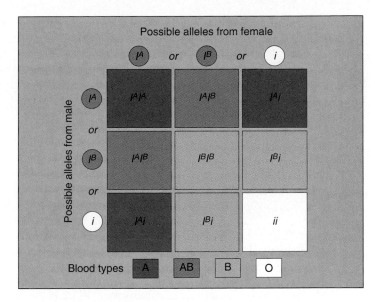

Figure 8.19 Multiple alleles controlling the ABO blood groups.
Three common alleles control the ABO blood groups. Different combinations of the three so-called I gene alleles occur in different individuals. An individual may be homozygous for any allele or heterozygous for any two. The different combinations of the three alleles result in four different blood type phenotypes: type A (either $I^A I^A$ homozygotes or $I^A i$ heterozygotes), type B (either $I^B I^B$ homozygotes or $I^B i$ heterozygotes), type AB ($I^A I^B$ heterozygotes), and type O (ii homozygotes).

blood groups, after the man who first described them. As Karl Landsteiner noted, a person's immune system can distinguish between these four phenotypes. If a type A individual receives a transfusion of type B blood, the recipient's immune system recognizes that the type B blood cells possess a "foreign" antigen (galactose) and attacks the donated blood cells, causing the cells to clump or agglutinate. This also happens if the donated blood is type AB. However, if the donated blood is type O, no immune attack will occur, as there are no galactose antigens on the surfaces of blood cells produced by the type O donor. In general, any individual's immune system will tolerate a transfusion of type O blood. Because neither galactose nor galactosamine is foreign to type AB individuals (whose red blood cells have both sugars), those individuals may receive any type of blood.

8.6 A variety of factors can disguise the Mendelian segregation of alleles. Among them are continuous variation, which results when many genes contribute to a trait; pleiotropic effects where one allele affects many phenotypes; incomplete dominance, which produces heterozygotes unlike either parent; environmental influences on the expression of phenotypes; and the interaction of more than one allele as seen in epistasis and codominance.

Mendelian Genetics Problems

These genetics problems will help you to see the far-reaching effects of Mendel's experiments. If you need help, the answers appear at www.mhhe.com/tlwessentials/geneticsanswers.

1. The annual plant *Haplopappus gracilis* has two pairs of chromosomes (1 and 2). In this species, the probability that two characters a and b selected at random will be on the same chromosome is equal to the probability that they will both be on chromosome 1 ($1/2 \times 1/2 = 1/4$, or 0.25), plus the probability that they will both be on chromosome 2 (also $1/2 \times 1/2 = 1/4$, or 0.25), for an overall probability of 1/2, or 0.5. In general, the probability that two randomly selected characters will be on the same chromosome is equal to $1/n$ where n is the number of chromosome pairs. Humans have 23 pairs of chromosomes. What is the probability that any two human characters selected at random will be on the same chromosome?

2. Among Hereford cattle there is a dominant allele called *polled;* the individuals that have this allele lack horns. Suppose you acquire a herd consisting entirely of polled cattle, and you carefully determine that no cow in the herd has horns. Some of the calves born that year, however, grow horns. You remove them from the herd and make certain that no horned adult has gotten into your pasture. Despite your efforts, more horned calves are born the next year. What is the reason for the appearance of the horned calves? If your goal is to maintain a herd consisting entirely of polled cattle, what should you do?

3. An inherited trait among humans in Norway causes affected individuals to have very wavy hair, not unlike that of a sheep. The trait, called *woolly,* is very evident when it occurs in families; no child possesses woolly hair unless at least one parent does. Imagine you are a Norwegian judge, and you have before you a woolly haired man suing his normal-haired wife for divorce because their first child has woolly hair but their second child has normal hair. The husband claims this constitutes evidence of his wife's infidelity. Do you accept his claim? Justify your decision.

4. In human beings, Down syndrome, a serious developmental abnormality, results from the presence of three copies of chromosome 21 rather than the usual two copies. If a female exhibiting Down syndrome mates with a normal male, what proportion of her offspring would you expect to be affected?

5. Many animals and plants bear recessive alleles for *albinism,* a condition in which homozygous individuals lack certain pigments. An albino plant, for example, lacks chlorophyll and is white, and an albino human lacks melanin. If two normally pigmented persons heterozygous for the same albinism allele marry, what proportion of their children would you expect to be albino?

6. You inherit a racehorse and decide to put him out to stud. In looking over the stud book, however, you discover that the horse's grandfather exhibited a rare disorder that causes brittle bones. The disorder is hereditary and results from homozygosity for a recessive allele. If your horse is heterozygous for the allele, it will not be possible to use him for stud because the genetic defect may be passed on. How would you determine whether your horse carries this allele?

7. Your instructor presents you with a *Drosophila* (fruitfly) with red eyes, as well as a stock of white-eyed flies and another stock of flies homozygous for the red-eye allele. You know that the presence of white eyes in *Drosophila* is caused by homozygosity for a recessive allele. How would you determine whether the single red-eyed fly was heterozygous for the white-eye allele?

8. Hemophilia is a recessive sex-linked human blood disease (see page 190) that leads to failure of blood to clot normally. One form of hemophilia has been traced to the royal family of England, from which it spread throughout the royal families of Europe. For the purposes of this problem, assume that it originated as a mutation either in Prince Albert or in his wife, Queen Victoria.
   a. Prince Albert did not have hemophilia. If the disease is a sex-linked recessive abnormality, how could it have originated in Prince Albert, a male, who would have been expected to exhibit sex-linked recessive traits?
   b. Alexis, the son of Czar Nicholas II of Russia and Empress Alexandra (a granddaughter of Victoria), had hemophilia, but their daughter Anastasia did not. Anastasia died, a victim of the Russian revolution, before she had any children. Can we assume that Anastasia would have been a carrier of the disease? Would your answer be different if the disease had been present in Nicholas II or in Alexandra?

9. A normally pigmented man marries an albino woman. They have three children, one of whom is an albino. What is the genotype of the father?

10. A man works in an atomic energy plant, and he is exposed daily to low-level background radiation. After several years, he has a child who has Duchenne muscular dystrophy, a recessive genetic defect caused by a mutation on the X chromosome. Neither the parents nor the grandparents have the disease. The man sues the plant, claiming that the abnormality in their child is the direct result of radiation-induced mutation of his gametes, and that the company should have protected him from this radiation. Before reaching a decision, the judge hearing the case insists on knowing the sex of the child. Which sex would be more likely to result in an award of damages, and why?

8.7 Chromosomes Are the Vehicles of Mendelian Inheritance

Chromosomes are not the only kinds of structures that segregate regularly when eukaryotic cells divide. Centrioles also divide and segregate in a regular fashion, as do the mitochondria and chloroplasts (when present) in the cytoplasm. Therefore, in the early twentieth century it was by no means obvious that chromosomes were the vehicles of hereditary information.

The Chromosomal Theory of Inheritance

A central role for chromosomes in heredity was first suggested in 1900 by the German geneticist Karl Correns, in one of the papers announcing the rediscovery of Mendel's work. Soon after, observations that similar chromosomes paired with one another during meiosis led directly to the *chromosomal theory of inheritance*, first formulated by the American Walter Sutton in 1902.

Several pieces of evidence supported Sutton's theory. One was that reproduction involves the initial union of only two cells, egg and sperm. If Mendel's model was correct, then these two gametes must make equal hereditary contributions. Sperm, however, contain little cytoplasm, suggesting that the hereditary material must reside within the nuclei of the gametes. Furthermore, while diploid individuals have two copies of each pair of homologous chromosomes, gametes have only one. This observation was consistent with Mendel's model, in which diploid individuals have two copies of each heritable gene and gametes have one. Finally, chromosomes segregate during meiosis, and each pair of homologues orients on the metaphase plate independently of every other pair. Segregation and independent assortment were two characteristics of the genes in Mendel's model.

Problems with the Chromosomal Theory

Investigators soon pointed out one problem with this theory, however. If Mendelian traits are determined by genes located on the chromosomes, and if the independent assortment of Mendelian traits reflects the independent assortment of chromosomes in meiosis, why does the number of traits that assort independently in a given kind of organism often greatly exceed the number of chromosome pairs the organism possesses? This seemed a fatal objection, and it led many early researchers to have serious reservations about Sutton's theory.

Morgan's White-Eyed Fly

The essential correctness of the chromosomal theory of heredity was demonstrated long before this paradox was resolved. A single small fly provided the confirmation. In 1910 Thomas Hunt Morgan, studying the fruit fly *Drosophila melanogaster,* detected a mutant male fly, one that differed strikingly from normal flies of the same species: its eyes were white instead of red (figure 8.20).

Morgan immediately set out to determine if this new trait would be inherited in a Mendelian fashion. He first crossed the

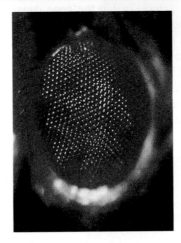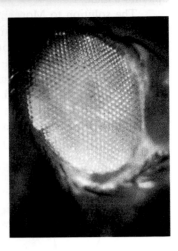

Figure 8.20   Red-eyed (wild type) and white-eyed (mutant) *Drosophila.*
The white-eye defect is hereditary, the result of a mutation in a gene located on the X chromosome. By studying this mutation, Morgan first demonstrated that genes are on chromosomes.

mutant male with a normal female to see if red or white eyes were dominant. All of the F_1 progeny had red eyes, so Morgan concluded that red eye color was dominant over white. Following the experimental procedure that Mendel had established long ago, Morgan then crossed the red-eyed flies from the F_1 generation with each other. Of the 4,252 F_2 progeny Morgan examined, 782 (18%) had white eyes. Although the ratio of red eyes to white eyes in the F_2 progeny was greater than 3:1, the results of the cross nevertheless provided clear evidence that eye color segregates. However, there was something about the outcome that was strange and totally unpredicted by Mendel's theory—*all of the white-eyed F_2 flies were males!*

How could this result be explained? Perhaps it was impossible for a white-eyed female fly to exist; such individuals might not be viable for some unknown reason. To test this idea, Morgan testcrossed the female F_1 progeny with the original white-eyed male. He obtained white-eyed and red-eyed males and females in a 1:1:1:1 ratio, just as Mendelian theory predicted. Hence, a female could have white eyes. Why, then, were there no white-eyed females among the progeny of the original cross?

Sex Linkage Confirms the Chromosomal Theory

The solution to this puzzle involved sex. In *Drosophila,* the sex of an individual is determined by the number of copies of a particular chromosome, the **X chromosome,** that an individual possesses. A fly with two X chromosomes is a female, and a fly with only one X chromosome is a male. In males, the single X chromosome pairs in meiosis with a large, dissimilar partner called the **Y chromosome.** The female thus produces only X gametes, while the male produces both X and Y gametes. When fertilization involves an X sperm, the result is an XX zygote, which develops into a female; when fertilization

Figure 8.24 Down syndrome.

(a) In this karyotype of a male individual with Down syndrome, the trisomy at position 21 can be clearly seen. (b) A person with Down syndrome.

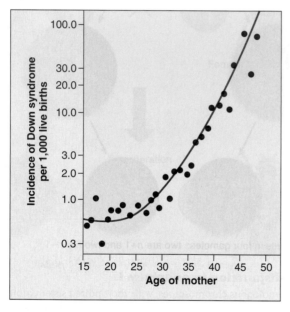

Figure 8.25 Correlation between maternal age and the incidence of Down syndrome.

As women age, the chances they will bear a child with Down syndrome increase. After a woman reaches age 35, the frequency of Down syndrome increases rapidly.

highly condensed and bears few functional genes in most organisms. Some of the active genes the Y chromosome does possess are responsible for the features associated with "maleness." Individuals who gain or lose a sex chromosome do not generally experience the severe developmental abnormalities caused by changes in autosomes. Such individuals may reach maturity, but with somewhat abnormal features.

Nondisjunction of the X Chromosome. When X chromosomes fail to separate during meiosis, some of the gametes that are produced possess both X chromosomes and so are XX gametes; the other gametes that result from such an event have no sex chromosome and are designated "O" (figure 8.26).

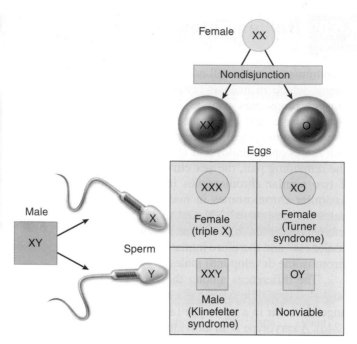

Figure 8.26 Nondisjunction of the X chromosome.

Nondisjunction of the X chromosome can produce sex chromosome aneuploidy—that is, abnormalities in the number of sex chromosomes.

If an XX gamete combines with an X gamete, the resulting XXX zygote develops into a female who is taller than average but other symptoms can vary greatly. Some are normal in most respects, others may have lower reading and verbal skills, and still others are mentally retarded. If an XX gamete combines with a Y gamete, the XXY zygote develops into a sterile male who has many female body characteristics and, in some cases, diminished mental capacity. This condition, called *Klinefelter syndrome,* occurs in about 1 in 500 male births.

If an O gamete fuses with a Y gamete, the OY zygote is nonviable and fails to develop further because humans cannot survive when they lack the genes on the X chromosome. If an O gamete fuses with an X gamete, the XO zygote develops into a sterile female of short stature, with a webbed neck and immature sex organs that do not undergo changes during puberty. The mental abilities of XO individuals are normal in verbal learning but lower in nonverbal/math-based problem solving. This condition, called *Turner syndrome,* occurs roughly once in every 5,000 female births.

Nondisjunction of the Y Chromosome. The Y chromosome can also fail to separate in meiosis, leading to the formation of YY gametes. When these gametes combine with X gametes, the XYY zygotes develop into fertile males of normal appearance. The frequency of the XYY genotype is about 1 per 1,000 newborn males. These males tend to be taller than average, and while a small percentage may be mildly retarded, in general they are phenotypically normal.

8.8 Autosome loss is always lethal, and an extra autosome is with few exceptions lethal too. Additional sex chromosomes have less serious consequences, although they can lead to sterility.

8.9 The Role of Mutations in Human Heredity

The proteins encoded by most of your genes must function in a very precise fashion for you to develop properly and for the many complex processes of your body to function correctly. Unfortunately, genes sometimes sustain damage or are copied incorrectly. We call these accidental changes in genes **mutations.** Mutations occur only rarely, because your cells police your genes and attempt to correct any damage they encounter. Still, some mutations get through. Many of them are bad for you in one way or another. It is easy to see why. Mutations hit genes at random—imagine that you randomly changed the number of a part on a design of a jet fighter. Sometimes it won't matter critically—a seatbelt becomes a radio, say. But what if a key rivet in the wing becomes a roll of toilet paper? The chance of a random mutation in a gene improving the performance of its protein is about the same as that of a randomly selected part making the jet fly faster.

Most mutations are rare in human populations. Almost all result in recessive alleles, and so they are not eliminated from the population by evolutionary forces—because they are not expressed in most individuals (heterozygotes) in which they occur. Do you see why they occur mostly in heterozygotes? Because mutant alleles are rare, it is unlikely that a person carrying a copy of the mutant allele will marry someone who also carries it. Instead, he or she will typically marry someone homozygous normal, and their children would not be homozygous for the mutant allele. While most mutations are harmful to normal functions and are usually recessive, that is not to say all mutations are undesirable; some mutations can lead to enhanced function. Nor are all mutations recessive; rarely they can occur as dominant alleles.

In some cases, particular mutant alleles have become more common in human populations. In these cases, the harmful effects that they produce are called genetic disorders. Some of the most common genetic disorders are listed in table 8.3. To study human heredity, scientists look at the results of crosses that have already been made. They study family trees, or **pedigrees,** to identify which relatives exhibit a trait (figure 8.27). Then they can often determine whether the gene

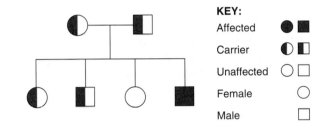

KEY:

Affected	● ■
Carrier	◐ ◧
Unaffected	○ □
Female	○
Male	□

Figure 8.27   A general pedigree.
This is a pedigree for the inheritance of a recessive trait. Males are depicted with squares, females with circles, unaffected persons with empty symbols, carriers (heterozygotes) with half-filled symbols, and affected persons (homozygotes) with solid symbols.

TABLE 8.3	SOME IMPORTANT GENETIC DISORDERS				
Disorder	**Symptom**	**Defect**	**Dominant/ Recessive**	**Frequency Among Human Births**	
Cystic fibrosis	Mucus clogs lungs, liver, and pancreas	Failure of chloride ion transport mechanism	Recessive	1/2,500 (Caucasians)	
Sickle-cell anemia	Poor blood circulation	Abnormal hemoglobin molecules	Recessive	1/625 (African Americans)	
Tay-Sachs disease	Deterioration of central nervous system in infancy	Defective enzyme (hexosaminidase A)	Recessive	1/3,500 (Ashkenazi Jews)	
Phenylketonuria	Brain fails to develop in infancy	Defective enzyme (phenylalanine hydroxylase)	Recessive	1/12,000	
Hemophilia	Blood fails to clot	Defective blood-clotting factor VIII	Sex-linked recessive	1/10,000 (Caucasian males)	
Huntington's disease	Brain tissue gradually deteriorates in middle age	Production of an inhibitor of brain cell metabolism	Dominant	1/24,000	
Muscular dystrophy (Duchenne)	Muscles waste away	Degradation of myelin coating of nerves stimulating muscles	Sex-linked recessive	1/3,700 (males)	
Congenital hypothyroidism	Increased birth weight, puffy face, constipation, lethargy	Failure of proper thyroid development	Recessive	1/1,000 (Hispanics) 1/700 (Native Americans)	
Hypercholesterolemia	Excessive cholesterol levels in blood, leading to heart disease	Abnormal form of cholesterol cell surface receptor	Dominant	1/500	

producing the trait is sex-linked or autosomal and whether the trait's phenotype is dominant or recessive. Frequently, they can infer which individuals are homozygous and which are heterozygous for the allele specifying the trait.

Hemophilia: A Sex-Linked Trait

Blood in a cut clots as a result of the polymerization of protein fibers circulating in the blood. A dozen proteins are involved in this process, and all must function properly for a blood clot to form. A mutation causing any of these proteins to lose their activity leads to a form of **hemophilia,** a hereditary condition in which the blood clots slowly or not at all.

Hemophilias are recessive disorders, expressed only when an individual does not possess any copy of the normal allele and so cannot produce one of the proteins necessary for clotting. Most of the genes that encode the blood-clotting proteins are on autosomes, but two (designated VIII and IX) are on the X chromosome. These two genes are sex-linked (see section 8.7): any male who inherits a mutant allele will develop hemophilia, because his other sex chromosome is a Y chromosome that lacks any alleles of those genes.

The most famous instance of hemophilia, often called the Royal hemophilia, is a sex-linked form that arose in the royal family of England. This hemophilia was caused by a mutation in gene IX that occurred in one of the parents of Queen Victoria of England (1819–1901; figure 8.28). In the five generations since Queen Victoria, 10 of her male descendants have had hemophilia. The present British royal family has escaped the disorder because Queen Victoria's son, King Edward VII, did not inherit the defective allele, and all the subsequent rulers of England are his descendants. Three of Victoria's nine children did receive the defective allele, however, and they carried it by marriage into many of the other royal families of Europe.

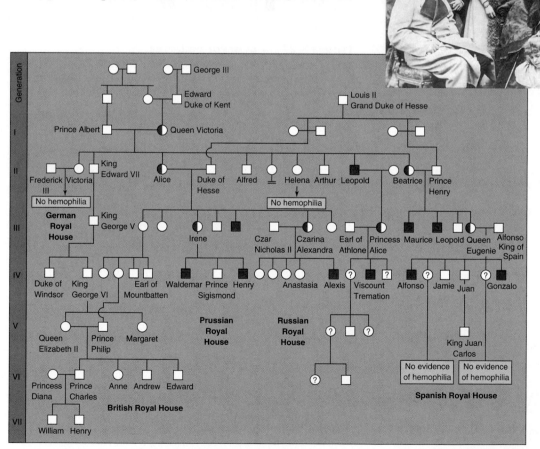

Figure 8.28   The Royal hemophilia pedigree.
Queen Victoria's daughter Alice introduced hemophilia into the Russian and Prussian royal houses, and her daughter Beatrice introduced it into the Spanish royal house. Victoria's son Leopold, himself a victim, also transmitted the disorder in a third line of descent. Half-shaded symbols represent carriers with one normal allele and one defective allele; fully shaded symbols represent affected individuals. Squares represent males; circles represent females. In this photo, Queen Victoria of England is surrounded by some of her descendants in 1894. Standing behind Victoria and wearing feathered boas are two of Victoria's granddaughters, Alice's daughter's: Princess Irene of Prussia (*right*), and Alexandra (*left*), who would soon become Czarina of Russia. Both Irene and Alexandra were also carriers of hemophilia.

Sickle-Cell Anemia: Recessive Trait

Sickle-cell anemia is a recessive hereditary disorder (figure 8.29) in which afflicted individuals have defective molecules of hemoglobin, the protein within red blood cells that carries oxygen. Consequently, these individuals are unable to properly transport oxygen to their tissues. The defective hemoglobin molecules stick to one another, forming stiff, rodlike structures and resulting in the formation of sickle-shaped red blood cells (figure 8.30). As a result of their stiffness and irregular shape, these cells have difficulty moving through the smallest blood vessels; they tend to accumulate in those vessels and form clots. People who have large proportions of sickle-shaped red blood cells tend to have intermittent illness and a shortened life span.

The hemoglobin in the defective red blood cells differs from that in normal red blood cells in only one of hemoglo-bin's 574 amino acid subunits. In the defective hemoglobin, the amino acid valine replaces a glutamic acid at a single position in the protein. Interestingly, the position of the change is far from the active site of hemoglobin where the iron-bearing heme group binds oxygen. Instead, the change occurs on the outer edge of the protein. Why then is the result so catastrophic? The sickle-cell mutation puts a very nonpolar amino acid on the surface of the hemoglobin protein, creating a "sticky patch" that sticks to other such patches—nonpolar amino acids tend to associate with one another in polar environments like water. As one hemoglobin adheres to another, chains of hemoglobin molecules form.

Individuals heterozygous for the sickle-cell allele are generally indistinguishable from normal persons. However, some of their red blood cells show the sickling characteristic when they are exposed to low levels of oxygen. The allele responsible for sickle-cell anemia is particularly common among people of African descent (figure 8.31); about 9% of African Americans are heterozygous for this allele,

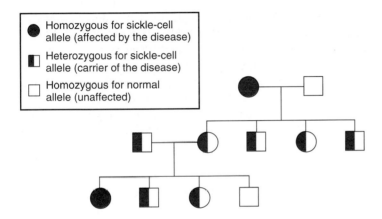

Figure 8.29 Inheritance of sickle-cell anemia.

Sickle-cell anemia is a recessive autosomal disorder. If one parent is homozygous for the recessive trait, all of the offspring will be carriers (heterozygotes) like the F_1 generation of Mendel's testcross.

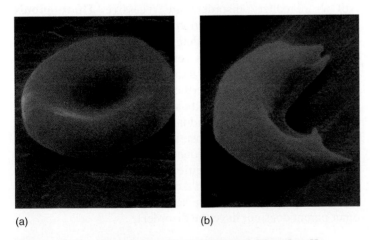

(a)   (b)

Figure 8.30 Normal and sickled red blood cells.

(a) A normal red blood cell is shaped like a flattened sphere. Its smooth edges pass easily through the body's tiny capillaries. In individuals homozygous for the sickle-cell trait, many of the red blood cells have sickle shapes (b). Their pointed edges get stuck while passing through capillaries, leading to internal bleeding and anemia.

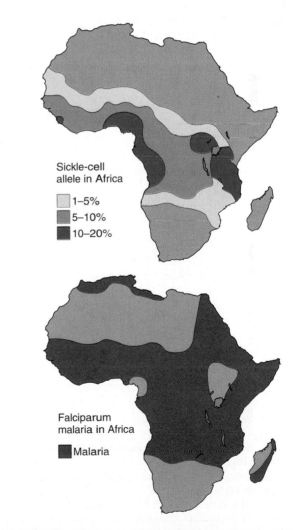

Figure 8.31 The sickle-cell allele confers resistance to malaria.

The distribution of sickle-cell anemia closely matches the occurrence of malaria in central Africa. This is not a coincidence. The sickle-cell allele, when heterozygous, confers resistance to malaria, a very serious disease.

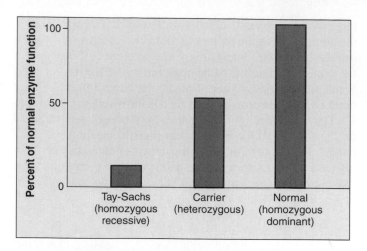

Figure 8.32 Tay-Sachs disease.
Homozygous individuals (*left bar*) typically have less than 10% of the normal level of hexosaminidase A (*right bar*), while heterozygous individuals (*middle bar*) have about 50% of the normal level—enough to prevent deterioration of the central nervous system.

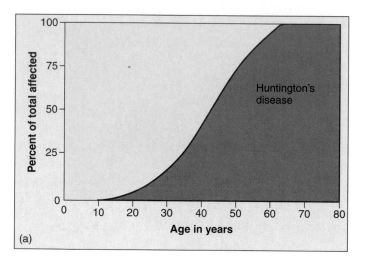

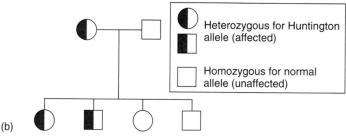

Figure 8.33 Huntington's disease is a dominant genetic disorder.
(*a*) It is because of the late age of onset of this disease that it persists despite the fact that it is dominant and fatal. (*b*) The pedigree shows how the dominant gene is inherited. All who inherit the gene are affected. Although the mother was affected, we can tell that she was heterozygous because if she were homozygous dominant, all of her children would have been affected. However, by the time she found out that she had the disease, she had probably already given birth to her children and so the trait passed on to the next generation even though it is fatal.

and about 0.2% are homozygous and therefore have the disorder. In some groups of people in Africa, up to 45% of all individuals are heterozygous for this allele, and fully 6% are homozygous and express the disorder. What factors determine the high frequency of sickle-cell anemia in Africa? It turns out that heterozygosity for the sickle-cell anemia allele increases resistance to malaria, a common and serious disease in central Africa. Sickle-cell anemia is discussed further in chapter 11.

Tay-Sachs Disease: Recessive Trait

Tay-Sachs disease is an incurable hereditary disorder in which the brain deteriorates. Affected children appear normal at birth and usually do not develop symptoms until about the eighth month, when signs of mental deterioration appear. The children are blind within a year after birth, and they rarely live past five years of age.

Tay-Sachs disease is rare in most human populations, occurring in only 1 in 300,000 births in the United States. However, the disease has a high incidence among Jews of Eastern and Central Europe (Ashkenazi) and among American Jews, 90% of whom trace their ancestry to Eastern and Central Europe. In these populations, it is estimated that 1 in 28 individuals is a heterozygous carrier of the disease, and approximately 1 in 3,500 infants has the disease. Because the disease is caused by a recessive allele, most of the people who carry the defective allele do not themselves develop symptoms of the disease (figure 8.32).

The Tay-Sachs allele produces the disease by encoding a nonfunctional form of the enzyme hexosaminidase A. This enzyme breaks down *gangliosides,* a class of lipids occurring within the lysosomes of brain cells. As a result, the lysosomes fill with gangliosides, swell, and eventually burst, releasing oxidative enzymes that kill the cells. There is no known cure for this disorder.

Huntington's Disease: Dominant Trait

Not all hereditary disorders are recessive. **Huntington's disease** is a hereditary condition caused by a dominant allele that causes the progressive deterioration of brain cells. Perhaps 1 in 24,000 individuals develops the disorder. Because the allele is dominant, every individual who carries the allele expresses the disorder. Nevertheless, the disorder persists in human populations because its symptoms usually do not develop until the affected individuals are more than 30 years old, and by that time most of those individuals have already had children (figure 8.33). Consequently, the allele is often transmitted before the lethal condition develops. A person who is heterozygous for Huntington's disease has a 50% chance of passing the disease to his or her children (even though the other parent does not have the disorder).

> **8.9 Many human hereditary disorders reflect the presence of rare (and sometimes not so rare) mutations within human populations.**

8.10 Genetic Counseling and Therapy

Although most genetic disorders cannot yet be cured, we are learning a great deal about them, and progress toward successful therapy is being made in many cases. However, in the absence of a cure, some parents may feel the only recourse is to try to avoid producing children with these conditions. The process of identifying parents at risk of producing children with genetic defects and of assessing the genetic state of early embryos is called **genetic counseling.** Genetic counseling can help prospective parents determine their risk of having a child with a genetic disorder and advise them on medical treatments or options if a genetic disorder is determined to exist in an unborn child.

High-Risk Pregnancies

If a genetic defect is caused by a recessive allele, how can potential parents determine the likelihood that they carry the allele? One way is through pedigree analysis, often employed as an aid in genetic counseling. By analyzing a person's pedigree, it is sometimes possible to estimate the likelihood that the person is a carrier for certain disorders. For example, if one of your relatives has been afflicted with a recessive genetic disorder such as cystic fibrosis, it is possible that you are a heterozygous carrier of the recessive allele for that disorder. When a pedigree analysis indicates that both parents of an expected child have a significant probability of being heterozygous carriers of a recessive allele responsible for a serious genetic disorder, the pregnancy is said to be a high-risk pregnancy. In such cases, there is a significant probability that the child will exhibit the clinical disorder.

Another class of high-risk pregnancies are those in which the mothers are more than 35 years old. As we have seen, the frequency of birth of infants with Down syndrome increases dramatically in the pregnancies of older women (see figure 8.25).

Genetic Screening

When a pregnancy is diagnosed as being high risk, many women elect to undergo **amniocentesis,** a procedure that permits the prenatal diagnosis of many genetic disorders (figure 8.34). In the fourth month of pregnancy, a sterile hypodermic needle is inserted into the expanded uterus of the mother, and a small sample of the amniotic fluid bathing the fetus is removed. Within the fluid are free-floating cells derived from the fetus; once removed, these cells can be grown in cultures in the laboratory. During amniocentesis, the position of the needle and that of the fetus are usually observed by means of **ultrasound** (figure 8.35). The sound waves used in ultrasound are not harmful to mother or fetus, and they permit the person withdrawing the amniotic fluid to do so without damaging the fetus. In addition, ultrasound can be used to examine the fetus for signs of major abnormalities.

In recent years, physicians have increasingly turned to a new procedure for genetic screening called **chorionic villus sampling.** In this procedure, the physician removes cells from the chorion, a membranous part of the placenta that nourishes the fetus. This procedure can be used earlier in pregnancy (by the eighth week) and yields results much more rapidly than does amniocentesis, but can increase the risk of miscarriage.

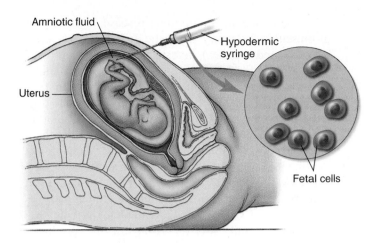

Figure 8.34 Amniocentesis.
A needle is inserted into the amniotic cavity, and a sample of amniotic fluid, containing some free cells derived from the fetus, is withdrawn into a syringe. The fetal cells are then grown in culture and their karyotype and many of their metabolic functions are examined.

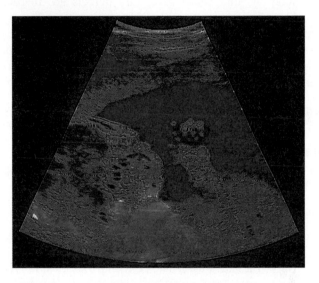

Figure 8.35 An ultrasound view of a fetus.
During the fourth month of pregnancy, when amniocentesis is normally performed, the fetus usually moves about actively. The head of the fetus (visualized in *green*) is to the *left*.

Genetic counselors look at three things in the cultures of cells obtained from amniocentesis or chorionic villus sampling:

1. **Chromosomal karyotype.** Analysis of the karyotype can reveal aneuploidy (extra or missing chromosomes) and gross chromosomal alterations.
2. **Enzyme activity.** In many cases, it is possible to test directly for the proper functioning of enzymes involved in genetic disorders. The lack of normal enzymatic activity signals the presence of the disorder. Thus, the lack of the enzyme responsible for breaking down phenylalanine signals PKU (phenylketonuria), the absence of the enzyme responsible for the breakdown of gangliosides indicates Tay-Sachs disease, and so forth.
3. **Genetic markers.** Genetic counselors can look for an association with known genetic markers. For

sickle-cell anemia, Huntington's disease, and one form of muscular dystrophy (a genetic disorder characterized by weakened muscles), investigators have found other mutations on the same chromosomes that, by chance, occur at about the same place as the mutations that cause those disorders. By testing for the presence of these other mutations, a genetic counselor can identify individuals with a high probability of possessing the disorder-causing mutations. Finding such mutations in the first place is a little like searching for a needle in a haystack, but persistent efforts have proved successful in these three disorders. The associated mutations are detectable because they alter the length of the DNA segments that DNA-cleaving enzymes produce when they cut strands of DNA at particular places, an approach described in more detail in chapter 10.

DNA Screening

The mutations that cause hereditary defects are frequently caused by alteration of a single DNA nucleotide within a key gene. Such spot differences between the version of a gene you have and the one another person has are called "single nucleotide polymorphisms," or SNPs. With the completion of the Human Genome Project, researchers have begun assembling a huge database of hundreds of thousands of SNPs. Each of us differs from the standard "type sequence" in several thousand gene-altering SNPs. Screening SNPs and comparing them to known SNP databases should soon allow genetic counselors to screen each patient for copies of genes leading to hereditary disorders such as cystic fibrosis and muscular dystrophy.

The possibility of convenient SNP profiling raises a number of important ethical issues. Your SNP profile, recorded on a DNA microchip no bigger than a postage stamp, can reveal if your profile correlates with any disease, with any tendency to excessive weight, or to any other medically significant condition. Should insurance companies or prospective employers have access to this information? Even more worrisome, with a big enough database it will become possible to search for multiple-SNP patterns that correlate with personality, intelligence, and other behavioral characteristics strongly influenced by genetic makeup. As these new technologies come online, society will need to give careful consideration to protecting the privacy of DNA profiles.

> **8.10** It has recently become possible to detect genetic defects early in pregnancy, allowing for appropriate planning by the prospective parents.

Exploring Current Issues

Additional Resources

Go to your campus library or look online to find the following articles, which further develop some of the concepts found in this chapter.

Bamshad, M. J. and S. E. Olson. (2003). Does race exist? *Scientific American,* 289(6), 78.

Lewis, R. (2003). A genetic checkup: lessons from Huntington disease and cystic fibrosis. *The Scientist,* 17(20), 24.

——. (2003). The bitter truth about PTC testing. *The Scientist,* 17(11), 23.

Nijhout, H. F. (2003). The importance of context in genetics. *American Scientist,* 91(5), 416.

U.S. Department of Energy Office of Science. (2003). Gene therapy. *Retrieved from www.ornl.gov/sci/techresources/Human_ Genome/medicine/genetherapy.shtml.*

In the News: Gene Therapy

Scientists have long sought a way to introduce "healthy" genes into humans that lack them. Such a therapy was actually achieved in 1990 for a rare blood disorder, but it has been difficult to advance the research further, as the adenovirus used to carry the healthy version of the gene proved inappropriate (it's a cold virus, and most people have antibodies directed against it). New virus vectors avoid these problems, and offer hope of cures for many hereditary disorders. Cystic fibrosis, for example, results from a defect in a single gene encoding a chloride ion transport protein, and gene therapy may finally offer a way to cure it.

Article 1. *Improvements in gene vectors renew hope for gene therapy.*
Article 2: *Altering ANDi: Inserting DNA into a primate.*
Article 3: *Gene therapy may allow us to combat debilitating Parkinson's disease.*

Find these articles, written by the author, online at www.mhhe.com/tlwessentials/exp8.

Summary

Mendel

8.1 Mendel and the Garden Pea

- Mendel studied inheritance using the garden pea (**figure 8.2**). He examined the patterns of inheritance of seven characteristics.

- Mendel used plants that were true-breeding for a particular characteristic as the P generation. He then crossed two P generation plants that expressed alternate forms of a characteristic. Their offspring were called the F_1 generation. He then allowed the F_1 plants to self-fertilize, giving rise to the F_2 generation (**figure 8.3**).

8.2 What Mendel Observed

- In Mendel's experiments, the F_1 generation plants all expressed the same alternative form, called the dominant trait. In the F_2 generation, 3/4 of the offspring expressed the dominant trait and 1/4 expressed the other trait, called the recessive trait. This is a 3:1 ratio (**table 8.1**). This 3:1 ratio was actually a 1:2:1 ratio— 1 true-breeding dominant, 2 not-true-breeding dominant, and 1 true-breeding recessive (**figure 8.5**).

8.3 Mendel Proposes a Theory

- Mendel's theory of heredity explains that characteristics are passed on from parent to offspring as alleles, one allele inherited from each parent (**figure 8.6**). If both of the alleles are the same (either both dominant or both recessive), the individual is homozygous for the trait. If the individual has one dominant and one recessive allele, it is heterozygous for the trait. An individual's alleles are referred to as its genotype and the expression of those alleles, what the individual looks like, is its phenotype.

- A Punnett square can be used to predict the genotypes and phenotypes of offspring of a cross (**figures 8.7 and 8.8**).

- A testcross is done to determine if a dominant individual is homozygous or heterozygous for a particular trait (**figure 8.9**).

8.4 Mendel's Laws

- Mendel's law of segregation states that alleles are distributed into gametes that combine randomly to produce offspring. Mendel's law of independent assortment states that genes located on different chromosomes are inherited independently of each other (**figure 8.10**).

From Genotype to Phenotype

8.5 How Genes Influence Traits

- Genes coded in DNA determine phenotype because DNA encodes the amino acid sequences of proteins and proteins are the outward expression of genes (**figure 8.11**). Alternative forms of a gene, alleles, result from mutations.

8.6 Why Some Traits Don't Show Mendelian Inheritance

- Not all traits follow the inheritance patterns outlined by Mendel. Continuous variation results when more than one gene contribute in a cumulative way to the phenotype (**figure 8.12**). Pleiotropic effects result when one gene influences more than one trait (**figure 8.13**).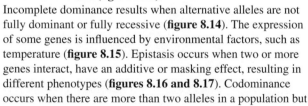
Incomplete dominance results when alternative alleles are not fully dominant or fully recessive (**figure 8.14**). The expression of some genes is influenced by environmental factors, such as temperature (**figure 8.15**). Epistasis occurs when two or more genes interact, have an additive or masking effect, resulting in different phenotypes (**figures 8.16 and 8.17**). Codominance occurs when there are more than two alleles in a population but

there isn't a dominant allele—two alleles are expressed resulting in phenotypic expression of both alleles (**figures 8.18 and 8.19**).

Chromosomes and Heredity

8.7 Chromosomes Are the Vehicles of Mendelian Inheritance

- Mendel explained that genes assort independently; this is because they are located on chromosomes that assort independently during meiosis. Morgan demonstrated this using an X-linked gene in fruit flies (**figure 8.21**).

8.8 Human Chromosomes

- Humans have 23 pairs of homologous chromosomes, for a total of 46 chromosomes. They have 22 pairs of autosomes and one pair of sex chromosomes. Nondisjunction occurs when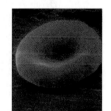
homologous pairs or sister chromatids fail to separate during meiosis, resulting in gametes with too many or too few chromosomes (**figure 8.23**).

- Nondisjunction of autosomes is usually fatal but the effects of nondisjunction of sex chromosomes is less severe (**figure 8.26**).

Human Hereditary Disorders

8.9 The Role of Mutations in Human Heredity

- Mutations can result in new alleles that can lead to genetic disorders. Pedigrees can be used to track and predict the inheritance of these disorders in families (**figure 8.27**). Disorders such as hemophilia (**figure 8.28**), sickle-cell anemia (**figures 8.29–8.31**), Tay-Sachs (**figure 8.32**), and Huntington's disease (**figure 8.33**) are all human genetic disorders.

8.10 Genetic Counseling and Therapy

- Some genetic disorders can be detected during pregnancy using methods such as amniocentesis (**figure 8.34**) and chorionic villus sampling. In these procedures, fetal DNA is examined by chromosomal karyotyping, enzyme activity analysis, and searching for genetic markers associated with genetic disorders.

Self-Test

1. Gregor Mendel studied the garden pea plants because
   a. pea plants are small, easy to grow, grow quickly, and produce lots of flowers and seeds.
   b. he knew about studies with the garden pea that had been done for hundreds of years, and wanted to continue them, using math—counting and recording differences.
   c. he knew that there were many varieties available with distinctive characteristics.
   d. All of these.
2. Mendel examined seven characteristics, such as flower color. He crossed plants with two different forms of a character (purple flowers and white flowers). In every case the first generation of offspring (F_1) were

   a. all purple flowers.
   b. half purple flowers and half white flowers.
   c. 3/4 purple and 1/4 white flowers.
   d. all white flowers.
3. Following question 2, if Mendel then randomly mated the offspring of that first cross, or F_1 generation, the offspring in the F_2 generation were
   a. all purple flowers.
   b. half purple flowers and half white flowers.
   c. 3/4 purple and 1/4 white flowers.
   d. all white flowers.

4. Mendel then studied his results, and proposed a set of hypotheses to explain them. The basis of these hypotheses is that parents transmit
   a. traits directly to their offspring and they are expressed.
   b. some factor, or information, about traits to their offspring and it may or may not be expressed.
   c. some factor, or information, about traits to their offspring and it will always be expressed sooner or later in some generation.
   d. some factor, or information, about traits to their offspring and it is expressed in every generation, perhaps in a "blended" form with information from the other parent.

5. Mendel's first law says that in one gamete (an egg or sperm)
   a. all of the parental information about a trait (both alleles) will be present.
   b. half of the parental information about a trait (one allele) will be present.
   c. 1/3 of the parental information about a trait (one allele) will be present.
   d. 1/4 of the parental information about a trait (one allele) will be present.

6. Each "chunk" of information, a gene, carries the directions for the body to produce
   a. an amino acid sequence that then folds into a protein that has a particular function.
   b. a protein, carbohydrate, or lipid that you need to build your organs.
   c. a trait directly.
   d. the nutrients your body needs to stay healthy.

7. Organisms don't show Mendelian inheritance in all traits (human eye color, for instance). Some of the factors involved include
   a. age and health of the organism.
   b. multiple traits all being expressed at once.
   c. multiple genes affecting one trait, environmental effects.
   d. one gene coding for multiple amino acid sequences with different effects.

8. One of the occurrences during meiosis that is of major importance to the study of inheritance is
   a. in telophase II when you finally have four nuclear envelopes.
   b. the breakdown and removal of some of the chromosomal material in gametes.
   c. the occasional mega-duplication of one chromosome into six or more chromosomes.
   d. the crossing over and information exchange between two nonsister chromatids.

9. There are a number of ways that genetic information can be changed, resulting in offspring that show characteristics not present in the genome of the parents. Some of the mechanisms are
   a. nondisjunction and mutation.
   b. amniocentesis and hemophilia.
   c. dominance and recession.
   d. All of these are mechanisms for changing genetic information.

10. If parents are concerned about their risk of producing children with serious genetic defects
    a. genetic infant design is now available.
    b. genetic correction of defects is now available.
    c. genetic modification of the parents is now available.
    d. genetic screening and prenatal diagnosis is now available.

Visual Understanding

1. **Figure 8.10** Using the four gametes shown from the F₁ generation, how many possible crosses are there? Remember that a gamete type could cross with another of the same type (such as *RY* × *RY*). Draw a Punnett square for each cross, and list the ratios of genotypes and phenotypes for the F₂ generation of that cross.

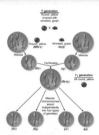

2. **Figure 8.19** Use Punnett squares to illustrate whether a type A female and a type B male can have a child with type O blood.

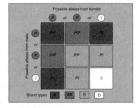

Challenge Questions

Mendel As Mendel struggled with understanding inheritance and formed his laws, how would the outcome have been different if he had only chosen five characteristics to study—flower position, pod shape, plant height, flower color, and seed color?

From Genotype to Phenotype Kim and Su-Ling are doing fruit fly crosses in their biology class. Kim wants to test whether the female they have is homozygous or heterozygous for red eyes by mating with a red-eyed male. How should Su-Ling explain the difficulty to him?

Chromosomes and Heredity Your biology class is collecting information on heredity. Michael realizes that he, along with three of his four brothers, are color blind, but his four sisters are not, and neither are his parents or his grandparents. Can you help Michael understand what happened?

Human Hereditary Disorders Kuzungu is a child orphaned by civil war in her country and raised in a group home. She has sickle-cell anemia, and type AB blood. Two couples who believe they are her grandparents ask you, a genetic counselor, to help them determine the truth. What do you suggest?

Online Learning Center

Visit the Online Learning Center for this chapter at www.mhhe.com/tlwessentials/ch8 for quizzes, animations, interactive learning exercises, and other study tools. At the site you will also find extended answers to the end-of-chapter questions.

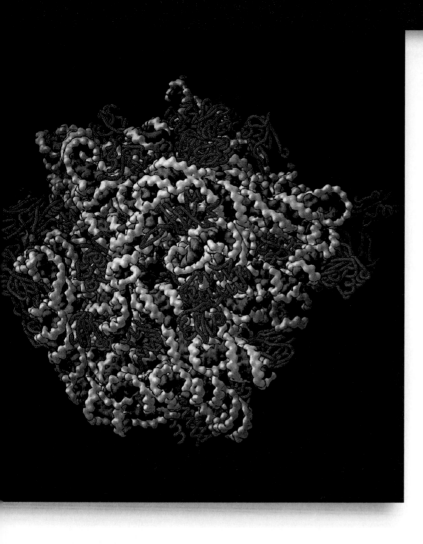

9

How Genes Work

R ibosomes like the one you see here are very complex cellular machines that assemble proteins, using information contributed from RNA transcripts of genes to determine the amino acid sequence of the new protein. Each ribosome is made of over 50 different proteins (shown here in gold), as well as three chains of RNA composed of some 3,000 nucleotides (shown here in gray). It has been traditionally assumed that the proteins in a ribosome act as enzymes to catalyze the amino acid assembly process, with the RNA acting as a scaffold to position the proteins. In the year 2000 we learned the reverse to be true. Powerful X-ray diffraction studies revealed the complete detailed structure of a ribosome at atomic resolution. Unexpectedly, the many proteins of a ribosome are scattered over its surface like decorations on a Christmas tree. The role of these proteins seems to be to stabilize the many bends and twists of the RNA chains, the proteins acting like spot-welds between the RNA strands they touch. Importantly, there are no proteins on the inside of the ribosome where the chemistry of protein synthesis takes place—just twists of RNA. Thus, it is the ribosome's RNA, not its proteins, that catalyzes the joining together of amino acids! Clearly, our knowledge of how genes work is still increasing, often adjusting what seem to be fundamental concepts.

9.1 The Griffith Experiment

As we learned in chapters 7 and 8, chromosomes contain genes, which, in turn, contain hereditary information. However, Mendel's work left a key question unanswered: What *is* a gene? When biologists began to examine chromosomes in their search for genes, they soon learned that chromosomes are made of two kinds of macromolecules, both of which you encountered in chapter 3: **proteins** (long chains of *amino acid subunits* linked together in a string) and **DNA** (deoxyribonucleic acid—long chains of *nucleotide* subunits linked together in a string). It was possible to imagine that either of the two was the stuff that genes are made of—information might be stored in a sequence of different amino acids, or in a sequence of different nucleotides. But which one is the stuff of genes, protein or DNA? This question was answered clearly in a variety of different experiments, all of which shared the same basic design: If you separate the DNA in an individual's chromosomes from the protein, which of the two materials is able to change another individual's genes?

In 1928, British microbiologist Frederick Griffith made a series of unexpected observations while experimenting with pathogenic (disease-causing) bacteria. When he infected mice with a virulent strain of *Streptococcus pneumoniae* bacteria (then known as *Pneumococcus*), the mice died of blood poisoning. However, when he infected similar mice with a mutant strain of *S. pneumoniae* that lacked the virulent strain's polysaccharide coat, the mice showed no ill effects. The coat was apparently necessary for infection. The normal pathogenic form of this bacterium is referred to as the S form because it forms smooth colonies in a culture dish. The mutant form, which lacks an enzyme needed to manufacture the polysaccharide coat, is called the R form because it forms rough colonies.

To determine whether the polysaccharide coat itself had a toxic effect, Griffith injected dead bacteria of the virulent S strain into mice; the mice remained perfectly healthy. Finally, he injected mice with a mixture containing dead S bacteria of the virulent strain and live, coatless R bacteria, each of which by itself did not harm the mice (figure 9.1). Unexpectedly, the mice developed disease symptoms and many of them died. The blood of the dead mice was found to contain high levels of live, virulent *Streptococcus* type S bacteria, which had surface proteins characteristic of the live (previously R) strain. Somehow, the information specifying the polysaccharide coat had passed from the dead, virulent S bacteria to the live, coatless R bacteria in the mixture, permanently transforming the coatless R bacteria into the virulent S variety.

> **9.1 Hereditary information can pass from dead cells to living ones and transform them.**

Transformation

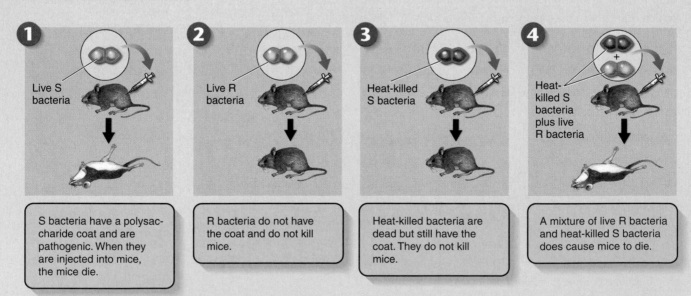

1 Live S bacteria
S bacteria have a polysaccharide coat and are pathogenic. When they are injected into mice, the mice die.

2 Live R bacteria
R bacteria do not have the coat and do not kill mice.

3 Heat-killed S bacteria
Heat-killed bacteria are dead but still have the coat. They do not kill mice.

4 Heat-killed S bacteria plus live R bacteria
A mixture of live R bacteria and heat-killed S bacteria does cause mice to die.

Figure 9.1 How Griffith discovered transformation.
Transformation, the movement of a gene from one organism to another, provided some of the key evidence that DNA is the genetic material. Griffith found that extracts of dead pathogenic strains of the bacterium *Streptococcus pneumoniae* can "transform" live harmless strains into live pathogenic strains.

9.2 The Avery and Hershey-Chase Experiments

The Avery Experiments

The agent responsible for transforming *Streptococcus* went undiscovered until 1944. In a classic series of experiments, Oswald Avery and his coworkers Colin MacLeod and Maclyn McCarty characterized what they referred to as the "transforming principle." Avery and his colleagues prepared the same mixture of dead S *Streptococcus* and live R *Streptococcus* that Griffith had used, but first they removed as much of the protein as they could from their preparation of dead S *Streptococcus,* eventually achieving 99.98% purity. Despite the removal of nearly all protein from the dead S *Streptococcus,* the transforming activity was not reduced. Moreover, the properties of the transforming principle resembled those of DNA in several ways:

Same chemistry as DNA.   When the purified principle was analyzed chemically, the array of elements agreed closely with DNA.

Same behavior as DNA.   In an ultracentrifuge, the transforming principle migrated like DNA; in electrophoresis and other chemical and physical procedures, it also acted like DNA.

Not affected by lipid and protein extraction.   Extracting the lipid and protein from the purified transforming principle did not reduce its activity.

Not destroyed by protein- or RNA-digesting enzymes.   Protein-digesting enzymes did not affect the principle's activity, nor did RNA-digesting enzymes.

Destroyed by DNA-digesting enzymes.   The DNA-digesting enzyme destroyed all transforming activity.

The evidence was overwhelming. They concluded that "a nucleic acid of the deoxyribose type is the fundamental unit of the transforming principle of *Pneumococcus* Type III"—in essence, that DNA is the hereditary material.

The Hershey-Chase Experiment

Avery's result was not widely appreciated at first, because most biologists still preferred to think that genes were made of proteins. In 1952, however, a simple experiment carried out by Alfred Hershey and Martha Chase was impossible to ignore. The team studied the genes of viruses that infect bacteria. These viruses attach themselves to the surface of bacterial cells and inject their genes into the interior; once inside, the genes take over the genetic machinery of the cell and order the manufacturing of hundreds of new viruses. When mature, the progeny viruses burst out to infect other cells. These bacteria-infecting viruses have a very simple structure: a core of DNA surrounded by a coat of protein.

In this experiment, Hershey and Chase used radioactive isotopes to "label" the DNA and protein of the viruses. In one preparation, the viruses were grown so that their DNA

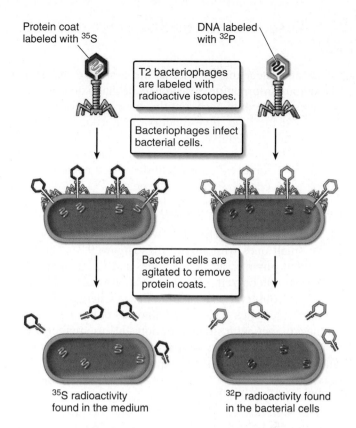

Figure 9.2   The Hershey-Chase experiment.
The experiment that convinced most biologists that DNA is the genetic material was carried out soon after World War II, when radioactive isotopes were first becoming commonly available to researchers. Hershey and Chase used different radioactive labels to "tag" and track protein and DNA. They found that when bacterial viruses inserted their genes into bacteria to guide the production of new viruses, it was DNA and not protein that was inserted. More specifically, ^{35}S radioactivity did not enter infected bacterial cells and ^{32}P radioactivity did. Clearly the virus DNA, not the virus protein, was responsible for directing the production of new viruses.

contained radioactive phosphorus (^{32}P); in another preparation, the viruses were grown so that their protein coats contained radioactive sulfur (^{35}S). After the labeled viruses were allowed to infect bacteria, Hershey and Chase shook the suspensions forcefully to dislodge attacking viruses from the surface of bacteria, used a rapidly spinning centrifuge to isolate the bacteria, and then asked a very simple question: What did the viruses inject into the bacterial cells, protein or DNA? They found that the bacterial cells infected by viruses containing the ^{32}P label had labeled tracer in their interiors; cells infected by viruses containing the ^{35}S labeled tracer did not. The conclusion was clear: the genes that viruses use to specify new viruses are made of DNA and not protein (figure 9.2).

> **9.2   Several key experiments demonstrated conclusively that DNA, not protein, is the hereditary material.**

9.3 Discovering the Structure of DNA

As it became clear that DNA was the molecule that stored the hereditary information, researchers began to question how this nucleic acid could carry out the complex function of inheritance. At the time, investigators did not know what the DNA molecule looked like.

We now know that DNA is a long, chainlike molecule made up of subunits called **nucleotides.** Each nucleotide has three parts: a central sugar, a phosphate (PO_4) group, and an organic base. The sugar and the phosphate group are the same in every nucleotide of DNA, but there are four different kinds of bases, two large ones with double-ring structures and two small ones with single rings (figure 9.3). The large bases, called **purines,** are **A** (adenine) and **G** (guanine). The small bases, called **pyrimidines,** are **C** (cytosine) and **T** (thymine). A key observation, made by Erwin Chargaff, was that DNA molecules always had equal amounts of purines and pyrimidines. In fact, with slight variations due to imprecision of measurement, the amount of A always equals the amount of T, and the amount of G always equals the amount of C. This observation (A = T, G = C), known as **Chargaff's rule,** suggested that DNA had a regular structure.

The significance of Chargaff's rule became clear in 1953 when the British chemist Rosalind Franklin carried out an X-ray diffraction experiment. In these experiments, DNA molecules are bombarded with X-ray beams, and when individual rays encounter atoms, their paths are bent or diffracted like a thrown ball bounces off or around an object. Each atomic encounter creates a pattern on photographic film that looks like the ripples created by tossing a rock into a smooth lake (figure 9.4a). Franklin's results suggested that the DNA molecule had the shape of a coiled spring or a corkscrew, a form called a **helix.**

Franklin's work was shared with two researchers at Cambridge University, Francis Crick and James Watson, before it was published. Using Tinkertoy-like models of the bases, Watson and Crick deduced the true structure of DNA (figure 9.4b): The DNA molecule is a **double helix,** a winding staircase of two strands whose bases face one another. Chargaff's rule is a direct reflection of this structure—every bulky purine on one strand is paired with a slender pyrimidine on the other strand. Specifically, A pairs with T, and G pairs with C (figure 9.4c). Because hydrogen bonds can form between the **base pairs,** the molecule keeps a constant thickness.

Watson and Crick continued on in the area of DNA research but Franklin's career was cut short due to her untimely death due to cancer at the age of 37.

9.3 The DNA molecule has two strands of nucleotides held together by hydrogen bonds between bases. The two strands wind into a double helix.

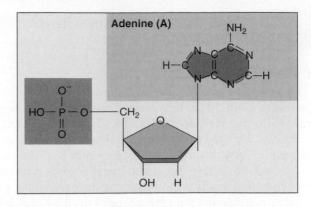

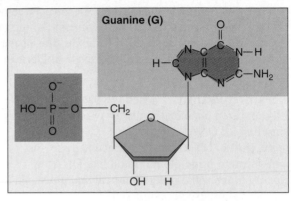

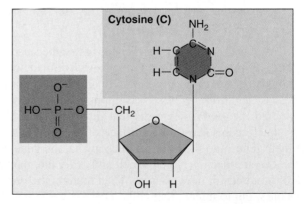

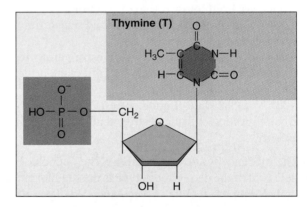

Figure 9.3 The four nucleotide subunits that make up DNA.

The nucleotide subunits of DNA are composed of three parts: a central five-carbon sugar, a phosphate group, and an organic, nitrogen-containing base.

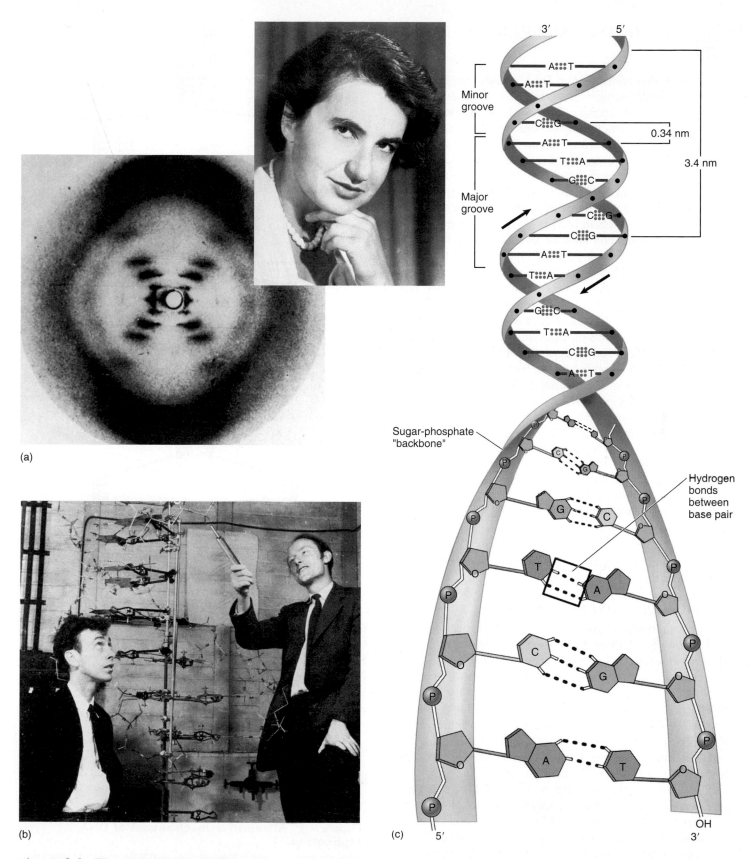

Figure 9.4 The DNA double helix.

(*a*) This X-ray diffraction photograph was made in 1953 by Rosalind Franklin (inset) in the laboratory of Maurice Wilkins. It suggested to Watson and Crick that the DNA molecule was a helix, like a winding staircase. (*b*) In 1953 Watson and Crick deduced the structure of DNA. James Watson (peering up at their homemade model of the DNA molecule) was a young American postdoctoral student, and Francis Crick (pointing) was an English scientist. (*c*) The dimensions of the double helix were suggested by the X-ray diffraction studies. In a DNA duplex molecule, only two base pairs are possible: adenine (A) with thymine (T) and guanine (G) with cytosine (C). A G–C base pair has three hydrogen bonds; an A–T base pair has only two.

9.4 How the DNA Molecule Replicates

The attraction that holds the two DNA strands together is the formation of weak hydrogen bonds between the bases that face each other from the two strands. That is why A pairs with T and not C—it can only form hydrogen bonds with T. Similarly, G can form hydrogen bonds with C but not T. In the Watson-Crick model of DNA, the two strands of the double helix are said to be *complementary* to each other. One chain of the helix can have any sequence of bases, of A, T, G, and C, but this sequence completely determines that of its partner in the helix. If the sequence of one chain is ATTGCAT, the sequence of its partner in the double helix must be TAACGTA. Each chain in the helix is a complementary mirror image of the other. This **complementarity** makes it possible for the DNA molecule to copy itself during cell division in a very direct manner. But, there are three possible alternatives as to how the DNA could serve as a template for the assembly of new DNA molecules.

First, the two strands of the double helix could separate and serve as templates for the assembly of two new strands by base pairing A with T and G with C, but then rejoin, preserving the original strand of DNA and forming an entirely new strand. This is called *conservative replication* (figure 9.5a). In the second alternative, the double helix need only "unzip" and assemble a new complementary chain along each single strand. This form of DNA replication is called *semiconservative replication* (figure 9.5b), because while the sequence of the original duplex is conserved after one round of replication, the duplex itself is not. Instead, each strand of the duplex becomes part of another duplex. In the third alternative, called *dispersive replication* (figure 9.5c), the original DNA would serve as a template for the formation of new DNA strands but the new and old DNA would be dispersed among the two daughter strands.

The Meselson-Stahl Experiment

The three alternative hypotheses of DNA replication were tested in 1958 by Matthew Meselson and Franklin Stahl of the California Institute of Technology. These two scientists grew bacteria in a medium containing the heavy isotope of nitrogen, ^{15}N, which became incorporated into the bases of the bacterial DNA (figure 9.6). After several generations, the DNA of these bacteria was denser than that of bacteria grown in a medium containing the lighter isotope of nitrogen, ^{14}N. Meselson and Stahl then transferred the bacteria from the ^{15}N medium to the ^{14}N medium and collected the DNA at various intervals.

By dissolving the DNA they had collected in a heavy salt called cesium chloride and then spinning the solution at very high speeds in an ultracentrifuge, Meselson and Stahl were able to separate DNA strands of different densities. The centrifugal forces caused the cesium ions to migrate toward the bottom of the centrifuge tube, creating a gradient of cesium concentration, and thus a gradation of density. Each DNA strand floats or sinks in the gradient until it reaches the position where its density exactly matches the density of the cesium there. Because ^{15}N strands are denser than ^{14}N strands, they migrate farther down the tube to a denser region of cesium.

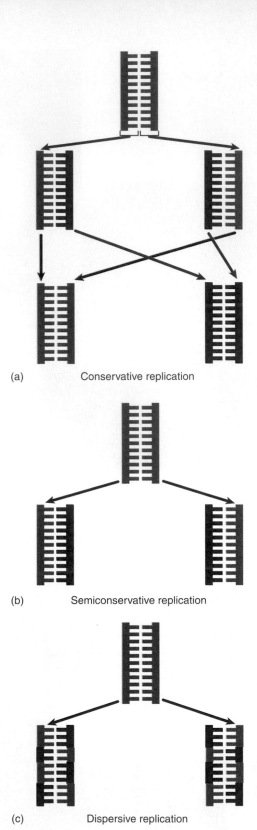

(a)   Conservative replication

(b)   Semiconservative replication

(c)   Dispersive replication

Figure 9.5 Alternative mechanisms of DNA replication.

(a) In conservative replication, the original DNA molecule (*blue*) is preserved, re-forming after serving as templates to make the new strands (*red*). (b) In semiconservative replication, the original DNA molecule separates and each strand serves as a template for forming a new strand so that the daughter DNA molecules contain one old and one new strand. (c) In dispersive replication, the original DNA molecule is used as a template to build new strands but the old and new strands are interspersed in the daughter DNA molecules.

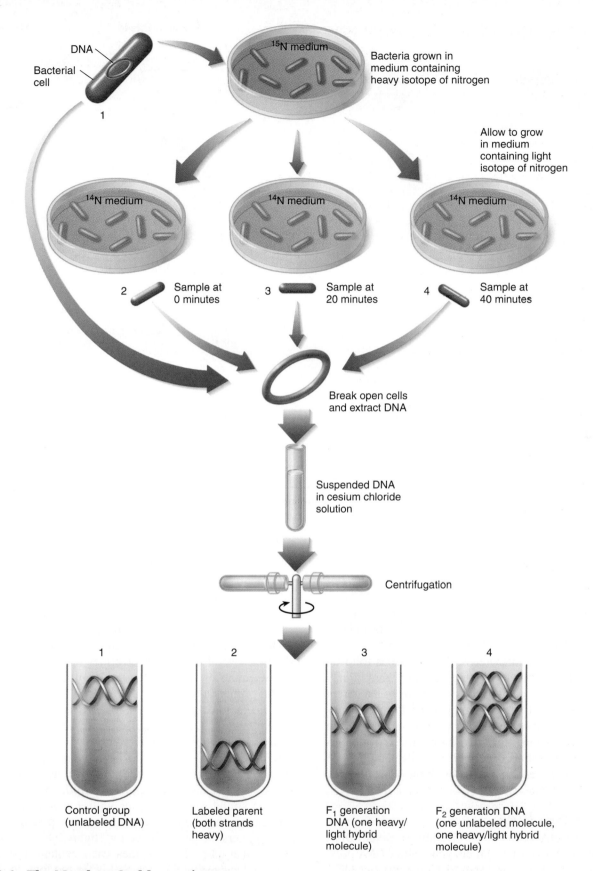

Figure 9.6 The Meselson-Stahl experiment.

Bacterial cells were grown for several generations in a medium containing a heavy isotope of nitrogen (^{15}N) and then were transferred to a new medium containing the normal lighter isotope (^{14}N). At various times thereafter, samples of the bacteria were collected, and their DNA was dissolved in a solution of cesium chloride, which was spun rapidly in a centrifuge. The labeled and unlabeled DNA settled in different areas of the tube because they differed in weight. The DNA with two heavy strands settled down toward the bottom of the tube. The DNA with two light strands settled higher up in the tube. The DNA with one heavy and one light strand settled in between the other two.

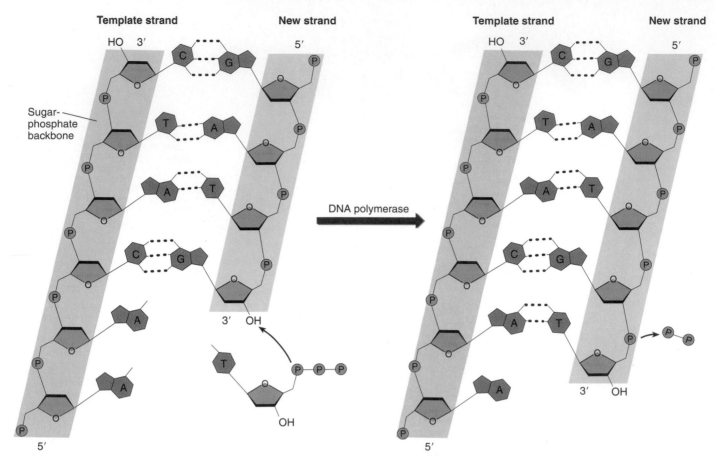

Figure 9.7 How nucleotides are added in DNA replication.

Nucleotides are added to the new growing strand of DNA by DNA polymerase. The addition of the nucleotides follows base pairing.

The DNA collected immediately after the transfer was all dense. However, after the bacteria completed their first round of DNA replication in the ^{14}N medium, the density of their DNA had decreased to a value intermediate between ^{14}N-DNA and ^{15}N-DNA. After the second round of replication, two density classes of DNA were observed, one intermediate and one equal to that of ^{14}N-DNA.

Meselson and Stahl interpreted their results as follows: after the first round of replication, each daughter DNA duplex was a hybrid possessing one of the heavy strands of the parent molecule and one light strand; when this hybrid duplex replicated, it contributed one heavy strand to form another hybrid duplex and one light strand to form a light duplex. Thus, this experiment clearly ruled out conservative and dispersive DNA replication, and confirmed the prediction of the Watson-Crick model that DNA replicates in a semiconservative manner.

How DNA Copies Itself

The copying of DNA before cell division is called **DNA replication** and is overseen by an enzyme called *DNA polymerase*. An enzyme called *helicase* first unwinds the DNA double helix, then DNA polymerase reads along each single strand and adds the correct complementary nucleotide (A pairs with T, G with C) at each position as it moves, creating a complementary strand (figure 9.7).

However, there are some limitations of the actions of DNA polymerase. First, it can only add to an existing strand; it cannot begin a strand. Another enzyme circumvents this difficulty by beginning the new strand with a section of nucleic acids called a *primer*. This happens at the place where the parent DNA molecule becomes unzipped, called the **replication fork.** At the replication fork, the polymerase very actively shuttles several hundred nucleotides up one strand, building a new strand of DNA called the **leading strand,** adding on to the primer in a continuous fashion. The phosphate group of a nucleotide, called the 5′ end, attaches to the sugar end, called the 3′ end, of the nucleotide at the end of the growing strand. The new strand assembles in a 5′ to 3′ direction. This reveals a second limitation; DNA polymerase can only build a strand of DNA in one direction, and so it assembles the other DNA strand, called the **lagging strand,** in segments. Each lagging strand segment begins with a primer, and the DNA polymerase then builds it away from the replication fork until it encounters the previous section (figure 9.8).

Eukaryotic chromosomes each contain a single, very long molecule of DNA, one that is far too long to copy conveniently all the way from one end to the other with a single replication fork. Each eukaryotic chromosome is instead copied in sections of about 100,000 nucleotides, each replication origin with its own replication fork.

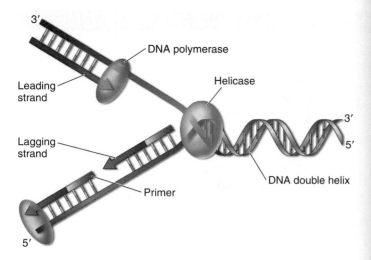

Figure 9.8   Building the leading and lagging strands.

DNA polymerase builds the leading strand as a continuous strand moving into the replication fork growing 5' to 3', but the lagging strand, also growing 5' to 3', is assembled moving away from the replication fork, in segments, each beginning with a primer.

Before the newly formed DNA molecules wind back into the double helix shape, the primers must be removed and the segments of DNA must be sealed together. Remember that the newly formed strand of DNA was assembled in sections on the lagging strand. These sections need to be covalently linked together. The enzyme that performs that function is *DNA ligase*. DNA ligase joins the ends of newly synthesized segments of DNA after the primers have been removed, resulting in one continuous strand of DNA (figure 9.9).

The enormous amount of DNA that resides within the 10 trillion cells of your body represents a long series of DNA replications, starting with the DNA of a single cell—the fertilized egg. Living cells have evolved many mechanisms to avoid errors during DNA replication and to preserve the DNA from damage. These mechanisms of **DNA repair** proofread the strands of each daughter cell against one another for accuracy and correct any mistakes. But the proofreading is not perfect. If it were, no mistakes such as mutations would occur, no variation in gene sequence would result, and evolution would come to a halt, for as we discussed in chapter 8, genetic variation that alters the phenotype is the raw material on which natural selection acts and evolution occurs. We will return to the subject of mutations and evolution in chapter 11.

9.4   The basis for the great accuracy of DNA replication is complementarity. DNA's two strands are complementary mirror images of each other, so either one can be used as a template to reconstruct the other.

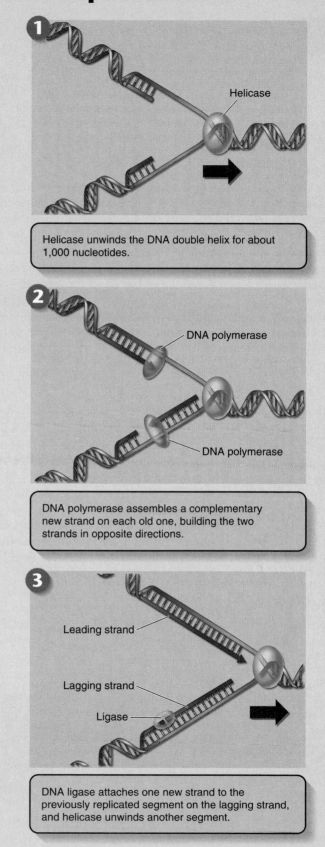

DNA Replication

Helicase unwinds the DNA double helix for about 1,000 nucleotides.

DNA polymerase assembles a complementary new strand on each old one, building the two strands in opposite directions.

DNA ligase attaches one new strand to the previously replicated segment on the lagging strand, and helicase unwinds another segment.

Figure 9.9   How DNA replication works.

9.5 Transcription

The discovery that genes are made of DNA left unanswered the question of how the information in DNA is used. How does a string of nucleotides in a spiral molecule determine if you have red hair? We now know that the information in DNA is arrayed in little blocks, like entries in a dictionary, each block a gene specifying a protein. These proteins determine what a particular cell will be like.

Just as an architect protects building plans from loss or damage by keeping them safe in a central place and issuing only blueprint copies to on-site workers, so your cells protect their DNA instructions by keeping them safe within a central DNA storage area, the nucleus. The DNA never leaves the nucleus. Instead, "blueprint" copies of particular genes within the DNA instructions are sent out into the cell to direct the assembly of proteins. These working copies of genes are made of ribonucleic acid (RNA) rather than DNA. Recall that RNA is the same as DNA except that the sugars in RNA have an extra oxygen atom and T is replaced by a similar pyrimidine base called uracil, U (see figure 3.27). The path of information is thus:

$$DNA \rightarrow RNA \rightarrow protein$$

This information path is often called the *central dogma,* because it describes the key organization used by your cells to express their genes (figure 9.10). A cell uses three kinds of RNA in the synthesis of proteins: messenger RNA (mRNA), ribosomal RNA (rRNA), and transfer RNA (tRNA).

The use of information in DNA to direct the production of particular proteins is called **gene expression.** Gene expression occurs in two stages: in the first stage, called **transcription,** an mRNA molecule is synthesized from a gene within the DNA; in the second stage, called **translation,** this mRNA is used to direct the production of a protein.

The Transcription Process

The RNA copy of a gene used in the cell to produce a protein is called **messenger RNA (mRNA)**—it is the messenger that

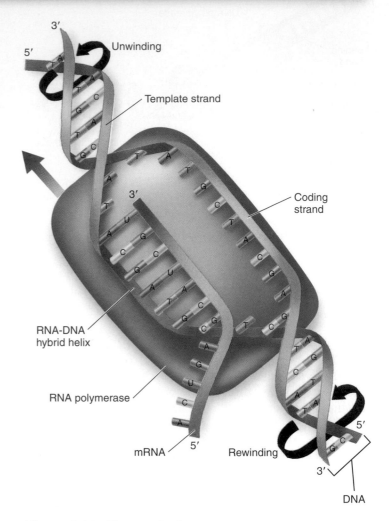

Figure 9.11   Transcription.
One of the strands of DNA functions as a template on which nucleotide building blocks are assembled into mRNA by RNA polymerase as it moves along the DNA strand.

conveys the information from the nucleus to the cytoplasm. The copying process that makes the mRNA is called transcription—just as monks in monasteries used to make copies of manuscripts by faithfully transcribing each letter, so enzymes within the nuclei of your cells make mRNA copies of your genes by faithfully complementing each nucleotide.

In your cells, the transcriber is a large and very sophisticated protein called **RNA polymerase.** It binds to one strand of a DNA double helix at a particular site called a *promoter* and then moves along the DNA strand like a train on a track. As it goes along the DNA "track," the polymerase pairs each nucleotide with its complementary RNA version (A with U, G with C), building an mRNA chain behind it as it moves along the DNA strand (figure 9.11).

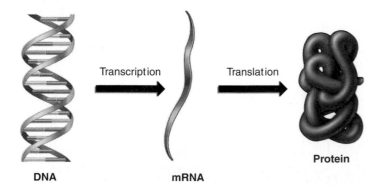

Transcription

Translation

Protein

DNA

mRNA

Figure 9.10   The central dogma of gene expression.
Through the production of mRNA (transcription) and the synthesis of proteins (translation), the information contained in DNA is expressed.

9.5   Transcription is the production of an mRNA copy of a gene by the enzyme RNA polymerase.

9.6 Translation

The Genetic Code

The essence of Mendelian genetics is that information determining hereditary traits, traits passed from parent to child, is encoded information. The information is written within the chromosomes in blocks called genes. Genes affect Mendelian traits by directing the production of particular proteins. The essence of gene expression, of using your genes, is reading the information encoded within DNA and using that information to direct the production of the correct protein.

To correctly read a gene, a cell must translate the information encoded in DNA into the language of proteins—that is, it must convert the *order* of the gene's nucleotides into the order of amino acids in a protein, a process called **translation.** The rules that govern this translation are called the **genetic code.**

The mRNA is transcribed from the gene in a linear sequence, one nucleotide following another, beginning at a fixed starting point on the gene called the promoter. There, the RNA polymerase binds to the DNA and begins its assembly of the mRNA. Transcription ends when the RNA polymerase reaches a certain nucleotide sequence that signals it to stop.

However, the mRNA is not translated in this same way. The mRNA is "read" by a ribosome in three-nucleotide units. Each three-nucleotide sequence of the mRNA corresponds to a particular amino acid, and is called a **codon.** Biologists worked out which codons correspond to which amino acids by trial-and-error experiments carried out in test tubes. In these experiments, investigators used artificial mRNAs to direct the synthesis of proteins in the tube, and then looked to see the sequence of amino acids in the newly formed proteins. An mRNA that was a string of UUUUUU …, for example, produced a protein that was a string of phenylalanine (PHE) amino acids, telling investigators that the codon UUU corresponded to the amino acid PHE. The entire **genetic code dictionary** is presented in figure 9.12. Because at each position of a three-letter codon, any of the four different nucleotides (U, C, A, G) may be used, there are 64 different possible three-letter codons ($4 \times 4 \times 4 = 64$) in the genetic code.

The genetic code is universal, the same in practically all organisms. GUC codes for valine in bacteria, in fruit flies, in eagles, and in your own cells. The only exception biologists have ever found to this rule is in the way in which cell organelles that contain DNA (mitochondria and chloroplasts) and a few microscopic protists read the "stop" codons. In every other instance, the same genetic code is employed by all living things.

The Genetic Code

First Letter	Second Letter								Third Letter
	U		**C**		**A**		**G**		
U	UUU UUC	Phenylalanine	UCU UCC	Serine	UAU UAC	Tyrosine	UGU UGC	Cysteine	U C
	UUA UUG	Leucine	UCA UCG		UAA UAG	Stop Stop	UGA UGG	Stop Tryptophan	A G
C	CUU CUC CUA CUG	Leucine	CCU CCC CCA CCG	Proline	CAU CAC	Histidine	CGU CGC CGA CGG	Arginine	U C A G
					CAA CAG	Glutamine			
A	AUU AUC AUA	Isoleucine	ACU ACC ACA ACG	Threonine	AAU AAC	Asparagine	AGU AGC	Serine	U C A G
	AUG	Methionine; Start			AAA AAG	Lysine	AGA AGG	Arginine	
G	GUU GUC GUA GUG	Valine	GCU GCC GCA GCG	Alanine	GAU GAC	Aspartate	GGU GGC GGA GGG	Glycine	U C A G
					GAA GAG	Glutamate			

Figure 9.12   The genetic code (RNA codons).

A codon consists of three nucleotides read in the sequence shown. For example, ACU codes for threonine. The first letter, A, is in the First Letter column; the second letter, C, is in the Second Letter column; and the third letter, U, is in the Third Letter column. Each of the mRNA codons is recognized by a corresponding anticodon sequence on a tRNA molecule. Some tRNA molecules recognize more than one codon in mRNA, but they always code for the same amino acid. In fact, most amino acids are specified by more than one codon. For example, threonine is specified by four codons, which differ only in the third nucleotide (ACU, ACC, ACA, and ACG).

Translating the RNA Message into Proteins

The final result of the transcription process is the production of an mRNA copy of a gene. Like a photocopy, the mRNA can be used without damage or wear and tear on the original. After transcription of a gene is finished, the mRNA passes out of the nucleus into the cytoplasm through pores in the nuclear membrane. There, translation of the genetic message occurs. In translation, organelles called **ribosomes** use the mRNA produced by transcription to direct the synthesis of a protein following the genetic code.

The Protein-Making Factory.   Ribosomes are the protein-making factories of the cell. Each is very complex, containing over 50 different proteins and several segments of **ribosomal RNA** (rRNA). Ribosomes use mRNA, the "blueprint" copies of nuclear genes, to direct the assembly of a protein.

Ribosomes are composed of two pieces, or subunits, one nested into the other like a fist in the palm of your hand (figure 9.13). The "fist" is the smaller of the two subunits. Its rRNA has a short nucleotide sequence exposed on the surface of the subunit. This exposed sequence is identical to a sequence called the leader region that occurs at the beginning of all genes. Because of this, an mRNA molecule binds to the exposed rRNA of the small subunit like a fly sticking to flypaper.

The Key Role of tRNA.   Directly adjacent to the exposed rRNA sequence are three small pockets or dents, called the A, P, and E sites (discussed shortly), in the surface of the ribosome that have just the right shape to bind yet a third kind of RNA molecule, **transfer RNA (tRNA).** It is tRNA molecules that bring amino acids to the ribosome to use in making proteins. tRNA molecules are chains about 80 nucleotides long, folded into a compact shape, with a three-nucleotide sequence

at one end and an amino acid attachment site on the other end (figure 9.14).

The three-nucleotide sequence, called the **anticodon,** is very important: it is the complementary sequence to 1 of the 64 codons of the genetic code! Special enzymes, called activating enzymes, match amino acids with their proper tRNAs, with the anticodon determining which amino acid will attach to a particular tRNA.

Because the first dent in the ribosome (the A site, which is the attachment site for tRNA) is directly adjacent to where the mRNA binds to the rRNA, three nucleotides of the mRNA are positioned directly facing the anticodon of the tRNA. Like the address on a letter, the anticodon ensures that an amino acid is delivered to its correct "address" on the mRNA where the ribosome is assembling the protein.

Making the Protein.   Once an mRNA molecule has bound to the small ribosomal subunit, the other larger ribosomal subunit binds as well, forming a complete ribosome. The ribosome then begins the process of translation (figure 9.15).

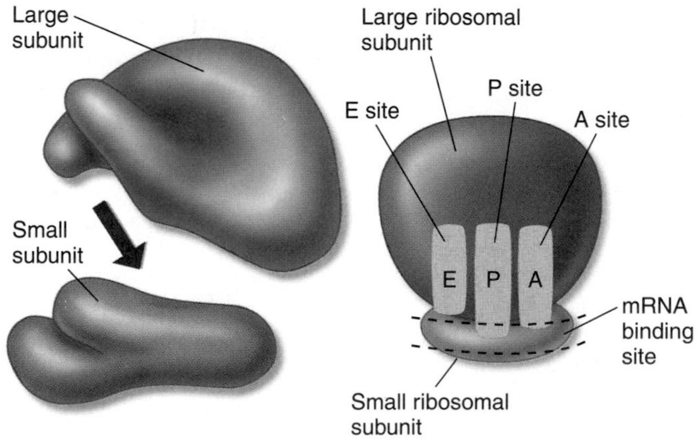

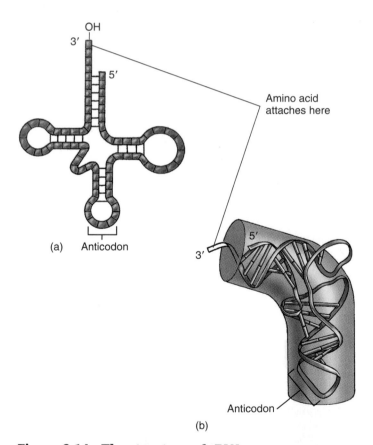

Figure 9.14   The structure of tRNA.

tRNA, like mRNA, is a long strand of nucleotides. However, unlike mRNA, hydrogen bonding occurs between its nucleotides, causing the strand to form hairpin loops, as seen in (a). The loops then fold up on each other to create the compact, three-dimensional shape seen in (b). Amino acids attach to the free, single-stranded —OH end of a tRNA molecule. A three-nucleotide sequence called the anticodon in the lower loop of tRNA interacts with a complementary codon on the mRNA.

Figure 9.13   A ribosome is composed of two subunits.

The smaller subunit fits into a depression on the surface of the larger one. The A, P, and E sites on the ribosome play key roles in protein synthesis.

Translation

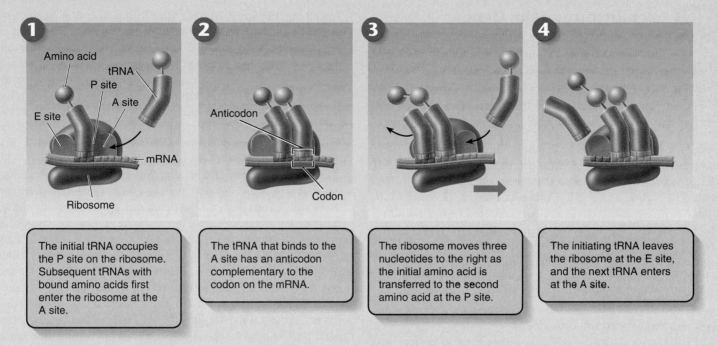

1 The initial tRNA occupies the P site on the ribosome. Subsequent tRNAs with bound amino acids first enter the ribosome at the A site.

2 The tRNA that binds to the A site has an anticodon complementary to the codon on the mRNA.

3 The ribosome moves three nucleotides to the right as the initial amino acid is transferred to the second amino acid at the P site.

4 The initiating tRNA leaves the ribosome at the E site, and the next tRNA enters at the A site.

Figure 9.15 How translation works.
The mRNA strand acts as a template for tRNA molecules. The appropriate tRNA is selected and positioned by the ribosome, which moves along the mRNA in three-nucleotide steps. tRNAs bring amino acids into the ribosome at the A site. A peptide bond is formed between the incoming amino acid and the growing polypeptide chain at the P site, and the empty tRNAs leave the ribosome at the E site.

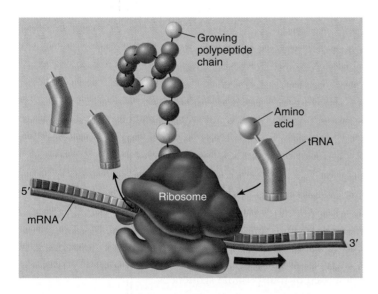

Figure 9.16 Ribosomes guide the translation process.
tRNA binds to an amino acid as determined by the anticodon sequence. Ribosomes bind the loaded tRNAs to their complementary sequences on the strand of mRNA. tRNA adds its amino acid to the growing polypeptide chain, which is released as the completed protein.

The mRNA begins to thread through the ribosome like a string passing through the hole in a donut. The mRNA passes through in short spurts, three nucleotides at a time, and at each burst of movement a new three-nucleotide codon is positioned opposite the A site in the ribosome, where tRNA molecules first bind.

As each new tRNA brings in an amino acid to each new codon presented at the A site, the old tRNA paired with the previous codon is passed over to the P site. The tRNA in the P site eventually shifts to the E site, as the amino acid it carried is attached to the end of a growing amino acid chain and the tRNA is released. So as the ribosome proceeds down the mRNA, one tRNA after another is selected to match the sequence of mRNA codons, until the end of the mRNA sequence is reached (figure 9.16). At this point, a "stop" codon is encountered which signals the end of the protein. The ribosome complex falls apart, and the newly made protein is released into the cell.

9.6 The genetic code dictates how a particular nucleotide sequence specifies a particular amino acid sequence. The genetic code is transcribed into mRNA and it is translated into a protein. Translation is the reading of mRNA by a ribosome, the sequence of mRNA codons dictating the assembly of a corresponding sequence of amino acids in a growing protein chain.

Introns

While it is tempting to think of a gene as simply the nucleotide version of a protein—an uninterrupted stretch of nucleotides that is read three at a time to make a chain of amino acids—this actually occurs only in prokaryotes. In eukaryotes, genes are fragmented. In these more complex genes, the DNA nucleotide sequences encoding the amino acid sequence of the protein (called **exons**) are interrupted frequently by extraneous "extra stuff" called **introns.** Imagine looking at an interstate highway from a satellite. Scattered randomly along the thread of concrete would be cars, some moving in clusters, others individually; most of the road would be bare. That is what a eukaryotic gene is like: scattered exons embedded within much longer sequences of introns. In humans, only 1% to 1.5% of the genome is devoted to the exons that encode proteins, while 24% is devoted to the noncoding introns within which these exons are embedded.

When a eukaryotic cell transcribes a gene, it first produces a **primary RNA transcript** of the entire gene. The primary transcript is then processed. First, enzyme-RNA complexes excise out the introns and join together the exons to form the shorter mature mRNA transcript that is actually translated into protein (figure 9.17). Because introns are excised from the RNA transcript before it is translated into protein, they do not affect the structure of the protein encoded by the gene in which they occur, despite the fact that introns represent over 90% of the nucleotide sequence of a typical human gene.

Protein synthesis in eukaryotes is more complex than in prokaryotes (figure 9.18). In addition to editing out intron sequences, enzymes add a *5' cap* (a modification of the first nucleotide) to the mRNA transcript that protects the 5' end of mRNA from being degraded during its long journey through the cytoplasm. The 5' cap also functions as a recognition site for the assembly of the ribosome, binding to the small ribosome. Another enzyme adds about 250 "A" nucleotides to the 3' end of the transcript. Called a *3' poly-A tail,* this long string of A nucleotides also protects the transcript from degradation and appears to make the transcript a better template for protein synthesis.

Why this crazy organization? It appears that many human genes can be spliced together in more than one way. In many instances, exons are not just random fragments, but rather functional modules. One exon encodes a straight stretch of protein, another a curve, yet another a flat place. Like mixing Tinkertoy parts, you can construct quite different assemblies by employing the same exons in different combinations and orders. With this sort of **alternative splicing,** the 25,000 genes of the human genome seem to encode as many as 120,000 different expressed messenger RNAs. It seems that added complexity in humans has been achieved not by gaining more gene parts (we have only about twice as many genes as a fruit fly), but rather by learning new ways to put them together.

Gene Families

Eukaryotic genes have other unique characteristics. For example, everything we have said in this chapter assumes that a chromosome carries one copy of a gene—one to make the enzyme that breaks down lactose, for instance, and one to encode the protein hemoglobin, which carries oxygen in your blood from lungs to tissues. However, most eukaryotic genes exist in multiple copies, clusters of almost identical sequences called **multigene families.** Multigene families may contain as few as three or as many as several hundred versions of a gene.

Transposons: Jumping Genes

Other DNA sequences are very unusual in that they are repeated hundreds of thousands of times, scattered randomly about on the chromosomes. These segments of DNA are called **transposable sequences** or **transposons.** Transposons have the remarkable ability to move about from one chromosomal location to another. Once every few thousand cell divisions, a transposon will copy itself and simply pick up and move elsewhere, jumping at random to a new location on the chromosome, taking the gene with it. Transposons appear to be molecular parasites. Fully 45% of the human genome is composed of transposable sequences.

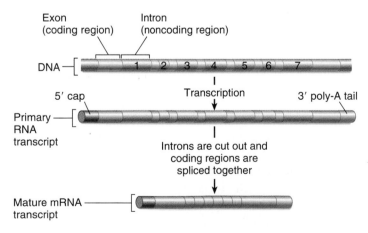

Figure 9.17   Processing eukaryotic RNA.
The gene shown here codes for a protein called ovalbumin. The ovalbumin gene and its primary transcript contain seven segments not present in the mRNA used by the ribosomes to direct the synthesis of the protein. These segments (introns) are removed by enzymes that cut them out and splice together the exons.

> **9.7**   The coding portions of most eukaryotic genes are embedded as exons within long sequences of noncoding introns. Many eukaryotic genes exist in multiple copies, some of which appear to have moved from one chromosomal location to another.

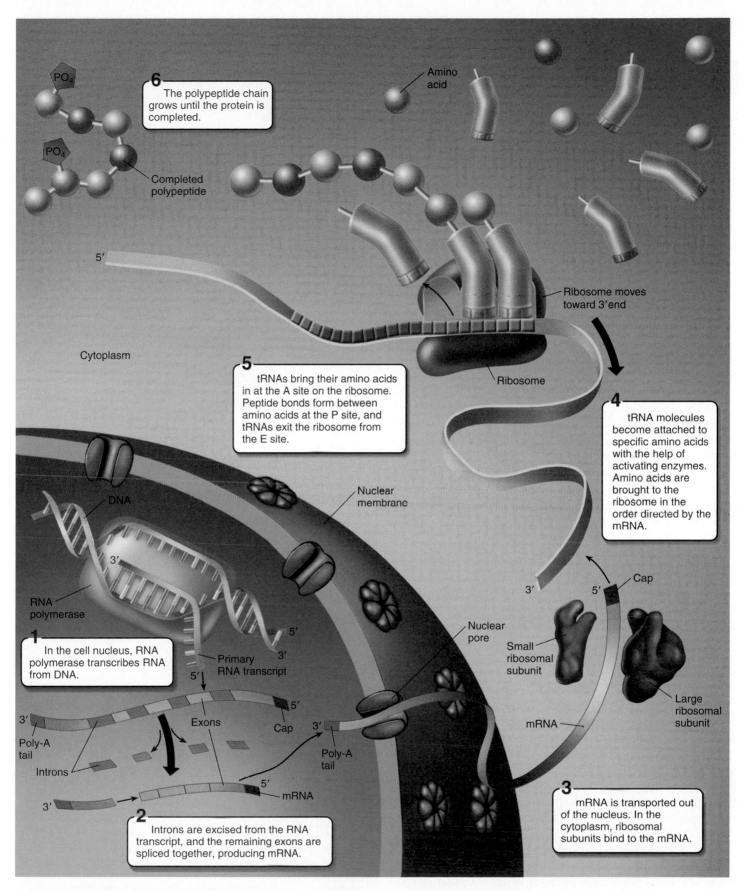

6 The polypeptide chain grows until the protein is completed.

PO_4

PO_4

Completed polypeptide

Amino acid

5′

Cytoplasm

Ribosome moves toward 3′end

Ribosome

5 tRNAs bring their amino acids in at the A site on the ribosome. Peptide bonds form between amino acids at the P site, and tRNAs exit the ribosome from the E site.

4 tRNA molecules become attached to specific amino acids with the help of activating enzymes. Amino acids are brought to the ribosome in the order directed by the mRNA.

Nuclear membrane

DNA

3′

RNA polymerase

1 In the cell nucleus, RNA polymerase transcribes RNA from DNA.

Primary RNA transcript

5′

3′

5′

3′

Exons

Cap

5′

Poly-A tail

Introns

3′

Poly-A tail

5′

mRNA

2 Introns are excised from the RNA transcript, and the remaining exons are spliced together, producing mRNA.

Nuclear pore

3′

5′

Cap

Small ribosomal subunit

Large ribosomal subunit

mRNA

3 mRNA is transported out of the nucleus. In the cytoplasm, ribosomal subunits bind to the mRNA.

Figure 9.18   How protein synthesis works in eukaryotes.

9.8 Turning Genes Off and On

Being able to translate a gene into a protein is only part of gene expression. Every cell must also be able to regulate when particular genes are used. Imagine if every instrument in a symphony played at full volume all the time, all the horns blowing full blast and each drum beating as fast and loudly as it could! No symphony plays that way, because music is more than noise—it is the controlled expression of sound. In the same way, growth and development are due to the controlled expression of genes, each brought into play at the proper moment to achieve precise and delicate effects.

Controlling Transcription

Cells control the expression of their genes by saying *when* individual genes are to be transcribed. At the beginning of each gene are special regulatory sites that act as points of control. Specific regulatory proteins within the cell bind to these sites, turning transcription of the gene off or on.

For a gene to be transcribed, the RNA polymerase has to bind to a **promoter,** a specific sequence of nucleotides that signals the beginning of a gene. In prokaryotes, gene expression is controlled by either blocking or allowing the RNA polymerase access to the promoter. Genes can be turned off by the binding of a **repressor,** a protein that binds to the DNA blocking the promoter. Genes can be turned on by the binding of an **activator,** a protein that makes the promoter more accessible to the RNA polymerase.

Repressors. Many genes are "negatively" controlled: they are turned off except when needed. In these genes, the regulatory site is located between the place where the RNA polymerase binds to the DNA (the promoter site) and the beginning edge of the gene. When a regulatory protein called a repressor is bound to its regulatory site, called the operator, its presence blocks the movement of the polymerase toward the gene. Imagine if you sat down to eat dinner and someone placed a brick wall between your chair and the table—you could not begin your meal until the wall was removed, any more than the polymerase can begin transcribing the gene until the repressor protein is removed.

To turn on a gene whose transcription is blocked by a repressor, all that is required is to remove the repressor. Cells do this by binding special "signal" molecules to the repressor protein; the binding causes the repressor protein to contort into a shape that doesn't fit DNA, and it falls off, removing the barrier to transcription. A specific example demonstrating how repressor proteins work is the set of genes called the *lac* operon in the bacterium *Escherichia coli.* An **operon** is a segment of DNA containing a cluster of genes that are transcribed as a unit. The *lac* operon consists of both protein-encoding genes (enzymes involved in breaking down the sugar lactose) and associated regulatory elements—the operator and promoter. When an *E. coli* encounters the sugar lactose, a metabolite of lactose called allolactose binds to the repressor protein

and induces a twist in its shape that causes it to fall from the DNA. RNA polymerase is no longer blocked, so it starts to transcribe the genes needed to break down the lactose to get energy (figure 9.19).

Activators. Because RNA polymerase binds to a specific promoter site on one strand of the DNA double helix, it is necessary that the DNA double helix unzip in the vicinity of this site for the polymerase protein to be able to sit down properly. In many genes, this unzipping cannot take place without the assistance of a regulatory protein called an activator that binds to the DNA in this region and helps it unwind. Just as in the case of the repressor protein described previously, cells can turn genes on and off by binding "signal" molecules to the activator protein. These molecules either prevent the activator from binding to the DNA or enable it to do so.

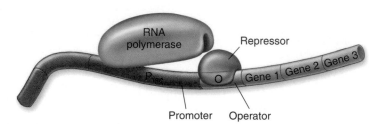

(a) *lac* operon is "repressed"

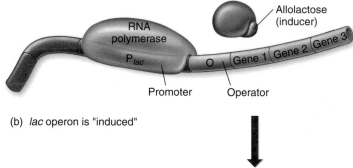

(b) *lac* operon is "induced"

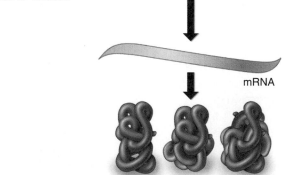

Figure 9.19 How the *lac* operon works.
(*a*) The *lac* operon is shut down ("repressed") when the repressor protein is bound to the operator site. Because promoter and operator sites overlap, RNA polymerase and the repressor cannot bind at the same time, any more than two people can sit in the same chair. (*b*) The *lac* operon is transcribed ("induced") when allolactose binds to the repressor protein changing its shape so that it can no longer sit on the operator site and block polymerase binding.

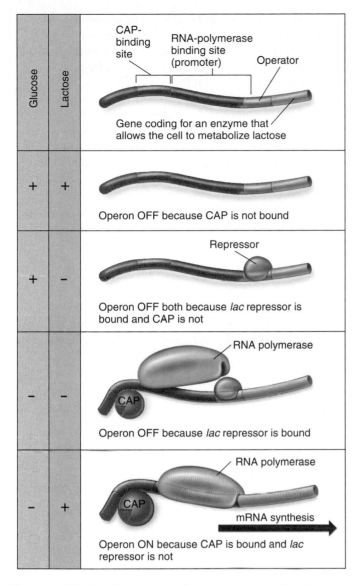

Figure 9.20   Activators and repressors at the *lac* operon.

Together, the *lac* repressor and the activator called CAP provide a very sensitive response to the cell's need for and ability to use lactose-metabolizing enzymes.

Why bother with activators? Imagine if you had to eat every time you encountered food! Activator proteins enable a cell to cope with this sort of problem. To understand how, let's return for a moment to the *lac* operon. When a bacterium encounters the sugar lactose, it may already have lots of energy and so does not need to break down more lactose. In such a bacterial cell, the activator protein is not able to bind to the DNA, and because the polymerase requires the activator to function, the *lac* operon is not expressed. A "low glucose" signal molecule needs to bind to the activator before the activator can "turn on" the polymerase. Without being prodded by this signal molecule, the activator protein, called CAP, cannot twist into the proper shape to fit the DNA unwinding site at the front of the operon; as a result, the genes are not transcribed, even though the repressor protein is not blocking the polymerase (figure 9.20).

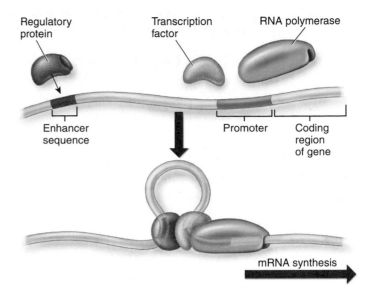

Figure 9.21   How enhancers work.

The enhancer site is located far away from the gene being regulated. Binding of a regulatory protein (*red*) to the enhancer allows the protein to interact with the transcription factors (*green*) associated with RNA polymerase, activating transcription.

Enhancers

A third level of control is exercised in eukaryotes by expanding access to the gene. To make the promoters of complexly-controlled genes accessible to regulatory proteins simultaneously, many eukaryotic genes possess special sequences called **enhancers.** These enhancer sequences bind specific regulatory proteins that interact with proteins called transcription factors, that help RNA polymerase find and attach to its promoter at the beginning of the structural gene. Unlike promoters and operators, which butt right up to the start of a gene, enhancers are usually located far away from the start of the gene, often thousands of nucleotides away. Although enhancers occur in exceptional instances in bacteria, they are the rule rather than the exception in eukaryotes.

How can regulatory proteins affect a promoter when they bind to the DNA at enhancer sites located far from the promoter? Apparently the DNA loops around so that the enhancer is positioned near the promoter. This brings the regulatory protein attached to the enhancer into direct contact with the transcription factor complex attached to the promoter (figure 9.21).

The enhancer mode of transcriptional control that has evolved in eukaryotes adds a great deal of flexibility to the control process. The positioning of regulatory sites at a distance permits a large number of different regulatory sequences scattered about the DNA to influence that particular gene.

> **9.8   Cells control the expression of genes by determining when they are transcribed. Some regulatory proteins block the binding of polymerase, and others facilitate it.**

A Closer Look

How Small RNAs Regulate Gene Expression

Thus far we have discussed gene regulation in terms of turning genes on and off by restricting RNA polymerase access to the gene so that it can't be transcribed. While this form of regulation, called transcriptional control, is how most genes are regulated, there are also other points in the process after transcription where gene expression is controlled. This level of control is called **posttranscriptional control.**

Posttranscriptional control acts on the primary RNA transcript. In some instances, RNA molecules, called *small RNAs,* regulate gene expression by interacting directly with primary RNA transcripts. The control may also occur during the splicing and editing of the primary transcript, resulting in the formation of different mRNAs by combining elements of the same primary transcript in different ways. Other controls involve regulating the process of moving the mRNA out of the nucleus and to the ribosome. Here we take a closer look at how small RNAs regulate gene expression.

Small RNAs form double-stranded loops.

These three miRNA molecules fold back to form hairpin loops because the sequences of the left and right halves are complementary, and form base pairs.

Recent experiments indicate that a class of RNA molecules loosely called **small RNAs** may play a major role in regulating gene expression by interacting directly with primary gene transcripts. Small RNAs are short segments of RNA ranging in length from 20 to 28 nucleotides. Researchers focusing on far larger messenger RNA (mRNA), transfer RNA (tRNA), and ribosomal RNA (rRNA) had not noticed these far smaller bits, tossing them out during experiments. The first hints of the existence of small RNAs emerged in 1993, when researchers reported in the nematode *Caenorhabditis elegans* the presence of tiny RNA molecules that don't encode any protein. These small RNAs appeared to regulate the activity of specific *C. elegans* genes.

Soon researchers found evidence of similar small RNAs in a wide range of other organisms. In the plant *Arabidopsis thaliana,* small RNAs seemed to be involved in the regulation of genes critical to early development, whereas in yeasts they were identified as the agents that silence genes in tightly packed regions of the genome. In the ciliated protozoan *Tetrahymena thermophila,* the loss of major blocks of DNA during development seems guided by small RNA molecules.

RNA interference.   What is going on here? How do small fragments of RNA act to regulate gene expression?

The first clue emerged in 1998, when researchers injected small stretches of double-stranded RNA into *C. elegans.* Double-stranded RNA forms when a single strand folds back in a hairpin loop, as shown in the figure on this page; this occurs because the two ends of the strand have a complementary nucleotide sequence, so that base pairing holds the strands together much as it does the strands of a DNA duplex. The result? The double-stranded RNA strongly inhibited the expression of the genes from which the double-stranded RNA had been generated. This kind of gene silencing, since seen in *Drosophila* and other organisms, is called **RNA interference.**

How small RNAs regulate gene expression.   In 2001 researchers identified an enzyme, dubbed "dicer," that appears to generate the small RNAs in the cell. Dicer chops double-stranded RNA molecules into little pieces. Two types of small RNA result: microRNAs (miRNAs) and small interfering RNAs (siRNAs).

miRNAs appear to act by binding directly to mRNAs and preventing their translation into protein. Researchers have identified over 100 different miRNAs, and are still trying to sort out how each functions, and which miRNAs occur in which species.

siRNAs appear to be the main agents of RNA interference, acting to degrade particular messenger RNAs after they have been transcribed but before they can be translated by the ribosomes. The exact way in which they achieve this degradation of selected gene transcripts is not yet known. Current data suggest that dicer delivers siRNAs to an enzyme complex called RISC, which searches out and degrades any mRNA molecules with a complementary sequence, as illustrated in the diagram on the facing page.

RNA interference appears to play a major role in epigenetic change. Epigenetic change is a change in gene expression that is passed from one generation to another but is not caused by changes in the DNA sequence of genes. Epigenetic regulation is in many cases the result of alterations in DNA packaging. As we saw in chapter 7, the DNA of eukaryotes is packaged in a highly compact form that enables it to fit into the cell nucleus. DNA is wrapped tightly around histone proteins to form nucleosomes and then the strand of nucleosomes is twisted into higher-order filaments. By altering how the strands are twisted, siRNAs can alter which genes are accessible for gene expression. Just how siRNAs achieve this change in chromatin shape is not known.

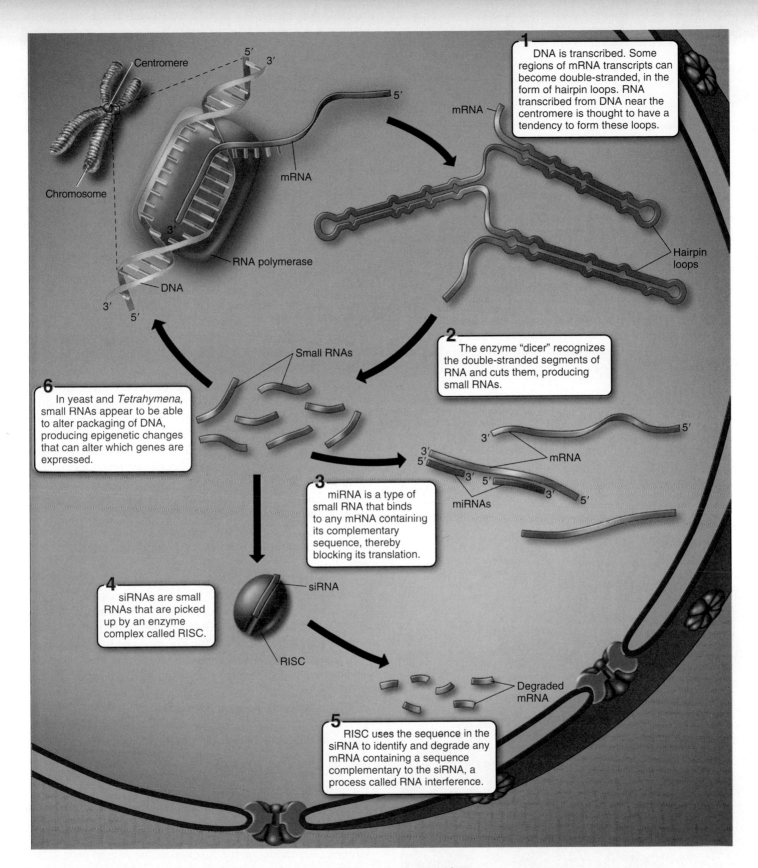

A closer look at how small RNAs may act to regulate gene expression.

Small RNAs are produced when double-stranded hairpin loops of an RNA transcript are cut. Although the details are not well understood, two types of small RNA, miRNA and siRNA, are thought to block gene expression within the nucleus at the level of the mRNA gene transcript, a process called RNA interference. As this diagram suggests, small RNAs are also thought to influence chromatin packaging in some organisms.

9.9 Mutation

There are two general ways in which the genetic message is altered: mutation and recombination. A change in the content of the genetic message—the base sequence of one or more genes—is referred to as a **mutation.** Return to our analogy about monks copying manuscripts. Their transcripts were for the most part accurate, but even an excellent transcriber could mistakenly transcribe a "d" instead of a "b." Some mutations alter the identity of a particular nucleotide, while others remove or add nucleotides to a gene. A change in the position of a portion of the genetic message is referred to as **recombination.** Some recombination events move a gene to a different chromosome; others alter the location of only part of a gene. The cells of eukaryotes contain an enormous amount of DNA, and the mechanisms that protect and proofread the DNA are not perfect. If they were, no variation would be generated.

Mistakes Happen

In fact, cells do make mistakes during replication, often causing a change in a cell's genetic message, or a mutation (figure 9.22). However, mutations are rare. Typically, a particular gene is altered in only one of a million gametes. If changes were common, the genetic instructions encoded in DNA would soon degrade into meaningless gibberish. Limited as it might seem, the steady trickle of change that does occur is the very stuff of evolution. Every difference in the genetic messages that specify different organisms arose as the result of genetic change.

The Importance of Genetic Change

All evolution begins with alterations in the genetic message: mutation creates new alleles, gene transfer and transposition alter gene location, reciprocal recombination shuffles and sorts these changes, and chromosomal rearrangement alters the organization of entire chromosomes. Some changes in germ-line tissue produce alterations that enable an organism to leave more offspring, and those changes tend to be preserved as the genetic endowment of future generations. An example of such a change was discussed in chapter 8, under "Environmental Effects." A genetic change occurred in the arctic fox that resulted in the production of fur pigments in warm weather but not in cold temperatures. This genetic change allowed the fox to be less conspicuous to predators, aiding its survival. This increased its ability to survive until the mating season, enabling them to leave more offspring, some of which inherited this same trait. Other changes reduce the ability of an organism to leave offspring. Those changes tend to be lost, as the organisms that carry them contribute fewer members to future generations. For example, a mutation that may cause an animal such as a zebra to lack speed will result in an early death at the jaws of a cheetah. However, the hunter is also affected by genetic mutations. A similar mutation in the cheetah will slow the animal down and result in an early death through starvation.

Evolution can be viewed as the selection of particular combinations of alleles from a pool of alternatives. The rate of evolution is ultimately limited by the rate at which these alternatives are generated. Genetic change through mutation and recombination provides the raw material for evolution.

Genetic changes in somatic cells do not pass on to offspring, and so they have less evolutionary consequence than germ-line changes. However, changes in the genes of somatic cells can have an important and immediate impact, particularly if the gene affects development or is involved with regulation of cell proliferation.

Kinds of Mutation

Because mutations can occur randomly in a cell's DNA, most mutations are detrimental, just as making a random change in a computer program usually worsens performance. The consequences of a detrimental mutation may be minor or catastrophic, depending on the function of the altered gene.

Mutations in Germ-Line Tissues. The effect of a mutation depends critically on the identity of the cell in which the mutation occurs. During the embryonic development of all multicellular organisms, there comes a point when cells destined to form gametes (germ-line cells) are segregated from those that will form the other cells of the body (somatic cells). Only

Figure 9.22 Mutation.
Fruit flies normally have one pair of wings, extending from the thorax. This fly is a *bithorax* mutant. Because of a mutation in a gene regulating a critical stage of development, it possesses *two* thorax segments and thus two sets of wings.

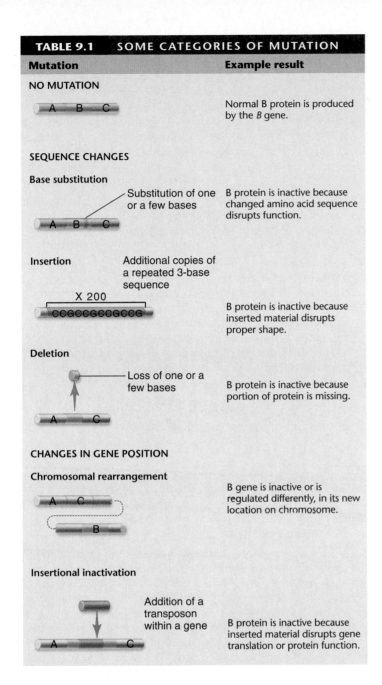

TABLE 9.1   SOME CATEGORIES OF MUTATION

Mutation	Example result
NO MUTATION	Normal B protein is produced by the *B* gene.
SEQUENCE CHANGES	
Base substitution — Substitution of one or a few bases	B protein is inactive because changed amino acid sequence disrupts function.
Insertion — Additional copies of a repeated 3-base sequence (X 200) CCGCCGCCGCCG	B protein is inactive because inserted material disrupts proper shape.
Deletion — Loss of one or a few bases	B protein is inactive because portion of protein is missing.
CHANGES IN GENE POSITION	
Chromosomal rearrangement	B gene is inactive or is regulated differently, in its new location on chromosome.
Insertional inactivation — Addition of a transposon within a gene	B protein is inactive because inserted material disrupts gene translation or protein function.

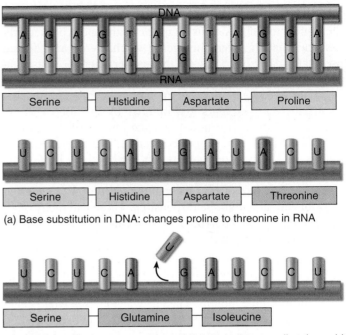

(a) Base substitution in DNA: changes proline to threonine in RNA

(b) Deletion in DNA: causes a frame-shift, which changes all amino acids after point of deletion in RNA

Figure 9.23   Two types of mutations.
Some changes in a DNA sequence can result in (*a*) a change in a single amino acid or (*b*) a frame-shift, which affects the entire amino acid sequence after the frame-shift.

when a mutation occurs within a germ-line cell is it passed to subsequent generations as part of the hereditary endowment of the gametes derived from that cell. Mutations in germ-line tissue are of enormous biological importance because they provide the raw material from which natural selection produces evolutionary change.

Mutations in Somatic Tissues.   Change can occur only if there are new, different allele combinations available to replace the old. Mutation produces new alleles, and recombination puts the alleles together in different combinations. In animals, it is the occurrence of these two processes in germ-line tissue that is important to evolution, because mutations in somatic cells (somatic mutations) are not passed from one generation to the next. However, a somatic mutation may have drastic effects on the individual organism in which it occurs, because it is passed on to all of the cells that are descended from the original mutant cell. Thus, if a mutant lung cell divides, all cells derived from it

will carry the mutation. Somatic mutations of lung cells are, as we shall see, the principal cause of lung cancer in humans.

Altering the Sequence of DNA.   One category of mutational changes affects the message itself, producing alterations in the sequence of DNA nucleotides (table 9.1). If alterations involve only one or a few base pairs in the coding sequence, they are called **point mutations.** Sometimes the identity of a base changes (**base substitution**), while other times one or a few bases are added (**insertion**) or lost (**deletion**). If an insertion or deletion throws the reading of the gene message out of register, a **frame-shift mutation** results (figure 9.23). While some point mutations arise due to spontaneous pairing errors that occur during DNA replication, others result from damage to the DNA caused by **mutagens,** usually radiation or chemicals. The latter class of mutations is of particular importance because modern industrial societies often release many chemical mutagens into the environment.

Changes in Gene Position.   Another category of mutations affects the way the genetic message is organized. In both prokaryotes and eukaryotes, individual genes may move from one place in the genome to another by **transposition** (see page 210). When a particular gene moves to a different location, its expression or the expression of neighboring genes may be altered. In addition, large segments of chromosomes in eukaryotes may change their relative locations or undergo duplication. Such **chromosomal rearrangements** often have drastic effects on the expression of the genetic message.

Mutation, Smoking, and Lung Cancer

The association of particular chemicals with cancer, particularly chemicals that are potent mutagens (see chapters 7 and 20), led researchers early on to suspect that cancer might be caused, at least in part, by the action of chemicals on the body.

The hypothesis that chemicals cause cancer was first advanced over 200 years ago in 1761 by Dr. John Hill, an English physician. Hill noted unusual tumors of the nose in heavy snuff users and suggested tobacco had produced these cancers. In 1775, a London surgeon, Sir Percivall Pott, made a similar observation, noting that men who had been chimney sweeps exhibited frequent cancer of the scrotum. He suggested that soot and tars might be responsible. These and many other observations led to the hypothesis that cancer results from the action of chemicals on the body.

It was over a century before this hypothesis was directly tested. In 1915, Japanese doctor Katsusaburo Yamagiwa applied extracts of coal tar to the skin of 137 rabbits every two or three days for three months. Then he waited to see what would happen. After a year, cancers appeared at the site of application in seven of the rabbits. Yamagiwa had induced cancer with the coal tar, the first direct demonstration of chemical carcinogenesis. In the decades that followed, this approach demonstrated that many chemicals were capable of causing cancer.

These were lab studies, and many did not accept that they applied to real people. Do tars in fact induce cancer in humans? In 1949, the American physician Ernst Winder and the British epidemiologist Richard Doll independently reported that lung cancer showed a strong link to the smoking of cigarettes, which introduces tars into the lungs. Winder interviewed 684 lung cancer patients and 600 normal controls, asking whether each had ever smoked. Cancer rates were 40 times higher in heavy smokers than in nonsmokers. From these studies, it seemed likely as long as 50 years ago that tars and other chemicals in cigarette smoke induce cancer in the lungs of persistent smokers. While this suggestion was resisted by the tobacco industry, the evidence that has accumulated since these pioneering studies makes a clear case, and there is no longer any real doubt. Chemicals in cigarette smoke cause cancer.

> **9.9** Rare changes in genes, called mutations, can have significant effects on the individual when they occur in somatic tissue, but they are inherited only if they occur in germ-line tissue. Inherited changes provide the raw material for evolution. Point mutations are changes in the hereditary message of an organism. Changes in gene position are changes in the organization of the genetic material. Chemicals that produce mutations in DNA, such as tars in cigarette smoke, are often potent carcinogens.

Exploring Current Issues

Additional Resources

Go to your campus library or look online to find the following articles, which further develop some of the concepts found in this chapter.

Jaffe, S. (2003). Alternative splicing goes mainstream: multilayered regulatory mechanisms at the RNA level increasingly show their stuff. *The Scientist,* 17(24), 28.

Johnston, M. and G. D. Stormo. (2003). Evolution: Heirlooms in the Attic. *Science,* 302(5647), 997.

Kreeger, K. Y. (2003). Asthma, genetics, and the environment. *The Scientist,* 17(7), 30.

Lewis, R. (2003). RNA calls the shots: the ultimate transcriber has a role in translation. *The Scientist,* 17(4), 29.

Lucentini, J. (2004). The body sleeps but the genes do not. *The Scientist,* 18(3), 24.

Biology and Society Lecture: Unraveling the Mystery of DNA

The realization that Mendel's patterns of heredity can be explained by the segregation of chromosomes in meiosis raised a question that occupied biologists for 50 years: What exactly is the nature of the connection between hereditary traits and chromosomes? In this lecture, we recount the chain of experiments that led to our current understanding of the molecular mechanisms of heredity. The experiments are among the most elegant in science. Just as in a good detective story, each conclusion has led to new questions. The intellectual path has not always been a straight one, the best questions are not always obvious. But however erratic and lurching the course of the experimental journey, our picture of heredity has become progressively clearer, the image more sharply defined. As we have mastered the details of what a gene is, and how genes do their job of dictating what we are like, DNA has become a household word, and its study the core of the new science of molecular biology.

Find this lecture, delivered by the author to his class at Washington University, online at www.mhhe.com/tlwessentials/exp9.

Summary

Genes Are Made of DNA

9.1 The Griffith Experiment

- Using *Streptococcus pneumoniae* bacteria, Griffith showed that information that controls physical characteristics can be passed from one bacterium to another, even from a dead bacterium. But Griffith did not determine whether this "transforming principle" was protein or DNA (**figure 9.1**).

9.2 The Avery and Hershey-Chase Experiments

- Avery showed that protein was not the source of this transformation; the virulent strain with its protein coat removed was still able to transform nonvirulent bacteria. This experimental result supported the hypothesis that DNA was the transforming principle.

- Using bacterial viruses, Hershey and Chase showed that genes were carried on DNA and not proteins. They used two different preparations, tagging DNA in one and protein in the other. They discovered that the DNA infected the bacteria, not the protein coat (**figure 9.2**).

9.3 Discovering the Structure of DNA

- Watson and Crick determined that DNA is a double helix, in which A on one strand pairs with T on the other, and G similarly pairs with C. They used research by Chargaff and Franklin to come to their conclusions (**figures 9.3 and 9.4**).

9.4 How the DNA Molecule Replicates

- Meselson and Stahl showed that DNA replicates semiconservatively, using each of the original strands as templates to form new strands (**figures 9.5 and 9.6**).

- Each DNA strand is copied by the actions of an enzyme called DNA polymerase. DNA polymerase adds nucleotides to the new DNA strand that are complementary to the original single strand, adding on to a primer (**figure 9.7**).

- DNA copies in a continuous manner on the leading strand and in a discontinuous manner on the lagging strand. DNA segments are linked together with another enzyme called DNA ligase (**figures 9.8 and 9.9**).

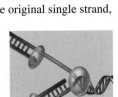

From Gene to Protein

9.5 Transcription

- DNA serves as a template on which mRNA is assembled by a protein called RNA polymerase, in a process called transcription (**figure 9.11**).

9.6 Translation

- The genetic information encoded in DNA is transcribed into three-nucleotide RNA units called codons, which correspond to particular amino acids (**figure 9.12**).

- A ribosome moves along mRNA while tRNA molecules that contain anticodon sequences that are complementary to the codons on the mRNA bring amino acids to the ribosome. The amino acids add to the end of a growing polypeptide chain (**figure 9.15**).

9.7 Architecture of the Gene

- Eukaryotic genes contain coding regions, called exons, and noncoding regions, called introns. The introns are spliced out before translation (**figure 9.17**).

- Some eukaryotic genes, called multigene families, are repeated on a chromosome. Genes can also move from one place to another in the genome as transposons.

Regulating Gene Expression

9.8 Turning Genes Off and On

- Most genes are regulated by controlling which genes are transcribed. Genes are turned on or shut off by attaching proteins that allow RNA polymerase access to the promoter or block the promoter, respectively (**figures 9.19–9.21**).

Altering the Genetic Message

9.9 Mutation

- A mutation is a change in the hereditary message (**figure 9.23**). Mutations that are passed on to offspring are the raw material on which natural selection acts to produce evolutionary changes.

- Mutations that change one or only a few nucleotides are called point mutations. They may arise as the result of errors in pairing during DNA replication, UV radiation, or chemical mutagens. Some mutations occur through the movement of sections of DNA from one place to another, a process called transposition (**table 9.1**).

Self-Test

1. As a result of the experiments performed by Frederick Griffith, we found that
   a. hereditary information within a cell cannot be changed.
   b. hereditary information can be added to cells from other cells.
   c. if hereditary information is added to a cell, it will kill the organism.
   d. hereditary information in the form of proteins can be added to cells.
2. The experiment performed by Alfred Hershey and Martha Chase showed that the molecule viruses use to specify new viruses is
   a. a protein.   c. ATP.
   b. a carbohydrate.   d. DNA.
3. Erwin Chargaff, Rosalind Franklin, Francis Crick, and James Watson all worked on pieces of information relating to the
   a. structure of DNA.   c. inheritance of DNA.
   b. function of DNA.   d. mutations of DNA.

4. Regarding the duplication of DNA, we now know that each double helix
   a. serves as a template to produce an identical double helix next to it.
   b. splits down the middle into two single helices, and each one then acts as a template to build its complement.
   c. fragments into small chunks that duplicate and reassemble.
   d. All of these are true for different types of DNA.
5. The process of obtaining a copy of the information in a gene as a strand of messenger RNA is called
   a. polymerase.   c. transcription.
   b. expression.   d. translation.
6. The process of taking the information on a strand of messenger RNA and building an amino acid chain, which will become all or part of a protein molecule, is called
   a. polymerase.   c. transcription.
   b. expression.   d. translation.

7. You can think of a gene as a recipe for an amino acid chain, like a recipe in a cookbook. However, you must keep in mind that it is like a

 a. straight version of a recipe, needing only to be read.
 b. coded version of a recipe, to be translated to mRNA and transcribed.
 c. version of a recipe that has been cut up and scattered into other information; all the bits need to be collected and read.
 d. coded version of a recipe that has been cut up and scattered into other information; all the bits need to be collected, transcribed to mRNA, and translated.

8. Regarding the activity level of genes

 a. all genes are on all the time in all cells, making the needed amino acid sequences.
 b. some genes are always off unless a promoter turns them on.
 c. some genes are always on unless a promoter turns them off.
 d. some genes are always off unless a repressor turns them on.

9. Genetic messages can be altered in two ways

 a. deliberately and accidentally.
 b. through the chromosome or through the protein.
 c. by mutation or by recombination.
 d. by activation or by repression.

10. Mutations can occur in

 a. germ-line tissues and be passed on to future generations.
 b. somatic tissues and be passed on to future generations.
 c. germ-line tissues and cause diseases such as lung cancer.
 d. somatic tissues and cause diseases such as cystic fibrosis.

Visual Understanding

1. **Figure 9.4c** What are some of the possible problems you see if the cytosine indicated with the red arrow is accidentally replaced with an adenine?

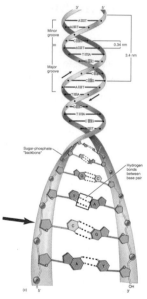

2. **Table 9.1** Your friend Gorinda wants to know if there are ever mutations that don't cause problems. What do you tell him?

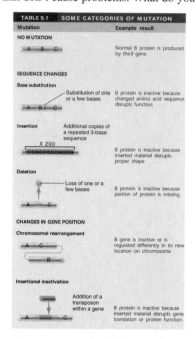

Challenge Questions

Genes are Made of DNA   Based on the experiments and discoveries discussed in chapters 8 and 9, defend the statement, attributed to Sir Isaac Newton in 1676 (though some say that Bernard of Chartres said it first, way back in about 1130!) that scientists build new ideas in science by "standing on the shoulders of giants."

From Gene to Protein   Compare DNA to a cookbook. The book is kept in a library and cannot be checked out (removed). Start with the letters and words in the cookbook compared with the bases and codons in DNA; end with the amino acid chain being folded into a protein, and a cake being baked.

Regulating Gene Expression   What would happen if all the genes in a cell were always active?

Altering the Genetic Message   Mutations can occur in somatic (body) cells or in germ-line cells. What problems or opportunities does each kind of cell face?

Online Learning Center

Visit the Online Learning Center for this chapter at www.mhhe.com/tlwessentials/ch9 for quizzes, animations, interactive learning exercises, and other study tools. At the site you will also find extended answers to the end-of-chapter questions.

10
The New Biology

T his sheep, Dolly, was the first animal to be cloned from a single adult cell. The lamb you see above beside her is her offspring, normal in every respect. From Dolly we learn that genes are not lost during development. If a single adult cell can be induced to switch the proper combination of genes on and off, that one cell can develop into a normal adult individual. Embryonic stem cells are like this—poised to become any cell of the body as the embryo develops. Controversial research suggests it may be possible to replace damaged tissues with healthy tissue grown from a patient's own embryonic stem cells. Another approach, when the damaged tissue results from a defective gene, is to repair rather than replace, using a virus to transfer a healthy gene into those tissues that lack it. In this chapter you will explore genomic screening, the application of gene technology to medicine and agriculture, reproductive cloning, stem cell tissue replacement, and gene therapy, all areas in which a revolution is reshaping biology.

10.1 Genomics

Recent years have seen an explosion of interest in comparing the entire DNA content of different organisms, a new field of biology called **genomics.** While initial successes focused on organisms with relatively small numbers of genes, researchers have recently completed the sequencing of several large eukaryotic genomes, including our own.

The full complement of genetic information of an organism—all of its genes and other DNA—is called its **genome.** The first genome to be sequenced was a very simple one: a small bacterial virus called ΦX174. Frederick Sanger, inventor of the first practical ways to sequence DNA, obtained the sequence of this 5,375-nucleotide genome in 1977. This was followed by the sequencing of dozens of prokaryotic genomes. The advent of automated DNA sequencing machines in recent years has made the DNA sequencing of much larger eukaryotic genomes practical (table 10.1).

Sequencing DNA

In sequencing DNA, a DNA fragment of unknown sequence is first amplified, so there are thousands of copies of the fragment. The DNA fragments are then mixed with copies of a primer, copies of DNA polymerase, a supply of the four nucleotide bases, and a supply of four different chain-terminating chemical tags that each act as one of the four nucleotide bases in DNA synthesis, undergoing complementary base pairing. First, heat is applied to denature the double-stranded DNA fragments. The solution is then allowed to cool, allowing the primer to bind to a single strand of the DNA, and synthesis of the complementary strand proceeds. Whenever a chemical

tag is added instead of a nucleotide base, the synthesis stops. Because of the relatively low concentration of the chemical tags compared with the nucleotides, a tag that binds to A on the DNA fragment, for example, will not necessarily be added to the first A site. Thus, the mixture will contain a series of double-stranded DNA fragments of different lengths, corresponding to the different distances the polymerase traveled from the primer before a chain-terminating tag was incorporated (figure 10.1a).

The series of fragments is then separated according to size using a technique called gel electrophoresis (see boxed feature: DNA Fingerprinting). The fragments become arrayed like the rungs of a ladder, each rung one base longer than the one below it (figure 10.1b). In automated DNA sequencing, fluorescently colored chemical tags are used to label the fragments, one color corresponding to each nucleotide. Computers read off the colors on the gel to determine the DNA sequence and display this sequence as a series of colored peaks (figure 10.1c, d). What made the attempt to sequence large eukaryotic genomes practical was the development in the mid-1990s of automated sequencers that perform electrophoresis of DNA fragments in capillary tubes instead of the traditional gel slabs. These systems can handle about 1,000 samples a day, with only 15 minutes of human attention. A research institute with several hundred such instruments can produce about 100 Mbp (million base pairs) every day.

> **10.1** Powerful automated DNA sequencing technology has begun to reveal the DNA sequences of entire genomes.

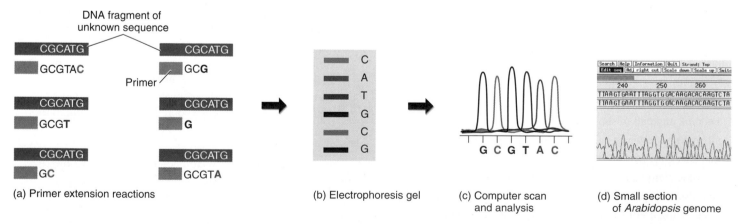

(a) Primer extension reactions   (b) Electrophoresis gel   (c) Computer scan and analysis   (d) Small section of *Arabidopsis* genome

Figure 10.1 How to sequence DNA.
(*a*) A DNA strand is sequenced by adding complementary bases to it. DNA synthesis stops when a chemical tag is inserted instead of a nucleotide, resulting in different sizes of DNA fragments. (*b*) The DNA fragments of varying lengths are separated by gel electrophoresis, the smaller fragments migrating farther down the gel. (*c*) Computers scan the gel, from smallest to largest fragments, and display the DNA sequence as a series of colored peaks. (*d*) Data from an automated DNA-sequencing run show the nucleotide sequence for a small section of the *Arabidopsis* (plant) genome.

TABLE 10.1 EUKARYOTIC GENOMES

Organism	Estimated Genome Size (Mbp)	Number of Genes (× 1,000)	Nature of Genome
VERTEBRATES			
Homo sapiens (human)	3,200	20-25	The first large genome to be sequenced; the number of transcribable genes is far less than expected; much of the genome is occupied by repeated DNA sequences.
Pan troglodytes (chimpanzee)	2,800	20-25	There are few base substitutions between chimp and human genomes, less than 2%, but many small sequences of DNA have been lost as the two species diverged, often with significant effects.
Mus musculus (mouse)	2,500	25	Roughly 80% of mouse genes have a functional equivalent in the human genome; importantly, large portions of the noncoding DNA of mouse and human have been conserved; overall, rodent genomes (mouse and rat) appear to be evolving more than 2X as fast as primate genomes (humans and chimpanzees).
Gallus gallus (chicken)	1,000	20-23	One-third the size of the human genome; genetic variation among domestic chickens seems much higher than in humans.
Fugu rubripes (pufferfish)	365	35	The Fugu genome is only one-ninth the size of the human genome, yet it contains ten thousand more genes.
INVERTEBRATES			
Caenorhabditis elegans (nematode)	97	21	The fact that every cell of C. elegans has been identified makes its genome a particularly powerful tool in developmental biology.
Drosophila melanogaster (fruit fly)	137	13	Drosophila telomere regions lack the simple repeated segments that are characteristic of most eukaryotic telomeres. About one-third of the genome consists of gene-poor centric heterochromatin.
Anopheles gambiae (mosquito)	278	15	The extent of similarity between Anopheles and Drosophila is approximately equal to that between human and pufferfish.
PLANTS			
Arabidopsis thaliana (wall cress)	115	26	A. thaliana was the first plant to have its genome fully sequenced. The evolution of its genome involved a whole-genome duplication followed by subsequent losses of genes and extensive local gene duplications.
Oryza sativa (rice)	430	33-50	The rice genome contains only 13% as much DNA as the human genome, but roughly twice as many genes; like the human genome, it is rich in repetitive DNA.
PROTISTS			
Plasmodium falciparum (malaria parasite)	23	5	The Plasmodium genome has an unusually high proportion of adenine and thymine. Scarcely 5,000 genes contain the bare essentials of the eukaryotic cell.
Dictyostelium discoideum (slime mold)	34	8-11	The genome for Dictyostelium discoideum has not yet been completely sequenced but partial sequencing reveals that D. discoideum is more closely related to animals than to plants or fungi.
FUNGI			
Saccharomyces cerevisiae (brewer's yeast)	13	6	S. cerevisiae was the first eukaryotic cell to have its genome fully sequenced.

10.2 The Human Genome

On June 26, 2000, geneticists reported that the entire human genome had been sequenced. This effort presented no small challenge, as the human genome is huge—more than 3 billion base pairs. To get an idea of the magnitude of the task, consider that if all 3.2 billion base pairs were written down on the pages of this book, the book would be 500,000 pages long and it would take you about 60 years, working eight hours a day, every day, at five bases a second, to read it all.

Geography of the Genome

The preliminary report of the human genome sequence, published in 2001, estimated the number of protein-encoding genes to be 30,000. The final report, published in 2004, lowered that estimate to 20,000–25,000 protein-encoding genes (figure 10.2). This is scarcely more than in nematodes, not quite double the number in *Drosophila,* and but a quarter of the number that had been anticipated by scientists counting unique messenger RNA (mRNA) molecules.

How can human cells contain four times as many kinds of mRNA as there are genes? Recall from chapter 9 that in a typical human gene, the sequence of DNA nucleotides that specifies a protein is broken into many bits called exons, scattered among much longer segments of nontranslated DNA called introns. Imagine this paragraph was a human gene; all the occurrences of the letter "e" could be considered exons, while the rest would be noncoding introns.

When a cell uses a human gene to make a protein, it first manufactures mRNA copies of all the exons (protein-specifying fragments) of the gene, then splices the exons together. Now here's the turn of events researchers had not anticipated: the transcripts of human genes are often spliced together in different ways, called alternative splicing. As we discussed in chapter 9, each exon is actually a module; one exon may code for one part of a protein, another for a different part of a protein. When the exon transcripts are mixed in different ways, very different protein shapes can be built.

With alternative mRNA splicing, it is easy to see how 25,000 genes can encode four times as many proteins. The added complexity of human proteins occurs because the gene parts are put together in new ways. Great music is made from simple tunes in much the same way.

In addition to the fragmenting of genes by the scattering of exons throughout the genome, there is another interesting "organizational" aspect of the genome. Genes are not distributed evenly over the genome. The small chromosome number 19 is packed densely with genes, transcription factors, and other functional elements. The much larger chromosome numbers 4 and 8, by contrast, have few genes. On most chromosomes, vast stretches of seemingly barren DNA fill the chromosomes between scattered clusters rich in genes.

DNA That Codes for Proteins

Four different classes of protein-encoding genes are found in the human genome, differing largely in gene copy number.

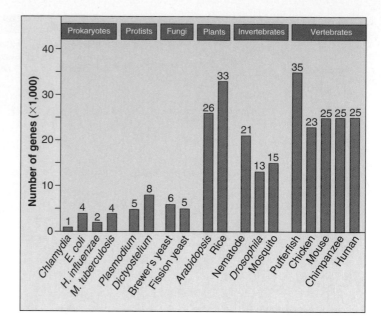

Figure 10.2   The human genome has an unexpectedly small number of protein-encoding genes.

The number of genes appears to be lower than expected in all higher eukaryotic organisms. The human genome has some 20,000–25,000 genes. This is the same as other mammals, and similar to the plant *Arabidopsis* and nematode worms.

Single-copy genes.   Many eukaryotic genes exist as single copies at a particular location on a chromosome. Mutations in these genes produce recessive Mendelian inheritance. Silent copies inactivated by mutation, called *pseudogenes,* are at least as common as protein-encoding genes.

Segmental duplications.   Human chromosomes contain many segmental duplications, whole blocks of genes that have been copied over from one chromosome to another. Blocks of similar genes in the same order are found throughout the genome. Chromosome 19 seems to have been the biggest borrower, with blocks of genes shared with 16 other chromosomes.

Multigene families.   Many genes exist as parts of multigene families, groups of related but distinctly different genes that often occur together in a cluster. Multigene families contain from three to several dozen genes. Although they differ from each other, the genes of a multigene family are clearly related in their sequences, making it likely that they arose from a single ancestral sequence.

Tandem clusters.   A second class of repeated genes consists of DNA sequences that are repeated many thousands of times, one copy following another in tandem array. By transcribing all of the copies in these tandem clusters simultaneously, a cell can rapidly obtain large amounts of the product they encode. For example, the genes encoding rRNA are typically present in clusters of several hundred copies.

Noncoding DNA

One of the most notable characteristics of the human genome is the startling amount of noncoding DNA it possesses. Only 1% to 1.5% of the human genome is coding DNA, devoted to genes encoding proteins. Each of your cells has about 6 feet of DNA stuffed into it, but of that, less than 1 inch is devoted to genes! Nearly 99% of the DNA in your cells seems to have little or nothing to do with the instructions that make you you. True genes are scattered about the human genome in clumps among the much larger amount of noncoding DNA, like isolated hamlets in a desert (table 10.2).

There are four major types of noncoding human DNA:

Noncoding DNA within genes. As we discussed on page 224, a human gene is made up of numerous fragments of protein-encoding information (exons) embedded within a much larger matrix of noncoding DNA (introns). Together, introns make up about 24% of the human genome, and exons about 1.5%.

Structural DNA. Some regions of the chromosomes remain highly condensed, tightly coiled, and untranscribed throughout the cell cycle. Called constitutive heterochromatin, these portions—about 20% of the DNA—tend to be localized around the centromere, or located near the ends of the chromosome.

Repeated sequences. Scattered about chromosomes are simple sequence repeats (SSRs). An SSR is a two- or three-nucleotide sequence like CA or CGG, repeated like a broken record thousands and thousands of times. SSRs make up about 3% of the human genome. An additional 7% is devoted to other sorts of duplicated sequences. Repetitive sequences with excess C and G tend to be found in the neighborhood of genes, while A- and T-rich repeats dominate the nongene deserts. The light bands on chromosome karyotypes now have an explanation—they are regions rich in GC and genes (see figure 8.24). Dark bands signal neighborhoods rich in AT and thin on genes. Chromosome 19, dense with genes, has few dark bands. Roughly 25% of the human genome has no genes at all.

Transposable elements. Fully 45% of the human genome consists of mobile bits of DNA called transposable elements. Discovered by Barbara McClintock in 1950 (she won the Nobel Prize for her discovery in 1983), transposable elements are bits of DNA that are able to jump from one location on a chromosome to another—tiny molecular versions of Mexican jumping beans.

Human chromosomes contain five sorts of transposable elements. Fully 20% of the genome consists of long interspersed elements (LINEs). An ancient and very successful element, LINEs are about 6 kb (6,000 DNA bases) long, and contain all the equipment needed for transposition, including genes for a DNA-loop-nicking enzyme and a reverse transcriptase.

Nested within the genome's LINEs are over half a million copies of a parasitic element called *Alu,* composing 10% of the human genome. *Alu* is only about 300 bases long, and has no transposition machinery of its own; like a flea on a dog, *Alu* moves with the LINE it resides within. Just as a flea sometimes jumps to a different dog, so *Alu* sometimes uses the enzymes of its LINE to move to a new chromosome location. Often jumping right into genes, *Alu* transpositions cause many harmful mutations.

Three other types of transposable elements are also present in the human genome. Eight percent of the genome is devoted to long terminal repeats (LTRs), also called "retrotransposons." Three percent is devoted to DNA transposons, which copy themselves as DNA rather than RNA. And, some 4% is devoted to dead transposons, elements that have lost the signals for replication and so can no longer jump.

10.2 The entire 3.2-billion-base-pair human genome has been sequenced. Gene sequences vary greatly in copy number, some occurring many thousands of times, others only once. Only about 1% of the human genome is devoted to protein-encoding genes. Much of the rest is composed of transposable elements.

TABLE 10.2	TYPES OF DNA SEQUENCES FOUND IN THE HUMAN GENOME	
Class	**Frequency**	**Description**
Protein-encoding genes	1%-1.5%	Translated exons, within some 25,000 genes scattered about the chromosomes
Introns	24%	Noncoding DNA comprising the great majority of most genes
Structural DNA	20%	Constitutive heterochromatin, localized near centromeres and telomeres
Repeated sequences	3%	Simple sequence repeats (SSRs) of a few nucleotides repeated millions of times
Duplicated sequences	7%	Duplicated sequences, other than the SSRs
Transposable elements	45%	20% long interspersed elements (LINEs), active transposons 15% other transposable elements, including long terminal repeats (LTRs) 10% the parasite sequence (*Alu*), present in half a million copies

10.3 A Scientific Revolution

In recent years the ability to manipulate genes and move them from one organism to another has led to great advances in medicine and agriculture. Moving genes from one organism to another is often called **genetic engineering.** Many of the gene transfers have placed eukaryotic genes into bacteria, converting the bacteria into tiny factories that produce prodigious amounts of the protein encoded by the eukaryotic gene.

Other gene transfers have moved genes from one animal or plant to another.

Genetic engineering is having a major impact on medicine and agriculture (figure 10.3). Most of the insulin used to treat diabetes is now obtained from bacteria that contain a human insulin gene. In late 1990, the first transfers of genes from one human to another were carried out in attempts to correct the effects of defective genes in a rare genetic disorder called *severe combined immune deficiency syndrome.* In addition,

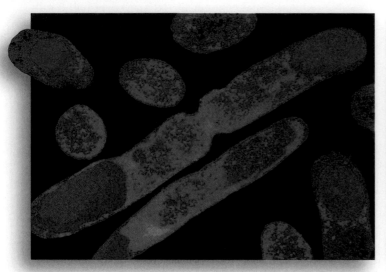

Producing insulin. The common bacteria *Escherichia coli* (*E. coli*) can be genetically engineered to contain the gene that codes for the protein insulin. The bacteria are turned into insulin-producing factories and can produce large quantities of insulin for diabetic patients. In the image above, insulin-producing sites inside genetically-altered *E. coli* cells are orange.

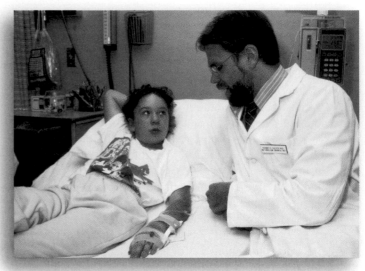

Curing disease. One of two young girls who were the first humans "cured" of a hereditary disorder by transferring into their bodies healthy versions of the gene they lacked. The transfer was successfully carried out in 1990, and the girls remain healthy.

Increasing yields. The genetically engineered salmon on the *right* have shortened production cycles and are heavier than the nontransgenic salmon on the *left*.

Figure 10.3 Examples of genetic engineering.

Pest-proofing plants. The genetically engineered cotton plants on the *right* have a gene that inhibits feeding by weevils; the cotton plants on the *left* lack this gene, and produce far fewer cotton bolls.

cultivated plants and animals can be genetically engineered to resist pests, grow bigger, or grow faster.

Restriction Enzymes

The first stage in any genetic engineering experiment is to chop up the "source" DNA to get a copy of the gene you wish to transfer. This first stage is the key to successful transfer of the gene, and learning how to do it is what has led to the genetic revolution. The trick is in how the DNA molecules are cut. The cutting must be done in such a way that the resulting DNA fragments have "sticky ends" that can later be inserted into another molecule of DNA.

This special form of molecular surgery is carried out by **restriction enzymes,** also called restriction endonucleases, which are special enzymes that bind to specific short sequences (typically four to six nucleotides long) on the DNA (figure 10.4a). These sequences are very unusual in that they are symmetrical—the two strands of the DNA duplex have the same nucleotide sequence, running in opposite directions! One of these sequences, for example, is GAATTC. Try writing down the sequence of the opposite strand: it is CTTAAG—the same sequence, written backward. This sequence is recognized by the restriction enzyme *Eco*RI.

What makes the DNA fragments "sticky" is that most restriction enzymes do not make their incision in the center of the sequence; rather, the cut is made to one side. In the sequence written above, for example, the cut is made after the first nucleotide, G/AATTC. This produces a break, with short, single strands of DNA dangling from each end. Because the two single-stranded ends are complementary in sequence, they could pair up and heal the break, with the aid of a sealing enzyme—*or they could pair with any other DNA fragment cut by the same enzyme,* because all would have the same single-stranded sticky ends (figure 10.4b)! Any gene in any organism cut by the enzyme that attacks GAATTC sequences can be joined to any other with the aid of a sealing enzyme called a **ligase,** which reforms the bonds between the sugars and phosphates of DNA (figure 10.4c).

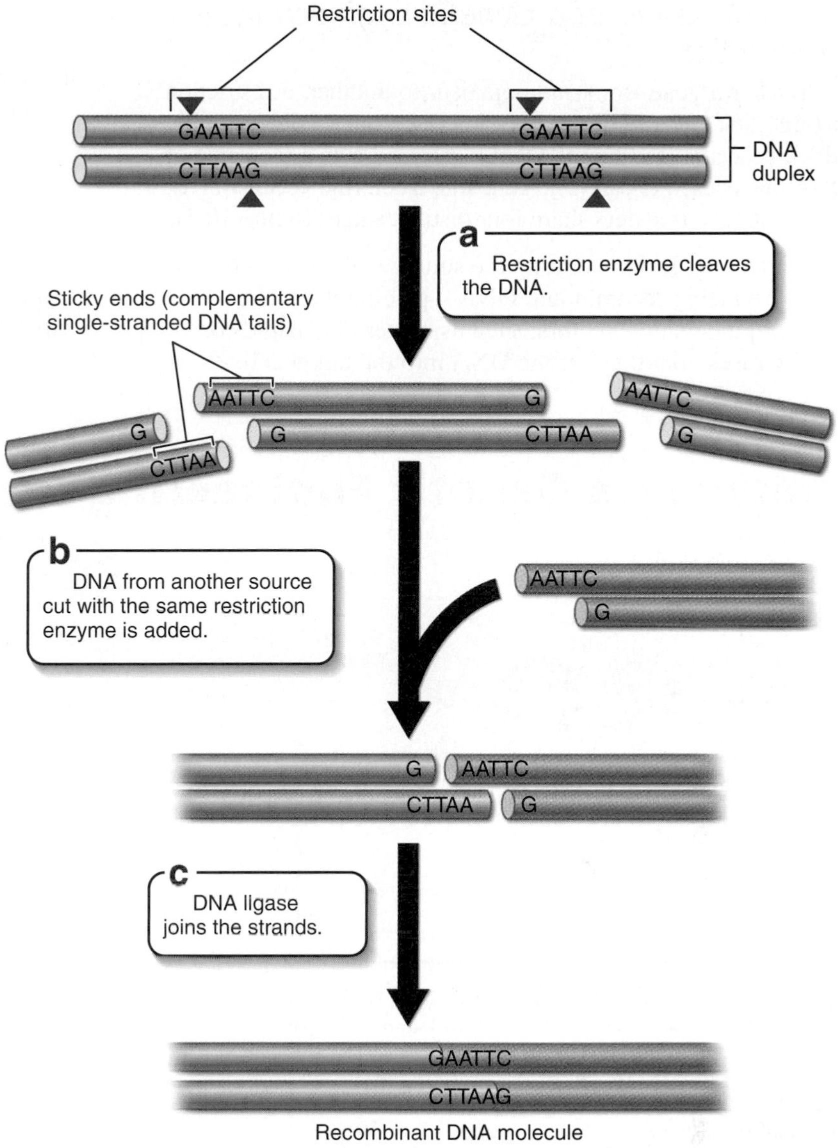

Restriction enzymes, bacterial enzymes that make cuts in double-stranded DNA, were discovered in the late 1960s by Werner Arber and Hamilton Smith (who were awarded the Nobel Prize for Medicine). Arber had observed that bacterial viruses could infect some cells but not others. Bacteria that exhibited this "host restriction" were found to contain enzymes that could cleave foreign DNA (their own DNA being protected from the enzyme action by chemical modifications of the DNA). Any viruses that attempt to infect bacterial cells

Figure 10.4   How restriction enzymes produce DNA fragments with sticky ends.

The restriction enzyme *Eco*RI always cleaves the sequence GAATTC between G and A. Because the same sequence occurs on both strands, both are cut. However, the two sequences run in opposite directions on the two strands. As a result, single-stranded tails are produced that are complementary to each other, or "sticky."

fail, because the bacterial restriction enzymes degrade the viral DNA (refer back to figure 9.2 to see how a bacterial virus infects a cell). Since their discovery, hundreds of different restriction enzymes have been identified, recognizing a wide variety of four- to six-nucleotide DNA sequences called restriction sites. Each kind of enzyme attacks only one sequence and always cuts at the same place. By trying one enzyme after another, biologists can almost always find an enzyme that cuts out the gene they seek, attacking a restriction site present by chance on both ends of the gene but not within it. Restriction enzymes are the basic tools of genetic engineering.

The Four Stages of a Genetic Engineering Experiment

To transfer a gene from one organism to another, a restriction enzyme first cuts out the gene of interest from the DNA within which it occurs (the source DNA), and also cuts a molecule of DNA that is used to carry the gene into a cell (the vector DNA).

All gene transfers share four distinct stages (figure 10.5):

1. Cleaving DNA—cutting the source and vector DNA.
2. Producing recombinant DNA—placing the DNA fragments into vectors, such as bacterial plasmids or viruses, that transfer the DNA into the target cells.
3. Cloning—infecting target cells with DNA-bearing vectors and allowing infected cells to reproduce.
4. Screening—selecting the particular infected cells that have received the gene of interest.

> **10.3** Restriction enzymes bind to specific short sequences of DNA and cut the DNA molecule there. This produces fragments with "sticky ends" that can be inserted into other DNA molecules.

Stages of a Genetic Engineering Experiment

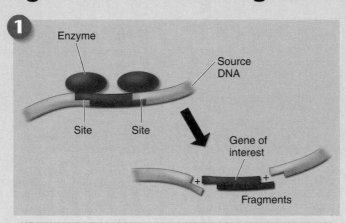

Cleaving DNA. Enzymes cut the source DNA at specific sites, cleaving the two strands short distances apart.

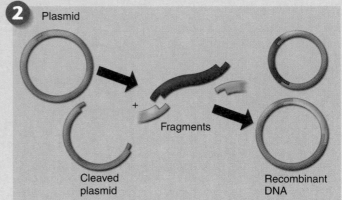

Producing recombinant DNA. A circular plasmid cut with the same enzyme is combined with the fragments of source DNA.

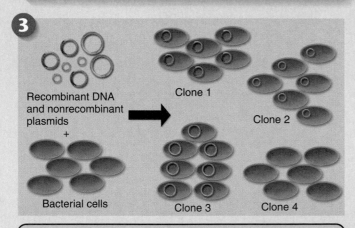

Cloning. A variety of recombinant plasmids are produced, some containing the gene of interest (*red*), others containing other fragments from the source DNA (*blue*), and still others containing no fragment. The plasmids are mixed with the bacterial cells. Some cells take up plasmids (clones 1–3) and some do not (clone 4). Each cell reproduces and forms a clone of bacterial cells, each clone containing one type of plasmid. All of the cells constitute a clone library.

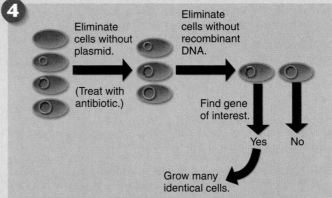

Screening. First, bacterial cells that did not take up the plasmid are screened out using an antibiotic for which the plasmid contains a resistance gene. Then those plasmid-containing cells that possess recombinant DNA are identified. Last, those cells containing the gene of interest are found using a probe sequence complementary to that gene.

Figure 10.5 How a genetic engineering experiment works.

≡❂≡ Science in Action

DNA Fingerprinting

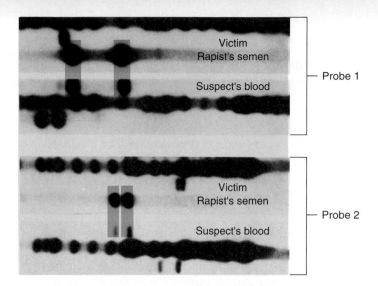

Two DNA profiles that led to conviction.
The two DNA probes seen here were used to characterize DNA isolated from the victim, the semen left by the rapist, and the suspect. The dark channels are multiband controls. There is a clear match between the suspect's DNA and the DNA of the rapist's semen, as indicated by the red boxes.

DNA fingerprinting is a process using probes to examine DNA samples that have been cut by restriction endonucleases. Because an endonuclease's recognition sequence is likely to occur many times within the source DNA, cleavage will produce a number of fragments of different sizes. For DNA fingerprinting, and a variety of other techniques used in genetic analyses, the fragments can be separated from each other according to their size by **electrophoresis.** In this procedure, solutions containing the fragments produced by restriction enzymes are loaded onto one side of a gel, and an electric current is applied. The DNA fragments will migrate different distances through the gel. As described in chapter 7, DNA is negatively charged because of the phosphate groups and thus moves toward the positively charged anode end of the gel. The distance a fragment moves relates to its size, larger fragments moving slower and thus traveling shorter distances during the period of time the current is applied. This produces a pattern of bands that can be visualized in a variety of ways.

DNA from different individuals rarely has exactly the same array of restriction sites and distances between sites, so the restriction fragment patterns for different individuals will be different. By cutting a DNA sample with a particular restriction endonuclease, separating the fragments according to length on an electrophoretic gel, and then using a radioactive probe to identify the fragments on the gel, one can obtain a pattern of bands often unique for each region of DNA analyzed. These "DNA fingerprints" are used in forensic analysis during criminal investigations.

The image here shows the DNA fingerprints a prosecuting attorney presented in a rape trial in 1987. This trial was the first time DNA evidence was used in a court of law. The lane with many bars represents a standardized control. Usually six to eight probes are used to identify the source of a DNA sample. The probes are unique DNA sequences found in noncoding regions of human DNA that vary much more frequently from one individual to the next than do coding regions of the DNA. The chances of any two individuals, other than identical twins, having the same restriction pattern for these noncoding sequences varies from 1 in 800,000 to 1 in 1 billion, depending on the number of probes used.

The results of two probes are shown here. A vaginal swab was taken from the rape victim within hours of her attack; from it, semen was collected, and the semen DNA analyzed for its restriction endonuclease patterns. Compare the restriction endonuclease patterns of the semen with that of the suspect, Tommie Lee Andrews.

You can see that the suspect's two patterns match that of the rapist (and are not at all like those of the victim). Although a DNA fingerprint doesn't prove with 100% certainty that the semen came from the suspect, it is at least as reliable as traditional fingerprinting when several probes are used. On November 6, 1987, the jury returned a verdict of guilty. Andrews became the first person in the United States to be convicted of a crime based on DNA evidence.

Since the Andrews verdict, DNA fingerprinting has been admitted as evidence in more than 2,000 court cases. While some probes contain DNA sequences shared by many people, others are quite rare. Using several probes, identity can be clearly established or ruled out.

Just as fingerprinting revolutionized forensic evidence in the early 1900s, so DNA fingerprinting is revolutionizing it today. A hair, a minute speck of blood, a drop of semen, all can serve as sources of DNA to convict or clear a suspect. As the man who analyzed Andrews' DNA says: "It's like leaving your name, address, and social security number at the scene of the crime. It's that precise." Of course, laboratory analyses of DNA samples must be carried out properly—sloppy procedures could lead to a wrongful conviction. After widely publicized instances of questionable lab procedures, national standards have been developed.

While the method of DNA fingerprinting described here has proven a valuable tool, increasingly it is being replaced by new more powerful methods of screening short segments of DNA.

Much of the excitement about genetic engineering has focused on its potential to improve medicine—to aid in curing and preventing illness. Major advances have been made in the production of proteins used to treat illness, in the creation of new vaccines to combat infections, and in the replacement of defective genes, or gene therapy, which will be discussed later in this chapter.

Making "Magic Bullets"

Many genetic defects occur because our bodies fail to make critical proteins. *Diabetes* is such an illness. The body is unable to control levels of sugar in the blood because a critical protein, **insulin,** cannot be made. These failures can be overcome if the body can be supplied with the protein it lacks. The donated protein is in a very real sense a "magic bullet" to combat the body's inability to regulate itself.

Until recently, the principal problem with using regulatory proteins as drugs was in manufacturing the protein. Proteins that regulate the body's functions are typically present in the body in very low amounts, and this makes them difficult and expensive to obtain in quantity. With genetic engineering techniques, the problem of obtaining large amounts of rare proteins has been largely overcome. The genes encoding the medically important proteins are now introduced into bacteria (table 10.3). Because the host bacteria can be grown cheaply, large amounts of the desired protein can be easily isolated. In 1982, the U.S. Food and Drug Administration approved the use of human insulin produced from genetically engineered bacteria, the first commercial product of genetic engineering.

Today hundreds of pharmaceutical companies around the world are busy producing other medically important proteins using these genetic engineering techniques. These products include growth hormone (figure 10.6), **anticoagulants** (proteins involved in dissolving blood clots), which are effective in treating heart attack patients, and **factor VIII,** a protein that promotes blood clotting. A deficiency in factor VIII leads to hemophilia, an inherited disorder discussed in chapter 8, which is characterized by prolonged bleeding. For a long time, hemophiliacs received blood factor VIII that had been isolated from donated blood. Unfortunately, some of the donated blood had been infected with viruses such as HIV and hepatitis B, which were then unknowingly transmitted to those people who received blood transfusions. Today the use of genetically engineered factor VIII eliminates the risks associated with blood products obtained from other individuals.

TABLE 10.3	GENETICALLY ENGINEERED DRUGS
Product	**Effects and Uses**
Anticoagulants	Involved in dissolving blood clots; used to treat heart attack patients
Colony-stimulating factors	Stimulate white blood cell production; used to treat infections and immune system deficiencies
Erythropoietin	Stimulates red blood cell production; used to treat anemia in individuals with kidney disorders
Factor VIII	Promotes blood clotting; used to treat hemophilia
Growth factors	Stimulate differentiation and growth of various cell types; used to aid wound healing
Human growth hormone	Used to treat dwarfism
Insulin	Involved in controlling blood sugar levels; used in treating diabetes
Interferons	Disrupt the reproduction of viruses; used to treat some cancers
Interleukins	Activate and stimulate white blood cells; used to treat wounds, HIV infections, cancer, immune deficiencies

Figure 10.6   Genetically engineered human growth hormone.
These two mice are genetically identical, but the large one has one extra gene: the gene encoding human growth hormone. The gene was added to the mouse's genome by genetic engineers and is now a stable part of the mouse's genetic endowment.

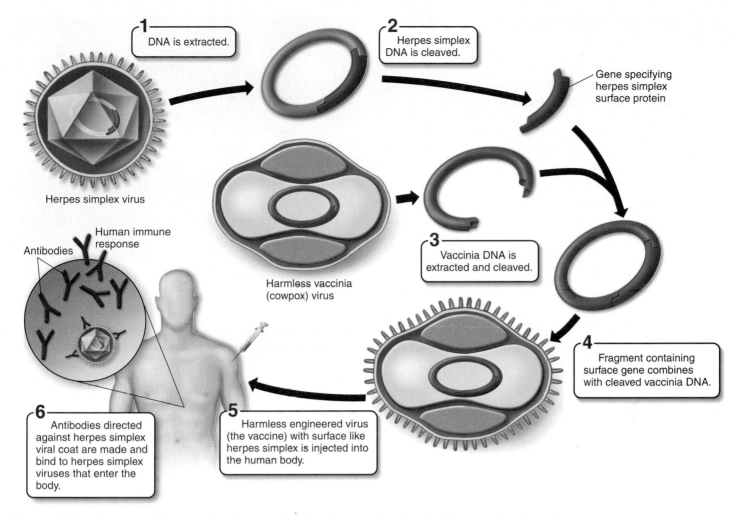

1 DNA is extracted.

2 Herpes simplex DNA is cleaved.

Gene specifying herpes simplex surface protein

Herpes simplex virus

Human immune response

Antibodies

Harmless vaccinia (cowpox) virus

3 Vaccinia DNA is extracted and cleaved.

4 Fragment containing surface gene combines with cleaved vaccinia DNA.

6 Antibodies directed against herpes simplex viral coat are made and bind to herpes simplex viruses that enter the body.

5 Harmless engineered virus (the vaccine) with surface like herpes simplex is injected into the human body.

Figure 10.7   Constructing a subunit, or piggyback, vaccine for the herpes simplex virus.

Piggyback Vaccines

Another area of potential significance involves the use of genetic engineering to produce **subunit vaccines** against viruses such as those that cause herpes and hepatitis. Genes encoding part of the protein-polysaccharide coat of the herpes simplex virus or hepatitis B virus are spliced into a fragment of the vaccinia (cowpox) virus genome (figure 10.7). The vaccinia virus, which British physician Edward Jenner used more than 200 years ago in his pioneering vaccinations against smallpox, is now used as a vector to carry the herpes or hepatitis viral coat gene into cultured mammalian cells. These cells produce many copies of the recombinant virus, which has the outside coat of a herpes or hepatitis virus. When this recombinant virus is injected into a mouse or rabbit, the immune system of the infected animal produces antibodies directed against the coat of the recombinant virus. It therefore develops an immunity to herpes or hepatitis virus. Vaccines produced in this way, also known as **piggyback vaccines,** are harmless because the vaccinia virus is benign and only a small fragment of the DNA from the disease-causing virus is introduced via the recombinant virus.

The great attraction of this approach is that it does not depend upon the nature of the viral disease. In the future,

similar recombinant viruses may be injected into humans to confer resistance to a wide variety of viral diseases.

In 1995, the first clinical trials began of a novel new kind of **DNA vaccine,** one that depends not on antibodies but rather on the second arm of the body's immune defense, the so-called cellular immune response, in which blood cells known as cytotoxic T cells attack infected cells. The infected cells are attacked and destroyed when they stick fragments of foreign proteins onto their outer surfaces that the T cells detect (the discovery by Peter Doherty and Rolf Zinkernagel that infected cells do so led to their receiving the Nobel Prize in Physiology or Medicine in 1996). The first DNA vaccines spliced an influenza virus gene encoding an internal nucleoprotein into a plasmid, which was then injected into mice. The mice developed strong cellular immune responses to influenza. New and controversial, the approach offers great promise.

10.4   Genetic engineering has facilitated the production of medically important proteins and led to novel vaccines.

Genetic Engineering and Agriculture

One of the greatest impacts of genetic engineering on society has been the successful manipulation of the genes of crop plants to make them more resistant to disease caused by insects, to make them resistant to herbicides (chemicals that kill plants), to improve their nutritional balance and protein content, and to make them hardier, able to resist stress caused by frost, drought, and other factors.

Pest Resistance

An important effort of genetic engineers in agriculture has involved making crops resistant to insect pests without spraying with pesticides, a great saving to the environment. Consider cotton. Its fibers are a major source of raw material for clothing throughout the world, yet the plant itself can hardly survive in a field because many insects attack it. Over 40% of the chemical insecticides used today are employed to kill insects that eat cotton plants. The world's environment would greatly benefit if these thousands of tons of insecticide were not needed. Biologists are now in the process of producing cotton plants that are resistant to attack by insects.

One successful approach uses a kind of soil bacterium *Bacillus thuringienses* (Bt) that produces a protein that is toxic when eaten by crop pests, such as larvae (caterpillars) of butterflies and other insects. When the gene producing the Bt protein is inserted into the chromosomes of tomatoes, the plants begin to manufacture Bt protein, which makes them highly toxic to tomato hornworms (one of the most serious pests of commercial tomato crops) but the Bt protein is not harmful to humans.

Many important plant pests also attack roots. To combat these pests, genetic engineers are introducing the *Bt* gene into

Figure 10.9 Genetically engineered herbicide resistance.

All four of these petunia plants were exposed to equal doses of an herbicide. The two on *top* were genetically engineered to be resistant to glyphosate, the active ingredient in the herbicide, whereas the two dead ones on the *bottom* were not.

different kinds of bacteria, ones that colonize the roots of crop plants. Any insect eating such roots consume the bacteria and so are lethally attacked by the enzyme.

Herbicide Resistance

A major advance has been the creation of crop plants that are resistant to the herbicide *glyphosate,* a powerful biodegradable herbicide that kills most actively growing plants. Glyphosate is used in orchards and agricultural fields to control weeds. Growing plants need to make a lot of protein, and glyphosate stops them from making protein by destroying an enzyme necessary for the manufacture of so-called aromatic amino acids (that is, amino acids that contain a ring structure like phenylalanine—see figure 3.19). Humans are unaffected by glyphosate because we don't make aromatic amino acids—we obtain them from plants we eat! To make crop plants resistant to this powerful plant killer, genetic engineers screened thousands of organisms until they found a species of bacteria that could make aromatic amino acids in the presence of glyphosate. They then isolated the gene encoding the resistant enzyme and, using plasmids or DNA particle guns (figure 10.8), successfully introduced the gene into plants (figure 10.9).

Agriculture is being revolutionized by glyphosate-resistant crops for two reasons. First, it lowers the cost of producing a crop, because a crop resistant to glyphosate does not need to be weeded. Second, the creation of glyphosate-tolerant crops is of major benefit to the environment. Glyphosate is quickly broken down in the environment, which makes its use a great improvement over long-lasting chemical herbicides. Perhaps even more important, not having to plow to remove weeds reduces the loss of fertile topsoil to erosion, one of the greatest environmental challenges facing our country today.

Figure 10.8 Shooting genes into cells.

This DNA particle gun fires tungsten pellets coated with DNA into plant cells such as the ones on the culture plate this experimenter is holding.

More Nutritious Crops

In the last 10 years the cultivation of genetically modified crops of corn, cotton, soybeans, and other plants (GM) has become commonplace in the United States. In 2003, 84% of soybeans in the United States were planted with seeds genetically modified to be herbicide resistant. The result has been that less tillage was needed and as a consequence soil erosion was greatly lessened. Pest-resistant GM corn in 2003 comprised 38% of all corn planted in the United States, and pest-resistant GM cotton comprised 81% of all cotton. In both cases, the change greatly lessens the amount of chemical pesticide used in raising the crops. These benefits of soil preservation and chemical pesticide reduction, while significant, have been largely bestowed upon farmers, making their cultivation of crops cheaper and more efficient.

Like the first act of a play, these developments have mainly set the stage for the real action, which is only now beginning to happen. The real promise of plant genetic engineering is to produce genetically modified plants with desirable traits that directly benefit the consumer.

One recent advance, nutritionally improved "golden" rice, gives us a hint of what is to come. In developing countries, large numbers of people live on simple diets that are poor sources of vitamins and minerals (what botanists call "micronutrients"). Worldwide, the two major micronutrient deficiencies are iron, which affects 1.4 billion women, 24% of the world population, and vitamin A, affecting 40 million children, 7% of the world population. The deficiencies are especially severe in developing countries where the major staple food is rice. In recent research, Swiss bioengineer Ingo Potrykus and his team at the Institute of Plant Sciences, Zurich, have gone a long way toward solving this problem. Supported by the Rockefeller Foundation and with results to be made free to developing countries, the work is a model of what plant genetic engineering can achieve.

To solve the problem of dietary iron deficiency among rice eaters, Potrykus first asked why rice is such a poor source of dietary iron. The problem, and the answer, proved to have three parts:

1. *Too little iron.* The proteins of rice endosperm have unusually low amounts of iron. To solve this problem, a ferritin gene was transferred into rice from beans (figure 10.10). Ferritin is a protein with an extraordinarily high iron content, and so greatly increased the iron content of the rice.
2. *Inhibition of iron absorption by the intestine.* Rice contains an unusually high concentration of a chemical called phytate, which inhibits iron reabsorption in the intestine—it stops your body from taking up the iron in the rice. To solve this problem, a gene encoding an enzyme that destroys phytate was transferred into rice from a fungus.
3. *Too little sulfur for efficient iron absorption.* The human body requires sulfur for the uptake of iron, and rice has very little of it. To solve this problem, a gene encoding a sulfur-rich metallothionin protein was transferred into rice from wild rice.

To solve the problem of vitamin A deficiency, the same approach was taken. First, the problem was identified. It turns out rice only goes partway toward making beta-carotene (provitamin A); there are no enzymes in rice to catalyze the last four steps. To solve the problem, genes encoding these four enzymes were added to rice from a flower, the daffodil.

The development of transgenic rice is only the first step in the battle to combat dietary deficiencies. The added nutritional value only makes up for half a person's requirements, and many years will be required to breed the genes into lines adapted to local conditions, but it is a promising start, representative of the very real promise of genetic engineering.

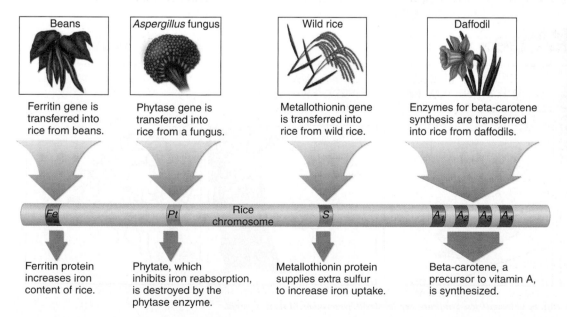

Beans	*Aspergillus* fungus	Wild rice	Daffodil
Ferritin gene is transferred into rice from beans.	Phytase gene is transferred into rice from a fungus.	Metallothionin gene is transferred into rice from wild rice.	Enzymes for beta-carotene synthesis are transferred into rice from daffodils.

Rice chromosome — Fe — Pt — S — A_1 A_2 A_3 A_4

Ferritin protein increases iron content of rice.	Phytate, which inhibits iron reabsorption, is destroyed by the phytase enzyme.	Metallothionin protein supplies extra sulfur to increase iron uptake.	Beta-carotene, a precursor to vitamin A, is synthesized.

Figure 10.10 Transgenic "golden" rice.

Developed by Swiss bioengineer Ingo Potrykus, transgenic rice offers the promise of improving the diets of people in rice-consuming developing countries, where iron and vitamin A deficiencies are a serious problem.

How Do We Measure the Potential Risks of Genetically Modified Crops?

The advantages afforded by genetic engineering are revolutionizing our lives. But what are the disadvantages, the potential costs and dangers, of genetic engineering? Many people, including influential activists and members of the scientific community, have expressed concern that genetic engineers are "playing God" by tampering with genetic material. Could genetically engineered products administered to plants or animals turn out to be dangerous for consumers after several generations? What kind of unforeseen impact on the ecosystem might "improved" crops have? Is it ethical to create "genetically superior" organisms, including humans?

While the promise of genetic engineering is very much in evidence, this same genetic engineering has been the cause of considerable controversy and protest. The intense feelings generated by this dispute point to the need to understand how we measure the risks associated with the genetic engineering of plants. Two sets of risks need to be considered. The first stems from eating genetically modified foods, the other concerns potential ecological effects.

Is Eating Genetically Modified Food Dangerous?
Bioengineers modify crops in two quite different ways. One class of gene modification makes the crop easier to grow; a second class of modification is intended to improve the food itself. Many consumers worry that either sort of genetically modified (GM) food may have been rendered somehow dangerous.

The introduction of glyphosate-resistant soybeans is an example of the first class of modification, producing a GM version that is easier to grow. Is the soybean that results nutritionally different? No. The gene that confers glyphosate resistance in soybeans does so by protecting the plant's ability to manufacture aromatic amino acids. In unprotected weeds, by contrast, glyphosate blocks this manufacturing process, killing the weed. As discussed earlier in this section, humans don't make any aromatic amino acids, so glyphosate doesn't hurt us.

The real issue, of course, is whether the gene modification that renders crop plants glyphosate resistant involves introducing novel proteins with potentially dangerous consequences. Thus in early GM crops like Flavr-Savr tomatoes, one of the issues raised was that the "selectable marker" (used in screening to ensure that the desired genes had been correctly introduced into the GM tomatoes) were still present in the modified plants, with unknown future consequences.

While new procedures eliminate this possibility in today's GM crops, a second issue raises a valid concern: Could introduced proteins like the enzyme making crops glyphosate tolerant become allergens, causing a potentially fatal immune reaction in some people? Because the potential danger of allergic reactions is quite real, every time a protein-encoding gene is introduced into a GM crop it is necessary to carry out extensive tests of the introduced protein's potential as an allergen.

Glyphosate-resistant corn has a single added protein, the enzyme EPSP synthetase, which catalyzes the first reaction in aromatic amino acid biosynthesis. How does this added enzyme make the plants tolerant to glyphosate? The introduced protein sequence, a version of the EPSP synthetase gene, is nearly identical to the "wild type" EPSP synthetase in conventional corn; the only difference is two amino acids out of 444 in the protein sequence. The change alters the EPSP synthetase's shape a little, so that glyphosate no longer can inhibit it.

The key safety question for consumers is whether the change in the EPSP synthetase is capable of inducing an allergic response in humans that consume corn containing it. The EPSP synthetase in glyphosate-resistant corn has passed extensive and stringent allergenicity tests, and so glyphosate-resistant corn is approved for human consumption by the Environmental Protection Agency (EPA).

This same issue arises each time a gene is added to a crop plant. The addition of the *Bt* gene to corn offers an illuminating example of the need for careful attention to this issue. The *Cry1A* gene isolated from the soil bacterium *Bacillus thuringiensis* creates a Bt protein that binds to specific receptors in the midgut of lepidopteran insects like the European corn borer, but is harmless to humans, wildlife, worms, and beneficial insects that can help control other pests. Thor-

Calvin and Hobbes

by Bill Watterson

CALVIN AND HOBBES © 1995 Watterson. Dist. by Universal Press Syndicate. Reprinted with permission. All rights reserved.

oughly tested for allergenicity, corn with this Bt protein is approved by the EPA for human consumption.

No GM crop currently being produced in the United States (see table 10.2) contains a protein that acts as an allergen to humans. On this score, then, the risk of bioengineering to the food supply seems to be slight, so long as adequate testing of new varieties continues.

Are GM Crops Harmful to the Environment?   Those concerned about the widespread use of GM crops raise three legitimate concerns that merit careful evaluation:

1. *Harm to Other Organisms.* The first concern is the possibility of unintentional harm to other organisms. This issue is seen most clearly in the case of crops like Bt corn. Results from a small laboratory experiment suggested that pollen from Bt corn could harm larvae from the Monarch butterfly, which is, like the corn borer, a lepidopteran insect. While this preliminary report received considerable publicity, subsequent studies suggest little possibility of harm. Monarch butterflies lay their eggs on milkweed, not corn, and field management ensures that there is little if any milkweed growing in or near cornfields. In field tests few Monarch larvae are found there. Additionally, field tests confirm that corn sheds its pollen at a different time than when the Monarch larval stage occurs. Finally, the amount of Bt pollen necessary to kill a Monarch caterpillar is far more than is encountered in field tests.

   Farmers focus on the fact that GM cornfields do not need to be sprayed with chemical pesticides to control the corn borer. An estimated $9 billion in damage is caused annually by the application of pesticides in the United States, and billions of insects and other animals, including an estimated 67 million birds, are killed each year.

   In a serious attempt to learn if GM crops are a danger to other organisms, the British government undertook a comprehensive three-year "farm-scale evaluation" of the effects of GM-herbicide-resistant beet, corn, and oilseed crops on biodiversity (in this instance, numbers of kinds of insects). In the study, 60 fields each of beets, corn, and rape (oilseed) were split between conventional varieties and genetically modified herbicide-tolerant strains. In a report released in 2003, the results "reveal significant differences in the effect on biodiversity when managing genetically herbicide-resistant crops as compared to conventional varieties." The report showed that weeds are important sources of food and shelter for insects; because GM crops get rid of weeds more effectively, they have a greater impact on insect populations.

2. *Resistance.* All insecticides used in agriculture share the problem that pests eventually evolve resistance to them, in much the same way that bacterial populations evolve resistance to antibiotics. Use of the insecticide creates a selective pressure favoring mutations that make the pest resistant to it. This is true of chemical pesticides, and also of insecticides produced by Bt corn.

Will pests eventually become resistant to the Bt toxin, just as many have become resistant to the high levels of chemical pesticide we sprayed on crops? This is certainly a possibility, as a few species of insects have developed resistance to Bt when it was sprayed directly on crops in years past. However, despite the widespread use of Bt crops like corn, soybeans, and cotton since 1996, there are as of yet no cases of insects developing resistance to Bt plants in the field. Still, because of this possibility farmers are required to plant at least 20% non-Bt crops alongside Bt crops in refugia to support insect populations that are not under selection pressure and so slow development of resistance. It is very important that this requirement be met. If ignored, Bt-resistant insect pests may appear in future fields.

3. *Gene Flow.* How about the possibility that introduced genes will pass from GM crops to their wild or weedy relatives? This sort of gene flow happens naturally all the time, and so this is a legitimate question. For the first round of major GM crops, there is usually no potential relative around to receive the modified gene from the GM crop. There are no wild relatives of soybeans in Europe, for example. Thus there can be no gene escape from GM soybeans in Europe, any more than genes can flow from you to your pet dog or cat. For secondary crops only now being gene modified, the risks are more tangible. In 2004, Environmental Protection Agency (EPA) scientists planted eight fields in central Oregon (400 acres in all) with genetically modified creeping bentgrass (*Agrostis stolonifera*). Bentgrass is widely planted on golf course greens because it can be closely mowed. But normal bentgrass is particularly susceptible to weeds, and so requires extensive chemical spraying. The GM form of creeping bentgrass being tested by the EPA had been made resistant to the herbicide glyphosate, which would eliminate the need for most of the spraying. Tracking the spread of GM pollen from these eight fields, the EPA scientists collected seeds from potted "sentinel" bentgrass plants and from natural grasses, grew them to seedling stage, and examined the DNA of the seedling plants for the glyphosate-resistance gene. The gene was found in the potted bentgrass plants up to 13 miles away, and in the wild relative grasses nearly nine miles away from the fields! Might the spread of the glyphosate-resistance gene to surrounding grasses create hard-to-kill "superweeds," or would it be easy to use other weedkillers to control any newly-resistant plants? However one interprets its findings, this study offers the strongest evidence yet that it will be difficult to control GM secondary crops from interbreeding with surrounding relatives to create new hybrids.

10.5 Genetic engineering affords great opportunities for progress in food production, although many are concerned about possible risks. On balance, the risks appear slight, and the potential benefits substantial.

One of the most active and exciting areas of biology involves recently developed approaches to manipulating animal cells. In this section, you will encounter three areas where landmark progress is being made in cell technology: reproductive cloning of farm animals, stem cell research, and gene therapy. Advances in cell technology hold the promise of literally revolutionizing our lives.

10.6 Reproductive Cloning

The idea of cloning animals was first suggested in 1938 by German embryologist Hans Spemann (called the "father of modern embryology"), who proposed what he called a "fantastical experiment": Remove the nucleus from an egg cell and put in its place a nucleus from another cell.

Early attempts to clone animals in this way failed, until researchers in Scotland suggested a novel idea: Maybe the egg and the donated nucleus needed to be at the same stage in the cell cycle. This proved to be a key insight. In 1994 researchers succeeded in cloning farm animals from advanced embryos by first starving the cells, so that they paused at the beginning of the cell cycle. Two starved cells are thus synchronized at the same point in the cell cycle.

Wilmut's Lamb

Reproductive biologist Ian Wilmut then attempted the key breakthrough, the experiment that had eluded researchers since Spemann proposed it 59 years before: He set out to transfer the nucleus from an adult differentiated cell into an enucleated egg, and to allow the resulting embryo to grow and develop in a surrogate mother, hopefully producing a healthy animal.

Wilmut removed mammary cells from the udder of a six-year-old sheep (figure 10.11). In preparation for cloning, Wilmut's team reduced for five days the concentration of nutrients on which the sheep mammary cells were growing. In parallel preparation, eggs obtained from a ewe were enucleated, the nucleus of each egg carefully removed with a micropipette.

Mammary cells and egg cells were then surgically combined in January of 1996, the mammary cells being inserted inside the covering around the egg cell. Wilmut then applied a brief electrical shock. A neat trick, this causes the plasma membranes surrounding the two cells to become leaky, so that the contents of the mammary cell pass into the egg cell. The shock also jump-starts the cell cycle, initiating cell division.

After six days, in 30 of 277 tries, the dividing embryo reached the hollow-ball "blastula" stage, and 29 of these were transplanted into surrogate mother sheep. Approximately five months later, on July 5, 1997, one sheep gave birth to a lamb. This lamb, "Dolly," was the first successful clone generated from a differentiated animal cell.

Wilmut's successful cloning of a fully differentiated sheep cell is a milestone event in cell technology. Even though his procedure proved inefficient (only 1 of 277 trials succeeded), it established the point beyond all doubt that cloning of adult animal cells, known as **reproductive cloning,** *can* be done.

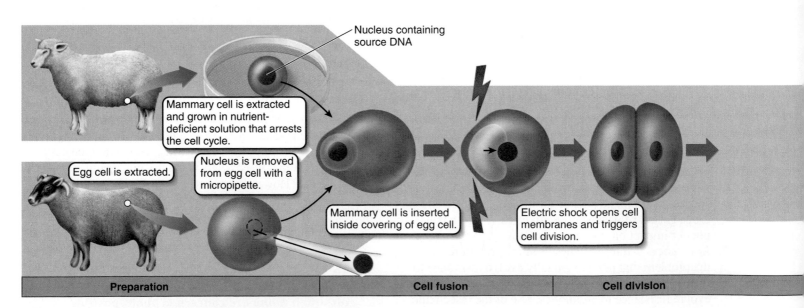

Figure 10.11 Wilmut's animal cloning experiment.
Wilmut combined a mammary cell and an egg cell (with its nucleus removed) to successfully clone a sheep.

Problems with Reproductive Cloning

Since Dolly's birth in 1997, scientists have successfully cloned sheep, mice, cattle, goats, and pigs. Only a small percentage of the transplanted embryos survive to term, however, most dying late in pregnancy. Those that survive to be born usually die soon thereafter. Many become oversized, a condition known as "large offspring syndrome."

The few cloned offspring that reach childhood face an uncertain future, as their development into adults tends to go unexpectedly haywire. Almost none survive to live a normal life span. Even Dolly died prematurely in 2002, having lived only half a normal sheep life span.

The Importance of Genomic Imprinting

What is going wrong? It turns out that as mammalian eggs and sperm mature, their DNA is conditioned by the parent female or male, a process called reprogramming. Chemical changes are made to the DNA that alter when particular genes are expressed without changing DNA sequences. In the years since Dolly, scientists have learned a lot about reprogramming. It appears to occur by a process called *genomic imprinting*. While the details are complex, the basic mechanism of genomic imprinting is simple.

Like a book, a gene can have no impact unless it is read. Genomic imprinting works by blocking the cell's ability to read certain genes. A gene is locked in the off position by chemically altering some of its cytosine DNA nucleotides. Because this involves adding a—CH_3 group (a methyl group), the process is called *methylation*. After a gene has been methylated, the polymerase protein that is supposed to "read" the gene can no longer recognize it. The gene has been shut off.

Genomic imprinting can also lock genes in the on position, permanently activating them. This process also uses methylation; in this case, however, it is not the gene that is blocked. Rather, a regulatory DNA sequence that normally would have prevented the gene from being read is blocked.

Why Cloning Often Fails

Normal animal development depends upon precise genomic imprinting. This chemical reprogramming of the DNA, which takes place in adult reproductive tissue, takes months for sperm and years for eggs.

During cloning, by contrast, the reprogramming of the donor DNA must occur within a few minutes. After the donor nucleus has been added to an enucleated egg, the reconstituted egg begins to divide within minutes, starting the process of making a new individual.

Cloning fails because there is simply not enough time in those few minutes to get the reprogramming job done properly. Thus, researchers find that in large offspring syndrome sheep, many genes have failed to become properly methylated.

Human reproductive cloning will not be practical until scientists figure out how to reprogram a donor nucleus, as occurs to DNA of sperm or eggs in our bodies. This reprogramming may be as simple as finding a way to postpone the onset of cell division after adding a donor nucleus to the enucleated egg, or may prove to be a much more complex process. The point is, we don't have a clue today as to how to do it. In light of this undeniable ignorance, any attempt to clone a human is simply throwing stones in the dark, hoping to hit a target you cannot see.

10.6 While recent experiments have demonstrated the possibility of cloning animals from adult tissue, the cloning of farm animals often fails, for lack of proper gene conditioning.

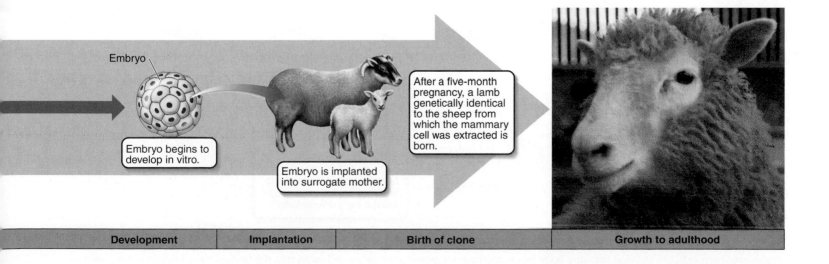

Embryo

Embryo begins to develop in vitro.

Embryo is implanted into surrogate mother.

After a five-month pregnancy, a lamb genetically identical to the sheep from which the mammary cell was extracted is born.

| Development | Implantation | Birth of clone | Growth to adulthood |

10.7 Embryonic Stem Cells

In 1981, mouse stem cells were first discovered to be **pluripotent**—to have the ability to form any body tissue, and even an adult animal—launching the era of stem cell research. After many years of failed attempts, human embryonic stem cells were isolated by James Thomson of the University of Wisconsin in 1998.

Embryonic Stem Cells Are Pluripotent

What is an embryonic stem cell, and why is it pluripotent? To answer this question, we need to consider for a moment where an embryo comes from. At the dawn of a human life, a sperm fertilizes an egg to create a single cell destined to become a child. As development commences, that cell begins to divide, producing after five or six days a small ball of a few hundred cells called a blastocyst. Described in more detail in chapter 25, a blastocyst consists of a protective outer layer of cells destined to form the placenta, enclosing an inner cell mass of **embryonic stem cells.** Each embryonic stem cell is capable by itself of developing into a healthy individual. In cattle breeding, for example, these cells are frequently separated by the breeder and used to produce multiple clones of valuable offspring.

As development proceeds, some of these embryonic stem cells become committed to forming neural tissues, and after this decision is taken, cannot ever produce any other kind of cell. They are then called *nerve stem cells.* Others become specialized to produce blood cells, others to produce muscle tissue, and still others to form the other tissues of the body. Each major tissue is formed from its own kind of tissue-specific **adult stem cell.** Because an adult stem cell forms only that one kind of tissue, it is not pluripotent.

Using Embryonic Stem Cells to Repair Damaged Tissues

Because they can develop into any tissue, embryonic stem cells offer the exciting possibility of restoring damaged tissues (figure 10.12). Experiments have already been tried successfully in mice. Heart muscle cells grown from mouse embryonic stem cells have been successfully integrated with the heart tissue of a living mouse. In other experiments, damaged spinal neurons have been partially repaired, suggesting a path to treating spinal injuries. DOPA-producing neurons of mouse brains, whose progressive loss is responsible for Parkinson's disease, have been successfully replaced with embryonic stem cells, as have the islet cells of the pancreas whose loss leads to juvenile diabetes, also called "type I" diabetes.

Because the course of development is broadly similar in all mammals, these experiments in mice suggest an exciting possibility, that embryonic stem cell therapy may allow the successful treatment of damaged or lost tissues in humans. Individuals with a severed spinal cord might have the injury repaired with embryonic stem cells that replace the damaged neurons. The damaged heart muscle cells of heart attack victims might be replaced with embryonic stem cells that form new heart muscle at the site of damage. Juvenile diabetes might be cured with new insulin-producing islet cells, Parkinson's disease with new DOPA-producing neurons.

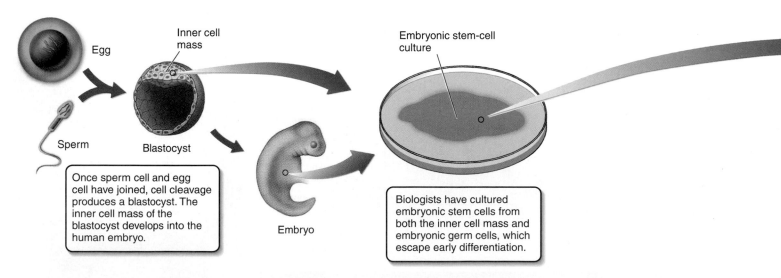

Figure 10.12 Using embryonic stem cells to restore damaged tissue.
Embryonic stem cells can develop into any body tissue. Methods for growing the tissue and using it to repair damaged tissue in adults, such as the brain cells of multiple sclerosis patients, heart muscle, and spinal nerves, are being developed.

Finding a Source for Human Embryonic Stem Cells

This research, however, is quite controversial. Human embryonic stem cells (figure 10.13) are typically isolated from discarded embryos of reproductive clinics, where *in vitro* fertilization procedures typically produce many more human embryos than can be successfully implanted in the prospective mother's womb. However, harvesting embryonic stem cells from these discarded embryos entails disruption and destruction of the embryos, raising serious ethical issues among some citizens who regard life as starting at fertilization, and these embryos as living individuals (see boxed feature later in this chapter).

New experimental results hint at a way around the ethical maze presented by the use of stem cells derived from embryos. Go back for a moment to our explanation of how a vertebrate develops. What happens to the embryonic stem cells? They start to take different developmental paths. For example, some become destined to form nerve tissue while others will go on to form muscle tissue. After this decision is taken, these cells will only give rise to cells of its specific tissue type. Each major tissue type in the body develops from a different tissue-specific stem cell. Now here's the key point: As development proceeds, these tissue-specific stem cells persist—even in adults. So why not use these adult stem cells, rather than embryonic stem cells?

The approach seems very straightforward and should apply to humans. Indeed, blood stem cells are already routinely used in humans to replenish the bone marrow of cancer patients after marrow-destroying therapy. The problem with extending the approach to other kinds of tissue-specific stem

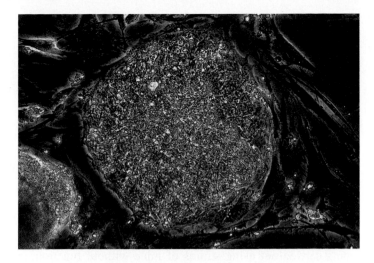

Figure 10.13 Human embryonic stem cells (20×).
Stem cells removed from a six-day blastocyst can be established in culture and then maintained indefinitely. This mass is a colony of undifferentiated human embryonic stem cells surrounded by fibroblasts (elongated cells) that serve as a "feeder layer."

cells is that it has not always been possible to find the kind of tissue-specific stem cell needed.

Therapeutic Cloning

The revolution in cell technology has been led by advances in stem cell research. By surgically transplanting embryonic stem cells, scientists can now routinely perform the remarkable feat of repairing disabled body tissues in mice. The basic strategy

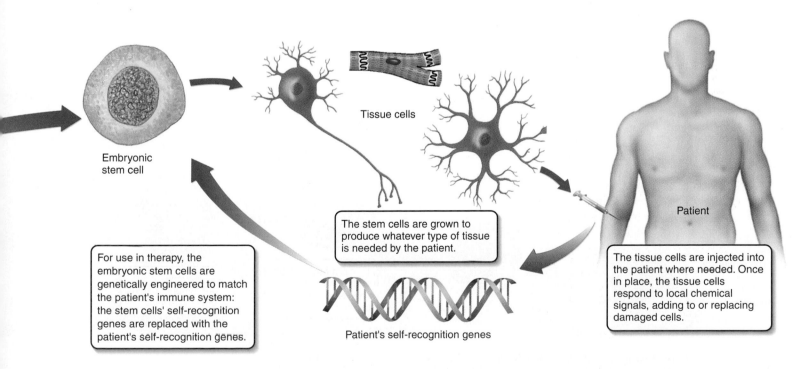

Embryonic stem cell

Tissue cells

Patient

For use in therapy, the embryonic stem cells are genetically engineered to match the patient's immune system: the stem cells' self-recognition genes are replaced with the patient's self-recognition genes.

The stem cells are grown to produce whatever type of tissue is needed by the patient.

Patient's self-recognition genes

The tissue cells are injected into the patient where needed. Once in place, the tissue cells respond to local chemical signals, adding to or replacing damaged cells.

for repairing damaged tissues is to surgically transfer embryonic stem cells to the damaged area, where the stem cells can form healthy replacement cells. Thus, embryonic stem cells transferred into the pancreas of a diabetic mouse that lacks the islet cells needed to produce insulin have been induced to become insulin-secreting islet cells. The new cells produce only about 2% as much insulin as normal islet cells do, so there is still plenty to learn, but the take-home message is clear: Transplanted embryonic stem cells offer a path to cure type I diabetes.

The Problem Posed by Immune Rejection

While exciting, these therapeutic uses of embryonic stem cells to cure type I diabetes, Parkinson's disease, damaged heart muscle, and injured nerve tissue were all achieved in experiments carried out using strains of mice without functioning immune systems. Why is this important? Because had these mice possessed fully functional immune systems, they almost certainly would have rejected the implanted stem cells as foreign. Humans with normal immune systems might well refuse to accept transplanted stem cells simply because they are from another individual. For such stem cell therapy to work in humans, this problem needs to be addressed and solved. The need to solve this problem has been the primary impetus behind the push to develop therapeutic cloning.

Using Cloning to Achieve Immune Acceptance

Early in 2001, a research team at the Rockefeller University reported a way around this potentially serious problem. Their solution? They isolated skin cells, then using the same procedure that created Dolly, they created a 120-cell embryo from them. The embryo was then destroyed, its embryonic stem cells harvested for transfer to injured tissue (figure 10.14).

Figure 10.14 Embryonic stem cells growing in tissue culture.

Embryonic stem cells derived from early human embryos will grow indefinitely in tissue culture. When transplanted, they can sometimes be induced to form new cells of the adult tissue into which they have been placed. This suggests exciting therapeutic uses.

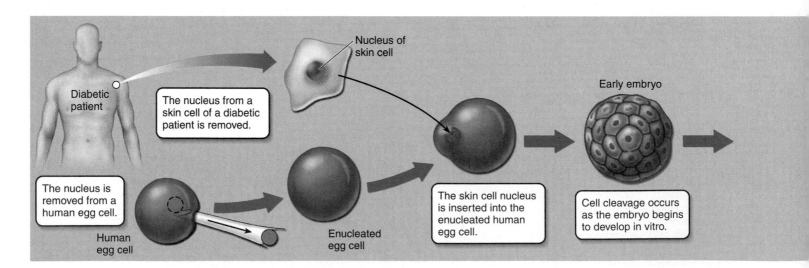

Figure 10.15 How human embryos might be used for therapeutic cloning.

Therapeutic cloning differs from reproductive cloning in that after the initial similar stages, the embryo is destroyed and its embryonic stem cells are extracted, grown in culture, and added to a tissue of the individual who provided the nucleus . In reproductive cloning, by contrast, the embryo is preserved to be implanted and grown to term in a surrogate mother. It is this latter procedure that was done in cloning Dolly, the sheep. Human cells were first cloned in November 2001, in a failed attempt to obtain stem cells for therapeutic cloning procedures such as outlined in this figure.

Using this procedure, which they called **therapeutic cloning,** or more technically, **somatic cell nuclear transfer,** the researchers succeeded in making cells from the tail of an immunologically normal mouse convert into the DOPA-producing cells of the brain that are lost in Parkinson's disease.

Therapeutic cloning successfully addresses the key problem that must be solved before embryonic stem cells can be used to repair damaged human tissues, which is immune acceptance. Because stem cells are cloned from the body's own tissues in therapeutic cloning, they pass the immune system's "self" identity check, and the body readily accepts them. The basic approach of therapeutic cloning is outlined in figure 10.15.

Therapeutic Cloning Is Controversial

It is important to draw a clear distinction between therapeutic cloning and reproductive cloning. In therapeutic cloning, the cloned embryo is destroyed to obtain embryonic stem cells—whereas in reproductive cloning, the embryo develops into an adult individual (figure 10.15). The possibility of therapeutic cloning is controversial, because some fear that a cloned embryo might be brought to term by inserting it into a human uterus. Actually, without first solving the gene imprinting problem, this would be impossible. In the only reported attempt, in 2001, the human clone embryos did not even survive long enough to make stem cells.

In 2003, researchers succeeded in converting mouse embryonic stem cells into what appeared to be oocytes, egg cells like those produced in the ovary. Unexpectedly, some of these eggs turned out to be able to develop parthenogenetically (that is, without being fertilized by sperm) into blastocyst embryos.

Use of stem cell–derived eggs might make therapeutic cloning ethically more acceptable to those who object to it for several reasons. First, it would provide a plentiful source of human eggs, which are now obtained from patients through an uncomfortable procedure requiring strong drugs and surgery. Second, the eggs could be genetically altered so as to be inviable in the womb, but still useful for therapeutic cloning, relieving the fear that the stem cell–derived egg might somehow be used to make a baby. Third, because the embryo used for therapeutic cloning is produced without fertilization, the approach has some theologians reconsidering their ideas about the nature of life as something that starts at conception with the union of egg and sperm. There is little doubt that the controversy over human therapeutic cloning will continue.

10.7 Human embryonic stem cells offer the possibility of replacing damaged or lost human tissues, although the procedures are controversial. Therapeutic cloning involves initiating blastocyst development from a patient's tissue using nuclear transplant procedures, then using these embryonic stem cells to replace the patient's damaged or lost tissue.

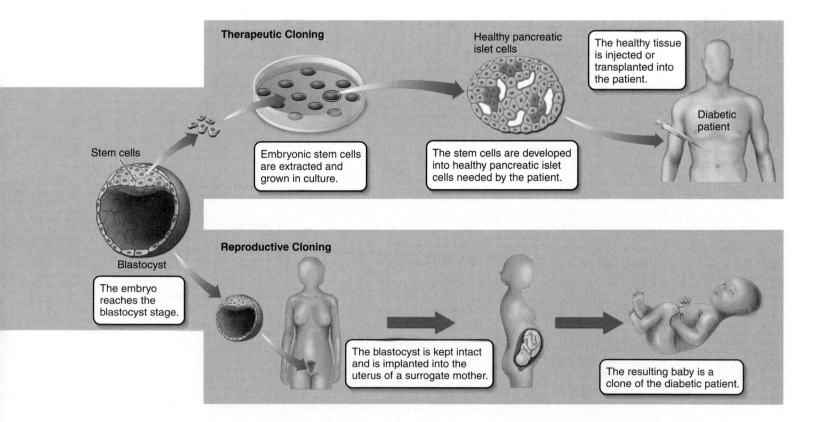

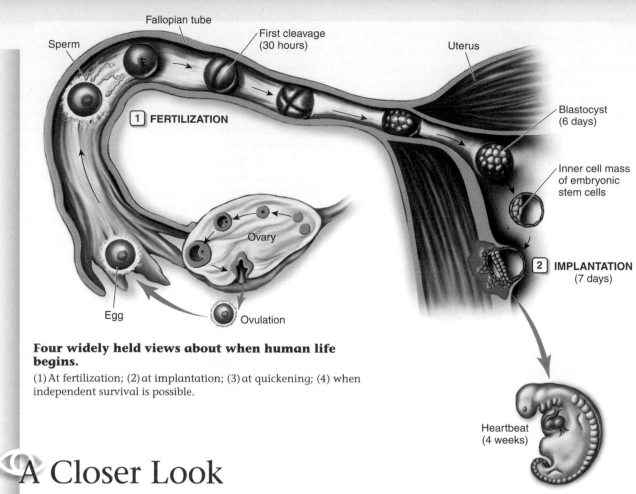

Fallopian tube

First cleavage
(30 hours)

Sperm

Uterus

1 FERTILIZATION

Blastocyst
(6 days)

Inner cell mass
of embryonic
stem cells

Ovary

Egg

Ovulation

2 IMPLANTATION
(7 days)

Four widely held views about when human life begins.

(1) At fertilization; (2) at implantation; (3) at quickening; (4) when independent survival is possible.

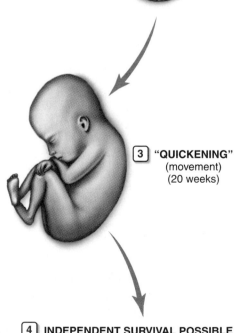

Heartbeat
(4 weeks)

A Closer Look

When Does Human Life Begin?

The story of when human life begins has a checkered past. Centuries before people knew of sperm and eggs, Aristotle argued that the fusion creating a new person did not exist until "quickening," the first noticeable movements in a woman's womb. He reckoned quickening occurred 40 days into pregnancy (18 to 20 weeks is the actual time). The 40-day rule was picked up by Jewish and Muslim religions. In 1591, Pope Gregory XIV supported this view of delayed animation and ensoulment. The Catholic Church did not reach its current conclusion that life begins at fertilization until 1896, when Pope Pius IX condemned abortion at any age after the moment of conception. Many Jewish theologians now argue that life begins seven days into pregnancy, with implantation of the embryo. Gene transcription starts even later (well after stem cells are harvested), and many scientists feel human individuality cannot be said to begin until then, when the embryo starts to actually use its genes. The U.S. Supreme Court takes the position that human life begins much later, when the fetus becomes capable of independent life if separated from the mother—roughly the third trimester.

3 "QUICKENING"
(movement)
(20 weeks)

4 INDEPENDENT SURVIVAL POSSIBLE
(third trimester)

Gene Therapy

The third major advance in cell technology involves introducing "healthy" genes into cells that lack them. For decades scientists have sought to cure often-fatal genetic disorders like cystic fibrosis, muscular dystrophy, and multiple sclerosis by replacing the defective gene with a functional one.

Early Success

That such **gene transfer therapy** can work was first demonstrated in 1990. Two girls were cured of a rare blood disorder due to a defective gene for the enzyme adenosine deaminase. Scientists isolated working copies of this gene and introduced them into bone marrow cells taken from the girls. The gene-modified bone marrow cells were allowed to proliferate, then were injected back into the girls. The girls recovered and stayed healthy. For the first time, a genetic disorder was cured by gene therapy.

The Rush to Cure Cystic Fibrosis

Like hounds to a hot scent, researchers set out to apply the new approach to one of the big killers, cystic fibrosis. The defective gene, labeled *cf,* had been isolated in 1989. Five years later, in 1994, researchers successfully transferred a healthy *cf* gene into a mouse with a defective one—they in effect had cured cystic fibrosis in a mouse. They achieved this remarkable result by adding the *cf* gene to a virus that infected the lungs of the mouse, carrying the gene with it "piggyback" into the lung cells. The virus chosen as the "vector" was adenovirus, a virus that causes colds and is very infective of lung cells. To avoid any complications, the lab mice used in the experiment had their immune systems disabled.

Very encouraged by these well-publicized preliminary trials with mice, several labs set out in 1995 to attempt to cure cystic fibrosis by transferring healthy copies of the *cf* gene into human patients. Confident of success, researchers added the human *cf* gene to adenovirus then squirted the gene-bearing virus into the lungs of cystic fibrosis patients. For eight weeks the gene therapy did seem successful, but then disaster struck. The gene-modified cells in the patients' lungs came under attack by the patients' own immune systems. The "healthy" *cf* genes were lost and with them any chance of a cure.

Problems with the Vector

Other attempts at gene therapy met with similar results, eight weeks of hope followed by failure. In retrospect, although it was not obvious then, the problem with these early attempts seems predictable. Adenovirus causes colds. Do you know anyone who has never had a cold? Due to previous colds, all of us have antibodies directed against adenovirus. We were introducing therapeutic genes in a vector our bodies are primed to destroy.

In 1995, the newly appointed head of the National Institutes of Health (the NIH), Nobel Prize winner Harold Varmus, held a comprehensive review of human gene therapy trials. Three problems became evident in the review: (1) The adenovirus vector being used in most trials elicits a strong immune response, leading to rejection of the added gene. (2) Adenovirus infection can, in rare instances, produce a very severe immune reaction, enough to kill. If many patients are treated, such instances can be expected to occur. (3) When the adenovirus infects a cell, it inserts its DNA into the human chromosome. Unfortunately, it does so at a random location. This means that the insertion events will cause mutations—by jumping into the middle of a gene, the virus inactivates that gene. Because the spot where adenovirus inserts is random, some of the mutations that result can be expected to cause cancer, certainly an unacceptable consequence.

Faced with these findings, Varmus called a halt to all further human clinical trials of gene therapy. "Go back to work in the laboratory," he told researchers, "until you get a vector that works."

More Promising Vectors

Within a few years, researchers had a much more promising vector. This new gene carrier is a tiny parvovirus called *adeno-associated virus* (AAV). It has only two genes and needs adenovirus to replicate. To create a vector for gene transfer, researchers remove both of the AAV genes. The shell that remains is still quite infective and can carry human genes into patients. More important, AAV enters human DNA far less frequently than adenovirus; thus, it is less likely to produce cancer-causing mutations. In addition, AAV does not elicit a strong immune response—cells infected with AAV are not eliminated by a patient's immune system. Finally, AAV never elicits a dangerously strong immune response, so it is safe to administer AAV to patients.

Success with the AAV Vector

In 1999, AAV successfully cured anemia in rhesus monkeys. In monkeys, humans, and other mammals, red blood cell production is stimulated by a protein called erythropoietin (EPO). People with anemia (that is, low red blood cell counts), like dialysis patients, get regular injections of EPO. Using AAV to carry a souped-up EPO gene into the monkeys, scientists were able to greatly elevate their red blood cell counts, curing the monkeys of anemia—and they stayed cured.

A similar experiment using AAV cured dogs of a hereditary disorder leading to retinal degeneration and blindness. These dogs had a defective gene that produced a mutant form of a protein associated with the retina of the eye, and were blind. Injection of AAV bearing the needed gene into the fluid-filled compartment behind the retina restored their sight (figure 10.16).

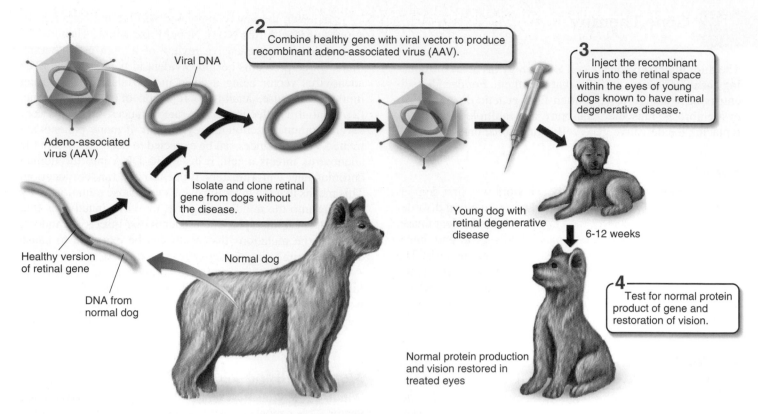

Figure 10.16   Using gene therapy to cure a retinal degenerative disease in dogs.
Recently, researchers were able to use genes from healthy dogs to restore vision in dogs blinded by an inherited retinal degenerative disease. This disease also occurs in human infants and is caused by a defective gene that leads to early vision loss, degeneration of the retinas, and blindness. In the gene therapy experiments, genes from dogs without the disease were inserted into 3-month-old dogs that were known to carry the defective gene and that had been blind since birth. Six weeks after the treatment, the dogs' eyes were producing the normal form of the gene's protein product, and by three months, tests showed that the dogs' vision was restored.

Human clinical trials are under way again. In 2000, scientists performed the first gene therapy experiment for muscular dystrophy, injecting genes into a 35-year-old South Dakota man. He is an early traveler on what is likely to become a well-traveled therapeutic highway. Trials are also under way for cystic fibrosis, rheumatoid arthritis, hemophilia, and a wide variety of cancers. The way seems open, and the possibility of progress is tantalizingly close.

> **10.8   In principle, it should be possible to cure hereditary disorders like cystic fibrosis by transferring a healthy gene into the cells of affected tissues. Early attempts using adenovirus vectors were not often successful. New virus vectors like AAV avoid the problems of earlier vectors and offer promise of gene transfer therapy cures.**

Exploring Current Issues

Additional Resources

Go to your campus library or look online to find the following articles, which further develop some of the concepts found in this chapter.

Coghlan, A. and N. Jones (2003). Landmark report aims to lay Frankenfood scares to rest. *New Scientist,* 179(2405), 6.

Henig, R. M. (2003). Pandora's baby. *Scientific American,* 288(6), 62.

U.S. Department of Energy Human Genome Program. (2003). Genomics and its impact on science and society: the human genome project and beyond. *Retrieved from: www.ornl.gov/ sci/techresources/Human_Genome/publicat/primer2001/index. shtml.*

U.S. Department of Energy Office of Science (2004). Cloning fact sheet. *Retrieved from: www.ornl.gov/sci/techresources/Human_ Genome/elsi/cloning.shtml.*

Biology and Society Lecture: **The Human Genome**

The sequencing of the human genome was announced on June 26, 2000. There were two quite unexpected results: There aren't as many genes as previously thought—only about the same number of genes contained in a mouse—and practically half of the entire genome is occupied by transposable elements.

Biology and Society Lecture: **Cloning and Stem Cells**

On July 5, 1996, the first mammal, a sheep dubbed Dolly, was cloned from an adult fully-differentiated cell. When the problem of properly "programming" DNA is solved, cloning offers a powerful opportunity to cure disorders and conditions involving lost tissue for which there is now no effective therapy.

Find these lectures, delivered by the author to his class at Washington University, online at www.mhhe.com/essentialstlw.com/exp10.

Sequencing Entire Genomes

10.1 Genomics

- The genetic information of an organism, its genes and other DNA, is called its genome. The sequencing and study of genomes is a new area of biology called genomics.

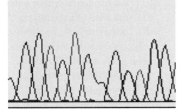

- The sequencing of entire genomes, a once long and tedious process, has been made faster and easier with automated systems (**figure 10.1**).

10.2 The Human Genome

- The human genome contains about 25,000 genes, not much more than other organisms (**table 10.1 and figure 10.2**) and far less than what was expected due to alternative splicing of exons.

- Genes are organized in different ways in the genome, with nearly 99% of the human genome containing noncoding segments (**table 10.2**).

Genetic Engineering

10.3 A Scientific Revolution

- Genetic engineering is the process of moving genes from one organism to another. It is having a major impact on medicine and agriculture (**figure 10.3**).

- Restriction enzymes are a special kind of enzyme that cuts DNA at specific sequences. When two different molecules of DNA are cut with the same restriction enzyme, sticky ends form, allowing the segments of different DNAs to be joined (**figure 10.4**). Cleaving DNA is the first of four stages of a genetic engineering experiment. The other three stages are: Producing recombinant DNA, cloning the DNA in a host cell, and screening the clones for the gene of interest (**figure 10.5**).

10.4 Genetic Engineering and Medicine

- Genetic engineering has been used in many medical applications. The first application was the production of medically important proteins used to treat illnesses (**table 10.3**). Genes encoding the proteins are inserted into bacteria that become small protein-producing factories.

- Vaccines are developed using genetic engineering. A gene that encodes a viral protein is put into the DNA of a harmless vector that expresses the protein. The vector is injected into the body where it directs the production of the viral protein. The body mounts an immune response against the viral protein. This protects the person from an infection by that virus in the future (**figure 10.7**).

10.5 Genetic Engineering and Agriculture

- Genetic engineering has been used in manipulating crop plants to make them more cost-effective to grow or more nutritious (**figures 10.8–10.10**).

- However, GM plants are a source of controversy because of potential dangers that may result from the genetic manipulation of crop plants.

The Revolution in Cell Technology

10.6 Reproductive Cloning

- Ian Wilmut succeeded in cloning a sheep by synchronizing the cell cycles of both the nucleus donor and host cells (**figure 10.11**).

- Other animals have been cloned but problems and complications usually arise, causing premature death. The problem with successful cloning appears to be caused by the lack of modifications that need to be made to the DNA, which turns genes "on" or "off", a process called genomic imprinting.

10.7 Stem Cells

- Embyronic stem cells are pluripotent cells, present in the early embryo, that are able to divide and develop into any cell in the body or develop into an entire individual. Because of their pluripotent nature, these cells could be used to replace lost or damaged tissues (**figure 10.12**). However, this procedure has problems because of immunological rejection and ethical issues concerning the destruction of human embryos.

- Therapeutic cloning is the process whereby a cell from an individual who has lost tissue function due to disease or accident is cloned, producing an embryo. Embryonic stem cells are harvested from the cloned embryo and injected into the same individual. The embryonic stem cells regrow the lost tissue without eliciting an immune response (**figure 10.15**). However, this procedure is also controversial.

10.8 Gene Therapy

- Using gene therapy, a patient with a genetic disorder is cured by replacing a defective gene with a "healthy" gene. In theory, this should work—but early attempts to cure cystic fibrosis failed because of immunological reactions to the adenovirus vector used to carry the healthy gene into the patient.

- The goal of researchers is to identify a vector that avoids the problems encountered with the adenovirus. Promising results in experiments using a parvovirus called adeno-associated virus (AAV) has scientists hopeful that this vector will eliminate the problems seen with adenovirus (**figure 10.16**).

1. The total amount of DNA in an organism, including all of its genes and other DNA, is its
   a. heredity.
   b. genetics.
   c. genome.
   d. genomics.

2. A possible reason why humans have such a small number of genes as opposed to what was anticipated by scientists is that
   a. humans don't need more than 25,000 genes to function.
   b. the exons used to make a specific mRNA can be rearranged to form genes for new proteins.
   c. the sample size used to determine the human genome was not big enough, so the number of genes estimated could be low.
   d. the estimate will increase as scientists find out what so-called "junk DNA" actually does.

3. A protein that can cut DNA at specific DNA base sequences is called a
   a. DNase.
   b. DNA ligase.
   c. restriction enzyme.
   d. DNA polymerase.

4. The four steps of a genetic engineering experiment are (in order)
   a. cleaving DNA, cloning, producing recombinant DNA, and screening.
   b. cleaving DNA, producing recombinant DNA, cloning, and screening.
   c. producing recombinant DNA, cleaving DNA, screening, and cloning.
   d. screening, producing recombinant DNA, cloning, and cleaving DNA.

5. Using drugs produced by genetically engineered bacteria allows
   a. the drug to be produced in far larger amounts than in the past.
   b. humans to permanently correct the effects of a missing gene from their own systems.
   c. humans to eliminate the chances of infection from blood transfusions.
   d. All of these answers are correct.

6. Some of the advantages to using genetically modified organisms in agriculture include
   a. increased yield.
   b. unchanged nutritive value.
   c. the ease of transferring the gene to other organisms.
   d. possibility of allergic reactions.

7. Which of the following is *not* a concern about the use of genetically modified crops?
   a. possible danger to humans after consumption
   b. insecticide resistance developing in pest species
   c. gene flow into natural relatives of GM crops
   d. harm to the crop itself from mutations

8. Genomic imprinting seems to involve
   a. protein signals that block transcription of a gene from its DNA.
   b. proteins that cause deformation of RNA polymerase.
   c. methylation of RNA polymerase.
   d. methylation of DNA.

9. One of the main biological problems with replacing damaged tissue through the use of embryonic stem cells is
   a. immunological rejection of the tissue by the patient.
   b. that stem cells may not target appropriate tissue.
   c. the time needed to grow sufficient amounts of tissue from stem cells.
   d. that genetic mutations of chosen stem cells may cause future problems

10. In gene therapy, healthy genes are placed into cells with defective genes by using
    a. bacteria.
    b. micropipettes (needles).
    c. viruses.
    d. Currently, cells are not modified genetically. Instead, healthy tissue is grown and transplanted into the patient.

1. **Figure 10.2** What can you say about the number of genes an organism contains and the complexity of an organism?

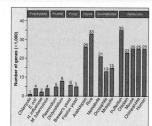

2. **Figure 10.11** Your friend Charlie wants to know why scientists can't just take the egg cell, with its own nucleus intact, and shock it to begin cell division? How do you answer him?

Sequencing Entire Genomes Your friend Marisa asks you why it is so difficult to sequence a genome; it seems as if it should simply be a matter of reading the DNA. What do you tell her?

Gene Engineering Many of the initial genetically engineered drugs have to do with the blood. What might be the reasons for this?

The Revolution in Cell Technology Why is it that we know the basics of how to clone, yet we cannot successfully clone our pets?

Visit the Online Learning Center for this chapter at www.mhhe.com/essentialstlw/ch10 for quizzes, animations, interactive learning exercises, and other study tools. At the site you will also find extended answers to the end-of-chapter questions.

11

Evolution and Natural Selection

T his photograph shows color variants of the peppered moth, *Biston betularia,* glued to the trunk of a soot-polluted tree. Historically, the dark form of the moth was rare, but with the onset of the Industrial Revolution the dark "melanic" forms became more common. To test the hypothesis that natural selection was favoring the dark form because of industrial pollution, biologists released equal numbers of the two variants in both polluted and nonpolluted areas. In polluted areas, twice as many dark individuals survived as light ones; in unpolluted areas, the reverse occurred, with more light-colored individuals surviving. This experiment clearly demonstrated that natural selection was favoring the dark form in polluted environments. How? The most likely hypothesis was that birds were eating the moths, selecting more often those that were most visible on tree trunks. As you can see in the photo, white moths are more visible on polluted tree trunks. As a test, moths were glued to dark tree trunks to see which the birds would eat. As you might expect, they ate the most visible ones. So is camouflage the reason natural selection favored the dark moths? Probably not. Further work showed these moths don't spend their days on tree trunks. Some other effect of the pollution seems to be at work. In this chapter, we will look closely at the action of natural selection, and at the evidence for evolution.

11.1 Evolution: Getting from There to Here

The word "evolution" is widely used in the natural and social sciences. It refers to how an entity—be it a social system, a gas, or a planet—changes through time. Although development of the modern concept of evolution in biology can be traced to Darwin, he initially used the phrase "descent with modification" to explain the concept of evolution. This phrase captures the essence of biological evolution: All species arise from other, pre-existing species. However, through time, they accumulate differences such that ancestral and descendant species are not identical. It is through this concept of evolution that we are able to explain the great paradox of biology, that in life there exists both unity and diversity.

The general concept of evolution was introduced in chapter 2, but it is important—before launching into a detailed discussion of evolution and natural selection—that we review how natural selection, the process that leads to evolution, works. The process of natural selection can be conveniently visualized as occurring in a series of steps:

1. There is gene variation among the individuals of a population.

2. This variation is often passed on to offspring, so that offspring have a tendency to look more like their parents than like other members of the population.

3. All populations overproduce young—only some of the offspring will survive to reproduce.

4. The likelihood that a particular individual will survive and reproduce is not random. The traits it inherits make it more or less likely to survive and reproduce.

5. Those individuals with traits that aid survival and reproduction will have a better chance of adding individuals to the next generation.

6. Over time, the population changes such that the traits of the more successful reproducers become more prevalent in the population.

Microevolution Leads to Macroevolution

When the word *evolution* is mentioned, it is difficult not to conjure up images of dinosaurs roaming the earth, woolly mammoths frozen in blocks of ice, or Darwin observing a monkey. Traces of ancient life-forms, now extinct, survive as fossils that help us piece together the evolutionary story. With such a background, we usually think of evolution in terms of changes that take place over long periods of time, changes in the kinds of animals and plants on earth as new forms replace old ones. This kind of evolution, **macroevolution,** is evolutionary change on a grand scale, encompassing the origins of novel kinds of organisms invading new habitats, and major episodes of extinction.

Much of the focus of Darwin's theory of natural selection, however, is directed not at the way in which new species are formed from old ones but rather at the way that changes occur within species. Natural selection is the process whereby some individuals in a population, those that possess certain inherited characteristics or adaptations, produce more surviving offspring than individuals lacking these characteristics. As a result, the population will gradually come to include more and more individuals with the advantageous characteristics. In this way the population evolves. Changes of this

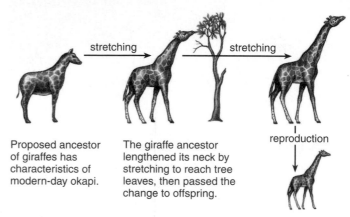

Proposed ancestor of giraffes has characteristics of modern-day okapi.

The giraffe ancestor lengthened its neck by stretching to reach tree leaves, then passed the change to offspring.

(a) **Lamarck's theory: variation is acquired.**

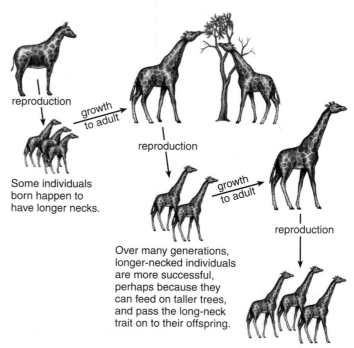

Some individuals born happen to have longer necks.

Over many generations, longer-necked individuals are more successful, perhaps because they can feed on taller trees, and pass the long-neck trait on to their offspring.

(b) **Darwin's theory: variation is inherited.**

Figure 11.1   How did long necks evolve in giraffes?
(*a*) Lamarck proposed that the trait for a longer neck was acquired and then passed on to offspring. (*b*) Darwin proposed that there existed some variation in the length of the neck in the population. Long-necked individuals may have survived better than shorter-necked individuals, producing more offspring. Over time, the long-necked trait became more prevalent in the population.

sort within populations—changes in gene frequencies—represent **microevolution.**

Adaptation results from microevolutionary changes that increase the likelihood of survival and reproduction of particular genetic traits in a population. In essence, Darwin's explanation of evolution is that adaptation by natural selection is responsible for evolutionary changes *within* a species (microevolution), and that the accumulation of these changes leads to the development of new species (macroevolution).

The Key Is the Source of the Variation

Darwin agreed with many earlier philosophers and naturalists who deduced that the many kinds of organisms around us were produced by a process of evolution.

A theory championed by a predecessor of Darwin, the prominent biologist Jean-Baptiste Lamarck, was that evolution occurred by the inheritance of acquired characteristics. According to Lamarck, individuals passed on to offspring body and behavior changes acquired during their lives. Thus, Lamarck proposed that ancestral giraffes with short necks tended to stretch their necks to feed on tree leaves, and this extension of the neck was passed on to subsequent generations, leading to the long-necked giraffe (figure 11.1). In Darwin's theory, by contrast, the variation is not created by experience but already exists when selection acts on it. Unlike his predecessors, Darwin proposed natural selection as the mechanism of evolution.

It is important to remember that natural selection can only act on the variation already present in a population. New gene variations are constantly being introduced into populations through random mutations. Of course, for the variation to be passed on to offspring, the mutation must occur in the germ cells. Mutations that only affect somatic cells (for example, mutations in lung cells that result from tobacco use and lead to lung cancer) cannot be passed on to offspring.

The Rate of Evolution

Different kinds of organisms evolve at different rates. Bacteria evolve very quickly compared with eukaryotic organisms, which in general have longer reproductive cycles and reproduce sexually. Some animal groups seem to evolve more rapidly than others. Thus, while mammals have evolved rapidly, lungfishes have changed little over the past 150 million years. On the basis of a relatively complete fossil record, it has been estimated that the duration of a typical mammal species, from the formation of the species to its extinction, is only a small fraction of that, about 200,000 years.

Not only does the rate of evolution differ greatly from group to group, but evolution within a group apparently

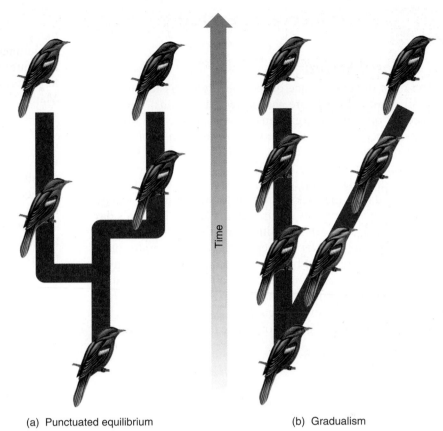

(a) Punctuated equilibrium   (b) Gradualism

Figure 11.2　Two views of the pace of macroevolution.
(*a*) Punctuated equilibrium surmises that species formation occurs in bursts, separated by long periods of stasis, while (*b*) gradualism surmises that species formation is constantly occurring by accumulating small differences.

proceeds rapidly during some periods and relatively slowly during others. The fossil record provides evidence for such variability in evolutionary rates, and evolutionists are very interested in understanding the factors that account for it. In some species, evolution appears to proceed in spurts. In times of environmental upheaval, evolutionary innovations give rise to new lines; these lines then persist unchanged for a long time, in "equilibrium," until a new spurt of evolution creates a "punctuation" in the fossil record. This theory of **punctuated equilibrium** contrasts with that of **gradualism,** or gradual evolutionary change, proposed by Darwin (figure 11.2).

Unfortunately, the distinctions are not as clear-cut as implied by this discussion. Some well-documented groups clearly have evolved gradually and not in spurts. Other groups seem to show the irregular pattern of evolutionary change the punctuated equilibrium model predicts.

11.1　Darwin proposed that natural selection on variants within populations leads to evolution. Viewed over long time periods, some groups evolve at a uniform rate, others in fits and starts.

11.2   The Evidence for Evolution

The Fossil Record

The most direct evidence of macroevolution is found in the fossil record. **Fossils** are the preserved remains, tracks, or traces of once-living organisms. Fossils are created when organisms become buried in sediment, the calcium in bone or other hard tissue mineralizes, and the surrounding sediment eventually hardens to form rock. The fossils contained in layers of sedimentary rock reveal a history of life on earth.

By dating the rocks in which fossils occur, we can get an accurate idea of how old the fossils are. In Darwin's day, rocks were dated by their position with respect to one another (relative dating); rocks in deeper strata are generally older. Today, rocks are dated by measuring the degree of decay of certain radioisotopes contained in the rock (absolute dating, see chapter 3); the older the rock, the more its isotopes have decayed. This is a more accurate way of dating rocks and provides dates stated in millions of years, rather than relative dates.

When fossils are arrayed according to their age, from oldest to youngest, they often provide evidence of successive evolutionary change. Among the hoofed mammals illustrated in figure 11.3, small, bony bumps on the nose can be seen to change continuously, until they become large, blunt horns. Many other examples illustrate a record of successive change and are one of the strongest lines of evidence that evolution has occurred.

While many gaps interrupted the fossil record in Darwin's era, even then, scientists knew of the *Archaeopteryx,* a transitional fossil between reptiles and birds. Today, the fossil record is far more complete, particularly among the vertebrates; fossils have been found linking all the major groups. The forms linking mammals to reptiles are particularly well known. A series of extinct marine mammals linking whales to terrestrial, four-legged hoofed ancestors filled in one of the last significant gaps (figure 11.4).

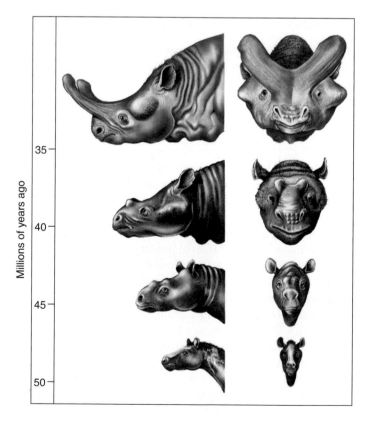

Figure 11.3   Evolution in the titanotheres.

Here you see illustrated changes in a group of hoofed mammals known as titanotheres between about 50 million and 35 million years ago. During this time, the small, bony protuberance located above the nose 50 million years ago evolved into relatively large, blunt horns.

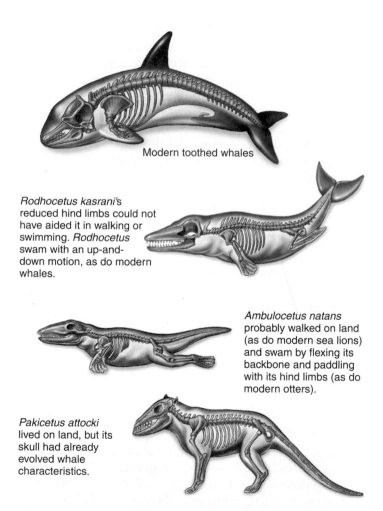

Modern toothed whales

Rodhocetus kasrani's reduced hind limbs could not have aided it in walking or swimming. *Rodhocetus* swam with an up-and-down motion, as do modern whales.

Ambulocetus natans probably walked on land (as do modern sea lions) and swam by flexing its backbone and paddling with its hind limbs (as do modern otters).

Pakicetus attocki lived on land, but its skull had already evolved whale characteristics.

Figure 11.4   Whale "missing links."

The recent discoveries of *Ambulocetus, Rodhocetus,* and *Pakicetus* have filled in the gaps between whales and their hoofed mammal ancestors. The features of *Pakicetus* illustrate that intermediate forms are not intermediate in all characteristics; rather, some traits evolve before others. All three fossil forms occurred in the Eocene period, 45 to 55 million years ago.

The Molecular Record

Traces of our evolutionary past are also evident at the molecular level. We possess color vision genes that have become more complex as vertebrates have evolved, and we employ pattern formation genes during early development that all animals share. If you think about it, the fact that organisms have evolved successively from relatively simple ancestors implies that a record of evolutionary change is present in the cells of each of us, in our DNA. According to evolutionary theory, new alleles arise from the old by mutation and come to predominance through favorable selection. Thus, a series of evolutionary changes involves a continual accumulation of genetic changes in the DNA. Organisms that are more distantly related will have accumulated a greater number of evolutionary differences, while two species that are more closely related will share a greater portion of their DNA. This pattern of divergence is clearly seen among vertebrates in a human hemoglobin polypeptide (figure 11.5): Macaques, primates closely related to humans, have fewer differences from humans in the 146-amino-acid hemoglobin β chain than do more distantly related mammals, like dogs. Nonmammalian terrestrial vertebrates differ even more, and marine vertebrates are the most different of all. In addition to analyzing specific chromosomal genes, researchers have also monitored changes in mitochondrial DNA as part of their study of human origins.

Molecular Clocks   This same pattern is seen when DNA sequences from various organisms are compared. For example, the longer the time since the organisms diverged, the greater the number of differences in the nucleotide sequence of the cytochrome *c* gene (figure 11.6), which codes for a protein that plays a key role in oxidative metabolism, as we learned in chapter 6. The changes appear to accumulate in cytochrome *c* at a constant rate, a phenomenon sometimes referred to as a **molecular clock.**

Proteins Evolve at Different Rates   All proteins for which data are available appear to accumulate changes over time. However, different proteins evolve at very different rates, which presents estimates of the number of "acceptable" point mutations (that is, changes successfully incorporated into the protein) per 100 amino acids per million years. The fastest rate of change appears to be in fibrinopeptides, while the most highly conserved protein is histone H4. Even faster rates of change are seen for pseudogenes, which are not transcribed, suggesting that molecular evolution proceeds more quickly when less constrained by selection.

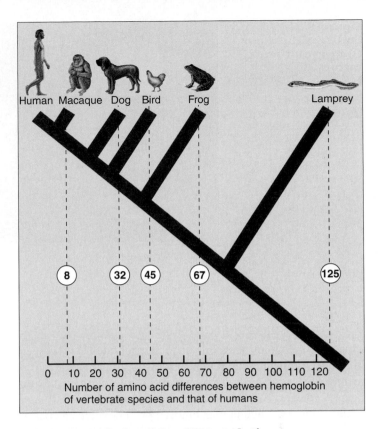

Figure 11.5   Molecules reflect evolutionary divergence.

The greater the evolutionary distance from humans (as revealed by the *blue* evolutionary tree, which is based on the fossil record), the greater the number of amino acid differences in the vertebrate hemoglobin polypeptide.

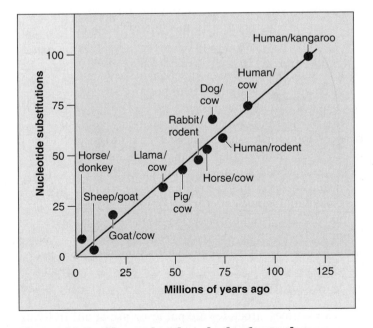

Figure 11.6   The molecular clock of cytochrome *c*.

When the time since each pair of organisms presumably diverged is plotted against the number of nucleotide differences in cytochrome *c*, the result is a straight line, suggesting that the cytochrome *c* gene is evolving at a constant rate.

**Figure 11.7
Embryos show our
early evolutionary
history.**

These embryos,
representing various
vertebrate animals, show
the primitive features
that all vertebrates
share early in their
development, such as gill
slits and a tail.

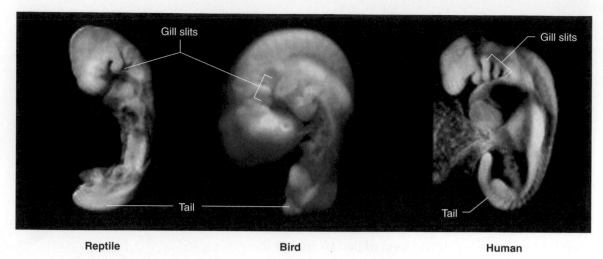

Reptile   Bird   Human

The Anatomical Record

Much of the evolutionary history of vertebrates can be seen
in the way in which their embryos develop. Early in development, all vertebrate embryos have gill slits like a fish; at a
later stage every vertebrate embryo has a long bony tail; later
in development, each human embryo even develops a coat of
fine fur! These relict developmental forms strongly suggest
that all vertebrates share a basic set of developmental instructions (figure 11.7).

As vertebrates have evolved, the same bones are sometimes put to different uses, yet they can still be seen, their
presence betraying their evolutionary past. For example, the
forelimbs of vertebrates are all **homologous structures;** that
is, although the structure and function of the bones have diverged, they are derived from the same body part present in a
common ancestor. You can see in figure 11.8 how the bones
of the forelimb have been modified in one way in the wings of
bats, in another way in the fins of porpoises, and in yet other
ways in horses and humans.

Not all similar features are homologous. Sometimes
features found in different lineages come to resemble each
other as a result of parallel evolutionary adaptations to similar
environments. This form of evolutionary change is referred
to as *convergent evolution* and these similar-looking features
are called **analogous structures.** For example, the marsupial
mammals of Australia evolved in isolation from placental
mammals, but similar selective pressures have generated very
similar kinds of animals (figure 11.9). Similarly, the wings of
birds, pterosaurs, and bats are analogous structures, modified
through natural selection to serve the same function and therefore look the same.

Sometimes structures are put to no use at all! In living
whales, which evolved from hoofed mammals, the bones of
the pelvis that formerly anchored the two hind limbs are all
that remain of the rear legs, unattached to any other bones
and serving no apparent purpose (see figure 11.4). Another example of a *vestigial organ* is the human appendix. In the great
apes, our closest relatives, we find an appendix much larger

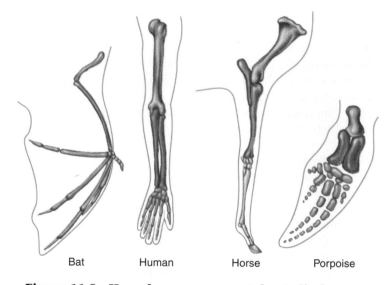

Bat   Human   Horse   Porpoise

Figure 11.8   Homology among vertebrate limbs.
Homologies among the forelimbs of four mammals show the ways
in which the proportions of the bones have changed in relation to
the particular way of life of each organism. Although considerable
differences can be seen in form and function, the same basic bones
are present in each forelimb.

than ours attached to the gut tube, which functions in digestion, holding bacteria used in digesting the cellulose cell walls
of the plants eaten by these primates. The human appendix is a
vestigial version of this structure that now serves no function
in digestion (although it may have acquired an alternate function in the lymphatic system).

**11.2   The fossil record provides a clear record
of continuous evolutionary change. The gene
record exhibits continual evolution, accumulating
increasing numbers of changes over time.
Comparative anatomy also offers evidence that
evolution has occurred.**

Taking Flight
To take to the air, three very different vertebrates lightened bones and transformed hands into wings.

Eastern bluebird

Pterosaur (extinct)

Samoan flying fox (fruitbat)

Mouse

Marsupial mouse

Wolf

Two Worlds
Marsupials evolved the same sort of adaptations in isolation in Australia that placental mammals did elsewhere.

Flying phalanger

Flying squirrel

Tasmanian wolf

Figure 11.9   Convergent evolution: many paths to one goal.

Over the course of evolution, form often follows function. Members of very different animal groups often adapt in similar fashions when challenged by similar opportunities. These are but a few of many examples of such convergent evolution. The flying vertebrates represent mammals (bat), reptiles (pterosaur), and birds (bluebird). The three pairs of terrestrial vertebrates each contrast a North American placental mammal with an Australian marsupial one.

"Taking Flight" image from The New York Times, *December 15, 1988.* The New York Times. *Reprinted with permission.*

11.8 The Biological Species Concept

A key aspect of Darwin's theory of evolution is his proposal that adaptation (microevolution) leads ultimately to species formation (macroevolution). The way natural selection leads to the formation of new species has been thoroughly documented by biologists, who have observed the stages of the species-forming process, or **speciation,** in many different plants, animals, and microorganisms. Speciation usually involves progressive change: First, local populations become increasingly specialized; then, if they become different enough, natural selection may act to keep them that way.

Before we can discuss how one species gives rise to another, we need to understand exactly what a species is. The evolutionary biologist Ernst Mayr coined the **biological species concept,** which defines species as "groups of actually or potentially interbreeding natural populations which are reproductively isolated from other such groups."

In other words, the biological species concept says that a species is composed of populations whose members mate with each other and produce fertile offspring—or would do so if they came into contact. Conversely, populations whose members do not mate with each other or who cannot produce fertile offspring are said to be **reproductively isolated** and, thus, members of different species.

What causes reproductive isolation? If organisms cannot interbreed or cannot produce fertile offspring, they clearly belong to different species. However, some populations that are considered to be separate species can interbreed and produce fertile offspring, but they ordinarily do not do so under natural conditions. They are still considered to be reproductively isolated in that genes from one species generally will not be able to enter the gene pool of the other species. Table 11.2 summarizes the steps at which barriers to successful reproduction may occur. Such barriers are termed **reproductive isolating mechanisms** because they prevent genetic exchange between species. We will discuss examples of these, beginning with **prezygotic isolating mechanisms,** those that prevent the formation of zygotes, followed by **postzygotic isolating mechanisms,** those that prevent the proper functioning of zygotes after they have formed.

Even though the definition of what constitutes a species is of fundamental importance to evolutionary biology, this issue has still not been completely settled and is currently the subject of considerable research and debate.

11.8 A species is generally defined as a group of similar organisms that does not exchange genes extensively with other groups in nature.

TABLE 11.2 ISOLATING MECHANISMS

Mechanism		Description
PREZYGOTIC ISOLATING MECHANISMS		
Geographic isolation		Species occur in different areas, which are often separated by a physical barrier such as a river or mountain range.
Ecological isolation		Species occur in the same area but they occupy different habitats. Survival of hybrids is low because they are not adapted to either environment of their parents.
Temporal isolation		Species reproduce in different seasons or at different times of the day.
Behavioral isolation		Species differ in their mating rituals.
Mechanical isolation		Structural differences between species prevent mating.
Prevention of gamete fusion	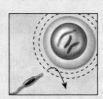	Gametes of one species function poorly with the gametes of another species or within the reproductive tract of another species.
POSTZYGOTIC ISOLATING MECHANISMS		
Hybrid inviability or infertility		Hybrid embryos do not develop properly, hybrid adults do not survive in nature, or hybrid adults are sterile or have reduced fertility.

11.9 Isolating Mechanisms

Prezygotic Isolating Mechanisms

Geographical Isolation. This mechanism is perhaps the easiest to understand; species that exist in different areas are not able to interbreed.

Ecological Isolation. Even if two species occur in the same area, they may utilize different portions of the environment and thus not hybridize because they do not encounter each other. For example, in India, the ranges of lions and tigers overlapped until about 150 years ago. Even when they did, however, there were no records of natural hybrids. Lions stayed mainly in the open grassland and hunted in groups called prides; tigers tended to be solitary creatures of the forest (figure 11.21). Because of their ecological and behavioral differences, lions and tigers rarely came into direct contact with each other, even though their ranges overlapped thousands of square kilometers.

Similar situations occur among many animals and plants. Two species of oaks occur widely in California: the valley oak, *Quercus lobata,* and the scrub oak, *Q. dumosa.* The valley oak, a graceful deciduous tree that can be as tall as 35 meters, occurs in the fertile soils of open grassland on gentle slopes and valley floors. In contrast, the scrub oak is an evergreen shrub, usually only 1 to 3 meters tall, which often forms the kind of dense scrub known as chaparral. The scrub oak is found on steep slopes in less fertile soils. Hybrids between these different oaks do occur and are fully fertile, but they are rare. The sharply distinct habitats of their parents limit their occurrence together, and there is no intermediate habitat where the hybrids might flourish.

Behavioral Isolation. In chapter 18, we will consider the often elaborate courtship and mating rituals of some groups of animals, which tend to keep these species distinct in nature even if they inhabit the same places. For example, mallard and pintail ducks are perhaps the two most common freshwater ducks in North America. In captivity, they produce completely fertile offspring, but in nature they nest side-by-side and rarely hybridize.

Temporal Isolation. *Lactuca graminifolia* and *L. canadensis,* two species of wild lettuce, grow together along roadsides throughout the southeastern United States. Hybrids between these two species are easily made experimentally and are completely fertile. But such hybrids are rare in nature because *L. graminifolia* flowers in early spring and *L. canadensis* flowers in summer. When their blooming periods overlap, as they do occasionally, the two species do form hybrids, which may become locally abundant.

Mechanical Isolation. Structural differences prevent mating between some related species of animals. Aside from such obvious features as size, the structure of the male and female copulatory organs may be incompatible. In many insect and other arthropod groups, the sexual organs, particularly those of the male, are so diverse that they are used as a primary basis for distinguishing species.

Similarly, flowers of related species of plants often differ significantly in their proportions and structures. Some of these differences limit the transfer of pollen from one plant species to another. For example, bees may pick up the pollen of one species on a certain place on their bodies; if this area

(a) (b) (c)

Figure 11.21 Lions and tigers are ecologically isolated.

The ranges of lions and tigers used to overlap in India. However, lions and tigers do not hybridize in the wild because they utilize different portions of the habitat. (*a*) Lions live in open grassland, whereas (*b*) tigers are solitary animals that live in the forest. (*c*) Hybrids, such as this tigon, have been successfully produced in captivity, but hybridization does not occur in the wild.

potatoes. Although much rarer than in plants, speciation by polyploidy is also known from a variety of animals, including insects, fish, and salamanders.

Problems with the Biological Species Concept

The biological species concept has proven to be an effective way of understanding the existence of species in nature. Nonetheless, it has a number of problems that have led some scientists to propose alternative species concepts.

One criticism concerns the extent to which all species truly are reproductively isolated. By definition, under the biological species concept, species should not interbreed and produce fertile offspring. Nonetheless, in recent years, biologists have detected much greater amounts of hybridization than previously realized between populations that seem to coexist as distinct biological entities. Botanists have always been aware that species can often experience substantial amounts of hybridization. For example, more than 50% of California plant species included in one study were not well defined by genetic isolation. Such coexistence without genetic isolation can be long-lasting: Fossil data show that balsam poplars and cottonwoods have been phenotypically distinct for 12 million years but have routinely produced hybrids throughout this time. Consequently, many botanists have long felt that the biological species concept only applies to animals.

What is becoming increasingly evident, however, is that hybridization is not all that uncommon in animals, either. One recent survey indicated that almost 10% of the world's 9,500 bird species are known to have hybridized in nature. Recent years have seen the documentation of many cases of substantial hybridization between animal species. Galápagos finches provide a particularly well-studied example. Three species on the island of Daphne Major—the medium ground finch, the cactus finch, and the small ground finch—are clearly distinct morphologically and occupy different ecological niches. Studies over the past 20 years by Peter and Rosemary Grant found that, on average, 2% of the medium ground finches and 1% of the cactus finches mated with other species every year. Furthermore, hybrid offspring appeared to be at no disadvantage in terms of survival or subsequent reproduction. This is not a trivial amount of genetic exchange, and one might expect to see the species coalesce into one genetically variable population, but the species are maintaining their distinctiveness.

This is not to say hybridization is rampant throughout the animal world. Most bird species do not hybridize, and even fewer probably experience significant amounts of hybridization. Still, it is common enough to cast doubt about whether reproductive isolation is the only force maintaining the integrity of species.

Natural Selection and the Ecological Species Concept

An alternative hypothesis is that the distinctions among species are maintained by natural selection. The idea is that each species has adapted to its own specific part of the environment. Stabilizing selection then maintains the species' adaptations; hybridization has little effect because alleles introduced into the gene pool from other species are quickly eliminated by natural selection.

Also, an interaction between gene flow and natural selection can have many outcomes. In some cases, strong selection can overwhelm any effects of gene flow, but in other situations, gene flow can prevent populations from eliminating less successful alleles from a population. As a general explanation, then, natural selection is not likely to have any fewer exceptions than the biological species concept, although it may prove more successful for certain types of organisms or habitats.

Other Problems with the Biological Species Concept

The biological species concept has been criticized for other reasons as well. For example, it can be difficult to apply the concept to populations that do not occur together in nature. Because individuals of these populations do not encounter each other, it is not possible to observe whether they would interbreed naturally. Although experiments can determine whether fertile hybrids can be produced, this information is not enough because many species that will coexist without interbreeding in nature will readily hybridize in the artificial settings of the laboratory or zoo. Consequently, evaluating whether such populations constitute different species is ultimately a judgment call. In addition, the concept is more limited than its name would imply. Many organisms are asexual and reproduce without mating; reproductive isolation has no meaning for such organisms.

For these reasons, a variety of other ideas have been put forward to establish criteria for defining species. Many of these are specific to a particular type of organism and none has universal applicability. In truth, it may be that there is no single explanation for what maintains the identity of species. Given the incredible variation evident in plants, animals, and microorganisms in all aspects of their biology, it is perhaps not surprising that different processes are operating in different organisms. In addition, some scientists have turned from emphasizing the processes that maintain species distinctions to examining the history of populations. These genealogical species concepts are currently a topic of great debate. The study of species concepts is thus an area of active research that demonstrates the dynamic nature of the field of evolutionary biology.

11.10   Speciation occurs much more readily in the absence of gene flow among populations. However, speciation can occur in sympatry by means of polyploidy. Because of the diversity of living organisms, no single definition of what constitutes a species may be universally applicable.

A Closer Look

Evolution's Critics

The theory that life on earth arose spontaneously and evolved into the forms living today is accepted by most, but not all, biologists. Some biologists, and many nonscientists, prefer a religious explanation of life's origin. Critics of evolution raise seven principal objections to evolution:

1. **Evolution is not solidly demonstrated.** "*Evolution is just a theory,*" critics point out, as if theory meant lack of knowledge, some kind of guess. Scientists, however, use the word theory in a very different sense than the general public does. Theories are the solid ground of science, that of which we are most certain. Few of us doubt the theory of gravity because it is "just a theory."

2. **There are no fossil intermediates.** "*No one ever saw a fin on the way to becoming a leg,*" critics claim, pointing to the many gaps in the fossil record in Darwin's day. Since then, however, most fossil intermediates in vertebrate evolution have indeed been found. A clear line of fossils now traces the transition between whales and hoofed mammals, between reptiles and mammals, and between apes and humans. The fossil evidence of evolution between major forms is compelling.

3. **The intelligent design argument.** "*The organs of living creatures are too complex for a random process to have produced—the existence of a clock is evidence of the existence of a clockmaker.*" Biologists do not agree. The intermediates in the evolution of the mammalian ear can be seen in fossils, and many intermediate "eyes" are known in various invertebrates. These intermediate forms arose because they have value—being able to detect light a little is better than not being able to detect it at all. Complex structures like eyes evolved as a progression of slight improvements.

4. **Evolution violates the second law of thermodynamics.** "*A jumble of soda cans doesn't by itself jump neatly into a stack—things become more disorganized due to random events, not more organized.*" Biologists point out that this argument ignores what the second law really says: disorder increases in a closed system, which the earth most certainly is not. Energy continually enters the biosphere from the sun, fueling life and all the processes that organize it.

5. **Proteins are too improbable.** "*Hemoglobin has 141 amino acids. The probability that the first one would be leucine is 1/20, and that all 141 would be the ones they are by chance is $(1/20)^{141}$, an impossibly rare event.*" This is statistical foolishness—you cannot use probability to argue backward. The probability that a student in a classroom has a particular birthday is 1/365; arguing this way, the probability that everyone in a class of 50 would have the birthdays they do is $(1/365)^{50}$, and yet there the class sits.

6. **Natural selection does not imply evolution.** "*No scientist has come up with an experiment where fish evolve into frogs and leap away from predators.*" Is microevolution (evolution within a species) the mechanism that has produced macroevolution (evolution among species)? Most biologists that have studied the problem think so. Some kinds of animals produced by human selection are remarkably distinctive, such as chihuahuas, dachshunds, and greyhounds. While all dogs are in fact the same species and can interbreed, laboratory selection experiments easily create forms that cannot interbreed and thus would in nature be considered different species. Thus, production of radically different forms has indeed been observed, repeatedly. To object that evolution still does not explain really major differences, like between fish and amphibians, simply takes us back to point 2—these changes take millions of years, and are seen clearly in the fossil record.

7. **The irreducible complexity argument.** "*The intricate molecular machinery of the cell cannot be explained by evolution from simpler stages. Because each part of a complex cellular process like blood clotting is essential to the overall process, how can natural selection fashion any one part?*" What's wrong with this argument is that each part of a complex molecular machine evolves as part of the system. Natural selection can act on a complex system because at every stage of its evolution, the system functions. Parts that improve function are added, and, because of later changes, become essential. The mammalian blood clotting system, for example, evolved from a much simpler system 600 million years ago, and is found today in lampreys, the most primitive fish. One hundred million years later, as vertebrates evolved, proteins were added to the clotting system, making it sensitive to substances released from damaged tissues. Fifty million years later, a third component was added, triggering clotting by contact with the jagged surfaces produced by injury. At each stage, the clotting system came to depend on the added elements, and thus has become "irreducibly complex."

11.11 The Pace of Evolution

We have discussed the manner in which speciation may occur, but we haven't yet considered the relationship between speciation and the evolutionary change that occurs within a species. For more than a century after the publication of *On the Origin of Species,* the standard view was that evolutionary change occurred extremely slowly. Such change would be nearly imperceptible from generation to generation, but would accumulate such that, over the course of thousands and millions of years, major changes could occur. This view is termed *gradualism* and was discussed earlier in this chapter.

Evolution in Spurts?

Gradualism was challenged in 1972 by paleontologists Niles Eldredge of the American Museum of Natural History in New York and Stephen Jay Gould of Harvard University, who argued that species experience long periods of little or no evolutionary change (termed **stasis**), punctuated by bursts of evolutionary change occurring over geologically short time intervals. Moreover, they argued that these periods of rapid change occurred only during the speciation process.

Eldredge and Gould's hypothesis of *punctuated equilibrium* (which was also discussed earlier in this chapter) prompted a great deal of research. Some well-documented groups such as African mammals clearly have evolved gradually, and not in spurts. Other groups, like marine bryozoa, seem to show the irregular pattern of evolutionary change the punctuated equilibrium model predicts. It appears, in fact, that gradualism and punctuated equilibrium are two ends of a continuum. Although some groups appear to have evolved solely in a gradual manner and others only in a punctuated mode, many other groups appear to show evidence of both gradual and punctuated episodes at different times in their evolutionary history. However, the idea that speciation is necessarily linked to phenotypic change has not been supported: It is now clear that speciation can occur without phenotypic change, and that phenotypic change can occur within species in the absence of speciation.

> **11.11** Evolutionary change can be slow and gradual, or rapid and discontinuous, separated by long periods of stasis, which may result from stabilizing selection.

Exploring Current Issues

Additional Resources

Go to your campus library or look online to find the following articles, which further develop some of the concepts found in this chapter.

Greener, M. (2004). A deadly selection. *The Scientist,* 18(5), 28.

Harris, C.R. (2004). The evolution of jealousy: did men and women, facing different selective pressures, evolve different "brands" of jealousy? *American Scientist,* 92(1), 62.

Hunter, P. (2004). Same tools, different boxes: convergent evolution plays out in plant and animal innate immunity. *The Scientist,* 18(3), 26.

Leonard, W.R. (2002). Food for thought: dietary change was a driving force in human evolution. *Scientific American,* 287(6), 106.

Russo, E. (2003). Microbial co-op in evolution. *The Scientist,* 17(19), 25.

Biology and Society Lecture: The Evidence for Evolution

At its core, the case for Darwin's theory of evolution by natural selection is built upon three pillars: First, evidence that natural selection can produce evolutionary change; second, evidence from the fossil record that evolution has occurred; and third, evidence of evolutionary descent in the DNA sequences of the genomes of different species. In addition, information from many different areas of biology—including areas as divergent as embryology, anatomy, and ecology—can only be interpreted sensibly as the outcome of evolution.

Biology and Society Lecture: Arguments Against Evolution

Of all the major ideas of biology, the theory that today's organisms evolved from now-extinct ancestors is perhaps the best known to the general public. This is not because the average person truly understands the basic facts of evolution, but rather because many people mistakenly believe that evolution represents a challenge to their religious beliefs. Although highly publicized criticisms of evolution have occurred ever since Darwin's time, the scientific community has found the objections to be without merit.

Find these lectures, delivered by the author to his class at Washington University, online at www.mhhe.com/tlwessentials/exp11.

Summary

The Theory of Evolution

11.1 Evolution: Getting from There to Here

- Darwin's proposal of evolution by natural selection, focuses primarily on microevolution. Microevolution occurs as changes in gene frequencies in a population. Over time microevolution leads to macroevolution, the formation of new species (**figure 11.1**).

- The rate of evolution isn't the same in all populations. Some evolve at a uniform rate, gradualism, while others evolve in spurts, punctuated equilibrium (**figure 11.2**).

11.2 The Evidence for Evolution

- The evidence for evolution includes the fossil record (**figures 11.3 and 11.4**), the molecular record that traces changes in the genomes of organisms (**figures 11.5 and 11.6**), and the anatomical record, which reveals similarities between species (**figures 11.7 and 11.8**).

How Populations Evolve

11.3 Genetic Change Within Populations: The Hardy-Weinberg Rule

- If a population follows the five assumptions of Hardy-Weinberg, the frequencies of alleles within the population will not change (**figure 11.10**). However, if a population is small, has selective mating, mutations, immigrations or emigrations, or is under the influence of natural selection, allele frequencies will be different in future populations.

11.4 Why Allele Frequencies Change

- There are five factors that act on populations to change their allele frequencies (**table 11.1**): mutations, changes in DNA; migrations, movement of individuals into or out of the population; genetic drift, the random loss of alleles in a population; nonrandom mating, individuals seeking out mates based on certain traits; and selection, individuals with certain traits leave more offspring.

11.5 Forms of Selection

- Selection can change a population such that certain traits are more common than others. Stabilizing selection tends to reduce extreme phenotypes, making the intermediate phenotype more common (**figures 11.12 and 11.13**). Disruptive selection tends to reduce intermediate phenotypes, leaving extreme phenotypes in the population (**figure 11.14**). Directional selection tends to reduce one extreme phenotype from the population (**figure 11.15**).

Adaptation Within Populations

11.6 Sickle-Cell Anemia

- Sickle-cell anemia is an example of heterozygous advantage, where individuals who are heterozygous for a trait tend to survive better than individuals with either of the two homozygous phenotypes (**figures 11.17 and 11.18**).

11.7 Selection on Color in Guppies

- Experimentation has been able to show evolutionary change due to natural selection on certain traits in a population. Although this isn't macroevolution, it is microevolution in action (**figures 11.19 and 11.20**).

How Species Form

11.8 The Biological Species Concept

- The biological species concept states that a species is a group of organisms that mate with each other and produce fertile offspring, or would do so if in contact with each other. If they cannot mate, or mate but cannot produce fertile offspring, they are said to be reproductively isolated. Many barriers to successful reproduction lead to reproductive isolation (**table 11.2**).

11.9 Isolating Mechanisms

- There are two types of isolating mechanisms, prezygotic and postzygotic. Prezygotic isolating mechanisms prevent the formation of a hybrid zygote. Postzygotic isolating mechanisms prevent normal development of a hybrid zygote or result in sterile offspring (**figure 11.22**).

11.10 Working with the Biological Species Concept

- Speciation is more likely to occur between populations that are geographically isolated from each other, called allopatric speciation (**figure 11.23**). However, sympatric speciation, where one species splits into two at a single location, is more common in plants than in animals.

- The biological species concept doesn't explain all situations of speciation, and so other concepts have been proposed, but no single concept is adequate.

11.11 The Pace of Evolution

- Examples of gradualism and punctuated equilibrium exist, and so speciation clearly occurs in different ways.

Self-Test

1. The study of changes in organisms over a long period of time, including the rise of novel organisms and extinction, is known as
   a. natural selection.
   b. microevolution.
   c. macroevolution.
   d. punctuated equilibrium.
2. One of the major sources of evidence for evolution is in the comparative anatomy of organisms. Features that have a similar look but different structural origin are called
   a. homologous structures.
   b. analogous structures.
   c. vestigial structures.
   d. equivalent structures.
3. Homologous structures in organisms are the result of
   a. divergence.
   b. convergent evolution.
   c. stasis.
   d. ontogeny.
4. The Hardy-Weinberg equilibrium allows scientists to study
   a. natural selection.
   b. microevolution.
   c. macroevolution.
   d. punctuated equilibrium.
5. A large group of organisms lives in a large, stable ecosystem that has an adequate carrying capacity. There is no competition for resources. Individuals mate at random. All organisms appear

to be identical except for a few individuals in the most recent generation of offspring that exhibit a different fur coat color and pattern. The ecosystem and population are geographically isolated from other populations of the same organism. Which Hardy-Weinberg assumption has been violated?

   a.   large population size
   b.   random mating within the population
   c.   no mutation within the population
   d.   no input of new alleles or loss of alleles

6. A population of 1,000 individuals has 200 individuals who show a homozygous recessive phenotype and 800 individuals who express the dominant phenotype. What is the frequency of homozygous dominant individuals (p^2) in this population?

   a. $p^2 = 0.20$ c. $p^2 = 0.45$
   b. $p^2 = 0.30$ d. $p^2 = 0.55$

7. A chance event occurs that causes a population to lose some individuals (they died)—hence, a loss of alleles to the population results in

   a.   mutation.   c.   selection.
   b.   migration.   d.   genetic drift.

8. Selection that causes an extreme phenotype to be more frequent in a population is an example of

   a.   disruptive selection.   c.   directional selection.
   b.   stabilizing selection.   d   equivalent selection.

9. A key element of Ernst Mayr's biological species is

   a.   homologous isolation.   c.   convergent isolation.
   b.   divergent isolation.   d.   reproductive isolation.

10. Sympatric species are populations of species that live in the same habitat. Sympatric species of deer mice (*Peromyscus* sp.) are externally identical. However, the males of different species have a differently shaped baculum or bone found in the penis. This is an example of which type of reproductive isolation?

   a.   temporal   c.   behavioral
   b.   mechanical   d.   ecological

Visual Understanding

1. **Figure 11.12** Because of prolonged drought, the trees on an island are producing nuts that are much smaller with thicker and harder shells. What will happen to the birds that depend on the nuts for food?

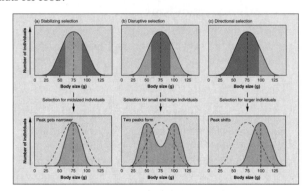

2. **Table 11.2** A very heavy rainstorm floods a mountain river, changing its course and digging a deep canyon through the soft soils of the meadow in the valley below. What happens to the mice on either side of the valley?

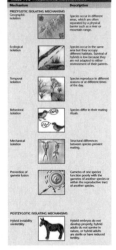

Challenge Questions

The Theory of Evolution How does microevolution differ from macroevolution?

How Populations Evolve How likely is it that any naturally occurring population on a continent (not on an island) will meet all the assumptions of the Hardy-Weinberg equations?

Adaptation Within Populations Use Hardy-Weinberg and the tenets of natural selection to explain why there are colorful, large guppies in pools without pike cichlids, and drab, small guppies when pike cichlids are present.

How Species Form A new volcanic island erupts from the ocean. Over millions of years, the black lava breaks down into black sand, seeds drift to the island and it becomes populated with plants, some flying insects, and a few birds, including a small hawk. In a storm, several palm trees are uprooted from the white sands mainland and float out to the island, carrying several insectivorous lizards of one species. Describe the subsequent changes that take place in the lizard population as they survive and reproduce.

Online Learning Center

Visit the Online Learning Center for this chapter at www.mhhe.com/tlwessentials/ch11 for quizzes, animations, interactive learning exercises, and other study tools. At the site you will also find extended answers to the end-of-chapter questions.

12
Exploring Biological Diversity

In 1799, the skin of a most unusual animal was sent to England by Captain John Hunter, governor of the British penal colony in New South Wales (Australia). Covered in soft fur, it was less than two feet long. As it had mammary glands with which to suckle its young, it was clearly a mammal, but in other ways it seemed very like a reptile. Males have internal testes, and females have a shared urinary and reproductive tract opening called a cloaca, lay eggs as reptiles do, and like reptilian eggs, the yolk of the fertilized egg does not divide. It thus seemed a confusing mixture of mammalian and reptilian traits. Adding to this impression was its appearance: It has a tail not unlike that of a beaver, a bill not unlike that of a duck, and webbed feet! It was as if a child had mixed together body parts at random—a most unusual animal. Individuals like the one pictured here are abundant in freshwater streams of eastern Australia today. What does one call such a beast? In its original 1799 description it was named *Platypus anatinus* (flatfooted ducklike animal), which was later changed to *Ornithorhynchus anatinus* (ducklike animal with a bird's snout)—informally, the duckbill platypus. How biologists assign names to the organisms they discover is the subject of this chapter. You will be surprised at how much information is crammed into the two words of a scientific name.

12.1 The Invention of the Linnaean System

Our world is populated by some 10 million different kinds of organisms. To talk about them and study them, it is necessary to give them names, just as it is necessary that people have names. Of course, no one can remember the name of every kind of organism, so biologists use a kind of multilevel grouping of individuals called **classification.**

Organisms were first classified more than 2,000 years ago by the Greek philosopher Aristotle, who categorized living things as either plants or animals. He classified animals as either land, water, or air dwellers, and he divided plants into three kinds based on stem differences. This simple classification system was expanded by the Greeks and Romans, who grouped animals and plants into basic units such as cats, horses, and oaks. Eventually, these units began to be called **genera** (singular, **genus**), the Latin word for "group." Starting in the Middle Ages, these names began to be systematically written down, using Latin, the language used by scholars at that time. Thus, cats were assigned to the genus *Felis,* horses to *Equus,* and oaks to *Quercus*—names that the Romans had applied to these groups. For genera that were not known to the Romans, new names were invented.

The classification system of the Middle Ages, called the polynomial system, was used virtually unchanged for hundreds of years, until it was replaced about 250 years ago by the **binomial system** introduced by Linnaeus.

The Polynomial System

Until the mid-1700s, biologists usually added a series of descriptive terms to the name of the genus when they wanted to refer to a particular kind of organism, which they called a **species.** These phrases, starting with the name of the genus, came to be known as **polynomials** (*poly,* many, and *nomial,* name), strings of Latin words and phrases consisting of up to 12 or more words. For example, the common wild briar rose was called *Rosa sylvestris inodora seu canina* by some and *Rosa sylvestris alba cum rubore, folio glabro* by others. This would be like the mayor of New York referring to a particular citizen as "Brooklyn resident: Democrat, male, Caucasian, middle income, Protestant, elderly, likely voter, short, bald, heavyset, wears glasses, works in the Bronx selling shoes." As you can imagine, these polynomial names were cumbersome. Even more worrisome, the names were altered at will by later authors, so that a given organism really did not have a single name that was its alone, as was the case with the briar rose.

The Binomial System

A much simpler system of naming animals, plants, and other organisms stems from the work of the Swedish biologist Carolus Linnaeus (1707–78). Linnaeus devoted his life to a challenge that had defeated many biologists before him—cataloging all the different kinds of organisms. Linnaeus, a botanist studying

(a) *Quercus phellos*
(Willow oak)

(b) *Quercus rubra*
(Red oak)

Figure 12.1 How Linnaeus named two species of oaks.

(a) Willow oak, *Quercus phellos.* (b) Red oak, *Quercus rubra.* Although they are clearly oaks (members of the genus *Quercus*), these two species differ sharply in the shapes and sizes of their leaves and in many other features, including their overall geographical distributions.

the plants of Sweden and from around the world, developed a plant classification system based on grouping plants based on their reproductive structures. This system resulted in some seemingly unnatural groupings and therefore was never universally accepted. However, in the 1750s he produced several major works that, like his earlier books, employed the polynomial system. But as a kind of shorthand, Linnaeus also included in these books a two-part name for each species (others had also occasionally done this, but Linnaeus used these shorthand names consistently). These two-part names, or **binomials** (*bi* is the Latin prefix for "two"), have become our standard way of designating species. For example, he designated the willow oak *Quercus phellos* and the red oak *Quercus rubra* (figure 12.1), even though he also included the polynomial name for these species. We also use binomial names for ourselves, our so-called given and family names. So, this naming system is like the mayor of New York calling the Brooklyn resident Sylvester Kingston.

Linnaeus took the naming of organisms a step further, grouping similar organisms into higher-level categories based on similar characteristics (discussed later). Although not intended to show evolutionary connections between different organisms, this hierarchical system acknowledged that there were broad similarities shared by groups of species that distinguished them from other groups.

> **12.1 Two-part (binomial) Latin names, first used by Linnaeus, are now universally employed by biologists to name particular organisms.**

12.2 Species Names

A group of organisms at a particular level in a classification system is called a **taxon** (plural, **taxa**), and the branch of biology that identifies and names such groups of organisms is called **taxonomy.** Taxonomists are in a real sense detectives, biologists who must use clues of appearance and behavior to identify and assign names to organisms.

By formal agreement among taxonomists throughout the world, no two organisms can have the same name. So that no one country is favored, a language spoken by no country—Latin—is used for the names. Because the scientific name of an organism is the same anywhere in the world, this system provides a standard and precise way of communicating, whether the language of a particular biologist is Chinese, Arabic, Spanish, or English. This is a great improvement over the use of common names, which often vary from one place to the next. As you can see in figure 12.2, corn in Europe refers to the plant Americans call wheat; a bear is a large placental omnivore in the United States but a koala (a vegetarian marsupial) in Australia; and a robin is a very different bird in Europe and North America.

By convention, the first word of the binomial name is the genus to which the organism belongs. This word is always capitalized. The second word, called the *epithet,* refers to the particular species and is not capitalized. The two words together are called the **scientific name,** or species name, and are written in italics. The system of naming animals, plants, and other organisms established by Linnaeus has served the science of biology well for nearly 250 years.

> **12.2** By convention, the first part of a binomial species name identifies the genus to which the species belongs, and the second part distinguishes that particular species from other species in the genus.

(a)

(b)

(c)

Figure 12.2   Common names make poor labels.
The common names corn (*a*), bear (*b*), and robin (*c*) bring clear images to our minds (photos on *top*), but the images would be very different to someone living in Europe or Australia (photos on *bottom*). There, the same common names are used to label very different species.

12.3 Higher Categories

Like the mayor of New York, a biologist needs more than two categories to classify all the world's living things. Taxonomists group the genera with similar properties into a cluster called a **family,** and place similar families into the same **order** (figure 12.3). Orders with common properties are placed into the same **class,** and classes with similar characteristics into the same **phylum** (plural, **phyla**). Botanists (that is, those who study plants) also call plant phyla "divisions." Finally, the phyla are assigned to one of several gigantic groups, the **kingdoms.** Biologists currently recognize six kingdoms: two kinds of prokaryotes (Archaea and Bacteria), a largely unicellular group of eukaryotes (Protista), and three multicellular groups (Fungi, Plantae, and Animalia). To remember the seven categories in their proper order, it may prove useful to memorize a phrase such as "**k**indly **p**ay **c**ash **o**r **f**urnish **g**ood **s**ecurity" or "**K**ing **P**hilip **c**ame **o**ver **f**or **g**reen **s**paghetti" (**k**ingdom–**p**hylum–**c**lass–**o**rder–**f**amily–**g**enus–**s**pecies).

In addition, an eighth level of classification, called *domains,* is sometimes used. Biologists recognize three domains, Bacteria, Archaea, and Eukarya, which are discussed later in this chapter.

Each of the categories in this **Linnaean system of classification** is loaded with information. For example, consider a honeybee:

Level 1: Its species name is *Apis mellifera.*

Level 2: Its genus name is *Apis.*

Level 3: Its family, Apidae, are all bees, some solitary, others living in hives as *A. mellifera* does.

Level 4: Its order, Hymenoptera, tells you that it is likely able to sting and may live in colonies.

Level 5: Its class, Insecta, says that *A. mellifera* has three major body segments, with wings and three pairs of legs attached to the middle segment.

Level 6: Its phylum, Arthropoda, tells us that it has a hard cuticle of chitin and jointed appendages.

Level 7: Its kingdom, Animalia, says that it is a multicellular heterotroph whose cells lack cell walls.

Level 8: An addition to the Linnaean system, its domain, Eukarya, says that its cells contain membrane-bounded organelles.

> **12.3** A hierarchical system is used to classify organisms, in which higher categories convey more general information about the group.

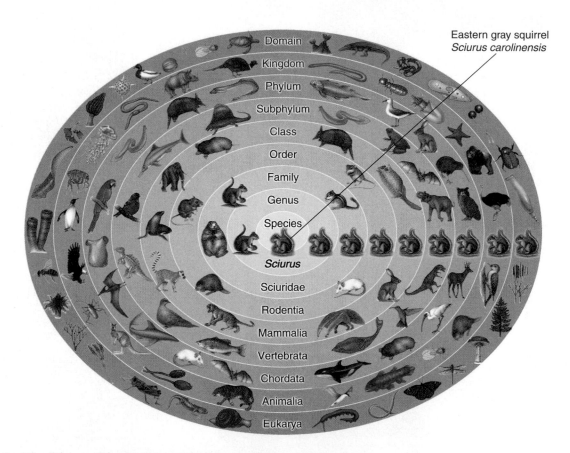

Figure 12.3 The hierarchical system used to classify an organism.

The organism is first recognized as a eukaryote (domain: Eukarya). Second, within this domain, it is an animal (kingdom: Animalia). Among the different phyla of animals, it is a vertebrate (phylum: Chordata, subphylum: Vertebrata). The organism's fur characterizes it as a mammal (class: Mammalia). Within this class, it is distinguished by its gnawing teeth (order: Rodentia). Next, because it has four front toes and five back toes, it is a squirrel (family: Sciuridae). Within this family, it is a tree squirrel (genus: *Sciurus*), with gray fur and white-tipped hairs on the tail (species: *Sciurus carolinensis,* the eastern gray squirrel).

12.4 What Is a Species?

The basic biological unit in the Linnaean system of classification is the species. John Ray (1627–1705), an English clergyman and scientist, was one of the first to propose a general definition of species. In about 1700, he suggested a simple way to recognize a species: All the individuals that belong to it can breed with one another and produce fertile offspring. By Ray's definition, the offspring of a single mating were all considered to belong to the same species, even if they contained different-looking individuals, as long as these individuals could interbreed. All domestic cats are one species (they can all interbreed), while carp are not the same species as goldfish (they cannot interbreed). The donkey you see in figure 12.4 is not the same species as the horse, because when they interbreed, the offspring—mules—are sterile.

Horse
Donkey
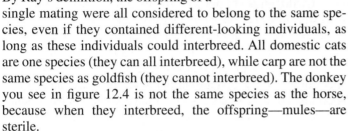
Mule

Figure 12.4 Ray's definition of a species.
According to Ray, donkeys and horses are not the same species. Even though they produce very robust offspring (mules) when they mate, the mules are sterile.

The Biological Species Concept

With Ray's observation, the species began to be regarded as an important biological unit that could be cataloged and understood, the task that Linnaeus set himself a generation later. Where information was available, Linnaeus used Ray's species concept, and it is still widely used today. When the evolutionary ideas of Darwin were joined to the genetic ideas of Mendel in the 1920s to form the field of population genetics, it became desirable to define the category species more precisely. The definition that emerged, the so-called *biological species concept* discussed in chapter 11, defined species as groups that are reproductively isolated. In other words, hybrids (offspring of different species that mate) occur rarely in nature, whereas individuals that belong to the same species are able to interbreed freely.

As we learned in chapter 11, the biological species concept works fairly well for animals, where strong barriers to hybridization between species exist, but very poorly for members of the other kingdoms. The problem is that the biological species concept assumes that organisms regularly outcross—that is, interbreed with individuals other than themselves, individuals with a different genetic makeup than their own but within their same species. Animals regularly outcross, and so the concept works well for animals. However, outcrossing is less common in the other five kingdoms. In prokaryotes and many protists, fungi, and some plants, asexual reproduction, reproduction without sex, predominates. These species clearly cannot be characterized in the same way as outcrossing animals and plants—they do not interbreed with one another, much less with the individuals of other species.

Complicating matters further, the reproductive barriers that are the key element in the biological species concept, although common among animal species, are not typical of other kinds of organisms. In fact, there are essentially no barriers to hybridization between the species in many groups of trees, such as oaks, and other plants, such as orchids. Even among animals, fish species are able to form fertile hybrids with one another, though they may not do so in nature.

In practice, biologists today recognize species in different groups in much the way they always did, as groups that differ from one another in their visible features. Within animals, the biological species concept is still widely employed, while among plants and other kingdoms it is not. Molecular data are causing a reevaluation of traditional classification systems, and, taking into account morphology, life cycles, metabolism, and other characteristics, they are changing the way scientists classify plants, protists, fungi, prokaryotes, and even animals.

How Many Kinds of Species Are There?

Since the time of Linnaeus, about 1.5 million species have been named. But the actual number of species in the world is undoubtedly much greater, judging from the very large numbers that are still being discovered. Some scientists estimate that at least 10 million species exist on earth, and at least two-thirds of these occur in the tropics.

12.4 Among animals, species are generally defined as reproductively isolated groups; among the other kingdoms such a definition is less useful, as their species typically have weaker barriers to hybridization.

12.5 How to Build a Family Tree

After naming and classifying some 1.5 million organisms, what have biologists learned? One very important advantage of being able to classify particular species of plants, animals, and other organisms is that individuals of species that are useful to humans as sources of food and medicine can be identified. For example, if you cannot tell the fungus *Penicillium* from *Aspergillus,* you have little chance of producing the antibiotic penicillin. In a thousand ways, just having names for organisms is of immense importance in our modern world.

Taxonomy also enables us to glimpse the evolutionary history of life on earth. The more similar two taxa are, the more closely related they are likely to be, for the same reason that you are more like your brothers and sisters than like strangers selected from a crowd. By looking at the differences and similarities between organisms, biologists can attempt to reconstruct the tree of life, inferring which organisms evolved from which other ones, in what order, and when. The evolutionary history of an organism and its relationship to other species is called **phylogeny** (see figure 12.7). The reconstruction and study of evolutionary trees, or **phylogenetic trees,** including the naming and classifying of organisms, is an area of study called **systematics.**

Cladistics

A simple and objective way to construct a phylogenetic tree is to focus on key characters that some organisms share because they have inherited them from a common ancestor. A **clade** is a group of organisms related by descent, and this approach to constructing a phylogeny is called **cladistics.** Cladistics infers phylogeny (that is, builds family trees) according to similarities derived from a common ancestor, so-called **derived characters.** The key to the approach is being able to identify morphological, physiological, or behavioral traits that differ among the organisms being studied and can be attributed to a common ancestor. By examining the distribution of these traits among the organisms, it is possible to construct a **cladogram** (figure 12.5), a branching diagram that represents the phylogeny.

Cladograms are not true family trees. They do not convey direct information about ancestors and descendants—who came from whom. Instead, they convey comparative information about *relative* relationships. Organisms that are closer together on a cladogram simply share a more recent common ancestor than those that are farther apart. Because the analysis is comparative, it is necessary to have something to anchor the comparison to, some solid ground against which the comparisons can be made. To achieve this, each cladogram must contain an **outgroup,** a rather different organism (but not *too* different) to serve as a baseline for comparisons among the other organisms being evaluated, the **ingroup.** For example, in figure 12.5, the lamprey is the outgroup to the clade of animals that have jaws.

Cladistics is a relatively new approach in biology and has become popular among students of evolution. This is because it does a very good job of portraying the *order* in which a series of evolutionary events have occurred. The great strength of a cladogram is that it can be completely objective

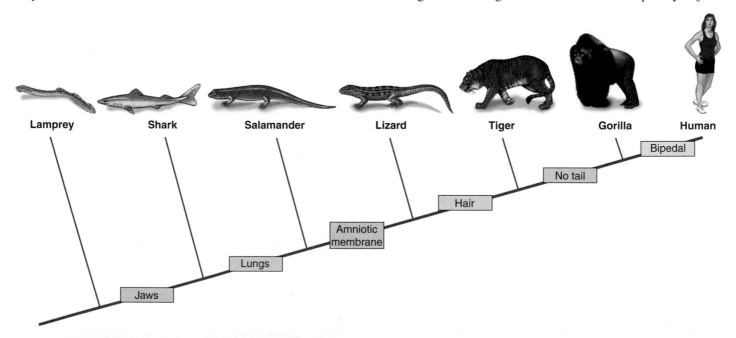

Figure 12.5   A cladogram of vertebrate animals.
The derived characters between the branch points are shared by all the animals to the right of each character and are not present in any organisms to the left of it.

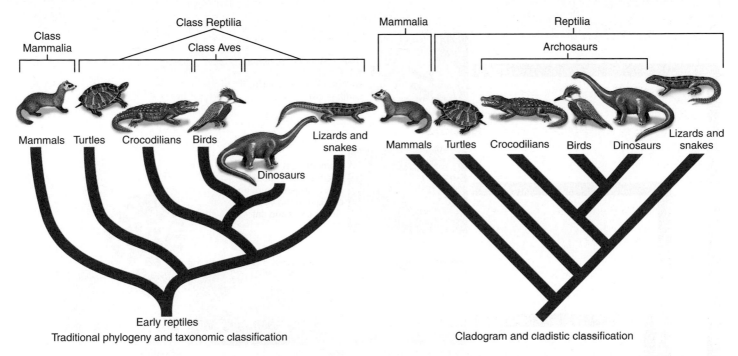

Figure 12.6 Two ways to classify terrestrial vertebrates.
Traditional taxonomic analyses place birds in their own class (Aves) because birds have evolved several unique adaptations that separate them from the reptiles. Cladistic analyses, however, place crocodiles, dinosaurs, and birds together (as archosaurs) because they share many derived characters.

(figure 12.6). A computer fed the data will generate exactly the same cladogram time and again. In fact, most cladistic analyses involve many characters, and computers are required to make the comparisons. Although objective, the phylogenetic trees are not absolute. Phylogenetic trees are hypotheses, proposed explanations of how organisms may have evolved (figure 12.7).

Sometimes it is necessary to "weight" characters, or take into account the variation in the "strength" (importance) of a character—the size or location of a fin, the effectiveness of a lung. For example, let's say that the following are five unique events that occurred on November 22, 1963: (1) my cat was declawed, (2) I had a wisdom tooth pulled, (3) I sold my first car, (4) President Kennedy was assassinated, and (5) I passed physics. Without weighting the events, each one is assigned equal importance. In a nonweighted cladistic sense, they are equal (all happened only once, and on that day), but in a practical, real-world sense, they certainly are not. One event, Kennedy's death, had a far greater impact and importance than the others. Because evolutionary success depends so critically on just such high-impact events, modern cladistics attempts to assign extra weight to the evolutionary significance of key characters.

Traditional Taxonomy

Weighting characters lies at the core of **traditional taxonomy.** In this approach, phylogenies are constructed based on a vast amount of information about the morphology and biology of the organism gathered over a long period of time. Tradition-

al taxonomists use both ancestral and derived characters to construct their trees, whereas cladists use only derived characters. The large amount of information used by traditional taxonomists permits a knowledgeable weighting of characters according to their biological significance. In traditional taxonomy, the full observational power and judgment of the biologist is brought to bear—and also any biases he or she may have. For example, in classifying the terrestrial vertebrates, traditional taxonomists place birds in their own class (Aves), giving great weight to the characters that made powered flight possible, such as feathers. However, a cladogram of vertebrate evolution lumps birds in among the reptiles with crocodiles and dinosaurs (see figure 12.6). This accurately reflects their true ancestry but ignores the immense evolutionary impact of a derived character such as feathers.

Overall, phylogenetic trees based on traditional taxonomy are information-rich, while cladograms do a better job of deciphering evolutionary histories. Thus, traditional taxonomy is the better approach when a great deal of information is available to guide character weighting, whereas cladistics is the better approach when little information is available about how the character affects the life of the organism.

12.5 A phylogeny may be represented as a cladogram based on the order in which groups evolved or as a traditional taxonomic tree that weights characters according to assumed importance.

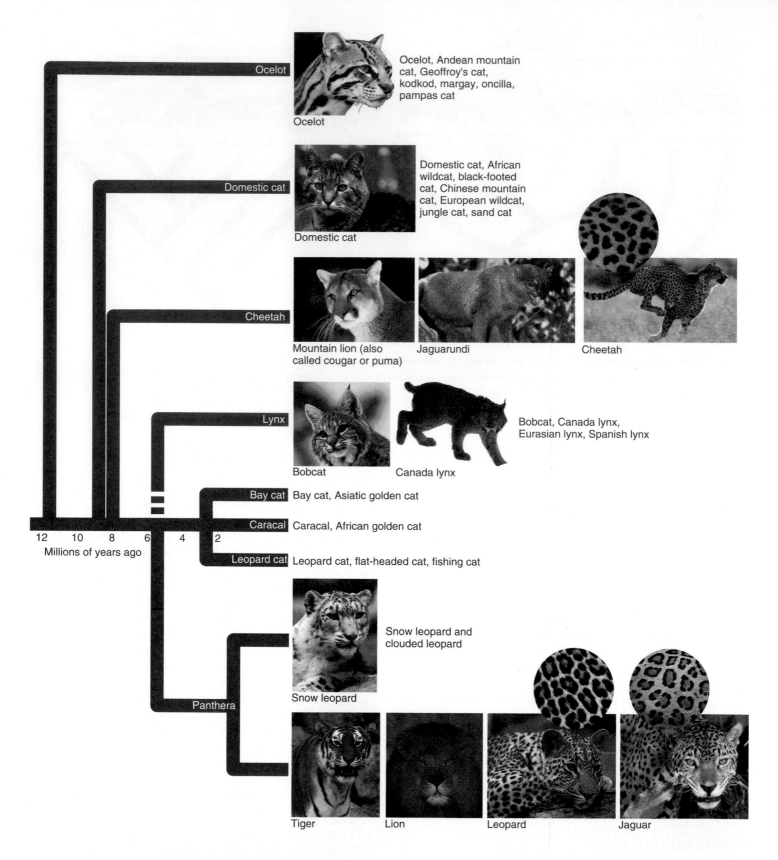

Figure 12.7 The cat family tree.

Recent studies of DNA, RNA, and protein similarities have allowed biologists to construct this feline family tree of the eight major cat lineages and their individual species. Among the oldest of all cats is the domestic cat. The most recently-evolved are the four big Panthers: tiger, lion, leopard, and jaguar. The other big cats, the cheetah and mountain lion, are members of a much older lineage and are not close relatives of the big four.

Author's Corner

Biodiversity Behind Bars

There is something about a child that doesn't like bars. When I was seven, I knew with a searing certainty that no person, no animal should have to live caged, peering out behind bars at a free world it cannot reach. And I acted on that certainty. I lived at 60 Hopkins Street, Hilton Village, a workers' suburb of the shipbuilding town of Newport News, Virginia. Across the street from me resided a public health nurse who maintained in her backyard a large colony of guinea pigs used in medical testing and research. Nowadays such an informal arrangement would surely violate some rule or regulation, but 1949 was a simpler time. I used to visit her guinea pigs in their outdoor cage, seeing in their liquid eyes all the despair that only a seven-year-old can feel. And one day I picked up a rock, broke the lock, and set the guinea pigs free. Out into the neighborhood they shot, 97 furry little lightning bolts, and I was proud to have struck this blow for freedom. I am amazed, looking back, that my parents kept me.

What they did was buy me a dog. That, and explain to me that some bars have a purpose. These guinea pigs were part of an effort to help sick people, they patiently explained, their freedom the price of a much greater good. It was only as I grew to adulthood that I came to understand that bitter truth. And now I find I must explain it to my children. At 13 years of age, my daughter Caitlin came to me, incandescent with anger, demanding to know why the apes at the Saint Louis Zoo weren't transported back to Africa and released to freedom. I looked into her eyes and saw myself 50 years ago (can it be so long?) and knew her question demanded an answer. So here it is.

Ape captivity, Caitlin, is no more justifiable than human captivity, for apes are our very close cousins, with loving families and complex cultures. But bars sometimes have a necessary purpose. The apes in the Saint Louis Zoo are captive for two reasons: to preserve them as a species, and to educate people about them.

Zoos as Conservators of a Disappearing World

There are only four kinds of living apes: gibbons, orangutans, gorillas, and chimpanzees. All are rare and highly endangered, living in relatively small areas, and their natural habitat is rapidly disappearing as human activities destroy the mountain forests of central Africa and Asia. In your lifetime, I told Caitlin, most of these wild ape populations will be gone, extinct. No amount of regret can change that sad fact. All that will be left of humanity's last link to its evolutionary past will be the captive ape populations living in zoos like ours.

That is the primary role of zoos today, not entertainment to distract children on hot summer days but preservation of an invaluable biological heritage that otherwise would be lost. Zoos are the conservators of a rapidly disappearing natural world, a living library in which we store biodiversity. Zoos take this role seriously. The breeding of endangered species is overseen collectively by the nation's zoos. Species preservation programs move animals from one zoo to another in carefully managed breeding programs that prevent unnecessary inbreeding.

Sometimes a captive species can be introduced back into the wild. Serious efforts are being made to establish large preserves where some of the natural habitat will be saved for the animals. For most endangered animals, however, there are no such rosy hopes, no wild habitat reserved for their return. Captivity, for better or worse, is their only future, surely a better choice than extinction.

Zoos as Educators of You and Me

The second role of zoos today, more subtle but equally important, is to make biodiversity immediate and real for the general public, for you and me. It was once wisely said that we will not preserve what we do not understand. Seeing with your own eyes a chimpanzee return your gesture, or watching with your own eyes a baby giraffe take its first awkward steps—no words or film have that impact, that immediate concrete contact with our animal relations. Like many zoos around the country, our Saint Louis Zoo has a very active education department, letting busloads of school children experience animals firsthand. But the bigger educational effort is simply our seeing the animals ourselves. Every zoo is a teaching institution; we the public are the students, and our response to the animals is the lesson we are being taught. The test we must pass, as the citizens upon whose support the world's zoos depends, will be graded severely. It is no less than the survival of this rich animal biodiversity for our children's children to see and share in turn.

12.6 The Kingdoms of Life

The earliest classification systems recognized only two kingdoms of living things: animals and plants (figure 12.8a). But as biologists discovered microorganisms and learned more about other organisms, they added kingdoms in recognition of fundamental differences (figure 12.8b). Most biologists now use a six-kingdom system first proposed by Carl Woese of the University of Illinois (figure 12.8c).

In this system, four kingdoms consist of eukaryotic organisms. The two most familiar kingdoms, **Animalia** and **Plantae,** contain only organisms that are multicellular during most of their life cycle. These groups of animals and plants are no doubt familiar to you. The kingdom **Fungi** contains multicellular forms, such as mushrooms and molds, and single-celled yeasts, which are thought to have multicellular ancestors. Fundamental differences divide these three kingdoms. Plants are mainly stationary, but some have motile sperm; fungi have no motile cells; animals are mainly motile. Animals ingest their food, plants manufacture it, and fungi digest it by means of secreted extracellular enzymes. Each of these kingdoms probably evolved from a different single-celled ancestor.

The large number of unicellular eukaryotes are arbitrarily grouped into a single kingdom called **Protista**. This kingdom includes the algae and microscopic aquatic organisms, all of which are unicellular during important parts of their life cycle.

The remaining two kingdoms, **Archaea** and **Bacteria,** consist of prokaryotic organisms, which are vastly different from all other living things. Most prokaryotes with which you are most familiar, those that cause disease or are used in industry, are members of the kingdom Bacteria. Archaea are a diverse group including the methanogens and extreme thermophiles, and they differ greatly from bacteria in many ways.

Domains

As biologists have learned more about the archaea, it has become increasingly clear that this ancient group is very different from all other organisms. When the full genomic DNA sequences of an archaea and a bacterium were first compared in 1996, the differences proved striking. Archaea are as different from bacteria as bacteria are from eukaryotes. Recognizing this, biologists are increasingly adopting a classification of living organisms that recognizes three **domains,** a taxonomic level higher than kingdom (figure 12.8d). Archaea are in one domain, bacteria in a second, and eukaryotes in the third. While the domain Eukarya contains four kingdoms of organisms, the domains Bacteria and Archaea contain only one kingdom in each.

> **12.6** Living organisms are grouped into three categories called domains. One of the domains, Eukarya, is divided into four kingdoms: Protista, Fungi, Plantae, and Animalia.

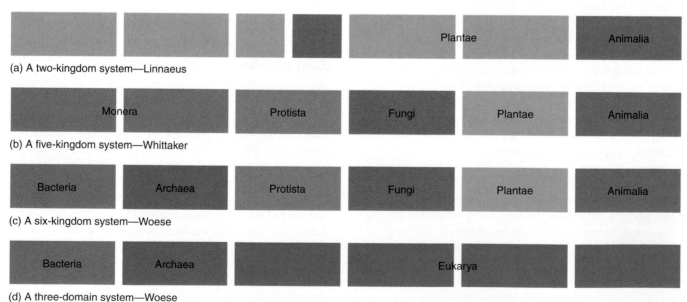

(a) A two-kingdom system—Linnaeus

(b) A five-kingdom system—Whittaker

(c) A six-kingdom system—Woese

(d) A three-domain system—Woese

Figure 12.8 Different approaches to classifying living organisms.
(a) Linnaeus popularized a two-kingdom approach, in which the fungi and the photosynthetic protists were classified as plants and the nonphotosynthetic protists as animals; when prokaryotes were described, they too were considered plants. (b) Whittaker in 1969 proposed a five-kingdom system that soon became widely accepted. (c) Woese has championed splitting the prokaryotes into two kingdoms for a total of six kingdoms or even assigning them separate domains, with a third domain containing the four eukaryotic kingdoms (d).

12.7 Domain Archaea

The domain Archaea contains one kingdom by the same name, the Archaea. The term *archaea* (Greek, *archaio,* ancient) refers to the ancient origin of this group of prokaryotes, which most likely diverged very early from the bacteria (figure 12.9). Today, archaea inhabit some of the most extreme environments on earth. Though a diverse group, all archaea share certain key characteristics. Their cell walls lack the peptidoglycan characteristic of the cell walls of bacteria. They possess very unusual lipids and characteristic ribosomal RNA (rRNA) sequences. Some of their genes possess introns, unlike those of bacteria.

Archaea are grouped into three general categories: methanogens, extremophiles, and nonextreme archaea.

Methanogens obtain their energy by using hydrogen gas (H_2) to reduce carbon dioxide (CO_2) to methane gas (CH_4). They are strict anaerobes, poisoned by even traces of oxygen. They live in swamps, marshes, and the intestines of mammals. Methanogens release about 2 billion tons of methane gas into the atmosphere each year.

Extremophiles are able to grow under conditions that seem extreme to us.

Thermophiles ("heat lovers") live in very hot places, typically from 60° to 80°C. Many thermophiles have metabolisms based on sulfur. Thus, the *Sulfolobus* inhabiting the hot sulfur springs of Yellowstone National Park at 70° to 75°C obtain their energy by oxidizing elemental sulfur to sulfuric acid. The recently described *Pyrolobus fumarii* holds the current record for heat stability, 106°C temperature optimum and 113°C maximum—it is so heat-tolerant that it is not killed by a one-hour treatment in an autoclave (121°C)!

Halophiles ("salt lovers") live in very salty places like the Great Salt Lake in Utah, Mono Lake in California, and the Dead Sea in Israel. Whereas the salinity of seawater is around 3%, these prokaryotes thrive in, and indeed require, water with a salinity of 15% to 20%.

pH-tolerant archaea grow in highly acidic (pH = 0.7) and very basic (pH = 11) environments.

Pressure-tolerant archaea have been isolated from ocean depths that require at least 300 atmospheres of pressure to survive, and tolerate up to 800 atmospheres!

Nonextreme archaea grow in the same environments bacteria do. As the genomes of archaea have become better known, microbiologists have been able to identify **signature sequences** of DNA present in all archaea and in no other organisms. When samples from soil or seawater are tested for genes matching these signature sequences, many of the prokaryotes living there prove to be archaea. Clearly, archaea are not restricted to extreme habitats, as microbiologists used to think.

> **12.7** Archaea are unique prokaryotes that inhabit diverse environments, some of them extreme.

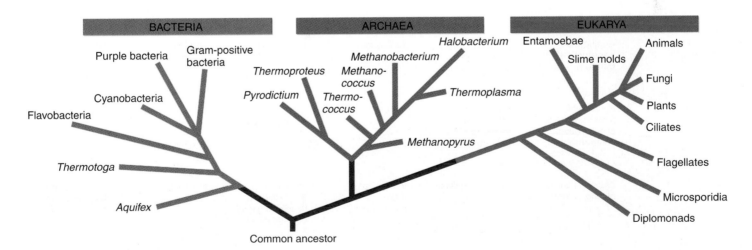

Figure 12.9 A tree of life.

This phylogeny, prepared from rRNA analyses, shows the evolutionary relationships among the three domains. The base of the tree was determined by examining genes that are duplicated in all three domains, the duplication presumably having occurred in the common ancestor. When one of the duplicates is used to construct the tree, the other can be used to root it. This approach clearly indicates that the root of the tree is within the bacterial domain. Archaea and eukaryotes diverged later and are more closely related to each other than either is to bacteria.

12.8 Domain Bacteria

The domain Bacteria contains one kingdom of the same name, Bacteria. The bacteria are the most abundant organisms on earth. There are more living bacteria in your mouth than there are mammals living on earth. Although too tiny to see with the unaided eye, bacteria play critical roles throughout the biosphere. They extract from the air all the nitrogen used by organisms, and they play key roles in cycling carbon and sulfur.

There are many different kinds of bacteria, and the evolutionary links between them are not well understood. Although there is considerable disagreement among taxonomists about the details of bacterial classification, most recognize 12 to 15 major groups of bacteria. Comparisons of the nucleotide sequences of rRNA molecules are beginning to reveal how these groups are related to each other and to the other two domains. The archaea and eukaryotes are more closely related to each other than to bacteria and are on a separate evolutionary branch of the tree, even though archaea and bacteria are both prokaryotes.

12.8 Bacteria are as different from archaea as they are from eukaryotes.

TABLE 12.1 CHARACTERISTICS OF THE SIX KINGDOMS

	Bacteria	Archaea	Protista	Plantae	Fungi	Animalia
Cell Type	Prokaryotic	Prokaryotic	Eukaryotic	Eukaryotic	Eukaryotic	Eukaryotic
Nuclear Envelope	Absent	Absent	Present	Present	Present	Present
Mitochondria	Absent	Absent	Present or absent	Present	Present or absent	Present
Chloroplasts	None (photosynthetic membranes in some types)	None (bacteriorhodopsin in one species)	Present in some forms	Present	Absent	Absent
Cell Wall	Present in most; peptidoglycan	Present in most; polysaccharide, glycoprotein, or protein	Present in some forms; various types	Cellulose and other polysaccharides	Chitin and other noncellulose polysaccharides	Absent
Means of Genetic Recombination, if present	Conjugation, transduction, transformation	Conjugation, transduction, transformation	Fertilization and meiosis	Fertilization and meiosis	Fertilization and meiosis	Fertilization and meiosis
Mode of Nutrition	Autotrophic (chemosynthetic, photosynthetic) or heterotrophic	Autotrophic (photosynthesis in one species) or heterotrophic	Photosynthetic or heterotrophic, or combination of both	Photosynthetic, chlorophylls *a* and *b*	Absorption	Digestion
Motility	Bacterial flagella, gliding, or nonmotile	Unique flagella in some	9 + 2 cilia and flagella; ameboid, contractile fibrils	None in most forms, 9 +2 cilia and flagella in gametes of some forms	Nonmotile	9 +2 cilia and flagella, contractile fibrils
Multicellularity	Absent	Absent	Absent in most forms	Present in all forms	Present in most forms	Present in all forms

Domain Eukarya

For at least 1 billion years, prokaryotes ruled the earth. No other organisms existed to eat them or compete with them, and their tiny cells formed the world's oldest fossils. The third great domain of life, the eukaryotes, appear in the fossil record much later, only about 1.5 billion years ago. Metabolically, eukaryotes are more uniform than prokaryotes. Each of the two domains of prokaryotic organisms has far more metabolic diversity than all eukaryotic organisms taken together. The characteristics of the six kingdoms are outlined in table 12.1.

Three Largely Multicellular Kingdoms

Fungi, plants, and animals are well-defined evolutionary groups, each of them clearly stemming from a different single-celled, eukaryotic ancestor. They are largely multicellular, each a distinct evolutionary line from an ancestor that would be classified in the kingdom Protista.

The amount of diversity among the protists, however, is much greater than that within or between the three largely multicellular kingdoms derived from the protists. Because of the size and ecological dominance of plants, animals, and fungi, and because they are predominantly multicellular, we recognize them as kingdoms distinct from Protista.

A Fourth Very Diverse Kingdom

When multicellularity evolved, the diverse kinds of single-celled organisms that existed at that time did not simply become extinct. A wide variety of unicellular eukaryotes and their relatives exists today, grouped together in the kingdom Protista solely because they are not fungi, plants, or animals. Protists are a fascinating group containing many organisms of intense interest and great biological significance.

Symbiosis and the Origin of Eukaryotes

The hallmark of eukaryotes is complex cellular organization, highlighted by an extensive endomembrane system that subdivides the eukaryotic cell into functional compartments called organelles (see chapter 4). Not all of these organelles however, are derived from the endomembrane system. Mitochondria and chloroplasts are both believed to have entered early eukaryotic cells by a process called endosymbiosis (*endo*, inside) where an organism such as a bacterium is taken into the cell and remains functional inside the cell (figure 12.10).

With few exceptions, all modern eukaryotic cells possess energy-producing organelles, the mitochondria. Mitochondria are about the size of bacteria and contain DNA. Comparison of the nucleotide sequence of this DNA with that of a variety of organisms indicates clearly that mitochondria are the descendants of purple bacteria that were incorporated

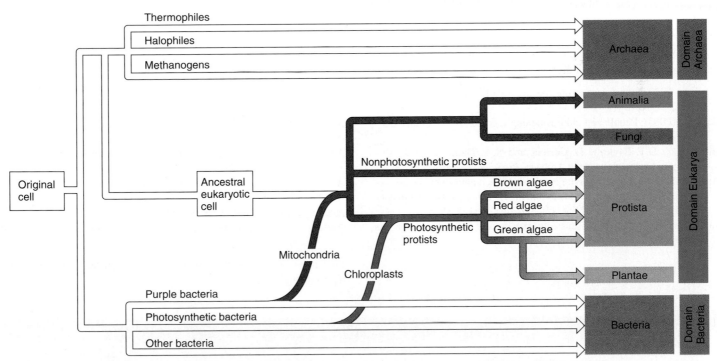

Figure 12.10 Diagram of the evolutionary relationship among the six kingdoms of organisms.
The colored lines indicate symbiotic events. Mitochondria are present in essentially all eukaryotic organisms. Chloroplasts are present in a subset of eukaryotic organisms, those that are photosynthetic.

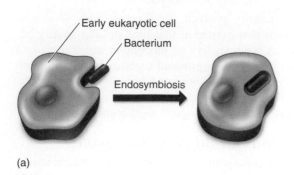

(a)

(b)

Figure 12.11 Endosymbiosis.

(*a*) This figure shows how an organelle could have arisen in early eukaryotic cells through a process called endosymbiosis. An organism, such as a bacterium, is taken into the cell through a process similar to endocytosis but remains functional inside the host cell. (*b*) Many corals contain endosymbionts, algae called zooxanthellae that carry out photosynthesis and provide the coral with nutrients. In this photograph, the zooxanthellae are the golden-brown spheres packed into the tentacles of a coral animal.

into eukaryotic cells early in the history of the group. Some protist phyla have in addition acquired chloroplasts during the course of their evolution and thus are photosynthetic (figure 12.11*a*). These chloroplasts are derived from cyanobacteria that became symbiotic in several groups of protists early in their history. Some of these photosynthetic protists gave rise to land plants. And, interestingly, others types of photosynthetic protists are endosymbionts of some eukaryotic organisms, such as certain species of sponges, jellyfish, corals

(figure 12.11*b*), octopuses, and others. We discussed the theory of the endosymbiotic origin of mitochondria and chloroplasts in chapter 4.

> **12.9 Eukaryotic cells acquired mitochrondia and chloroplasts by endosymbiosis, mitochondria being derived from purple bacteria and chloroplasts from cyanobacteria.**

Exploring Current Issues

Additional Resources

Go to your campus library or look online to find the following articles, which further develop some of the concepts found in this chapter.

Doolittle, W. F. (2000). Uprooting the tree of life. *Scientific American,* 282(2), 90.

Hecht, J. (2003). People and chimps belong together on the family tree. *New Scientist,* 178(2396), 15.

Jarrell, K. F., D. P. Bayley, J. D. Correia, and N. A. Thomas. (1999). Recent excitement about the Archaea. *BioScience,* 49(7), 530.

Pray, L. A. (2003). Modern phylogeneticists branch out: scientists use sequencing and analysis tools to find places for organisms, from bats to bacilli, in the tree of life. *The Scientist,* 17(11), 35.

Reydom, T. A. C. (2003). Discussion: species are individuals—or are they? *Philosophy of Science,* 70(1), 49.

In the News: *Biodiversity*

At its heart, biology is about critters—the great diversity of organisms that share the earth. Research into biodiversity, often carried out in the field, is a rich and rewarding activity. It can also be fun, funny, and intensely interesting.

Article 1. *A fierce argument has broken out among ecologists over biodiversity.*

Article 2. *Is the number of men in a female's life written in her genes? Polyandry in Galápagos hawks.*

Article 3. *Violence in Eden: Is Flipper a senseless killer?*

Article 4. *The killer bees are coming.*

Article 5. *How did Saint Patrick get the snakes out of Ireland?*

Article 6. *Going batty: One of the most successful mammals flies at night.*

Find these articles, written by the author, online at www.mhhe.com/tlwessentials/exp12.

Summary

The Classification of Organisms

12.1 The Invention of the Linnaean System

- Scientists use a system of grouping similar organisms together, called classification. Latin is used because it was the language used by earlier philosophers and scientists.

- The polynomial system of classification named an organism by using a list of adjectives that described the organism. The binomial system, using a two-part name, was originally developed as a "shorthand" reference to the polynomial name. Linnaeus used this two-part naming system consistently and its use became widespread (**figure 12.1**).

12.2 Species Names

- Taxonomy is the area of biology involved in identifying, naming, and grouping organisms. Scientific names that consist of two parts, the genus and species, are standardized, universal names that are less confusing than common names (**figure 12.2**).

12.3 Higher Categories

- In addition to the genus and species names, an organism is also assigned to higher levels of classification. The higher categories convey more general information about the organisms in a particular group. The most general category, domain, is the largest grouping followed by ever-increasingly specific information that is used to group organisms into a kingdom, phylum, class, order, family, genus, and species (**figure 12.3**).

12.4 What Is a Species?

- The biological species concept states that a species is a group of organisms that is reproductively isolated, meaning that the individuals mate and produce fertile offspring with each other but not with those of other species.

- This concept works well to define animal species because animals regularly outcross (mate with other individuals—**figure 12.4**). However, the concept does not apply to other organisms (fungi, protists, plants, and prokaryotes) that regularly reproduce without mating through asexual reproduction. The classification of these organisms relies more on physical, behavioral, and genetic characteristics.

Inferring Phylogeny

12.5 How to Build a Family Tree

- In addition to organizing a great number of organisms, the study of taxonomy also gives us a glimpse of the evolutionary history of life on earth. Organisms with similar characteristics are more likely to be related to each other. The evolutionary history of an organism and its relationship to other species is called phylogeny, and relationships are often mapped out using phylogenetic trees.

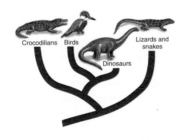

- Phylogenetic tress can be created using key characteristics that are shared by some organisms, presumably having been inherited from a common ancestor. A group of organisms that have shared characteristics is called a clade, and a phylogenetic tree organized in this manner is called a cladogram (**figure 12.5**). A cladogram suggests the order in which evolutionary changes occurred.

- Traditional taxonomy focuses more on the significance or evolutionary impact of a characteristic and not just on the commonality of the characteristic (**figures 12.6 and 12.7**).

Kingdoms and Domains

12.6 The Kingdoms of Life

- The designation of kingdoms, the second-highest category used in classification, has changed over the years as more and more information about organisms has been uncovered. Originally there were two kingdoms, Plantae and Animalia, but as more information was obtained, biologists began to classify organisms into other kingdoms: Fungi, Protista, Archaea, and Bacteria (**figure 12.8**). The domain level of classification was added in the mid-1990s, recognizing three fundamentally different types of cells: Eukarya (eukaryotic cells), Archaea (prokaryotic archaea), and Bacteria (prokaryotic bacteria) (**figure 12.9**).

12.7 Domain Archaea

- The domain Archaea contains prokaryotic organisms in the kingdom Archaebacteria. These single-celled organisms are found in diverse environments but most interestingly, in very extreme environments.

12.8 Domain Bacteria

- The domain Bacteria contains prokaryotic organisms in the kingdom Bacteria. These single-celled organisms play key roles in ecology (**table 12.1**).

12.9 Domain Eukarya

- The domain Eukarya contains very diverse organisms from four kingdoms but are similar in that they are all eukaryotes. They contain cellular organelles that were most likely acquired through endosymbiosis (**figures 12.10 and 12.11**).

Self-Test

1. The wolf, domestic dog, and red fox are all in the same family, Canidae. The scientific name for the wolf is *Canis lupus,* the domestic dog is *Canis familiar,* and the red fox is *Vulpes vulpes.* This means that
   a. the red fox is in the same family, but different genus than dogs and wolves.
   b. the dog is in the same family, but different genus than red fox and wolves.
   c. the wolf is in the same family, but different genus than dogs and red foxes.
   d. all three organisms are in different genera.

2. The evolutionary relationship of an organism, and its relationship to other species, are its
   a. taxonomy.
   b. phylogeny.
   c. ontogeny.
   d. systematics.

3. All classification systems for organisms are based on
   a. physical and chemical characteristics.
   b where the organism lives.
   c. what the organism eats.
   d. the size of the organism.

4. The six kingdoms of organisms can be organized into three domains based on
   a. where the organism lives.
   b. what the organism eats.
   c. cell structure.
   d. cell and DNA structure.
5. All of the extremophiles belong to the domain of
   a. Bacteria.
   b. Archaea.
   c. Prokarya.
   d. Eukarya.
6. Bacteria are also known as prokaryotic cells because they
   a. are chemosynthetic.
   b. are unicellular.
   c. cause disease.
   d. do not have an internal membrane system.
7. Organisms in the domain Bacteria are different than organisms in the domain Archaea because they
   a. are prokaryotes.
   b. have a nucleus.

   c. have cell walls that are made of different materials.
   d. have mitochondria.
8. It is theorized that the ancestral organism for the domain Eukarya came from the phylum
   a. Animalia.
   b. Plantae.
   c. Protista.
   d. Fungi.
9. One difference between the kingdom Protista and the other three kingdoms in the domain Eukarya is that the other kingdoms are
   a. chemosynthetic.
   b. multicellular.
   c. eukaryotic.
   d. unicellular.
10. It is thought that two important organelles of eukaryotic cells came from
    a. development of the internal membrane system.
    b. ingestion of endosymbiotic protists.
    c. mutation.
    d. ingestion of endosymbiotic bacteria.

Visual Understanding

1. **Figure 12.5** Which of the organisms shown have amniotic membranes that surround the fetus?

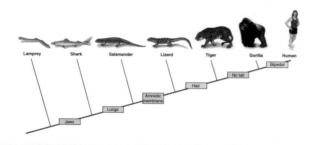

2. **Figure 12.9** Compared to the different types of bacteria and archaea, how similar are animals and plants?

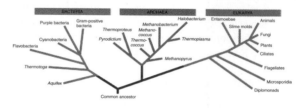

Challenge Questions

The Classification of Organisms Your friend, Julio, wants to know what the big deal is—everyone knows that a rose is a rose, why bother with all the fancy Latin stuff, like *Rosa odorata?* What do you tell him?

Inferring Phylogeny Why are birds classified so differently in traditional phylogeny and in cladistics?

Kingdoms and Domains If we already have things divided into kingdoms, why do we also need domains?

Online Learning Center

Visit the Online Learning Center for this chapter at www.mhhe.com/tlwessentials/ch12 for quizzes, animations, interactive learning exercises, and other study tools. At the site you will also find extended answers to the end-of-chapter questions.

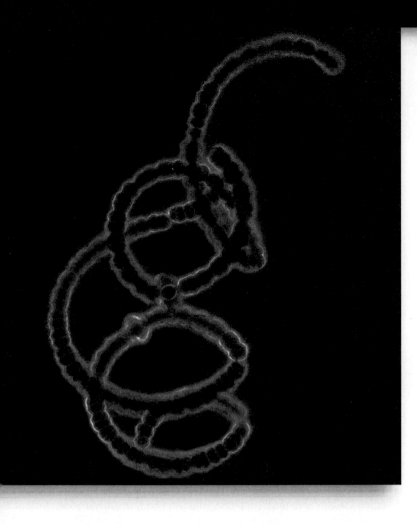

13

Evolution of Microbial Life

Much of the living world is composed of organisms too small for you to see without a microscope to magnify them. The beautiful jewel-like creature you see above is a prokaryote, the cyanobacterium *Anabaena,* in which cells adhere in filaments. The larger cells (areas on the filament that seem to be bulging) are heterocysts, specialized cells in which nitrogen fixation occurs. These organisms exhibit one of the closest approaches to multicellularity among the prokaryotes, and in the nineteenth century some biologists considered *Anabaena* to be a very simple sort of plant. Today, however, most biologists consider *Anabaena* to be a colonial prokaryote. In this chapter we will explore microbes—the sorts of creatures too small to see with the unaided eye. Most living organisms are microbes, and they have an enormous impact on human life. Some of the other creatures you will encounter in this chapter are near relatives of microbes that are multicellular and large. Others will not be organisms at all, but viruses, which are renegade segments of genomes that infect living cells. All of these creatures—microbes, multicellular relatives, and viruses—are important components of the living world.

13.1 How Cells Arose

All living organisms are constructed of the same four kinds of macromolecules discussed in chapter 3, the bricks and mortar of cells. Where the first macromolecules came from and how they came to be assembled together into cells are among the least understood questions in biology—questions that address the very origin of life itself.

No one knows for sure where the first organisms (thought to be like today's bacteria) came from. It is not possible to go back in time and watch how life originated, nor are there any witnesses. Nevertheless, it is difficult to avoid being curious about the origin of life, about what, or who, is responsible for the appearance of the first living organisms on earth. There are, in principle, at least three possibilities:

1. **Extraterrestrial origin.** Life may not have originated on earth at all but instead may have been carried to it, perhaps as an extraterrestrial infection of spores originating on a planet of a distant star. How life came to exist on that planet is a question we cannot hope to answer soon.
2. **Special creation.** Life-forms may have been put on earth by supernatural or divine forces. This viewpoint, called creationism, is common to most Western religions and is the oldest hypothesis. However, almost all scientists reject creationism, preferring evolution as a scientific explanation of life's diversity.
3. **Evolution.** Life may have evolved from inanimate matter, with associations among molecules becoming more and more complex. In this view, the force leading to life was selection; changes in molecules that increased their stability caused the molecules to persist longer.

In this text, we focus on the third possibility and attempt to understand whether the forces of evolution could have led to the origin of life and, if so, how the process might have occurred. This is not to say that the third possibility, evolution, is definitely the correct one. Any one of the three possibilities might be true. Nor does the third possibility preclude religion: A divine agency might have acted via evolution. Rather, we are limiting the scope of our inquiry to scientific matters. Of the three possibilities, only the third permits testable hypotheses to be constructed and so provides the only scientific explanation—that is, one that could potentially be disproved by experiment.

Forming Life's Building Blocks

If we look at the presence of living organisms on earth as a 24-hour clock of biological time (figure 13.1), with the formation of the earth 4.5 billion years ago being midnight, the first cells are seen just before mid-day, while humans do not appear until the day is almost all over, only minutes before its end. How can we learn about the origin of the first cells, so

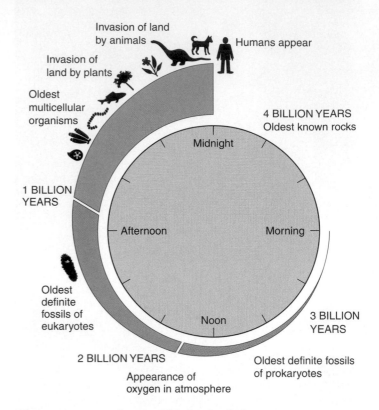

Figure 13.1 A clock of biological time.
A billion seconds ago, most students using this text had not yet been born. A billion minutes ago, Jesus was alive and walking in Galilee. A billion hours ago, the first human had not been born. A billion days ago, no biped walked on earth. A billion months ago, the last dinosaurs had not yet been hatched. A billion years ago, no creature had ever walked on the surface of the earth.

long ago? One way is to try to reconstruct what the earth was like long ago. We know from rocks that there was little or no oxygen in the earth's atmosphere then, and considerably more of the hydrogen-rich gases hydrogen sulfide (H_2S), ammonia (NH_3), and methane (CH_4). Electrons in these gases would have been frequently pushed to higher energy levels by photons crashing into them from the sun or by electrical energy in lightning . Today, high-energy electrons are quickly soaked up by the oxygen in earth's atmosphere (air is 21% oxygen, all of it contributed by photosynthesis) because oxygen atoms have a great "thirst" for such electrons. But in the absence of oxygen, high-energy electrons would have been free to help form biological molecules.

When the scientists Stanley Miller and Harold Urey reconstructed the oxygen-free atmosphere of the early earth in their laboratory and subjected it to the lightning and UV radiation it would have experienced then, they found that many of the building blocks of organisms, such as amino acids and nucleotides, formed spontaneously. They concluded that life may have evolved in a "primordial soup" of biological molecules formed in the ancient earth's oceans.

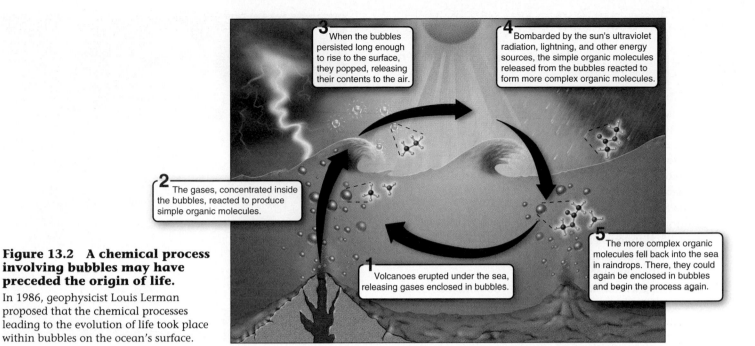

Figure 13.2 A chemical process involving bubbles may have preceded the origin of life.

In 1986, geophysicist Louis Lerman proposed that the chemical processes leading to the evolution of life took place within bubbles on the ocean's surface.

Text in figure:

3 When the bubbles persisted long enough to rise to the surface, they popped, releasing their contents to the air.

4 Bombarded by the sun's ultraviolet radiation, lightning, and other energy sources, the simple organic molecules released from the bubbles reacted to form more complex organic molecules.

2 The gases, concentrated inside the bubbles, reacted to produce simple organic molecules.

5 The more complex organic molecules fell back into the sea in raindrops. There, they could again be enclosed in bubbles and begin the process again.

1 Volcanoes erupted under the sea, releasing gases enclosed in bubbles.

Recently, concerns have been raised regarding the "primordial soup" hypothesis as the origin of life on earth. If the earth's atmosphere had no oxygen soon after it was formed, as Miller and Urey assumed (and most evidence supports this assumption), then there would have been no protective layer of ozone to shield the earth's surface from the sun's damaging UV radiation. Without an ozone layer, scientists think UV radiation would have destroyed any ammonia and methane present in the atmosphere. When these gases are missing, the **Miller-Urey experiment** does not produce key biological molecules such as amino acids. If the necessary ammonia and methane were not in the atmosphere, where were they?

In the last two decades, support has grown among scientists for what has been called the **bubble model.** The bubble model proposes that the key chemical processes generating the building blocks of life took place not in a primordial soup but rather within bubbles on the ocean's surface (figure 13.2). Bubbles produced by wind, wave action, the impact of raindrops, and the eruption of volcanoes cover about 5% of the ocean's surface at any given time. Because water molecules are polar, water bubbles tend to attract other polar molecules, in effect concentrating them within the bubbles. Chemical reactions would proceed much faster in bubbles, where polar reactants would be concentrated. The bubble model solves a key problem with the primordial soup hypothesis. Inside the bubbles, the methane and ammonia required to produce amino acids would have been protected from destruction by UV radiation.

The First Cells

We don't know how the first cells formed, but most scientists suspect they aggregated spontaneously. When complex carbon-containing macromolecules are present in water, they tend to gather together, much as the people from the same foreign country tend to aggregate within a large city. Sometimes the aggregations form a cluster big enough to see. Try vigorously shaking a bottle of oil-and-vinegar salad dressing—tiny bubbles called **microspheres** form spontaneously, suspended in the vinegar. Similar microspheres might have represented the first step in the evolution of cellular organization. Such microspheres have many cell-like properties—their outer boundary resembles the membranes of a cell in that it has two layers (see chapter 4), and the microspheres can increase in size and divide. Over millions of years, those microspheres better able to incorporate molecules and energy would have tended to persist longer than others.

As we learned earlier, scientists suspect that the first macromolecules to form were RNA molecules, and with the recent discovery that RNA molecules can behave as enzymes, catalyzing their own assembly, this provides a possible early mechanism of inheritance. Eventually DNA may have taken the place of RNA as the storage molecule for genetic information because the double-stranded DNA would have been more stable than single-stranded RNA.

As you can see, the scientific vision of life's origin is at best a hazy outline. Many different scenarios seem possible, and some have solid support from experiments. Deep-sea hydrothermal vents are an interesting possibility; the prokaryotes populating these vents are among the most primitive of living organisms. Other researchers have proposed that life originated deep in the earth's crust. How life might have originated naturally and spontaneously remains a subject of intense interest, research, and discussion among scientists.

> **13.1** Life appeared on earth 2.5 billion years ago. It may have arisen spontaneously, although the nature of the process is not clearly understood. Little is known about how the first cells originated.

A Closer Look

Has Life Evolved Elsewhere?

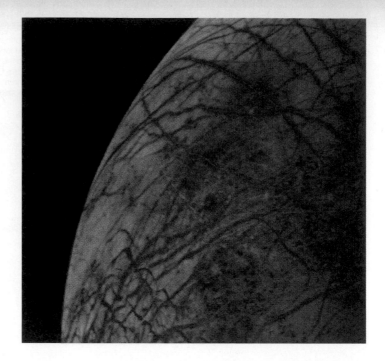

We should not overlook the possibility that life processes might have evolved in different ways on other planets. A functional genetic system, capable of accumulating and replicating changes and thus of adaptation and evolution, could theoretically evolve from molecules other than carbon, hydrogen, nitrogen, and oxygen in a different environment. Silicon, like carbon, needs four electrons to fill its outer energy level, and ammonia is even more polar than water. Perhaps under radically different temperatures and pressures, these elements might form molecules as diverse and flexible as those carbon has formed on earth.

The universe has 10^{20} (100,000,000,000,000,000,000) stars similar to our sun. We don't know how many of these stars have planets, but it seems increasingly likely that many do. Since 1996, astronomers have been detecting planets orbiting distant stars. At least 10% of stars are thought to have planetary systems. If only 1 in 10,000 of these planets is the right size and at the right distance from its star to duplicate the conditions in which life originated on earth, the "life experiment" will have been repeated 10^{15} times (that is, a million billion times). It does not seem likely that we are alone.

A dull gray chunk of rock collected in 1984 in Antarctica ignited an uproar about ancient life on Mars with the report that the rock contains evidence of possible life. Analysis of gases trapped within small pockets of the rock indicate it is a meteorite from Mars. It is, in fact, the oldest rock known to science—fully 4.5 billion years old. Evidence collected by the 2004 NASA Mars mission (the photo below of the martian surface was taken by the rover Spirit) suggests that the surface, now cold and arid, was much warmer when the Antarctic meteorite formed 4.5 billion years ago, that water flowed over its surface, and that it had a carbon dioxide atmosphere—conditions not too different from those that spawned life on earth.

When examined with powerful electron microscopes, carbonate patches within the meteorite exhibit what look like microfossils, some 20 to 100 nanometers in length. As they are one hundred times smaller than any known bacteria, it is not clear they actually are fossils, but the resemblance to bacteria is striking.

Viewed as a whole, the evidence of bacterial life associated with the Mars meteorite is not compelling. Clearly, more painstaking research remains to be done before the discovery can claim a scientific consensus. However, while there is no conclusive evidence of bacterial life associated with this meteorite, it seems very possible that life has evolved on other worlds in addition to our own.

There are planets other than ancient Mars with conditions not unlike those on earth. Europa, a large moon of Jupiter, is a promising candidate (photo *above*). Europa is covered with ice, and photos taken in close orbit in the winter of 1998 reveal seas of liquid water beneath a thin skin of ice. Additional satellite photos taken in 1999 suggest that a few miles under the ice lies a liquid ocean of water larger than earth's, warmed by the push and pull of the gravitational attraction of Jupiter's many large satellite moons. The conditions on Europa now are far less hostile to life than the conditions that existed in the oceans of the primitive earth. In coming decades, satellite missions are scheduled to explore this ocean for life.

13.2 The Simplest Organisms

Judging from fossils in ancient rocks, prokaryotes have been plentiful on earth for over 2.5 billion years. From the diverse array of early living forms, a few became the ancestors of the great majority of organisms alive today. Several ancient forms including cyanobacteria have survived or have given rise to members of the domain Bacteria; others gave rise to the second great group of prokaryotes, members of the domain Archaea. Still others probably became extinct millions or even billions of years ago. The fossil record indicates that eukaryotic cells, being much larger than prokaryotes and exhibiting elaborate shapes in some cases, did not appear until about 1.5 billion years ago. Therefore, for at least 1 billion years prokaryotes were the only organisms that existed.

Today prokaryotes are the simplest and most abundant form of life on earth. In a spoonful of farmland soil, 2.5 billion bacteria may be present. In 1 hectare (about 2.5 acres) of wheat land in England, the weight of bacteria in the soil is approximately equal to that of 100 sheep!

It is not surprising, then, that prokaryotes occupy a very important place in the web of life on earth. They play a key role in cycling minerals within the earth's ecosystems. In fact, photosynthetic bacteria were in large measure responsible for the introduction of oxygen into the earth's atmosphere. Bacteria are responsible for some of the most deadly animal and plant diseases, including many humans diseases. Bacteria and archaea are our constant companions, present in everything we eat and on everything we touch.

The Structure of a Prokaryote

The essential character of prokaryotes can be conveyed in a simple sentence: **Prokaryotes** are small, simply organized, single cells that lack an organized nucleus. Therefore, bacteria and archaea are prokaryotes; their single circle of DNA is not confined by a nuclear membrane in a nucleus (see figure 4.8), as in the cells of eukaryotes. Too tiny to see with the naked eye, a bacterial cell is usually simple in form, either rod-shaped (bacilli), spherical (cocci), or spirally coiled (spirilla) (see figure 4.9). A few kinds of bacteria aggregate into stalked structures or filaments.

The prokaryotic cell's plasma membrane is encased within a cell wall. The cell wall of bacteria is made of peptidoglycan, a network of polysaccharide molecules linked together by protein cross-links. The cell walls of archaea lack peptidoglycan and are made of proteins or polysaccharides or both. Many species of bacteria have a cell wall composed of layers of peptidoglycan, but in others, an outer membrane composed of large molecules of lipopolysaccharide (a chain of sugars with lipids attached to it) covers a thinner peptidoglycan cell wall (figure 13.3). Bacteria are commonly classified by the

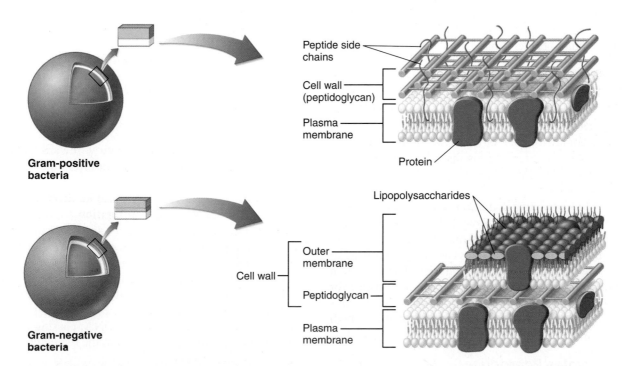

Figure 13.3 The structures of bacterial cell walls.
The peptidoglycan layer encasing gram-positive bacteria traps crystal violet dye, so the bacteria appear purple in a Gram-stained smear (named after Hans Christian Gram, who developed the technique). Because gram-negative bacteria have much less peptidoglycan (located between the plasma membrane and an outer membrane), they do not retain the crystal violet dye and so exhibit the red counterstain (usually a safranin dye) in a Gram-stained smear.

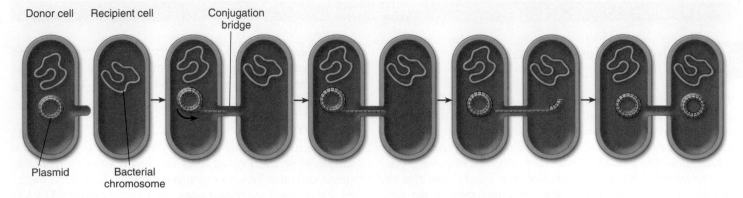

Donor cell   Recipient cell   Conjugation
                                 bridge

Plasmid   Bacterial
               chromosome

Figure 13.4   Bacterial conjugation.
Donor cells contain a plasmid that recipient cells lack. The plasmid replicates itself and transfers the copy across a conjugation bridge. The remaining strand of the plasmid serves as a template to build a replacement. When the single strand enters the recipient cell, it serves as a template to assemble a double-stranded plasmid. When the process is complete, both cells contain a complete copy of the plasmid.

presence or absence of this membrane as **gram-positive** (no outer membrane) or **gram-negative** (possess an outer membrane). The name refers to the Danish microbiologist Hans Gram, who developed a staining process that stains the cell types differently. A purple dye is retained in the peptidoglycan layer in the cell walls of gram-positive bacteria and they stain purple. In bacteria with an outer membrane, the peptidoglycan layer is thinner and does not retain the purple dye, which is washed away easily. A counter stain with a red dye is retained and so the cells stain red, not purple. The outer membranes of gram-negative bacteria make them resistant to antibiotics that attack the bacterial cell wall. That is why penicillin, which targets the protein cross-links of the bacterial cell wall, is effective only against gram-positive bacteria. Outside the cell wall and membrane, many bacteria have a gelatinous layer called a **capsule.**

Many kinds of bacteria possess threadlike **flagella,** long strands of protein that may extend out several times the length of the cell body. Bacteria swim by twisting these flagella in a corkscrew motion. Flagella may be distributed all over the cell body or be confined to one or both ends of the cell. Some bacteria also possess shorter outgrowths called **pili** (singular, **pilus**), which act as docking cables, helping the cell to attach to surfaces or other cells.

When exposed to harsh conditions (dryness or high temperature), some bacteria form thick-walled **endospores** around their DNA and a small bit of cytoplasm. These endospores are highly resistant to environmental stress and may germinate to form new active bacteria even after centuries. The formation of endospores is the reason that the bacterium responsible for botulism, *Clostridium botulinum,* sometimes persists in cans and bottles if the containers have not been heated at a high enough temperature to kill the spores.

How Prokaryotes Reproduce

Prokaryotes, like all other living cells, grow and divide. Prokaryotes reproduce using a process called **binary fission,** in which an individual cell simply increases in size and divides in two. Following replication of the prokaryote DNA, the

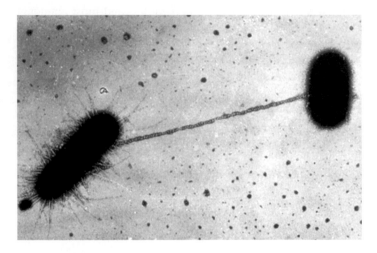

Figure 13.5   Contact by a pilus.
The pilus of a donor bacterial cell connects to a recipient bacterial cell and draws the two cells close together so that DNA transfer can occur.

plasma membrane and cell wall grow inward and eventually divide the cell by forming a new wall from the outside toward the center of the old cell.

Some bacteria can pass plasmids from one cell to another in a process called **conjugation** (figure 13.4). Recall from chapter 10 that a plasmid is a small, circular fragment of DNA that replicates outside the main bacterial chromosome. In bacterial conjugation, the pilus of one cell, called the donor cell, contacts a recipient cell (figure 13.5) and draws the two cells close together. The plasmid within the donor cell is then transferred across a conjugation bridge. The plasmid in the donor cell begins to replicate its DNA, passing the replicated copy out across the bridge and into the recipient cell.

13.2   Prokaryotes are the smallest and simplest organisms, composed of a single cell with no internal compartments or organelles. They divide by binary fission.

13.3 Comparing Prokaryotes to Eukaryotes

Prokaryotes differ from eukaryotes in many respects: the cytoplasm of prokaryotes has very little internal organization, prokaryotes are unicellular and much smaller than eukaryotes, the prokaryotic chromosome is a single circle of DNA, cell division and flagella are simple, and prokaryotes are far more metabolically diverse than eukaryotes (table 13.1).

Prokaryotic Metabolism

Prokaryotes have evolved many more ways than eukaryotes to acquire the carbon atoms and energy necessary for growth and reproduction. Many are **autotrophs,** organisms that obtain their carbon from inorganic CO_2. Autotrophs that obtain their energy from sunlight are called *photoautotrophs,* while those that harvest energy from inorganic chemicals are called *chemoautotrophs.* Other prokaryotes are **heterotrophs,** organisms that obtain at least some of their carbon from organic molecules like glucose. Heterotrophs that obtain their energy from sunlight are called *photoheterotrophs,* while those that harvest energy from organic molecules are called *chemoheterotrophs.*

Photoautotrophs. Many prokaryotes carry out photosynthesis, using the energy of sunlight to build organic molecules from carbon dioxide. The cyanobacteria use chlorophyll *a* as the key light-capturing pigment and use H_2O as an electron donor, leaving oxygen gas as a by-product. Other prokaryotes use bacteriochlorophyll as their pigment and H_2S as an electron donor, leaving elemental sulfur as the by-product.

Chemoautotrophs. Some prokaryotes obtain their energy by oxidizing inorganic substances. Nitrifiers, for example, oxidize ammonia or nitrite to form the nitrate that is taken up by plants. Other prokaryotes oxidize sulfur, hydrogen gas, and other inorganic molecules. On the dark ocean floor at depths of 2,500 meters, entire ecosystems subsist on prokaryotes that oxidize hydrogen sulfide as it escapes from volcanic vents.

Photoheterotrophs. The so-called purple nonsulfur bacteria use light as their source of energy but obtain carbon from organic molecules such as carbohydrates or alcohols that have been produced by other organisms.

Chemoheterotrophs. Most prokaryotes obtain both carbon atoms and energy from organic molecules. These include decomposers and most pathogens.

> **13.3** Prokaryotes differ from eukaryotes in having no nucleus or other interior compartments, in being far more metabolically diverse, and in many other fundamental respects.

TABLE 13.1	PROKARYOTES COMPARED TO EUKARYOTES
Feature	**Example**

Internal compartmentalization. Unlike eukaryotic cells, prokaryotic cells contain no internal compartments, no internal membrane system, and no cell nucleus.

Prokaryotic cell

Cell size. Most prokaryotic cells are only about 1 micrometer in diameter, whereas most eukaryotic cells are well over 10 times that size.

Prokaryotic cell   Eukaryotic cell

Unicellularity. All prokaryotes are fundamentally single-celled. Even though some may adhere together in a matrix or form filaments, their cytoplasms are not directly interconnected, and their activities are not integrated and coordinated, as is the case in multicellular eukaryotes.

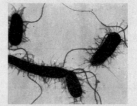
Unicellular bacteria

Chromosomes. Prokaryotes do not possess chromosomes in which proteins are complexed with the DNA, as eukaryotes do. Instead, their DNA exists as a single circle in the cytoplasm.

Prokaryotic chromosome   Eukaryotic chromosomes

Cell division. Cell division in prokaryotes takes place by binary fission (see chapter 7). The cells simply pinch in two. In eukaryotes, microtubules pull chromosomes to opposite poles during the cell division process, called mitosis.

Binary fission in prokaryotes

Mitosis in eukaryotes

Flagella. Prokaryotic flagella are simple, composed of a single fiber of protein that is spun like a propeller. Eukaryotic flagella are more complex structures, with a 9 + 2 arrangement of microtubules, that whip back and forth rather than rotate.

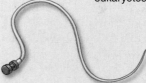
Simple bacterial flagellum

Metabolic diversity. Prokaryotes possess many metabolic abilities that eukaryotes do not: Prokaryotes perform several different kinds of anaerobic and aerobic photosynthesis; prokaryotes can obtain their energy from oxidizing inorganic compounds (so-called chemoautotrophs); and prokaryotes can fix atmospheric nitrogen.

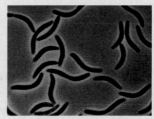
Chemoautotrophs

13.4 Viruses Infect Organisms

The border between the living and the nonliving is very clear to a biologist. Living organisms are cellular and able to grow and reproduce independently, guided by information encoded within DNA. As discussed earlier in this chapter, the simplest creatures living on earth today that satisfy these criteria are prokaryotes. Viruses, on the other hand, do not satisfy the criteria for "living" because they possess only a portion of the properties of organisms. **Viruses** are literally "parasitic" chemicals, segments of DNA (or sometimes RNA) wrapped in a protein coat. They cannot reproduce on their own, and for this reason they are not considered alive by biologists. They can, however, reproduce within cells, often with disastrous results to the host organism. For this reason, viruses have a major impact on the living world.

Viruses are very small. The smallest viruses are only about 17 nanometers in diameter; the largest ones measure up to 1,000 nanometers, big enough to be barely visible with the light microscope. Viruses are so small that they are smaller than many of the molecules in a cell. Most viruses can be detected only by using the higher resolution of an electron microscope.

The true nature of viruses was discovered in 1935, when the biologist Wendell Stanley prepared an extract of a plant virus called *tobacco mosaic virus* (*TMV*) and attempted to purify it. To his great surprise, the purified TMV preparation precipitated (that is, separated from solution) in the form of crystals. This was surprising because precipitation is something that only chemicals do—the TMV virus was acting like a chemical rather than an organism. Stanley concluded that TMV is best regarded as just that—a chemical matter rather than a living organism.

Each particle of TMV virus is in fact a mixture of two chemicals: RNA and protein. The TMV virus has the structure of a Twinkie, a tube made of an RNA core surrounded by a coat of protein. Later workers were able to separate the RNA from the protein and purify and store each chemical. Then, when they reassembled the two components, the reconstructed TMV particles were fully able to infect healthy tobacco plants and so clearly *were* the virus itself, not merely chemicals derived from it.

Viruses occur in all organisms, from bacteria to humans, and in every case the basic structure of the virus is the same, a core of nucleic acid surrounded by protein. There is considerable difference, however, in the details. In figure 13.6 you can compare the structure of bacterial, plant, and animal viruses—they are clearly quite different from one another. Many plant viruses like TMV have a core of RNA, and some animal viruses like HIV do too. Several different segments of DNA or RNA may be present in animal virus particles, along with many different kinds of protein. Like TMV, most viruses form a protein sheath, or **capsid,** around their nucleic acid core. In addition, many viruses form a membranelike **envelope,** rich in proteins, lipids, and glycoprotein molecules, around the capsid.

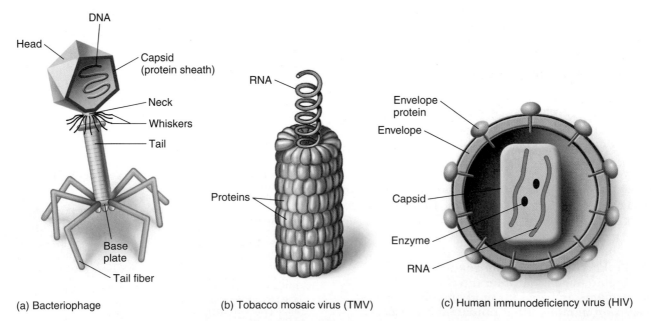

(a) Bacteriophage   (b) Tobacco mosaic virus (TMV)   (c) Human immunodeficiency virus (HIV)

Figure 13.6   The structure of bacterial, plant, and animal viruses.

(a) Bacterial viruses, called bacteriophages, often have a complex structure. (b) TMV infects plants and consists of 2,130 identical protein molecules (*purple*) that form a cylindrical coat around the single strand of RNA (*green*). The RNA backbone determines the shape of the virus and is protected by the identical protein molecules packed tightly around it. (c) In the human immunodeficiency virus (HIV), the RNA core is held within a capsid that is encased by a protein envelope.

The Origin of Viral Diseases

Sometimes viruses that originate in one organism pass to another, causing a disease in the new host. Thus, influenza is fundamentally a bird virus, and smallpox is thought to have passed from cattle to humans when cows were first domesticated.

New pathogens arising in this way, called *emerging viruses,* represent a greater threat today than in the past, as air travel and world trade in animals allow infected individuals and animals to move about the world quickly, spreading an infection. The widespread conversion of tropical forests into agricultural land has greatly increased the contact between people and wild animals, amplifying the opportunity for the introduction into people of novel viruses.

Influenza. Perhaps the most lethal virus in human history has been the influenza virus. Over 20 million worldwide died of flu within 18 months in 1918 and 1919—an astonishing number. The natural reservoir of influenza virus is in ducks (and pigs) in central Asia. Major flu pandemics (that is, worldwide epidemics) arise in Asian ducks through recombination within multiple infected individuals, putting together novel combinations of virus surface proteins unrecognizable by human immune defenses. The Asian flu of 1957 killed over 100,000 Americans. The Hong Kong flu of 1968 infected 50 million people in the United States alone, of which 70,000 died.

AIDS (HIV virus). The virus that causes AIDS first entered humans from chimpanzees somewhere in Central Africa, probably between 1910 and 1950. The chimpanzee virus, called simian immunodeficiency virus, or SIV, mutates rapidly, at a rate of 1% a year, and in humans it continued to do so, soon becoming what we now know as HIV and spreading widely, mostly through sexual contact with an infected person. Where did chimpanzees acquire SIV? SIV viruses are rampant in African monkeys, and chimpanzees eat monkeys. Study of the nucleotide sequences of monkey SIVs revealed in 2001 that one end of the chimp virus RNA closely resembles the SIV found in red-capped mangabey monkeys, while the other end resembles the virus from the greater spot-nosed monkey. It thus seems certain that chimpanzees acquired SIV from monkeys they ate.

Ebola virus. Among the most lethal of emerging viruses is a collection of filamentous viruses arising in Central Africa that attack human connective tissue. With lethality rates in excess of 50%, these so-called filoviruses cause the most lethal infectious diseases known. One, Ebola virus, has exhibited lethality rates in excess of 90% in isolated outbreaks in Central Africa. Luckily, victims die too fast to spread the disease very far. The natural host of Ebola is unknown. Extensive searches among wild and domestic animals have failed to reveal the identity of its natural reservoir, although there must be one, as isolated outbreaks continue to be reported in Africa every few years.

Hantavirus. A sudden outbreak of a highly fatal hemorrhagic infection in the southwestern United States in 1993 was soon attributed to a species of hantavirus, an RNA virus associated with rodents. This species was eventually traced to deer mice. The deer mouse hantavirus is transmitted to humans through fecal contamination in areas of human habitation. Control of deer mouse populations has limited the disease.

SARS. A recently emerged species of coronavirus was responsible for a worldwide outbreak in 2003 of *severe acute respiratory syndrome* (SARS), a respiratory infection with pneumonia-like symptoms that in over 8% of cases is fatal. When the 29,751-nucleotide RNA genome of the SARS virus was sequenced, it proved to be a completely new form of coronavirus, not closely related to any of the three previously described forms. Virologists suspect that the SARS coronavirus most likely came from civets (a weasel-like mammal) and other wild animals that are eaten as delicacies in China. If the SARS virus indeed exists in natural animal populations, it will be difficult to prevent future outbreaks.

West Nile Virus. A mosquito-borne virus, West Nile virus, first infected people in North America in 1999, killing four people in Queens, New York. Carried by infected crows and other birds, the virus proceeded to spread across the country, from 62 severe cases in 1999 to 4,156 cases in 2002, 284 of whom died. By 2003 it had been reported in every state. First detected in humans in Uganda, Africa, in 1937, the virus is common among birds, and is thought to have been transmitted to humans by mosquitos that had previously bitten infected birds, much as it is being transmitted now. Earlier spread of the virus through Europe abated after several years.

13.4 Viruses are genomes of DNA or RNA, encased in a protein shell, that can infect cells and replicate within them. They are chemical assemblies, not cells, and are not alive. Viruses are responsible for some of the most lethal diseases of humans.

13.5 The Origin of Eukaryotic Cells

The First Eukaryotic Cells

All fossils more than 1.7 billion years old are small, simple cells, similar to the bacteria of today. In rocks about 1.7 billion years old, we begin to see the first microfossils that are noticeably larger than bacteria and have internal membranes and thicker walls. A new kind of organism had appeared, called a **eukaryote,** from the Greek words for "true" and "nucleus," because eukaryotic cells possess an internal structure called a nucleus.

Many bacteria have infoldings of their outer membranes extending into the interior that serve as passageways to the surface. The network of internal membranes in eukaryotes called the endoplasmic reticulum (ER) is thought to have evolved from such infoldings, as is the nuclear envelope.

Endosymbiosis

The **endosymbiotic theory,** now widely accepted, suggests that at a critical stage in the evolution of eukaryotic cells, energy-producing bacteria came to reside symbiotically (that is, cooperatively) within larger early eukaryotic cells, eventually evolving into the cell organelles we now know as mitochondria. Similarly, photosynthetic bacteria came to live within some of these early eukaryotic cells, leading to the evolution of chloroplasts, the photosynthetic organelles of plants and algae (figure 13.7). Present-day mitochondria and chloroplasts still contain their own DNA, which is remarkably similar to the DNA of bacteria in size and character.

Mitochondria. Mitochondria, the energy-generating organelles in eukaryotic cells, are sausage-shaped organelles about 1 to 3 micrometers long, about the same size as most bacteria. Mitochondria are bounded by *two* membranes. The outer membrane is smooth and was apparently derived from the host cell. The inner membrane is folded into numerous layers, resembling the folded membranes of nonsulfur purple bacteria; embedded within this membrane are the proteins that carry out oxidative metabolism.

During the billion-and-a-half years in which mitochondria have existed as endosymbionts within eukaryotic cells, most of their genes have been transferred to the chromosomes of the host cells—but not all. Each mitochondrion still has its own genome, a circular, closed molecule of DNA similar to that found in bacteria, on which is located genes encoding some of the essential proteins of oxidative metabolism. Mitochondria divide by simple fission, just as bacteria do, and can divide on their own, without the cell nucleus dividing. However, the cell's nuclear genes direct the process, and mitochondria cannot be grown outside of the eukaryotic cell, in cell-free culture.

Chloroplasts. Many eukaryotic cells contain other endosymbiotic bacteria in addition to mitochondria. Plants and algae contain chloroplasts, bacteria-like organelles that were apparently derived from symbiotic photosynthetic bacteria. Chloroplasts have a complex system of inner membranes and a circle of DNA.

While all mitochondria are thought to have arisen from a single symbiotic event, it is difficult to be sure with chloroplasts. Three biochemically distinct classes of chloroplasts exist, but all appear to have their origin in the cyanobacteria.

> **13.5** The theory of endosymbiosis proposes that mitochondria originated as symbiotic aerobic bacteria and chloroplasts originated from a second endosymbiotic event.

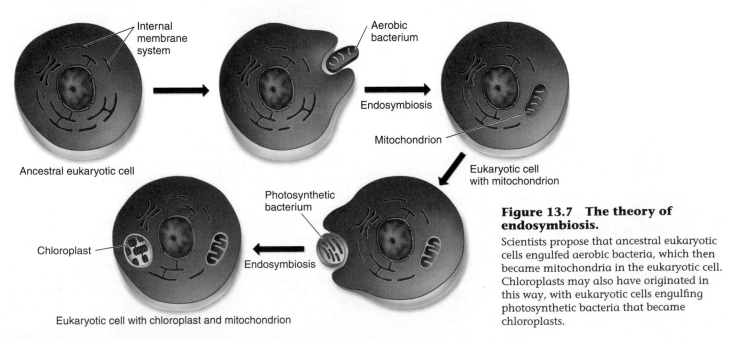

Figure 13.7 The theory of endosymbiosis.

Scientists propose that ancestral eukaryotic cells engulfed aerobic bacteria, which then became mitochondria in the eukaryotic cell. Chloroplasts may also have originated in this way, with eukaryotic cells engulfing photosynthetic bacteria that became chloroplasts.

13.6 General Biology of Protists

Protists are eukaryotes united on the basis of a single negative characteristic: They are not fungi, plants, or animals. In all other respects, they are highly variable with no uniting features. Many are unicellular (figure 13.8), but there are numerous colonial and multicellular groups. Most are microscopic, but some are as large as trees.

The Cell Surface

Protists possess varied types of cell surfaces. All protists have plasma membranes. But, some protists, like algae and molds, are additionally encased within strong cell walls. Still others, like diatoms and radiolarians, secrete glassy shells of silica.

Locomotor Organelles

Movement in protists is also accomplished by diverse mechanisms. Protists move by cilia, flagella, pseudopods, or gliding mechanisms. Many protists wave one or more flagella to propel themselves through the water, whereas others use banks of short, flagella-like structures called cilia to create water currents for their feeding or propulsion. Pseudopodia are the chief means of locomotion among amoeba, whose pseudopods are large, blunt extensions of the cell body called lobopodia.

Cyst Formation

Many protists with delicate surfaces are successful in quite harsh habitats. How do they manage to survive so well? They survive inhospitable conditions by forming **cysts.** A cyst is a dormant form of a cell with a resistant outer covering in which cell metabolism is more or less completely shut down. Amoeba parasites in vertebrates, for example, form cysts that are quite resistant to gastric acidity (although they will not tolerate desiccation or high temperature).

Nutrition

Protists employ every form of nutritional acquisition except chemoautotrophy, which has so far been observed only in prokaryotes. Some protists are photosynthetic autotrophs and are called **phototrophs.** Others are heterotrophs that obtain energy from organic molecules synthesized by other organisms. Among heterotrophic protists, those that ingest visible particles of food are called **phagotrophs,** or **holozoic feeders.** Those ingesting food in soluble form are called **osmotrophs,** or **saprozoic feeders.**

Phagotrophs ingest food particles into intracellular vesicles called **food vacuoles** or **phagosomes.** Lysosomes fuse with the food vacuoles, introducing enzymes that digest the food particles within. As the digested molecules are absorbed across the vacuolar membrane, the food vacuole becomes progressively smaller.

Reproduction

Protists typically reproduce asexually, most reproducing sexually only in times of stress. Asexual reproduction involves mitosis, but the process is often somewhat different from the mitosis that occurs in multicellular animals. The nuclear membrane, for example, often persists throughout mitosis, with the microtubular spindle forming within it. In some groups, asexual reproduction involves spore formation, in others fission. The most common type of fission is **binary,** in which a cell simply splits into nearly equal halves. When the progeny cell is considerably smaller than its parent, and then grows to adult size, the fission is called **budding.**

Sexual reproduction also takes place in many forms among the protists. In ciliates and some flagellates, **gametic meiosis** occurs just before gamete formation, as it does in most animals. In the sporozoans, **zygotic meiosis** occurs directly *after* fertilization, and all the individuals that are produced are haploid until the next zygote is formed. In algae, there is **sporic meiosis,** producing an alternation of generations similar to that seen in plants, with significant portions of the life cycle spent as haploid as well as diploid.

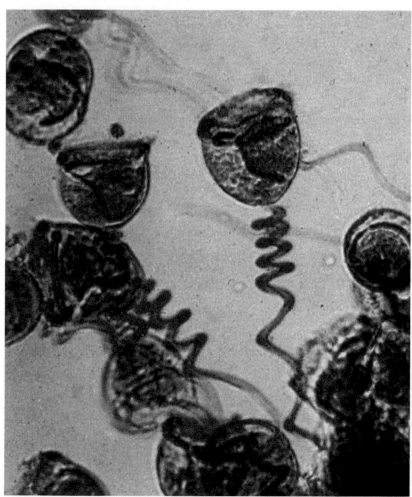

Figure 13.8 A unicellular protist.
The protist kingdom is a catchall kingdom for many different groups of unicellular organisms, such as this *Vorticella* (phylum Ciliophora), which is heterotrophic, feeding on bacteria, and has a retractable stalk.

Multicellularity

The algae are structurally simple multicellular organisms that fill the evolutionary gap between unicellular protists, colonial protists (figure 13.9), and more complexly multicellular organisms (fungi, plants, and animals). In **complex multicellular organisms,** individuals are composed of many highly specialized kinds of cells that coordinate their activities. There are three kingdoms that exhibit more complex multicellularity:

1. **Plants.** Multicellular green algae were almost certainly the direct ancestors of the plants (see chapter 14) and were themselves considered plants in the nineteenth century. However, green algae are basically aquatic and much simpler in structure than plants and are considered protists in the six-kingdom system used widely today.

2. **Animals.** Animals arose from a unicellular protist ancestor. Several groups of animal-like protists have been considered to be tiny animals in the past ("protozoa"), including flagellates, ciliates, and amoebas. The simplest (and seemingly most primitive) animals today, the sponges, seem clearly to have evolved from a kind of flagellate.

3. **Fungi.** Fungi also arose from a unicellular protist ancestor, one different from the ancestor of animals. Certain protists, including slime molds and water molds, have been considered fungi ("molds"), although they are usually classified as protists and are not thought to resemble ancestors of fungi. The true protist ancestor of fungi is as yet unknown. This is one of the great unsolved problems of taxonomy.

Two key characteristics of complex multicellular organisms distinguish them from simple multicellular organisms like marine algae: **cell specialization** and **intercellular coordination.** In a fungus, plant, or animal, the body of an individual possesses different kinds of cells that have very different structures and are coordinated in complex ways.

Perhaps the most important characteristic of complex multicellular organisms is cell specialization. If you think about it, having a variety of different sorts of cells within the same individual implies something very important about the genes of the individual: *Different cells are using different genes!* The process whereby a single cell (in humans, a fertilized egg) becomes a multicellular individual with many different kinds of cells is called **development.** The cell specialization that is the hallmark of complex multicellular life is the direct result of cells developing in different ways by activating different genes.

A second key characteristic of complex multicellular organisms is intercellular coordination, the adjustment of a cell's activity in response to what other cells are doing. The cells of all complex multicellular organisms communicate with one another with chemical signals called hormones. In some organisms like sponges, there is relatively little coor-

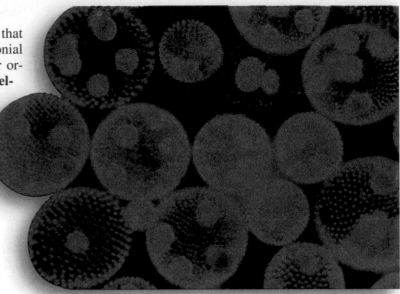

Figure 13.9   A colonial protist.
Individual, motile, unicellular green algae are united in the protist *Volvox* as a hollow colony of cells that moves by the beating of the flagella of its individual cells. Some species of *Volvox* have cytoplasmic connections between the cells that help coordinate colony activities. *Volvox* is a highly complex form of colony that has many of the properties of multicellular life.

dination between the cells; in other organisms like humans, almost every cell is under complex coordination.

The bodies of most fungi are far more complex than simple multicellular protists. Fungi contain cells that are joined end to end, forming long microscopic filaments called hyphae. The hyphae associate with each other forming larger, macroscopic structures. Fungal cells exhibit a high degree of communication and can become quite specialized in forming reproductive structures. The cells of some hyphae are separated by cross-walls called septa. The septa rarely form a complete barrier, allowing the flow of cytoplasm throughout the hyphae. Because of this cytoplasmic streaming, proteins synthesized throughout the hyphae can be carried to the hyphal tips. As a result, fungi can respond quickly to environmental changes, growing very rapidly when food and water are plentiful and the temperature is optimal.

13.6   Protists exhibit a wide range of forms, locomotion, nutrition, and reproduction. They mostly exhibit asexual reproduction but often undergo sexual reproduction in times of stress. Within the protists, only the algae are truly multicellular, and complex multicellularity is exhibited by plants, animals, and fungi.

Protists are the most diverse of the four kingdoms in the domain Eukarya. The kingdom Protista contains many unicellular, colonial, and multicellular groups. Probably the most important statement we can make about the kingdom Protista is that it is an artificial group; as a matter of convenience, single-celled eukaryotic organisms have typically been grouped together into this kingdom. This lumps many very different and only distantly related forms together. The "single-kingdom" classification of the Protista is not representative of any evolutionary relationships. The phyla of protists are, with very few exceptions, only distantly related to one another.

New applications of a wide variety of molecular methods are providing important insights into the relationships among the protists. Of all the groups of organisms biologists study, protists are probably in the greatest state of flux when it comes to classification. There is little consensus, even among experts, as to how the different kinds of protists should be classified. Are they a single, very diverse kingdom, or are they better considered as several different kingdoms, each of equal rank with animals, plants, and fungi?

Because the Protista are still predominantly considered part of one diverse, ununified group, that is how we will treat them in this chapter, bearing in mind that biologists are rapidly gaining a better understanding of the evolutionary relationships among members of the kingdom Protista. It seems likely that within a few years, the traditional kingdom Protista will be replaced by another more illuminating arrangement.

Five Groups of Protists

There are some 15 distinct phyla of protists. It is difficult to encompass their great diversity with any simple scheme. Traditionally, texts have grouped them artificially (as was done in the nineteenth century) into photosynthesizers (algae), heterotrophs (protozoa), and absorbers (funguslike protists).

In this text, we group the protists into five general groups according to some of the major shared characteristics (figure 13.10). These are characteristics that taxonomists are using today in broad attempts to classify the kingdom Protista. These include (1) the presence or absence and type of cilia or flagella, (2) the presence and kinds of pigments, (3) the type of mitosis, (4) the kinds of cristae present in the mitochondria, (5) the molecular genetics of the ribosomal "S" subunit, (6) the kind of inclusions the protist may have, (7) overall body form (amoeboid, coccoid, and so forth), (8) whether the protist has any kind of shell or other body "armor," and (9) modes of nutrition and movement. These represent only some of the characters used to define phylogenetic relationships.

The criteria we have chosen to define the five general groups are not the only ones that might be chosen, and there

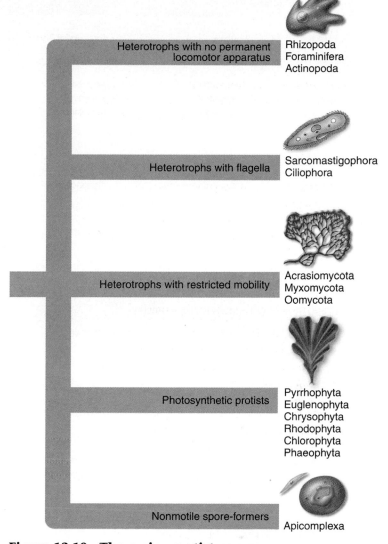

Heterotrophs with no permanent locomotor apparatus — Rhizopoda / Foraminifera / Actinopoda

Heterotrophs with flagella — Sarcomastigophora / Ciliophora

Heterotrophs with restricted mobility — Acrasiomycota / Myxomycota / Oomycota

Photosynthetic protists — Pyrrhophyta / Euglenophyta / Chrysophyta / Rhodophyta / Chlorophyta / Phaeophyta

Nonmotile spore-formers — Apicomplexa

Figure 13.10 The major protist groups.

is no broad agreement among biologists as to which set of criteria is preferable. As molecular analysis gives us a clearer picture of the phylogenetic relationships among the protists, more evolutionarily suitable groupings will without a doubt replace the one represented here. Table 13.2 summarizes some of the general characteristics and groupings of the 15 major phyla of protists. It is important to remember that while the phyla of protists discussed here are generally accepted taxa, the larger groupings of phyla presented are functional groupings.

13.7 The 15 protist phyla can be grouped into five categories according to major shared characteristics.

TABLE 13.2   KINDS OF PROTISTS

Group	Phylum	Typical Examples		Key Characteristics
HETEROTROPHS WITH NO PERMANENT LOCOMOTOR APPARATUS				
Amoebas	Rhizopoda	*Amoeba*		Move by pseudopodia
Forams	Foraminifera	Forams		Rigid shells; move by protoplasmic streaming
Radiolarians	Actinopoda	Radiolarians		Glassy skeletons; needlelike pseudopods
HETEROTROPHS WITH FLAGELLA				
Zoomastigotes	Sarcomastigophora	Trypanosomes		Heterotrophic; unicellular
Ciliates	Ciliophora	*Paramecium*		Heterotrophic unicellular protists with cells of fixed shape possessing two nuclei and many cilia; many cells also contain highly complex and specialized organelles
HETEROTROPHS WITH RESTRICTED MOBILITY				
Cellular slime molds	Acrasiomycota	*Dictyostelium*		Colonial aggregations of individual cells; most closely related to amoebas
Plasmodial slime molds	Myxomycota	*Fuligo*		Stream along as a multinucleate mass of cytoplasm
Water molds	Oomycota	Water molds and downy mildew		Terrestrial and freshwater
PHOTOSYNTHETIC PROTISTS				
Dinoflagellates	Pyrrhophyta	Red tides		Unicellular; two flagella; contain chlorophylls *a* and *b*
Euglenoids	Euglenophyta	*Euglena*		Unicellular; some photosynthetic; others heterotrophic; contain chlorophylls *a* and *b* or none
Diatoms	Chrysophyta	*Diatoma*		Unicellular; manufacture the carbohydrate chrysolaminarin; unique double shells of silica; contain chlorophylls *a* and *c*
Golden algae	Chrysophyta	*Dinobryon*		Unicellular, but often colonial; manufacture the carbohydrate chrysolaminarin; contain chlorophylls *a* and *c*
Red algae	Rhodophyta	Coralline algae		Most multicellular; contain chlorophyll *a* and a red pigment
Green algae	Chlorophyta	*Chlamydomonas*		Unicellular or multicellular; contain chlorophylls *a* and *b*
Brown algae	Phaeophyta	Kelp		Multicellular; contain chlorophylls *a* and *c*
NONMOTILE SPORE-FORMERS				
Sporozoans	Apicomplexa	*Plasmodium*		Nonmotile; unicellular; the apical end of the spores contains a complex mass of organelles

13.8 A Fungus Is Not a Plant

The fungi are a distinct kingdom of organisms, comprising about 74,000 named species. **Mycologists,** scientists who study fungi, believe there may be many more species in existence. Although fungi were at one time included in the plant kingdom, they lack chlorophyll and resemble plants only in their general appearance and lack of mobility. Significant differences between fungi and plants include the following:

Fungi are heterotrophs. Perhaps most obviously, a mushroom is not green. Virtually all plants are photosynthesizers, whereas no fungi carry out photosynthesis. Instead, fungi obtain their food by secreting digestive enzymes onto whatever they are attached to and then absorbing into their bodies the organic molecules that are released by the enzymes.

Fungi have filamentous bodies. Fungi are basically filamentous in their growth form (that is, their body consists of long, slender filaments), even though these filaments may be packed together to form complex structures like the mushroom.

Fungi have nonmotile sperm. Some plants have motile sperm with flagella. The majority of fungi do not.

Fungi have cell walls made of chitin. The cell walls of fungi contain chitin, the same tough material that a crab shell is made of. The cell walls of plants are made of cellulose, also a strong building material. Chitin, however, is far more resistant to microbial degradation than is cellulose.

Fungi have nuclear mitosis. Mitosis in fungi is different from plants or most other eukaryotes in one key respect: the nuclear envelope does not break down and re-form. Instead, all of mitosis takes place *within* the nucleus. A spindle apparatus forms there, dragging chromosomes to opposite poles of the *nucleus* (not the cell, as in all other eukaryotes).

You could build a much longer list, but already the take-home lesson is clear: Fungi are not like plants at all! Their many unique features are strong evidence that fungi are not closely related to any other group of organisms.

The Body of a Fungus

Fungi exist mainly in the form of slender filaments, barely visible with the naked eye, called **hyphae** (singular, **hypha**). A hypha is basically a long string of cells. The walls dividing one cell from another are called **septa** (singular, **septum**). The presence of septa is another of the ways in which fungi differ fundamentally from all other multicellular organisms, for the septa rarely form a complete barrier! From one fungal cell to the next, cytoplasm flows, streaming freely down the hypha through openings in the septa.

The main body of a fungus is not the familiar mushroom (figure 13.11), which is a temporary reproductive structure, but rather the extensive network of fine hyphae that penetrate the soil, wood, or flesh in which the fungus is growing. A mass of hyphae is called a **mycelium** (plural, **mycelia**) and may contain many meters of individual hyphae. This body organization creates a unique relationship between the fungus and its environment. All parts of the fungal body are metabolically active, secreting digestive enzymes and actively attempting to digest and absorb any organic material with which the fungus comes in contact.

Because of cytoplasmic streaming, many nuclei may be connected by the shared cytoplasm of a fungal mycelium. None of them (except for reproductive cells) are isolated in any one cell; all of them are linked cytoplasmically with every cell of the mycelium. Indeed, the entire concept of multicellularity takes on a new meaning among the fungi, the ultimate communal sharers among the multicellular organisms.

(a)   (b)

Figure 13.11 Mushrooms.

Most of the body of a fungus is below ground, a network of fine threads that penetrate the ground, often over great distances. Mushrooms are above-ground reproductive structures that release spores into the air, allowing the fungus to invade new habitats. (*a*) Yellow morel. (*b*) *Amanita* mushroom.

How Fungi Reproduce

Fungi reproduce both asexually and sexually. All fungal nuclei except for the zygote are haploid. Often in the sexual reproduction of fungi, individuals of different "mating type" must participate, much as two sexes are required for human reproduction. Sexual reproduction is initiated when two hyphae of genetically different mating types come in contact, and the hyphae fuse. What happens next? In animals and plants, when the two haploid gametes fuse, the two haploid nuclei immediately fuse to form the diploid nucleus of the zygote. As you might by now expect, fungi handle things differently. In most fungi, the two nuclei do not fuse immediately. Instead, they remain unmarried inhabitants of the same house, coexisting in a common cytoplasm for most of the life of the fungus! A fungal hypha that has two nuclei is called **dikaryotic.** If the nuclei are derived from two genetically different individuals, the hypha is called a **heterokaryon** (Greek, *heteros,* other, and *karyon,* kernel or nucleus). A fungal hypha in which all the nuclei are genetically similar is said to be a **homokaryon** (Greek, *homo,* one).

When reproductive structures are formed in fungi, complete septa form between cells, the only exception to the free flow of cytoplasm between cells of the fungal body. There are three kinds of reproductive structures: (1) **gametangia** form haploid gametes, which fuse to give rise to a zygote that undergoes meiosis; (2) **sporangia** produce haploid spores that can be dispersed; and (3) **conidiophores** produce asexual spores called **conidia,** which can be produced quickly and allow for the rapid colonization of a new food source.

Spores are a common means of reproduction among the fungi (figure 13.12). They are well suited to the needs of an organism anchored to one place. They are so small and light

Figure 13.13   The oyster mushroom.
This species, *Pleurotus ostreatus,* immobilizes nematodes, which the fungus uses as a source of food.

that they may remain suspended in the air for long periods of time and may be carried great distances. When a spore lands in a suitable place, it germinates and begins to divide, soon giving rise to a new fungal hypha.

How Fungi Obtain Nutrients

All fungi obtain their food by secreting digestive enzymes into their surroundings and then absorbing back into the fungus the organic molecules produced by this **external digestion.** Many fungi are able to break down the cellulose in wood, cleaving the linkages between glucose subunits and then absorbing the glucose molecules as food. That is why fungi are so often seen growing on trees.

Just as some plants like the Venus's-flytrap are active carnivores, so some fungi are active predators. For example, the edible oyster fungus *Pleurotus ostreatus* attracts tiny roundworms known as nematodes that feed on it—and secretes a substance that anesthetizes the nematodes (figure 13.13). When the worms become sluggish and inactive, the fungal hyphae envelop and penetrate their bodies and absorb their contents, a rich source of nitrogen (always in short supply in natural ecosystems). Other fungi are even more active predators, snaring or trapping prey or firing projectiles into nematodes, rotifers, and other small animals that come near.

Figure 13.12   Many fungi produce spores.
Spores explode from the surface of a puffball fungus.

> **13.8   Fungi are not at all like plants. The fungal body is basically long strings of cells, often interconnected. Fungi reproduce both asexually and sexually. They obtain their nutrients by secreting digestive enzymes into their surroundings and then absorbing the digested molecules back into the fungal body.**

13.9 Kinds of Fungi

Fungi are an ancient group of organisms at least 400 million years old. There are nearly 74,000 described species, in five groups (table 13.3), and many more awaiting discovery. Many fungi are harmful because they decay, rot, and spoil many different materials as they obtain food and because they cause serious diseases in animals and particularly in plants. Other fungi, however, are extremely useful. The manufacture of both bread and beer depends on the biochemical activities of yeasts, single-celled fungi that produce abundant quantities of carbon dioxide and ethanol. Fungi are used on a major scale in industry to convert one complex organic molecule into another; many commercially important steroids are synthesized in this way.

The four fungal phyla, distinguished from one another primarily by their mode of sexual reproduction, are the zygomycetes, the ascomycetes, the basidiomycetes, and the chytridiomycetes. A fifth group, the imperfect fungi, is an artificial grouping of fungi in which sexual reproduction has not been observed; these organisms are assigned to an appropriate group once their mode of sexual reproduction is identified. Molecular data are contributing to our understanding of the fungal phylogeny and as additional molecular evidence is acquired, a new fungal phylogeny is likely to appear. However, it already seems clear that fungi are more closely related to animals than to plants, both diverging from a common ancestor.

Ecological Roles of Fungi as Decomposers

Fungi, together with bacteria, are the principal decomposers in the biosphere. They break down organic materials and return the substances locked in those molecules to circulation in the ecosystem. Fungi are virtually the only organisms capable of breaking down lignin, one of the major constituents of wood. By breaking down such substances, fungi release critical building blocks, such as carbon, nitrogen, and phosphorus, from the bodies of dead organisms and make them available to other organisms.

TABLE 13.3 FUNGI

Phylum	Typical Examples	Key Characteristics	Approximate Number of Living Species
Zygomycota	*Rhizopus* (black bread mold)	Reproduce sexually and asexually; multinucleate hyphae lack septa, except for reproductive structures; fusion of hyphae leads directly to formation of a zygote, in which meiosis occurs just before it germinates	1,050
Ascomycota	Yeasts, truffles, morels	Reproduce by sexual means; ascospores are formed inside a sac called an ascus; asexual reproduction is also common	32,000
Basidiomycota	Mushrooms, toadstools, rusts	Reproduce by sexual means; basidiospores are borne on club-shaped structures called basidia; the terminal hyphal cell that produces spores is called a basidium; asexual reproduction occurs occasionally	22,000
Chytridiomycota	*Allomyces*	Produce flagellated gametes (zoospores); predominately aquatic, some freshwater and some marine; oldest group of fungi	1,500
Imperfect fungi (not a phylum)	*Aspergillus*, *Penicillium*	Sexual reproduction has not been observed; most are thought to be ascomycetes that have lost the ability to reproduce sexually	17,000

In breaking down organic matter, some fungi attack living plants and animals as a source of organic molecules, whereas others attack dead ones. Fungi often act as disease-causing organisms for both plants and animals, and they are responsible for billions of dollars in agricultural losses every year. Not only are fungi the most harmful pests of living plants, but they also attack food products once they have been harvested and stored. In addition, fungi often secrete substances into the foods they are attacking that make these foods unpalatable or poisonous.

Commercial Uses

The same aggressive metabolism that makes fungi ecologically important has been put to commercial use in many ways. The manufacture of both bread and beer depends on the biochemical activities of yeasts, single-celled fungi that produce abundant quantities of ethanol and carbon dioxide. Cheese and wine achieve their delicate flavors because of the metabolic processes of certain fungi, and others make possible the manufacture of soy sauce. Vast industries depend on the biochemical manufacture of organic substances such as citric acid by fungi in culture, and yeasts are now used on a large scale to produce protein for the enrichment of animal food. Many antibiotics, including the first one that was used on a wide scale, penicillin, are derived from fungi.

Some fungi are used to convert one complex organic molecule into another, cleaning up toxic substances in the environment. For example, at least three species of fungi have been isolated that combine selenium, accumulated at the San Luis National Wildlife Refuge in California's San Joaquin Valley, with harmless volatile chemicals—thus removing it from the soil.

Fungal Associations

Two kinds of mutualistic associations between fungi and autotrophic organisms are ecologically important: mycorrhizae and lichens. In each case, a photosynthetic organism fixes atmo-

Figure 13.14   Lichens growing on a rock.

spheric carbon dioxide and thus makes organic material available to the fungi. The metabolic activities of the fungi, in turn, enhance the overall ability of the symbiotic association to exist in a particular habitat. Mycorrhizae are symbiotic associations between fungi and the roots of plants, where the fungal partner expedites the plant's absorption of essential nutrients such as phosphorus. Lichens are symbiotic associations between fungi and either green algae or cyanobacteria (figure 13.14). They are prominent nearly everywhere in the world, especially in unusually harsh habitats such as bare rock.

> **13.9   The fungal phyla are distinguished primarily by their modes of sexual reproduction. Fungi are key decomposers within almost all terrestrial ecosystems. In mycorrhizae and lichens, fungi form ecologically important associations with autotrophic organisms.**

Exploring Current Issues

Additional Resources

Go to your campus library or look online to find the following articles, which further develop some of the concepts found in this chapter.

Amabile-Cuevas, C.F. (2003). New antibiotics and new resistance: in many ways, the fight against antibiotic resistance is already lost; preventing bacterial diseases requires thoughtful new approaches. *American Scientist,* 91(2), 138.

Bhattacharya, D., H.S. Yoon, and J.D. Hacket. (2004). Photosynthetic eukaryotes unite: endosymbiosis connects the dots. *BioEssays,* 26(1), 50.

Pray, L. (2003). Microbial multicellularity: many species of bacteria live in a social, coordinated fashion, and they'll even die to keep it that way. *The Scientist,* 17(23), 20.

Smith, D.J. (2003). Partners for life. *World and I,* 18(4), 36.

Biology and Society Lecture: Investigating the Origin of Life

Few questions have fascinated humankind so intensively as the origin of life. There are both religious and scientific views about how life arose on earth. We limit ourselves to discussing scientific ones, proposals that are at least in principle subject to test and rejection. It seems clear that the earth itself was formed about 4.5 billion years ago. The oldest clear evidence of life, microfossils in ancient rock, are 2.5 billion years old. Life arose at a time when chemically rich oceans covered much of the earth. One scenario for the origin of life is that it originated spontaneously in this dilute hot chemical soup; another is that it arose in hydrothermal deep-sea vents. Yet another is that life arrived on earth from an extraterrestrial source. While hypotheses abound, little is known for certain about exactly how the first cells originated.

Find this lecture, delivered by the author to his class at Washington University, online at www.mhhe.com/tlwessentials/exp13.

Origin of Life

13.1 How Cells Arose

- Life on earth may have arisen in bubbles in the ocean. The "bubble model" suggests that biological molecules were captured in bubbles where they underwent chemical reactions leading to the origin of life (**figure 13.2**). The first cells may have formed from bubble-enclosed molecules, such as RNA.

Prokaryotes

13.2 The Simplest Organisms

- Prokaryotes have very simple internal structures, lacking nuclei and other membrane-bound compartments. The plasma membrane of prokaryotes is encased in a cell wall.  The cell wall of bacteria is made of peptidoglycan, and the cell wall of archaea lacks peptidoglycan, being made of protein and/or polysaccharides. Bacteria are divided into two groups, gram-positive and gram-negative, based on the construction of their cell walls (**figure 13.3**).

- Bacteria reproduce by splitting in two, called binary fission, and may exchange genetic information through conjugation (**figure 13.4**).

13.3 Comparing Prokaryotes to Eukaryotes

- Prokaryotes are different from eukaryotes in many ways, including that they lack interior compartments, such as nuclei, and are more metabolically diverse (**table 13.1**).

Viruses

13.4 Viruses Infect Organisms

- Viruses are not living organisms, but are parasitic chemicals that enter and replicate inside cells. They contain a nucleic acid core surrounded by a protein coat. Some viruses have an outer protein capsid or membranelike envelope (**figure 13.6**). Viruses cause many diseases, often spreading from animals to humans.

The Protists

13.5 The Origin of Eukaryotic Cells

- The endosymbiont theory proposes that energy-producing bacteria became incorporated into the cells of early eukaryotic cells, giving rise to mitochondria. Similarly, photosynthetic bacteria became incorporated, giving rise to chloroplasts (**figure 13.7**).

13.6 General Biology of Protists

- Protists are a very diverse group of eukaryotes, having diverse cell surfaces, modes of locomotion, and modes of acquiring nutrients. Some protists form cysts and most reproduce asexually except for times of stress. Some exist as single cells (**figure 13.8**), whereas others form colonies (**figure 13.9**) or aggregates, but only a few exhibit true multicellularity like that seen in plants, fungi, and animals.

13.7 Kinds of Protists

- The classification of the organisms in the kingdom Protista is in a state of flux. The 15 phyla of protists are often grouped into five categories based on modes of locomotion, photosynthetic capabilities, and spore-formation (**figure 13.10 and table 13.2**).

Fungi

13.8 A Fungus Is Not a Plant

- Fungi are nonmobile heterotrophs. The body of a fungus is composed of long slender filaments, called hyphae, that pack together to form a mycelium (**figure 13.11**). Most fungi have nonmotile sperm, unlike some plants. They have cell walls made of chitin, which is different from plant cell walls, and they undergo nuclear mitosis, where the nuclei divide but not the cell. Most fungal cells are separated by an incomplete wall called a septum, which allows cytoplasm to pass between cells.

- Sexual reproduction takes place between two genetically different "mating types" where two haploid hyphae fuse. Their nuclei can either remain separate, forming a heterokaryon, or they may fuse, forming a zygote, in a specialized reproductive structure which include gametangia, sporangia, or conidiophores.

- Fungi obtain nutrients through external digestion. They grow on their food, releasing enzymes that digest the food. The products of digestion are then absorbed by the fungus.

13.9 Kinds of Fungi

- There are four phyla of fungi— Zygomycota, Ascomycota, Basidiomycota, and Chytridiomycota— classified according to their mode of sexual reproduction. A fifth group, the imperfect fungi, is a "catch-all" category of fungi in which sexual reproduction has not been observed (**table 13.3**).

- Fungi play key roles in the environment. They are decomposers. They are involved in symbiotic associations with the roots of some plants, called mycorrhizae, and with cyanobacteria or algae, called lichens (**figure 13.14**).

1. While it is still unknown how the first cells formed, scientists suspect that the first active biological macromolecule was
   a. protein.
   c. RNA.
   b. DNA.
   d. carbohydrates.

2. Bacteria
   a. are prokaryotic.
   b. have been on the earth for at least 2.5 billion years.
   c. are the most abundant life-form on earth.
   d. All answers are correct.

3. Some species of prokaryotes are able to obtain energy by oxidizing inorganic chemicals. These organisms are called
   a. photoautotrophs.
   c. photoheterotrophs.
   b. chemoautotrophs.
   d. chemoheterotrophs.

4. Viruses are
   a. protein shells that contain DNA or RNA.
   b. simple eukaryotic cells.
   c. simple prokaryotic cells.
   d. living organisms.

5. One piece of supporting evidence for the endosymbiotic theory for the origin of eukaryotic cells is that
   a. eukaryotic cells have internal membranes.
   b. mitochondria and chloroplasts have their own DNA.
   c. Golgi bodies and endoplasmic reticulum represent ancestral cells.
   d. the nuclear membrane could only have come from another cell.

6. Many protists survive unfavorable environmental conditions by forming
   a. gametes.
   c. cysts.
   b. zygotes.
   d. aggregates.

7. The main problem in classifying protists is that
   a. all have a common lifestyle.
   b. all are unicellular.
   c. all are photoautotrophic.
   d. any organism that is not a plant, animal, or fungus is a protist.

8. The main body of a fungus is the
   a. hyphae.
   b. septa.
   c. mushroom.
   d. mycelium.

9. Fungi reproduce
   a. both sexually and asexually.
   b. only sexually.
   c. only asexually.
   d. by fragmentation.

10. Lichens are mutualistic associations between
    a. plants and fungi.
    b. algae and fungi.
    c. termites and fungi.
    d. coral and fungi.

1. **Figure 13.6** Viruses have dramatically different forms and shapes. What are the consistent features of all of the viruses shown?

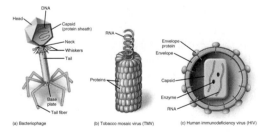

2. **Figure 13.9** Are the *Volvox* shown here multicellular organisms, or numerous single-celled organisms working together?

Origin of Life   Why can't we prove how life began on earth?

Prokaryotes   How do prokaryotes obtain the energy for life?

Viruses   Why do we say that bacteria are alive but viruses are not?

The Protists   Compare the methods of movement found in five major groups of protists.

Fungi   Describe three ways that fungi cooperate with other organisms.

Visit the Online Learning Center for this chapter at www.mhhe.com/tlwessentials/ch13 for quizzes, animations, interactive learning exercises, and other study tools. At the site you will also find extended answers to the end-of-chapter questions.

14
Evolution of Plants

Plants are thought to be descendants of green algae. Plants first invaded the land over 455 million years ago in partnership with fungi—the most ancient surviving plants have mycorrhizal roots. One of the many challenges posed by the terrestrial environment was the difficulty of finding a mate when anchored to one spot. The solution adopted by most early plants was for male individuals to cast their gametes—pollen—into the wind, and let air currents carry pollen grains to nearby female plants. This strategy works particularly well in dense stands, in which there are many nearby individuals of the same species. The massive redwood tree seen here, one of the largest individual organisms living on land, grows in dense stands and is wind pollinated. An alternative solution proved even better, however. Plants evolved flowers, devices to attract insects. When insects visit the flower to obtain nectar, they become coated with pollen. When they then visit another flower seeking more nectar, they deposit some of this pollen, pollinating that flower. It doesn't matter how far apart the two plants are from each other—the insect will seek them out. In this chapter, you will explore this and other evolutionary challenges met by plants as they colonized the land.

14.1 Adapting to Terrestrial Living

Plants are complex multicellular organisms that are terrestrial **autotrophs**—that is, they occur almost exclusively on land and feed themselves by photosynthesis. The name *autotroph* comes from the Greek, *autos,* self, and *trophos,* feeder. Today, plants are the dominant organisms on the surface of the earth (figure 14.1). An estimated 263,500 species are now in existence, covering every part of the terrestrial landscape except the extreme polar regions and the highest mountaintops. In this chapter, we examine how plants adapted to life on land.

The green algae that were probably the ancestors of today's plants are aquatic organisms that are not well adapted to living on land. Before their descendants could live on land, they had to overcome many environmental challenges. For example, they had to absorb minerals from the rocky surface. They had to find a means of conserving water. They had to develop a way to reproduce on land.

Absorbing Minerals

Plants require relatively large amounts of six inorganic minerals: nitrogen, potassium, calcium, phosphorus, magnesium, and sulfur. Each of these minerals constitutes 1% or more of a plant's dry weight. Algae absorb these minerals from water, but where is a plant on land to get them? From the soil. Soil is the weathered outer layer of the earth's crust. It is composed of a mixture of ingredients, which may include sand, rocks, clay, silt, humus (partly decayed organic material), and various other forms of mineral and organic matter. The soil is also rich in microorganisms that

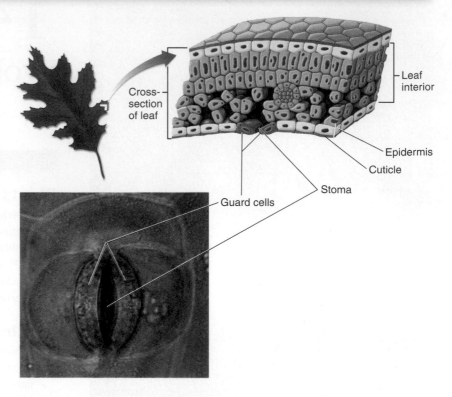

Figure 14.2 A stoma.

A stoma is a passage through the cuticle covering the epidermis of a leaf (400×). Water and oxygen pass out through the stoma, and carbon dioxide enters by the same portal. The cells flanking the stoma are called guard cells.

break down and recycle organic debris. Plants absorb these materials, along with water, through their *roots* (described in chapter 26). Most roots are found in topsoil, which is a mixture of mineral particles, living organisms, and humus. When topsoil is lost due to erosion, the soil loses its ability to hold water and nutrients.

The first plants seem to have developed a special relationship with fungi, which was a key factor in their ability to absorb minerals in terrestrial habitats. Within the roots of many early fossil plants like *Cooksonia* (see figure 14.7) and *Rhynia,* fungi can be seen living intimately within and among the root cells. These kinds of symbiotic associations are called **mycorrhizae.** In plants with mycorrhizae, the fungi enable the plant to take up phosphorus, zinc, copper, and other nutrients from rocky soil, while the plant supplies organic molecules to the fungus.

Conserving Water

One of the key challenges to living on land is to avoid drying out. To solve this problem, plants have a watertight outer covering called a **cuticle.** The covering is formed from a waxy substance that is impermeable to water. Like the wax on a shiny car, the cuticle prevents water from entering or leaving the stem or leaves. Water enters the plant only from the roots, while the cuticle prevents water loss to the air. Passages do ex-

Figure 14.1 There are many kinds of plants.

The plant kingdom is astonishingly diverse, often displaying remarkable adaptations.

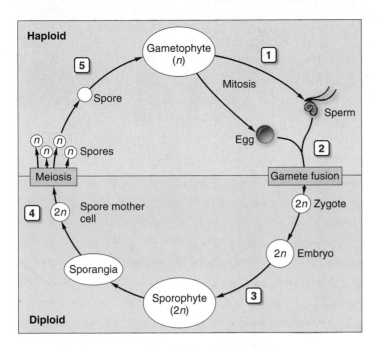

Figure 14.3 Generalized plant life cycle.
In a plant life cycle, there is alternation of generations, in which a diploid generation alternates with a haploid one. Gametophytes, which are haploid (*n*), alternate with sporophytes, which are diploid (2*n*). (1) Gametophytes give rise by mitosis to sperm and eggs. (2) The sperm and egg ultimately come together to produce the first diploid cell of the sporophyte generation, the zygote. (3) The zygote undergoes cell division, ultimately forming the sporophyte. (4) Meiosis takes place within the sporangia, the spore-producing organs of the sporophyte, resulting in the production of the spores, which are haploid and are the first cells (5) of gametophyte generations.

ist through the cuticle, in the form of specialized pores called **stomata** (singular, **stoma**) in the leaves and sometimes the green portions of the stems (figure 14.2). Stomata, which occur on at least some portions of all plants except liverworts, allow carbon dioxide to pass into the plant bodies for photosynthesis and allow water and oxygen gas to pass out of them. The guard cells that border stomata swell (opening the stoma) and shrink (closing the stoma) with the movement of water into and out of them, thus controlling the loss of water from the leaf while allowing the entrance of carbon dioxide. In most plants, water enters through the roots as a liquid and exits through the underside of the leaves as water vapor.

Reproducing on Land

To reproduce sexually on land, it is necessary to pass gametes from one individual to another, and because plants cannot move about, it is necessary that the gametes avoid drying out while they are transferred by wind or insects. In the first plants, the eggs were surrounded by a jacket of cells, and a film of water was required for the sperm to swim to the egg and fertilize it. Today, mosses still reproduce this way. However, soon after mosses evolved, changes occurred in the plant life cycle that favored the development of **spores,** which are reproductive cells very resistant to drying out.

(a)

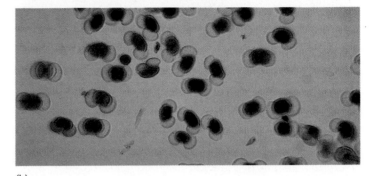

(b)

Figure 14.4 Two types of gametophytes.
(*a*) This gametophyte of a moss (a nonvascular plant) is green and free-living. (*b*) Male gametophytes (pollen) of a pine (a vascular plant) are barely large enough to be visible to the naked eye (200×).

Changing the Life Cycle. Among many algae, haploid cells occupy the major portion of the life cycle. The zygote formed by the fusion of gametes is the only diploid cell, and it immediately undergoes meiosis to form haploid cells again. In early plants, meiosis became delayed, so that for a significant portion of the life cycle, cells were diploid. This resulted in an **alternation of generations,** in which a diploid generation alternates with a haploid one (figure 14.3). Botanists call the diploid generation the **sporophyte** because it forms haploid spores by meiosis. The haploid generation is called the **gametophyte** because it forms haploid gametes by mitosis. When you look at primitive plants such as mosses and liverworts, you see mostly gametophyte tissue—the sporophytes are smaller brown structures attached to or enclosed within the tissues of the larger gametophyte (figure 14.4*a*). When you look at plants that evolved later, you see mostly sporophyte tissue. The gametophytes of these plants are always much smaller than the sporophytes (figure 14.4*b*) and are often enclosed within sporophyte tissues.

> **14.1** Plants are multicellular terrestrial photosynthesizers, evolved from green algae, that adapted to life on land by developing ways to absorb minerals in partnership with fungi, to conserve water with watertight coverings, and to reproduce with spores and seeds.

14.2 Plant Evolution

Once plants became established on land, they gradually developed many other features that aided their evolutionary success in this new, demanding habitat. For example, among the first plants there was no fundamental difference between the aboveground and the belowground parts. Later, roots and shoots with specialized structures evolved, each suited to its particular below- or aboveground environment. The evolution of specialized *vascular tissue* allowed plants to grow larger. Compare the size of a moss, an early plant, with a more highly evolved tree on which it grows.

As we explore plant diversity, we will examine several key evolutionary innovations that have given rise to the wide variety of plants we see today (table 14.1). While other interesting and important changes also arose, the key innovations discussed here serve to highlight the evolutionary trends exhibited by the plant kingdom (figure 14.5).

Figure 14.5   The evolution of plants.

TABLE 14.1	PLANT PHYLA			
Phylum	**Typical Examples**		**Key Characteristics**	**Approximate Number of Living Species**
NONVASCULAR PLANTS				
Hepaticophyta (liverworts)	*Marchantia*		Without true vascular tissues; lack true roots and leaves; live in moist habitats and obtain water and nutrients by osmosis and diffusion; require water for fertilization; gametophyte is dominant structure in the life cycle; the three phyla were once grouped together	15,600
Anthocerophyta (hornworts)	*Anthoceros*			
Bryophyta (mosses)	*Polytrichum, Sphagnum* (hairy cap and peat moss)			
SEEDLESS VASCULAR PLANTS				
Lycophyta (lycopods)	*Lycopodium* (club mosses)		Seedless vascular plants similar in appearance to mosses but diploid; require water for fertilization; sporophyte is dominant structure in life cycle; found in moist woodland habitats	1,150
Pterophyta (ferns)	*Azolla, Sphaeropteris* (water and tree ferns) *Equisetum* (horsetails) *Psilotum* (whisk ferns)		Seedless vascular plants; require water for fertilization; sporophytes diverse in form and dominate the life cycle	11,000

Four key evolutionary innovations serve to trace the evolution of the plant kingdom:

1. **Alternation of generations.** Although algae exhibit a haploid and diploid phase, the diploid phase is not a significant portion of their life cycle. By contrast, even in early plants the diploid sporophyte is a larger structure and offers protection for the egg and developing embryo. The dominance of the sporophyte, both in size and the proportion of time devoted to it in the life cycle, becomes greater throughout the evolutionary history of plants.

2. **Vascular tissue.** A second key innovation was the emergence of vascular tissue. Vascular tissue transports water and nutrients throughout the plant body. With the evolution of vascular tissue, plants were able to supply the upper portions of their bodies with water absorbed from the soil, allowing the plants to grow larger and in drier conditions.

3. **Seeds.** The evolution of seeds was a key innovation that allowed plants to dominate their terrestrial environments. Seeds provide nutrients and a tough, durable cover that protects the embryo until it encounters favorable growing conditions.

4. **Flowers and fruits.** The evolution of flowers and fruits were key innovations that improved the chances of successful mating in sedentary organisms and facilitated the dispersal of their seeds. Flowers both protected the egg and improved the odds of its fertilization, allowing plants that were located at considerable distances to mate successfully. Fruit, which surrounds the seed and aids in its dispersal, allows plant species to better invade new and possibly more favorable environments.

14.2 Plants evolved from freshwater green algae and eventually developed more dominant diploid phases of the life cycle, conducting systems of vascular tissue, seeds that protected the embryo, and flowers and fruits that aided in fertilization and distribution of the seeds.

TABLE 14.1	(CONTINUED)			
Phylum	**Typical Examples**		**Key Characteristics**	**Approximate Number of Living Species**
SEED PLANTS				
Coniferophyta (conifers)	Pines, spruce, fir, redwood, cedar		Gymnosperms; wind pollinated; ovules partially exposed at time of pollination; flowerless; seeds are dispersed by the wind; sperm lack flagella; sporophyte is dominant structure in life cycle; leaves are needlelike or scalelike; most species are evergreens and live in dense stands; among the most common trees on earth	550
Cycadophyta (cycads)	Cycads, sago palms		Gymnosperms; wind pollination or possibly insect-pollination; very slow growing, palmlike trees; sperm have flagella but reach vicinity of egg by a pollen tube; sporophyte dominant in the life cycle	140
Gnetophyta (shrub teas)	Mormon tea, *Welwitschia*		Gymnosperms; nonmotile sperm; shrubs and vines; wind pollination and possibly insect pollination; sporophyte is dominant in the life cycle	70
Ginkgophyta (ginkgo)	Ginkgo trees		Gymnosperms; fanlike leaves that are dropped in winter (deciduous); seeds fleshy and ill-scented; motile sperm; trees are either male or female; sporophyte is dominant in the life cycle	1
Anthophyta (flowering plants, also called angiosperms)	Oak trees, corn, wheat, roses		Flowering; pollination by wind, animal, and water; characterized by ovules that are fully enclosed by the carpel; fertilization involves two sperm nuclei; one forms the embryo, the other fuses with the polar body to form endosperm for the seed; after fertilization, carpels and the fertilized ovules (now seeds) mature to become fruit; sporophyte is dominant in life cycle	235,000

14.3 Nonvascular Plants

Liverworts and Hornworts

The first successful land plants had no vascular system—no tubes or pipes to transport water and nutrients throughout the plant. This greatly limited the maximum size of the plant body because all materials had to be transported by osmosis and diffusion. Only two phyla of living plants, the **liverworts** (phylum Hepaticophyta) and the **hornworts** (phylum Anthocerophyta), completely lack a vascular system. The word *wort* meant *herb* in medieval Anglo-Saxon when these plants were named. Liverworts are the simplest of all living plants. About 6,000 species of liverworts and 100 species of hornworts survive today, usually growing in moist and shady places.

Primitive Conducting Systems: Mosses

Another phylum of plants, the **mosses** (phylum Bryophyta), were the first plants to evolve strands of specialized cells that conduct water and carbohydrates up the stem of the gametophyte. The conducting cells do not have specialized wall thickenings, instead they are like nonrigid pipes and cannot carry water very high. Because these conducting cells could at the most be considered a primitive vascular system, mosses are usually grouped by botanists with the liverworts and hornworts as "nonvascular" plants. Today about 9,500 species of mosses grow in moist places all over the world. The life cycle of a moss is illustrated in figure 14.6.

> **14.3** While liverworts and hornworts totally lack a vascular system, mosses have simple soft strands of conducting cells.

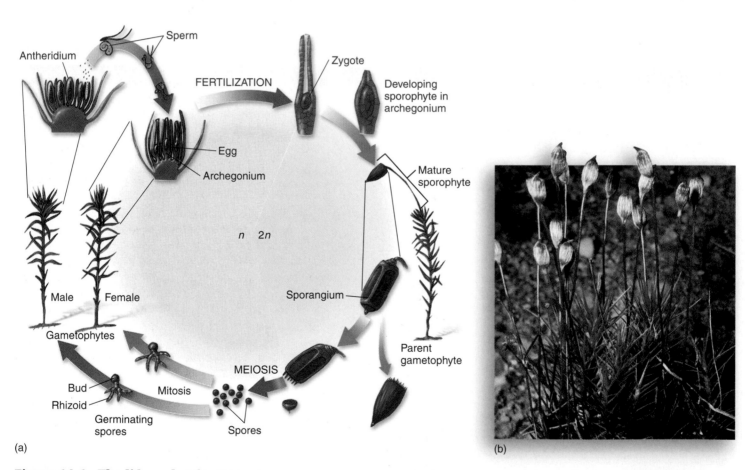

(a)

(b)

Figure 14.6 The life cycle of a moss.

(a) On the gametophytes, which are haploid, sperm are released from each antheridium (sperm-producing structure). They then swim through free water to an archegonium (egg-producing structure) and down its neck to the egg. Fertilization takes place there; the resulting zygote develops into a sporophyte, which is diploid. The sporophyte grows out of the archegonium and differentiates into a slender, basal stalk with a swollen capsule, the sporangium, at its apex. The capsule is covered, at least at first, with a cap formed from the swollen archegonium. The sporophyte grows on the gametophyte and eventually produces spores as a result of meiosis. The spores are shed from the capsule. The spores germinate, giving rise to gametophytes. The gametophytes initially are threadlike; they grow along the ground. Ultimately, buds form on them, from which leafy gametophytes arise. (b) In this hair-cup moss, *Polytrichum,* the small, green leaf-like projections at the base belong to the gametophyte, while each of the large, yellowish brown stalks, with the capsule at its summit, is a sporophyte. Although moss sporophytes may be green and carry out a limited amount of photosynthesis when they are immature, they are soon completely dependent nutritionally on the gametophyte.

14.4 The Evolution of Vascular Tissue

The remaining seven phyla of plants, which have efficient vascular systems made of highly specialized cells, are called **vascular plants.** The first vascular plant appeared approximately 430 million years ago, but only incomplete fossils have been found. The first vascular plants for which we have relatively complete fossils, the extinct phylum Rhyniophyta, lived 410 million years ago. Among them is the oldest known vascular plant, *Cooksonia,* with branched, leafless shoots that formed spores at their tips (figure 14.7).

Cooksonia and the other early plants that followed it became successful colonizers of the land through the development of efficient water- and food-conducting systems known as **vascular tissues** (Latin, *vasculum,* vessel or duct) (figure 14.8). These tissues, discussed in detail in chapter 26, consist of strands of specialized cylindrical or elongated cells that form a network throughout a plant, extending from near the tips of the roots, through the stems, and into the leaves. The presence of a cuticle and stomata are also characteristic of vascular plants.

Early vascular plants grew by cell division at the tips of the stem and roots. Imagine stacking dishes—the stack can get taller but not wider! This sort of growth is called **primary growth** and was quite successful. During the so-called Coal Age (between 250 and 215 million years ago), when much of the world's fossil fuel was formed, the lowland swamps

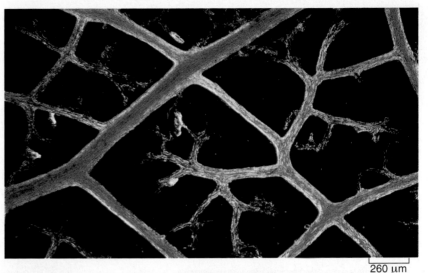

Figure 14.8 The vascular system of a leaf.

The veins of a vascular plant contain strands of specialized cells for conducting carbohydrates, as well as water containing dissolved minerals. The veins run from the tips of the roots to the tips of the shoots and throughout the leaves, as shown in this greatly magnified photomicrograph of the veins of a leaf.

that covered Europe and North America were dominated by an early form of seedless tree called a lycophyte. Lycophyte trees grew to heights of 10 to 35 meters (33 to 115 ft), and their trunks did not branch until they attained most of their total height. The pace of evolution was rapid during this period, for the world's climate was changing, growing dryer and colder. As the world's swamplands began to dry up, the lycophyte trees vanished, disappearing abruptly from the fossil record. They were replaced by tree-sized ferns, a form of vascular plant that will be described in detail in the next section. Tree ferns grew to heights of more than 20 meters (66 ft) with trunks 30 centimeters (12 in) thick. Like the lycophytes, the trunks of tree ferns were formed entirely by primary growth.

About 380 million years ago, vascular plants developed a new pattern of growth, in which a cylinder of cells beneath the bark divides, producing new cells in regions around the plant's periphery. This growth is called **secondary growth.** Secondary growth makes it possible for a plant stem to increase in diameter. Only after the evolution of secondary growth could vascular plants become thick-trunked and therefore tall. Redwood trees today reach heights of up to 117 meters (384 ft) and trunk diameters in excess of 11 meters (36 ft). This evolutionary advance made possible the dominance of the tall forests that today cover northern North America. You are familiar with the product of plant secondary growth as **wood.** The growth rings that are so visible in cross sections of trees are zones of secondary growth (spring–summer) spaced by zones of little growth (fall–winter).

Sporangia

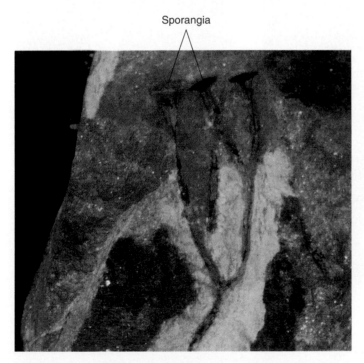

Figure 14.7 The earliest vascular plant.

The earliest vascular plant of which we have complete fossils is *Cooksonia.* This fossil shows a plant that lived some 410 million years ago; its upright branched stems terminated in spore-producing sporangia at the tips.

14.4 Vascular plants have specialized vascular tissue composed of hollow tubes that conduct water to the leaves and cells that form cylinders that conduct food from them.

14.5 Seedless Vascular Plants

The earliest vascular plants lacked seeds, and two of the seven phyla of modern-day vascular plants do not have them. The two phyla of living seedless vascular plants include the ferns, phylum Pterophyta, which includes the horsetails and the whisk ferns, and the club mosses, phylum Lycophyta (figure 14.9). These phyla have free-swimming sperm that require the presence of free water for fertilization.

By far the most abundant of seedless vascular plants are the **ferns,** with about 11,000 living species. Ferns are found throughout the world, although they are much more abundant in the tropics than elsewhere. Many are small, only a few centimeters in diameter, but some of the largest plants that live today are also ferns. Descendants of ancient tree ferns, they can have trunks more than 24 meters (79 ft) tall and leaves up to 5 meters (16 ft) long!

The Life of a Fern

In ferns, the life cycle of plants begins a revolutionary change that culminates later with seed plants. Nonvascular plants like mosses are made largely of gametophyte (haploid) tissue. Vascular seedless plants like ferns have both gametophyte and sporophyte individuals, each independent and self-sufficient. The gametophyte produces eggs and sperm; after sperm swim through water and fertilize the egg, the zygote grows into a sporophyte. The sporophyte bears and releases haploid **spores** that float to the ground and germinate,

Figure 14.9   Seedless vascular plants.

(a) A tree fern in the forests of Malaysia (phylum Pterophyta). The ferns are by far the largest group of spore-producing vascular plants. (b) Ferns on the floor of a redwood forest. (c) A whisk fern. Whisk ferns have no roots or leaves. (d) A horsetail, *Equisetum telmateia.* This species forms two kinds of erect stems; one is green and photosynthetic, and the other, which terminates in a spore-producing "cone," is mostly light brown. (e) The club moss *Lycopodium lucidulum,* recently renamed *Huperzia lucidula* (phylum Lycophyta). Although superficially similar to the gametophytes of mosses, the conspicuous club moss plants shown here are sporophytes.

growing into haploid gametophytes. The fern gametophytes are small, thin, heart-shaped photosynthetic plants, usually no more than a centimeter in length, that live in moist places. The fern sporophytes are much larger and more complex, with long vertical leaves called **fronds.** When you see a fern, you are almost always looking at the sporophyte. The fern life cycle is described in figure 14.10.

14.5 Ferns are among the vascular plants that lack seeds, reproducing with spores as nonvascular plants do.

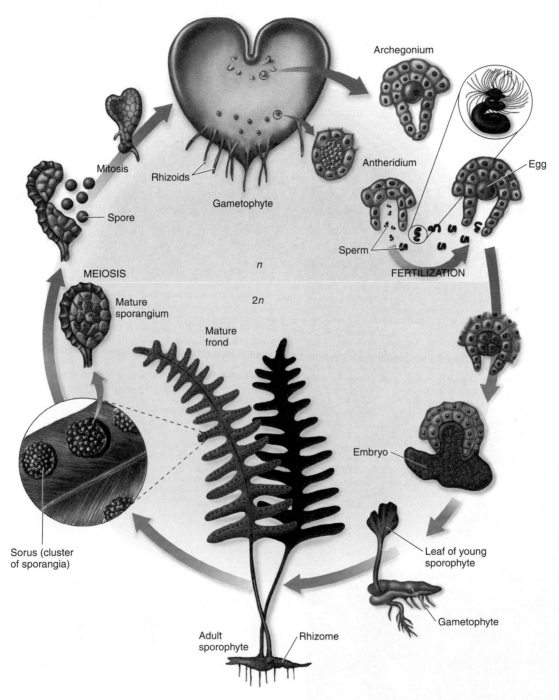

Figure 14.10 Fern life cycle.

The gametophytes, which are haploid, grow in moist places. Rhizoids (anchoring structures) project from their lower surface. Eggs develop in an archegonium, and sperm develop in an antheridium, both located on the gametophyte's lower surface. The sperm, when released, swim through free water to the mouth of the archegonium, entering and fertilizing the single egg. Following the fusion of egg and sperm to form a zygote—the first cell of the diploid sporophyte generation—the zygote starts to grow within the archegonium. Eventually, the sporophyte, the fern plant, becomes much larger than the gametophyte. Most ferns have more or less horizontal stems, called rhizomes, that creep along below ground. On the sporophyte's leaves, called fronds, occur clusters (called sori; singular, sorus) of sporangia, within which meiosis occurs and spores are formed. The release of these spores, which is explosive in many ferns, and their germination lead to the development of new gametophytes.

14.6 Evolution of Seed Plants

A key evolutionary advance among the vascular plants was the development of a protective cover for the embryo called a **seed** (figure 14.11). The seed is a crucial adaptation to life on land because it protects the embryonic plant when it is at its most vulnerable stage. The evolution of the seed was a critical step in the domination of the land by plants.

The change in the life cycle of vascular plants in favor of the sporophyte (diploid) generation reaches its full force with the advent of the seed plants. Seed plants produce two kinds of gametophytes—male and female, each of which consists of just a few cells. Both kinds of gametophytes develop separately within the sporophyte and are completely dependent on it for their nutrition. Male gametophytes, commonly referred to as **pollen grains,** arise from **microspores.** The pollen grains become mature when sperm are produced. The sperm are carried to the egg in the female gametophyte without using free water in the environment. A female gametophyte contains the egg and develops from a **megaspore** produced within an **ovule.** The transfer of pollen to an ovule by insects, wind, or other agents is referred to as **pollination.** The pollen grain then cracks open and sprouts, or germinates, and the pollen tube, containing the sperm cells, grows out, transporting the sperm directly to the egg. Thus there is no need for free water in the pollination and fertilization process.

Botanists generally agree that all seed plants are derived from a single common ancestor. There are five living phyla. In four of them, collectively called the **gymnosperms** (Greek, *gymnos,* naked, and *sperma,* seed), the ovules are not completely enclosed by sporophyte tissue at the time of pollination. Gymnosperms were the first seed plants. From gymnosperms evolved the fifth group of seed plants, called **angiosperms** (Greek, *angion,* vessel, and *sperma,* seed), phylum Anthophyta. Angiosperms, or flowering plants, are the most recently evolved of all the plant phyla. Angiosperms differ from all gymnosperms in that their ovules are completely enclosed by a vessel of sporophyte tissue in the flower called the **carpel** at the time they are pollinated.

The Structure of a Seed

A seed has three parts: (1) a sporophyte plant embryo, (2) a source of food for the developing embryo called **endosperm,** and (3) a drought-resistant protective cover (figure 14.12). In some seeds, the endosperm is used up during the development of the embryo and food is stored in thick "leaflike" structures called **cotyledons.** Seeds are one way in which plants, anchored by their roots to one place in the ground, solve the

Figure 14.11   A seed plant.
The seeds of this cycad, like all seeds, consist of a plant embryo and a protective covering. A cycad is a gymnosperm (naked-seeded plant), and its seeds develop out in the open on the edges of the cone scales.

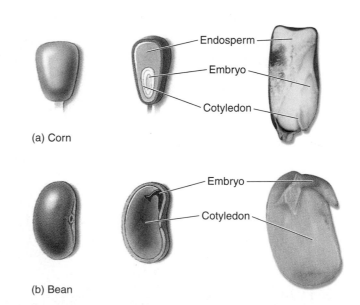

(a) Corn

(b) Bean

Figure 14.12   Basic structure of seeds.
A seed contains a sporophyte (diploid) embryo and a source of food, either endosperm (*a*) or food stored in the cotyledons (*b*). A seed coat, formed of sporophytic tissue from the parent, surrounds the seed and protects the embryo.

Figure 14.13   Seeds allow plants to bypass the dry season.
When it does rain, seeds can germinate, and plants can grow rapidly to take advantage of the relatively short periods when water is available. This palo verde desert tree (*Cercidium floridum*) has tough seeds (*inset*) that germinate only after they are cracked. Rains leach out the chemicals in the seed coats that inhibit germination, and the hard coats of the seeds may be cracked when they are washed down gullies in temporary floods.

problem of dispersing their progeny to new locations. The hard cover of the seed (formed from the tissue of the parent plant) protects the seed while it travels to a new location. The seed travels by many means such as by air, water, and animals. Many airborne seeds have devices to aid in carrying them farther. Most species of pine, for example, have seeds with thin flat wings attached. These wings help catch air currents, which carry the seeds to new areas. Seed dispersal will be discussed further in chapter 27.

Once a seed has fallen to the ground, it may lie there, dormant, for many years. When conditions are favorable, however, and particularly when moisture is present, the seed germinates and begins to grow into a young plant (figure 14.13). Most seeds have abundant food stored in them to provide a ready source of energy for the new plant as it starts its growth.

The advent of seeds had an enormous influence on the evolution of plants. Seeds have greatly improved the adaptation of plants to living on land in at least four respects:

1. **Dispersal.** Most important, seeds facilitate the migration and dispersal of plant offspring into new habitats.

2. **Dormancy.** Seeds permit plants to postpone development when conditions are unfavorable, as during a drought, and to remain dormant until conditions improve.

3. **Germination.** By making the reinitiation of development dependent upon environmental factors such as temperature, seeds permit the course of embryonic development to be synchronized with critical aspects of the plant's habitat, such as the season of the year.

4. **Nourishment.** The seed offers the young plant nourishment during the critical period just after germination, when the seedling must establish itself.

14.6   A seed is a dormant diploid embryo encased with food reserves in a hard protective coat. Seeds play critical roles in improving a plant's chances of successfully reproducing in a varied environment.

14.7 Gymnosperms

Four phyla constitute the gymnosperms: the conifers (Coniferophyta), the cycads (Cycadophyta), the gnetophytes (Gnetophyta), and the ginkgo (Ginkgophyta). The conifers are the most familiar of the four phyla of gymnosperms and include pine, spruce, hemlock, cedar, redwood, yew, cypress, and fir trees (figure 14.14). **Conifers** are trees that produce their seeds in **cones.** The seeds (ovules) of conifers develop on scales within the cones and are exposed at the time of pollination. Most of the conifers have needlelike leaves, an evolutionary adaptation for retarding water loss. Conifers are often found growing in moderately dry regions of the world, including the vast taiga forests of the northern latitudes. Many are very important as sources of timber and pulp.

There are about 550 living species of conifers. The tallest living vascular plant, the coastal sequoia (*Sequoia sempervirens*), found in coastal California and Oregon, is a conifer and reaches over 100 meters (328 ft). The biggest redwood, however, is the mountain sequoia redwood species (*Sequoiadendron gigantea*) of the Sierra Nevadas. The largest individual tree is nicknamed after General Sherman of the Civil War, and it stands more than 83 meters (274 ft) tall while measuring 31 meters (102 ft) around its base. Another much smaller type of conifer, the bristlecone pines in Nevada, may be the oldest trees in the world—about 5,000 years old.

The other three gymnosperm phyla (figure 14.15) are much less widespread. Cycads, the predominant land plant in the golden age of dinosaurs, the Jurassic period (213–144

Figure 14.14 A familiar group of gymnosperms: conifers (phylum Coniferophyta).

These douglas fir trees, a type of conifer, often occur in vast forests.

(a) (b) (c)

Figure 14.15 Three more phyla of gymnosperms.

(a) An African cycad, *Encephalartos transvenosus*, phylum Cycadophyta. The cycads have fernlike leaves and seed-forming cones, like the ones shown here. (b) *Welwitschia mirabilis*, phylum Gnetophyta, is found in the extremely dry deserts of southwestern Africa. In *Welwitschia*, two enormous, beltlike leaves grow from a circular zone of cell division that surrounds the apex of the carrot-shaped root. (c) Maidenhair tree, *Ginkgo biloba*, the only living representative of the phylum Ginkgophyta, a group of plants that was abundant 200 million years ago. Among living seed plants, only the cycads and ginkgo have swimming sperm.

million years ago), have short stems and palmlike leaves. They are still widespread throughout the tropics. There is only one living species of ginkgo, the maidenhair tree, which has fan-shaped leaves shed in the autumn. Because ginkgos are resistant to air pollution, they are commonly planted along city streets. The gnetophytes are thought to be the most closely related to angiosperms. This phylum contains only three kinds of plants, all unusual. One of them is perhaps the most bizarre of all plants, *Welwitschia*, which grows on the exposed sands of the harsh Namibian Desert of southwestern Africa. *Welwitschia* acts like a plant standing on its head! Its two beltlike, leathery leaves are generated continuously from their base, splitting as they grow out over the desert sands.

The fossil record indicates that members of the ginkgo phylum were once widely distributed, particularly in the Northern Hemisphere; today, only one living species, the maidenhair tree (*Ginkgo biloba*), remains. The reproductive structures of ginkgos are produced on separate trees. The fleshy outer coverings of the seeds of female ginkgo plants exude the foul smell of rancid butter caused by butyric and isobutyric acids. In many Asian countries, however, the seeds are considered a delicacy. In Western countries, because of the seed odor, male plants vegetatively propagated are preferred for cultivation.

The Life of a Gymnosperm

We will examine conifers as typical gymnosperms. The conifer life cycle is illustrated in figure 14.16. Conifer trees form two kinds of cones. Seed cones contain the female gametophytes, with their egg cells; pollen cones contain pollen grains. Conifer pollen grains are small and light and are carried by the wind to seed cones. Each pollen grain has a pair of air sacs to help carry it in the wind. Because it is very unlikely that any particular pollen grain will succeed in being carried to a seed cone (the wind can take it anywhere), a great many pollen grains are needed to be sure that at least a few succeed in pollinating seed cones. For this reason, pollen grains are shed from their cones in huge quantities, often appearing as a sticky yellow layer on the surfaces of ponds and lakes—and even on windshields.

When a grain of pollen settles down on a scale of a female cone, a slender tube grows out of the pollen cell into the scale, delivering the male gamete (the sperm cell) to the female gametophyte containing the egg, or ovum. Fertilization occurs when the sperm cell fuses with the egg, forming a zygote. This zygote is the beginning of the sporophyte generation. What happens next is the essential improvement in

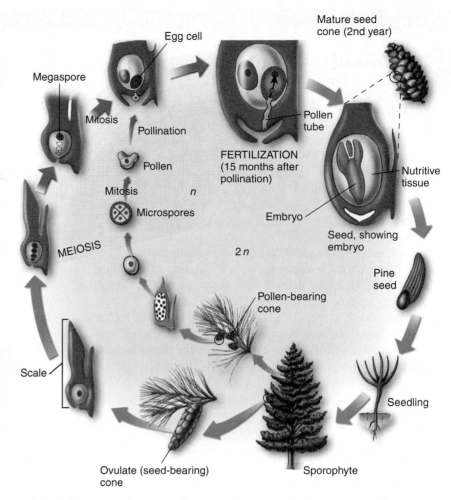

Figure 14.16   Life cycle of a conifer.
In all seed plants, the gametophyte generation is greatly reduced. In conifers such as pines, the relatively delicate pollen-bearing cones contain microspores, which give rise to pollen grains, the male gametophytes. The familiar seed-bearing cones of pines are much heavier and more substantial structures than the pollen-bearing cones. Two ovules, and ultimately two seeds, are borne on the upper surface of each scale, which contains the megaspores that give rise to the female gametophytes. After a pollen grain has reached a scale, it germinates, and a slender pollen tube grows toward the egg. When the pollen tube grows to the vicinity of the female gametophyte, sperm are released, fertilizing the egg and producing a zygote there. The development of the zygote into an embryo takes place within the ovule, which matures into a seed. Eventually, the seed falls from the cone and germinates, the embryo resuming growth and becoming a new pine tree.

reproduction achieved by seed plants. Instead of the zygote growing immediately into an adult sporophyte—just as you grow directly into an adult from a fertilized zygote—the fertilized ovule forms a seed. The seed can then be dispersed into new habitats.

> **14.7   Gymnosperms are seed plants in which the ovules are not completely enclosed by diploid tissue at pollination. Gymnosperms do not have flowers.**

14.8 Rise of the Angiosperms

Angiosperms, plants in which the ovule is completely enclosed by sporophyte tissue when it is fertilized, are the most successful of all plants. Ninety percent of all living plants are angiosperms, over 235,000 species, including many trees, shrubs, herbs, grasses, vegetables, and grains—in short, nearly all of the plants that we see every day. Virtually all of our food is derived, directly or indirectly, from angiosperms. In fact, more than half of the calories we consume come from just three species: rice, corn, and wheat.

In a very real sense, the remarkable evolutionary success of the angiosperms is the apparent culmination of the plant kingdom's adaptation to life on land, although plants continue to evolve. Angiosperms successfully meet the last difficult challenge posed by terrestrial living: the inherent conflict between the need to obtain nutrients (solved by roots, which anchor the plant to one place) and the need to find mates (solved by accessing plants of the same species). This challenge has never really been overcome by gymnosperms, whose pollen grains are carried passively by the wind on the chance that they might by luck encounter a female cone. Think about how inefficient this is! Angiosperms are also able to deliver their pollen directly, as if in an addressed envelope, from one individual of a species to another. How? *By inducing insects and other animals to carry it for them!* The tool that makes this animal-dictated pollination possible, the great advance of the angiosperms, is the flower. While some very successful later-evolving angiosperms like grasses have reverted to wind pollination, the directed pollination of flowering plants has led to phenomenal evolutionary success.

The Flower

Flowers are the reproductive organs of angiosperm plants. A flower is a sophisticated pollination machine. It employs bright colors to attract the attention of insects (or birds or small mammals), nectar to induce the insect to enter the flower, and structures that coat the insect with pollen grains while it is visiting. Then, when the insect visits another flower, it carries the pollen with it into that flower.

The basic structure of a flower consists of four concentric circles (figure 14.17), or **whorls,** connected to a base called the **receptacle:**

1. The outermost whorl, called the **sepals** of the flower, typically serves to protect the flower from physical damage. Sepals are in effect modified leaves that protect the flower while it is a bud.
2. The second whorl, called the **petals** of the flower, serves to attract particular pollinators. Petals have particular pigments, often vividly colored.
3. The third whorl, called the **stamens** of the flower, contains the "male" parts that produce the pollen

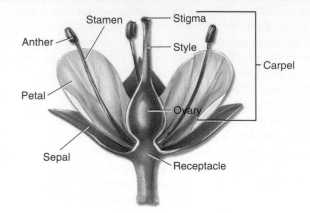

(a)

(b)

Figure 14.17 An angiosperm flower.
(*a*) The basic structure of a flower is a series of four concentric circles, or whorls: the sepals, petals, stamens, and the carpel (ovary, style, and stigma). (*b*) This flower of the wild woodland plant *Geranium* shows the five free petals, 10 stamens, and a fused carpel typical of angiosperm flowers.

grains. Stamens are slender, threadlike filaments with a swollen **anther** at the tip containing pollen.
4. The fourth and innermost whorl, called the **carpel** of the flower, contains the "female" parts that produce eggs. The carpel is sporophyte tissue that completely encases the ovules within which the egg cell develops. The ovules occur in the bulging lower portion of the carpel, called the **ovary;** usually there is a slender stalk rising from the ovary called the **style,** with a sticky tip called a **stigma,** which receives pollen. When the flower is pollinated, a pollen tube grows down from the pollen grain on the stigma through the style to the ovary to fertilize the egg.

14.8 Angiosperms are seed plants in which the ovule is completely enclosed by diploid tissue at pollination.

14.9 Why Are There Different Kinds of Flowers?

If you were to watch insects visiting flowers, you would quickly discover that the visits are not random. Instead, certain insects are attracted by particular flowers. Insects recognize a particular color pattern and odor and search for flowers that look similar. Insects and plants have coevolved (see chapters 27 and 17) so that certain insects specialize in visiting particular kinds of flowers. As a result, a particular insect carries pollen from one individual flower to another *of the same species.* It is this keying in on particular species that makes insect pollination so effective.

Of all insect pollinators, the most numerous are bees. Bees evolved soon after flowering plants, some 125 million years ago. Today there are over 20,000 species. Bees locate sources of nectar largely by odor at first (that is why flowers smell sweet) and then focus in on the flower's color and shape. Bee-pollinated flowers are usually yellow or blue, and frequently they have guiding stripes or lines of dots to indicate the position in the flower of the nectar, usually in the throat of the flower (figure 14.18*a, b*). While inside the flower, the bee becomes coated with pollen (figure 14.18*c*).

Many other insects pollinate flowers. Butterflies tend to visit flowers of plants like phlox that have "landing platforms" on which they can perch. These flowers typically have long, slender floral tubes filled with nectar that a butterfly can reach by uncoiling its long proboscis (a hoselike tube extending out from the mouth). Moths, which visit flowers at night, are attracted to white or very pale-colored flowers, often heavily scented, that are easy to locate in dim light. Flowers pollinated by flies, such as members of the milkweed family, are usually brownish in color and foul smelling.

Red flowers, interestingly, are not typically visited by insects, most of which cannot "see" red as a distinct color. Who pollinates these flowers? Hummingbirds and sunbirds (figure 14.19)! To these birds, red is a very conspicuous color, just as it is to us. Birds do not have a well-developed sense of smell, and do not orient to odor, which is why red flowers often do not have a strong smell.

Some angiosperms have reverted to the wind pollination practiced by their ancestors, notably oaks, birches, and, most important, the grasses. The flowers of these plants are small, greenish, and odorless. Other angiosperm species are aquatic, and while some have developed specialized pollination systems where the pollen is transported underwater or floats from one plant to another, most aquatic angiosperms are either wind pollinated or insect pollinated like their terrestrial ancestors. Their flowers extend up above the surface of the water.

14.9 Flowers can be viewed as pollinator-attracting devices, with different kinds of pollinators attracted to different kinds of flowers.

(a) (b) (c)

Figure 14.18 How a bee sees a flower.
(a) The yellow flower of *Ludwigia peruviana* (Onagraceae) photographed in normal light and (b) with a filter that selectively transmits ultraviolet light. The outer sections of the petals reflect both yellow and ultraviolet, a mixture of colors called "bee's purple"; the inner portions of the petals reflect yellow only and therefore appear dark in the photograph that emphasizes ultraviolet reflection. To a bee, this flower appears as if it has a conspicuous central bull's-eye. (c) When inside the flower, the bee becomes covered in pollen, which it takes to a neighboring flower.

Figure 14.19 Red flowers are pollinated by hummingbirds.
This long-tailed hermit hummingbird is extracting nectar from the red flowers of *Heliconia imbricata* in the forests of Costa Rica. Note the pollen on the bird's beak. Hummingbirds of this group obtain nectar primarily at long, curved flowers that more or less match the length and shape of their beaks.

14.10 Improving Seeds: Double Fertilization

The seeds of gymnosperms often contain food to nourish the developing plant in the critical time immediately after germination, but the seeds of angiosperms have greatly improved on this aspect of seed function. Angiosperms produce a special, highly nutritious tissue called **endosperm** within their seeds. Here is how it happens. In the angiosperm life cycle, the male gametophyte (with haploid gametes) produces two sperm, not one (figure 14.20). Upon adhering to the stigma, the pollen begins to form a pollen tube, which grows until it reaches the ovule in the ovary. The two sperm travel down the pollen tube and into the ovary. The first sperm fuses with the egg, as in all sexually reproducing organisms, forming the zygote. The other fuses with two other products of meiosis, called polar nuclei, to form a triploid (three copies of the chromosomes, 3*n*) endosperm cell. This cell divides much more rapidly than the zygote, giving rise to the nutritive endosperm tissue within the seed. This process of fertilization with two sperm to produce

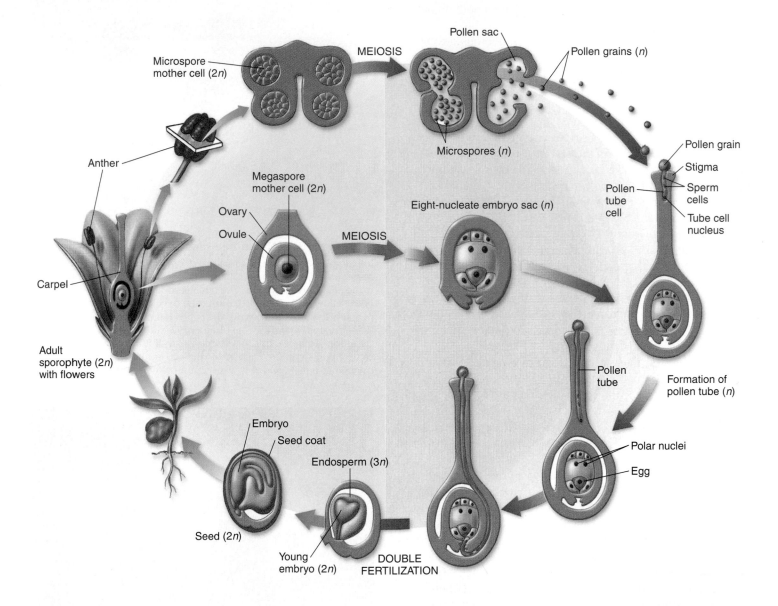

Figure 14.20   Life cycle of an angiosperm.

In angiosperms, as in gymnosperms, the sporophyte is the dominant generation. Eggs form within the embryo sac, inside the ovules, which, in turn, are enclosed in the carpels. The carpel is differentiated in most angiosperms into a slender portion, or style, ending in a stigma, the surface on which the pollen grains germinate. The pollen grains, meanwhile, are formed within the microsporangia of the anthers and complete their differentiation to the mature, three-celled stage either before or after grains are shed. Fertilization is distinctive in angiosperms, being a double process. A sperm and an egg come together, producing a zygote; at the same time, another sperm fuses with the two polar nuclei, producing the primary endosperm nucleus, which is triploid. The zygote and the primary endosperm nucleus divide mitotically, giving rise, respectively, to the embryo and the endosperm. The endosperm is the tissue, almost exclusive to angiosperms, that nourishes the embryo and young plant.

both a zygote and endosperm is called **double fertilization.** Double fertilization is almost exclusive to angiosperms; it is found in only one group of gymnosperms, the gnetophytes.

In some angiosperms, such as the common bean or pea, the endosperm is fully used up by the time the seed is mature. Food reserves are stored in swollen, fleshy leaves called cotyledons, or seed leaves. In other angiosperms, such as corn, the mature seed contains abundant endosperm, which is used after germination.

Some angiosperm embryos have two cotyledons, and are called dicotyledons, or **dicots.** The first angiosperms were like this. Dicots typically have leaves with netlike branching of veins and flowers with four to five parts per whorl. Oak and maple trees are dicots, as are many shrubs.

The embryos of other angiosperms, which evolved somewhat later, have a single cotyledon and are called monocotyledons, or **monocots.** Monocots typically have leaves with parallel veins and flowers with three parts per whorl. Grasses, one of the most abundant of all plants, are wind-pollinated monocots. The basic characteristics of monocots and dicots are outlined in figure 14.21. There are also differences in the organization of vascular tissue in the stems of monocots and dicots, which will be compared in more detail in chapter 26.

14.10 Two sperm fertilize each angiosperm ovule. One fuses with the egg to form the zygote, the other with two polar nuclei to form triploid (3*n*) nutritious endosperm.

(a) Monocots

(b) Dicots

Figure 14.21 Dicots and monocots.
(*a*) Monocots are characterized by one cotyledon, parallel veins, and the occurrence of flower parts in threes (or multiples of three). (*b*) Dicots have two cotyledons and netlike (reticulate) veins. Their flower parts occur in fours and fives.

(a)

(b)

(c)

Figure 14.22 Different ways of dispersing fruits.
(a) Berries are fruits that are dispersed by animals. (b) Coconuts are fruits dispersed by water, where they are carried off to new island habitats. (c) The fruits of maples are dry and winged, carried by the wind like small helicopters floating through the air.

14.11 Improving Seed Dispersal: Fruits

Just as a mature ovule becomes a seed, so a mature ovary that surrounds the ovule becomes all or a part of the **fruit.** A fruit is a mature ripened ovary containing fertilized seeds. Fruits provide angiosperms with a far better way of dispersing their progeny than simply sending their seeds off on the wind. Instead, just as in pollination, they employ animals. By making fruits fleshy and tasty to animals, angiosperms encourage animals to eat them. The seeds within the fruit are resistant to chewing and digestion. They pass out of the animal with the feces, undamaged and ready to germinate at a new location far from the parent plant (figure 14.22a).

Although many fruits are dispersed by animals, some fruits are dispersed by water (figure 14.22b), and many plant fruits are specialized for wind dispersal. The small, nonfleshy fruits of the dandelion, for example, have a plumelike structure that allows them be carried long distances on wind currents. The fruits of many grasses are small particles, so light wind bears them easily. Maples have long wings attached to the fruit (figure 14.22c). In tumbleweeds, the whole plant breaks off and is blown across open country by the wind, scattering seeds as it moves. Fruits will be discussed in more detail in chapter 27.

> **14.11** A fruit is a mature ovary containing fertilized seeds, often specialized to aid in seed dispersal.

Exploring Current Issues

Additional Resources

Go to your campus library or look online to find the following articles, which further develop some of the concepts found in this chapter.

Ananthaswamy, A. (2002). Inside the hanging gardens of Arcata. *New Scientist,* 176(2368), 46.

Blackwell, M. (2000). Terrestrial life: fungal from the start? *Science,* 289(5486), 1884.

Fleming, T. H. (2000). Pollination on cacti in the Sonoran Desert. *American Scientist,* 88(5), 432.

Raghavan, V. (2003). Some reflections on double fertilization, from its discovery to the present. *New Phytologist,* 159(3), 565.

Stuessey, T. F. (2004). A transitional-combinational theory for the origin of angiosperms. *Taxon,* 53(1), 3.

Biology and Society Lecture: Unearthing the Roots of the Tree of Life

Organisms were first classified more than 2,000 years ago by the Greek philosopher Aristotle, who categorized all living things as being either animals or plants. Biologists now call each of these great groups of organisms a "kingdom" and recognize six kingdoms of life: animals, plants, fungi, and protists (eukaryotes), and bacteria and archaea (prokaryotes). When genomic DNA sequences of the two prokaryotic kingdoms were first compared in 1996, the differences were astonishingly great, as great as either kingdom is from any eukaryote, and so they were placed into separate domains. How the three domains relate to each other is proving difficult to sort out, as there appears to have been considerable exchange of genes early in the history of life between the three groups. The base of the tree of life seems to be less a trunk than a thicket.

Find this lecture, delivered by the author to his class at Washington University, online at www.mhhe.com/tlwessentials/exp14.

Plants

14.1 Adapting to Terrestrial Living

- Plants are complex multicellular autotrophs, producing their own food through photosynthesis. Plants acquire water and minerals from the soil through their roots. They control water loss through a watertight layer called  the cuticle, with openings called stomata, which allow for gas exchange with the air (**figure 14.2**).

- Plant life cycles involve an alternation of generations where a haploid gametophyte alternates with a diploid sporophyte, with the sporophyte dominating the life cycle more and more as plants evolved (**figure 14.3**).

14.2 Plant Evolution

- Plants evolved from green algae. Four innovations illustrate the challenges that were overcome to invade the land. These innovations include an alternation of generation life cycle, the evolution of vascular tissue, seeds, and flower and fruits (**table 14.1 and figure 14.5**).

Seedless Plants

14.3 Nonvascular Plants

- Liverworts and hornworts don't have vascular tissue, which limits how large they can grow. Mosses have specialized cells for conducting water and carbohydrates in the plant, but not rigid vascular tissue. These plants are grouped together as nonvascular plants. Their life cycles are dominated by the haploid gametophyte (**figure 14.6**).

14.4 The Evolution of Vascular Tissue

- Vascular systems consist of specialized cylindrical or elongated cells that form a network throughout the plant. These cells conduct water from the roots throughout the plant and carbohydrates manufactured in the leaves throughout the plant (**figure 14.8**).

14.5 Seedless Vascular Plants

- Among the most primitive vascular plants are the Pterophyta (ferns) and Lycophyta (**figure 14.9**). These plants can grow very tall due to vascularization but lack seeds and require water for fertilization. The gametophyte and sporophyte both make up significant portions of the life cycle (**figure 14.10**).

The Advent of Seeds

14.6 Evolution of Seed Plants

- The seed was an evolutionary innovation that provided protection for the plant embryo. The seed contains the embryo and food for the developing embryo, inside a drought-resistant cover (**figure 14.12**). In seed plants, the sporophyte also becomes the dominant structure in the life cycle.

- Seeds are a key adaptation to land-dwelling plants by improving the dispersal of embryos, providing dormancy when needed, allowing for germination when conditions are favorable, and nourishment during germination.

14.7 Gymnosperms

- Gymnosperms are nonflowering seed plants and include conifers, cycads, gnetophytes, and the ginkgo. The seeds are produced in cones where pollination occurs. Pollen, produced in  smaller, pollen-bearing cones, travels by wind to the seed-bearing cones containing the egg cells and fertilizes the egg. The ovules are not completely enclosed by diploid tissue as is the case with flowering plants. The seed forms and is released from the cone for dispersal. (**figure 14.16**).

The Evolution of Flowers

14.8 Rise of the Angiosperms

- Angiosperms, flowering plants, are plants in which the ovule is completely enclosed by sporophyte tissue. The flowers are the reproductive structures, containing the pollen on the anthers and the egg within the ovary (**figure 14.17**). They improved the efficiency of mating.

14.9 Why Are There Different Kinds of Flowers?

- Flowers vary greatly in size, shape, and color so that they are identifiable by a particular pollinator. This improves the likelihood that the pollen will be carried to the appropriate mate (**figures 14.18 and 14.19**).

14.10 Improving Seeds: Double Fertilization

- Double fertilization provides food for the germinating plant. The pollen grain produces two sperm cells, one fuses with the egg and the other fuses with two polar nuclei, producing endosperm (**figure 14.20**). The angiosperms are divided into two groups: monocots and dicots (**figure 14.21**).

14.11 Improving Seed Dispersal: Fruits

- A further evolutionary advancement in angiosperms was the development of the ovary into fruit tissue. Fruits aid in the dispersal of seeds to new habitats (**figure 14.22**).

1. A major consideration for the evolution of terrestrial plants is the problem of
   a. lack of nutrients.
   c. dehydration.
   b predators.
   d. not enough carbon.

2. Which of the following structures or systems does *not* give the plants that have them an evolutionary advantage?
   a. chloroplasts
   c. seeds
   b. vascular tissue
   d. flowers and fruits

3. Mosses, liverworts, and hornworts do not reach a large size because
   a. they lack chlorophyll.
   b. they do not have specialized vascular tissue to transport water very high.
   c. photosynthesis does not take place at a very fast rate.
   d. alternation of generations does not allow the plants to grow very tall before reproduction.

4. One characteristic that separates ferns from complex vascular plants is that ferns do not have
   a. a vascular system.
   b. chloroplasts.
   c. alternation of generations in their life cycle.
   d. seeds.

5. As seed plants evolved, the _____ form became more visible.
   a. gametophyte
   c. sporophyte
   b gymnosperm
   d. angiosperm

6. In seeds, the seed coat aids in
   a. germination.
   c. nourishment.
   b. photosynthesis.
   d. dispersal.

7. What separates the gymnosperms from the rest of the vascular plants is/are
   a. a vascular system.
   b. ovules completely covered by the gametophyte.
   c. ovules not completely covered by the sporophyte.
   d. fruits and flowers.

8. What separates the angiosperms from the rest of the vascular plants is/are
   a. a vascular system.
   b. ovules completely covered by the gametophyte.
   c. fruits and flowers.
   d ovules not completely covered by the sporophyte.

9. Flower shape and color can be linked to the process of
   a. pollination.
   c. senescence.
   b. photosynthesis.
   d. secondary growth.

10. If the seeds of a plant are encased in a fleshy fruit, then the most likely form of dispersal is
   a. to attach to an animal's fur or skin.
   b. wind.
   c. an animal's digestive system and processes.
   d. water.

1. **Figure 14.12** Why do you think these two seeds were the most important foods of the Native Americans of this country? *Hint:* Think about the purpose of a seed.

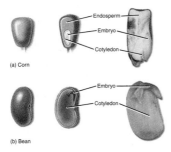

(a) Corn

(b) Bean

2. **Figure 14.13** These seeds fall from the tree and are scattered about on the ground nearby. What are the benefits or drawbacks to having (a) chemicals in the seed to inhibit sprouting, and (b) hard seed coats?

Plants Why do most plant pollinators have wings?

Seedless Plants Why are mosses limited to moist places?

The Advent of Seeds Why does a pine tree, which might live more than 100 years, produce hundreds of pine cones each year, each with 30 to 50 seeds?

The Evolution of Flowers Your friend, Jessica, is admiring the wildflowers one spring day, and wonders aloud why in the world plants evolved such beautiful colors and shapes and produce, such wonderful scents. What do you tell her?

Visit the Online Learning Center for this chapter at www.mhhe.com/tlwessentials/ch14 for quizzes, animations, interactive learning exercises, and other study tools. At the site you will also find extended answers to the end-of-chapter questions.

15
Evolution of Animals

A nimals are the most diverse in appearance of all eukaryotes. *Polistes,* the common paper wasp you see here, is a member of the most diverse of all animal groups, the insects. It has been a major challenge for biologists to sort out the millions of kinds of animals. This wasp has a segmented external skeleton and jointed appendages, and so is clearly an arthropod. But how are arthropods related to mollusks such as snails, and to segmented worms like earthworms? Until recently, biologists grouped all three kinds of animals together, as they all share a true body cavity called a coelom, a character assumed to be so fundamental it could have evolved only once. Now, molecular analyses suggest this assumption may be wrong. Instead, mollusks and segmented worms are grouped together with other animals that grow the same way you do, by adding additional mass to an existing body, while arthropods are grouped with other molting animals. These animals increase in size by molting their external skeletons, an ability that seems to have evolved only once. Thus we learn that even in a long-established field like taxonomy, biology is constantly changing. In this chapter we will explore the most diverse of all kingdoms, the animals.

15.1 General Features of Animals

We now explore the great diversity of animals, the result of a long evolutionary history. Animals are the eaters of the earth. All are heterotrophs and depend directly or indirectly for nourishment on plants, photosynthetic protists (algae), or autotrophic bacteria. Many animals are able to move from place to place in search of food (figure 15.1). In most, ingestion of food is followed by digestion in an internal cavity.

Multicellular Heterotrophs

All animals are multicellular heterotrophs that require oxygen for respiration. Several decades ago, a biologist would not have made that statement. In recent years, the generally accepted scientific definition of the kingdom Animalia, which contains the animals, has been changed in one respect. The unicellular, heterotrophic organisms called protozoa, which were at one time regarded as simple animals, are now considered to be members of the kingdom Protista, the large and diverse group we discussed in chapter 13.

Diverse in Form

Almost all animals (99%) are **invertebrates,** animal species that lack a backbone. Of the estimated 10 million living species, only 54,000 have a backbone and are referred to as **vertebrates.** Animals are very diverse in form, ranging in size from ones too small to see with the naked eye to enormous whales and giant squids (see table 15.1). The animal kingdom includes about 36 phyla, most of which occur in the sea. Members of three phyla, Arthropoda (spiders, insects), Mollusca (snails), and Chordata (vertebrates), dominate animal life on land.

No Cell Walls

Animal cells are distinct among multicellular organisms because they lack rigid cell walls and are usually quite flexible. As we shall learn, the cells of all animals except sponges are organized into structural and functional units called **tissues.** A tissue is a collection of cells that have joined together and are specialized to perform a specific function.

Active Movement

The ability of animals to move more rapidly and in more complex ways than members of other kingdoms is perhaps their most striking characteristic and one that is directly related to the flexibility of their cells. A remarkable form of movement unique to animals is flying, an ability that is well developed among both insects and vertebrates. Among vertebrates, birds, bats, and pterosaurs (now-extinct flying reptiles) are or were all strong fliers. The only terrestrial vertebrate group never to have had flying representatives is amphibians.

Figure 15.1 Animals on the move.

These large antelopes, known as wildebeests, move about in large groups consuming grass and finding new areas to feed.

Sexual Reproduction

Most animals reproduce sexually. Animal eggs, which are nonmotile, are much larger than the small, usually flagellated sperm. In animals, cells formed in meiosis function directly as gametes. The haploid cells do not divide by mitosis first, as they do in plants and fungi, but rather fuse directly with one another to form the zygote. Consequently, with a few exceptions, there is no counterpart among animals to the alternation of haploid (gametophyte) and diploid (sporophyte) generations characteristic of plants (see chapter 14).

Embryonic Development

The complex form of a given animal develops from a zygote by a characteristic process of embryonic development. The zygote first undergoes a series of mitotic divisions and becomes a solid ball of cells, the **morula,** and then a hollow ball of cells, the **blastula,** a developmental stage that occurs in all animals. In most, the blastula folds inward at one point to form a hollow sac with an opening at one end called the **blastopore.** An embryo at this stage is called a **gastrula.** The subsequent growth and movement of cells of the gastrula produce the digestive system, also called the gut or intestine. The details of embryonic development differ widely from one phylum of animals to another and often provide important clues to the evolutionary relationships among the animal phyla (figure 15.2). Taken as a whole, the pattern of embryology is characteristic of the animal kingdom.

> **15.1** Animals are complex multicellular organisms typically characterized by mobility and heterotrophy. Most animals also exhibit sexual reproduction and tissue-level organization.

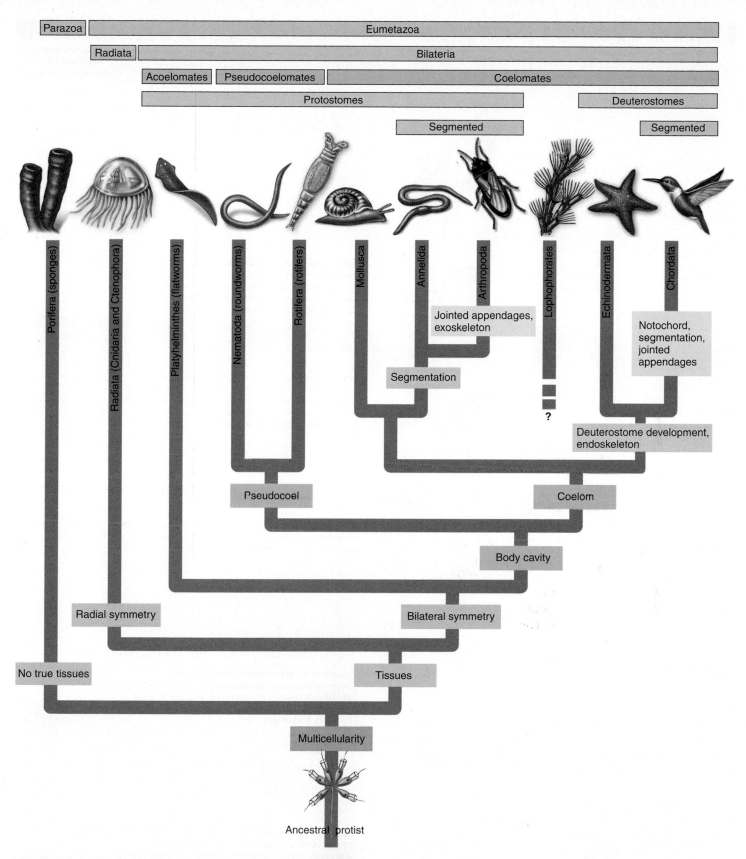

Figure 15.2 Evolutionary trends among the animals.

In this chapter we examine a series of key evolutionary innovations in the animal body plan, shown here along the branches.

TABLE 15.1 THE MAJOR ANIMAL PHYLA

Phylum	Typical Examples		Key Characteristics	Approximate Number of Named Species
Arthropoda (arthropods)	Insects, crabs, spiders, millipedes		Most successful of all animal phyla; chitinous exoskeleton covering segmented bodies with paired, jointed appendages; most insect groups have wings; nearly all are freshwater or terrestrial	1,000,000
Mollusca (mollusks)	Snails, clams, octopuses, nudibranchs		Soft-bodied coelomates whose bodies are divided into three parts: head-foot, visceral mass, and mantle; many have shells; almost all possess a unique rasping tongue called a radula; most are marine or freshwater but 35,000 species are terrestrial	110,000
Chordata (chordates)	Mammals, fish, reptiles, birds, amphibians		Segmented coelomates with a notochord; possess a dorsal nerve cord, pharyngeal slits, and a tail at some stage of life; in vertebrates, the notochord is replaced during development by the spinal column; most are marine, many are freshwater, and 20,000 species are terrestrial	56,000
Platyhelminthes (flatworms)	*Planaria*, tapeworms, flukes		Solid, unsegmented, bilaterally symmetrical worms; no body cavity; digestive cavity, if present, has only one opening; marine, freshwater, or parasitic	20,000
Nematoda (roundworms)	*Ascaris*, pinworms, hookworms, *Filaria*		Pseudocoelomate, unsegmented, bilaterally symmetrical worms; tubular digestive tract passing from mouth to anus; tiny; without cilia; live in great numbers in soil and aquatic sediments; some are important animal parasites	20,000
Annelida (segmented worms)	Earthworms, marine worms, leeches		Coelomate, serially segmented, bilaterally symmetrical worms; complete digestive tract; most have bristles called setae on each segment that anchor them during crawling; marine, freshwater, and terrestrial	12,000

TABLE 15.1   CONTINUED

Phylum	Typical Examples		Key Characteristics	Approximate Number of Named Species
Cnidaria (cnidarians)	Jellyfish, hydra, corals, sea anemones		Soft, gelatinous, radially symmetrical bodies whose digestive cavity has a single opening; possess tentacles armed with stinging cells called cnidocytes that shoot sharp harpoons called nematocysts; almost entirely marine	10,000
Echinodermata (echinoderms)	Sea stars, sea urchins, sand dollars, sea cucumbers		Deuterostomes with radially symmetrical adult bodies; endoskeleton of calcium plates; pentamerous (five-part) body plan and unique water vascular system with tube feet; able to regenerate lost body parts; all are marine	6,000
Porifera (sponges)	Barrel sponges, boring sponges, basket sponges, vase sponges		Asymmetrical bodies without distinct tissues or organs; saclike body consists of two layers breached by many pores; internal cavity lined with food-filtering cells called choanocytes; most are marine (150 species live in freshwater)	5,150
Lophophorates (moss animals, also called Bryoza)	*Bowerbankia, Plumatella,* sea mats, sea moss		Microscopic, aquatic deuterostomes that form branching colonies, possess circular or U-shaped row of ciliated tentacles for feeding called a lophophore that usually protrudes through pores in a hard exoskeleton; also called Ectoprocta because the anus or proct is external to the lophophore; marine or freshwater	4,000
Rotifera (wheel animals)	Rotifers		Small, aquatic pseudocoelomates with a crown of cilia around the mouth resembling a wheel; almost all live in freshwater	2,000

15.2 Sponges and Cnidarians: The Simplest Animals

The kingdom Animalia consists of two subkingdoms: (1) *Parazoa,* animals that lack a definite symmetry and possess neither tissues nor organs, and (2) *Eumetazoa,* animals that have a definite shape and symmetry, and in most cases tissues organized into organs. The subkingdom Parazoa consists primarily of the sponges, phylum Porifera. The other animals, comprising about 35 phyla, belong to the subkingdom Eumetazoa.

Sponges

Sponges, members of the phylum Porifera, are the simplest animals. Most sponges completely lack symmetry, and although some of their cells are highly specialized, they are not organized into tissues. The bodies of sponges consist of little more than masses of specialized cells embedded in a gel-like matrix, like chopped fruit in Jell-O. However, sponge cells do possess a key property of animal cells: cell recognition. For example, when a sponge is passed through a fine silk mesh, individual cells separate and then reaggregate on the other side to re-form the sponge. Clumps of cells disassociated from a sponge can give rise to entirely new sponges.

About 5,000 species exist, almost all in the sea (a few live in freshwater). Some are tiny, and others are more than 2 meters in diameter (figure 15.3). The body of an adult sponge is anchored in place on the seafloor and is shaped like a vase. The outside of the sponge is covered with a skin of flattened cells called epithelial cells that protect the sponge.

The body of the sponge is perforated by tiny holes. The name of the phylum, Porifera, refers to this system of pores. Unique flagellated cells called **choanocytes,** or collar cells, line the body cavity of the sponge. The beating of the flagella of the many choanocytes draws water in through the pores and drives it through the cavity. One cubic centimeter of sponge tissue can propel more than 20 liters of water a day in and out of the sponge body! Why all this moving of water? The sponge is a "filter-feeder." The beating of each choanocyte's flagellum draws water down through its collar, made of small hairlike projections resembling a picket fence. Any food particles in the water, such as protists and tiny animals, are trapped in the fence and later ingested by the choanocyte or other cells of the sponge.

The choanocytes of sponges very closely resemble a kind of protist called choanoflagellates, which seem almost certain to have been the ancestors of sponges. Indeed, they may be the ancestors of *all* animals, although it is difficult to be certain that sponges are the direct ancestors of the other more complex phyla of animals.

All animals other than sponges have both symmetry and tissues and thus are eumetazoans.

(a)

(b)

Figure 15.3 Sponges.
These two marine sponges are barrel sponges. They are among the largest of sponges, with well-organized forms. Many are more than 2 meters in diameter (*a*), while others are smaller (*b*).

Cnidarians

All eumetazoans form distinct embryonic layers. The **radially symmetrical** (that is, with body parts arranged around a central axis) eumetazoans have two embryonic layers, an outer **ectoderm**, which gives rise to the epidermis, and an inner **endoderm**, which gives rise to the gastrodermis. A jellylike layer called the *mesoglea* forms between the epidermis and gastrodermis. These layers give rise to the basic body plan, differentiating into the many tissues of the body. No such layers are present in sponges.

The most primitive eumetazoans to exhibit symmetry and tissues are two radially symmetrical phyla, whose bodies are organized around an oral-aboral axis, like the petals of a daisy. The oral side of the animal contains the "mouth." Radial symmetry offers advantages to animals that either remain attached or closely associated to the surface or to animals that are free-floating. These animals don't pass through their environment, but rather they interact with their environment on all sides. These two phyla are Cnidaria (pronounced ni-DAH-ree-ah), which includes hydra (figure 15.4), jellyfish, sea anemones, and corals, and Ctenophora (pronounced tea-NO-fo-rah), a minor phylum that includes the comb jellies. These two phyla together are called the Radiata. The bodies of all other eumetazoans, the Bilateria, are marked by a fundamental bilateral symmetry (discussed in next section). Even sea stars, which exhibit radial symmetry as adults, are bilaterally symmetrical when young.

A major evolutionary innovation among the radiates is the **extracellular digestion** of food. In sponges, food trapped by a choanocyte is taken directly into that cell, or into a circulating amoeboid cell, by endocytosis. In radiates, digestion begins *outside of cells*, in a gut cavity, called the gastrovascular cavity. After the food is broken down into smaller pieces, cells lining the gut cavity will complete digestion intracellularly. Extracellular digestion is the same heterotrophic strategy pursued by fungi, except that fungi digest food outside their bodies, while animals digest it within their bodies, in a cavity. This evolutionary advance has been retained by all of the more advanced groups of animals.

Cnidarians (phylum Cnidaria) are carnivores that capture their prey, such as fishes and shellfish, with tentacles that ring their mouths. These tentacles, and sometimes the body surface cells, bear unique stinging cells called **cnidocytes** that occur in no other organism and give the phylum its name. Within each cnidocyte is a small but powerful harpoon called a **nematocyst** that cnidarians use to spear their prey and then draw the harpooned prey back to the tentacle containing the cnidocyte. The cnidocyte builds up a very high internal osmotic pressure and uses it to push the nematocyst outward so explosively that the barb can penetrate even the hard shell of a crab.

Cnidarians have two basic body forms, **medusae** and **polyps** (figure 15.5). Some cnidarians exist only as medusae, others only as polyps, and still others alternate between these two phases during the course of their life cycles. Medusae are free-floating, gelatinous, and often umbrella-shaped. Their mouths point downward, with a ring of tentacles hanging down around the edges (hence the radial symmetry). Medusae are commonly called "jellyfish" because of their gelatinous

Figure 15.4   A cnidarian.
Hydroids are a group of cnidarians that are mostly marine and colonial. However, *Hydra,* shown above, is a freshwater genus whose members exist as solitary polyps.

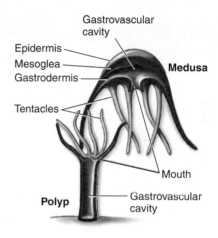

Figure 15.5   Two basic body forms of cnidarians.
The medusa (*top*) and the polyp (*bottom*) are the two phases that alternate in the life cycles of many cnidarians, but several species (corals and sea anemones, for example) exist only as polyps.

interior or "stinging nettles" because of their nematocysts. Polyps are cylindrical, pipe-shaped animals that are usually attached to a rock. They also exhibit radial symmetry. The *Hydra*, sea anemones, and corals are examples of polyps. In polyps, the mouth faces away from the rock and therefore is often directed upward. Polyps are primarily sessile and although *Hydra* are able to move around using a "cartwheel-type" motion, many polyps remain fixed and exposed to predators. For shelter and protection, corals deposit an external "skeleton" of calcium carbonate within which they live. This is the structure usually identified as coral.

> **15.2   Sponges have a multicellular body with specialized cells but lack definite symmetry and organized tissues. Cnidarians possess radial symmetry and specialized tissues and carry out extracellular digestion.**

The Advent of Bilateral Symmetry

All eumetazoans other than cnidarians and ctenophores are **bilaterally symmetrical**—that is, they have a right half and a left half that are mirror images of each other. In looking at a bilaterally symmetrical animal, you refer to the top half of the animal as **dorsal** and the bottom half as **ventral** (figure 15.6). The front is called **anterior** and the back **posterior.**

Bilateral symmetry was a major evolutionary advance among the animals because it allows different parts of the body to become specialized in different ways. For example, most bilaterally symmetrical animals have evolved a definite head end, a process called **cephalization.** Animals that have heads are often active and mobile, moving through their environment headfirst, with sensory organs concentrated in front so the animal can test for food, danger, and mates, as it enters new surroundings.

The bilaterally symmetrical eumetazoans produce three embryonic layers: an outer **ectoderm,** an inner **endoderm,** and a third layer, the **mesoderm,** between the ectoderm and endoderm. In general, the outer coverings of the body and the nervous system develop from the ectoderm, the digestive organs and intestines develop from the endoderm, and the skeleton and muscles develop from the mesoderm (figure 15.7).

The simplest of all bilaterally symmetrical animals are the **solid worms.** By far the largest phylum of these, with about 20,000 species, is Platyhelminthes (pronounced plat-ee-hel-MIN-theeze), which includes the flatworms. Flatworms are the simplest animals in which organs occur. An organ is a collection of different tissues that function as a unit. The testes and uterus of flatworms are reproductive organs, for example. The dark spots on the head are eyespots that can detect light, although they cannot focus an image like your eyes can.

Solid worms lack any internal cavity other than the digestive tract. Flatworms are soft-bodied animals flattened from top to bottom, like a piece of tape or ribbon. If you were to cut a flatworm in half across its body, you would see that the gut is completely surrounded by tissues and organs. This solid body construction is called **acoelomate,** meaning without a body cavity.

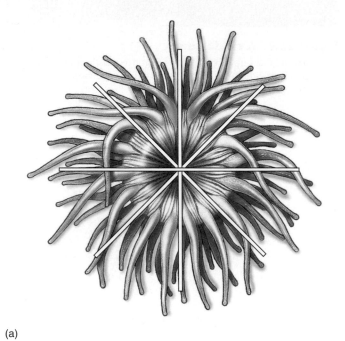
(a)

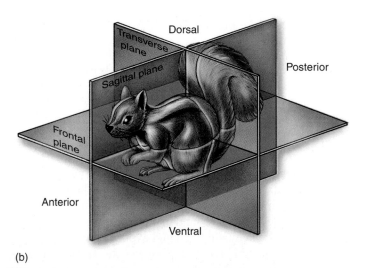
(b)

Figure 15.6 How radial and bilateral symmetry differ.

(*a*) Radial symmetry is the regular arrangement of parts around a central axis, so that any plane passing through the central axis divides the organism into halves that are approximate mirror images. (*b*) Bilateral symmetry is reflected in a body form that can only be bisected into halves that are approximate mirror images by one plane (the sagittal plane).

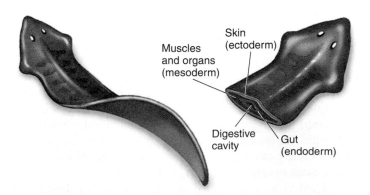

Figure 15.7 Body plan of a solid worm.

All bilaterally symmetrical eumetazoans produce three layers during embryonic development: an outer ectoderm, a middle mesoderm, and an inner endoderm. These layers differentiate to form the skin, muscles and organs, and gut, respectively, in the adult animal.

Flatworms

Although flatworms have a simple body design, they do have a definite head at the anterior end and they do possess organs. Flatworms range in size from a millimeter or less to many meters long, as in some tapeworms. Most species of flatworms are parasitic, occurring within the bodies of many other kinds of animals. Other flatworms are free-living, occurring in a wide variety of marine and freshwater habitats, as well as moist places on land (figure 15.8). Free-living flatworms are carnivores and scavengers; they eat various small animals and bits of organic debris. They move from place to place by means of ciliated epithelial cells, which are particularly concentrated on their ventral surfaces.

There are two classes of parasitic flatworms, which live within the bodies of other animals: flukes and tapeworms. Both groups of worms have epithelial layers resistant to the digestive enzymes and immune defenses produced by their hosts—an important feature in their parasitic way of life. Some parasitic flatworms require only one host, but many flukes require two or more hosts to complete their life cycles. The parasitic lifestyle has resulted in the eventual loss of features not used or needed by the parasite. Parasitic flatworms lack certain features of the free-living flatworms, such as cilia in the adult stage, eyespots, and other sensory organs that lack adaptive significance for an organism that lives within the body of another animal, a loss sometimes dubbed "degenerative evolution."

Characteristics of Flatworms

Those flatworms that have a digestive cavity have an incomplete gut, one with only one opening. As a result, they cannot eat, digest, and eliminate undigested particles of food simultaneously. Thus, flatworms cannot feed continuously, as more advanced animals can. The gut is branched and extends throughout the body, functioning in both digestion and transport of food. Cells that line the gut engulf most of the food particles by phagocytosis and digest them; but, as in the cnidarians, some of these particles are partly digested extracellularly. Tapeworms, which are parasitic flatworms, lack digestive systems. They absorb their food directly through their body walls.

Unlike cnidarians, flatworms have an excretory system, which consists of a network of fine tubules (little tubes) that runs throughout the body. Cilia line the hollow centers of bulblike **flame cells,** which are located on the side branches of the tubules. Cilia in the flame cells move water and excretory substances into the tubules and then to exit pores located between the epidermal cells. Flame cells were named because of the flickering movements of the tuft of cilia within them. They primarily regulate the water balance of the organism. The excretory function of flame cells appears to be a secondary one. A large proportion of the metabolic wastes excreted by flatworms probably diffuses directly into the gut and is eliminated through the mouth.

Like sponges, cnidarians, and ctenophorans, flatworms lack a **circulatory system,** a network of vessels that carries

 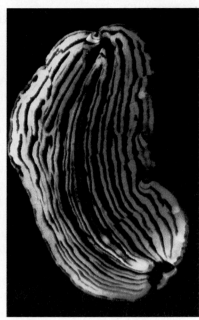

Figure 15.8 Flatworms.
(a) A common flatworm, *Planaria.* (b) A marine free-living flatworm.

fluids, oxygen, and food molecules to parts of the body. Consequently, all flatworm cells must be within diffusion distance of oxygen and food. Flatworms have thin bodies and highly branched digestive cavities that make such a relationship possible.

The nervous system of flatworms is very simple. Some primitive flatworms have only a loosely organized nerve net. However, most members of this phylum have longitudinal nerve cords that constitute a simple central nervous system. Between the longitudinal cords are cross-connections, so that the flatworm nervous system resembles a ladder.

Free-living flatworms use sensory pits or tentacles along the sides of their heads to detect food, chemicals, or movements of the fluid in which they are living. Free-living members of this phylum also have eyespots on their heads. These are inverted, pigmented cups containing light-sensitive cells connected to the nervous system. These eyespots enable the worms to distinguish light from dark. Flatworms are far more active than cnidarians or ctenophores. Such activity is characteristic of bilaterally symmetrical animals.

The reproductive systems of flatworms are complex. Most flatworms are **hermaphroditic,** with each individual containing both male and female sexual structures. Some genera of flatworms are also capable of asexual regeneration; when a single individual is divided into two or more parts, each part can regenerate an entirely new flatworm.

15.3 Flatworms have internal organs, bilateral symmetry, and a distinct head. They do not have a body cavity.

15.4 The Advent of a Body Cavity

A key transition in the evolution of the animal body plan was the evolution of the body cavity. All bilaterally symmetrical animals other than solid worms have a cavity within their body. The evolution of an internal body cavity was an important improvement in animal body design for three reasons:

1. **Circulation.** Fluids that move within the body cavity can serve as a circulatory system, permitting the rapid passage of materials from one part of the body to another and opening the way to larger bodies.
2. **Movement.** Fluid in the cavity makes the animal's body rigid, permitting resistance to muscle contraction and thus opening the way to muscle-driven body movement.
3. **Organ function.** In a fluid-filled enclosure, body organs can function without being deformed by surrounding muscles. For example, food can pass freely through a gut suspended within a cavity, at a rate not controlled by when the animal moves.

Kinds of Body Cavities

There are three basic kinds of body plans found in bilaterally symmetrical animals (figure 15.9). Acoelomates, such as solid worms, have no body cavity. **Pseudocoelomates** have a body cavity called the **pseudocoel** located between the mesoderm and endoderm. A third way of organizing the body is one in which the fluid-filled body cavity develops not between endoderm and mesoderm but rather entirely within the mesoderm. Such a body cavity is called a **coelom,** and animals that possess such a cavity are called **coelomates.** In coelomates, the gut is suspended, along with other organ systems of the animal, within the coelom; the coelom, in turn, is surrounded by a layer of epithelial cells entirely derived from the mesoderm.

The development of a body cavity poses a problem—circulation—solved in pseudocoelomates by churning the fluid within the body cavity. In coelomates, the gut is again surrounded by tissue that presents a barrier to diffusion, just as it was in solid worms. This problem is solved among coelomates by the development of a circulatory system. The circulating fluid, or blood, carries nutrients and oxygen to the tissues and removes wastes and carbon dioxide. Blood is usually pushed through the circulatory system by contraction of one or more muscular hearts. In an **open circulatory system,** the blood passes from vessels into sinuses, mixes with body fluid, and then reenters the vessels later in another location. In a **closed circulatory system,** the blood is separate from the body fluid and can be separately controlled. Also, blood moves through a closed circulatory system faster and more efficiently than it does through an open system.

The evolutionary relationship among coelomates, pseudocoelomates, and acoelomates is not clear. Acoelomates, for example, could have given rise to coelomates, but scientists

Acoelomate
— Ectoderm
— Mesoderm
— Endoderm

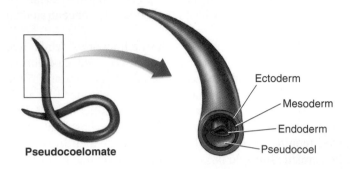

Pseudocoelomate
— Ectoderm
— Mesoderm
— Endoderm
— Pseudocoel

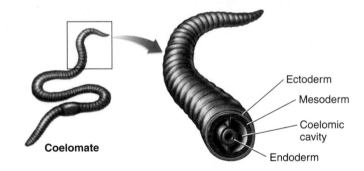

Coelomate
— Ectoderm
— Mesoderm
— Coelomic cavity
— Endoderm

Figure 15.9 Three body plans for bilaterally symmetrical animals.

Acoelomates, such as flatworms, have no body cavity between the digestive tract (endoderm) and the outer body layer (ectoderm). Pseudocoelomates have a body cavity, the pseudocoel, between the endoderm and the mesoderm. Coelomates have a body cavity, the coelom, that develops entirely within the mesoderm, and so is lined on both sides by mesoderm tissue.

also cannot rule out the possibility that acoelomates were derived from coelomates. The different phyla of pseudocoelomates form two groups that do not appear to be closely related.

Pseudocoelomates

As we have noted, all bilaterally symmetrical animals except solid worms possess an internal body cavity. Among them, seven phyla are characterized by their possession of a pseudocoel. In all pseudocoelomates, the pseudocoel serves as a hydrostatic skeleton—one that gains its rigidity from being filled with fluid under pressure. The animal's muscles can work against this "skeleton," thus making the movements of pseudocoelomates far more efficient than those of the acoelomates.

Only one of the seven pseudocoelomate phyla includes a large number of species. This phylum, Nematoda, includes some 20,000 recognized species of **nematodes,** eelworms, and other roundworms. Scientists estimate that the actual number might approach 100 times that many. Members of this phylum are found everywhere. Nematodes are abundant and diverse in marine and freshwater habitats, and many members of this phylum are parasites of animals and plants (figure 15.10*a*). Many nematodes are microscopic and live in soil. It has been estimated that a spadeful of fertile soil may contain, on the average, a million nematodes.

A second phylum consisting of animals with a pseudocoelomate body plan is Rotifera, the rotifers. **Rotifers** are common, small, basically aquatic animals that have a crown of cilia at their heads; they range from 0.04 to 2 millimeters long (figure 15.10*b*). About 2,000 species exist throughout the world. Bilaterally symmetrical and covered with chitin, rotifers depend on their cilia for both locomotion and feeding, ingesting bacteria, protists, and small animals.

All pseudocoelomates lack a defined circulatory system; this role is performed by the fluids that move within the pseudocoel. Most pseudocoelomates have a complete, one-way digestive tract that acts like an assembly line. Food is broken down, then absorbed, and then treated and stored.

Phylum Nematoda: The Roundworms

Nematodes are bilaterally symmetrical, cylindrical, unsegmented worms. They are covered by a flexible, thick cuticle, which is molted as they grow. Their muscles constitute a layer beneath the epidermis and extend along the length of the worm, rather than encircling its body. These longitudinal muscles pull both against the cuticle and the pseudocoel, which forms a hydrostatic skeleton. When nematodes move, their bodies whip about from side to side.

Near the mouth of a nematode, at its anterior end, are usually 16 raised, hairlike sensory organs. The mouth is often equipped with piercing organs called **stylets.** Food passes through the mouth as a result of the sucking action of a muscular chamber called the **pharynx.** After passing through a short corridor into the pharynx, food continues through the other portions of the digestive tract, where it is broken down and then digested. Some of the water with which the food has been mixed is reabsorbed near the end of the digestive tract, and material that has not been digested is eliminated through the anus.

Nematodes completely lack flagella or cilia, even on sperm cells. Reproduction in nematodes is sexual, with sexes usually separate. Their development is simple, and the adults consist of very few cells. For this reason, nematodes have become extremely important subjects for genetic and developmental studies. The 1-millimeter-long *Caenorhabditis elegans* matures in only three days, its body is transparent, and

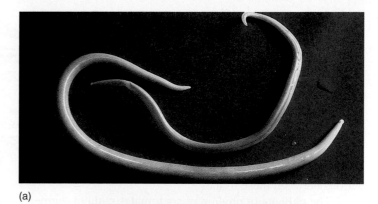

(a)

(b)

Figure 15.10   Pseudocoelomates.
(*a*) These nematodes (phylum Nematoda) are intestinal roundworms that infect humans and some other animals. Their fertilized eggs pass out with feces and can remain viable in soil for years. (*b*) Rotifers (phylum Rotifera) are common aquatic animals that depend on their crown of cilia for feeding and locomotion.

it has only 959 cells. It is the only animal whose complete developmental cellular anatomy is known, and the first animal whose genome (97 million DNA bases encoding over 21,000 different genes) was fully sequenced.

About 50 species of nematodes, including several that are rather common in the United States, regularly parasitize human beings. Trichinosis, a nematode-caused disease in temperate regions, is caused by worms of the genus *Trichinella*. These worms live in the small intestine of pigs, where fertilized female worms burrow into the intestinal wall. Once it has penetrated these tissues, each female produces about 1,500 live young. The young enter the lymph channels and travel to muscle tissue throughout the body, where they mature and form cysts. Infection in human beings or other animals arises from eating undercooked or raw pork in which the cysts of *Trichinella* are present. If the worms are abundant, a fatal infection can result, but such infections are rare.

Mollusks

Even though acoelomates and pseudocoelomates have proven very successful, the bulk of the animal kingdom consists of coelomates. A major advantage of the coelomate body plan is that it allows contact between mesoderm and endoderm during development, permitting localized portions of the digestive tract to develop into complex, highly specialized regions like the stomach. In pseudocoelomates, mesoderm and endoderm are separated by the body cavity, limiting developmental interactions between these tissues.

The only major phylum of coelomates without segmented bodies is phylum Mollusca. The **mollusks** are the largest animal phylum, except for the arthropods, with over 110,000 species. Mollusks are mostly marine, but occur almost everywhere.

Mollusks include three general groups with outwardly different body plans. However, the seeming differences hide a basically similar body design. The body of mollusks is composed of three distinct parts: a head-foot, a central section called the visceral mass that contains the body's organs, and a mantle. The foot of a mollusk is muscular and may be adapted for locomotion, attachment, food capture (in squids and octopuses), or various combinations of these functions. The **mantle** is a heavy fold of tissue wrapped around the visceral mass like a cape, with the gills positioned on its inner surface like the lining of a coat. The **gills** are filamentous projections of tissue, rich in blood vessels, that capture oxygen from the water circulating between the mantle and visceral mass and release carbon dioxide.

The three major groups of mollusks, all different variations upon this same basic design, are gastropods, bivalves, and cephalopods (figure 15.11).

1. **Gastropods** (snails and slugs) use the muscular foot to crawl, and their mantle often secretes a single, hard, protective shell. All terrestrial mollusks are gastropods.
2. **Bivalves** (clams, oysters, and scallops) secrete a two-part shell with a hinge, as their name implies. They filter feed by drawing water into their shell.
3. **Cephalopods** (octopuses and squids) have modified the mantle cavity to create a jet propulsion system that can propel them rapidly through the water. Their shell is greatly reduced to an internal structure or is absent.

One of the most characteristic features of mollusks is the **radula,** a rasping, tonguelike organ. With rows of pointed, backward-curving teeth, the radula is used by some snails to scrape algae off rocks. The small holes often seen in oyster shells are produced by gastropods that have bored holes to kill the oyster and extract its body.

In most mollusks, the outer surface of the mantle also secretes a protective shell. Its shell consists of a horny outer layer, rich in protein, which protects the two underlying calcium-rich

(a)

(b)

(c)

Figure 15.11   Three major groups of mollusks.
(a) A gastropod, (b) a bivalve, (c) a cephalopod.

layers from erosion. The inner layer is pearly and is used as mother-of-pearl. Pearls themselves are formed when a foreign object, such as a grain of sand, becomes lodged between the mantle and the inner shell layer of a bivalve, including clams and oysters. The mantle coats the foreign object with layer upon layer of shell material to reduce irritation. The shell serves primarily for protection with some mollusks withdrawing into their shell when threatened.

Annelids

One of the early key innovations in body plan to arise among the coelomates was **segmentation,** the building of a body from a series of similar segments. The first segmented animals to evolve were the **annelid worms,** phylum Annelida. These advanced coelomates are assembled as a chain of nearly identical segments, like the boxcars of a train. The great advantage of such segmentation is the evolutionary flexibility it offers—a small change in an existing segment can produce a new kind of segment with a different function. Thus, some segments are modified for reproduction, some for feeding, and others for eliminating wastes.

Two-thirds of all annelids live in the sea (about 8,000 species) but some live in freshwater, and most of the rest—some 3,100 species—are earthworms (figure 15.12). The basic body plan of an annelid is a tube within a tube: The digestive tract is a tube suspended within the coelom, which is itself a tube running from mouth to anus. There are three characteristics of this organization:

1. **Repeated segments.** The body segments of an annelid are visible as a series of ringlike structures running the length of the body, looking like a stack of donuts. The segments are divided internally from one another by partitions, just as walls separate the rooms of a building. In each of the cylindrical segments, the excretory and locomotor organs are repeated. The body fluid within the coelom of each segment creates a hydrostatic (liquid-supported) skeleton that gives the segment rigidity, like an inflated balloon. Muscles within each segment play against the fluid in the coelom. Because each segment is separate, each is able to expand or contract independently. This lets the worm body move in ways that are quite complex. When an earthworm crawls on a flat surface, for example, it lengthens some parts of its body while shortening others.

2. **Specialized segments.** The anterior (front) segments of annelids contain the sensory organs of the worm. Some of these are organs sensitive to light, and elaborate eyes with lenses and retinas have evolved in some annelids. One anterior segment contains a well-developed cerebral ganglion, or brain.

3. **Connections.** Because partitions separate the segments, it is necessary to provide ways for materials and information to pass between segments. A circulatory system carries blood from one segment to another, while nerve cords connect the nerve centers or ganglia located in each segment with each other and the brain. The brain can then coordinate the worm's activities.

Segmentation underlies the body organization of all complex coelomate animals, not only annelids but also arthropods (crustaceans, spiders, and insects) and chordates (mostly vertebrates). For example, vertebrate muscles develop from repeated blocks of tissue called somites that occur in the embryo. Vertebrate segmentation is also seen in the vertebral column, which is a stack of very similar vertebrae.

(a)   (b)

Figure 15.12 Representative annelids.
(a) Earthworms are the terrestrial annelids. This night crawler, *Lumbricus terrestris,* is in its burrow. (b) Shiny bristle worm, *Oenone fulgida,* is an aquatic annelid, a polychaete.

Arthropods

The evolution of segmentation among the annelids marked a major innovation in body structure among the coelomates. An even more profound innovation marks the origin of the body plan characteristic of the most successful of all animal groups, the **arthropods,** phylum Arthropoda (figure 15.13). This innovation was the development of jointed appendages.

Jointed Appendages. The name *arthropod* comes from two Greek words, *arthros,* jointed, and *podes,* feet. All arthropods have jointed appendages (figure 15.14). Some are legs, and others may be modified for other uses. To gain some idea of the importance of jointed appendages, imagine yourself without them—no hips, knees, ankles, shoulders, elbows, wrists, or knuckles. Without jointed appendages, you could not walk or grasp an object. Arthropods use jointed appendages as legs and wings for moving, as antennae to sense their environment, and as mouthparts for sucking, ripping, and chewing prey. A scorpion, for example, seizes and tears apart its prey with mouthpart appendages modified as large pincers.

Rigid Exoskeleton. The arthropod body plan has a second great innovation: Arthropods have a rigid external skeleton, or **exoskeleton,** made of chitin. In any animal, a key function of the skeleton is to provide places for muscle attachment, and in arthropods the muscles attach to the interior surface of the hard chitin shell, which also protects the animal from predators and impedes water loss.

However, while chitin is hard and tough, it is also brittle and cannot support great weight. As a result, the exoskeleton must be much thicker to bear the pull of the muscles in large insects than in small ones, so there is a limit to how big an arthropod body can be. That is why you don't see beetles as big as birds or crabs the size of a cow—the exoskeleton would be so thick the animal couldn't move its great weight. In fact, the great majority of arthropod species consist of small animals—mostly about a millimeter in length—but members of the phylum range in adult size from about 80 micrometers long (some parasitic mites) to 3.6 meters across (a gigantic crab found in the sea off Japan). Some lobsters are nearly a meter in length. In these cases, their aquatic environments help support the weight of their exoskeletons. The largest living insects are about 33 centimeters long, but the giant dragonflies that lived 300 million years ago had wingspans of as much as 60 centimeters (2 feet)!

Because this size limitation is inherent in the body design of arthropods, few arthropods have ever grown to great size. As we will see, to overcome this limitation, a strong, flexible endoskeleton is required.

Arthropods have proven very successful. About two-thirds of all named species on earth are arthropods. Among the major kinds are arachnids (which include spiders, ticks, mites, scorpions, and daddy longlegs), crustaceans (which include lobsters, crabs, shrimps, and barnacles), cen-

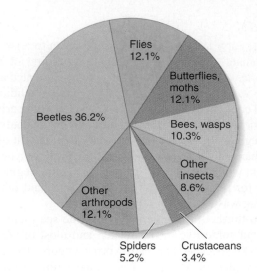

Figure 15.13 Arthropods are a successful group.
About two-thirds of all named species are arthropods. About 80% of all arthropods are insects, and about half of the named species of insects are beetles.

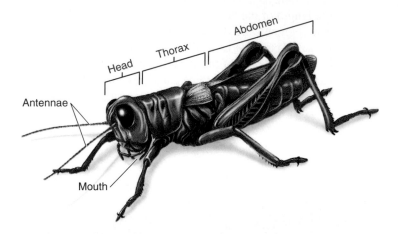

Figure 15.14 The body plan of an insect.
This grasshopper illustrates the body segmentation found in adult insects. The many segments found in most larval stages of insects become fused in the adult, giving rise to three adult body segments: the head, thorax, and abdomen. The appendages—legs, wings, mouthparts, antennae—are jointed.

tipedes and millipedes, and insects (figure 15.15). Scientists estimate that a quintillion (a billion billion) insects are alive at any one time—200 million insects for each living human!

> **15.4** Roundworms have a pseudocoel body cavity, while mollusks have a coelom body cavity; neither is segmented. Annelids are mostly marine segmented worms. Arthropods, the most successful animal phylum, have jointed appendages, a rigid exoskeleton, and, in the case of insects, wings.

Figure 15.15 Insect diversity.

(*a*) Some insects have a tough exoskeleton, like this South American scarab beetle, *Dilobderus abderus* (order Coleoptera). (*b*) Human flea, *Pulex irritans* (order Siphonaptera), in California. Fleas are flattened laterally, slipping easily through hair. (*c*) The honeybee, *Apis mellifera* (order Hymenoptera), is a widely domesticated and efficient pollinator of flowering plants. (*d*) This pot dragonfly (order Odonata) has a fragile exoskeleton. (*e*) A true bug, *Edessa rufomarginata* (order Hemiptera), in Panama. (*f*) Copulating grasshoppers (order Orthoptera). (*g*) Luna moth, *Actias luna,* in Virginia. Luna moths and their relatives are among the most spectacular insects (order Lepidoptera).

15.5 Redesigning the Embryo

There are two major kinds of coelomate animals representing two distinct evolutionary lines. All the coelomates we have met so far have essentially the same kind of embryo, starting as a hollow ball of cells, a blastula, which indents to form a two-layer-thick ball with a blastopore opening to the outside. In mollusks, annelids, and arthropods, the mouth (stoma) develops from or near the blastopore. This same pattern of development, in a general sense, is seen in all noncoelomate animals. An animal whose mouth develops in this way is called a **protostome** (from the Greek words *protos,* first, and *stoma,* mouth).

A second distinct pattern of embryological development occurs in the echinoderms and chordates. In these animals, the anus forms from or near the blastopore, and the mouth forms subsequently on another part of the blastula (figure 15.16). This group of phyla consists of animals that are called the **deuterostomes** (Greek, *deuteros,* second, and *stoma,* mouth).

Deuterostomes represent a revolution in embryonic development. In addition to the pattern of blastopore formation, deuterostomes differ from protostomes in three other fundamental embryological features:

1. The progressive division of cells during embryonic growth is called **cleavage.** The cleavage pattern relative to the embryo's polar axis determines how the cells array. In nearly all protostomes, each new cell buds off at an angle oblique to the polar axis in a pattern called **spiral cleavage.** In deuterostomes, the cells divide parallel to and at right angles to the polar axis. As a result, the pairs of cells from each division are positioned directly above and below one another; this process gives rise to a loosely packed array of cells. This pattern is called **radial cleavage.**

2. In protostomes, the developmental fate of each cell in the embryo is fixed when that cell first appears. Even at the four-celled stage, each cell is different, and no one cell, if separated from the others, can develop into a complete animal because the chemicals that act as developmental signals have already been localized in different parts of the egg. In deuterostomes, on the other hand, the first cleavage divisions of the fertilized embryo produce identical daughter cells, and any single cell, if separated, can develop into a complete organism.

3. In all coelomates, the coelom originates from mesoderm. In protostomes, this occurs simply and directly: The cells simply move away from one another as the coelomic cavity expands within the mesoderm. However, in deuterostomes, whole groups of cells usually move around to form new tissue associations. The coelom is normally produced by an evagination of the **archenteron**—the main cavity, also called the primitive gut. This cavity, lined with endoderm, opens to the outside via the blastopore and eventually becomes the gut cavity.

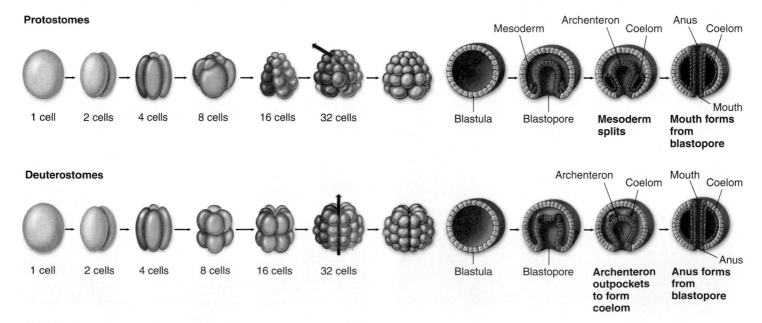

Figure 15.16 Embryonic development in protostomes and deuterostomes.
Cleavage of the egg produces a hollow ball of cells called the blastula. Invagination, or infolding, of the blastula produces the blastopore. In protostomes, embryonic cells cleave in a spiral pattern and become tightly packed. The blastopore becomes the animal's mouth, and the coelom originates from a mesodermal split. In deuterostomes, embryonic cells cleave radially and form a loosely packed array. The blastopore becomes the animal's anus, and the mouth develops at the other end. The coelom originates from an evagination, or outpouching, of the archenteron in deuterostomes.

Echinoderms

The first deuterostomes, marine animals called **echinoderms** in the phylum Echinodermata, appeared more than 650 million years ago. The term *echinoderm* means "spiny skin" and refers to an **endoskeleton** composed of hard, calcium-rich plates called ossicles just beneath the delicate skin. When they are first formed, the plates are enclosed in living tissue, and so are truly an endoskeleton, although in adults they fuse, forming a hard shell. About 6,000 species of echinoderms are living today, almost all of them on the ocean bottom (figure 15.17). Many of the most familiar animals seen along the seashore are echinoderms, including sea stars (starfish), sea urchins, sand dollars, and sea cucumbers.

The body plan of echinoderms undergoes a fundamental shift during development: All echinoderms are bilaterally symmetrical as larvae but become radially symmetrical as adults. Many biologists believe that early echinoderms were sessile and evolved adult radiality as an adaptation to the sessile existence. Bilaterality is of adaptive value to an animal that travels through its environment, whereas radiality is of value to an animal whose environment meets it on all sides. Adult echinoderms have a five-part body plan, easily seen in the five arms of a sea star. Its nervous system consists of a central ring of nerves from which five branches arise—while the animal is capable of complex response patterns, there is no centralization of function, no "brain." Some echinoderms like feather stars have 10 or 15 arms, but always multiples of five.

A key evolutionary innovation of echinoderms is the development of a hydraulic system to aid movement. Called a **water vascular system,** this fluid-filled system is composed of a central ring canal from which five radial canals extend out into the arms. From each radial canal, tiny vessels extend out through short side branches into thousands of tiny, hollow **tube feet.** At the base of each tube foot is a fluid-filled muscular sac that acts as a valve. When a sac contracts, its fluid is prevented from reentering the radial canal and instead is forced into the tube foot, thus extending it. When extended, the tube foot attaches itself to the ocean bottom, often aided by suckers. The sea star can then pull against these tube feet and so haul itself over the seafloor.

Most echinoderms reproduce sexually but they have the ability to regenerate lost parts, which can lead to asexual reproduction. In a few sea stars, asexual reproduction takes place by splitting, and the broken parts of the sea star can sometimes regenerate whole animals.

(a)

(b)   (c)

(d)   (e)

(f)

Figure 15.17   Diversity in echinoderms.

(a) Warty sea cucumber, *Parastichopus parvimensis*. (b) Feather star (class Crinoidea) on the Great Barrier Reef in Australia. (c) Brittle star, *Ophiothrix* (class Ophiuroidea). (d) Sand dollar, *Echinarachnius parma*. (e) Giant red sea urchin, *Strongylocentrotus franciscanus*. (f) Sea star, *Oreaster occidentalis* (class Asteroidea), in the Gulf of California, Mexico.

Chordates

Chordates (phylum Chordata) are deuterostomes that employ an endoskeleton that is truly internal, characterized by a flexible rod called a **notochord** that develops along the back of the embryo. Muscles attached to this rod allowed early chordates to swing their backs back and forth, swimming through the water. This key evolutionary innovation, attaching muscles to an internal element, started chordates along an evolutionary path that leads to the vertebrates and for the first time to truly large animals.

Chordates are distinguished by four principal features:

1. **Notochord.** A stiff, but flexible, rod that forms beneath the nerve cord, between it and the developing gut in the early embryo.
2. **Nerve cord.** A single dorsal (along the back) hollow nerve cord, to which the nerves that reach the different parts of the body are attached.
3. **Pharyngeal slits.** A series of slits behind the mouth into the pharynx, which is a muscular tube that connects the mouth to the digestive tract and windpipe.
4. **Postanal tail.** Chordates have a postanal tail, a tail that extends beyond the anus. A postanal tail is present at least during their embryonic development if not in the adult. Nearly all other animals have a terminal anus.

All chordates have all four of these characteristics at some time in their lives. For example, the tunicate that appears in figure 15.18 looks more like a sponge than a chordate, but its larval stage resembles a tadpole. All four features just listed are present in the tunicate larva. Likewise, human embryos have pharyngeal slits, a nerve cord, a notochord, and a postanal tail as embryos. The pharyngeal slits and postanal tail disappear during human development, and the notochord is replaced with the vertebral column.

Vertebrates

With the exception of tunicates and a group of fishlike marine animals, the lancelets, all chordates are **vertebrates.** Vertebrates differ from tunicates and lancelets in two important respects:

1. **Backbone.** The notochord becomes surrounded and then replaced during the course of the embryo's development by a bony vertebral column, a tube of hollow bones called vertebrae that encloses the dorsal nerve cord like a sleeve and protects it.
2. **Head.** All living vertebrates have a distinct and well-differentiated head, with a skull and brain.

All vertebrates have an internal skeleton made of bone or cartilage against which the muscles work. This endoskeleton makes possible the great size and extraordinary powers of movement that characterize the vertebrates.

> **15.5** Echinoderms and chordates are deuterostomes. Their eggs cleave radially, and the blastopore of their embryo becomes the animal's anus. Chordates have a notochord at some stage of their development. In adult vertebrates, the notochord is replaced by a backbone.

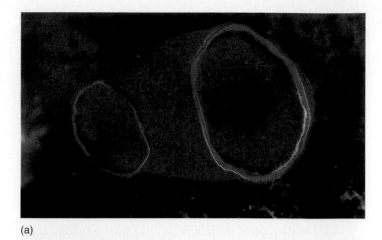

(a)

(b)

(c)

Figure 15.18 Diversity in chordates.

(a) A beautiful blue and gold tunicate. (b) Two lancelets, *Branchiostoma lanceolatum,* partly buried in shell gravel, with their anterior ends protruding. (c) This lion, stalking a potential dinner, is both large and very mobile, two traits characteristic of many terrestrial vertebrates.

15.6 Overview of Vertebrate Evolution

When scientists first began to study and date fossils, they had to find some way to organize the different time periods from which the fossils came. They divided the earth's past into large blocks of time called **eras.** Eras are further subdivided into smaller blocks of time called **periods,** and some periods, in turn, are subdivided into **epochs,** which can be divided into **ages.**

Virtually all of the major groups of animals that survive at the present time originated in the sea at the beginning of the **Paleozoic era,** during or soon after the Cambrian period (545–490 M.Y.A.). Thus, the diversification of animal life on earth is basically a marine record, and the fossils from the Paleozoic era all originated in the sea.

Many of the animal phyla that appeared in the Cambrian period have no living relatives. Their fossils indicate that this was a period of experimentation with different body forms and ways of life, some of which ultimately led to the contemporary phyla of animals and others to extinction. For example, the trilobites appear to be the ancestors of at least one living group, the horseshoe crabs, whereas the ammonites, which were abundant 100 million years ago, have no surviving descendants.

The first vertebrates evolved about 470 million years ago in the oceans—fishes without jaws. They didn't have paired fins either—many of them looked something like a flat hotdog with a hole at one end and a fin at the other. For 100 million years, a parade of different kinds of fishes were the only vertebrates on earth. They became the dominant creatures in the sea, some bigger than cars.

Invasion of the Land

Only a few of the animal phyla that evolved in the Cambrian seas have invaded the land successfully; most others have remained exclusively marine. The first organisms to colonize the land were fungi and plants, over 500 million years ago. The ancestors of plants were specialized members of a group of photosynthetic protists known as the green algae. It seems probable that plants first occupied the land in symbiotic association with fungi, as discussed in chapter 13.

The first invasion of the land by animals, and perhaps the most successful invasion of the land, was accomplished by the arthropods, a phylum of hard-shelled animals with jointed legs and segmented bodies. This invasion occurred about 410 million years ago.

Vertebrates invaded the land during the Carboniferous period (360–280 M.Y.A.). The first vertebrates to live on land were the amphibians, represented today by frogs, toads, salamanders, and caecilians (legless amphibians). The earliest amphibians known are from the Devonian period, and among their descendants are the reptiles, which became the ancestors of the dinosaurs, birds, and mammals.

Mass Extinctions

The history of life on earth has been marked by periodic episodes of extinction, where the loss of species outpaces the formation of new species. Particularly sharp declines in species diversity are called **mass extinctions.** Five mass extinctions have occurred, the first of them near the end of the Ordovician period about 438 million years ago. At that time, most of the existing families of trilobites, a very common type of marine arthropod, became extinct. Another mass extinction occurred about 360 million years ago at the end of the Devonian period.

The third and most drastic mass extinction in the history of life on earth happened during the last 10 million years of the Permian period, marking the end of the Paleozoic era. It is estimated that 96% of all species of marine animals that were living at that time became extinct! All of the trilobites disappeared forever. Brachiopods, marine animals resembling mollusks but with a different filter-feeding system, were extremely diverse and widespread during the Permian; only a few species survived. Bryozoans, marine filter feeders that formed coral-like colonies in oceans throughout the world in the Permian, became rare afterward.

Mass extinctions left vacant many ecological opportunities, and for this reason they were followed by rapid evolution among the relatively few plants, animals, and other organisms that survived the extinction. Little is known about the causes of major extinctions. In the case of the Permian mass extinction, some scientists argue that the extinction was brought on by a gradual accumulation of carbon dioxide in ocean waters, the result of large-scale volcanism due to the collision of the earth's landmasses during the formation of the single large "supercontinent" of Pangaea. Such an increase in carbon dioxide would have severely disrupted the ability of animals to carry out metabolism and form their shells.

The most famous and well-studied extinction, though not as drastic, occurred at the end of the Cretaceous period (65 million years ago), at which time the dinosaurs and a variety of other organisms went extinct. Recent findings have supported the hypothesis that this fifth mass extinction event was triggered when a large asteroid slammed into the earth, perhaps causing global forest fires and obscuring the sun for months by throwing particles into the air.

We are living during a new sixth mass extinction event. The number of species in the world is greater today than it has ever been. Unfortunately, that number is decreasing at an alarming rate due to human activity. Some estimate that as many as one-fourth of all species will become extinct in the near future, a rate of extinction not seen on earth since the Cretaceous mass extinction.

> **15.6** The diversification of the animal phyla occurred in the sea. Only two animal phyla have invaded the land successfully: arthropods and chordates (the vertebrates).

15.7 Fishes Dominate the Sea

A series of key evolutionary advances allowed vertebrates to first conquer the sea and then the land (figure 15.19). About half of all vertebrates are **fishes.** The most diverse and successful vertebrate group, they provided the evolutionary base for invasion of land by amphibians.

Characteristics of Fishes

From whale sharks that are 12 meters long to tiny cichlids no larger than your fingernail, fishes vary considerably in size, shape, color, and appearance. However varied, all fishes have four important characteristics in common:

1. **Gills**. Fish are water-dwelling creatures, and they use gills to extract dissolved oxygen gas from the water around them. Gills are fine filaments of tissue rich in blood vessels. When water passes over the gills in the back of the mouth, oxygen gas diffuses from the water into the fish's blood.
2. **Vertebral column.** All fishes have an internal skeleton with a spine surrounding the dorsal nerve cord, although it may not necessarily be made of bone. The

brain is fully encased within a protective box, the skull or cranium, made of bone or cartilage.

3. **Single-loop blood circulation.** Blood is pumped from the heart to the gills. From the gills, the oxygenated blood passes to the rest of the body and then returns to the heart.
4. **Nutritional deficiencies.** Fishes are unable to synthesize the aromatic amino acids and must consume them in their diet. This inability has been inherited by all their vertebrate descendants.

The earliest fishes are now extinct. Only their head-shields were made of bone; their elaborate internal skeletons were constructed of cartilage. Wriggling through the water, jawless and toothless, these fishes sucked up small food particles from the ocean floor. One group of jawless fishes, the agnathans, survive today as hagfish and parasitic lampreys. These first fishes were eventually replaced by larger, heavier, jawed fishes that were predators. Jaws seem to have evolved from the front-most of a series of arch supports made of cartilage that were used to reinforce the tissue between gill slits, holding the slits open. These heavier, jawed fish were eventually replaced by fishes that moved through the water faster—the sharks and the bony fishes.

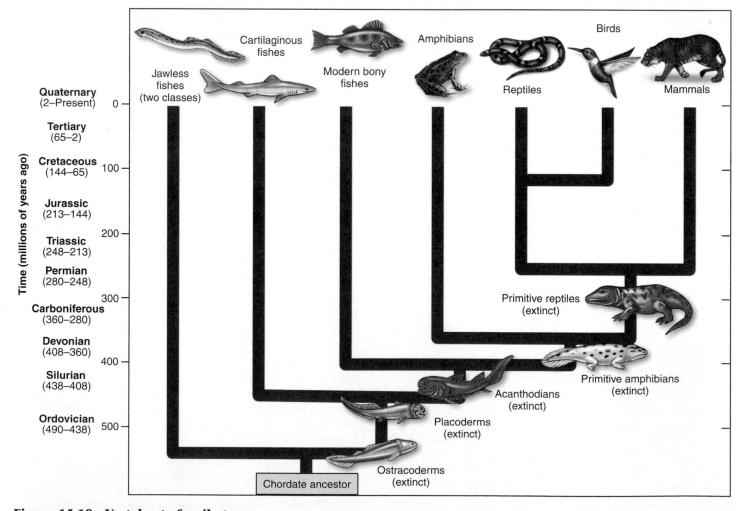

Figure 15.19 Vertebrate family tree.

Primitive amphibians arose from fishes. Primitive reptiles arose from amphibians and gave rise to mammals and to dinosaurs, which survive today as birds.

Figure 15.20 Chondrichthyes.
The Galápagos shark is a member of the class Chondrichthyes, which is composed of mainly predators or scavengers who spend most of their time in graceful motion. As they move, they create a flow of water past their gills, from which they extract oxygen.

Sharks

Sharks are fast and maneuverable swimmers due to their light and flexible cartilaginous skeleton. Members of this group, the class Chondrichthyes, consist of sharks, skates, and rays. Sharks are very powerful swimmers, with a back fin, a tail fin, and two sets of paired side fins for controlled thrusting through the water (figure 15.20). Skates and rays are flattened sharks that are bottom-dwellers. Today there are about 750 species of sharks, skates, and rays.

Reproduction among the Chondrichthyes is the most advanced of any fish. Shark eggs are fertilized internally. During mating, the male grasps the female with modified fins called claspers. Sperm pass from the male into the female through grooves in the claspers. About 40% of sharks lay fertilized eggs. The eggs of other species develop within the female's body, and the pups are born alive.

Bony Fishes

Instead of gaining speed through lightness, as sharks did, bony fishes adopted a heavy internal skeleton made completely of bone. Such an internal skeleton is very strong, providing a base against which very strong muscles could pull. Bony fishes are still buoyant though because they possess a **swim bladder,** a gas-filled sac that allows them to regulate their buoyant density and so remain effortlessly suspended at any depth in the water (figure 15.21). By adjusting the amount of gas in its swim bladder, a bony fish can rise up and down in the water the same way a submarine does. Sharks, by contrast, increase buoyancy with oil in their liver, but still must move through the water or sink, because their bodies are denser than water. The swim bladder solution to the challenge of swimming has proven to be a great success.

Bony fishes are the most successful of all fishes, indeed of all vertebrates. Of the nearly 30,800 living species of fishes in the world today, about 30,000 species are bony fishes with swim bladders. That's more species than all other kinds of vertebrates combined!

The remarkable success of the bony fishes has resulted from a series of significant adaptations. In addition to the swim bladder, bony fishes have a highly developed **lateral line system,** a sensory system that enables them to detect changes in water pressure and thus the movement of predators and prey in the water. Also, most bony fishes have a hard plate called the **operculum** that covers the gills on each side of the head. Flexing the operculum permits bony fishes to pump water over their gills. Using the operculum as very efficient bellows, bony fishes can pass water over the gills while stationary in the water. That is what a goldfish in a fishtank is doing when it seems to be gulping.

> **15.7** Fishes are characterized by gills, a simple, single-loop circulatory system, and a vertebral column. Sharks are fast swimmers, while the very successful bony fishes have a unique swim bladder.

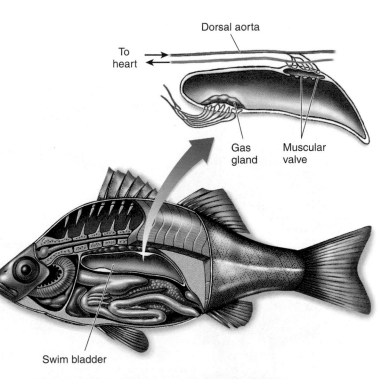

Figure 15.21 Diagram of a swim bladder.
The bony fishes use this structure, which evolved as a dorsal outpocketing of the pharynx, to control their buoyancy in water. The swim bladder can be filled with or drained of gas to allow the fish to control buoyancy. Gases are taken from the blood, and the gas gland secretes the gases into the swim bladder; gas is released from the bladder by a muscular valve.

Amphibians and Reptiles Invade the Land

Frogs, salamanders, and caecilians, the damp-skinned verte-brates, are direct descendants of fishes. They are the sole sur-vivors of a very successful group, the **amphibians,** the first vertebrates to walk on land (figure 15.22). Amphibians almost certainly evolved from the lobe-finned fishes, fish with paired fins that consist of a long fleshy muscular lobe, supported by a central core of bones that form fully articulated joints with one another.

Characteristics of Amphibians

Amphibians have five key characteristics that allowed them to successfully invade the land.

1. **Legs.** Frogs and salamanders have four legs and can move about on land quite well. Legs, which evolved from fins (figure 15.23), were one of the key adaptations to life on land. Caecilians have lost their legs during the course of adapting to a burrowing existence.

2. **Lungs.** Most amphibians possess a pair of lungs, although the internal surfaces are poorly developed. Lungs were necessary because the delicate structure of fish gills requires the buoyancy of water to support it. Also, there is far more oxygen in air than water and so lungs provide a more efficient means of respiration.

Figure 15.22 A representative amphibian.
This red-eyed tree frog, *Agalychnis callidryas,* is a member of the group of amphibians that includes frogs and toads (order Anura).

3. **Cutaneous respiration.** Frogs, salamanders, and caecilians all supplement the use of lungs by respiring directly across their skin, which is kept moist and provides an extensive surface area. This mode of respiration limits the body size of amphibians, because it is only efficient for a high surface-to-volume ratio.

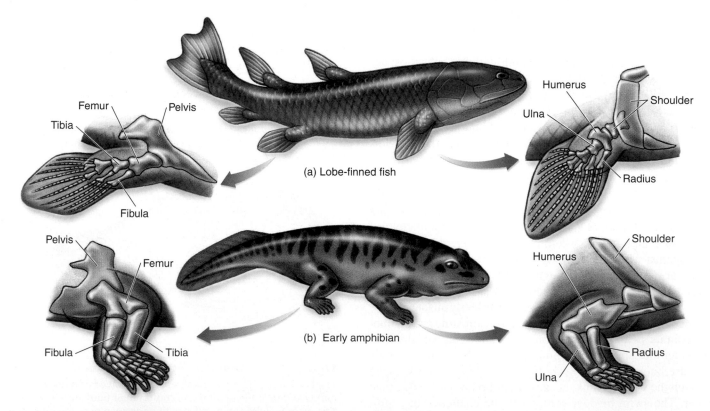

(a) Lobe-finned fish

(b) Early amphibian

Figure 15.23 A key adaptation of amphibians: the evolution of legs.
(a) The limbs of a lobe-finned fish. Some lobe-finned fishes could move out onto land. (b) The limbs of an early amphibian. As illustrated by their skeletal structure, the legs of primitive amphibians could clearly function on land better than could the fins of lobe-finned fishes.

4. **Pulmonary veins.** After blood is pumped through the lungs, two large veins called pulmonary veins return the aerated blood to the heart for repumping. This allows the aerated blood to be pumped to the tissues at a much higher pressure than when it leaves the lungs.

5. **Partially divided heart.** The heart evolved to deliver greater amounts of oxygen to the amphibian tissues, because greater amounts of oxygen are required by muscles for walking. The initial chamber of the fish heart is absent in amphibians, and the second and last chambers are separated by a dividing wall that helps prevent aerated blood from the lungs from mixing with nonaerated blood being returned to the heart from the rest of the body. This separates the blood circulation into two separate paths, pulmonary and systemic. The separation is imperfect; the third chamber has no dividing wall.

History of Amphibians

Amphibians were the dominant land vertebrate for 100 million years. They first became common in the late Paleozoic era, when much of North America was covered by lowland tropical swamps. Amphibians reached their greatest diversity during the mid-Permian period, when 40 families existed. Sixty percent of them were fully terrestrial, with bony plates and armor covering their bodies, and many grew to be very large—some as big as a pony! After the great Permian extinction, the terrestrial forms began to decline, and by the time dinosaurs evolved, only 15 families remained, all aquatic. Only two of these families survived the Age of Dinosaurs, both aquatic: the anurans (frogs and toads) and the urodeles (salamanders).

Approximately 4,850 species of amphibians exist today, in 37 different families, all aquatic and all descended from the two aquatic families that survived the Age of Dinosaurs. Three orders constitute the class Amphibia: Anura, frogs and toads; Urodela, salamanders and newts; and Apoda, caecilians. Most of today's amphibians must reproduce in water and live the early part of their lives there, so amphibians are not completely terrestrial. However, in most habitats, particularly in the tropics, they are often today the most abundant and successful vertebrates to be found.

Reptiles

If we think of amphibians as the "first draft" of a manuscript about survival on land, then **reptiles** were the finished book. All 7,000 species of living reptiles share certain fundamental characteristics, features they retained from the time when they replaced amphibians as the dominant terrestrial vertebrates. Among the most important are:

1. **Amniotic egg.** Amphibians never succeeded in becoming fully terrestrial because amphibian eggs must be laid in water to avoid drying out. Most reptiles lay watertight eggs that contain a food source

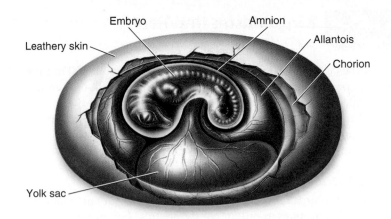

Figure 15.24 A key adaptation of reptiles: watertight eggs.
The watertight amniotic egg allows reptiles to live in a wide variety of terrestrial habitats.

(the yolk) and a series of four membranes—the chorion, the amnion, the yolk sac, and the allantois (figure 15.24). Each membrane plays a role in making the egg an independent life-support system. The outermost membrane of the egg, the **chorion,** allows oxygen to enter the porous shell but retains water within the egg. The **amnion** encases the developing embryo within a fluid-filled cavity. The **yolk sac** provides food from the yolk for the embryo via blood vessels connecting to the embryo's gut. The **allantois** surrounds a cavity into which waste products from the embryo are excreted.

2. **Dry skin.** Amphibians have a moist skin and must remain in moist places to avoid drying out. Like their ancestors, reptiles have dry skin. A layer of scales or armor covers their bodies, preventing water loss.

3. **Thoracic breathing.** Amphibians breathe by squeezing their throats to pump air into their lungs; this limits their breathing capacity to the volume of their mouths. Reptiles developed pulmonary breathing, expanding and contracting the rib cage to suck air into the lungs and then force it out.

In addition, reptiles improved on the innovations first attempted by amphibians. Legs were arranged to more effectively support the body's weight, allowing reptile bodies to be bigger and to run. Also, the lungs and heart were altered to make them far more efficient.

15.8 **Amphibians were the first vertebrates to successfully invade land, helped by legs, lungs, and the pulmonary vein. Reptiles have three characteristics that suit them well for life on land: a watertight (amniotic) egg, dry skin, and thoracic breathing.**

15.9 Birds Master the Air

Birds evolved from small bipedal dinosaurs about 150 million years ago, but they were not common until the flying reptiles called pterosaurs became extinct along with the dinosaurs. Unlike pterosaurs, birds are insulated with feathers. Birds are so structurally similar to dinosaurs in all other respects that many scientists consider birds to be simply feathered dinosaurs.

Characteristics of Birds

Modern birds lack teeth and have only vestigial tails, but they still retain many reptilian characteristics. For instance, birds lay amniotic eggs, although the shells of bird eggs are hard rather than leathery. Also, reptilian scales are present on the feet and lower legs of birds. What makes birds unique? What distinguishes them from living reptiles?

1. **Feathers.** Derived from reptilian scales, feathers are the ideal adaptation for flight, lightweight and easily replaced if damaged (figure 15.25).
2. **Flight skeleton.** The bones of birds are thin and hollow. Many of the bones are fused, making the bird skeleton more rigid than a reptilian skeleton, forming a sturdy frame that anchors muscles during flight. The power for active flight comes from large breast muscles that can make up 30% of a bird's total body weight. They stretch down from the wing and attach to the breastbone, which is greatly enlarged and bears a prominent keel for muscle attachment. They also attach to the fused collarbones that form the so-called wishbone. No other living vertebrates have a fused collarbone or a keeled breastbone.

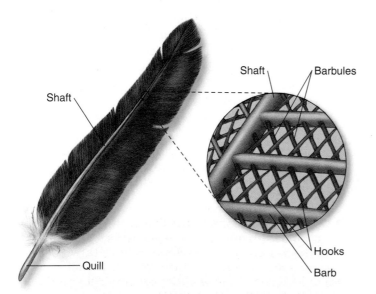

Figure 15.25   A key adaptation of birds: feathers.
The barbs off the main shaft of a feather have secondary branches called barbules. The barbules of adjacent barbs are attached to one another by microscopic hooks.

Figure 15.26   *Archaeopteryx*.
An artist's reconstruction of *Archaeopteryx* based on fossil records.

Birds, like mammals, are endothermic. They generate enough heat through metabolism to maintain a high body temperature. Birds maintain body temperatures significantly higher than most mammals. The high body temperature permits a faster metabolism, necessary to satisfy the large energy requirements of flight.

History of Birds

The oldest bird of which there is a clear fossil is ***Archaeopteryx*** (meaning "ancient wing"), which was about the size of a crow and shared many features with small, bipedal, carnivorous dinosaurs (figure 15.26). For example, it had teeth and a long reptilian tail. And unlike the hollow bones of today's birds, its bones were solid. Because of its many dinosaur features, several *Archaeopteryx* fossils were originally classified as *Compsognathus,* a small theropod dinosaur of similar size—until feathers were discovered on the fossils. What makes *Archaeopteryx* distinctly avian is the presence of feathers on its wings and tail. It also has other birdlike features, notably the presence of a wishbone. Dinosaurs lack a wishbone, although thecodonts had them.

By the early Cretaceous, only a few million years after *Archaeopteryx,* a diverse array of birds had evolved, with many of the features of modern birds. Fossils in Mongolia, Spain, and China discovered within the last few years reveal a diverse collection of toothed birds with the hollow bones and breastbones necessary for sustained flight. The diverse birds of the Cretaceous shared the skies with pterosaurs for 70 million years.

Today about 8,600 species of birds (class Aves) occupy a variety of habitats all over the world. You can tell a great deal about birds by examining their beaks. For example, carnivorous birds such as hawks have a sharp beak for tearing apart meat, the beaks of ducks are flat for shoveling through mud, and the beaks of finches are short and thick for crushing seeds.

> **15.9   Birds are essentially dinosaurs with feathers. Feathers and a strong, light skeleton make flight possible.**

15.10 Mammals Adapt to Colder Times

Characteristics of Mammals

The **mammals** (class Mammalia) that evolved about 220 million years ago would look strange to you, not at all like modern-day lions and tigers and bears. They share three key characteristics with mammals today:

1. **Mammary glands.** Female mammals have mammary glands, which produce milk to nurse the newborns. Even baby whales are nursed by their mother's milk. Milk is a very-high-calorie food (human milk has 750 kcal per liter), important because of the high energy needs of a rapidly growing newborn mammal.

2. **Hair.** Among living vertebrates, only mammals have hair, and all mammals do (even whales and dolphins have a few sensitive bristles on their snout). A hair is a filament composed of dead cells filled with the protein keratin. The primary function of hair is insulation. The insulation provided by fur may have ensured the survival of mammals when the dinosaurs perished.

3. **Middle ear.** All mammals have three middle ear bones, which amplify vibrations created by sound waves that beat upon the eardrum.

History of Mammals

Mammals have been around since the time of the dinosaurs, although they were always small until dinosaurs disappeared. We have learned a lot about the evolutionary history of mammals from their fossils. The first mammals were tiny, shrewlike creatures that lived in trees chasing insects. For 155 million years, all the time the dinosaurs flourished, mammals were a minor group that changed little.

Today, over 4,500 species of mammals occupy all the large-body niches that dinosaurs once claimed, among many others. They range in size from 1.5-gram shrews to 100-ton whales. Almost half of all mammals are rodents—mice and their relatives. Almost one-quarter of all mammals are bats! Mammals have even invaded the seas, as did some types of extinct reptiles—79 species of whales and dolphins live in today's oceans. The placental mammals that walked the earth during the ice ages were even larger than today's versions; the world's climate has warmed again in recent times, favoring smaller bodies, which are easier to cool.

Primates, the order to which we belong, are not a particularly large group; there are only 233 known species. Human beings evolved only very recently, less than 2 million years ago. There have been at least three species of humans, but our species, *Homo sapiens,* is the only one that survives today. We are notable among primates for having less hair, walking upright, making complex tools and having complicated language.

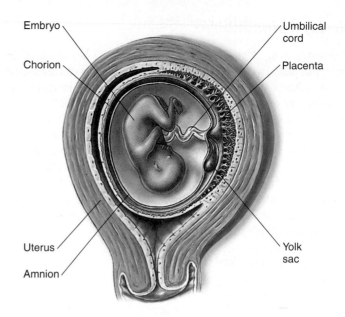

Figure 15.27   The placenta.
The placenta is characteristic of the largest group of mammals, the placental mammals. It evolved from membranes in the amniotic egg. The umbilical cord evolved from the allantois. The chorion, or outermost part of the amniotic egg, forms most of the placenta itself. The placenta serves as the provisional lungs, intestine, and kidneys of the embryo, without ever mixing maternal and fetal blood.

Other Characteristics of Modern Mammals

Endothermy.   Mammals are endothermic, a crucial adaptation that allows them to be active at any time of the day or night and to colonize severe environments, from deserts to ice fields. Many characteristics, such as hair, which provides insulation, played important roles in making endothermy possible. Also, the more efficient blood circulation provided by a four-chambered heart and the more efficient breathing provided by the *diaphragm* (a special sheet of muscles below the rib cage that aids breathing) make possible the higher metabolic rate upon which endothermy depends.

Placenta.   In most mammal species, females carry their young in the uterus during development, nourishing them by a placenta, and give birth to live young. The placenta is a specialized organ within the womb of the mother that brings the bloodstream of the fetus into close contact with the bloodstream of the mother (figure 15.27). Food, water, and oxygen can pass across from mother to child, and wastes can pass over to the mother's blood and be carried away.

Teeth.   Reptiles have homodont dentition: Their teeth are all the same. However, mammals have heterodont dentition, with different types of teeth that are highly specialized to match particular eating habits.

Hooves and Horns.   Keratin, the protein of hair, is also the structural building material in claws, fingernails, and hooves. Hooves are specialized keratin pads on the toes of horses, cows, sheep, antelopes, and other running mammals. The pads are hard and horny, protecting the toe and cushioning it from impact.

Today's Mammals

Monotremes: Egg-Laying Mammals. The duck-billed platypus and two species of echidna, or spiny anteater (figure 15.28a), are the only living monotremes. The monotremes have many reptilian features, including laying shelled eggs, but they also have both of the defining mammalian features: hair and functioning mammary glands. Females lack well-developed nipples, so the newly hatched babies cannot suckle. Instead, the milk oozes onto the mother's fur, and the babies lap it off with their tongues. The platypus, found only in Australia, lives much of its life in the water and is a good swimmer. It uses its bill much as a duck does, rooting in the mud for worms and other small animals.

Marsupials: Pouched Mammals. The major difference between marsupials (figure 15.28b) and other mammals is their pattern of embryonic development. In marsupials, a fertilized egg is surrounded by chorion and amniotic membranes, but no shell forms around the egg as it does in monotremes. The marsupial embryo is nourished by an abundant yolk within the shell-less egg. Shortly before birth, a short-lived placenta forms from the chorion membrane. After the embryo is born, tiny and hairless, it crawls into the marsupial pouch where it latches onto a nipple and continues its development.

Placental Mammals. As stated earlier, mammals that produce a true placenta, which nourishes the embryo throughout its entire development, are called placental mammals (figure 15.28c). Most species of mammals living today, including humans, are in this group. Early in the course of embryonic development, the placenta forms from both fetal and maternal tissues. Unlike marsupials, the young undergo a considerable period of development before they are born.

Figure 15.28 The three kinds of living mammals.
(a) This echidna, *Tachyglossus aculeatus,* is a monotreme.
(b) Marsupials include kangaroos, like this adult with young in its pouch. (c) This female African lion, *Panthera leo* (order Carnivora), is a placental mammal.

15.10 Mammals are endotherms that nurse their young with milk and exhibit a variety of different kinds of teeth. All mammals have at least some hair.

Exploring Current Issues

Additional Resources

Go to your campus library or look online to find the following articles, which further develop some of the concepts found in this chapter.

Erwin, D., J. Valentine, and D. Jablonski. (1997). The origin of animal body plans. *American Scientist,* 85(2), 126.

Fox, D. (1999). Why we don't lay eggs: mammals have perfected the art of incubating their young in their bodies. *New Scientist,* 162(2190), 26.

Hendrix, P.F. and P.J. Bohen. (2002). Exotic earthworm invasions in North America: ecological and policy implications. *BioScience,* 52(9), 801.

Johnson, C.C. (2002). The rise and fall of rudist reefs: reefs of the dinosaur era were dominated not by corals but by odd mollusks, which died off at the end of the Cretaceous from causes yet to be discovered. *American Scientist,* 90(2), 128.

Wong, K. (2002). Taking wing. *Scientific American,* 286(1), 16.

Biology and Society Lecture: The Challenge of Invading Land

The first vertebrates to invade the land were descendants of bony fishes. To successfully invade land, they had to solve many problems including: 1. how to get around out of water; 2. how to avoid drying out; 3. how to acquire oxygen from air; 4. how to get more oxygen to muscles; and 5. how to reproduce out of water.

Biology and Society Lecture: A Parade of Phyla

Animals, millions of species of them, are among the most abundant living things. Found in every conceivable habitat, they bewilder us with their diversity. The evolution of the animal phyla involved five key transitions in body plan: 1. evolution of tissues; 2. evolution of bilateral symmetry; 3. evolution of body cavity; 4. evolution of segmentation; and 5. evolution of deuterostome development.

Find these lectures, delivered by the author to his class at Washington University, online at www.mhhe.com/tlwessentials/exp15.

Introduction to the Animals

15.1 General Features of Animals

- Animals are complex multicellular heterotrophs. They are mobile and reproduce sexually. Animals share common characteristics, but the evolution of key innovations allows us to see evolutionary relationships (**figure 15.2 and table 15.1**).

Evolution of the Animal Phyla

15.2 Sponges and Cnidarians: The Simplest Animals

- Sponges, in the subkingdom Parazoa, are aquatic, have specialized cells, but lack tissues. The vaselike shaped adult is anchored to the substrate (**figure 15.3**).

- Cnidarians have radially symmetrical bodies and two tissue layers, an ectoderm and an endoderm. They capture prey and digest it extracellularly in the gastrovascular cavity (**figure 15.5**).

15.3 The Advent of Bilateral Symmetry

- The simplest bilaterally symmetrical animals are the solid worms, including flatworms (**figure 15.7**). They have three embryonic tissue layers, ectoderm, mesoderm, and endoderm. They lack a body cavity and so are called acoelomates.

15.4 The Advent of a Body Cavity

- The evolution of a body cavity improved body functions. The roundworms have a body cavity, but it is not a true body cavity, so they are called pseudocoelomates (**figure 15.9**).

- Mollusks have a true body cavity and so are called coelomates. Although a diverse group, they all contain a head-foot, a visceral mass, and a mantle (**figure 15.11**).

- Segmentation first evolved in annelid worms, which allowed for specialization, providing flexibility (**figure 15.12**).

- Arthropods are the most successful animal phylum. Jointed appendages first evolved in this group, as well as the evolution of a rigid exoskeleton, that improved protection and mobility (**figure 15.14**).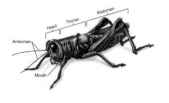

15.5 Redesigning the Embryo

- In the coelomates, there are two different developmental patterns. A deuterostome developmental pattern evolved in echinoderms and chordates, suggesting that they share a common ancestor. All other coelomates are protostomes (**figure 15.16**).

- Echinoderms have an endoskeleton made up of bony plates that lie under the skin. The adults are radially symmetrical, but that appears to be an adaptation to their environment (**figure 15.17**).

- The chordates have a truly internal endoskeleton and are distinguished by the presence of a notochord, a dorsal nerve cord, pharyngeal slits, and a postanal tail (**figure 15.18**). The notochord is replaced with a backbone in vertebrates.

The Parade of Vertebrates

15.6 Overview of Vertebrate Evolution

- Animal life began in the sea primarily during the Paleozoic era with only two animal groups having successfully invaded the land, arthropods and vertebrates.

- Mass extinctions, where the loss of species outpaces new species formation, have occurred throughout earth's history.

15.7 Fishes Dominate the Sea

- Fishes are the ancestors of all vertebrates (**figure 15.19**). Although a diverse group, all fishes have gills, a vertebral column, a single-loop circulatory system, and nutritional deficiencies.

- Sharks have a flexible cartilaginous skeleton and are very fast swimmers. Bony fish have swim bladders for buoyancy (**figure 15.21**) and are the most successful group of vertebrates.

15.8 Amphibians and Reptiles Invade the Land

- Amphibians were the first vertebrates to invade the land, having evolved from lobe-finned fishes (**figure 15.23**). Adaptations to a terrestrial environment included the development of legs, lungs, cutaneous respiration, pulmonary veins, and a partially divided heart.

- The key adaptations found in reptiles that made them better suited to life on land were the evolution of a watertight amniotic egg (**figure 15.24**), watertight skin, and thoracic breathing.

15.9 Birds Master the Air

- Birds evolved from dinosaurs but were a minor group until the mass extinction of dinosaurs, leaving open large niches for birds to fill. Two key adaptations allowed birds to dominate the skies: feathers (**figure 15.25**) and a skeleton modified for flight, with its lightweight, hollow bones.

15.10 Mammals Adapt to Colder Times

- Mammals expanded and diversified to fill niches left vacant after the mass extinction of dinosaurs. Mammals are distinguished by the presence of mammary glands to nurse their young, hair for insulation, and bones in the middle ear to amplify sound. They are endothermic. The evolution of the placenta offered protection to the fetus (**figure 15.27**) and distinguishes the placental mammals from other mammals, the monotremes and marsupials.

1. Sponges possess unique, collared flagellated cells called
   a. cnidocytes.
   c. choanoflagellates.
   b. choanocytes.
   d. epithelial cells.
2. Which of the following characteristics is *not* seen in the flatworms?
   a. cephalization.
   c. a body cavity.
   b. the presence of a mesoderm.
   d. bilateral symmetry.
3. One difference between the pseudocoel found in the roundworms and the coelom found in the segmented worms is
   a. the coelom develops between the mesoderm and ectoderm in segmented worms and the pseudocoel in the mesoderm in roundworms.
   b. the pseudocoel develops between the mesoderm and the endoderm in roundworms and the coelom develops in the mesoderm in segmented worms.
   c. the coelom develops between the endoderm and ectoderm in segmented worms, and the pseudocoel develops in the mesoderm of roundworms.
   d. the pseudocoel develops between the ectoderm and endoderm in roundworms, and the coelom develops in the mesoderm in segmented worms.
4. The main limiting factor of arthropod size is the
   a. inefficiency of open circulatory systems.
   b. weight of the muscles needed to move the organism.
   c. weight of the thick exoskeleton needed to support a large insect.
   d. entire weight of the organism that would crush the soft body during molting.
5. Echinoderms take on radial symmetry as adults. Some biologists explain this change in symmetry as an adaptation for
   a. an animal traveling through its environment, rather than a sessile lifestyle.
   b. an animal with a sessile lifestyle, rather than one that moves through its environment.
   c. a predator, rather than a filterfeeder.
   d. an animal living in a marine environment, rather than a freshwater environment.
6. All fish species share all of the following characteristics *except*
   a. gills.
   c. endoskeleton with dorsal nerve cord.
   b. jaws.
   d. single loop circulatory system.
7. The two fish families, the sharks and bony fish, have evolved adaptations for swimming. Which modification do these two vertebrate groups have in common?
   a. dorsal (back) fin, tail fins, and paired ventral fins
   b. buoyancy control through swim bladders
   c. endoskeleton made of cartilage
   d. use of oil in the liver to prevent sinking
8. Adaptations that evolved in reptiles that allow them to overcome dehydration do *not* include
   a. an amniotic egg.
   b. a layer of scales on the skin.
   c. more effective arrangement of appendages.
   d. modifications to the respiratory system.
9. Characteristics that evolved in birds to allow for flight include
   a. reptilian-like scales on the legs.
   b. a hard-shelled amniotic egg.
   c. internal fertilization.
   d. thin, hollow bones in the skeleton.
10. A characteristic unique to almost all mammals and no other vertebrate is
    a. thoracic breathing.
    b. skin covering for insulation and protection from dehydration.
    c. a placenta and internal development of offspring.
    d. endothermy, able to control body temperature.

1. **Figure 15.13** About two-thirds of all the named species on our planet are arthropods. From the arthropod pie chart, estimate about what percentage of the named species on our planet are beetles?

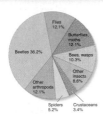

2. **Figure 15.24** What advantages do reptiles have by covering the reptile egg with a leathery outer shell?

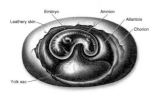

Introduction to the Animals You have discovered a new organism deep in the jungle. You are trying to decide if it is a slow-moving animal or a plant that responds to light touch. What characteristics can you investigate to help you decide?

Evolution of the Animal Phyla Why is the body cavity such a useful innovation in the animal kingdom?

The Parade of Vertebrates Your friend, Ahmed, tells you that his father can't understand why, since there have already been five mass extinctions on our planet, people get so upset about the possibility of a few more species dying out because of global warming or rainforest destruction. What do you tell him?

Visit the Online Learning Center for this chapter at www.mhhe.com/tlwessentials/ch.15 for quizzes, animations, interactive learning exercises, and other study tools. At the site you will also find extended answers to the end-of-chapter questions.

16
Ecosystems

The earth provides living organisms with much more than a place to stand or swim. Many chemicals cycle between our bodies and the physical environment around us. We live in a delicate balance with our physical surroundings, one easily disturbed by human activities. The collection of organisms that live in a place, and all the physical aspects of the environment that affect how they live, operate as a fundamental biological unit, the ecological system or ecosystem. This mountain meadow, in the High Sierras of California, is an ecosystem, and so is the desert of Death Valley. All of the earth's surface—mountains and deserts and deep-sea floors—is teeming with life, although it may not always appear so. The same ecological principles apply to the organization of all the earth's communities, both on land and in the sea, although the details may differ greatly. Ecology is the study of ecosystems and how they function. In this chapter, we focus on principles that govern the functioning of communities of organisms living together, and on the physical and biological factors that determine why particular kinds of organisms can be found living in one place and not in another. A proper understanding of how ecosystems function will be critical to preserving the living world in this new century.

The ecosystem is the most complex level of biological organization. The biosphere includes all the ecosystems on earth. The earth is a closed system with respect to chemicals but an open system in terms of energy. Collectively, the organisms in ecosystems regulate the capture and expenditure of energy and the cycling of chemicals. All organisms depend on the ability of other organisms—plants, algae, and some bacteria—to recycle the basic components of life.

16.1 Energy Flows Through Ecosystems

What Is an Ecosystem?

Ecology is the study of the interactions of living organisms with one another and with their physical environment (soil, water, weather, and so on). Ecologists, the scientists who study ecology, view the world as a patchwork quilt of different environments, all bordering on and interacting with one another. Consider for a moment a patch of forest, the sort of place a deer might live. Ecologists call the collection of creatures that live in a particular place a **community**—all the animals, plants, fungi, and microorganisms that live together in a forest, for example, are the forest community. Ecologists call the place where a community lives its **habitat**—the soil, and the water flowing through it, are key components of the forest habitat. The sum of these two, community and habitat, is an ecological system, or **ecosystem.** An ecosystem is a largely self-sustaining collection of organisms and their physical environment. An ecosystem can be as large as a forest or as small as a tidepool.

The Path of Energy: Who Eats Whom in Ecosystems

Energy flows into the biological world from the sun, which shines a constant beam of light on our earth. Life exists on earth because some of that continual flow of light energy can be captured and transformed into chemical energy through the process of photosynthesis and used to make organic molecules such as carbohydrates, proteins, and fats. These organic molecules, which also include nucleic acids, are what we call food. Living organisms use the energy in food to make new materials for growth, to repair damaged tissues, to reproduce, and to do myriad other things that require energy, like turning the pages of this text.

You can think of all the organisms in an ecosystem as chemical machines fueled by energy captured in photosynthesis. The organisms that first capture the energy, the **producers,** are plants (and algae and some bacteria), which produce their own energy-storing molecules by carrying out photosynthesis. They are also referred to as *autotrophs*. All other organisms in an ecosystem are **consumers,** obtaining their energy-storing molecules by consuming plants or other animals, and are referred to as *heterotrophs*. Ecologists assign every organism in an ecosystem to a trophic, or feeding, level,

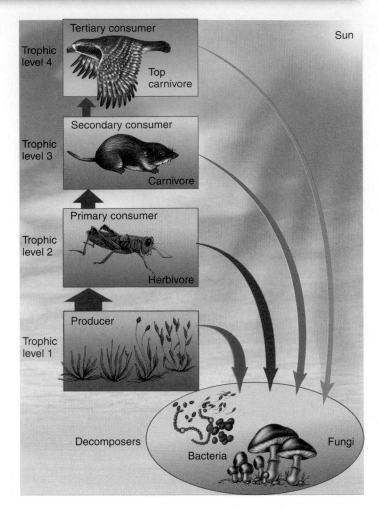

Figure 16.1 Trophic levels within an ecosystem.
Ecologists assign all the members of a community to various trophic levels based on feeding relationships. Producers, such as photosynthetic plants, obtain their energy directly from the sun and are assigned to trophic level 1. Animals that eat plants (herbivores) are at trophic level 2. Animals that eat plant-eating animals (carnivores) are at trophic level 3 and higher. Decomposers, organisms that break down organic substances including dead organisms, use all trophic levels for food.

depending on the source of its energy. A **trophic level** is composed of those organisms within an ecosystem whose source of energy is the same number of consumption "steps" away from the sun. Thus, a plant's trophic level is 1, while animals that graze on plants are in trophic level 2, and animals that eat these grazers are in trophic level 3 (figure 16.1). Higher trophic levels exist for animals that eat higher on the food chain. Food energy passes through an ecosystem from one trophic level to another. When the path is a simple linear progression, like the links of a chain, it is called a **food chain.** In most ecosystems, however, the path of energy is not a simple linear one, because individual animals often feed at several trophic levels. This creates a complicated path of energy flow called a **food web** (figure 16.2).

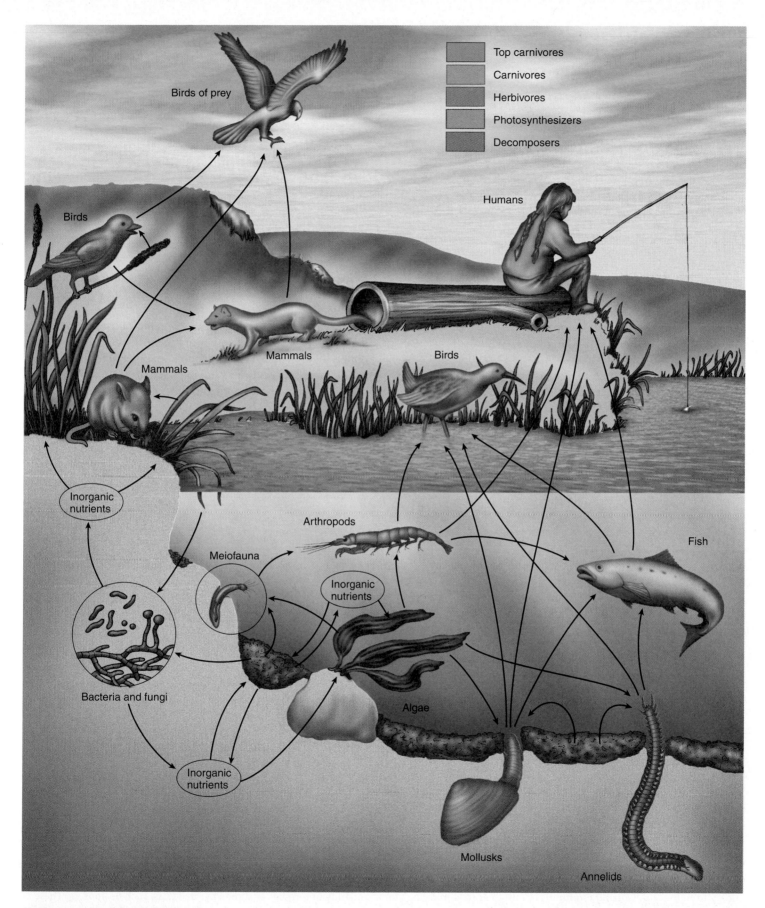

Figure 16.2 A food web.

A food web is much more complicated than a linear food chain. The path of energy passes from one trophic level to another and back again in complex ways.

Producers

The lowest trophic level of any ecosystem is occupied by the producers (figure 16.3a)—green plants in most land ecosystems (and, usually, algae in freshwater). Plants use the energy of the sun to build energy-rich sugar molecules. They also absorb carbon dioxide from the air, and nitrogen and other key substances from the soil and build them into biological molecules. It is important to realize that plants consume as well as produce. The roots of a plant, for example, do not carry out photosynthesis—there is no sunlight underground. Roots obtain their energy the same way you do, by using energy-storing molecules produced elsewhere (in this case, in the leaves of the plant).

Herbivores

At the second trophic level are **herbivores,** animals that eat plants (figure 16.3a). They are the *primary consumers* of ecosystems. Deer and horses are herbivores, and so are rhinoceroses, chickens (primarily herbivores), and caterpillars. Most herbivores rely on "helpers" to aid in the digestion of cellulose, a structural material found in plants. A cow, for instance, has a thriving colony of bacteria in its gut that digests cellulose for it. So does a termite. Humans cannot digest cellulose, because we lack these bacteria—that is why a cow can live on a diet of grass and you cannot.

Carnivores

At the third trophic level are animals that eat herbivores, called **carnivores** (meat eaters). They are the *secondary consumers* of ecosystems. Tigers and wolves are carnivores (figure 16.3b), and so are mosquitoes and blue jays. Some animals, like bears and humans, eat both plants and animals and are called **omnivores** (figure 16.3c). They use the simple sugars and starches stored in plants as food and not the cellulose. Many complex ecosystems contain a fourth trophic level, composed of animals that consume other carnivores. They are called tertiary consumers, or top carnivores. A weasel that eats a blue jay is a tertiary consumer. Only rarely do ecosystems contain more than four trophic levels, for reasons we will discuss later.

Detritivores and Decomposers

In every ecosystem there is a special class of consumers that include **detritivores** (figure 16.3d), organisms that eat dead organisms (also referred to as scavengers) and **decomposers,** organisms that break down organic substances making the nutrients available to other organisms (figure 16.3e). They obtain their energy from all trophic levels. Bacteria and fungi are the principal decomposers in land ecosystems. Worms, arthropods, and vultures are examples of detritivores.

(a) Producers and herbivores

(b) Carnivores

(c) Omnivore

(d) Detritivore

(e) Decomposer

Figure 16.3   Members of the food chain.

(a) The East African grasslands are covered by a dense growth of grasses, with interspersed trees; these plants are primary producers, capturing energy from the sun. Grazing mammals obtain their food from the plants and may in turn be consumed by predators, such as lions. (b) These wolves are carnivores that roam the forests of North America searching for prey. (c) This grizzly bear is an omnivore, this one fishing for salmon, but bears also eat berries and other plant materials. (d) This crab, *Geocarcinus quadratus,* photographed on the beach at Mazatlán, Mexico, is a detritivore, playing the same role that vultures and similar animals do in other ecosystems. (e) Fungi, such as the basidiomycete whose mycelium is shown here growing through the soil in Costa Rica, are, together with bacteria, the primary decomposers of terrestrial ecosystems.

Energy Flows Through Trophic Levels

How much energy passes through an ecosystem? **Primary productivity** is the total amount of light energy converted by photosynthetic organisms into organic compounds in a given area per unit of time. An ecosystem's **net primary productivity** is the total amount of energy fixed by photosynthesis per unit of time, minus that which is expended by photosynthetic organisms to fuel metabolic activities. In short, it is the energy stored in organic compounds that is available to heterotrophs. The total weight of all ecosystem organisms, called the ecosystem's **biomass,** increases as a result of the ecosystem's net productivity. Some ecosystems, such as cattail swamps, which are wetlands, have a high net primary productivity. Others, such as tropical rain forests, also have a relatively high net primary productivity, but a rain forest has a much larger biomass than a wetlands area. Consequently, a rain forest's net primary productivity is much lower in relation to its biomass.

When a plant uses the energy from sunlight to make structural molecules such as cellulose, it loses a lot of the energy as heat. In fact, only about half of the energy captured by the plant ends up stored in its molecules. The other half of the energy is lost. This is the first of many such losses as the energy passes through the ecosystem. When the energy flow through an ecosystem is measured at each trophic level, we find that 80% to 95% of the energy available at one trophic level is not transferred to the next. In other words, there is one order of magnitude drop in available energy from one trophic level to the next. For example, the amount of energy that ends up in the herbivore's body is approximately an order of magnitude less than the energy present in the plant molecules it eats (figure 16.4). Similarly, when a carnivore eats the herbivore, an order of magnitude is lost from the amount of energy present in the herbivore's molecules. This is why food chains generally consist of only three or four steps. So much energy

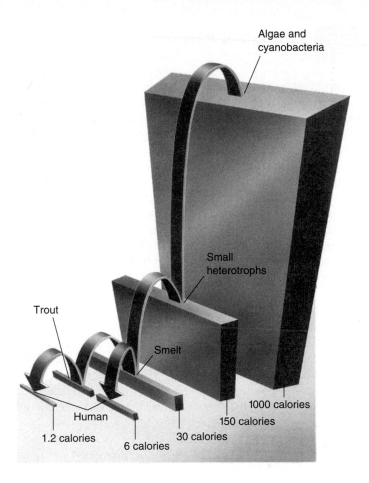

Figure 16.5   Energy loss in an ecosystem.

In a classic study of Cayuga Lake in New York, the path of energy was measured precisely at all points in the food web. This diagram summarizes the results. Photosynthetic algae and cyanobacteria fix the energy of the sun; animal plankton (small heterotrophs) feed on them; and both are consumed by smelt (small fish less than six inches in length). The smelt are eaten by trout or humans, with about a 10-fold loss in fixed energy.

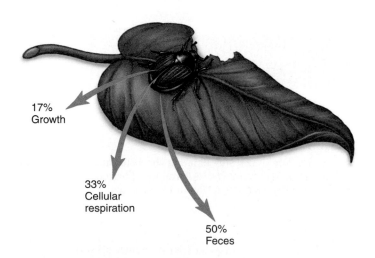

Figure 16.4   How heterotrophs use food energy.

A heterotroph assimilates only a fraction of the energy it consumes. For example, if a "bite" is composed of 500 Joules of energy (1 Joule = 0.239 calories), about 50%, 250 J, is lost in feces, about 33%, 165 J, is used to fuel cellular respiration, and about 17%, 85 J, is converted into consumer biomass. Only this 85 J is available to the next trophic level.

is lost at each step that very little usable energy remains in the system after it has been incorporated into the bodies of organisms at four successive trophic levels.

Lamont Cole of Cornell University studied the flow of energy in a freshwater ecosystem in Cayuga Lake in upstate New York. He calculated that about 150 of each 1,000 calories of potential energy fixed by algae and cyanobacteria are transferred into the bodies of small heterotrophs. Of these, about 30 calories are incorporated into the bodies of smelt, the principal secondary consumers of the system. If humans eat the smelt, they gain about 6 of the 1,000 calories that originally entered the system. If trout eat the smelt and humans eat the trout, humans gain only about 1.2 calories of the 1,000 (figure 16.5).

16.1   Energy moves through ecosystems from producers, to herbivores, to carnivores, and finally to detritivores and decomposers, which consume the dead bodies of all the others. Much energy is lost as heat at each stage of a food chain.

16.2 Ecological Pyramids

As previously mentioned, a plant fixes about 1% of the sun's energy that falls on its green parts. The successive members of a food chain, in turn, process into their own bodies about 10% of the energy available in the organisms on which they feed. For this reason, there are generally far more individuals at the lower trophic levels of any ecosystem than at the higher levels. Similarly, the biomass of the primary producers present in a given ecosystem is greater than the biomass of the primary consumers, with successive trophic levels having a lower and lower biomass and correspondingly less potential energy. Larger animals are characteristically members of the higher trophic levels. To some extent, they must be larger to be able to capture enough prey in the lower trophic levels.

These relationships, if shown diagrammatically, appear as pyramids (figure 16.6). Ecologists speak of "pyramids of numbers," where the sizes of the blocks reflect the number of individuals at each trophic level; "pyramids of biomass," where the sizes of the blocks reflect the total weight of all the organisms at each trophic level; and "pyramids of energy," where the sizes of the blocks reflect the amount of energy stored at each trophic level.

Inverted Pyramids

Some aquatic ecosystems have inverted biomass pyramids. For example, in a planktonic ecosystem—dominated by small organisms floating in water—the turnover of photosynthetic phytoplankton at the lowest level is very rapid, with zooplankton consuming phytoplankton so quickly that the phytoplankton (the producers at the base of the food chain) can never develop a large population size. Because the phytoplankton reproduce very rapidly, the community can support a population of heterotrophs that is larger in biomass and more numerous than the phytoplankton (figure 16.6b). However, don't confuse the sizes of these bars with the energy present at each level. The zooplankton that eat the phytoplankton are present in greater numbers but they contain about only 10% of the energy.

Top Carnivores

The loss of energy that occurs at each trophic level places a limit on how many top-level carnivores a community can support. As we have seen, only about one-thousandth of the energy captured by photosynthesis passes all the way through a three-stage food chain to a tertiary consumer such as a snake or hawk. This explains why there are no predators that subsist on lions or eagles—the biomass of these animals is simply insufficient to support another trophic level.

In the pyramid of numbers, top-level predators tend to be fairly large animals. Thus, the small residual biomass available at the top of the pyramid is concentrated in a relatively small number of individuals.

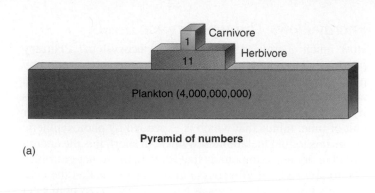

Pyramid of numbers

(a)

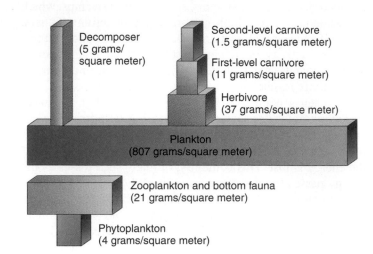

Pyramid of biomass

(b)

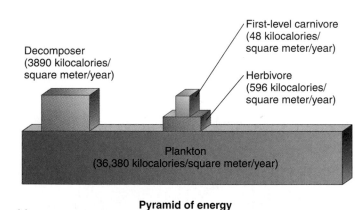

Pyramid of energy

(c)

Figure 16.6 Ecological pyramids.
Ecological pyramids measure different characteristics of each trophic level. (*a*) Pyramid of numbers. (*b*) Pyramids of biomass, both normal (*top*) and inverted (*bottom*). (*c*) Pyramid of energy.

> **16.2** Because energy is lost at every step of a food chain, the biomass of primary producers (photosynthesizers) tends to be greater than that of the herbivores that consume them, and herbivore biomass tends to be greater than the biomass of the predators that consume them.

16.3 The Water Cycle

Unlike energy, which flows through the earth's ecosystems in one direction (from the sun to producers to consumers), the physical components of ecosystems are passed around and reused within ecosystems. Ecologists speak of such constant reuse as recycling or, more commonly, **cycling.** Materials that are constantly recycled include all the chemicals that make up the soil, water, and air. While many are important and will be considered later, the proper cycling of four materials is particularly critical to the health of any ecosystem: water, carbon, and the soil nutrients nitrogen and phosphorus.

The paths of water, carbon, and soil nutrients as they pass from the environment to living organisms and back form closed circles, or cycles. In each cycle, the chemical resides for a time in an organism and then returns to the nonliving environment, often referred to as a *biogeochemical cycle.*

Of all the nonliving components of an ecosystem, water has the greatest influence on the living portion. The availability of water and the way in which it cycles in an ecosystem system in large measure determines the biological richness of that ecosystem—how many different kinds of creatures live there and how many of each.

Water cycles within an ecosystem in two ways: the environmental water cycle and the organismic water cycle.

The Environmental Water Cycle

In the environmental water cycle, water vapor in the atmosphere condenses and falls to the earth's surface as rain or snow. Heated there by the sun, it reenters the atmosphere by **evaporation** from lakes, rivers, and oceans (figure 16.7).

The Organismic Water Cycle

In the organismic water cycle, surface water does not return directly to the atmosphere. Instead, it is taken up by the roots of plants. After passing through the plant, the water reenters the atmosphere through tiny openings in the leaves, evaporating from their surface. This evaporation from leaf surfaces is called **transpiration.** Transpiration is also driven by the sun: The sun's heat creates wind currents that draw moisture from the plant by passing air over the leaves.

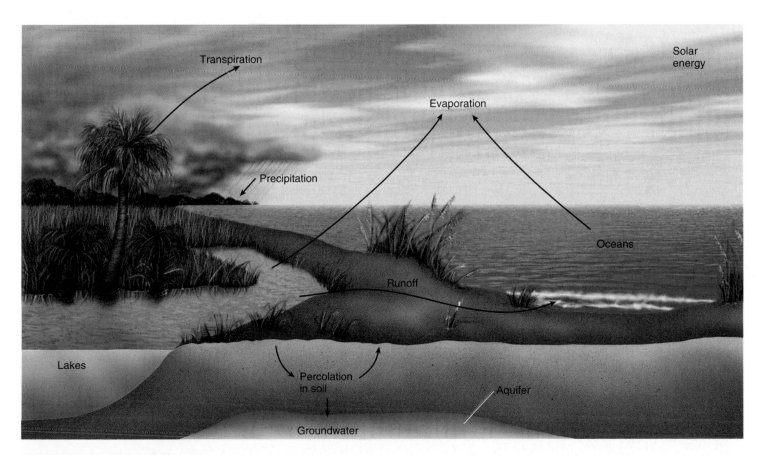

Figure 16.7   The water cycle.
Precipitation on land eventually makes its way to the ocean via groundwater, lakes, and finally, rivers. Solar energy causes evaporation, adding water to the sky. Plants give off excess water through transpiration, also adding water to the atmosphere. Atmospheric water falls as rain or snow over land and oceans, completing the water cycle.

Breaking the Cycle

In very dense forest ecosystems, such as tropical rain forests, more than 90% of the moisture in the ecosystem is taken up by plants and then transpired back into the air. Because so many plants in a rain forest are doing this, the vegetation is the primary source of local rainfall. In a very real sense, these plants create their own rain: The moisture that travels up from the plants into the atmosphere falls back to earth as rain.

Where forests are cut down, the organismic water cycle is broken, and moisture is not returned to the atmosphere. Water drains off to the sea instead of rising to the clouds and falling again on the forest. During his expeditions from 1799–1805, the great German explorer Alexander von Humboldt reported that stripping the trees from a tropical rain forest in Colombia prevented water from returning to the atmosphere and created a semiarid desert. It is a tragedy of our time that just such a transformation is occurring in many tropical areas, as tropical and temperate rain forests are being clear-cut or burned in the name of "development" (figure 16.8).

Groundwater

Much less obvious than the surface waters seen in streams, lakes, and ponds is the groundwater, which occurs in perme-able, saturated, underground layers of rock, sand, and gravel called aquifers. In many areas, groundwater is the most important water reservoir; for example, in the United States, more than 96% of all freshwater is groundwater. Groundwater flows much more slowly than surface water, anywhere from a few millimeters to as much as a meter or so per day. In the United States, groundwater provides about 25% of the water used for all purposes and provides about 50% of the population with drinking water. Rural areas tend to depend on groundwater almost exclusively, and its use is growing at about twice the rate of surface water use.

Because of the greater rate at which groundwater is being used, the increasing chemical pollution of groundwater is a very serious problem. Pesticides, herbicides, and fertilizers are key sources of groundwater pollution. Because of the large volume of water, its slow rate of turnover, and its inaccessibility, removing pollutants from aquifers is virtually impossible.

> **16.3** Water cycles through ecosystems in the atmosphere via precipitation and evaporation, some of it passing through plants on the way.

Figure 16.8 Burning or clear-cutting forests breaks the water cycle.
The high density and large size of plants in a forest translate into great quantities of water being transpired to the atmosphere, creating rain over the forests. In this way rain forests perpetuate the wet climate that supports them. Tropical deforestation permanently alters the climate in these areas, creating arid zones.

16.4 The Carbon Cycle

The earth's atmosphere contains plentiful carbon, present as carbon dioxide (CO_2) gas. This carbon cycles between the atmosphere and living organisms, often being locked up for long periods of time in fossil organisms. The cycle is begun by plants that use CO_2 in photosynthesis to build organic molecules—in effect, they trap the carbon atoms of CO_2 within the living world. The carbon atoms are returned to the atmosphere's pool of CO_2 through respiration, combustion, and erosion (figure 16.9).

Respiration

Most of the organisms in ecosystems respire—that is, they extract energy from organic food molecules by stripping away the carbon atoms and combining them with oxygen to form CO_2. Plants respire, as do the herbivores, which eat the plants, and the carnivores, which eat the herbivores. All of these organisms use oxygen to extract energy from food, and CO_2 is what is left when they are done.

Combustion

A lot of carbon is tied up in wood, and it may stay trapped there for many years, only returning to the atmosphere when the wood is burned. Sometimes the duration of the carbon's visit to the organic world is long indeed. Plants that become buried in sediment, for example, may be gradually transformed by pressure into coal or oil. The carbon originally trapped by these plants is only released back into the atmosphere when the coal or oil (called **fossil fuels**) is burned.

Erosion

Very large amounts of carbon are present in seawater as dissolved CO_2. Substantial amounts of this carbon are extracted from the water by marine organisms, which use it to build their calcium carbonate shells. When these marine organisms die, their shells sink to the ocean floor, become covered with sediments, and form limestone. Eventually, the ocean recedes and the limestone becomes exposed to weather and erodes, the carbon washes back to the oceans where it is returned to the cycle.

> **16.4** Carbon captured from the atmosphere by photosynthesis is returned to it through respiration, combustion, and erosion.

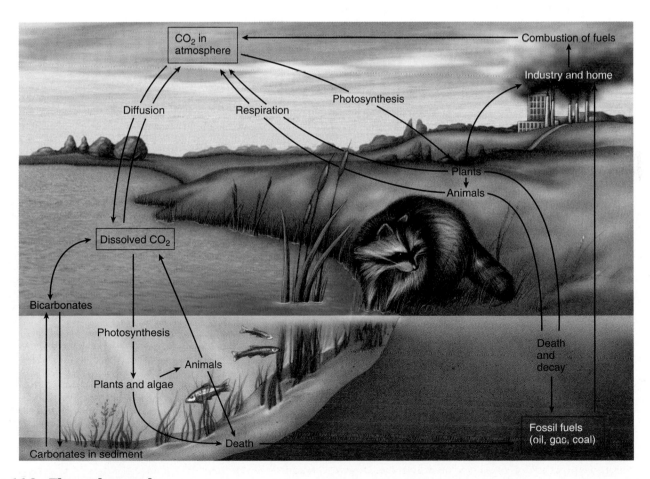

Figure 16.9 The carbon cycle.
Carbon from the atmosphere and from water is fixed by photosynthetic organisms and returned through respiration, combustion, and erosion.

16.5 Soil Nutrients and Other Chemical Cycles

The Nitrogen Cycle

Organisms contain a lot of nitrogen (a principal component of protein) and so does the atmosphere, which is 78.08% nitrogen gas (N_2). However, the chemical connection between these two reservoirs is very delicate, because most living organisms are unable to use the N_2 so plentifully available in the air surrounding them. The two nitrogen atoms of N_2 are bound together by a particularly strong "triple" covalent bond that is very difficult to break. Luckily, a few kinds of bacteria can break the nitrogen triple bond and bind its nitrogen atoms to hydrogen (forming "fixed" nitrogen, ammonia [NH_3], which becomes ammonium ion [NH_4^+]) in a process called **nitrogen fixation.**

Bacteria evolved the ability to fix nitrogen early in the history of life, before photosynthesis had introduced oxygen gas into the earth's atmosphere, and that is still the only way the bacteria are able to do it—even a trace of oxygen poisons the process. In today's world, awash with oxygen, these bacteria live encased within bubbles called cysts that admit no oxygen or within special airtight cells in nodules of tissue on the roots of beans, aspen trees, and a few other plants. Figure 16.10 shows how bacteria make needed nitrogen available to other organisms, a process called the nitrogen cycle. Following the uptake of nitrogen by plants and animals and the death and excretion of plants and animals, decomposing bacteria and then ammonifying bacteria return the nitrogen to ammonia and ammonium ion forms. Continuing the cycle, nitrifying bacteria can convert ammonium ion into nitrate (NO_3^-), and denitrifying bacteria are able to convert nitrate back into atmospheric nitrogen (N_2).

The growth of plants in ecosystems is often severely limited by the availability of "fixed" nitrogen in the soil, which is why farmers fertilize fields. This agricultural practice is a very old one, known even to primitive societies—the American Indians instructed the pilgrims to bury fish, a rich source of fixed nitrogen, with their corn seeds. Today most fixed nitrogen added to soils by farmers is not organic but instead is produced in factories by industrial rather than bacterial nitrogen fixation, a process that accounts for a prodigious 30% of the entire nitrogen cycle.

The Phosphorus Cycle

Phosphorus is an essential element in all living organisms, a key part of both ATP and DNA. Phosphorus is often in very limited supply in the soil of particular ecosystems, and because phosphorus does not form a gas, none is available in the atmosphere. Most phosphorus exists in soil and rock as the mineral calcium phosphate, which dissolves in water to form phosphate ions (Coca-Cola is a sweetened solution of phosphate ions). These phosphate ions are absorbed by the roots of plants and used by them to build organic molecules like ATP and DNA. When the plants and animals die and decay, bacteria in the soil convert the organic phosphorus back into phosphorus ions, completing the cycle (figure 16.11).

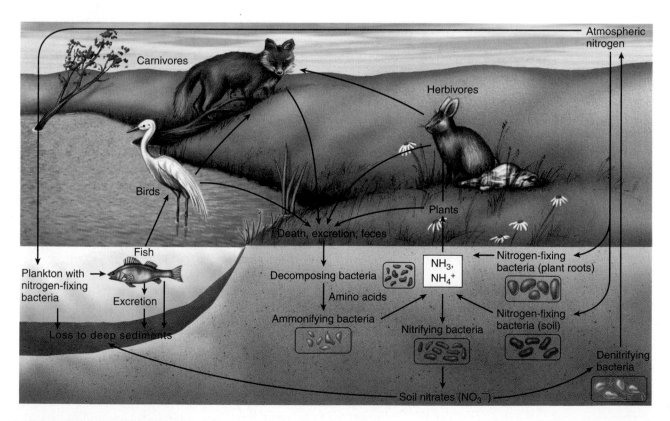

Figure 16.10 The nitrogen cycle.
Relatively few kinds of organisms—all of them bacteria—can convert atmospheric nitrogen into forms that can be used for biological processes.

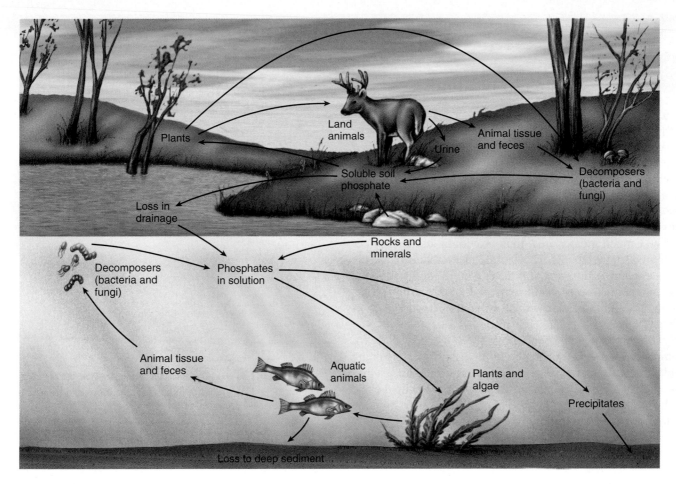

Figure 16.11   The phosphorus cycle.
Phosphorus plays a critical role in plant nutrition; next to nitrogen, phosphorus is the element most likely to be so scarce that it limits plant growth.

The phosphorus level in freshwater lake ecosystems is often quite low, preventing much growth of photosynthetic algae in these systems. Such ecosystems are particularly vulnerable to the inadvertent addition of phosphorus by human activity. For example, agricultural fertilizers and many commercial detergents are rich in phosphorus. Pollution of a lake by the addition of phosphorus to its waters first produces a green scum of algal growth on the surface of the lake and then, if the pollution continues, proceeds to "kill" the lake. After the initial bloom of rapid algal growth, aging algae die, and bacteria feeding on the dead algae cells use up so much of the lake's dissolved oxygen that fish and invertebrate animals suffocate. Such rapid, uncontrolled growth caused by excessive nutrients in an aquatic ecosystem is called **eutrophication.**

The Cycling of Other Chemicals

Many other chemicals cycle through an ecosystem and must be maintained in a balanced state for the ecosystem to be healthy. Proper balance is important. Some chemicals can become harmful when their concentrations exceed normal levels for cycling, as we saw with phosphorus. Other chemicals, when in excess of normal cycling levels, can have similar devastating effects on an ecosystem.

Sulfur, a chemical that cycles through the atmosphere, can harm an ecosystem when large amounts of it are pumped into the atmosphere through coal-burning power plants. The excess sulfur combines with water vapor and oxygen, producing sulfuric acid: $SO_2 + 1/2 \ O_2 + H_2O = H_2SO_4$. This acid then reenters the ecosystem as precipitation.

Heavy metals, which include mercury, cadmium, and lead, are particularly damaging as they cycle through biological food chains, as they tend to progressively accumulate in organisms of higher trophic levels. This process is called *biological magnification*.

16.5 Most of the earth's atmosphere is diatomic nitrogen gas that cannot be used by most organisms. Certain bacteria are able to convert this nitrogen gas into ammonia through nitrogen fixation. These nitrogen atoms then cycle through the earth's ecosystem. Phosphorus, critical to organisms, is available in soil and dissolved in water. It cycles between organisms and the environment and is often the limiting factor in determining what organisms are able to live in an ecosystem.

16.6 The Sun and Atmospheric Circulation

The world contains a great diversity of ecosystems because its climate varies a great deal from place to place. On a given day, Miami and Boston often have very different weather. There is no mystery about this. The tropics are warmer than the temperate regions because the sun's rays arrive almost perpendicular (that is, dead on) at regions near the equator. As you move from the equator into temperate latitudes, sunlight strikes the earth at more oblique angles, which spreads it out over a much greater area, thus providing less energy per unit of area (figure 16.12). This simple fact—that because the earth is a sphere some parts of it receive more energy from the sun than others—is responsible for much of the earth's different climates and thus, indirectly, for much of the diversity of its ecosystems.

The earth's annual orbit around the sun and its daily rotation on its own axis are also both important in determining world climate. Because of the daily cycle, the climate at a given latitude is relatively constant. Because of the annual cycle and the inclination of the earth's axis, all parts away from the equator experience a progression of seasons.

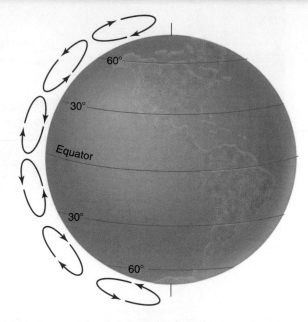

Figure 16.13   Air rises at the equator and then falls.
The pattern of air movement out from and back to the earth's surface forms three pairs of great cycles. At 30 degrees and at the poles, zones of very dry climate are created by descending dry air. Wet climates occur near the equator and at 60 degrees.

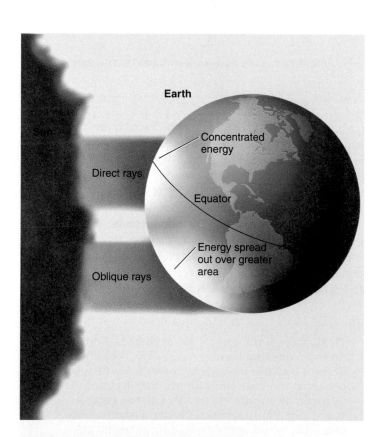

Figure 16.12   Latitude affects climate.
The relationship between the earth and sun is critical in determining the nature and distribution of life on earth. The tropics are warmer than the temperate regions because the sun's rays strike at a direct angle, providing more energy per unit of area.

The major atmospheric circulation patterns result from the interactions between six large air masses. These great air masses occur in pairs, with one air mass of the pair occurring in the northern latitudes and the other occurring in the southern latitudes. These air masses affect climate because the rising and falling of an air mass influence its temperature, which, in turn, influences its moisture-holding capacity.

Near the equator, warm air rises and flows toward the poles (figure 16.13). As it rises and cools, this air loses most of its moisture because cool air holds less water vapor than warm air. (This explains why it rains so much in the tropics.) When this air has traveled to about 30 degrees north and south latitudes, the cool, dry air sinks and becomes reheated, soaking up water like a sponge as it warms and producing a broad zone of low rainfall. It is no accident that all of the great deserts of the world lie near 30 degrees north or 30 degrees south latitude. Air at these latitudes is still warmer than it is in the polar regions, and thus it continues to flow toward the poles. At about 60 degrees north and south latitudes, air rises and cools and sheds its moisture, and such are the locations of the great temperate forests of the world. Finally, this rising air descends near the poles, producing zones of very low temperatures and precipitation.

16.6   The sun drives circulation of the atmosphere, causing rain in the tropics and a band of deserts at 30 degrees latitude.

16.7 Latitude and Elevation

Temperatures are higher in tropical ecosystems for a simple reason: More sunlight per unit area falls on tropical latitudes. Solar radiation is most intense when the sun is directly overhead, and this occurs only in the tropics, where sunlight strikes the equator perpendicularly. Temperature also varies with elevation, with higher altitudes becoming progressively colder. At any given latitude, air temperature falls about 6°C for every 1,000-meter increase in elevation. The ecological consequences of temperature varying with elevation are the same as temperature varying with latitude (figure 16.14). Thus, in North America a 1,000-meter increase in elevation results in a temperature drop equal to that of a 880-kilometer increase in latitude. This is why "timberline" (the elevation above which trees do not grow) occurs at progressively lower elevations as one moves farther from the equator.

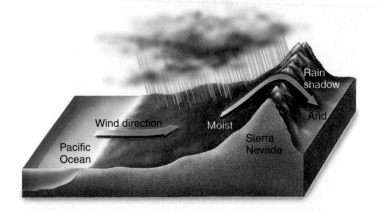

Figure 16.15   The rain shadow effect.
Moisture-laden winds from the Pacific Ocean rise and are cooled when they encounter the Sierra Nevada. As they cool, their moisture-holding capacity decreases and precipitation occurs. As the air descends on the east side of the range, it warms, its moisture-holding capacity increases, and the air picks up moisture from its surroundings. As a result, arid conditions prevail on the east side of these mountains.

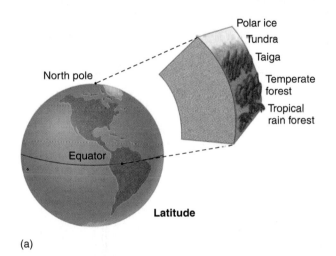

(a)

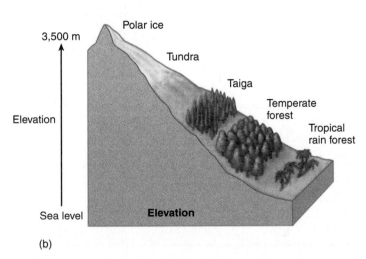

(b)

Figure 16.14   How elevation affects ecosystems.
The same land ecosystems that normally occur north and south of the equator at sea level (a) can occur in the tropics as elevation increases (b). Thus, on a tall mountain in southern Mexico or Guatemala, you might see a sequence of ecosystems such as is illustrated here.

Rain Shadows

When a moving body of air encounters a mountain, it is forced upward, and as it is cooled at higher elevations the air's moisture-holding capacity decreases, producing rain on the windward side of the mountains—the side from which the wind is blowing. The effect on the other side of the mountain—the leeward side—is quite different. As the air passes the peak and descends on the far side of the mountains, it is warmed, so its moisture-holding capacity increases. Sucking up all available moisture, the air dries the surrounding landscape, often producing a desert. This effect, called a **rain shadow,** is responsible for deserts such as Death Valley, which is in the rain shadow of Mount Whitney, the tallest mountain in the Sierra Nevada (figure 16.15).

Similar effects can occur on a larger scale. Regional climates are areas that are located on different parts of the globe but share similar climates because of similar geography. A so-called Mediterranean climate results when winds blow from a cool ocean onto a warm land during the summer. As a result, the air's moisture-holding capacity is increased and precipitation is blocked, similar to what occurs on the leeward side of mountains. This effect accounts for dry, hot summers and cool, moist winters in areas with a Mediterranean climate such as portions of southern California or Oregon, central Chile, southwestern Australia, and the Cape region of South Africa. Such a climate is unusual on a world scale. In the regions where it occurs, many unusual kinds of endemic (local in distribution) plants and animals have evolved.

> **16.7   Temperatures fall with increasing latitude and also with increasing altitude. Rainfall is higher on the windward side of mountains, with air losing its moisture as it rises up the mountain; descending on the far side, the dry air warms and sucks up moisture, creating deserts.**

16.8 Patterns of Circulation in the Ocean

Patterns of ocean circulation are determined by the patterns of atmospheric circulation, but they are modified by the locations of land masses. Oceanic circulation is dominated by huge surface gyres (figure 16.16), which move around the subtropical zones of high pressure between approximately 30 degrees north and south latitudes. These gyres move clockwise in the Northern Hemisphere and counterclockwise in the Southern Hemisphere. The ways they redistribute heat profoundly affects life not only in the oceans but also on coastal lands. For example, the Gulf Stream, in the North Atlantic, swings away from North America near Cape Hatteras, North Carolina, and reaches Europe near the southern British Isles. Because of the Gulf Stream, western Europe is much warmer and more temperate than eastern North America at similar latitudes. As a general principle, western sides of continents in temperate zones of the Northern Hemisphere are warmer than their eastern sides; the opposite is true of the Southern Hemisphere. In addition, winds passing over cold water onto warm land increase their moisture-holding capacity, limiting precipitation.

In South America, the Humboldt current carries phosphorus-rich cold water northward up the west coast. Phosphorus is brought up from the ocean depths by the upwelling of cool water that occurs as offshore winds blow from the mountainous slopes that border the Pacific Ocean. This nutrient-rich current helps make possible the abundance of marine life that supports the fisheries of Peru and northern Chile. Marine birds, which feed on these organisms, are responsible for the commercially important, phosphorus-rich guano deposits on the seacoasts of these countries.

El Niño Southern Oscillations and Ocean Ecology

Every Christmas a tepid current sweeps down the coast of Peru and Ecuador from the tropics, reducing the fish population slightly and giving local fishers some time off. The local fishers named this Christmas current *El Niño* ("The Christ Child"). Now, though, the term is reserved for a catastrophic version of the same phenomenon, one that occurs every two to seven years and is not only felt locally but on a global scale.

Scientists now have a pretty good idea of what goes on in an El Niño. Normally the Pacific Ocean is fanned by con-

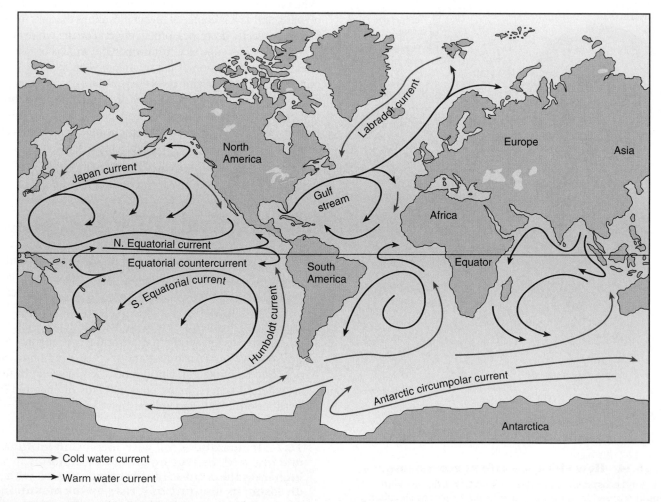

Figure 16.16 Oceanic circulation.

The circulation in the oceans moves in great surface spiral patterns called gyres; oceanic circulation profoundly affects the climate on adjacent lands.

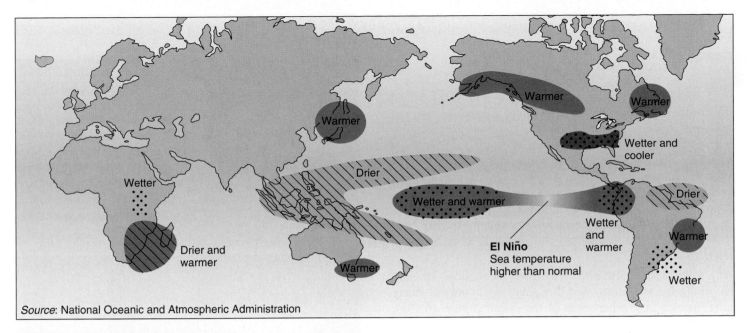

Figure 16.17 An El Niño winter.
El Niño currents produce unusual weather patterns all over the world as warm waters from the western Pacific move eastward.

Source: National Oceanic and Atmospheric Administration

stantly blowing east-to-west trade winds. These winds push warm surface water away from the ocean's eastern side (Peru, Ecuador, and Chile) and allow cold water to well up from the depths in its place, carrying nutrients that feed plankton and hence fish. This surface water piles up in the west, around Australia and the Philippines, making it several degrees warmer and a meter or so higher than the eastern side of the ocean. But if the winds slacken briefly, warm water begins to slosh back across the ocean.

Once this happens, ocean and atmosphere conspire to ensure it keeps happening. The warmer the eastern ocean gets, the warmer and lighter the air above it becomes, and hence more similar to the air on the western side. This reduces the difference in pressure across the ocean. Because a pressure difference is what makes winds blow, the easterly trades weaken further, letting the warm water continue its eastward advance.

The end result is to shift the weather systems of the western Pacific Ocean 6,000 kilometers eastward. The tropical rainstorms that usually drench Indonesia and the Philippines are caused when warm seawater abutting these islands causes the air above it to rise, cool, and condense its moisture into clouds. When the warm water moves east, so do the clouds, leaving the previously rainy areas in drought. Conversely, the western edge of South America, its coastal waters usually too cold to trigger much rain, gets a soaking, while the upwelling slows down. During an El Niño, commercial fish stocks virtually disappear from the waters of Peru and northern Chile, and plankton drop to a twentieth of their normal abundance. The commercially valuable anchovy fisheries of Peru were essentially destroyed by the 1972 and 1997 El Niños.

That is just the beginning. El Niño's effects are propagated across the world's weather systems (figure 16.17). Violent winter storms lash the coast of California, accompanied by flooding, and El Niño produces colder and wetter winters than normal in Florida and along the Gulf Coast. The U.S. Midwest experiences heavier-than-normal rains.

Though the effects of an El Niño are now fairly clear, what triggers them still remains a mystery. Models of these weather disturbances suggest that the climatic change that triggers an El Niño is "chaotic." Wind and ocean currents return again and again to the same condition, but never in a regular pattern, and small nudges can send them off in many different directions—including an El Niño.

La Niña. El Niño is an extreme phase of a naturally occurring climatic cycle, but as in all cycles, there is an opposite side to it. While El Niño is characterized by unusually warm ocean temperatures in the eastern Pacific, *La Niña* is characterized by unusually cold ocean temperatures in the eastern Pacific. The strengthening of the east-to-west trade winds intensifies the cold upwelling along the eastern Pacific, with coastal water temperatures along the South American coast falling as much as 7°F below normal. Although not as well known as El Niño, La Niña causes equally extreme effects that are nearly opposite to those of El Niño. In the United States, the effects of La Niña are most apparent during the winter months. La Niña brings wetter than normal conditions across the Pacific Northwest and dryer, warmer conditions across much of the southern United States. Winter temperatures are warmer than normal in the Southeast and cooler than normal in the Northwest in La Niña years.

16.8 The world's oceans circulate in huge gyres deflected by continental land masses. Disturbances in ocean currents like an El Niño and La Niña can have profound influences on world climate.

16.9 Ocean Ecosystems

Most of the earth's surface—nearly three-quarters—is covered by water. The seas have an average depth of more than 3 kilometers, and they are, for the most part, cold and dark. Photosynthetic organisms are confined to the upper few hundred meters, because light does not penetrate any deeper. Almost all organisms that live below this level feed on organic debris that rains downward. The three main kinds of ecosystems are shallow waters, open-sea surface, and deep-sea waters (figure 16.18).

Shallow Waters

Very little of the earth's ocean surface is shallow—mostly that along the shoreline—but this small area contains many more species than other parts of the ocean (figure 16.19*a*). The world's great commercial fisheries occur on banks in the coastal zones, where nutrients derived from the land are more abundant than in the open ocean. Part of this zone consists of the **intertidal region,** which is exposed to the air whenever the tides recede. Partly enclosed bodies of water, such as those that often form at river mouths and in coastal bays, where the salinity is intermediate between that of seawater and freshwater, are called **estuaries.** Estuaries are among the most naturally fertile areas in the world, often containing rich stands of submerged and emergent plants, algae, and microscopic organisms. They provide the breeding grounds for most of the coastal fish and shellfish that are harvested both in the estuaries and in open water.

(a)

(b)

Figure 16.19   Shallow waters and open sea surface.
(*a*) Fishes and many other kinds of animals find food and shelter among the coral in the coastal waters of some regions. (*b*) The upper layers of the open ocean contain plankton and large schools of fish, like these yellow tail grunts.

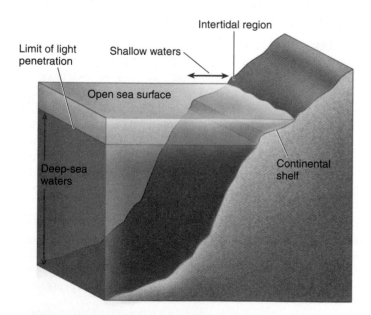

Figure 16.18   Ocean ecosystems.
There are three primary ecosystems found in the earth's oceans. Shallow water ecosystems occur along the shoreline and at areas of coral reefs. Open sea surface ecosystems occur in the upper 100–200 meters where light can penetrate and support photosynthetic organisms. Finally, deep-sea water ecosystems are areas below 300 meters where relatively few organisms live.

Open-Sea Surface

Drifting freely in the upper, better-illuminated waters of the ocean is a diverse biological community of microscopic organisms: phytoplankton and zooplankton. Most of the plankton occurs in the top 100 meters of the sea. Many fishes swim in these waters as well, feeding on the plankton and one another (figure 16.19*b*). Some members of the plankton, including algae and some bacteria, are photosynthetic. Collectively, these organisms are responsible for about 40% of all photosynthesis that takes place on earth. Over half of this is carried out by organisms less than 10 micrometers in diameter—at the lower limits of size for organisms—and almost all of it near the surface of the sea, in the zone into which light from the surface penetrates freely.

Populations of organisms that make up the plankton are able to increase rapidly, and the turnover of nutrients in the plankton is much more rapid than in most other ecosystems, although the total amounts of nutrients in the sea are very low.

Deep-Sea Waters

In the deep waters of the sea, below the top 300 meters, little light penetrates. Very few organisms live there, compared to the rest of the ocean, but those that do include some of the most bizarre organisms found anywhere on earth. Many deep-sea inhabitants have bioluminescent (light-producing) body parts that they use to communicate or to attract prey (figure 16.20*a*).

The supply of oxygen can often be critical in the deep ocean, and as water temperatures become warmer, the water holds less oxygen. For this reason, the amount of available oxygen becomes an important limiting factor for deep-sea organisms in warmer marine regions of the globe. Carbon dioxide, in contrast, is almost never limited in the deep ocean. The distribution of minerals is much more uniform in the ocean than it is on land, where individual soils reflect the composition of the parent rocks from which they have weathered.

Frigid and bare, the floors of the deep sea have long been considered a biological desert. Recent close-up looks taken by marine biologists, however, paint a different picture (figure 16.20*c*). The ocean floor is teeming with life. Often kilometers deep, thriving in pitch darkness under enormous pressure, crowds of marine invertebrates have been found in hundreds of deep samples from the Atlantic and Pacific. Rough estimates of deep-sea diversity have soared to millions of species. Many appear endemic (local). The diversity of species is so high it may rival that of tropical rain forests! This profusion is unexpected. New species usually require some kind of barrier to diverge (see chapter 11), and the ocean floor seems boringly uniform. However, little migration occurs among deep populations, and this lack of movement may encourage local specialization and species formation. A patchy environment may also contribute to species formation there; deep-sea ecologists find evidence that fine but nonetheless formidable resource barriers arise in the deep sea.

No light falls in the deep ocean. From where do deep-sea organisms obtain their energy? While some utilize energy falling to the ocean floor as debris from above, other deep-sea organisms are autotrophic, gaining their energy from **hydrothermal vent systems,** areas in which seawater circulates through porous rock surrounding fissures where molten material from beneath the earth's crust comes close to the surface. Hydrothermal vent systems, also called deep-sea vents, support a broad array of heterotrophic life (figure 16.20*b*). Water in the area of these hydrothermal vents is heated to temperatures in excess of 350°C, and contains high concentrations of hydrogen sulfide. Prokaryotes that live by these deep-sea vents obtain energy and produce carbohydrates through chemosynthesis instead of photosynthesis. Like plants, they are autotrophs; they extract energy from hydrogen sulfide to manufacture food, much as a plant extracts energy from the sun to manufacture its food. These prokaryotes live symbiotically within the tissues of heterotrophs that live around the deep-sea vents. The animals provide a place for the prokaryotes to live and obtain nutrients, and in turn the prokaryotes supply the animal with organic compounds to use as food.

Despite the many new forms of small invertebrates now being discovered on the seafloor, and the huge biomass that occurs in the sea, more than 90% of all *described* species of organisms occur on land. Each of the largest groups of organisms, including insects, mites, nematodes, fungi, and plants, has marine representatives, but they constitute only a very small fraction of the total number of described species.

(a)   (b)

(c)

Figure 16.20   Deep-sea waters.

(*a*) The luminous spot below the eye of this deep-sea fish results from the presence of a symbiotic colony of luminous bacteria.
(*b*) These giant beardworms live along vents where water jets from fissures at 350°C and then cools to the 2°C of the surrounding water.
(*c*) Looking for all the world like some undersea sunflower, these two sea anemones (actually animals) use a glass-sponge stalk to catch "marine snow," food particles raining down on the ocean floor from the ocean surface several kilometers above.

16.9   The three principal ocean ecosystems occur in shallow water, in the open-sea surface, and along the deep-sea bottom. Both intertidal shallows and deep-sea communities are very diverse.

16.10 Freshwater Ecosystems

Freshwater ecosystems (lakes, ponds, rivers, and wetlands) are distinct from both ocean and land ecosystems, and they are very limited in area. Inland lakes cover about 1.8% of the earth's surface and rivers, streams, and wetlands about 0.4%. All freshwater habitats are strongly connected to land habitats, with marshes and swamps (wetlands) constituting intermediate habitats. In addition, a large amount of organic and inorganic material continually enters bodies of freshwater from communities growing on the land nearby (figure 16.21). Many kinds of organisms are restricted to freshwater habitats (figure 16.22). When they occur in rivers and streams, they must be able to attach themselves in such a way as to resist or avoid the effects of current or risk being swept away.

Like the ocean, ponds and lakes have three zones in which organisms live: a shallow "edge" zone, an open-water surface zone, and a deep-water zone where light does not penetrate (figure 16.23). **Thermal stratification,** characteristic of the larger lakes in temperate regions, is the process whereby water at a temperature of 4°C (which is when water is most dense) sinks beneath water that is either warmer or cooler. In winter, water at 4°C sinks beneath cooler water that freezes at the surface at 0°C. Below the ice, the water remains between 0° and 4°C, and plants and animals survive there. In spring, as the ice melts, the surface water is warmed to 4°C and sinks be-

low the cooler water, bringing the cooler water to the top with nutrients from the lake's lower regions. This process is known as the **spring overturn** (figure 16.24).

In summer, warmer water forms a layer over the cooler water (about 4°C) that lies below. In the area between these two layers, called the *thermocline,* temperature changes abruptly. Depending on the climate of the particular area, the warm upper layer may become as much as 20 meters thick during the summer. In autumn, its temperature drops until it

(a)

(b)

Figure 16.22 Freshwater organisms.
(a) This speckled darter and (b) this giant waterbug with eggs on its back can only live in freshwater habitats.

Figure 16.21 A nutrient-rich stream.
In this stream in the northern coastal mountains of California, as in all streams, much organic material falls or seeps into the water from communities along the edges. This input is responsible for much of the stream's biological productivity.

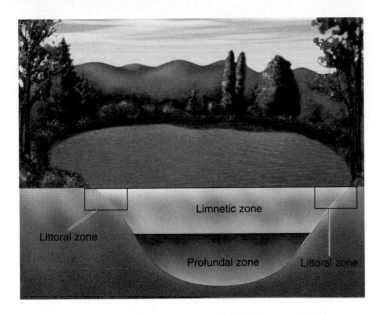

Figure 16.23   The three zones in ponds and lakes.
A shallow "edge" (littoral) zone lines the periphery of the lake where attached algae and their insect herbivores live. An open-water surface (limnetic) zone lies across the entire lake and is inhabited by floating algae, zooplankton, and fish. A dark, deep-water (profundal) zone overlies the sediments at the bottom of the lake. The profundal zone contains numerous bacteria and wormlike organisms that consume dead debris settling at the bottom of the lake.

reaches that of the cooler layer underneath—4°C. When this occurs, the upper and lower layers mix—a process called the **fall overturn.** Therefore, colder waters reach the surfaces of lakes in the spring and fall, bringing up fresh supplies of dissolved nutrients.

Lakes can be divided into two categories, based on their production of organic material. **Eutrophic lakes** have an abundant supply of minerals and organic matter. Oxygen is depleted below the thermocline in the summer because of the abundant organic material and high rate at which aerobic decomposers in the lower layer use oxygen. These stagnant waters again reach the surface after the fall overturn and are then infused with more oxygen. In **oligotrophic lakes,** on the other hand, organic matter and nutrients are relatively scarce. Such lakes are often deeper than eutrophic ones, and their deep waters are always rich in oxygen. Oligotrophic lakes are highly susceptible to pollution from excess phosphorus from such sources as fertilizer runoff, sewage, and detergents.

> **16.10**   Freshwater ecosystems cover only about 2% of the earth's surface; all are strongly tied to adjacent terrestrial ecosystems. In some, organic materials are common, and in others, scarce. The temperature zones in lakes overturn twice a year, in spring and fall.

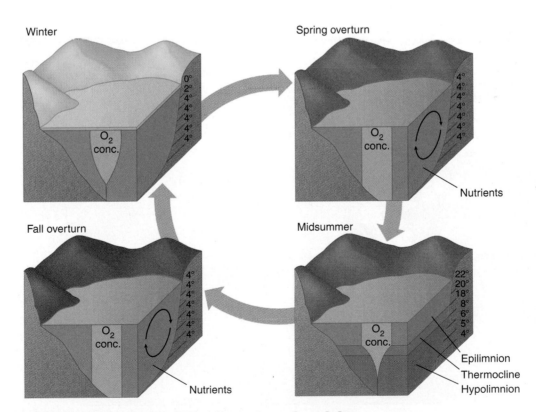

Figure 16.24   Spring and fall overturns in freshwater ponds or lakes.
The pattern of stratification in a large pond or lake in temperate regions is upset in the spring and fall overturns. Of the three layers of water shown in midsummer (*lower right*), the densest water occurs at 4°C (the hypolimnion). The warmer water at the surface is less dense (the epilimnion). The thermocline is the zone of abrupt change in temperature that lies between them. If you have dived into a pond in temperate regions in the summer, you have experienced the existence of these layers directly.

16.11 Land Ecosystems

Living on land ourselves, we humans tend to focus much of our attention on terrestrial ecosystems. A **biome** is a terrestrial ecosystem that occurs over a broad area. Each biome is characterized by a particular climate and a defined group of organisms.

While biomes can be classified in a number of ways, the seven most widely occurring biomes are (1) tropical rain forest, (2) savanna, (3) desert, (4) temperate grassland, (5) temperate deciduous forest, (6) taiga, and (7) tundra. The reason that there are seven primary biomes, and not one or 80, is that they have evolved to suit the climate of the region, and the earth has seven principal climates. The seven biomes differ remarkably from one another but are consistent within; a particular biome looks the same, with the same types of creatures living there, wherever it occurs on earth.

There are seven other less widespread biomes: chaparral; polar ice; mountain zone; temperate evergreen forest; warm, moist evergreen forest; tropical monsoon forest; and semidesert. Figure 16.25 shows the geographical distribution of all 14 biomes, both the 7 principal ones and the 7 less common ones.

If there were no mountains and no climatic effects caused by the irregular outlines of the continents and by different sea temperatures, each biome would form an even belt around the globe. In fact, their distribution is greatly affected by these factors, especially by elevation. Thus, the summits of the Rocky Mountains are covered with a vegetation type that resembles tundra, whereas other forest types that resemble taiga occur farther down. It is for reasons such as these that the distributions of the biomes are so irregular. One trend that is apparent is that those biomes that normally occur at high latitudes also follow an altitudinal gradient along mountains. That is, biomes found far north and far south of the equator at sea level also occur in the tropics but at high mountain elevations (see figure 16.14).

Distinctive features of the seven major biomes—tropical rain forest, savanna, desert, temperate grassland, temperate deciduous forest, taiga, and tundra—along with several of the less widespread biomes are now discussed in more detail.

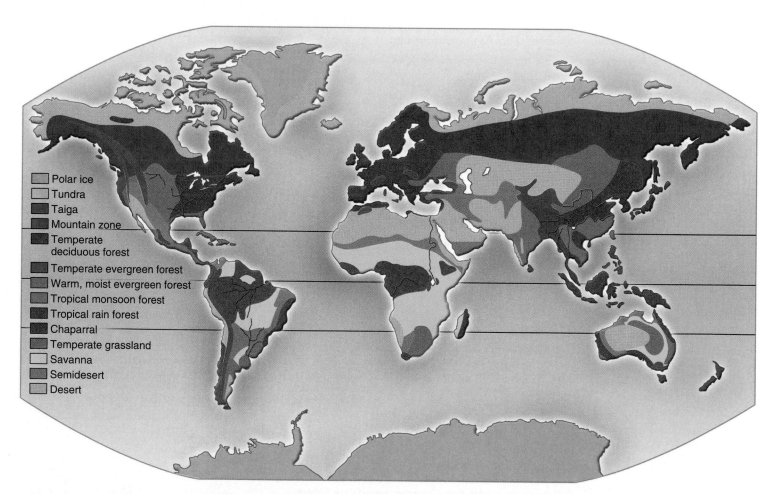

Figure 16.25 Distribution of the earth's biomes.

The seven primary types of biomes are tropical rain forest, savanna, desert, temperate grassland, temperate deciduous forest, taiga, and tundra. In addition, seven less widespread biomes are shown.

Lush Tropical Rain Forests

Rain forests, which experience over 250 centimeters of rain a year, are the richest ecosystems on earth (figure 16.26). They contain at least half of the earth's species of terrestrial plants and animals—more than 2 million species! In a single square mile of tropical forest in Rondonia, Brazil, there are 1,200 species of butterflies—twice the total number found in the United States and Canada combined. The communities that make up tropical rain forests are diverse in that each kind of animal, plant, or microorganism is often represented in a given area by very few individuals. There are extensive tropical rain forests in South America, Africa, and Southeast Asia. But the world's tropical rain forests are being destroyed, and with them, countless species, many of them never seen by humans. Perhaps a quarter of the world's species will disappear with the rain forests during the lifetime of many of us.

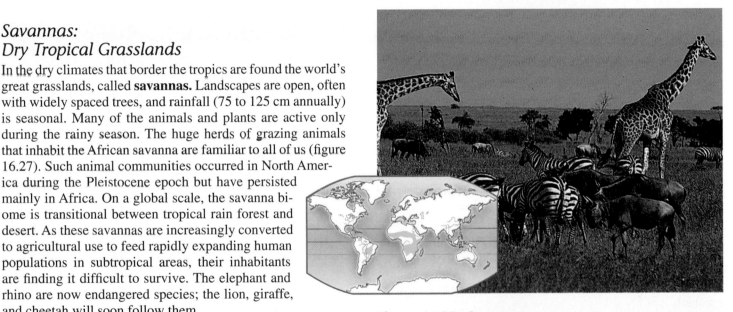

Figure 16.26   Tropical rain forest.

Savannas: Dry Tropical Grasslands

In the dry climates that border the tropics are found the world's great grasslands, called **savannas.** Landscapes are open, often with widely spaced trees, and rainfall (75 to 125 cm annually) is seasonal. Many of the animals and plants are active only during the rainy season. The huge herds of grazing animals that inhabit the African savanna are familiar to all of us (figure 16.27). Such animal communities occurred in North America during the Pleistocene epoch but have persisted mainly in Africa. On a global scale, the savanna biome is transitional between tropical rain forest and desert. As these savannas are increasingly converted to agricultural use to feed rapidly expanding human populations in subtropical areas, their inhabitants are finding it difficult to survive. The elephant and rhino are now endangered species; the lion, giraffe, and cheetah will soon follow them.

Figure 16.27   Savanna.

Deserts: Burning Hot Sands

In the interior of continents are found the world's great deserts, especially in Africa (the Sahara), Asia (the Gobi), and Australia (the Great Sandy Desert). **Deserts** are dry places where less than 25 centimeters of rain falls in a year—an amount so low that vegetation is sparse and survival depends on water conservation (figure 16.28). Plants and animals may restrict their activity to favorable times of the year, when water is present. To avoid high temperatures, most desert vertebrates live in deep, cool, and sometimes even somewhat moist burrows. Those that are active over a greater portion of the year emerge only at night, when temperatures are relatively cool. Some, such as camels, can drink large quantities of water when it is available and then survive long, dry periods. Many animals simply migrate to or through the desert, where they exploit food that may be abundant seasonally.

Figure 16.28   Desert.

Grasslands: Seas of Grass

Halfway between the equator and the poles are temperate regions where rich **grasslands** grow. These grasslands once covered much of the interior of North America, and they were widespread in Eurasia and South America as well. Such grasslands are often highly productive when converted to agriculture. Many of the rich agricultural lands in the United States and southern Canada were originally occupied by **prairies,** another name for temperate grasslands. The roots of perennial grasses characteristically penetrate far into the soil, and grassland soils tend to be deep and fertile. Temperate grasslands are often populated by herds of grazing mammals. In North America, the prairies were once inhabited by huge herds of bison and pronghorns (figure 16.29). The herds are almost all gone now, with most of the prairies having been converted to the richest agricultural region on earth.

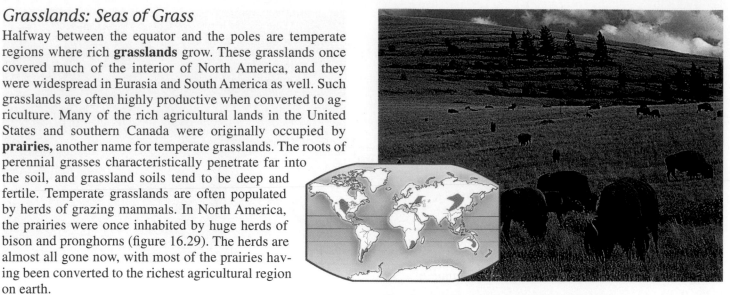

Figure 16.29 Temperate grassland.

Deciduous Forests: Rich Hardwood Forests

Mild climates (warm summers and cool winters) and plentiful rains promote the growth of **deciduous** ("hardwood") **forests** in Eurasia, the northeastern United States, and eastern Canada (figure 16.30). A deciduous tree is one that drops its leaves in the winter. Deer, bears, beavers, and raccoons are the familiar animals of the temperate regions. Because the temperate deciduous forests represent the remnants of more extensive forests that stretched across North America and Eurasia several million years ago, these remaining areas—especially those in eastern Asia and eastern North America—share animals and plants that were once more widespread. Alligators, for example, are found only in China and in the southeastern United States. The deciduous forest in eastern Asia is rich in species because climatic conditions have remained constant.

Figure 16.30 Temperate deciduous forest.

Taiga: Trackless Conifer Forests

A great ring of northern forests of coniferous trees (spruce, hemlock, larch, and fir) extends across vast areas of Asia and North America. Coniferous trees are ones with leaves like needles that are kept all year long. This ecosystem, called **taiga,** is one of the largest on earth (figure 16.31). Here, the winters are long and cold, and most of the limited amount of precipitation falls in the summer. Because it has too short a growing season for farming, few people live there. Many large mammals, including elk, moose, deer, and such carnivores as wolves, bears, lynx, and wolverines, live in the taiga. Traditionally, fur trapping has been extensive in this region. Lumber production is also important. Marshes, lakes, and ponds are common and are often fringed by willows or birches. Most of the trees occur in dense stands of one or a few species.

Figure 16.31 Taiga.

Tundra: Cold Boggy Plains

In the far north, above the great coniferous forests and below the polar ice, there are few trees. There the grassland, called **tundra,** is open, windswept, and often boggy (figure 16.32). Enormous in extent, this ecosystem covers one-fifth of the earth's land surface. Very little rain or snow falls. When rain does fall during the brief arctic summer, it sits on frozen ground, creating a sea of boggy ground. **Permafrost,** or permanent ice, usually exists within a meter of the surface. Trees are small and are mostly confined to the margins of streams and lakes. Large grazing mammals, including musk-oxen, caribou, reindeer, and carnivores such as wolves, foxes, and lynx, live in the tundra. Lemming populations rise and fall on a long-term cycle, with important effects on the animals that prey on them.

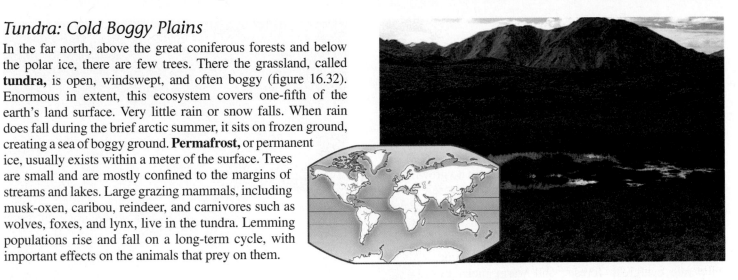

Figure 16.32   Tundra.

Chaparral

Chaparral consists of evergreen, often spiny shrubs and low trees that form communities in regions with a Mediterranean, dry summer climate: the Mediterranean area itself, California, central Chile, the Cape region of South Africa, and southwestern Australia (figure 16.33). Many plant species found in chaparral can germinate only when they have been exposed to the hot temperatures generated during a fire. The chaparral of California and adjacent regions is historically derived from deciduous forests.

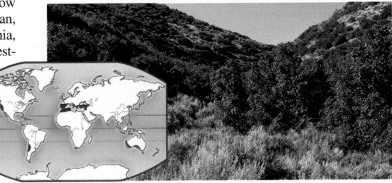

Figure 16.33   Chaparral.

Polar Ice Caps

Polar ice caps lie over the Arctic Ocean in the north and Antarctica in the south (figure 16.34). The poles receive almost no precipitation, so although ice is abundant, freshwater is scarce. The sun barely rises in the winter months. Life in Antarctica is largely limited to the coasts. Because the Antarctic ice cap lies over a landmass, it is not warmed by the latent heat of circulating ocean water and becomes very cold. As a result, only prokaryotes, algae, and some small insects inhabit the vast Antarctic interior.

Figure 16.34   Polar ice.

Tropical Monsoon Forest

Tropical monsoon forests, also called tropical upland forests (figure 16.35), occur in the tropics and semitropics at slightly higher latitudes than rain forests or where local climates are drier. Most trees in these forests are deciduous, losing many of their leaves during the dry season. This loss of leaves allows sunlight to penetrate to the understory and ground levels of the forest, where a dense layer of shrubs and small trees grow rapidly. Leaves are smaller, and photosynthesis is depressed because of lower temperatures. Rainfall is typically very seasonal, measuring several inches daily in the monsoon season and approaching drought conditions in the dry season, particularly in locations far from oceans, such as in central India. Tropical monsoon forests are also found in Bangladesh, southeastern Asia, China, the West Indies, Central and South America, and Australia.

Figure 16.35   Tropical monsoon forest.

Semidesert

Semidesert areas occur in regions with less rain than monsoon forests but more rain than savannas (figure 16.36). Vegetation is dominated by bushes and trees with thorns and spikes, which is why these regions are also known as thornwood forests. Little or no rain falls for eight or nine months during the winter. Plants survive on one or a few short heavy rains received during the summer wet season, growing intensively in response to the moisture. The brief rain is followed by the long dry season, when leaves fall from plants and little or no growth occurs. Semideserts are found along the edges of desert biomes such as in Africa, Asia, and Australia.

Other Biomes

Other biomes include the mountain (alpine) zone, temperate evergreen forests, and warm moist evergreen forests. **Mountain zone** areas are similar to tundras because the increasing altitude produces many of the same changes in temperature and moisture as seen with increasing latitudes (see figure 16.14). The tops of mountains have a typical windswept vegetation similar in many respects to tundra. Few if any trees are able to grow in this alpine zone, which, like the polar tundra, is alive with life in the warm summer months. During the harsh winters, little grows. **Temperate evergreen forests** occur in regions where winters are cold and there is a strong, seasonal dry period. The pine forests of the western United States, the California oak woodlands, and the Australian eucalyptus forests are typical temperate evergreen forests. Many of these forests are endangered by overlogging, particularly in the western United States. **Warm, moist evergreen forests** occur in temperate regions

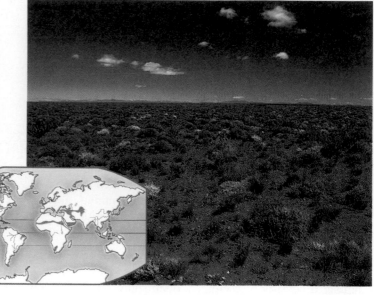

Figure 16.36 Semidesert.

where winters are mild and moisture is plentiful. These can be seen in central China, in the pine forests covering much of the southeastern United States, and in the coastal redwood forest of northern California.

16.11 Biomes are major terrestrial communities defined largely by temperature and rainfall patterns.

Exploring Current Issues

Additional Resources

Go to your campus library or look online to find the following articles, which further develop some of the concepts found in this chapter.

Gleik, P.H. (2001). Making every drop count. *Scientific American,* 284(2), 40.

Levin, L.A. (2002). Deep-ocean life where oxygen is scarce: oxygen-deprived zones are common and might become more so with climate change. Here life hangs on with some unusual adaptations. *American Scientist,* 90(5), 436.

McNulty, S.G. and J.D. Aber. (2001). U.S. national climate change assessment on forest ecosystems: an introduction. *BioScience,* 51(9), 720.

Pearce, F. (2003). Oceans raped of their former riches: ten times as many whales as thought may have once swum the seas. Is it the latest sign that our plundering has irrevocably upset the world's marine ecosystems? *New Scientist,* 179(2406), 4.

Walker, G. (2000). Wild weather. *New Scientist,* 167(2256), 26.

Biology and Society Lecture: **The Global Environmental Challenge**

The current world population of more than 6.3 billion people is placing severe strains on our earth's ability to sustain and support so many people. While the worldwide average birthrate has dropped only slightly in the last 300 years, the death rate has fallen dramatically. This difference between birthrate and death rate produces an annual worldwide population growth rate of 1.3%. The rate may seem small, but it doubles the world's population in 53 years! The environmental problems that haunt this new century—global warming, ozone depletion, acid rain, chemical pollution—are all a direct result of this explosion of human population growth, as well as the "development" of third world resources, and the environmental destruction this economic activity fosters. Perhaps the most serious harm concerns loss of nonrenewable resources, such as topsoil, groundwater, and biodiversity. Current levels of consumption of these resources are not sustainable, and replenishment can take many centuries.

Find this lecture, delivered by the author to his class at Washington University, online at www.mhhe.com/tlwessentials/exp16.

The Energy in Ecosystems

16.1 Energy Flows Through Ecosystems

- Energy flows through an ecosystem by way of food chains. Energy from the sun is captured by photosynthetic producers, which are eaten by herbivores, which are in turn eaten by carnivores. Organisms at all trophic levels die and are consumed by detritivores and decomposers (**figure 16.1**). Energy is lost at every level, such that only about 10% of available energy in an organism is passed on to organisms at the next trophic level (**figures 16.4 and 16.5**).

16.2 Ecological Pyramids

- Because energy is lost as it passes up through the trophic levels of the food chain, there tends to be more individuals at the lower trophic levels (that is, there are more producers than herbivores and more herbivores than carnivores). Ecological pyramids illustrate this distribution of energy and biomass (**figure 16.6**).

Materials Cycle Within Ecosystems

16.3 The Water Cycle

- Physical components of the ecosystem cycle through the ecosystem, being used then recycled then reused. This cycling of materials often involves living organisms and is referred to as a biochemical cycle.

- Water availability determines how many and what kinds of organisms can be supported in the ecosystem. Water cycles from the atmosphere as precipitation, where it falls to the earth and is heated. Upon heating, water reenters the atmosphere through evaporation. Water also cycles through plants, entering through the roots and leaving as water vapor in transpiration (**figure 16.7**).

16.4 The Carbon Cycle

- Carbon cycles through plants—via carbon fixation in photosynthesis—which are then eaten by animals. It returns to the atmosphere as CO_2 from cellular respiration. Carbon also cycles through the ecosystem by diffusion and by the burning of fossil fuels (**figure 16.9**).

16.5 Soil Nutrients and Other Chemical Cycles

- Nitrogen gas in the atmosphere cannot be readily used by organisms and needs to be fixed by certain types of bacteria into ammonia. Animals eat plants that have taken up the fixed nitrogen. Nitrogen reenters the ecosystem through animal excretion and decomposition through detritivores and decomposers (**figure 16.10**).

- Phosphorus also cycles through the ecosystem and may limit growth when not available (**figure 16.11**).

How Weather Shapes Ecosystems

16.6 The Sun and Atmospheric Circulation

- The heating power of the sun affects evaporation and air currents, causing certain parts of the globe, such as the tropics, to have larger amounts of precipitation (**figures 16.12 and 16.13**).

16.7 Latitude and Elevation

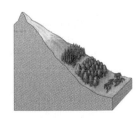

- Temperature and precipitation are similarly affected by altitude and latitude. Changes in ecosystems seen going from the equator to the poles are similarly reflected in the changes in ecosystems seen going from sea level to mountaintops (**figure 16.14**). The changes in temperature also cause the rainshadow effect, where precipitation is deposited on the windward side of mountains (**figure 16.15**).

16.8 Patterns of Circulation in the Ocean

- The earth's oceans circulate in patterns that distribute warmer and cooler waters to different areas of the world. These ocean patterns affect climates across the globe (**figures 16.16 and 16.17**).

Major Kinds of Ecosystems

16.9 Ocean Ecosystems

- There are three primary ocean ecosystems: shallow waters, open-sea surfaces, and deep-sea bottoms (**figure 16.18**). Each is affected by light and temperature.

16.10 Freshwater Ecosystems

- Freshwater ecosystems are closely tied to and affected by the terrestrial environments that surround them. Freshwater ecosystems are affected by light, temperature, and nutrients (**figures 16.23 and 16.24**).

16.11 Land Ecosystems

- Biomes are terrestrial communities found throughout the world. Each biome contains its own characteristic representation of plants and animals based on temperature and rainfall patterns (**figure 16.25**).

1. Energy from light is converted into chemical energy for organisms by
   - a. herbivores.
   - b. carnivores.
   - c. producers.
   - d. detritivores.

2. As energy is transferred from one trophic level to the next, substantial amounts of energy are lost to
   - a. undigestible biomass.
   - b. heat.
   - c. metabolism.
   - d. All answers are correct.

3. The number of carnivores found at the top of an ecological pyramid is limited by the
   a. number of organisms below the top carnivores.
   b. number of trophic levels below the top carnivores.
   c. amount of biomass below the top carnivores.
   d. amount of energy transferred to the top carnivores.
4. Hydrologists, scientists who study the movements and cycles of water, refer to the return of water from the ground to the air as *evapotranspiration*. The first part of the word refers to evaporation. The second part of the word refers to transpiration, which is evaporation of water
   a. from plants.
   b. through animal perspiration.
   c. off the ground shaded by plants.
   d. from the surface of rivers.
5. The carbon cycle includes the gathering of
   a. materials to make proteins.
   b. light energy through photosynthesis.
   c. phosphorus to make DNA.
   d. No answer is correct.
6. The element phosphorus is needed in organisms to build
   a. proteins.
   c. ATP.
   b. lipids.
   d. steroids.
7. A rain shadow results in
   a. extremely wet conditions due to loss of moisture from winds rising over a mountain range.
   b. dry air moving toward the poles that cools and sinks in regions 15° to 30° north/south latitude.
   c. global polar regions that rarely receive moisture from the warmer, tropical regions, and are therefore dryer.
   d. desert conditions on the down-wind side of a mountain due to increased moisture-holding capacity of the winds as the air heats up.
8. As one travels from northern Canada south to the United States, the timberline increases in elevation, but never disappears. This is because as elevation
   a. increases, temperature decreases.
   b. decreases, temperature increases.
   c. increases, humidity decreases.
   d. decreases, humidity increases.
9. In freshwater lakes during the summer, layers of sudden temperature change called _____ form.
   a. eutrophy   c. oligotrophy
   b. thermal restratification   d. thermocline
10. A temperate evergreen forest can exist in the southwest deserts of the United States if the forest exists
    a. in desert riparian (stream or river) areas.
    b. near natural springs.
    c. near the top of a desert mountain range.
    d. in the desert flatlands.

Visual Understanding

1. **Figure 16.5** A. One thousand calories of energy is produced by a certain amount of algae. Explain the probable efficiency of having humans eat, respectively, (1) the algae itself, (2) the heterotrophs, (3) the smelt, or (4) the trout. B. How many more people could be fed if we all ate algae? C. Why can't we survive just eating algae?

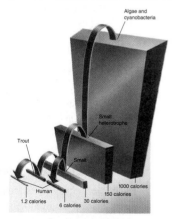

2. **Figure 16.7** Imagine that you are a molecule of water. Create a journey for yourself, starting with falling as part of a drop of rain onto the earth, and ending in a cloud ready to fall once again. Take a trip through a plant along the way.

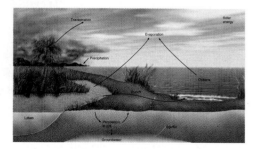

Challenge Questions

The Energy in Ecosystems   Given the amount of sunlight that hits the plants on our planet, and the ability of plants for rapid growth and reproduction, how come we aren't all hip deep in dead plants?

Materials Cycle Within Ecosystem   Plants need both carbon (C) and nitrogen (N) in large amounts. Compare and contrast how plants obtain these two important elements.

How Weather Shapes Ecosystems   How is it that the amount of sunlight hitting a particular place on the earth affects the plants that grow there in almost exactly the same way as the elevation of that place above sea level?

Major Kinds of Ecosystems   Pick two very different land ecosystems. Compare and contrast the sunshine, rainfall, major temperature features, and something about the plants and animals.

Online Learning Center

Visit the Online Learning Center for this chapter at www.mhhe.com/tlwessentials/ch16 for quizzes, animations, interactive learning exercises, and other study tools. At the site you will also find extended answers to the end-of-chapter questions.

17

Populations and Communities

O ften the most significant events that occur in an ecosystem involve the organisms that inhabit it. The swarming insects you see here are migratory locusts, *Locusta migratoria,* moving across farmland in North Africa in 1988. In most years, the locusts are not plentiful and do not swarm. In particularly favorable years, however, when food is plentiful and the weather mild, the abundance of resources leads to greater-than-usual growth of locust populations. When high population densities are reached, the locusts exhibit different hormonal and physical characteristics and take off as a swarm. Moving over the landscape, the swarm eats every available plant, denuding the landscape. Swarming locusts, although not common in North America, are a legendary plague of large areas of Africa and Eurasia. In this chapter, we examine how natural populations grow, and what factors limit this growth. The organisms of the living world have evolved many accommodations to facilitate living together, creating complex evolutionary arrangements. When these arrangements are disturbed by unusual weather—or human intervention—the consequences can be catastrophic.

Successfully reproducing is the very essence of evolutionary success, and few organisms do it alone. Rather, organisms live as members of **populations,** groups of individuals of a species that live together and influence each other's survival. In this chapter, we will explore the properties of populations, focusing on the factors that influence whether a population will grow or shrink, and at what rate. Although we humans picture ourselves as different from populations of animals living in the wild, factors that affect wild populations, such as population densities, dispersion, growth, competition, and sharing resources, affect human populations in similar ways.

17.1   Population Growth

One of the critical properties of any population is its **population size**—the number of individuals in the population. For example, if an entire species consists of only one or a few small populations, that species is likely to become extinct, especially if it occurs in areas that have been or are being radically changed. In addition to population size, **population density**—the number of individuals that occur in a unit area, such as per square kilometer—is often an important characteristic. The density of a population, how closely individuals associate with each other, is an indication of how they live. Animals that live in large groups, such as herds of wildebeests or zebras, may find safety in numbers. A third significant property is **population dispersion,** the scatter of individual organisms within the population's range. Individuals may be spaced randomly (less common in nature), in clumps (due to uneven distribution of resources or in response to social interaction), or uniformly (as a result of competition for resources) (figure 17.1). In addition to size, density, and dispersion, another key characteristic of any population is its capacity to grow. To understand populations, we must consider how they grow and what factors in nature limit **population growth.**

The Exponential Growth Model

The simplest model of population growth assumes a population growing without limits at its maximal rate. This rate, symbolized *r* and called the **biotic potential,** is the rate at which a population of a given species will increase when no limits are placed on its rate of growth. In mathematical terms, this is defined by the following formula:

$$growth\ rate = dN/dt = r_iN$$

where N is the number of individuals in the population, dN/dt is the rate of change in its numbers over time, and r_i is the *intrinsic* rate of natural increase for that population—its innate capacity for growth.

The *actual* rate of population increase, *r,* is defined as the difference between the birthrate and the death rate cor-

(b)

Figure 17.1   Population dispersion.
(*a*) Different arrangements of bacterial colonies, and (*b*) starlings distributed uniformly along telephone wires.

rected for any movement of individuals in or out of the population, whether net emigration (movement out of the area) or net immigration (movement into the area). Thus,

$$r = (b - d) + (i - e)$$

Movements of individuals can have a major impact on population growth rates. For example, the increase in human population in the United States during the closing decades of the twentieth century was mostly due to immigrants. Less than half of the increase came from the reproduction of the people already living there.

The innate capacity for growth of any population is exponential. Even when the *rate* of increase remains constant, the actual increase in the *number* of individuals accelerates rapidly as the size of the population grows. This sort of growth pattern is similar to that obtained by compounding interest on an investment. In practice, such patterns prevail only for short periods, usually when an organism reaches a new habitat with abundant resources. Natural examples include dandelions reaching the fields, lawns, and meadows of North America from Europe for the first time; algae colonizing a newly formed pond; or the first terrestrial immigrants arriving on an island recently thrust up from the sea.

Carrying Capacity

No matter how rapidly populations grow, they eventually reach a limit imposed by shortages of important environmental factors such as space, light, water, or nutrients. A population ultimately stabilizes at a certain size, called the **carrying capacity** of the particular place where it lives. The carrying capacity, symbolized by *K,* is the maximum number of individuals that an area can support, a dynamic rather than static measure as the characteristics of an area change.

The Logistic Growth Model

As a population approaches its carrying capacity, its rate of growth slows greatly, because fewer resources remain for each new individual to use. The growth curve of such a population, which is always limited by one or more factors in the environment, can be approximated by the following **logistic growth equation** that adjusts the growth rate to account for the lessening availability of limiting factors:

$$dN/dt = rN \left(\frac{K - N}{K} \right)$$

In this logistic model of population growth, the growth rate of the population (*dN/dt*) equals its rate of increase (*r* multiplied by *N*, the number of individuals present at any one time), adjusted for the amount of resources available. The adjustment is made by multiplying *rN* by the fraction of *K* still unused (*K* minus *N*, divided by *K*). As *N* increases (the population grows in size), the fraction by which *r* is multiplied (the remaining resources) becomes smaller and smaller, and the rate of increase of the population declines.

In mathematical terms, as *N* approaches *K,* the rate of population growth (*dN/dt*) begins to slow, until it reaches 0 when *N = K* (figure 17.2). In practical terms, factors such as increasing competition among more individuals for a given set of resources, the buildup of waste, or an increased rate of predation causes the decline in the rate of population growth.

Graphically, if you plot *N* versus *t* (time), you obtain an S-shaped **sigmoid growth curve** characteristic of most biological populations. The curve is called "sigmoid" because its shape has a double curve like the letter S. As the size of a population stabilizes at the carrying capacity, its rate of growth slows down, eventually coming to a halt (figure 17.3).

Processes such as competition for resources, emigration, and the accumulation of toxic wastes all tend to increase as a population approaches its carrying capacity for a particular habitat. The resources for which the members of the population are competing may be food, shelter, light, mating sites, mates, or any other factor the species needs to carry out its life cycle and reproduce.

> **17.1** The size at which a population stabilizes in a particular place is defined as the carrying capacity of that place for that species.
> Populations grow to the carrying capacity of their environment.

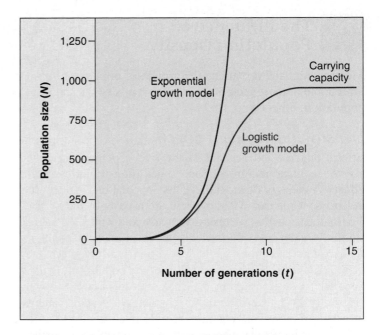

Figure 17.2   Two models of population growth.
The *red line* illustrates the exponential growth model for a population with an *r* of 1.0. The *blue line* illustrates the logistic growth model in a population with *r* = 1.0 and *K* = 1,000 individuals. At first, logistic growth accelerates exponentially, and then, as resources become limiting, the death rate increases and growth slows. Growth ceases when the death rate equals the birthrate. The carrying capacity (*K*) ultimately depends on the resources available in the environment.

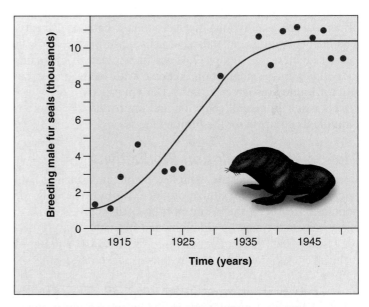

Figure 17.3   Most natural populations exhibit logistic growth.
These data present the history of a fur seal (*Callorhinus ursinus*) population on St. Paul Island, Alaska. Driven almost to extinction by hunting in the late 1800s, the fur seal made a comeback after hunting was banned in 1911. Today the number of breeding males with "harems" oscillates around 10,000 individuals, presumably the carrying capacity of the island for fur seals.

17.2 The Influence of Population Density

Many factors act to regulate the growth of populations in nature. Some of these factors act independently of the size of the population; others do not.

Density-Independent Effects

Effects that are independent of the size of a population and act to regulate its growth are called **density-independent effects.** A variety of factors may affect populations in a density-independent manner. Most of these are aspects of the external environment, such as weather (extremely cold winters, droughts, storms, floods) and physical disruptions (volcanic eruptions, fire, road construction). Individuals often will be affected by these activities regardless of the size of the population. Populations that occur in areas in which such events occur relatively frequently will display erratic population growth patterns, increasing rapidly when conditions are relatively good, but suffering extreme reductions whenever the environment turns hostile.

Density-Dependent Effects

Effects that are dependent on the size of the population and act to regulate its growth are called **density-dependent effects.** Among animals, these effects may be accompanied by hormonal changes that can alter behavior that will directly affect the ultimate size of the population. One striking example occurs in migratory locusts ("short-horned" grasshoppers). When they become crowded, the locusts produce hormones that cause them to enter a migratory phase; the locusts take off as a swarm and fly long distances to new habitats. Density-dependent effects, in general, have an increasing effect as population size increases (figure 17.4). As the population grows, the individuals in the population compete with increasing intensity for limited resources. Charles Darwin proposed that these effects result in natural selection and improved adaptation as individuals compete for the limiting factors.

Maximizing Population Productivity

In natural systems that are exploited by humans, such as fisheries, the aim is to maximize productivity by exploiting the population early in the rising portion of its sigmoid growth curve. At such times, populations and individuals are growing rapidly, and net productivity—in terms of the amount of material incorporated into the bodies of these organisms—is highest.

Commercial fisheries attempt to operate so that they are always harvesting populations in the steep, rapidly growing parts of the curve. The point of *maximal sustainable yield* lies partway up the sigmoid curve (figure 17.5). Harvesting the population of an economically desirable species near this point will result in the best sustained yields. Overharvesting a population that is smaller than this critical size can destroy its productivity for many years or even drive it to extinction. This evidently happened in the Peruvian anchovy fishery after the populations had been depressed by the 1972 El Niño. It is often difficult to determine population levels of commercially

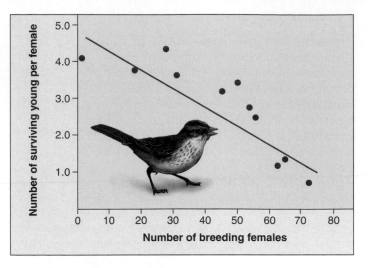

Figure 17.4   Density-dependent effects.
Reproductive success of the song sparrow (*Melospiza melodia*) decreases as population size increases.

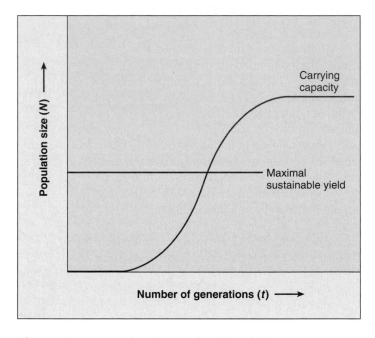

Figure 17.5   Maximal sustainable yield.
The goal of harvesting organisms for commercial purposes is to harvest just enough organisms to maximize current yields but also to sustain the population for future yields. Harvesting the organisms when the population is in the rapid growth phase of the sigmoidal curve, but not overharvesting, will result in sustained yields.

valuable species, and without this information it is equally difficult to determine the yield most suitable for long-term, productive harvesting.

> **17.2**   Density-independent effects are controlled by factors that operate regardless of population size; density-dependent effects are caused by factors that come into play particularly when the population size is larger.

17.3 Life History Adaptations

Populations of any species, including annual plants, some insects, and most bacteria, can have very fast rates of growth when not limited by dwindling environmental resources. Habitats with more available resources than the population requires favor very rapid reproduction rates, which often approximate the exponential growth model discussed earlier.

Populations of most animals have much slower rates of growth with numbers limited by available resources. Growth slows as available resources become limiting, producing a sigmoid growth curve approximating the logistic growth model discussed earlier. Habitats with limited resources lead to more intense competition for resources, and favor individuals that can survive and successfully reproduce more efficiently. The number of individuals that can survive at this limit is the carrying capacity of the environment, or K.

The complete life cycle of an organism constitutes its *life history*. Life histories are very diverse, with different organisms having different adaptations in response to their environments. Some life history adaptations of a population favor very rapid growth in a habitat with unlimited resources, or in unpredictable or volatile environments where organisms have to take advantage of the resources when they are available, among them reproducing early, producing many small offspring that mature quickly, and engaging in other aspects of "big bang" reproduction. Using the terms of the exponential model, these adaptations, all favoring a high rate of increase, r, are called **r-selected adaptations.** Examples of organisms displaying r-selected life history adaptations include dandelions, aphids, mice, and cockroaches (figure 17.6).

Other life history adaptations favor survival in an environment where individuals are competing for limited resources, among them reproducing late, having small numbers of large offspring that mature slowly and receive intensive parental care, and other aspects of "carrying capacity" reproduction. In terms of the logistic model, these adaptations, all favoring reproduction near the carrying capacity of the environment, K, are called **K-selected adaptations.** Examples of organisms displaying K-selected life history adaptations include coconut palms, whooping cranes, and whales.

Although the r/K concept of life histories is often used to compare different taxa, it also provides a powerful way to examine more closely related organisms living in different types of habitats. In general, populations living in rapidly changing habitats tend to exhibit r-selected adaptations, whereas populations of closely related organisms living in more stable but highly competitive habitats exhibit more K-selected adaptations.

Most natural populations show life history adaptations that exist along a continuum, ranging from completely r-selected traits to completely K-selected traits. Table 17.1 outlines the adaptations at the extreme ends of the continuum.

17.3 Some life history adaptations favor near-exponential growth, others the more competitive logistic growth. Most natural populations exhibit a combination of the two.

Figure 17.6   The consequences of exponential growth.

All organisms have the potential to produce populations larger than those that actually occur in nature. The German cockroach (*Blatella germanica*), a major household pest, produces 80 young every six months. If every cockroach that hatched survived for three generations, kitchens might look like this theoretical culinary nightmare concocted by the Smithsonian Museum of Natural History.

TABLE 17.1	r-SELECTED AND K-SELECTED LIFE HISTORY ADAPTATIONS	
Adaptation	**r-Selected Populations**	**K-Selected Populations**
Age at first reproduction	Early	Late
Homeostatic capability	Limited	Often extensive
Life span	Short	Long
Maturation time	Short	Long
Mortality rate	Often high	Usually low
Number of offspring produced per reproductive episode	Many	Few
Number of reproductions per lifetime	Usually one	Often several
Parental care	None	Often extensive
Size of offspring or eggs	Small	Large

Source: After E. R. Pianka. Evolutionary Ecology, 4th ed. New York: Harper & Row, 1987.

17.4 Population Demography

Demography is the statistical study of populations. The term comes from two Greek words: *demos,* "the people" (the same root we see in the word *democracy*), and *graphos,* "measurement." Demography therefore means measurement of people, or, by extension, of the characteristics of populations. Demography is the science that helps predict how population sizes will change in the future. Populations grow if births outnumber deaths and shrink if deaths outnumber births. Because birth and death rates depend significantly on age and sex, the future size of a population depends on its present age structure and sex ratio.

Age Structure

Many annual plants and insects time their reproduction to particular seasons of the year and then die. All members of these populations are the same age. Perennial plants and longer-lived animals contain individuals of more than one generation, so that in any given year individuals of different ages are reproducing within the population. A group of individuals of the same age is referred to as a **cohort.**

Within a population, every cohort has a characteristic birthrate, or **fecundity,** defined as the number of offspring produced in a standard time (for example, per year), and a characteristic death rate, or **mortality,** the number of individuals that die in that period. The rate of a population's growth depends directly on the difference between these two rates.

The relative number of individuals in each cohort defines a population's age structure (refer back to figure 2.23 in relation to human populations). Because individuals of different ages have different fecundity and death rates, age structure has a critical impact on a population's growth rate. A population with a large proportion of young individuals, for example, tends to grow rapidly because an increasing proportion of its individuals are reproductive.

Sex Ratio

The proportion of males and females in a population is its **sex ratio.** The number of births is usually directly related to the number of females, but it may not be as closely related to the number of males in species where a single male can mate with several females. In deer, elk, lions, and many other animals, a reproductive male guards a "harem" of females with which he mates, while preventing other males from mating with them. In such species, a reduction in the number of males simply changes the identities of the reproductive males without reducing the number of births. Among monogamous species like many birds, by contrast, where pairs form long-lasting reproductive relationships, a reduction in the number of males can directly reduce the number of births.

Mortality and Survivorship Curves

A population's intrinsic rate of increase depends on the ages of the organisms in it and the reproductive performance of

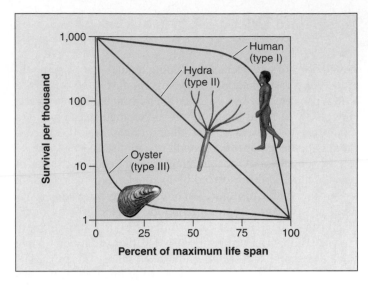

Figure 17.7 Survivorship curves.
By convention, survival (the *vertical axis*) is plotted on a log scale. Humans have a type I life cycle, the hydra (an animal related to jellyfish) type II, and oysters type III.

the individuals in the various age groups. When a population lives in a constant environment for a few generations, its **age distribution**—the proportion of individuals in different age categories—tends to stabilize. This distribution differs greatly from species to species and even, to some extent, from population to population within a given species. Depending on the mating system of the species, sex ratio and generation time can also have a significant effect on population growth. A population whose size remains fairly constant through time is called a stable population. In such a population, births plus immigration must balance deaths plus emigration.

One way to express the age distribution characteristics of populations is through a **survivorship curve.** Survivorship is defined as the percentage of an original population that survives to a given age. Examples of different kinds of survivorship curves are shown in figure 17.7. In hydra, animals related to jellyfish, individuals are equally likely to die at any age, as indicated by the straight survivorship curve (type II). Oysters, like plants, produce vast numbers of offspring, only a few of which live to reproduce. However, once they become established and grow into reproductive individuals, their mortality is extremely low (type III survivorship curve). Finally, even though human babies are susceptible to death at relatively high rates, mortality in humans, as in many animals and protists, rises in the postreproductive years (type I survivorship curve).

> **17.4** The growth rate of a population is a sensitive function of its age structure. In some species, mortality is focused among the young, and in others, among the old; in only a few is mortality independent of age.

A Closer Look

Invasion of the Killer Bees

One of the harshest lessons of environmental biology is that the unexpected does happen. Precisely because science operates at the edge of what we know, nosing ahead to learn more, researchers sometimes stumble over the unexpected. This lesson has been brought clearly to mind in recent years, with reports of killer bees being discovered east of the Mississippi River.

In the continental United States, we are used to the mild-mannered European honeybee, a subspecies called *mellifera*. The African variety, subspecies *scutelar*, look very much like them, but they are hardly mild mannered. They are in fact very aggressive critters with a chip on their shoulder, and when swarming do not have to be provoked to start trouble. A few individuals may see you at a distance, "lose it," and lead thousands of bees in a concerted attempt to do you in. The only thing you can do is run—fast. They will keep after you for up to a mile. It doesn't do any good to duck under water, as they just wait for you.

They are nicknamed "killer" bees not because any one sting is worse than the European kind, but rather because so many of the bees try to sting you. An average human can survive no more than 300 bee stings. A horrible total of more than 8,000 bee stings is not unusual for a killer bee attack. An American graduate student attacked and killed in a Costa Rican jungle had 10,000 stings.

The reason killer bees remind us of the "law of unintended consequences" is that their invasion of this continent is the direct result of researchers stumbling over the unexpected.

Killer bees were brought from Africa to Brazil 47 years ago by a prominent Brazilian scientist, Warwick Estevan Kerr. A famous geneticist, Kerr is the only Brazilian to be a member of the U.S. National Academy of Sciences. What he was doing in 1956, at the request of the Brazilian government, was attempting to establish tropical bees in Brazil to expand the commercial bee industry (bees pollinate crops and make commercial honey in the process). African bees seemed ideal candidates, better adapted to the tropics than European bees and more prolific honey producers.

Kerr established a quarantined colony of African bees at a remote field station outside the city of Rio Claro, several hundred miles from his university at São Paulo. Although he had brought back many queens from South Africa, the colony came to be dominated by the offspring of a single very productive queen from Tanzania. Kerr noted at the time that she appeared unusually aggressive.

In the fall of 1957, the field station was visited by a beekeeper. As no one else was around that day, the visitor performed the routine courtesy of tending the hives. A hive with a queen in it has a set of bars across the door so the queen can't get out. Called a "queen excluder," the bars are far enough apart that the smaller worker bees can squeeze through. Now once a queen starts to lay eggs, she'll never leave the hive, so there is little point in slowing down entry of workers to the hive, and the queen excluder is routinely removed. On that day, the visitor saw the Tanzanian queen laying eggs in the African colony hive, and so removed the queen excluder.

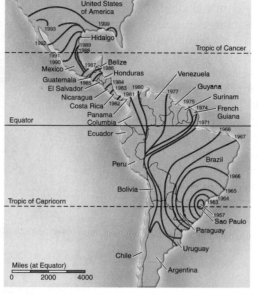

One and a half days later, when a staff member inspected the African colony, the Tanzanian queen and 26 of her daughter queens had decamped. Out into the neighboring forest they went. And that's what I mean by being bushwhacked by the unexpected. European queen bees never leave the hive after they have started to lay eggs. No one could have guessed that the Tanzanian queen would behave differently. But she did.

By 1970 the superaggressive African bees had blanketed Brazil, totally replacing local colonies. They reached Central America by 1980, Mexico by 1986, and Texas by 1990, having conquered 5 million square miles in 33 years. In the process they killed an estimated 1,000 people and over 100,000 cows. The first American to be killed, a rancher named Lino Lopez, died of multiple stings in Texas in 1993.

All during the 1990s, the bees continued their invasion of the United States. All of Arizona and much of Texas has been occupied. A few years ago, Los Angeles county was officially declared colonized, and it looks like African bees will eventually move at least halfway up the state of California.

Soon, however, the invasion is predicted to cease, on a line roughly from San Francisco, California, to Richmond, Virginia. Winter cold is expected to limit any further northward advance of the invading hoards.

For the states below this line, sure as the sun rises, the bees are coming, not caring one bit that they were unanticipated. The deep lesson—that unexpected things do happen—is being driven home by millions of tiny aggressive teachers.

17.5 Communities

All organisms that live together in an area are called a **community.** The countless number of species that inhabit a rain forest or the sparse number of species that live in the boiling waters of a hot spring make up a community. Indeed, every inhabited place on earth supports its own particular array of organisms. Over time, the different species have made many complex adjustments to community living, evolving together and forging relationships that give the community character and stability. Both competition and cooperation have played key roles in molding these communities.

The magnificent redwood forest that extends along the coast of central and northern California and into the southwestern corner of Oregon is an example of a community. Within it, the most obvious organisms are the redwood trees, *Sequoia sempervirens.* These trees are the sole survivors of a genus that was once distributed throughout much of the Northern Hemisphere. A number of other plants and animals are regularly associated with redwood trees (figure 17.8). Their coexistence is in part made possible by the special conditions the redwood trees themselves create, providing shade, water (dripping from the branches), and relatively cool temperatures. This particular distinctive assemblage of organisms is called the redwood community. The organisms characteristic of this community have each had a complex and unique evolutionary history. They evolved at different times in the past and then came to be associated with the redwoods.

We recognize this community mainly because of the redwood trees, and its boundaries are determined by the redwood's distribution. The distributions of the other organisms in the redwood community may differ a good deal. Some organisms may not be distributed as widely as the redwoods, and some may be distributed over a broader range. In the redwood community or any other community, the ranges of the different organisms overlap; that is why they occur together.

Many communities are very similar in species composition and appearance over wide areas. For example, the open savanna that stretches across much of Africa includes many plant and animal species that coexist over thousands of square kilometers. Interactions between these organisms occur in a similar manner throughout these grassland communities, and some interactions have evolved over millions of years.

> **17.5** We recognize a community largely because of the presence of its dominant species, but many other kinds of organisms are also characteristic of each community. A community exists in a place because the ranges of its species overlap there.

(a)

(b)

(c)

(d)

Figure 17.8 The redwood community.
(*a*) The redwood forest of coastal California and southwestern Oregon is dominated by the redwoods (*Sequoia sempervirens*) themselves. Other organisms in the redwood community include (*b*) redwood sorrel (*Oxalis oregana*), (*c*) sword ferns (*Polystichum munitum*), and (*d*) ground beetles (*Scaphinotus velutinus*), this one feeding on a slug on a sword fern leaf. The ecological requirements of each of these organisms differ, but they overlap enough that the organisms occur together in the redwood community.

17.6 The Niche and Competition

Within a community, each organism occupies a particular biological role, or **niche.** As we discussed in chapter 2, the niche an organism occupies is the sum total of all the ways it uses the resources of its environment, including space, food, and many other factors of the environment. A niche is a pattern of living.

Sometimes organisms are not able to occupy their entire niche because some other organism is using it. We call such situations when two organisms attempt to use the same resource **competition.** Competition is the struggle of two organisms to use the same resource when there is not enough of the resource to satisfy both.

Interspecific competition refers to the interactions between individuals of different species when both require the same scarce resource. Interspecific competition is often greatest between organisms that obtain their food in similar ways; thus, green plants compete mainly with other green plants, herbivores with other herbivores, and carnivores with other carnivores. In addition, competition is more acute between similar organisms than between those that are less similar. Although interspecific competition occurs between members of different species, it is to be distinguished from **intraspecific competition,** which occurs between individuals of a single species.

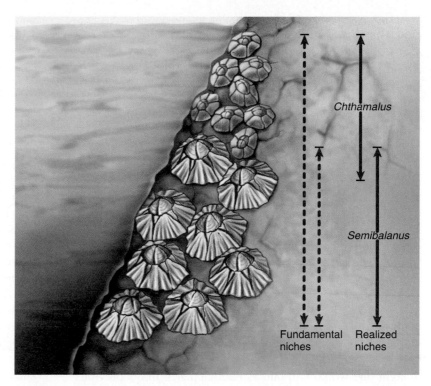

Figure 17.9 Competition among two species of barnacles limits niche use.

Chthamalus can live in both deep and shallow zones (its fundamental niche), but *Semibalanus* forces *Chthamalus* out of the part of its fundamental niche that overlaps the realized niche of *Semibalanus.*

The Realized Niche

Because of competition, organisms may not be able to occupy the entire niche they are theoretically capable of using, called the **fundamental niche** (or theoretical niche). The actual niche the organism is able to occupy in the presence of competitors is called its **realized niche.**

In a classic study, J. H. Connell of the University of California, Santa Barbara, investigated competitive interactions between two species of barnacles that grow together on rocks along the coast of Scotland. Barnacles are marine animals (crustaceans) that have free-swimming larvae. The larvae eventually settle down, cementing themselves to rocks and remaining attached for the rest of their lives. Of the two species Connell studied, *Chthamalus stellatus* lives in shallower water, where tidal action often exposes it to air, and *Semibalanus balanoides* lives lower down, where it is rarely exposed to the atmosphere (figure 17.9). In the deeper zone, *Semibalanus* could always outcompete *Chthamalus* by crowding it off the rocks, undercutting it, and replacing it even where it had begun to grow. When Connell removed *Semibalanus* from the area, however, *Chthamalus* was easily able to occupy the deeper zone, indicating that no physiological or other general obstacles prevented it from becoming established there. In contrast, *Semibalanus* could not survive in the shallow-water habitats where *Chthamalus* normally occurs; it evidently does not have the special physiological and morphological adapta-

tions that allow *Chthamalus* to occupy this zone. Thus, the fundamental niche of the barnacle *Chthamalus* in Connell's experiments in Scotland included that of *Semibalanus,* but its realized niche was much narrower because *Chthamalus* was outcompeted by *Semibalanus* in its fundamental niche.

In the previous example, *Chthamalus* was able to fully occupy its fundamental niche when there was no competition. This is often the case when a species first enters a new, very favorable habitat that presents it with adequate resources, little or no competition, and no predators. However, once resources begin to become limiting as the population approaches carrying capacity and other species begin to compete for them, and as predators begin to more frequently recognize them, the population will be forced into its realized niche. For example, a plant called the St. John's-wort was introduced and became widespread in open rangeland habitats in California. It occupied all of its fundamental niche until a species of beetle that feeds on the plant was introduced into the habitat. Populations of the plant then quickly decreased, and it is now only found in shady sites where the beetle cannot thrive. In this case, the presence of a predator limits the realized niche of a plant.

Competitive Exclusion

In classic experiments carried out between 1934 and 1935, Russian ecologist G. F. Gause studied competition among three species of *Paramecium,* a tiny protist. All three species grew well alone in culture tubes, preying on bacteria and yeasts that fed on oatmeal suspended in the culture fluid. However, when Gause grew *P. aurelia* together with *P. caudatum* in the same culture tube, the numbers of *P. caudatum* always declined to extinction, leaving *P. aurelia* the only survivor. Why? Gause found *P. aurelia* was able to grow six times faster than its competitor, *P. caudatum,* because it was able to better use the limited available resources.

From experiments such as this, Gause formulated what is now called the *principle of competitive exclusion.* This principle states that if two species are competing for a resource, the species that uses the resource more efficiently will eventually eliminate the other locally—no two species with the same niche can coexist. This overlapping of niches leads to competition, but is one competitor always eliminated? No, as we shall soon see.

Niche Overlap

In a revealing experiment, Gause challenged *P. caudatum*—the defeated species in his earlier experiments—with a third species, *P. bursaria.* Because he expected these two species to also compete for the limited bacterial food supply, Gause thought one would win out, as had happened in his previous experiments. But that's not what happened. Instead, both species survived in the culture tubes; the paramecia found a way to divide the food resources. How did they do it? In the upper part of the culture tubes, where the oxygen concentration and bacterial density were high, *P. caudatum* dominated because it was better able to feed on bacteria. However, in the lower part of the tubes, the lower oxygen concentration favored the growth of a different potential food, yeast, and *P. bursaria* was better able to eat this food. The fundamental niche of each

species was the whole culture tube, but the realized niche of each species was only a portion of the tube. Figure 17.10 summarizes Gause's experiments. The graph also demonstrates the negative effect competition had on the participants: Competition was always detrimental to both species involved. Both species reach about twice the density when grown without a competitor as when grown together.

Gause's principle of competitive exclusion can be restated to say that no two species can occupy the same niche indefinitely. Certainly species can and do coexist while competing for the same resources; we have seen many examples of such relationships. Nevertheless, Gause's theory predicts that when two species are able to coexist on a long-term basis, their niches always differ in one or more features; otherwise, one species outcompetes the other and the extinction of the second species inevitably results, a process referred to as **competitive exclusion.**

Niche is a complex concept, involving all facets of the environment that are important to individual species. In recent years, a vigorous debate has arisen concerning the role of competitive exclusion, not only in determining the structure of communities but also in setting the course of evolution. When one or more resources suddenly become sharply limiting, as in periods of drought, the role of competition becomes much more obvious. However, in less immediately stressful situations, species act to avoid competition whenever possible. When their niches overlap, two outcomes are possible: competitive exclusion (winner takes all), or **resource partitioning** (dividing up resources to create two realized niches). It is only through resource partitioning, the topic of the next section, that the two species can continue to coexist over long periods.

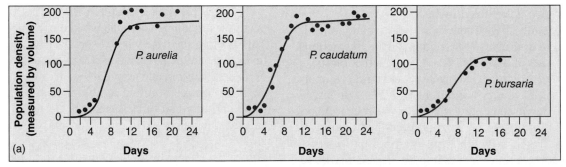

Figure 17.10 Competitive exclusion among three species of *Paramecium*.

In the microscopic world, *Paramecium* is a ferocious predator. *Paramecia* eat by ingesting their prey; their plasma membranes surround bacterial or yeast cells, forming a food vacuole containing the prey cell. In his experiments, (*a*) Gause found that three species of *Paramecium* grew well alone in culture tubes. (*b*) However, *P. caudatum* declined to extinction when grown with *P. aurelia* because they shared the same realized niche, and *P. aurelia* outcompeted *P. caudatum* for food resources. (*c*) However, *P. caudatum* and *P. bursaria* were able to coexist, although in smaller populations, because the two have different realized niches and thus avoid competition.

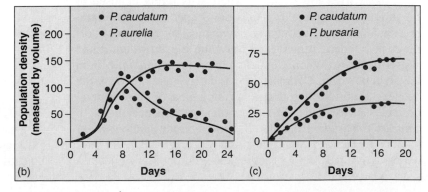

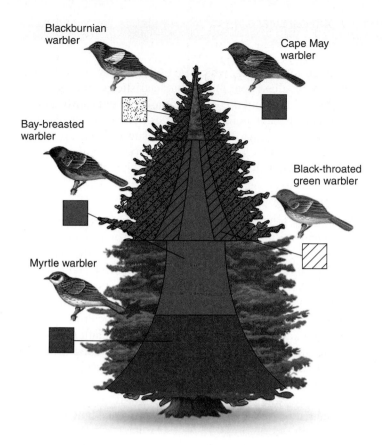

Figure 17.11 Resource partitioning in warblers.
Five different species of warblers appear to be competing for the same resources, but they actually feed in different parts of spruce trees.

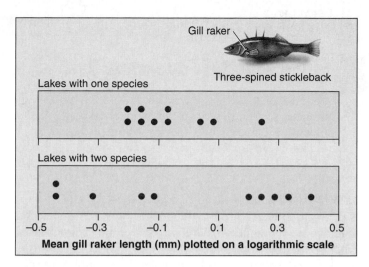

Figure 17.12 Character displacement in stickleback fish.
In lakes that have two species of three-spined stickleback (*lower* graph), one species (*blue* dots) feeds on plankton in open water and has long gill rakers (structures on the gills of fish) that act like a sieve to filter the plankton out. The other species (*red* dots) is more of a bottom-dweller, has shorter gill rakers, and feeds on larger prey. However, in lakes with only one species of stickleback (*upper* graph), the fish take advantage of both food resources, and the gill rakers are intermediate in length.

Resource Partitioning

Gause's exclusion principle has a very important consequence: Persistent competition between two species is rare in natural communities. Either one species drives the other to extinction, or natural selection reduces the competition between them. When the late Princeton ecologist Robert MacArthur studied five species of warblers, small insect-eating forest songbirds, he found that they all appeared to be competing for the same resources. However, when he studied them more carefully, he found that each species actually fed in a different part of spruce trees and so ate different subsets of insects. One species fed on insects near the tips of branches, a second within the dense foliage, a third on the lower branches, a fourth high on the trees, and a fifth at the very apex of the trees. Thus, the species of warblers had *subdivided the niche,* partitioning the available resource so as to avoid direct competition with one another (figure 17.11).

Resource partitioning can often be seen in similar species that occupy the same geographical area. Called **sympatric species** (Greek, *syn,* same, and *patria,* country), these species avoid competition by living in different portions of the habitat or by using different food or other resources. For example, as we saw in chapter 2, sympatric *Anolis* lizards species are able to partition the tree habitat (see figure 2.15). Species that do not live in the same geographical area, called **allopatric species** (Greek, *allos,* other, and *patria,* country), often use the same habitat locations and food resources—because they are not in competition, natural selection does not favor evolutionary changes that subdivide their niche.

When a pair of species occupy the same habitat (that is, when they are sympatric), they tend to exhibit greater differences in morphology and behavior than the same two species do when living in different habitats (that is, when they are allopatric). Called **character displacement,** the differences evident between sympatric species are thought to have been favored by natural selection as a mechanism to facilitate habitat partitioning and thus reduce competition. Thus, the stickleback fish in figure 17.12 have different-sized gill rakers (structures of the fish gill) where the fish are sympatric, two species sharing the same lake but partitioning the food resources. But the gill rakers are intermediate in size in lakes with only one species of stickleback, the fish taking advantage of all the available food resources. In chapter 2, we also discussed the difference in bill size seen in sympatric Darwin's finches, another example of character displacement (see figure 2.16).

17.6 A niche may be defined as the way in which an organism uses its environment. No two species can occupy the same niche indefinitely without competition driving one to extinction. Sympatric species partition available resources, reducing competition between them.

17.7 Coevolution and Symbiosis

The previous section described the "winner take all" results of competition between two species whose niches overlap. Other relationships in nature are less competitive and more cooperative.

Coevolution

The plants, animals, protists, fungi, and prokaryotes that live together in communities have changed and adjusted to one another continually over millions of years. For example, many features of flowering plants have evolved in relation to the dispersal of the plant's gametes by animals (figure 17.13). These animals, in turn, have evolved a number of special traits that enable them to obtain food or other resources efficiently from the plants they visit, often from their flowers. In addition, the seeds of many flowering plants have features that make them more likely to be dispersed to new areas of favorable habitat.

Figure 17.13 Pollination by bat.
Many flowers have coevolved with other species to facilitate pollen transfer. Insects are widely known as pollinators, but they're not the only ones. Notice the cargo of pollen on the bat's snout.

Such interactions, which involve the long-term, mutual evolutionary adjustment of the characteristics of the members of biological communities, are examples of **coevolution.** Coevolution is the adaptation of a species not only to its physical environment but also to the other organisms that share it. In this section, we consider some examples of coevolution, including symbiotic relationships and predator-prey interactions.

Symbiosis Is Widespread

In symbiotic relationships, two or more kinds of organisms live together in often elaborate and more or less permanent relationships. All symbiotic relationships carry the potential for coevolution between the organisms involved, and in many instances the results of this coevolution are fascinating. Examples of symbiosis include lichens, which are associations of certain fungi with green algae or cyanobacteria. Another important example are mycorrhizae, the association between fungi and the roots of most kinds of plants. The fungi expedite the plant's absorption of certain nutrients, and the plants in turn provide the fungi with carbohydrates. Similarly, root nodules that occur in legumes and certain other kinds of plants contain bacteria that fix atmospheric nitrogen and make it available to their host plants.

In the tropics, leaf-cutter ants are often so abundant that they can remove a quarter or more of the total leaf surface of the plants in a given area. They do not eat these leaves directly; rather, they take them to underground nests, where they chew them up and inoculate them with the spores of particular fungi. These fungi are cultivated by the ants and brought from one specially prepared bed to another, where they grow and reproduce. In turn, the fungi constitute the primary food of the ants and their larvae. The relationship between leaf-cutter ants and these fungi is an excellent example of symbiosis.

The major kinds of symbiotic relationships include (1) mutualism, in which both participating species benefit; (2) parasitism, in which one species benefits but the other is harmed; and (3) commensalism, in which one species benefits while the other neither benefits nor is harmed. Parasitism can also be viewed as a form of predation, although the organism that is preyed upon does not necessarily die.

Mutualism

Mutualism is a symbiotic relationship among organisms in which both species benefit. Examples of mutualism are of fundamental importance in determining the structure of biological communities. Some of the most spectacular examples of mutualism occur among flowering plants and their animal visitors, including insects, birds, and bats. As we discussed in chapter 14, during the course of their evolution, the characteristics of flowers have evolved in large part in relation to the characteristics of the animals that visit them for food and, in doing so, spread their pollen from individual to individual. At the same time, characteristics of the animals have changed,

Figure 17.14 Mutualism: ants and aphids.
These ants are tending to aphids (small oval structures), feeding on the "honeydew" that the aphids excrete continuously, moving the aphids from place to place, and protecting them from potential predators.

Figure 17.15 Mutualism: ants and acacias.
Ants of the genus *Pseudomyrmex* live within the hollow thorns of certain species of acacia trees. The nectaries at the bases of the leaves and the Beltian bodies at the ends of the leaflets (yellow structures) provide food for the ants. The ants, in turn, supply the acacias with organic nutrients, protect the trees from herbivores, and reduce shading by other plants.

increasing their specialization for obtaining food or other substances from particular kinds of flowers.

Another example of mutualism involves ants and aphids. Aphids, also called greenflies, are small insects that suck fluids with their piercing mouthparts from the phloem of living plants. They extract a certain amount of the sucrose and other nutrients from this fluid, but they excrete much of it in an altered form through their anus. Certain ants have taken advantage of this—in effect, domesticating the aphids (figure 17.14). The ants carry the aphids to new plants, where they come into contact with new sources of food, and then consume as food the "honeydew" that the aphids excrete.

Ants and Acacias.   A particularly striking example of mutualism involves ants and certain Latin American species of the plant genus *Acacia*. In these species, certain leaf parts, called stipules, are modified as paired, hollow thorns; these particular species are called "bull's horn acacias." The thorns are inhabited by stinging ants of the genus *Pseudomyrmex,* which do not nest anywhere else. Like all thorns that occur on plants, the acacia horns serve to deter herbivores.

At the tip of the leaflets of these acacias are unique, protein-rich bodies called Beltian bodies, named after Thomas Belt, a nineteenth-century British naturalist who first wrote about them after seeing them in Nicaragua. Beltian bodies do not occur in species of *Acacia* that are not inhabited by ants, and their role is clear: They serve as a primary food for the ants (figure 17.15). In addition, the plants secrete nectar from glands near the bases of their leaves. The ants consume this nectar as well, feeding it and the Beltian bodies to their larvae.

Obviously, this association is beneficial to the ants, and one can readily see why they inhabit acacias of this group. The ants and their larvae are protected within the swollen thorns, and the trees provide a balanced diet, including the sugar-rich nectar and the protein-rich Beltian bodies. What, if anything, do the ants do for the plants? This question had fascinated observers for nearly a century until it was answered by Daniel Janzen, then a graduate student at the University of California, Berkeley, in a beautifully conceived and executed series of field experiments.

Whenever any herbivore lands on the branches or leaves of an acacia inhabited by ants, the ants immediately attack and devour the herbivore. Thus, the ants protect the acacias from being eaten, and the herbivore also provides additional food for the ants, which continually patrol the acacia's branches. Related species of acacias that do not have the special features of the bull's horn acacias and are not protected by ants have bitter-tasting substances in their leaves that the bull's horn acacias lack. Evidently, these bitter-tasting substances protect the acacias in which they occur from herbivores in a different way.

The ants that live in the bull's horn acacias also help their hosts to compete with other plants. The ants cut away any branches of other plants that touch the bull's horn acacia in which they are living. They create, in effect, a tunnel of light through which the acacia can grow, even in the lush deciduous forests of lowland Central America. Without the ants, as Janzen showed experimentally by poisoning the ant colonies that inhabited individual plants, the acacia is unable to compete successfully in this habitat. Finally, the ants bring organic material into their nests. The parts they do not consume, together with their excretions, provide the acacias with an abundant source of nitrogen.

Parasitism

Parasitism is a symbiotic relationship that may be regarded as a special form of predator/prey relationship, discussed later. In this symbiotic relationship the predator, or parasite, is much smaller than the prey, or host, and remains closely associated with it. Parasitism is harmful to the host organism and beneficial to the parasite, but unlike a predator/prey relationship, a good parasite does not kill its host. The concept of parasitism seems obvious, but individual instances are often surprisingly difficult to distinguish from predation and from other kinds of symbiosis.

External Parasites. Parasites that feed on the exterior surface of an organism are external parasites, or **ectoparasites.** Many instances of external parasitism are known (figure 17.16*a*). Lice, which live their entire lives on the bodies of vertebrates—mainly birds and mammals—are normally considered parasites. Mosquitoes are not considered parasites, even though they draw food from birds and mammals in a similar manner to lice, because their interaction with their host is so brief.

Parasitoids are insects that lay eggs on living hosts. This behavior is common among wasps, whose larvae feed on the body of the unfortunate host, often killing it.

Internal Parasites. Vertebrates are parasitized internally by **endoparasites,** members of many different phyla of animals and protists (figure 17.16*b*). Invertebrates also have many kinds of parasites that live within their bodies. Bacteria and viruses are not usually considered parasites, even though they fit our definition precisely.

Internal parasitism is generally marked by much more extreme specialization than external parasitism, as shown by the many protist and invertebrate parasites that infect humans. The more closely the life of the parasite is linked with that of its host, the more its morphology and behavior are likely to have been modified during the course of its evolution. The same is true of symbiotic relationships of all sorts. Conditions within the body of an organism are different from those encountered outside and are apt to be much more constant. Consequently, the structure of an internal parasite is often simplified, and unnecessary armaments and structures are lost as it evolves.

Many parasites have complex life cycles that require several different hosts for growth to adulthood and reproduction.

Brood Parasitism. Not all parasites consume the body of their host. In brood parasitism, birds like cowbirds and European cuckoos lay their eggs in the nests of other species (figure 17.16*c*). The host parents raise the brood parasite as if it were one of their own clutch, in many cases investing more in feeding the imposter than in feeding their own offspring. The brood parasite reduces the reproductive success of the foster parent hosts, so it is not surprising that evolution has fostered the hosts' ability to detect parasite eggs and reject them.

(a)

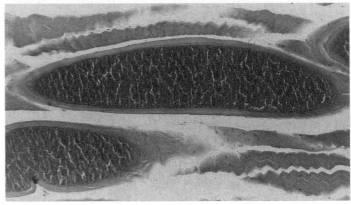

(b)

(c)

Figure 17.16 Parasitism.

(*a*) Dodder (*Cuscuta*) is a parasitic plant that has lost its chlorophyll and its leaves in the course of its evolution and is heterotrophic—unable to manufacture its own food. Instead, it obtains its food from the host plants it grows on. (*b*) *Sarcocystis* is a protistan endoparasite that infects a number of different kinds of vertebrates and is seen here in the muscle of a sheep. (*c*) Brood parasitism. Cuckoos lay their eggs in the nests of other species of birds. Because the young cuckoos (large bird to the *right*) are raised by a different species (like this meadow pipit, smaller bird to the *left*), they have no opportunity to *learn* the cuckoo song; the cuckoo song they later sing is innate.

Commensalism

Commensalism is a symbiotic relationship that benefits one species and neither hurts nor helps the other. In nature, individuals of one species are often physically attached to members of another. For example, epiphytes are plants that grow on the branches of other plants. In general, the host plant is unharmed, and the epiphyte that grows on it benefits. Similarly, various marine animals, such as barnacles, grow on other, often actively moving, sea animals like whales and thus are carried passively from place to place. These "passengers" presumably gain more protection from predation than they would if they were fixed in one place, and they also reach new sources of food. The increased water circulation that such animals receive as their host moves around may be of great importance, particularly if the passengers are filter-feeders.

Examples of Commensalism. The best-known examples of commensalism involve the relationships between certain small tropical fishes and sea anemones, marine animals that have stinging tentacles. These fish have evolved the ability to live among the tentacles of sea anemones, even though these tentacles would quickly paralyze other fishes that touched them (figure 17.17). The anemone fishes feed on the detritus left from the meals of the host anemone, remaining uninjured under remarkable circumstances.

On land, an analogous relationship exists between birds called oxpeckers and grazing animals such as cattle or rhinoc-

Figure 17.18 Commensalism between cattle egrets and an African cape buffalo.
The egrets eat insects off the buffalo.

eroses. The birds spend most of their time clinging to the animals, picking off parasites and other insects, carrying out their entire life cycles in close association with the host animals. Cattle egrets, which have extended their range greatly during the past few decades, engage in a loosely coupled relationship of this kind (figure 17.18).

When Is Commensalism Commensalism? In each of these instances, it is difficult to be certain whether the second partner receives a benefit or not; there is no clear-cut boundary between commensalism and mutualism. For instance, it may be advantageous to the sea anemone to have particles of food removed from its tentacles; it may then be better able to catch other prey. Similarly, the grazing animals may benefit from the relationship with the oxpeckers or cattle egrets if the bugs that are picked off of them are harmful, such as ticks or fleas. If so, then the relationship is mutualism. However, if the birds also pick at scabs, causing bleeding and possibly infections, the relationship may be parasitic. In true commensalism, only one of the partners benefits and the other neither benefits nor is harmed. If the grazing animals are not harmed by either the ticks that are eaten off of their bodies or by the oxpeckers that feed off of them, then it is an example of commensalism.

Figure 17.17 Commensalism in the sea.
Clownfishes, such as this *Amphiprion perideraion* in Guam, often form symbiotic associations with sea anemones, gaining protection by remaining among their tentacles and gleaning scraps from their food. Different species of anemones secrete different chemical mediators; these attract particular species of fishes and may be toxic to the fish species that occur symbiotically with other species of anemones in the same habitat. There are 26 species of clownfishes, all found only in association with sea anemones; 10 species of anemones are involved in such associations, so that some of the anemone species are host to more than one species of clownfish.

17.7 Coevolution is a term that describes the long-term evolutionary adjustments of species to one another. In symbiosis, two or more species live together. Mutualism involves cooperation between species, to the mutual benefit of both. In parasitism, one organism serves as a host to another organism, usually to the host's disadvantage. Commensalism is the benign use of one organism by another.

17.8 Predator-Prey Interactions

In the previous section, we considered parasitism, a symbiotic relationship, a specialized form of **predator/prey interaction** in which the predator is much smaller than its prey and does not generally kill it. **Predation** is the consuming of one organism by another, usually of a similar or larger size. In this sense, predation includes everything from a leopard capturing and eating an antelope to a whale grazing on microscopic ocean plankton.

The relationships between large carnivores and grazing mammals have a major impact on biological communities in many parts of the world. Appearances, however, are sometimes deceiving. On Isle Royale in Lake Superior, moose reached the island by crossing over ice in an unusually cold winter and multiplied freely there in isolation. When wolves later reached the island by crossing over the ice, naturalists widely assumed that the wolves were playing a key role in controlling the moose population. More careful studies have demonstrated that this is not in fact the case. The moose that the wolves eat are, for the most part, old or diseased animals that would not survive long anyway. In general, the moose are controlled by food availability, disease, and other factors rather than by the wolves (figure 17.19).

Figure 17.19   Wolves chasing a moose—what will the outcome be?
On Isle Royale, Michigan, a large pack of wolves pursue a moose. They chased this moose for almost 2 kilometers; it then turned and faced the wolves, who by that time were exhausted from running through chest-deep snow. The wolves lay down, and the moose walked away.

Predator-Prey Cycles

When experimental populations are set up under simple laboratory conditions, the predator often exterminates its prey and then becomes extinct itself, having nothing left to eat (figure 17.20). However, if refuges are provided for the prey, its population drops to low levels but not to extinction. Low prey population levels then provide inadequate food for the predators, causing the predator population to decrease. When this occurs, the prey population can recover. In this situation the predator and prey populations may continue in this cyclical pattern for some time.

Cycles in Hare Populations: A Case Study

Population cycles are characteristic of some species of small mammals, such as lemmings, and they appear to be stimulated, at least in some situations, by their predators. Ecologists have studied cycles in hare populations since the 1920s. They have found that the North American snowshoe hare, *Lepus americanus,* follows a "10-year cycle" (in reality, it varies from 8 to 11 years). Its numbers fall 10-fold to 30-fold in a typical cycle, and 100-fold changes can occur. Two factors appear to be generating the cycle: food plants and predators.

1. **Food plants.** The preferred foods of snowshoe hares are willow and birch twigs. As hare density increases, the quantity of these twigs decreases, forcing the

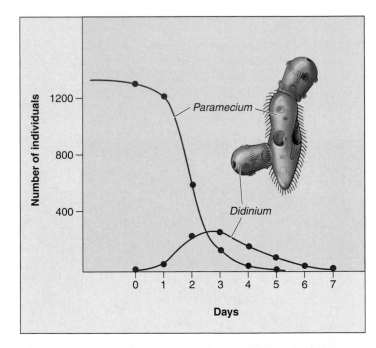

Figure 17.20   Predator-prey in the microscopic world.
When the predatory *Didinium* is added to a *Paramecium* population, the numbers of *Didinium* initially rise, while the numbers of *Paramecium* steadily fall. As the *Paramecium* population is depleted, however, the *Didinium* individuals also die.

hares to feed on high-fiber (low-quality) food. Lower birthrates, low juvenile survivorship, and low growth rates follow. The result is a precipitous decline in willow and birch twig abundance, and a corresponding fall in hare abundance. It takes two to three years for the quantity of mature twigs to recover.

2. **Predators.** A key predator of the snowshoe hare is the Canada lynx, *Lynx canadensis.* The Canada lynx shows a "10-year cycle" of abundance that seems remarkably

entrained to the hare abundance cycle (figure 17.21). As hare numbers increase, lynx numbers do, too, rising in response to the increased availability of lynx food. When hare numbers fall, so do lynx numbers, their food supply depleted.

Which factor is responsible for the predator-prey oscillations? Do increasing numbers of hares lead to overharvesting of plants (a hare-plant cycle), or do increasing numbers of lynx lead to overharvesting of hares (a hare-lynx cycle)? Field experiments carried out by C. Krebs and coworkers in 1992 provide an answer. Krebs set up experimental plots in Canada's Yukon containing hare populations. If food is added (no food effect) and predators excluded (no predator effect) from an experimental area, hare numbers increase 10-fold and stay there—the cycle is lost. However, the cycle is retained if either of the factors is allowed to operate alone: exclude predators but don't add food (food effect alone), or add food in presence of predators (predator effect alone). Thus, both factors can affect the cycle, which, in practice, seems to be generated by the interaction between the two factors.

Predation Reduces Competition

Predator-prey interactions are an essential factor in the maintenance of communities that are rich and diverse in species. The predators prevent or greatly reduce competitive exclusion by reducing the numbers of individuals of competing species. For example, in preying selectively on bivalves in marine intertidal habitats, sea stars prevent bivalves from monopolizing such habitats (figure 17.22), opening up space for many other organisms. When sea stars are removed, species diversity falls precipitously, the seafloor community coming to be dominated by a few species of bivalves. Because predation tends to reduce competition in natural communities, it is usually a mistake to attempt to eliminate a major predator such as sea stars, wolves, or mountain lions from a community. The result is to decrease rather than increase the biological diversity of the community, the opposite of what is intended. Predators such as these are examples of **keystone species,** species that play key roles in their communities.

A given predator may often feed on two, three, or more kinds of plants or animals in a given community. The predator's choice depends partly on the relative abundance of the prey options. A given prey species may be a primary source of food for increasing numbers of species as it becomes more abundant, which tends to limit the size of its population automatically. Such feedback is a key factor in structuring many natural communities.

17.8 Predators and their prey often show similar cyclic oscillations, often promoted by refuges that prevent prey populations from being driven to extinction.

(a)

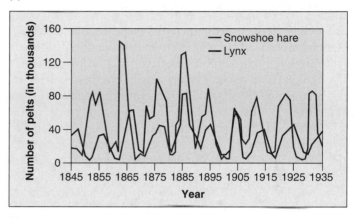

(b)

Figure 17.21 A predator-prey cycle.

(a) A snowshoe hare being chased by a lynx. (b) The numbers of lynxes and snowshoe hares oscillate in tune with each other in northern Canada. The data are based on numbers of animal pelts from 1845 to 1935. As the number of hares grows, so does the number of lynxes, with the cycle repeating about every nine years. Both predators (lynxes) and available food resources control the number of hares. The number of lynxes is controlled by the availability of prey (snowshoe hares).

Figure 17.22 Sea stars prevent bivalves from dominating intertidal habitats.

Sea stars prey on bivalves, such as clams, keeping their numbers low enough for other types of organisms to share the habitat.

17.9 Plant and Animal Defenses

Plant Defenses

As we have noted, predator-prey interactions are interactions between organisms in which one organism uses the other for food. Plants have evolved many mechanisms to defend themselves from herbivores. The most obvious are morphological (structural) defenses: Thorns, spines, and prickles play an important role in discouraging browsers, and plant hairs, especially those that have a glandular, sticky tip, deter insect herbivores. Some plants, such as grasses, deposit silica in their leaves, both strengthening and protecting themselves. If enough silica is present in their cells, these plants are simply too tough to eat.

Chemical Defenses. Significant as these morphological adaptations are, the chemical defenses that occur so widely in plants are even more crucial. Best known and perhaps most important in the defenses of plants against herbivores are secondary chemical compounds. These are distinguished from primary compounds, which are regular components of the major metabolic pathways, such as respiration. Virtually all plants, and apparently many algae as well, contain very structurally diverse secondary compounds that are either toxic to most herbivores or disturb their metabolism so greatly that they are unable to complete normal development. Consequently, most herbivores tend to avoid the plants that possess these compounds.

The mustard family (Brassicaceae) is characterized by a group of chemicals known as mustard oils. These are the substances that give the pungent aromas and tastes to such plants as mustard, cabbage, watercress, radish, and horseradish. The same tastes we enjoy signal the presence of toxic chemicals to many groups of insects. Similarly, plants of the milkweed family (Asclepiadaceae) and the related dogbane family (Apocynaceae) produce a milky sap that deters herbivores from eating them. In addition, these plants usually contain cardiac glycosides, molecules named for their drastic effect on heart function in vertebrates.

The Evolutionary Response of Herbivores. Certain groups of herbivores are associated with each family or group of plants protected by a particular kind of secondary compound. These herbivores are able to feed on these plants without harm, often as their exclusive food source. For example, cabbage butterfly caterpillars (subfamily Pierinae) feed almost exclusively on plants of the mustard and caper families, as well as on a few other small families of plants that also contain mustard oils (figure 17.23). Similarly, caterpillars of monarch butterflies and their relatives (subfamily Danainae) feed on plants of the milkweed and dogbane families. How do these animals manage to avoid the chemical defenses of the plants, and what are the evolutionary precursors and ecological consequences of such patterns of specialization?

We can offer a potential explanation for the evolution of these particular patterns. Once the ability to manufacture mustard oils evolved in the ancestors of the caper and mustard

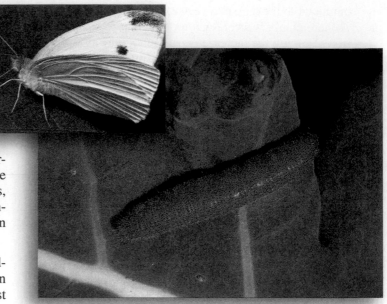

Figure 17.23 Insect herbivores are well suited to their hosts.

The green caterpillars of the cabbage butterfly, *Pieris rapae*, are camouflaged on the leaves of cabbage and other plants on which they feed. Although mustard oils protect these plants against most herbivores, the cabbage butterfly caterpillars are able to break down the mustard oil compounds. (*Inset*) An adult cabbage butterfly.

families, the plants were protected for a time against most or all herbivores that were feeding on other plants in their area. At some point, certain groups of insects—for example, the caterpillar of the cabbage butterfly—developed the ability to break down mustard oils and thus feed on these plants without harming themselves. Having evolved this ability, the butterflies were able to use a new resource without competing with other herbivores for it. Often, in groups of insects such as cabbage butterflies, sense organs have evolved that are able to detect the secondary compounds that their food plants produce. Clearly, the relationship that has formed between cabbage butterflies and the plants of the mustard and caper families is an example of coevolution.

Animal Defenses

Some animals that feed on plants rich in secondary chemical compounds receive an extra benefit. When the caterpillars of monarch butterflies feed on plants of the milkweed family, they do not break down the cardiac glycosides that protect these plants from herbivores. Instead, the caterpillars concentrate and store the cardiac glycosides in fat bodies; they then pass them through the chrysalis stage to the adult and even to the eggs of the next generation. The incorporation of cardiac glycosides thus protects all stages of the monarch life cycle from predators. A bird that eats a monarch butterfly quickly regurgitates it and in the future avoids the conspicuous orange-and-black pattern that characterizes the adult monarch (figure 17.24). Some birds, however, appear to have acquired the ability to tolerate the protective chemicals. These birds eat the monarchs.

Defensive Coloration. Many insects that feed on milkweed plants are brightly colored; they advertise their poisonous nature using an ecological strategy known as **warning coloration** or **aposematic coloration.** Showy coloration is characteristic of animals that use poisons (see figure 17.26) and stings to repel predators, whereas organisms that lack specific chemical defenses are seldom brightly colored. In fact, many have **cryptic coloration**—color that blends with the surroundings and thus hides the individual from predators (figure 17.25). Camouflaged animals usually do not live together in groups because a predator that discovers one individual gains a valuable clue to the presence of others.

Chemical Defenses. Animals also manufacture and use a startling array of substances to perform a variety of defensive functions. Bees, wasps, predatory bugs, scorpions, spiders, and many other arthropods use chemicals to defend themselves and to kill their prey. In addition, various chemical defenses have evolved among marine animals and the vertebrates, including jellyfish, venomous snakes, lizards, fishes,

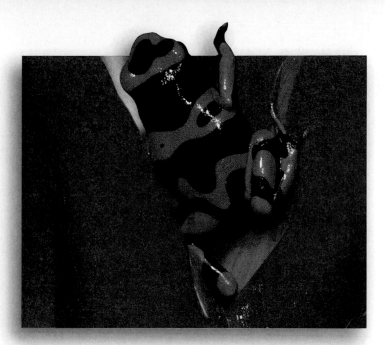

Figure 17.26 Vertebrate chemical defenses.
Dendrobatid frogs advertise their toxicity with aposematic coloration.

 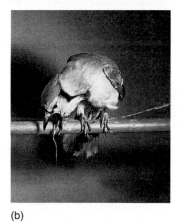

(a)   (b)

Figure 17.24 A blue jay learns that monarch butterflies taste bad.

(a) This cage-reared jay, which had never seen a monarch butterfly before, tried eating one. (b) The same jay regurgitated the butterfly a few minutes later. This bird is not likely to attempt to eat an orange-and-black insect again.

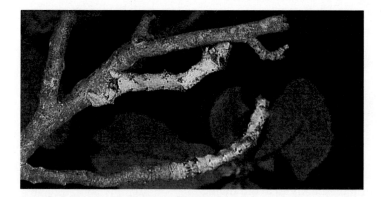

Figure 17.25 Cryptic coloration.
An inchworm caterpillar (*Necophora quernaria*) closely resembles a twig.

and some birds. The poison-dart frogs of the family Dendrobatidae produce toxic alkaloids in the mucus that covers their brightly colored skin (figure 17.26). Some of these toxins are so powerful that a few micrograms will kill a person if injected into the bloodstream. More than 200 different alkaloids have been isolated from these frogs, and some are playing important roles in neuromuscular research. There is an intensive investigation of marine animals, algae, and flowering plants for new drugs to fight cancer and other diseases, or as sources of antibiotics.

Coevolution of Predator and Prey

Predation can exert strong selective pressures on prey populations. Any feature that acts to decrease the probability of capture should thus be strongly favored by natural selection. In turn, the evolution of such features by prey would be expected to encourage natural selection to then favor counteradaptations in their predator populations. In this way, a coevolutionary arms race may ensue, in which predators and prey are continually evolving better defenses and better means of circumventing these defenses. Herbivores that are able to feed on mustard, milkweed, and dogbane are predators that have acquired counteradaptations to the chemicals produced by the plants. And, birds that are able to eat monarch butterflies seem to be counteradapting to the defense of their animal prey.

17.9 The members of many groups of plants are protected from most herbivores by their secondary compounds. Once the members of a particular herbivore group evolve the ability to feed on them, these herbivores gain access to a new resource. Animals defend themselves against predators with warning coloration, camouflage, and chemical defenses such as poisons and stings.

Mimicry

During the course of their evolution, many unprotected (non-poisonous) species have come to resemble distasteful ones that exhibit aposematic coloration. The unprotected mimic gains an advantage by looking like the distasteful model. Two types of mimicry have been identified: Batesian mimicry, as described here, and Müllerian mimicry, which is a type of "group" protection.

Batesian Mimicry

Batesian mimicry is named for Henry Bates, the nineteenth-century British naturalist who first brought this type of mimicry to general attention in 1857. In his journeys to the Amazon region of South America, Bates discovered many instances of palatable insects that resembled brightly colored, distasteful

(a) Model

(b) Batesian mimic

Figure 17.27   A Batesian mimic.

(a) The model. Monarch butterflies (*Danaus plexippus*) are protected from birds and other predators by the cardiac glycosides they incorporate from the milkweeds and dogbanes they feed on as larvae. Adult monarch butterflies advertise their poisonous nature with warning coloration. (b) The mimic. Viceroy butterflies, *Limenitis archippus,* are Batesian mimics of the poisonous monarch. Although the viceroy is not related to the monarch, it looks a lot like it, so predators that have learned not to eat distasteful monarchs avoid viceroys, too.

species. He reasoned that if the unprotected animals (mimics) are present in numbers that are low relative to those of the distasteful species that they resemble (the model), the mimics are avoided by predators, who are fooled by the disguise into thinking the mimic actually is the distasteful model.

It is important that mimics be rare relative to models. If the unprotected mimic animals are too common, or if they do not live among their models, many are eaten by predators that have not yet learned to avoid model individuals with those particular characteristics. As a result, the predator learns that animals with that coloration are good rather than bad to eat.

Many of the best-known examples of Batesian mimicry occur among butterflies and moths. Obviously, predators in systems of this kind must use visual cues to hunt for their prey; otherwise, similar color patterns would not matter to potential predators. There is also increasing evidence indicating that Batesian mimicry can also involve nonvisual cues, such as olfaction, although such examples are less obvious to humans.

The kinds of butterflies that provide the models in Batesian mimicry are, not surprisingly, members of groups whose caterpillars feed only on one or a few closely related plant families. The plant families on which they feed are strongly protected by toxic chemicals. The model butterflies incorporate the poisonous molecules from these plants into their bodies. The mimic butterflies, in contrast, belong to groups in which the feeding habits of the caterpillars are not so restricted. As caterpillars, these butterflies feed on a number of different plant families unprotected by toxic chemicals.

One often-studied mimic among North American butterflies is the viceroy, *Limenitis archippus* (figure 17.27). This butterfly, which resembles the poisonous monarch, ranges from central Canada through much of the United States and into Mexico. The caterpillars feed on willows and cottonwoods, and neither caterpillars nor adults were thought to be distasteful to birds, although recent findings may dispute this. Interestingly, the Batesian mimicry seen in the adult viceroy butterfly does not extend to the caterpillars: Viceroy caterpillars are camouflaged on leaves, resembling bird droppings, whereas the monarch's distasteful caterpillars are very conspicuous.

Müllerian Mimicry

Another kind of mimicry, **Müllerian mimicry,** was named for German biologist Fritz Müller, who first described it in 1878. In Müllerian mimicry, several unrelated but protected animal species come to resemble one another (figure 17.28). Thus, different kinds of stinging wasps have yellow-and-black-striped abdomens, but they may not all be descended from a common yellow-and-black-striped ancestor. In general, yellow-and-black and bright red tend to be common color patterns that warn predators relying on vision. If animals that resemble one another are all poisonous or dangerous, they gain an advantage because a predator learns more quickly to avoid them.

In both Batesian and Müllerian mimicry, mimic and model must not only look alike but also act alike if predators are to be deceived. For example, the members of several families of insects that resemble wasps behave surprisingly like

(a)

(b)

(c)

(d)

Figure 17.28 Müllerian mimics.
Because the color patterns of these insects are very similar, and because they all sting, they are Müllerian mimics. The yellow jacket (a), the masarid wasp (b), the sand wasp (c), and the anthidiine bee (d) all act as models for each other, strongly reinforcing the color pattern that they all share.

the wasps they mimic, flying often and actively from place to place.

Self Mimicry

Another type of mimicry, called **self mimicry,** involves adaptations where one animal body part comes to resemble another body part. This type of mimicry is used by both prey and predators. In prey, it is used to increase survival during an attack. For example, many moths, butterflies, and fish appear to have "eye-spots," which are large dark circular markings. In some cases, these eye-spots startle a predator, allowing the prey time to escape, or they present a false target for the predator's attack, attacking the tail where the eye-spots appear instead of the head. Predators may use mimicry to simulate

bait to lure prey in. For example, some predators have accessory body parts that look like food, such as a wriggly worm, that attract prey and bring them in close enough for a successful attack.

17.10 In Batesian mimicry, unprotected species resemble others that are distasteful. Both species exhibit aposematic coloration. If they are relatively scarce, the unprotected mimics are avoided by predators. In Müllerian mimicry, two or more unrelated but protected species resemble one another, thus achieving a kind of group defense.

17.11 Ecological Succession

Competition and cooperation often produce dramatic changes in communities. This results in changes in ecosystems, by the orderly replacement of one community with another, from simple to complex in a process known as **succession.** This process is familiar to anyone who has seen a vacant lot or cleared woods slowly become occupied by an increasing number of plants, or a pond become dry land as it is filled with vegetation encroaching from the sides.

Secondary Succession

If a wooded area is cleared and left alone, plants slowly reclaim the area. Eventually, traces of the clearing disappear and the area is again woods. Similarly, a stream that experiences intense flooding may clear the stream bed of all organisms, leaving sand and rock; the bed is progressively reinhabited by protists, invertebrates, and other aquatic organisms. This kind of succession, which occurs in areas where an existing community has been disturbed, is called **secondary succession.** Humans are often responsible for initiating secondary succession, which may also take place when a fire has burned off an area or in abandoned agricultural fields.

Primary Succession

In contrast, **primary succession** occurs on bare, lifeless substrate, such as rocks. Primary succession occurs in lakes left behind after the retreat of glaciers, on volcanic islands that rise above the sea, and on land exposed by retreating glaciers (figure 17.29). Primary succession on glacial moraines provides an example (figure 17.30). On bare, mineral-poor soil, lichens grow first, forming small pockets of soil. Acidic secretions from the lichens help to break down the substrate and add to the accumulation of soil. Mosses then colonize these pockets of soil, eventually building up enough nutrients in the soil for alder shrubs to take hold. These first plants to appear form a **pioneering community.** Over 100 years, the alders build up the soil nitrogen levels until spruce are able to thrive, eventually crowding out the alder and forming a dense spruce forest.

Primary successions end with a community called a **climax community,** whose populations remain relatively stable, and that is characteristic of the region as a whole. However, because local climate keeps changing, the process of succession is often very slow, and human activities have a major impact, so many successions do not reach climax.

Why Succession Happens

Succession happens because species alter the habitat and the resources available in it, often in ways that favor other species. Three dynamic concepts are of critical importance in the process:

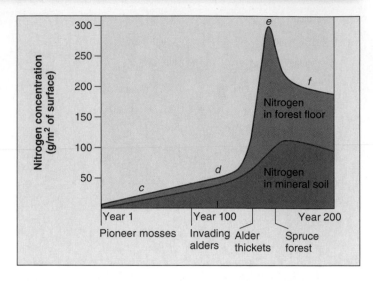

Figure 17.29 Plant succession produces progressive changes in the soil.

Initially the glacial moraine at Glacier Bay, Alaska, portrayed in figure 17.30, had little soil nitrogen, but nitrogen-fixing alders led to a buildup of nitrogen in the soil, encouraging the subsequent growth of the conifer forest. Letters in the graph correspond to photographs in figure 17.30 c–f.

1. **Tolerance.** Early successional stages are characterized by weedy r-selected species that do not compete well in established communities but are tolerant of the harsh, abiotic conditions in barren areas.

2. **Facilitation.** The weedy early successional stages introduce local changes in the habitat that favor other, less weedy species. Thus, the mosses in the Glacier Bay succession of figure 17.30 fix nitrogen, which allows alders to invade. The alders in turn lower soil pH as their fallen leaves decompose, and spruce and hemlock, which require acidic soil, are able to invade.

3. **Inhibition.** Sometimes the changes in the habitat caused by one species, while favoring other species, inhibit the growth of the species that caused them. Alders, for example, do not grow as well in acidic soil as the spruce and hemlock that replace them.

As ecosystems mature, and more K-selected species replace r-selected ones, species richness (see next section) and total biomass increase but net productivity decreases. Because earlier successional stages are more productive than later ones, agricultural systems are intentionally maintained in early successional stages to keep net productivity high.

> **17.11 Communities evolve to have greater total biomass and species richness in a process called succession.**

(a) Retreating Glacier

(b) Barren Moraine

(c) Pioneering Mosses

(d) Invading Alders

(e) Alder Thickets

(f) Spruce Forest

Figure 17.30   Primary succession at Alaska's Glacier Bay.
The sides of the glacier (a) have been retreating at a rate of some 8 meters a year, leaving behind exposed soil (b) from which nitrogen and other minerals have been leached out. The first invaders of these exposed sites are pioneer moss species (c) with nitrogen-fixing mutualistic microbes. Within 20 years, young alder shrubs take hold (d). Rapidly fixing nitrogen, they soon form dense thickets (e). As soil nitrogen levels rise, spruce crowd out the mature alders, forming a forest (f).

Mature ecosystems in nature persist because they are able to cope with changes in climate and with the many conflicting demands of their inhabitants. Climate, the coevolution of species, and competition—these driving forces of evolutionary change have molded ecosystems over long periods of time to create the world we see today. However, the most important single influence on natural ecosystems today—human activity—is not acting slowly. Human influence now extends to all parts of the globe, to every blade of grass growing anywhere on our world. The fate of every ecosystem on earth in this century will be influenced more by its stability in the face of human disruption than by any other factor. Thus, it is essential that we consider how ecosystems respond to disruption and try to gain an understanding of why some ecosystems are more stable than others.

Ecologists have long sought to understand why some ecosystems are more stable than others—better able to avoid permanent change and return to normal after disturbances like land clearing, fire, invasion by plagues of insects, or severe storm damage. Most ecologists now agree that biologically diverse ecosystems are in general more stable than simple ones. Ecosystems with more kinds of different organisms support a more complex web of interactions, and as a result, an alternative niche is more likely to be able to compensate for the effect of a disruption. The number of species in an ecosystem, called **species richness,** is the quantity usually measured by biologists in attempting to characterize an ecosystem's **biodiversity** (figure 17.31).

However, the very complexity of highly diverse ecosystems, while buffering them from everyday insult, also makes them more likely to have particular points of vulnerability that, if damaged, can have far-reaching consequences. A species that interacts with many other elements of an ecosystem in critical ways is called a keystone species. Because the loss of a keystone species may affect many other organisms in a diverse community, such species represent points of particular sensitivity in the ecosystem—places where the complex ecological machine can be easily broken. Think of how you would design a computer to protect it from "crashing"—like a complex ecosystem, you would make it with a diverse array of redundant (repeated) circuits that could take over the function of others if the need arose. However, this added complexity has a price: The advanced chip is uniquely vulnerable to damage at those points where functional crossover takes place—a single flaw can destroy the operation of the entire computer, while a simple chip would have lost only one circuit.

Thus, if we wish to preserve natural ecosystems, we should do all we can to preserve their biological diversity, while at the same time realizing that diverse ecosystems can be damaged too.

No ecosystem, however complex or simple, can be stable in the face of massive disruption. Evolution has provided no mechanism to buffer ecosystems from total destruction by human activity. Unhappily, much of today's disruption of the world's ecosystems is of this drastic sort—replacing forests with grazing land, with "pure stands" of a single kind of tree, with plowed fields for farming, or with asphalt.

Figure 17.31 How many species are there?
Scientists are attempting to determine the species richness in the canopy of moist tropical forests, the most diverse biological community on earth. In these experiments, the insects and other animals living in the treetops are sprayed with insecticide and then sampled as they drop onto the sheets below. By such methods, Terry L. Erwin, Jr., of the Smithsonian Institution, has estimated that there may be as many as 30 million kinds of organisms in the world. Taxonomists have thus far recognized only about 1.75 million of these.

What Promotes Biodiversity?

If diversity is such a good thing, why doesn't evolution drive all ecosystems toward high species richness? Why isn't tundra as diverse as tropical rain forest? Although many factors contribute to biodiversity, including predation, competition, and disturbances, among others, if we examine all the species-rich ecosystems on earth, and ask what they have in common that species-poor ecosystems lack, two general trends emerge: large size and tropical latitude.

Ecosystem Size

Larger ecosystems, because they contain a more varied array of physical habitats, are able to support a greater number of different species. Because they offer a more diverse array of potential niches, more species can be packed into them.

This dependence of diversity on the physical size of an ecosystem has a very important consequence: If you reduce the size of the ecosystem, you reduce the number of species it can support. In today's world, this happens with increasing frequency. A new road cut across a forest divides the ecosystem in two for the many animals that cannot or do not cross it. In extreme cases, reduction in ecosystem area can produce *faunal collapse,* a situation in which many of the animal species living in the ecosystem become extinct there because the smaller ecosystem is simply unable to divide its resources into that many pieces.

For exactly this reason, in all but the largest of our national parks, a high proportion of the mammals once there have become extinct. Different animals have been lost from different parks, so nothing is entirely lost, but the pattern of species loss is clear. When the parks were formed at the beginning of the twentieth century, their animals were part of much larger communities, but, walled off from their surroundings by society's growing development around the parks, the animals have been restricted to much smaller areas. The loss of species richness has been the result. If we want to maintain animal diversity for future generations, the parks must be managed carefully, and lost species must be reintroduced. In today's world, there is no longer any true wilderness. Our national parks, like gardens, have to be tended.

Latitude

Compared to the arctic region, tropical areas have many more species (figure 17.32), estimated at more than 6 million species of organisms. Why should this be so? There are two principal reasons:

1. **Length of growing season.** As a general rule, ecosystems with more resources are able to subdivide them into more niches, and so maintain a greater number of species. In the tropics, with ample sunlight, warm temperatures, and generous rainfall throughout the year, the growing season never stops. Progressing toward the poles, however, growing seasons shorten, so that polar ecosystems have much less energy to work with in weaving their food webs.

2. **Climatic stability.** The climate in the tropics has not been subjected to glaciers and other major disruptions during the evolutionary past. Thus, the unchanging physical conditions have provided a longer evolutionary window in the tropics for specialized relationships to develop.

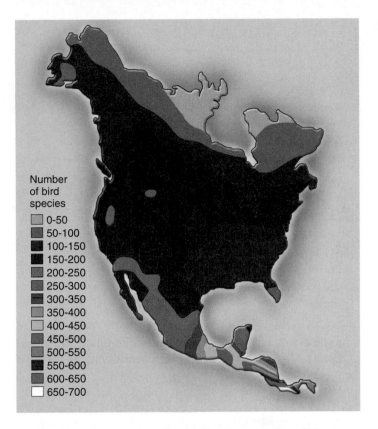

Number of bird species
- [ ] 0-50
- 50-100
- 100-150
- 150-200
- 200-250
- 250-300
- 300-350
- 350-400
- 400-450
- 450-500
- 500-550
- 550-600
- 600-650
- [ ] 650-700

Figure 17.32   A latitudinal cline in species richness. Among North and Central American birds, a marked increase in the number of species occurs as one moves toward the tropics. Fewer than 100 species are found at arctic latitudes, while more than 600 species live in southern Central America.

Island Biodiversity

Oceanic islands provide a natural laboratory for studying the factors that promote species richness. In 1967, Robert MacArthur of Princeton University and Edward O. Wilson of Harvard University proposed that the number of species on such islands is related to the size of the island and the distance between the island and the source of colonization.

The Equilibrium Model.   MacArthur and Wilson reasoned that species are constantly being dispersed to islands, so islands have a tendency to accumulate more and more species. At the same time that new species are added, however, other species are lost by extinction. Once the number of species fills the capacity of that island, no more species can be established unless one of the species already there becomes extinct. Every island of a given size, then, has a characteristic equilibrium number of species that tends to persist through time (the intersection point in figure 17.33*a*), although the individual species may change.

MacArthur and Wilson's equilibrium theory proposes that island species richness is a dynamic equilibrium between colonization and extinction. Both island size and distance from the mainland would play important roles. We would expect smaller islands to have higher rates of extinction because their population sizes would, on average, be smaller. Also, we would expect fewer colonizers to reach islands that lie farther

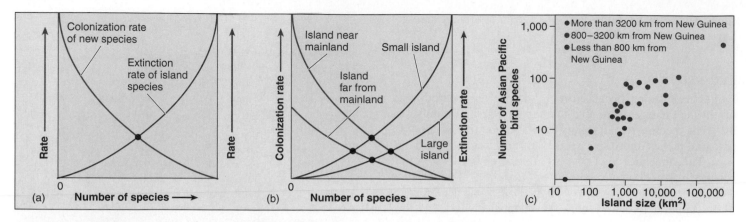

Figure 17.33 The equilibrium model of island biogeography.

(*a*) Island species richness reaches an equilibrium (*black dot*) when the colonization rate of new species equals the extinction rate of species on the island. (*b*) The equilibrium shifts when the colonization rate is affected by distance from the mainland and when the extinction rate is affected by size of the island. Species richness is positively correlated with island size and inversely correlated with distance from the mainland. (*c*) Small distant Asian islands have fewer bird species, bearing out the principles of the equilibrium model.

from the mainland. Thus, small islands far from the mainland have the fewest species; large islands near the mainland have the most (figure 17.33*b*).

The predictions of this simple model bear out well in field data. Asian Pacific bird species (figure 17.33*c*) exhibit a positive correlation of species richness with island size but a negative correlation of species richness with distance from the mainland (New Guinea).

> **17.12 Ecosystems with biodiversity are more stable, while, at the same time, they may also have points of vulnerability. Species richness on islands is a dynamic equilibrium between colonization and extinction.**

Exploring Current Issues

Additional Resources

Go to your campus library or look online to find the following articles, which further develop some of the concepts found in this chapter.

Barnett, A. (2000). Copy your neighbour. *New Scientist,* 167(2255), 34.

de Waal, F. B. M., F. Aureli, and P. G. Judge. (2000). Coping with crowding. *Scientific American,* 282(5), 76.

Doyle, R. (2003). Fertility volatility: fluctuating U.S. birth rates elude definitive explanation. *Scientific American,* 289(3), 38.

Levy, S. (2002). Top dogs: high in the rocky mountains a top predator has returned after an absence of seventy years. What lessons does this have for reintroducing species to other habitats? *New Scientist,* 176(2367), 46.

Lovejoy, T. (2002). Biodiversity: dismissing scientific process. *Scientific American,* 286(1), 69.

In the News: Science in the New Millennium

We have entered a new century, one that presents unique challenges. The ability of modern society to sustain itself will depend in important ways on the challenge presented by problems involving biology, challenges that we all must come to understand if we as a society are to meet them.

Article 1. *In the new millennium, science must become more than a tool for growth.*

Article 2. *Was Malthus mistaken? A 150-year-old warning about population growth may have been off only in timing.*

Article 3. *Seniors graduating from college enter a century of challenging responsibilities.*

Article 4. *Two minor science stories—both about ice—are likely to profoundly affect the new century.*

Find these articles, written by the author, online at www.mhhe.com/tlwessentials/exp17.

Summary

Population Dynamics

17.1 Population Growth

- Several factors affect population growth including population size, density, and dispersion (**figure 17.1**). A population often grows exponentially when no factors are

limiting its growth. As resources are used up, its growth slows and stabilizes at a size called the carrying capacity (**figures 17.2 and 17.3**).

17.2 The Influence of Population Density

- Factors, such as weather and physical disruptions, are density-independent effects and act on population growth, regardless of population size. Density-dependent effects are factors, such as resources, that are affected by increases in population size (**figure 17.4**).

17.3 Life History Adaptations

- Populations whose resources are limitless experience little competition and reproduce rapidly; these organisms exhibit *r*-selected adaptations. Populations that experience competition over limited resources tend to be more reproductively efficient and exhibit *K*-selected adaptations (**table 17.1**).

17.4 Population Demography

- Survivorship curves illustrate the impact of mortality rates among different age groups in a population (**figure 17.7**).

How Competition Shapes Communities

17.5 Communities

- The array of organisms that live together in an area is called a community. Different species in a community compete and cooperate with each other to make the community stable. Communities are often identified by the dominant species (**figure 17.8**).

17.6 The Niche and Competition

- A niche is the way an organism uses all available resources in its environment. Competition limits an organism from using its entire niche (**figure 17.9**). Two species cannot use the same niche; one will either outcompete the other, driving it to extinction, called competitive exclusion (**figure 17.10*a, b***), or they will divide the niche into two smaller niches, called resource partitioning (**figures 17.10*c* and 17.11**). Resource partitioning can affect morphological characteristics, as each species adapts to its portion of the niche, called character displacement (**figure 17.12**).

How Coevolution Shapes Communities

17.7 Coevolution and Symbiosis

- Coevolution is the process whereby one species changes as a consequence of a relationship with another, and the other species changes in response. Symbiotic relationships such as mutualism (**figures 17.14 and 17.15**), parasitism (**figure 17.16**), and commensalism (**figures 17.17 and 17.18**) can lead to coevolution.

17.8 Predator-Prey Interactions

- In predatory-prey relationships, the predator kills and consumes the prey. Predator and prey populations often exhibit cyclic oscillations, the prey population being hunted to a low number, which begins to negatively affect the predator population. When the predator population decreases in numbers, the prey population rebounds (**figure 17.21**).

17.9 Plant and Animal Defenses

- Plants have evolved defense mechanisms, such as toxic or distasteful chemical compounds, but this in turn led to counteradaptations in certain herbivores that allow them to consume the plants (**figure 17.23**). Animals have also evolved chemical defense mechanisms and warning coloration in order to let potential predators know that they should "stay away" (**figures 17.24 and 17.26**). Cryptic coloration is also a defense mechanism in animals to camouflage them to avoid predation (**figure 17.25**).

17.10 Mimicry

- Mimicry is where one organism takes advantage of the warning coloration of another organism. Batesian mimicry is where a harmless species has come to resemble a harmful species (**figure 17.27**). Müllerian mimicry is where a group of harmful species has a similar warning coloration pattern (**figure 17.28**).

Community Stability

17.11 Ecological Succession

- Succession is the ordered replacement of one community with another as in secondary succession. Primary succession is the emergence of a pioneering community where no life existed before (**figures 17.29 and 17.30**).

17.12 Biodiversity

- A large variety of species, called biodiversity, increases the stability within an ecosystem. A smaller ecosystem, such as an island, is more susceptible to periods of colonization and extinction (**figure 17.33**).

Self-Test

1. When the number of organisms remains more or less the same over time in the specific place where these organisms live, it is said that this population of organisms has reached the _____ of that place.
   a. dispersion
   c. carrying capacity
   b. biotic potential
   d. population density
2. Which of the following is a density-dependent effect on population?
   a. earthquake
   b. increased competition for food
   c. habitat destruction by humans
   d. seasonal flooding
3. Which of the following traits is *not* a characteristic of an organism that has *K*-selected adaptations?

   a. short life span
   b. few offspring per breeding season
   c. extensive parental care of offspring
   d. low mortality rate
4. If the age structure of a population shows more older organisms than younger organisms, then the fecundity
   a. will increase, and the mortality will decrease.
   b. will decrease, and the mortality will increase.
   c. and the mortality will be equal.
   d. and the mortality will not change.
5. All the animals and plants that live in the same location make up a(n)
   a. biome.
   c. ecosystem.
   b. population.
   d. community.

6. For similar species to occupy the same space, their niches must be different in some way. One way for these species to both survive is
   a. competitive exclusion.
   b. interspecific competition.
   c. resource partitioning.
   d. intraspecific competition.
7. A relationship between two species where one species benefits and the other is neither hurt nor helped is known as
   a. parasitism.
   b. commensalism.
   c. mutualism.
   d. competition.
8. Predators can assist in maintaining the species diversity of an area by
   a. increasing competitive exclusion between prey species.
   b. decreasing competitive exclusion between prey species.
   c. not affecting competitive exclusion between prey species.
   d. decreasing resource partitioning between prey species.
9. The bright colors of poison-dart frogs and Gila monsters are examples of
   a. aposematic coloration.
   b. Müllerian mimicry.
   c. cryptic coloration.
   d. Batesian mimicry.
10. One of the factors that will increase biodiversity of an area is the size of the area. This factor is crucial to having a high biodiversity because a large area
    a. results in fewer contacts with predators.
    b. contains more water.
    c. has fewer niches but they are big in size.
    d. has a wide variety of niches.

Visual Understanding

1. **Figure 17.2** Speculate on what happens with populations that follow the exponential growth model.

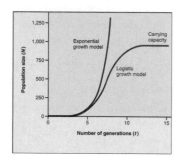

2. **Figure 17.33b** This graph predicts the equilibrium number of species on an island according to the island's size and distance from the mainland. Briefly explain the relationship.

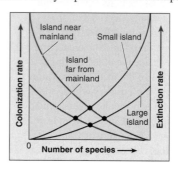

Challenge Questions

Population Dynamics   Give at least two examples, from your own area, of limiting factors on a population that are density-dependent factors, and two that are density-independent factors on the same population.

How Competition Shapes Communities   Use the terms from this section: interspecific competition, fundamental niche, realized niche, niche overlap, competitive exclusion, and resource partitioning to discuss Robert MacArthur's warbler research.

How Coevolution Shapes Communities   Briefly discuss and give examples of the three types of symbiosis.

Community Stability   Your friend, Susan, asks you why the rain forests are so diverse, when most other biomes are not. Explain the reasons for high or low biodiversity of an area.

Online Learning Center

Visit the Online Learning Center for this chapter at www.mhhe.com/tlwessentials/ch17 for quizzes, animations, interactive learning exercises, and other study tools. At the site you will also find extended answers to the end-of-chapter questions.

18

Planet Under Stress

The girl gazing from this now-classic *National Geographic* photo faces an uncertain future. An Afghani refugee, the whims of war destroyed her home, her family, and all that was familiar to her. Her expression carries a message about our own future: The problems humanity faces on an increasingly unstable, overcrowded, and polluted earth are no longer hypothetical. They are with us today and demand solutions. This chapter provides an overview of the problems and then focuses on solutions—on what can be done to address very real problems. As a concerned citizen, your first task must be to clearly understand the nature of the problem. You cannot hope to preserve what you do not understand. The world's environmental problems are acute, and a knowledge of biology is an essential tool you will need to contribute to the effort to solve them. It has been said that we do not inherit the earth from our parents—we borrow it from our children. We must preserve for them a world in which they can live. That is our challenge for the future, and it is a challenge that must be met soon. In many parts of the world, the future is happening right now.

18.1 Pollution

Our world is one ecological continent, one highly interactive biosphere, and damage done to any one ecosystem can have ill effects on many others. Burning high-sulfur coal in Illinois kills trees in Vermont, while dumping refrigerator coolants in New York destroys atmospheric ozone over Antarctica and leads to increased skin cancer in Madrid. Biologists call such wide-spread effects on the worldwide ecosystem **global change.** The pattern of global change that has become evident within recent years, including chemical pollution, acid precipitation, the ozone hole, the greenhouse effect, and the loss of biodiversity, is one of the most serious problems facing humanity's future.

Chemical Pollution

The problem posed by chemical pollution has grown very serious in recent years, both because of the growth of heavy industry and because of an overly casual attitude in industrialized countries. In one example, a poorly piloted oil tanker named the *Exxon Valdez* ran aground in Alaska in 1989 and heavily polluted many kilometers of North American coastline, and the organisms that live there, with oil. If the tanker had been loaded no higher than the waterline, little oil would have been lost, but it was loaded far higher than that, and the weight of the above-waterline oil forced thousands of tons of oil out the hole in the ship's hull. Why do policies permit overloading like this?

Chemicals are released into both the air and into water—therefore, their effects are far-reaching.

Air Pollution. Air pollution is a major problem in the world's cities. In Mexico City, oxygen is sold routinely on corners for patrons to inhale. Cities such as New York, Boston, and Philadelphia are known as gray-air cities because the pollutants in the air are usually sulfur oxides emitted by industry. Cities such as Los Angeles, however, are called brown-air cities because the pollutants in the air undergo chemical reactions in the sunlight.

Water Pollution. Water pollution is a very serious consequence of our casual attitude about pollution. "Flushing it down the sink" doesn't work in today's crowded world. There is simply not enough water available to dilute the many substances that the enormous human population produces continuously. Despite improved methods of sewage treatment, lakes and rivers throughout the world are becoming increasingly polluted. Recall from chapter 16 that excess phosphorus in lakes can cause rapid algal growth that eventually kills the lake, a process called eutrophication.

Agricultural Chemicals

The spread of "modern" agriculture, and particularly the Green Revolution, which brought high-intensity farming to developing countries, has caused very large amounts of many kinds of new chemicals to be introduced into the global ecosystem,

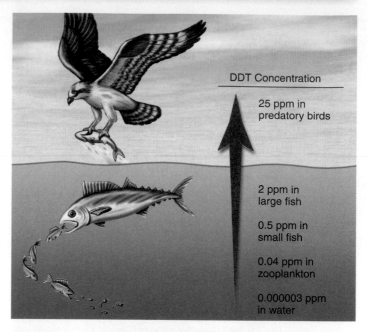

Figure 18.1 Biological magnification of DDT.
Because DDT accumulates in animal fat, the compound becomes increasingly concentrated in higher levels of the food chain. Before DDT was banned in the United States, predatory bird populations drastically declined because DDT made their eggshells thin and fragile enough to break during incubation.

particularly pesticides, herbicides, and fertilizers. Industrialized countries like the United States now attempt to carefully monitor side effects of these chemicals. Unfortunately, large quantities of many toxic chemicals, although no longer manufactured, still circulate in the ecosystem.

For example, the chlorinated hydrocarbons, a class of compounds that includes DDT, chlordane, lindane, and dieldrin, have all been banned for normal use in the United States, where they were once widely used. They are still manufactured in the United States and exported to other countries, where their use continues. Chlorinated hydrocarbon molecules break down slowly and accumulate in animal fat tissue. Furthermore, as they pass through a food chain, they become increasingly concentrated in a process called **biological magnification** (figure 18.1). DDT caused serious ecological problems by leading to the production of thin, fragile eggshells in many predatory bird species, such as peregrine falcons, bald eagles, osprey, and brown pelicans, in the United States and elsewhere until the late 1960s, when it was banned in time to save the birds from extinction. Chlorinated compounds have other undesirable side effects and exhibit hormonelike activities in the bodies of animals.

18.1 All over the globe, increasing industrialization is leading to higher levels of pollution.

18.2 Acid Precipitation

The smokestacks you see in figure 18.2 are those of the Four Corners power plant in New Mexico. This facility burns coal, sending the smoke high into the atmosphere through these stacks, each of which is over 65 meters tall. The smoke the stacks belch out contains high concentrations of sulfur, because the cheap coal that the plant burns is rich in sulfur. The sulfur-rich smoke is dispersed and diluted by winds and air currents. The first tall stacks were introduced in Britain in the mid-1950s. Such tall stacks rapidly became popular in the United States and Europe—there are now over 800 of them in the United States alone.

In the 1970s, 20 years after the stacks were introduced, ecologists began to report evidence that the tall stacks were not eliminating the problems associated with the sulfur, just exporting the ill effects elsewhere. Throughout northern Europe, lakes were reported to have suffered drastic drops in biodiversity, some even becoming devoid of life. The trees of the great Black Forest of Germany seemed to be dying—and the damage was not limited to Europe. It turned out that the sulfur introduced into the upper atmosphere by high smokestacks combines with water vapor to produce sulfuric acid. When this water later falls back to earth as rain or snow, it carries the sulfuric acid with it. Drifting to the northeast in the United States, the acid is taken far from its source, but the result is now clear. When schoolchildren measured the pH of natural rainwater as part of a nationwide project in 1989, locations around the United States rarely had a pH lower than 5.6, except in the northeastern United States. Rain and snow in the Northeast now have a pH of about 4.3, about 100 times more acidic than the rest of the country. This pollution-acidified precipitation is called **acid rain** (but the term acid precipitation is actually more correct).

Acid precipitation destroys life. Many of the forests of the northeastern United States and Canada have been seriously damaged. In fact, it is now estimated that at least 1.4

Figure 18.2 Tall stacks export pollution.

Tall stacks like those of the Four Corners coal-burning power plant in New Mexico send pollution far up into the atmosphere.

Figure 18.3 Acid precipitation.

Acid precipitation is killing many of the trees in North American and European forests. Much of the damage is done to the mycorrhizae, fungi growing within the cells of the tree roots. Trees need mycorrhizae in order to extract nutrients from the soil.

million acres of forests in the Northern Hemisphere have been adversely affected (figure 18.3). Not just trees are affected. Thousands of lakes in Sweden and Norway no longer support fish—these lakes are now eerily clear. In the northeastern United States and Canada, too, tens of thousands of lakes are dying biologically as their pH levels fall to below 5.0.

The solution at first seems obvious: Capture and remove the emissions instead of releasing them into the atmosphere. But there have been serious problems with implementing this solution. First, it is expensive. Estimates of the cost of installing and maintaining the necessary "scrubbers" in the United States are on the order of $5 billion a year. An additional difficulty is that the polluter and the recipient of the pollution are far from one another, and neither wants to pay so much for what they view as someone else's problem. The Clean Air Act revisions of 1990 have begun to address this problem by mandating some cleaning of emissions in the United States, although much still remains to be done worldwide.

> **18.2** Pollution-acidified precipitation—loosely called acid rain—is destroying forest and lake ecosystems in Europe and North America. The solution is to clean up the emissions.

18.3 The Ozone Hole

Living things were able to leave the oceans and colonize the surface of the earth only after a protective shield of ozone had been added to the atmosphere by photosynthesis. For 2 billion years before that, life was trapped in the oceans because radiation from the sun seared the earth's surface unchecked. Nothing could survive that bath of destructive energy. Imagine if that shield were taken away. Alarmingly, it appears that we are destroying it ourselves. Starting in 1975, the earth's ozone shield began to disintegrate. Over the South Pole in September of that year, satellite photos revealed that the ozone concentration was unexpectedly less than elsewhere in the earth's atmosphere. It was as if some "ozone eater" were chewing it up in the Antarctic sky, leaving a mysterious zone of lower-than-normal ozone concentration, an **ozone hole.** Every year after that, more of the ozone has been depleted, and the hole grows bigger and deeper (figure 18.4).

What is eating the ozone? Scientists soon discovered that the culprit was a class of chemicals that everyone had thought to be harmless: **chlorofluorocarbons (CFCs).** CFCs were invented in the 1920s, a miracle chemical that was stable, harmless, and a near-ideal heat exchanger. Throughout the world, CFCs are used in large amounts as coolants in refrigerators and air conditioners, as the gas in aerosol dispensers, and as the foaming agent in Styrofoam containers. All of these CFCs eventually escape into the atmosphere, but no one worried about this until recently, both because CFCs were thought to be chemically inert and because everyone tends to think of the atmosphere as limitless. But CFCs are very stable chemicals, so over many years they have accumulated in the atmosphere.

It turned out that the CFCs were causing mischief the chemists had not imagined. High over the South and North Poles, nearly 50 kilometers up, where it was very, very cold, the CFCs stuck to frozen water vapor and began to act as catalysts of a chemical reaction. Just as an enzyme carries out a reaction in your cells without being changed itself, so the CFCs began to catalyze the conversion of ozone (O_3) into oxygen (O_2) without being used up themselves. Very stable, the CFCs in the atmosphere just kept at it—little machines that never stop. They are still there, still doing it, today. The drop in ozone worldwide is now over 3%.

Ultraviolet radiation is a serious human health concern. Every 1% drop in the atmospheric ozone content is estimated to lead to a 6% increase in the incidence of skin cancers. At middle latitudes, the drop of approximately 3% that has occurred worldwide is estimated to have led to an increase of perhaps as much as 20% in lethal melanoma skin cancers. Importantly, photosynthetic plankton species, critically important to global productivity, are apparently much more highly susceptible than humans.

International agreements to lower the level of CFC production worldwide over the next several decades have been signed, but no one knows if this gradual phase-out will be adequate, or too little too late. The vast majority of the CFCs that have already been manufactured have not yet reached the upper atmosphere, and an enormous additional amount is scheduled to be produced in the coming decades. We can only guess how much of the earth's ozone will be destroyed.

> **18.3** CFCs and other chemicals are catalytically destroying the ozone in the upper atmosphere, exposing earth's surface to dangerous radiation.

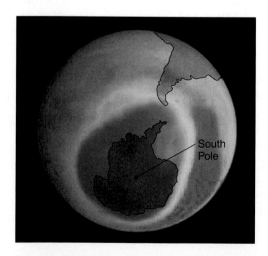

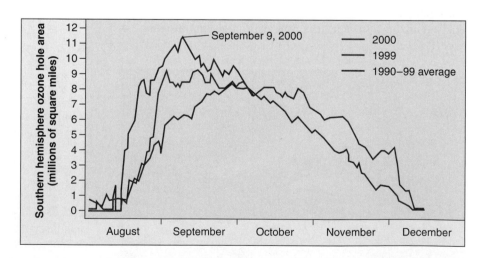

Figure 18.4   The ozone hole over Antarctica is still growing.

For decades NASA satellites have tracked the extent of ozone depletion over Antarctica. Every year since 1975 an ozone "hole" has appeared in August when sunlight triggers chemical reactions in cold air trapped over the South Pole during Antarctic winter. The hole intensifies during September before tailing off as temperatures rise in November–December. In 2000, the 11.4-million-square-mile hole (purple in the satellite image) covered an area larger than the United States, Canada, and Mexico combined, the largest hole ever recorded. In September 2000, the hole extended over Punta Arenas, a city of about 120,000 people in southern Chile, exposing residents to very high levels of UV radiation.

18.4 The Greenhouse Effect

For over 150 years, the growth of our industrial society has been fueled by cheap energy, much of it obtained by burning fossil fuels—coal, oil, and gas. Coal, oil, and gas are the remains of ancient plants, transformed by pressure and time into carbon-rich "fossil fuels." When such fossil fuels are burned, this carbon is combined with oxygen atoms, producing carbon dioxide (CO_2). Industrial society's burning of fossil fuels has released huge amounts of carbon dioxide into the atmosphere. As with CFCs, no one paid any attention to this because the carbon dioxide was thought to be harmless and because the atmosphere was thought to be a limitless reservoir, able to absorb and disperse any amount. It turns out neither assumption was true, and in recent decades, the levels of carbon dioxide in the atmosphere have risen sharply and continue to rise.

What is alarming is that the carbon dioxide doesn't just sit in the air doing nothing. The chemical bonds in carbon dioxide molecules transmit radiant energy from the sun but trap the longer wavelengths of infrared light, or heat, and prevent them from radiating into space. This creates what is known as the **greenhouse effect.** Planets that lack this type of "trapping" atmosphere are much colder than those that possess one. If the earth did not have a "trapping" atmosphere, the average earth temperature would be about –20°C, instead of the actual +15°C.

Global Warming

The earth's greenhouse effect is intensifying with increased fossil fuel combustion and certain types of waste disposal. These activities are increasing the amounts of carbon dioxide, CFCs, nitrogen oxides, and methane—all "greenhouse gases"—in the atmosphere. The rise in average global temperatures during recent decades is consistent with increased carbon dioxide concentrations in the atmosphere (figure 18.5). The idea of **global warming** due to accumulation of greenhouse gases in the earth's atmosphere has been controversial, because correlations do not prove a cause-and-effect relationship. However, as more data become available, a growing consensus of scientists accept global warming as an unwelcome reality.

Increases in the amounts of greenhouse gases could increase average global temperatures from 1° to 4°C, which could have serious associated effects in alterations in rain patterns, in prime agricultural lands, and in changes in sea levels.

Effects on Rain Patterns. Global warming is predicted to have a major effect on rainfall patterns. Areas that have already been experiencing droughts may see even less rain, contributing to even greater water shortages. Recent increases in the frequency of El Niño events (see chapter 16) may indicate that these global warming climatic changes are already beginning to occur.

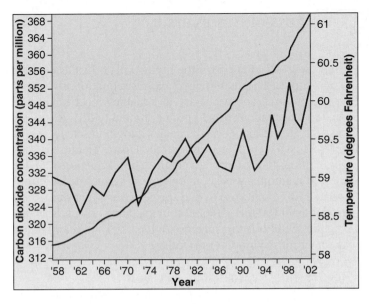

Figure 18.5 The greenhouse effect.
The concentration of carbon dioxide in the atmosphere has shown a steady increase for many years (*blue line*). The *red line* shows the average global temperature for the same period of time. Note the general increase in temperature since the 1950s and, specifically, the sharp rise beginning in the 1980s. Data from Geophysical Monograph, American Geophysical Union, National Academy of Sciences, and National Center for Atmospheric Research.

Effects on Agriculture. Both positive and negative effects of global warming on agriculture are predicted. Warmer temperatures and increased levels of carbon dioxide in the atmosphere would be expected to increase the yields of some crops, while having a negative impact on others. Droughts that may result from global warming will also negatively affect crops. Plants in the tropics are growing at maximal temperature limits; any further increases in temperature will probably begin to have a negative impact on agricultural yields of tropical farms.

Rising Sea Levels. Much of the water on earth is locked into ice in glaciers and polar ice caps. As global temperatures increase, these large stores of ice have begun to melt. Most of the water from the melted ice ends up in the oceans, causing water levels to rise (because the Arctic ice cap floats, its melting will not raise sea levels, any more than melting ice raises the level of water in a glass). Higher water levels can be expected to cause increased flooding of low-lying lands.

There is considerable disagreement among governments about what ought to be done about global warming. The Clean Air Act of 1990 and the as-yet-unratified Kyoto Treaty have established goals for reducing the emission of greenhouse gases. Countries across the globe are making progress toward reducing emission, but much more needs to be done.

> **18.4** Humanity's burning of fossil fuels has greatly increased atmospheric levels of CO_2, leading to global warming.

18.5 Loss of Biodiversity

Extinction is a fact of life in the living world. Just as death is as necessary to a normal life cycle as reproduction, so extinction is as normal and necessary to a stable world ecosystem as species formation. Most species, probably all, go extinct eventually. More than 99% of species known to science (most from the fossil record) are now extinct. However, current rates of extinctions are alarmingly high. The extinction rate for birds and mammals was about one species every decade from 1600 to 1700, but it rose to one species every year during the period from 1850 to 1950, and four species per year between 1986 and 1990. It is this increase in the rate of extinction that is the heart of the **biodiversity** crisis.

Factors Responsible for Extinction

What factors are responsible for extinction? Studying a wide array of recorded extinctions, and many species currently threatened with extinction, conservation biologists have identified three factors that seem to play a key role in many extinctions: habitat loss, species overexploitation, and introduced species (figure 18.6).

Habitat Loss.   Habitat loss is the single most important cause of extinction. Given the tremendous amounts of ongoing destruction of all types of habitat, from rain forest to ocean floor, this should come as no surprise. Natural habitats may be adversely affected by human influences in four ways: (1) destruction (figure 18.7), (2) pollution, (3) human disrup-

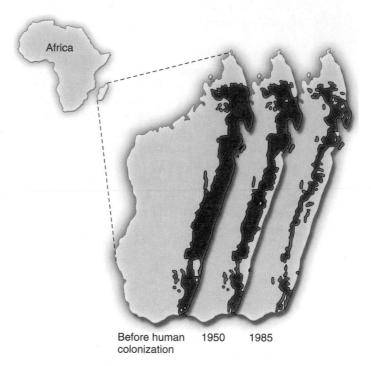

Before human colonization   1950   1985

Figure 18.7   Extinction and habitat destruction.
The rain forest covering the eastern coast of Madagascar, an island off the coast of East Africa, has been progressively destroyed as the island's human population has grown. Ninety percent of the original forest cover is now gone. Many species have become extinct, and many others are threatened, including 16 of Madagascar's 31 primate species.

tion, and (4) habitat fragmentation (dividing up the habitat into small isolated areas).

Species Overexploitation.   Species that are hunted or harvested by humans have historically been at grave risk of extinction, even when the species populations are initially very abundant. There are many examples in our recent history of overexploitation: passenger pigeons, bison, many species of whales, commercial fish such as cod and Atlantic bluefin tuna, and mahogany trees in the West Indies are but a few.

Introduced Species.   Occasionally, a new species will enter a habitat and colonize it, usually at the expense of native species. Colonization occurs in nature, but it is rare; however, humans have made this process more common with devastating ecological consequence. The introduction of exotic species has wiped out or threatened many native populations. African bees are an obvious example. Species introductions occur in many ways, usually unintentionally. Plants and animals can be transported in the ballast of large ocean vessels, in nursery plants, as stowaways in boats, cars, and planes, and as beetle larvae within wood products. These species enter new environments where they have no native predators to keep their population sizes in check. Free to populate the habitat, they crowd out native species.

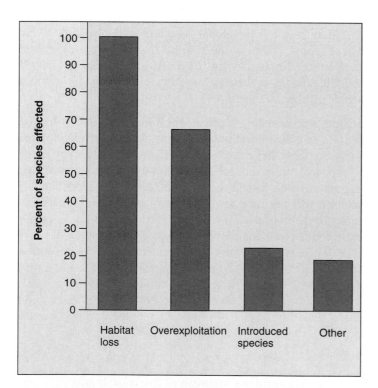

Figure 18.6   Factors responsible for animal extinction.
These data represent known extinctions of mammals in Australia, Asia, and the Americas.

18.5   The loss of biodiversity can usually be attributed to one of a few main causes, including habitat loss, overexploitation, and introduced species.

A Closer Look

The Global Decline in Amphibians

Sometimes important things happen, right under our eyes, without anyone noticing. That thought occurred to David Bradford as he stood looking at a quiet lake high in the Sierra Nevada Mountains of California in the summer of 1988. Bradford, a biologist, had hiked all day to get to the lake, and when he got there his worst fears were confirmed. The lake was on a list of mountain lakes that Bradford had been visiting that summer in Sequoia-Kings Canyon National Parks while looking for a little frog with yellow legs. The frog's scientific name was *Rana muscosa,* and it had lived in the lakes of the parks for as long as anyone had kept records. But this silent summer evening, the little frog was gone. The last major census of the frog's populations within the parks had been taken in the mid-1970s, and *R. muscosa* had been everywhere, a common inhabitant of the many freshwater ponds and lakes within the parks. Now, for some reason Bradford did not understand, the frogs had disappeared from 98% of the ponds that had been their homes.

After Bradford reported this puzzling disappearance to other biologists, an alarming pattern soon became evident. Throughout the world, local populations of amphibians (frogs, toads, and salamanders) were becoming extinct. Waves of extinction have swept through high-elevation amphibian populations in the western United States, and have also cut through the frogs of Central America and coastal Australia.

Amphibians have been around for 350 million years, since long before the dinosaurs. Their sudden disappearance from so many of their natural homes sounded an alarm among biologists. What are we doing to our world? If amphibians cannot survive the world we are making, can we?

In 1998 the U.S. National Research Council brought scientists together from many disciplines in a serious attempt to address the problem. After years of intensive investigation, they have begun to sort out the reasons for the global decline in amphibians. Like many important questions in science, this one does not have a simple answer.

Four factors seem to be contributing in a major way to the worldwide amphibian decline: (1) habitat deterioration and destruction, particularly clear-cutting of forests, which drastically lowers the humidity (water in the air) that amphibians require; (2) the introduction of exotic species that outcompete local amphibian populations; (3) chemical pollutants that are toxic to amphibians; and (4) fatal infections by pathogens.

Infection by parasites appears to have played a particularly important role in the western United States and coastal Australia. Amphibian ecology expert James Collins of Arizona State University has reported one clear instance of infection leading to amphibian decline. When Collins examined populations of salamanders living on the Kaibab Plateau along the Grand Canyon rim, he found many sick salamanders. Their skin was covered with white pustules, and most infected ones died, their hearts and spleens collapsed. The infectious agent proved to be a virus common in fish called a ranavirus. Ranavirus isolated by Collins from one sick salamander would cause the disease in a healthy salamander, so there was no doubt that ranavirus was the culprit responsible for the salamander decline on the Kaibab Plateau.

Ranavirus outbreaks eliminate small populations, but in larger ones a few individuals survive infection, sloughing off their pustule-laden skin. These populations slowly recover.

A second kind of infection, very common in Australia but also seen in the United States, is the real species killer. Populations infected with this microbe, a kind of fungus called a chytrid (pronounced "kit-rid"), do not recover. Usually a harmless soil fungus that decomposes plant material, this particular chytrid (with the Latin name of *Batrachochytrium dendrobatidis*) is far from harmless to amphibians. It dissolves and absorbs the chitinous mouthparts of amphibian larvae, killing them.

This killer chytrid was introduced to Australia near Melbourne in the early 1980s. Now almost all Australia is affected. How did the disease spread so rapidly? Apparently it traveled by truck. Infected frogs moved all across Australia in wooden boxes with bunches of bananas. In one year, 5,000 frogs were collected from banana crates in one Melbourne market alone.

In other parts of the world, infection does not seem to play as important a role as acid precipitation, habitat loss, and introduction of exotic species. This complex pattern of cause and effect only serves to emphasize the take-home lesson: Worldwide amphibian decline has no one culprit. Instead, all four factors play important roles. It is their total impact that has shifted the worldwide balance toward extinction.

To reverse the trend toward extinction, we must work to lessen the impact of all four factors. It is important that we not get discouraged at the size of the job, however. Any progress we make on any one factor will help shift the balance back toward survival. Extinction is only inevitable if we let it be.

18.6 Reducing Pollution

The pattern of global change that is overtaking our world is very disturbing. Human activities are placing a severe stress on the biosphere, and we must quickly find ways to reduce the harmful impact. There are four key areas in which it will be particularly important to meet the challenge successfully: reducing pollution, finding other sources of energy, preserving nonreplaceable resources, and curbing population growth.

To solve the problem of industrial pollution, it is first necessary to understand the cause of the problem. In essence, it is a failure of our economy to set a proper price on environmental health. To understand how this happens, we must think for a moment about money. The economy of the United States (and much of the rest of the industrial world) is based on a simple feedback system of supply and demand. As a commodity gets scarce, its price goes up, and this added profit acts as an incentive for more of the item to be produced; if too much is produced, the price falls and less of it is made because it is no longer so profitable to produce it.

This system works very well and is responsible for the economic strength of our nation, but it has one great weakness. If demand is set by price, then it is very important that all the costs be included in the price. Imagine that the person selling the item were able to pass off part of the production cost to a third person. The seller would then be able to set a lower price and sell more of the item! Driven by the lower price, the buyer would purchase more than if all the costs had been added into the price.

Unfortunately, that sort of pricing error is what has driven the pollution of the environment by industry. The true costs of energy and of the many products of industry are composed of direct production costs, such as materials and wages, and of indirect costs, such as pollution of the ecosystem. Economists have identified an "optimum" amount of pollution based on how much it costs to reduce pollution versus the social and environmental cost of allowing pollution (figure 18.8).

The indirect costs of pollution are usually not taken into account. However, the indirect costs do not disappear because we ignore them. They are simply passed on to future generations, which must pay the bill in terms of damage to the ecosystems on which we all depend. Increasingly, the future is now. Our world, unable to support more damage, is demanding that something be done—that we finally pay up.

Antipollution Laws

Two effective approaches have been devised to curb pollution in this country. The first is to pass laws forbidding it. In the last 20 years, laws have begun to significantly curb the spread of pollution by setting stiff standards for what can be released into the environment. For example, all cars are required to have effective catalytic converters to eliminate

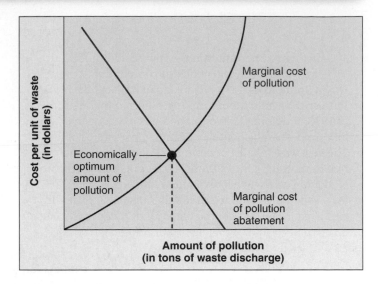

Figure 18.8 Is there an optimum amount of pollution?
Economists identify the "optimum" amount of pollution as the amount whose marginal cost equals the marginal cost of pollution abatement (the point where the two curves intersect). If more pollution than the optimum is allowed, the social cost is unacceptably high. If less than the optimum amount of pollution is allowed, the economic cost is unacceptably high.

automobile smog. Similarly, the Clean Air Act of 1990 requires that power plants eliminate sulfur emissions. They can accomplish this by either installing scrubbers on their smokestacks or by burning low-sulfur coal (clean-coal technology), which is more expensive. The effect is that the consumer pays to avoid polluting the environment. The cost of the converters makes cars more expensive, and the cost of the scrubbers increases the price of the energy. The new, higher costs are closer to the true costs, lowering consumption to more appropriate levels.

Pollution Taxes

A second approach to curbing pollution has been to increase the consumer costs directly by placing a tax on the pollution, in effect an artificial price hike imposed by the government as a tax added to the price of production. This added cost lowers consumption too, but by adjusting the tax, the government can attempt to balance the conflicting demands of environmental safety and economic growth. Such taxes, often imposed as "pollution permits," are becoming an increasingly important part of antipollution laws.

18.6 Free market economies often foster pollution when prices do not include environmental costs. Laws and taxes are being designed in an attempt to compensate.

18.7 Finding Other Sources of Energy

The pollution generated by burning coal and oil, the increasing scarcity of oil, and the potential contributions of carbon dioxide to global warming all make it desirable to find alternative energy sources. Many countries are turning to nuclear power for their growing energy needs. In less than 50 years, nuclear power has become a leading source of energy. In 1995, more than 500 nuclear reactors were producing power worldwide. Over 70% of France's electricity is now produced by nuclear power plants.

Nuclear power plants have not been as popular in this country as in the rest of the world, because we have ample access to cheap coal and because the public fears the consequences of an accident. A reactor partial meltdown at the Three Mile Island nuclear plant in Pennsylvania in 1979 released little radiation into the environment but galvanized these fears. There has been little nuclear power development in this country since then (figure 18.9).

In theory, nuclear power can provide plentiful, cheap energy, but the reality is less encouraging. Nuclear power presents several problems that must be overcome if it is to provide a significant portion of the energy that will fuel our future world.

Safe operation. Because of the potential for vast radioactive contamination such as occurred when a reactor exploded at Chernobyl in the Ukraine in 1986

Figure 18.10 A nuclear disaster.
The Russian scientists in this helicopter measure radiation levels over the Chernobyl reactor days after a 1986 explosion blew it apart.

Figure 18.9 Three Mile Island nuclear power plant.
Since a nuclear accident here in 1979, the building of nuclear power stations in the United States has slowed dramatically.

(figure 18.10), it is crucial that nuclear power plants be designed "fail-safe" and be sited away from densely populated areas. Although new power plant designs are far safer than those of Chernobyl, there is still much room for improvement.

Waste disposal. Spent nuclear fuel remains very radioactive for thousands of years, and the plant itself quickly wears out—after about 25 years, the intense radioactivity makes its metal pipes brittle. As yet, no safe way has been devised to dispose of this growing mountain of highly radioactive material, although a few nuclear power plants in France and the United States are in the process of decommissioning and dismantling. Their dangerously radioactive components are to be disposed of by long-term storage.

Security. Some of the uranium in nuclear fuel is converted to plutonium during its decay, and this plutonium could be recovered from spent fuel and used to make nuclear weapons. It will be very important to find ways to effectively guard nuclear power plants against terrorism.

> **18.7** Safety, security, and particularly waste disposal remain serious obstacles to the widespread use of nuclear power.

18.8 Preserving Nonreplaceable Resources

Among the many ways ecosystems are being damaged, one class of problem stands out as more serious than all the rest: consuming or destroying resources that we all share in common (figure 18.11) but cannot replace in the future. Although a polluted stream can be cleaned, no one can restore an extinct species. In the United States, three sorts of nonreplaceable resources are being reduced at alarming rates: topsoil, groundwater, and biodiversity.

Topsoil

Soil is composed of a mixture of rocks and minerals with partially decayed organic matter called humus. Plant growth is strongly affected by soil composition. Minerals like nitrogen and phosphorus are critical to plant growth, and are abundant in humus-rich soils.

The United States is one of the most productive agricultural countries on earth, largely because much of it is covered with particularly fertile soils. Our midwestern farm belt sits astride what was once a great prairie. The soil of that ecosystem accumulated bit by bit from countless generations of animals and plants until, by the time humans came to plow, the humus-rich soil extended down several feet.

We can never replace this rich **topsoil,** the capital upon which our country's greatness is built, yet we are allowing it to be lost at a rate of centimeters every decade. Our country has lost one-quarter of its topsoil since 1950! By repeatedly tilling (turning the soil over) to eliminate weeds, we permit rain to wash more and more of the topsoil away, into rivers and eventually out to sea. New approaches are desperately needed to lessen the reliance on intensive cultivation. Some possible solutions include using genetic engineering to make crops resistant to weed-killing herbicides and terracing to recapture lost topsoil.

Groundwater

A second resource that we cannot replace is **groundwater,** water trapped beneath the soil within porous rock reservoirs called aquifers. This water seeped into its underground reservoir very slowly during the last ice age over 12,000 years ago. We should not waste this treasure, for we cannot replace it.

In most areas of the United States, local governments exert relatively little control over the use of groundwater. As a result, a very large portion is wasted watering lawns, washing cars, and running fountains. A great deal more is inadvertently being polluted by poor disposal of chemical wastes—and once pollution enters the groundwater, there is no effective means of removing it.

"Freedom in a Commons Brings Ruin to All"

The essence of Hardin's original essay:

Picture a pasture open to all. It is expected that each herdsman will try to keep as many cattle as possible on [this] commons....What is the utility...of adding one more animal?...Since the herdsman receives all the proceeds from the sale of the additional animal, the positive utility [to the herdsman] is nearly +1.... Since, however, the effects of overgrazing are shared by all the herdsmen, the negative utility for any particular decision-making herdsman is only a fraction of -1. Adding together the...partial utilities, the rational herdsman concludes that the only sensible course for him to pursue is to add another animal to [the] herd. And another; and another.... Therein is the tragedy. Each man is locked into a system that [causes] him to increase his herd without limit—in a world that is limited....Freedom in a commons brings ruin to all.

—G. Hardin, "The Tragedy of the Commons," *Science* **162,** 1243 (1968), p. 1244

Figure 18.11　The tragedy of the commons.
In a now-famous essay, ecologist Garrett Hardin argues that destruction of the environment is driven by freedom without responsibility.

(a)

(b)

(c)

Figure 18.12 Tropical rain forest destruction.

(a) These fires are destroying the rain forest in Brazil, which is being cleared for cattle pasture. (b) The flames are so widespread and so high that their smoke can be viewed from space. This satellite photo shows a plume of smoke generated from the burning of the rain forest. (c) The consequences of deforestation can be seen on these middle-elevation slopes in Ecuador. The slopes now support only low-grade pastures where they used to support highly productive forest, which protected the watersheds of the area.

Biodiversity

The number of species in danger of extinction during your lifetime is far greater than the number that became extinct with the dinosaurs. This disastrous loss of biodiversity is important to every one of us, because as these species disappear, so does our chance to learn about them and their possible benefits for ourselves. The fact that our entire supply of food is based on 20 kinds of plants, out of the 250,000 available, should give us pause. Like burning a library without reading the books, we don't know what it is we waste. All we can be sure of is that we cannot retrieve it. Extinct is forever.

Over the last 20 years, about half of the world's tropical rain forests have been either burned to make pasture land or cut for timber (figure 18.12). Over 6 million square kilometers have been destroyed. Every year the rate of loss increases as the human population of the tropics grows. About 160,000 square kilometers were cut each year in the 1990s, a rate greater than 0.6 hectares (1.5 acres) per second! At this rate, all the rain forests of the world will be gone in your lifetime. In the process, it is estimated that one-fifth or more of the world's species of animals and plants will become extinct— more than a million species. This would be an extinction event unparalleled for at least 65 million years, since the age of the dinosaurs.

You should not be lulled into thinking that loss of biodiversity is a problem limited to the tropics. The ancient forests of the Pacific Northwest are being cut at a ferocious rate today, largely to supply jobs (the lumber is exported), with much of the cost of cutting it down subsidized by our government (the Forest Service builds the necessary access roads, for example). At the current rate, very little will remain in a decade. Nor is the problem restricted to one area. Throughout our country, natural forests are being "clear-cut," replaced by pure stands of lumber trees planted in rows like so many lines of corn. It is difficult to scold those living in the tropics when we ourselves do such a poor job of preserving our own country's biodiversity.

But what is so bad about losing species? What is the value of biodiversity? Loss of a species entails three costs: (1) the direct economic value of the products we might have obtained from species; (2) the indirect economic value of benefits produced by species without our consuming them, such as nutrient recycling in ecosystems; and (3) their ethical and aesthetic value. It is not difficult to see the value in protecting species that we use to obtain food, medicine, clothing, energy, and shelter, but other species are vitally important to maintaining healthy ecosystems; by destroying biodiversity, we are creating conditions of instability and lessened productivity. Other species add beauty to the living world, no less crucial because it is hard to set a price upon.

18.8 Nonreplaceable resources are being consumed at an alarming rate all over the world, key among them topsoil, groundwater, and biodiversity.

18.9 Curbing Population Growth

If we were to solve all the problems mentioned in this chapter, we would merely buy time to address the fundamental problem: There are getting to be too many of us.

Humans first reached North America at least 12,000 to 13,000 years ago, crossing the narrow straits between Siberia and Alaska and moving swiftly to the southern tip of South America. By 10,000 years ago, when the continental ice sheets withdrew and agriculture first developed, about 5 million people lived on earth, distributed over all the continents except Antarctica. With the new and much more dependable sources of food that became available through agriculture, the human population began to grow more rapidly. By the time of Christ, 2,000 years ago, an estimated 130 million people lived on earth. By the year 1650, the world's population had doubled, and doubled again, reaching 500 million.

As we discussed in chapter 2, the human population has grown explosively for the last 300 years (figure 18.13). The average human birthrate has stabilized at about 22 births per year per 1,000 people worldwide. However, with the spread of better sanitation and improved medical techniques, the death rate has fallen steadily, to its present level of 9 per 1,000 per year. The difference between birth and death rates amounts to a population growth rate of 1.3% per year, which seems like a small number, but it is not, given the large population size.

The world population reached 6.3 billion people in 2003, and the annual increase now amounts to about 80 million people, which leads to a doubling of the world population in about 53 years. Put another way, more than 216,000 people are added to the world population each day, or more than 150 every minute. At this rate, the world's population will continue to grow over 6.3 billion, and perhaps stabilize at a figure between 8.5 billion and 20 billion. Such growth cannot continue, because our world cannot support it. Just as a cancer cannot grow unabated in your body without eventually killing you, so humanity cannot continue to grow unchecked in the biosphere without killing it.

One of the most alarming trends taking place in developing countries is the massive movement to urban centers. For example, Mexico City, one of the largest cities in the world, is plagued by smog, traffic, inadequate waste disposal, and other problems; it has a population of about 26 million people (figure 18.14). The prospects of supplying adequate food, water, and sanitation to this

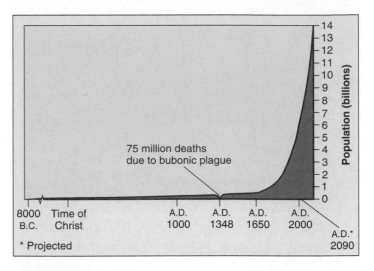

Figure 18.13 Growth curve of the human population.

Over the past 300 years, the world population has been growing steadily. Currently, there are over 6 billion people on the earth.

Figure 18.14 The world's population is centered in mega-cities.

Mexico City, one of the world's largest cities, has about 26 million inhabitants.

city's people are almost unimaginable. The lot of the rural poor, mainly farmers, in Mexico is even worse.

In view of the limited resources available to the human population, and the need to learn how to manage those resources well, the first and most necessary step toward global prosperity is to stabilize the human population. One of the surest signs of the pressure being placed on the environment is human use of about 40% of the total net global photosynthetic productivity on land. Given that statistic, a doubling of the human population in 53 years poses extraordinarily severe problems. The facts virtually demand restraint in population growth. If and when technology is developed that would allow greater numbers of people to inhabit the earth in a stable condition, the human population can be increased to whatever level might be appropriate.

A key element in the world's population growth is its uneven distribution among countries. Of the billion people added to the world's population in the 1990s, 80% to 90% live in developing countries (figure 18.15) and of that number, about 60% of the people in the world live in countries that are at least partly tropical or subtropical. An additional 20% live in China. The remaining 20% live in the so-called developed, or industrialized, countries: Europe, Russia, Japan, the United States, Canada, Australia, and New Zealand. Whereas the populations of the developed countries are growing at an annual rate of only about 0.1%, those of the less developed, mostly tropical countries (excluding China) are growing at an annual rate estimated to be about 1.9%.

Most countries are devoting considerable attention to slowing the growth rate of their populations, and there are genuine signs of progress. If it continues, the United Nations estimates that the world's population may stabilize by the close of this century at a level of 13 to 15 billion people, nearly three times the number living today. No one knows whether the world can support so many people indefinitely. Finding a way to do so is the greatest task facing humanity. The quality of life that will be available for your children in this new century will depend to a large extent on our success.

Population Growth Rate Has Been Declining

The world population growth rate has been declining, from a high of 2.0% in the period 1965–1970 to 1.3% in 2003. Nonetheless, because of the larger population, this amounts to an increase of 80 million people per year to the world population, compared to 53 million per year in the 1960s.

The United Nations attributes the decline to increased family planning efforts and the increased economic power and social status of women. Although the United Nations applauds the United States for leading the world in funding family planning programs abroad, some oppose spending money on international family planning. The opposition states that money is better spent on improving education and the economy in other countries, leading to an increased awareness and lowered fertility rates. The United Nations certainly supports the improvement of education programs in developing countries, but, interestingly, it has reported increased education

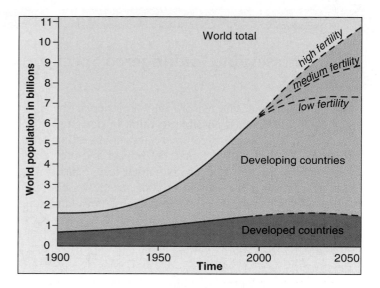

Figure 18.15   Distribution of population growth.
Most of the worldwide increase in population since 1950 has occurred in developing countries. This trend will likely increase in the near future. World population in 2050 is estimated to be between 7.3 and 10.7 billion, according to a recent United Nations study. Depending on fertility rates, the population at that time will either be increasing rapidly or slightly, or in the best case, declining slightly.

levels *following* a decrease in family size as a result of family planning.

Slowing population growth will help sustain the world's resources, but per capita consumption is also important. Even though the vast majority of the world's population is in developing countries, the vast majority of resource consumption occurs in the developed world—the wealthiest 20% of the world's population accounts for 80% of the world's consumption of resources, whereas the poorest 20% is responsible for only 1.3% of consumption. The developed world must lessen the impact each of us makes.

No one knows whether the world can sustain today's population of 6.3 billion people, much less the far greater numbers expected in the future. We cannot reasonably expect to expand the world's carrying capacity indefinitely. The population will begin scaling back in size, as predicted by logistic growth models; indeed it is already happening. In the sub-Saharan area of Africa, population projections for the year 2025 have been scaled back from 1.33 billion to 1.05 billion because of the impact of AIDS. If we are to avoid catastrophic increases in death rates, the birthrates must continue to fall dramatically.

18.9   The problem at the core of all other environmental concerns is the rapid growth of the world's human population. Serious efforts are being made to slow its growth.

18.10 Preserving Endangered Species

Once you understand the reasons why a particular species is endangered, it becomes possible to think of designing a recovery plan. If the cause is commercial overharvesting, regulations can be designed to lessen the impact and protect the threatened species. If the cause is habitat loss, plans can be instituted to restore lost habitat. Loss of genetic variability in isolated subpopulations can be countered by transplanting individuals from genetically different populations. Populations in immediate danger of extinction can be captured, introduced into a captive breeding program, and later reintroduced to other suitable habitats.

Of course, all of these solutions are extremely expensive. As Bruce Babbitt, Interior Secretary in the Clinton administration, noted, it is much more economical to prevent such "environmental trainwrecks" from occurring than it is to clean them up afterward. Preserving ecosystems and monitoring species before they are threatened is the most effective means of protecting the environment and preventing extinctions.

Habitat Restoration

Conservation biology typically concerns itself with preserving populations and species in danger of decline or extinction. Conservation, however, requires that there be something left to preserve, while in many situations, conservation is no longer an option. Species, and in some cases whole communities, have disappeared or have been irretrievably modified. The clear-cutting of the temperate forests of Washington State leaves little behind to conserve; nor does converting a piece of land into a wheat field or an asphalt parking lot. Redeeming these situations requires restoration rather than conservation.

Three quite different sorts of habitat restoration programs might be undertaken, depending very much on the cause of the habitat loss.

Pristine Restoration. In situations where all species have been effectively removed, one might attempt to restore the plants and animals that are believed to be the natural inhabitants of the area, when such information is available. When abandoned farmland is to be restored to prairie (figure 18.16), how do you know what to plant? Although it is in principle possible to reestablish each of the original species in their original proportions, rebuilding a community requires that you know the identity of all of the original inhabitants, and the ecologies of each of the species. We rarely ever have this much information, so no restoration is truly pristine.

Removing Introduced Species. Sometimes the habitat of a species has been destroyed by a single introduced species. In such a case, habitat restoration involves removal of the introduced species. For example, Lake Victoria, Africa, was home to over 300 species of cichlid fishes, small perchlike fishes that display incredible diversity. However, in 1954, the Nile perch, a

(a)

(b)

Figure 18.16 Habitat restoration.
The University of Wisconsin-Madison Arboretum has pioneered restoration ecology. (*a*) The restoration of the prairie was at an early stage in November, 1935. (*b*) The prairie as it looks today. This picture was taken at approximately the same location as the 1935 photograph.

commercial fish with a voracious appetite, was introduced into Lake Victoria. For decades, these perch did not seem to have a significant impact, and then something happened to cause the Nile perch to explode and spread rapidly through the lake, eating their way through the cichlids. By 1986, over 70% of cichlid species had disappeared, including all open-water species.

So what happened to kick-start the mass extinction of the cichlids? The trigger seems to have been eutrophication. High inputs of nutrients from agricultural runoff and sewage from towns and villages led to algal blooms that severely depleted oxygen levels in deeper parts of the lake. This is thought to have led to an increase in cichlids that feed on algae, and a subsequent explosion of Nile perch numbers. The situation has been compounded by a second factor, the introduction into Lake Victoria of a floating water weed from South America, the water hyacinth *Eichornia crassipes*. Extremely prolific under eutrophic conditions, thick mats of water hyacinth soon covered entire bays and inlets, choking off the coastal habitats of non-open-water cichlids.

Restoration of the once-diverse cichlid fishes to Lake Victoria will require more than breeding and restocking the endangered species. Eutrophication will have to be reversed, and the introduced water hyacinth and Nile perch populations brought under control or removed.

It is important to act quickly if an introduced species is to be removed. When aggressive African bees (the so-called "killer bees") were inadvertently released in Brazil. They remained in the local area only one season. Now they occupy much of the Western Hemisphere.

Cleanup and Rehabilitation. Habitats seriously degraded by chemical pollution cannot be restored until the pollution is cleaned up. The successful restoration of the Nashua River in New England, discussed later in this chapter, is one example of how a concerted effort can succeed in restoring a heavily polluted habitat to a relatively pristine condition.

Captive Propagation

Recovery programs, particularly those focused on one or a few species, often must involve direct intervention in natural populations to avoid an immediate threat of extinction. Introducing wild-caught individuals into captive breeding programs is being used in an attempt to save ferret and prairie chicken populations in immediate danger of disappearing. Several other such captive propagation programs have had significant success.

Case History: The Peregrine Falcon. U.S. populations of birds of prey such as the Peregrine falcon (*Falco peregrinus*) began an abrupt decline shortly after World War II. Of the approximately 350 breeding pairs east of the Mississippi River in 1942, all had disappeared by 1960. The culprit proved to be the chemical pesticide DDT (dichlorodiphenyltrichloroethane) and related organochlorine pesticides. Birds of prey are particularly vulnerable to DDT because they feed at the top of the food chain, where DDT becomes concentrated. DDT interferes with the deposition of calcium in the bird's eggshells, causing most of the eggs to break before they hatch.

The use of DDT was banned by federal law in 1972, causing levels in the eastern United States to fall quickly. There were no peregrine falcons left in the eastern United States to reestablish a natural population, however. Falcons from other parts of the country were used to establish a captive breeding program at Cornell University in 1970, with the intent of reestablishing the peregrine falcon in the eastern United States by releasing offspring of these birds. By the end of 1986, over 850 birds had been released in 13 eastern states, producing an astonishingly strong recovery.

Sustaining Genetic Diversity

One of the chief obstacles to a successful species recovery program is that a species is generally in serious trouble by the time a recovery program is instituted. When populations become very small, much of their genetic diversity is lost. If a program is to have any chance of success, every effort must be made to sustain as much genetic diversity as possible.

Case History: The Black Rhino. All five species of rhinoceros are critically endangered. The three Asian species live in a forest habitat that is rapidly being destroyed, while the two African species are illegally killed for their horns. Fewer than 11,000 individuals of all five species survive today. The problem is intensified by the fact that many of the remaining animals live in very small, isolated populations. The 2,400 wild-living individuals of the black rhino, *Diceros bicornis* (figure 18.17), live in approximately 75 small, widely separated groups consisting of six subspecies adapted to local conditions throughout the species' range. All of these subspecies appear to have low genetic variability; in three of the subspecies, only a few dozen animals remain. Analysis of mitochondrial DNA suggests that in these populations most individuals are genetically very similar.

This lack of genetic variability represents the greatest challenge to the future of the species. Much of the range of the black rhino is still open and not yet subject to human encroachment. To have any significant chance of success, a species recovery program will have to find a way to sustain the genetic diversity that remains in this species. Heterozygosity could be best maintained by bringing all black rhinos together in a single breeding population, but this is not a practical possibility. A more feasible solution would be to move individuals between populations. Managing the black rhino populations for genetic diversity could fully restore the species to its original numbers and much of its range.

Placing black rhinos from a number of different locations together in a sanctuary to increase genetic diversity raises a potential problem: Local subspecies may be adapted in different ways to their immediate habitats–what if these local adaptations are crucial to their survival? Homogenizing the black

Figure 18.17 Sustaining genetic diversity.
The black rhino is highly endangered, living in 75 small, widely separated populations. Only about 2400 individuals survive in the wild. Conservation biologists have the difficult job of finding ways to preserve genetic diversity in small, isolated populations.

rhino populations by pooling their genes risks destroying such local adaptations, if they exist, perhaps at great cost to survival.

Preserving Keystone Species

Keystone species are species that exert a particularly strong influence on the structure and functioning of a particular ecosystem. Their removal can have disastrous consequences.

Case History: Flying Foxes. The severe decline of many species of pteropodid bats, or "flying foxes," in the Old World tropics is an example of how the loss of a keystone species can have dramatic effects on the other species living within an ecosystem, sometimes even leading to a cascade of further extinctions (figure 18.18). These bats have very close relationships with important plant species on the islands of the Pacific and Indian Oceans. The family Pteropodidae contains nearly 200 species, approximately a quarter of them in the genus *Pteropus,* and is widespread on the islands of the South Pacific, where they are the most important— and often the only—pollinators and seed dispersers. A study in Samoa found that 80% to 100% of the seeds landing on the ground during the dry season were deposited by flying foxes. Many species are entirely dependent on these bats for pollination.

In Guam, where the two local species of flying fox have recently been driven extinct or nearly so, the impact on the ecosystem appears to be substantial. Many plant species are not fruiting, or are doing so only marginally, with fewer fruits than normal. Fruits are not being dispersed away from parent plants, so offspring shoots are being crowded out by the adults.

Flying foxes are being driven to extinction by human hunting. They are hunted for food, for sport, and by orchard farmers, who consider them pests. Flying foxes are particularly vulnerable because they live in large, easily seen groups of up to a million individuals. Because they move in regular and predictable patterns and can be easily tracked to their home roost, hunters can easily bag thousands at a time.

Species preservation programs aimed at preserving particular species of flying foxes are only just beginning. One particularly successful example is the program to save the Rodrigues fruit bat, *Pteropus rodricensis,* which occurs only on Rodrigues Island in the Indian Ocean near Madagascar. The population dropped from about 1,000 individuals in 1955 to fewer than 100 by 1974, the drop reflecting largely the loss of the fruit bat's forest habitat to farming. Since 1974 the species has been legally protected, and the forest area of the island is being increased through a tree-planting program. Eleven captive breeding colonies have been established, and the bat population is now increasing rapidly. The combination of legal protection, habitat restoration, and captive breeding has in this instance produced a very effective preservation program.

Figure 18.18   Preserving keystone species.

The flying fox is a keystone species in many Old World tropical islands. It pollinates many of the plants, and is a key disperser of seeds. Its elimination by hunting and habitat loss is having a devastating effect on the ecosystems of many South Pacific islands.

Conservation of Ecosystems

Habitat fragmentation is one of the most pervasive enemies of biodiversity conservation efforts. Some species simply require large patches of habitat to thrive, and conservation efforts that cannot provide suitable habitat of such a size are doomed to failure. As it has become clear that isolated patches of habitat lose species far more rapidly than large preserves do, conservation biologists have promoted the creation, particularly in the tropics, of so-called megareserves, large areas of land containing a core of one or more undisturbed habitats.

In addition to this focus on maintaining large enough reserves, in recent years, conservation biologists also have recognized that the best way to preserve biodiversity is to focus on preserving intact ecosystems, rather than focusing on particular species. For this reason, attention in many cases is turning to identifying those ecosystems most in need of preservation and devising the means to protect not only the species within the ecosystem, but the functioning of the ecosystem itself.

> **18.10**   Recovery programs at the species level must deal with habitat loss and fragmentation, and often with a marked reduction in genetic diversity. Captive breeding programs that stabilize genetic diversity and pay careful attention to habitat preservation and restoration are typically involved in successful recoveries.

18.11 Individuals Can Make the Difference

The development of appropriate solutions to the world's environmental problems must rest partly on the shoulders of politicians, economists, bankers, engineers—many kinds of public and commercial activity will be required. However, it is important not to lose sight of the key role often played by informed individuals in solving environmental problems. Often one person has made the difference; two examples serve to illustrate the point.

The Nashua River

Running through the heart of New England, the Nashua River was severely polluted by mills established in Massachusetts in the early 1900s. By the 1960s, the river was clogged with pollution and declared ecologically dead. When Marion Stoddart moved to a town along the river in 1962, she was appalled. She approached the state about setting aside a "greenway" (trees running the length of the river on both sides), but the state wasn't interested in buying land along a filthy river. So Stoddart organized the Nashua River Cleanup Committee and began a campaign to ban the dumping of chemicals and wastes into the river. The committee presented bottles of dirty river water to politicians, spoke at town meetings, recruited businesspeople to help finance a waste treatment plant, and began to clean garbage from the Nashua's banks. This citizen's campaign, coordinated by Stoddart, greatly aided passage of the Massachusetts Clean Water Act of 1966. Industrial dumping into the river is now banned, and the river has largely recovered (figure 18.19).

Lake Washington

A large, 86-square-kilometer freshwater lake east of Seattle, Lake Washington became surrounded by Seattle suburbs in the building boom following the Second World War. Between 1940 and 1953, a ring of 10 municipal sewage plants discharged their treated effluent into the lake. Safe enough to drink, the effluent was believed "harmless." By the mid-1950s a great deal of effluent had been dumped into the lake (try multiplying 80 million liters/day × 365 days/year × 10 years). In 1954, an ecology professor at the University of Washington in Seattle, W. T. Edmondson, noted that his research students were reporting filamentous blue-green algae growing in the lake. Such algae require plentiful nutrients, which deep freshwater lakes usually lack—the sewage had been fertilizing the lake! Edmondson, alarmed, began a campaign in 1956 to educate public officials to the danger: Bacteria decomposing dead algae would soon so deplete the lake's oxygen that the lake would die. After five years, joint municipal taxes financed the building of a sewer to carry the effluent out to sea. The lake is now clean (figure 18.20).

(a)   (b)

(c)

Figure 18.19   Cleaning up the Nashua River.
(a) The Nashua River, seen here in the 1960s, was severely polluted because factories set up along its banks dumped their wastes directly into the river. Today, the river is mostly clean (b), largely due to the efforts of Marion Stoddart (c), who began a campaign forty years ago to clean up the river.

Figure 18.20   Lake Washington, Seattle.
Lake Washington in Seattle is surrounded by residences, businesses, and industries. By the 1950s, the dumping of sewage and the runoff of fertilizers had caused an algal bloom in the lake, which would eventually deplete the lake's oxygen. Efforts to reverse this effect and clean up the lake were started by W. T. Edmondson of the University of Washington in 1956. The lake is now clean.

Solving Environmental Problems

It is easy to become discouraged when considering the world's many environmental problems, but do not lose track of the single most important conclusion that emerges from our examination of these problems—the fact that each is solvable. A polluted lake can be cleaned; a dirty smokestack can be altered to remove noxious gas; waste of key resources can be stopped. What is required is a clear understanding of the problem and a commitment to doing something about it. The extent to which U.S. families **recycle** aluminum cans and newspapers is evidence of the degree to which people want to become part of the solution, rather than part of the problem.

If we look at how success was achieved in those instances where environmental problems have been overcome, a simple pattern emerges. Viewed simply, five components are involved in successfully solving any environmental problem:

1. **Assessment.** The first stage of addressing any environmental problem is scientific analysis. Data must be collected and experiments performed in order to construct a "model" of the ecosystem that describes how it is responding to the situation. Such a model can then be used to make predictions about the future course of events in the ecosystem.

2. **Risk analysis.** Using the results of the scientific investigation as a tool, it is possible to predict the consequences of environmental intervention. It is necessary to evaluate not only the potential for solving the environmental problem but also any adverse effects the action plan might create.

3. **Public education.** When a clear choice can be made among alternative courses of action, the public must be informed. This involves explaining the problem in terms people can understand, presenting the alternative actions available, and describing the probable costs and results of the different choices.

4. **Political education.** The public, through its elected officials, selects a course of action and implements it. Individuals can have a major impact at this stage by exercising their right to vote and be heard. Many voters do not understand the magnitude of what they can achieve by writing letters and supporting special-interest groups.

5. **Follow-through.** The results of any action should be carefully monitored to see if the environmental problem is being solved and, more basically, to evaluate and improve the initial assessment and modeling of the problem. We learn by doing.

> **18.11** In solving environmental problems, the commitment of one person can make a critical difference. Biological literacy is no longer a luxury for scientists—it has become a necessity for all of us.

Exploring Current Issues

Additional Resources

Go to your campus library or look online to find the following articles, which further develop some of the concepts found in this chapter.

Blaustein, A. R. and P. T. J. Johnson. (2003). Explaining frog deformities: an eight-year investigation into the cause of a shocking increase in deformed amphibians has sorted out the roles of three prime suspects. *Scientific American,* 288(2), 60.

Brimblecombe, P. (2002). Acid drops. *New Scientist,* 174(2343), 1.

Herzog, H., B. Eliasson, O. Kaarstad, D. Martindale, D.W. Keith, and E.A. Parson. (2000). Capturing greenhouse gases. *Scientific American,* 282(2), 72.

Mukerjee, M. (2003). Greenhouse suits: litigation becomes a tool against global warming. *Scientific American,* 288(2), 14.

West, K. (2002). Lion versus lamb: in New Mexico a battle brews between two rare species. *Scientific American,* 286(5), 20.

Biology and Society Lecture: Eco-economics and Population Growth

The world faces two serious environmental challenges. The most immediate is degradation of the global environment. This can be seen in acid rain, atmospheric ozone depletion, and global warming. At least as serious a challenge is posed by overutilization of resources that cannot be replaced. At the core of both challenges is the deeper issue of population growth, and the resulting encouragement of economic "development."

Biology and Society Lecture: Solving Environmental Problems

Every successful solution of an environmental problem involves the same key elements: assessment, risk analysis, public education, political action, and follow-through. Very often success results from the determined actions of one individual such as the cleaning up of Lake Washington near Seattle and the Nashua River in New England.

Find these lectures, delivered by the author to his class at Washington University, online at www.mhhe.com/tlwessentials/exp18.

Global Change

18.1 Pollution

- Pollution leads to global change because its effects can spread far from the source. Air and water become polluted when chemicals that are harmful to organisms are released into the ecosystem. The use of agricultural chemicals, such as pesticides, herbicides, and fertilizers, has been widespread with devastating effects on animals through biological magnification, where the chemicals become more concentrated as they pass up through the food chain (**figure 18.1**).

18.2 Acid Precipitation

- The burning of coal releases sulfur into the atmosphere, where it combines with water vapor, forming sulfuric acid. This acid falls back to earth in rain and snow, commonly called acid rain, far from the source of the pollution, killing animals and vegetation (**figures 18.2 and 18.3**).

18.3 The Ozone Hole

- Ozone (O_3) forms a protective shield in the earth's upper atmosphere that blocks out harmful UV rays from the sun. In the mid-1970s, scientists determined that ozone was being depleted. The culprit, chlorofluorocarbons (CFCs), used in refrigeration systems, react with ozone, converting it to oxygen gas (O_2), which doesn't block UV rays. Termed the ozone hole, this reduction in ozone over and extending from the South Pole is resulting in dangerously high levels of radiation reaching the earth (**figure 18.4**).

18.4 The Greenhouse Effect

- The burning of fossil fuels releases carbon dioxide into the atmosphere, more carbon dioxide than can be cycled back through the ecosystem. It remains in the atmosphere, where it traps infrared light (heat) from the sun, a phenomenon known as the greenhouse effect. As a result, the average global temperatures have been steadily increasing, a process known as global warming (**figure 18.5**). Global warming is predicted to have major impacts on global rain patterns, agriculture, and rising sea levels.

18.5 Loss of Biodiversity

- Extinction is a fact of life, but the current rate of the loss of species is alarmingly high. Three factors most responsible for present-day extinctions include loss of habitat, overexploitation, and the introduction of exotic species; habitat loss being the most devastating (**figures 18.6 and 18.7**).

Saving Our Environment

18.6 Reducing Pollution

- Human activities are placing severe stress on the biosphere. Reducing pollution requires examining the costs associated with pollution, which usually aren't factored into industrial development plans (**figure 18.8**). Antipollution laws and pollution taxes are ways to begin factoring in the costs of pollution.

18.7 Finding Other Sources of Energy

- The burning of fossil fuels leads to pollution, depletion of valuable resources, and global warming. New sources of energy are needed, and some countries have looked to nuclear power, but nuclear power has its own drawbacks such as safety issues, the problems of waste disposal, and security issues associated with protecting the plutonium produced as a by-product.

18.8 Preserving Nonreplaceable Resources

- The consumption or destruction of nonreplaceable resources is perhaps the most serious problem humans face. Topsoil, necessary for agriculture, is being depleted rapidly. Groundwater, which percolates through the soil to underground  reservoirs, is our primary source of drinking water, but it is being wasted and polluted. Biodiversity is being reduced through extinctions. Extinctions are greatly outpacing the formation of new species.

18.9 Curbing Population Growth

- The root of all environmental problems is the rapid growth of the human population. More people means more resources are depleted, more land is developed, and more pollution is created (**figures 18.13–18.15**).

Solving Environmental Problems

18.10 Preserving Endangered Species

- In an attempt to slow the loss of biodiversity, recovery programs are under way, designed to save endangered species. These programs include habitat restoration (**figure 18.16**), breeding in captivity, and conservation of ecosystems.

18.11 Individuals Can Make the Difference

- There are environmental success stories, where one or a few individuals made a difference and reversed an ecological disaster (**figures 18.19 and 18.20**).

1. "Gray-air cities" are the result of
   a. biological magnification of air pollutants.
   b. chlorinated hydrocarbons as a major air pollutant.
   c. pesticides as a major air pollutant.
   d. sulfur oxides as a major air pollutant.

2. The main cause of acid rain is
   a. car and truck exhaust.
   b. coal-powered industry.
   c. chlorofluorocarbons.
   d. chlorinated hydrocarbons.

3. Destruction of the ozone layer is due to
   a. car and truck exhaust.
   b. coal-powered industry.
   c. chlorofluorocarbons.
   d. chlorinated hydrocarbons.
4. Which of the following is *not* a gas that contributes to the greenhouse effect?
   a. ammonia
   b. nitrogen oxides
   c. methane
   d. CFCs
5. Introduced species can reduce the biodiversity of an area because they
   a. may be better competitors for resources than native species.
   b. have not coevolved with native predators and diseases.
   c. may have more effective adaptations for the environment than native organisms do.
   d. All answers are correct.
6. Free market economies often promote pollution. This is because
   a. environmental costs are hardly ever recognized as part of the economy.
   b. supply never keeps up with demand, so industry must increase output to address the demand.
   c. the cost of energy and raw materials are so variable.
   d. laws about pollution are unenforceable.

7. Preserving biodiversity is
   a. needed to preserve possible direct value from species, such as new medicines.
   b. not needed as extinction is a "natural" cycle and should not be disturbed.
   c. needed to make sure all niches are filled.
   d. not needed as it interferes with industrial development.
8. Which factor is *not* responsible for the large increase in the human population over the last 300 or so years?
   a. larger and more reliable food reserves from the modernization of farming techniques
   b. decreasing mortality rate due to improvements in medicine
   c. increasing amounts of open space as countries develop
   d. increased sanitation practices
9. Removal of the endangered black-footed ferret and California condor populations from the wild for breeding programs in zoos and field laboratories preserves endangered species through
   a. pristine restoration.   c. habitat rehabilitation.
   b. habitat restoration.   d. captive propagation.
10. If the removal of a species causes an ecosystem to collapse, that species is known as a(n)
   a. keystone species.   c. threatened species.
   b. endangered species.   d. No answer is correct.

Visual Understanding

1. **Figure 18.6**   Your friends, Ron and Barbara, tell you that their father complains that environmentalists are trying to save "every confounded bug and weed on the planet." He says that things have always gone extinct, why should now be any different. Use the graph to help you respond to your friends.

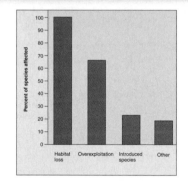

2. **Figures 18.2 and 18.13**   Discuss the human growth curve from this chapter in terms of the graph from chapter 17.

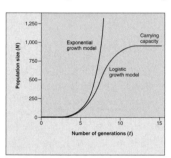
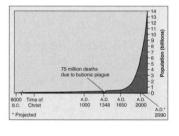

Challenge Questions

Global Change   Explain why being exposed to even very tiny amounts of some chemical pollutants can, over time, be hazardous to your health.

Saving Our Environment   Economists tend to look at the world in terms of costs. They say, "If you increase pollution, then you do increase the social costs to people's health, but if you try to decrease pollution, then you increase the economic costs of cleaning it up." Discuss whether you think these two costs are paid equally by the same set of people.

Solving Environmental Problems   Consider the stories of Marion Stoddart, W. T. Edmondson, and the following statement: "Never doubt that a small group of thoughtful, committed citizens can change the world, indeed it's the only thing that ever has."— Margaret Mead, Anthropologist. What *could* you do to make your neighborhood/area/community a better, healthier place? What *will* you do?

Online Learning Center

Visit the Online Learning Center for this chapter at www.mhhe.com/tlwessentials/ch18 for quizzes, animations, interactive learning exercises, and other study tools. At the site you will also find extended answers to the end-of-chapter questions.

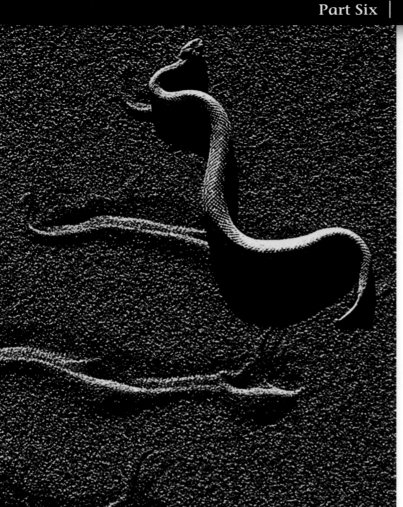

19

The Animal Body and How It Moves

A nimals are unrivaled among the inhabitants of the living world in their ability to move about from one place to another. Under water, on land, and in the air, animals swim, burrow, crawl, slither, slide, walk, jump, run, glide, soar, and fly. This sidewinder rattlesnake can move surprisingly rapidly over the desert sand by a coordinated series of muscle contractions, throwing its long body into a series of sinuous curves. Propulsion is not by a wave of contraction undulating the body, but by a simultaneous lateral thrust in all segments of the body in contact with the ground. For the snake's body to go forward, it is necessary that its muscles push against the ground opposite the direction of movement, so that the thrust of the snake's looping motion tends to occur on the inward-curving side of the loop of the snake's body, at the outside edge. On land, only vertebrates and arthropods have developed a means of rapid surface locomotion. In both groups, the body is raised above the ground and moved forward by pushing against the ground with a series of jointed appendages, the legs. How muscles and nerves produce such movements is the subject of this chapter.

19.1 Innovations in Body Design

Several evolutionary innovations in the design of animals bodies have led to the diversity seen in the phylum Animalia. We will review some of those innovations here and in table 19.1.

Radial Versus Bilateral Symmetry

The simplest animals, the sponges, mostly lack any definite symmetry. They grow asymmetrically as irregular masses. Symmetrical bodies first evolved in cnidarians (jellyfish, sea anemones, and corals) and ctenophores (comb jellies). The bodies of these two types of animals exhibit **radial symmetry,** in which the parts of the body are arranged around a central axis in such a way that any plane passing through the central axis divides the organism into halves that are approximate mirror images.

The bodies of all other animals are marked by a fundamental **bilateral symmetry,** in which the body has a right and a left half that are mirror images of each other. Bilateral symmetry constitutes a major evolutionary advance. This unique form of organization allows parts of the body to evolve in different ways, permitting different organs to be located in different parts of the body. For example, the nervous system in bilaterally symmetrical animals is in the form of major longitudinal nerve cords with nerve cells grouped in the anterior portion of the body. This trend ultimately led to the evolution of a definite head and brain area, a process called **cephalization.** In some higher animals like echinoderms (sea stars, also called starfish), the adults are radially symmetrical, but even in them, the larvae are bilaterally symmetrical.

No Body Cavity Versus Body Cavity

A second key transition in the evolution of the animal body plan was the evolution of the body cavity. The evolution of efficient organ systems within the animal body depended critically upon a body cavity for supporting organs, distributing materials, and fostering complex developmental interactions.

In the animal kingdom, we see three body arrangements: Some animals have no body cavity, whereas others have either of two different types of body cavities, distinguished primarily by where they develop within the three embryonic layers.

1. Animals with no body cavity are called *acoelomates.* Sponges are acoelomates, and so are jellyfish and flatworms.
2. A body cavity that forms between the endoderm and the mesoderm is called a pseudocoel, and the animals in which it occurs are called *pseudocoelomates.* Nematodes are pseudocoelomates.
3. A body cavity that forms entirely within the mesoderm is called a coelom, and animals in which it occurs are called *coelomates.* Mollusks, arthropods, echinoderms, and vertebrates are all coelomates.

Nonsegmented Versus Segmented Bodies

The third key transition in animal body plan involved the subdivision of the body into **segments.** Just as it is efficient for workers to construct a tunnel from a series of identical prefabricated parts, so segmented animals are "assembled" from a succession of identical segments. During the animal's early development, these segments become most obvious in the mesoderm but later are reflected in the ectoderm and endoderm as well. Two advantages result from early embryonic segmentation:

1. In annelids and other highly segmented animals, each segment may go on to develop a more or less complete set of adult organ systems. Damage to any one segment need not be fatal to the individual, because the other segments duplicate that segment's functions.
2. Locomotion is far more effective when individual segments can move independently because the animal as a whole has more flexibility of movement. Because the separations isolate each segment into an individual hydrostatic skeletal unit, each is able to contract or expand autonomously. Therefore, a long body can move in ways that are often quite complex. When an earthworm crawls on a flat surface, it lengthens some parts of its body while shortening others. The elaborate and obvious segmentation of the annelid body clearly represents an evolutionary specialization. Many scientists believe that segmentation is an adaptation for burrowing; it makes possible the production of strong peristaltic waves along the length of the worm, thus enabling vigorous digging.

Segmentation underlies the organization of all advanced animals. Segmentation can be seen in the bodies of all annelids, arthropods, and chordates, although it is not always obvious.

Protostomes Versus Deuterostomes

Two outwardly dissimilar phyla of animals, the echinoderms (sea stars) and the chordates (vertebrates), together with two smaller phyla of animals, have a series of key embryological features different from those shared by the other animal phyla. Because it is extremely unlikely that these features evolved more than once, it is believed that these four phyla share a common ancestry. They are the members of a group called the **deuterostomes.** All other coelomate animals are called **protostomes.** Deuterostomes evolved from protostomes more than 630 million years ago. Deuterostomes, like protostomes, are coelomates. They differ fundamentally from protostomes, however, in three aspects of embryonic growth.

1. **How cleavage forms a hollow ball of cells.** Deuterostomes differ from protostomes in one of the earliest steps of development, the divisions that determine the plane in which the cells divide.

Most protostomes undergo spiral cleavage, whereas deuterostomes undergo radial cleavage.

2. **How the blastopore determines the body axis.** Deuterostomes differ from protostomes in the way in which the embryo grows. The blastopore of a protostome becomes the animal's mouth, and the anus develops at the other end. In a deuterostome, by contrast, the blastopore becomes the animal's anus, and the mouth develops at the other end.

3. **How the developmental fate of the embryo is fixed.** Most protostomes undergo determinate cleavage, which rigidly fixes the developmental fate of each cell very early. No one cell isolated at even the four-cell stage can go on to form a normal individual. In marked contrast, deuterostomes undergo indeterminate cleavage, with each cell retaining the capacity to develop into a complete individual.

19.1 Several key innovations in animal body design set the stage for a great diversity in the phylum.

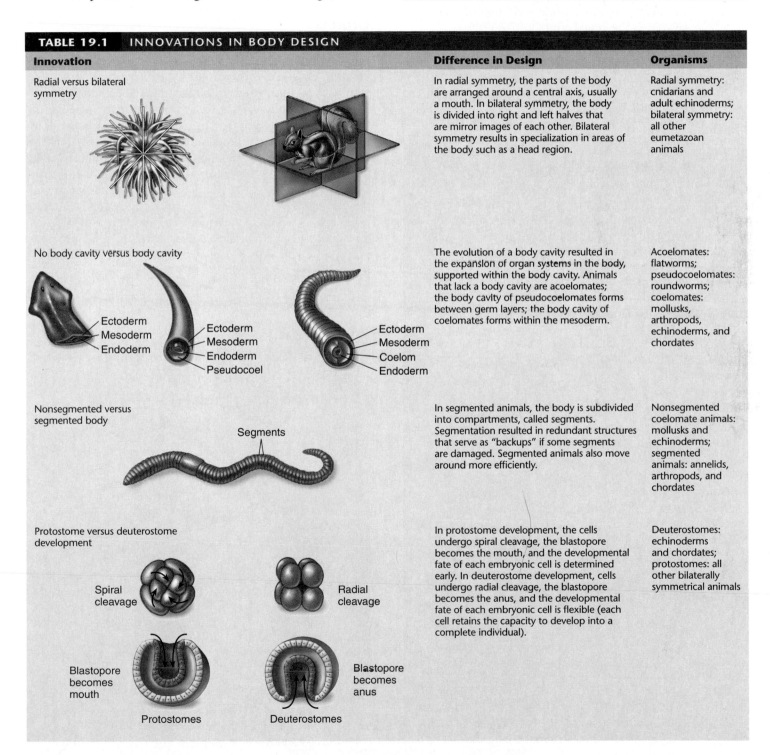

TABLE 19.1	INNOVATIONS IN BODY DESIGN		
Innovation		**Difference in Design**	**Organisms**
Radial versus bilateral symmetry		In radial symmetry, the parts of the body are arranged around a central axis, usually a mouth. In bilateral symmetry, the body is divided into right and left halves that are mirror images of each other. Bilateral symmetry results in specialization in areas of the body such as a head region.	Radial symmetry: cnidarians and adult echinoderms; bilateral symmetry: all other eumetazoan animals
No body cavity versus body cavity		The evolution of a body cavity resulted in the expansion of organ systems in the body, supported within the body cavity. Animals that lack a body cavity are acoelomates; the body cavity of pseudocoelomates forms between germ layers; the body cavity of coelomates forms within the mesoderm.	Acoelomates: flatworms; pseudocoelomates: roundworms; coelomates: mollusks, arthropods, echinoderms, and chordates
Nonsegmented versus segmented body		In segmented animals, the body is subdivided into compartments, called segments. Segmentation resulted in redundant structures that serve as "backups" if some segments are damaged. Segmented animals also move around more efficiently.	Nonsegmented coelomate animals: mollusks and echinoderms; segmented animals: annelids, arthropods, and chordates
Protostome versus deuterostome development		In protostome development, the cells undergo spiral cleavage, the blastopore becomes the mouth, and the developmental fate of each embryonic cell is determined early. In deuterostome development, cells undergo radial cleavage, the blastopore becomes the anus, and the developmental fate of each embryonic cell is flexible (each cell retains the capacity to develop into a complete individual).	Deuterostomes: echinoderms and chordates; protostomes: all other bilaterally symmetrical animals

Organization of the Vertebrate Body

All vertebrates have the same general architecture: A long internal tube that extends from mouth to anus, which is suspended within an internal body cavity called the *coelom*. The coelom of many terrestrial vertebrates is divided into two parts: the *thoracic cavity,* which contains the heart and lungs, and the *abdominal cavity,* which contains the stomach, intestines, and liver. The vertebrate body is supported by an internal scaffold, or skeleton, made up of jointed bones. A bony skull surrounds and protects the brain, while a column of bones, the vertebrae, surrounds the spinal cord.

Like all animals, the vertebrate body is composed of cells—over 100 trillion of them in your body. It's difficult to picture how many 100 trillion actually is. A line with 100 trillion cars in it would stretch from the earth to the sun and back 50 million times! Not all of these 100 trillion cells are the same, of course. If they were, we would not be bodies but amorphous blobs. Vertebrate bodies contain over 100 different kinds of cells.

Tissues

Groups of cells of the same type are organized within the body into **tissues,** which are the structural and functional units of the vertebrate body. A tissue is a group of cells of the same type that performs a particular function in the body.

Tissues form as the vertebrate body develops. Early in development, the growing mass of cells that will become a mature animal differentiates into three fundamental layers of cells: endoderm, mesoderm, and ectoderm. These three kinds of embryonic cell layers, in turn, differentiate into the more than 100 different kinds of cells in the adult body.

It is possible to assemble many different kinds of tissue from 100 cell types, but biologists have traditionally grouped adult tissues into four general classes: *epithelial, connective, muscle,* and *nerve tissue* (figure 19.1). Of these, connective tissues are particularly diverse.

Organs

Organs are body structures composed of several different tissues grouped together into a larger structural and functional unit, just as a factory is a group of people with different jobs who work together to make something. The heart is an organ. It contains cardiac muscle tissue wrapped in connective tissue and joined to many nerves. All of these tissues work together to pump blood through the body: The cardiac muscles squeeze to push the blood; the connective tissues act as a bag to hold the heart in the proper shape and ensure that the different chambers of the heart contract in the proper order; and the nerves control the rate at which the heart beats. No single tissue can do the job of the heart, any more than one piston can do the job of an automobile engine.

Figure 19.1
Vertebrate tissue types.

The four basic classes of tissue are epithelial, connective, muscle, and nerve.

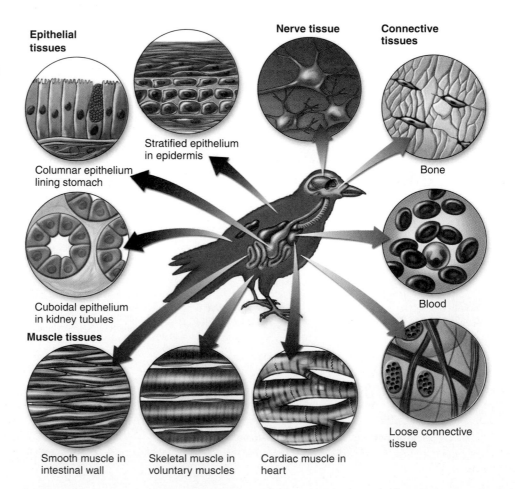

Epithelial tissues

Stratified epithelium in epidermis

Columnar epithelium lining stomach

Cuboidal epithelium in kidney tubules

Muscle tissues

Smooth muscle in intestinal wall

Skeletal muscle in voluntary muscles

Cardiac muscle in heart

Nerve tissue

Connective tissues

Bone

Blood

Loose connective tissue

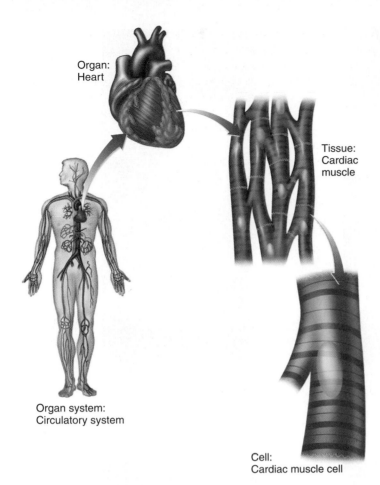

Organ:
Heart

Tissue:
Cardiac
muscle

Organ system:
Circulatory system

Cell:
Cardiac muscle cell

Figure 19.2 Levels of organization within the vertebrate body.
Similar cell types operate together and form tissues. Tissues functioning together form organs. Several organs working together to carry out a function for the body are called an organ system. The circulatory system is an example of an organ system.

You are probably familiar with many of the major organs of a vertebrate body. Lungs are organs that terrestrial vertebrates use to extract oxygen from the air. Fish use gills to accomplish the same task from water. The stomach is an organ that digests food, and the liver an organ that controls the level of sugar and other chemicals in the blood. Organs are the machines of the vertebrate body, each built from several different tissues and each doing a particular job. How many others can you name?

Organ Systems

An **organ system** is a group of organs that work together to carry out an important function. For example, the vertebrate digestive system is an organ system composed of individual organs that break up the food (beaks or teeth), pass the food to the stomach (esophagus), break down the food (stomach), absorb the food (intestine), and expel the solid residue (rectum). If all of these organs do their job right, the body obtains energy and necessary building materials from food. Figure 19.2 illustrates the relationship between cells, tissues, organs, and organ systems.

The vertebrate body contains 11 principal organ systems (figure 19.3):

1. **Skeletal.** Perhaps the most important feature of the vertebrate body is its bony internal skeleton. The skeletal system protects the body and provides support for locomotion and movement. Its principal components are bones, skull, cartilage, and ligaments. Like arthropods, vertebrates have jointed appendages—the arms, hands, legs, and feet.

2. **Circulatory.** The circulatory system transports oxygen, nutrients, and chemical signals to the cells of the body and removes carbon dioxide, chemical wastes, and water. Its principal components are the heart, blood vessels, and blood.

3. **Endocrine.** The endocrine system coordinates and integrates the activities of the body. Its principal components are the pituitary, adrenal, thyroid, and other ductless glands.

4. **Nervous.** The activities of the body are coordinated by the nervous system. Its principal components are the nerves, sense organs, brain, and spinal cord.

5. **Respiratory.** The respiratory system captures oxygen and exchanges gases and is composed of the lungs, trachea, and other air passageways.

6. **Immune and lymphatic.** The immune system removes foreign bodies from the bloodstream using special cells, such as lymphocytes, macrophages, and antibodies. The lymphatic system provides vessels that transport extracellular fluid and fats to the circulatory system but also provides sites (lymph nodes and thymus, tonsils, and spleen) for the storage of immune cells.

7. **Digestive.** The digestive system captures soluble nutrients from ingested food. Its principal components are the mouth, esophagus, stomach, intestines, liver, and pancreas.

8. **Urinary.** The urinary system removes metabolic wastes from the bloodstream. Its principal components are the kidneys, bladder, and associated ducts.

9. **Muscular.** The muscular system produces movement, both within the body and of its limbs. Its principal components are skeletal muscle, cardiac muscle, and smooth muscle.

10. **Reproductive.** The reproductive system carries out reproduction. Its principal components are the testes in males, ovaries in females, and associated reproductive structures.

11. **Integumentary.** The integumentary system covers and protects the body. Its principal components are the skin, hair, nails, and sweat glands.

19.2 Groups of cells of the same type are organized in the vertebrate body into tissues. Organs are body structures composed of several different tissues. An organ system is a group of organs that work together to carry out an important function.

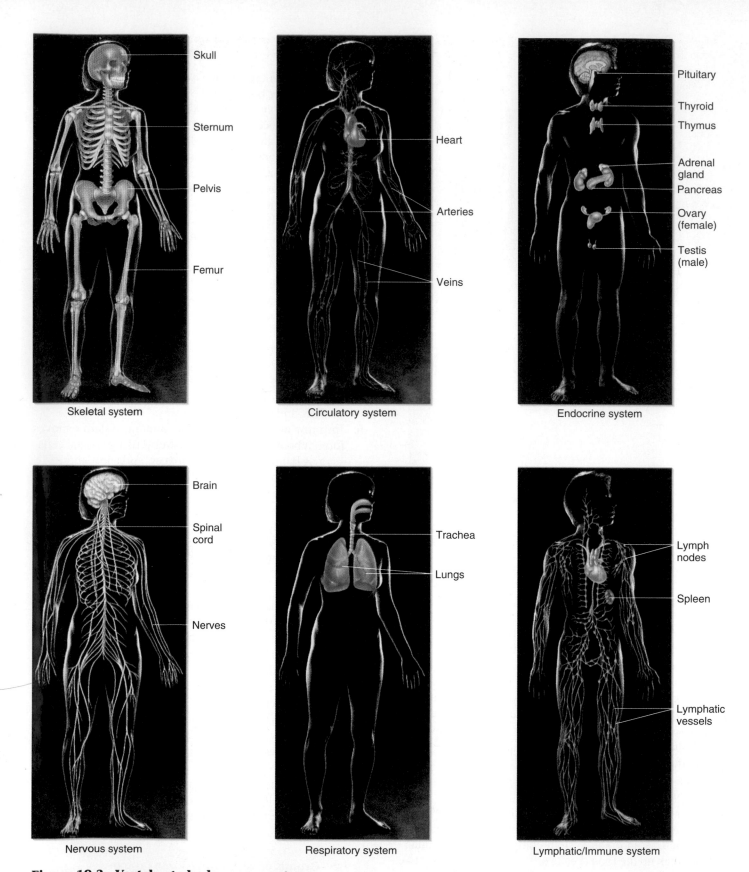

Figure 19.3 Vertebrate body organ systems.

The 11 principal organ systems of the human body are shown, including both male and female reproductive systems.

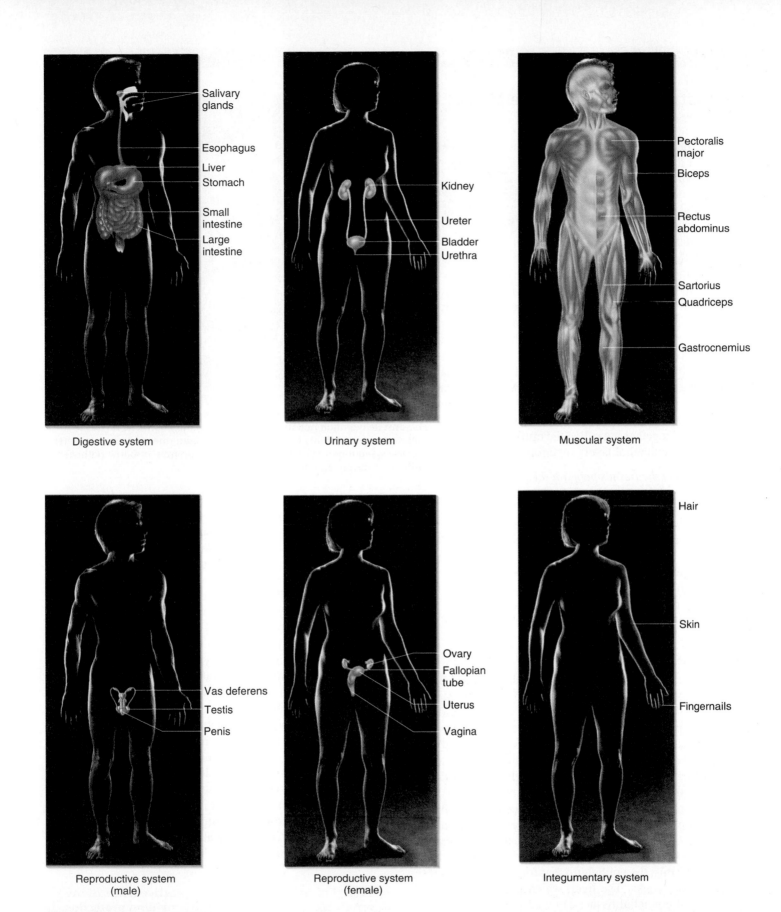

Digestive system

- Salivary glands
- Esophagus
- Liver
- Stomach
- Small intestine
- Large intestine

Urinary system

- Kidney
- Ureter
- Bladder
- Urethra

Muscular system

- Pectoralis major
- Biceps
- Rectus abdominus
- Sartorius
- Quadriceps
- Gastrocnemius

Reproductive system (male)

- Vas deferens
- Testis
- Penis

Reproductive system (female)

- Ovary
- Fallopian tube
- Uterus
- Vagina

Integumentary system

- Hair
- Skin
- Fingernails

Figure 19.3   (continued)

19.3 Epithelium Is Protective Tissue

As we have stated, the vertebrate body is composed of over 100 kinds of cells, traditionally grouped into four types of tissues: epithelial, connective, muscle, and nerve.

Epithelial cells are the guards and protectors of the body. They cover its surface and determine which substances enter it and which do not. The organization of the vertebrate body is fundamentally tubular, with one tube (the digestive tract) suspended inside another (the body cavity or coelom) like an inner tube inside a tire. The outside of the body is covered with cells (skin) that develop from embryonic *ectoderm* tissue; the body cavity is lined with cells that develop from embryonic *mesoderm* tissue; and the hollow inner core of the digestive tract (the gut) is lined with cells that develop from embryonic *endoderm* tissue. All three germ layers give rise to epithelial cells. Although different in embryonic origin, all epithelial cells are broadly similar in form and function and together are called the **epithelium.**

The body's epithelial layers function in three ways:

1. They *protect the tissues beneath them* from dehydration (water loss) and mechanical damage. Because epithelium encases all the body's surfaces, every substance that enters or leaves the body must cross an epithelial layer (figure 19.4).
2. They *provide sensory surfaces.* Many of a vertebrate's sense organs are in fact modified epithelial cells.
3. They *secrete materials.* Most secretory glands are derived from pockets of epithelial cells that pinch together during embryonic development.

Types of Epithelial Cells and Epithelial Tissues

Epithelial cells are classified into three types according to their shapes: squamous, cuboidal, and columnar. Layers of epithelial tissue are usually only one or a few cells thick. Individual epithelial cells possess only a small amount of cytoplasm and have a relatively low metabolic rate. A characteristic of all epithelia is that sheets of cells are tightly bound together, with very little space between them. This forms the barrier that is key to the functioning of the epithelium.

Epithelium possesses remarkable regenerative abilities. The cells of epithelial layers are constantly being replaced throughout the life of the organism. The cells lining the digestive tract, for example, are continuously replaced every few days. The epidermis, the epithelium that forms the skin, is renewed every two weeks. The liver, which is a football-sized gland formed of epithelial tissue, can readily regenerate substantial portions of itself removed during surgery.

There are two general kinds of epithelial tissue. First, the membranes that line the lungs and the major cavities of the

Figure 19.4   The epithelium prevents dehydration.
The tough, scaly skin of this gila monster provides a layer of protection against dehydration and injury. For all land-dwelling vertebrates, the relative impermeability of the surface epithelium (the epidermis) to water offers essential protection from dehydration and from airborne pathogens (disease-causing organisms).

body are a **simple epithelium** only a single cell layer thick. You can see why—these are surfaces across which many materials must pass, entering and leaving the body's compartments, and it is important that the "road" into and out of the body not be too long. Second, the skin, or epidermis, is a **stratified epithelium** composed of more complex epithelial cells several layers thick. Several layers are necessary to provide adequate cushioning and protection and to enable the skin to continuously replace its cells. Table 19.2 summarizes the characteristics of these two types of epithelial tissues.

A type of simple epithelial tissue that has a secretory function is cuboidal epithelium, which is found in the **glands** of the body. Endocrine glands secrete hormones into the blood. Exocrine glands (those with ducts that open to the body's outside) secrete sweat, milk, saliva, and digestive enzymes out of the body. Exocrine glands also secrete digestive enzymes into the stomach. If you think about it, the stomach and digestive tract are *outside* the body, because they are the inner canal that passes right through the body. It is possible for a substance to pass all the way through this digestive tract, from mouth to anus, and never enter the body at all. A substance must cross an epithelial layer to truly enter the body.

> **19.3**   Epithelial tissue is the protective tissue of the vertebrate body. In addition to providing protection and support, vertebrate epithelial tissues provide sensory surfaces and secrete key materials.

TABLE 19.2   EPITHELIAL TISSUE

Tissue		Typical Location	Tissue Function

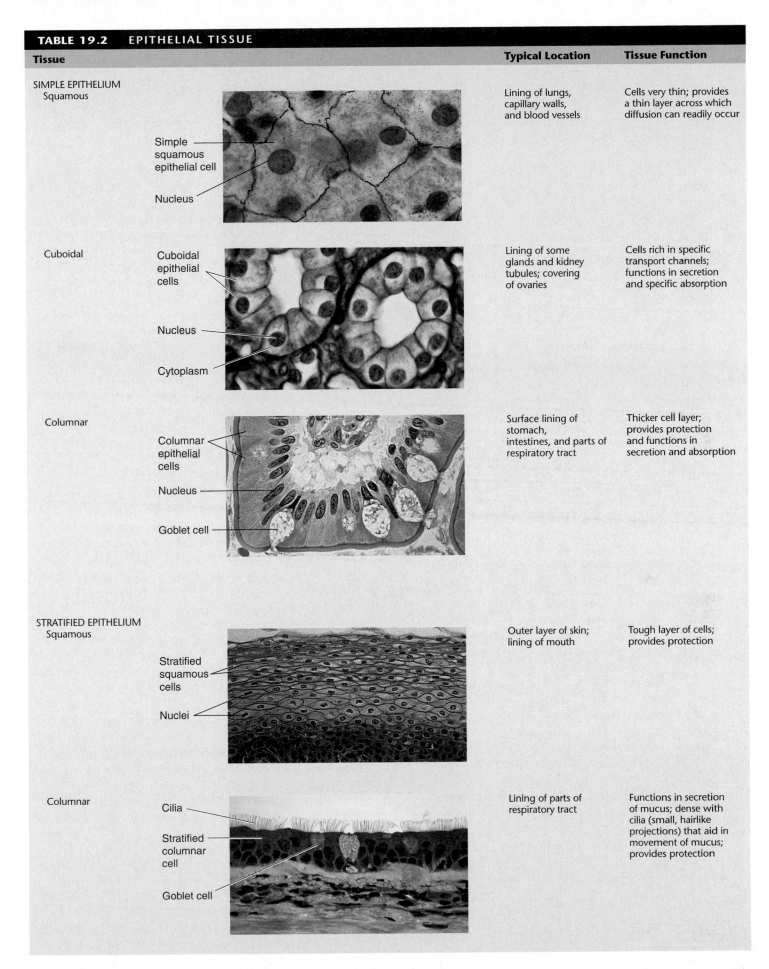

SIMPLE EPITHELIUM
Squamous

Simple squamous epithelial cell

Nucleus

Lining of lungs, capillary walls, and blood vessels

Cells very thin; provides a thin layer across which diffusion can readily occur

Cuboidal

Cuboidal epithelial cells

Nucleus

Cytoplasm

Lining of some glands and kidney tubules; covering of ovaries

Cells rich in specific transport channels; functions in secretion and specific absorption

Columnar

Columnar epithelial cells

Nucleus

Goblet cell

Surface lining of stomach, intestines, and parts of respiratory tract

Thicker cell layer; provides protection and functions in secretion and absorption

STRATIFIED EPITHELIUM
Squamous

Stratified squamous cells

Nuclei

Outer layer of skin; lining of mouth

Tough layer of cells; provides protection

Columnar

Cilia

Stratified columnar cell

Goblet cell

Lining of parts of respiratory tract

Functions in secretion of mucus; dense with cilia (small, hairlike projections) that aid in movement of mucus; provides protection

19.4 Connective Tissue Supports the Body

The cells of connective tissue provide the vertebrate body with its structural building blocks and also with its most potent defenses. Derived from the mesoderm, these cells are sometimes densely packed together, and sometimes widely dispersed, just as the soldiers of an army are sometimes massed together in a formation and sometimes widely scattered as guerrillas. **Connective tissue** cells fall into three functional categories: (1) the cells of the immune system, which act to defend the body; (2) the cells of the skeletal system, which support the body; and (3) the blood and fat cells, which store and distribute substances throughout the body. The grouping of these very diverse types of cells may seem odd but all connective tissues do share a common structural feature: They all have abundant extracellular material, known as the matrix, between widely spaced cells.

Immune Connective Tissue

The cells of the immune system roam the body within the bloodstream. They are mobile hunters of invading microorganisms and cancer cells. The two principal kinds of immune system cells (called white blood cells) are **macrophages,** which engulf and digest invading microorganisms, and **lymphocytes**, which make antibodies or attack virus-infected cells. Immune cells are carried through the body in a fluid matrix, called plasma, that will be discussed later.

Skeletal Connective Tissue

Three kinds of connective tissue are the principal components of the skeletal system: fibroblasts, cartilage, and bone. Although composed of similar cells, they differ in the nature of the material, the matrix, that is laid down between individual cells.

1. **Fibroblasts.** The most common kind of connective tissue in the vertebrate body consists of flat, irregularly branching cells called fibroblasts that secrete structurally strong proteins into the spaces between the cells. The many types of proteins give tissues different strengths. The most commonly secreted protein, collagen, is the most abundant protein in the human body—in fact, one-quarter of all the protein in your body is collagen! Fibroblasts are active in wound healing; scar tissue, for example, possesses a collagen matrix.

2. **Cartilage.** In cartilage, the collagen matrix between cells forms in long parallel arrays along the lines of mechanical stress. What results is a firm and flexible tissue of great strength, just as strands of nylon molecules laid down in long, parallel arrays produce strong, flexible ropes. Cartilage makes up the entire skeletal system of the modern agnathans and cartilaginous fishes. In most adult vertebrates, however, cartilage is restricted to the articular (joint) surfaces of bones that form freely movable joints and to other specific locations.

3. **Bone.** Bone is similar to cartilage, except that the collagen fibers are coated with a calcium phosphate salt, making the tissue rigid. The structure of bone and the way it is formed are discussed shortly.

Storage and Transport Connective Tissue

The third general class of connective tissue is made up of cells that are specialized to accumulate and transport particular molecules. They include the fat-accumulating cells of **adipose tissue.** They also include red blood cells, called **erythrocytes.** About 5 billion erythrocytes are present in every milliliter of your blood. Erythrocytes transport oxygen and carbon dioxide in blood. They are unusual in that during their maturation they lose most of their organelles, including the nucleus, mitochondria, and endoplasmic reticulum. Instead, occupying the interior of each erythrocyte are about 300 million molecules of hemoglobin, the protein that carries oxygen.

The fluid, or **plasma,** in which erythrocytes move is both the "banquet table" and the "refuse heap" of the vertebrate body. Practically every substance used by cells is dissolved in plasma, including inorganic salts like sodium and calcium, body wastes, and food molecules like sugars, lipids, and amino acids. Plasma also contains a wide variety of proteins, including antibodies and albumin, which gives the blood its viscosity. The various types of connective tissue are summarized in table 19.3.

A Closer Look at Bone

The vertebrate endoskeleton is strong because of the structural nature of bone. Bone is produced by coating collagen fibers with a calcium phosphate salt; the result is a material that is strong without being brittle. To understand how coating collagen fibers with calcium salts makes such an ideal structural material, consider fiberglass. Fiberglass is composed of glass fibers embedded in epoxy glue. The individual fibers are rigid, giving great strength, but they are also brittle. The epoxy glue, on the other hand, is flexible but weak. The composite, fiberglass, is both strong and flexible because when stress causes an individual fiber to break, the crack runs into glue before it reaches another fiber. The glue distorts and reduces the concentration of the stress—in effect, the glue spreads the stress over many fibers.

The construction of bone is similar to that of fiberglass: Small, needle-shaped crystals of a calcium phosphate mineral, hydroxyapatite, surround and impregnate collagen fibers within bone. No crack can penetrate far into bone because any stress that breaks a hard hydroxyapatite crystal passes into the collagenous matrix, which dissipates the stress. The hydroxyapatite mineral provides rigidity, whereas the collagen "glue" provides flexibility.

Most of us think of bones as solid and rocklike. But actually, bone is a dynamic tissue that is constantly being reconstructed. The outer layer of bone is very dense and compact and so is called **compact bone.** The interior is less compact, with a more open lattice structure, and is called **spongy bone.** Red blood cells form in the red marrow of spongy bone. New

TABLE 19.3 **CONNECTIVE TISSUE**

Tissue		Typical Location	Tissue Function	Characteristic Cell Types
IMMUNE White blood cells	White blood cell / Invading micro-organism .15 μm	Circulatory system	Attack invading microorganisms and virus-infected cells	Macrophages; lymphocytes; mast cells
SKELETAL Fibroblasts Loose	Loose connective tissue (fibroblasts)	Beneath skin and other epithelial tissues	Support; provide a fluid reservoir for epithelium	Fibroblasts
Dense		Tendons; sheath around muscles; kidney; liver; dermis of skin	Provide flexible, strong connections	Fibroblasts
Elastic		Ligaments; large arteries; lung tissue; skin	Enable tissues to expand and then return to normal size	Fibroblasts
Cartilage	Chondrocytes (cartilage cells)	Spinal disks; knees and other joints; ear; nose; tracheal rings	Provides flexible support; functions in shock absorption and reduction of friction on load-bearing surfaces	Chondrocytes (specialized fibroblast-like cells)
Bone		Most of skeleton	Protects internal organs; provides rigid support for muscle attachment	Osteocytes (specialized fibroblast-like cells)
STORAGE AND TRANSPORT Red blood cells	Red blood cells	In plasma	Transport oxygen	Red blood cells (erythrocytes)
Adipose tissue	Adipose tissue	Beneath skin	Stores fat	Specialized fibroblasts (adipocytes)

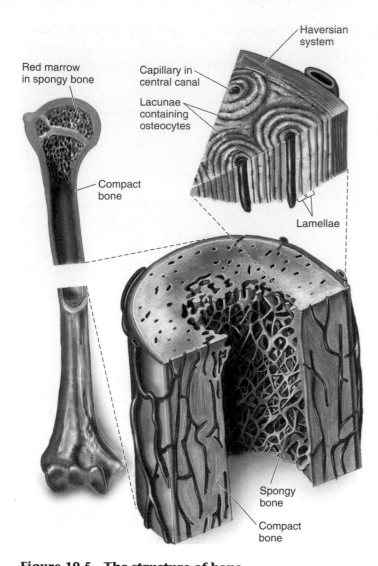

Figure 19.5   The structure of bone.

Some parts of bones are dense and compact, giving the bone strength. Other parts are spongy, with a more open lattice; red blood cells form in the bone marrow.

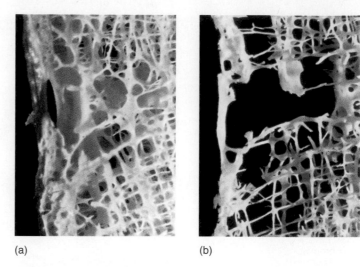

(a)   (b)

Figure 19.6   Osteoporosis.

Common in older women, osteoporosis is a bone disorder in which bones progressively lose minerals. (*a*) Normal bone tissue, in which the rate of bone deposit by osteoblasts equals the rate of bone withdrawal by osteoclasts. (*b*) Advanced osteoporosis, in which an excess of osteoclast activity over a considerable period of time has led to significant bone loss.

bone is formed in two stages: First, collagen is secreted by cells called **osteoblasts,** which lay down a matrix of fibers along lines of stress. Then calcium minerals impregnate the fibers. Bone is laid down in thin, concentric layers, like layers of paint on an old pipe. The layers form as a series of tubes around a narrow central channel called a **central canal,** also called a *Haversian canal,* which runs parallel to the length of the bone (figure 19.5). The many central canals within a bone, all interconnected, contain blood vessels and nerves that provide a lifeline to its living, bone-forming cells.

When bone is first formed in the embryo, osteoblasts use the cartilage skeleton as a template for bone formation. During childhood, bones grow actively. The total bone mass in a healthy young adult, by contrast, does not change much from one year to the next. This does not mean change is not occurring. Large amounts of calcium and thousands of *osteocytes* (former osteoblasts) are constantly being removed and replaced, but total bone mass does not change because deposit and removal take place at about the same rate.

Two cell types are responsible for this dynamic bone "remodeling": *osteoblasts* deposit bone, and *osteoclasts* secrete enzymes that digest the organic matrix of bone, liberating calcium for reabsorption by the bloodstream. The dynamic remodeling of bone adjusts bone strength to workload, new bone being formed along lines of stress. When a bone is subjected to compression, mineral deposition by osteoblasts exceeds withdrawals by osteoclasts. That is why long-distance runners must slowly increase the distances they attempt, to allow their bones to strengthen along lines of stress; otherwise, stress fractures can cripple them.

As a person ages, the backbone and other bones tend to decline in mass. Excessive bone loss is a condition called **osteoporosis** (figure 19.6). After the onset of osteoporosis, the replacement of calcium and other minerals lags behind withdrawal, causing the bone tissue to gradually erode. Eventually the bones become brittle and easily broken.

The best defense against osteoporosis is to maximize calcium deposition in bone during the years when active bone remodeling is occurring. High levels of dietary calcium consumed in the first decades of life stimulate osteoblasts to lay down more minerals, producing dense bones. Later in life, when osteoblast activity decreases and mineral withdrawal predominates, there is a much higher mineral reservoir from which to draw.

> **19.4   Connective tissues support the vertebrate body and consist of cells embedded in an extracellular matrix. They include cells of the immune system, cells of the skeletal system, and cells found throughout the body like blood and fat cells. Bone is a type of connective tissue.**

19.5 Muscle Tissue Lets the Body Move

Muscle cells are the motors of the vertebrate body. The distinguishing characteristic of muscle cells, the thing that makes them unique, is the abundance of contractible protein fibers within them. These fibers, called **microfilaments,** are made of the proteins actin and myosin. Vertebrate cells have a fine network of these microfilaments, but muscle cells have many more than other cells. Crammed in like the fibers of a rope, they take up practically the entire volume of the muscle cell. When actin and myosin slide past each other, shortening the fibers, the muscle contracts. Like slamming a spring-loaded door, the shortening of all of these fibers together within a muscle cell can produce considerable force. The process of muscle contraction will be discussed later in this chapter.

The vertebrate body possesses three different kinds of muscle cells: **smooth muscle, skeletal muscle,** and **cardiac muscle** (table 19.4). In smooth muscle, the microfilaments are only loosely organized. In skeletal and cardiac muscle, the microfilaments are bunched together into fibers called *myofibrils.* Each myofibril contains many thousands of microfilaments, all aligned to provide maximum force when they simultaneously shorten. Skeletal and cardiac muscle are often called striated muscles because the alignment of so many microfilaments gives the muscle myofibril a banded appearance.

Smooth Muscle

Smooth muscle cells are long and spindle-shaped, each containing a single nucleus. However, the individual myofilaments are not aligned into orderly assemblies as they are in skeletal and cardiac muscles. Smooth muscle tissue is organized into

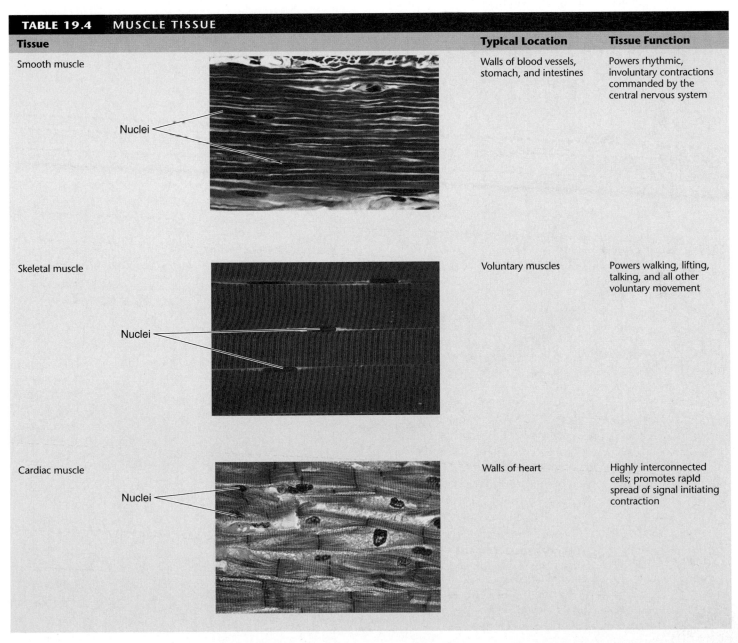

TABLE 19.4	MUSCLE TISSUE		
Tissue		**Typical Location**	**Tissue Function**
Smooth muscle	Nuclei	Walls of blood vessels, stomach, and intestines	Powers rhythmic, involuntary contractions commanded by the central nervous system
Skeletal muscle	Nuclei	Voluntary muscles	Powers walking, lifting, talking, and all other voluntary movement
Cardiac muscle	Nuclei	Walls of heart	Highly interconnected cells; promotes rapid spread of signal initiating contraction

sheets of cells. In some tissues, smooth muscle cells contract only when they are stimulated by a nerve or hormone. Examples are the muscles that line the walls of many blood vessels and those that make up the iris of the vertebrate eye. In other smooth muscle tissue, such as that found in the wall of the gut, the individual cells contract spontaneously, leading to a slow, steady contraction of the tissue.

Skeletal Muscle

Skeletal muscles move the bones of the skeleton. Skeletal muscle cells are produced during development by the fusion of several cells at their ends to form a very long fiber. Each of these muscle cell fibers still contains all the original nuclei, pushed out to the periphery of the cytoplasm. Each **muscle fiber** consists of many elongated **myofibrils,** and each myofibril is, in turn, composed of many myofilaments (figure 19.7). Furthermore, each myofilament contains the protein filaments actin and myosin. The key property of muscle cells is the relative abundance of actin and myosin within them, which enable a muscle cell to contract. These protein filaments are present as part of the cytoskeleton of all eukaryotic cells, but they are far more abundant and more highly organized in muscle cells.

Cardiac Muscle

The vertebrate heart is composed of striated muscle fibers arranged very differently from the fibers of skeletal muscle. Instead of very long, multinucleate cells running the length of the muscle, heart muscle is composed of chains of single cells, each with its own nucleus. Chains of cells are organized into fibers that branch and interconnect, forming a lattice-work. This lattice structure is critical to the way heart muscle functions. Each heart muscle cell is coupled to its neighbors electrically by tiny holes called *gap junctions* that pierce the plasma membranes in regions where the cells touch each other. Heart contraction is initiated at one location by the opening of transmembrane channels that conduct ions across the membrane. This changes the electrical properties of the membrane, causing it to depolarize. A wave of electrical depolarization then passes from cell to cell across the gap junctions, causing the heart to contract in an orderly pulsation.

> **19.5** Muscle tissue is the tool the vertebrate body uses to move its limbs, contract its organs, and pump the blood through its circulatory system.

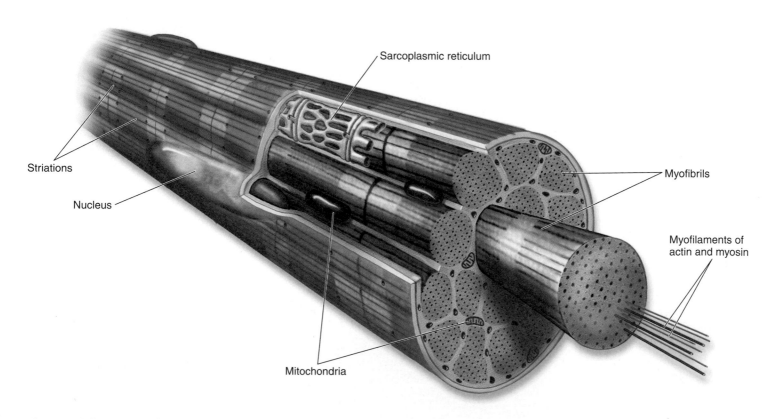

Striations

Nucleus

Mitochondria

Sarcoplasmic reticulum

Myofibrils

Myofilaments of actin and myosin

Figure 19.7 A muscle fiber, or muscle cell.
Each muscle is composed of bundles of muscle cells, or fibers. Each fiber is composed of many myofibrils, which are each, in turn, composed of myofilaments. Muscle cells have a modified endoplasmic reticulum called the sarcoplasmic reticulum that is involved in the regulation of calcium ions in muscles.

19.6 Nerve Tissue Conducts Signals Rapidly

Nerve cells carry information rapidly from one vertebrate organ to another. Nerve tissue, the fourth major class of vertebrate tissue, is composed of two kinds of cells: (1) **neurons,** which are specialized for the transmission of nerve impulses, and (2) supporting **glial cells,** which supply the neurons with nutrients, support, and insulation.

Neurons have a highly specialized cell architecture that enables them to conduct signals rapidly throughout the body. Their plasma membranes are rich in ion-selective channels that maintain a voltage difference between the interior and the exterior of the cell, the equivalent of a battery. When ion channels in a local area of the membrane open, ions flood in from the exterior, temporarily wiping out the charge difference. This process, called depolarization, tends to open nearby voltage-sensitive channels in the neuron membrane, resulting in a wave of electrical activity that travels down the entire length of the neuron as a nerve impulse.

Each neuron is composed of three parts: (1) a **cell body,** which contains the nucleus; (2) threadlike extensions called **dendrites,** which act as antennae, bringing nerve impulses to the cell body from other cells or sensory systems; and (3) a single, long extension called an **axon,** which carries nerve impulses away from the cell body (figure 19.8). Axons often carry nerve impulses for considerable distances: The axons that extend from the skull to the pelvis in a giraffe are about 3 meters long!

The body contains many different kinds of neurons (table 19.5). Some neurons are tiny and have only a few projections, others are bushy and have more projections, and still others have extensions that are meters long. Neurons are not normally in direct contact with one another. Instead, a tiny gap called a **synapse** separates them. Neurons communicate with other neurons by passing chemical signals called **neurotransmitters** across the gap. Special receptor proteins present on the far side of the synapse respond to the arrival of the neurotransmitters by starting a new nerve impulse in the receiving neuron.

Vertebrate nerves appear as fine white threads when viewed with the naked eye, but they are actually composed of bundles of axons. Like a telephone trunk cable, nerves include large numbers of independent communication channels—bundles composed of hundreds of axons, each connecting a nerve cell to a muscle fiber. In addition, the nerve contains numerous supporting glial cells bunched around the axons. It is important not to confuse a nerve with a neuron. A nerve is made up of the axons of many neurons, just as a cable is made of many wires.

> **19.6** Nerve tissue provides the vertebrate body with a means of communication and coordination.

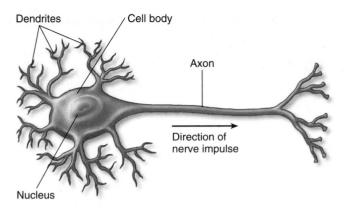

Figure 19.8   Neurons carry nerve impulses.
Neurons carry nerve impulses, which are electrical signals, from their initiation in dendrites, to the cell body, and down the length of the axon, where they may pass the signal to a neighboring cell.

TABLE 19.5	NERVE TISSUE		
Tissue		**Typical Location**	**Tissue Function**
Sensory neurons		Eyes; ears; surface of skin	Receive information about body's condition and external environment; send impulses from sensory receptors to CNS
Motor neurons		Brain and spinal cord	Stimulate muscles and glands; conduct impulses out of CNS toward muscles and glands
Association neurons		Brain and spinal cord	Integrate information; conduct impulses between neurons within CNS

19.7 Types of Skeletons

With muscles alone, the animal body could not move—it would simply pulsate as its muscles contracted and relaxed in futile cycles. For a muscle to produce movement, it must direct its force against another object. Animals are able to move because the opposite ends of their muscles are attached to a rigid scaffold, or **skeleton,** so that the muscles have something to pull against. There are three types of skeletal systems in the animal kingdom: hydraulic skeletons, exoskeletons, and endoskeletons.

Hydraulic skeletons are found in soft-bodied invertebrates such as earthworms and jellyfish. In this case, a fluid-filled cavity is encircled by muscle fibers that raise the pressure of the fluid when they contract. In an earthworm, for example, a wave of contractions of circular muscles begins anteriorly and compresses the body, so that the fluid pressure pushes it forward (figure 19.9). Contractions of longitudinal muscles then pull the rest of the body.

Figure 19.10   Crustaceans have an exoskeleton.

The exoskeleton of this rock crab is bright orange.

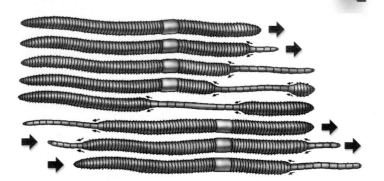

Figure 19.9   Earthworms have a hydraulic skeleton.

When an earthworm's circular muscles contract, the internal fluid presses on the longitudinal muscles, which then stretch to elongate segments of the earthworm. A wave of contractions down the body of the earthworm produces forward movement.

Exoskeletons surround the body as a rigid hard case to which muscles attach internally. Arthropods, such as crustaceans (figure 19.10) and insects, have exoskeletons made of the polysaccharide *chitin.* An animal with an exoskeleton cannot get too large because its exoskeleton would have to become thicker and heavier to prevent collapse. If an insect were the size of an elephant, its exoskeleton would have to be so thick and heavy it would hardly be able to move.

Endoskeletons, found in vertebrates and echinoderms, are rigid internal skeletons to which muscles are attached. Vertebrates have a soft, flexible exterior that stretches to accommodate the movements of their skeleton. The endoskeleton of vertebrates is composed of bone (figure 19.11). Unlike chitin, bone is a cellular, living tissue capable of growth, self-repair, and remodeling in response to physical stresses.

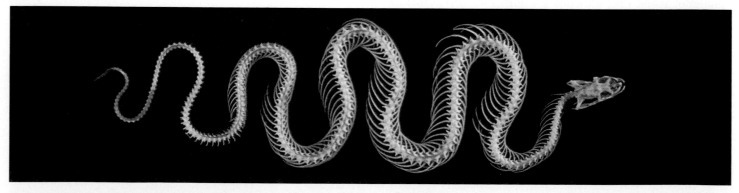

Figure 19.11   Snakes have an endoskeleton.

The endoskeleton of most vertebrates is made of bone. A snake's skeleton is specialized for quick lateral movement because it has numerous vertebrae and ribs and no pelvis (pelvic girdle is vestigial in some snakes).

A Vertebrate Endoskeleton: The Human Skeleton

The human skeleton is made up of 206 individual bones. If you saw them as a pile of bones jumbled together, it would be hard to make any sense of them. To understand the skeleton, it is necessary to group the 206 bones according to their function and position in the body. The 80 bones of the **axial skeleton** support the main body axis, while the remaining 126 bones of the **appendicular skeleton** support the arms and legs (figure 19.12). These two skeletons function more or less independently—that is, the muscles controlling the axial skeleton (postural muscles) are managed by the brain separately from those controlling the appendages (manipulatory muscles).

The Axial Skeleton

The axial skeleton is made up of the skull, backbone, and rib cage. Of the skull's 28 bones, only 8 form the cranium, which encases the brain; the rest are facial bones and middle ear bones. An additional bone, the hyoid bone, supports the tongue but is not really part of the skull.

The skull is attached to the upper end of the backbone, which is also called the *spine,* or **vertebral column.** The spine is made up of 26 vertebrae, stacked one on top of the other to provide a flexible column surrounding and protecting the spinal cord. Curving forward from the vertebrae are 12 pairs of ribs, attached at the front to the breastbone, or sternum, and forming a protective cage around the heart and lungs.

The Appendicular Skeleton

The 126 bones of the appendicular skeleton are attached to the axial skeleton at the shoulders and hips. The shoulder, or **pectoral girdle,** is composed of two large, flat shoulder blades, each connected to the top of the breastbone by a slender, curved collarbone. The arms are attached to the pectoral girdle; each arm and hand contains 30 bones. The collarbone is the most frequently broken bone of the body. Can you guess why? Because if you fall on an outstretched arm, a large component of the force is transmitted to the collarbone.

The **pelvic girdle** forms a bowl that provides strong connections for the legs, which must bear the weight of the body. Each leg and foot contains a total of 30 bones.

Joints, points where two bones come together, confer flexibility to the rigid endoskeleton, allowing a range of motion determined by the type of joint. Slightly movable joints bridged by cartilage allow the bones of the axial skeleton's spine some movement, whereas the freely movable joints of the appendicular skeleton's limbs are encased within fibrous capsules filled with a lubricating fluid.

19.7 **The animal skeletal system provides a framework against which the body's muscles can pull. Many soft-bodied invertebrates employ a hydraulic skeleton, whereas arthropods have a rigid, hard exoskeleton surrounding their body. Echinoderms and vertebrates have an internal endoskeleton to which muscles attach.**

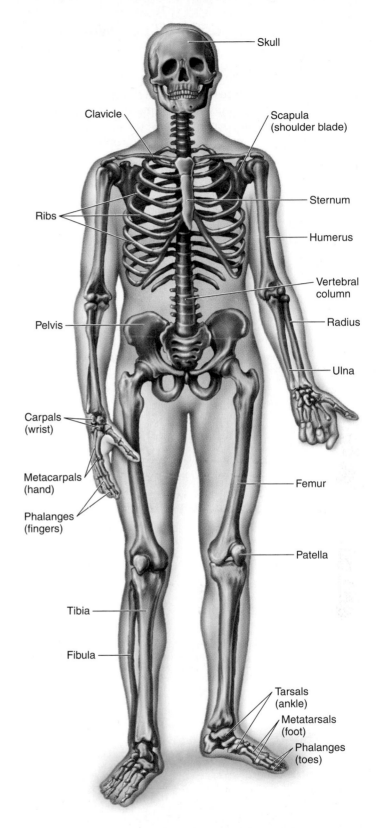

Figure 19.12 Axial and appendicular skeletons.
The axial skeleton is shown in *purple,* and the appendicular skeleton is shown in *tan.*

Muscles and How They Work

Three kinds of muscle together form the vertebrate muscular system. As we have discussed, the vertebrate body is able to move because *skeletal muscles* pull the bones with considerable force. The heart pumps because of the contraction of *cardiac muscle.* Food moves through the intestines because of the rhythmic contractions of *smooth muscle.*

Actions of Skeletal Muscle

Skeletal muscles move the bones of the skeleton. Some of the major human muscles are shown in figure 19.13. Muscles are attached to bones by straps of dense connective tissue called **tendons.** Bones pivot about flexible connections called *joints,* pulled back and forth by the muscles attached to them. Each muscle pulls on a specific bone. One end of the muscle, the *origin,* is attached by a tendon to a bone that remains stationary during a contraction. This provides an object against which the muscle can pull. The other end of the muscle, the *insertion,* is attached to a bone that moves if the muscle contracts.

Muscles can only pull, not push, because their myofibrils contract rather than expand. For this reason, the muscles in the movable joints of vertebrate are attached in opposing pairs, called flexors and extensors, which when contracted, move the bones in different directions. When the **flexor** muscle of your upper leg contracts, the lower leg is moved closer to the thigh. When the **extensor** muscle of your upper leg contracts, the lower leg is moved in the opposite direction, farther away (figure 19.14).

Just as there are two types of skeletal muscles, so there are also two types of muscle contractions, isotonic and isometric contractions. In **isotonic contractions,** the muscle shortens, moving the bones as just described. In **isometric contractions,** a force is exerted by the muscle, but the muscle does not shorten. This occurs when you try to lift something very heavy. Eventually, if your muscles generate enough force and you are able to lift the object, the isometric contraction becomes isotonic.

Muscle Contraction

Recall from figure 19.7 that myofibrils are composed of bundles of myofilaments. Far too fine to see with the naked eye, the individual myofilaments of vertebrate muscles are only 6 nanometers thick. Each is composed of long, threadlike filaments of the proteins actin and myosin. An **actin filament** consists of two strings of actin molecules wrapped around one another, like two strands of pearls loosely wound together. A **myosin filament** is also composed of two strings of protein wound about each other, but a myosin filament is 10 times longer than an actin filament, and the myosin strings have a

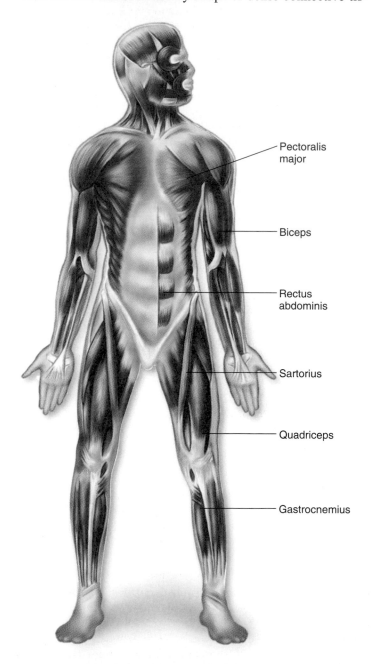

Figure 19.13   The muscular system.

Some of the major muscles in the human body are labeled.

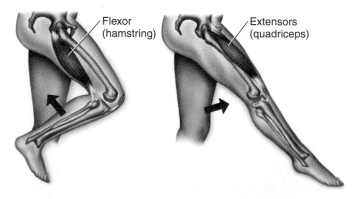

Figure 19.14   Flexor and extensor muscles.

Limb movement is always the result of muscle contraction, never muscle extension. Muscles that retract limbs are called flexors; those that extend limbs are called extensors. Thus, you bend your leg back by contracting your hamstring muscle, a flexor muscle, and you extend your leg by contracting your quadriceps, an extensor muscle.

Author's Corner

Running Improperly Provides a Painful Lesson in the Biology of Bones and Muscles

No one seeing the ring of fat decorating my middle would take me for a runner. Only in my memory do I get up with the robins, lace on my running shoes, bounce out the front door, and run the streets around Washington University before going to work. Now my 5-K runs are 25-year-old memories. Any mention I make of my running in a race only evokes screams of laughter from my daughters, and an arch look from my wife. Memory is cruelest when it is accurate.

I remember clearly the day I stopped running. It was a cool fall morning in 1978, and I was part of a mob running a 5 K (that's 5 kilometers for the uninitiated) race, winding around the hills near the university. I started to get flashes of pain in my legs below the knees—like shin splints, but much worse.

Imagine fire pouring on your bones. Did I stop running? No. Like a bonehead I kept going, "working through the pain," and finished the race. I have never run a race since.

I had pulled a muscle in my thigh, which caused part of the pain. But that wasn't all. The pain in my lower legs wasn't shin splints, and didn't go away. A trip to the doctor revealed compound stress fractures in both legs. The X rays of my legs looked like tiny threads had been wrapped around the shaft of each bone, like the red stripe on a barber's pole. It was summer before I could walk without pain.

What went wrong? Isn't running supposed to be GOOD for you? Not if you run improperly. In my enthusiasm to be healthy, I ignored some simple rules and paid the price. The biology lesson I ignored had to do with how bones grow. The long bones of your legs are not made of stone, solid and permanent. They are dynamic structures, constantly being re-formed and strengthened in response to the stresses to which you subject them.

To understand how bone grows, we first need to know a bit about what bone is like. Bone is made of fibers of a flexible protein called collagen stuck together to form cartilage. While an embryo, all your bones are made of cartilage. As your adult body develops, the collagen fibers become impregnated with tiny needle-shaped crystals of calcium phosphate, turning the cartilage into bone. The

crystals are brittle but rigid, giving bone great strength. Collagen is flexible but weak, but like the epoxy of fiberglass, it acts to spread any stress over many crystals, making bone resistant to fracture. As a result, bone is both strong and flexible.

When you subject a bone in your body to stress—say, by running—the bone grows so as to withstand the greater workload. How does the bone "know" just where to add more material? When stress deforms the collagen fibers of a leg bone, the interior of the collagen fibers becomes exposed, like opening your jacket and exposing your shirt. The fiber interior has a minute electrical charge. Cells called fibroblasts are attracted to the electricity like bugs to night lights, and secrete more collagen there. As a result, new collagen fibers are laid down on a bone along the lines of stress. Slowly, over months, calcium phosphate crystals convert the new collagen to new bone. In your legs, the new bone forms along the long stress lines that curve down along the shank of the bone.

Now go back 25 years, and visualize me pounding happily down the concrete pavement each morning. I had only recently begun to run on the sidewalk, and for an hour or more at a stretch. Every stride I took those mornings was a blow to my shinbones, a stress to which my bones no doubt began to respond by forming collagen along the spiral lines of stress. Had I run on a softer surface, the daily stress would have been far less severe. Had I gradually increased my running, new bone would have had time to form properly in response to the added stress. I gave my leg bones a lot of stress, and no time to respond to it. I pushed them too hard, too fast, and they gave way.

Nor was my improper running limited to overstressed leg bones. Remember that pulled thigh muscle? In my excessive enthusiasm, I never warmed up before I ran. I was having too much fun to worry about such details. Wiser now, I am sure the pulled thigh muscle was a direct result of failing to properly stretch before running.

I was reminded of that pulled muscle recently, listening to a good friend of my wife's describe how she sets out early each morning for a long run without stretching or warming up. I can see her in my mind's eye, bundled up warmly on the cooler mornings, an enthusiastic gazelle pounding down the pavement in search of health. Unless she uses more sense than I did, she may fail to find it.

very unusual shape. One end of a myosin filament consists of a very long rod, while the other end consists of a double-headed globular region, or "head." In electron micrographs, a myosin filament looks like a two-headed snake. This odd structure is the key to how muscles work.

How Myofilaments Contract

Look at the diagram of myosin and actin in a myofilament (figure 19.15), and focus on the myosin heads. When a myofilament contracts, the heads of the myosin filaments move first. Like flexing your hand downward at the wrist, the heads bend backward and inward. This moves them closer to their rodlike backbones and several nanometers in the direction of the flex. In itself, this myosin head-flex accomplishes nothing—but the myosin head is attached to the actin filament! As a result, the actin filament is pulled along with the myosin head as it flexes, causing the actin filament to slide by the myosin filament in the direction of the flex. As one after another myosin head flexes, the myosin in effect "walks" step by step along the actin. Each step uses a molecule of ATP to recock the myosin head before each flex.

How does this sliding of actin past myosin lead to myofilament contraction and muscle cell movement? Within each myofilament, the actin is anchored at one end, at a position in striated muscle called the Z line (figure 19.16). Because it is tethered like this, the actin cannot simply move off. Instead, the actin pulls the anchor with it! As actin moves past myosin, it drags the Z line toward the myosin. The secret of muscle contraction is that each myosin is interposed between two pairs of actin filaments—the myofilament is attached at *both* ends to Z lines by actin. One moving to the left and the other to the right, the two pairs of actin molecules drag the Z lines toward each other as they slide past the myosin core. As the Z lines are pulled closer together, the plasma membranes to which they are attached move toward one another, and the cell contracts.

How Nerves Signal Muscles to Contract

In vertebrate skeletal muscle, contraction is initiated by a nerve impulse. The nerve fiber is embedded in the surface of the muscle fiber, forming a **neuromuscular junction.** When a signal reaches the end of a neuron, at the point where the neuron almost touches the muscle cell (the **motor end plate**), the neuron releases the chemical acetylcholine into the tiny gap separating neuron from muscle. Acetylcholine passes across the gap to the muscle plasma membrane, binds to receptor proteins there, and so causes ion channels in the motor end plate to open. The muscle membrane is said to be depolarized, because opening these channels allows ions to move freely in and out, wiping out any differences there might be in ion concentration between inside and outside.

Myofilament Contraction

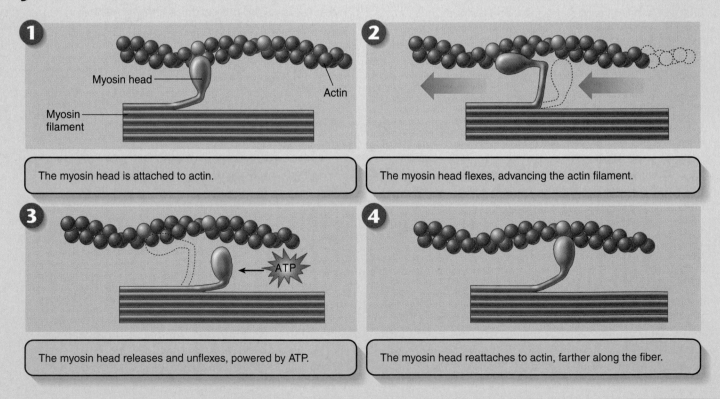

Figure 19.15 How myofilament contraction works.

The Sliding Filament Model of Muscle Contraction

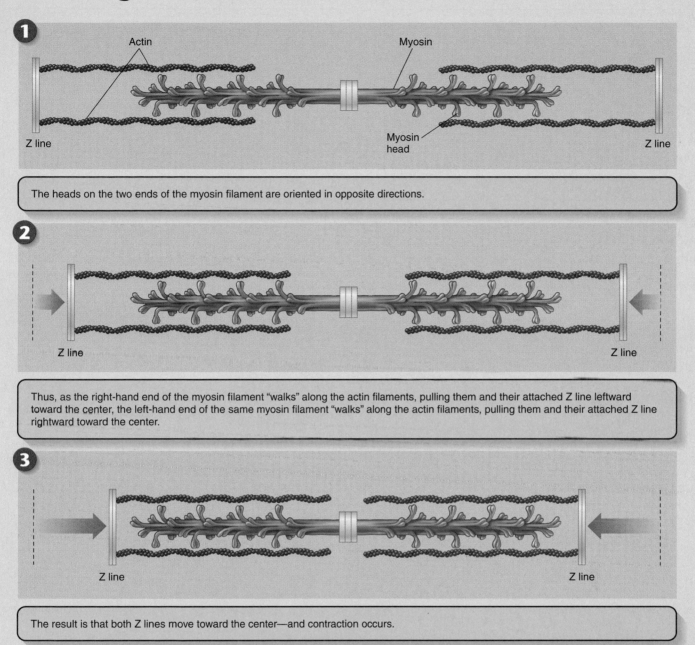

1

Actin   Myosin

Myosin head

Z line   Z line

The heads on the two ends of the myosin filament are oriented in opposite directions.

2

Z line   Z line

Thus, as the right-hand end of the myosin filament "walks" along the actin filaments, pulling them and their attached Z line leftward toward the center, the left-hand end of the same myosin filament "walks" along the actin filaments, pulling them and their attached Z line rightward toward the center.

3

Z line   Z line

The result is that both Z lines move toward the center—and contraction occurs.

Figure 19.16   How actin and myosin filaments interact.

The Role of Calcium Ions in Contraction

When a muscle is relaxed, its myosin heads are "cocked" and ready, through the splitting of ATP, but are unable to bind to actin. This is because the attachment sites for the myosin heads on the actin are physically blocked by another protein, known as **tropomyosin.** Myosin heads therefore cannot bind to actin in the relaxed muscle, and the filaments cannot slide.

To contract a muscle, the tropomyosin must be moved out of the way so that the myosin heads can bind to actin. This requires the function of **troponin,** a regulatory protein that binds to the tropomyosin. The troponin and tropomyosin form a complex that is regulated by the calcium ion (Ca^{++}) concentration of the muscle cell cytoplasm.

When the Ca^{++} concentration of the muscle cell cytoplasm is low, tropomyosin inhibits myosin binding, and the muscle is relaxed (figure 19.17). When the Ca^{++} concentration is raised, Ca^{++} binds to troponin. This causes the troponin-tropomyosin complex to be shifted away from the attachment sites for the myosin heads on the actin. When this repositioning has occurred, the myosin heads attach to actin and, using ATP energy, move along the actin in a stepwise fashion to shorten the myofibril.

Where does the Ca^{++} come from? Muscle fibers store Ca^{++} in a modified endoplasmic reticulum called the **sarcoplasmic reticulum** (figure 19.18). When a muscle fiber is stimulated to contract, Ca^{++} is released from the sarcoplasmic

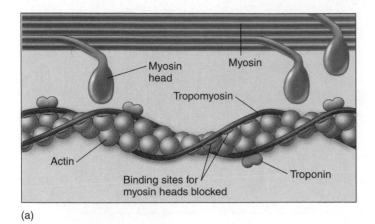

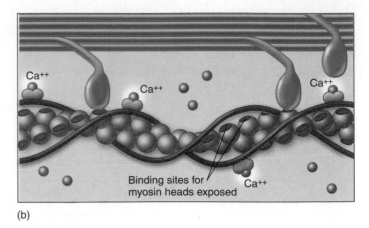

(a)   (b)

Figure 19.17   How calcium controls muscle contraction.

(a) When the muscle is at rest, a long filament composed of the molecule tropomyosin blocks the myosin binding sites of the actin molecule so that muscle contraction cannot occur. (b) When calcium ions bind to another protein, troponin, the resulting complex displaces the filament of tropomyosin, exposing the myosin binding sites of actin. Myosin heads can bind to the actin, and contraction occurs.

reticulum and diffuses into the myofibrils, where it binds to troponin and causes contraction. The contraction of muscles is regulated by nerve activity, and so nerves must influence the distribution of Ca^{++} in the muscle fiber.

> **19.8**   Muscles are made of many tiny threadlike filaments of actin and myosin called myofilaments. Muscles work by using ATP to power the sliding of myosin along actin, causing the myofilaments to contract.

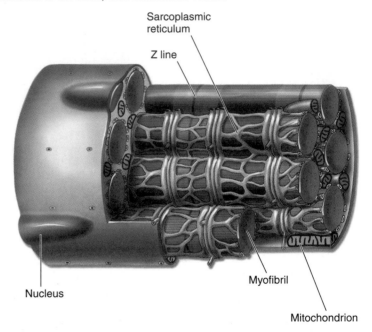

Figure 19.18   The sarcoplasmic reticulum.

The sarcoplasmic reticulum is a system of membranes that wraps around the individual myofibrils of a muscle fiber.

Exploring Current Issues

Additional Resources

Go to your campus library or look online to find the following articles, which further develop some of the concepts found in this chapter.

Andersen, J. L., P. Schjerling, and B. Saltin. (2000). Muscles, genes and athletic performance: the cellular biology of muscles explains why a particular athlete wins and suggests what future athletes might do to better their odds. *Scientific American,* 283(3), 48.

Constans, A. (2003). Body by science: tissue engineers advance in bid to build organs in the lab, but significant roadblocks remain. *The Scientist,* 17(19), 34.

Kreeger, K. Y. (2003). Hoping to mend their sporting ways: researchers turn to gene therapy, tissue engineering, and more, to heal sports stars' popped knees and torn ligaments. *The Scientist,* 17(1), 32.

Rosen, C. J. (2003). Restoring aging bones: the bone decay of osteoporosis can cripple, but an improved understanding of how the body builds and loses bone is leading to ever better prevention and treatment options. *Scientific American,* 288(3), 70.

Biology and Society Lecture: The Biology of Jogging

Jogging is an activity that many Americans do to get or stay in shape. Although moderate exercise is beneficial, two important things can go wrong if you get overly enthusiastic too soon: (1) Tendons attach muscles to bones, and while they are somewhat elastic, a good jerk can result in a "pulled muscle." Stretching before jogging helps alleviate this. (2) Bones are living tissue and grow like other tissues in the body. As you run, you create areas of stress on the bone, and the bone over time strengthens around areas of stress. Push too hard, too fast and bones give way, forming painful cracks called stress fractures.

Find this lecture, delivered by the author to his class at Washington University, online at www.mhhe.com/tlwessentials/exp19.

The Animal Body Plan

19.1 Innovations in Body Design

- Four innovations have been key to the diversity seen in the animal kingdom: radial versus bilateral symmetry, no body cavity versus a body cavity, nonsegmentation versus segmentation, and protostome versus deuterostome development (**table 19.1**).

19.2 Organization of the Vertebrate Body

- 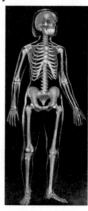 The animal body is composed of cells that exhibit specialization of function and intercellular communication. Animal bodies have increasing levels of structural and functional complexity. Cells with similar functions group together into tissues, which act as functional units (**figure 19.1**). Organs of the body are composed of several different tissues that act together to perform a higher level of function. Organs work together in an organ system to perform larger-scale body functions (**figures 19.2 and 19.3**).

Tissues of the Vertebrate Body

19.3 Epithelium Is Protective Tissue

- Epithelial tissue is composed of different types of epithelial cells (**table 19.2**). It covers surfaces of the body, both internal and external surfaces, providing protection. Many sensory surfaces and glands are lined with a layer of epithelium and, in these cases, the epithelial tissue has sensory or secretory functions.

19.4 Connective Tissue Supports the Body

- The connective tissues of the body are very diverse in structure and function, but all are composed of cells embedded in an extracellular matrix (**table 19.3**). The matrix may be hard, as in bone (**figure 19.5**), or may be fluid, as in blood. Connective tissues function in providing the body structural support, storage, and transportation of nutrients, gases, and immune cells.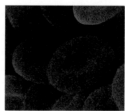

19.5 Muscle Tissue Lets the Body Move

- Muscle tissue is dynamic tissue; it contracts, causing the body to move. There are three types of muscle tissue: smooth, skeletal, and cardiac muscle (**table 19.4**). All three types of muscle contain actin and myosin microfilaments but differ in the organization of the microfilaments. Smooth muscle cells are organized into sheets of cells with little alignment of the cells. This results in muscle contractions that are less coordinated.

- Smooth muscle is found in the walls of blood vessels, the digestive system, and so forth. Skeletal muscle cells are aligned but independent of each other (**figure 19.7**). They contract as small units when stimulated by nerves. Skeletal muscle is attached to the skeleton, so when the muscle contracts, the skeleton moves. Cells of cardiac muscle, found in the heart, are aligned and interconnected so that they contract together as one unit.

19.6 Nerve Tissue Conducts Signals Rapidly

- Nerve tissue is composed of neurons and supporting glial cells (**table 19.5**). Neurons carry electrical impulses from one area of the body to another (**figure 19.8**), providing the body a means of communication and coordination.

The Skeletal and Muscular Systems

19.7 Types of Skeletons

- The skeletal system provides a framework on which muscles act to move the body. Soft bodied invertebrates have hydraulic skeletons, where muscles act on a fluid-filled cavity (**figure 19.9**). Arthropods have exoskeletons, where muscles attach from within to the hard outer covering of the body (**figure 19.10**). Vertebrates and echinoderms have endoskeletons, where muscles attach to bones or cartilage inside the body (**figures 19.11** and **19.12**).

19.8 Muscles and How They Work

- Skeletal muscles attach to bone at two points and cause the skeleton to move at joints. The end of the muscle that attaches to the stationary bone is called the origin. The muscle passes over a joint and attaches to another bone that moves, called the insertion. As the muscle contracts and shortens, the insertion is brought closer to the origin and the joint flexes. Muscles act to flex or extend a joint (**figure 19.14**).

- The contraction of a muscle is due to molecular forces between actin and myosin microfilament within the muscle cells. Actin and myosin filaments stack on top of each other,  creating an orderly array. Myosin attaches to actin and pulls the actin along its length, causing the microfilaments to slide past each other. Energy from ATP causes the microfilaments to dissociate and reattach, triggering the sliding motion again. This is called the sliding filament model (**figure 19.16**). This process is controlled by Ca^{++} concentrations in the muscle cells (**figures 19.17 and 19.18**).

1. One of the four innovations in animal body design, segmentation, allowed for
   a. development of efficient internal organ systems.
   b. more flexible movement as individual segments can move independently of each other.
   c. locating organs in different areas of the body.
   d. embryonic cells to not have a fixed fate.

2. Which of the following is the correct organization sequence from smallest to largest in animals?
   a. cells, tissues, organs, organ systems, organism
   b. organism, organ systems, organs, tissues, cells
   c. tissues, organs, cells, organ systems, organism
   d. organs, tissues, cells, organism, organ systems

3. Which of the following is *not* a function of the epithelial tissue?
   a. secrete materials
   b. provide sensory surfaces
   c. move the body
   d. protect underlying tissue from damage and dehydration
4. An example of connective tissue is
   a. nerve cells in your fingers.
   b. skin cells.
   c. brain cells.
   d. red blood cells.
5. When a person has osteoporosis, the work of _____ falls behind the work of _____.
   a. osteoclasts, osteoblasts
   b. osteoclasts, collagen
   c. osteoblasts, osteoclasts
   d. osteoblasts, collagen
6. Nerve impulses jump between nerve cells through the use of
   a. hormones.
   b. neurotransmitters.
   c. pheromones.
   d. calcium ions.
7. The type of muscle used for voluntary body motion is
   a. skeletal.
   b. cardiac.
   c. smooth.
   d. squamous.

8. The vertebral column is part of the
   a. appendicular skeleton.
   b. axial skeleton.
   c. hydrostatic skeleton.
   d. exoskeleton.
9. Movement requires a pair of muscles because
   a. a single muscle can only pull and not push.
   b. a single muscle can only push and not pull.
   c. both muscles are required to move a limb in any single direction.
   d. No answer is correct.
10. The role of calcium in the process of muscle contraction is to
   a. gather ATP for the myosin to use.
   b. cause the myosin head to shift position, contracting the myofibril.
   c. cause the myosin head to detach from the actin, causing the muscle to relax.
   d. expose myosin attachment sites on actin.

Visual Understanding

1. **Figure 19.6a** Normal bone is composed of an open lattice framework of minerals, including calcium. Why wouldn't it be more sensible to have bones be almost solid mineral, and be more resistant to breakage?

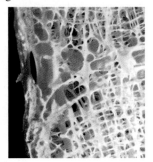

2. **Figure 19.12** The bones of your head and trunk form three cagelike structures; these are the skull, the rib cage, and the pelvic girdle. What is the importance of these structures?

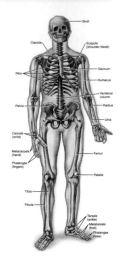

Challenge Questions

The Animal Body Plan Your friends are having a friendly debate about which organ systems are the most important to them. Lucas says it's his muscles and bones so that he can play soccer. Lakshmi insists that her nervous system and respiratory system are the most important so that she can think, feel, and talk. Joseph is sure that it's his digestive system; otherwise he couldn't enjoy foods like pizza, tacos, hamburgers, and curry. Pick one or two systems and explain why you think they are the most important to your life.

Tissues of the Vertebrate Body Imagine that you are designing a living organism, some type of vertebrate. Explain briefly how the four types of tissue are all necessary to your design.

The Skeletal and Muscular Systems When you are exercising rapidly, such as playing tennis, dancing to fast music, or doing aerobics, you begin to breathe rapidly and your heart rate increases. If you continue, you might even say that you are "out of breath." Why is rapid breathing and heart rate important when you are giving your muscles a lot of work to do?

Online Learning Center

Visit the Online Learning Center for this chapter at www.mhhe.com/tlwessentials/ch19 for quizzes, animations, interactive learning exercises, and other study tools. At the site you will also find extended answers to the end-of-chapter questions.

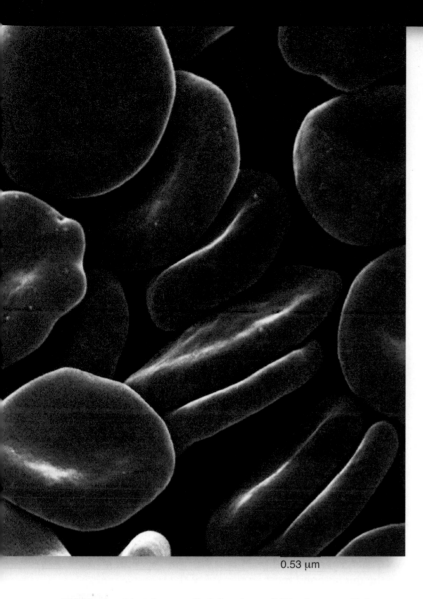

0.53 µm

20

Circulation and Respiration

B lood has been called the river of life. Among all the vertebrate body's tissues, blood is the only liquid tissue, a fluid highway transporting gases, nutrients, hormones, antibodies, and wastes throughout the body. The material that makes blood a liquid is a protein-rich fluid called plasma that makes up approximately 55% of blood. The other 45% is made mostly of red and white blood cells. Red blood cells like those seen above are the oxygen transporters of the vertebrate circulatory system. There are approximately 5 million of them in each cubic millimeter (1 µl) of blood! Each red blood cell is shaped like a rounded cushion, squashed in the center, and is crammed full of a protein called hemoglobin, an iron-containing molecule that gives blood its red color. Oxygen binds easily to the iron in hemoglobin, making red blood cells efficient oxygen carriers. A single red blood cell contains about 250 million hemoglobin molecules, and each red blood cell can carry about 1 billion molecules of oxygen at one time. The average life span of a red blood cell is only 120 days—about 2 million new ones are produced in the bone marrow every second to replace those that die or are worn out.

Every cell in the vertebrate body must acquire the energy it needs for living from organic molecules outside the body. Like residents of a city whose food is imported from farms in the countryside, the cells of the body need trucks to carry the food, highways for the trucks to travel on, and a means to cook the food when it arrives. In the vertebrate body, the trucks are blood, the highways are blood vessels, and oxygen molecules are used to cook the food. Remember from chapter 5 that cells obtain energy by "burning" sugars like glucose, using up oxygen and generating carbon dioxide. In animals, the organ system that provides the trucks and highways is called the *circulatory system,* while the organ system that acquires the oxygen fuel and disposes of the carbon dioxide waste is called the *respiratory system.* We discuss the functions of these two organ systems in this chapter.

20.1 Open and Closed Circulatory Systems

Among the unicellular protists, oxygen and nutrients are obtained directly by simple diffusion from the aqueous external environment. Cnidarians, such as *Hydra,* and flatworms, such as *Planaria,* have cells that are directly exposed to either the external environment or to a body cavity that functions in both digestion and circulation called the **gastrovascular cavity** (figure 20.1a). The gastrovascular cavity of *Hydra* extends even into the tentacles, and that of *Planaria* branches extensively to supply every cell with oxygen and the nourishment obtained by digestion. Larger animals, however, have tissues that are several cell layers thick, so that many cells are too far away from the body surface or digestive cavity to exchange materials directly with the environment. Instead, oxygen and nutrients are transported from the environment and digestive cavity to the body cells by an internal fluid within a **circulatory system.**

There are two main types of circulatory systems: *open* or *closed.* In an **open circulatory system,** such as that found in mollusks and arthropods (figure 20.1b), there is no distinction between the circulating fluid (blood) and the extracellular fluid of the body tissues (interstitial fluid or lymph). This fluid is thus called **hemolymph.** Insects have a muscular tube that serves as a heart to pump the hemolymph through a network of open-ended channels that empty into the cavities in the body. There, the hemolymph delivers nutrients to the cells of the body. It then reenters the circulatory system through pores in the heart. The pores close when the heart pumps to keep the hemolymph from flowing back out into the body cavity.

In a **closed circulatory system,** the circulating fluid, or *blood,* is always enclosed within blood vessels that transport blood away from and back to a pump, the *heart.* Annelids and all vertebrates have a closed circulatory system. In annelids such as an earthworm, a dorsal blood vessel contracts rhythmically to function as a pump. Blood is pumped through five small connecting vessels called lateral hearts, which also function as pumps, to a ventral blood vessel, which transports the blood posteriorly until it eventually reenters the dorsal blood vessel. Smaller vessels branch between the ventral and dorsal blood vessels to supply the tissues of the earthworm with oxygen and nutrients and to transport waste products (figure 20.1c).

In vertebrates, blood vessels form a tubular network that permits blood to flow from the heart to all the cells of the body and then back to the heart. *Arteries* carry blood away from the heart, whereas *veins* return blood to the heart. Blood passes from the arterial to the venous system in *capillaries,* which are the thinnest and most numerous of the blood vessels.

As blood plasma passes through capillaries, the pressure of the blood forces some of this fluid out of the capillary walls. Fluid derived this way is called **interstitial fluid.** Some

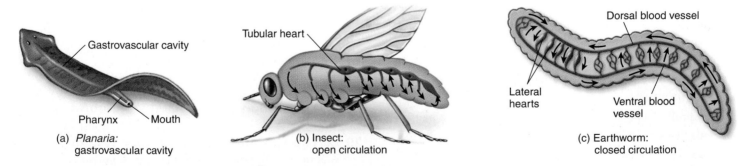

(a) *Planaria:* gastrovascular cavity — Gastrovascular cavity, Pharynx, Mouth

(b) Insect: open circulation — Tubular heart

(c) Earthworm: closed circulation — Dorsal blood vessel, Lateral hearts, Ventral blood vessel

Figure 20.1 Three types of circulatory systems found in the animal kingdom.
(a) The gastrovascular cavity of *Planaria* serves as both a digestive and circulatory system, delivering nutrients directly to the tissue cells by diffusion from the digestive cavity. (b) In the open circulation of an insect, hemolymph is pumped from a tubular heart into cavities in the insect's body; the hemolymph then returns to the blood vessels so that it can be recirculated. (c) In the closed circulation of the earthworm, blood pumped from the hearts remains within a system of vessels that returns it to the hearts. All vertebrates also have closed circulatory systems.

of this fluid returns directly to capillaries, and some enters into **lymph vessels,** located in the connective tissues around the blood vessels. This fluid, now called *lymph,* is returned to the venous blood at specific sites. The lymphatic system is considered a part of the circulatory system and is discussed later in this chapter.

The Functions of Vertebrate Circulatory Systems

The functions of the circulatory system can be divided into three areas: transportation, regulation, and protection.

1. **Transportation.** All of the substances essential for cellular metabolism are transported by the circulatory system. These substances can be categorized as follows:

   Respiratory. Red blood cells, or erythrocytes, transport oxygen to the tissue cells. In the capillaries of the lungs or gills, oxygen attaches to hemoglobin molecules within the erythrocytes and is transported to the cells for aerobic respiration. Carbon dioxide produced by cell respiration is carried by the blood to the lungs or gills for elimination.

   Nutritive. The digestive system is responsible for the breakdown of food so that nutrients can be absorbed through the intestinal wall and into the blood vessels of the circulatory system. The blood then carries these absorbed products of digestion through the liver and to the cells of the body.

   Excretory. Metabolic wastes, excessive water and ions, and other molecules in plasma (the fluid portion of blood) are filtered through the capillaries of the kidneys and excreted in urine.

2. **Regulation.** The cardiovascular system transports hormones and participates in temperature regulation.

   Hormone transport. The blood carries hormones from the endocrine glands, where they are secreted, to the distant target organs they regulate.

   Temperature regulation. In warm-blooded vertebrates, or homeotherms, a constant body temperature is maintained, regardless of the surrounding temperature. This is accomplished in part by blood vessels located just under the epidermis. When the ambient temperature is cold, the superficial vessels constrict to divert the warm blood to deeper vessels. When the ambient temperature is warm, the superficial vessels dilate so that the warmth of the blood can be lost by radiation.

   Some vertebrates also retain heat in a cold environment by using a **countercurrent heat exchange.** In this process, a vessel carrying warm blood from deep within the body passes next to a vessel carrying cold blood from the surface of the body (figure 20.2). The warm blood going out heats the cold blood

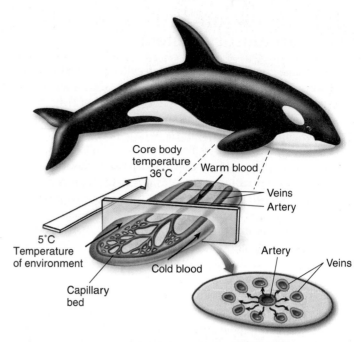

Figure 20.2   Countercurrent heat exchange.
Many marine mammals, such as this killer whale, limit heat loss in cold water by countercurrent flow that allows heat exchange between arteries and veins. The warm blood pumped from within the body in arteries warms the cold blood returning from the skin in veins, so that the core body temperature can remain constant in cold water. The cut-away portion in the figure shows how the veins surround the artery, maximizing the heat exchange between the artery and the veins.

returning from the body surface, so that this blood is no longer cold when it reaches the interior of the body, helping to maintain a stable core body temperature.

3. **Protection.** The circulatory system protects against injury and foreign microbes or toxins introduced into the body.

   Blood clotting. The clotting mechanism protects against blood loss when vessels are damaged. This clotting mechanism involves both proteins from the blood plasma and cell structures called platelets (discussed in section 20.4).

   Immune defense. The blood contains white blood cells, or leukocytes, that provide immunity against many disease-causing agents. Some white blood cells are phagocytic, some produce antibodies, and some act by other mechanisms to protect the body.

20.1   Circulatory systems may be open or closed. All vertebrates have a closed circulatory system, in which blood circulates away from the heart in arteries and back to the heart in veins. The circulatory system serves a variety of functions, including transportation, regulation, and protection.

20.2 Architecture of the Vertebrate Circulatory System

The vertebrate circulatory system, also called the **cardiovascular system,** is made up of three elements: (1) the **heart,** a muscular pump that pushes blood through the body; (2) the **blood vessels,** a network of tubes through which the blood moves; and (3) the **blood,** which circulates within these vessels.

Blood leaves the heart through vessels known as **arteries.** From the arteries, the blood passes into a network of smaller arteries called **arterioles.** From these, it is eventually forced through a capillary bed, a fine latticework of very narrow tubes called **capillaries** (from the Latin, *capillus,* "a hair"). While passing through the capillaries, the blood exchanges gases and metabolites (glucose, vitamins, hormones) with the cells of the body. Capillary beds can be opened or closed, based on the physiological needs of the tissues, by the relaxation or contraction of small circular muscles called *precapillary sphincters.* After traversing the capillaries, the blood passes into a fourth kind of vessel, the **venules,** or small veins (figure 20.3). A network of venules empties into larger **veins** that collect the circulating blood and carry it back to the heart.

The capillaries have a much smaller diameter than the other blood vessels of the body. Blood leaves the mammalian heart through a large artery, the aorta, a tube that has a diameter of about 2 centimeters (about the same as your thumb), but when it reaches the capillaries it passes through vessels with an average diameter of only 8 micrometers, a reduction in radius of some 1,250 times! This decrease in size of blood vessels has a very important consequence.

Although each capillary is very narrow, there are so many of them that the capillaries have the greatest *total* cross-sectional area of any other type of vessel. Consequently, this allows more time for blood to exchange materials with the surrounding extracellular fluid. By the time the blood reaches the end of a capillary, it has released some of its oxygen and nutrients and picked up carbon dioxide and other waste products. Blood loses most of its pressure and velocity in passing through the vast capillary networks, and so is under very low pressure when it enters the veins. Like water flowing out of the sprinkler head of a watering can, the stream of blood spreads out to many small streams. These smaller streams don't flow with as much force, nor as quickly, as the larger stream that entered the capillary bed.

Arteries: Highways from the Heart

The arterial system, composed of arteries and arterioles, carries blood away from the heart. An artery is more than simply a pipe. Blood comes from the heart not in a smooth flow but rather in pulses, slammed into the artery in great big slugs as the heart forcefully ejects its contents with each contraction. The artery has to be able to *expand* to withstand the pressure caused by each contraction of the heart. An artery, then, is designed as an expandable tube, with its walls made up of three layers of tissue (figure 20.4). The innermost thin layer is composed of endothelial cells. Surrounding them is a thick layer of smooth muscle and elastic fibers, which in turn is encased within an envelope of protective connective tissue. Because this sheath and envelope are elastic, the artery is able to expand its volume considerably when the heart contracts, shoving a new volume of blood into the artery—just as a tu-

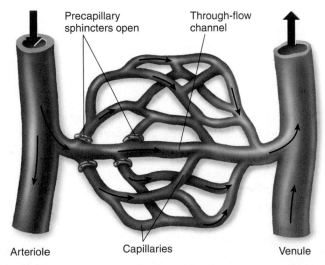

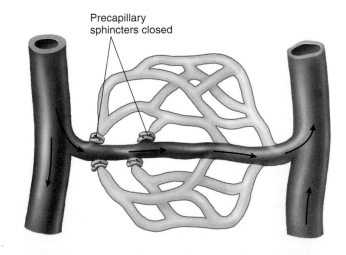

(a) Blood flows through capillary network

(b) Blood flow in capillary network is limited

Figure 20.3   The capillary network connects arteries with veins.
Through-flow channels connect arterioles directly to venules. Branching from these through-flow channels is a network of finer channels, the capillaries. Most of the exchange between the body tissues and the red blood cells occurs while they are in this capillary network. The flow of blood into the capillaries is controlled by bands of muscle called precapillary sphincters located at the entrance to each capillary. (a) When a sphincter is open, blood flows through that capillary. (b) When a sphincter contracts, it closes off the capillary. By contracting these sphincters, the body can limit the amount of blood in the capillary network of a particular tissue; through-flow channels then allow the blood to bypass the capillary network when it is needed elsewhere.

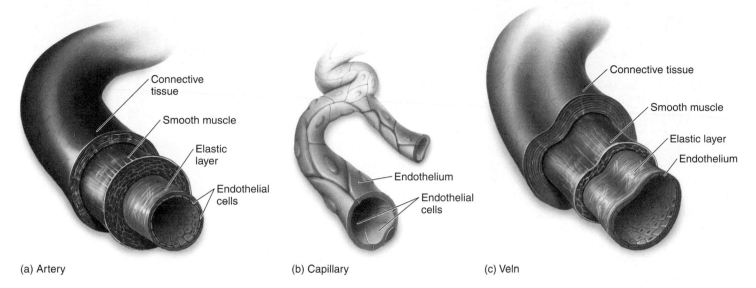

| (a) Artery | (b) Capillary | (c) Vein |

Figure 20.4   The structure of blood vessels.
(*a*) Arteries, which carry blood away from the heart, are expandable and are composed of layers of tissue. (*b*) Capillaries are simple tubes whose thin walls facilitate the exchange of materials between the blood and the cells of the body. (*c*) Veins, which transport blood back to the heart, do not need to be as sturdy as arteries. The walls of veins have thinner muscle layers than arteries, and they collapse when empty. Note, this drawing is not to scale; as stated in the text, arteries can be up to two centimeters in diameter, the capillaries are only about eight micrometers in diameter, and the largest veins can be up to three centimeters in diameter.

bular balloon expands when you blow more air into it. The steady contraction of the smooth muscle layer strengthens the wall of the vessel against overexpansion.

Arterioles differ from arteries in two ways. They are smaller in diameter, and the muscle layer that surrounds an arteriole can be relaxed under the influence of hormones to enlarge the diameter. When the diameter increases, the blood flow also increases, an advantage during times of high body activity. Most arterioles are also in contact with nerve fibers. When stimulated by these nerves, the muscle lining of the arteriole contracts, constricting the diameter of the vessel. Such contraction limits the flow of blood to the extremities during periods of low temperature or stress. You turn pale when you are scared or cold because the arterioles in your skin are constricting. You blush for just the opposite reason. When you overheat or are embarrassed, the nerve fibers connected to muscles surrounding the arterioles are inhibited, which relaxes the smooth muscle and causes the arterioles in the skin to expand, bringing heat to the surface for escape.

Capillaries: Where Exchange Takes Place

Capillaries are where oxygen and food molecules are transferred from the blood to the body's cells and where waste carbon dioxide is picked up. To facilitate this back-and-forth traffic, capillaries have thin walls across which gases and metabolites pass easily. Capillaries have the simplest structure of any element in the cardiovascular system. They are built like a soft-drink straw, simple tubes with walls only one cell thick (see figure 20.4*b*). The average capillary is about 1 millimeter long and connects an arteriole with a venule. All capillaries are very narrow, with an internal diameter of about

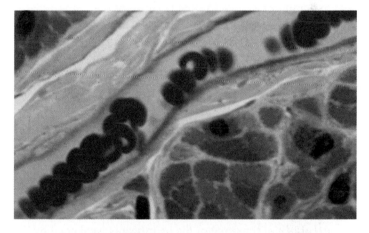

Figure 20.5   Red blood cells within a capillary.
The red blood cells in this capillary pass along in single file. Many capillaries are even narrower than the one shown here. However, red blood cells can pass through capillaries narrower than their own diameter, pushed along by the pressure of the pumping heart.

8 micrometers, just bigger than the diameter of a red blood cell (5 to 7 micrometers). This design is critical to the function of capillaries. By bumping against the sides of the vessel as they pass through, the red blood cells are forced into close contact with the capillary walls, making exchange easier. Examining figure 20.5, you can better understand the complications that result from sickle-cell anemia, discussed in chapters 8 and 11. The red blood cells of people with sickle-cell anemia take on an elongated shape, and do not easily flow through the narrow passageways of the capillaries, causing the capillaries to become blocked.

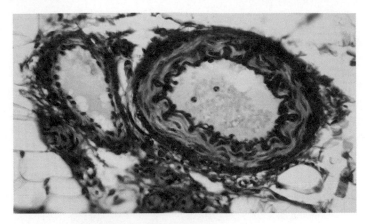

Figure 20.6 Veins and arteries.
The vein (*left*) has the same general structure as the artery (*right*) but much thinner layers of muscle and elastic fiber. An artery retains its shape when empty, but a vein collapses.

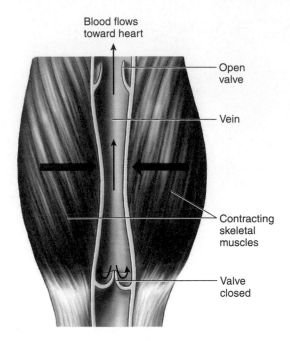

Figure 20.7 Flow of blood through veins.
Venous valves ensure that blood moves through the veins in only one direction back to the heart. This movement of blood is aided by the contraction of skeletal muscles surrounding the veins.

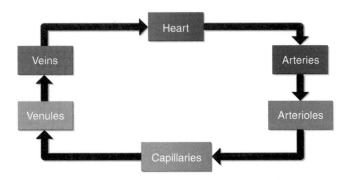

Figure 20.8 The flow of blood through the circulatory system.

Almost all cells of the vertebrate body are no more than 100 micrometers from a capillary. At any one moment, about 5% of the circulating blood is in capillaries, a network that amounts to several thousand miles in overall length. If all the capillaries in your body were laid end to end, they would extend across the United States! Individual capillaries have high resistance to flow because of their small diameters. However, the total cross-sectional area of the extensive capillary network (that is, the sum of all the diameters of all the capillaries, expressed as area) is greater than that of the arteries leading to it. As a result, the blood pressure is actually far lower in the capillaries than in the arteries. This is important, because the walls of capillaries are not strong, and they would burst if exposed to the pressures that arteries routinely withstand.

Veins: Returning Blood to the Heart

Veins are vessels that return blood to the heart. Veins do not have to accommodate the pulsing pressures that arteries do, because much of the force of the heartbeat is weakened by the high resistance and great cross-sectional area of the capillary network. For this reason, the walls of veins have much thinner layers of muscle and elastic fiber. An empty artery is still a hollow tube, like a pipe, but when a vein is empty, its walls collapse like an empty balloon (figure 20.6).

Because the pressure of the blood flowing within veins is low, it becomes important to avoid any further resistance to flow, lest there not be enough pressure to get the blood back to the heart. Because a wide tube presents much less resistance to flow than a narrow one, the internal passageway of veins is often quite large, requiring only a small pressure difference to return blood to the heart. The diameters of the largest veins in the human body, the venae cavae, which lead into the heart, are fully 3 centimeters; this is wider than your thumb! Pressure alone cannot force the blood in the veins back to the heart but several features provide help. Most significantly, when skeletal muscles surrounding the veins contract, they

move blood by squeezing the veins (figure 20.7). Veins have unidirectional valves that ensure the return of this blood by preventing it from flowing backward. These structural features keep the blood flowing in a cycle through the circulatory system (figure 20.8).

20.2 The vertebrate circulatory system is composed of arteries and arterioles, which are elastic and carry blood away from the heart; a fine network of capillaries across whose thin walls the exchange of gases and food molecules takes place; and venules and veins, which return blood from the capillaries to the heart.

20.3 The Lymphatic System: Recovering Lost Fluid

The cardiovascular system is very leaky. Fluids are forced out across the thin walls of the capillaries by the pumping pressure of the heart. Although this loss is unavoidable—the circulatory system could not do its job of gas and metabolite exchange without tiny vessels with thin walls—it is important that the loss be made up. In your body, about 4 liters of fluid leave your cardiovascular system in this way each day, more than half the body's total supply of about 5.6 liters of blood! To collect and recycle this fluid, the body uses a second circulatory system called the **lymphatic system** (figure 20.9).

Fluid is filtered from the capillaries near their arteriole ends where the blood pressure is higher. Most of the fluid returns to the blood capillaries by osmosis; however, excess fluid enters the open-ended lymphatic capillaries (figure 20.10). These capillaries gather up liquid from the spaces surrounding cells and carry it through a series of progressively larger vessels to two large lymphatic vessels, which drain into veins in the lower part of the neck through one-way valves. Once within the lymphatic system, this fluid is called **lymph**.

Fluid is driven through the lymphatic system when its vessels are squeezed by the movements of the body's muscles.

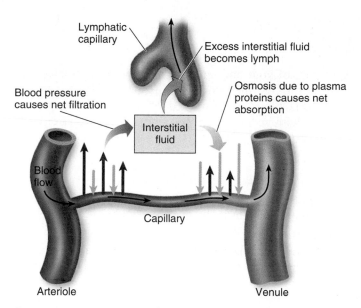

Figure 20.10 Lymphatic capillaries reclaim fluid from interstitial fluid.

Blood pressure forces fluid from capillaries into the interstitial fluid surrounding cells. Much of this fluid is returned to the blood capillaries by osmosis but the excess interstitial fluid is drained into lymphatic capillaries.

The lymphatic vessels contain a series of one-way valves that permit movement only in the direction of the neck. In some cases, the lymphatic vessels also contract rhythmically. In many fishes, all amphibians and reptiles, bird embryos, and some adult birds, movement of lymph is propelled by **lymph hearts.**

The lymphatic system has three other important functions:

1. **It returns proteins to the circulation.** So much blood flows through the capillaries that there is significant leakage of blood proteins, even though the capillary walls are not very permeable to proteins (proteins are very large molecules). By recapturing the lost fluid, the lymphatic system also returns this lost protein. Whenever this protein is not returned to the blood, the body becomes swollen, a condition known as *edema.*

2. **It transports fats absorbed from the intestine.** Lymph capillaries called *lacteals* penetrate the lining of the small intestine. These lacteals absorb fats from the digestive tract and eventually transport them to the circulatory system by way of the lymphatic system.

3. **It aids in the body's defense.** Along a lymph vessel are swellings called **lymph nodes.** Inside each node is a honeycomb of spaces filled with white blood cells specialized for defense. The lymphatic system carries bacteria and dead blood cells to the lymph nodes and spleen for destruction.

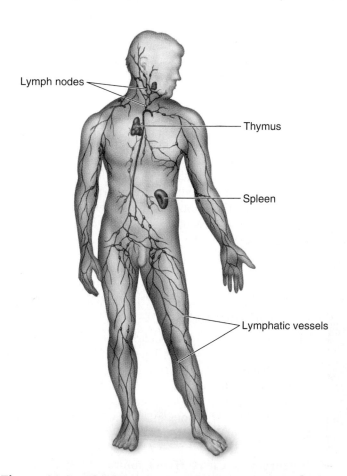

Figure 20.9 The human lymphatic system.

The lymphatic system consists of lymphatic vessels and capillaries, lymph nodes, and lymphatic organs, including the spleen and thymus gland.

20.3 The lymphatic system returns to the circulatory system fluids and proteins that leak through the capillaries.

20.4 Blood

About 5% of your body mass is composed of the blood circulating through the arteries, veins, and capillaries of your body. This blood is composed of a fluid called **plasma**, together with several different kinds of cells that circulate within that fluid.

Blood Plasma: The Blood's Fluid

Blood plasma is a complex solution of water with three very different sorts of substances dissolved within it:

1. **Metabolites and wastes.** If the circulatory system is the highway of the vertebrate body, the blood contains the traffic traveling on that highway. Dissolved within its plasma are glucose, vitamins, hormones, and wastes that circulate between the cells of the body.

2. **Salts and ions.** Like the seas in which life arose, plasma is a dilute salt solution. The chief plasma ions are sodium, chloride, and bicarbonate. In addition, trace amounts of other salts, such as calcium and magnesium, as well as metallic ions, including copper, potassium, and zinc, are present in plasma. The composition of the plasma is not unlike that of seawater.

3. **Proteins.** Blood plasma is 90% water. Passing by all the cells of the body, blood would soon lose most of its water to them by osmosis if it did not contain as high a concentration of proteins as the cells it passes. Some of the proteins blood plasma contains are antibodies that are active in the immune system. More than half the amount of protein that is necessary to balance the protein content of the cells of the body consists of a single protein, **serum albumin,** which circulates in the blood as an osmotic counterforce. Human blood contains 46 grams of serum albumin per liter—that's over half a pound of it in your body. Starvation and protein deficiency result in reduced levels of protein in the blood. This lack of plasma proteins produces swelling of the body because the body's cells, which now have a higher level of solutes than the blood, take up water from the albumin-deficient blood. A symptom of protein deficiency diseases such as kwashiorkor is edema, a swelling of tissues, although other factors can also result in edema.

The liver produces most of the plasma proteins, including albumin; the alpha and beta globulins, which serve as carriers of lipids and steroid hormones; and *fibrinogen,* which is required for blood clotting. When blood in a test tube clots, the fibrinogen is converted into insoluble threads of *fibrin* that become part of the clot (figure 20.11). The fluid that's left, which lacks fibrinogen and so cannot clot, is called serum.

Blood Cells: Cells That Circulate Through the Body

Although blood is liquid, nearly half of its volume is actually occupied by cells. The fraction of the total volume of the blood that is occupied by cells is referred to as the blood's **hematocrit.** In humans, the hematocrit is usually about 45%. The three principal types of cells in the blood are erythrocytes, leukocytes, and cell fragments called platelets (figure 20.12).

Erythrocytes Carry Hemoglobin. Each cubic millimeter (1 µl) of blood contains about 5 million **erythrocytes,** also called **red blood cells.** Each erythrocyte is a flat disk with a central depression on both sides,

Figure 20.11 Threads of fibrin.

This scanning electron micrograph (×1,430) shows fibrin threads among red blood cells. Fibrin is formed from a soluble protein, fibrinogen, in the plasma to produce a blood clot when a blood vessel is damaged.

Blood cell	Life span in blood	Function
 Erythrocyte	120 days	O$_2$ and CO$_2$ transport
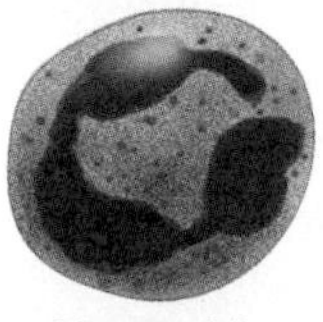 Neutrophil	7 hours	Immune defenses
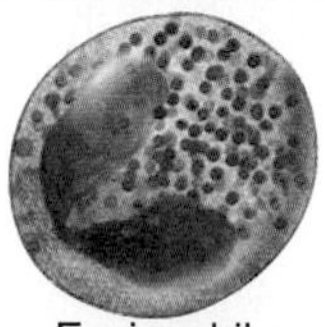 Eosinophil	Unknown	Defense against parasites
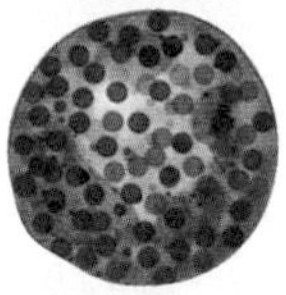 Basophil	Unknown	Inflammatory response
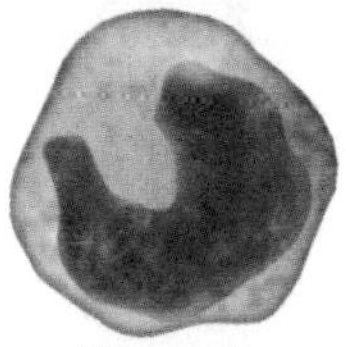 Monocyte	3 days	Immune surveillance (precursor of tissue macrophage)
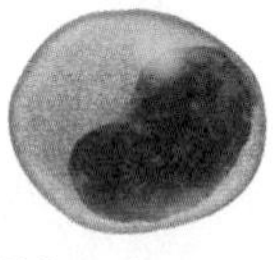 B lymphocyte	Unknown	Antibody production (precursor of plasma cells)
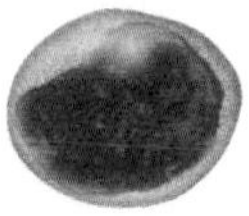 T lymphocyte	Unknown	Cellular immune response
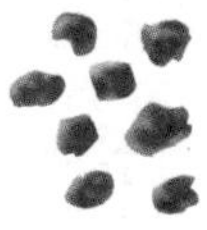 Platelets	7- 8 days	Blood clotting

Figure 20.12   Types of blood cells.
Erythrocytes, leukocytes (neutrophils, eosinophils, basophils, monocytes, and lymphocytes), and platelets are the three principal types of blood cells in vertebrates.

something like a doughnut with a hole that doesn't go all the way through. Erythrocytes carry oxygen to the cells of the body. Almost the entire interior of an erythrocyte is packed with hemoglobin, a protein that binds oxygen in the lungs and delivers it to the cells of the body.

Mature mammalian erythrocytes function like boxcars rather than trucks. Like a vehicle without an engine, erythrocytes contain neither a nucleus nor the machinery to make proteins. Because they lack a nucleus, these cells are unable to repair themselves and therefore have a rather short life; any one erythrocyte lives only about four months. New erythrocytes are constantly being synthesized and released into the blood by cells within the soft interior marrow of bones.

Leukocytes Defend the Body.   Less than 1% of the cells in mammalian blood are **leukocytes,** also called **white blood cells.** Leukocytes are larger than red blood cells. They contain no hemoglobin and are essentially colorless. There are several kinds of leukocytes, each with a different function. Neutrophils are the most numerous of the leukocytes, followed in order by lymphocytes, monocytes, eosinophils, and basophils. *Neutrophils* attack like kamikazes, responding to foreign cells by releasing chemicals that kill all the cells in the neighborhood—including the neutrophils. Monocytes give rise to *macrophages,* which attack and kill foreign cells by ingesting them (the name means "large eater"). Lymphocytes include *B cells,* which produce antibodies, and *T cells,* which literally drill holes in invading bacteria.

All of these white blood cell types, and others, help defend the body against invading microorganisms and other foreign substances, as you will see in chapter 22. Unlike other blood cells, leukocytes are not confined to the bloodstream; they are mobile soldiers that also migrate out into the fluid surrounding cells.

Platelets Help Blood to Clot.   Certain large cells within the bone marrow, called **megakaryocytes,** regularly pinch off bits of their cytoplasm. These cell fragments, called **platelets,** contain no nuclei. Entering the bloodstream, they play a key role in blood clotting. In a clot, a gluey mesh of fibrin protein fibers sticks platelets together to form a mass that plugs the rupture in the blood vessel. The clot provides a tight, strong seal, much as the inner lining of a tubeless tire seals punctures. The fibrin that forms the clot is made in a series of reactions that start when circulating platelets first encounter the site of an injury. Responding to chemicals released by the damaged blood vessel, platelets release a protein factor into the blood that starts the clotting process.

20.4   Blood is a collection of cells that circulate within a protein-rich salty fluid called plasma. Some of the cells circulating in the blood carry out gas transport; others are engaged in defending the body from infection and cancer.

In chapter 19, we described one of the greatest evolutionary achievements of animals—locomotion. For vertebrates, the combination of large body size and locomotion was possible because of the "coevolution" of the circulatory and respiratory systems. As the body size and physiological abilities of animals increased, so did the need for more efficient mechanisms to both deliver nutrients and oxygen and remove wastes and carbon dioxide from the growing mass of tissues. Animals have evolved a remarkable set of adaptations intricately linking circulation and respiration. In this context, the remainder of this chapter examines the evolution of the heart, the evolution of separate pulmonary and systemic circulations, and the principles of gas exchange.

20.5 Fish Circulation

The chordates that were ancestral to the vertebrates are thought to have had simple tubular hearts, similar to those now seen in lancelets. The heart was little more than a specialized zone of the ventral artery, more heavily muscled than the rest of the arteries, which contracted in simple peristaltic waves. A pumping action results because the uncontracted portions of the vessel have a larger diameter than the contracted portion, and thus present less resistance to blood flow.

The development of gills by fishes required a more efficient pump, and in fishes we see the evolution of a true chamber-pump heart. The fish heart is, in essence, a tube with four chambers arrayed one after the other (figure 20.13*a*). The first two chambers—the **sinus venosus** and **atrium**—are collection chambers, while the second two—the **ventricle** and **conus arteriosus**—are pumping chambers.

As might be expected from the early chordate hearts from which the fish heart evolved, the sequence of the heartbeat in fishes is a peristaltic sequence, starting at the rear and moving to the front. The first of the four chambers to contract is the sinus venosus, followed by the atrium, the ventricle, and finally the conus arteriosus. Despite shifts in the relative positions of the chambers in the vertebrates that evolved later, this heartbeat sequence is maintained in all vertebrates. In fish, the electrical impulse that produces the contraction is initiated in the sinus venosus; in other vertebrates, the electrical impulse is initiated by their equivalent of the sinus venosus.

The fish heart is remarkably well suited to the gill respiratory apparatus and represents one of the major evolutionary innovations in the vertebrates. Perhaps its greatest advantage is that the blood it delivers to the tissues of the body is fully oxygenated. Blood is pumped first through the gills, where it becomes oxygenated; from the gills, it flows through a network of arteries and capillaries to the rest of the body; then it returns to the heart through the veins (figure 20.13*b*). This arrangement has one great limitation, however. In passing through the capillaries in the gills, the blood loses much of the pressure developed by the contraction of the heart, so the circulation from the gills through the rest of the body is sluggish. This feature limits the rate of oxygen delivery to the rest of the body.

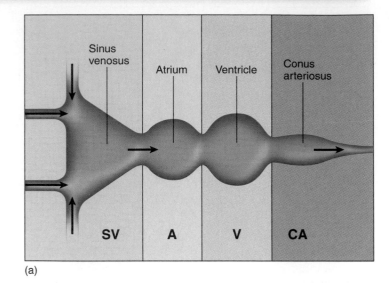

(a)

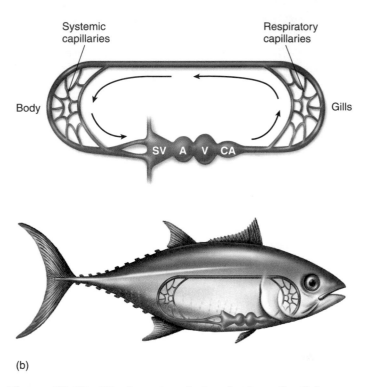

(b)

Figure 20.13 The heart and circulation of a fish.
(*a*) Diagram of a fish heart, showing the chambers in series with each other. (*b*) Diagram of the fish circulation, showing that blood is pumped by the ventricle to the gills, and then blood flows directly to the body, resulting in sluggish circulation. Blood rich in oxygen (oxygenated) is shown in *red;* blood low in oxygen (deoxygenated) is shown in *blue.*

> **20.5** The fish heart is a modified tube, consisting of a series of four chambers. Blood first enters the heart at the sinus venosus, where the wavelike contraction of the heart begins.

20.6 Amphibian and Reptile Circulation

The advent of lungs involved a major change in the pattern of circulation. After blood is pumped by the heart to the lungs, it does not go directly to the tissues of the body but instead returns to the heart. This results in two circulations: one that goes to and from the heart to the lungs, called the **pulmonary circulation,** and one that goes to and from the heart to the rest of the body, called the **systemic circulation.**

If no changes had occurred in the structure of the heart, the oxygenated blood from the lungs would be mixed in the heart with the deoxygenated blood returning from the rest of the body. Consequently, the heart would pump a mixture of oxygenated and deoxygenated blood rather than fully oxygenated blood. The amphibian heart has two structural features that help reduce this mixing (figure 20.14a). First, the atrium is divided by a *septum,* or dividing wall, into two chambers: The right atrium receives deoxygenated blood from the systemic circulation, and the left atrium receives oxygenated blood from the lungs. The septum prevents the two stores of blood from mixing in the atria, but some mixing might be expected when the contents of each atrium enter the single, common ventricle. Surprisingly, however, little mixing actually occurs. Two factors contribute to the separation of the two blood supplies. First, the ventricle in some amphibians is lined with folds that help direct the flow of blood from the atria. Second, the conus arteriosus is partially separated by another septum that directs deoxygenated blood into the *pulmonary arteries* toward the lungs, and oxygenated blood into the *aorta,* the major artery of the systemic circulation to the body.

Amphibians in water supplement the oxygenation of the blood by obtaining additional oxygen by diffusion through their skin. This process is called **cutaneous respiration.**

Among reptiles, additional modifications have reduced the mixing of blood in the heart still further. In addition to having two separate atria, reptiles have a septum that partially subdivides the ventricle (figure 20.14b). This results in an even greater separation of oxygenated and deoxygenated blood within the heart. The separation is complete in one order of reptiles, the crocodiles, which have two separate ventricles divided by a complete septum. Crocodiles therefore have a completely divided pulmonary and systemic circulation. Another change in the circulation of reptiles is that the conus arteriosus has become incorporated into the trunks of the large arteries leaving the heart.

> **20.6** Amphibians and reptiles have two circulations, pulmonary and systemic, that deliver blood to the lungs and to the rest of the body, respectively.

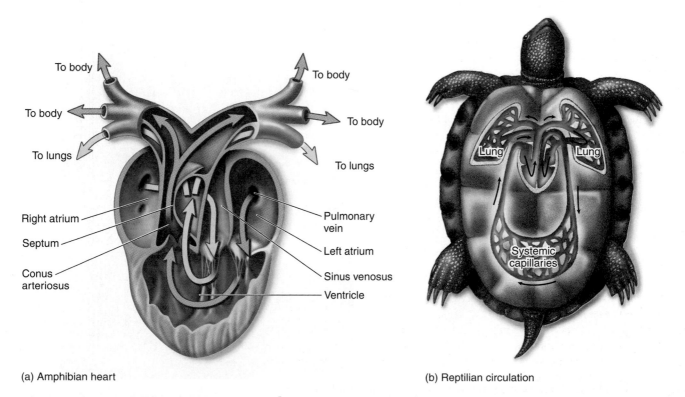

(a) Amphibian heart

(b) Reptilian circulation

Figure 20.14 The amphibian heart and circulation in reptiles.
(a) The frog heart has two atria but only one ventricle, which pumps blood both to the lungs and to the body. Despite the potential for mixing, the oxygenated and deoxygenated bloods (*red* and *blue,* respectively) mix very little as they are pumped to the body and lungs. A valve and septum separate the blood coming from the ventricle to either the arteries taking blood to the lungs or the arteries taking blood to the rest of the body. The slight mixing is shown in *purple.* (b) In reptiles, not only are there two separate atria, but the ventricle is also partially divided. In amphibians and reptiles, blood is returned to the heart after leaving the lungs and is repumped, resulting in much more vigorous circulation to the body than in fishes.

20.7 Mammalian and Bird Circulation

Mammals, birds, and crocodiles have a four-chambered heart that is really two separate pumping systems operating together within a single unit. One of these pumps blood to the lungs, while the other pumps blood to the rest of the body. The left side has two connected chambers, and so does the right, but the two sides are not connected with one another. The increased efficiency of the double circulatory system in mammals and birds is thought to have been important in the evolution of endothermy (warm-bloodedness), because a more efficient circulation is necessary to support the high metabolic rate required.

Circulation Through the Heart

Let's follow the journey of blood through the mammalian heart, starting with the entry of oxygen-rich blood into the heart from the lungs (figure 20.15). Oxygenated blood from the lungs enters the left side of the heart, emptying directly into the **left atrium** through large vessels called the **pulmonary veins.** From the atrium, blood flows through an opening into the adjoining chamber, the **left ventricle.** Most of this flow, roughly 80%, occurs while the heart is relaxed. When the heart starts to contract, the atrium contracts first, pushing the remaining 20% of its blood into the ventricle.

After a slight delay, the ventricle contracts. The walls of the ventricle are far more muscular than those of the atrium, and thus this contraction is much stronger. It forces most of the blood out of the ventricle in a single strong pulse. The blood is prevented from going back into the atrium by a large, one-way valve, the **bicuspid (mitral) valve,** or left atrioventricular valve, whose two flaps are pushed shut as the ventricle contracts.

Prevented from reentering the atrium, the blood within the contracting left ventricle takes the only other passage out. It moves through a second opening that leads into a large

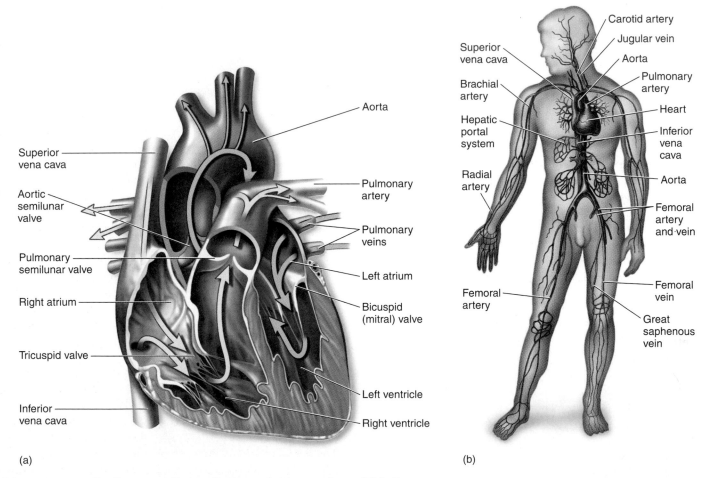

(a)

(b)

Figure 20.15 The heart and circulation of mammals and birds.

(a) Unlike the amphibian heart, this heart has a septum dividing the ventricle into left and right ventricles. Oxygenated blood from the lungs enters the left atrium of the heart by way of the pulmonary veins. This blood then enters the left ventricle, from which it passes into the aorta to circulate throughout the body and deliver oxygen to the tissues. When gas exchange has taken place at the tissues, veins return blood to the heart. After entering the right atrium by way of the superior and inferior venae cavae, deoxygenated blood passes into the right ventricle and then through the pulmonary valve to the lungs by way of the pulmonary artery. (b) Some of the major arteries and veins in the human circulatory system are shown.

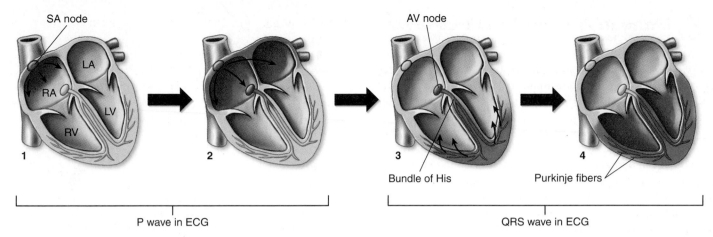

P wave in ECG   QRS wave in ECG

Figure 20.16 How the mammalian heart contracts.
(1) Contraction of the mammalian heart is initiated by a wave of depolarization that begins at the SA node; the contraction can be recorded using an electrocardiogram (ECG). (2) After passing over the right and left atria and causing their contraction (forming the P wave on the ECG), (3) the wave of depolarization reaches the AV node, from which it passes to the ventricles by the bundle of His. (4) The depolarization is then conducted rapidly over the surface of the ventricles by a set of finer fibers called Purkinje fibers, which branch throughout each ventricle causing them to contract (forming the QRS wave on the ECG). The T wave on the ECG corresponds to the repolarization of the ventricles.

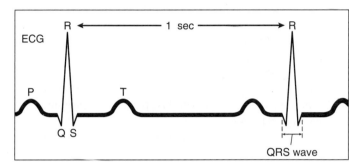

blood vessel called the **aorta.** The aorta is separated from the left ventricle by a one-way valve, the **aortic semilunar valve.** Unlike the bicuspid valve, the aortic valve is oriented to permit the flow of the blood *out* of the ventricle. Once this outward flow has occurred, the aortic valve closes, preventing the reentry of blood from the aorta into the heart. The aorta and its many branches are systemic arteries and carry oxygen-rich blood to all parts of the body.

Eventually, this blood returns to the heart after delivering its cargo of oxygen to the cells of the body. In returning, it passes through a series of progressively larger veins, ending in two large veins that empty into the right atrium of the heart. The **superior vena cava** drains the upper body, and the **inferior vena cava** drains the lower body.

The right side of the heart is similar in organization to the left side. Blood passes from the **right atrium** into the **right ventricle** through a one-way valve, the **tricuspid valve** or right atrioventricular valve. It passes out of the contracting right ventricle through a second valve, the **pulmonary semilunar valve,** into the **pulmonary arteries,** which carry the deoxygenated blood to the lungs. The blood then returns from the lungs to the left side of the heart with a new cargo of oxygen, which is pumped to the rest of the body.

How the Heart Contracts

Throughout the evolutionary history of the vertebrate heart, the sinus venosus has served as a pacemaker, the site where the impulses that produce the heartbeat originate. Although it constitutes a major chamber in the fish heart, it is reduced in

size in amphibians and further reduced in reptiles. In mammals and birds, the sinus venosus is no longer a separate chamber, but some of its tissue remains in the wall of the right atrium, near the point where the systemic veins empty into the atrium. This tissue, which is called the **sinoatrial (SA) node,** is still the site where each heartbeat originates.

The contraction of the heart consists of a carefully orchestrated series of muscle contractions. These contractions generate electrical potentials that can be recorded as described on page 468. First, the atria contract together, followed by the ventricles. Contraction is initiated by the sinoatrial (SA) node (figure 20.16). Its membranes spontaneously depolarize with a regular rhythm that determines the rhythm of the heart's beating. Each depolarization initiated within this pacemaker region passes quickly from one heart muscle cell to another in a wave that envelops the left and the right atria almost simultaneously.

But the wave of depolarization does not immediately spread to the ventricles. There is a pause before the lower half of the heart starts to contract. The reason for the delay is that the atria of the heart are separated from the ventricles by connective tissue that does not propagate a depolarization wave. The depolarization would not pass to the ventricles at all except for a slender connection of cardiac muscle cells known as the **atrioventricular (AV) node,** which connects across the gap to a strand of specialized muscle known as the atrioventricular bundle, or **bundle of His.** Bundle branches divide into fast-conducting **Purkinje fibers,** which initiate the almost simultaneous contraction of all the cells of the right and left ventricles about 0.1 seconds after the atria contract. This delay permits the atria to finish emptying their contents into the corresponding ventricles before those ventricles start to contract. The contraction of the ventricles begins at the apex (the bottom of the heart) where the depolarization of Purkinje fibers begins. The contraction then spreads up toward the atria. This results in a "wringing out" of the ventricle, forcing the blood up and out of the heart.

Because the vertebrate body basically consists of water, it conducts electrical currents rather well. A wave of membrane depolarization passing over the surface of the heart generates an electrical current that passes in a wave throughout the body. The magnitude of this electrical pulse is tiny, but it can be detected by sensors placed on the skin. The recording, called an **electrocardiogram** (**ECG** or **EKG**), shows how the cells of the heart depolarize and repolarize during the cardiac cycle (figure 20.16). Depolarization causes contraction of the heart, while repolarization causes relaxation. The first peak in the recording, P, is produced by the depolarization of the atria. The second, larger peak, QRS, is produced by ventricular depolarization; during this time, the ventricles contract and eject blood into the arteries. The last peak, T, is produced by ventricular repolarization.

Because the overall circulatory system is closed, the same volume of blood must move through the pulmonary circulation as through the much larger systemic circulation with each heartbeat. Therefore, the right and left ventricles must pump the same amount of blood each time they contract. If the output of one ventricle did not match that of the other, fluid would accumulate and pressure would increase in one of the circuits. The result would be increased filtration out of the capillaries and edema (as occurs in congestive heart failure, for example). Although the volume of blood pumped by the two ventricles is the same, the pressure they generate is not. The left ventricle, which pumps blood through the systemic pathway, is more muscular and generates more pressure than does the right ventricle.

Monitoring the Heart's Performance

As you can see, the heartbeat is not simply a squeeze-release, squeeze-release cycle but rather a little play in which a series of events occur in a predictable order. The simplest way to monitor heartbeat is to listen to the heart at work, using a stethoscope. The first sound you hear, a low-pitched *lub,* is the closing of the bicuspid and tricuspid valves at the start of ventricular contraction. A little later, you hear a higher-pitched *dub,* the closing of the pulmonary and aortic valves at the end of ventricular contraction. If the valves are not closing fully, or if they open incompletely, turbulence is created within the heart. This turbulence can be heard as a **heart murmur.** It often sounds like liquid sloshing.

A second way to examine the events of the heartbeat is to monitor the blood pressure. This is done using a device called a sphygmomanometer that measures the blood pressure in the brachial artery found on the inside part of the arm, at the elbow (figure 20.17). A cuff wrapped around the upper arm is tightened enough to stop the flow of blood

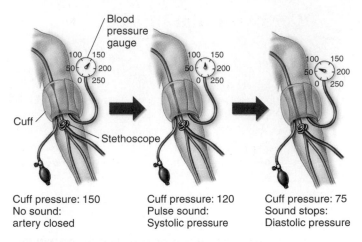

Cuff pressure: 150
No sound:
artery closed

Cuff pressure: 120
Pulse sound:
Systolic pressure

Cuff pressure: 75
Sound stops:
Diastolic pressure

Figure 20.17   Measuring blood pressure.
The blood pressure cuff is tightened to stop the blood flow through the brachial artery. As the cuff is loosened, the systolic pressure is recorded as the pressure at which a pulse is heard through a stethoscope. The diastolic pressure is recorded as the pressure at which a sound is no longer heard.

to the lower part of the arm. As the cuff is loosened, blood begins pulsating through the artery and can be detected using a stethoscope. Two measurements are recorded: the systolic and the diastolic pressures. During the first part of the heartbeat, the atria are filling. At this time the pressure in the arteries leading from the left side of the heart out to the tissues of the body decreases slightly. This low pressure is referred to as the **diastolic** pressure. During the contraction of the left ventricle, a pulse of blood is forced into the systemic arterial system, immediately raising the blood pressure within these vessels. The high blood pressure produced in this pushing period, which ends with the closing of the aortic valve, is referred to as the **systolic** pressure. Normal blood pressure values are 70 to 90 diastolic and 110 to 130 systolic. When the inner walls of the arteries accumulate fats, as they do in the condition known as *atherosclerosis,* the diameters of the passageways are narrowed. If this occurs, the systolic blood pressure is elevated.

The evolution of multicellular organisms has depended critically on the ability to circulate materials throughout the body efficiently. Vertebrates carefully regulate the operation of their circulatory systems and are able to integrate their body activities. Indeed, the metabolic demands of different vertebrates have shaped the evolution of circulatory systems.

20.7   The mammalian heart is a two-cycle pump. The left side pumps oxygenated blood to the body's tissues, while the right side pumps O$_2$-depleted blood to the lung. The performance of the heart can be monitored in many ways.

Cardiovascular Diseases

Cardiovascular diseases are the leading cause of death in the United States, accounting for nearly 40% of all deaths in 2002; more than 61 million people have some form of cardiovascular disease. Heart attacks are the main cause of cardiovascular deaths in the United States. They result from an insufficient supply of blood reaching one or more parts of the heart muscle, which causes myocardial cells in those parts to die. Heart attacks, also called myocardial infarctions, may be caused by a blood clot forming somewhere in the coronary arteries (the arteries that supply the heart muscle with blood) and blocking the passage of blood through those vessels. They may also result if an artery is blocked by atherosclerosis (discussed later). Recovery from a heart attack is possible if the portion of the heart that was damaged is small. **Angina pectoris,** which literally means "chest pain," occurs for reasons similar to those that cause heart attacks. The pain may occur in the heart and often also in the left arm and shoulder. Angina pectoris is a warning sign that the blood supply to the heart is inadequate but still sufficient to avoid myocardial cell death.

Strokes are caused by an interference with the blood supply to the brain. They may occur when a blood vessel bursts in the brain or when blood flow in a cerebral artery is blocked by a thrombus (blood clot) or by atherosclerosis. The effects of a stroke depend on how severe the damage is and where in the brain the stroke occurs.

Atherosclerosis is an accumulation within the arteries of fatty materials, abnormal amounts of smooth muscle, deposits of cholesterol or fibrin, or various kinds of cellular debris. These accumulations cause blood flow to be reduced (figure 20.18). The lumen (interior) of the artery may be further reduced in size by a clot that forms as a result of the atherosclerosis. In the severest cases, the artery may be blocked completely. Atherosclerosis is promoted by genetic factors, smoking, hypertension (high blood pressure), and high blood cholesterol levels. Diets low in cholesterol and saturated fats (from which cholesterol can be made) can help lower the level of blood cholesterol, and therapy for hypertension can reduce that risk factor. Stopping smoking, however, is the single most effective action a smoker can take to reduce the risk of atherosclerosis.

Arteriosclerosis, or hardening of the arteries, occurs when calcium is deposited in arterial walls. It tends to occur when atherosclerosis is severe. Not only do such arteries have restricted blood flow, but they also lack the ability to expand as normal arteries do to accommodate the volume of blood pumped out by the heart. This inflexibility forces the heart to work harder.

Treatments of Blocked Coronary Arteries

Atherosclerosis is treated both with medication and with invasive procedures. Medications include enzymes, which help dissolve clots; anticoagulants, which prevent clots from forming (aspirin works as a weak anticoagulant); and nitroglycerin, which dilates blood vessels.

Invasive treatments include *heart transplants,* where the damaged heart is replaced by a donor heart; *coronary bypass surgery,* where healthy segments of blood vessels are patched into a coronary artery, which diverts the flow of blood around a blocked section of artery; and *angioplasty,* where the blockage is reduced. Angioplasty is a procedure where a small balloon is threaded into a partially blocked coronary artery. Once in the blocked artery, the balloon is inflated, flattening the atherosclerosis deposit against the side of the artery. In some cases, a small metal mesh sleeve called a stent may also be inserted to prop the artery open.

> **20.8** Humans are subject to a variety of cardiovascular diseases, many of them associated with the accumulation of fatty materials on the inner surfaces of arteries.

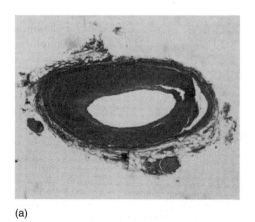

(a)

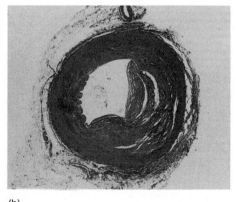

(b)

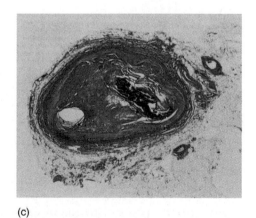

(c)

Figure 20.18   The path to a heart attack.

(a) The coronary artery shows only minor blockage. (b) The artery exhibits severe atherosclerosis—much of the passage is blocked by buildup on the interior walls of the artery. (c) The coronary artery is essentially completely blocked.

20.9 Types of Respiratory Systems

All animals obtain the energy that powers their lives by oxidizing molecules rich in energy-laden carbon–hydrogen bonds. This oxidative metabolism requires a ready supply of oxygen. The uptake of oxygen and the simultaneous release of carbon dioxide together constitute a form of gas exchange called **respiration.**

Most of the primitive phyla of organisms obtain oxygen by direct diffusion from their aquatic environments, which contains about 10 milliliters of dissolved oxygen per liter (figure 20.19*a*). Sponges, cnidarians, many flatworms and roundworms, and some annelid worms all obtain their oxygen by diffusion from surrounding water. Similarly, some members of the vertebrate class Amphibia conduct gas exchange by direct diffusion through their moist skin.

The more advanced marine invertebrates (mollusks, arthropods, and echinoderms) possess special respiratory organs called gills that increase the surface area available for diffusion of oxygen. A **gill** is basically a thin sheet of tissue that waves through the water. Gills can be simple, as in the papulae of echinoderms, or complex, as in the highly convoluted gills of fish (figure 20.19*c*). Terrestrial arthropods do not have a single major respiratory organ like a gill. Instead, a network of air ducts called **tracheae,** branching into smaller and smaller tubes, carries air to every part of the body (figure 20.19*b*). The openings of tracheae to the outside are through special structures called **spiracles,** which can be closed and opened. Terrestrial vertebrates, except for some amphibians, do have a single respiratory organ, called the **lung** (figure 20.19*d*).

> **20.9** Aquatic animals extract oxygen dissolved in water, some by direct diffusion, others with gills. Terrestrial animals use tracheae or lungs.

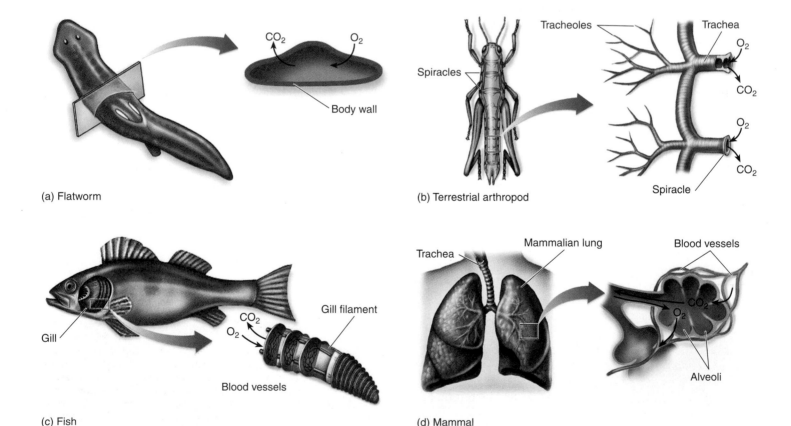

(a) Flatworm

(b) Terrestrial arthropod

(c) Fish

(d) Mammal

Figure 20.19   Gas exchange in animals.

(*a*) Gases diffuse directly across the body wall in many invertebrates, including flatworms, and in some species of amphibians. (*b*) Terrestrial arthropods respire through tracheae, which open to the outside through spiracles. (*c*) Fish gills provide a very large respiratory surface and employ countercurrent flow, which will be discussed later.(*d*) Mammalian lungs provide a large respiratory surface but do not permit countercurrent flow.

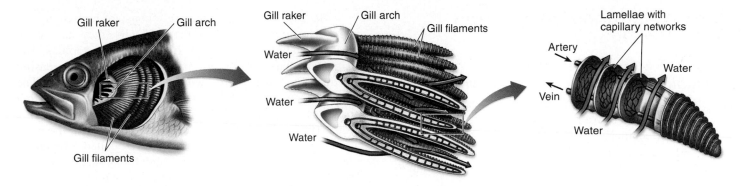

Figure 20.20   Structure of a fish gill.

Water passes from the gill arch over the filaments (from *left* to *right* in the diagram). Water always passes the lamellae in the same direction, which is opposite to the direction the blood circulates across the lamellae. This opposite orientation of blood and water flow is critical to the success of the gill's operation.

20.10 Respiration in Aquatic Vertebrates

Have you ever seen the face of a swimming fish up close? A fish swimming in water continuously opens and closes its mouth, pushing water through the mouth cavity and out a slit at the rear of the mouth—and (this is the whole point) past the gills on its one-way journey.

This swallowing process, which seems so awkward, is at the heart of a great advance in gill design achieved by the fishes. What is important about the swallowing is that it causes the water to always move past the fish's gills *in the same direction.* Moving the water past the gills in the same direction permits **countercurrent flow,** which is a supremely efficient way of extracting oxygen. Here is how it works:

Each gill is composed of two rows of gill filaments, thin membranous plates stacked one on top of the other and projecting out into the flow of water (figure 20.20). As water flows past the filaments from front to back, oxygen diffuses from the water into blood circulating within the gill filament. Within each filament the blood circulation is arranged so that the blood is carried in the direction opposite the movement of the water, from the back of the filament to the front. The advantage of the countercurrent flow system is that blood in the blood vessels of the gill filaments always encounters water that has a higher oxygen concentration, resulting in the diffusion of oxygen into the blood vessels. The water with the highest concentration of oxygen is passing by blood vessels that also have higher oxygen concentrations, but the oxygen concentration in the water is still higher (figure 20.21). Thus, countercurrent flow ensures that a concentration gradient remains between blood and water throughout the flow, permitting oxygen diffusion along the entire length of the filament.

Because of the countercurrent flow, the blood in the fish's gills can build up oxygen concentrations as high as those of the water entering the gills. The gills of bony fishes are the most efficient respiratory machines that have ever evolved among organisms. They are able to extract up to 85% of the available oxygen from water.

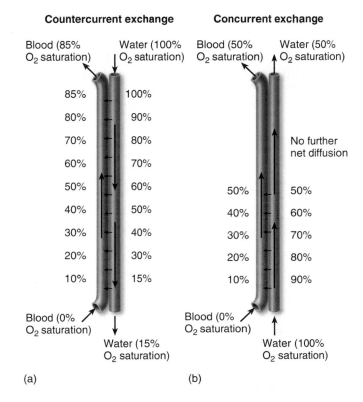

Figure 20.21   Countercurrent flow.

When blood and water flow in *opposite* directions (*a*), the initial oxygen concentration difference between the two is not large, but it is sufficient for oxygen to diffuse from water to blood. As the blood oxygen concentration rises, the blood encounters water with ever higher oxygen concentrations. At every point, the oxygen concentration is higher in the water, so that diffusion continues. In this example, blood attains an oxygen concentration of 85%. When blood and water flow in the *same* direction (*b*), oxygen can diffuse from the water into the blood rapidly at first, but the diffusion rate slows as more oxygen diffuses from the water into the blood, until finally the concentrations of oxygen in water and blood are equal. In this example, blood's oxygen concentration cannot exceed 50%.

20.10   Fish gills achieve countercurrent flow, making them very efficient at extracting oxygen.

20.11 Respiration in Terrestrial Vertebrates

Amphibians Get Oxygen from Air with Lungs

One of the major challenges facing the first land vertebrates was obtaining oxygen from air. Fish gills, which are superb oxygen-gathering machines in water, don't work in air. The gill's system of delicate membranes has no means of support in air, and the membranes collapse on top of one another—that's why a fish dies when kept out of water, literally suffocating in air for lack of oxygen.

Unlike a fish, if you lift a frog out of water and place it on dry ground, it doesn't suffocate. Partly this is because the frog is able to respire through its moist skin, but mainly it is because the frog has lungs. A **lung** is a respiratory organ designed like a bag. The amphibian lung is hardly more than a sac with a convoluted internal membrane (figure 20.22a). The air moves into the sac through a tubular passage from the head and then back out again through the same passage. Lungs are not as efficient as gills because new air that is inhaled mixes with old air already in the lung. But, air contains about 210 milliliters of oxygen per liter, over 20 times as much as seawater. So, because there is so much more oxygen *in* air, the lung doesn't have to be as efficient as the gill.

Reptiles and Mammals Increase the Lung Surface

Reptiles are far more active than amphibians, so they need more oxygen. But reptiles cannot rely on their skin for respiration the way amphibians can; their dry scaly skin is "watertight" to avoid water loss. Instead, the lungs of reptiles contain many small air chambers, which greatly increase the surface area of the lung available for diffusion of oxygen (figure 20.22b).

Because mammals maintain a constant body temperature by heating their bodies metabolically, they have even greater metabolic demands for oxygen than do reptiles. The problem of harvesting more oxygen is solved by increasing the diffusion surface area within the lung even more. The lungs of mammals possess on their inner surface many small chambers called **alveoli,** which are clustered together like grapes (figure 20.22c). Each cluster is connected to the main air sac in the lung by a short passageway called a **bronchiole.** Air within the lung passes through the bronchioles to the alveoli, where all oxygen uptake and carbon dioxide disposal takes place. In more active mammals, the individual alveoli are smaller and more numerous, increasing the diffusion surface area even more. Humans have about 300 million alveoli in each of their lungs, for a total surface area devoted to diffusion of about 80 square meters (about 42 times the surface area of the body)!

Birds Perfect the Lung

There is a limit to how much efficiency can be improved by increasing the surface area of the lung, a limit that has already been reached by the more active mammals. This efficiency is not enough for the metabolic needs of birds. Flying creates a respiratory demand for oxygen that exceeds the capacity of the saclike lungs of even the most active mammal. Unlike bats, whose flight involves considerable gliding, most birds beat their wings rapidly as they fly, often for quite a long time. This intensive wing beating uses up a lot of energy quickly, because the wing muscles must contract very frequently. Flying birds thus must carry out very active oxidative respiration within their cells to replenish the ATP expended by their flight muscles, and this requires a great deal of oxygen.

A novel way to improve the efficiency of the lung, one that does not involve further increases in its surface area, evolved in birds' lungs. This higher-efficiency lung copes with

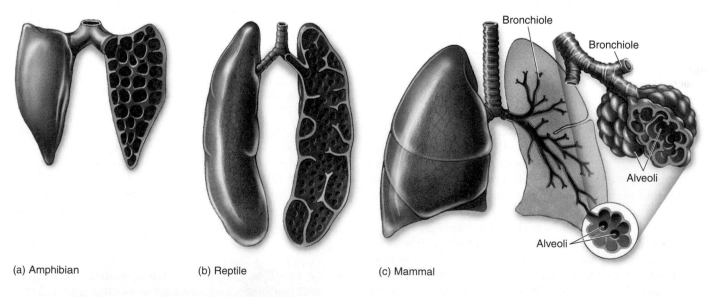

(a) Amphibian (b) Reptile (c) Mammal

Figure 20.22 Evolution of the vertebrate lung.
(*a*) Amphibian lung. (*b*) Reptile lung. (*c*) Mammalian lung.

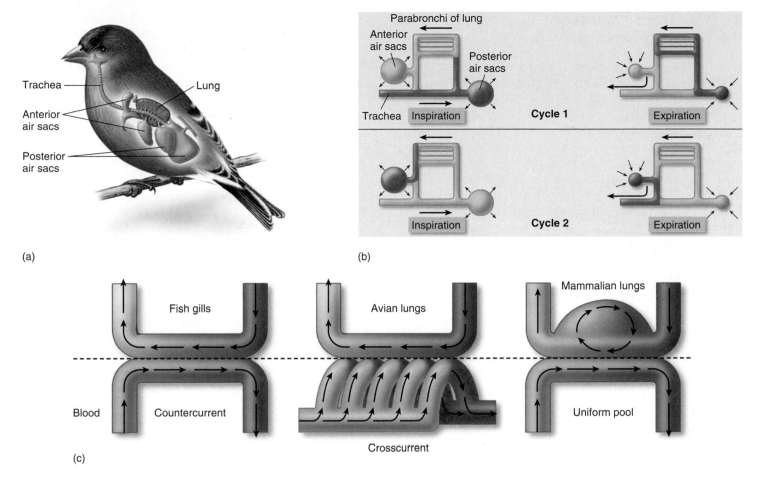

(a)

(b)

(c)

Figure 20.23 How a bird breathes.

(a) A bird's respiratory system is composed of the trachea, anterior air sacs, lungs, and posterior air sacs. (b) Breathing occurs in two cycles: in cycle 1, air is drawn from the trachea into the posterior air sacs and then is exhaled into a lung; in cycle 2, the air is drawn from the lung into the anterior air sacs and then is exhaled through the trachea. Passage of air through the lungs is always in the same direction, from posterior to anterior (*right* to *left* here). (c) Because blood circulates in the lung from anterior to posterior, the lung achieves a type of crosscurrent flow that is very efficient in picking up oxygen from the air.

the demands of flight. Can you guess what it is? In effect, they do what fishes do! An avian lung is connected to a series of air sacs behind the lung. When a bird inhales, the air passes directly to these posterior air sacs, which act as holding tanks (figure 20.23*a*). When the bird exhales, the air flows from these air sacs forward into the lung and then on through another set of anterior air sacs in front of the lungs and out of the body (figure 20.23*b*). What is the advantage of this complicated passage? It creates a unidirectional flow of air through the lungs.

Air flows through the lungs of birds in one direction only, from back to front. This one-way air flow results in two significant improvements: (1) There is no dead volume, as in the mammalian lung, so the air passing across the diffusion surface of the bird lung is always fully oxygenated. (2) Just as in the gills of fishes, the flow of blood past the lung runs in a direction different from that of the unidirectional air flow. It is not opposite, as in fish; instead, the latticework of capillaries is arranged across the air flow, at a 90-degree angle, called a **crosscurrent flow** (figure 20.23*c*). This is not as efficient as the 180-degree arrangement of fishes, but the blood leaving the lung can still contain more oxygen than exhaled air,

which no mammalian lung can do. That is why a sparrow has no trouble flying at an altitude of 6,000 meters on an Andean mountain peak, whereas a mouse of the same body mass and with a similar high metabolic rate will stand panting, unable even to walk. The sparrow is simply getting more oxygen than the mouse.

Just as fish gills are the most efficient aquatic respiratory machines, so bird lungs are the most efficient atmospheric ones. Both achieve high efficiency by using forms of countercurrent flow. No one knows what the lungs of dinosaurs were like, but if dinosaurs were indeed endothermic, it is likely that they evolved lungs more efficient than those of today's reptiles. As dinosaurs evolved at about the same time as mammals, they may have adopted the same evolutionary strategy of expanding surface area or adopted the approach seen in the birds.

20.11 Terrestrial vertebrates employ lungs to extract oxygen from air. Bird lungs are the most efficient atmospheric respiratory machines, achieving a form of crosscurrent flow.

20.12 The Mammalian Respiratory System

The oxygen-gathering mechanism of mammals, although less efficient than birds, adapts them well to their terrestrial habitat. Mammals, like all terrestrial vertebrates, obtain the oxygen they need for metabolism from air, which is about 21% oxygen gas. A pair of lungs is located in the chest, or **thoracic,** cavity. The two lungs hang free within the cavity, connected to the rest of the body only at one position, where the lung's blood vessels and air tube enter. This air tube is called a **bronchus.** It connects each lung to a long tube called the **trachea,** which passes upward and opens into the rear of the mouth (figure 20.24).

Air normally enters through the nostrils into the nasal cavity where it is moistened and warmed. In addition, the nostrils are lined with hairs that filter out dust and other particles. As the air passes through the nasal cavity, an extensive ar-ray of cilia further filters it. Mucous secretory ciliated cells in the bronchi also trap foreign particles and carry them upward, where they can be swallowed. The air then passes through the back of the mouth, entering first the **larynx** (voice box) and then the trachea. Because the air crosses the path of food at the back of the throat, a special flap called the epiglottis covers the trachea whenever food is swallowed, to keep it from "going down the wrong pipe." From there, it passes down through the bronchus to the lungs. The lungs contain millions of **alveoli,** tiny sacs clustered like grapes. The alveoli are surrounded by an extremely extensive capillary network. All gas exchange between the air and blood takes place across the walls of the alveoli.

The mammal respiratory apparatus is simple in structure and functions as a one-cycle pump. The thoracic cavity is bounded on its sides by the ribs and on the bottom by a thick layer of muscle, the **diaphragm,** which separates the thoracic cavity from the abdominal cavity. Each lung is covered by a very thin, smooth membrane called the **pleural membrane.** A second pleural membrane lines the interior of the thoracic cavity, into which the lungs hang. The space between these two membranes is very small and filled with fluid. This fluid causes the two membranes to adhere to each other in the same way a thin film of water can hold two plates of glass together, effectively coupling the lungs to the thoracic cavity.

Air is drawn into the lungs by the creation of negative pressure—that is, pressure in the lungs that is less than atmospheric pressure. The pressure in the lungs is reduced when

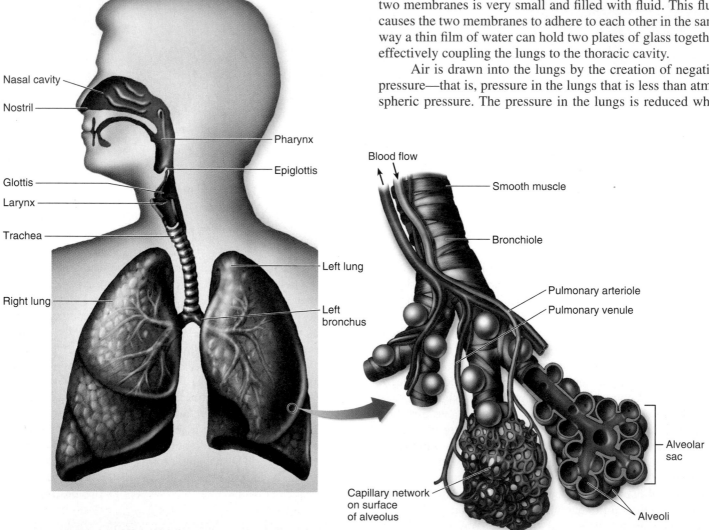

Figure 20.24 The human respiratory system.
The respiratory system consists of the lungs and the passages that lead to them.

Breathing

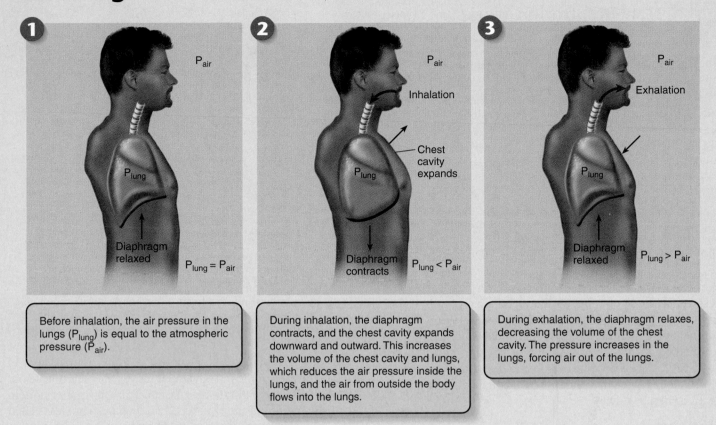

1 Before inhalation, the air pressure in the lungs (P_{lung}) is equal to the atmospheric pressure (P_{air}).

P_{air}

P_{lung}

Diaphragm relaxed

$P_{lung} = P_{air}$

2 During inhalation, the diaphragm contracts, and the chest cavity expands downward and outward. This increases the volume of the chest cavity and lungs, which reduces the air pressure inside the lungs, and the air from outside the body flows into the lungs.

P_{air}

Inhalation

Chest cavity expands

P_{lung}

Diaphragm contracts

$P_{lung} < P_{air}$

3 During exhalation, the diaphragm relaxes, decreasing the volume of the chest cavity. The pressure increases in the lungs, forcing air out of the lungs.

P_{air}

Exhalation

P_{lung}

Diaphragm relaxed

$P_{lung} > P_{air}$

Figure 20.25 How breathing works.

the volume of the lungs is increased. This is similar to how a bellow pump or accordion works. In both cases, when the bellow is extended the volume inside increases, causing air to rush in. How does this occur in the lungs? The volume in the lungs increases when muscles that surround the thoracic cavity contract, causing the thoracic cavity to increase in size. Because the lungs adhere to the thoracic cavity, they also expand in size. This creates negative pressure in the lungs, and air rushes in.

The Mechanics of Breathing

The active pumping of air in and out through the lungs is called **breathing.** During *inhalation,* muscular contraction causes the walls of the chest cavity to expand so that the rib cage moves outward and upward. The diaphragm is dome-shaped when relaxed but moves downward as it flattens during this contraction (figure 20.25). In effect, we have enlarged the bellow in the accordion.

During *exhalation,* the ribs and diaphragm return to their original resting position. In doing so, they exert pressure on the lungs. This pressure is transmitted uniformly over the entire surface of the lung, forcing air from the inner cavity back out to the atmosphere. In a human, a typical breath at rest moves about 0.5 liters of air, called the **tidal volume.** The ex-

tra amount that can be forced into and out of the lung is about 4.5 liters in men and 3.1 liters in women. The air remaining in the lung after such a maximal expiration is the residual volume, or dead volume, typically about 1.2 liters.

When each breath is completed, the lung still contains a volume of air, the **residual volume.** In human lungs, this volume is about 1,200 milliliters. Each inhalation adds from 500 milliliters (resting) to 3,000 milliliters (exercising) of additional air. Each exhalation removes approximately the same volume as inhalation added, reducing the air volume in the lungs once more to about 1,200 milliliters. Because the diffusion surfaces of the lungs are not exposed to fully oxygenated air, but rather to a mixture of fresh and partly oxygenated air, the respiratory efficiency of mammalian lungs is far from maximal. A bird, for example, whose lungs do not retain a residual volume, is able to achieve far greater respiratory efficiency.

20.12 In mammals, the lungs are located within a thoracic cavity bounded on the bottom by a muscular diaphragm. By contracting and relaxing, the diaphragm expands or reduces the volume of the cavity, drawing air into the lungs or forcing it out.

20.13 How Respiration Works: Gas Exchange

When oxygen has diffused from the air into the moist cells lining the inner surface of the lung, its journey has just begun. Passing from these cells into the bloodstream, the oxygen is carried throughout the body by the circulatory system, described earlier in this chapter. It has been estimated that it would take a molecule of oxygen three years to diffuse from your lung to your toe if transport depended only on diffusion, unassisted by a circulatory system.

Oxygen moves within the circulatory system carried piggyback on the protein **hemoglobin.** Hemoglobin molecules contain iron, which combines with oxygen in a reversible way (figure 20.26). Hemoglobin is manufactured within red blood cells and gives them their color. Hemoglobin never leaves these cells, which circulate in the bloodstream like ships bearing cargo. Oxygen binds to hemoglobin within these cells as they pass through the capillaries surrounding the alveoli of the lungs. Carried away within red blood cells, the oxygen-bearing hemoglobin molecules later release their oxygen molecules to metabolizing cells at distant locations in the body.

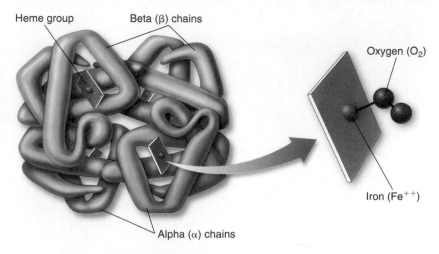

Figure 20.26 The hemoglobin molecule.
The hemoglobin molecule is actually composed of four protein chain subunits: two copies of the "alpha chain" and two copies of the "beta chain." Each chain is associated with a heme group, and each heme group has a central iron atom, which can bind to a molecule of oxygen.

O_2 Transport

Hemoglobin molecules act like little oxygen (O_2) sponges, soaking up oxygen within red blood cells and causing more to diffuse in from the blood plasma. At the high O_2 levels that occur in the blood supply of the lung, most hemoglobin molecules carry a full load of oxygen atoms. Later, in the tissues, the O_2 levels are much lower, so that hemoglobin gives up its bound oxygen.

In tissue, the presence of carbon dioxide (CO_2) causes the hemoglobin molecule to assume a different shape, one that gives up its oxygen more easily. This speeds up the unloading of oxygen from hemoglobin even more. The effect of CO_2 on oxygen unloading is of real importance, because CO_2 is produced by the tissues at the site of cell metabolism. For this reason the blood unloads oxygen more readily within those tissues undergoing metabolism and generating CO_2.

CO_2 Transport

At the same time the red blood cells are unloading oxygen, they are also absorbing CO_2 from the tissue (figure 20.27). About 8% of the CO_2 in blood is simply dissolved in plasma; another 20% is bound to hemoglobin; however, it does not bind to the heme group but rather to another site on the hemoglobin molecule, and so it does not compete with oxygen binding. The remaining 72% of the CO_2 diffuses into the red blood cells. To keep CO_2 from diffusing out of the red blood cells back into the plasma where CO_2 levels are low, the enzyme *carbonic anhydrase* combines CO_2 molecules with water molecules to form carbonic acid (H_2CO_3). This acid dissociates into **bicarbonate (HCO_3^-)** and hydrogen (H^+) ions. The H^+ binds to hemoglobin. A transporter protein in the membrane of the red blood cell moves the bicarbonate out of the red blood cell into the plasma. This transporter protein exchanges one chloride ion for a bicarbonate, a process called the "chloride shift." This reaction removes large amounts of CO_2 from the blood plasma, facilitating the diffusion of more CO_2 into it from the surrounding tissue. The facilitation is critical to CO_2 removal, because the difference in CO_2 concentration between blood and tissue is not large (only 5%). The formation of carbonic acid is also important in maintaining the acid-base balance of the blood, because bicarbonate serves as the major buffer of the blood plasma. You'll recall from chapter 3 that a buffer is a substance that can adjust the levels of hydrogen ions (H^+) in a solution, thereby controlling the pH of a solution.

The red blood cells carry their cargo of bicarbonate ions back to the lungs. The lower CO_2 concentration in the air inside the lungs causes the carbonic anhydrase reaction to proceed in the reverse direction, releasing gaseous CO_2, which diffuses outward from the blood into the alveoli. With the next exhalation, this CO_2 leaves the body. The one-fifth of the CO_2 bound to hemoglobin also leaves because hemoglobin has a greater affinity for oxygen than for CO_2 at low CO_2 concentrations. The diffusion of CO_2 outward from the red blood cells causes the hemoglobin within these cells to release its bound CO_2 and take up oxygen instead. The red blood cells, with their newly bound oxygen, then start the next respiratory journey.

NO Transport

Hemoglobin also has the ability to hold and release the gas nitric oxide (NO). Nitric oxide, although a noxious gas in the atmosphere, has an important physiological role in the body, acting on many kinds of cells to change their shape and function. For example, in blood vessels the presence of NO causes

Gas Exchange During Respiration

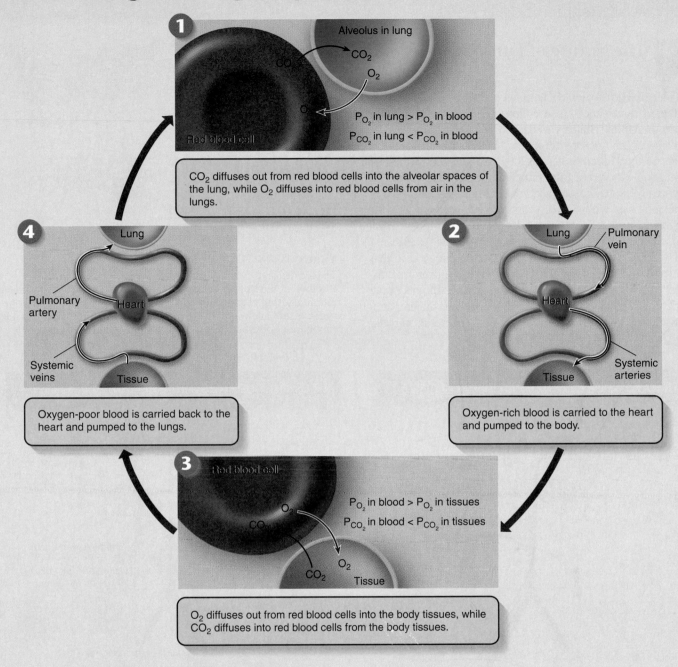

1 Alveolus in lung

CO_2
O_2

P_{O_2} in lung > P_{O_2} in blood
P_{CO_2} in lung < P_{CO_2} in blood

Red blood cell

CO_2 diffuses out from red blood cells into the alveolar spaces of the lung, while O_2 diffuses into red blood cells from air in the lungs.

4 Lung

Pulmonary artery

Heart

Systemic veins

Tissue

Oxygen-poor blood is carried back to the heart and pumped to the lungs.

2 Lung   Pulmonary vein

Heart

Systemic arteries

Tissue

Oxygen-rich blood is carried to the heart and pumped to the body.

3 Red blood cell

O_2
CO_2

P_{O_2} in blood > P_{O_2} in tissues
P_{CO_2} in blood < P_{CO_2} in tissues

O_2
CO_2 Tissue

O_2 diffuses out from red blood cells into the body tissues, while CO_2 diffuses into red blood cells from the body tissues.

Figure 20.27   How respiratory gas exchange works.

the blood vessels to expand because it relaxes the surrounding muscle cells. Thus, blood flow and blood pressure are regulated by the amount of NO released into the bloodstream.

Hemoglobin carries NO in a special form called super nitric oxide. In this form, NO has acquired an extra electron and is able to bind to an amino acid, called cysteine, present in hemoglobin. In the lungs, hemoglobin that is dumping CO_2 and picking up O_2 also picks up NO as super nitric oxide. In tissues, hemoglobin that is releasing its O_2 and picking up CO_2 may also release super nitric oxide as NO into the blood, making blood vessels expand, thereby increasing blood flow to the tissue. Alternatively, hemoglobin may trap any excesses

of NO on its iron atoms left vacant by the release of oxygen, causing blood vessels to constrict, which reduces blood flow to the tissue. When the red blood cells return to the lungs, hemoglobin dumps its CO_2 and the regular form of NO bound to the iron atoms. It is then ready to pick up O_2 and super nitric oxide and continue the cycle.

20.13   Oxygen and NO move through the circulatory system carried by the protein hemoglobin within red blood cells. Most CO_2 moves dissolved in the cytoplasm of red blood cells.

20.14 The Nature of Lung Cancer

Of all the diseases to which humans are susceptible, none is more feared than cancer (see chapter 7). One in every four deaths in the United States was caused by cancer in 2003. The American Cancer Society estimates that 556,500 people died of cancer in the United States in 2003. About 28% of these—157,200 people—died of **lung cancer.** About 140,000 cases of lung cancer were diagnosed each year in the 1980s, and 90% of these persons died within three years. Lung cancer is one of the leading causes of death among adults in the world today. What has caused lung cancer to become a major killer of Americans?

The search for a cause of cancers such as lung cancer has uncovered a host of environmental factors that appear to be associated with cancer. For example, the incidence of cancer per 1,000 people is not uniform throughout the United States. Rather, it is centered in cities and in the Mississippi Delta, suggest-

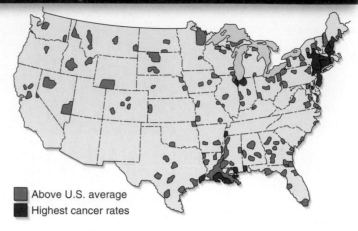

Above U.S. average

Highest cancer rates

Figure 20.28   Cancer in the United States.

The incidence of cancer per 1,000 people is not uniform throughout the United States. It is centered in cities where chemical manufacturing is common and in the Mississippi Delta. This suggests that environmental factors such as pollution and pesticide runoff may contribute to the development of cancer.

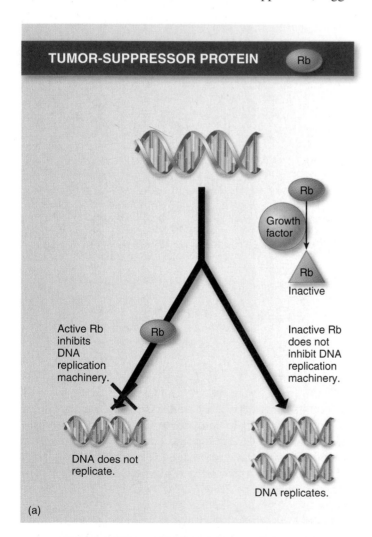

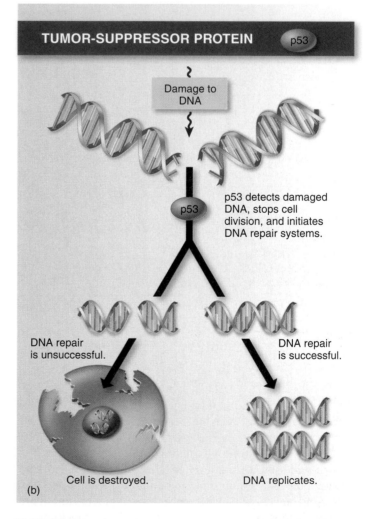

Figure 20.29   The roles Rb and p53 play in controlling cell division.

(*a*) The retinoblastoma protein, Rb, slows down cell division by inhibiting DNA replication. A normal cell can only reinitiate DNA replication when a growth factor renders Rb inactive. Lung cancer cells often have no Rb protein, and their cells will divide at a rapid rate. (*b*) p53 checks DNA for damage as it gets ready to replicate. If p53 detects damage, cell division is halted, and the cell's DNA repair systems are activated. If the damage does not get repaired, p53 initiates the destruction of the cancer cell.

ing that environmental factors such as pollution and pesticide runoff may contribute to cancer (figure 20.28). When the many environmental factors associated with cancer are analyzed, a clear pattern emerges: Most cancer-causing agents, or carcinogens, share the property of being potent mutagens. Recall from chapter 9 that a mutagen is a chemical or radiation that damages DNA, destroying or changing genes (a change in a gene is called a mutation). The conclusion that cancer is caused by mutation is now supported by an overwhelming body of evidence.

What sort of genes are being mutated? In the last several years, researchers have found that mutation of only a few genes is all that is needed to transform normally dividing cells into cancerous ones. Identifying and isolating these cancer-causing genes, investigators have learned that all are involved with regulating cell proliferation (how fast cells grow and divide). A key element in this regulation are so-called tumor suppressors, genes that actively prevent tumors from forming. Two of the most important tumor-suppressor genes are called *Rb* and *p53,* and the proteins they produce are Rb and p53, respectively (recall from chapter 7 that genes are usually indicated in italics and proteins in regular type).

The Rb Protein

The **Rb protein** (named after retinoblastoma, the rare eye cancer in which it was first discovered) acts as a brake on cell division, attaching itself to the machinery the cell uses to replicate its DNA, and preventing it from doing so. When the cell wants to divide, a growth signal molecule ties up Rb so that it is not available to act as a brake on the division process (figure 20.29*a*). If the gene that produces Rb is disabled, there are no brakes to prevent the cell from replicating its DNA and dividing. The control switch is locked in the "ON" position.

The p53 Protein

The **p53 protein,** a tumor suppressor sometimes called the "guardian angel" of the cell, inspects the DNA to ensure it is ready to divide. When p53 detects damaged or foreign DNA, it stops cell division and activates the cell's DNA repair systems. If the damage doesn't get repaired in a reasonable time, p53 pulls the plug, triggering events that kill the cell (figure 20.29*b*). In this way, mutations such as those that cause cancer are either repaired or the cells containing them eliminated. If the gene that produces p53 is itself destroyed by mutation, future damage accumulates unrepaired. Among this damage are mutations that lead to cancer, mutations that would have been repaired by healthy p53. Fifty percent of all cancers have a disabled *p53* gene.

Smoking Causes Lung Cancer

If cancer is caused by damage to growth-regulating genes, what then has led to the rapid increase in lung cancer in the United States? Two lines of evidence are particularly telling. The first consists of detailed information about cancer rates among smokers. The annual incidence of lung cancer among nonsmokers is only a few per 100,000 but increases with the number of cigarettes smoked per day to a staggering 300 per 100,000 for those smoking 30 cigarettes a day.

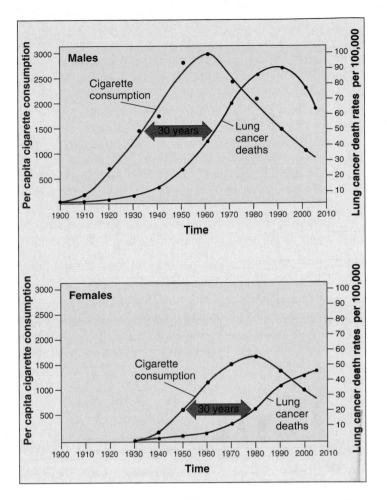

Figure 20.30 Incidence of lung cancer in men and women.

Lung cancer was a rare disease a century ago. As men in the United States increased smoking in the early 1900s, the incidence of lung cancer also increased. Women followed suit years later, and in 2004, more than 68,500 women died of lung cancer, a rate of 46 per 100,000. In that year, about 20% of women smoked and about 25% of men.

A second line of evidence consists of changes in the incidence of lung cancer that mirror changes in smoking habits. Look carefully at the data presented in figure 20.30. The upper curves are compiled from data on men and show the incidence of smoking and of lung cancer in the United States since 1900. As late as 1920, lung cancer was a rare disease. About 30 years after the incidence of smoking began to increase among men, lung cancer also started to become more common. Now look at the lower curves, which present data on women. Because of social mores, significant numbers of women in the United States did not smoke until after World War II, when many social conventions changed. As late as 1963, only 6,588 women had died of lung cancer. But as women's frequency of smoking has increased, so has their incidence of lung cancer, again with a lag of about 30 years. Women today have achieved equality with men in the number of cigarettes they smoke, and their lung cancer death rates are now rapidly approaching those for men. In 2004, an estimated 68,500 women died of lung cancer in the United States.

How does smoking cause cancer? Cigarette smoke contains many powerful mutagens, among them benzo[*a*]pyrene, and smoking introduces these mutagens to the lung tissues. Benzo[*a*]pyrene, for example, binds to three sites on the *p53* gene and causes mutations at these sites that inactivate the gene. In 1997, scientists studying this tumor-suppressor gene demonstrated a direct link between cigarettes and lung cancer. They found that the *p53* gene is inactivated in 70% of all lung cancers. When these inactivated *p53* genes are examined, they prove to have mutations at just the three sites where benzo[*a*]pyrene binds! Clearly, the chemical in cigarette smoke is responsible for the lung cancer.

In the face of these facts, why do so many people continue to smoke? Because the nicotine in cigarette smoke is an addictive drug. Researchers have identified the receptor on central nervous system neurons that it binds to, and they have demonstrated that smoking leads to a decrease in the brain's population of these receptors, leading to a craving for more cigarettes. This mechanism of nicotine addiction is very similar to that of cocaine addiction; once a person becomes addicted, it is difficult to quit. About half of those who try eventually succeed. Because your life is at stake, it is well worth the effort.

Clearly, an effective way to avoid lung cancer is not to smoke. Life insurance companies have computed that, on a statistical basis, smoking a single cigarette lowers your life expectancy 10.7 minutes (more than the time it takes to smoke the cigarette!). Every pack of 20 cigarettes bears an unwritten label: *The price of smoking this pack of cigarettes is 3 1/2 hours of your life.* Smoking a cigarette is very much like going into a totally dark room with a person who has a gun and standing still. The person with the gun cannot see you, does not know where you are, and shoots once in a random direction. A hit is unlikely, and most shots miss. As the person keeps shooting, however, the chance of eventually scoring a hit becomes more likely. Every time an individual smokes a cigarette, mutagens are being shot at his or her genes. Nor do statistics protect any one individual: Nothing says the first shot will not hit. Older people are not the only ones who die of lung cancer.

> **20.14** Cancer results from the destruction of genes by mutations that, when healthy, enable the cell to regulate cell division. To avoid lung cancer, don't smoke.

Exploring Current Issues

Additional Resources

Go to your campus library or look online to find the following articles, which further develop some of the concepts found in this chapter.

Gorman, C. (2003). The no. 1 killer of women: no, it's not breast cancer. More women die of heart disease than of all cancers combined. What you should know about the latest research and how you can protect yourself. *Time,* 161(17), 60.

Libby, P. (2002). Atherosclerosis: the new view, it causes chest pain, heart attack, and stroke, leading to more deaths every year than cancer. The long-held conception of how the disease develops turns out to be wrong. *Scientific American,* 286(5), 47.

Noonan, D. (2003). A healthy heart: cardiovascular disease is still the no. 1 killer of American men. New screening tests may help millions avoid the operating room. *Newsweek,* 141(24), 48.

Petit-Zeman, S. (2003). Smoke gets in your mind: it's bad enough that cigarettes give you lung cancer and heart disease. Could they really cause mental illness too? *The Scientist,* 174(2338), 30.

Wilson, J. F. (2003). Angiogenesis research moves past cancer: abnormal blood vessel growth affects many diseases, including psoriasis, Crohn disease, and rheumatoid arthritis. *The Scientist,* 17(8), 30.

In the News: **Cancer and Smoking**

One of the great tragedies of modern life is the widespread use of tobacco, particularly smoked in cigarettes. Tobacco contains the chemical nicotine, which is very addictive. Other chemicals in cigarette smoke are powerful mutagens, producing lung cancer with great regularity. Smokers are thus addicted to a habit that will kill them. The morality of such legalized deadly addictions has attracted national attention.

Article 1. *Does smoking really cause cancer?*
Article 2. *Is smoking cigarettes addictive?*
Article 3. *Nicotine addiction from cigarettes may begin in a few days of smoking.*
Article 4. *Should the FDA regulate nicotine?*
Article 5. *Curing lung cancer is a step closer.*

Find these articles, written by the author, online at www.mhhe.com/tlwessentials/exp20.

Summary

Circulation

20.1 Open and Closed Circulatory Systems

- Cnidarians and flatworms have a gastrovascular cavity that functions in digestion and circulation. Mollusks and arthropods have open circulatory systems, and annelids and all vertebrates have closed circulatory systems (**figure 20.1**).

20.2 Architecture of the Vertebrate Circulatory System

- In the vertebrate circulatory system, blood circulates from the heart through arteries and arterioles to capillaries. The capillary is the site of gas, food, and waste exchange. Blood flows from the capillaries back to the heart through venules and larger veins (**figures 20.4 and 20.8**).

20.3 The Lymphatic System: Recovering Lost Fluid

- Needed fluid is lost to the body tissues through leaky capillary walls. The fluid is recovered by lymphatic capillaries that drain into lymphatic vessels. Lymphatic vessels return the fluid back to the circulatory system (**figure 20.10**).

20.4 Blood

- Blood is a salty, protein-rich fluid that circulates through the body in the circulatory system. Blood contains red blood cells that are involved in gas exchange, and various types of white blood cells, called leukocytes, that defend the body against infection and cancer. Platelets are fragments of cells involved in blood clotting (**figure 20.12**).

Evolution of Vertebrate Circulatory Systems

20.5 Fish Circulation

- The fish heart consists of a series of chambers. Blood enters the heart and is collected in the first two chambers, the sinus venosus and the atrium. Blood is then pumped from the third and fourth chambers, the ventricle and conus arteriosus, to the body. It passes to the gills, where gas exchange occurs, and then oxygenated blood travels throughout the body (**figure 20.13**).

20.6 Amphibian and Reptile Circulation

- Amphibians and reptiles have lungs and so the blood pumps in two cycles: between the heart and lungs (the pulmonary circulation) and between the heart and the body (the systemic circulation). Blood enters the heart through the atria and then passes to the ventricle where it is pumped out to the lungs and body (**figure 20.14**).

20.7 Mammalian and Bird Circulation

- Mammalian and bird hearts are two-cycle pumps but, unlike amphibians and reptiles, the ventricle is separated so that the heart has four chambers (**figure 20.15**). Deoxygenated blood enters the right side of the heart, where it is pumped to the lungs. Oxygenated blood from the lungs enters the left side of the heart and is pumped throughout the body.

20.8 Cardiovascular Diseases

- Cardiovascular diseases such as heart attacks, angina pectoris, strokes, atherosclerosis, and arteriosclerosis are primarily caused by fatty deposits in arteries that interfere with blood flow (**figure 20.18**).

Respiration

20.9 Types of Respiratory Systems

- Some aquatic animals extract oxygen across the skin and others use gills. Terrestrial animals extract oxygen from the air with tracheae or lungs (**figure 20.19**).

20.10 Respiration in Aquatic Vertebrates

- Fishes' gills (**figure 20.20**) are very efficient by using a countercurrent flow system. Oxygenated water flows through the gills in a direction opposite of blood flow through the gills such that the water always has a higher concentration of oxygen driving the diffusion of oxygen into the blood (**figure 20.21**).

20.11 Respiration in Terrestrial Vertebrates

- Terrestrial vertebrate lungs (**figure 20.22**) are less efficient than a fish's gills, but bird lungs use a crosscurrent flow system that is more efficient than other terrestrial animals' lungs (**figure 20.23**).

20.12 The Mammalian Respiratory System

- Mammals breathe by contracting muscles that line the thoracic cavity that holds the lungs. Contraction of these muscles, including the diaphragm, expands the space in the lungs, causing air to rush into the lungs. Relaxation of the muscles causes air to be exhaled (**figure 20.25**).

20.13 How Respiration Works: Gas Exchange

- Hemoglobin carries oxygen and nitric oxide. Carbon dioxide is primarily carried as bicarbonate and hydrogen ions in the cytoplasm of red blood cells (**figure 20.27**).

Lung Cancer and Smoking

20.14 The Nature of Lung Cancer

- Lung cancer results from harmful mutations caused by chemicals in cigarette smoke (**figure 20.30**).

Self-Test

1. Which of the following is *not* a function of the circulatory system?
   a. regulation of some body processes and characteristics
   b. protection against injury, foreign toxins, and microbes
   c. transportation of materials in the body
   d. All of the above are functions of the circulatory system.
2. Exchange of waste material, oxygen, carbon dioxide, and metabolites such as salts and food molecules, takes place in the
   a. capillaries. c. arterioles.
   b. venules. d. arteries.
3. The most numerous blood cell is the
   a. macrophage. c. platelet.
   b. leukocyte. d. erythrocyte.

4. Additional septa in the amphibian and reptile heart allows for
   a. higher blood pressure to move blood faster.
   b. better separation of oxygenated and deoxygenated blood.
   c. better body temperature regulation.
   d. better transport of food to needy tissues.
5. The four-chambered heart and double loop vessel system is thought to be important in the evolution of
   a. locomotion. c. exothermy.
   b. ectothermy. d. endothermy.
6. In the gills of aquatic animals, countercurrent flow allows
   a. the animal's blood to be continually exposed to water of lower oxygen concentration.

b the animal's blood to be continually exposed to water of equal oxygen concentration.
c. the animal's blood to be continually exposed to water of higher oxygen concentration.
d. the animal to be endothermic.

7. The purpose of gill filaments and the lung alveoli is to
a. decrease the surface area available for gas exchange.
b. increase the surface area available for gas exchange.
c. decrease the volume available for gas exchange.
d. increase the volume available for gas exchange.

8. In general, the need for alveoli increases as the
a. energy need for different vertebrate classes increases.
b. energy need for different vertebrate classes decreases.

c. habitat need for different vertebrate classes increases.
d. nutrition need for different vertebrate classes increases.

9. Oxygen is transported by
a. hemoglobin in red blood cells.
b. dissolving it in the blood plasma.
c. blood proteins in the blood plasma.
d. platelets in the blood plasma.

10. Most of the carbon dioxide is transported by
a. hemoglobin in red blood cells.
b. dissolving it in the blood plasma.
c. blood proteins in the blood plasma.
d. platelets in the blood plasma.

Visual Understanding

1. **Figure 20.7** Does being a dedicated "couch potato" or "computer geek" ever cause circulatory problems?

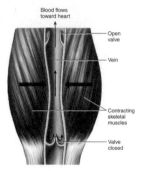

2. **Figure 20.30** What conclusions can you draw about cigarette smoking and women's health? Explain.

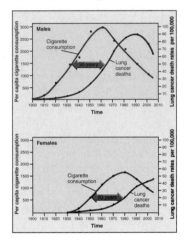

Challenge Questions

Circulation Your friend, Masaji, has a painful paper cut on his finger and it's bleeding. Disgusted, because it already stained his shirt, he asks you why the blood is so important, anyway. What do you tell him?

Evolution of Vertebrate Circulatory Systems Explain how the mnemonic "VAVA lung, VAVA body" describes the path of the blood in humans.

Respiration Sometimes when people are eating, they take a bite that is too big, or is not completely chewed, and when they

swallow it becomes stuck partway down the esophagus. Because the esophagus is a soft, muscular tube that lies just behind the trachea, a somewhat stiffer tube, this bulge of food in the esophagus can sometimes push hard enough on the trachea to close it off. In this case, people have been trained to do the Heimlich maneuver, which is a method of pushing up rapidly on the diaphragm, compressing the lungs. Why might this help?

Lung Cancer and Smoking How is it that cigarette smoking can be linked to an increased incidence of many kinds of cancer?

Online Learning Center

Visit the Online Learning Center for this chapter at www.mhhe.com/tlwessentials/ch20 for quizzes, animations, interactive learning exercises, and other study tools. At the site you will also find extended answers to the end-of-chapter questions.

21

The Path of Food Through the Animal Body

There are no photosynthetic animals. Animals are heterotrophs, gaining the energy to power their lives by consuming and oxidizing organic molecules present in other organisms. All animals must continuously consume plant material or other animals in order to live. The nuts in this chipmunk's cheeks will be consumed and converted within its cells to body tissue, energy, and refuse. Most of the molecules in the nuts are far too large to be conveniently absorbed by the chipmunk's cells, and so they are first broken down into smaller pieces: carbohydrates are broken down into sugars, proteins into amino acids, fats into fatty acids. This process, called digestion, is the focus of this chapter. The chipmunk's digestive system is a long tube passing from mouth to anus, with specialized segments for digestion and for the subsequent absorption into its body of the resulting sugars, amino acids, and fatty acids. Whatever is left is excreted from the body as feces. The water it drinks is also absorbed by the chipmunk's digestive tube; metabolic wastes are filtered from the chipmunk's body fluids by its kidneys and excreted as a concentrated solution called urine.

21.1   Food for Energy and Growth

The food animals eat provides both a source of energy and essential molecules such as certain amino acids and fats that the animal body is not able to manufacture for itself. An optimal diet contains more carbohydrates than fats and also a significant amount of protein, as the "pyramid of nutrition" recommended by the federal government indicates (figure 21.1). Fats have a far greater number of energy-rich carbon–hydrogen bonds and thus a much higher energy content per gram than carbohydrates or proteins, which contain more carbon–oxygen bonds that are already oxidized. For this reason, fats are a very efficient way to store energy. When food is consumed, it is either metabolized by muscles and other cells of the body, or it is converted into fat and stored in fat cells.

Carbohydrates are obtained primarily from cereals and grains, breads, fruits, and vegetables. On the average, carbohydrates contain 4.1 **calories** per gram; fats, by comparison, contain 9.3 calories per gram, over twice as much. Dietary fats are obtained from oils, margarine, and butter and are abundant in fried foods, meats, and processed snack foods, such as potato chips and crackers. Protein, like carbohy-

Fats and sweets
(Use sparingly)

Milk group
(2–3 servings):
Milk, cheese,
yogurt

Meat group
(2–3 servings):
Meat, poultry,
fish, dry beans,
eggs, nuts

Vegetable group
(3–5 servings)

Fruit group
(2–4 servings)

Bread and cereal group (6–11 servings):
Whole grains, complex carbohydrates

Figure 21.1   The pyramid of nutrition.
A healthy diet uses more foods at the base of the pyramid and fewer at the top.
Source: U.S. Department of Agriculture, U.S. Department of Health and Human Services.

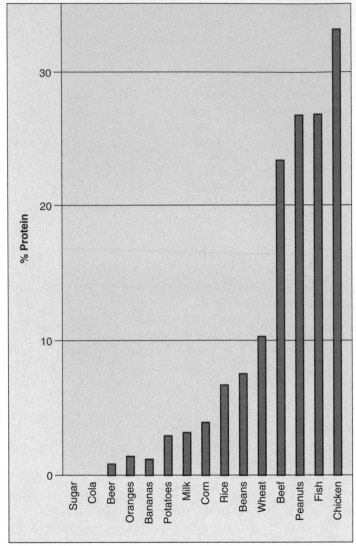

Figure 21.2   The protein content of a variety of common foods.
Humans and many other vertebrates must eat protein to obtain the amino acids they are unable to synthesize.

drates, has 4.1 calories per gram and can be obtained from many foods including dairy products, poultry, fish, meat, and grains (figure 21.2).

The body uses carbohydrates for energy and fats to construct cell membranes and other cell structures, to insulate nervous tissue, and to provide energy. Fat-soluble vitamins that are essential for proper health are also absorbed with fats. Proteins are used for energy and as building materials for cell structures, enzymes, hemoglobin, hormones, and muscle and bone tissue.

In wealthy countries such as those of North America and Europe, being significantly overweight is common, the result of habitual overeating and high-fat diets, in which fats con-

WEIGHT	100	105	110	115	120	125	130	135	140	145	150	155	160	165	170	175	180	185	190	195	200	205
HEIGHT																						
5' 0"	20	21	21	22	23	24	25	26	27	28	29	30	31	32	33	34	35	36	37	38	39	40
5' 1"	19	20	21	22	23	24	25	26	26	27	28	29	30	31	32	33	34	35	36	37	38	39
5' 2"	18	19	20	21	22	23	24	25	26	27	27	28	29	30	31	32	33	34	35	36	37	37
5' 3"	18	19	19	20	21	22	23	24	25	26	27	27	28	29	30	31	32	33	34	35	35	36
5' 4"	17	18	19	20	21	21	22	23	24	25	26	27	27	28	29	30	31	32	33	33	34	35
5' 5"	17	17	18	19	20	21	22	22	23	24	25	26	27	27	28	29	30	31	32	32	33	34
5' 6"	16	17	18	19	19	20	21	22	23	23	24	25	26	27	27	28	29	30	31	31	32	33
5' 7"	16	16	17	18	19	20	20	21	22	23	23	24	25	26	27	27	28	29	30	31	31	32
5' 8"	15	16	17	17	18	19	20	21	21	22	23	24	24	25	26	27	27	28	29	30	30	31
5' 9"	15	16	16	17	18	18	19	20	21	21	22	23	24	24	25	26	27	27	28	29	30	30
5' 10"	14	15	16	17	17	18	19	19	20	21	22	22	23	24	24	25	26	27	27	28	29	29
5' 11"	14	15	15	16	17	17	18	19	20	20	21	22	22	23	24	24	25	26	26	27	28	29
6' 0"	14	14	15	16	16	17	18	18	19	20	20	21	22	22	23	24	24	25	26	26	27	28
6' 1"	13	14	15	15	16	16	17	18	18	19	20	20	21	22	22	23	24	24	25	26	26	27
6' 2"	13	13	14	15	15	16	17	17	18	19	19	20	21	21	22	22	23	24	24	25	26	26
6' 3"	12	13	14	14	15	16	16	17	17	18	19	19	20	21	21	22	22	23	24	24	25	26
6' 4"	12	13	13	14	15	16	16	16	17	18	18	19	19	20	21	21	22	23	23	24	24	25

Figure 21.3 Are you overweight?

This chart presents the body mass index (BMI) values used by federal health authorities to determine who is overweight. Your body mass index is at the intersection of your height and weight. BMI is calculated by dividing body weight in kilograms by height in meters squared. For pounds and feet, first multiply weight in pounds by 703, and then divide by height in inches squared. A BMI value of 25 or over is considered overweight and 30 or over is considered obese.

Source: "Shape Up America" National Institutes of Health.

stitute over 35% of the total caloric intake. The international standard measure of appropriate body weight is the body mass index (BMI), estimated as your body weight in kilograms, divided by your height in meters squared (figure 21.3). In the United States, the National Institutes of Health estimated in 2000 that 64.5% of adults, 129.6 million Americans, were overweight, with a body mass index of 25 or more. Of those individuals, 61.3 million were considered obese with a body mass index of 30 or greater. Being overweight is highly correlated with coronary heart disease, diabetes, and many other disorders. A BMI of less than 18.5 is also unhealthy, often resulting from eating disorders including anorexia nervosa.

One essential characteristic of food is its fiber content. Fiber is the part of plant food that cannot be digested by humans and is found in fruits, vegetables, and grains such as in breads and cereals. Other animals, however, have evolved many different ways to process food that has a relatively high fiber content. Diets that are low in fiber, now common in the United States, result in a slower passage of food through the colon. This low dietary fiber content is thought to be associated with incidences of colon cancer in the United States, which are among the highest levels in the world.

Essential Substances for Growth

Over the course of their evolution, many animals have lost the ability to manufacture certain substances they need, substances that often play critical roles in their metabolism. Mosquitoes and many other blood-sucking insects, for example, cannot manufacture cholesterol, but they obtain it in their diet because human blood is rich in cholesterol. Many vertebrates are unable to manufacture one or more of the 20 amino acids used to make proteins. Humans are unable to synthesize eight amino acids: lysine, tryptophan, threonine, methionine, phenylalanine, leucine, isoleucine, and valine. These amino acids, called **essential amino acids,** must therefore be obtained from proteins in the food we eat. For this reason, it is important to eat so-called complete proteins—that is, ones containing all the essential amino acids. In addition, all vertebrates have also lost the ability to synthesize certain polyunsaturated fats that provide backbones for the many kinds of fats their bodies manufacture.

Trace Elements.   In addition to supplying energy, food that is consumed must also supply the body with essential minerals such as calcium and phosphorus, as well as a wide variety of **trace elements,** which are minerals required in very small amounts. Among the trace elements are iodine (a component of thyroid hormone), cobalt (a component of vitamin B_{12}), zinc and molybdenum (components of enzymes), manganese, and selenium. All of these, with the possible exception of selenium, are also essential for plant growth; animals obtain them directly from plants that they eat or indirectly from animals that have eaten plants, or plant eaters.

Vitamins. Essential organic substances that are used in trace amounts are called **vitamins.** Humans require at least 13 different vitamins (table 21.1). Many vitamins are required cofactors for cellular enzymes. Humans, monkeys, and guinea pigs, for example, have lost the ability to synthesize ascorbic acid (vitamin C) and will develop the potentially fatal disease called scurvy—characterized by weakness, spongy gums, and bleeding of the skin and mucous membranes—if vitamin C is not supplied in their diets. All other mammals are able to synthesize ascorbic acid.

21.1 Food is an essential source of calories. It is important to maintain a proper balance of carbohydrate, protein, and fat. Individuals with a body mass index of 25 or more are considered overweight. Food also provides key amino acids that the body cannot manufacture for itself, as well as necessary trace elements and vitamins.

TABLE 21.1 MAJOR VITAMINS

Vitamin	Function	Dietary Source	Recommended Daily Allowance (milligrams)	Deficiency Symptoms
Vitamin A	Used in making visual pigments, maintenance of epithelial tissues	Green vegetables, carrots, milk products, liver	1	Night blindness, flaky skin
B-complex vitamins B_1	Coenzyme in CO_2 removal during cellular respiration	Meat, grains, legumes	1.5	Beriberi, weakening of heart, edema
B_2 (riboflavin)	Part of coenzymes FAD and FMN, which play metabolic roles	In many different kinds of foods	1.8	Inflammation and breakdown of skin, eye irritation
B_3 (niacin)	Part of coenzymes NAD^+ and $NADP^+$	Liver, lean meats, grains	20	Pellagra, inflammation of nerves, mental disorders
B_5 (pantothenic acid)	Part of coenzyme A, a key connection between carbohydrate and fat metabolism	In many different kinds of foods	5 to 10	Rare: fatigue, loss of coordination
B_6 (pyridoxine)	Coenzyme in many phases of amino acid metabolism	Cereals, vegetables, meats	2	Anemia, convulsions, irritability
B_{12} (cyanocobalamin)	Coenzyme in the production of nucleic acids	Red meat, dairy products	0.003	Pernicious anemia
Biotin	Coenzyme in fat synthesis and amino acid metabolism	Meat, vegetables	Minute	Rare: depression, nausea
Folic acid	Coenzyme in amino acid and nucleic acid metabolism	Green leafy vegetables, whole-grain products	0.4	Anemia, diarrhea
Vitamin C	Important in forming collagen, cement of bone, teeth, connective tissue of blood vessels; may help maintain resistance to infection	Fruit, green leafy vegetables	45	Scurvy, breakdown of skin, blood vessels
Vitamin D (calciferol)	Increases absorption of calcium and promotes bone formation	Dairy products, cod liver oil	0.01	Rickets, bone deformities
Vitamin E (tocopherol)	Protects fatty acids and cell membranes from oxidation	Margarine, seeds, green leafy vegetables	15	Rare
Vitamin K	Essential to blood clotting	Green leafy vegetables	0.03	Severe bleeding

A Closer Look

Closing in on the Long-Sought Link Between Diabetes and Obesity

We Americans love to eat, but in 2004 the Centers for Disease Control and Prevention released a report warning we are eating ourselves into a diabetes epidemic. Diabetes affected 7 million Americans in 1991. At the close of 2001, the number was 17 million, over 7% of all Americans. This represents an increase of 60% in 10 years! Over that same period, the obesity rate increased from 12% to nearly 21%, reflecting a 74% increase since 1991.

Diabetes is a disorder in which the body's cells fail to take up glucose from the blood. Tissues waste away as glucose-starved cells are forced to consume their own proteins. Diabetes is the leading cause of kidney failure, blindness, and amputation in adults. Almost all the increase in diabetes in the last decade is in the 85% of diabetics who suffer from type II, or "adult-onset" diabetes. These individuals lack the ability to use the hormone insulin.

Your body manufactures insulin after a meal as a way to alert cells that higher levels of glucose are coming soon. The insulin signal attaches to special receptors on the cell surfaces, which respond by causing the cell to turn on its glucose-transporting machinery.

Individuals who suffer from type II diabetes have normal or even elevated levels of insulin in their blood, and normal insulin receptors, but for some reason the binding of insulin to their cell receptors does not turn on the glucose-transporting machinery like it is supposed to do. For 30 years researchers have been trying to figure out why not.

How does insulin act to turn on a normal cell's glucose transporting machinery? Proteins called IRS proteins (the names refer not to tax collectors, but to *insulin receptor substrate*) snuggle up against the insulin receptor inside the cell. When insulin attaches to the receptor protein, the receptor responds by adding a phosphate group onto the IRS molecules. Like being touched by a red-hot poker, this galvanizes the IRS molecules into action. Dashing about, they activate a variety of processes, including an enzyme that turns on the glucose-transporting machinery.

When the IRS genes are deliberately taken out of action in so-called "knockout" mice, type II diabetes results. Are defects in the genes for IRS proteins responsible for type II diabetes? Probably not. When researchers look for IRS gene mutations in inherited type II diabetes, they don't find them. The IRS genes are normal.

This suggests that in type II diabetes something is interfering with the action of the IRS proteins. What might it be? An estimated 80% of those who develop type II diabetes are obese, a tantalizing clue.

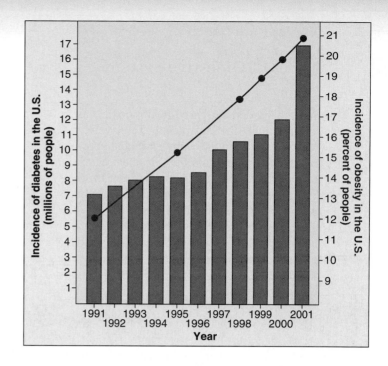

What is the link between diabetes and obesity? Recent research suggests an answer to this key question. A team of scientists at the University of Pennsylvania School of Medicine had been investigating why a class of drugs called thiazolidinediones (TZDs) helped combat diabetes. They found that TZDs cause the body's cells to use insulin more effectively, and this suggested to them that the TZD drug might be targeting a hormone.

The researchers then set out to see if they could find such a hormone in mice. In search of a clue, they started by looking to see which mouse genes were activated or deactivated by TZD. Several were. Examining them, they were able to zero in on the hormone they sought. Dubbed *resistin,* the hormone is produced by fat cells and prompts tissues to resist insulin. The same resistin gene is present in humans too. The researchers speculate that resistin may have evolved to help the body deal with periods of famine.

Mice given resistin by the researchers lost much of their ability to take up blood sugar. When given a drug that lowers resistin levels, these mice recovered the lost glucose-transporting ability.

Researchers don't yet know how resistin acts to lower insulin sensitivity, although blocking the action of IRS proteins seems a likely possibility.

Importantly, dramatically high levels of the hormone were found in mice obese from overeating. Finding this sort of result is like ringing a dinner bell to diabetes researchers. If obesity is causing high resistin levels in humans, leading to type II diabetes, then resistin-lowering drugs might offer a diabetes cure!

On the scent of something important, resistin researchers are now shifting their efforts from mice to humans. Much needs to be checked, as there are no guarantees that what works in a mouse will do so in the same way in a human. Still, the excitement is tangible.

21.2 Types of Digestive Systems

Heterotrophs are divided into three groups on the basis of their food sources. Animals that eat plants exclusively are classified as **herbivores;** common examples include cows, horses, rabbits, and sparrows. Animals that are meat eaters, such as cats, eagles, trout, and frogs, are **carnivores. Omnivores** are animals that eat both plants and other animals. We humans are omnivores, as are pigs, bears, and crows.

Single-celled organisms (as well as sponges) digest their food intracellularly. Other animals digest their food extracellularly, within a digestive cavity. In this case, the digestive enzymes are released into a cavity that is continuous with the animal's external environment. In cnidarians and flatworms (such as *Planaria*), the digestive cavity has only one opening that serves as both mouth and anus. There can be no specialization within this type of digestive system, called a gastrovascular cavity, because every cell is exposed to all stages of food digestion (figure 21.4).

Specialization occurs when the digestive tract, or alimentary canal, has a separate mouth and anus, so that transport of food is one way. The most primitive digestive tract

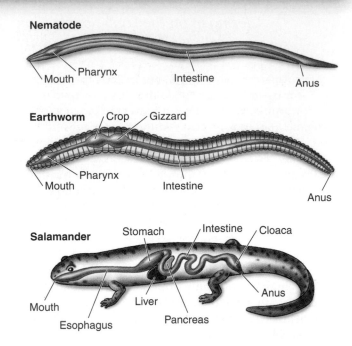

Figure 21.5 One-way digestive tracts.
One-way movement through the digestive tract allows different regions of the digestive system to become specialized for different functions.

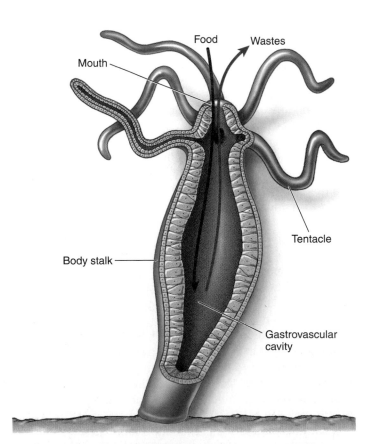

Figure 21.4 The gastrovascular cavity of *Hydra*.
Because there is only one opening, the mouth is also the anus, and no specialization is possible in the different regions that participate in extracellular digestion.

is seen in nematodes (phylum Nematoda), where it is simply a tubular *gut* lined by an epithelial membrane. Earthworms (phylum Annelida) have a digestive tract specialized in different regions for the ingestion, storage, fragmentation, digestion, and absorption of food. All higher animals show similar specializations (figure 21.5).

The ingested food may be stored in a specialized region of the digestive tract or may first be subjected to physical fragmentation through the chewing action of teeth (in the mouth of many vertebrates) or the grinding action of pebbles (in the gizzard of earthworms and birds). Chemical **digestion** then occurs primarily in the intestine, breaking down the larger food molecules of polysaccharides, fats, and proteins into smaller subunits. Carbohydrate digestion begins in the mouth of some animals, and protein digestion begins in the stomach in some animals. Chemical digestion involves hydrolysis reactions that liberate the subunits—primarily monosaccharides, amino acids, and fatty acids—from the food. These products of chemical digestion pass through the epithelial lining of the gut into the blood, in a process known as absorption. Any molecules in the food that are not absorbed cannot be used by the animal. These wastes are excreted from the anus.

> **21.2 Most animals digest their food extracellularly. A digestive tract with a one-way transport of food allows specialization of regions for different functions.**

21.3 Vertebrate Digestive Systems

In humans and other vertebrates, the digestive system consists of a tubular gastrointestinal tract and accessory digestive organs (figure 21.6). The initial components of the gastrointestinal tract are the mouth and the pharynx, which is the common passage of the oral and nasal cavities. The pharynx leads to the esophagus, a muscular tube that delivers food to the stomach, where some preliminary digestion occurs. From the stomach, food passes to the first part of the small intestine, where a battery of digestive enzymes continues the digestive process. The products of digestion then pass across the wall of the small intestine into the bloodstream. The small intestine empties what remains into the large intestine, where water and minerals are absorbed. In most vertebrates other than mammals, the waste products emerge from the large intestine into a cavity called the cloaca (see figure 21.5), which also receives the products of the urinary and reproductive systems. In mammals, the urogenital products are separated from the fecal material in the large intestine, also called the colon; the fecal material enters the rectum and is expelled through the anus.

In general, carnivores have shorter intestines for their size than do herbivores. A short intestine is adequate for a carnivore, but herbivores ingest a large amount of plant cellulose, which resists digestion. These animals have a long, convoluted small intestine. In addition, mammals called *ruminants* (such as cows) that consume grass and other vegetation have stomachs with multiple chambers, where bacteria aid in the digestion of cellulose. Other herbivores, including rabbits and horses, digest cellulose (with the aid of bacteria) in a blind pouch called the **cecum** located at the beginning of the large intestine. Accessory digestive organs described later in this chapter include the liver, the gallbladder, and the pancreas.

The tubular gastrointestinal tract of a vertebrate has a characteristic layered structure (figure 21.7). The innermost layer is the mucosa, an epithelium that lines the interior of the tract (the lumen). The next major tissue layer, composed of connective tissue, is called the submucosa. Just outside the submucosa is the muscularis, which consists of a double layer of smooth muscles. The muscles in the inner layer have a circular orientation, and those in the outer layer are arranged longitudinally. Another connective tissue layer, the serosa, covers the external surface of the tract. Nerves, intertwined in regions called *plexuses,* are located in the submucosa and help regulate the gastrointestinal activities.

21.3 The vertebrate digestive system consists of a tubular gastrointestinal tract, which is modified in different animals, composed of a series of tissue layers.

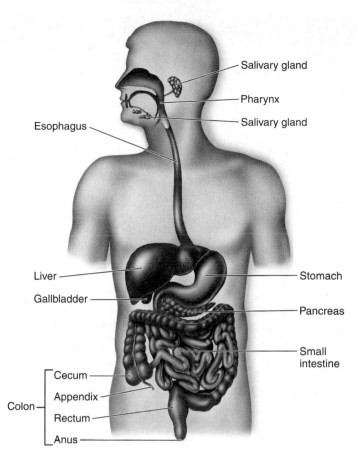

Figure 21.6 The human digestive system.
The tubular gastrointestinal tract and accessory digestive organs are shown. The colon extends from the cecum to the anus.

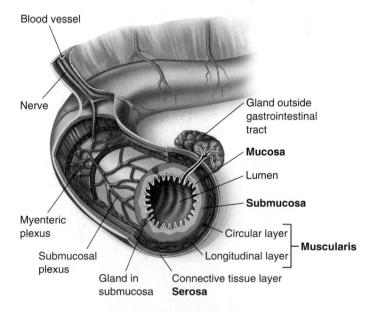

Figure 21.7 The layers of the gastrointestinal tract.
The mucosa contains a lining epithelium, the submucosa is composed of connective tissue (as is the serosa), and the muscularis consists of smooth muscles.

Specializations of the digestive systems in different kinds of vertebrates reflect differences in the way these animals live. Fishes have a large pharynx with gill slits, whereas air-breathing vertebrates have a greatly reduced pharynx. Many vertebrates have teeth (figure 21.8), and chewing (*mastication*) breaks up food into small particles and mixes it with fluid secretions. Birds, which lack teeth, break up food in their two-chambered stomachs (figure 21.9). In one of these chambers, the gizzard, small pebbles ingested by the bird are churned together with the food by muscular action. This churning grinds up the seeds and other hard plant material into smaller chunks that can be digested more easily in the second chamber of the stomach.

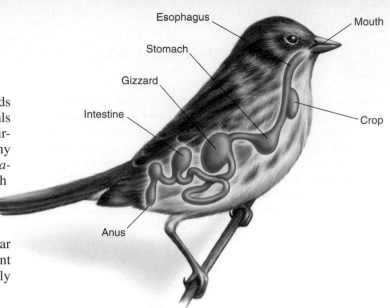

Figure 21.9 The digestive tract of birds.

Birds, which lack teeth, have a muscular chamber called the gizzard that works to break down food. Birds swallow gritty objects or pebbles that lodge in the gizzard and pulverize food before it passes into the intestine. Food is stored in the crop.

Vertebrate Teeth

Carnivorous mammals have pointed teeth that lack flat grinding surfaces. Such teeth are adapted for cutting and shearing. Carnivores often tear off pieces of their prey but have little need to chew them, because digestive enzymes can act directly on animal cells. (Recall how a cat or dog gulps down its food.) By contrast, grass-eating herbivores, such as cows and horses, must pulverize the cellulose cell walls of plant tissue before digesting it. These animals have large, flat teeth with complex ridges well suited to grinding.

Humans are omnivores, and human teeth are specialized for eating both plant and animal food. Viewed simply, humans are carnivores in the front of the mouth and herbivores in the back. The four front teeth in the upper and lower jaws are sharp, chisel-shaped incisors used for biting. On each side of the incisors are sharp, pointed teeth called cuspids (sometimes

referred to as "canine" teeth), which are used for tearing food. Behind the canines are two premolars and three molars, all with flattened, ridged surfaces for grinding and crushing food. Children have only 20 teeth, but these deciduous teeth are lost during childhood and are replaced by 32 adult teeth. The third molars are the wisdom teeth, which usually grow in during the late teen or early twenties years, when a person is assumed to have gained a little "wisdom."

The tooth is a living organ, composed of connective tissue, nerves, and blood vessels, held in place by cementum, a bonelike substance that anchors the tooth in the jaw (figure 21.10). The interior of the tooth contains connective tissue called pulp that extends into the root canals and contains nerves and blood vessels. A layer of calcified tissue called dentin surrounds the pulp cavity. The portion of the tooth that projects above the gums is called the crown and is covered with an extremely hard, nonliving substance called enamel. Enamel protects the tooth against abrasion and acids that are produced by bacteria living in the mouth.

Processing Food in the Mouth

Inside the mouth, the tongue mixes food with a mucous solution, **saliva.** In humans, three pairs of salivary glands secrete saliva into the mouth through ducts in the mouth's mucosal lining. Saliva moistens and lubricates the food so that it is easier to swallow and does not abrade the tissue it passes on its way through the esophagus. Saliva also contains the hydrolytic enzyme salivary **amylase,** which initiates the breakdown of the polysaccharide starch into the disaccharide maltose. This digestion is usually minimal in humans, however, because most people don't chew their food very long.

The secretions of the salivary glands are controlled by the nervous system, which in humans maintains a constant

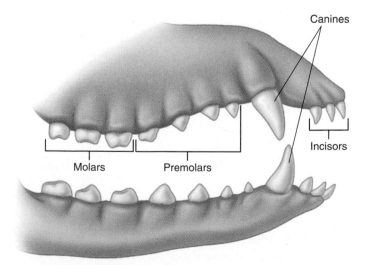

Figure 21.8 Diagram of generalized vertebrate dentition.

Different vertebrates have specific variations from this generalized pattern, depending on whether the vertebrate is an herbivore, carnivore, or omnivore.

flow of about half a milliliter per minute when the mouth is empty of food. This continuous secretion keeps the mouth moist. The presence of food in the mouth triggers an increased rate of secretion, as taste-sensitive neurons in the mouth send impulses to the brain, which responds by stimulating the salivary glands. The most potent stimuli are acidic solutions; lemon juice, for example, can increase the rate of salivation eightfold. The sight, sound, or smell of food can stimulate salivation markedly in dogs, but in humans, these stimuli are much less effective than thinking or talking about food.

Swallowing

When food is ready to be swallowed, the tongue moves it to the back of the mouth. In mammals, the process of swallowing begins when the soft palate elevates, pushing against the back wall of the pharynx (figure 21.11). Elevation of the soft palate seals off the nasal cavity and prevents food from entering it. Pressure against the pharynx stimulates neurons within its walls, which send impulses to the swallowing center in the brain. In response, muscles are stimulated to contract and raise the *larynx* (voice box). This pushes the *glottis*, the opening from the larynx into the trachea (windpipe), against a flap of tissue called the *epiglottis*. These actions keep food out of the respiratory tract, directing it instead into the esophagus.

21.4 In many vertebrates, ingested food is fragmented through the tearing or grinding action of specialized teeth. In birds, this is accomplished through the grinding action of pebbles in the gizzard. Food mixed with saliva is swallowed and enters the esophagus.

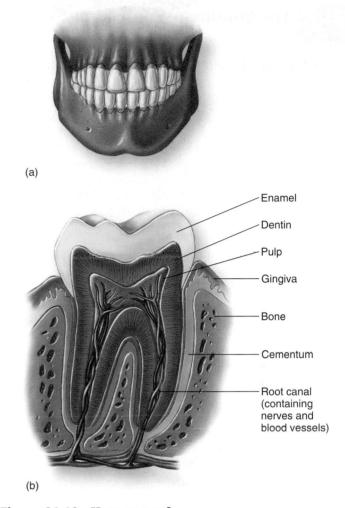

(a)

(b)

Figure 21.10 Human teeth.
(*a*) The front six teeth on the upper and lower jaws are cuspids and incisors. The remaining teeth, running along the sides of the mouth, are grinders called premolars and molars. Hence, humans are carnivores in the front of their mouths and herbivores in the back. (*b*) Each tooth is alive, with a central pulp containing nerves and blood vessels. The actual chewing surface is a hard enamel layered over the softer dentin, which forms the body of the tooth.

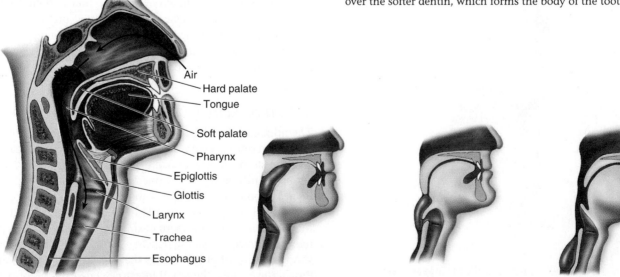

Figure 21.11 The human pharynx, palate, and larynx.
Food that enters the pharynx is prevented from entering the nasal cavity by elevation of the soft palate and from entering the larynx and trachea (the airways of the respiratory system) by elevation of the larynx against the epiglottis.

The Esophagus and Stomach

Structure and Function of the Esophagus

Swallowed food enters a muscular tube called the **esophagus,** which connects the pharynx to the stomach. In adult humans, the esophagus is about 25 centimeters long; the upper third is enveloped in skeletal muscle, for voluntary control of swallowing, while the lower two-thirds is surrounded by involuntary smooth muscle. The swallowing center stimulates successive waves of contraction in these muscles that move food along the esophagus to the stomach. These rhythmic waves of muscular contraction are called **peristalsis** (figure 21.12); they enable humans and other vertebrates to swallow even if they are upside down.

In many vertebrates, the movement of food from the esophagus into the stomach is controlled by a ring of circular smooth muscle, the **sphincter,** that opens in response to the pressure exerted by the food. Contraction of this sphincter prevents food in the stomach from moving back into the esophagus. Rodents and horses have a true sphincter at this site and thus cannot regurgitate, while humans lack a true sphincter and so are able to regurgitate. Regurgitation occurs during vomiting, when the sphincter between the stomach and esophagus is relaxed and the contents of the stomach are forcefully expelled through the mouth. The relaxing of this sphincter can also result in the movement of stomach acid into the esophagus, causing an irritation called heartburn. Chronic and severe heartburn is a condition known as acid reflux.

Structure and Function of the Stomach

The **stomach** is a saclike portion of the digestive tract. Its inner surface is highly convoluted, enabling it to fold up when empty and open out like an expanding balloon as it fills with food. Thus, while the human stomach has a volume of only about 50 milliliters when empty, it may expand to contain 2 to 4 liters of food when full. Carnivores that engage in sporadic gorging as an important survival strategy possess stomachs that are able to distend much more than that.

The stomach contains an extra layer of smooth muscle for churning food and mixing it with *gastric juice,* an acidic secretion of the tubular gastric glands of the mucosa (figure 21.13). These exocrine glands contain two kinds of secretory cells: *parietal cells,* which secrete hydrochloric acid (HCl); and *chief cells,* which secrete pepsinogen, a weak protease (protein-digesting enzyme) that requires a very low pH to be active. This low pH is provided by the HCl. Activated pepsinogen molecules then cleave each other at specific sites, producing a much more active protease, pepsin. This process of secreting a relatively inactive enzyme that is then converted into a more active enzyme outside the cell prevents the chief cells from digesting themselves. It should be noted that only proteins are partially digested in the stomach—there is no significant digestion of carbohydrates or fats.

Action of Acid

The human stomach produces about 2 liters of HCl and other gastric secretions every day, creating a very acidic solution inside the stomach. The concentration of HCl in this solution is about 10 millimolar, corresponding to a pH of 2. Thus, gastric juice is about 250,000 times more acidic than blood, whose normal pH is 7.4. The low pH in the stomach helps denature food proteins, making them easier to digest, and keeps pepsin maximally active. Active pepsin hydrolyzes food proteins into shorter chains of polypeptides that are not fully digested until the mixture enters the small intestine. The mixture of partially digested food and gastric juice is called **chyme.**

The acidic solution within the stomach also kills most of the bacteria that are ingested with the food. The few bacteria

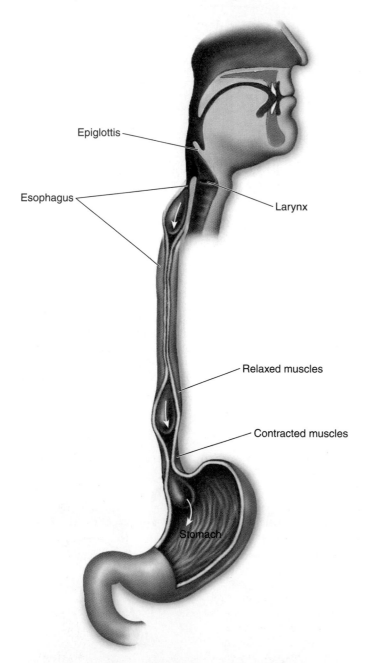

Epiglottis

Esophagus

Larynx

Relaxed muscles

Contracted muscles

Stomach

Figure 21.12   The esophagus and peristalsis.

After food has entered the esophagus, rhythmic waves of muscular contraction, called peristalsis, move the food down to the stomach.

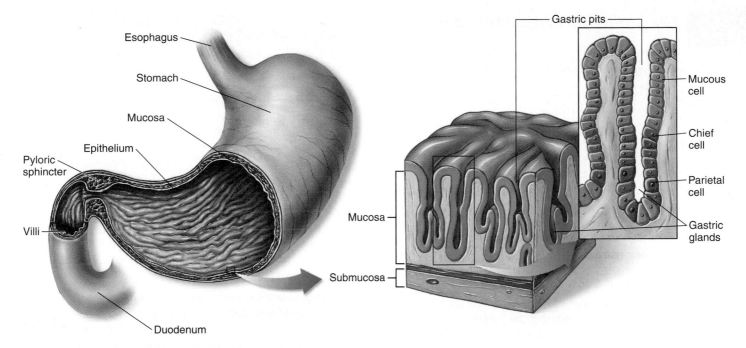

Figure 21.13 The stomach and gastric glands.
Food enters the stomach from the esophagus. A band of smooth muscle called the pyloric sphincter controls the entrance to the duodenum, the upper part of the small intestine. The epithelial walls of the stomach are dotted with gastric pits, which contain glands that secrete hydrochloric acid (HCl) and the enzyme pepsinogen. The gastric glands consist of mucous cells, chief cells that secrete pepsinogen, and parietal cells that secrete HCl. Gastric pits are the openings of the gastric glands.

that survive the stomach and enter the intestine intact are able to grow and multiply there, particularly in the large intestine. In fact, most vertebrates harbor thriving colonies of bacteria within their intestines, and bacteria are a major component of feces. As we discuss later, bacteria that live within the digestive tract of cows and other ruminants play a key role in the ability of these mammals to digest cellulose.

Ulcers

It is important that the stomach not produce too much acid. If it did, the body could not neutralize the acid later in the small intestine, a step essential for the final stage of digestion. Production of acid is controlled by hormones. These hormones are produced by endocrine cells scattered within the walls of the stomach. The hormone **gastrin** regulates the synthesis of HCl by the parietal cells of the gastric pits, permitting HCl to be made only when the pH of the stomach is higher than about 1.5.

Overproduction of gastric acid can occasionally eat a hole through the wall of the stomach. Such gastric **ulcers** are rare, however, because epithelial cells in the mucosa of the stomach are protected somewhat by a layer of alkaline mucus, and because those cells are rapidly replaced by cell division if they become damaged (gastric epithelial cells are replaced every two to three days). Over 90% of gastrointestinal ulcers are duodenal ulcers, which are ulcers of the small intestine. These may be produced when excessive amounts of acidic chyme are delivered into the duodenum, so that the acid cannot be properly neutralized through the action of alkaline pancreatic juice

(described later). Susceptibility to ulcers is increased when the mucosal barriers to self-digestion are weakened by an infection of the bacterium *Helicobacter pylori*. Indeed, modern antibiotic treatments of this infection can reduce symptoms and often even cure the ulcer.

In addition to producing HCl, the parietal cells of the stomach also secrete intrinsic factor, a polypeptide needed for the intestinal absorption of vitamin B_{12}. Because this vitamin is required for the production of red blood cells, persons who lack sufficient intrinsic factor develop a type of anemia (low red blood cell count) called *pernicious anemia*.

Leaving the Stomach

Chyme leaves the stomach through the *pyloric sphincter* (figure 21.13) to enter the small intestine. This is where all terminal digestion of carbohydrates, fats, and proteins occurs, and where the products of digestion—amino acids, glucose, and fatty acids—are absorbed into the blood. Only some of the water in chyme and a few substances such as aspirin and alcohol are absorbed through the wall of the stomach.

21.5 Peristaltic waves of contraction propel food along the esophagus to the stomach. Gastric juice contains strong hydrochloric acid and the protein-digesting enzyme pepsin, which begins the digestion of proteins into shorter polypeptides. The acidic chyme is then transferred through the pyloric sphincter to the small intestine.

Digestion and Absorption: The Small Intestine

The digestive tract exits from the stomach into the **small intestine,** where the breaking down of large molecules into small ones occurs. Only relatively small portions of food are introduced into the small intestine at one time, to allow time for acid to be neutralized and enzymes to act. The small intestine is the true digestive vat of the body. Within it, carbohydrates are broken down into sugars, proteins into amino acids, and fats into fatty acids. Once these small molecules have been produced, they pass across the epithelial wall of the small intestine into the bloodstream.

Some of the enzymes necessary for these digestive processes are secreted by the cells of the intestinal wall. Most, however, are made in a large gland called the *pancreas* (discussed in section 21.8), situated near the junction of the stomach and the small intestine. It is one of the body's major exocrine (secreting through ducts) glands. The pancreas sends its secretions into the small intestine through a duct that empties into its initial segment, the **duodenum.** Your small intestine is approximately 6 meters long—unwound and stood on its end, it would be far taller than you are! Only the first 25 centimeters, about 4% of the total length, is the duodenum. It is within this initial segment, where the pancreatic enzymes enter the small intestine, that digestion occurs.

Much of the food energy the vertebrate body harvests is obtained from fats. The digestion of fats is carried out by a collection of molecules known as *bile salts* secreted into the duodenum from the *liver* (discussed in section 21.8). Because fats are insoluble in water, they enter the intestine as drops within the watery chyme. The bile salts, which are partly lipid-soluble and partly water-soluble, work like detergents. They combine with fats to form microscopic droplets in a process called emulsification. These tiny droplets have greater surface areas upon which the enzyme that breaks down fats, called lipase, can work. This allows the digestion of fats to proceed more rapidly.

All the rest of the small intestine (96% of its length) is called the **ileum.** The ileum is devoted to absorbing water and the products of digestion into the bloodstream. The lining of the small intestine is covered with fine fingerlike projections called **villi** (singular, **villus**), each too small to see with the naked eye (figure 21.14*a*). In turn, each of the cells covering a villus is covered on its outer surface by a field of cytoplasmic projections called **microvilli** (figure 21.14*b,c*). Both kinds of projections greatly increase the absorptive surface of the lining of the small intestine. The average surface area of the small intestine of an adult human is about 300 square meters, more than the surface of many swimming pools!

The amount of material passing through the small intestine is startlingly large. Per day, an average human consumes about 800 grams of solid food, and 1,200 milliliters of water, for a total volume of about 2 liters. To this amount is added about 1.5 liters of fluid from the salivary glands, 2 liters from the gastric secretions of the stomach, 1.5 liters from the pancreas, 0.5 liters from the liver, and 1.5 liters of intestinal secretions. The total adds up to a remarkable 9 liters—more than 10% of the total volume of your body! However, although the flux is great, the *net* passage is small. Almost all these fluids and solids are reabsorbed during their passage through the small intestine—about 8.5 liters across the walls of the small intestine and 0.35 liters across the wall of the large intestine. Of the 800 grams of solids and 9 liters of liquids that enter the digestive tract each day, only about 50 grams of solids and 100 milliliters of liquids leave the body as feces. The fluid absorption efficiency of the digestive tract thus approaches 99%, very high indeed.

Concentration of Solids: The Large Intestine

The **large intestine,** or **colon,** is much shorter than the small intestine, approximately 1 meter long, but it is called the large intestine because of its larger diameter. The small intestine empties directly into the large intestine at a junction where the cecum and the appendix are located, which are two structures no longer actively used in humans (see figure 21.6). No digestion takes place within the large intestine, and only about 6% to 7% of fluid absorption occurs there. The large intestine is not convoluted, lying instead in three relatively straight segments, and its inner surface does not possess villi. As a consequence, the large intestine has only one-thirtieth the absorptive surface area of the small intestine. Although some water, sodium, and vitamin K are absorbed across its walls, the primary function of the large intestine is to act as a refuse dump. Within it, undigested material, including large amounts of plant fiber and cellulose, is compacted and stored. Many bacteria live and actively divide within the large intestine, where they play a role in the processing of undigested material into the final excretory product, **feces.** Bacterial fermentation produces gas within the colon at a rate of about 500 milliliters per day. This rate increases greatly after the consumption of beans or other vegetable matter because the passage of undigested plant material (fiber) into the large intestine provides substrates for fermentation.

The final segment of the digestive tract is a short extension of the large intestine called the **rectum.** Compact solids within the colon pass into the rectum as a result of the peristaltic contractions of the muscles encasing the large intestine. From the rectum, the solid material passes out of the body through the **anus.**

21.6 Most digestion occurs in the initial upper portion of the small intestine, called the duodenum. The rest of the small intestine is devoted to absorption of water and the products of digestion. The large intestine compacts residual solid wastes.

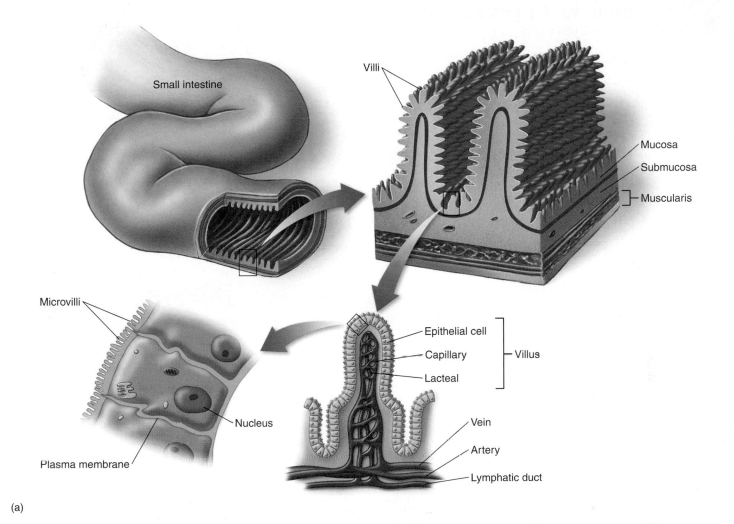

(a)

(b)

(c)

Figure 21.14   The small intestine.

(a) Cross section of the small intestine with details showing villi structure. (b) Microvilli, shown in a scanning electron micrograph, are very densely clustered, giving the small intestine an enormous surface area, which is very important for efficient absorption. (c) Intestinal microvilli as shown in a transmission electron micrograph.

21.7 Variations in Vertebrate Digestive Systems

Most animals lack the enzymes necessary to digest cellulose, the carbohydrate that functions as the chief structural component of plants. The digestive tracts of some animals, however, contain prokaryotes and protists that convert cellulose into substances the host can digest. Although digestion by gastrointestinal microorganisms plays a relatively small role in human nutrition, it is an essential element in the nutrition of many other kinds of animals, including insects like termites and cockroaches and a few groups of herbivorous mammals. The relationships between these microorganisms and their animal hosts are mutually beneficial and provide an excellent example of symbiosis.

Cows, deer, and other herbivores called ruminants have large, divided stomachs (figure 21.15). The first portion consists of the rumen and a smaller chamber, the reticulum; the second portion consists of two additional chambers, the omasum and abomasum. The rumen, which may hold up to 50 gallons, serves as a fermentation vat in which prokaryotes and protists convert cellulose and other molecules into a variety of simpler compounds. The location of the rumen at the front of the four chambers is important because it allows the animal to regurgitate and rechew the contents of the rumen, an activity called *rumination*, or "chewing the cud." The cud is then swallowed and enters the reticulum, from which it passes to the omasum and then the abomasum, where it is finally mixed with gastric juice. Hence, only the abomasum is equivalent to the human stomach in its function. This process leads to a far more efficient digestion of cellulose in ruminants than in mammals that lack a rumen, such as horses.

In some animals such as rodents, horses, and lagomorphs (rabbits and hares), the digestion of cellulose by microorganisms takes place in the cecum, which is greatly enlarged (figure 21.16). Because the cecum is located beyond the stomach, regurgitation of its contents is impossible. However, rodents and lagomorphs have evolved another way to digest cellulose that achieves a degree of efficiency similar to that of ruminant digestion. They do this by eating their feces, thus passing the food through their digestive tract a second time. The second passage makes it possible for the animal to absorb the nutrients produced by the microorganisms in its cecum. Animals that engage in this practice of **coprophagy** (from the Greek words *copros,* excrement, and *phagein,* eat) cannot remain healthy if they are prevented from eating their feces.

Cellulose is not the only plant product that vertebrates can use as a food source because of the digestive activities of intestinal microorganisms. Wax, a substance indigestible by most terrestrial animals, is digested by symbiotic bacteria liv-

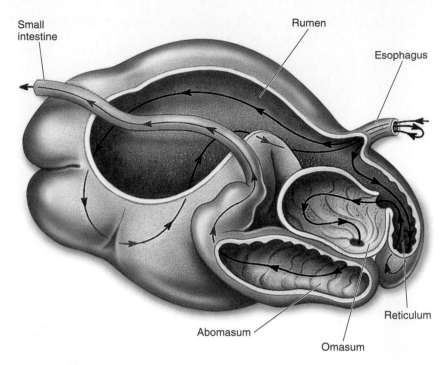

Figure 21.15 Four-chambered stomach of a ruminant.
The grass and other plants that a ruminant, such as a cow, eats enter the rumen, where they are partially digested. Before moving into a second chamber, the reticulum, the food may be regurgitated and rechewed. The food is then transferred to the rear two chambers, the omasum and abomasum. Only the abomasum is equivalent to the human stomach in its function of secreting gastric juice.

ing in the gut of honeyguides, African birds that eat the wax in bees' nests. In the marine food chain, wax is a major constituent of copepods (crustaceans in the plankton), and many marine fish and birds appear to be able to digest wax with the aid of symbiotic microorganisms.

Another example of the way intestinal microorganisms function in the metabolism of their animal hosts is provided by the synthesis of vitamin K. All mammals rely on intestinal bacteria to synthesize this vitamin, which is necessary for the clotting of blood. Birds, which lack these bacteria, must consume the required quantities of vitamin K in their food. In humans, prolonged treatment with antibiotics greatly reduces the populations of bacteria in the intestine; under such circumstances, it may be necessary to provide supplementary vitamin K.

> **21.7** Much of the food value of plants is tied up in cellulose, and the digestive tract of many animals harbors colonies of cellulose-digestive microorganisms. Intestinal microorganisms also produce molecules such as vitamin K that are important to the well-being of their vertebrate hosts.

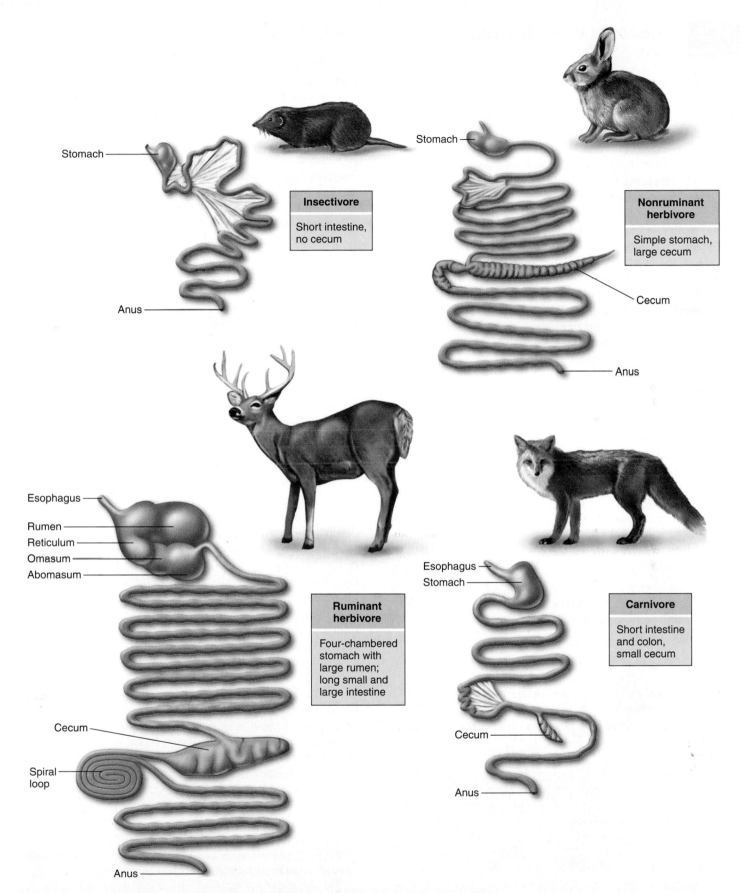

Figure 21.16 The digestive systems of different mammals reflect their diets.
Herbivores require long digestive tracts with specialized compartments for the breakdown of plant matter. Protein diets are more easily digested; thus, insectivorous and carnivorous mammals have short digestive tracts with few specialized pouches.

Labels within figure:

Insectivore — Short intestine, no cecum
Stomach
Anus

Nonruminant herbivore — Simple stomach, large cecum
Stomach
Cecum
Anus

Ruminant herbivore — Four-chambered stomach with large rumen; long small and large intestine
Esophagus
Rumen
Reticulum
Omasum
Abomasum
Cecum
Spiral loop
Anus

Carnivore — Short intestine and colon, small cecum
Esophagus
Stomach
Cecum
Anus

Accessory Digestive Organs

The Pancreas

The **pancreas,** a large gland situated near the junction of the stomach and the small intestine, is one of the accessory organs that contribute secretions to the digestive tract. Fluid from the pancreas is secreted into the duodenum through the *pancreatic duct* (figure 21.17). This fluid contains a host of enzymes, including trypsin and chymotrypsin, which digest proteins. Inactive forms of these enzymes are released into the duodenum and are then activated by the enzymes of the intestine. Pancreatic fluid also contains pancreatic amylase, which digests starch; and lipase, which digests fats. Pancreatic enzymes digest proteins into smaller polypeptides, polysaccharides into shorter chains of sugars, and fat into free fatty acids and other products. The digestion of these molecules is then completed by the intestinal enzymes.

Pancreatic fluid also contains bicarbonate, which neutralizes the HCl from the stomach and gives the chyme in the duodenum a slightly alkaline pH. The digestive enzymes and bicarbonate are produced by clusters of secretory cells known as *acini.*

In addition to its exocrine role in digestion, the pancreas also functions as an endocrine gland, secreting several hormones into the blood that control the blood levels of glucose and other nutrients. These hormones are produced in the **islets of Langerhans,** clusters of endocrine cells scattered throughout the pancreas. The two most important pancreatic hormones, insulin and glucagon, are discussed later in this chapter and in chapter 24.

The Liver and Gallbladder

The **liver** is the largest internal organ of the body. In an adult human, the liver weighs about 1.5 kilograms and is the size of a football. The main exocrine secretion of the liver is **bile,** a fluid mixture consisting of *bile pigments* and *bile salts*

that is delivered into the duodenum during the digestion of a meal.

The bile salts play a very important role in the digestion of fats. Because fats are insoluble in water, they enter the intestine as drops within the watery chyme. The bile salts, which are partly fat-soluble and partly water-soluble, work like detergents, dispersing the large drops of fat into a fine suspension of smaller droplets. This emulsification process produces a greater surface area of fat upon which the lipase enzymes can act, and thus allows the digestion of fat to proceed more rapidly.

After it is produced in the liver, bile is stored and concentrated in the **gallbladder** (see figure 21.17). The arrival of fatty food in the duodenum triggers a neural and endocrine reflex that stimulates the gallbladder to contract, causing bile to be transported through the common bile duct and injected into the duodenum. If the bile duct is blocked by a *gallstone* (formed from a hardened precipitate of cholesterol), contraction of the gallbladder causes pain that is generally felt under the right scapula (shoulder blade).

Regulatory Functions of the Liver

Because a large vein carries blood from the stomach and intestine directly to the liver, the liver is in a position to chemically modify the substances absorbed in the gastrointestinal tract before they reach the rest of the body. For example, ingested alcohol and other drugs are taken into liver cells and metabolized; this is why the liver is often damaged as a result of alcohol and drug abuse. The liver also removes toxins, pesticides, carcinogens, and other poisons, converting them into less toxic forms (figure 21.18). Also, excess amino acids that may be present in the blood are converted to glucose by liver enzymes. The first step in this conversion is the removal of the amino group ($-NH_2$) from the amino acid, a process called *deamination.* Unlike plants, animals cannot reuse the nitrogen from these amino groups and must excrete it as nitrogenous waste. The product of amino acid deamination, ammonia (NH_3), combines with carbon dioxide to form urea. The urea is released by the liver into the bloodstream, where—as you will learn later in this chapter—the kidneys subsequently remove it.

The liver also produces most of the proteins found in blood plasma. The total concentration of plasma proteins is significant because it must be kept within normal limits to maintain osmotic balance between blood and interstitial (tissue) fluid. If the concentration of plasma proteins drops too low, as can happen as a result of liver disease such as cirrhosis, fluid accumulates in the tissues; this condition is called *edema.*

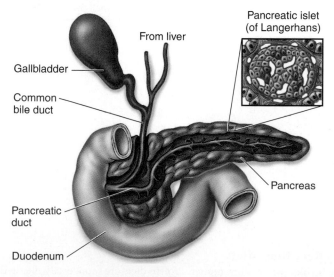

Figure 21.17 **The pancreatic and bile ducts empty into the duodenum.**

> **21.8** The pancreas secretes digestive enzymes and bicarbonate into the pancreatic duct.
> The liver produces bile, which is stored and concentrated in the gallbladder. The liver and the pancreatic hormones regulate blood glucose concentration.

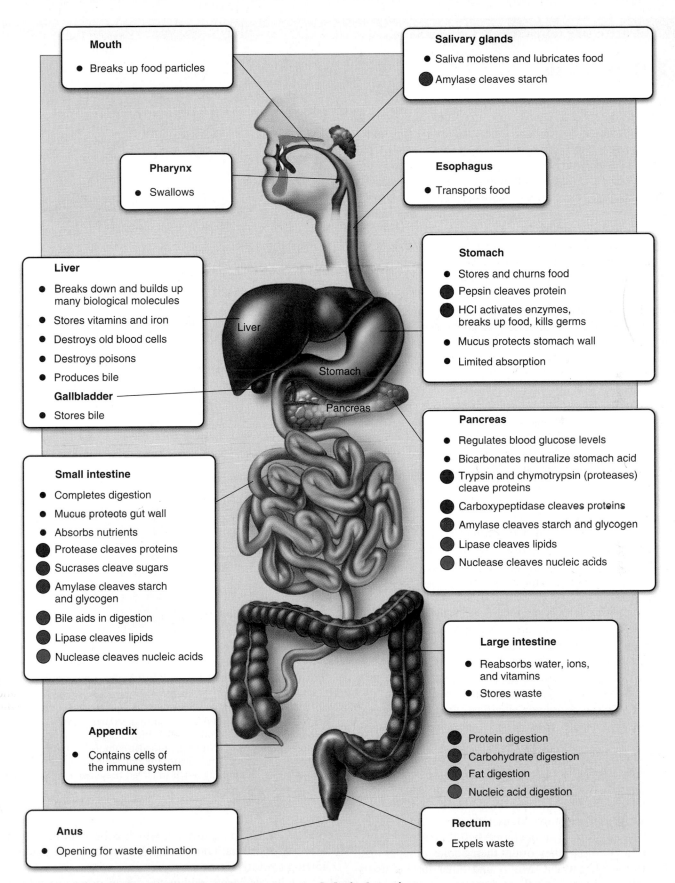

Mouth
- Breaks up food particles

Salivary glands
- Saliva moistens and lubricates food
- Amylase cleaves starch

Pharynx
- Swallows

Esophagus
- Transports food

Liver
- Breaks down and builds up many biological molecules
- Stores vitamins and iron
- Destroys old blood cells
- Destroys poisons
- Produces bile

Gallbladder
- Stores bile

Stomach
- Stores and churns food
- Pepsin cleaves protein
- HCl activates enzymes, breaks up food, kills germs
- Mucus protects stomach wall
- Limited absorption

Pancreas
- Regulates blood glucose levels
- Bicarbonates neutralize stomach acid
- Trypsin and chymotrypsin (proteases) cleave proteins
- Carboxypeptidase cleaves proteins
- Amylase cleaves starch and glycogen
- Lipase cleaves lipids
- Nuclease cleaves nucleic acids

Small intestine
- Completes digestion
- Mucus protects gut wall
- Absorbs nutrients
- Protease cleaves proteins
- Sucrases cleave sugars
- Amylase cleaves starch and glycogen
- Bile aids in digestion
- Lipase cleaves lipids
- Nuclease cleaves nucleic acids

Large intestine
- Reabsorbs water, ions, and vitamins
- Stores waste

Appendix
- Contains cells of the immune system

- Protein digestion
- Carbohydrate digestion
- Fat digestion
- Nucleic acid digestion

Anus
- Opening for waste elimination

Rectum
- Expels waste

Liver

Stomach

Pancreas

Figure 21.18   The organs of the digestive system and their functions.
The digestive system contains some dozen different organs that act on the food that is consumed, starting with the mouth and ending with the anus. All of these organs must work properly for the body to effectively obtain nutrients.

21.9 Homeostasis

As the animal body has evolved, specialization has increased. Each cell is a sophisticated machine, finely tuned to carry out a precise role within the body. Such specialization of cell function is possible only when extracellular conditions are kept within narrow limits. Temperature, pH, the concentrations of glucose and oxygen, and many other factors must be held fairly constant for cells to function efficiently and interact properly with one another.

Homeostasis may be defined as the dynamic constancy of the internal environment. The term *dynamic* is used because conditions are never absolutely constant but fluctuate continuously within narrow limits. Homeostasis is essential for life, and most of the regulatory mechanisms of the vertebrate body that are not devoted to reproduction are concerned with maintaining homeostasis.

Regulating Body Temperature

Humans, together with other mammals and with birds, are endothermic; they can maintain relatively constant body temperatures independent of the environmental temperature. When the temperature of your blood exceeds 37°C (98.6°F), neurons in a part of the brain called the hypothalamus (see chapters 23 and 24) detect the temperature change. Acting through the control of neurons, the hypothalamus responds by promoting the dissipation of heat through sweating, dilation of blood vessels in the skin, and other mechanisms. These responses tend to counteract the rise in body temperature. When body temperature falls, the hypothalamus coordinates a different set of responses, such as shivering and the constriction of blood vessels in the skin, which help to raise body temperature and correct the initial challenge to homeostasis.

Vertebrates other than mammals and birds are ectothermic; their body temperatures are more or less dependent on the environmental temperature. However, to the extent that it is possible, many ectothermic vertebrates attempt to maintain some degree of temperature homeostasis. Certain large fish, including tuna, swordfish, and some sharks, for example, can maintain parts of their body at a significantly higher temperature than that of the water. Reptiles attempt to maintain a constant body temperature through behavioral means—by placing themselves in varying locations of sun and shade. That's why you frequently see lizards basking in the sun. Sick lizards even give themselves a "fever" by seeking warmer locations!

Most invertebrates, like reptiles, modify behaviors to adjust their body temperature. Many butterflies, for example, must reach a certain body temperature before they can fly. In the cool of the morning, they orient their bodies to maximize their absorption of sunlight. Moths and other insects use a shivering reflex to warm their flight muscles.

Regulating Blood Glucose

When you digest a carbohydrate-containing meal, you absorb glucose into your blood. This causes a temporary rise in the

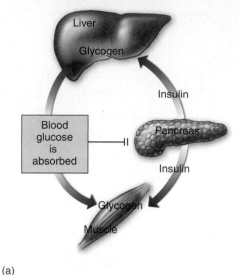

HIGH BLOOD SUGAR

(a)

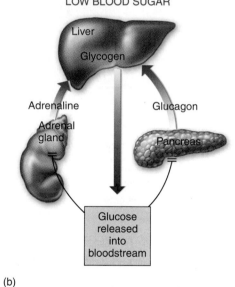

LOW BLOOD SUGAR

(b)

Figure 21.19   Control of blood glucose levels.

(a) When blood glucose levels are high, cells within the pancreas produce the hormone insulin, which stimulates the liver and muscles to convert blood glucose into glycogen. (b) When blood glucose levels are low, other cells within the pancreas release the hormone glucagon into the bloodstream; in addition, cells within the adrenal gland release the hormone adrenaline into the bloodstream. When they reach the liver, these two hormones act to increase the liver's breakdown of glycogen to glucose.

blood glucose concentration, which is brought back down in a few hours. What counteracts the rise in blood glucose following a meal?

Glucose levels within the blood are constantly monitored by a sensor, cells called the islets of Langerhans in the pancreas. When levels increase, the islets secrete the hormone insulin, which stimulates the uptake of blood glucose into muscles, liver, and adipose tissue. The muscles and liver can

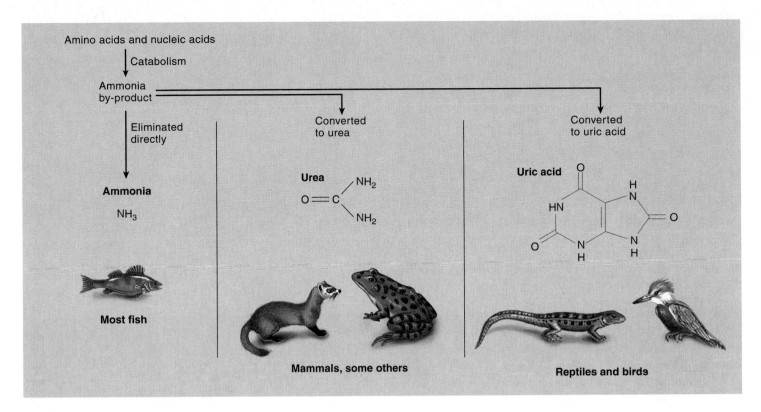

Figure 21.20 Nitrogenous wastes.
When amino acids and nucleic acids are metabolized, the immediate nitrogen by-product is ammonia, which is quite toxic but can be eliminated through the gills of bony fish. Mammals convert ammonia into urea, which is less toxic. Birds and terrestrial reptiles convert it instead into uric acid, which is insoluble in water.

convert the glucose into the polysaccharide glycogen (figure 21.19); adipose cells can convert glucose into fat. These actions lower the blood glucose and help to store energy in forms that the body can use later. When blood glucose levels decrease, as they do between meals, during periods of fasting, and during exercise, the liver secretes glucose into the blood. This glucose is obtained in part from the breakdown of liver glycogen. This breakdown of liver glycogen is stimulated by another hormone, *glucagon,* which is also secreted by the islets of Langerhans.

Eliminating Nitrogenous Wastes

Amino acids and nucleic acids are nitrogen-containing molecules. When animals catabolize these molecules for energy or convert them into carbohydrates or lipids, they produce nitrogen-containing by-products called **nitrogenous wastes** (figure 21.20) that must be eliminated from the body.

The first step in the metabolism of amino acids and nucleic acids is the removal of the amino ($-NH_2$) group and its combination with H^+ to form **ammonia** (NH_3) in the liver. Ammonia is quite toxic to cells and therefore is safe only in very dilute concentrations. The excretion of ammonia is not a problem for the bony fish and tadpoles, which eliminate most of it by diffusion through the gills and the rest by excretion in very dilute urine. In sharks, adult amphibians, and mammals, the nitrogenous wastes are eliminated in the far less toxic form of **urea.** Urea is water-soluble and so can be excreted in large amounts in the urine. It is carried in the bloodstream from its place of synthesis in the liver to the kidneys, where it is excreted in the urine.

Reptiles, birds, and insects excrete nitrogenous wastes in the form of **uric acid,** which is only slightly soluble in water. As a result of its low solubility, uric acid precipitates and thus can be excreted using very little water. Uric acid forms the pasty white material in bird droppings called *guano.* The ability to synthesize uric acid in these groups of animals is also important because their eggs are encased within shells, and nitrogenous wastes build up as the embryo grows within the egg. The formation of uric acid, while a lengthy process that requires considerable energy, produces a compound that crystallizes and precipitates. As a precipitate, it is unable to affect the embryo's development even though it is still inside the egg.

Mammals also produce some uric acid, but it is a waste product of the degradation of purine nucleotides (see chapter 3), not of amino acids. Most mammals have an enzyme called *uricase,* which converts uric acid into a more soluble derivative, **allantoin.** Only humans, apes, and the dalmatian dog lack this enzyme and so must excrete the uric acid. In humans, excessive accumulation of uric acid in the joints produces a condition known as *gout.*

> **21.9** Animals tend to maintain a relatively constant internal environment, a condition called homeostasis. Body temperature, glucose levels, and nitrogenous waste levels are all carefully regulated.

Osmoregulatory Organs

Animals must also carefully monitor the water content of their bodies. The first animals evolved in seawater, and the physiology of all animals reflects this origin. Approximately two-thirds of every vertebrate's body is water. If the amount of water in the body of a vertebrate falls much lower than this, the animal dies. Animals use various mechanisms for **osmoregulation,** the regulation of the body's osmotic composition, or how much water and salt it contains. The proper operation of many vertebrate organ systems of the body requires that the osmotic concentration of the blood—the concentration of solutes dissolved within it—be kept within narrow bounds.

Animals have evolved a variety of mechanisms to cope with problems of water balance. In many animals, the removal of water or salts from the body is coupled with the removal of metabolic wastes through the excretory system. Protists employ contractile vacuoles for this purpose (figure 21.21), as do sponges. Other multicellular animals have a system of excretory tubules (little tubes) that expel fluid and wastes from the body.

In flatworms, these tubules are called *protonephridia,* and they branch throughout the body into bulblike **flame cells** (figure 21.22). While these simple excretory structures open to the outside of the body, they do not open to the inside of the body. Rather, cilia within the flame cells must draw in fluid from the body. Water and metabolites are then reabsorbed, and the substances to be excreted are expelled through excretory pores.

Other invertebrates have a system of tubules that open both to the inside and to the outside of the body. In the earthworm, these tubules are known as *nephridia* (figure 21.23).

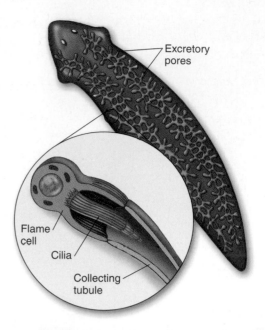

Figure 21.22   The protonephridia of flatworms.

A branching system of tubules, bulblike flame cells, and excretory pores make up the protonephridia of flatworms. Cilia inside the flame cells draw in fluids from the body by their beating action. Substances are then expelled through pores that open to the outside of the body.

The nephridia obtain fluid from the body cavity through a process of filtration into funnel-shaped structures called *nephrostomes.* The term *filtration* is used because the fluid is formed under pressure and passes through small openings, so that molecules larger than a certain size are excluded. This filtered fluid is isotonic (having the same osmotic concentration) to the fluid in the coelom, but as it passes through the tubules of the nephridia, NaCl is removed by active transport processes. A general term for transport out of the tubule and into the surrounding body fluids is *reabsorption.* Because salt is reabsorbed from the filtrate, the urine excreted is more dilute than the body fluids (is hypotonic). The kidneys of mollusks and the excretory organs of crustaceans (called *antennal glands*) also produce urine by filtration and reclaim certain ions by reabsorption.

The excretory organs in insects are called **Malpighian tubules** (figure 21.24), extensions of the digestive tract that branch off anterior to the hindgut. Urine is not formed by filtration in these tubules, because there is no pressure difference between the blood in the body cavity and the tubule. Instead, waste molecules and potassium ions (K^+) are secreted into the tubules by active transport. *Secretion* is the opposite of reabsorption—ions or molecules are transported from the body fluid into the tubule. The secretion of K^+ creates an osmotic gradient that causes water to enter the tubules by osmosis from the body's open circulatory system. Most of the water and K^+ is then reabsorbed into the circulatory system through the epithelium of the hindgut, leaving only small molecules and waste products to be excreted from the rectum along with feces. Malpighian tubules thus provide a very efficient means of water conservation.

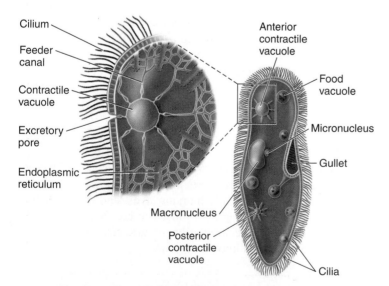

Figure 21.21   Contractile vacuoles.

Protists, such as *Paramecium,* use contractile vacuoles to eliminate excess water and metabolic wastes from their bodies. Water and wastes collected by the endoplasmic reticulum flow through feeder canals to the contractile vacuole, where they are expelled to the outside.

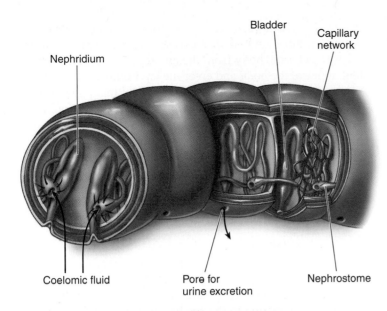

Figure 21.23 **The nephridia of annelids.**
Most invertebrates, such as the annelid shown here, have nephridia. These consist of tubules that receive a filtrate of coelomic fluid, which enters the funnel-like nephrostomes. Salt can be reabsorbed from these tubules, and the fluid that remains, urine, is released from pores into the external environment.

Kidneys are the excretory organs in vertebrates, and unlike the Malpighian tubules of insects, kidneys create a tubular fluid by filtration of the blood under pressure. In addition to containing waste products and water, the filtrate contains many small molecules that are of value to the animal, including glucose, amino acids, and vitamins. These molecules and most of the water are reabsorbed from the tubules into the blood, while wastes remain in the filtrate. Additional wastes may be secreted by the tubules and added to the filtrate, and the final waste product, urine, is eliminated from the body.

It may seem odd that the vertebrate kidney should filter out almost everything from blood plasma (except proteins, which are too large to be filtered) and then spend energy to take back or reabsorb what the body needs. But selective reabsorption provides great flexibility; various vertebrate groups have evolved the ability to reabsorb different molecules that are especially valuable in particular habitats. This flexibility is a key factor underlying the successful colonization of many diverse environments by the vertebrates.

21.10 **Many invertebrates filter fluid into a system of tubules and then reabsorb ions and water, leaving waste products for excretion. Insects create an excretory fluid by secreting K⁺ into tubules, which draws water osmotically. The vertebrate kidney produces a filtrate that enters tubules and is modified to become urine.**

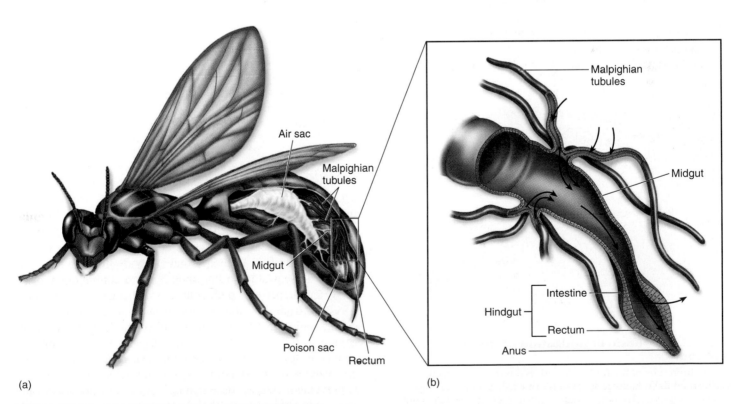

(a) (b)

Figure 21.24 **The Malpighian tubules of insects.**
(a) The Malpighian tubules of insects are extensions of the digestive tract that collect water and wastes from the body's circulatory system. (b) K⁺ is secreted into these tubules, drawing water with it osmotically. Much of this water (*see arrows*) is reabsorbed across the wall of the hindgut.

21.11 Evolution of the Vertebrate Kidney

The kidney is a complex organ made up of up to a million repeating disposal units called **nephrons,** each with the structure of a bent tube (figure 21.25). Blood pressure forces the fluid in blood through a capillary bed, called the *glomerulus,* at the top of each nephron. The glomerulus retains blood cells, proteins, and other useful large molecules in the blood but allows the water, and the small molecules and wastes dissolved in it, to pass through and into the bent tube part of the nephron. As the filtered fluid passes through the nephron tube, useful sugars and ions are recovered from it by active transport, leaving the water and metabolic wastes dissolved in it behind in a fluid urine.

Although the same basic design has been retained in all vertebrate kidneys, there have been a few modifications. Because the original glomerular filtrate is isotonic to blood, all vertebrates can produce a urine that is isotonic to (by reabsorbing ions) or hypotonic to (more dilute than) blood. Only birds and mammals can reabsorb water from their glomerular filtrate to produce a urine that is hypertonic to (more concentrated than) blood.

Freshwater Fish

Kidneys are thought to have evolved first among the freshwater teleosts, or bony fish. Because the body fluids of a freshwater fish have a greater osmotic concentration than the surrounding water, these animals face two serious problems: (1) water tends to enter the body from the environment; and (2) solutes tend to leave the body and enter the environment. Freshwater fish address the first problem by *not* drinking water and by excreting a large volume of dilute urine, which is hypotonic to their body fluids (figure 21.26, *top*). They address the second problem by reabsorbing ions across the nephron tubules, from the glomerular filtrate back into the blood. In addition, they actively transport ions across their gills from the surrounding water into the blood.

Marine Bony Fish

Although most groups of animals seem to have evolved first in the sea, marine bony fish (teleosts) probably evolved from freshwater ancestors. They faced significant new problems in making the transition to the sea because their body fluids are hypotonic to the surrounding seawater. Consequently, water tends to leave their bodies by osmosis across their gills, and they also lose water in their urine. To compensate for this continuous water loss, marine fish drink large amounts of seawater.

Many of the divalent cations (principally Ca^{++} and Mg^{++}) in the seawater that a marine fish drinks remain in the digestive tract and are eliminated through the anus. Some, however, are absorbed into the blood, as are the monovalent ions K^+, Na^+, and Cl^-. Most of the monovalent ions are actively transported out of the blood across the gills, while the divalent ions that enter the blood are secreted into the nephron tubules and excreted in the urine (figure 21.26, *bottom*). In these two ways, marine bony fish eliminate the ions they get from the seawater they drink. The urine they excrete is isotonic to their body fluids. It is more concentrated than the urine of freshwater fish but not as concentrated as that of birds and mammals.

Cartilaginous Fish

The elasmobranchs—sharks, rays, and skates—are by far the most common subclass in the class Chondrichthyes (cartilaginous fish). Elasmobranchs have solved the osmotic problem posed by their seawater environment in a different way than have the bony fish. Instead of having body fluids that are hypotonic to seawater, so that they have to continuously drink seawater and actively pump out ions, the elasmobranchs reabsorb urea from the nephron tubules and maintain a blood urea concentration that is 100 times higher than that of mammals (figure 21.27). This added urea makes their blood approximately isotonic to the surrounding sea. Because there is no net water movement between isotonic solutions, water loss is prevented. Hence, these fish do not need to drink seawater for osmotic balance, and their kidneys and gills do not have to remove large amounts of ions from their bodies. The enzymes and tissues of the cartilaginous fish have evolved to tolerate the high urea concentrations.

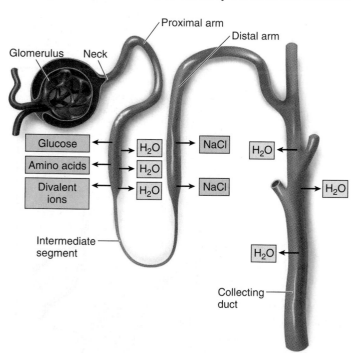

Figure 21.25 Basic organization of the vertebrate nephron.

The nephron tube of the freshwater fish is a basic design that has been retained in the kidneys of marine fishes and terrestrial vertebrates that evolved later. Sugars, small proteins, and divalent ions such as Ca^{++} are recovered at the beginning of the tube (the so-called proximal arm). Ions such as Na^+ and Cl^- are recovered after the initial bend (the so-called distal arm). Water is reabsorbed by the body from the collecting duct.

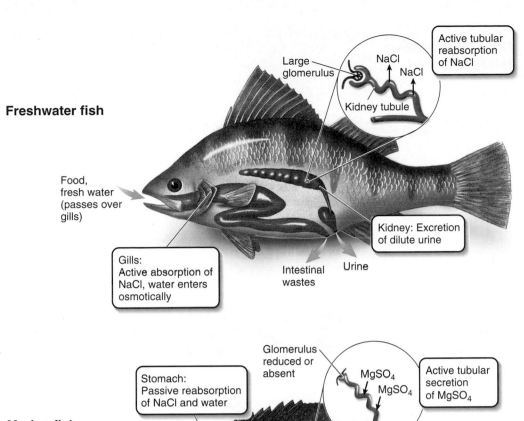

Freshwater fish

Large glomerulus

NaCl

NaCl

Active tubular reabsorption of NaCl

Kidney tubule

Food, fresh water (passes over gills)

Gills: Active absorption of NaCl, water enters osmotically

Kidney: Excretion of dilute urine

Intestinal wastes

Urine

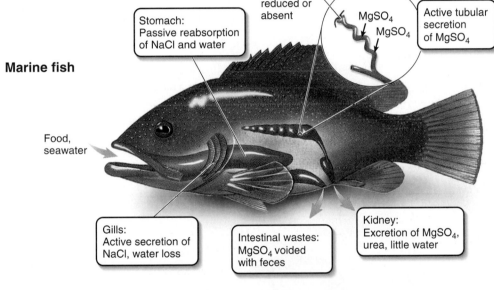

Glomerulus reduced or absent

MgSO₄

MgSO₄

Active tubular secretion of MgSO₄

Stomach: Passive reabsorption of NaCl and water

Marine fish

Food, seawater

Gills: Active secretion of NaCl, water loss

Intestinal wastes: MgSO₄ voided with feces

Kidney: Excretion of MgSO₄, urea, little water

Figure 21.26 Freshwater and marine teleosts (bony fish) face different osmotic problems.

Whereas the freshwater teleost is hypertonic to its environment, the marine teleost is hypotonic to seawater. To compensate for its tendency to take in water and lose ions, a freshwater fish excretes dilute urine, avoids drinking water, and reabsorbs ions across the nephron tubules. To compensate for its osmotic loss of water, the marine teleost drinks seawater and eliminates the excess ions through transport across epithelia of the gills and kidney tubules.

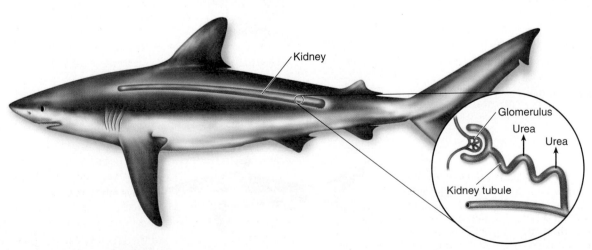

Kidney

Glomerulus

Urea

Urea

Kidney tubule

Figure 21.27 Osmoregulation in elasmobranchs.

The elasmobranchs control osmotic balance in seawater by reabsorbing urea from the nephron, thereby maintaining an internal osmotic concentration that is the same as the surrounding seawater.

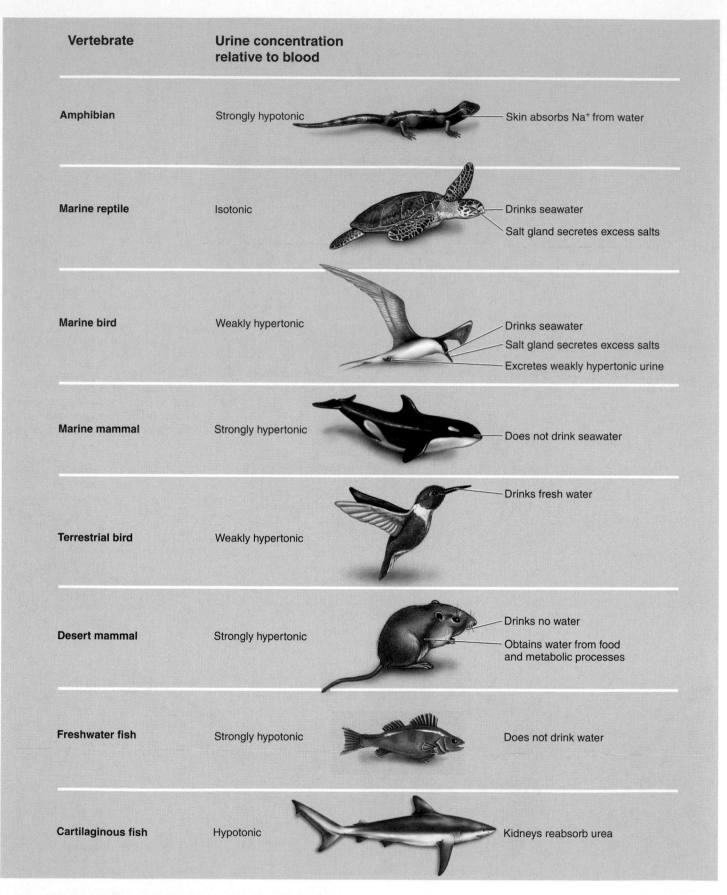

Vertebrate	Urine concentration relative to blood		
Amphibian	Strongly hypotonic		Skin absorbs Na⁺ from water
Marine reptile	Isotonic		Drinks seawater
			Salt gland secretes excess salts
Marine bird	Weakly hypertonic		Drinks seawater
			Salt gland secretes excess salts
			Excretes weakly hypertonic urine
Marine mammal	Strongly hypertonic		Does not drink seawater
Terrestrial bird	Weakly hypertonic		Drinks fresh water
Desert mammal	Strongly hypertonic		Drinks no water
			Obtains water from food and metabolic processes
Freshwater fish	Strongly hypotonic		Does not drink water
Cartilaginous fish	Hypotonic		Kidneys reabsorb urea

Figure 21.28   Osmoregulation by some vertebrates.

Only birds and mammals can produce a hypertonic urine and thereby retain water efficiently, but marine reptiles and birds can drink seawater and excrete the excess salt through salt glands.

Amphibians and Reptiles

The first terrestrial vertebrates were the amphibians, and the amphibian kidney is identical to that of freshwater fish. This is not surprising because amphibians spend a significant portion of their time in freshwater, and when on land, they generally stay in wet places. Amphibians produce a very dilute urine and compensate for their loss of Na^+ by actively transporting Na^+ across their skin from the surrounding water.

Reptiles, on the other hand, live in diverse habitats. Those living mainly in freshwater, like some of the crocodilians, occupy a habitat in many ways similar to that of the freshwater fish and amphibians, and thus have similar kidneys. Marine reptiles, which consist of other crocodilians, turtles, sea snakes, and one lizard, possess kidneys similar to those of their freshwater relatives but face opposite problems; they tend to lose water and take in salts. Like marine teleosts (bony fish), they drink the seawater and excrete an isotonic urine. Marine teleosts eliminate the excess salt by transport across their gills, while marine reptiles eliminate excess salt through salt glands located near the nose or the eye (figure 21.28).

The kidneys of terrestrial reptiles also reabsorb much of the salt and water in the nephron tubules, helping somewhat to conserve blood volume in dry environments. Like fish and amphibians, they cannot produce urine that is more concentrated than the blood plasma. However, when their urine enters their cloaca (the common exit of the digestive and urinary tracts), additional water can be reabsorbed.

Mammals and Birds

Mammals and birds are the only vertebrates able to produce urine with a higher osmotic concentration than their body fluids. This allows these vertebrates to excrete their waste products in a small volume of water, so that more water can be retained in the body. Human kidneys can produce urine that

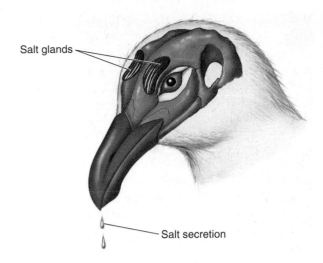

Figure 21.30 Marine birds drink seawater and then excrete the salt through salt glands.
The salty fluid excreted by these glands can then dribble down the beak.

is as much as 4.2 times as concentrated as blood plasma, but the kidneys of some other mammals are even more efficient at conserving water. For example, the camel, gerbil, and pocket mouse, *Perognathus,* can excrete urine 8, 14, and 22 times as concentrated as their blood plasma, respectively. The kidneys of the kangaroo rat (figure 21.29) are so efficient it never has to drink water; it can obtain all the water it needs from its food and from water produced in aerobic cell respiration!

The production of hypertonic urine is accomplished by the *loop of Henle* portion of the nephron (see figure 21.31c), found only in mammals and birds. A nephron with a long loop of Henle extends deeper into the renal medulla and can produce more concentrated urine. Most mammals have some nephrons with short loops and other nephrons with loops that are much longer. Birds, however, have relatively few or no nephrons with long loops, so they cannot produce urine that is as concentrated as that of mammals. At most, they can only reabsorb enough water to produce a urine that is about twice the concentration of their blood. Marine birds solve the problem of water loss by drinking seawater and then excreting the excess salt from salt glands near the eyes (figure 21.30).

The moderately hypertonic urine of a bird is delivered to its cloaca, along with the fecal material from its digestive tract. If needed, additional water can be absorbed across the wall of the cloaca to produce a semisolid white paste or pellet, which is excreted.

Figure 21.29 The kangaroo rat, *Dipodomys panamintensis.*
This mammal has very efficient kidneys that can concentrate urine to a high degree by reabsorbing water, thereby minimizing water loss from the body. This feature is extremely important to the kangaroo rat's survival in dry or desert habitats.

21.11 The kidneys of freshwater fish must excrete copious amounts of very dilute urine, whereas marine teleosts drink seawater and excrete an isotonic urine. The basic design and function of the nephron of freshwater fish have been retained in the terrestrial vertebrates. Modifications, particularly the presence of a loop of Henle, allow mammals and birds to reabsorb water and produce a hypertonic urine.

The Mammalian Kidney

In humans, the kidneys are fist-sized organs located in the region of the lower back (figure 21.31a). Each kidney receives blood from a renal artery, and it is from this blood that urine is produced. Urine drains from each kidney through a **ureter,** which carries the urine to a **urinary bladder.** Within the kidney, the mouth of the ureter flares open to form a funnel-like structure, the *renal pelvis.* The renal pelvis, in turn, has cup-shaped extensions that receive urine from the renal tissue. This tissue is divided into an outer **renal cortex** and an inner **renal medulla** (figure 21.31b). Together, these structures perform filtration, reabsorption, secretion, and excretion.

The mammalian kidney is composed of roughly 1 million nephrons (figure 21.31c), each of which is composed of three regions:

1. **Filter.** The filtration device at the top of each nephron is called a **Bowman's capsule.** Within each capsule an arteriole enters and splits into a fine network of vessels called a **glomerulus** (figure 21.32). The walls of these capillaries act as a filtration device. Blood pressure forces fluid through the capillary walls. These walls withhold proteins and other large molecules in the blood, while passing water, small molecules, ions, and urea, the primary waste product of metabolism.

2. **Tube.** The Bowman's capsule is connected to a long narrow tube called a renal tubule, which is bent back on itself in its center, called the **loop of Henle.** This long hairpin loop is a reabsorption device. Like the mammalian small intestine, it extracts from the filtrate passing through the tube molecules useful to the body, such as glucose and a variety of ions.

3. **Duct.** The tube empties into a large collection tube called a **collecting duct.** The collecting duct operates as a water conservation device, reclaiming water from the urine so that it is not lost from the body. Human urine is four times as concentrated as blood plasma—that is, the collecting ducts remove much of the water from the filtrate passing through the kidney. Your kidneys achieve this remarkable degree of water conservation by a simple but superbly designed mechanism: They bend the duct back alongside the nephron tube and make the duct permeable to urea. This greatly increases the local salt (urea) concentration in the tissue surrounding the tube, causing water in urine to pass out of the tube by osmosis. The salty tissue sucks up water from the urine like blotting paper, passing it on to blood vessels that carry it out of the kidneys and back to the bloodstream.

The Kidney at Work

The formation of urine within the mammalian kidney involves the movement of several kinds of molecules between nephrons and the capillaries that surround them. Five steps are involved: pressure filtration, reabsorption of water, selective reabsorption of ions and nutrients, tubular excretion, and further reabsorption of water.

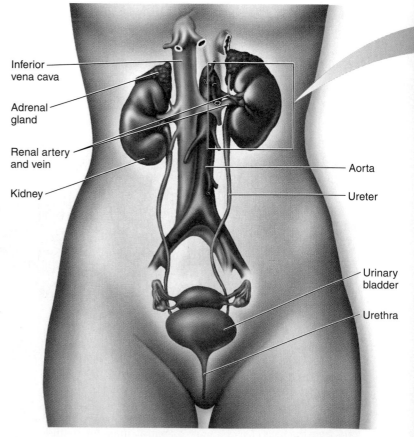

Figure 21.31 The mammalian urinary system contains two kidneys, each of which contain about a million nephrons that lie in the renal cortex and renal medulla.

(a) The urinary system consists of the kidneys, the ureter, which transports urine from the kidneys to the urinary bladder, and the urethra. (b) The kidney is a bean-shaped reddish brown organ and contains about 1 million nephrons. (c) The glomerulus is enclosed within a filtration device called a Bowman's capsule. Blood pressure forces liquid from blood through the glomerulus and into the proximal tubule of the nephron, where glucose and small proteins are reabsorbed from the filtrate. The filtrate then passes through a double-loop arrangement consisting of the loop of Henle and the collecting duct, which act to remove water from the filtrate. The water is then collected by blood vessels and transported out of the kidney to the systemic (body) circulation.

Labels on figure: Inferior vena cava; Adrenal gland; Renal artery and vein; Kidney; Aorta; Ureter; Urinary bladder; Urethra

(a)

Pressure Filtration. Driven by the blood pressure, small molecules are pushed across the thin walls of the glomerulus to the inside of the Bowman's capsule. Blood cells and large molecules like proteins cannot pass through, and as a result the blood that enters the glomerulus is divided into two paths: nonfilterable blood components that are retained and leave the glomerulus in the bloodstream and filterable components that pass across and leave the glomerulus in the urine. This filterable stream is called the **glomerular filtrate.** It contains water, nitrogenous wastes (principally urea), nutrients (principally glucose and amino acids), and a variety of ions.

Reabsorption of Water. Filtrate from the glomerulus passes down the descending arm of the loop of Henle. The walls of this portion of the tube are impermeable to either salts or urea but are freely permeable to water. Because (for reasons we discuss later) the surrounding tissue has a high concentration of urea, water passes out of the descending arm by osmosis, leaving behind a more concentrated filtrate.

Selective Reabsorption. At the turn in the loop, the walls of the tubule become permeable to salts and other nutrients, like sugars and amino acids, but much less permeable to water. As the concentrated filtrate passes up this ascending arm, these nutrients pass out into the surrounding tissue, where they are carried away by blood vessels. In the upper region of

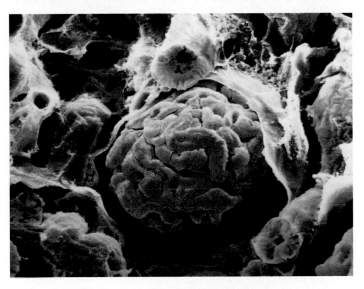

Figure 21.32   A glomerulus in the kidney.

The red spherical structure in this micrograph is a glomerulus, a fine network of capillaries associated with a nephron tubule and connected to a pair of arterioles.

From R. G. Kessel and R. H. Kardon, Tissues and Organs: A Text Atlas of Scanning Electron Microscopy, *1979, W. H. Freeman Co.*

the ascending arm are active transport channels that pump out salt (NaCl). Left behind in the filtrate is the urea that initially passed through the glomerulus as nitrogenous waste. The urea concentration is becoming very high within the tubule.

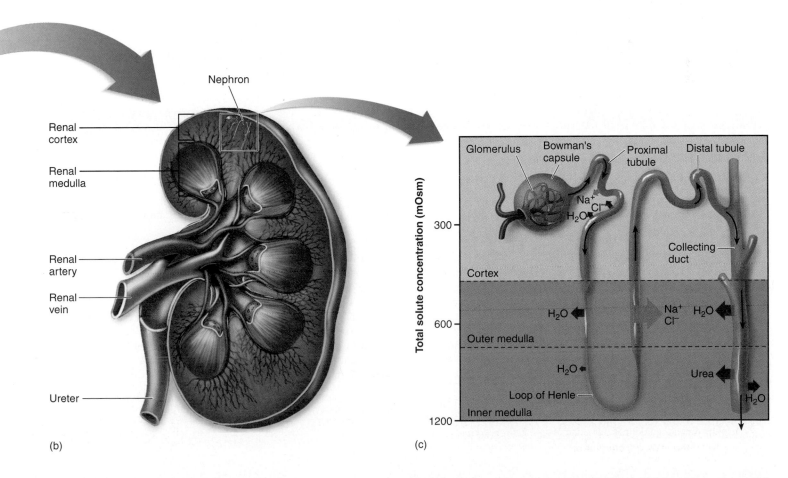

(b)

(c)

Reabsorption of Water in the Kidneys

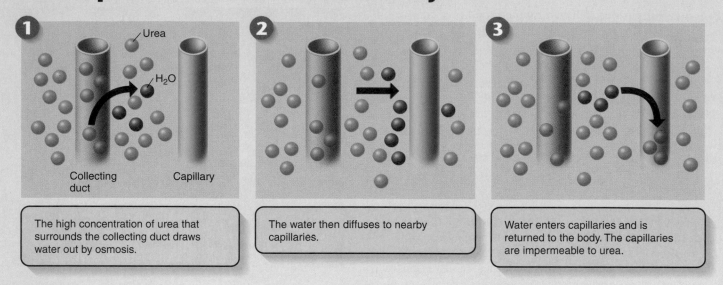

Figure 21.33   How the kidneys reabsorb water.

Tubular Excretion.   In the ascending loop, substances are also added to the urine by a process called tubular excretion. This active transport process excretes into the urine other nitrogenous wastes such as uric acid and ammonia, as well as excess hydrogen ions.

Further Reabsorption of Water.   The tubule then empties into a collecting duct that passes back through the tissue of the kidney. Unlike the tubule, the lower portions of the collecting duct are permeable to urea, some of which diffuses out into the surrounding tissue (that is why the tissue surrounding the descending arm of the loop of Henle has a high urea concentration). A high urea concentration in the tissue results, causing even more water to pass outward from the filtrate by osmosis (figure 21.33). The filtrate that is left after salts, nutrients, and water have been removed is urine.

> **21.12   The mammalian kidney pushes waste molecules through a filter and then reclaims water and useful metabolites and ions from the filtrate before eliminating the residual urine.**

Exploring Current Issues

Additional Resources

Go to your campus library or look online to find the following articles, which further develop some of the concepts found in this chapter.

Cowley, G. (2002). Hepatitis C: the insidious spread of a killer virus: this stealthy virus can incubate for decades. Now thousands of people are getting sick. By 2010 it may strike down more Americans each year than AIDS. *Newsweek,* 139(16), 46.

Guarner, F. and J. R. Malagelada. (2003). Gut flora in health and disease. *The Lancet,* 361(9356), 512.

Hamilton, G. (2001). Dead man walking: Barry Marshall's research into *Helicobacter pylori,* ulcers, and stomach cancer. *New Scientist,* 171(2303), 31.

Peplow, M. (2002). Full of goodness: giving violent young offenders a cocktail of vitamins, minerals, and fatty acids seems to transform them into well behaved kids. Can better nutrition tackle crime? *New Scientist,* 176(2369), 38.

United States Department of Health and Human Services and United States Department of Agriculture. (2000). Nutrition and your health: dietary guidelines for Americans. *Retrieved from www. health.gov/dietaryguidelines/dga2000/document/frontcover.htm.*

In the News: Dieting

Almost all of us at one time or another worry about our weight. It turns out there's a lot of interesting biology involved.

Article 1. Impossible dreams: Fad diets and our futile search for an easy way to lose weight.

Article 2. Why fat people are hungrier.

Article 3. New drugs to reduce cholesterol by inhibiting a key enzyme used in manufacturing it.

Article 4. Diet drugs may be good news for those of us who are overweight.

Article 5. Despite earlier hopes, high-fiber diets don't seem to protect against colon cancer.

Article 6. The turkey's revenge: Why eating Thanksgiving dinner puts on the pounds.

Article 7. In the battle to lose weight, I seem to be losing.

Find these articles, written by the author, online at www.mhhe.com/tlwessentials/exp21.

Food Energy and Essential Nutrients

21.1 Food for Energy and Growth

- Animals consume food as a source of energy and as a source of essential molecules and minerals. When food is consumed, it is either used up in metabolic processes or stored as fat. A balanced diet that is higher in complex carbohydrates, fruits, and vegetables and low in fats and sweets is recommended (**figure 21.1**). Consuming excess calories that the body doesn't use results in gaining weight and obesity, a major health problem (**figure 21.3**).

- Many animals must consume proteins, fruits, and vegetables to obtain essential amino acids and minerals that the body needs but cannot produce itself (**table 21.1**).

Digestion

21.2 Types of Digestive Systems

- Most animals digest food extracellularly, in a cavity or tract that contains digestive enzymes (**figures 21.4 and 21.5**). The products of digestion are then absorbed by the body.

21.3 Vertebrate Digestive Systems

- Vertebrate digestion occurs in a gastrointestinal tract (**figure 21.6**). Areas of the tract are specialized for different digestive functions.

21.4 The Mouth and Teeth

- Digestion begins in the mouth with the chewing of food, where it mixes with saliva and is swallowed (**figures 21.8, 21.10, and 21.11**).

- Birds don't have teeth but break up food in the gizzard, where the food is churned and ground up with pebbles that the bird has swallowed (**figure 21.9**).

21.5 The Esophagus and Stomach

- Once food is swallowed, it passes into the esophagus, where peristaltic waves of muscle contractions move the food down the esophagus to the stomach (**figure 21.12**).

- In the stomach, muscle contractions churn up the food with gastric juice, which contains hydrochloric acid and pepsin, a protein-digesting enzyme activated by HCl (**figure 21.13**). Proteins are partially digested in the stomach.

21.6 The Small and Large Intestines

- The acidic chyme from the stomach passes into the upper portion of the small intestine, where it is neutralized and mixed with other digestive enzymes (**figure 21.14**). Some enzymes are secreted by the cells that line the walls of the intestine, but most enzymes and other digestive substances are produced in the pancreas or other accessory organs. The rest of the small intestine is involved in absorption of food molecules and water.

- The large intestine collects and compacts solid waste and releases the waste from the body through the rectum and anus.

21.7 Variations in Vertebrate Digestive Systems

- In ruminants, cellulose-digesting microorganisms live in a chamber of the stomach called the rumen (**figure 21.15**), and the products of cellulose digestion are absorbed in the small intestine. In others, cellulose digestion occurs in the cecum. To gain the nutritional value of cellulose digestion, these animals eat their feces. The digestive systems of animals differ, based on their diets (**figure 21.16**).

21.8 Accessory Digestive Organs

- The pancreas produces the protein-digesting enzymes trypsin and chymotrypsin, the starch-digesting enzyme pancreatic amylase, and the fat-digesting enzyme lipase, which are released into the small intestine. The liver produces bile (a mixture of bile pigments and bile salts), which breaks down fats. Bile is stored in the gallbladder and released into the small intestine (**figure 21.17**). All of the organs of digestion work together (**figure 21.18**).

Maintaining the Internal Environment

21.9 Homeostasis

- Animals maintain relatively constant internal conditions, a process called homeostasis. Examples of homeostasis include body temperature, blood glucose levels (**figure 21.19**), and nitrogenous wastes (**figure 21.20**).

21.10 Osmoregulatory Organs

- Osmotic balance in the body is important, and animals have evolved various mechanisms to control water balance. Many invertebrates use systems of tubules that collect fluid and reabsorb ions and water (**figures 21.22, 21.23 and 21.24**). The kidneys serve this excretory function in vertebrates.

21.11 Evolution of the Vertebrate Kidney

- The kidney has evolved in different animals, adapting to different environmental conditions. The nephron is the basic unit of the kidney (**figure 21.25**). The kidneys of freshwater fish excrete dilute urine (**figure 21.26**), whereas marine animals drink seawater and either excrete an isotonic urine or excrete salt through salt glands (**figures 21.28 and 21.30**).

21.12 The Mammalian Kidney

- In mammals, the loop of Henle is longer (**figure 21.31**), allowing the formation of a hypertonic urine (**figure 21.33**).

1. One-way passage of food through the digestive system of many animal groups allows
   a. intracellular digestion.
   b. specialization of different regions of the digestive system.
   c. release of digestive enzymes into the gut.
   d. extracellular digestion.

2. Organisms with longer digestive systems, which help break down difficult to digest food, are usually
   a. herbivores.
   c. omnivores.
   b. carnivores.
   d. detritivores.

3. The purpose of a gizzard, like teeth, is to
   a. hold on to prey.
   b. begin the chemical digestion of food.
   c. release enzymes.
   d. begin the physical digestion of food.

4. Most of the absorption of food molecules takes place in the
   a. stomach.
   c. small intestine.
   b. liver.
   d. large intestine.

5. The purpose of the villi in the small intestine is to
   a. neutralize stomach acid.
   b. produce bile.
   c. produce digestive enzymes.
   d. increase the surface area of the small intestine for absorption of nutrients.

6. The monitoring and adjusting of the body's condition, such as temperature and pH, is also known as
   a. exothermy.
   c. osmoregulation.
   b. homeostasis.
   d. ectothermy.

7. If your blood sugar is too low, the hormone glucagon is released by the pancreas. This hormone will cause
   a. the release of insulin.
   c. glycogen to be formed.
   b. glycogen to break down.
   d. fat to be formed.

8. One of the challenges to osmoregulatory tissues and organs is
   a. retaining metabolites and water while excreting soluble wastes.
   b. getting rid of glucose.
   c. getting rid of ammonia-based waste products.
   d. reducing the amount of solutes in the bloodstream while retaining water.

9. To keep the proper concentrations of water and solutes in their blood, freshwater bony fish must drink
   a. lots of water and excrete large volumes of urine that are hypotonic to body fluids.
   b. no water and excrete large volumes of urine that are hypotonic to body fluids.
   c. lots of water and excrete large volumes of urine that are isotonic to body fluids.
   d. no water and excrete large volumes of urine that are isotonic to body fluids.

10. Water is removed from kidney filtrate by the process of
    a. diffusion.
    c. facilitated diffusion.
    b. active transport.
    d. osmosis.

Visual Understanding

1. **Figure 21.2** People throughout the world have existed for thousands of years on diets of various types of beans and nuts, along with corn (New World), wheat (Europe), rice (Asia), or other grains (Africa), supplemented by the occasional mammal, fish, or bird. How did they get enough protein to build strong bodies?

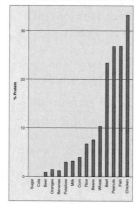

2. **A Closer Look** Discuss the relationship of obesity and diabetes in the United States, and our ability to make headway against the disease.

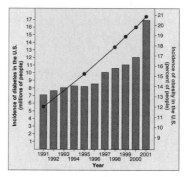

Challenge Questions

Food Energy and Essential Nutrients Why is it so important for Popeye, and you, to eat your green, leafy vegetables (spinach, chard, turnip and mustard greens, bok choi, broccoli, cauliflower, cabbage)?

Digestion You're going on a backpacking trip with three friends. Lisa wants to pack trail snacks of chocolate cupcakes; Andre wants to take beef jerky. Antje insists that you should all pack GORP (*g*ood *o*ld *r*aisins and *p*eanuts—sometimes known as trail mix). They turn to you to decide. Explain which one is best for a quick snack to keep up your energy on a long hike, and why.

Maintaining the Internal Environment Hummingbirds, pocket mice, and young Chiricahua leopard frogs are all about the same size, and all live in the Southwestern deserts. Why is there such a difference in their urine?

Online Learning Center

Visit the Online Learning Center for this chapter at www.mhhe.com/tlwessentials/ch21 for quizzes, animations, interactive learning exercises, and other study tools. At the site you will also find extended answers to the end-of-chapter questions.

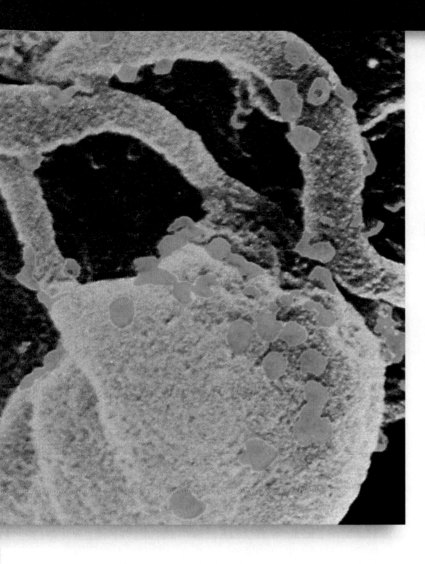

22

How the Animal Body Defends Itself

A ll animals are constantly at war with bacteria and viruses that attempt to use the rich resources of the cellular environment to fuel their own reproduction. Another war is fought on a very different front by animals against their own cells when cells become cancerous and begin to grow without restraint. Both of these wars, against invading microbes and against cancer, are fought with the same defensive weapons, the immune system. This chapter focuses on the vertebrate immune system and how it defends the body in the face of these onslaughts. Sometimes the immune system itself is the target of infection, leading to a loss of the body's ability to defend itself. AIDS results from just this sort of infection by a virus called HIV (human immunodeficiency virus). The cell you see here is a human immune system cell called a lymphocyte that has been infected by HIV. Progeny HIV viruses are being released from the infected cell, budding out from its surface. These viruses soon spread to neighboring lymphocytes, infecting them in turn. Eventually a large majority of lymphocytes become infected, and the immune defense is destroyed. The individual viruses, in buds colored blue in this scanning electron micrograph, are extremely small; over 200 million would fit on the period at the end of this sentence.

22.1 Skin: The First Line of Defense

Multicellular bodies offer a feast of nutrients for tiny, single-celled creatures, as well as a warm, sheltered environment in which they can grow and reproduce. We live in a world awash with microbes, and no animal can long withstand their onslaught unprotected. Animals survive because they have a variety of very effective defenses against this constant attack.

Overview of the Three Lines of Defense

The vertebrate is defended from infection the same way knights defended medieval cities. "Walls and moats" make entry difficult; "roaming patrols" attack strangers; and "sentries" challenge anyone wandering about and call patrols if a proper "ID" is not presented.

Figure 22.1 Skin is the vertebrate body's first line of defense. This young elephant has tough, leathery skin thicker than a belt, allowing it to follow the herd through dense thickets without injury.

1. **Walls and moats.** The outermost layer of the vertebrate body, the skin, is the first barrier to penetration by microbes. Mucous membranes in the respiratory and digestive tracts are also important barriers that protect the body from invasion.

2. **Roaming patrols.** If the first line of defense is penetrated, the response of the body is to mount a **cellular counterattack,** using a battery of cells and chemicals that kill microbes. These defenses act very rapidly after the onset of infection.

3. **Sentries.** Lastly, the body is also guarded by mobile cells that patrol the bloodstream, scanning the surfaces of every cell they encounter. They are part of the **specific immune response.** One kind of immune cell aggressively attacks and kills any cell identified as foreign, whereas the other type marks the foreign cell or virus for elimination by the roaming patrols.

The Skin

Skin is the outermost layer of the vertebrate body (figure 22.1) and provides the first defense against invasion by microbes. Skin is our largest organ, comprising some 15% of our total weight. One square centimeter of skin from your forearm (about the size of a dime) contains 200 nerve endings, 10 hairs and muscles, 100 sweat glands, 15 oil glands, 3 blood vessels, 12 heat-sensing organs, 2 cold-sensing organs, and 25 pressure-sensing organs. Skin has two distinct layers: an outer **epidermis** and a lower **dermis.** A **subcutaneous layer** lies underneath the dermis. Cells of the outer epidermis are continually being worn away and replaced by cells moving up from below—in one hour your body loses and replaces approximately 1.5 million skin cells!

The epidermis of skin is from 10 to 30 cells thick, about as thick as this page. The outer layer, called the **stratum corneum,** is the one you see when you look at your arm or face. Cells from this layer are continuously subjected to damage. They are abraded, injured, and worn by friction and stress during the body's many activities. They also lose moisture and dry out. The body deals with this damage not by repairing cells but by replacing them. Cells from the stratum corneum are shed continuously, replaced by new cells produced deep within the epidermis. The cells of the innermost **basal layer** are among the most actively dividing cells of the vertebrate body. New cells formed there migrate upward, and as they move they manufacture keratin protein, which makes them tough. Each cell eventually arrives at the outer surface and takes its turn in the stratum corneum, residing there for about a month before it is shed and replaced by a newer cell. Persistent dandruff (psoriasis) is a chronic skin disorder in which new cells reach the epidermal surface every three or four days, about eight times faster than normal.

The dermis of the skin is from 15 to 40 times thicker than the epidermis. It provides structural support for the epidermis, as well as a matrix for the many specialized cells residing within the skin. The wrinkling that occurs as we grow older occurs here. The leather used to manufacture belts and shoes is derived from thick animal dermis. The layer of subcutaneous tissue below the dermis is composed of fat-rich cells that act as shock absorbers and provide insulation, which conserves body heat.

The skin not only defends the body by providing a nearly impermeable barrier, but it also reinforces this defense with chemical weapons. The oil and sweat glands, for example (figure 22.2), make the skin's surface very acidic, which inhibits the growth of many microbes. Sweat also contains the enzyme lysozyme, which attacks and digests the cell walls of many bacteria.

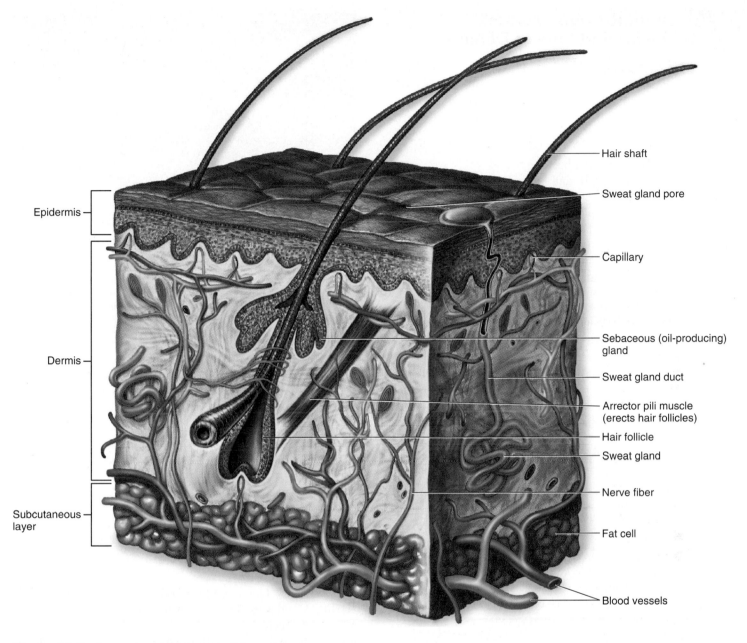

Figure 22.2 A section of human skin.
The skin defends the body by providing a barrier and sweat and oil glands whose secretions make the skin's surface acidic enough to inhibit the growth of microorganisms.

Other External Surfaces

In addition to the skin, two other potential routes of entry by viruses and microorganisms must be guarded: the *digestive tract* and the *respiratory tract*. Microbes are present in food, but many are killed by saliva (which also contains lysozyme), by the very acidic environment of the stomach, and by digestive enzymes in the intestine. Microorganisms are also present in inhaled air. The cells lining the smaller bronchi and bronchioles secrete a layer of sticky mucus that traps most microorganisms before they can reach the warm, moist lungs, which would provide ideal breeding grounds for them. Other cells lining these passages have cilia that continually sweep the mucus up toward the glottis in the throat. There it can be

swallowed, carrying potential invaders out of the lungs and into the digestive tract.

The surface defenses are very effective, but they are occasionally breached. Through breathing, eating, or cuts and nicks, bacteria and viruses now and then enter our bodies. When these invaders reach deeper tissue, a second line of defense comes into play, a cellular counterattack.

> **22.1 Skin and the mucous membranes lining the digestive and respiratory tracts are the body's first defenses.**

22.2 Cellular Counterattack: The Second Line of Defense

When the body's interior is invaded, a host of cellular and chemical defenses swing into action. Four are of particular importance: (1) cells that kill invading microbes; (2) proteins that kill invading microbes; (3) the inflammatory response, which speeds defending cells to the point of infection; and (4) the temperature response, which elevates body temperature to slow the growth of invading bacteria.

Although these cells and proteins roam through the body, there is a central location for their storage and distribution; it is called the **lymphatic system** (see chapter 20). The lymphatic system consists of a network of lymphatic capillaries, ducts, nodes, and lymphatic organs (figure 22.3), and although it has other functions involved with circulation, it is also involved in the immune response.

Cells That Kill Invading Microbes

The most important counterattack to infection is mounted by white blood cells, which attack invading microbes. These cells patrol the bloodstream and await invaders within the tissues.

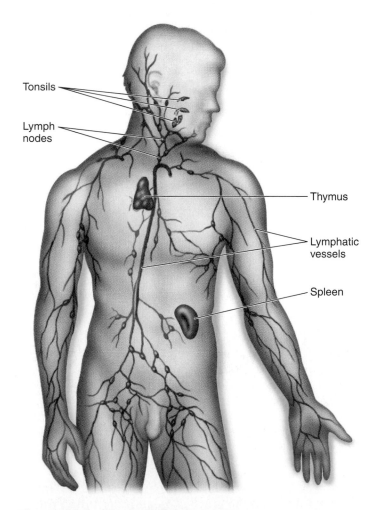

Figure 22.3 The lymphatic system.
The major lymphatic vessels, organs, and nodes are shown.

Figure 22.4 A macrophage in action.
In this scanning electron micrograph, a macrophage is "fishing" with long, sticky cytoplasmic extensions. Bacterial cells unfortunate enough to come in contact with the extensions are drawn back to the macrophage and engulfed.

The three basic kinds of killing cells are macrophages and neutrophils, which are phagocytes, and natural killer cells. Each uses a different tactic to kill invading microbes.

Macrophages. White blood cells called **macrophages** (Greek, "big eaters") kill bacteria by ingesting them (figure 22.4), much as an amoeba ingests a food particle. Although some macrophages are anchored within particular organs, particularly the spleen, most patrol the byways of the body, circulating as precursor cells called **monocytes** in the blood, lymph, and fluid between cells. Macrophages are among the most actively mobile cells of the body.

Neutrophils. Other white blood cells called **neutrophils** act like kamikazes. In addition to ingesting microbes, they release chemicals (identical to household bleach) to "neutralize" the entire area, killing any bacteria in the neighborhood—and themselves in the process. A neutrophil is like a grenade tossed into an infection. It kills everything in the vicinity. Macrophages, by contrast, kill only one invading cell at a time but live to keep on doing it.

Natural Killer Cells. A third kind of white blood cell, called a **natural killer cell,** does not attack invading microbes but rather the body cells that are infected by them. Natural killer cells puncture the membrane of the infected target cell with a molecule called *perforin,* allowing water to rush in and causing the cell to swell and burst (figure 22.5). Natural killer cells are particularly effective at detecting and attacking body cells that have been infected with viruses. They are also one of the body's most potent defenses against cancer, which they kill before the cancer cell has a chance to develop into a tumor (figure 22.6).

These three kinds of cells can distinguish the body's cells (self) from foreign cells (nonself) because the body's

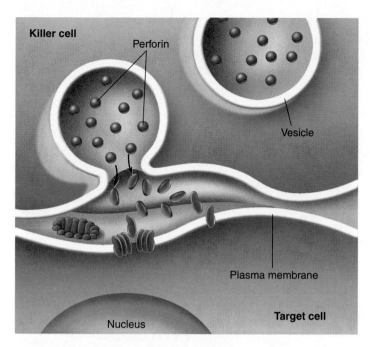

Figure 22.5 How natural killer cells attack target cells.

The initial event is the tight binding of the natural killer cell to the target cell. Binding initiates a chain of events within the killer cell in which vesicles loaded with perforin molecules move to the outer plasma membrane and expel their contents into the intercellular space over the target. The perforin molecules insert into the membrane like boards on a picket fence to form a pore that admits water and ruptures the cell.

cells contain self-identifying *MHC proteins* (discussed later in this chapter). When the body's defensive cells fail to make the self versus nonself distinction correctly, they may attack the body's own tissues. Diseases resulting from this failure are known as *autoimmune diseases.*

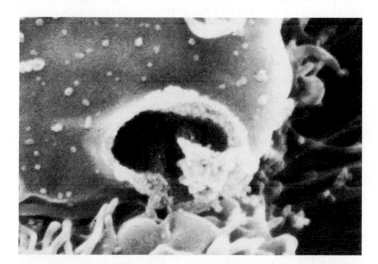

Figure 22.6 Death of a tumor cell.

A natural killer cell has attacked this cancer cell, punching a hole in its plasma membrane. Water has rushed in, making it balloon out. Soon it will burst.

Proteins That Kill Invading Microbes

The cellular defenses of vertebrates are complemented by a very effective chemical defense called the **complement system.** This system consists of approximately 20 different proteins that circulate freely in the blood plasma in an inactive state. Their defensive activity is triggered when they encounter the cell walls of bacteria or fungi. The complement proteins then aggregate to form a *membrane attack complex* that inserts itself into the foreign cell's plasma membrane, forming a pore like that produced by natural killer cells (figure 22.7). Water enters the foreign cell through this pore, causing the cell to swell and burst. Aggregation of the complement proteins is also triggered by the binding of antibodies to invading microbes, as we'll see in a later section.

The proteins of the complement system can augment the effects of other body defenses. Some amplify the inflammatory response (discussed next) by stimulating histamine release; others attract phagocytes (monocytes and neutrophils) to the area of infection; and still others coat invading microbes, roughening the microbes' surfaces so that phagocytes may attach to them more readily.

Another class of proteins that play a key role in body defense are interferons. There are three major categories of interferons: *alpha, beta,* and *gamma*. Almost all cells in the body make alpha and beta interferons. These polypeptides act as messengers that protect other cells in the vicinity from viral infection. The viruses are still able to penetrate the other cells, but the ability of the viruses to replicate and assemble new

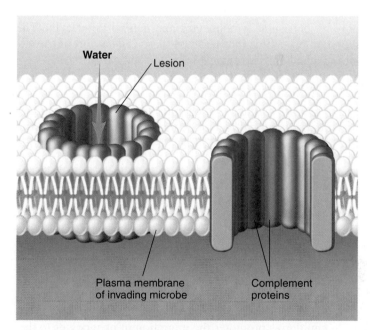

Figure 22.7 Complement creates a hole in a cell.

The complement proteins form a transmembrane channel resembling the perforin-lined lesion made by natural killer cells, although there are some differences between the two: The complement proteins are free-floating in blood plasma, and they attach to the invading microbe directly; the perforin molecules are produced in natural killer cells and poke holes in infected body cells, not in the actual microbe.

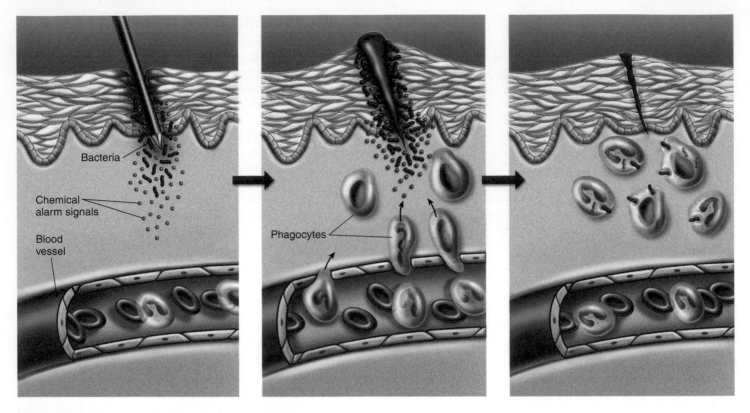

Figure 22.8 The events in a local inflammation.
When an invading microbe has penetrated the skin, chemicals, such as histamine and prostaglandins, act as alarm signals that cause nearby blood vessels to dilate. Increased blood flow brings a wave of phagocytic cells that attack and engulf invading bacteria.

virus particles is inhibited. Gamma interferon is produced only by particular lymphocytes and natural killer cells. The secretion of gamma interferon by these cells is part of the immunological defense against infection and cancer.

The Inflammatory Response

The aggressive cellular and chemical counterattacks to infection are made more effective by the **inflammatory response.** The inflammatory response can be broken down into three stages:

1. Infected or injured cells release chemical alarm signals, most notably histamine and prostaglandins.
2. These chemical alarm signals cause blood vessels to expand, both increasing the flow of blood to the site of infection or injury and, by stretching their thin walls, making the capillaries more permeable (figure 22.8). This produces the redness and swelling so often associated with infection.
3. The increased blood flow through larger, leakier capillaries promotes the migration of phagocytes to the site of infection. Neutrophils arrive first, spilling out chemicals that kill the microbes (as well as tissue cells in the vicinity and themselves). Monocytes follow and become macrophages, which engulf the pathogens and remains of all the dead cells. This counterattack takes a considerable toll; the pus associated with infections is a mixture of dead or dying neutrophils, tissue cells, and pathogens.

The Temperature Response

Human pathogenic bacteria do not grow well at high temperatures. Thus, when macrophages initiate their counterattack, they increase the odds in their favor by sending a message to the brain to raise the body's temperature. The cluster of brain cells that serves as the body's thermostat responds to the chemical signal by boosting the body temperature several degrees above the normal value of 37°C (98.6°F). The higher-than-normal temperature that results is called a **fever.** Although fever is quite effective at inhibiting microbial growth, very high fevers are dangerous because excessive heat can inactivate critical cellular enzymes. In general, temperatures greater than 39.4°C (103°F) are considered dangerous; those greater than 40.6°C (105°F) are often fatal.

The second line of defense, with both chemical and cellular weapons, provides a sophisticated defense against microbial infection. Only occasionally do bacteria or viruses overwhelm this defense. When this happens, they face yet a third line of defense, more difficult to evade than any they have encountered. It is the specific immune response, the most elaborate of the body's defenses.

> **22.2 Vertebrates respond to infection with a battery of cellular and chemical weapons, including cells and proteins that kill invading microbes, and inflammatory and temperature responses.**

22.3 Specific Immunity: The Third Line of Defense

Specific immune defense mechanisms of the body involve the actions of white blood cells, or leukocytes. They are very numerous—of the 100 trillion cells of your body, two in every 100 are white blood cells! Macrophages are white blood cells, as are neutrophils and natural killer cells. In addition, there are T cells, B cells, plasma cells, mast cells, and monocytes (table 22.1). T cells and B cells are called **lymphocytes** and are critical to the specific immune response.

After their origin in the bone marrow, **T cells** migrate to the thymus (hence the designation "T"), a gland just above the heart (see figure 22.3). There they develop the ability to identify microorganisms and viruses by the antigens exposed on their surfaces. An **antigen** is a molecule that provokes a specific immune response. Antigens are large, complex molecules, such as proteins, and they are generally foreign to the body, usually belonging to bacteria and viruses. Tens of millions of different T cells are made, each specializing in the recognition of one particular antigen. No invader can escape being recognized by at least a few T cells. There are four principal kinds of T cells: helper T cells (often symbolized T_H) initiate the immune response; memory T cells (T_M) provide a quick response to a previously encountered antigen; cytotoxic ("cell-poisoning") T cells (T_C) lyse cells that have been infected by viruses; and suppressor T cells (T_S) terminate the immune response.

Unlike T cells, **B cells** do not travel to the thymus; they complete their maturation in the bone marrow. (B cells are so named because they were originally characterized in a region of chickens called the bursa.) From the bone marrow, B cells are released to circulate in the blood and lymph. Individual B cells, like T cells, are specialized to recognize particular foreign antigens. When a B cell encounters the antigen to which it is targeted, it begins to divide rapidly, and its progeny differentiate into plasma cells and memory cells. Each plasma cell is a miniature factory producing markers called *antibodies*. These antibodies stick like flags to that antigen wherever it occurs in the body, marking any cell bearing the antigen for destruction.

B cells and T cells also produce memory cells that provide the body with the ability to recall a previous exposure to an antigen and mount an attack against that antigen very quickly. As described later in this chapter, the initial specific immune response to an antigen encountered for the first time is delayed, which allows the pathogen time to infect the body. A second infection is halted much earlier due to the presence of memory cells, which respond to the pathogen more quickly.

> **22.3** T cells develop in the thymus, whereas B cells develop in the bone marrow. When a B cell encounters a specific antigen, it gives rise to plasma cells that produce antibodies.

TABLE 22.1 CELLS OF THE IMMUNE SYSTEM

Cell Type		Function
Helper T cell		Commander of the immune response; detects infection and sounds the alarm, initiating both T cell and B cell responses
Memory T cell		Provides a quick and effective response to an antigen previously encountered by the body
Cytotoxic T cell		Detects and kills infected body cells; recruited by helper T cells
Suppressor T cell		Dampens the activity of T and B cells, scaling back the defense after the infection has been checked
B cell		Precursor of plasma cell; specialized to recognize specific foreign antigens
Neutrophil		Engulfs invading bacteria and releases chemicals that kill neighboring bacteria
Plasma cell		Biochemical factory devoted to the production of antibodies directed against specific foreign antigens
Mast cell		Initiator of the inflammatory response, which aids the arrival of leukocytes at a site of infection; secretes histamine and is important in allergic responses
Monocyte		Precursor of macrophage
Macrophage		The body's first cellular line of defense; also serves as antigen-presenting cell to B and T cells and engulfs antibody-covered cells
Natural killer cell		Recognizes and kills infected body cells; natural killer cell detects and kills cells infected by a broad range of invaders

22.4 Initiating the Immune Response

To understand how this third line of defense works, imagine you have just come down with the flu. Influenza viruses enter your body in small water droplets inhaled into your respiratory system. If they avoid becoming ensnared in the mucus lining the respiratory membranes (first line of defense), and avoid consumption by macrophages (second line of defense), the viruses infect and kill mucous membrane cells.

At this point, macrophages initiate the immune defense. Macrophages inspect the surfaces of all cells they encounter. Every cell in the body carries special marker proteins on its surface, called major histocompatibility proteins, or **MHC proteins.** The MHC proteins are different for each individual, much as fingerprints are. As a result, the MHC proteins on the tissue cells serve as "self" markers that enable the individual's immune system to distinguish its cells from foreign cells. This distinction is critical for the function of T cells because, unlike B cells, T cells cannot bind to free antigens. T cells can only bind to antigens that are presented to them on the surface of cells.

When a foreign particle, such as a virus, infects the body, it is taken in by cells and partially digested. Within the cells, the viral antigens are processed and moved to the surface of the plasma membrane. The cells that perform this function are called **antigen-presenting cells** (figure 22.9). At the membrane, the processed antigens are complexed with the MHC proteins. T cell receptors can only interact with cells that have this combination of MHC proteins and antigens.

Macrophages that encounter pathogens—either a foreign cell such as a bacterial one, which lacks proper MHC proteins, or a virus-infected body cell with telltale virus proteins stuck to its surface—respond by secreting a chemical alarm signal. The alarm signal is a protein called **interleukin-1** (Latin for "between white blood cells"). This protein stimulates **helper T cells.** The helper T cells respond to the interleukin-1 alarm by simultaneously initiating two different parallel lines of immune system defense: the cellular immune response carried out by T cells and the antibody or humoral response carried out by B cells. The immune response carried out by T cells is called the cellular response because the T lymphocytes attack the cells that carry antigens. The B cell response is called the humoral response because antibodies are secreted into the blood and body fluids (*humor* refers to a body fluid).

> **22.4** When macrophages encounter cells without the proper MHC proteins, they secrete a chemical alarm that initiates the immune defense.

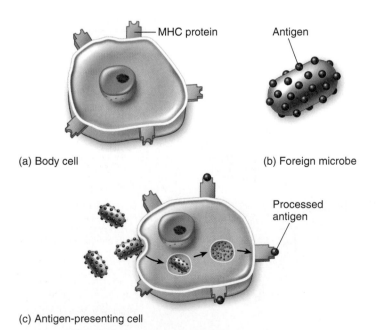

(a) Body cell

(b) Foreign microbe

(c) Antigen-presenting cell

(d)

Figure 22.9 How antigens are presented.
(*a*) Cells of the body have MHC proteins on their surfaces that identify them as "self" cells. Immune system cells do not attack these cells. (*b*) Foreign cells or microbes have antigens on their surfaces. B lymphocytes are able to bind directly to free antigens in the body to initiate an attack on a foreign invader. (*c*) T cells can bind to the antigens to initiate an attack only after the antigens are processed and complexed with MHC proteins on the surface of an antigen-presenting cell. (*d*) In this electron micrograph, a lymphocyte (*right*) contacts a macrophage (*left*), an antigen-presenting cell.

22.5 T Cells: The Cellular Response

When macrophages process the foreign antigens, they secrete interleukin-1, which stimulates cell division and proliferation of T cells, the **cellular immune response.** Helper T cells become activated when they bind to the complex of MHC proteins and antigens presented to them by the macrophages. The helper T cells then secrete **interleukin-2,** which stimulates the proliferation of **cytotoxic T cells,** which recognize and destroy infected body cells (figure 22.10). Cytotoxic T cells can destroy infected cells only if those cells display the foreign antigen together with their MHC proteins.

The body makes millions of different types of T cells. Each type bears a single, unique kind of receptor protein on its membrane, a receptor able to bind to a particular antigen-MHC protein complex on the surface of an antigen-presenting cell. Any cytotoxic T cell whose receptor fits the particular antigen-MHC protein complex present in the body begins to multiply rapidly, soon forming large numbers of T cells capable of recognizing the complex containing the particular foreign antigen. Large numbers of infected cells can be quickly eliminated, because the single T cell able to recognize the invading virus is amplified in number to form a large clone of identical T cells, all able to carry out the attack. Any of the body's cells that bear traces of viral infection are destroyed. The method used by cytotoxic T cells to kill infected body cells is similar to that used by natural killer cells and complement—they puncture the plasma membrane of the infected cell.

Cytotoxic T cells will also attack any foreign version of the MHC proteins. Thus even though vertebrates did not evolve the immune system as a defense against tissue transplants, their immune systems will attack transplanted tissue and cause graft rejection. It is for this reason that relatives are often sought for kidney transplants—their MHC proteins are genetically closer to the recipient. The drug cyclosporin is often given to transplant patients because it inactivates cytotoxic T cells.

Cancer cells also exhibit foreign antigens on their surfaces that activate an immune response. The concept of immunological surveillance against cancer was introduced in the early 1970s to describe the role of the immune system in fighting cancer.

> **22.5** The cellular immune response is carried out by T cells, which mount an immediate attack on infecting and infected cells, killing any that present unusual surface antigens.

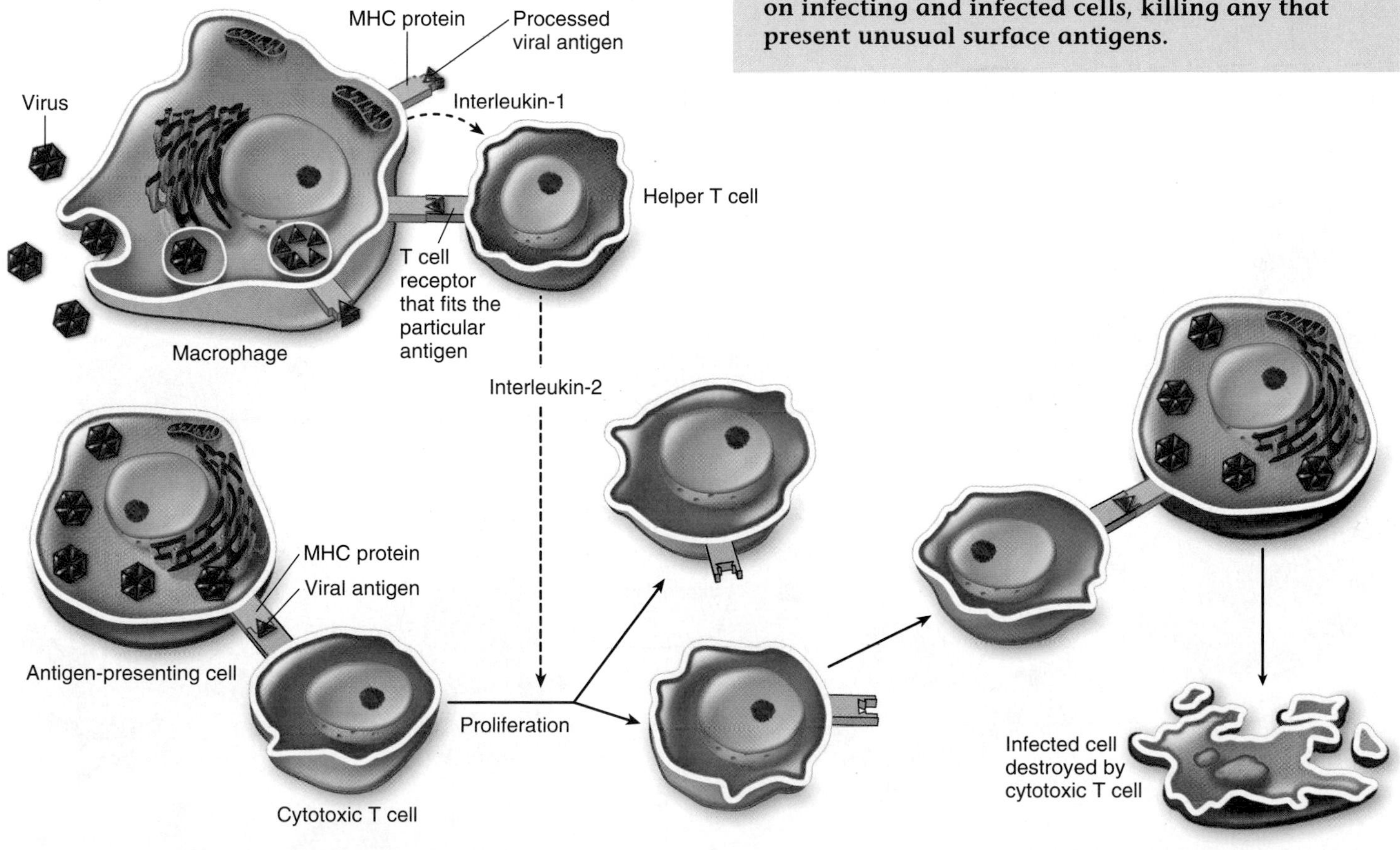

Figure 22.10 The T cell immune defense.
After a macrophage has processed an antigen, it releases interleukin-1, signaling helper T cells to bind to the antigen-MHC protein complex. This triggers the helper T cell to release interleukin-2, which stimulates the multiplication of cytotoxic T cells. In addition, proliferation of cytotoxic T cells is stimulated when a T cell with a receptor that fits the antigen displayed by an antigen-presenting cell binds to the antigen-MHC protein complex. Body cells that have been infected by the antigen are destroyed by the cytotoxic T cells. As the infection subsides, suppressor T cells "turn off" the immune response.

22.6 B Cells: The Humoral Response

B cells also respond to helper T cells activated by interleukin-1. Like cytotoxic T cells, B cells have receptor proteins on their surfaces, one type of receptor for each type of B cell. B cells recognize invading microbes much as cytotoxic T cells recognize infected cells, but unlike cytotoxic T cells, they do not go on the attack themselves. Rather, they mark the pathogen for destruction by mechanisms that have no "ID check" system of their own. Early in the immune response known as the **humoral immune response,** the markers placed by B cells alert complement proteins to attack the cells carrying them. Later in the response, the markers placed by B cells activate macrophages and natural killer cells.

The way B cells do their marking is simple and foolproof. Unlike the receptors on T cells, which bind only to antigen-MHC protein complexes on antigen-presenting cells, B cell receptors can bind to free, unprocessed antigens. When a B cell encounters an antigen, antigen particles enter the B cell by endocytosis and get processed. Helper T cells that are able to recognize the specific antigen bind to the antigen-MHC protein complex on the B cell and release interleukin-2, which stimulates the B cell to divide. In addition, free, unprocessed antigens stick to antibodies on the B cell surface. This antigen exposure triggers even more B cell proliferation. B cells divide to produce long-lived memory B cells and plasma cells that serve as short-lived antibody factories (figure 22.11).

Antibodies are proteins in a class called **immunoglobulins** (abbreviated *Ig*), which is divided into subclasses based on the structures and functions of the antibodies. The five different immunoglobulin subclasses are as follows:

1. **IgM.** This is the first type of antibody to be secreted into the blood during the primary response and to serve as a receptor on the B cell surface. These antibodies also promote agglutination reactions (causing antigen-containing particles to stick together, or agglutinate).
2. **IgG.** This is the major form of antibody secreted in a secondary response and the major one in the blood plasma.
3. **IgD.** These antibodies serve as receptors for antigens on the B cell surface. Their other functions are unknown.
4. **IgA.** This is the form of antibody in external secretions, such as saliva and mother's milk.
5. **IgE.** This form of antibody promotes the release of histamine and other agents that produce allergic symptoms, such as those of hay fever.

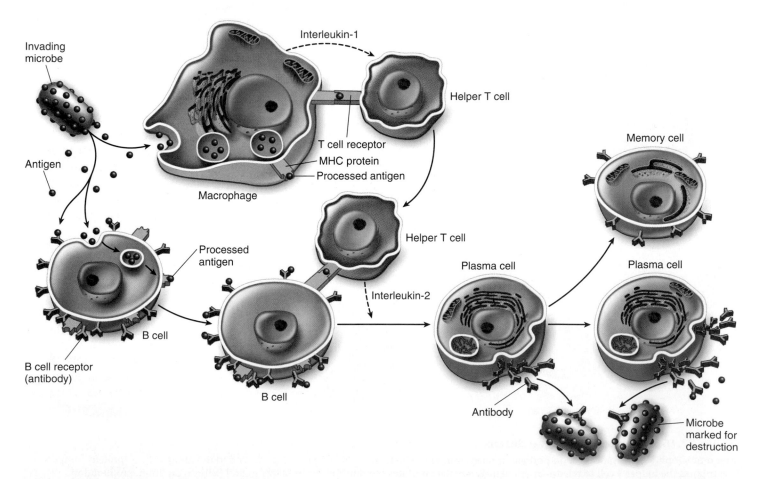

Figure 22.11   The B cell immune defense.

Invading particles are bound by B cells, which interact with helper T cells and are activated to divide. The multiplying B cells produce either memory B cells or plasma cells that secrete antibodies that bind to invading microbes and tag them for destruction by macrophages.

B cells divide, giving rise to **plasma cells** that produce lots of the same particular antibody that was able to bind the antigen in the initial immune response. Flooding through the bloodstream, these antibody proteins (figure 22.12) are able to stick to antigens on any cells or microbes that present them, flagging those cells and microbes for destruction. Complement proteins, macrophages, or natural killer cells then destroy the antibody-displaying cells and microbes.

The B cell defense is very powerful because it amplifies the reaction to an initial pathogen encounter a millionfold. It is also a very long-lived defense in that a few of the multiplying B cells do not become antibody producers. Instead they become a line of **memory B cells** that continue to patrol your body's tissues, circulating through your blood and lymph for a long time—sometimes for the rest of your life.

Antibody Diversity

The vertebrate immune system is capable of recognizing as foreign practically any nonself molecule presented to it—literally millions of different antigens. Although vertebrate chromosomes contain only a few hundred receptor-encoding genes, it is estimated that human B cells can make between 10^6 and 10^9 different antibody molecules. How do vertebrates generate millions of different antigen receptors when their chromosomes contain only a few hundred versions of the genes encoding those receptors?

The answer to this question is that the millions of immune receptor genes do not exist as single sequences of nucleotides. Rather, they are assembled by stitching together three or four DNA segments that code for different parts of the receptor molecule. When an antibody is assembled, these different sequences of DNA are brought together to form a composite gene (figure 22.13). This process is called **somatic rearrangement.**

Two other processes generate even more sequences. First, the DNA segments are often joined together with one

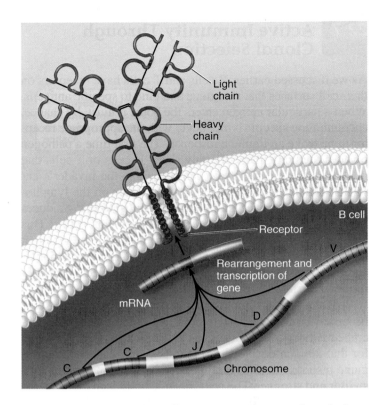

Figure 22.13 The lymphocyte receptor molecule is produced by a composite gene.

Different regions of the DNA code for different regions of the receptor structure (*C*, constant regions; *J*, joining regions; *D*, diversity regions; and *V*, variable regions) and are brought together to make a composite gene that codes for the receptor. Through different somatic rearrangements of these DNA segments, an enormous number of different receptor molecules can be produced.

or two nucleotides off-register, shifting the reading frame during gene transcription and so generating a totally different sequence of amino acids in the protein. Second, random mistakes occur during successive DNA replications as the lymphocytes divide during clonal expansion. Both mutational processes produce changes in amino acid sequences, a phenomenon known as **somatic mutation** because it takes place in a somatic cell rather than in a gamete.

Because a cell may end up with any heavy chain gene and any light chain gene during its maturation, the total number of different antibodies possible is staggering: 16,000 heavy chain combinations $\times$ 1,200 light chain combinations = 19 million different possible antibodies. If one also takes into account the changes induced by somatic mutation, the total can exceed 200 million! It should be understood that although this discussion has centered on B cells and their receptors, the receptors on T cells are as diverse as those on B cells because they also are subject to similar somatic rearrangements and mutations.

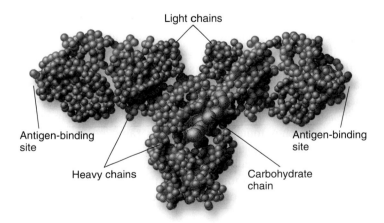

Figure 22.12 An antibody molecule.

In this molecular model of an antibody molecule, each amino acid is represented by a small sphere. Each molecule consists of four protein chains, two identical "light" (*red*) and two identical "heavy" (*blue*). The four protein chains wind around one another to form a Y shape. Foreign molecules, called antigens, bind to the arms of the Y.

22.6 In the humoral immune response, B cells label infecting and infected cells with antibodies for destruction by complement proteins, natural killer cells, and macrophages.

HIV's attack on CD4$^+$ T cells progressively cripples the immune system, because HIV-infected cells die only after releasing replicated viruses that proceed to infect other CD4$^+$ T cells, until the body's entire population of CD4$^+$ T cells is destroyed (figure 22.26). In a normal individual, CD4$^+$ T cells make up 60% to 80% of circulating T cells; in AIDS patients, CD4$^+$ T cells often become too rare to detect, wiping out the human immune defense. With no defense against infection, any of a variety of otherwise commonplace infections proves fatal. With no ability to recognize and destroy cancer cells when they arise, death by cancer becomes far more likely. Indeed, AIDS was first recognized because of a cluster of cases of a rare cancer, Karposi's sarcoma. More AIDS victims die of cancer than from any other cause.

The fatality rate of AIDS is 100%; no patient exhibiting the symptoms of AIDS has ever been known to survive more than a few years. However, the disease is *not* highly contagious, because it is only transmitted from one individual to another through the transfer of internal body fluids, typically in semen during sexual intercourse and in blood transmitted by needles during drug use.

A variety of drugs inhibit HIV in the test tube. These include AZT and its analogs (which inhibit virus nucleic acid replication) and protease inhibitors (which inhibit the cleavage of the large virus proteins into functional segments). A combination of a protease inhibitor and two AZT analog drugs entirely eliminates the HIV virus from many patients' bloodstreams. Widespread use of this **combination therapy** has cut the U.S. AIDS death rate by almost two-thirds since its introduction in the mid-1990s, from 43,000 AIDS deaths in 1995 to 31,000 in 1996, and six years later in 2002, deaths remained low, at just over 16,000.

Unfortunately, this sort of combination therapy does not appear to actually succeed in eliminating HIV from the body. While the virus disappears from the bloodstream, traces of it can still be detected in lymph tissue of the patients. When combination therapy is discontinued, virus levels in the blood-

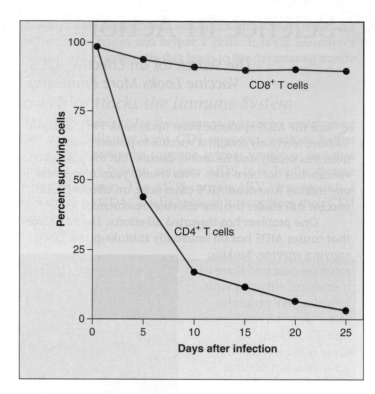

Figure 22.26   Survival of T cells in culture after exposure to HIV.

The virus has little effect on the number of CD8$^+$ T cells, T cells with CD8 cell surface receptors. But HIV causes the number of CD4$^+$ T cells (this group includes helper T cells) to decline dramatically.

stream once again rise. Because of demanding therapy schedules and many side effects, long-term combination therapy does not seem a promising approach.

> **22.12   HIV cripples the vertebrate immune defense by infecting and killing key lymphocytes.**

Exploring Current Issues

Additional Resources

Go to your campus library or look online to find the following articles, which further develop some of the concepts found in this chapter.

Bunk, S. (2003). Cancer's other conduit: evidence suggests that some tumors sprout lymph vessels for metastasis. *New Scientist,* 17(17), 33.

Fitzgerald, D. (2003). Antibody drug development: on target: as researchers grow ever more clever in their manipulations, monoclonal antibody therapeutics stage a comeback. *The Scientist,* 17(22), 29.

Gorman, C., A. Park, K. Dell, and D. Cray. (2004). The fires within: inflammation is the body's first defense against disease, but when it goes awry, it can lead to heart attacks, colon cancer, Alzheimer's, and a host of other diseases. *Time,* 163(8), 38.

Hodgkinson, N. (2003). AIDS: scientific or viral catastrophe? *Journal of Scientific Exploration,* 17(1), 87.

Russo, E. (2003). A new approach to autoimmune diseases: instead of studying disorders in isolation, some investigators are seeking common denominators among them. *The Scientist,* 17(9), 30.

In the News: AIDS

Since the AIDS epidemic started, over 24 million people have died of AIDS. Researchers have worked feverishly to find a way to halt the epidemic, and after many disappointments seem to be having some success.

Article 1. *Combination drug therapy buys time, not a cure.*
Article 2. *The search for an AIDS vaccine just got harder.*
Article 3. *Looking for novel ways to hinder the sexual transmission of AIDS.*
Article 4. *Did a contaminated polio vaccine bring about the AIDS epidemic?*
Article 5. *The battle against AIDS proceeds on two fronts.*
Article 6. *AIDS at 20: New approaches to combating HIV.*

Find these articles, written by the author, online at www.mhhe.com/tlwessentials/exp22.

Three Lines of Defense

22.1 Skin: The First Line of Defense

- The body has three lines of defense against infection, the first being skin (**figure 22.2**) and the mucous membranes that line the digestive and respiratory tracts. Skin and mucous membranes form barriers to pathogens.

22.2 Cellular Counterattack: The Second Line of Defense

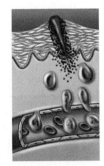

- The second line of defense is a nonspecific cellular attack. The cells and chemicals used in this line of defense attack all foreign agents that they encounter. Macrophages and neutrophils attack the invading pathogen, while natural killer cells attack infected cells, killing them before the pathogen can spread to other cells (**figures 22.4 and 22.5**). Free-floating proteins in the blood, called complement, insert into the membranes of foreign cells, killing them (**figure 22.7**). The inflammatory and temperature responses are also part of the second line of defense (**figure 22.8**).

22.3 Specific Immunity: The Third Line of Defense

- The third line of defense is a specific cellular attack (**table 22.1**). T cells and B cells are "programmed" by exposure to specific antigens and once programmed, will seek out the antigens or cells carrying those antigens and destroy them.

The Immune Response

22.4 Initiating the Immune Response

- Macrophages survey cells for "self" MHC proteins. A cell that exhibits "nonself" MHC proteins is engulfed by a macrophage which secretes interleukin-1. Interleukin-1 stimulates helper T cells that trigger the cellular and humoral immune responses. These macrophages also insert foreign antigens in their membranes and present these antigens to T cells, activating the T cell response. These macrophages are called antigen-presenting cells (**figure 22.9**).

22.5 T Cells: The Cellular Response

- The cellular response involves T cells. Antigen-presenting cells activate helper T cells that release interleukin-2. Interleukin-2 stimulates the cloning of cytotoxic T cells that recognize and kill cells with the specific antigen found on the antigen-presenting cells (**figure 22.10**).

22.6 B Cells: The Humoral Response

- The humoral response involves B cells. B cells are able to recognize foreign antigens on the surfaces of pathogens and when activated by interleukin-2, released from helper T cells,

they "mark" the foreign invaders with specific antibody proteins. The "marked" cells are then attacked by the nonspecific immune response. Activated B cells divide to form plasma cells, that produce and release the antibodies, and memory cells, that circulate in the blood. Memory cells quickly become plasma cells in the event of a second infection by the same antigen (**figure 22.11**).

22.7 Active Immunity Through Clonal Selection

- The initial humoral response triggered by infection is called the primary response. It is a delayed and somewhat weak response. But through clonal selection, a large population of memory cells is present in the body, such that a second infection by the same antigen will trigger a quicker and larger humoral response, called the secondary response (**figures 22.14–22.16**).

22.8 Evolution of the Immune System

- The immune system evolved from a nonspecific immune response, involving phagocytic cells only, present in invertebrates, to a two-part immune response involving B cells and T cells, present in vertebrates (**figure 22.18**).

22.9 Vaccination

- Vaccination takes advantage of the mechanism of the primary and secondary immune responses. Vaccines introduce harmless antigens present on the pathogenic cell into the body. The body produces memory B cells toward the antigen, such that with an actual infection, the body elicits a swift and large immune response (**figure 22.21**).

22.10 Antibodies in Medical Diagnosis

- Antibodies are keen detectors of antigens, and so are used in various medical diagnostic applications such as blood typing and monoclonal antibody assays (**figure 22.22**).

Defeat of the Immune System

22.11 Overactive Immune System

- Sometimes the immune system attacks antigens that are not foreign or pathogenic. In autoimmune responses, the body attacks its own cells. In allergic reactions, the body mounts an attack against a harmless substance (**figure 22.24**).

22.12 AIDS: Immune System Collapse

- AIDS is a fatal disease caused by infection with the HIV virus (**figure 22.25**). HIV attacks macrophages and helper T cells, eventually destroying the cells that protect the body from other infections (**figure 22.26**).

1. Besides acting as a physical barrier, the skin also provides chemical protection from
   a. sweat and oil glands.   c. the skin's basal layer.
   b. mucus.   d. the stratum corneum of the skin.

2. The immune system can identify foreign cells in the bloodstream because these foreign cells
   a. are observed destroying other cells by the immune system.
   b. do not have the proper cell surface proteins that identify the cell as the body's own.

c.   are observed causing inflammatory response.

d.   are observed causing temperature response.

3. The purpose of the inflammatory response is to
   a.   increase the temperature of an infected area.
   b.   reduce pain of an infected area.
   c.   increase the number of immune system cells in an infected area.
   d.   increase physical protection of an infected area.

4. Increasing human body temperature—that is, causing a fever—assists the immune system because
   a.   increased temperature speeds up the chemical reactions used by the immune system.
   b.   pathogenic bacteria do not grow well at high temperatures.
   c.   increased temperature causes foreign proteins to denature.
   d.   All of these answers are correct.

5. Immune responses tailored to specific pathogens involve
   a.   T cells.   c.   monocytes.
   b.   macrophages.   d.   neutrophils.

6. Antibody production takes place in
   a.   T cells.   c.   B cells.
   b.   natural killer cells. d.   mast cells.

7. Cytotoxic T cells
   a.   produce antibodies.
   b.   destroy pathogens directly.
   c.   destroy foreign antigens floating freely in the bloodstream.
   d.   destroy cells infected by pathogens.

8. Immunity to future invasion of a specific pathogen is accomplished by production of
   a.   plasma cells.   c.   helper T cells.
   b.   memory T and B cells. d.   monocytes.

9. When a body's own immune system attacks the body's own cells, this is called
   a.   an inflammatory response.
   b.   a temperature response.
   c.   an autoimmune response.
   d.   an allergic response.

10. HIV-infected people with advanced AIDS usually die of an infectious disease such as pneumonia. This is because HIV attacks
    a.   helper T cells.   c.   memory T and B cells.
    b.   neutrophils.   d.   mast cells.

Visual Understanding

1. **Figure 22.14**   Certain diseases are considered to be primarily "childhood" diseases—measles, mumps, chicken pox, for instance—and once someone has had these illnesses as a child, they don't catch them again when, as parents, they take care of their own children who are sick. Use the graph to help explain the reason.

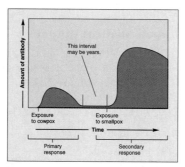

2. **Figure 22.25**   What might account for the large first curve, from 1981 to 1992, then the decline until 1997, the increase once again in 1998, and now the essentially steady state or flat line, neither growing nor diminishing by very much, for the past five years?

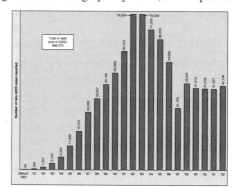

Challenge Questions

Three Lines of Defense   Your friend, Sadako, fell yesterday while she was skateboarding and has a deep cut on her forearm from a piece of wire. She shows it to you and complains about how sore it is. You can see the skin is red and swollen around the small puncture wound. Explain to her what is happening.

The Immune Response   They give you an antibiotic after you already have a bacterial infection, but they give you a vaccination before a virus (and some bacteria) has made you ill. What is the difference between these approaches?

Defeat of the Immune System   What are two major reasons why we have been unable to develop a good vaccine for AIDS/HIV?

Online Learning Center

Visit the Online Learning Center for this chapter at www.mhhe.com/tlwessentials/ch22 for quizzes, animations, interactive learning exercises, and other study tools. At the site you will also find extended answers to the end-of-chapter questions.

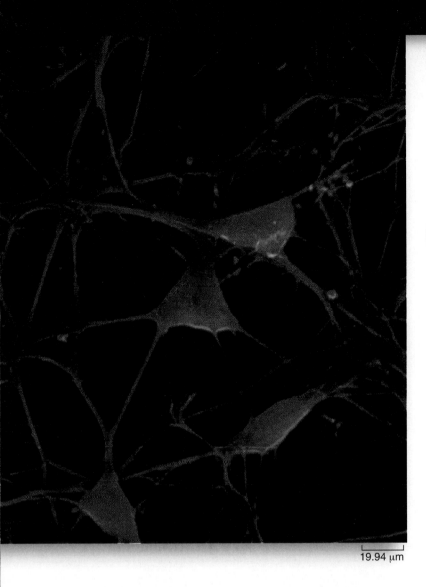

19.94 µm

23

The Nervous System

In vertebrates, the central nervous system coordinates and regulates the diverse activities of the body, using a network of specialized cells called neurons to direct the voluntary muscles, and a second network not under voluntary control to direct cardiac and smooth muscles. All sensory information is acquired through depolarization of sensory nerve endings. From a knowledge of which neurons are sending signals, and how often they are doing so, the brain builds a picture of the body's internal condition and of the external environment. The network of neurons (nerve cells) seen here, magnified over a thousand times, is transmitting signals within the portion of the brain called the cerebral cortex. The vertebrate brain contains a staggering number of neurons—the human brain contains an estimated 100 billion. The cerebral cortex is a layer of gray matter only a few millimeters thick on the brain's outer surface. Densely packed with neurons and highly convoluted, it is the site of higher mental activities.

23.1 Evolution of the Animal Nervous System

An animal must be able to respond to environmental stimuli. To do this, it must have sensory receptors that can detect the stimulus and motor *effectors* that can respond to it. In most invertebrate phyla and in all vertebrate classes, sensory receptors and motor effectors are linked by way of the **nervous system.** As described in chapter 19, the nervous system consists of neurons and supporting cells. One type of neuron, called **association neurons** (or **interneurons**), is present in the nervous systems of most invertebrates and all vertebrates. These neurons are located in the brain and spinal cord of vertebrates, together called the **central nervous system (CNS).** They help provide more complex reflexes and in the case of the brain, higher associative functions, including learning and memory, which require integration of many sensory inputs (figure 23.1).

There are two other types of neurons. **Motor** (or **efferent**) **neurons** carry impulses away from the CNS to effectors—muscles and glands. **Sensory** (or **afferent**) **neurons** carry impulses from sensory receptors to the CNS (figure 23.2). Together, motor and sensory neurons constitute the **peripheral nervous system (PNS)** of vertebrates.

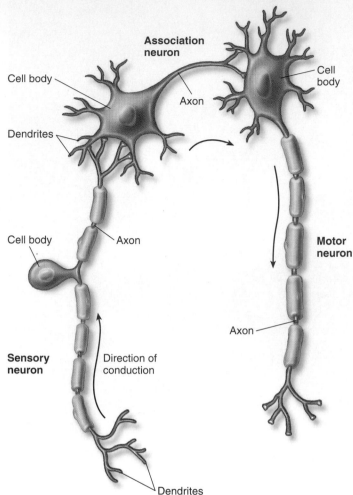

Figure 23.2   Three types of neurons.
Sensory neurons carry information about the environment to the brain and spinal cord. *Association neurons* are found in the brain and spinal cord and often provide links between sensory and motor neurons. *Motor neurons* carry impulses to muscles and glands (effectors).

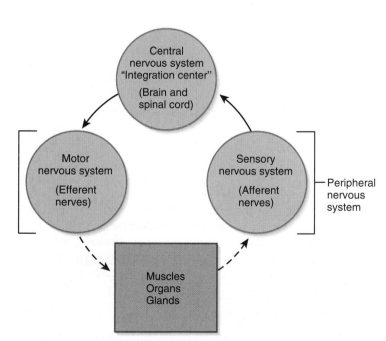

Figure 23.1   Organization of the vertebrate nervous system.
The central nervous system, consisting of the brain and spinal cord, issues commands via the motor nervous system and receives information from the sensory nervous system. The motor and sensory nervous systems together make up the peripheral nervous system.

Invertebrate Nervous Systems

Sponges are the only major phylum of multicellular animals that lack nerves. If you prick a sponge, the nearby surface contracts slowly. The cytoplasm of each individual cell conducts an impulse that fades within a few millimeters. No messages dart from one part of the sponge body to another, as they do in all other multicellular animals.

The Simplest Nervous Systems: Reflexes.   The simplest nervous systems occur among cnidarians (figure 23.3): all neurons are similar, each having fibers of approximately equal length. Cnidarian neurons are linked to one another in a web, or *nerve net,* dispersed through the body. Although conduction

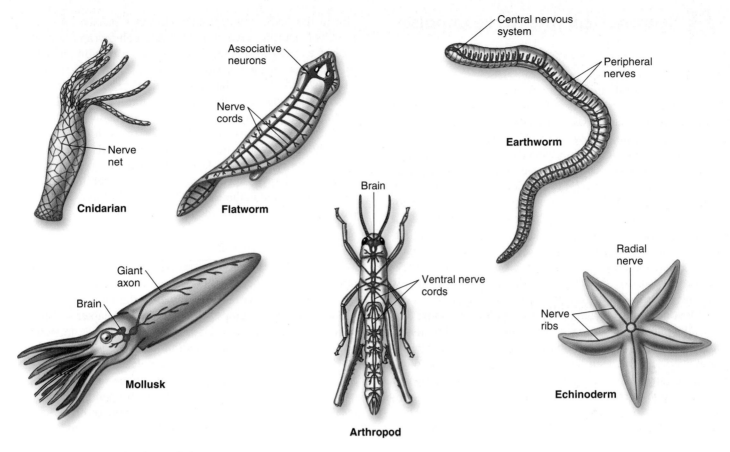

Figure 23.3 Evolution of the nervous system.
Invertebrates exhibit a progressive elaboration of organized nerve cords and the centralization of complex responses in the front end of the nerve cord.

is slow, a stimulus anywhere can eventually spread through the whole net. There is no associative activity, no control of complex actions, and little coordination. Any motion that results is called a **reflex** because it is an automatic consequence of the nerve stimulation.

More Complex Nervous Systems: Associative Activities.
The first associative activity in nervous systems is seen in the free-living flatworms, phylum Platyhelminthes. Running down the bodies of these flatworms are two nerve cords; peripheral nerves extend outward to the muscles of the body. The two nerve cords converge at the front end of the body, forming an enlarged mass of nervous tissue that also contains associative neurons that connect neurons to one another. This primitive "brain" is a rudimentary central nervous system and permits a far more complex control of muscular responses than is possible in cnidarians.

The Evolutionary Path to the Vertebrates. All of the subsequent evolutionary changes in nervous systems can be viewed as a series of elaborations on the characteristics already present in flatworms. Five trends can be identified, each becoming progressively more pronounced as nervous systems evolved greater complexity.

1. *More sophisticated sensory mechanisms.* Particularly among the vertebrates, sensory systems become highly complex.
2. *Differentiation into central and peripheral nervous systems.* For example, earthworms exhibit a central nervous system that is connected to all other parts of the body by peripheral nerves.
3. *Differentiation of sensory and motor nerves.* Neurons operating in particular directions (sensory signals traveling to the brain, or motor signals traveling from the brain) become increasingly specialized.
4. *Increased complexity of association.* Central nervous systems with more numerous interneurons evolved, increasing association capabilities dramatically.
5. *Elaboration of the brain.* Coordination of body activities became increasingly localized in arthropods, mollusks, and vertebrates in the front end of the nerve cord, which evolved into the vertebrate brain discussed later in the chapter.

23.1 As nervous systems became more complex, there was a progressive increase in associative activity, increasingly localized in a brain.

Neurons Generate Nerve Impulses

Neurons

The basic structural unit of the nervous system, whether central, motor, or sensory, is the nerve cell, or **neuron.** All neurons have the same basic structure (figure 23.4a). The **cell body** is an enlarged region of the neuron containing the nucleus. Short, slender branches called **dendrites** extend from one end of a neuron's cell body. Dendrites are input channels. Nerve impulses travel inward along them, toward the cell body. Motor and association neurons possess a profusion of highly branched dendrites, enabling those cells to receive information from many different sources simultaneously. Projecting out from the other end of the cell body is a single, long, tubelike extension called an **axon.** Axons are output channels. Nerve impulses travel outward along them, away from the cell body, toward other neurons or to muscles or glands.

Most neurons are unable to survive alone for long; they require the nutritional support provided by companion **neuroglial cells.** More than half the volume of the human nervous system is composed of supporting neuroglial cells. Two of the most important kinds of supporting neurons are the **Schwann cells** and **oligodendrocytes,** which envelop the axons of many neurons with a sheath of fatty material called myelin, which acts as an electrical insulator. Schwann cells produce my-elin in the PNS, while oligodendrocytes produce myelin in the CNS. During development, these cells wrap themselves around each axon several times to form a **myelin sheath,** an insulating covering consisting of multiple layers of membrane (figure 23.4b). Axons that have myelin sheaths are said to be myelinated, and those that don't are unmyelinated. The myelin sheath is interrupted at intervals, leaving uninsulated gaps called **nodes of Ranvier** where the axon is in direct contact with the surrounding fluid. The nerve impulse jumps from node to node, speeding its travel down the axon. Multiple sclerosis and Tay-Sachs (see chapter 8) are debilitating clinical disorders, and in the case of Tay-Sachs is fatal. They result from the degeneration of the myelin sheath.

The Nerve Impulse

When a neuron is "at rest," not carrying an impulse, active transport channels in the neuron's plasma membrane transport sodium ions (Na^+) out of the cell and potassium ions (K^+) in. This sodium-potassium pump was described in chapter 4. Sodium ions cannot easily move back into the cell once they are pumped out, so the concentration of sodium ions builds up outside the cell. Similarly, potassium ions accumulate inside the cell, although they are not as highly concentrated because many potassium ions are able to diffuse out through open channels. The result is to make the outside of the neuron more positive than the inside. The plasma membrane is said to be "polarized."

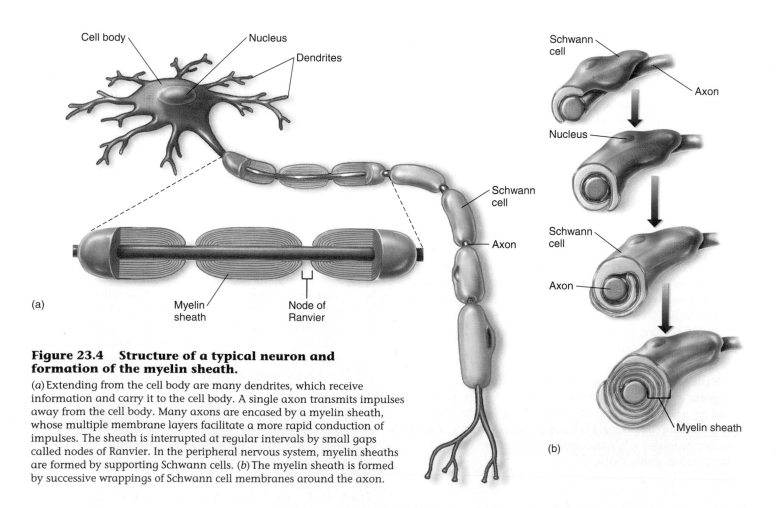

Figure 23.4 Structure of a typical neuron and formation of the myelin sheath.

(a) Extending from the cell body are many dendrites, which receive information and carry it to the cell body. A single axon transmits impulses away from the cell body. Many axons are encased by a myelin sheath, whose multiple membrane layers facilitate a more rapid conduction of impulses. The sheath is interrupted at regular intervals by small gaps called nodes of Ranvier. In the peripheral nervous system, myelin sheaths are formed by supporting Schwann cells. (b) The myelin sheath is formed by successive wrappings of Schwann cell membranes around the axon.

Neurons are constantly expending energy to pump sodium ions out of the cell, in order to maintain a positive charge on the exterior of the cell and a negative charge on the interior. The net negative charge of most proteins within the cell also adds to this charge difference. This charge separation is called the **resting membrane potential.** Using sophisticated instruments, scientists have been able to measure the voltage difference between the neuron interior and exterior as -70 millivolts (thousandth of a volt). The resting membrane potential is the starting point for a nerve impulse.

A nerve impulse travels along the axon and dendrites as an electrical current gathered by ions moving in and out of the neuron through **voltage-gated channels** (that is, protein channels in the neuron membrane that open and close in response to an electrical voltage). The impulse starts when pressure or other sensory inputs disturb a neuron's plasma membrane, causing sodium channels on a dendrite to open. As a result, sodium ions flood into the neuron from outside, down its concentration gradient, and for a brief moment a localized area inside of the membrane is "depolarized," becoming more positive than the outside in that immediate area of the dendrite.

The sodium channels in the small patch of depolarized membrane remain open for only about a half a millisecond. However, if the change in voltage is large enough, it causes nearby voltage-gated sodium and potassium channels to open. The sodium channels open first, which starts a wave of depolarization moving down the neuron. The opening of the gated channels causes nearby voltage-gated channels to open, like a chain of falling dominoes. This moving local reversal of voltage is called an **action potential.** An action potential follows an all-or-none law: A large enough depolarization produces either a full action potential or none at all, because the voltage-gated Na^+ channels open completely or not at all. Once they open, an action potential occurs. After a slight delay, potassium voltage-gated channels also open, and K^+ flows out of the cells down its concentration gradient, making the inside of the cell more negative. The increasingly negative membrane potential causes the voltage-gated sodium channels to snap closed again. The resting potential is restored by the actions of the sodium-potassium pump (figure 23.5).

The depolarization and restoration of the resting membrane potential takes only about 5 milliseconds. Fully 100 such cycles could occur, one after another, in the time it takes to say the word *nerve.*

23.2 **Neurons are cells specialized to conduct impulses. Signals typically arrive along any of numerous dendrites, pass over the cell body's surface, and travel outward on a single long axon. Nerve impulses result from ion movements across the neuron plasma membrane through special protein channels that open and close in response to chemical or electrical stimulation.**

The Nerve Impulse

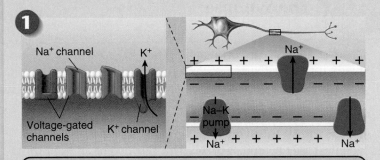

At the resting membrane potential, the inside of the axon is negatively charged because the sodium-potassium pump keeps a higher concentration of Na^+ outside. Voltage-gated ion channels are closed, but there is some leakage of K^+.

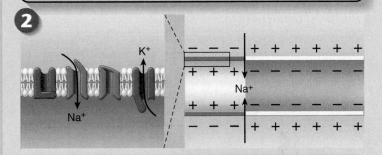

In response to a stimulus, the membrane depolarizes: voltage-gated Na^+ channels open, Na^+ flows into the cell, and the inside becomes more positive.

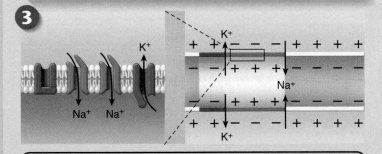

The local change in voltage opens adjacent voltage-gated Na^+ channels, and an action potential is produced.

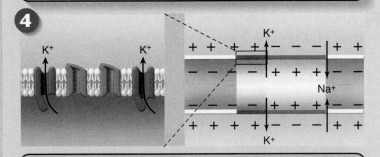

As the action potential travels farther down the axon, voltage-gated Na^+ channels close and K^+ channels open, allowing K^+ to flow out of the cell and restoring the negative charge inside the cell. Ultimately, the sodium-potassium pump restores the resting membrane potential.

Figure 23.5 How an action potential works.

23.3 The Synapse

A nerve impulse can travel only so far along a neuron. Eventually it reaches the end of the axon, usually positioned very close to another neuron or to a muscle cell or gland. Axons, however, do not actually make direct contact with other neurons or with other cells. Instead, a narrow gap, 10 to 20 nanometers across, called the synaptic cleft, separates the axon tip and the target neuron or tissue. This junction of an axon with another cell is called a **synapse.** The membrane on the near (axon) side of the synapse is called the **presynaptic membrane;** the membrane on the far (receiving) side of the synapse is called the **postsynaptic membrane** (figure 23.6).

Neurotransmitters

When a nerve impulse gets to the end of an axon, its message must cross the synapse if it is to continue. Messages do not "jump" across synapses. Instead, they are carried across by chemical messengers called **neurotransmitters.** These chemicals are packaged in tiny sacs, or vesicles, at the tip of the axon. When a nerve impulse arrives at the tip, it causes the sacs to release their contents into the synapse. The neurotransmitters diffuse across the synaptic cleft and bind to receptors in the membrane of the cell on the other side, passing the signal to that cell by causing special ion channels in the postsynaptic membrane to open (figure 23.7). Because these channels open when stimulated by a chemical (in this case, a neurotransmitter), they are said to be *chemically gated.*

Why go to all this trouble? Why not just wire the neurons directly together? For the same reason that the wires of your house are not all connected but instead are separated by a host of switches. When you turn on one light switch, you don't want every light in the house to go on, the toaster to start heating, and the television to come on! If every neuron in your body were connected to every other neuron, it would be impossible to move your hand without moving every other part of your body at the same time. Synapses are the control switches of the nervous system. However, the control switch must be turned off at some point by getting rid of the neurotransmitter, or the postsynaptic cell would keep firing action potentials. In some cases, the neurotransmitter molecules diffuse away from the synapse. In other cases, the neurotransmitter molecules are either reabsorbed by the presynaptic cell, or are degraded in the synaptic cleft.

Kinds of Synapses

The vertebrate nervous system uses dozens of different kinds of neurotransmitters, each recognized by specific receptors on receiving cells. They fall into two classes, depending on whether they excite or inhibit the postsynaptic cell.

In an *excitatory synapse,* the receptor protein is usually a chemically-gated sodium channel, meaning that a sodium

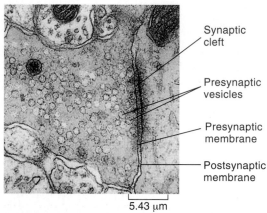

Figure 23.6   A synapse between two neurons.

This micrograph clearly shows the space between the presynaptic and postsynaptic membranes, which is called the synaptic cleft.

Synaptic cleft

Presynaptic vesicles

Presynaptic membrane

Postsynaptic membrane

5.43 μm

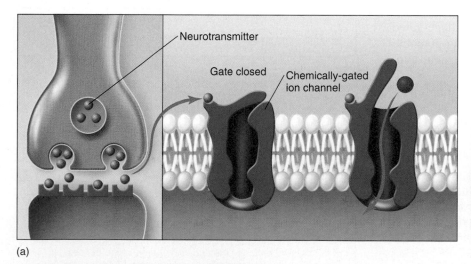

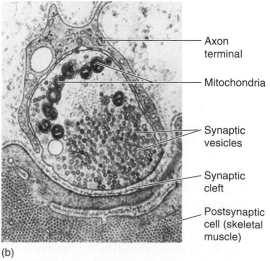

Neurotransmitter

Gate closed

Chemically-gated ion channel

(a)

Axon terminal

Mitochondria

Synaptic vesicles

Synaptic cleft

Postsynaptic cell (skeletal muscle)

(b)

Figure 23.7   Events at the synapse.

(a) When a nerve impulse reaches the end of an axon, it releases a neurotransmitter into the synaptic cleft. The neurotransmitter molecules diffuse across the synapse and bind to receptors on the postsynaptic cell, a neuron in this case, passing the signal to that cell. (b) A transmission electron micrograph of the tip of an axon filled with synaptic vesicles.

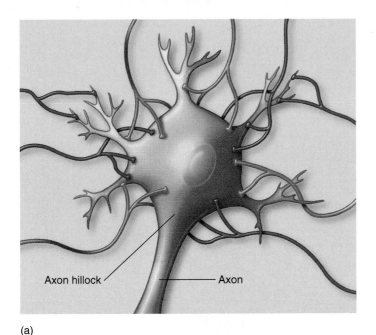

(a)

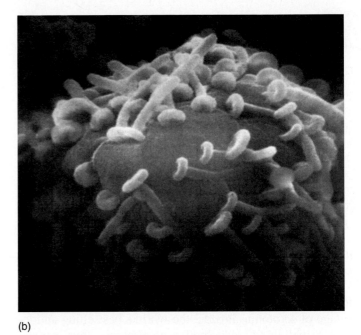

(b)

Figure 23.8   Integration.
(a) Many different axons synapse with the cell body of the postsynaptic neuron illustrated here. Excitatory synapses are shown in *red* and inhibitory synapses are shown in *blue*. The summed influence of their input at the axon hillock determines whether or not a large enough negative potential will be realized and a nerve impulse will be sent down the axon extending below. (b) Scanning electron micrograph of a neuronal cell body with numerous synapses.

channel through the membrane is opened by the neurotransmitter. On binding with a neurotransmitter whose shape fits it, the sodium channel opens, allowing sodium ions to flood inward. If enough sodium ion channels are opened by neurotransmitters, an action potential begins.

In an *inhibitory synapse,* the receptor protein is a chemically-gated potassium or chloride channel. Binding with its neurotransmitter opens these channels, leading to the exit of positively charged potassium ions or the influx of negatively charged chloride ions, resulting in a more negative interior in the receiving cell. This inhibits the start of an action potential, because the negative voltage change inside means that even more sodium ion channels must be opened to get a domino effect started among voltage-gated sodium channels, and so start an action potential.

An individual nerve cell can possess both kinds of synaptic connections to other nerve cells. When signals from both excitatory and inhibitory synapses reach the cell body of a neuron, the excitatory effects (which cause less internal negative charge) and the inhibitory effects (which cause more internal negative charge) interact with one another. The result is a process of **integration** in which the various excitatory and inhibitory electrical effects tend to cancel or reinforce one another. An area at the base of the axon, called the **axon hillock,** is the site of this integration process (figure 23.8). Neurons often receive many inputs. A single motor neuron in the spinal cord may have as many as 50,000 synapses on it!

Neurotransmitters and Their Functions

Acetylcholine (ACh) is the neurotransmitter released at the neuromuscular junction, the synapse that forms between a neuron and a muscle fiber. ACh forms an excitatory synapse with skeletal muscle but has the opposite effect on cardiac muscle, causing an inhibitory synapse.

Glycine and *GABA* are inhibitory neurotransmitters. This inhibitory effect is very important for neural control of body movements and other brain functions. Interestingly, the drug diazepam (Valium) causes its sedative and other effects by enhancing the binding of GABA to its receptors.

Biogenic amines are a group of neurotransmitters that include *dopamine, norepinephrine, serotonin,* and the hormone *epinephrine.* These neurotransmitters have various effects on the body: Dopamine is important in controlling body movements, norepinephrine and epinephrine are involved in the autonomic nervous system, which will be discussed later, and, serotonin is involved in sleep regulation and other emotional states. The drug PCP (angel dust) elicits its actions by blocking the elimination of biogenic amines from the synapse. The symptoms vary depending on the dosage.

23.3   A synapse is the junction of an axon with another cell, a gap across which neurotransmitters carry a signal either facilitating or inhibiting transmission of a signal, depending on which ion channels they open.

Addictive Drugs Act on Chemical Synapses

Neuromodulators

The body sometimes deliberately prolongs the transmission of a signal across a synapse by slowing the destruction of neurotransmitters. It does this by releasing into the synapse special long-lasting chemicals called **neuromodulators.** Some neuromodulators aid the release of neurotransmitters into the synapse; others inhibit the reabsorption of neurotransmitters so that they remain in the synapse; still others delay the breakdown of neurotransmitters after their reabsorption, leaving them in the tip to be released back into the synapse when the next signal arrives.

Mood, pleasure, pain, and other mental states are determined by particular groups of neurons in the brain that use special sets of neurotransmitters and neuromodulators. Mood, for example, is strongly influenced by the neurotransmitter serotonin. Many researchers think that depression results from a shortage of serotonin. Prozac, the world's bestselling antidepressant, inhibits the reabsorption of serotonin, thus increasing the amount in the synapse (figure 23.9).

Drug Addiction

When a cell of the body is exposed to a chemical signal for a prolonged period, it tends to lose its ability to respond to the stimulus with its original intensity. (You are familiar with this loss of sensitivity—when you sit in a chair, how long are you aware of the chair?) Nerve cells are particularly prone to this loss of sensitivity. If receptor proteins within synapses are exposed to high levels of neurotransmitter molecules for prolonged periods, that nerve cell often responds by inserting fewer receptor proteins into the membrane. This feedback is a normal part of the functioning of all neurons, a simple mechanism that has evolved to make the cell more efficient by adjusting the number of "tools" (receptor proteins) in the membrane "workshop" to suit the workload.

Cocaine.   The drug cocaine is a neuromodulator that causes abnormally large amounts of neurotransmitters to remain in the synapses for long periods of time. Cocaine affects nerve cells in the brain's pleasure pathways (the so-called limbic system). These cells transmit pleasure messages using the neurotransmitter dopamine. Using radioactively labeled cocaine molecules, investigators found that cocaine binds tightly to the transporter proteins on presynaptic membranes. These proteins normally remove the neurotransmitter dopamine after it has acted. Like a game of musical chairs in which all the chairs become occupied, eventually there are no unoccupied transporter proteins available to the dopamine molecules, so the dopamine stays in the synapse, firing the receptors again and again. As new signals arrive, more and more dopamine is added, firing the pleasure pathway more and more often.

When receptor proteins on limbic system nerve cells are exposed to high levels of dopamine neurotransmitter molecules for prolonged periods of time, the nerve cells "turn down the volume" of the signal by lowering the number of receptor proteins on their surfaces. They respond to the greater number of neurotransmitter molecules by simply reducing the

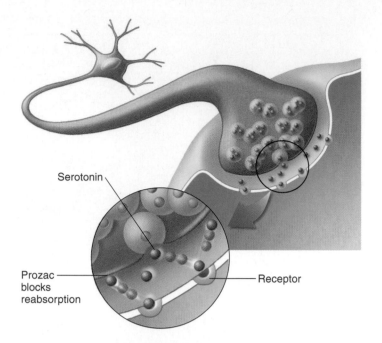

Figure 23.9   Drugs alter transmission of impulses across the synapse.

Depression can result from a shortage of the neurotransmitter serotonin. The antidepressant drug Prozac works by blocking reabsorption of serotonin, making up for the shortage by keeping serotonin in the synapse longer.

number of targets available for these molecules to hit. The cocaine user is now addicted (figure 23.10). **Addiction** occurs when chronic exposure to a drug induces the nervous system to adapt physiologically. With so few receptors, the user needs the drug to maintain even normal levels of limbic activity.

Is Addiction to Smoking Cigarettes Drug Addiction?

Investigators attempting to explore the habit-forming nature of smoking cigarettes used what had been learned about cocaine to carry out what seems a reasonable experiment—they introduced radioactively-labeled nicotine from tobacco into the brain and looked to see what sort of transporter protein it attached itself to. To their great surprise, the nicotine ignored proteins on the presynaptic membrane and instead bound directly to a specific receptor on the postsynaptic cell! This was totally unexpected, as nicotine does not normally occur in the brain—why should it have a receptor there?

Intensive research followed, and researchers soon learned that the "nicotine receptors" normally served to bind the neurotransmitter acetylcholine. It was just an accident of nature that nicotine, an obscure chemical from a tobacco plant, was also able to bind to them. What, then, is the normal function of these receptors? The target of considerable research, these receptors turned out to be one of the brain's most important tools. The brain uses them to coordinate the activities of many other kinds of receptors, acting to "fine-tune" the sensitivity of a wide variety of behaviors.

When neurobiologists compare the limbic system nerve cells of smokers to those of nonsmokers, they find changes in both the number of nicotine receptors and in the levels of

Drug Addiction

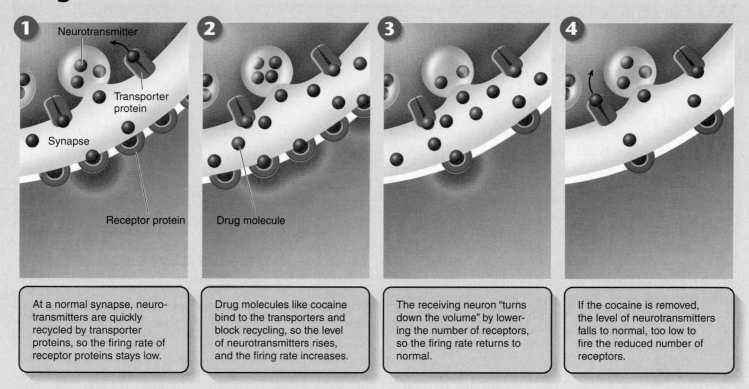

1 At a normal synapse, neurotransmitters are quickly recycled by transporter proteins, so the firing rate of receptor proteins stays low.

2 Drug molecules like cocaine bind to the transporters and block recycling, so the level of neurotransmitters rises, and the firing rate increases.

3 The receiving neuron "turns down the volume" by lowering the number of receptors, so the firing rate returns to normal.

4 If the cocaine is removed, the level of neurotransmitters falls to normal, too low to fire the reduced number of receptors.

Labels in figure: Neurotransmitter, Transporter protein, Synapse, Receptor protein, Drug molecule

Figure 23.10   How drug addiction works.

RNA used to make the receptors. They have found that the brain adjusts to prolonged exposure to nicotine by "turning down the volume" in two ways: (1) by making fewer receptor proteins to which nicotine can bind; and (2) by altering the pattern of *activation* of the nicotine receptors (that is, their sensitivity to neurotransmitters).

It is this second adjustment that is responsible for the profound effect smoking has on the brain's activities. By overriding the normal system used by the brain to coordinate its many activities, nicotine alters the pattern of release of many neurotransmitters into synaptic clefts, including acetylcholine, dopamine, serotonin, and many others. As a result, changes in level of activity occur in a wide variety of nerve pathways within the brain.

Addiction to nicotine occurs because the brain compensates for the many changes nicotine induces by making other changes. Adjustments are made to the numbers and sensitivities of many kinds of receptors within the brain, restoring an appropriate balance of activity.

Now what happens if you stop smoking? Everything is out of whack! The newly coordinated system *requires* nicotine to achieve an appropriate balance of nerve pathway activities. This is addiction in any sensible use of the term. The body's physiological response is profound and unavoidable. There is no way to prevent addiction to nicotine with willpower, any more than willpower can stop a bullet when playing Russian roulette with a loaded gun. If you smoke cigarettes for a prolonged period, you will become addicted.

What do you do if you are addicted to smoking cigarettes and you want to stop? When use of an addictive drug like nicotine is stopped, the level of signaling changes to levels far from normal. If the drug is not reintroduced, the altered level of signaling eventually induces the nerve cells to once again make compensatory changes that restore an appropriate balance of activities within the brain. Over time, receptor numbers, their sensitivity, and patterns of release of neurotransmitters all revert to normal, once again producing normal levels of signaling along the pathways. There is no way to avoid the down side of addiction. The pleasure pathways will not function at normal levels until the number of receptors on the affected nerve cells has time to readjust.

Many people attempt to quit smoking by using patches containing nicotine; the idea is that by providing gradually smaller doses of nicotine, the smoker can be weaned of his or her craving for cigarettes. The patches do reduce the craving for cigarettes—as long as you keep using the patches! Actually, using such patches simply substitutes one (admittedly less dangerous) nicotine source for another. If you are going to quit smoking, there is no way to avoid the necessity of eliminating the drug to which you are addicted. Hard as it is to hear the bad news, there is no easy way out. The only way to quit is to quit.

23.4   Cigarette smokers find it difficult to quit because they have become addicted to nicotine, a powerful neuromodulator.

23.5 Evolution of the Vertebrate Brain

The structure and function of the vertebrate brain have long been the subject of scientific inquiry. Despite ongoing research, scientists are still not sure how the brain performs many of its functions. For instance, scientists continue to look for the mechanism the brain employs to store memories, and they do not understand how some memories can be "locked away," only to surface in times of stress. The brain is the most complex vertebrate organ ever to evolve, and it can perform a bewildering variety of complex functions (figure 23.11).

Casts of the interior braincases of fossil agnathans, fishes that swam 500 million years ago, have revealed much about the early evolutionary stages of the vertebrate brain. Although small, these brains already had the three divisions that characterize the brains of all contemporary vertebrates: (1) the hindbrain, or rhombencephalon; (2) the midbrain, or mesencephalon; and (3) the forebrain, or prosencephalon (figure 23.12).

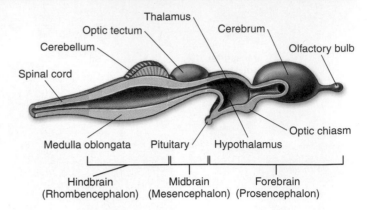

Figure 23.12   The brain of a primitive fish.

The basic organization of the vertebrate brain can be seen in the brains of primitive fishes. The brain is divided into three regions that are found in differing proportions in all vertebrates: the hindbrain, which is the largest portion of the brain in fishes; the midbrain, which in fishes is devoted primarily to processing visual information; and the forebrain, which is concerned mainly with olfaction (the sense of smell) in fishes. In terrestrial vertebrates, the forebrain plays a far more dominant role in neural processing than it does in fishes.

Figure 23.11   Singing well takes practice.

This baby coyote is greeting the approaching evening. His howling is not as impressive as his dad's—a good performance takes practice. His brain is learning by repetition how to control the vocal cords properly.

The hindbrain was the major component of these early brains, as it still is in fishes today. Composed of the *cerebellum, pons,* and *medulla oblongata,* the hindbrain may be considered an extension of the spinal cord devoted primarily to coordinating motor reflexes. Tracts containing large numbers of axons run like cables up and down the spinal cord to the hindbrain. The hindbrain, in turn, integrates the many sensory signals coming from the muscles and coordinates the pattern of motor responses.

Much of this coordination is carried on within a small extension of the hindbrain called the cerebellum ("little cerebrum"). In more advanced vertebrates, the cerebellum plays an increasingly important role as a coordinating center and is correspondingly larger than it is in the fishes. In all vertebrates, the cerebellum processes data on the current position and movement of each limb, the state of relaxation or contraction of the muscles involved, and the general position of the body and its relation to the outside world. These data are gathered in the cerebellum and synthesized, and the resulting commands are issued to efferent pathways.

In fishes, the remainder of the brain is devoted to the reception and processing of sensory information. The midbrain is composed primarily of the **optic lobes** (also called the optic tectum), which receive and process visual information, while the forebrain is devoted to the processing of *olfactory* (smell) information. The brains of fishes continue growing throughout their lives. This continued growth is in marked contrast to the brains of other classes of vertebrates, which generally complete their development by infancy. The human brain continues to develop through early childhood, but no new neurons are produced once development has ceased, except in the hippocampus, involved in long-term memory.

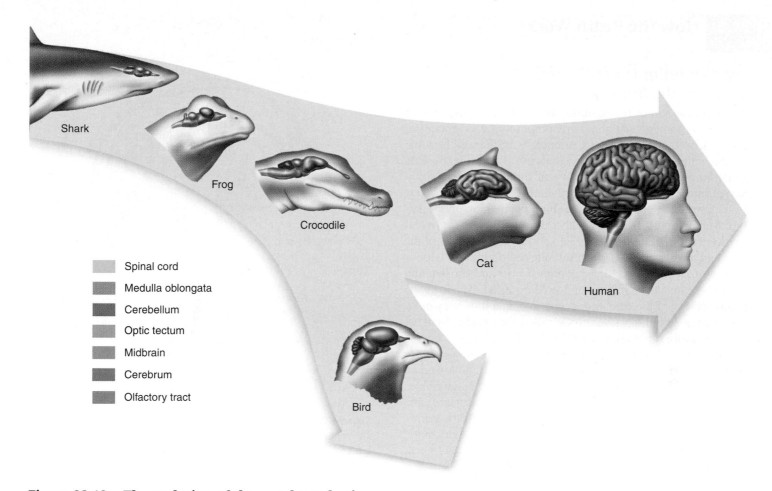

Figure 23.13 The evolution of the vertebrate brain.

In sharks and other fishes, the hindbrain is predominant, and the rest of the brain serves primarily to process sensory information. In amphibians and reptiles, the forebrain is far larger, and it contains a larger cerebrum devoted to associative activity. In birds, which evolved from reptiles, the cerebrum is even more pronounced. In mammals, the cerebrum covers the optic tectum and is the largest portion of the brain. The dominance of the cerebrum is greatest in humans, where it envelops much of the rest of the brain.

Legend:
- Spinal cord
- Medulla oblongata
- Cerebellum
- Optic tectum
- Midbrain
- Cerebrum
- Olfactory tract

Labels: Shark, Frog, Crocodile, Cat, Human, Bird

The Dominant Forebrain

Starting with the amphibians and continuing more prominently in the reptiles, sensory information is increasingly centered in the forebrain. This pattern was the dominant evolutionary trend in the further development of the vertebrate brain (figure 23.13).

The forebrain in reptiles, amphibians, birds, and mammals is composed of two elements that have distinct functions. The *diencephalon* (Greek, *dia,* between) consists of the thalamus and hypothalamus. The **thalamus** is an integrating and relay center between incoming sensory information and the cerebrum. The **hypothalamus** participates in basic drives and emotions and controls the secretions of the pituitary gland, which in turn regulates many of the other endocrine glands of the body (see chapter 24). Through its connections with the nervous system and endocrine system, the hypothalamus helps coordinate the neural and hormonal responses to many internal stimuli and emotions. The *telencephalon,* or "end brain" (Greek, *telos,* end), is located at the front of the forebrain and is devoted largely to associative activity. In mammals, the telencephalon is called the **cerebrum.**

The Expansion of the Cerebrum

If you examine the relationship between brain size and body size in the animals pictured in figure 23.13, you can see a remarkable difference between fishes, amphibians, and reptiles (to the left of the branch point in the figure), and birds and mammals (to the right of the branch point in the figure). Mammals in particular have brains that are particularly large relative to their body size. This is especially true of porpoises and humans. The increase in brain size in the mammals largely reflects the great enlargement of the cerebrum, the dominant part of the mammalian brain. The cerebrum is the center for correlation, association, and learning in the mammalian brain. It receives sensory data from throughout the body and issues motor commands to the body.

23.5 In fishes, the hindbrain forms much of the brain; as terrestrial vertebrates evolved, the forebrain became increasingly more prominent.

The Cerebrum Is the Control Center of the Brain

Although vertebrate brains differ in the relative importance of different components, the human brain is a good model of how vertebrate brains function. About 85% of the weight of the human brain is made up of the cerebrum (figure 23.14). The cerebrum is the large rounded area of the brain divided by a groove into right and left halves called cerebral hemispheres. It functions in language, conscious thought, memory, personality development, vision, and a host of other activities we call "thinking and feeling." Figure 23.15 shows general areas of the brain and the functions they control. The cerebrum, which looks like a wrinkled mushroom, is positioned over and surrounding the rest of the brain, like a hand holding a fist. Much of the neural activity of the cerebrum occurs within a thin, gray outer layer only a few millimeters thick called the **cerebral cortex** (*cortex* is Latin for "bark of a tree"). This layer is gray because it is densely packed with neuron cell bodies. The human cerebral cortex contains the cell bodies of more than 10 billion nerve cells, roughly 10% of all the neurons in the brain. The wrinkles in the surface of the cerebral cortex increase its surface area (and number of cell bodies) threefold. Underneath the cortex is a solid white region of myelinated nerve fibers that shuttle information between the cortex and the rest of the brain.

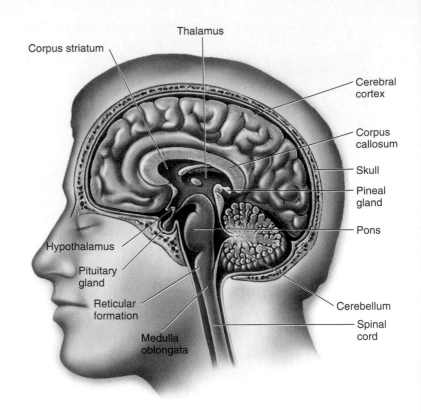

Figure 23.14   A section through the human brain.
The cerebrum occupies most of the brain. Only its outer layer, the cerebral cortex, is visible on the surface.

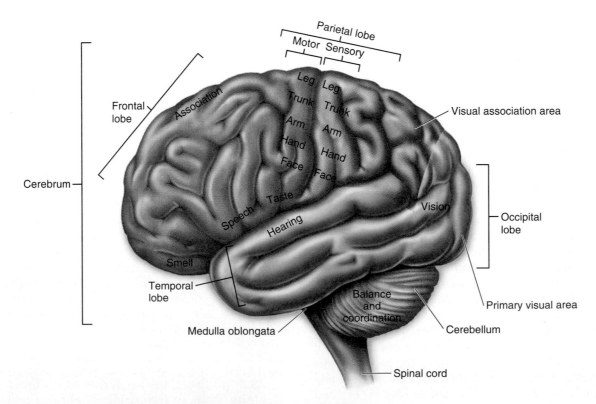

Figure 23.15   The major functional regions of the human brain.
Specific areas of the cerebral cortex are associated with different regions and functions of the body.

The right and left cerebral hemispheres are linked by bundles of neurons called **tracts.** These tracts serve as information highways, telling each half of the brain what the other half is doing. Because these tracts cross over, in the area of the brain called the *corpus callosum,* each half of the brain controls muscles and glands on the opposite side of the body. In general, the left brain is associated with language, speech, and mathematical abilities, whereas the right brain is associated with intuitive, musical, and artistic abilities.

Researchers have found that the two sides of the cerebrum can operate as two different brains. For instance, in some people the tract between the two hemispheres has been cut by accident or surgery. In laboratory experiments, one eye of an individual with such a "split brain" is covered and a stranger is introduced. If the other eye is then covered instead, the person does not recognize the stranger who was just introduced!

Sometimes blood vessels in the brain are blocked by blood clots, causing a disorder called a **stroke.** During a stroke, circulation to an area in the brain is blocked and the brain tissue dies. A severe stroke in one side of the cerebrum may cause paralysis of the other side of the body.

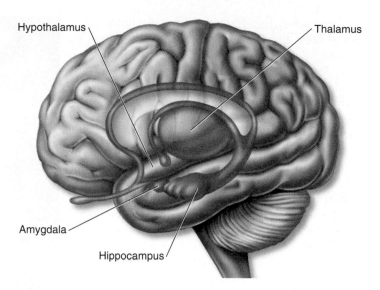

Figure 23.16   The limbic system.
The hippocampus and the amygdala are the major components of the limbic system, which controls our most deep-seated drives and emotions.

The Thalamus and Hypothalamus Process Information

Beneath the cerebrum are the thalamus and hypothalamus, important centers for information processing. The **thalamus** is the major site of sensory processing in the brain. Auditory (sound), visual, and other information from sensory receptors enter the thalamus and then are passed to the sensory areas of the cerebral cortex. The thalamus also controls balance. Information about posture, derived from the muscles, and information about orientation, derived from sensors within the ear, combine with information from the cerebellum and pass to the thalamus. The thalamus processes the information and channels it to the appropriate motor center on the cerebral cortex.

The **hypothalamus** integrates all the internal activities. It controls centers in the brain stem that in turn regulate body temperature, blood pressure, respiration, and heartbeat. It also directs the secretions of the brain's major hormone-producing gland, the pituitary gland. The hypothalamus is linked by an extensive network of neurons to some areas of the cerebral cortex. This network, along with parts of the hypothalamus and areas of the brain called the hippocampus and amygdala, make up the **limbic system** (figure 23.16). The operations of the limbic system are responsible for many of the most deep-seated drives and emotions of vertebrates, including pain, anger, sex, hunger, thirst, and pleasure, centered in the amygdala. It is also involved in memory, centered in the hippocampus.

The Cerebellum Coordinates Muscle Movements

Extending back from the base of the brain is a structure known as the **cerebellum.** The cerebellum controls balance, posture, and muscular coordination. This small, cauliflower-shaped structure, while well developed in humans and other mammals, is even better developed in birds. Birds perform more complicated feats of balance than we do, because they move through the air in three dimensions. Imagine the kind of balance and coordination needed for a bird to land on a branch, stopping at precisely the right moment without crashing into it.

The Brain Stem Controls Vital Body Processes

The **brain stem,** a term used to collectively refer to the midbrain, pons, and **medulla oblongata,** connects the rest of the brain to the spinal cord. This stalklike structure contains nerves that control your breathing, swallowing, and digestive processes, as well as the beating of your heart and the diameter of your blood vessels. A network of nerves called the **reticular formation** runs through the brain stem and connects to other parts of the brain. Their widespread connections make these nerves essential to consciousness, awareness, and sleep. One part of the reticular formation filters sensory input, enabling you to sleep through repetitive noises such as traffic yet awaken instantly when a telephone rings.

Language and Other Higher Functions

Although the two cerebral hemispheres seem structurally similar, they are responsible for different activities. The most thoroughly investigated example of this lateralization of function is language. The left hemisphere is the "dominant" hemisphere for language—the hemisphere in which most neural processing related to language is performed—in 90% of right-handed people and nearly two-thirds of left-handed people. There are two language areas in the dominant hemisphere: One is important for language comprehension and the formulation of thoughts into speech, and the other is responsible for the generation of motor output needed for language communication (figure 23.17).

While the dominant hemisphere for language is adept at sequential reasoning, like that needed to formulate a sentence, the nondominant hemisphere (the right hemisphere in most people) is adept at spatial reasoning, the type of reasoning needed to assemble a puzzle or draw a picture. It is also the hemisphere primarily involved in musical ability—a person with damage to the speech area in the left hemisphere may not be able to speak but may retain the ability to sing! Damage to the nondominant hemisphere may lead to an inability to appreciate spatial relationships and may impair musical activities such as singing. Reading, writing, and oral comprehension remain normal. The nondominant hemisphere is also important for the consolidation of memories of nonverbal experiences.

One of the great mysteries of the brain is the basis of memory and learning. There is no one part of the brain in which all aspects of a memory appear to reside. Although memory is impaired if portions of the brain, particularly the temporal lobes, are removed, it is not lost entirely. Many memories persist in spite of the damage, and the ability to access them gradually recovers with time. Therefore, investigators who have tried to probe the physical mechanisms underlying memory often have felt that they were grasping at a shadow. Although we still do not have a complete understanding of these mechanisms, we have learned a good deal about the basic processes in which memories are formed.

There appear to be fundamental differences between short-term and long-term memory. Short-term memory is transient, lasting only a few moments. Such memories can readily be erased by the application of an electrical shock, leaving previously stored long-term memories intact. This result suggests that short-term memories are stored electrically in the form of a transient neural excitation. Long-term memory, in contrast, appears to involve structural changes in certain neural connections within the brain. Two parts of the temporal lobes, the hippocampus and the amygdala, are involved in both short-term memory and its consolidation into long-term memory. Damage to these structures impairs the ability to process recent events into long-term memories.

The Mechanism of Alzheimer Disease Still a Mystery

In the past, little was known about Alzheimer disease, a condition in which the memory and thought processes of the brain become dysfunctional. Drug companies are eager to develop new products for the treatment of Alzheimer disease, but they have little concrete evidence to go on. Scientists disagree about the biological nature of the disease and its cause. Two hypotheses have been proposed: One suggests that nerve cells in the brain are killed from the outside in, and the other that the cells are killed from the inside out.

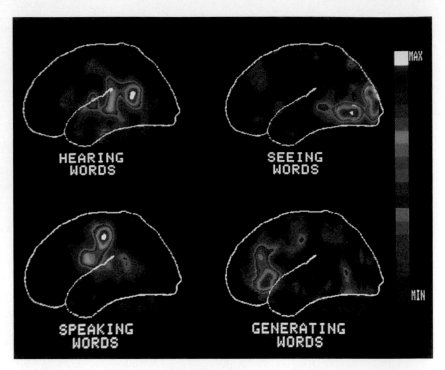

Figure 23.17   Different brain regions control various activities.
This illustration shows how the brain reacts in human subjects asked to listen to a spoken word, to read that same word silently, to repeat the word out loud, and then to speak a word related to the first. Regions of white, red, and yellow show the greatest activity. Compare this to Figure 23.15 to see how regions of the brain are mapped.

In the first hypothesis, external proteins called β-amyloid peptides kill nerve cells. A mistake in protein processing produces an abnormal form of the peptide, which then forms aggregates, or plaques. The plaques begin to fill-in the brain and then damage and kill nerve cells. However, these amyloid plaques have been found in autopsies of people who did not have Alzheimer disease.

The second hypothesis maintains that the nerve cells are killed by an abnormal form of an internal protein. This protein, called tau (τ), normally functions to maintain protein transport microtubules. Abnormal forms of τ assemble into helical segments that form tangles, which interfere with the normal functioning of the nerve cells. Researchers continue to study whether tangles and plaques are causes or effects of Alzheimer disease.

Progress has been made in identifying genes that increase the likelihood of developing Alzheimer disease and genes that, when mutated, can cause the disorder. Most Alzheimer patients do not have these mutated genes, but for those that do, the symptoms of Alzheimer are expressed much earlier in life.

23.6   The associative activity of the brain is centered in the wrinkled cerebral cortex, which lies over the cerebrum. Beneath, the thalamus and hypothalamus process information and integrate body activities. At the base of the brain, the cerebellum coordinates muscle movements.

23.7 The Spinal Cord

The **spinal cord** is a cable of neurons extending from the brain down through the backbone (figure 23.18). The gray neuron cell bodies form a column in the center of the cord, surrounded by a sheath of axons and dendrites, which make the outer edges of the cord white because they are coated with myelin. The spinal cord is surrounded and protected by the vertebrae, through which spinal nerves pass out to the body (figure 23.19). Messages from the body and the brain run up and down the spinal cord, an information highway.

In each segment of the spine, motor nerves extend out of the spinal cord to the muscles. Motor nerves from the spine control most of the muscles below the head. This is why injuries to the spinal cord often paralyze the lower part of the body. A muscle is paralyzed and cannot move if its motor neurons are damaged.

Spinal Cord Regeneration

In the past, scientists have tried to repair severed spinal cords by installing nerves from another part of the body to bridge the gap and act as guides for the spinal cord to regenerate. But most of these experiments have failed because the nerve bridges did not go from white matter to gray matter. Also, there is a factor that inhibits nerve growth in the spinal cord. After discovering that fibroblast growth factor stimulates nerve growth, neurobiologists tried gluing on the nerves, from white to gray matter, with fibrin that had been mixed with the fibroblast growth factor.

Three months later, rats with the nerve bridges began to show movement in their lower bodies. In further analyses of

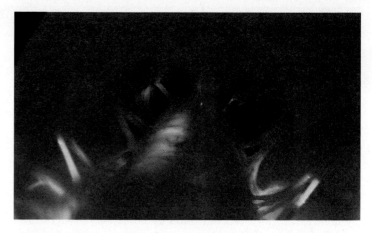

Figure 23.18   A view down the human spinal cord.
Pairs of spinal nerves can be seen extending out from the spinal cord. Along these nerves, the brain and spinal cord communicate with the body.

the experimental animals, dye tests indicated that the spinal cord nerves had regrown from both sides of the gap. Many scientists are encouraged by the potential to use a similar treatment in human medicine. However, most spinal cord injuries in humans do not involve a completely severed spinal cord; often, nerves are crushed. Also, although the rats with nerve bridges did regain some locomotory ability, tests indicated that they were barely able to walk or stand.

> **23.7**   The spinal cord, protected in vertebrates by a backbone, extends motor nerves to the muscles below the head.

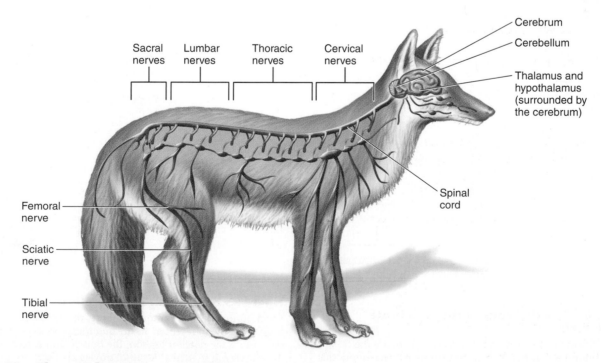

Figure 23.19   The vertebrate nervous system.
The brain is colored *tan* and the spinal cord and nerves are colored *yellow*.

23.8 Voluntary and Autonomic Nervous Systems

The motor pathways of a vertebrate can be subdivided into the **voluntary nervous system,** which relays commands to skeletal muscles, and the **autonomic nervous system,** which stimulates glands and relays commands to the smooth muscles of the body (figure 23.20). The voluntary nervous system can be controlled by conscious thought. You can, for example, command your hand to move. The autonomic nervous system, by contrast, cannot be controlled by conscious thought. You cannot, for example, tell the smooth muscles in your digestive tract to speed up their action. The central nervous system issues commands over both voluntary and autonomic systems, but you are conscious of only the voluntary commands.

Voluntary Nervous System

Motor neurons carry information from the central nervous system to muscles and glands. For example, if your eyes see a runaway car speeding toward you, the CNS sends messages through motor neurons to glands that secrete the hormone adrenaline. The adrenaline increases your heartbeat and breathing rate. The CNS also sends messages through motor neurons to many muscles, which contract and get your body out of there—fast!

Reflexes Enable Quick Action. The motor neurons of the body have been wired to enable the body to act particularly quickly in time of danger—even before the animal is consciously aware of the threat. These sudden, involuntary movements are called reflexes. A **reflex** produces a rapid motor response to a stimulus because the sensory neuron bringing information about the threat passes the information directly to a motor neuron. The escape reaction of a fly about to be swatted is a reflex. One of the most frequently used reflexes in your body is blinking, a reflex that protects your eyes. If anything, such as an insect or a cloud of dust, approaches your eye, the eyelid blinks closed even before you realize what has happened. The reflex occurs before the cerebrum is aware the eye is in danger.

Because they involve passing information between few neurons, reflexes are very fast. Many reflexes never reach the brain. The "danger" nerve impulse travels only as far as the spinal cord and then comes right back as a motor response. Most reflexes involve a single connecting interneuron between the sensory neuron and the motor neuron. A few, like the knee-jerk reflex (figure 23.21), are monosynaptic reflex arcs. In these, the sensory neuron synapses directly with a motor neuron in the spinal cord—there is no interneuron intermediary between them. If you step on something sharp, your leg jerks away from the danger: The prick causes nerve impulses in sensory neurons, which pass up the spinal cord directly to motor neurons, which cause your leg muscles to contract, jerking your leg up.

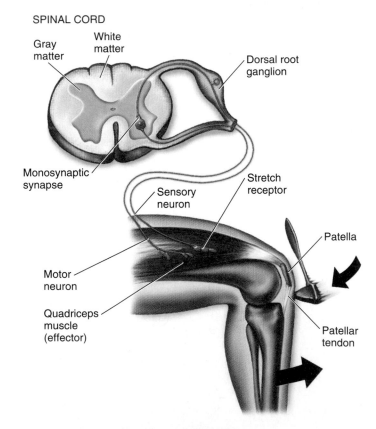

Figure 23.21   The knee-jerk reflex.
The most famous involuntary response, the knee jerk, is produced by activating stretch receptors in the quadriceps muscle. When a rubber mallet taps the patellar tendon, the muscle and stretch receptors in the muscle are stretched. A signal travels up a sensory neuron to the spinal cord, where the sensory neuron stimulates a motor neuron, which sends a signal to the quadriceps muscle to contract.

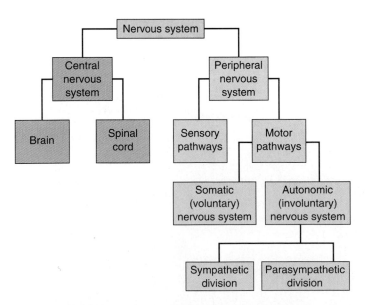

Figure 23.20   The divisions of the vertebrate nervous system.
The motor pathways of the peripheral nervous system are the somatic (voluntary) and autonomic nervous systems. The autonomic nervous system is divided into the sympathetic and parasympathetic divisions.

Autonomic Nervous System

Some motor neurons are active all the time, even during sleep. These neurons carry messages from the CNS that keep the body going even when it is not active. These neurons are called the **autonomic nervous system.** The word *autonomic* means involuntary. The autonomic nervous system carries messages to muscles and glands that usually work without the animal noticing.

The autonomic nervous system is the command network the CNS uses to maintain the body's homeostasis. Using it, the CNS regulates heartbeat and controls muscle contractions in the walls of the blood vessels. It directs the muscles that control blood pressure, breathing, and the movement of food through the digestive system. It also carries messages that help stimulate glands to secrete tears, mucus, and digestive enzymes.

The autonomic nervous system is composed of two elements that act in opposition to one another. One division, the **sympathetic nervous system,** dominates in times of stress. It controls the "fight-or-flight" reaction, increasing blood pressure, heart rate, breathing rate, and blood flow to the muscles. It consists of a network of short motor axons extending out from the spine to clusters of neuron cell bodies, called **gan-** **glia,** located near the spine. It also consists of long motor neurons extending from the ganglia directly to each target organ. Another division, the **parasympathetic nervous system,** has the opposite effect. It conserves energy by slowing the heartbeat and breathing rate and by promoting digestion and elimination. It consists of a network of long axons extending out from motor neurons within the spine; these axons extend to ganglia in the immediate vicinity of an organ. It also consists of short motor neurons extending from the ganglia to the nearby organ.

Most glands, smooth muscles, and cardiac muscles constantly get input from *both* the sympathetic and parasympathetic systems. The CNS controls activity by varying the ratio of the two signals to either stimulate or inhibit the organ (figure 23.22).

> **23.8** The voluntary nervous system relays commands to skeletal muscles and can be controlled by conscious thought. The autonomic nervous system relays commands to muscles and glands that cannot be controlled by conscious thought.

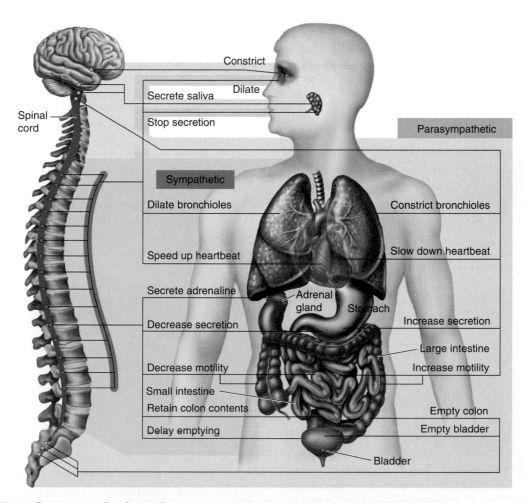

Figure 23.22 How the sympathetic and parasympathetic nervous systems interact.
A nerve path runs from both of the systems to every organ indicated except the adrenal gland, which is only innervated by the sympathetic nervous system.

23.9 Sensory Perception

Did you ever wonder what it would be like not to know anything about what is going on around you? Imagine if you couldn't hear, or see, or feel, or smell. After a while, a human goes mad if completely deprived of sensory input. The senses are the bridge to experience and perceive the way the body relates to everything around it.

Sensory Receptors

The **sensory nervous system** tells the central nervous system what is happening. Sensory neurons carry impulses to the CNS from more than a dozen different types of sensory cells that detect changes outside and inside the body. Called **sensory receptors,** these specialized sensory cells detect many different things, including changes in blood pressure, strain on ligaments, and smells in the air. Particularly complex sensory receptors, made up of many cell and tissue types, are called **sensory organs.** The eyes and ears (figure 23.23) are sensory organs, and so are the taste buds in your mouth.

How does the brain know whether an incoming nerve impulse is light, sound, or pain? This information is built into the "wiring"—into which neurons interact while passing the information to the CNS and into the location in the brain where the information is sent. The brain "knows" it is responding to light because the message from a sensory neuron is wired to light receptor cells. That is why when you press your fingertips gently against the corners of your eyes, you "see stars"—the brain treats any impulse from the eyes as light, even though the eye received no light.

The Path of Sensory Information

The path of sensory information to the CNS is a simple one, composed of three stages:

1. **Stimulation.** A physical stimulus impinges on a sensory receptor.
2. **Transduction.** The sensory receptor initiates the opening or closing of ion channels in a sensory neuron.
3. **Transmission.** The sensory neuron conducts a nerve impulse along an afferent pathway to the CNS.

All sensory receptors are able to initiate nerve impulses by opening or closing **stimulus-gated channels** within sensory neuron membranes. Except for visual photoreceptors, these channels are sodium ion channels that depolarize the membrane and so start an electrical signal. If the stimulus is large enough, the depolarization will trigger an action potential. The greater the sensory stimulus, the greater the depolarization of the sensory receptor and the higher the frequency of action potentials. The channels are opened by chemical or mechanical stimulation, often a disturbance such as touch, heat, or cold. The receptors differ from one another in the nature of the environmental input that triggers the opening of the channel. The body contains many sorts of receptors, each sensitive to a different aspect of the body's condition or to a different quality of the external environment.

Exteroceptors are receptors that sense stimuli that arise in the external environment. Almost all of a vertebrate's exterior senses evolved in water before vertebrates invaded the land. Consequently, many senses of terrestrial vertebrates emphasize stimuli that travel well in water, using receptors that have been retained in the transition from the sea to the land. Hearing, for example, converts an airborne stimulus into a waterborne one, using receptors similar to those that originally evolved in aquatic animals. A few vertebrate sensory systems that function well in the water, such as the electric organs of fish, cannot function in the air and are not found among terrestrial vertebrates. On the other hand, some land dwellers have sensory systems, such as infrared receptors, that could not function in the sea.

Interoceptors sense stimuli that arise from within the body. These internal receptors detect stimuli related to muscle length and tension, limb position, pain, blood chemistry, blood volume and pressure, and body temperature. Many of these receptors are simpler than those that monitor the external environment and are believed to bear a closer resemblance to primitive sensory receptors.

Figure 23.23 Kangaroo rats have specialized ears.
The ears of kangaroo rats (*Dipodomys*) are adapted to nocturnal life and allow them to hear the low-frequency sounds of their predators, such as an owl's wingbeats or a sidewinder rattlesnake's scales rubbing against the ground. Also, the ears seem to be adapted to the poor sound-carrying quality of dry, desert air.

Sensing the Internal Environment

Sensory receptors inside the body inform the CNS about the condition of the body. Much of this information passes to a coordinating center in the brain, the hypothalamus, the part of the brain responsible for maintaining the body's homeostasis—that is, keeping the body's internal environment constant. The vertebrate body uses a variety of different sensory receptors to respond to different aspects of its internal environment.

Temperature change. Two kinds of nerve endings in the skin are sensitive to changes in temperature, one stimulated by cold, the other by warmth. By comparing information from the two, the CNS can learn what the temperature is and if it is changing.

Blood chemistry. Receptors in the walls of arteries sense CO_2 levels in the blood. The brain uses this information to regulate the body's respiration rate, increasing it when CO_2 levels rise above normal.

Pain. Damage to tissue is detected by special nerve endings within tissues, usually near the surface, where damage is most likely to occur. When these nerve endings are physically damaged or deformed, the CNS responds by reflexively withdrawing the body segment and often by changing heartbeat and blood pressure as well.

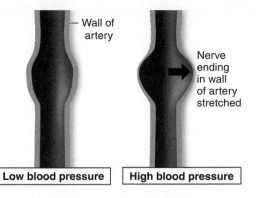

Low blood pressure | High blood pressure

Figure 23.25   How a baroreceptor works.
A network of nerve endings covers a region where the wall of the artery is thin. High blood pressure causes the wall to balloon out there, stretching the nerve endings and causing them to fire impulses.

Muscle contraction. Buried deep within muscles are sensory receptors called stretch receptors. In each, the end of a sensory neuron is wrapped around a muscle fiber (figure 23.24): When the muscle is stretched, the fiber elongates, stretching the spiral nerve ending and causing repeated signals to be sent to the brain. From these signals the brain can determine the rate of change of muscle length at any given moment. The CNS uses this information to control movements that require the combined action of several muscles, such as those that carry out breathing or locomotion.

Blood pressure. Blood pressure is sensed by neurons called baroreceptors with highly branched nerve endings within the walls of major arteries. When blood pressure increases, the stretching of the arterial wall causes the sensory neuron to increase the rate at which it sends signals to the CNS, while when it decreases, the rate of firing of the sensory neuron goes down (figure 23.25). Thus, the frequency of impulses provides the CNS with a continuous measure of blood pressure.

Touch. Touch is sensed by pressure receptors buried below the surface of the skin. There are a variety of different types, some specialized to detect rapid changes in pressure, others to measure the duration and extent to which pressure is applied, and still others sensitive to vibrations.

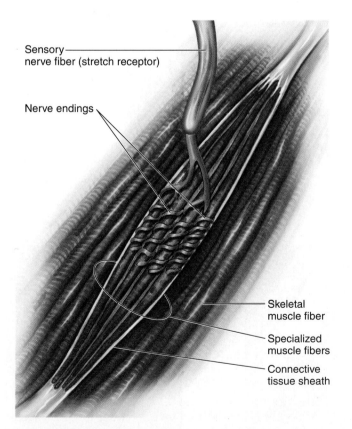

Figure 23.24   A stretch receptor embedded within skeletal muscle.
Stretching the muscle elongates the specialized muscle fibers, which deforms the nerve endings, causing them to send a nerve impulse out along the nerve fiber.

23.9 Sensory receptors initiate nerve impulses in response to stimulation. All sensory nerve impulses are the same, differing only in the stimulus that fires them and their destination in the brain. A variety of different sensory receptors inform the hypothalamus about different aspects of the body's internal environment, enabling it to maintain the body's homeostasis.

23.10 Sensing Gravity and Motion

Receptors in the ear (figure 23.26a) inform the brain where the body is in three dimensions. This knowledge enables an animal to move freely and maintain its balance.

Balance. To keep the body's balance the brain needs a frame of reference, and the reference point it uses is gravity. The sensory receptors that detect gravity are hair cells that project into a gelatinous matrix with embedded particles called **otoliths** (figure 23.26b). The receptors are located in two chambers, the utricle and the saccule, within the inner ear. To illustrate how these receptors work, imagine a pencil standing in a glass. No matter which way you tip the glass, the pencil rolls along the rim, applying pressure to the lip of the glass. If you want to know the direction the glass is tipped, you need only ask where on the rim pressure is being applied. Similarly, the otoliths will shift in the matrix in response to the pull of gravity stimulating hair cells, which the brain uses to determine vertical positioning.

Motion. The brain senses motion by employing a receptor in which fluid deflects cilia in a direction opposite that of the motion. Within the inner ear are three fluid-filled **semicircular canals,** each oriented in a different plane at right angles to the other two so that motion in any direction can be detected. Protruding into the canal are groups of cilia from sensory cells. The cilia from each cell are arranged in a tentlike assembly called a *cupula* (figure 23.26c), which is pushed by moving ear fluid in a direction opposite that of the head's movement. Because the three canals are oriented in all three planes, movement in any plane is sensed by at least one of them, and the brain is able to analyze complex movements by comparing the sensory inputs from each canal.

The semicircular canals do not react if the body moves in a straight line because the fluid in the canals does not move. That is why traveling in a car or airplane at a constant speed in one direction gives no sense of motion.

> **23.10** The body senses gravity and acceleration by the deflection of cilia by moving objects or fluid. The body cannot sense motion at a constant velocity and direction.

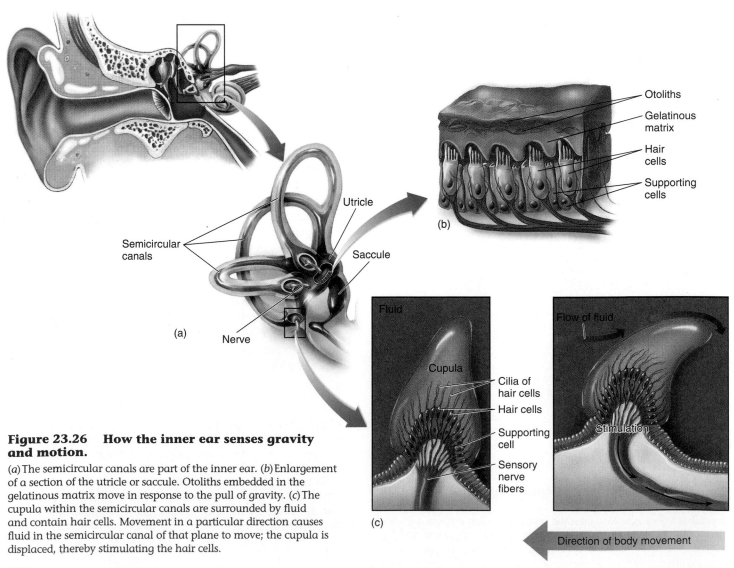

Figure 23.26 How the inner ear senses gravity and motion.

(a) The semicircular canals are part of the inner ear. (b) Enlargement of a section of the utricle or saccule. Otoliths embedded in the gelatinous matrix move in response to the pull of gravity. (c) The cupula within the semicircular canals are surrounded by fluid and contain hair cells. Movement in a particular direction causes fluid in the semicircular canal of that plane to move; the cupula is displaced, thereby stimulating the hair cells.

23.11 Sensing Chemicals: Taste and Smell

Vertebrates are able to detect many of the chemicals in air and in food.

Taste. Embedded within the surface of the tongue are *taste buds*, which are located within raised areas called *papillae* (figure 23.27). Taste buds are onion-shaped structures that contain many taste receptor cells, each of which has fingerlike microvilli that project into an opening called the taste pore. Chemicals from food dissolve in saliva and contact the taste cells through the taste pore. Salty, sour, sweet, bitter, and umami (a "meaty" taste) are perceived because chemicals in food are detected in different ways by taste buds. When the tongue encounters a chemical, information from the taste cells passes to sensory neurons, which transmit the signals to the brain.

Smell. In the nose are chemically sensitive neurons whose cell bodies are embedded within the epithelium of the nasal passage (figure 23.28). When they detect chemicals, these sensory neurons transmit information to a location in the brain where smell information is processed and analyzed. For discovering how this works, American scientists Richard Axel and Linda Buck were awarded the Noble Prize in Physiology or Medicine in 2004. In many vertebrates (dogs are a familiar example), these neurons are far more sensitive than in humans.

Smell as well as taste is very important in telling an animal about its food. That is why when you have a bad cold and your nose is stuffed up, your food has little taste. Other receptors also play a role. Thus the "hot" sensation of foods such as chili peppers is detected by pain receptors, not chemical receptors.

> **23.11** **Taste and smell are chemical senses. In many vertebrates, the sense of smell is very well developed.**

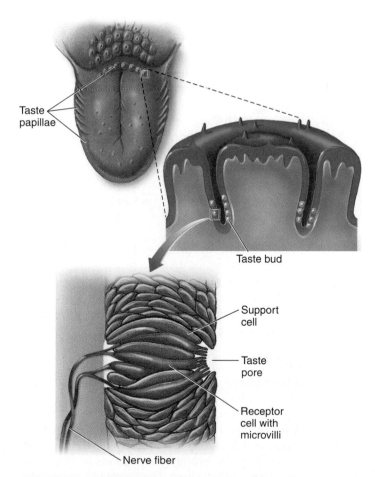

Figure 23.27 Taste.
Taste buds on the human tongue are typically grouped into projections called papillae. Individual taste buds are bulb-shaped collections of taste receptor cells that open out into the mouth through a taste pore. Taste buds in humans can detect bitter, sour, salty, sweet, and umami (a meaty taste) chemicals in food.

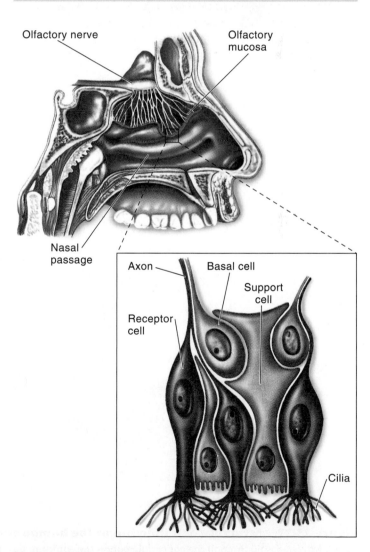

Figure 23.28 Smell.
Humans smell by using receptor cells located in the lining of the nasal passage. The receptor cells are neurons. Axons from these sensory neurons project back through the olfactory nerve directly to the brain.

23.12 Sensing Sounds: Hearing

When you hear a sound, you are detecting the air vibrating—waves of pressure in the air beating against your ear, pushing a membrane called the **eardrum** in and out. On the other side of the eardrum are three small bones, called ossicles, that act as a lever system to increase the force of the vibration. They transfer the amplified vibration across a second membrane to fluid within the inner ear. The chamber of the inner ear is shaped like a tightly coiled snail shell and is called the **cochlea** (figure 23.29), from the Latin name for "snail." The middle ear is connected to the throat by the eustachian tube in such a way that there is no difference in air pressure between the middle ear and the outside. That is why your ears sometimes "pop" when landing in an airplane—the pressure is equalizing between the two sides of the eardrum.

The sound receptors within the cochlea are hair cells that rest on a membrane that runs up and down the middle of the spiraling chamber, separating it into two halves like a wall. The hair cells do not project into the fluid-filled canals of the cochlea; instead, they are covered by a second membrane. When a sound enters the cochlea, the sound waves cause this membrane "sandwich" to vibrate, bending the hairs pressed against the outer membrane and causing them to send nerve impulses to sensory neurons that travel to the brain.

Sounds of different frequencies cause different parts of the membrane to vibrate, and thus fire different sensory neurons—the identity of the sensory neuron being fired tells the CNS what the frequency of the sound is. Sound waves of higher frequencies, about 20,000 vibrations (or cycles) per second, also called hertz (Hz), move the membrane in the area closest to the middle ear. Medium-length frequencies, about 2,000 Hz, move the membrane in the area about midway down the length of the cochlea. The lowest-frequency sound waves, about 500 Hz, move the membrane in the area near the tip of the cochlea.

The intensity of the sound is determined by how *often* the neurons fire. Our ability to hear depends upon the flexibility of the membranes within the cochlea. Humans cannot hear low-pitched sounds, below 20 vibrations (or cycles) per second, although some vertebrates can. As children, we can hear high-pitched sounds, up to 20,000 cycles per second, but this ability decreases as we get older. Other vertebrates can hear sounds at far higher frequencies. Dogs readily hear sounds of 40,000 cycles per second and so can respond to a high-pitched dog whistle when it seems silent to a human observer.

Frequent or prolonged exposure to loud noises can result in damage of the hair cells and membrane, especially in the high-frequency area of the cochlea. The loss of the ability to detect high-frequency sounds affects a person's ability to hear certain sounds, especially in a noisy setting.

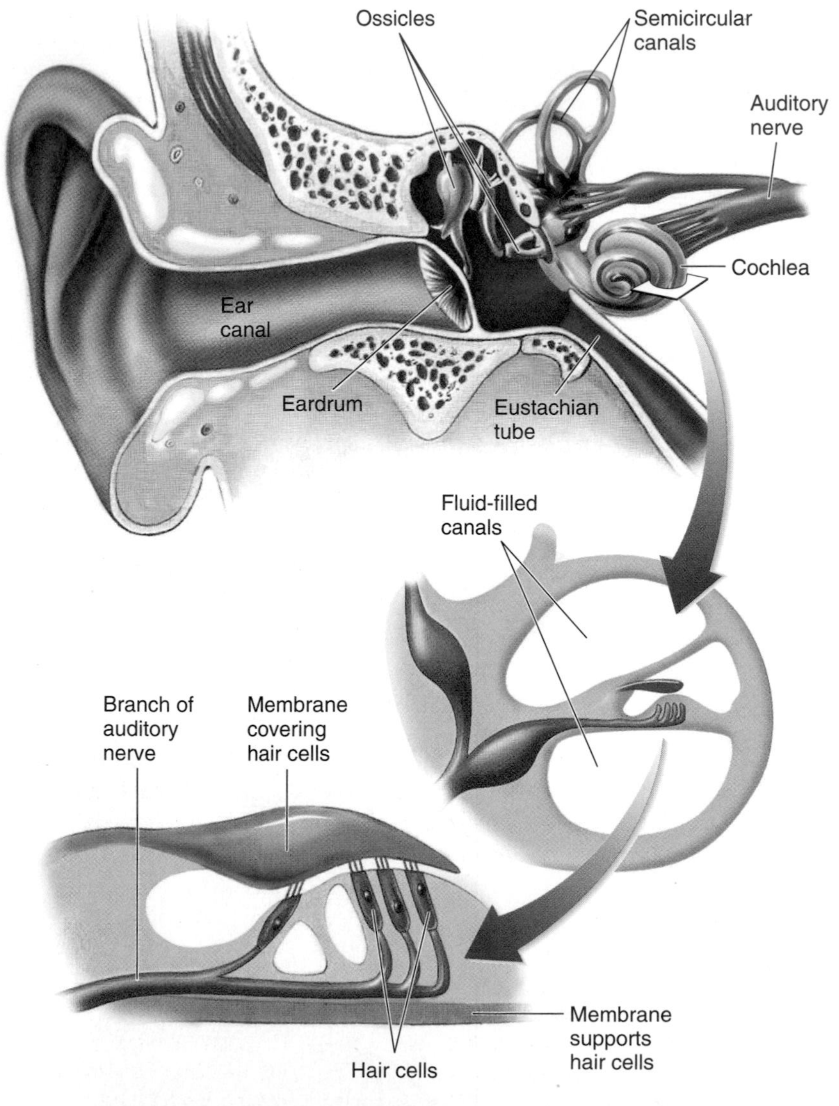

The Lateral Line System

A lateral line system supplements the fish's sense of hearing, which is performed by a different sensory structure, and provides a sense of "distant touch." A fish is able to sense objects that reflect pressure waves and low-frequency vibrations and thus can detect prey, for example, and swim in synchrony with the rest of its school. The lateral line system also enables a blind cave fish to sense its environment by monitoring changes in the patterns of water flow past the lateral line receptors. The same system is found in amphibian larvae, but it is lost at metamorphosis and is not present in any terrestrial vertebrate.

Figure 23.29 Structure and function of the human ear.
Sound waves passing through the ear canal beat on the eardrum, pushing a set of three small bones, or ossicles, against an inner membrane. This sets up a wave motion in the fluid filling the canals within the cochlea. When the sound wave beats against the sides of the canals, the membrane covering the hair cells moves back and forth against the hair cells, which causes associated neurons to fire impulses.

The lateral line system consists of sensory structures within a longitudinal canal in the fish's skin that extends along each side of the body and within several canals in the head (figure 23.30). The sensory structures are known as hair cells because they have hairlike processes at their surface that project into a gelatinous membrane called a *cupula* (Latin, "little cup"). The hair cells are innervated by sensory neurons that transmit impulses to the brain. Vibrations carried through the fish's environment produce movements of the cupula, which cause the hairs to bend. When the hair cells bend, the associated sensory neurons are stimulated and generate a receptor potential that is transmitted to the brain.

Figure 23.31   Using ultrasound to locate a moth.

This bat is emitting high-frequency "chirps" as it flies. It then listens for the sound's reflection against the moth. By timing how long it takes for a sound to return, the bat can "see" the moth even in total darkness.

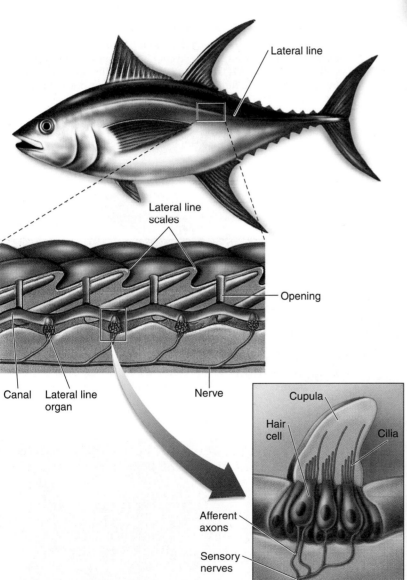

Figure 23.30   The lateral line system.

This system consists of canals running the length of the fish's body beneath the surface of the skin. Within these canals are sensory structures containing hair cells with cilia that project into a gelatinous cupula. Pressure waves traveling through the water in the canals deflect the cilia and depolarize the sensory neurons associated with the hair cells.

Sonar

A few groups of mammals that live and obtain their food in dark environments have circumvented the limitations of darkness. A bat flying in a completely dark room easily avoids objects that are placed in its path—even a wire less than a millimeter in diameter (figure 23.31). Shrews use a similar form of "lightless vision" beneath the ground, as do whales and dolphins beneath the sea. All of these mammals perceive distance by means of sonar. They emit sounds and then determine the time it takes these sounds to reach an object and return to the animal. This process is called **echolocation.** A bat, for example, produces clicks that last 2 to 3 milliseconds and are repeated several hundred times per second. The three-dimensional imaging achieved with such an auditory sonar system can be quite sophisticated.

23.12   Sound receptors detect the air vibrating as waves of pressure push against the membrane covering the ear. Inside, these waves are amplified and press down hair cells that send signals to the brain. Fish sense pressure waves in water much as an ear senses sound. Many vertebrates sense distant objects by bouncing sounds off of them.

23.13 Sensing Light: Vision

No other stimulus provides as much detailed information about the environment as light. Vision, the perception of light, is carried out by a special sensory apparatus called an eye. All the sensory receptors described to this point have been chemical or mechanical ones. Eyes contain sensory receptors called rods and cones that respond to photons of light. The light energy is absorbed by pigments in the rods and cones, which respond by triggering nerve impulses in sensory neurons.

Evolution of the Eye

Vision begins with the capture of light energy by photoreceptors. Because light travels in a straight line and arrives virtually instantaneously, visual information can be used to determine both the direction and the distance of an object. No other stimulus provides as much detailed information.

Many invertebrates have simple visual systems with photoreceptors clustered in an eyespot (figure 23.32). Although an eyespot can perceive the direction of light, it cannot be used to construct a visual image. The members of four phyla—

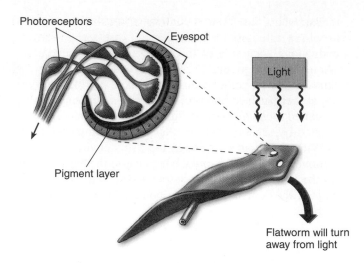

Figure 23.32   Simple eyespots in the flatworm.
Eyespots will detect the direction of light because a pigmented layer on one side of the eyespot screens out light coming from the back of the animal. Light is thus detected more readily coming from the front of the animal; flatworms will respond by turning away from the light.

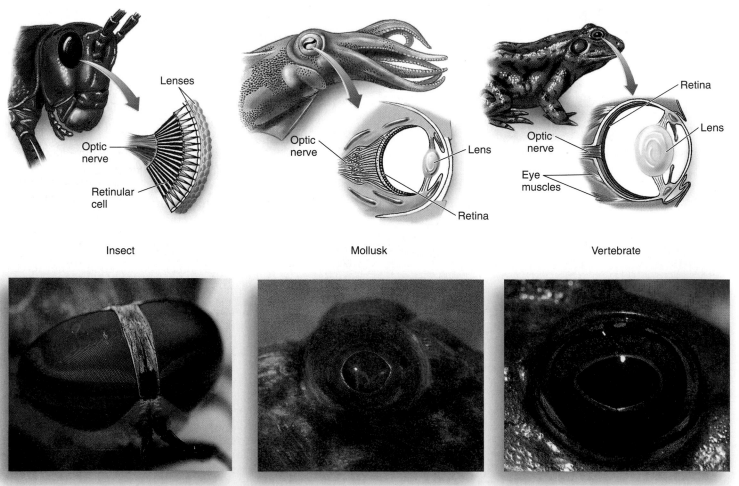

Insect

Mollusk

Vertebrate

Figure 23.33   Eyes in three phyla of animals.
Although they are superficially similar, these eyes differ greatly in structure and are not homologous. Each has evolved separately and, despite the apparent structural complexity, has done so from simpler structures.

annelids, mollusks, arthropods, and vertebrates—have evolved well-developed, image-forming eyes. True image-forming eyes in these phyla, though strikingly similar in structure, are believed to have evolved independently (figure 23.33). Interestingly, the photoreceptors in all of them use the same light-capturing molecule, suggesting that not many alternative molecules are able to play this role.

Structure of the Vertebrate Eye

The vertebrate eye works like a lens-focused camera (figure 23.34). Light first passes through a transparent protective covering, the **cornea,** which begins to focus the light onto the rear of the eye. The beam of light then passes through the **lens,** which completes the focusing. The lens is a fat disc, somewhat resembling a flattened balloon. It is attached by suspending ligaments to **ciliary muscles.** When these muscles contract, they change the shape of the lens and thus the point of focus on the rear of the eye. The amount of light entering the eye and reaching the lens is controlled by a shutter, called the **iris,** between the cornea and the lens. The transparent zone in the middle of the iris, the **pupil,** gets larger in dim light and smaller in bright light.

The light that passes through the pupil is focused by the lens onto the back of the eye. An array of light-sensitive receptor cells line the back surface of the eye, called the **retina.** The retina is the light-sensing portion of the eye. The vertebrate retina contains two kinds of photoreceptors, called **rods** and **cones** (figure 23.35), which, when stimulated by light, generate nerve impulses that travel to the brain along a short, thick nerve pathway called the optic nerve. Rods are receptor cells that are extremely sensitive to light, and they can detect various shades of gray even in dim light. However, they can-

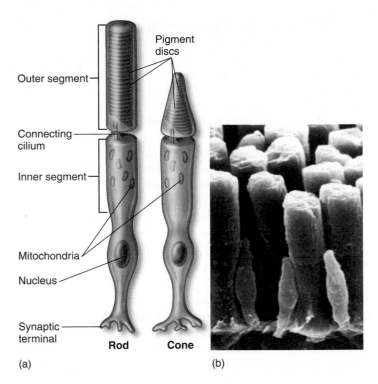

Figure 23.35   Rods and cones.
(a) The broad tubular cell on the *left* is a rod. The shorter, tapered cell next to it is a cone. (b) Electron micrograph of rods and cones.

not distinguish colors, and because they do not detect edges well, they produce poorly defined images. Cones are receptor cells that detect color and are sensitive to edges so that they produce sharp images. The center of the vertebrate retina contains a tiny pit, called the **fovea,** densely packed with some 3 million cones. This area produces the sharpest image, which is why we tend to move our eyes so that the image of an object we want to see clearly falls on this area.

The lens of the vertebrate eye is constructed to filter out short-wavelength light. This solves a difficult optical problem: Any uniform lens bends short wavelengths more than it does longer ones, a phenomenon known as chromatic aberration. Consequently, these short wavelengths cannot be brought into focus simultaneously with longer wavelengths. Unable to focus the short wavelengths, the vertebrate eye eliminates them. Insects, whose eyes do not focus light, are able to see these lower, ultraviolet wavelengths quite well and often use them to locate food or mates.

How Rods and Cones Work

A rod or cone cell in the eye is able to detect a single photon of light. How can it be so sensitive? The primary sensing event of vision is the absorption of a photon of light by a pigment. The pigments in rods and cones are made from plant pigments called carotenoids. That is why eating carrots is said to be good for night vision—the orange color of carrots is due to the presence of carotenoids called carotenes. The visual pigment in the human eye is a fragment of carotene called *cis*-**retinal.** The pigment is attached to a protein called **opsin** to form a light-detecting complex called **rhodopsin.**

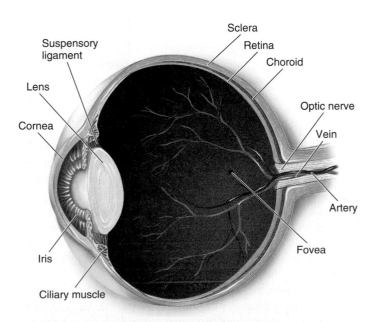

Figure 23.34   The structure of the human eye.
Light passes through the transparent cornea and is focused by the lens on the rear surface of the eye, the retina, at a particular location called the fovea. The retina is rich in photoreceptors.

When it receives a photon of light, the pigment undergoes a change in shape. This change in shape must be large enough to alter the shape of the opsin protein attached to it. When light is absorbed by the *cis*-retinal pigment, the linear end of the molecule rotates sharply upward, straightening out that end of the molecule (figure 23.36). The new form of the pigment is referred to as **trans-retinal.** This radical change in the pigment's shape induces a change in the shape of the protein opsin to which the pigment is bound, initiating a chain of events that leads to the generation of a nerve impulse.

Each rhodopsin activates several hundred molecules of a protein called transducin. Each of these activates several hundred molecules of an enzyme whose product stimulates sodium channels in the photoreceptor membrane at a rate of about 1,000 per second. This cascade of events allows a single photon to have a large effect on the receptor.

Color Vision

Three kinds of **cone cells** provide us with color vision. Each possesses a different version of the opsin protein (that is, one with a distinctive amino acid sequence and thus a different shape). These differences in shape affect the flexibility of the attached retinal pigment, shifting the wavelength at which it absorbs light (figure 23.37). In rods, light is absorbed at 500 nanometers. In cones, the three versions of opsin absorb light at 420 nanometers (blue-absorbing), 530 nanometers (green-absorbing), or 560 nanometers (red-absorbing). By comparing the relative intensities of the signals from the three cones, the brain can calculate the intensity of other colors.

Some people are not able to see all three colors, a condition referred to as *color blindness*. Color blindness is typically due to an inherited lack of one or more types of cones. People with normal vision have all three types of cones. People with only two types of cones lack the ability to detect the third color. For example, people with red color blindness lack red

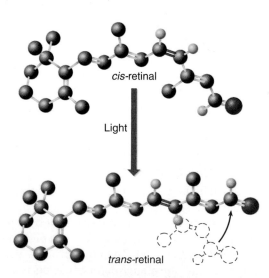

Figure 23.36   Absorption of light.

When light is absorbed by *cis*-retinal, the pigment undergoes a change in shape and becomes *trans*-retinal. This change in shape initiates a chain of events that leads to the generation of a nerve impulse.

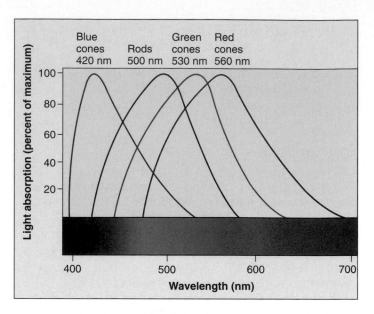

Figure 23.37   Color vision.

The absorption spectrum of *cis*-retinal is shifted in cone cells from the 500 nanometers characteristic of rod cells. The amount of the shift determines what color the cone absorbs: a shift down to 420 nanometers yields blue absorption; a shift up to 530 nanometers yields green absorption; and a shift farther up to 560 nanometers yields red absorption. Red cones do not peak in the red part of the spectrum, but they are the only cones that absorb red light.

cones and have difficulty distinguishing red from green (figure 23.38). Color blindness is a sex-linked trait (see chapter 8) and so men are far more likely to be color blind than women.

Most vertebrates, particularly those that are diurnal (active during the day), have color vision, as do many insects. Indeed, honeybees can see light in the near-ultraviolet range, which is invisible to the human eye. Color vision requires the presence of more than one photopigment in different receptor cells, but not all animals with color vision have the three-cone system characteristic of humans and other primates. Fish, turtles, and birds, for example, have four or five kinds of cones; the "extra" cones enable these animals to see near-ultraviolet light. Many mammals (such as squirrels), on the other hand, have only two types of cones.

Conveying the Light Information to the Brain

The path of light through each eye is the reverse of what you might expect. The rods and cones are at the rear of the retina, not the front. Light passes through several layers of ganglion and bipolar cells before it reaches the rods and cones (figure 23.39). Once the photoreceptors are activated, they stimulate bipolar cells, which in turn stimulate ganglion cells. The direction of nerve impulses in the retina is thus opposite to the direction of light.

Action potentials propagated along the axons of ganglion cells are relayed through structures called the *lateral geniculate nuclei* of the thalamus and projected to the occipital lobe of the cerebral cortex. There the brain interprets this information as light in a specific region of the eye's receptive field. The pattern of activity among the ganglion cells across the retina encodes a point-to-point map of the receptive field, allowing the retina and brain to image objects in visual space. In addition, the frequency of impulses in each ganglion cell provides information about the light intensity at each point, while the relative activity of ganglion cells connected (through bipolar cells) with the three types of cones provides color information.

Binocular Vision

Primates (including humans) and most predators have two eyes, one located on each side of the face. When both eyes are trained on the same object, the image that each sees is slightly different because each eye views the object from a different angle. This slight displacement of the images permits **binocular vision,** the ability to perceive three-dimensional images and to sense depth or the distance to an object. Having their eyes facing forward maximizes the field of overlap in which this stereoscopic vision occurs (figure 23.40).

In contrast, prey animals generally have eyes located to the sides of the head, preventing binocular vision but enlarging the overall receptive field. Depth perception is less important to prey than detection of potential enemies from any quarter. The eyes of the American woodcock, for example, are located at exactly opposite sides of its skull so that it has a 360-degree field of view without turning its head! Most birds have laterally placed eyes and, as an adaptation, have two foveas in each retina. One fovea provides sharp frontal vision, like the single fovea in the retina of mammals, and the other fovea provides sharper lateral vision.

23.13 Vision receptors detect reflected light; binocular vision allows the brain to form three-dimensional images of objects.

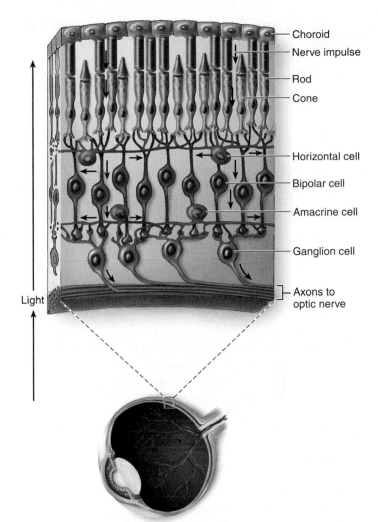

Figure 23.39 Structure of the retina.
The rods and cones are at the rear of the retina, not the front. Light passes over four other types of cells in the retina before it reaches the rods and cones. Nerve impulses then travel through the bipolar cells to the ganglion cells and on to the optic nerve.

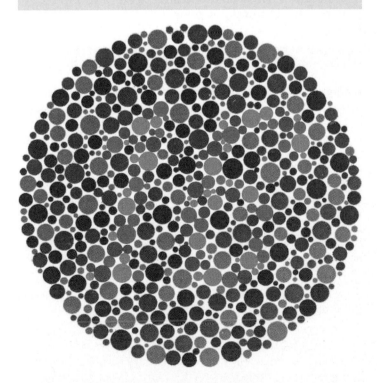

Figure 23.38 Test for color blindness.
People with normal color vision see the number 16, but people that are red-green color blind see just spots and no discernible number.

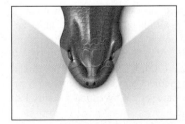 

Figure 23.40 Binocular vision.
When the eyes are located on the sides of the head (as on the *left*), the two vision fields do not overlap and binocular vision does not occur. When both eyes are located toward the front of the head (as on the *right*) so that the two fields of vision overlap, depth can be perceived.

23.14 Other Types of Sensory Reception

Although vision is the primary sense used by all vertebrates that live in a light-filled environment, other parts of the electromagnetic spectrum are also used by some vertebrates.

Heat

Infrared ("below red") radiation, what we normally think of as radiant heat, is used by certain snakes called pit vipers to sense their environment. Pit vipers possess a pair of heat-detecting pit organs located on either side of the head between the eye and the nostril (figure 23.41). The pit organs permit a blindfolded rattlesnake to accurately strike at a warm, dead rat. Each pit organ is composed of two chambers separated by a membrane. The infrared radiation falls on the membrane and warms it. The organ apparently operates by comparing the temperatures of the two chambers.

Electricity

While air does not readily conduct an electrical current, water is a good conductor. All aquatic animals generate electrical currents from contractions of their muscles. A number of different groups of fishes can detect these electrical currents. The *electric fish* even have the ability to produce electrical discharges from specialized electric organs. Electric fish use these weak discharges to locate their prey and mates and to construct a three-dimensional image of their environment even in murky water.

Magnetism

Eels, sharks, and many birds appear to navigate along the magnetic field lines of the earth. Even some bacteria use such forces to orient themselves. Birds kept in blind cages, with no

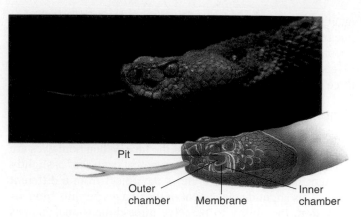

Figure 23.41   "Seeing" heat.
The depression between the nostril and the eye of this rattlesnake opens into the pit organ. In the cutaway portion of the diagram, you can see that the organ is composed of two chambers separated by a membrane.

visual cues to guide them, will peck and attempt to move in the direction in which they would normally migrate at the appropriate time of the year. They will not do so, however, if the cage is shielded from magnetic fields by steel. There has been much speculation about the nature of the magnetic receptors in these vertebrates, but the mechanism is still very poorly understood (see chapter 18).

> **23.14 Pit vipers can locate warm prey by infrared radiation (heat), and many aquatic vertebrates can locate prey and ascertain the contours of their environment by means of electroreceptors. Some vertebrates can even orient themselves with respect to the earth's magnetic field.**

Exploring Current Issues

Additional Resources

Go to your campus library or look online to find the following articles, which further develop some of the concepts found in this chapter.

Adler, J., A. Murr, A. Underwood, and J. Raymond. (2004). The war on strokes; they strike out of the blue, insidious and deadly, killing brain cells, destroying lives. Now a new wave of research offers hope to millions. Inside the search for treatments that work. *Newsweek,* 143(10), 42.

Gage, F. H. (2003). Brain, repair yourself: how do you fix a broken brain? The answers may lie literally within our heads. The same approaches might also boost the power of an already healthy brain. *Scientific American,* 289(3), 46.

Gold, P. E., L. Cahill, and G. L. Wenk. (2003). The lowdown on ginkgo biloba: this popular herbal supplement may slightly improve your memory, but you can get the same effect by eating a candy bar. *Scientific American,* 288(4), 86.

Ingram, V. (2003). Alzheimer's disease: the molecular origins of this disease are coming to light, suggesting several novel therapies. *American Scientist,* 91(4), 312.

Singer, E. (2004). The master switch: the brain's central circuits were once a no-go area for drug treatments. But not anymore, and there could be a medical revolution in the making. *New Scientist,* 181(2437), 34.

Biology and Society Lecture: **Doin' Drugs**

Addictive drugs like cocaine or nicotine act at the level of individual synapses within the brain. Typically, the drug produces euphoria by stimulating the synapses of pleasure-producing pathways. Cocaine, for example, blocks the reabsorption of the neurotransmitter dopamine in nerve synapses of the limbic system, with the result that more and more dopamine builds up in the synapse. Higher levels of dopamine cause the postsynaptic neurons to fire more often, increasing pleasure. Addiction results when the brain attempts to "turn down the volume" by removing dopamine receptors from the postsynaptic membrane. So later, when cocaine isn't added to the synapse, there are not enough postsynaptic dopamine receptors to fire the pleasure-producing pathway. To feel normal, you take cocaine to reelevate dopamine's concentration. You are addicted.

Find this lecture, delivered by the author to his class at Washington University, online at www.mhhe.com/tlwessentials/exp23.

Neurons and How They Work

23.1 Evolution of the Animal Nervous System

- The nervous system is the communication network in the body (**figure 23.1**), evolving from nerve nets to more and more complex systems, with specialized cell types and localization of integration centers in the brain (**figure 23.3**).

23.2 Neurons Generate Nerve Impulses

- Neurons are cells that conduct electrical impulses (**figure 23.4**). Electrical signals begin in dendrites and travel down an axon. Nerve impulses result from the movement of ions across the plasma membranes through specialized channels. The movement of ions in one area of the membrane causes a change in electrical properties, called depolarization, which causes the opening of adjacent ion channels, spreading the electrical impulse down the length of the axon (**figure 23.5**).

23.3 The Synapse

- When a nerve impulse reaches the end of the axon, it triggers the release of neurotransmitters that pass across a small gap, called a synaptic cleft, between the neuron and another cell. Neurotransmitter molecules bind to receptors on the postsynaptic cell, causing ion channels to open, creating electrical impulses in the postsynatic cell (**figure 23.7**). All neural inputs are integrated in the postsynaptic cell, producing an overall positive or negative change in membrane potential (**figure 23.8**).

23.4 Addictive Drugs Act on Chemical Synapses

- Molecules called neuromodulators increase or decrease the effects of neurotransmitters at a synapse (**figure 23.9**). Many addictive drugs act as neuromodulators (**figure 23.10**).

The Central Nervous System

23.5 Evolution of the Vertebrate Brain

- The evolution of the brain resulted in a larger forebrain, leading to more complex functions (**figure 23.13**).

23.6 How the Brain Works

- The cerebral cortex lies over the cerebrum and is the site of neural activities such as language, conscious thought, memory, personality development, vision, and many other higher-level activities (**figure 23.15**). The thalamus and hypothalamus integrate bodily functions and lie underneath the cerebrum (**figure 23.16**). The brain stem controls vital functions and the cerebellum controls muscular coordination.

23.7 The Spinal Cord

- The spinal cord is a cable of neurons that extends down the back and is encased in the bony vertebrae of the backbone

(**figure 23.19**). Motor nerves carry impulses from the spinal cord and brain out to the body, and sensory nerves carry impulses from the body to the brain and spinal cord.

The Peripheral Nervous System

23.8 Voluntary and Autonomic Nervous Systems

- The voluntary nervous system relays commands between the CNS and skeletal muscles and can be consciously controlled; however, reflexes work without conscious control (**figure 23.21**). The autonomic nervous system consists of opposing sympathetic and parasympathetic divisions that unconsciously relay commands between the CNS and muscles and glands (**figure 23.22**).

The Sensory Nervous System

23.9 Sensory Perception

- Neurons called sensory receptors initiate and carry nerve impulses to the CNS. Different sensory cells are stimulated by different stimuli. Exteroceptors sense stimuli from the external environment, and interoceptors sense stimuli within the body (**figures 23.24 and 23.25**).

23.10 Sensing Gravity and Motion

- Sensory receptors in the ear sense gravity and acceleration. The otolith sensory receptors detect gravity by the deflection of hair cells caused by the movement of otoliths in a gelatin-like matrix (**figure 23.26b**). Motion is detected by the deflection of hair cells in the cupula of the semicircular canals (**figure 23.26c**).

23.11 Sensing Chemicals: Taste and Smell

- Chemicals are detected through the senses of taste, using taste buds on the tongue and smell, using olfactory receptors that line the nasal passages (**figures 23.27 and 23.28**).

23.12 Sensing Sound: Hearing

- Sound receptors detect vibrations of air and water through the deflection of hair cells. These senses include hearing in terrestrial vertebrates (**figure 23.29**) and the lateral line system in fishes (**figure 23.30**).

23.13 Sensing Light: Vision

- Sensory receptors in the eye detect light. Light receptors evolved in several different animal phyla (**figures 23.32–23.34**). Rod cells detect the intensity of light, while cone cells detect different colors of light (**figures 23.35–23.39**).

23.14 Other Types of Sensory Reception

- Other sensory receptors, for specialized functions, detect heat, electricity, and magnetic fields.

1. Complexity in animal nervous systems evolved with an increase in
   a. animal body size.
   b. animal body nutritive requirements.
   c. the amount of associative neurons that eventually formed the "brain."
   d. types of animal behavior.

2. A nerve impulse is caused by a quick reversal of polarity in a nerve cell resulting from the
   a. exchange of sodium and potassium ions.
   b. outflow of sodium ions.
   c. inflow of potassium ions.
   d. exchange of sodium and chloride ions.

3. Excitatory neurotransmitters initiate an action potential in a postsynaptic neuron by opening
   a. sodium ion gates in the postsynaptic cell.
   b. potassium ion gates in the postsynaptic cell.
   c. chloride ion gates in the postsynaptic cell.
   d. calcium ion gates in the postsynaptic cell.

4. The purpose of the hindbrain is to
   a. coordinate olfactory information from the nose with the eyes.
   b. process critical thought and learning.
   c. coordinate optic information with the muscles.
   d. coordinate sensory information from muscles and motor responses.

5. Involuntary body activities such as breathing are controlled by the
   a. cerebrum.   c. hypothalamus.
   b. cerebellum.   d. brain stem.

6. The optic nerve, the nerve that connects the eye to the brain, is part of the
   a. central nervous system.
   b. peripheral nervous system.
   c. autonomic nervous system.
   d. sympathetic nervous division of the peripheral nervous system.

7. The purpose of the autonomic nervous system is to
   a. control involuntary muscle movement.
   b. control voluntary muscle movement.
   c. process external and internal environmental information.
   d. regulate the body's homeostasis.

8. When arm muscles hurt after heavy exercise, the pain comes from
   a. neurotransmitters.   c. associative neurons.
   b. interoceptors.   d. exteroceptors.

9. Motion and orientation to gravity are sensed in the ear, specifically in the
   a. pinnae.
   b. ear bones (hammer, anvil, stirrup).
   c. semicircular canals.
   d. Eustachian tubes.

10. An animal that sees well in the dark has
    a. a larger than average pupil.
    b. more rod cells in the retina of the eye.
    c. more cone cells in the retina of the eye.
    d. an oddly shaped lens.

Visual Understanding

1. **Figure 23.10** Describe the qualitative feelings that might be experienced by someone who goes through the four stages shown in the drawings.

2. **Figure 23.13** Look at the comparative size of the olfactory tract and the cerebrum in the various animals. What does the figure tell you about the possible importance of olfaction in humans?

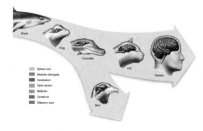

Challenge Questions

Neurons and How They Work There is a lot of concern today about high sodium (Na) levels in our diet, and the resulting high blood pressure for many people. Your friend, Henry, suggests that if a low-sodium diet is healthy, then a no-sodium diet should be even better. What do you tell him?

The Central Nervous System When people become very angry, they are said to be operating from their "dinosaur brain"—not thinking clearly. Explain what this means.

The Peripheral Nervous System Why is it important that the neurons in the autonomic nervous system are always active, not shutting down at night while you're asleep?

The Sensory Nervous System If you only taste a few things (salt, bitter, sweet, sour, umami), why do foods all seem so distinct?

Online Learning Center

Visit the Online Learning Center for this chapter at www.mhhe.com/tlwessentials/ch23 for quizzes, animations, interactive learning exercises, and other study tools. At the site you will also find extended answers to the end-of-chapter questions.

24
Chemical Signaling Within the Animal Body

In vertebrates and most other animals, the central nervous system coordinates and regulates the diverse activities of the body by using chemical signals called hormones to effect changes in physiological activities. Many hormones—adrenaline (epinephrine), estrogen, testosterone, insulin, thyroid hormone—are probably familiar to you. Some of these hormones have very different roles in other animals, however. For example, thyroid hormone is needed in amphibians for the metamorphosis of larvae into adults. If the thyroid gland is removed from a tadpole, it will not change into a frog. Conversely, if an immature tadpole is fed pieces of a thyroid gland, it will undergo premature metamorphosis and become a miniature frog! Similarly, melanocyte-stimulating hormone, a peptide hormone, is present in mammals, but we don't know what it does. In reptiles and amphibians, this hormone stimulates color changes. When the green anole (*Anolis carolinensis*) shown here moves from a light background to a dark background, its color darkens. The color change is the result of dispersal of pigment-containing granules from the center of skin cells into extensions of the cells, darkening the skin. Triggered by melanocyte-stimulating hormone, the reversible color change takes 5 to 10 minutes.

24.1 Hormones

A **hormone** is a chemical signal produced in one part of the body that is stable enough to be transported in active form far from where it is produced and that typically acts at a distant site. There are three big advantages to using chemical messengers rather than electrical signals (nerves) to control body organs. First, chemical molecules can spread to all tissues via the blood and are usually required in only small amounts. (Imagine trying to wire every cell with its own nerve!) Second, chemical signals can persist much longer than electrical ones, a great advantage for hormones controlling slow processes like growth and development. Third, many different kinds of chemicals can act as hormones, so different hormone molecules can be targeted at different tissues.

Hormones, in general, are produced by glands, most of which are controlled by the central nervous system. Because these glands are completely enclosed in tissue rather than having ducts that empty to the outside, they are called **endocrine glands** (from the Greek, *endon,* within). Hormones are secreted from them directly into the bloodstream (this is in contrast to **exocrine glands,** which, like sweat glands, have ducts). Your body has a dozen principal endocrine glands that together make up the endocrine system (figure 24.1).

The *endocrine system* and the *motor nervous system* are the two main routes the central nervous system (CNS) uses to issue commands to the organs of the body. The two are so closely linked that they are often considered a single system—the **neuroendocrine system.** The **hypothalamus** can be considered the main switchboard of the neuroendocrine system. The hypothalamus is continually checking conditions inside the body to maintain a constant internal environment, a condition known as homeostasis. Is the body too hot or too cold? Is it running out of fuel? Is the blood pressure too high? If homeostasis is no longer maintained, the hypothalamus has several ways to set things right again. For example, if the hypothalamus needs to speed up the heart rate, it can send a nerve signal to the medulla, or it can use a chemical command, causing the adrenal gland to produce the hormone adrenaline, which also speeds up the heart rate. Which command the hypothalamus uses depends on the desired duration of the effect. A chemical message is typically far longer lasting than a nerve signal.

The Chain of Command

The hypothalamus issues commands to a nearby gland, the pituitary, which in turn sends chemical signals to the various hormone-producing glands of the body. The pituitary is suspended from the hypothalamus by a short stalk, across which chemical messages pass from the hypothalamus to the pituitary. The first of these chemical messages to be discovered was a short peptide called thyrotropin-releasing hormone

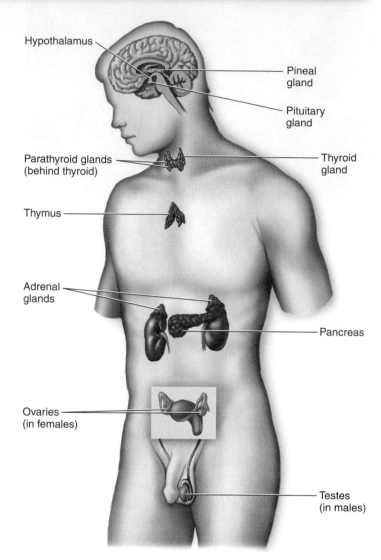

Figure 24.1 Major glands of the human endocrine system.

Hormone-secreting cells are clustered in endocrine glands. The pituitary and adrenal glands are each composed of two glands.

(TRH), which was isolated in 1969. The release of TRH from the hypothalamus triggers the pituitary to release a hormone called thyrotropin, or thyroid-stimulating hormone (TSH), which travels to the thyroid and causes the thyroid gland to release thyroid hormones.

Seven other hypothalamic hormones have since been isolated, which together govern the pituitary. Thus, the CNS regulates the body's hormones through a chain of command. Each of the seven "releasing" hormones made by the hypothalamus causes the pituitary to synthesize a corresponding pituitary hormone, which travels to a distant endocrine gland and causes that gland to begin producing its particular endocrine hormone.

Hormonal Communication

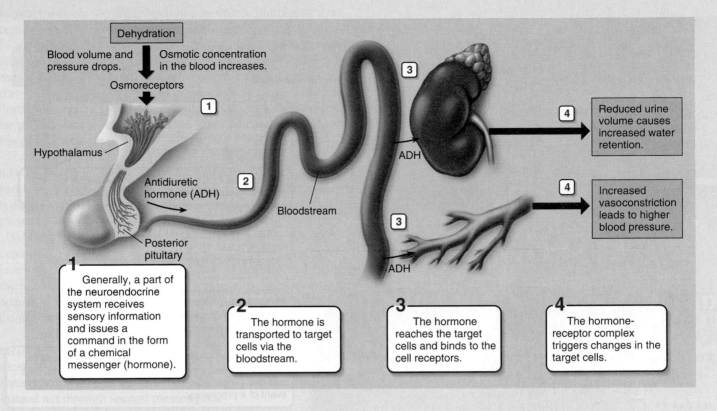

Figure 24.2 How hormonal communication works.

How Hormones Work

The key reason why hormones are effective messengers within the body is because a particular hormone can influence a specific target cell. How does the target cell recognize that hormone, ignoring all others? Embedded in the plasma membrane or within the target cell are receptor proteins that match the shape of the potential signal hormone like a hand fits a glove. As you recall from chapter 23, nerve cells have highly specific receptors within their synapses, each receptor shaped to "respond" to a different neurotransmitter molecule. Cells that the body has targeted to respond to a particular hormone have receptor proteins shaped to fit that hormone and no other. Thus, chemical communication within the body involves *two* elements: a molecular signal (the hormone) and a protein receptor on or in target cells. The system is highly specific because each protein receptor has a shape that only a particular hormone fits.

Hormones secreted by endocrine glands belong to four different chemical categories:

1. **Polypeptides** are composed of chains of amino acids that are shorter than about 100 amino acids. Some important examples include insulin and antidiuretic hormone (ADH).
2. **Glycoproteins** are composed of polypeptides significantly longer than 100 amino acids to which are attached carbohydrates. Examples include follicle-stimulating hormone (FSH) and luteinizing hormone (LH).
3. **Amines,** derived from the amino acids tyrosine and tryptophan, include hormones secreted by the adrenal medulla, thyroid, and pineal glands.
4. **Steroids** are lipids derived from cholesterol, and include the hormones testosterone, estrogen, progesterone, aldosterone, and cortisol.

The path of communication taken by a hormonal signal can be visualized as a series of simple steps (figure 24.2):

1. **Issuing the command.** The hypothalamus of the CNS controls the release of many hormones. It chemically signals the pituitary gland, which is located very close to it, to release hormones into the bloodstream.
2. **Transporting the signal.** While hormones can act on an adjacent cell, most are transported throughout the body by the bloodstream.
3. **Hitting the target.** When a hormone encounters a cell with a matching receptor, called a target cell, the hormone binds to that receptor.
4. **Having an effect.** When the hormone binds to the receptor protein, the protein responds by changing shape, which triggers a change in cell activity.

24.1 Hormones are effective because they are recognized by specific receptors. Thus, only cells possessing the appropriate receptor will respond to a particular hormone.

24.3 The Hypothalamus and the Pituitary

As discussed earlier, the hypothalamus is the "control center" of the neuroendocrine system. It exerts control over the pituitary gland. The **pituitary gland,** located in a bony recess in the brain below the hypothalamus, is where nine major hormones are produced. These hormones act principally to influence other endocrine glands. The pituitary is actually two glands. The back portion, or *posterior lobe,* regulates water conservation, milk letdown, and uterine contraction in women; the front portion, or *anterior lobe,* regulates the other endocrine glands.

The Posterior Pituitary

The posterior pituitary appears fibrous because it contains axons that originate in cell bodies within the hypothalamus and extend along the stalk of the pituitary as a tract of fibers. This anatomical relationship results from the way that the posterior pituitary is formed in embryonic development. As the floor of the third ventricle of the brain forms the hypothalamus, part of this neural tissue grows downward to produce the posterior pituitary. The hypothalamus and posterior pituitary thus become interconnected by a tract of axons. The hormones released from the posterior pituitary are actually produced by neuron cell bodies located in the hypothalamus. The hormones are transported to the posterior pituitary through the axon tracts and are stored and released from the posterior pituitary (figure 24.6).

The role of the posterior pituitary first became evident in 1912, when a remarkable medical case was reported: A man who had been shot in the head developed a surprising disorder—he began to urinate every 30 minutes, unceasingly. The bullet had lodged in his pituitary gland, and subsequent research demonstrated that surgical removal of the pituitary also produces these unusual symptoms. Pituitary extracts were shown to contain a substance that makes the kidneys conserve water, and in the early 1950s the peptide hormone **vasopressin** (also called antidiuretic hormone, **ADH**) was isolated. Vasopressin regulates the kidney's retention of water. When vasopressin is missing, the kidneys cannot retain water, which is why the bullet caused excessive urination (and why excessive alcohol and caffeine, which inhibit vasopressin secretion, have the same effect).

The posterior pituitary also releases a second hormone of very similar structure—both are short peptides composed of nine amino acids—but very different function, called **oxytocin.** Oxytocin initiates uterine contraction during childbirth and milk release in mothers. Here is how milk release works: Sensory receptors in the mother's nipples, when stimulated by sucking, send messages to the hypothalamus causing the hypothalamus to stimulate the release of oxytocin from the posterior pituitary. The oxytocin travels in the bloodstream to the breasts, where it stimulates contraction of the muscles around the ducts into which the mammary glands secrete

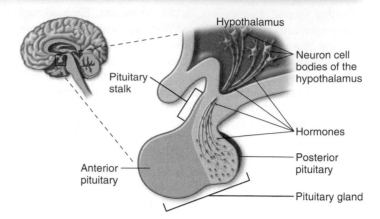

Figure 24.6   The posterior pituitary contains cells that originate in the hypothalamus.

A tract of nerve cells originates in the hypothalamus and extends down along the pituitary stalk and ends in the posterior pituitary. The cell bodies of the neurons produce hormones, which travel down the axons and are stored in the posterior pituitary. Thus, the hormones released from the posterior pituitary are not produced by the gland, but are actually synthesized in neurons in the hypothalamus.

milk. Oxytocin and vasopressin are both synthesized inside neurons within the hypothalamus, and they are transported down nerve axons and stored in the posterior pituitary.

The Anterior Pituitary

The anterior pituitary, unlike the posterior pituitary, does not develop from a downgrowth of the brain; instead, it develops from a pouch of epithelial tissue that pinches off from the roof of the embryo's mouth. Because it is epithelial tissue, the anterior pituitary is a complete gland—it produces the hormones it secretes. The key role of the anterior pituitary first became understood in 1909, when a 38-year-old South Dakota farmer was cured of the growth disorder called acromegaly by the surgical removal of a pituitary tumor. Acromegaly is a form of giantism in which the jaw begins to protrude and the features thicken. It turned out that giantism is almost always associated with pituitary tumors. Robert Wadlow, born in Alton, Illinois, in 1928, grew to a height of 8 feet, 11 inches, and weighed 475 pounds before he died from infection at age 22—the tallest human being ever recorded. Skull X rays showed he had a pituitary tumor. Pituitary hormones have also proven to be the cause of several other well-known cases of giantism, such as the 8-foot, 2-inch Irish giant Charles Byrne, born in 1761; his skeleton, preserved in the Royal College of Surgeons, London, shows the effects of a pituitary tumor.

Why did removal of the pituitary tumor cure the South Dakota farmer? Pituitary tumors produce giants because the tumor cells produce enormous amounts of a growth-promoting hormone. This **growth hormone (GH),** a long peptide of 191 amino acids, is normally produced in only minute amounts by the anterior pituitary gland and usually only during periods of body growth, such as infancy and puberty.

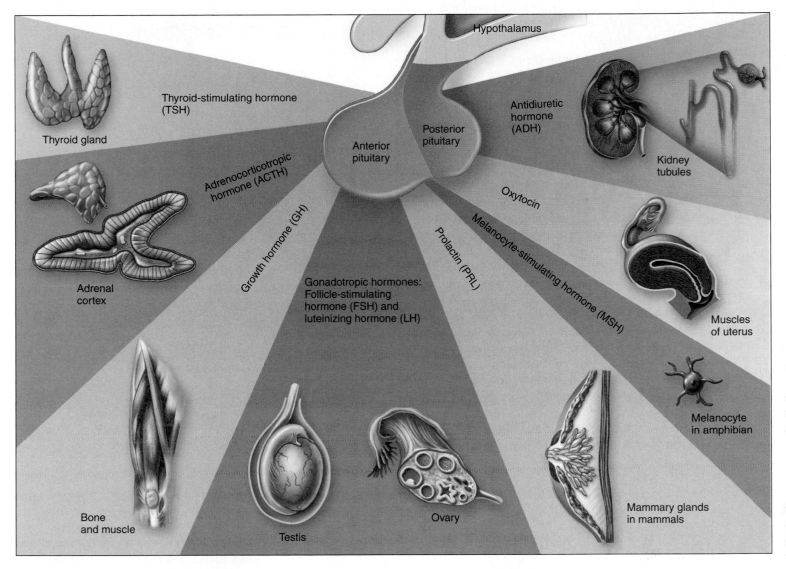

Figure 24.7 The role of the pituitary.
Interactions between the anterior and posterior lobes of the pituitary (two distinct glands) and various organs of the human body are shown in this diagram.

The anterior pituitary gland produces seven major peptide hormones (figure 24.7), each controlled by a particular releasing signal secreted from cells in the hypothalamus. All seven of these hormones have major endocrine roles, as well as poorly understood roles in the central nervous system.

1. **Thyroid-stimulating hormone (TSH).** TSH stimulates the thyroid gland to produce the thyroid hormone thyroxine, which in turn stimulates oxidative respiration.
2. **Luteinizing hormone (LH).** LH plays an important role in the female menstrual cycle by triggering ovulation, the release of a mature egg. It also stimulates the male gonads to produce testosterone, which initiates and maintains the development of male secondary sexual characteristics not involved directly in reproduction.
3. **Follicle-stimulating hormone (FSH).** FSH is significant in the female menstrual cycle by triggering

the maturation of egg cells and stimulating the release of estrogen. In males, it stimulates cells in the testes to produce a hormone that regulates development of the sperm.
4. **Adrenocorticotropic hormone (ACTH).** ACTH stimulates the adrenal gland to produce a variety of steroid hormones. Some regulate the production of glucose from fat; others regulate the balance of sodium and potassium ions in the blood.
5. **Growth hormone (GH).** GH stimulates the growth of muscle and bone throughout the body.
6. **Prolactin (PRL).** Prolactin stimulates the breasts to produce milk.
7. **Melanocyte-stimulating hormone (MSH).** In reptiles and amphibians, melatonin stimulates color changes in the epidermis. The function of this hormone in humans is still poorly understood.

How the Hypothalamus Controls the Anterior Pituitary

As noted earlier, the hypothalamus controls production and secretion of the anterior pituitary hormones by means of a family of special hormones. Neurons in the hypothalamus secrete these releasing and inhibiting hormones into blood capillaries at the base of the hypothalamus (figure 24.8). These capillaries drain into small veins that run within the stalk of the pituitary to a second bed of capillaries in the anterior pituitary. This unusual system of vessels is known as the *hypothalamo-hypophyseal portal system.* It is called a portal system because it has a second capillary bed downstream from the first; the only other body location with a similar system is the liver.

Each releasing hormone delivered to the anterior pituitary by this portal system regulates the secretion of a specific hormone. For example, thyrotropin-releasing hormone (TRH) stimulates the release of TSH, corticotropin-releasing hormone (CRH) stimulates the release of ACTH, gonadotropin-releasing hormone (GnRH) stimulates the release of FSH and LH, and prolactin-releasing factor (PRF) stimulates the release of prolactin, however, this factor has not yet been chemically characterized and may actually be a chemical similar to thyrotropin-releasing hormone. A releasing hormone for growth hormone, called growth-hormone-releasing hormone (GHRH) has also been discovered.

The hypothalamus also secretes hormones that inhibit the release of certain anterior pituitary hormones. To date, three such hormones have been discovered: somatostatin inhibits the secretion of GH; prolactin-inhibiting hormone

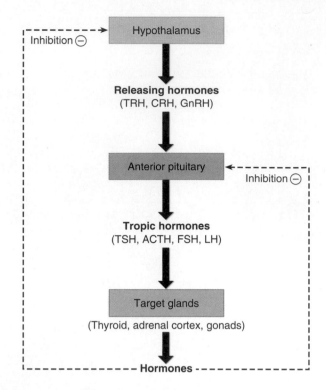

Figure 24.9   Negative feedback.
The hormones secreted by some endocrine glands feed back to inhibit the secretion of hypothalamic-releasing hormones and anterior pituitary tropic hormones.

(PIH), possibly dopamine, inhibits the secretion of prolactin; and melanotropin-inhibiting hormone (MIH) inhibits the secretion of MSH.

Because hypothalamic hormones control the secretions of the anterior pituitary gland, and the anterior pituitary hormones control the secretions of some other endocrine glands, it may seem that the hypothalamus functions as a "master gland," in charge of hormonal secretion in the body. This idea is not generally valid, however, for two reasons. First, a number of endocrine organs, such as the adrenal medulla and the pancreas, are not directly regulated by this control system. Second, the hypothalamus and the anterior pituitary gland are themselves controlled by the very hormones whose secretion they stimulate! In most cases this is an inhibitory control, where the target gland hormones inhibit the secretions of the hypothalamus and anterior pituitary (figure 24.9). This type of control system is an example of **negative feedback (or feedback inhibition)** with specific examples given throughout this chapter.

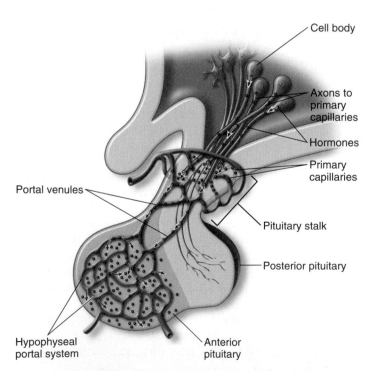

Figure 24.8   Hormonal control of the anterior pituitary gland by the hypothalamus.
Neurons in the hypothalamus secrete hormones that are carried by short blood vessels directly to the anterior pituitary gland, where they either stimulate or inhibit the secretion of anterior pituitary hormones.

24.3   The posterior pituitary gland contains axons originating from neurons in the hypothalamus. The anterior pituitary responds to hormonal signals from the hypothalamus by itself producing a family of pituitary hormones that are carried to distant glands and that induce them to produce specific hormones.

24.4 The Pancreas

The **pancreas** gland is located behind the stomach and is connected to the front end of the small intestine by a narrow tube. It secretes a variety of digestive enzymes into the digestive tract through this tube, and for a long time it was thought to be solely an exocrine gland. In 1869, however, a German medical student named Paul Langerhans described some unusual clusters of cells scattered throughout the pancreas. In 1893, doctors concluded that these clusters of cells, which came to be called islets of Langerhans, produced a substance that prevented diabetes mellitus. **Diabetes mellitus** is a serious disorder in which affected individuals are unable to take up glucose from the blood, even though their levels of blood glucose become very high. Such individuals lose weight and literally starve. Breaking down their fat reserves results in the production of acids called ketones, which lower the pH of the blood. Affected individuals may eventually suffer brain damage and even death. Diabetes is the leading cause of blindness among adults, and it accounts for one-third of all kidney failures. It is the seventh leading cause of death in the United States.

The substance produced by the islets of Langerhans, which we now know to be the peptide hormone *insulin,* was not isolated until 1922. Two young doctors working in a Toronto hospital injected an extract purified from beef pancreas glands into a 13-year-old boy, a diabetic whose weight had fallen to 29 kilograms (65 lb) and who was not expected to survive. The hospital record gives no indication of the historic importance of the trial, only stating, "15 cc of MacLeod's serum. 7-1/2 cc into each buttock." With this single injection, the glucose level in the boy's blood fell 25%—his cells were taking up glucose. A more potent extract soon brought levels down to near normal.

This was the first instance of successful insulin therapy. The islets of Langerhans in the pancreas produce two hormones that interact to govern the levels of glucose in the blood. These hormones are *insulin* and *glucagon.* Insulin is a storage hormone, designed to put away nutrients for leaner times. It promotes the accumulation of glycogen in the liver and triglycerides in fat cells. When food is consumed, beta cells in the islets of Langerhans secrete insulin, causing the body to store glucose to be used later (figure 24.10). When body activity causes the level of glucose in the blood to fall as it is used as fuel, other cells in the islets of Langerhans, called alpha cells, secrete glucagon, which causes liver cells to release stored glucose and fat cells to break down triglycerides. The two hormones work together to keep glucose levels in the blood within a narrow range.

Eighteen million people in the United States, and over 194 million people worldwide, have **diabetes.** There are *two* kinds of diabetes mellitus. About 5% to 10% of affected individuals suffer from type I diabetes, a hereditary autoimmune disease in which the immune system attacks the islets of Langerhans, resulting in abnormally low insulin secretion. Called juvenile-onset diabetes, this type usually develops before age 20. Affected individuals can be treated by daily injections of insulin. Active research on the possibility of transplanting is-

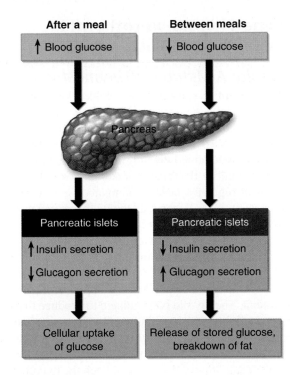

Figure 24.10   Insulin and glucagon secreted by the pancreas regulate blood glucose levels.

After a meal, an increased secretion of insulin by the beta cells of the pancreatic islets of Langerhans promotes the movement of glucose from blood into tissue cells. Between meals, an increased secretion of glucagon by the alpha cells of the pancreatic islets and decreased secretion of insulin cause the release of stored glucose and the breakdown of fat.

lets of Langerhans holds much promise as a lasting treatment for this form of diabetes.

In type II diabetes, the number of insulin receptors on the target tissue is abnormally low, while the level of insulin in the blood is often higher than normal. This form of diabetes usually develops in people over 40 years of age. It is almost always a consequence of excessive weight; in the United States, 90% of those who develop type II diabetes are obese. Cells of these individuals, overwhelmed with food, adjust their appetite for glucose downward by reducing their sensitivity to insulin. As a drug addict's neurons reduce their number of neurotransmitter receptors after continued exposure to a drug, the obese individual's cells reduce their number of insulin receptors. To compensate, the pancreas pumps out ever-more insulin, and, in some people, the insulin-producing cells are unable to keep up with the ever-heavier workload and stop functioning. Type II diabetes is usually treatable by diet and exercise, and most affected individuals do not need daily injections of insulin.

24.4   Clusters of cells within the pancreas secrete the hormones insulin and glucagon. Insulin stimulates the storage of glucose as glycogen, while glucagon stimulates glycogen breakdown to glucose. Working together, these hormones keep glucose levels within a narrow range.

24.5 The Thyroid, Parathyroid, and Adrenal Glands

The Thyroid: A Metabolic Thermostat

The **thyroid gland** is shaped like a shield (its name comes from *thyros*, the Greek word for "shield"). It lies just below the Adam's apple in the front of the neck. The thyroid makes several hormones, the two most important of which are **thyroxine,** which increases metabolic rate and promotes growth, and **calcitonin,** which inhibits the release of calcium from bones.

Thyroxine regulates the level of metabolism in the body in several important ways. Without adequate thyroxine, growth is retarded. For example, children with underactive thyroid glands are not able to carry out carbohydrate breakdown and protein synthesis at normal rates, a condition called cretinism, which results in stunted growth. Mental retardation can also result, because thyroxine is needed for normal development of the central nervous system. The thyroid is stimulated to produce thyroxine by the hypothalamus, which is inhibited by thyroxine via negative feedback (figure 24.11*a*). Thyroxine contains iodine, and if the amount of iodine in the diet is too low, the thyroid cannot make adequate amounts of thyroxine to keep the hypothalamus inhibited. The hypothalamus will then continue to stimulate the thyroid, which will grow larger in a futile attempt to manufacture more thyroxine. The greatly enlarged thyroid gland that results is called a goiter (figure 24.11*b*). This need for iodine in the diet is why iodine is added to table salt.

Calcitonin plays a key role in maintaining proper calcium levels in the body. If levels of calcium in the blood become too high, calcitonin reduces calcium release from bone, thus lowering calcium levels in the bloodstream. As you will see, another hormone has an even more critical role in maintaining the body's levels of calcium.

The Parathyroids: Builders of Bone

The **parathyroid glands** are four small glands attached to the thyroid. Small and unobtrusive, they were ignored by researchers until well into the last century. The first suggestion that the parathyroids produce a hormone came from experiments in which they were removed from dogs: The concentration of calcium in the dogs' blood plummeted to less than half the normal level. However, if an extract of the parathyroid gland was administered, calcium levels returned to normal. If an excess was administered, calcium levels in the blood became *too* high, and the bones of the dogs were literally dismantled by the extract. It was clear that the parathyroid glands were producing a hormone that acted on calcium uptake into and out of the bones.

The hormone produced by the parathyroids is **parathyroid hormone (PTH).** It is one of only two hormones in the body that is absolutely essential for survival (the other is aldosterone, a hormone produced by the adrenal glands, discussed on page 577). PTH regulates the level of calcium in blood. Recall that calcium ions are the key actors in muscle contraction—by initiating calcium release, nerve impulses cause muscles to contract. A vertebrate cannot live without the muscles that pump the heart and drive the body, and these

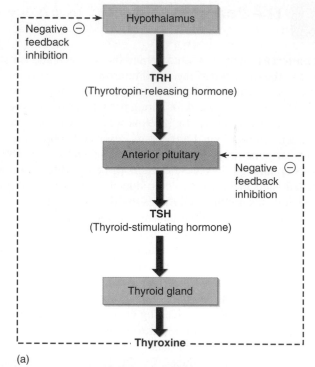

(a)

(b)

Figure 24.11   The thyroid gland secretes thyroxine.

(*a*) The hypothalamus secretes TRH, which stimulates the anterior pituitary to secrete TSH. The TSH then stimulates the thyroid to secrete thyroxine, which exerts its negative feedback control of the hypothalamus and anterior pituitary. (*b*) A goiter is caused by a lack of iodine in the diet. As a result, thyroxine secretion is low, so there is less negative feedback inhibition of TSH. The elevated TSH secretion, in turn, stimulates the thyroid to swell, producing a goiter.

muscles cannot function if calcium levels are not kept within narrow limits.

PTH acts as a fail-safe to make sure calcium levels never fall too low (figure 24.12). PTH is released into the bloodstream, where it travels to the bones and acts on the osteoclast cells within bones, stimulating them to dismantle bone tissue and release calcium into the bloodstream. PTH also acts on the kidneys to reabsorb calcium ions from the urine and leads to activation of vitamin D, necessary for calcium absorption by the intestine. A diet deficient in vitamin D leads to poor bone formation, a condition called rickets. When PTH is synthesized by the parathyroids in response to falling levels of calcium in the blood, the body is essentially sacrificing bone to keep calcium levels within the narrow limits necessary for proper functioning of muscle and nerve tissues.

The Adrenals: Two Glands in One

Mammals have two **adrenal glands,** one located just above each kidney. Each adrenal gland is composed of two parts: an inner core called the **medulla,** which produces the hormones adrenaline (also called epinephrine) and norepinephrine, and an outer shell called the **cortex,** which produces the steroid hormones cortisol and aldosterone. Similar to the pituitary gland, the adrenal gland is formed from different embryonic tissues. The adrenal medulla originates from nervous tissue and behaves as an extension of the autonomic nervous system. The adrenal cortex originates from glandular mesoderm and is entirely steroid-producing tissue.

The Adrenal Medulla: Emergency Warning Siren

The medulla releases adrenaline (epinephrine) and norepinephrine in times of stress. These hormones act as emergency signals that stimulate rapid deployment of body fuel. The "alarm" response these hormones produce throughout the body is identical to the individual effects achieved by the sympathetic nervous system, but it is much longer lasting. Among the effects of these hormones are an accelerated heartbeat, increased blood pressure, higher levels of blood sugar, and increased blood flow to the heart and lungs. These hormones can thus be thought of as extensions of the sympathetic nervous system.

The Adrenal Cortex: Maintaining the Proper Amount of Salt

The adrenal cortex produces the steroid hormone cortisol. Cortisol (also called hydrocortisone) acts on many different cells in the body to maintain nutritional well-being. It stimulates carbohydrate metabolism and reduces inflammation. Synthetic derivatives of this hormone, such as prednisone, have widespread medical use as anti-inflammatory agents.

The adrenal cortex also produces aldosterone. Aldosterone acts primarily in the kidney to promote the uptake of sodium and other salts from the urine. Sodium ions play crucial roles in nerve conduction and many other body functions. Aldosterone is, with PTH, one of the two endocrine hormones essential for survival. That is why removal of the adrenal glands is invariably fatal.

Table 24.1 (see page 578) summarizes the actions of the principal endocrine glands, including the major ones discussed so far and some others discussed in the next section.

> **24.5** **The thyroid acts as a metabolic thermostat, secreting hormones that adjust metabolic rate. Parathyroid hormone (PTH) regulates calcium levels in the blood. The adrenal hormone aldosterone promotes the uptake of sodium and other salts in the kidney.**

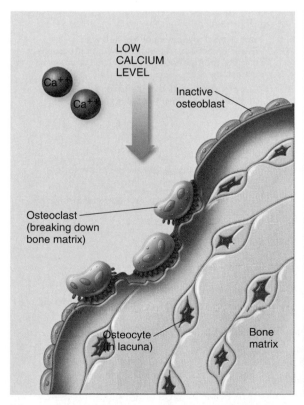

(a)

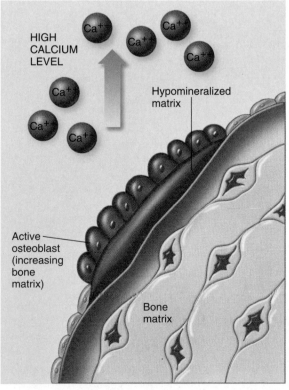

(b)

Figure 24.12 Maintenance of proper calcium levels in the blood.

(a) When calcium levels in the blood become too low, the parathyroid gland produces additional amounts of PTH, a hormone that stimulates the osteoclast cells to break down bone, releasing calcium. This raises the calcium levels in the blood. (b) Conversely, abnormally high levels of calcium in the blood trigger the thyroid gland to secrete calcitonin, which inhibits the release of calcium from bone, and promotes the activity of osteoblasts to remove calcium from the blood and deposit it in bone.

TABLE 24.1	THE PRINCIPAL ENDOCRINE GLANDS		
Endocrine Gland and Hormone		**Target**	**Principal Actions**
Adrenal Cortex			
Aldosterone		Kidney tubules	Maintains proper balance of sodium and potassium ions
Cortisol		General	Adaptation to long-term stress; raises blood glucose level; mobilizes fat
Adrenal Medulla			
Epinephrine (adrenaline) and norepinephrine (noradrenaline)		Smooth muscle, cardiac muscle, blood vessels, skeletal muscle	Initiate stress responses; increase heart rate, blood pressure, metabolic rate; dilate blood vessels; mobilize fat; raise blood glucose level
Hypothalamus			
Thyrotropin-releasing hormone (TRH)		Anterior pituitary	Stimulates TSH release from anterior pituitary
Corticotropin-releasing hormone (CRH)		Anterior pituitary	Stimulates ACTH release from anterior pituitary
Gonadotropin-releasing hormone (GnRH)		Anterior pituitary	Stimulates FSH and LH release from anterior pituitary
Prolactin-releasing factor (PRF)		Anterior pituitary	Stimulates PRL release from anterior pituitary
Growth-hormone-releasing hormone (GHRH)		Anterior pituitary	Stimulates GH release from anterior pituitary
Prolactin-inhibiting hormone (PIH)		Anterior pituitary	Inhibits PRL release from anterior pituitary
Growth-hormone-inhibiting hormone (somatostatin)		Anterior pituitary	Inhibits GH release from anterior pituitary
Melanotropin-inhibiting hormone (MIH)		Anterior pituitary	Inhibits MSH release from anterior pituitary
Ovary			
Estrogen		General; female reproductive structures	Stimulates development of secondary sex characteristics in females and growth of sex organs at puberty; prompts monthly preparation of uterus for pregnancy
Progesterone		Uterus, breasts	Completes preparation of uterus for pregnancy; stimulates development of breasts
Pancreas			
Insulin		General	Lowers blood glucose level; increases storage of glycogen in liver
Glucagon		Liver, adipose tissue	Raises blood glucose level; stimulates breakdown of glycogen in liver
Parathyroid Glands			
Parathyroid hormone (PTH)		Bone, kidneys, digestive tract	Increases blood calcium level by stimulating bone breakdown; stimulates calcium reabsorption in kidneys; activates vitamin D

TABLE 24.1 (CONTINUED)

Endocrine Gland and Hormone	Target	Principal Actions
Pineal Gland		
Melatonin	Hypothalamus	Function not well understood; may help control onset of puberty in humans
Posterior Lobe of Pituitary		
Oxytocin (OT)	Uterus	Stimulates contraction of uterus
	Mammary glands	Stimulates ejection of milk
Vasopressin (antidiuretic hormone, ADH)	Kidneys	Conserves water; increases blood pressure
Anterior Lobe of Pituitary		
Growth hormone (GH)	General	Stimulates growth by promoting protein synthesis and breakdown of fatty acids
Prolactin (PRL)	Mammary glands	Sustains milk production after birth
Thyroid-stimulating hormone (TSH)	Thyroid gland	Stimulates secretion of thyroid hormones
Adrenocorticotropic hormone (ACTH)	Adrenal cortex	Stimulates secretion of adrenal cortical hormones
Follicle-stimulating hormone (FSH)	Gonads	Stimulates ovarian follicle growth and secretion of estrogen in females; stimulates production of sperm cells in males
Luteinizing hormone (LH)	Ovaries and testes	Stimulates ovulation and corpus luteum formation in females; stimulates secretion of testosterone in males
Melanocyte-stimulating hormone (MSH)	Skin	Stimulates color change in reptiles and amphibians; unknown function in mammals
Testes		
Testosterone	General; male reproductive structures	Stimulates development of secondary sex characteristics in males and growth spurt at puberty; stimulates development of sex organs; stimulates sperm production
Thyroid Gland		
Thyroid hormone (thyroxine, T_4, and triiodothyronine, T_3)	General	Stimulates metabolic rate; essential to normal growth and development
Calcitonin	Bone	Lowers blood calcium level by inhibiting release of calcium from bone
Thymus		
Thymosin	White blood cells	Promotes production and maturation of white blood cells

Sexual Development

The ovaries and testes are important endocrine glands, producing the steroid sex hormones (including estrogen, progesterone, and testosterone), to be described in detail in chapter 25.

Other Hormones

There are a variety of hormones secreted by organs whose primary functions are not endocrine in nature. The thymus is the site of production of particular lymphocytes called T cells, and it secretes a number of hormones that function in the regulation of the immune system. The right atrium of the heart secretes atrial natriuretic hormone, which stimulates the kidneys to excrete salt and water in the urine. The kidneys secrete erythropoietin, a hormone that stimulates the bone marrow to produce red blood cells. Even the skin has an endocrine function, because it secretes vitamin D.

Molting and Metamorphosis in Insects

In insects, hormonal secretions influence both metamorphosis and molting. Prior to molting, neurosecretory cells on the surface of the brain secrete brain hormone, which in turn stimulates a gland in the thorax called the prothoracic gland to produce **molting hormone,** or *ecdysone* (figure 24.13). Another hormone, called juvenile hormone, is produced by structures near the brain called the corpora allata. For insects to molt, both ecdysone and juvenile hormone must be present, but the level of juvenile hormone determines the result of a particular molt. When juvenile hormone levels are high, the molt produces another larva. At the late stages of metamorphosis, juvenile hormone levels decrease, and the molt produces a pupa and eventually an adult insect.

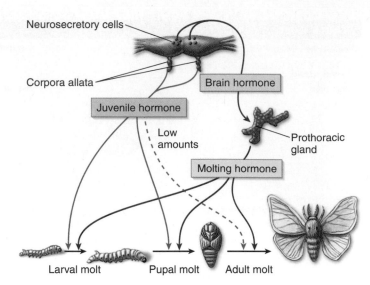

Figure 24.13   The hormonal control of metamorphosis in the silkworm moth, *Bombyx mori.*
While molting hormone (ecdysone), produced by the prothoracic gland, triggers when molting occurs, juvenile hormone, produced by structures near the brain called the corpora allata, determines the result of a particular molt. High levels of juvenile hormone inhibit the formation of the pupa and adult forms. At the late stages of metamorphosis, therefore, it is important that the corpora allata not produce large amounts of juvenile hormone.

> **24.6**   Sex steroid hormones from the gonads regulate reproduction. Other hormones are released from nontraditional endocrine glands. Molting hormone, or ecdysone, and juvenile hormone regulate metamorphosis and molting in insects.

Exploring Current Issues

Additional Resources

Go to your campus library or look online to find the following articles, which further develop some of the concepts found in this chapter.

American Diabetes Association. (2004). All about diabetes. *Retrieved from www.diabetes.org/about-diabetes.jsp.*

Blackman, S. (2003). The hunger hormone unharnessed: ghrelin as a source of drugs for combating eating disorders remains in doubt. *The Scientist,* 17(19), 30.

Greener, M. (2003). Steroid action gets a rewrite: nontranscriptional roles and new tissue sources add to signaling complexity. *The Scientist,* 17(17), 31.

McDowell, N. (2003). How to flick the switch that makes liver cells produce insulin. *New Scientist,* 177(2380), 16.

Wakefield, J. (2002). Boys won't be boys. *New Scientist,* 174(2349), 42.

Biology and Society Lecture: *Searching for a Cure for Diabetes*

Diabetes is a disorder affecting 18 million people in the United States in which the body's cells fail to take up glucose from the blood. Tissues waste away as glucose-starved cells are forced to consume their own protein. Some 10% of diabetics suffer from type I or juvenile diabetes, where their bodies do not produce insulin, a hormone needed for glucose transport. Ninety percent of diabetics suffer from type II, or late-onset, diabetes. Typically overweight, these individuals have normal levels of insulin, but have elevated levels of a hormone called resistin produced by fat cells that renders the body cells insensitive to insulin. Resistin appears to block the information pathway within the cell between the insulin receptor protein and the cell's glucose-transporting machinery.

Find this lecture, delivered by the author to his class at Washington University, online at www.mhhe.com/tlwessentials/exp24.

The Neuroendocrine System

24.1 Hormones

- Hormones are chemical signals produced in glands or other endocrine tissues (**figure 24.1**) and transported to distant sites in the body. Endocrine glands produce hormones and release the hormones into the bloodstream.

- Endocrine glands and tissues are under the control of the central nervous system, primarily the hypothalamus. A command from the hypothalamus causes the release of a hormone from an endocrine gland. Only cells that have receptors for the hormone respond and are called "target cells." The hormone binds to the receptor and elicits a response in the cell, often a change in cellular activity or genetic expression (**figure 24.2**).

24.2 How Hormones Target Cells

- Steroid hormones are lipid-soluble molecules, transported through the bloodstream by carrier proteins. They pass through the plasma membrane of the target cell and bind to receptors in the cytoplasm or nucleus. The hormone-receptor complex binds to DNA, causing a change in gene expression that alters cell function (**figure 24.3**).

- Peptide hormones are not lipid-soluble and therefore travel through the bloodstream unaided but are not able to pass through the plasma membrane. Instead, they bind to transmembrane protein receptors. The binding of the hormone causes a change in the internal side of the receptor, triggering a change in cellular activity (**figure 24.4**). This change is facilitated by a second messenger system, which is a cascade of reactions that amplifies the signal (**figure 24.5**).

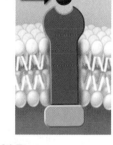

The Major Endocrine Glands

24.3 The Hypothalamus and the Pituitary

- The pituitary is actually two glands: the anterior and posterior pituitary glands. The posterior pituitary develops as an extension of the hypothalamus and contains axons that extend from cell bodies in the hypothalamus (**figure 24.6**). The hormones released from the posterior pituitary are actually produced in the hypothalamus and transported by the axons to the posterior pituitary for storage and release. The hormones of the posterior pituitary include antidiuretic hormone (ADH), which regulates water retention in the kidneys, and oxytocin, which initiates uterine contractions during childbirth and milk release in the mother (**figure 24.7**).

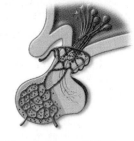

- The anterior pituitary originates from epithelial tissue and produces the hormones it releases, although it is controlled by the hypothalamus. Hormones released from the hypothalamus control the release of anterior pituitary hormones (**figure 24.8**). Seven hormones are released from the anterior pituitary; some have direct effects on target cells, such as LH, FSH, prolactin, and MSH, while others, such as TSH and ACTH, stimulate target cells to release hormones (**figure 24.7**). Many hormones are controlled by negative feedback, where the hormone itself feeds back to shut down the process (**figure 24.9**).

24.4 The Pancreas

- The pancreas secretes two hormones—insulin and glucagon—that interact to maintain stable blood sugar levels. Insulin stimulates cell uptake of glucose from the blood. Glucagon stimulates the breakdown of glycogen to glucose. These hormones are controlled through negative feedback (**figure 24.10**). When insulin is not available or cells fail to respond to insulin, diabetes can result.

24.5 The Thyroid, Parathyroid, and Adrenal Glands

- The two most important hormones produced by the thyroid are thyroxine, which increases metabolism and growth, and calcitonin, which stimulates calcium uptake by bones. Thyroxine is controlled by negative feedback, and the over- or underproduction of thyroxine can lead to serious health problems (**figure 24.11**).

- The parathyroid glands produce parathyroid hormone, a hormone necessary for survival. PTH regulates the levels of calcium in the blood (**figure 24.12**).

- The adrenal gland is actually two glands: The adrenal medulla is the inner core, and the adrenal cortex is the outer shell. The adrenal medulla secretes epinephrine and norepinephrine, and the adrenal cortex secretes aldosterone, which, like PTH, is necessary for survival. It promotes the uptake of sodium from urine.

24.6 A Host of Other Hormones

- The sex steroid hormones, including estrogen, progesterone, and testosterone, are released from the gonads and regulates sexual development and reproduction. Hormones are also released from other nontraditional endocrine tissues. In insects, molting hormone, ecdysone, and juvenile hormone regulate metamorphosis (**figure 24.13**).

1. One advantage chemical signaling has over nervous signaling is that
   a. reaction to stimuli can happen very quickly.
   b. although it takes large amounts of chemicals, the chemical signals are efficient.
   c. chemical signals stick around longer than nervous signals and can be used for slow processes.
   d. chemical signals are used in reaction to external and internal stimuli.

2. A coordination center for some of the endocrine system is the
   a. hypothalamus.   c. thymus.
   b. pituitary gland.   d. pineal gland.
3. Hormones are very similar to neurotransmitters and antibodies because they
   a. all fit into receptors specifically shaped for them.
   b. are all proteins.
   c. are all involved with maintaining homeostasis.
   d. appear in response to a signal from the brain.
4. The action of steroid hormones is different from peptide hormones because
   a. peptide hormones must enter the cell to begin action, whereas steroid hormones must begin action on the external surface of the cell membrane.
   b. steroid hormones must enter the cell to begin action, whereas peptide hormones must begin action on the external surface of the cell membrane.
   c. peptide hormones produce a hormone receptor complex that works directly on the DNA, whereas steroid hormones cause the release of a secondary messenger that triggers enzymes.
   d. No answer is correct.
5. Regulation of kidney function is done by the
   a. thyroid gland.   c. anterior pituitary gland.
   b. thymus.   d. posterior pituitary gland.

6. _____ is the hormone that stimulates the adrenal gland to produce a number of steroid hormones.
   a. ACTH   c. TSH
   b. LH   d. MSH
7. Diabetes mellitus is caused by an abnormality in endocrine cells of the
   a. pancreas.   c. adrenal glands.
   b. thymus.   d. hypothalamus.
8. The release of thyroid hormone calcitonin is determined by
   a. too much glucose in the blood.
   b. too much sodium in the blood.
   c. too much calcium in the blood.
   d. too little calcium in the blood.
9. Epinephrine directly stimulates the
   a. autonomic nervous system.
   b. central nervous system.
   c. parasympathetic nervous system.
   d. sympathetic nervous system.
10. A person afflicted with acromegaly, a form of giantism, may have too much
   a. GH.   c oxytocin.
   b. GnRH.   d. thyroxine.

Visual Understanding

1. **Figure 24.1** Some of the body systems are located primarily in one area of the body, or are obviously connected. The respiratory system, for instance, is located in the head and upper portion of the body. The skeletal system is articulated, every bone connected to others. The endocrine system, however, is spread out, a batch of glands that do not appear connected with one another. Speculate on why this is so.

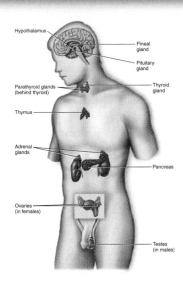

2. **Figure 24.7** Make a table that shows the hormones that are released from the anterior and posterior pituitary, and briefly explain their actions.

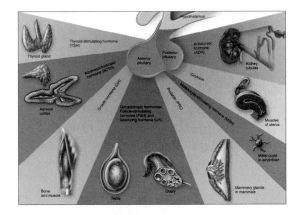

Challenge Questions

The Neuroendocrine System   Tadeusz, the younger brother of your friend, Zofia, wants to be a sports star in high school. Only in the eighth grade, he brags that he is taking steroids he gets from a friend in order to "bulk up" for next year. He asks if you have ever heard of any problems for kids his age—he only wants to take them "a couple of years" to get a football scholarship so his family can afford for him to go to college. How would you advise him?

The Major Endocrine Glands   Since your bones are so very important, why would you ever need a system that includes osteoclasts, which literally break down your bone tissue?

Online Learning Center

Visit the Online Learning Center for this chapter at www.mhhe.com/tlwessentials/ch24 for quizzes, animations, interactive learning exercises, and other study tools. At the site you will also find extended answers to the end-of-chapter questions.

25

Reproduction and Development

F ew subjects pervade our everyday thinking more than sex; few urges are more insistent. They are no accident, these strong feelings. They are a natural part of being human. All animals share them. The cry of a cat in heat, insects chirping outside the windows, frogs croaking in swamps, wolves howling in a frozen northern scene—all these are the sounds of the living world's essential act, an urgent desire to reproduce that has been patterned by a long history of evolution. It is a pattern that each of us shares. The reproduction of our families spontaneously elicits in us a sense of rightness and fulfillment. It is difficult not to return the smile of a new infant, not to feel warmed by it and by the look of wonder and delight to be seen on the faces of parents like this nursing mother. This chapter deals with sex and reproduction among the vertebrates, of which we human beings are one kind. Few subjects are of more direct concern to students than sex. Because many students must make important decisions about sex, the subject is of far more than academic interest, and is one about which all students need to be well informed.

25.1 Asexual and Sexual Reproduction

Not all sex involves two parents. Asexual reproduction, in which the offspring are genetically identical to one parent, is the primary means of reproduction among protists, cnidarians, and tunicates, and also occurs in some of the more complex animals.

Through mitosis, genetically identical cells are produced from a single parent cell. This permits asexual reproduction to occur in protists by division of the organism, or **fission** (figure 25.1). Cnidaria commonly reproduce by **budding,** where a part of the parent's body becomes separated from the rest and differentiates into a new individual. The new individual may become an independent animal or may remain attached to the parent, forming a colony.

Unlike asexual reproduction, sexual reproduction occurs when a new individual is formed by the union of *two* cells. These cells are called **gametes,** and the two kinds that combine are generally called *sperm* and *eggs* (or ova). The union of a sperm and an egg produces a fertilized egg, or **zygote,** that develops by mitotic division into a new multicellular organism. The zygote and the cells it forms by mitosis are diploid; they contain both members of each homologous pair of chromosomes. The gametes, formed by meiosis in the sex organs, or **gonads**—the *testes* and *ovaries*—are haploid (see chapter 7). The processes of spermatogenesis (sperm formation) and oogenesis (egg formation) are described in later sections.

Different Approaches to Sex

Parthenogenesis, a type of reproduction in which offspring are produced from unfertilized eggs, is common in many species of arthropods; some species are exclusively parthenogenic, whereas others switch between sexual reproduction and parthenogenesis in different generations. In honeybees, for example, a queen bee mates only once and stores the sperm. She then can control the release of sperm. If no sperm are released, the eggs develop parthenogenetically into drones, which are males; if sperm are allowed to fertilize the eggs, the fertilized eggs develop into other queens or worker bees, which are female.

The Russian biologist Ilya Darevsky reported in 1958 one of the first cases of unusual modes of reproduction among vertebrates. He observed that some populations of small lizards of the genus *Lacerta* were exclusively female, and he suggested that these lizards could lay eggs that were viable even if they were not fertilized. In other words, they were capable of asexual reproduction in the absence of sperm, a type of parthenogenesis. Further work has shown that parthenogenesis occurs among populations of other lizard genera.

Another variation in reproductive strategies is **hermaphroditism,** when one individual has both testes and ovaries and so can produce both sperm and eggs (figure 25.2a). A tapeworm is hermaphroditic and can fertilize itself as well as cross fertilize, a useful strategy because it is unlikely to

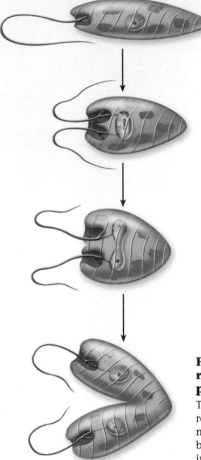

Figure 25.1 Asexual reproduction in protists.

This protist, *Euglena*, reproduces asexually: a mature individual divides by fission, and two complete individuals result.

encounter another tapeworm. Most hermaphroditic animals, however, require another individual to reproduce. Two earthworms, for example, are required for reproduction—each functions as both male and female, and each leaves the encounter with fertilized eggs.

There are some deep-sea fish that are hermaphrodites—both male and female at the same time. Numerous fish genera include species in which individuals can change their sex, a process called *sequential hermaphroditism.* Among coral reef fish, for example, both **protogyny** ("first female," a change from female to male) and **protandry** ("first male," a change from male to female) occur. In fish that practice protogyny (figure 25.2b), the sex change appears to be under social control. These fish commonly live in large groups, or schools, where successful reproduction is typically limited to one or a few large, dominant males. If those males are removed, the largest female rapidly changes sex and becomes a dominant male.

Sex Determination

Among the fish just described, and in some species of reptiles, environmental changes can cause changes in the sex of the animal. In mammals, the sex is determined early in embryonic development. The reproductive systems of human males and

(a)

(b)

Figure 25.2 Hermaphroditism and protogyny.
(a) The hamlet bass (genus *Hypoplectrus*) is a deep-sea fish that is a hermaphrodite—both male and female at the same time. In the course of a single pair-mating, one fish may switch sexual roles as many as four times, alternately offering eggs to be fertilized and fertilizing its partner's eggs. Here the fish, acting as a male, curves around its motionless partner, fertilizing the upward-floating eggs. (b) The bluehead wrasse, *Thalassoma bifasciatium,* is protogynous—females sometimes turn into males. Here a large male, or sex-changed female, is seen among females, typically much smaller.

females appear similar for the first 40 days after conception. During this time, the cells that will give rise to ova or sperm migrate from the yolk sac to the embryonic gonads, which have the potential to become either ovaries in females or testes in males. For this reason, the embryonic gonads are said to be "indifferent." If the embryo is XY it is a male, and will carry a gene on the Y chromosome whose product converts the indifferent gonads into testes. In females, who are XX, this Y chromosome gene and the protein it encodes are absent, and the gonads become ovaries. Recent evidence suggests that the sex-determining gene may be one known as **SRY** (for "*sex-determining region of the Y* chromosome") (figure 25.3). The *SRY* gene appears to have been highly conserved during the evolution of different vertebrate groups.

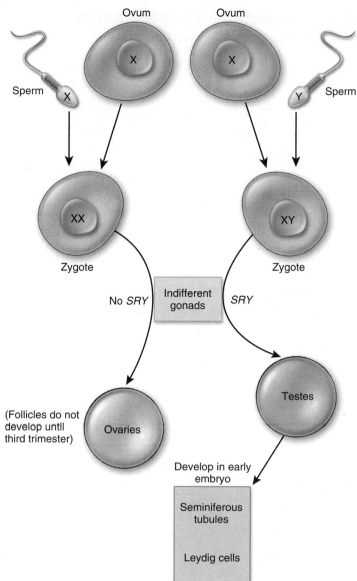

Figure 25.3 Sex determination.
Sex determination in mammals is made by a gene of the Y chromosome designated *SRY*. Testes are formed when the Y chromosome and *SRY* are present; ovaries are formed when they are absent.

Once testes form in the embryo, they secrete testosterone and other hormones that promote the development of the male external genitalia and accessory reproductive organs. If testes do not form (the ovaries are nonfunctional at this stage), the embryo develops female external genitalia and accessory reproductive organs. In other words, all mammalian embryos will develop female sex accessory organs and external genitalia by default unless they are masculinized by the secretions of the testes.

> **25.1 Sexual reproduction is most common among animals, but many reproduce asexually by fission, budding, or parthenogenesis. Sexual reproduction generally involves the fusion of gametes derived from different individuals of a species, but some species are hermaphroditic.**

25.2 Evolution of Reproduction Among the Vertebrates

Vertebrate sexual reproduction evolved in the ocean before vertebrates colonized the land. The females of most species of marine bony fish produce eggs or ova in batches and release them into the water. The males generally release their sperm into the water containing the eggs, where the union of the free gametes occurs. This process is known as **external fertilization.**

Although seawater is not a hostile environment for gametes, it does cause the gametes to disperse rapidly, so their release by females and males must be almost simultaneous. Thus, most marine fish restrict the release of their eggs and sperm to a few brief and well-defined periods. Some reproduce just once a year, while others do so more frequently. There are few seasonal cues in the ocean that organisms can use as signals for synchronizing reproduction, but one all-pervasive signal is the cycle of the moon. Once each month, the moon approaches closer to the earth than usual, and when it does, its increased gravitational attraction causes somewhat higher tides. Many marine organisms sense the tidal changes and entrain the production and release of their gametes to the lunar cycle.

Fertilization is external in most fish but internal in most other vertebrates. The invasion of land posed the new danger of desiccation (drying out), a problem that was especially severe for the small and vulnerable gametes. On land, the gametes could not simply be released near each other, because they would soon dry up and perish. Consequently, there was intense selective pressure for terrestrial vertebrates (as well as some groups of fish) to evolve **internal fertilization,** that is, the introduction of male gametes into the female reproductive tract. By this means, fertilization still occurs in a nondesiccating environment, even when the adult animals are fully terrestrial. The vertebrates that practice internal fertilization have three strategies for embryonic and fetal development. Depending upon the relationship of the developing embryo to the mother and egg, those vertebrates with internal fertilization may be classified as oviparous, ovoviviparous, or viviparous.

Oviparity.   In oviparity, the eggs, after being fertilized internally, are deposited outside the mother's body to complete their development. This is found in some bony fish, most reptiles, some cartilaginous fish, some amphibians, a few mammals, and all birds.

Ovoviviparity.   In ovoviviparity, the fertilized eggs are retained within the mother to complete their development, but the embryos still obtain all of their nourishment from the egg yolk. The young are fully developed when they are hatched and released from the mother. This is found in some bony fish (including mollies, guppies, and mosquito fish), some cartilaginous fish, and many reptiles.

Viviparity.   In viviparity, the young develop within the mother and obtain nourishment directly from their mother's blood, rather than from the egg yolk (figure 25.4). This is found in most cartilaginous fish, some amphibians, a few reptiles, and almost all mammals.

Fish and Amphibians

Most fish and amphibians, unlike other vertebrates, reproduce by means of external fertilization.

Fish.   The eggs of most bony fish and amphibians are fertilized externally. Fertilization in most species of bony fish (teleosts), for example, is external, and the eggs contain only

Figure 25.4 Viviparous vertebrates carry live, mobile young within their bodies.
The young complete their development within the body of the mother and are then released as small but competent adults. Here a lemon shark has just given birth to a young shark, which is still attached by the umbilical cord.

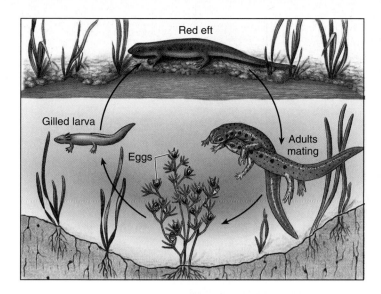

Figure 25.5   Life cycle of the red-spotted newt.
Many salamanders have both aquatic and terrestrial stages in their life cycle. In the red-spotted newt (*Notophthalmus viridescens*), eggs are laid in water and hatch into aquatic larvae with external gills and a finlike tail. After a period of growth, the larvae can metamorphose into a terrestrial "red eft" stage that later metamorphose again to produce aquatic, breeding adults. In some environments, the "red eft" stage is skipped, and the populations remain entirely aquatic.

enough yolk to sustain the developing embryo for a short time. After the initial supply of yolk has been exhausted, the young fish must seek its food from the waters around it. Development is speedy, and the young that survive mature rapidly. Although thousands of eggs are fertilized in a single mating, many of the resulting individuals succumb to microbial infection or predation, and few grow to maturity.

In marked contrast to the bony fish, fertilization in most cartilaginous fish is internal. The male introduces sperm into the female through a modified pelvic fin. Development of the young in these vertebrates is generally viviparous.

Amphibians.   The amphibians invaded the land without fully adapting to the terrestrial environment, and their life cycle is still tied to the water (figure 25.5). Fertilization is external in most amphibians, just as it is in most species of bony fish. Gametes from both males and females are released through the cloaca, a common opening used by the digestive, reproductive, and urinary systems. Among the frogs and toads, the male grasps the female and discharges fluid containing the sperm onto the eggs as they are released into the water (figure 25.6). Although the eggs of most amphibians develop in the water, there are some interesting exceptions (figure 25.7). In two species of frogs, for example, the eggs develop in the vocal sacs and the stomach, and the young frogs leave through their parent's mouth!

The time required for development of most amphibians is much longer than that for fish, but amphibian eggs do not include a significantly greater amount of yolk. Instead, the process of development in most amphibians is divided into embryonic, larval, and adult stages, in a way reminiscent of the life cycles found in some insects. The embryo develops within

Figure 25.6   The eggs of frogs are fertilized externally.
When frogs mate, as these two are doing, the clasp of the male induces the female to release a large mass of mature eggs, over which the male discharges his sperm.

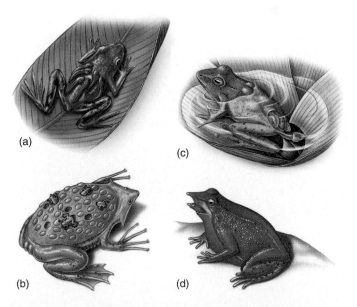

Figure 25.7   Different ways young develop in frogs.
(*a*) In the poison arrow frog, the male carries the tadpoles on his back. (*b*) In the female Surinam frog, froglets develop from eggs in special brooding pouches on the back. (*c*) In the South American pygmy marsupial frog, the female carries the developing larvae in a pouch on her back. (*d*) Tadpoles of the Darwin's frog develop into froglets in the vocal pouch of the male and emerge from the mouth.

the egg, obtaining nutrients from the yolk. After hatching from the egg, the aquatic larva then functions as a free-swimming, food-gathering machine, often for a considerable period of time. The larvae may increase in size rapidly; some tadpoles, which are the larvae of frogs and toads, grow in a matter of weeks from creatures no bigger than the tip of a pencil into individuals as big as a goldfish. When the larva has grown to a sufficient size, it undergoes a developmental transition, or metamorphosis, into the terrestrial adult form.

Reptiles and Birds

Most reptiles and all birds are oviparous, laying amniotic eggs that are protected by watertight membranes from desiccation. After the eggs are fertilized internally, they are deposited outside of the mother's body to complete their development. Like most vertebrates that fertilize internally, most male reptiles use a cylindrical organ, the penis, to inject sperm into the female, a process called copulation (figure 25.8). The penis, containing erectile tissue, can become quite rigid and penetrate far into the female reproductive tract. Most reptiles are oviparous, laying eggs and then abandoning them. These eggs are surrounded by a leathery shell that is deposited as the egg passes through the oviduct, the part of the female reproductive tract leading from the ovary. Other species of reptiles are ovoviviparous or viviparous, forming eggs that develop into embryos within the body of the mother.

All birds practice internal fertilization, though most male birds lack a penis. In some of the larger birds (including swans, geese, and ostriches), however, the male cloaca extends to form a false penis. As the egg passes along the oviduct, glands secrete albumin proteins (the egg white) and the hard, calcareous shell that distinguishes bird eggs from reptilian eggs (figure 25.9a). While modern reptiles are poikilotherms (ani-

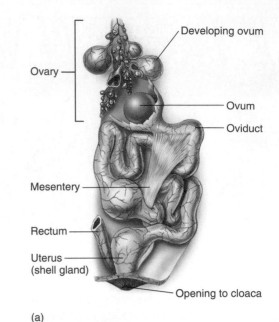

(a)

Figure 25.8 The introduction of sperm by the male into the female's body is called copulation.
Reptiles, such as these tortoises, were the first terrestrial vertebrates to develop this form of reproduction, called copulation, which is particularly suited to a terrestrial environment.

Figure 25.9 Egg formation and incubation in birds.
(a) In birds, fertilization of the egg (ovum) takes place within the female, in the upper portion of the oviduct. As the fertilized egg passes down the oviduct, albumin (egg white), shell membranes, and the shell is secreted around the egg. (b) This nesting pair of crested penguins is changing the parental guard in a stylized ritual.

mals whose body temperature varies with the temperature of their environment), birds are homeotherms (animals that maintain a relatively constant body temperature independent of environmental temperatures). Hence, most birds incubate their eggs after laying them to keep them warm (figure 25.9b).

The shelled eggs of reptiles and birds constitute one of the most important adaptations of these vertebrates to life on land, because shelled eggs can be laid in dry places. Such eggs are known as amniotic eggs because the embryo develops within a fluid-filled cavity surrounded by a membrane called the amnion. The amnion is an extraembryonic membrane—that is, a membrane formed from embryonic cells but located outside the body of the embryo. Other extraembryonic mem-

(a) Monotremes

(b) Marsupials

(c) Placentals

Figure 25.10 Reproduction in mammals.
(*a*) Monotremes, like the duck-billed platypus shown here, lay eggs in a nest. (*b*) Marsupials, such as this kangaroo, give birth to small fetuses that complete their development in a pouch. (*c*) In placental mammals, such as this doe nursing her fawn, the young remain inside the mother's uterus for a longer period of time and are born relatively more developed.

branes in amniotic eggs include the chorion, which lines the inside of the eggshell, the yolk sac, and the allantois. In contrast, the eggs of fish and amphibians contain only one extra-embryonic membrane, the yolk sac. The viviparous mammals, including humans, also have extraembryonic membranes that are described later in this chapter.

Mammals

Some mammals are seasonal breeders, reproducing only once a year, such as dogs, foxes, and bears, whereas others, such as horses and sheep, have multiple short reproductive cycles throughout a given time of the year. Among the latter, the females generally undergo the reproductive cycles, whereas the males are more constant in their reproductive activity. Cycling in females involves the periodic release of a mature ovum from the ovary in a process known as ovulation. Most female mammals are "in heat," or sexually receptive to males, only around the time of ovulation. This period of sexual receptivity is called **estrus,** and the reproductive cycle is therefore called an **estrous cycle.** The females continue to cycle until they become pregnant.

In the estrous cycle of most mammals, changes in the secretion of follicle-stimulating hormone (FSH) and luteinizing hormone (LH) by the anterior pituitary gland cause changes in egg cell development and hormone secretion in the ovaries. Humans and apes have menstrual cycles that are similar to the estrous cycles of other mammals in their cyclic pattern of hormone secretion and ovulation. Unlike mammals with estrous cycles, however, human and some ape females bleed when they shed the inner lining of their uterus, a process called menstruation, and may engage in copulation at any time during the cycle.

Rabbits and cats differ from most other mammals in that they are **induced ovulators.** Instead of ovulating in a cyclic fashion regardless of sexual activity, the females ovulate only after copulation as a result of a reflex stimulation of LH secretion (described later). This makes these animals extremely fertile.

The most primitive mammals, the **monotremes** (consisting solely of the duck-billed platypus and the echidna), are oviparous, like the reptiles from which they evolved. They incubate their eggs in a nest (figure 25.10*a*) or specialized pouch, and the young hatchlings obtain milk from their mother's mammary glands by licking her skin, because monotremes lack nipples. All other mammals are viviparous and are divided into two subcategories based on how they nourish their young. The **marsupials,** a group that includes opossums and kangaroos, give birth to fetuses that are incompletely developed. The fetuses complete their development in a pouch of their mother's skin, where they can obtain nourishment from nipples of the mammary glands (figure 25.10*b*). The **placental mammals** (figure 25.10*c*) retain their young for a much longer period of development within the mother's uterus. The fetuses are nourished by a structure known as the placenta, which is derived from both an extraembryonic membrane (the chorion) and from the mother's uterine lining. Because the fetal and maternal blood vessels are in very close proximity in the placenta, the fetus can obtain nutrients by diffusion from the mother's blood. The functioning of the placenta is discussed in more detail later in this chapter.

25.2 Fertilization is external in frogs and most bony fish and internal in other vertebrates. Birds and most reptiles lay watertight eggs, as do monotreme mammals. All other mammals are viviparous, giving birth to live young.

25.3 Males

The human male gamete, or **sperm,** is highly specialized for its role as a carrier of genetic information. Produced after meiosis, sperm cells have 23 chromosomes instead of the 46 found in other cells of the male body. Sperm do not successfully complete their development at 37°C (98.6°F), the normal human body temperature. The sperm-producing organs, the **testes** (singular, **testis**), move during the course of fetal development into a sac called the **scrotum** (figure 25.11), which hangs between the legs of the male, maintaining the testes at a temperature about 3°C cooler than the rest of the body.

Male Gametes Are Formed in the Testes

The two testes are composed of several hundred compartments, each packed with large numbers of tightly coiled tubes called **seminiferous tubules.** Within these tubes sperm production, *spermatogenesis,* takes place (figure 25.12). The number of sperm produced is truly incredible. A typical adult male produces several hundred million sperm each day of his life! Those that are not ejaculated from the body are broken down, and their materials are resorbed and recycled.

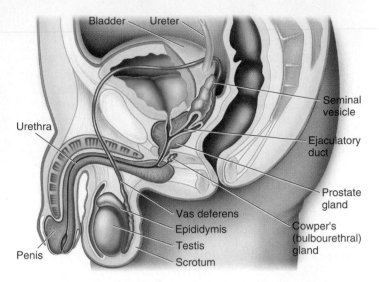

Figure 25.11   The male reproductive organs.

The testis is where sperm are formed. Cupped above the testis is the epididymis, a highly coiled passageway within which sperm complete their maturation. Extending away from the epididymis is a long tube, the vas deferens.

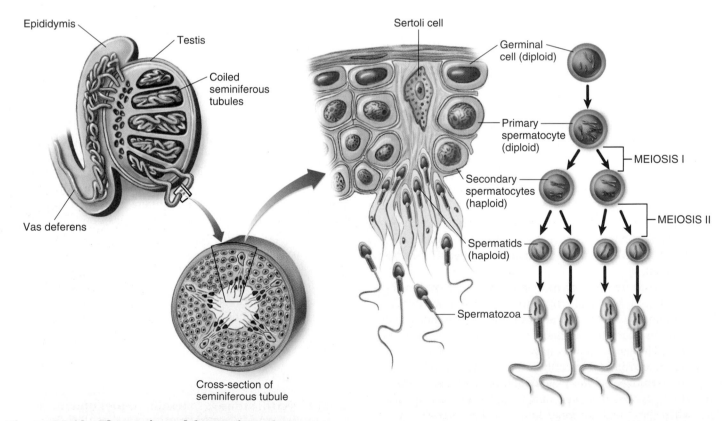

Figure 25.12   The testis and formation of sperm.

Inside the testis, the seminiferous tubules are the sites of sperm formation. Germinal cells in the seminiferous tubules give rise to primary spermatocytes (diploid), which undergo meiosis to form haploid spermatids. Spermatids develop into mobile spermatozoa, or sperm. Sertoli cells are nongerminal cells within the walls of the seminiferous tubules. They assist spermatogenesis in several ways, such as helping to convert spermatids into spermatozoa.

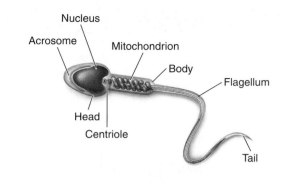

(a)

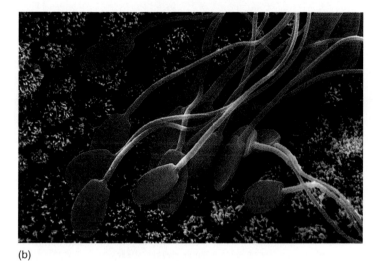

(b)

Figure 25.13 Human sperm cells.
(*a*) Each sperm possesses a long tail that propels the sperm and a head that contains the nucleus. The tip, or acrosome, contains enzymes to help the sperm cell digest a passageway into the egg for fertilization. (*b*) Scanning electron micrograph of sperm.

The testes contain cells that secrete the male sex hormone **testosterone.** Sperm and all the cells of the testes also require a combination of the pituitary hormones FSH and LH for their normal function.

After a sperm cell is manufactured within the testes through intermediate stages of meiosis, it is delivered to a long, coiled tube called the **epididymis,** where it matures. The sperm cell (figure 25.13) is not motile when it arrives in the epididymis, and it must remain there for at least 18 hours before its motility develops. From there, the sperm is delivered to another long tube, the **vas deferens.** When sperm are released during intercourse, they travel through a tube from the vas deferens to the **urethra,** where the reproductive and urinary tracts join, emptying through the penis.

Male Gametes Are Delivered by the Penis

In the case of humans and some other mammals, the **penis** is an external tube containing two long cylinders of spongy tissue side by side. Below and between them runs a third cylinder of spongy tissue that contains in its center a tube called the urethra, through which both semen (during ejaculation) and urine (during urination) pass (figure 25.14). Why this unusual design? The penis is designed to inflate. The spongy tissues

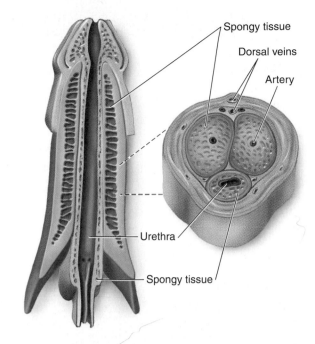

Figure 25.14 Structure of the penis.
(*Left*) Longitudinal section; (*right*) cross section.

that make up the three cylinders are riddled with small spaces between the cells, and when nerve impulses from the CNS cause the arterioles leading into this tissue to expand, blood collects within these spaces. Like blowing up a balloon, this causes the penis to become erect and rigid. Continued stimulation by the CNS is required for this erection to be maintained. Nitric oxide (NO) as described in chapter 20 elicits this response.

Erection can be achieved without any physical stimulation of the penis, but physical stimulation is required for semen to be delivered. Stimulation of the penis, as by repeated thrusts into the vagina of a female, leads first to the mobilization of the sperm. In this process, muscles encircling the vas deferens contract, moving the sperm along the vas deferens into the urethra. The stimulation then leads to the strong contraction of the muscles at the base of the penis. The result is **ejaculation,** the forceful ejection of 2 to 5 milliliters of semen. Semen contains sperm and a collection of secretions from the prostate and other glands that provides metabolic energy sources for the sperm. Within this small 5-milliliter volume are several hundred million sperm. Because the odds against any one individual sperm cell successfully completing the long journey to the egg and fertilizing it are extraordinarily high, successful fertilization requires a high sperm count. Males with fewer than 20 million sperm per milliliter are generally considered sterile.

> **25.3** Male testes continuously produce large numbers of male gametes, sperm, which mature in the epididymis, are stored in the vas deferens, and are delivered through the penis into the female.

25.4 Females

In females, eggs develop from cells called **oocytes,** located in the outer layer of compact masses of cells called **ovaries** within the abdominal cavity (figure 25.15). Recall that in males the gamete-producing cells are constantly dividing. In females all of the oocytes needed for a lifetime are already present at birth. During each reproductive cycle, one or a few of these oocytes are initiated to continue their development in a process called **ovulation;** the others remain in a developmental holding pattern.

Only One Female Gamete Matures Each Month

At birth, a female's ovaries contain some 2 million oocytes, all of which have begun the first meiotic division. At this stage they are called **primary oocytes** (figure 25.16). Each primary oocyte is poised to develop further, but it does not continue on with meiosis. Instead, it waits to receive the proper developmental "go" signal, and until a primary oocyte receives this signal, its meiosis is arrested in prophase of the first meiotic division. Very few ever receive the awaited signal, which

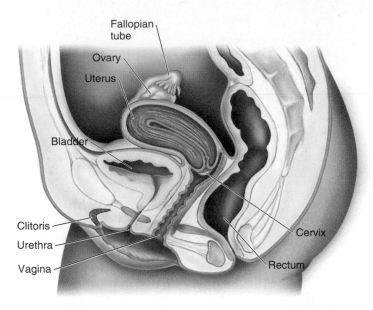

Figure 25.15   The female reproductive system.
The organs of the female reproductive system are specialized to produce gametes and to provide a site for embryonic development if the gamete is fertilized.

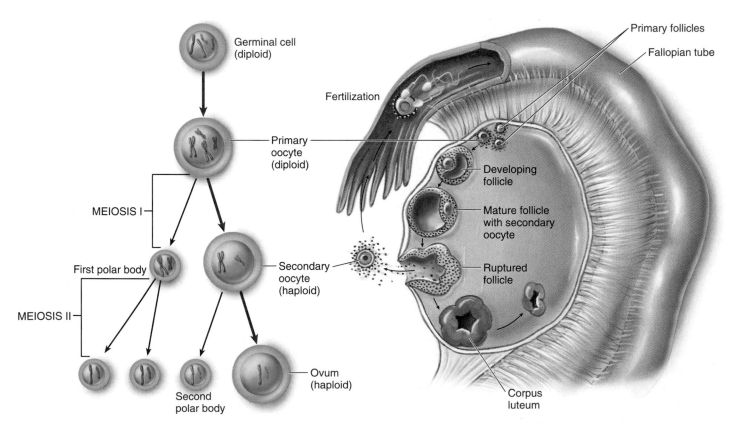

Figure 25.16   The ovary and formation of an ovum.
At birth, a human female's ovaries contain about 2 million egg-forming cells called oocytes, which have begun the first meiotic division and stopped. At this stage, they are called primary oocytes, and their further development is halted until they receive the proper developmental signal, which is the hormone FSH. At the onset of puberty in females, the release of FSH causes the development of approximately one oocyte every 28 days, with only about 400 of the 2 million oocytes maturing during a woman's lifetime. When FSH is released, meiosis resumes in a few oocytes, but only one oocyte usually continues to mature while the others regress. The primary oocyte (diploid) completes the first meiotic division, and one division product is eliminated as a nonfunctional polar body. The other product, the secondary oocyte, is released during ovulation. The secondary oocyte does not complete the second meiotic division until after fertilization; that division yields a second nonfunctional polar body and a single haploid egg, or ovum. Fusion of the haploid egg with a haploid sperm during fertilization produces a diploid zygote.

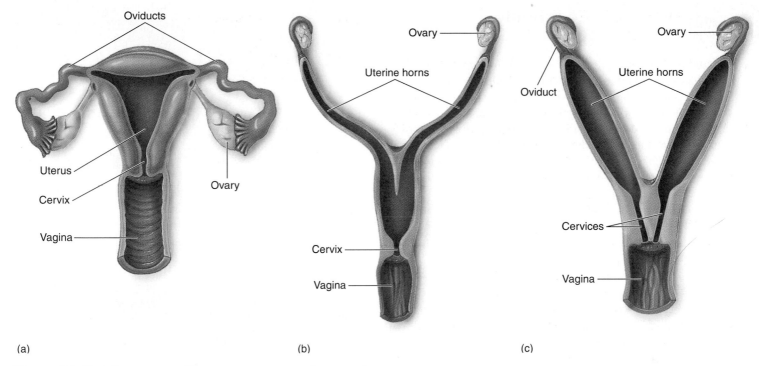

Figure 25.17 The mammalian uterus—several examples.
(a) Humans and other primates; (b) cats, dogs, and cows; and (c) rats, mice, and rabbits.

turns out to be the hormone FSH, which was discussed in chapter 24.

With the onset of puberty, females mature sexually. At this time, the release of FSH initiates the resumption of the first meiotic division in a few oocytes, but in humans, usually only a single oocyte continues to mature and becomes an ovum, and the others regress. Approximately every 28 days after that, another oocyte matures and is ovulated, although the exact timing may vary from month to month. Only about 400 of the approximately 2 million oocytes a woman is born with mature and are ovulated during her lifetime. When they mature, the egg cells are called **ova** (singular, **ovum**), the Latin word for "egg."

Fertilization Occurs in the Oviducts

The **fallopian tubes** (also called uterine tubes or **oviducts**) transport ova from the ovaries to the **uterus.** In humans, the uterus is a muscular, pear-shaped organ about the size of a fist that narrows to a muscular ring called the **cervix,** which leads to the vagina (figure 25.17a). The uterus is lined with a stratified epithelial membrane called the **endometrium.** The surface of the endometrium is shed during menstruation, while the underlying portion remains to generate a new surface during the next cycle.

Mammals other than primates have more complex female reproductive tracts, where part of the uterus divides to form uterine "horns," each of which leads to an oviduct (figure 25.17b, c). In cats, dogs, and cows, for example, there is one cervix but two uterine horns separated by a septum, or wall. Marsupials, such as opossums, carry the split even further, with two unconnected uterine horns, two cervices, and two vaginas. A male marsupial has a forked penis that can enter both vaginas simultaneously.

Smooth muscles lining the fallopian tubes contract rhythmically, moving the egg down the tube to the uterus in much the same way that food is moved down through your intestines, pushing it along by squeezing the tube behind it. The journey of the egg through the fallopian tube is a slow one, taking from five to seven days to complete. If the egg is unfertilized, it loses its capacity to develop within a few days. Any egg that arrives at the uterus unfertilized can never become so. For this reason the sperm cannot simply lie in wait within the uterus. To fertilize an egg successfully, a sperm must make its way far up the fallopian tube, a long passage that few survive.

Sperm are deposited within the vagina, a thin-walled muscular tube about 7 centimeters long that leads to the mouth of the uterus. Sperm entering the uterus swim up to and enter the fallopian tube. They swim upward against the current generated by the tube's contractions, which are carrying the ovum downward toward the uterus.

When a sperm succeeds in fertilizing an egg high in the fallopian tube, the fertilized egg—now a zygote—continues on its journey down the fallopian tube. When it reaches the uterus, the zygote attaches itself to the endometrial lining and starts the long developmental journey that eventually leads to the birth of a child.

> **25.4 In human females, hormones trigger the maturation of one or a few oocytes each 28 days. When mature, the egg cells travel down the fallopian tubes and, if fertilized during their journey, implant in the wall of the uterus and initiate embryonic development.**

25.5 Hormones Coordinate the Reproductive Cycle

The female reproductive cycle, called a **menstrual cycle,** is composed of two distinct phases, the follicular phase, in which an egg reaches maturation and is ovulated, and the luteal phase, where the body continues to prepare for pregnancy. These phases are coordinated by a family of hormones. Hormones play many roles in human reproduction. Sexual development is initiated by hormones, released from the anterior pituitary and ovary, that coordinate simultaneous sexual development in many kinds of tissues. The production of gametes is another closely orchestrated process, involving a series of carefully timed developmental events. Successful fertilization initiates yet another developmental "program," in which the female body continues its preparation for the many changes of pregnancy.

Production of the sex hormones that coordinate all these processes is coordinated by the hypothalamus, which sends releasing hormones to the pituitary, directing it to produce particular sex hormones. Negative feedback, discussed in chapter 24, plays a key role in regulating these activities of the hypothalamus. When target organs receive a pituitary hormone, they begin to produce a hormone of their own, which circulates back to the hypothalamus, shutting down production of the pituitary hormone.

Triggering the Maturation of an Egg

The first, or **follicular, phase** of the reproductive cycle is when the egg develops within the ovary. This development is carefully regulated by hormones. The anterior pituitary, after receiving a chemical signal (GnRH) from the hypothalamus, starts the cycle by secreting **follicle-stimulating hormone (FSH),** which binds to receptors on the surface of cells surrounding the egg (the oocyte and its surrounding mass of tissue is called a **follicle**) and triggers resumption of meiosis (figure 25.18). Several follicles are stimulated to grow under FSH stimulation, but only one achieves full maturity. FSH levels then fall, which keep more than just a few eggs from being stimulated to grow.

The fall of FSH levels is achieved by a feedback command to the pituitary. FSH does not itself carry out this feedback—instead, the ovary sends another hormone as a messenger. FSH not only triggers final egg development, it also causes the ovary to start producing the female sex hormone **estrogen** (see figure 25.18). Rising levels of estrogen in the bloodstream feed back to the hypothalamus, which responds to the rising estrogen by commanding the anterior pituitary to cut off the further production of FSH. This "shuts the door" on further FSH-induced egg development. The rise in estrogen levels signals the completion of the follicular phase of the reproductive cycle.

Preparing the Body for Fertilization

The second, or **luteal, phase** of the cycle follows smoothly from the first (see figure 25.18). The hypothalamus responds to

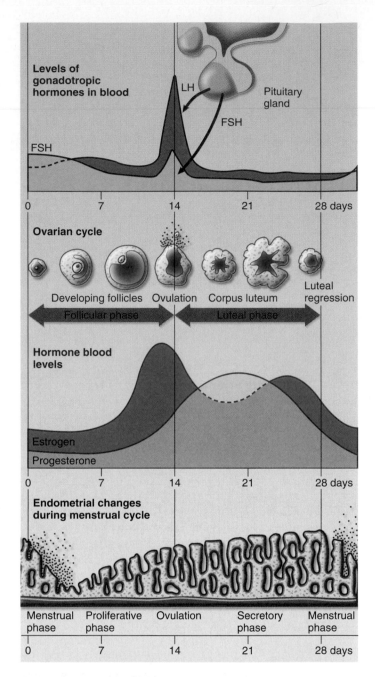

Figure 25.18 The human menstrual cycle.
The growth and thickening of the uterine (endometrial) lining is governed by increasing levels of the hormone progesterone; menstruation, the sloughing off of this blood-rich tissue, is initiated by lower levels of progesterone.

estrogen not only by shutting down the pituitary's FSH production but also by causing the anterior pituitary to begin secreting a second hormone, called **luteinizing hormone (LH).** LH is the hormone that causes ovulation, sending the now-mature egg on its journey toward fertilization (figure 25.19). LH is carried in the bloodstream to the developing follicle, where it inhibits further estrogen production and causes the wall of the follicle to burst. The egg within the follicle is released into one of the fallopian tubes extending from the ovary to the uterus.

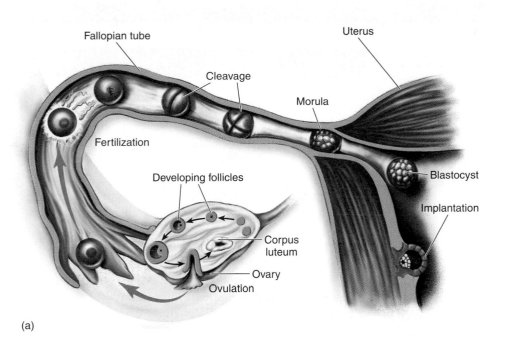

Fallopian tube

Uterus

Cleavage

Morula

Fertilization

Developing follicles

Blastocyst

Implantation

Corpus luteum

Ovary

Ovulation

(a)

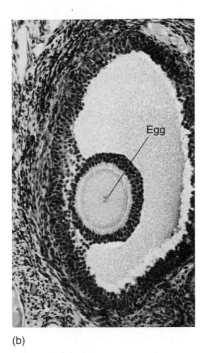

Egg

(b)

Figure 25.19   The journey of an ovum.

(a) Produced within a follicle and released at ovulation, an ovum is swept up into a fallopian tube and carried down by waves of contraction of the tube walls. Fertilization occurs within the tube, by sperm journeying upward. Several mitotic divisions occur while the fertilized ovum (a zygote) continues its journey down the fallopian tube. The zygote implants itself within the wall of the uterus, where it continues its development. (b) A mature egg within an ovarian follicle. In each menstrual cycle, a few follicles are stimulated to grow under the influence of FSH, but usually only one achieves full maturity and ovulation.

After the egg's release and departure, LH directs the repair of the ruptured follicle that fills in and becomes yellowish. In this condition, it is called the **corpus luteum,** which is simply the Latin phrase for "yellow body." The corpus luteum soon begins to secrete a hormone, **progesterone,** which inhibits FSH (a backup for estrogen in preventing further ovulations). Progesterone completes the body's preparation of the uterus for fertilization including the thickening of the endometrium. If fertilization does *not* occur soon after ovulation, however, production of progesterone slows and eventually ceases, marking the end of the luteal phase. The decreasing levels of progesterone cause the thickened layer of blood-rich tissue to be sloughed off, a process that results in the bleeding associated with menstruation. **Menstruation,** or "having a period," usually occurs about midway between successive ovulations, although its timing varies widely for individual females.

At the end of the luteal phase, neither estrogen nor progesterone is being produced. In their absence, the anterior pituitary can again initiate production of FSH, thus starting another reproductive cycle. Each cycle begins immediately after the preceding one ends. A cycle usually occurs every 28 days, or a little more frequently than once a month, although this varies in individual cases. The Latin word for "month" is *mens,* which is why the reproductive cycle is called the menstrual cycle, or monthly cycle.

If fertilization does occur, the corpus luteum is maintained and subsequent menstruation does not occur because of a hormone released from the embryo. The tiny embryo secretes human chorionic gonadotropin (hCG), an LH-like hormone. By maintaining the corpus luteum, hCG keeps the levels of estrogen and progesterone high, thereby preventing menstruation, which would terminate the pregnancy. Because hCG comes from the embryo and not from the mother, it is the hormone that is tested for in all pregnancy tests.

Two other hormones, both secreted by the pituitary, are important in the female reproductive system. For the first couple of days after childbirth, the mammary glands produce a fluid called colostrum, which contains protein and lactose but little fat. Then milk production is stimulated by the anterior pituitary hormone **prolactin,** usually by the third day after delivery. When the infant suckles at the breast, the posterior pituitary hormone **oxytocin** is released, initiating milk release. Earlier, in combination with chemicals released from the uterus, oxytocin initiates labor and delivery.

25.5   Humans and apes have menstrual cycles that are similar to the estrous cycle of other mammals in that they are driven by cyclic patterns of hormone secretion and ovulation. The cycle is composed of two distinct phases, follicular and luteal, coordinated by a family of four sex hormones.

25.6 Embryonic Development

TABLE 25.1	FATES OF THE PRIMARY GERM LAYERS
Ectoderm	Epidermis, central nervous system, sense organs, neural crest
Mesoderm	Skeleton, muscles, blood vessels, heart, gonads
Endoderm	Lining of digestive and respiratory tracts; liver, pancreas

Cleavage: Setting the Stage for Development

The first major event in human embryonic development is the rapid division of the zygote into a larger and larger number of smaller and smaller cells, becoming first 2 cells, then 4, then 8, and so on. The first of these divisions occurs about 30 hours after union of the egg and the sperm, and the second, 30 hours later. During this period of division, called **cleavage,** the overall size does not increase from that of the zygote. The resulting tightly packed mass of about 32 cells is called a **morula,** and each individual cell in the morula is referred to as a **blastomere.** The cells of the morula continue to divide, each cell secreting a fluid into the center of the cell mass. Eventually, a hollow ball of 500 to 2,000 cells is formed. This is the **blastocyst,** which contains a fluid-filled cavity called the **blastocoel.** Within the ball is an *inner cell mass* concentrated at one pole that goes on to form the developing embryo. The outer sphere of cells, called the *trophoblast,* releases the hCG hormone, discussed earlier.

During this period, the zygote continues its journey down the mother's fallopian tube. On about the sixth day, the blastocyst reaches the uterus, attaches to the uterine lining, and penetrates into the tissue of the lining. The zygote now begins to grow rapidly, initiating the formation of the membranes that will later surround, protect, and nourish it. One of these membranes, the **amnion,** will enclose the developing embryo, whereas another, the **chorion,** which forms from the trophoblast, will interact with uterine tissue to form the **placenta,** which will nourish the growing embryo. The placenta connects the developing embryo to the blood supply of the mother. Fully 61 of the cells at the 64-celled stage develop into the trophoblast and only 3 into the embryo proper.

Gastrulation: The Onset of Developmental Change

Ten to 11 days after fertilization, certain groups of cells move inward from the surface of the cell mass in a carefully orchestrated migration called **gastrulation.** First, the lower cell layer of the blastocyst cell mass differentiates into **endoderm,** one of the three primary embryonic tissues, and the upper layer into **ectoderm.** Just after this differentiation, much of the **mesoderm** and endoderm arise by the invagination of cells that move from the upper layer of the cell mass *inward,* along the edges of a furrow that appears at the midline of the embryo. The site of this invagination appears as a slit in the embryo, called the **primitive streak** (see table 25.2).

During gastrulation, about half of the cells of the blastocyst cell mass move into the interior of the human embryo. This movement largely determines the future development of the embryo. By the end of gastrulation, distribution of cells into the three primary germ layers has been completed. The ectoderm is destined to form the epidermis and neural tissue. The mesoderm is destined to form the connective tissue,

muscle, and vascular elements. The endoderm forms the lining of the gut and its derivative organs (table 25.1).

Neurulation: Determination of Body Architecture

In the third week of embryonic development, the three primary cell types begin their development into the tissues and organs of the body. As in all vertebrates, this begins with the formation of two characteristic vertebrate features, the notochord and the hollow neural tube. This stage in development is called **neurulation.**

The first of these two structures to form is the **notochord,** a flexible rod. Soon after gastrulation is complete, it forms from mesoderm tissue along the midline of the embryo, below its dorsal surface. After the notochord has been formed, the **neural tube** forms from the region of the ectoderm that is located above the notochord and later differentiates into the spinal cord and brain (table 25.2). Just before the neural tube closes, two strips of cells break away and form the **neural crest.** These neural crest cells give rise to neural structures found in the vertebrate body.

While the neural tube is forming from ectoderm, the rest of the basic architecture of the human body is being rapidly determined by changes in the mesoderm. On either side of the developing notochord, segmented blocks of tissue form. Ultimately, these blocks, or **somites,** give rise to the muscles, vertebrae, and connective tissues. As development continues, more and more somites are formed. Within another strip of mesoderm that runs alongside the somites, many of the significant glands of the body, including the kidneys, adrenal glands, and gonads, develop. The remainder of the mesoderm layer moves out and around the inner endoderm layer of cells and eventually surrounds it entirely. As a result, the mesoderm forms two layers. The outer layer is associated with the body wall and the inner layer is associated with the gut. Between these two layers of mesoderm is the **coelom,** which becomes the body cavity of the adult.

By the end of the third week, over a dozen somites are evident, and the blood vessels and gut have begun to develop. At this point the embryo is about 2 millimeters (less than a tenth of an inch) long.

25.6 The vertebrate embryo develops in three stages. In cleavage, a hollow ball of cells forms. In gastrulation, cells move into the interior, forming the primary tissues. In neurulation, the organs of the body form.

TABLE 25.2 STAGES OF MAMMALIAN DEVELOPMENT

	Stage	Description
Sperm — Ovum — (Fertilization illustration)	Fertilization	The haploid male and female gametes fuse to form a diploid zygote.
Blastocyst — Inner cell mass — Trophoblast — Blastocoel (Cleavage illustration)	Cleavage	The zygote rapidly divides into many cells, with no overall increase in size. These divisions affect future development, because different cells receive different portions of the egg cytoplasm and, hence, different regulatory signals.
Amniotic cavity — Ectoderm — Endoderm (Gastrulation illustration)	Gastrulation	The cells of the embryo move, forming three primary germ layers: ectoderm and endoderm form first, followed by the formation of mesoderm.
Primitive streak — Ectoderm — Mcsoderm — Endoderm — Formation of extraembryonic membranes		
Neural groove — Notochord	Neurulation	In all chordates, the first organ to form is the notochord; the second is the neural tube.
Neural crest — Neural tube — Notochord		During neurulation, the neural crest is produced as the neural tube is formed. The neural crest gives rise to several uniquely vertebrate structures such as sensory neurons, sympathetic neurons, Schwann cells, and other cell types.
(embryo illustration)	Organogenesis	Cells from the three primary cell layers combine in various ways to produce the organs of the body.

25.7 Fetal Development

The Fourth Week: Organ Formation

In the fourth week of pregnancy, the body organs begin to form, a process called **organogenesis** (figure 25.20*a*). The eyes form, and the heart begins a rhythmic beating and develops four chambers. At 70 beats per minute, the little heart is destined to beat more than 2.5 billion times during a lifetime of about 70 years. More than 30 pairs of somites are visible by the end of the fourth week, and the arm and leg buds have begun to form. The embryo more than doubles in length during this week, reaching about 5 millimeters.

By the end of the fourth week, the developmental scenario is far advanced, although most women are not yet aware that they are pregnant. This is a crucial time in development because the proper course of events can be interrupted easily. For example, alcohol use by pregnant women during the first

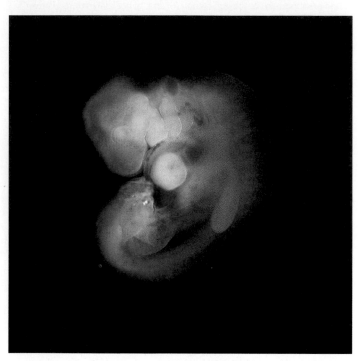

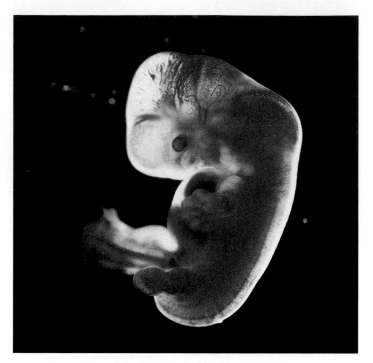

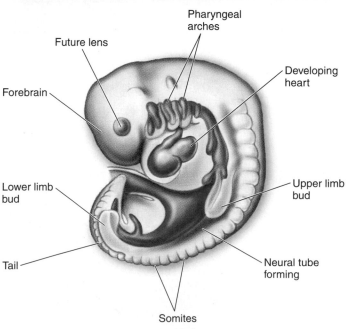

(a)

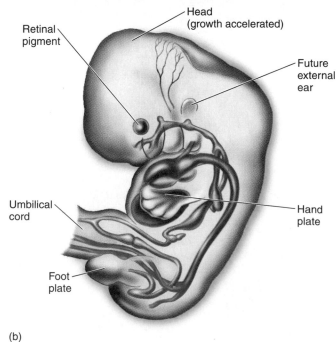

(b)

Figure 25.20 The developing human.

(*a*) Four weeks; (*b*) seven weeks; (*c*) three months; and (*d*) four months.

months of pregnancy is one of the leading causes of birth defects, producing **fetal alcohol syndrome,** in which the baby is born with a deformed face and often severe mental retardation. One in 250 newborns in the United States is affected with fetal alcohol syndrome. Also, most spontaneous abortions (miscarriages) occur during this period.

The Second Month: The Embryo Takes Shape

During the second month of pregnancy, great changes in morphology occur as the embryo takes shape (figure 25.20*b*). The miniature limbs of the embryo assume their adult shapes. The arms, legs, knees, elbows, fingers, and toes can all be seen—

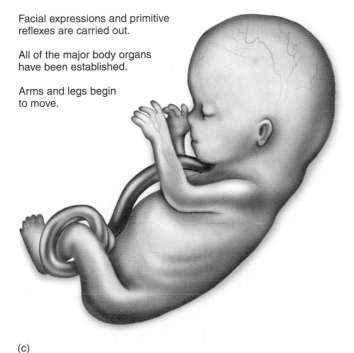

Development is essentially complete.

The developing human is now referred to as a fetus.

Facial expressions and primitive reflexes are carried out.

All of the major body organs have been established.

Arms and legs begin to move.

(c)

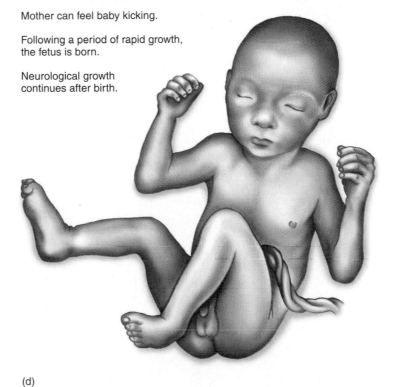

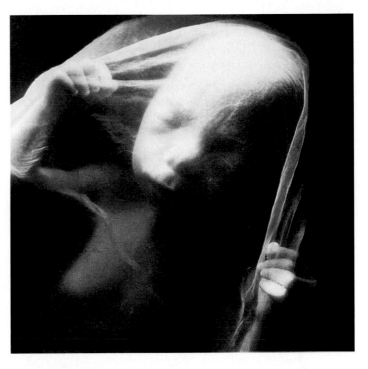

Bones actively enlarge.

Mother can feel baby kicking.

Following a period of rapid growth, the fetus is born.

Neurological growth continues after birth.

(d)

Figure 25.20 (continued)

as well as a short, bony tail. The bones of the embryonic tail, an evolutionary reminder of our past, later fuse to form the coccyx, or tailbone. Within the body cavity, the major internal organs are evident, including the liver and pancreas. By the end of the second month, the embryo has grown to about 25 millimeters in length—it is 1 inch long. It weighs perhaps a gram and is beginning to look distinctly human.

The Third Month:
Completion of Development

Development of the embryo is essentially complete except for the lungs and brain. The lungs don't complete development until the third trimester and the brain continues to develop even after birth. From this point on, the developing human is referred to as a **fetus** rather than an embryo. What remains is essentially growth. The nervous system and sense organs develop during the third month. The fetus begins to show facial expressions and carries out primitive reflexes such as the startle reflex and sucking. By the end of the third month, all of the major organs of the body have been established and the arms and legs begin to move (figure 25.20c).

The Second Trimester:
The Fetus Grows in Earnest

The second trimester is a time of growth. In the fourth (figure 25.20d) and fifth months of pregnancy, the fetus grows to about 175 millimeters in length (almost 7 in. long), with a body weight of about 225 grams. Bone formation occurs actively during the fourth month. During the fifth month, the head and body become covered with fine hair. This downy body hair, called **lanugo,** is

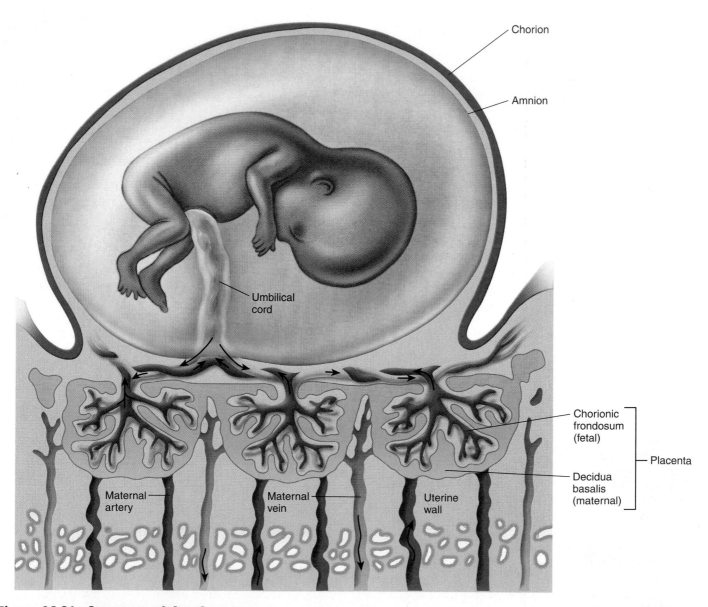

Figure 25.21 Structure of the placenta.

The placenta contains a fetal component, the chorionic frondosum, and a maternal component, the decidua basalis. Oxygen and nutrients enter the fetal blood from the maternal blood by diffusion. Waste substances enter the maternal blood from the fetal blood, also by diffusion.

another evolutionary relic and is lost later in development. By the end of the fourth month, the mother can feel the baby kicking; by the end of the fifth month, she can hear its rapid heartbeat with a stethoscope.

In the sixth month, growth accelerates. By the end of the sixth month, the baby is over 0.3 meters (1 ft) long and weighs 0.6 kilograms (about 1.5 lb)—and most of its prebirth growth is still to come. At this stage, the fetus cannot yet survive outside the uterus without special medical intervention.

The Third Trimester: The Pace of Growth Accelerates

The third trimester is a period of rapid growth. In the seventh, eighth, and ninth months of pregnancy, the weight of the fetus more than doubles. This increase in bulk is not the only kind of growth that occurs. Most of the major nerve tracts are formed within the brain during this period, as are new brain cells.

All of this growth is fueled by nutrients provided by the mother's bloodstream, passing into the fetal blood supply within the placenta (figure 25.21). The undernourishment of the fetus by a malnourished mother can adversely affect this growth and result in severe retardation of the infant. Retardation resulting from fetal malnourishment is a severe problem in many underdeveloped countries, where poverty and hunger are common.

By the end of the third trimester, the neurological growth of the fetus is far from complete and, in fact, continues long after birth. But by this time the fetus is able to exist on its own. Why doesn't the fetus continue to develop within the uterus until its neurological development is complete? What's the rush to get out and be born? Because physical growth is continuing as well, and the fetus is about as large as it can get and still be delivered through the pelvis without damage to mother or child. As any woman who has had a baby can testify, it is a tight fit. Birth takes place as soon as the probability of survival is high.

Postnatal Development

Growth continues rapidly after birth. Babies typically double their birth weight within a few months. Different organs grow at different rates, however. The reason that adult body proportions are different from infant ones is that different parts of the body grow or cease growing at different times, a growth pattern called **allometric growth** (figure 25.22). At birth, the developing nervous system is generating new nerve cells at an average rate of more than 250,000 per minute. Then, about six months after birth, this astonishing production of new neurons ceases permanently. Because both jaw and skull continue to grow at the same rate, the proportions of the head do not change after birth. That is why a young human fetus seems so incredibly adultlike. The fact that the human brain continues to grow significantly for the first few years of postnatal life means that adequate nutrition and a safe environment are particularly crucial during this period for the full development of a person's intellectual potential.

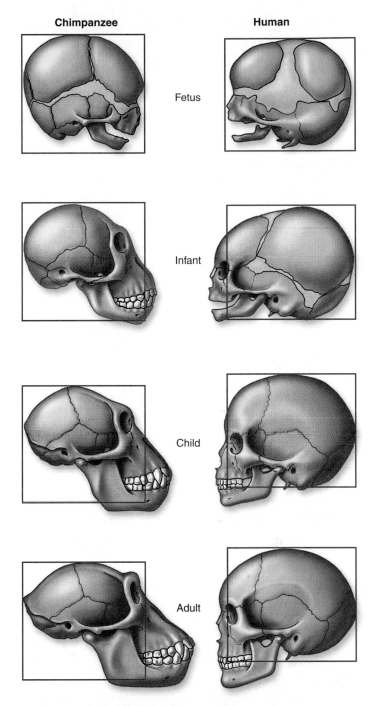

Chimpanzee   Human

Fetus

Infant

Child

Adult

Figure 25.22   Allometric growth.

In young chimpanzees, the jaw grows at a faster rate than the rest of the head. As a result, the adult chimpanzee head shape differs greatly from its head shape as a newborn. In humans, the difference in growth between the jaw and the rest of the head is much smaller, and the adult head shape is similar to that of the newborn.

25.7   Most of the key events in fetal development occur early. Organs begin to form in the fourth week, and by the end of the second month the developing body looks distinctly human. While many women are not yet aware they are pregnant, development of the embryo is essentially complete. What remains is primarily growth.

⚛ Science in Action

Why You Age and Cancer Cells Don't

Death is an integral part of the life cycle, its inevitable end. All organisms die. However, while each of us knows we shall die someday, few of us can escape wishing we could delay the process. Some succeed. The oldest documented living person, Marie-Louise Febronie Meilleur of Ontario, Canada, reached the age of 117 years in 1997. The tantalizing possibility of long life that she represents is one reason why there is such interest in the aging process—if we knew enough about it, perhaps we could slow it. A wide variety of theories have been advanced to explain why we age. In recent years, scientists have come a long way toward unraveling one aspect of the puzzle.

The first clue was the discovery that cells appear to die on schedule, as if following a script. In a famous experiment carried out in 1961, geneticist Leonard Hayflick demonstrated that fibroblast cells growing in tissue culture will divide only a certain number of times. After about 50 population doublings, cell division stops, the cell cycle blocked just before DNA replication. If a cell sample is taken after 20 doublings and frozen, when thawed it resumes growth for 30 more doublings, then stops.

An explanation of the "Hayflick limit" was suggested in 1986 when cell biologist Howard Cooke first glimpsed an extra length of DNA at the ends of chromosomes. These so-called telomeric regions, about 5,000 nucleotides long, are each composed of several thousand repeats of the sequence TTAGGG. Cooke found the telomere region to be substantially shorter in body tissue chromosomes than in those of germ-line cells, eggs and sperm. He speculated that in body cells a portion of the telomere cap was lost by a chromosome during each cycle of DNA replication.

Cooke was right. The cell machinery that replicates the DNA of each chromosome sits on the last 100 units of DNA at the chromosome's tip, and so cannot copy that bit. As a consequence, each time the cell divides, its chromosomes get a little shorter. Eventually, after some 50 replication cycles, the telomeric cap is used up, and the cell line then enters senescence (old age), no longer able to proliferate.

How do our sperm and egg avoid this trap, dividing continuously for decades? Scientists have recently learned that all human cells contain an enzyme, telomerase, which lengthens telomeres. This enzyme is active in sperm and egg cells, maintaining their chromosome telomeres at a constant length of 5,000 nucleotides. In body cells, by contrast, the telomerase gene is silent.

Research published in the last few years has confirmed Cooke's hypothesis, providing direct evidence of a causal relationship between telomeric shortening and cell aging. Using genetic engineering, teams of researchers from California and Texas transferred into human body cell cultures a DNA fragment that unleashes each cell's telomerase gene. The result was unequivocal. New telomeric caps were added to the chromosomes of the cells, and the cells with the artificially elongated telomeres did not senesce at the Hayflick limit, continuing to divide in a healthy and vigorous manner for more than 20 additional generations.

This research confirms the hypothesis that loss of telomere DNA eventually restrains the ability of human cells to proliferate—and yet every human cell possesses a copy of the telomerase gene that, if expressed, would rebuild the telomere. Why do our cells accept aging, if they need not? The answer, it seems, is to avoid cancer. By limiting the number of cell divisions allotted to human cell lines, the body ensures that no cell can continue to

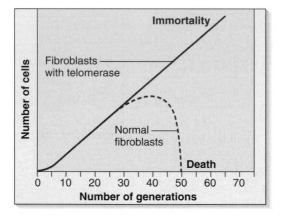

divide indefinitely. Most adult cells are but a few cell generations from the Hayflick limit. Uncontrolled proliferation of a normal cell would soon grind to a halt. When scientists examine cancer cells, they commonly find that the telomerase genes in the cancer cells have been activated, and are maintaining telomeres at full length. Thus, the telomere shortening that occurs in normal cells is a tumor-suppressing mechanism and one of your body's key safeguards against cancer.

Aging, then, is at least partly a strategy to avoid the inevitable consequence of wear and tear to our genes, producing mutations that sooner or later lead to cancer. A powerful source of such mutations are free radicals, highly reactive fragments of atoms with unpaired electrons that shear through DNA like a shotgun blast. Free radicals are an unavoidable by-product of how our bodies use oxygen to metabolize food. Because free radicals are so destructive, every cell has numerous mechanisms to control and eliminate them. If all else fails to eliminate free radicals, the cell activates a second key safeguard against cancer. A special fail-safe gene called *p66* pulls the plug, causing the cell to commit suicide rather than permit free radicals to damage the DNA and produce cancer.

Interestingly, Italian researchers have found that mice with disabled *p66* genes lived longer than normal mice. Their cells were no longer being zapped by the hair-trigger *p66* self-destruct mechanism. Inhibitory drugs of the p66 class of proteins are known. If you are willing to accept the added risk of cancer, perhaps taking them would, like the Italian mice, let you live longer.

25.8 Contraception and Sexually Transmitted Diseases

Contraception

Not all couples want to initiate a pregnancy every time they have sex, yet sexual intercourse may be a necessary and important part of their emotional lives together. The solution to this dilemma is to find a way to avoid reproduction without avoiding sexual intercourse, an approach that is commonly called **birth control,** or contraception. Several different birth control methods are currently available (figure 25.23). These methods differ from one another in their effectiveness and in their acceptability to different couples.

Abstinence. The simplest and most reliable way to avoid pregnancy is not to have sex at all. Of all methods of birth control, this is the most certain—and the most limiting, because it denies a couple the emotional support of a sexual relationship. A variant of this approach is to avoid sex only on the two days preceding and following ovulation, because this is the period during which successful fertilization is likely to occur. The rest of the sexual cycle is considered relatively "safe" for intercourse. This approach, called the rhythm method, or **natural family planning** when other indicators are also monitored, is satisfactory in principle but difficult in application because ovulation is not easy to predict and may occur unexpectedly. Failure rate can be as high as 20% to 30%.

Prevention of Egg Maturation. Since about 1960, a widespread form of birth control in the United States has been the daily ingestion of hormones, or **birth control pills.** These pills contain estrogen and progesterone. These hormones shut down production of the pituitary hormones FSH and LH, fooling the body into acting as if ovulation had already occurred, when in fact it has not. The ovarian follicles do not ripen in the absence of FSH, and ovulation does not occur in the absence of LH. Other methods of delivering hormones that prevent egg maturation include medroxy progesterone (Depo-Provera), which is injected every one to three months, the seven-day birth control patch, which releases the hormones through the skin, and surgically implanted capsules that release hormones. Failure rates of these methods are less than 2%.

Prevention of Embryo Implantation. The insertion of a coil or other irregularly shaped object into the uterus is an effective means of birth control because the irritation in the uterus prevents the implantation of the descending embryo within the uterine wall. Such **intrauterine devices (IUDs)** are very effective, with a failure rate of less than 2%. Their high degree of effectiveness, like surgically implanted hormones, undoubtedly reflects their being "no-brainers"—once they are inserted, they can be forgotten.

A chemical means of preventing embryo implantation or ending a early pregnancy is RU486, or the "morning after pill." This pill, taken after sexual intercourse, blocks the action of progesterone, causing the endometrium to slough off.

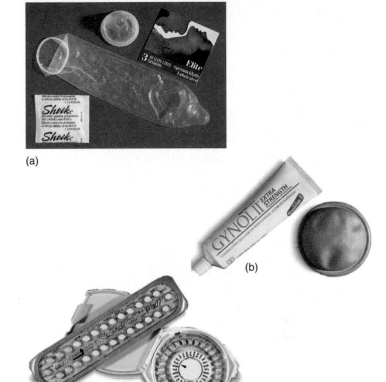

(a)

(b)

(c)

Figure 25.23 Three common birth control methods.
(a) Condom; (b) diaphragm and spermicidal jelly; and (c) oral contraceptives.

RU486 must be administered under a doctor's care because of potentially serious side effects.

Sperm Blockage. If sperm is not delivered to the uterus, fertilization cannot occur. One way to prevent the delivery of sperm is to encase the penis within a thin rubber bag, or **condom.** In principle, this method is easy to apply and foolproof, but in practice, it proves to be less effective than you might expect, with a failure rate of up to 15%. A second way to prevent the entry of sperm is to cover the cervix with a rubber dome called a **diaphragm,** inserted immediately before intercourse. Because the dimensions of individual cervices vary, diaphragms must be fitted by a physician. Failure rates average 20%.

Sperm Destruction. A third general approach to birth control is to remove or destroy the sperm after ejaculation. Sperm can be destroyed within the vagina with **spermicidal jellies, suppositories,** and **foams.** These require application immediately before intercourse. The failure rate varies widely, from 3% to 22%.

Sterilization. Sterilization is the surgical removal of portions of the tubes that transport the gametes from the testes or ovaries. A vasectomy in males involves removing a segment of the vas deferens from each testis (see figure 25.11). A tubal ligation in females involves removal of a section of each fallopian tube (see figure 25.16).

Sexually Transmitted Diseases

Sexually transmitted diseases (STDs) are diseases that spread from one person to another through sexual contact. AIDS, discussed in chapter 22, is a deadly viral STD. Other significant sexually transmitted diseases include:

Gonorrhea. Caused by the bacterium *Neisseria gonorrhoeae,* this disease is on the increase worldwide. The primary symptom is discharge from the penis or vagina. It can be treated with antibiotics. If left untreated in women, gonorrhea can cause pelvic inflammatory disease (PID), a condition in which the fallopian tubes become scarred and blocked. PID can eventually lead to sterility.

Chlamydia. Caused by the bacterium *Chlamydia trachomatis,* this disease is sometimes called the "silent STD" because women usually experience no symptoms until after the infection has become established. Like gonorrhea, chlamydia can cause PID in women if left untreated with antibiotics.

Syphilis. Caused by the bacterium *Treponema pallidum,* this disease is one of the most potentially devastating STDs. The disease progresses in four stages. In the first stage, a small, painless lesion called a chancre appears on the site of infection, usually on the penis or hidden in the vagina. In the second stage, a pink rash appears over the body. In the third stage, symptoms may come and go for several years with the person being contagious even when symptoms aren't present. The fourth stage is the most debilitating, with heart disease, mental deficiency, and nerve damage that may include loss of motor function or blindness.

Genital herpes. Caused by the herpes simplex virus type 2 (HSV-2), this disease is the most common STD in the United States. The virus causes red blisters on the penis or on the labia, vagina, or cervix that rupture and scab over. The lesions heal, but the virus travels to the dorsal root ganglion by way of the sensory neurons. They become dormant in the dorsal root ganglion, from which they can spread out to other parts of the body forming similar lesions. An infected person is always contagious when the lesions are present, but can also be contagious even when not exhibiting lesions. Medication can reduce the symptoms but cannot prevent the spread or recurrences of outbreaks.

This summary of STDs may give the impression that sexual activity is fraught with danger, and when practiced with unknown partners it is. In light of this, it is folly not to take precautions to avoid STDs. The best way to avoid STDs is to avoid sex. If sexually active, it is important to know one's sexual partner well enough to discuss the possible presence of an STD. Condom use can also prevent transmission of most of the diseases. Anyone responsible enough to have sex should be responsible enough to protect themselves.

> **25.8** A variety of birth control methods are available, many of them quite effective. Sexually transmitted diseases are spread through sexual contact.

Exploring Current Issues

Additional Resources

Go to your campus library or look online to find the following articles, which further develop some of the concepts found in this chapter.

Centers for Disease Control and Prevention, Department of Health and Human Services. (2004). Sexually transmitted diseases. *Retrieved from www.cdc.gov/node.do/id/0900f3ec80009a98.*

Herndon, E. J. and M. Zieman. (2004). New contraceptive options. *American Family Physician,* 69(4), 853.

Hodgkinson, N. (2003). AIDS: scientific or viral catastrophe? *Journal of Scientific Exploration,* 17(1), 87.

Kingsland, J. (2004). Sperm warfare: forget everything you have heard about the male pill. A new generation of smart pills in on the drawing board, says James Kingsland, and they will target sperm with military precision. *New Scientist,* 181(2429), 38.

Westphal, S. P. (2003). The next IVF revolution? We could be on the verge of creating children from artificial eggs and sperm made in a lab dish, if work on mice can be repeated with human cells. *New Scientist,* 178(2394), 4.

In the News: Stem Cells and Cloning

One of the ethically most controversial and scientifically promising areas of research concerns embryonic stem cells and their potential use to repair damaged tissues. Therapeutic cloning, or somatic cell nuclear transfer, where the embryonic stem cells are isolated from a clone of the patient's own tissue, is particularly exciting.

Article 1. *Stem cell research is a new frontier full of ethical and political questions.*

Article 2. *Scientists use transplanted stem cells to reverse juvenile diabetes in mice.*

Article 3. *Ethics and science collide over the therapeutic cloning of human stem cells.*

Article 4. *Cloning humans is not going to work until scientists solve a key problem.*

Article 5. *Newspaper reports of genetically modified humans are misleading.*

Find these articles, written by the author, online at www.mhhe.com/tlwessentials/exp25.

Vertebrate Reproduction

25.1 Asexual and Sexual Reproduction

- Asexual reproduction through fission (**figure 25.1**) or budding is the primary means of reproduction among protists and some animals, but most animals reproduce sexually.

- Although sexual reproduction is most common in animals, parthenogenesis and hermaphroditism are variations seen in the animal kingdom. Parthenogenesis is reproduction where offspring are produced from unfertilized eggs. Hermaphroditism is a reproductive strategy where an individual has both testes and ovaries, producing both sperm and eggs. Some hermaphrodites can self-fertilize.

25.2 Evolution of Reproduction Among the Vertebrates

- Fertilization in most fish and amphibians occurs externally (**figures 25.5 and 25.6**), but internal fertilization occurs in most other vertebrates. Even with internal fertilization, development may occur inside or outside of the mother (**figure 25.7**). Birds and reptiles lay watertight eggs, whereas almost all mammals are viviparous, giving birth to live young.

The Human Reproductive System

25.3 Males

- Sperm is produced in the testes (**figure 25.11**). The testes contain a large number of tightly coiled tubes called seminiferous tubules, where sperm develop in a process called spermatogenesis (**figure 25.12**). As sperm develop, they move to the center of the tubules, where they pass into the epididymis. Once matured, they are stored in the vas deferens. During sexual intercourse, sperm are delivered through the penis (**figure 25.14**) into the female.

25.4 Females

- Female gametes, eggs or ova, develop in the ovary from oocytes (**figure 25.15**). A female is born with some 2 million oocytes, all arrested in meiosis. The hormone FSH initiates the resumption of meiosis in a few oocytes, but usually only one completes meiosis and becomes a mature egg each month. The egg ruptures from the ovary, called ovulation, and enters the fallopian tube (**figure 25.16**). Sperm deposited in the vagina travel up into the fallopian tube, where fertilization occurs.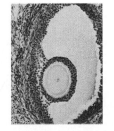

25.5 Hormones Coordinate the Reproductive Cycle

- The follicular phase begins with the secretion of FSH, which triggers oocyte maturation and secretion of estrogen. The luteal phase begins with the secretion of LH, which causes ovulation and the formation of the corpus luteum. The corpus luteum begins secreting progesterone, which acts in preparing the uterus for implantation of the zygote (**figures 25.18 and 25.19**).

- If a zygote implants in the lining of the uterus (**figure 25.19**), estrogen and progesterone levels remain high due to the release of human chorionic gonadotropin (hCG) from the embryo. The uterus is maintained and no further egg maturation occurs. If fertilization does not occur, estrogen and progesterone levels drop and the endometrial lining of the uterus sloughs off (menstruation).

The Course of Development

25.6 Embryonic Development

- The vertebrate embryo develops in three stages. The first stage, called cleavage, involves hundreds of cell divisions, eventually producing a hollow ball of cells called a blastocyst. The second stage, called gastrulation, involves the orchestrated movement of cells, forming the three germ layers. The third stage is neurulation, where the notochord and neural tube form (**table 25.2**).

25.7 Fetal Development

- Organs begin forming by the fourth week, and by the end of the second month the embryo looks distinctly human (**figure 25.20**). By the end of the third month, all major organs except the brain and lungs are developed. The second and third trimesters are periods of considerable growth.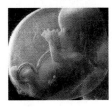

Birth Control and Sexually Transmitted Diseases

25.8 Contraception and Sexually Transmitted Diseases

- Various birth control methods are available and work by preventing egg maturation, preventing embryo implantation, blocking or killing sperm, or sterilization.

- Sexually transmitted diseases are spread through sexual contact. AIDS is a deadly STD, but there are other STDs that may not be as fatal as AIDS but are quite destructive, especially if left untreated.

1. If the offspring are genetically similar but not identical to each other and the parent, then the organism reproduces through
   a. parthenogenesis. c. budding.
   b. sexual reproduction. d. fission.

2. For terrestrial vertebrates, dehydration was a selection pressure that caused the evolution of
   a. parthenogenesis. c. budding.
   b. external fertilization. d. internal fertilization.

3. Kangaroos show
   a. viviparity.   c. oviparity.
   b. ovoviviparity.   d. parthenogenesis.
4 Temperature regulation of spermatogenesis in human males is controlled by the position of the
   a. seminiferous tubules. c. vas deferens.
   b. epididymis.   d. scrotum.
5. Gametogenesis in human males and females requires the hormones
   a. estrogen and testosterone.
   b. FSH and LH.
   c. progesterone and testosterone.
   d. oxytocin and prolactin.
6. When pregnancy occurs, the endometrium is maintained by the
   a. embryo releasing hCG.
   b. corpus luteum releasing progesterone.
   c. hypothalamus releasing GnRH.
   d. ovary releasing estrogen.

7. Initiation of labor and childbirth is controlled by the hormone
   a. estrogen.   c. oxytocin.
   b prolactin.   d. progesterone.
8. A human embryo has formed the three germ layers from which all tissues form by the time
   a. the blastula forms.
   b. neurulation is complete.
   c. the blastocyst forms.
   d. gastrulation is complete.
9. In a developing human, the first tissues to begin forming are the
   a. skeletal.   c. nervous.
   b. muscular.   d. digestive.
10. Premature childbirth is dangerous for a human baby because the
   a. respiratory system is not fully formed.
   b. circulatory system is not fully formed.
   c. digestive system is not fully formed.
   d. excretory system is not fully formed.

Visual Understanding

1. **Figures 25.7 and 25.10b** These figures represent some of the more unusual methods that some organisms use to ensure the survival of their offspring. Briefly describe them.

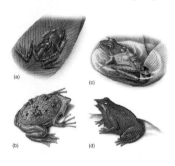

2. **Figure 25.17** Observe the shapes of the uteri shown in the figure. Speculate on the number of offspring the individuals are able to gestate at one time.

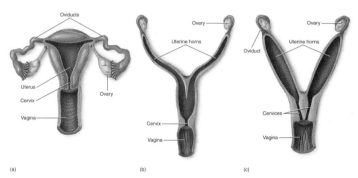

Challenge Questions

Vertebrate Reproduction Compare the advantages and disadvantages of external and internal fertilization.

The Human Reproductive System Speculate on why males produce so many sperm and females have, comparatively, so few eggs.

The Course of Development Explain some of the ways that a woman's use of drugs—alcohol, cocaine, prescription drugs, over-the-counter drugs—can affect the embryo or fetus that she is carrying.

Birth Control and Sexually Transmitted Diseases Your friend, Paloma, tells you that the most reliable method of contraception and prevention of sexually transmitted disease is communication. Explain what she means.

Online Learning Center

Visit the Online Learning Center for this chapter at www.mhhe.com/tlwessentials/ch25 for quizzes, animations, interactive learning exercises, and other study tools. At the site you will also find extended answers to the end-of-chapter questions.

26

Plant Form and Function

Of all the many kinds of plants, trees are the largest, rising high above the surrounding vegetation to capture the sun's rays. This hardwood forest, green with the leaves of spring, can be thought of as an enormous photosynthesis machine, each of its trees competing with its neighbors for light and soil nutrients, and putting the raw materials it captures to work producing the organic molecules necessary for growth and reproduction. Typical of plants, a tree captures light with the green pigment chlorophyll, which gives its leaves their characteristic color. A tree captures soil nutrients with its roots, which spread out in a fine network through the surrounding soil. Connecting the leaves of a tree with its roots is the stem, the massive, tall woody cylinder that makes up most of the mass of the tree. The stem of a tall tree is an engineering marvel, piping water and dissolved soil nutrients to the leaves many meters higher up in the air, and sending back down to the roots the carbohydrates produced by photosynthesis in the leaves. In this chapter, we will journey through the plant body, one of nature's most interesting creations.

26.1 Organization of a Vascular Plant

The similarities between a cactus, an orchid, and a pine tree may not at first be obvious. However, most plants possess the same fundamental architecture and the same three major groups of organs—roots, stems, and leaves. The organs of vascular plants have an outer covering of protective tissue and an inner matrix of tissue. Within the inner matrix is embedded vascular tissue, which conducts water, minerals, and food throughout the plant. The cells and tissues of vascular plants, and how the plant body carries out the functions of living, are the focus of this chapter. We discuss the funda-mental differences between the roots, which are usually be-lowground, and the shoot, which is typically aboveground. We also examine the structural and functional relationships between them.

A vascular plant is organized along a vertical axis. The part belowground is called the **root,** and the part above-ground is called the **shoot** (although in some instances roots may extend above the ground, and some shoots can extend below it) (figure 26.1). Although roots and shoots differ in their basic structure, growth at the tips throughout the life of the individual is characteristic of both. The root penetrates the soil and absorbs water and various ions, which are cru-cial for plant nutrition. It also anchors the plant. The shoot consists of stem and leaves. The **stem** serves as a framework for the positioning of the **leaves,** where most photosynthesis takes place. The arrangement, size, and other characteristics of the leaves are critically important in the plant's produc-tion of food. Flowers, and ultimately fruits and seeds, are also formed on the shoot.

Meristems

Animals grow all over, but plants do not. As children grow into adults, their torsos grow at the same time their legs do. If, instead, children grew in only one place, with only their legs getting longer and longer, they would be growing in a way similar to the way plants grow.

Plants contain growth zones of unspecialized cells called **meristems,** clumps of small cells with dense cytoplasm and proportionately large nuclei that act as stem cells do in animals. That is, one cell divides to give rise to two cells. One remains meristematic, while the other is free to differentiate and contrib-ute to the plant body. In this way, the population of meristem cells is continually renewed. Molecular genetic evidence sup-ports the hypothesis that stem cells and meristem cells may also share some common pathways of gene expression.

In plants, **primary growth** is initiated at the tips by the **apical meristems,** regions of active cell division that occur at the tips of roots and shoots. The growth of these meristems results primarily in the extension of the plant body. As it elon-gates, it forms what is known as the primary plant body, which is made up of the primary tissues.

Growth in thickness, **secondary growth,** involves the activity of the **lateral meristems,** which are cylinders of mer-istematic tissue. The continued division of their cells results primarily in the thickening of the plant body. There are two kinds of lateral meristems: the **vascular cambium,** which gives rise to ultimately thick accumulations of secondary xy-lem and phloem, and the **cork cambium,** from which arise the outer layers of bark on both roots and shoots.

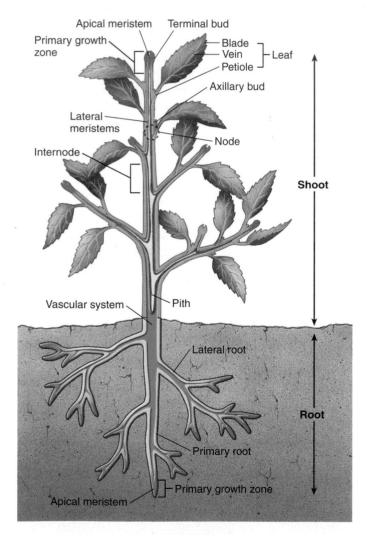

Figure 26.1 The body of a plant.
The plant body consists of an aboveground portion called the shoot (stems and leaves) and a belowground portion called the root. Elongation of the plant, so-called primary growth, takes place when clusters of cells called the apical meristems (*lime green areas*) divide at the ends of the roots and the stems. Thickening of the plant, so-called secondary growth, takes place in the lateral meristems of the stem, allowing the plant to enlarge in girth like letting out a belt.

26.1 The body of a vascular plant is a continuous structure, a grouping of tubes connecting roots to leaves, with growth zones called meristems.

26.2 Plant Tissue Types

The organs of a plant—the roots, stem, leaves, and in some cases, flowers and fruits—are composed of different combinations of tissues, just as your legs are composed of bone, muscle, and connective tissue. A tissue is a group of similar cells—cells that are specialized in the same way—organized into a structural and functional unit (see figure 1.4). Most plants have three major tissue types: (1) *ground tissue,* in which the vascular tissue is embedded; (2) *dermal tissue,* the outer protective covering of the plant; and (3) *vascular tissue,* which conducts water and dissolved minerals up the plant and conducts the products of photosynthesis throughout.

Each major tissue type is composed of distinctive kinds of cells, whose structures are related to the functions of the tissues in which they occur. For example, vascular tissue is composed of *xylem,* which conducts water and dissolved minerals, and *phloem,* which conducts carbohydrates (mostly sugars), which the plant uses as food.

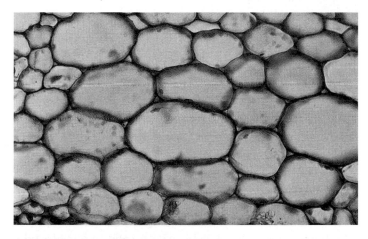

Figure 26.2   Parenchyma cells.
Cross section of parenchyma cells from grass. Only thin primary cell walls are seen in this living tissue.

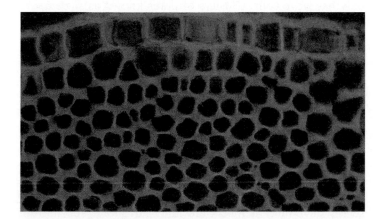

Figure 26.3   Collenchyma cells.
Cross section of collenchyma cells, with thickened side walls, from a young branch of elderberry (*Sambucus*). In other kinds of collenchyma cells, the thickened areas may occur at the corners of the cells or in other kinds of strips.

Ground Tissue

Parenchyma cells are the least specialized and the most common of all plant cell types (figure 26.2); they form masses in leaves, stems, and roots. Parenchyma cells, unlike some other cell types, are characteristically alive at maturity, with fully functional cytoplasm and a nucleus. They carry out the functions of photosynthesis and food and water storage. The edible parts of most fruits and vegetables are composed of parenchyma cells. They are capable of cell division and are important in cell regeneration and wound healing. Most parenchyma cells have only thin cell walls called **primary cell walls,** which are mostly cellulose that is laid down while the cells are still growing.

Collenchyma cells, which are also living at maturity, form strands or continuous cylinders beneath the epidermis of stems or leaf stalks and along veins in leaves. They are usually elongated, with unevenly thickened primary walls, which are their distinguishing feature. Strands of collenchyma provide much of the support for plant organs in which secondary growth has not occurred. These cells also provide the plant with flexibility due to the uneven nature of the walls, the thinner areas becoming flex points (figure 26.3).

In contrast to parenchyma and collenchyma cells, **sclerenchyma cells** have tough, thick cell walls called **secondary cell walls;** they usually do not contain living cytoplasm when mature. The secondary cell wall is laid down inside of the primary cell wall after the cell has stopped growing and expanding in size. The secondary cell wall provides cells with strength and rigidity. There are two types of sclerenchyma: **fibers,** which are long, slender cells that usually form strands, and **sclereids,** which are variable in shape but often branched. Sclereids are sometimes called "stone cells" because they make up the bulk of the stones of peaches and other "stone" fruits (figure 26.4), as well as that of nut shells. Both fibers and sclereids are thick-walled and strengthen the tissues in which they occur.

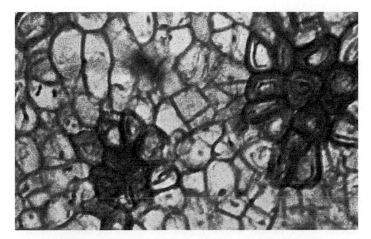

Figure 26.4   Sclerenchyma cells in sclereids.
Clusters of sclereids ("stone cells"), stained *red* in this preparation, in the pulp of a pear. The surrounding thin-walled cells, stained *bluish-green,* are parenchyma cells. These sclereid clusters give pears their gritty texture.

Dermal Tissue

All parts of the outer layer of a primary plant body are covered by flattened epidermal cells, which are often covered with a thick, waxy layer called the **cuticle.** These are the most abundant cells in the plant epidermis, or outer covering. They protect the plant and provide an effective barrier against water loss. One type of specialized cell that occurs among the epidermal cells is the guard cell.

Guard cells are paired cells with openings that lie between them called **stomata** (singular, stoma). Guard cells and stomata occur frequently in the epidermis of leaves and occasionally on other parts of the shoot, such as on stems or fruits (figure 26.5*a*). Oxygen, carbon dioxide, and water pass into and out of the leaves almost exclusively through the stomata, which open and shut in response to such external factors as supply of moisture and light.

Trichomes are outgrowths of the epidermis that occur on the shoot, on the surfaces of stems and leaves (figure 26.5*b*). Trichomes vary greatly in form in different kinds of plants. A "fuzzy" or "woolly" leaf is covered with trichomes, which when viewed under the microscope look like a thicket of fibers. Trichomes play an important role in regulating the heat and water balance of the leaf, much as the hairs of a fur coat provide insulation. Other trichomes are glandular, secreting sticky or toxic substances that may deter potential herbivores.

Other outgrowths of the epidermis occur belowground, on the surface of roots near their tips. Called **root hairs,** these tubular extensions of single epidermal cells keep the root in intimate contact with the particles of soil (see figure 26.22). Root hairs play an important role in the absorption of water and minerals from the soil, by increasing the surface area of the root.

Vascular Tissue

Vascular plants contain two kinds of conducting, or vascular, tissue: the xylem and the phloem. **Xylem** is the plant's principal water-conducting tissue, forming a continuous system that runs throughout the plant body. Within this system, water (and minerals dissolved in it) passes from the roots up through the shoot in an unbroken stream. When water reaches the leaves, much of it passes into the air as water vapor, through the stomata.

The two principal types of conducting cells in the xylem are **tracheids** and **vessel elements,** both of which have thick secondary walls that are laid down inside the primary cell wall, are elongated, and have no living cytoplasm (they are dead) at maturity. In conducting elements composed of tracheids, water flows from tracheid to tracheid through openings called pits in the secondary walls. In contrast, vessel elements have not only pits but also definite openings, or perforations, in their end walls by which they are linked together and through which water flows. A linked row of vessel elements forms a vessel (figure 26.6). Primitive angiosperms and other vascular plants have only tracheids, but the majority of angiosperms have vessels. Vessels conduct water much more efficiently than do strands of tracheids. In addition to conducting cells, xylem includes fibers and parenchyma cells.

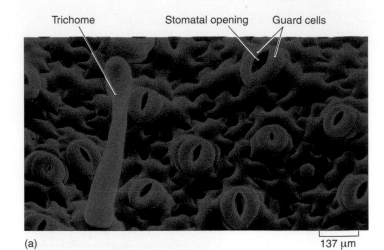

Trichome   Stomatal opening   Guard cells

(a)   137 μm

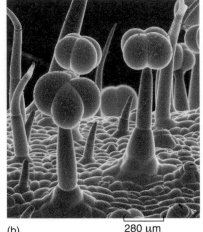

Figure 26.5   Guard cells and trichomes.
(*a*) Numerous stomata occur among the leaf epidermal cells of this tobacco (*Nicotiana tabacum*) leaf. A trichome is also visible. (*b*) Trichomes on a tomato plant (*Lycopersicon lycopersicum*).

(b)   280 μm

Phloem is the principal food-conducting tissue in vascular plants. Food conduction in phloem is carried out through two kinds of elongated cells: **sieve cells** and **sieve-tube members** (figure 26.7). Seedless vascular plants and gymnosperms have only sieve cells; most angiosperms have sieve-tube members, but at least one primitive angiosperm only has sieve cells. Clusters of pores known as *sieve areas* occur on both kinds of cells and connect the cytoplasms of adjoining sieve cells and sieve-tube members. Both cell types are living, but their nuclei are lost during maturation.

In sieve-tube members, some sieve areas have larger pores and are called *sieve plates.* Sieve-tube members occur end to end, forming longitudinal series called **sieve tubes.** Specialized parenchyma cells known as **companion cells** occur regularly in association with sieve-tube members. Companion cells apparently carry out some of the metabolic functions that are needed to maintain the associated sieve-tube member; their cytoplasms are connected to the sieve-tube members through openings called *plasmodesmata.* In an evolutionary sense, sieve-tube members are more advanced, specialized, and presumably more efficient than sieve cells.

26.2   Plants contain a variety of ground, dermal, and vascular tissues.

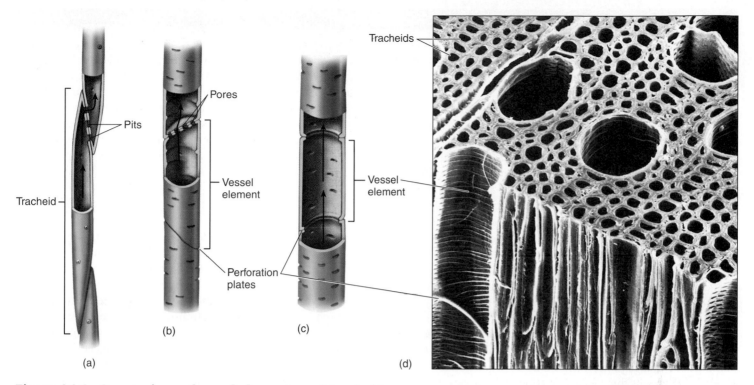

Figure 26.6   Comparison of vessel elements and tracheids.
(*a*) In tracheids the water passes from cell to cell by means of pits. (*b*) In vessel elements, water moves by way of perforation plates, which may be simple or interrupted by bars. (*c*) Open-ended vessel elements. (*d*) A scanning electron micrograph of the red maple (*Acer rubrum*) wood, showing the xylem.

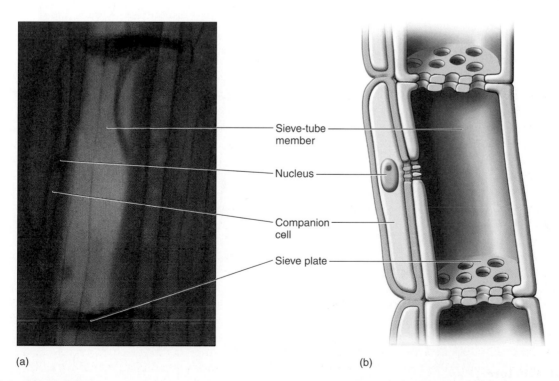

Figure 26.7   Sieve tubes.
(*a*) Sieve-tube member from the phloem of squash (*Cucurbita*), connected with the cells above and below to form a sieve tube. (*b*) In this diagram, note the thickened end walls, which are at right angles to the sieve tube. The narrow cell with the nucleus at the *left* of the sieve-tube member is a companion cell.

26.3   Roots

We now consider the three kinds of vegetative organs that form the body of a plant: **roots, stems,** and **leaves.** Roots have a simpler pattern of organization and development than do stems, and we will examine them first. Although different patterns exist, the kind of root described here and shown in figure 26.8*a* is found in many dicots. Roots contain xylem and phloem. Roots have a central column of xylem with radiating arms. Alternating with the radiating arms of xylem are strands of primary phloem. Surrounding the column of vascular tissue, and forming its outer boundary, is a cylinder of cells one or more cell layers thick called the **pericycle.** Branch, or lateral, roots are formed from cells of the pericycle. The outer layer of the root, as in the shoot, is the epidermis. The mass of parenchyma in which the root's vascular tissue is located is the cortex. Its innermost layer—the endodermis—consists of specialized cells that regulate the flow of water between the vascular tissues and the root's outer portion (figure 26.9). The endodermis is a single layer of cells that lies just outside of the pericycle. Endodermis cells are encircled by a thickened, waxy band called the **Casparian strip.** The Casparian strip directs the movement of water through the endodermis cells, regulating the plant's supply of minerals by controlling the passage of minerals into the xylem through the endodermis cells (figure 26.9*c*).

The apical meristem of the root divides and produces cells both inwardly, back toward the body of the plant, and outwardly. The three primary meristems are the **protoderm,** which becomes the epidermis; the **procambium,** which produces primary vascular tissues (primary xylem and primary phloem); and the **ground meristem,** which differentiates further into ground tissue, which is composed of parenchyma cells. Outward cell division results in the formation of a thimblelike mass of relatively unorganized cells, the **root cap,** that covers and protects the root's apical meristem as it grows through the soil.

The root elongates relatively rapidly just behind its tip in the area called the *zone of elongation.* Abundant root hairs (see figure 26.22), extensions of single epidermal cells, form above that zone, in the area called the *zone of differentiation.* Virtually all water and minerals are absorbed from the soil through the root hairs, which greatly increase the root's surface area and absorptive powers. In plants with symbiotic mycorrhizae, the root hairs are often greatly reduced in number, and the fungal filaments of the mycorrhizae play a role similar to that of the root hairs, increasing the surface area for absorption. Another symbiotic relationship involving plants is often key to the health of ecosystems. The roots of some plants, specifically plants of the pea family, called legumes, form symbiotic relationships with bacteria that are able to break down atmospheric nitrogen into a source that can be taken up

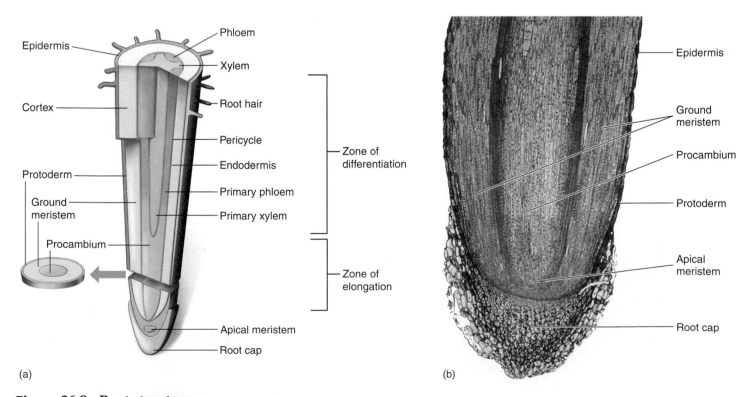

(a)

(b)

Figure 26.8   Root structure.

(*a*) Diagram of primary meristems in a dicot root, showing their relation to the apical meristem. The three primary meristems are the protoderm, which differentiates further into epidermis; the procambium, which differentiates further into primary vascular strands; and the ground meristem, which differentiates further into ground tissue. (*b*) Median longitudinal section of a monocot root tip in corn, *Zea mays,* showing the differentiation of protoderm, procambium, and ground meristem.

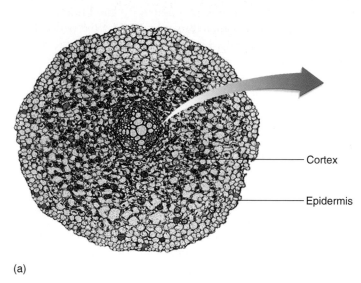

(a)

- Cortex
- Epidermis

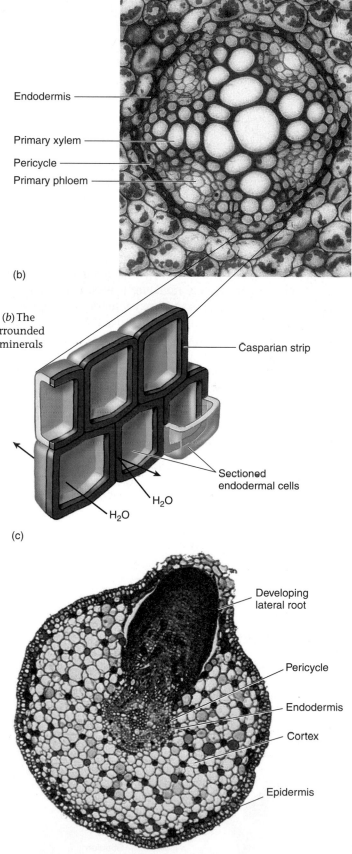

- Endodermis
- Primary xylem
- Pericycle
- Primary phloem

(b)

- Casparian strip
- Sectioned endodermal cells
- H_2O
- H_2O

(c)

- Developing lateral root
- Pericycle
- Endodermis
- Cortex
- Epidermis

Figure 26.9   A root cross section.

(a) Cross section through a root of a buttercup (*Ranunculus*), a dicot (40×). (b) The enlargement shows the various tissues present. (c) Endodermal cells are surrounded by a water-proofing band, called a Casparian strip, that forces water and minerals to pass through the cell membranes rather than between two cells.

and used by plants. These plants are a key component of the nitrogen cycle, cycling nitrogen back into the ecosystem in a form that can be used by other organisms.

One of the fundamental differences between roots and shoots has to do with the nature of their branching. In stems, branching occurs from buds on the stem surface; in roots, branching is initiated well back of the root tip as a result of cell divisions in the pericycle. The developing lateral roots grow out through the cortex toward the surface of the root, eventually breaking through and becoming established as lateral roots (figure 26.10). In some plants, roots may arise along a stem, or in some place other than the root of the plants. These roots are called *adventitious roots*. Adventitious roots occur in ivy, in bulb plants such as onions, and in perennial grasses, irises, and other plants that produce rhizomes, which are horizontal stems that grow underground.

Secondary growth occurs in some roots. In dicots and other plants with secondary growth, part of the pericycle and the parenchyma cells between the phloem patches and the xylem arms become the root vascular cambium, which starts producing secondary xylem to the inside and secondary phloem to the outside. Eventually, the secondary tissues acquire the form of concentric cylinders and expand thickness of the root. Cork cambium is also present in the roots of woody plants, producing cork cells to the outside of the pericycle. Secondary growth will be discussed in more detail in the next section on stems, where the process is more extensive than that found in roots.

26.3   Roots, the belowground portion of the plant body, are adapted to absorb water and minerals from the soil.

Figure 26.10   Lateral roots.

A lateral root growing out through the cortex of the black willow, *Salix nigra*. Lateral roots originate beneath the surface of the main root, whereas lateral stems originate at the surface.

Stems serve as the main structural support of the plant and the framework for the positioning of the leaves. Often experiencing both primary and secondary growth, stems are the source of an economically important product—wood.

Primary Growth

In the primary growth of a shoot, leaves first appear as leaf primordia (singular, primordium), rudimentary young leaves that cluster around the apical meristem, unfolding and growing as the stem itself elongates. The places on the stem at which leaves form are called nodes (figure 26.11). The portions of the stem between these attachment points are called the internodes. As the leaves expand to maturity, a bud, a tiny undeveloped side shoot, develops in the **axil** of each leaf, the angle between a leaf and the stem from which it arises. These buds, which have their own immature leaves, may elongate and form lateral branches, or they may remain small and dormant. A hormone moving downward from the terminal bud of the shoot continuously suppresses the expansion of the lateral buds.

Within the soft, young stems, the strands of vascular tissue, xylem and phloem, are either arranged around the outside of the stem as a cylinder (dicots), or scattered through it (monocots) (figure 26.12). This difference in vascular tissue organization, in addition to other characteristics discussed in chapter 14, illustrates the differences between these two major groups of angiosperms. The vascular bundles contain both primary xylem and primary phloem. At the stage when only primary growth has occurred, the inner portion of the ground tissue of a dicot stem is called the **pith,** and the outer portion is the **cortex.**

Secondary Growth

In stems, secondary growth is initiated by the differentiation of the **vascular cambium,** a thin cylinder of actively dividing cells located between the bark and the main stem in woody plants. The vascular cambium develops from cells within the vascular bundles of the stem, between the primary xylem and the primary phloem (figure 26.13). The cylindrical form of the vascular cambium is completed by the differentiation of some of the parenchyma cells that lie between the bundles. Once established, the vascular cambium consists of elongated, somewhat flattened cells with large vacuoles. The cells that divide from the vascular cambium outwardly, toward the bark, become secondary phloem; those that divide from it inwardly become secondary xylem.

While the vascular cambium is becoming established, a second kind of lateral cambium, the cork cambium, develops in the stem's outer layers. The cork cambium usually consists

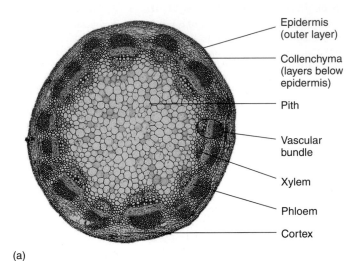

(a)

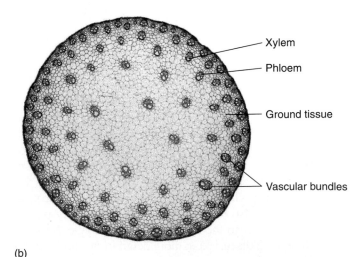

(b)

Figure 26.12   A comparison of dicot and monocot stems.

(a) Transection of a young stem of a dicot, the common sunflower, *Helianthus annuus,* in which the vascular bundles are arranged around the outside of the stem. (b) Transection of a monocot stem, corn, *Zea mays,* with the scattered vascular bundles characteristic of the class.

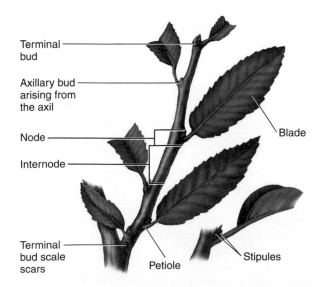

Figure 26.11   A woody twig.

This twig shows key stem structures, including the node and internode areas, the axillary bud in the axil, and leaves.

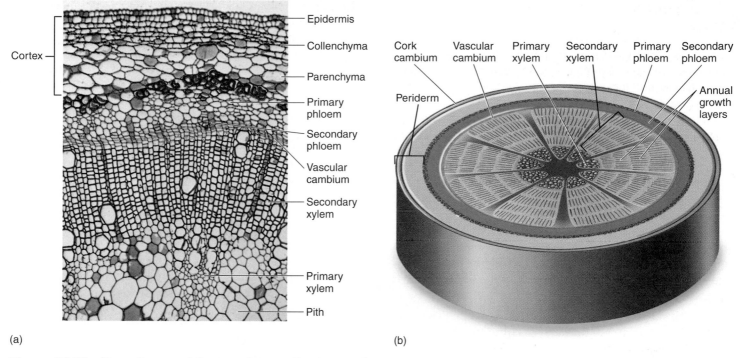

Cortex — Epidermis
Collenchyma
Parenchyma
Primary phloem
Secondary phloem
Vascular cambium
Secondary xylem
Primary xylem
Pith

(a)

Cork cambium · Vascular cambium · Primary xylem · Secondary xylem · Primary phloem · Secondary phloem · Periderm · Annual growth layers

(b)

Figure 26.13 Vascular cambium and secondary growth.

(a) Early stage in the differentiation of the vascular cambium in the castor bean, *Ricirus.* (b) Vascular cambium and cork cambium (lateral meristems) produce secondary tissues, causing the stem's girth to increase. Each year, a new layer of secondary tissue is laid down, forming rings in the wood.

Figure 26.14 Annual rings in a section of pine.

The rings that you see in this section of pine reflect the fact that the vascular cambium of trees divides more actively in the spring and summer, when water is plentiful and temperatures are suitable for growth, than in the fall and winter, when water is scarce and the weather is cold. As a result, layers of larger, thinner-walled cells formed during the growing season alternate with the smaller, darker layers of thick-walled cells formed during the rest of the year. A count of such annual rings in a tree trunk can be used to calculate the tree's age, and the width of rings can reveal information about environmental factors. For example, the region of thinner rings could indicate a period of prolonged drought conditions that was followed by wetter years.

of plates of dividing cells that move deeper and deeper into the stem as they divide. Outwardly, the cork cambium splits off densely packed **cork cells;** they contain a fatty substance and are nearly impermeable to water. Cork cells are dead at maturity. Inwardly, the cork cambium divides to produce a layer of parenchyma cells. The cork, the cork cambium that produces it, and this layer of parenchyma cells make up a layer called the **periderm,** which is the plant's outer protective covering.

Cork covers the surfaces of mature stems or roots. The term **bark** refers to all of the tissues of a mature stem or root outside of the vascular cambium. Because the vascular cambium has the thinnest-walled cells that occur anywhere in a secondary plant body, it is the layer at which bark breaks away from the accumulated secondary xylem.

Wood is one of the most useful, economically important, and beautiful products obtained from plants. Anatomically, wood is accumulated secondary xylem. As the secondary xylem ages, its cells become infiltrated with gums and resins, and the wood becomes darker. For this reason, the wood located nearer the central regions of a given trunk, called heartwood, is often darker and denser than the wood nearer the vascular cambium, called sapwood, which is still actively involved in transport within the plant. Because of the way it is accumulated, wood often displays rings. In temperate regions, these rings are annual rings (figure 26.14).

26.4 Stems, the aboveground framework of the plant body, grow both at their tips and in circumference.

26.5 | Leaves

Leaves are usually the most prominent shoot organs and are structurally diverse (figure 26.15). As outgrowths of the shoot apex, leaves are the major light-capturing organs of most plants. Most of the chloroplast-containing cells of a plant are within its leaves, and it is there where the bulk of photosynthesis occurs (see chapter 6). Exceptions to this are found in some plants, such as cacti, whose green stems have largely taken over the function of photosynthesis for the plant. Photosynthesis is conducted mainly by the "greener" parts of plants because they contain more chlorophyll, the most efficient photosynthetic pigment. In some plants, other pigments may also be present, giving the leaves a color other than green. Recall in chapter 6, we described accessory pigments that absorb light of other wavelengths. Thus, although coleus plants and red maple trees have leaves that are reddish in color, these leaves still contain chlorophyll and are the primary sites of photosynthetic activity in the plant.

The apical meristems of stems and roots are capable of growing indefinitely under appropriate conditions. Leaves, in contrast, grow by means of **marginal meristems,** which flank their thick central portions. These marginal meristems grow outward and ultimately form the **blade** (flattened portion) of the leaf, while the central portion becomes the midrib. Once a leaf is fully expanded, its marginal meristems cease to grow.

In addition to the flattened blade, most leaves have a slender stalk, the **petiole.** Two leaflike organs, the **stipules,** may flank the base of the petiole where it joins the stem (see figure 26.11). Veins, consisting of both xylem and phloem, run through the leaves. As mentioned in chapter 14, in most dicots, the pattern is net or reticulate venation; in many monocots, the veins are parallel (figure 26.16).

Leaf blades come in a variety of forms from oval to deeply lobed to having separate leaflets (the blade being divided but attached to a

Figure 26.15   Leaves.

Angiosperm leaves are stunningly variable. (*a*) *Simple leaves* from a gray birch, in which there is a single blade. (*b*) A simple leaf, its margin lobed, from the vine maple. (*c*) A *pinnately compound leaf* of a black walnut tree, where leaflets occur in pairs along the central axis of the main vein. (*d*) *Palmately compound leaves* of a California buckeye, in which the leaflets radiate out from a single point. (*e*) The leaves of pine trees are tough and needlelike. (*f*) Many unusual types of modified leaves occur in different kinds of plants. For example, some plants produce floral leaves or bracts; the most conspicuous parts of this poinsettia flower are the red bracts, which are modified leaves that surround the small yellowish true flowers in the center.

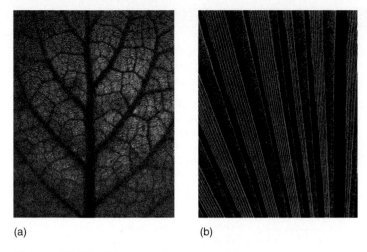

(a)   (b)

Figure 26.16   Dicot and monocot leaves.

The leaves of dicots, such as this (a) African violet relative from Sri Lanka, have netted, or reticulate, veins; (b) those of monocots, like this cabbage palmetto, have parallel veins. The dicot leaf has been stained to make the veins show up more clearly.

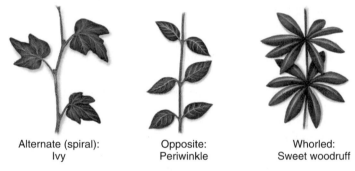

| Alternate (spiral): | Opposite: | Whorled: |
| Ivy | Periwinkle | Sweet woodruff |

Figure 26.17   Types of leaf arrangements.

The three common types of leaf arrangements are alternate, opposite, and whorled.

single petiole). In **simple leaves** (see figure 26.15*a, b*), such as those of lilacs or birch trees, there is a single blade, undivided, but some simple leaves may have teeth, indentations, or lobes, such as the leaves of maples and oaks. In **compound leaves,** such as those of ashes, box elders, and walnuts, the blade is divided into leaflets. If the leaflets are arranged in pairs along a common axis—the equivalent of the main central vein, or *midrib,* in simple leaves—the leaf is **pinnately compound** (see figure 26.15*c*), such as in the black walnut. If, however, the leaflets radiate out from a common point at the blade end of the petiole, the leaf is **palmately compound** (see figure 26.15*d*), such as in buckeyes and Virginia creepers. Leaves may be **alternately** arranged (alternate leaves usually spiral around a shoot) or they may be in **opposite** pairs. Less often, three or more leaves may be in a **whorl,** a circle of leaves at the same level at a node (figure 26.17).

A typical leaf contains masses of parenchyma, called **mesophyll** ("middle leaf"), through which the vascular bundles, or veins, run. Beneath the upper epidermis of a leaf are one or more layers of closely packed, columnlike parenchyma cells called **palisade mesophyll.** The rest of the leaf interior, except for the veins, consists of a tissue called **spongy mesophyll** (figure 26.18). Between the spongy mesophyll cells are large intercellular spaces that function in gas exchange and particularly in the passage of carbon dioxide from the atmosphere to the mesophyll cells. These intercellular spaces are connected, directly or indirectly, with the stomata.

26.5   Leaves, the photosynthetic organs of the plant body, are varied in shape and arrangement.

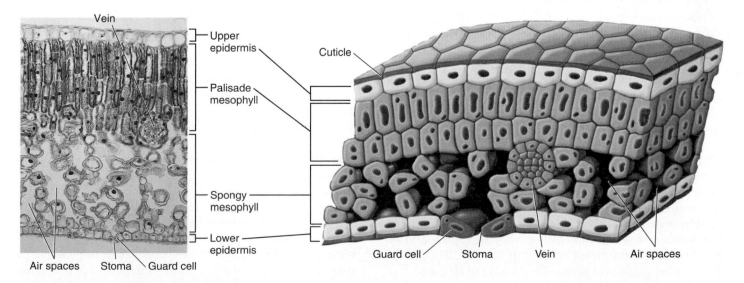

Figure 26.18   A leaf in cross section.

Cross section of a leaf, showing the arrangement of palisade and spongy mesophyll, a vascular bundle or vein, and the epidermis, with paired guard cells flanking the stoma.

26.6 | Water Movement

Vascular plants have a conducting system, as humans do, for transporting fluids and nutrients from one part to another. Functionally, a plant is essentially a bundle of tubes with its base embedded in the ground. At the base of the tubes are roots, and at their tops are leaves. For a plant to function, two kinds of transport processes must occur: first, the carbohydrate molecules produced in the leaves by photosynthesis must be carried to all of the other living plant cells. To accomplish this, liquid, with these carbohydrate molecules dissolved in it, must move both up and down the tubes. Second, minerals and water in the ground must be taken up by the roots and ferried to the leaves and other plant cells. In this process, liquid moves up the tubes. Plants accomplish these two processes by using chains of specialized cells: those of the phloem transport photosynthetically produced carbohydrates up and down the plant, and those of the xylem carry water and minerals upward.

Cohesion-Adhesion-Tension Theory

Many of the leaves of a large tree may be more than 10 stories off the ground. How does a tree manage to raise water so high? Several factors are at work to move water up the height of a plant. The initial movement of water into the roots of a plant involves osmosis. Water moves into the cells of the root because the fluid in the xylem contains more solutes than the surroundings—recall from chapter 4 that water will move across a membrane from an area of lower solute concentration to an area of higher solute concentration. However, this force, called *root pressure,* is not by itself strong enough to "push" water up a plant's stem.

Capillary action adds a "pull." *Capillary action* results from the tiny electrical attractions of polar water molecules to surfaces that carry an electrical charge, a process called *adhesion.* In the laboratory, a column of water rises up a tube of glass because the attraction of the water molecules to the charged molecules on the interior surface of the glass tube "pull" the water up in the tube (figure 26.19).

However, although capillary action can produce enough force to raise water a meter or two, it cannot account for the movement of water to the tops of tall trees. A second very strong "pull" accomplishes this, provided by transpiration. Opening up the tube and blowing air across its upper end demonstrates how transpiration draws water up a plant stem. The stream of relatively dry air causes water molecules at the water column's exposed top surface to evaporate from the tube. The water level in the tube does not fall, because as water molecules are drawn from the top, they are replenished by new water molecules pulled up from the bottom. This, in essence, is what happens in plants. The passage of air across leaf surfaces results in the loss of water by evaporation, creating

Figure 26.19 Capillary action.
Capillary action causes the water within a narrow tube to rise above the surrounding water; the attraction of the water molecules to the glass surface, which draws water upwards, is stronger than the force of gravity, which tends to draw it down. The narrower the tube, the greater the surface area available for adhesion for a given volume of water, and the higher the water rises in the tube.

a "pull" at the open upper end of the plant. New water molecules that enter the roots are pulled up the plant. Adhesion of water molecules to the walls of the narrow vessels in plants also helps to maintain water flow to the tops of plants.

A column of water in a tall tree does not collapse simply because of its weight because water molecules have an inherent strength that arises from their tendency to form hydrogen bonds with one another. These hydrogen bonds cause *cohesion* of the water molecules; in other words, a column of water resists separation. The beading of water droplets illustrates the property of cohesion. This resistance, called *tensile strength,* varies inversely with the diameter of the column; that is, the smaller the diameter of the column, the greater the tensile strength. Therefore, plants must have very narrow transporting vessels to take advantage of tensile strength.

How the combination of gravity, tensile strength, and cohesion affect water movement in plants is called the **cohesion-adhesion-tension theory.** It is important to note that the movement of water up through a plant is a passive process and requires no expenditure of energy on the part of the plant.

Transpiration

The process by which water leaves a plant is called **transpiration.** More than 90% of the water taken in by plant roots is ultimately lost to the atmosphere, almost all of it from the leaves. It passes out primarily through the stomata in the evaporation of water vapor. On its journey from the plant's interior to the outside, a molecule of water first passes into the pockets of air within the leaf by evaporating from the walls of the spongy mesophyll that lines the intercellular spaces. These intercellular spaces open to the outside of the leaf by way of the stomata. The water that evaporates from these surfaces of the spongy mesophyll cells is continuously replenished from the tips of the veinlets in the leaves. Because the strands of xylem conduct water within the plant in an unbroken stream all the way from the roots to the leaves, when a portion of the water vapor in the intercellular spaces passes out through the stomata, the supply of water vapor in these spaces is continually renewed (figure 26.20). Because the process of transpira-

Transpiration

1 Dry air passes across the leaves and causes water vapor to evaporate out of the stomata.

2 The loss of water from the leaves creates a type of "suction" that draws water up the stem through the xylem.

3 New water enters the plant through the roots to replace the water moving up the stem.

Figure 26.20   How transpiration works.
Water evaporating from the leaves through the stomata causes the movement of water upward in the xylem and the entrance of water through the roots.

tion is dependent upon evaporation, factors that affect evaporation also affect transpiration. In addition to the movement of air across the stomata, mentioned earlier, humidity levels in the air will affect the rate of evaporation, high humidity reducing it and low humidity increasing it. Temperature will also affect the rate of evaporation, high temperatures increasing it and lower temperatures reducing it. This temperature effect is especially important because evaporation also acts to cool plant tissues.

Structural features such as the stomata, the cuticle, and the intercellular spaces in leaves have evolved in response to one or both of two contradictory requirements: minimizing the loss of water to the atmosphere, on the one hand, and admitting carbon dioxide, which is essential for photosynthesis, on the other. How plants resolve this problem is discussed next.

Regulation of Transpiration: Open and Closed Stomata

The only way plants can control water loss on a short-term basis is to close their stomata. Many plants can do this when subjected to water stress. But the stomata must be open at least part of the time so that carbon dioxide, which is necessary for photosynthesis, can enter the plant. In its pattern of opening or closing its stomata, a plant must respond to both the need to conserve water and the need to admit carbon dioxide.

The stomata open and close because of changes in the water pressure of their guard cells (figure 26.21). Stomatal guard cells are long, sausage-shaped cells attached at their ends. The cellulose microfibrils of their cell wall wrap around the cell such that when the guard cells are **turgid** (plump and swollen with water), they expand in length, causing the cells to bow, thus opening the stomata as wide as possible.

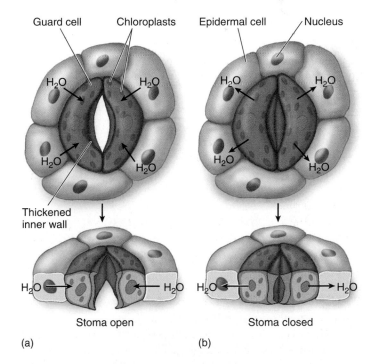

Figure 26.21   How guard cells regulate the opening and closing of stomata.
(a) When guard cells contain a high level of solutes, water enters the guard cells, causing them to swell and bow outward. This bowing opens the stoma. (b) When guard cells contain a low level of solutes, water leaves the guard cells, causing them to become flaccid. This flaccidity closes the stoma.

A number of environmental factors affect the opening and closing of stomata. The most important is water loss. The stomata of plants that are wilted because of a lack of water tend to close. An increase in carbon dioxide concentration also causes the stomata of most species to close. In most plant species, stomata open in the light and close in the dark. Very high temperatures (above 30° to 35°C) also tend to cause stomata to close.

Water Absorption by Roots

Most of the water absorbed by plants comes in through the root hairs (figure 26.22). These root hairs greatly increase the surface area and therefore the absorptive powers of the roots. In plants that have ectomycorrhizae, the root hairs often are greatly reduced in number; the fungal filaments take their place in promoting absorption. Root hairs are turgid—plump and swollen with water—because they contain a higher concentration of dissolved minerals and other solutes than does the water in the soil solution; water, therefore, tends to move into them steadily. Once inside the roots, water passes inward to the conducting elements of the xylem.

Water is not the only substance that enters the roots by passing into the cells of root hairs. Minerals also enter the root. Membranes of root hair cells contain a variety of ion transport channels that actively pump specific ions into the plant, even against large concentration gradients. These ions, many of which are plant nutrients, are then transported throughout the plant as a component of the water flowing through the xylem (figure 26.23).

Figure 26.22   Root hairs.

Abundant fine root hairs can be seen in the back of the root apex of this germinating seedling of radish, *Raphanus sativus*.

26.6   Water is drawn up the plant stem from the roots by transpiration from the leaves.

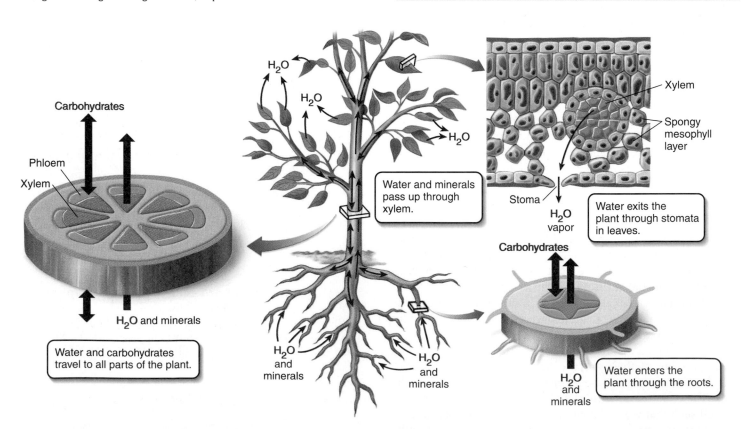

Figure 26.23   The flow of materials into, out of, and within a plant.

Water and minerals enter through the roots of a plant and are transported through the xylem to all parts of the plant body (*blue arrows*). Water leaves the plant through the stomata in the leaves. Carbohydrates synthesized in the leaves are circulated throughout the plant by the phloem (*red arrows*).

26.7 Carbohydrate Transport

Most of the carbohydrates manufactured in plant leaves and other green parts are moved through the phloem to other parts of the plant. This process, known as **translocation,** makes suitable carbohydrate building blocks available at the plant's actively growing regions. The carbohydrates concentrated in storage organs such as underground stems (potatoes), roots (carrots), and leaves (onions and cabbage), often in the form of starch, are also converted into transportable molecules, such as sucrose, and moved through the phloem.

The pathway that sugars and other substances travel within the plant has been demonstrated precisely by using radioactive isotopes and aphids, a group of insects that suck the sap of plants. Aphids thrust their piercing mouthparts into the phloem cells of leaves and stems to obtain the abundant sugars there. When the aphids are cut off of the leaf, the liquid continues to flow from the detached mouthparts protruding from the plant tissue and is thus available in pure form for analysis. The liquid in the phloem contains 10% to 25% dissolved solid matter, almost all of which is sucrose. The harvesting of sap (the liquid removed from phloem) from maple trees uses a similar process. A hole is drilled in the tree and the phloem is drained from the tree using tubing and collected in buckets. The sap is then processed into maple syrup.

Using aphids to obtain the critical samples and radioactive tracers to mark them, researchers have learned that movement of substances in the phloem can be remarkably fast—rates of 50 to 100 centimeters per hour have been measured. This translocation movement is a passive process that does not require the expenditure of energy by the plant. The **mass flow** of materials transported in the phloem occurs because of water pressure, which develops as a result of osmosis. First, sucrose produced as a result of photosynthesis is actively "loaded" into the sieve tubes (or sieve cells) of the vascular bundles. This loading increases the solute concentration of the sieve tubes, so water passes into them by osmosis. An area where the sucrose is made is called a *source;* an area where sucrose is delivered from the sieve tubes is called a *sink.* Sinks include the roots and other regions of the plant that are not photosynthetic, such as young leaves and fruits. The sucrose is unloaded and stored in these areas. There the solute concentration of the sieve tubes is decreased as the sucrose is removed. As a result of these processes, water moves through the sieve tubes from the areas where sucrose is being added into those areas where it is being withdrawn, and the sucrose moves passively with the water (figure 26.24). This is called the *pressure-flow hypothesis.*

26.7 Carbohydrates move through the plant by the passive osmotic process of translocation.

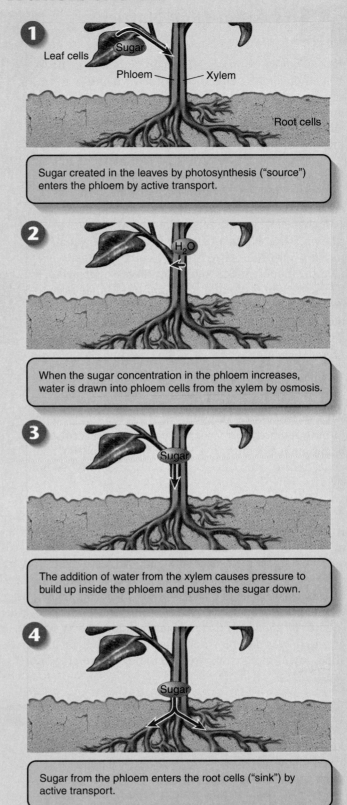

Translocation

1 Sugar created in the leaves by photosynthesis ("source") enters the phloem by active transport.

2 When the sugar concentration in the phloem increases, water is drawn into phloem cells from the xylem by osmosis.

3 The addition of water from the xylem causes pressure to build up inside the phloem and pushes the sugar down.

4 Sugar from the phloem enters the root cells ("sink") by active transport.

Figure 26.24   How translocation works.

26.8 Essential Plant Nutrients

Just as human beings need certain nutrients, such as carbohydrates, amino acids, and vitamins, to survive, plants also need various nutrients to remain alive and healthy. Lack of an important nutrient may slow a plant's growth or make the plant more susceptible to disease or even death.

Minerals are involved in plant metabolism in many ways. *Nitrogen* (N), acquired from the soil with the help of nitrogen-fixing bacteria, is an essential part of proteins and nucleic acids. *Potassium* (K) ions regulate the **turgor pressure** (the pressure within a cell that results from water moving into the cell) of guard cells and therefore the rate at which the plant loses water and takes in carbon dioxide. *Calcium* (Ca) is an essential component of the middle lamellae, the structural elements laid down between plant cell walls, and it also helps to maintain the physical integrity of membranes. *Magnesium* (Mg) is a part of the chlorophyll molecule. The presence of *phosphorus* (P) in many key biological molecules such as nucleic acids and ATP has been explored in detail in earlier chapters. *Sulfur* (S) is a key component of an amino acid (cysteine), essential in building proteins. Other essential minerals for plant health include chlorine (Cl), iron (Fe), boron (B), manganese (Mn), zinc (Zn), copper (Cu), and molybdenum (Mo).

Most plants acquire minerals from the soil, although some carnivorous plants are able to use other organisms directly as sources of nitrogen, just as animals do. Carnivorous plants lure and trap insects and other small animals and then digest their prey with enzymes secreted from various kinds of glands. The Venus's-flytrap (*Dionaea muscipula*) has sensitive hairs on each side of each leaf, which, when touched, trigger the two halves of the leaf to snap together (figure 26.25*a, b*). Pitcher plants attract insects with their bright, flowerlike colors; once inside the pitchers, the insects slide down into a cavity containing digestive enzymes (figure 26.25*c*).

Figure 26.25   Carnivorous plants.
(*a*) Venus's-flytrap, *Dionaea muscipula,* which inhabits low boggy ground in North and South Carolina. (*b*) A Venus's-flytrap leaf has snapped together, imprisoning a fly. (*c*) A tropical Asian pitcher plant, *Nepenthes*. Insects seeking nectar enter the pitchers, which are modified leaves, and are trapped and digested. Complex communities of invertebrate animals and protists inhabit the pitchers.

26.8   Plants require ample supplies of nitrogen, phosphorus, and potassium, and smaller amounts of many other nutrients.

Exploring Current Issues

Additional Resources

Go to your campus library or look online to find the following articles, which further develop some of the concepts found in this chapter.

Coen, E. (2003). Way to grow: how do plants and animals translate their genes into graceful curves of leaf and limb? *New Scientist,* 180(2420), 44.

Fresco, L. O. (2003). Fertilize the plants, not the soil—dispelling myths about fertilizers and plant nutrients. *UN Chronicle,* 40(3), 62.

Kozela, C. and S. Regan, (2003). How plants make tubes. *Trends in Plant Science,* 8(4), 159.

Sperry, J. S. (2003). Evolution of water transport and xylem structure. *International Journal of Plant Sciences,* 164(3), 115.

Vines, G. (2000). Follow that food. *New Scientist,* 166(2240), 28.

Biology and Society Lecture: History of Agriculture

Among plant, fungal, and animal species, only a small minority are edible by humans. The few species of plants and animals that we can eat were domesticated early after human agriculture began. The oldest clear evidence of domestication is the dog, found in both Asia and North America. Other organisms followed: wheat, sheep, and goats in Southwest Asia; rice and pigs in China; cows in India; horses in the Ukraine; corn and beans in Middle America; and potatoes in the Andes. Why domestication of these organisms and not others? Much has to do with the ease of domestication—there just aren't many organisms suitable for agriculture.

Find this lecture, delivered by the author to his class at Washington University, online at www.mhhe.com/tlwessentials/exp26.

Structure and Function of Plant Tissues

26.1 Organization of a Vascular Plant

- Most plants possess the same three major groups of organs: roots, stems, and leaves, although they may appear very different in different types of plants (**figure 26.1**). Vascular tissue extends throughout the plant, connecting roots, stems, and leaves.

- Growth occurs in regions called meristems. The tips of the roots and shoots contain apical meristem, which is the site of primary growth, extending the plant body in vertical directions. Extending the thickness or girth of the plant, called secondary growth, occurs at the lateral meristem, cylinders of meristematic tissue.

26.2 Plant Tissue Types

- Plants contain different types of tissue. A tissue is a group of similar cells that functions as a unit. There are three main types of tissues in plants: ground tissue, dermal tissue, and vascular tissue.

- Ground tissue makes up the main body of the plant and contains several different cell types. Parenchyma cells form the masses of tissue in leaves, stems, and roots. They carry out functions such as photosynthesis, food, and water storage. The edible parts of fruits and vegetables are primarily parenchyma cells (**figure 26.2**). Collenchyma cells form strands or cylinders that provide support (**figure 26.3**). Sclerenchyma cells have thick, secondary cell walls. They provide strength and rigidity to the plant body. They are organized into long fibers or branched structures called sclereids or "stone cells" (**figure 26.4**).

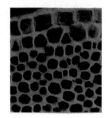

- Dermal tissue makes up the outer layer of the plant body, which consists of epidermal cells covered with a waxy layer called the cuticle (roots have epidermal cells but no cuticle). This outer layer protects the plant and the cuticle provides a barrier to water loss. Guard cells are specialized epidermal cells that provide openings for gas exchange. Trichomes and root hairs are extensions of epidermal cells (**figure 26.5**).

- Vascular tissue is composed of xylem and phloem. Xylem contains water-conducting cells, tracheids and vessel elements, that are dead at maturity (**figure 26.6**). Phloem contains food-conduction cells, sieve cells and sieve-tube members, both of which are living but lack nuclei when mature. Companion cells associated with the phloem cells carry out metabolic functions needed to maintain the sieve-tube members (**figure 26.7**).

The Plant Body

26.3 Roots

- Roots are organs adapted to absorb water and minerals from the soil (**figure 26.8**). A waxy Casparian strip keeps water from passing between endodermal cells and forces the water to pass through the cells into the xylem cells (**figure 26.9**).

26.4 Stems

- The stem serves as a framework for positioning the leaves. Primary growth occurs at the apical meristem laying down the stem tissue behind the meristem. Leaves grow out of the stems at node areas (**figure 26.11**). Secondary growth occurs at the lateral meristems with the differentiation of vascular cambium, which gives rise to xylem and phloem and cork cambium, which gives rise to the layers of cork inside the bark (**figures 26.12–26.14**).

26.5 Leaves

- Leaves are the primary site for photosynthesis. They vary in size, shape, and arrangement (**figures 26.15–26.18**).

Plant Transport and Nutrition

26.6 Water Movement

- Water enters the plant through the roots by osmosis. Root pressure and capillary action cause the water to pass up farther into the tissues (**figure 26.19**). However, for water to travel up the length of the stem, it requires a stronger force, the combination of cohesion and adhesion, known as the cohesion-adhesion-tension theory. Transpiration, the evaporation of water vapor from the leaves, creates the "pull" that raises the water through the plant (**figures 26.20 and 26.21**). Minerals follow the flow of water into the plant (**figure 26.23**).

26.7 Carbohydrate Transport

- Carbohydrates produced in the leaves through photosynthesis travel throughout the plant in phloem tissue. The process, called translocation, involves osmotic movement of water into the phloem cells, forcing the sugars to "sinks" for carbohydrate storage (**figure 26.24**).

26.8 Essential Plant Nutrients

- Plants require minerals for metabolic functions that they obtain from the soil. However, some carnivorous plants obtain nitrogen directly from animals that are digested by plant enzymes (**figure 26.25**).

1. Growth in vascular plants is regulated and coordinated by
   a   photosynthetic tissue.
   b. root tissue.
   c. meristematic tissue.
   d. leaf epidermal tissue.

2. In vascular plants, phloem tissue
   a. transports water.
   b. transports carbohydrates.
   c. transports minerals.
   d. supports the plant.

3. In vascular plant leaves, gases enter and leave the plant through
   a. stomata.   c. chloroplasts.
   b. guard cells.   d. trichomes.
4. In roots, growth of lateral branches begins
   a. on the root epidermis.   c. at the ground meristem.
   b. on the root hairs.   d. at the pericycle.
5. In stems, the tissue responsible for secondary growth is the
   a. collenchyma.   c. cambium.
   b. pith.   d. cortex.
6. One difference between monocot and dicot plant stems is the
   a. absence of buds in monocots.
   b. organization of vascular tissue.
   c. presence of guard cells.
   d. absence of stomata.

7. In leaves, gas exchange and transport takes place in the
   a. cuticle.   c. spongy mesophyll.
   b. palisade mesophyll.   d. guard cells.
8. Which of the following is *not* a process that directly assists in water movement from the roots to the leaves?
   a. photosynthesis   c. capillary action
   b. root pressure   d. transpiration
9. The passive process of moving carbohydrates throughout a plant is called
   a. transpiration.   c. translation.
   b. translocation.   d. evaporation.
10. Bacteria are needed to aid the plant in obtaining which nutrient from the soil?
   a. potassium   c. magnesium
   b. calcium   d. nitrogen

Visual Understanding

1. **Figure 26.8**   What is the purpose of the root cap covering the apical meristem of the root?

2. **Figure 26.14**   Besides wet and dry years, and the age of the tree, what else might the study of tree rings tell us?

Challenge Questions

Structure and Function of Plant Tissues   Why do land plants need schlerenchyma cells?

The Plant Body   Some plants in cold climates lose their leaves in the winter. In desert climates some plants, such as palo verde and ocotillo, may lose some or all of their leaves in the hot, dry summer to minimize water loss. These plants have green stems. Why?

Plant Transport and Nutrition   Your friend, Alex, just returned from a family trip to northern Michigan, where he visited a maple tree farm where they made maple syrup. On the maple trees, they made just one small hole (or two small holes on larger trees) and hung a bucket beneath to catch the sap. Why, he asks you, wouldn't they just make a cut completely around the tree on the diagonal and collect much more sap, much faster? Why isn't this feasible?

Online Learning Center

Visit the Online Learning Center for this chapter at www.mhhe.com/tlwessentials/ch26 for quizzes, animations, interactive learning exercises, and other study tools. At the site you will also find extended answers to the end-of-chapter questions.

27

Plant Reproduction and Growth

S eeds are one of the cleverest adaptations of plants. Carried by wind or animals, they are capable of transporting the next generation to distant locations and so ensuring the plant an opportunity to occupy any available habitats. A seed is a protected package of genetic information, an embryonic individual kept in a dormant state by a variety of mechanisms such as a watertight covering that keeps the seed's interior free of water. When the seed is deposited in suitable soil, the watertight covering splits open and the embryo begins to grow, a process called germination. These seeds of a soybean plant are germinating, with leaves thrusting upward and roots downward toward the soil. For many seeds, moisture and moderate temperatures are sufficient to trigger germination. For some seeds, however, more extreme cues are required. The seeds of many species of pine tree, for example, will not germinate unless exposed to extreme temperatures of the sort experienced in forest fires; periodic forest fires provide openings for sunlight to reach the seedlings, and abundant nutrients enter the soil from the tissues of fire-killed trees.

27.1 Angiosperm Reproduction

Although reproduction varies greatly among the members of the plant kingdom, we focus in this chapter on reproduction among flowering plants. While the evolution of their unique sexual reproductive features, flowers and fruits, have contributed to their success, angiosperms also reproduce asexually.

Asexual Reproduction

In **asexual reproduction,** an individual inherits all of its chromosomes from a single parent and is, therefore, genetically identical to that parent. Asexual reproduction produces a "clone" of the parent. In a stable environment, asexual reproduction may prove more advantageous than sexual reproduction because it allows individuals to reproduce with a lower investment of energy than sexual reproduction. A common type of asexual reproduction called *vegetative reproduction* results when new individuals are simply cloned from parts of the parent (figure 27.1).

The forms of vegetative reproduction in plants are many and varied:

Runners. Some plants reproduce by means of runners— long, slender stems that grow along the surface of the soil. At node regions on the stem, adventitious roots form, extending into the soil. Leaves and flowers form, and a new shoot is sent out, continuing the runner. Strawberry plants reproduce by runners.

Rhizomes. Rhizomes are underground horizontal stems that create a network underground. Like runners, nodes give rise to new flowering shoots. The noxious character of many weeds results from this type of growth pattern, but so do grasses and many garden plants such as irises. Other specialized stems, for example, tubers, function in food storage and reproduction. White potatoes are specialized underground stems that store food, and the "eyes" give rise to new plants.

Suckers. The roots of some plants produce "suckers," or sprouts, which give rise to new plants, such as found in cherry, apple, raspberry, and blackberry plants. When the root of a dandelion is broken, which may occur if one tries to pull it from the ground, each root fragment may give rise to a new plant.

Adventitious Plantlets. In a few species, even the leaves are reproductive. One example is the house plant *Kalanchoë daigremontiana,* familiar to many people as the "maternity plant," or "mother of thousands." The common names of this plant are based on the fact that numerous plantlets arise from meristematic tissue

Figure 20.1   Vegetative reproduction.
Small plants arise from notches along the leaves of the house plant *Kalanchoë daigremontiana.*

located in notches along the leaves. The maternity plant is ordinarily propagated by means of these small plants, which, when they mature, drop to the soil and take root (figure 27.1).

Sexual Reproduction

Plant sexual life cycles are characterized by an alternation of generations, in which a diploid sporophyte generation gives rise to a haploid gametophyte generation. In angiosperms, the developing gametophyte generation is completely enclosed within the tissues of the parent sporophyte (see figure 14.20). The male gametophytes are **pollen grains,** and they develop from *microspores.* The female gametophyte is the **embryo sac,** which develops from a *megaspore.* Pollen grains and the embryo sac both are produced in separate, specialized structures of the angiosperm flower.

Like animals, angiosperms have separate structures for producing male and female gametes, but the reproductive organs of angiosperms are different from those of animals in two ways: First, in angiosperms, both male and female structures usually occur together in the same individual flower. Second, angiosperm reproductive structures are not permanent parts of the adult individual. Angiosperm flowers and reproductive organs develop seasonally; these flowering seasons correspond to times of the year most favorable for pollination.

Structure of the Flower. Flowers contain male and female parts, called *stamens* and *carpels,* respectively (see figure 14.17). Often flowers contain both stamens and carpels, but there are some exceptions. In various species of flowering plants, for example, willows and some mulberries, flowers containing only male or only female parts, called *imperfect*

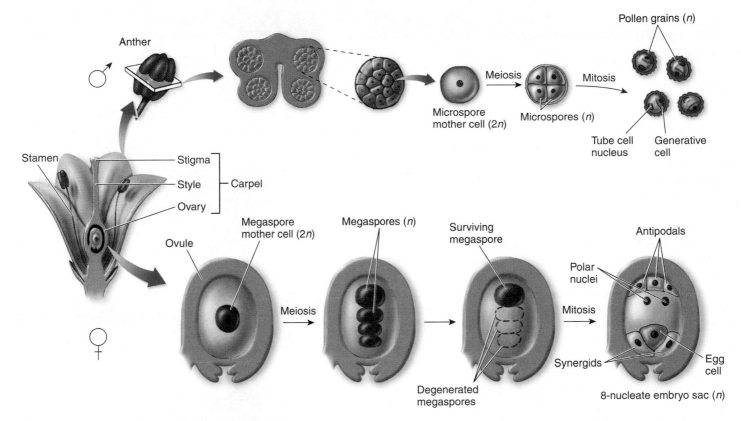

Figure 27.2   Formation of pollen and egg.
Diploid microspore mother cells are housed in the anther and divide by meiosis to form four haploid (*n*) microspores. Each microspore develops by mitosis into a pollen grain with a generative cell and a tube cell nucleus. The generative cell will later divide to form two sperm cells. Within the ovule, one diploid megaspore mother cell divides by meiosis to produce four haploid megaspores. Usually only one megaspore survives, and the other three degenerate. The surviving megaspore divides by mitosis to produce an embryo sac with eight nuclei. Upon fertilization, the egg cell becomes the embryo and the polar nuclei become the endosperm.

flowers, occur on separate plants. Plants that contain imperfect flowers that produce only ovules or only pollen are called *dioecious,* from the Greek words for "two houses." In other plants, there are separate male and female flowers, but they occur on the same plant. These plants are called *monoecious,* meaning "one house." In monoecious plants, the male and female flowers may mature at different times, which keeps the plant from pollinating itself. Even if, as is usually the case, functional stamens and carpels are both present in each flower of a particular plant, these organs, as is the case with separate flowers, may reach maturity at different times, which also keeps the plant from pollinating itself.

Pollen Formation.   Pollen grains develop from microspores formed in the four pollen sacs located in the flower's anther. Each pollen sac contains specialized chambers in which the microspore mother cells are enclosed and protected. Each microspore mother cell undergoes meiosis to form four haploid microspores. Subsequently, mitotic divisions form pollen grains that contain a generative cell and a tube cell nucleus. The tube cell nucleus forms the pollen tube; the generative cell will later divide to form two sperm cells (figure 27.2).

Pollination.   **Pollination** is the process by which pollen is transferred from the anther to the stigma. The pollen may be carried to the flower by wind or by animals, or it may originate within the individual flower itself. When pollen from a flower's anther pollinates the same flower's stigma, the process is called *self-pollination,* which can lead to *self-fertilization.* For some plants, self-pollination and self-fertilization occur because self-pollination eliminates the need for animal pollinators and maintains beneficial phenotypes in stable environments. However, other plants are adapted to *outcrossing,* the crossing of two different plants. As described earlier, the presence of only male or female flowers on a plant requires outcrossing, as does the different timing of appearance of the male and female parts on a flower. Even when a flower's stamen and stigma mature at the same time, some plants exhibit **self-incompatibility.** Self-incompatibility results when the pollen and stigma recognize each other as being genetically related, and respond by blocking the fertilization of the flower.

In many angiosperms, the pollen grains are carried from flower to flower by insects and other animals that visit the flowers for food or other rewards or are deceived into doing so because the flower's characteristics suggest such rewards.

A liquid called **nectar,** which is rich in sugar as well as amino acids and other substances, is often the reward sought by animals. Successful pollination depends on the plants attracting insects and other animals regularly enough that the pollen is carried from one flower of that particular species to another.

The relationship between such animals, known as *pollinators,* and flowering plants has been important to the evolution of both groups, a process called **coevolution.** By using insects to transfer pollen, the flowering plants can disperse their gametes on a regular and more or less controlled basis, despite their being anchored to the ground. The more attractive the plant is to the pollinator, the more frequently the plant will be visited. Therefore, any changes in the phenotype of the plant that result in more visits by pollinators offer a selective advantage. This has resulted in the evolution of a wide variety of angiosperm species. Furthermore, animals that could obtain food from the flowers became more abundant and diverse as their food supply increased and diversified.

For pollination by animals to be effective, a particular insect or other animal must visit plant individuals of the same species. A flower's color and form have been shaped by evolution to promote such specialization. Yellow flowers are particularly attractive to bees (figure 27.3*a*), whereas red flowers attract birds but are not particularly noticed by insects. Some flowers have very long floral tubes with the nectar produced deep within them; only the long, slender beaks of hummingbirds or the long, coiled proboscis of moths or butterflies can reach such nectar supplies (figure 27.3*b*).

In certain angiosperms and all gymnosperms, pollen is blown about by the wind and reaches the stigmas passively. For such a system to operate efficiently, the individuals of a given plant species must grow relatively close together because wind does not carry pollen very far or very precisely, compared to transport by insects or other animals. Because gymnosperms, such as spruces or pines, grow in dense stands, wind pollination is very effective. Wind-pollinated angiosperms, such as birches, grasses, and ragweed, also tend to grow in dense stands. The flowers of wind-pollinated angiosperms are usually small, greenish, and odorless, and their petals are either reduced in size or absent altogether. They typically produce large quantities of pollen.

Egg Formation. Eggs develop in the **ovules** of the angiosperm flower. Within each ovule is a megaspore mother cell. Each megaspore mother cell undergoes meiosis to produce four haploid megaspores. In most plants, only one of these megaspores, however, survives; the rest are absorbed by the ovule. The lone remaining megaspore undergoes repeated mitotic divisions to produce eight haploid nuclei, which are enclosed within a seven-celled *embryo sac* (see figure 27.2). Within the embryo sac, the eight nuclei are arranged in precise positions. One nucleus is located near the opening of the embryo sac in the egg cell. Two nuclei are located in a single cell in the middle of the embryo sac and are called polar nuclei. Two nuclei reside in cells that flank the egg cell and are called synergids; and the other three nuclei are located in cells at the end of the embryo sac, opposite the egg cell, and are called antipodals.

Figure 27.3   Insect pollination.
(*a*) Bees are usually attracted to yellow flowers. (*b*) This alfalfa butterfly (*Colias eurytheme*) has a long proboscis that allows it to feed on nectar deep in the flower.

Fertilization.   Once a pollen grain has been spread by wind, an animal, or self-pollination, it adheres to the sticky, sugary substance that covers the stigma and begins to grow a **pollen tube,** which pierces the style. The pollen tube, nourished by the sugary substance, grows until it reaches the ovule in the ovary.

When the pollen tube reaches the entry to the embryo sac in the ovule, the tip of the tube bursts and releases the two sperm cells that form from the generative cell. One of the sperm cells fertilizes the egg cell, forming a zygote. The other sperm cell fuses with the two polar nuclei located at the center of the embryo sac, forming the triploid (3*n*) primary endosperm nucleus (see figure 14.20). This process of fertilization in angiosperms in which two sperm cells are used is called **double fertilization.** The primary endosperm nucleus eventually develops into the endosperm, which nourishes the embryo.

27.1   **Reproduction in angiosperms involves asexual and sexual reproduction. In sexual reproduction, pollen is transferred to the female stigma. Double fertilization leads to the development of an embryo and endosperm.**

The entire series of events that occurs between fertilization and maturity is called *development*. During development, cells become progressively more specialized, or differentiated. The first stage in the development of a plant is active cell division to form an organized mass of cells, the embryo. In angiosperms, the differentiation of cell types within the embryo begins almost immediately after fertilization (figure 27.4). By the fifth day, the principal tissue systems can be detected within the embryo mass, and within another day, the root and shoot apical meristems can be detected.

Early in the development of an angiosperm embryo, a profoundly significant event occurs: The embryo simply stops developing and becomes dormant as a result of drying. In many plants, embryo development is arrested soon after apical meristems and the seed leaves, or **cotyledons,** are differentiated. The integuments—the outermost covering of the ovule—develop into a relatively impermeable seed coat, which encloses the dormant embryo within the seed, together with a source of stored food.

Once the seed coat fully develops around the embryo, most of the embryo's metabolic activities cease; a mature seed contains only about 10% water. Under these conditions, the seed and the young plant within it are very stable.

Germination, or the resumption of metabolic activities that leads to the growth of a mature plant, cannot take place until water and oxygen reach the embryo, a process that sometimes involves cracking the seed. Seeds of some plants have been known to remain viable for hundreds and in some cases thousands of years. The seed will germinate when conditions are favorable for the plant's survival.

> **27.2** A seed contains a dormant embryo and substantial food reserves, encased within a tough drought-resistant coat.

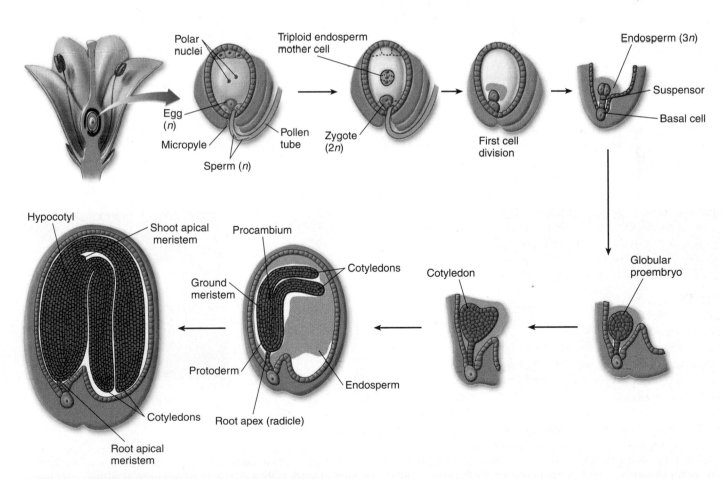

Figure 27.4 Development in an angiosperm embryo.
After the zygote forms, the first cell division is asymmetric. After another division, the basal cell, the one nearest the opening through which the pollen tube entered, undergoes a series of divisions and forms a narrow column of cells called the suspensor. The other three cells continue to divide and form a mass of cells arranged in layers. By about the fifth day of cell division, the principal tissue systems of the developing plant can be detected within this mass.

27.3 Fruit

During seed formation, the flower ovary begins to develop into fruit. The evolution of flowers was key to the success and diversification of the angiosperms. But of equal importance to angiosperm success has been the evolution of these fruits in response to their modes of dispersal. Fruits form in many ways and exhibit a wide array of modes of specialization.

Three layers of ovary wall can have distinct fates and account for the diversity of fruit types, from fleshy to dry and hard. There are three main kinds of fleshy fruits (figure 27.5): berries, drupes, and pomes. In *berries*—such as grapes, tomatoes, and peppers—which are typically many-seeded, the inner layers of the ovary wall are fleshy. In *drupes*—peaches, olives, plums, and cherries—the inner layer of the fruit is stony and adheres tightly to a single seed. In *pomes*—apples and pears—the fleshy portion of the fruit forms from the ovary, which is embedded in the receptacle (the swollen end of the flower stem that holds the petals and sepals).

Fruits that have fleshy coverings, often black, bright blue, or red, are normally dispersed by birds and other vertebrates (figure 27.5*d*). By feeding on these fruits, the birds and other animals carry seeds from place to place before excreting the seeds as solid waste. The seeds, not harmed by the animal digestive system, thus are transferred from one suitable habitat to another. Other fruits that are dispersed by wind (figure 27.5*e*), or by attaching themselves to the fur of mammals or the feathers of birds (figure 27.5*f*), are called dry fruits because they lack the fleshy tissue of edible fruits. Still other fruits, such as those of mangroves, coconuts, and certain other plants that characteristically occur on or near beaches, swamps, or other bodies of water, are regularly spread from place to place by water.

> **27.3** Fruits are specialized to achieve widespread dispersal by wind, by water, by attachment to animals, or, in the case of fleshy fruits, by being eaten.

(a) Berries   (b) Drupes   (c) Pomes

(d) Eaten by animals   (e) Dispersed by wind   (f) Dispersed by attaching to animals

Figure 27.5 Types of fruits and common modes of dispersion.
(*a*) Tomatoes are a type of fleshy fruit called berries that have multiple seeds. (*b*) Peaches are a type of fleshy fruit called drupes that contain a single large seed. (*c*) Apples are a type of fleshy fruit that contains multiple seeds. (*d*) The bright red berries of this honeysuckle, *Lonicera*, are highly attractive to birds. Birds may carry the berry seeds either internally or stuck to their feet for great distances. (*e*) The seeds of this dandelion, *Taraxacum officinale*, are enclosed in a dry fruit with a "parachute" structure that aids its dispersal by wind. (*f*) The spiny fruits of this cocklebur, *Xanthium strumarium*, adhere readily to any passing animal.

27.4 Germination

What happens to a seed when it encounters conditions suitable for its germination? First, it absorbs water. Seed tissues are so dry at the start of germination that the seed takes up water with great force, after which metabolism resumes. Initially, the metabolism may be anaerobic, but when the seed coat ruptures, aerobic metabolism takes over. At this point, oxygen must be available to the developing embryo because plants, which drown for the same reason people do, require oxygen for active growth (see chapter 6). Few plants produce seeds that germinate successfully underwater, although some, such as rice, have evolved a tolerance of anaerobic conditions and can initially respire anaerobically. Figure 27.6 shows the development of a dicot and monocot (see chapter 14) from germination through early stages.

> **27.4** Germination is the resumption of a seed's growth and reproduction, triggered by water.

Figure 27.6 Development of angiosperms.

Dicot development in a soybean. The first structure to emerge is the embryonic root followed by the two cotyledons of the dicot. The cotyledons are pulled up through the soil along with the hypocotyl (the stem below the cotyledons). The cotyledons are the seed leaves that provide nutrients to the growing plant. As other leaves develop, they provide nutrients through photosynthesis, and the cotyledons shrivel and fall off the stem. Flowers develop in buds at the nodes.
Monocot development in corn. The first structure to emerge is the radicle or primary root. Monocots have one cotyledon, which does not emerge from underground. The coleoptile is a tubular sheath; it encloses and protects the shoot and leaves as they push their way up through the soil.

Dicot

Monocot

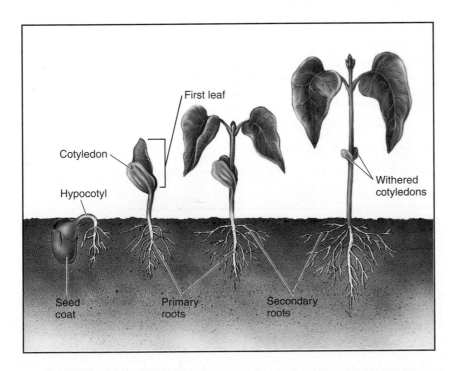

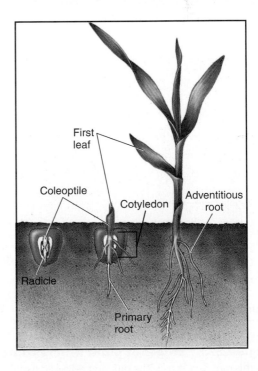

27.5 Plant Hormones

After a seed germinates, the pattern of growth and differentiation that was established in the embryo is repeated indefinitely until the plant dies. But differentiation in plants, unlike that in animals, is largely reversible. Botanists first demonstrated in the 1950s that individual differentiated cells isolated from mature individuals could give rise to entire individuals. F.C. Steward was able to induce isolated bits of phloem tissue taken from carrots to form new plants, plants that were normal in appearance and fully fertile (figure 27.7). Regeneration of entire plants from differentiated tissue has since been carried out in many plants, including cotton, tomatoes, and cherries. These experiments clearly demonstrate that the original differentiated phloem tissue still contains cells that retain all of the genetic potential needed for the differentiation of entire plants. No information is lost during plant tissue differentiation in these cells, and no irreversible steps are taken.

Once a seed has germinated, the plant's further development depends on the activities of the meristematic tissues, which interact with the environment. The shoot and root apical meristems give rise to all of the other cells of the adult plant. Differentiation, or the formation of specialized tissues, occurs in five stages in plants (figure 27.8).

The tissue regeneration experiments of Steward and many others have led to the general conclusion that some nucleated cells in differentiated plant tissue are capable of expressing their hidden genetic information when provided with suitable environmental signals. What halts the expression of genetic potential when the same kinds of cells are

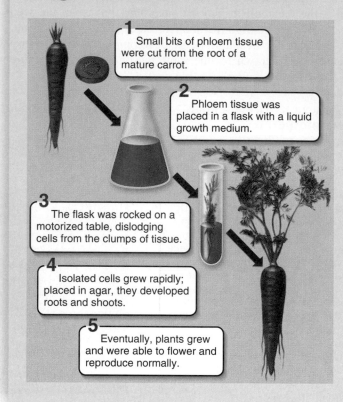

Regeneration in Plants

1 Small bits of phloem tissue were cut from the root of a mature carrot.

2 Phloem tissue was placed in a flask with a liquid growth medium.

3 The flask was rocked on a motorized table, dislodging cells from the clumps of tissue.

4 Isolated cells grew rapidly; placed in agar, they developed roots and shoots.

5 Eventually, plants grew and were able to flower and reproduce normally.

Figure 27.7   How Steward regenerated a plant from differentiated tissue.

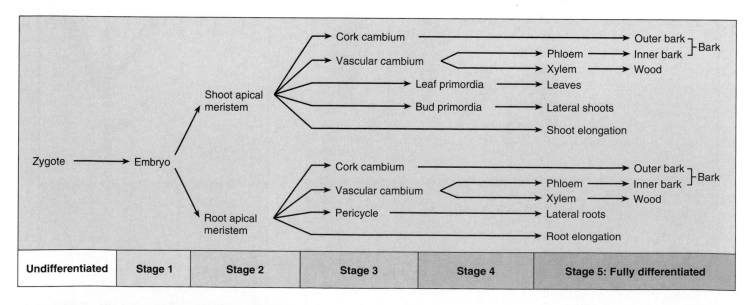

Figure 27.8   Stages of plant differentiation.

As this diagram shows, the different cells and tissues in a plant all originate from the shoot and root apical meristems. It is important to remember, however, that this is showing the origin of the tissue, not the location of the tissue in the plant. For example, the vascular tissues of xylem and phloem arise from the vascular cambium, but these tissues are present throughout the plant, in the leaves, shoots, and roots.

incorporated into normal, growing plants? As we will see, the expression of some of these genes is controlled by plant hormones.

Hormones are chemical substances produced in small (often minute) quantities in one part of an organism and then transported to another part of the organism, where they stimulate certain physiological processes and inhibit others. How they act in a particular instance is influenced both by which hormones are present and by how they affect the particular tissue that receives their message.

In animals, there are several organs, called endocrine glands, that are solely involved with hormone production (hormones are produced in other organs as well). In plants, on the other hand, all hormones are produced in tissues that are not specialized for that purpose and carry out many other functions.

At least five major kinds of hormones are found in plants: auxin, gibberellins, cytokinins, ethylene, and abscisic acid (table 27.1). Other kinds of plant hormones certainly exist but are less well understood. The study of plant hormones, especially how hormones produce their effects, is today an active and important field of research.

27.5   The development of plant tissues is controlled by the actions of hormones. They act on the plant by regulating the expression of key genes.

TABLE 27.1 FUNCTIONS OF THE MAJOR PLANT HORMONES			
Hormone	**Major Functions**	**Where Produced or Found in Plant**	**Practical Applications**
Auxin (IAA)	Promotes stem elongation and growth; forms adventitious roots; inhibits leaf abscission; promotes cell division (with cytokinins); induces ethylene production; promotes lateral bud dormancy	Apical meristems; other immature parts of plants	Seedless fruit production; synthetic auxins act as herbicides
Gibberellins (GA_1, GA_2, GA_3, etc.)	Promotes stem elongation; stimulates enzyme production in germinating seeds	Root and shoot tips; young leaves; seeds	Uniform seed germination for production of barley malt used in brewing; early seed production of biennial plants; increasing size of grapes by allowing more space for growth
Cytokinins	Stimulates cell division, but only in the presence of auxin; promotes chloroplast development; delays leaf aging; promotes bud formation	Root apical meristems; immature fruits	Tissue culture and biotechnology; pruning trees and shrubs, which cause them to "fill out"
Ethylene	Controls leaf, flower, and fruit abscission; promotes fruit ripening	Roots, shoot apical meristems; leaf nodes; aging flowers; ripening fruits	Fruit ripening of agricultural products that are picked early to retain freshness
Abscisic acid (ABA)	Controls stomatal closure; some control of seed dormancy; inhibits effects of other hormones	Leaves, fruits, root caps, seeds	Research on stress tolerance in plants, specifically drought tolerance

27.6 Auxin

In his later years, the great evolutionist Charles Darwin became increasingly devoted to the study of plants. In 1881, he and his son Francis published a book called *The Power of Movement in Plants,* in which they reported their systematic experiments concerning the way in which growing plants bend toward light, a phenomenon known as **phototropism.**

After conducting a series of experiments (figure 27.9), the Darwins hypothesized that when plant shoots were illuminated from one side, an "influence" that arose in the uppermost part of the shoot was then transmitted downward, causing the shoot to bend. Later, several botanists conducted a series of experiments that demonstrated that the substance causing the shoots to bend was a chemical we call **auxin.** Auxin is now known to regulate cell growth in plants.

How auxin controls plant growth was discovered in 1926 by Frits Went, a Dutch plant physiologist, in the course of studies for his doctoral dissertation. From his experiments, described in figure 27.10, Went was able to show that the substance that flowed into the agar from the tips of the light-grown grass seedlings enhanced cell elongation. This chemical messenger caused the tissues on the side of the seedling into which it flowed to grow more than those on the opposite side. He named the substance that he had discovered auxin, from the Greek word *auxin,* meaning "to increase."

Went's experiments provided a basis for understanding the responses the Darwins had obtained some 45 years earlier: Grass seedlings bend toward the light because the auxin contents on the two sides of the shoot differ. The side of the shoot that is in the shade has more auxin; therefore, its cells elongate more than those on the lighted side, bending the plant toward the light (figure 27.11). Later experiments by other investigators showed that auxin in normal plants migrates away from the illuminated side toward the dark side in response to light and thus causes the plant to bend toward the light.

Auxin appears to act by increasing the stretchability of the plant cell wall within minutes of its application. Researchers speculate that the covalent bonds linking the polysaccharides of the cell wall to one another change extensively in response to auxin. This in turn allows auxin-treated cells to take up water and thus enlarge.

The Discovery of Auxin

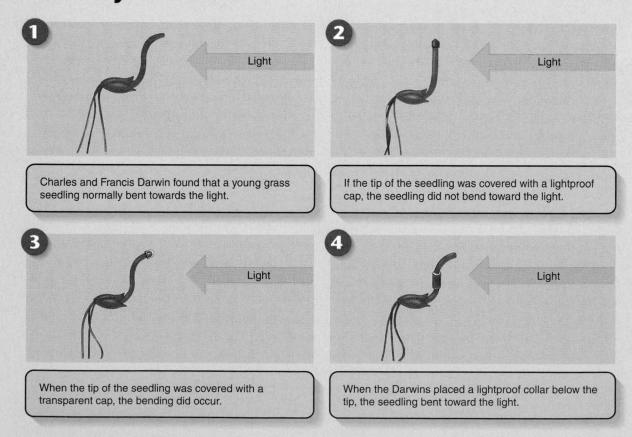

1 Charles and Francis Darwin found that a young grass seedling normally bent towards the light. Light

2 If the tip of the seedling was covered with a lightproof cap, the seedling did not bend toward the light. Light

3 When the tip of the seedling was covered with a transparent cap, the bending did occur. Light

4 When the Darwins placed a lightproof collar below the tip, the seedling bent toward the light. Light

Figure 27.9 The Darwins' experiment with phototropism.
From these experiments, the Darwins concluded that, in response to light, an "influence" that causes bending was transmitted from the tip of the seedling to the area below the tip, where bending usually occurs.

Auxin Promotes Plant Growth

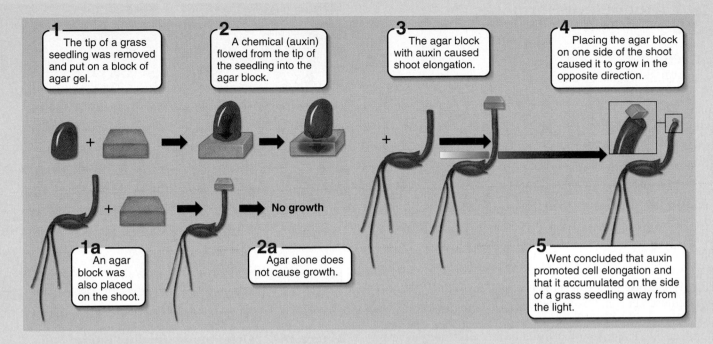

1 The tip of a grass seedling was removed and put on a block of agar gel.

2 A chemical (auxin) flowed from the tip of the seedling into the agar block.

3 The agar block with auxin caused shoot elongation.

4 Placing the agar block on one side of the shoot caused it to grow in the opposite direction.

1a An agar block was also placed on the shoot.

No growth

2a Agar alone does not cause growth.

5 Went concluded that auxin promoted cell elongation and that it accumulated on the side of a grass seedling away from the light.

Figure 27.10 How Went demonstrated the effects of auxin on plant growth.
The experiment showed how a chemical at the tip of the seedling caused the shoot to elongate and to bend. Steps 1a and 2a show the control experiment.

Synthetic auxins are routinely used to control weeds. When applied as herbicides, they are used in higher concentrations than those at which auxin normally occurs in plants. One of the most important of the synthetic auxins used in this way is 2,4-dichlorophenoxyacetic acid, usually known as 2,4-D. It kills weeds in lawns without harming the grass because 2,4-D affects only broad-leaved dicots. When treated, the weeds literally "grow to death," rapidly reducing ATP production so that no source of energy remains for transport or other essential functions.

Closely related to 2,4-D is the herbicide 2,4,5-trichlorophenoxyacetic acid (2,4,5-T), which is widely used to kill woody seedlings and weeds. Notorious as the Agent Orange of the Vietnam War, 2,4,5-T is easily contaminated with a by-product of its manufacture, dioxin. Dioxin is harmful to people because it is an **endocrine disrupter,** a chemical that interferes with the course of human development. The growing release of endocrine disrupters as by-products of modern chemical manufacturing is a subject of great environmental concern.

27.6 The primary growth-promoting hormone of plants is auxin, which increases the plasticity of plant cell walls, allowing growth in specific directions.

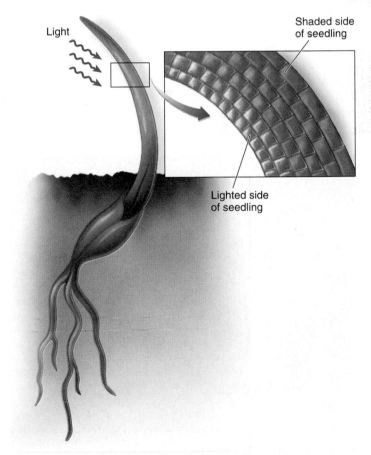

Light

Shaded side of seedling

Lighted side of seedling

Figure 27.11 Auxin causes cells to elongate.
Plant cells that are in the shade have more auxin and grow faster, elongating more, than cells on the lighted side, causing the plant to bend toward light.

Other Plant Hormones

Gibberellins

Gibberellins are a large class of over 100 naturally occurring hormones, abbreviated GA and numbered. Synthesized in the apical portions of both shoots and roots, gibberellins have important effects on stem elongation in plants and play the leading role in controlling this process in mature trees and shrubs (figure 27.12). In these plants, the application of gibberellins characteristically promotes elongation within the spaces between leaf nodes on stems, and this effect is enhanced if auxin is also present. Gibberellins also affect a number of other aspects of plant growth and development. The application of gibberellins can often induce biennial plants (plants that live for two years) to flower early during their first year of growth. These hormones also hasten seed germination, apparently because they can substitute for the effects of cold or light requirements in this process. Gibberellins are used commercially to space out the flowers of grape vines by extending the internode lengths so the fruits have more room to grow and become larger.

Cytokinins

A **cytokinin** is a plant hormone that, in combination with auxin, stimulates cell division in plants and determines the course of differentiation. Substances with these properties are widespread, both in bacteria and in eukaryotes. In vascular plants, most cytokinins seem to be produced in the roots, from which they are then transported throughout the rest of the plant. Cytokinins apparently stimulate cell division by influencing the synthesis or activation of proteins specifically required for mitosis and are therefore used in tissue culture.

Figure 27.12 The effect of a gibberellin.
This rosette mutant (*left*) of the mustard family plant (*Brassica rapa*) is defective in producing gibberellins. It can be rescued by applying gibberellin, which will cause it to flower and grow taller like the plant on the *right*.

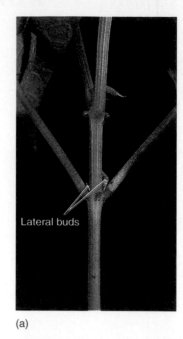

(a)   (b)

Figure 27.13 Cytokinins stimulate lateral bud growth in the absence of auxin.
(*a*) When the apical meristem of a plant is intact, auxin from the apical bud will inhibit the growth of lateral buds. (*b*) When the apical bud is removed, cytokinins are able to promote the growth of lateral buds into branches. This is why pruning bushes and trees makes the plants fuller.

Cytokinins promote growth of lateral buds into branches (figure 27.13); thus, along with auxin and ethylene, they play a role in the control of apical dominance and lateral bud growth. Cytokinins inhibit formation of lateral roots, while auxins promote their formation. As a consequence of these relationships, the balance between cytokinins and auxin, along with other factors, determines the appearance of a mature plant. In addition, the application of cytokinins to leaves detached from a plant retards their yellowing.

Ethylene

Ethylene is a gas that is produced in relatively large quantities during a certain phase of fruit ripening, when the fruit's respiration is proceeding at its most rapid rate. At this phase, complex carbohydrates are broken down into simple sugars, cell walls become soft, and the volatile compounds associated with flavor and scent in the ripe fruits are produced. When applied to fruits, ethylene hastens their ripening.

One of the first lines of evidence that led to the recognition of ethylene as a plant hormone was the observation that gases from oranges caused premature ripening in bananas. Such relationships have led to major commercial uses. Tomatoes are often picked green and then artificially ripened as desired by the application of ethylene. Ethylene is widely used to speed the color formation of lemons and oranges as well.

Genetic engineers have placed genes that interfere with the synthesis of ethylene into tomatoes to slow the ripening process. Until now, commercial tomatoes had to be picked

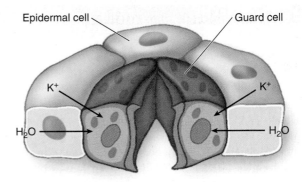

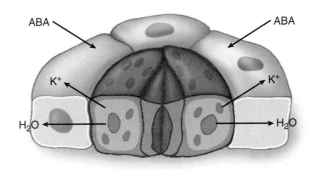

In the absence of ABA—stoma is open

In the presence of ABA—stoma is closed

Figure 27.15   Effects of abscisic acid.
Abscisic acid (ABA) affects the closing of stomata by influencing the movement of potassium ions out of guard cells. As potassium ions are transported out of the guard cells, water also passes out of the cells due to osmosis. The efflux of water causes the guard cells to shrink, resulting in the closing of the stomata.

Figure 27.14   The effects of ethylene.
A holly twig was placed under the glass jar on the *left* for a week. Under the jar on the *right*, a holly twig spent a week with a ripe apple. Ethylene produced by the apple caused abscission of the holly leaves.

very early in order to get them to market before they become overripe, and store-bought tomatoes typically lacked the taste of "homegrown" tomatoes. However, in the genetically engineered tomatoes ripening is delayed, and the tomatoes can be left on the vine longer, improving their taste.

Ethylene also plays an important ecological role. Ethylene production increases rapidly when a plant is exposed to ozone and other toxic chemicals, temperature extremes, drought, attack by pathogens or herbivores, and other stresses. The increased production of ethylene that occurs can accelerate the abscission of leaves (figure 27.14) or fruits that have been damaged by these stresses. It now appears that some of the damage associated with exposure to ozone is due to the ethylene produced by the plants. Some studies suggest that the production of ethylene by plants subjected to attack by herbivores or infected with diseases may be a signal to activate the defense mechanisms of the plants. Such mechanisms may include the production of molecules toxic to the animals or pests attacking them. A full understanding of these relationships is obviously important for agriculture and forestry.

Abscisic Acid

Abscisic acid (ABA) is a naturally occurring plant hormone that is synthesized mainly in mature green leaves, fruits, and root caps. The hormone was given its name because it was thought that it stimulated leaves to age rapidly and fall off (the process of abscission), but evidence that abscisic acid plays an important natural role in this process is scant. In fact, it is believed that abscisic acid may cause ethylene synthesis, and that it is actually the ethylene that promotes senescence and abscission. When abscisic acid is applied to a green leaf, the areas of contact turn yellow. Thus, abscisic acid has the exact opposite effect on a leaf from that of the cytokinins; a yellowing leaf remains green in an area where cytokinins are applied.

Abscisic acid was also initially thought to induce the formation of winter buds—dormant buds that remain through the winter—by suppressing growth, but recent evidence does not support this. Abscisic acid levels increase during seed development and decrease during germination, and so it is likely that ABA plays a role in causing the dormancy of many seeds. ABA may also function in transpiration. During drought conditions, leaves produce large amounts of ABA, which induces the closing of stomata. ABA's effects on stomata occur on the order of minutes, suggesting that this action does not involve turning genes on and off, but rather occurs by influencing the membrane permeability of guard cells. Stomata open when water flows into guard cells, causing them to swell. Water flows into the guard cells osmotically, driven by the influx of potassium ions into the guard cells. ABA most likely stimulates the transport of potassium ions out of the guard cells, causing water to pass out of the guard cells by osmosis. This loss of water causes the stomata to close (figure 27.15).

27.7   Other plant hormones work with auxin and each other to control growth. These include gibberellins, cytokinins, ethylene, and abscisic acid.

27.8 Photoperiodism and Dormancy

Plants respond to different environmental stimuli in a variety of ways. As discussed earlier in the chapter, plants bend toward light as they grow in response to this environmental stimulus. A host of other plant responses, including flowering, dropping of leaves, and yellowing of leaves due to loss of chlorophyll, are also prompted by various environmental stimuli.

Photoperiodism

Essentially all eukaryotic organisms are affected by the cycle of night and day, and many features of plant growth and development are keyed to changes in the proportions of light and dark in the daily 24-hour cycle. Such responses constitute **photoperiodism,** a mechanism by which organisms measure seasonal changes in relative day and night length. One of the most obvious of these photoperiodic reactions concerns angiosperm flower production.

Day length changes with the seasons; the farther from the equator you are, the greater the variation. Plants' flowering responses fall into three basic categories in relation to day length: long-day plants, short-day plants, and day-neutral plants. Long-day plants initiate flowers when nights become shorter than a certain length (and days become longer). Short-day plants, on the other hand, begin to form flowers when nights become longer than a critical length (and days become shorter). Thus, many spring and early summer flowers are long-day plants, and many fall flowers are short-day plants (figure 27.16). The "interrupted night" experiment of figure 27.16 makes it clear that it is the length of uninterrupted dark that is the flowering trigger.

In addition to long-day and short-day plants, a number of plants are described as day-neutral. Day-neutral plants produce flowers without regard to day length.

Photoperiodism

1 Midnight / 6 P.M. / 6 A.M. / Noon

Long-day plants   Short-day plants

Early summer. Short periods of darkness induce flowering in long-day plants, such as iris, but not in short-day plants, such as goldenrod.

2 Midnight / 6 P.M. / 6 A.M. / Noon

Long-day plants   Short-day plants

Late fall. Long periods of darkness induce flowering in short-day plants, such as goldenrod, but not in long-day plants, such as iris.

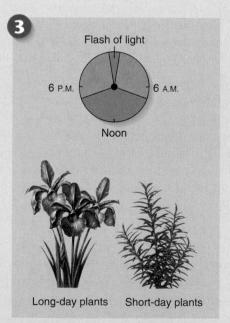

3 Flash of light / 6 P.M. / 6 A.M. / Noon

Long-day plants   Short-day plants

Interrupted night. If the long night of winter is artificially interrupted by a flash of light, the goldenrod will not bloom and the iris will.

Figure 27.16 How photoperiodism works in plants.
In each case, the duration of uninterrupted darkness determines when flowering occurs.

The Chemical Basis of Photoperiodism

Flowering responses to daylight and darkness are controlled by several chemicals that interact in complex ways. Although the nature of some of these chemicals has been deduced, how the various chemicals work together to promote or inhibit flowering responses is still being debated.

Phytochromes. Plants contain a pigment, **phytochrome**, that exists in two interconvertible forms, P_r and P_{fr}. In the first form, phytochrome absorbs red light; in the second, it absorbs far-red light. When a molecule of P_r absorbs a photon of red light (660 nm), it is instantly converted into a molecule of P_{fr}, and when a molecule of P_{fr} absorbs a photon of far-red light (730 nm), it is instantly converted to P_r. P_{fr} is biologically active and P_r is biologically inactive (figure 27.17). In other words, when P_{fr} is present, a given biological reaction that is affected by phytochrome occurs. When most of the P_{fr} has been replaced by P_r, the reaction does not occur.

Phytochrome is a light receptor, but it does not act directly to bring about reactions to light. In short-day plants, the presence of P_{fr} leads to a biological reaction that suppresses flowering. The amount of P_{fr} steadily declines in darkness, the molecules converting to P_r. When the period of darkness is long enough, the suppression reaction ceases and the flowering response is triggered. However, a single flash of red light at a wavelength of about 660 nanometers converts most of the molecules of P_r to P_{fr}, and the flowering reaction is blocked. Still, because most of the P_{fr} is converted to P_r within the first three to four hours of darkness, the conversion of P_r to P_{fr} cannot fully explain the flowering responses of short-day plants; other factors, still not understood, must also be involved.

Phytochrome is also involved in many other plant growth responses. For example, seed germination is inhibited by far-red light and stimulated by red light in many plants. Because chlorophyll absorbs red light strongly but does not absorb far-red light, light passing through green leaves inhibits seed germination. Consequently, seeds on the ground under deciduous plants that lose their leaves in winter are more apt to germinate in the spring after the leaves have decomposed and the seedlings are exposed to direct sunlight. This greatly improves the chances the seedlings will become established.

The Flowering Hormone. Working with long-day and short-day plants, some investigators have gathered evidence for the existence of a flowering hormone. It has been shown that plants will not flower in response to day-length stimuli if their leaves have been removed before exposure to the light. However, the presence of a single leaf or exposure of a single leaf to the appropriate stimuli usually initiates flowering. If the leaf is removed immediately after exposure, the plant will not produce flowers; but if it is left on the plant for a few hours and then removed, flowering occurs normally. These results indicate that a substance passes from the leaves to the apices of the plant, where it induces flowering. Other experiments have shown that, unlike auxin, the substance cannot be transmitted through agar but actually requires a connection through living plant parts.

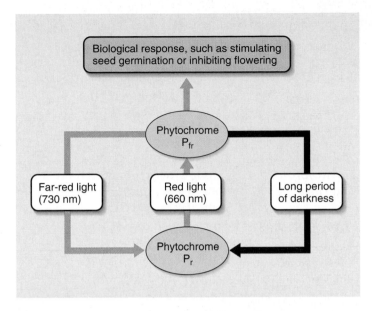

Figure 27.17 How phytochrome works.
When exposed to red light, the phytochrome molecule P_r is converted to P_{fr}, which is the active form that elicits a response in plants such as stimulating seed germination or inhibiting flowering. P_{fr} is converted to P_r when exposed to far-red light, as well as in darkness.

Scientists have searched for a flowering hormone for more than 50 years, but their quest has been unsuccessful. A considerable amount of evidence demonstrates the existence of substances that promote flowering and substances that inhibit it. These poorly understood substances appear to interact in complex ways. The complexity of their interactions, as well as the fact that multiple chemical messengers are evidently involved, has made this scientifically and commercially interesting search very difficult, and to this day, the existence of a flowering hormone remains strictly hypothetical.

Dormancy

Plants respond to their external environment largely by changes in growth rate. Plants' ability to stop growing altogether when conditions are not favorable—to become dormant—is critical to their survival.

In temperate regions, dormancy is generally associated with winter, when low temperatures and the unavailability of water because of freezing make it impossible for plants to grow. During this season, the buds of deciduous trees and shrubs remain dormant, and the apical meristems remain well protected inside enfolding scales. Perennial herbs spend the winter underground as stout stems or roots packed with stored food. Many other kinds of plants, including most annuals, pass the winter as seeds.

27.8 Plant growth and reproduction are sensitive to photoperiod, using chemicals to link flowering to season.

27.9 Tropisms

Tropisms are directional and irreversible growth responses of plants to external stimuli. They control patterns of plant growth and thus plant appearance. Three major classes of plant tropisms include phototropism (figure 27.18*a*, introduced in the discussion of auxin), gravitropism, and thigmotropism.

Gravitropism

Gravitropism causes stems to grow upward and roots downward (figure 27.18*b*). Both of these responses clearly have adaptive significance: Stems that grow upward are apt to receive more light than those that do not; roots that grow downward are more apt to encounter a more favorable environment than those that do not. The phenomenon is called gravitropism because it is clearly a response to gravity.

Thigmotropism

Still another commonly observed response of plants is **thigmotropism,** a name derived from the Greek root *thigma,* meaning "touch." Thigmotropism is defined as the response of plants to touch. Examples include plant tendrils, which rapidly curl around and cling to stems or other objects, and twining plants, such as bindweed, which also coil around objects (figure 27.18*c*). These behaviors result from rapid growth responses to touch. Specialized groups of cells in the plant epidermis appear to be concerned with thigmotropic reactions, but again, their exact mode of action is not well understood.

> **27.9** Growth of the plant body is often sensitive to light, gravity, or touch.

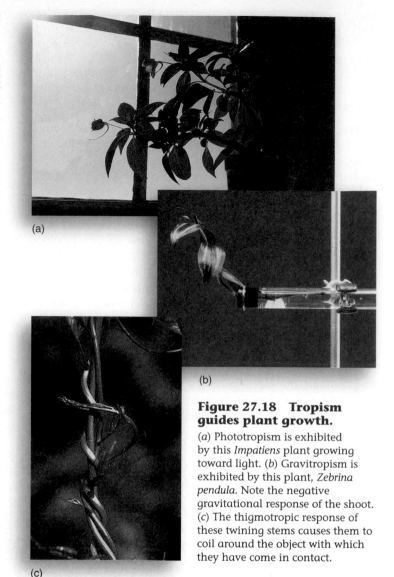

(a)

(b)

(c)

Figure 27.18 Tropism guides plant growth.

(*a*) Phototropism is exhibited by this *Impatiens* plant growing toward light. (*b*) Gravitropism is exhibited by this plant, *Zebrina pendula.* Note the negative gravitational response of the shoot. (*c*) The thigmotropic response of these twining stems causes them to coil around the object with which they have come in contact.

Exploring Current Issues

Additional Resources

Go to your campus library or look online to find the following articles, which further develop some of the concepts found in this chapter.

Carlson, S. (2001). Geotropism, one last time. *Scientific American,* 284(3), 78.

Ma, H. (2003). Plant reproduction: GABA gradient, guidance and growth. *Current Biology,* 13(21), 834.

Moore, I. (2002). Gravitropism: lateral thinking in auxin transport. *Current Biology,* 12(13), 452.

Phillips, H. (2002). Not just a pretty face: they may be green, but plants ain't stupid. *New Scientist,* 175(2353), 40.

Voesenek, L.A.C.J., J.H.G.M. Rijinders, A.J.M. Peeters, H.M. van de Steeg, and H. deKroon. (2004). Plant hormones regulate fast shooting elongation under water: from genes to community. *Ecology,* 85(1), 16.

Biology and Society Lecture: The Promise of GM Crops

In the last decade, the United States has undergone a revolution in agriculture. Genetically modified crops of corn, cotton, and soybeans are now commonplace. Soybeans have been genetically modified to be herbicide resistant, with the result that less tillage has been needed, lessening soil erosion. Bioengineers have modified corn to be pesticide resistant and rice to be more nutritious, containing genes from beans for high-iron protein and other genes to promote the body's absorption of iron and production of beta-carotene. The real promise of GM crops lies in their ability to improve the human condition.

Find this lecture, delivered by the author to his class at Washington University, online at www.mhhe.com/tlwessentials/exp27.

Flowering Plant Reproduction

27.1 Angiosperm Reproduction

- Angiosperms reproduce sexually and asexually. In asexual reproduction, offspring are genetically identical to the parent; this often involves cloning of the parent, a process called vegetative reproduction. Examples of vegetative reproduction include runners, rhizomes, suckers, and adventitious plants (**figure 27.1**).

- Sexual reproduction involves the pollination and fertilization of flowers. Flowers have male and female parts. The male parts include the stamen and the pollen-producing anthers. The female parts include the stigma, style, and ovary, which make up the carpel. The egg cell and two polar nuclei are produced in the ovule, within the ovary (**figure 27.2**).

- Pollen grains are carried by animals, insects, or wind to a flower of the same species (**figure 27.3**). The pollen, after landing on the stigma, extends a pollen tube through the style to the base of the ovule. Two sperm cells travel down the pollen tube, into the ovule. One sperm fertilizes the egg, giving rise to the zygote, and the other fuses with the two polar nuclei, giving rise to the endosperm. This process is called double fertilization.

27.2 Seeds

- Development begins after fertilization. The fertilized egg begins dividing, forming the embryo. After a few days, the embryo stops growing and becomes dormant. The endosperm may remain intact or may be consumed by the embryo, forming cotyledons (seed leaves) that becomes the food source. The outer layer of the ovule becomes the seed coat, enclosing the embryo and food source (**figure 27.4**).

27.3 Fruit

- The evolution of fruit was another key advancement in the angiosperms. During seed formation, the flower's ovary begins to develop into fruit that surrounds the seed. Fruits vary from fleshy to dry, depending on their means of dispersal. Fleshy fruits are usually eaten by animals and then dispersed. Dry fruits are usually dispersed by wind, water, or carried on animals (**figure 27.5**).

27.4 Germination

- When a seed encounters favorable conditions, it will resume growth, a process called germination. The seed absorbs water and uses the endosperm or cotyledons as a food source. Eventually, the seed coat cracks open and the plant begins to grow, sending out roots and shoots (**figure 27.6**).

Regulating Plant Growth

27.5 Plant Hormones

- As a plant grows, its cells differentiate, but differentiation in plants is largely reversible. New plants can be grown from parts of adult plants through regeneration (**figure 27.7**). Differentiation results from the activation or suppression of particular genes, controlled by hormones (**table 27.1**).

27.6 Auxin

- Early researchers, including Darwin, described a process, now called phototropism, where plants grow toward light (**figure 27.9**). Went identified a chemical he called auxin as the hormone involved in phototropism (**figure 27.10**). When exposed to light on one side, auxin is released from the tip of the shoot and causes cells on the shady side of the plant to elongate, causing the plant to grow toward the light (**figure 27.11**).

27.7 Other Plant Hormones

- Other plant hormones include gibberellins, cytokinins, ethylene, and abscisic acid. Gibberellins affect stem elongation, promoting elongation between the node regions (**figure 27.12**). Cytokinin stimulates cell division and acts in combination with auxin. Cytokinin stimulates lateral bud growth in the absence of auxin (**figure 27.13**). Ethylene is released as a gas during fruit ripening. It also accelerates abscission (falling off) of leaves (**figure 27.14**). Abscisic acid (ABA) works in combination with other hormones but seems to play a role in seed dormancy and the closing of stomata (**figure 27.15**).

Plant Responses to Environmental Stimuli

27.8 Photoperiodism and Dormancy

- The length of daylight affects flowering, a process called photoperiodism (**figure 27.16**) and dormancy. A plant pigment, phytochrome, exists in two forms converted by darkness. P_{fr} inhibits flowering, while P_r allows flowering to occur (**figure 27.17**). Some plants flower in response to short days, some to long days, and some are day-neutral.

27.9 Tropisms

- Tropisms are directed growth responses in plants that are irreversible. Phototropism is a growth response toward light; gravitropism is growth in response to the pull of gravity; thigmotropism is growth in response to touch (**figure 27.18**).

1. Sexual reproduction in angiosperms requires
   a. pollen.
   b. runners.
   c. rhizomes.
   d. suckers.

2. Angiosperm flowers that contain both male and female structures are called
   a. dioecious.
   b. monoecious.
   c. parthenogenetic.
   d. incomplete.

3. The flower shape, scent, color, and presence of nectar in the flowers of some angiosperms are related to the plant's
   a. predators.
   b. animal pollinators.
   c. insect pests.
   d. symbionts.
4. For a seed to germinate, the dormant plant embryo must get
   a. carbon dioxide and water.
   b. nitrogen and water.
   c. oxygen and nitrogen.
   d. oxygen and water.
5. Fruit forms from a flower's
   a. ovary.
   b. sepals.
   c. carpels.
   d. stigma.
6. Meristematic tissue regulates plant growth and development through the use of
   a. carbohydrates.
   b. minerals.
   c. hormones.
   d. magnesium.
7. The hormone auxin causes
   a. plant stem cells to shorten by releasing water.
   b. fruit to ripen.
   c. plant stem cells to elongate through absorption of water.
   d. increase of plant diameter through growth of more lateral branches.
8. The hormone ethylene causes
   a. plant stem cells to shorten by releasing water.
   b. fruit to ripen.
   c. increase in fruit size.
   d. increase of plant diameter through growth of more lateral branches.
9. Angiosperm flower production is controlled by
   a. temperature.
   b. amount of available water.
   c. photoperiod.
   d. phototropism.
10. Sensitivity of plants to touch is known as
    a. thigmotropism.
    b. photoperiodism.
    c. phototropism.
    d. gravitropism.

Visual Understanding

1. **Figure 27.3** Bees tend to pollinate yellow flowers; hummingbirds cue in on red flowers. Why are there so many white flowers?

2. **Figure 27.5** If fruits are produced by plants to encourage organisms, especially birds and mammals, to eat them and carry the seeds elsewhere, then speculate on the purpose of fruits such as peaches, shown here, mangoes, and avocados.

(a) Berries (b) Drupes (c) Pomes
(d) Eaten by animals (e) Dispersed by wind (f) Dispersed by attaching to animals

Challenge Questions

Flowering Plant Reproduction Your friend, Shonille, says her teacher made the comment that "nectar is to pollination as fruit is to dispersal." She asks you to explain.

Regulating Plant Growth An old saying is "one bad apple spoils the whole barrel." We even refer to troublemakers as "a bad apple." Can you explain the scientific basis for these statements?

Plant Responses to Environmental Stimuli If it is out in the open, trailing ivy grows along the ground and can form dense mats of ground cover. If planted in a hanging pot, the stems dangle down, forming a graceful flow. If ivy is planted next to a building, however, its stems attach to the side of the building and, in time, it will cover the building. What causes this?

Online Learning Center

Visit the Online Learning Center for this chapter at www.mhhe.com/tlwessentials/ch27 for quizzes, animations, interactive learning exercises, and other study tools. At the site you will also find extended answers to the end-of-chapter questions.

Answers to Self-Test Questions

Chapter 1

1. a. kingdoms 2. c. cellular organization
3. b. atom, molecule, organelle, cell, tissue, organ, organ system, organism, population, species, community, ecosystem 4. d. emergent properties
5. b. evolution, energy flow, cooperation, structure determines function, and homeostasis 6. a. inductive reasoning 7. b. test each hypothesis, using appropriate controls, to rule out as many as possible 8. c. After sufficient testing, you can accept it as probable, being aware that it may be revised or rejected in the future 9. d. all living organisms consist of cells, and all cells come from other cells 10. c. is contained in a long molecule called DNA

Chapter 2

1. c. populations can change over time, sometimes forming new species 2. b. the characteristics of a species varied in different places; there were geographic patterns 3. c. populations are capable of geometric increase, yet remain at constant levels 4. a. natural selection 5. b. seems to agree with Darwin's original ideas 6. d. niche 7. a. a population 8. d. energy flows through once and is lost, while materials cycle and recycle 9. d. competitive exclusion 10. c. the rate at which the population can grow and reproduce if there are no limits

Chapter 3

1. b. an atom 2. c. an ion 3. d. ionic, covalent, and hydrogen 4. a. hydrogen bonds between the individual water molecules 5. a. (1) acids and (2) bases 6. b. proteins, carbohydrates, lipids, and nucleic acids 7. c. structure, function 8. d. are information storage devices found in every cell in the body 9. a. structure and for energy 10. c. energy storage and for hormones

Chapter 4

1. a. cells are the smallest living things. Nothing smaller than a cell is considered alive. 2. b. a double lipid layer with proteins inserted in it, which surrounds every cell individually 3. d. prokaryotes, eukaryotes 4. a. a nucleolus 5. c. the endoplasmic reticulum and the Golgi complex 6. d. mitochondria and the chloroplasts 7. b. Eukaryotic cells in plants and fungi, and all prokaryotes, have a cell wall 8. d. slowly disperse throughout the water; this is because of diffusion 9. b. endocytosis and phagocytosis 10. c. energy and specialized pumps or channels

Chapter 5

1. c. energy 2. d. says that energy can change forms, but cannot be made nor destroyed 3. b. says that entropy, or disorder, continually increases in a closed system 4. a. exergonic and release energy
5. b. enzymes 6. c. temperature and pH
7. d. All of these are true 8. d. produces repressor molecules that alter the enzyme shape 9. c. an inhibitor molecule competes with the substrate for the same binding site on the enzyme 10. a. ATP molecules

Chapter 6

1. c. photosynthesis 2. b. with molecules called pigments that absorb photons and use their energy 3. d. All of these are true 4. c. only go through the system once; they are obtained by splitting a water molecule 5. a. obtain the energy to make organic molecules from carbon dioxide 6. b. use C_4 photosynthesis or CAM 7. a. breaking down the organic molecules that were consumed 8. b. makes ATP by splitting glucose and capturing the energy 9. c. mitochondria of the cell and are broken down in the presence of O_2 to make more ATP 10. d. from the electron transport chain

Chapter 7

1. a. copying DNA then undergoing binary fission 2. c. and most eukaryotes have between 10 and 50 pairs of chromosomes 3. b. metaphase 4. a. a tumor 5. d. germ cells went through meiosis; the egg and sperm only have half the parental chromosomes 6. a. $1n$ gametes (haploid), then next there are $2n$ zygotes (diploid) 7. b. randomly separate the homologous pairs, called independent assortment 8. c. separate the duplicated sister chromatids 9. d. make new cells, and all body cells do it; make eggs or sperm, and only germ-line cells do it 10. c. has a lot of genetic reassortment due to processes in meiosis I

Chapter 8

1. d. All of these 2. a. all purple flowers 3. c. 3/4 purple and 1/4 white flowers 4. b. some factor, or information, about traits to their offspring and it may or may not be expressed 5. b. half of the parental information about a trait (one allele) will be present 6. a. an amino acid sequence that then fold into a protein that has a particular function 7. c. multiple genes affecting one trait, environmental effects 8. d. the crossing over and information exchange between two nonsister chromatids 9. a. nondisjunction and mutation 10. d. genetic screening and prenatal diagnosis is now available

Chapter 9

1. b. hereditary information can be added to cells from other cells 2. d. DNA 3. a. the structure of DNA 4. b. splits down the middle into two single helices, and each one then acts as a template to build its complement 5. c. transcription 6. d. translation 7. d. coded version of a recipe that has been cut up and scattered into other information; all the bits need to be collected, transcribed to mRNA and translated 8. b. some genes are always off unless a promoter turns them on 9. c. by mutation or by recombination 10. a. germ-line tissues and be passed on to future generations

Chapter 10

1. c. genome 2. b. the exons used to make a specific mRNA can be rearranged to form genes for new proteins 3. c. restriction enzyme 4. b. cleaving DNA, producing recombinant DNA, cloning, and screening 5. a. the drug to be produced in far larger amounts than in the past 6. a. increased yield 7. d. harm to the crop itself from mutations 8. d. methylation or demethylation of DNA 9. a. immunological rejection of the tissue by the patient 10. c. viruses

Chapter 11

1. c. macroevolution 2. b. analogous structures 3. a. divergence 4. b. microevolution 5. c. no mutation within the population 6. b. $p^2 = 0.30$ 7. d. genetic drift 8. c. directional selection 9. d. reproductive isolation 10. b. mechanical

Chapter 12

1. a. the red fox is in the same family, but different genus than dogs and wolves 2. b. phylogeny 3. a. physical and chemical characteristics 4. d. cell and DNA structure 5. b. Archaea 6. d. they do not have an internal membrane system 7. c. have cell walls that are made of different materials 8. c. Protista 9. b. multicellular 10. d. ingestion of endosymbiotic bacteria

Chapter 13

1. c. RNA 2. d. All answers are correct 3. b. chemoautotrophs 4. a. protein shells that contain DNA or RNA 5. b. mitochondria and chloroplasts have their own DNA 6. c. cysts 7. d. any organism that is not plant, animal or fungi is a protist 8. d. mycelium 9. a. both sexually and asexually 10. b. algae and fungi

Chapter 14

1. c. dehydration 2. a. chloroplasts 3. b. they do not have specialized vascular tissue to transport water very high 4. d. seeds 5. c. sporophyte 6. d. dispersal 7. c. ovules not completely covered by the sporophyte 8. c. fruits and flowers 9. a. pollination 10. c. an animal's digestive system and processes

Chapter 15

1. b. choanocytes 2. b. specialization of digestive tract 3. b. the pseudocoel develops between the mesoderm and the endoderm in roundworms and the coelom develops in the mesoderm in segmented worms 4. c. the weight of the thick exoskeleton needed to support very large insects 5. a. an animal traveling through the environment to a form more appropriate to a sessile lifestyle 6. b. jaws 7. a. dorsal (back) fin, tail fins and paired fins on the ventral side 8. c. more effective arrangement of appendages 9. d. thin, hollow bones in the skeleton 10. c. placenta and internal development of offspring

Chapter 16

1. c. producers 2. d. All answers are correct
3. d. amount of energy transferred to the top
carnivores 4. a. from plants 5. b. light energy
through photosynthesis 6. c. ATP 7. d. desert
conditions on the down-wind side of a mountain
due to increased moisture-holding capacity of the
winds as the air heats up 8. a. increases, temperature
decreases 9. d. thermocline 10. c. near the top of a
desert mountain range

Chapter 17

1. c. carrying capacity 2. b. increased competition
for food 3. a. short life span 4. b. will decrease,
and the mortality will increase 5. d. community
6. c. resource partitioning 7. b. commensalisms
8. b. decreasing competitive exclusion between prey
species 9. a. aposematic coloration 10. d. has a wide
variety of niches

Chapter 18

1. d. sulfur oxides as a major air pollutant
2. b. coal-powered industry 3. c. chlorofluorocarbons
4. a. ammonia 5. d. All answers are correct
6. a. environmental costs are hardly ever recognized
as part of the economy 7. a. needed to preserve
possible direct value from species, such as new
medicines 8. c. increasing amounts of open space
as countries develop 9. d. captive propagation
10. a. keystone species

Chapter 19

1. b. more flexible movement as individual
segments can move independently of each other
2. a. cells, tissues, organs, organ systems,
organism 3. c. move the body 4. d. red blood cells
5. c. osteoblasts, osteoclasts 6. b. neurotransmitters
7. a. skeletal 8. b. axial skeleton 9. a. a single
muscle can only pull and not push 10. d. expose
myosin attachment sites on actin

Chapter 20

1. d. All of the above are functions of the circulatory
system 2. a. capillaries 3. d. erythrocyte 4. b. better
separation of oxygenated and deoxygenated blood
5. d. endothermy 6. c. the animal's blood to be
continually exposed to water of higher oxygen
concentration 7. b. increase the surface area available
for gas exchange 8. a. energy need for different
vertebrate classes increases 9. a. hemoglobin in red
blood cells 10. b. dissolving it in the blood plasma

Chapter 21

1. b. specialization of different regions of digestive
system 2. a. herbivores 3. d. begin the physical
digestion of food 4. c. small intestine 5. d. increase
the surface area of the small intestine for absorption
of nutrients 6. b. homeostasis 7. b. glycogen to
break down 8. a. retaining metabolites and water
while excreting soluble wastes 9. b. no water and
excrete large volumes of urine that are hypotonic to
body fluids 10. d. osmosis

Chapter 22

1. a. sweat and oil glands 2. b. do not have the
proper cell surface proteins that identify the cell
as the body's own 3. c. increase the number of
immune system cells in an infected area 4. d. All
of the answers are correct 5. a. T cells 6. c. B cells
7. d. destroy cells infected by pathogens
8. b. memory T and B cells 9. c. an autoimmune
response 10. a. helper T cells

Chapter 23

1. c. the amount of associative neurons which
eventually formed the "brain" 2. a. exchange of
sodium and potassium ions. 3. a. sodium ion gates
in the postsynaptic cell 4. d. coordinate sensory
information from muscles and motor responses
5. d. brain stem 6. b. peripheral nervous system
7. d. regulate the body's homeostasis
8. b. interoceptors 9. c. semicircular canals
10. b. more rod cells in retina of the eye

Chapter 24

1. c. chemical signals stick around longer than
nervous signals and can be used for slow processes
2. a. hypothalamus 3. a. all fit into receptors
specifically shaped for them 4. b. steroid hormones
must enter the cell to begin action, whereas peptide
hormones must begin action on the external surface
of the cell membrane 5. d. posterior pituitary gland
6. a. ACTH 7. a. pancreas 8. c. too much calcium
in the blood 9. d. sympathetic nervous system
10. a. GH

Chapter 25

1. b. sexual reproduction 2. d. internal fertilization
3. a. viviparity 4. d. scrotum 5. b. FSH and LH
6. a. embryo releasing hCG 7. c. oxytocin
8. d. gastrulation is complete 9. c. nervous
10. a. respiratory system is not fully formed

Chapter 26

1. c. meristematic tissue 2. b. transports
carbohydrates 3. a. stomata 4. d. at the pericycle
5. c. cambium 6. b. organization of vascular tissue
7. c. spongy mesophyll 8. a. photosynthesis
9. b. translocation 10. d. nitrogen

Chapter 27

1. a. pollen 2. b. monoecious 3. b. animal pollinators
4. d. oxygen and water 5. a. ovary 6. c. hormones
7. c. plant stem cells to elongate through absorption
of water 8. b. fruit to ripen 9. c. photoperiod
10. a. thigmotropism

Glossary

A

absorption (L. *absorbere,* to swallow down) The movement of water and substances dissolved in water into a cell, tissue, or organism.

acid Any substance that dissociates to form H$^+$ ions when dissolved in water. Having a pH value less than 7.

acoelomate (Gr. *a,* not + *koiloma,* cavity) A bilaterally symmetrical animal not possessing a body cavity, such as a flatworm.

actin (Gr. *actis,* ray) One of the two major proteins that make up myofilaments (the other is myosin). It provides the cell with mechanical support and plays major roles in determining cell shape and cell movement.

action potential A single nerve impulse. A transient all-or-none reversal of the electrical potential across a neuron membrane. Because it can activate nearby voltage-sensitive channels, an action potential propagates along a nerve cell.

activation energy The energy a molecule must acquire to undergo a specific chemical reaction.

active transport The transport of a solute across a membrane by protein carrier molecules to a region of higher concentration by the expenditure of chemical energy. One of the most important functions of any cell.

adaptation (L. *adaptare,* to fit) Any peculiarity of structure, physiology, or behavior that promotes the likelihood of an organism's survival and reproduction in a particular environment.

adenosine triphosphate (ATP) A molecule composed of ribose, adenine, and a triphosphate group. ATP is the chief energy currency of all cells. Cells focus all of their energy resources on the manufacture of ATP from ADP and phosphate, which requires the cell to supply 7 kilocalories of energy obtained from photosynthesis or from electrons stripped from foodstuffs to form 1 mole of ATP. Cells then use this ATP to drive endergonic reactions.

adhesion (L. *adhaerere,* to stick to) The molecular attraction exerted between the surfaces of unlike bodies in contact, as water molecules to the walls of the narrow tubes that occur in plants.

aerobic (Gr. *aer,* air + *bios,* life) Oxygen-requiring.

allele (Gr. *allelon,* of one another) One of two or more alternative forms of a gene.

allele frequency The relative proportion of a particular allele among individuals of a population. Not equivalent to gene frequency, although the two terms are sometimes confused.

allosteric interaction (Gr. *allos,* other + *stereos,* shape) The change in shape that occurs when an activator or inhibitor binds to an enzyme. These changes result when specific, small molecules bind to the enzyme, molecules that are not substrates of that enzyme.

alternation of generations A reproductive life cycle in which the multicellular diploid phase produces spores that give rise to the multicellular haploid phase and the multicellular haploid phase produces gametes that fuse to give rise to the zygote. The zygote is the first cell of the multicellular diploid phase.

alveolus, *pl.* alveoli (L. *alveus,* a small cavity) One of the many small, thin-walled air sacs within the lungs in which the bronchioles terminate.

amniotic egg An egg that is isolated and protected from the environment by a more or less impervious shell. The shell protects the embryo from drying out, nourishes it, and enables it to develop outside of water.

anaerobic (Gr. *an,* without + *aer,* air + *bios,* life) Any process that can occur without oxygen. Includes glycolysis and fermentation. Anaerobic organisms can live without free oxygen.

anaphase In mitosis and meiosis II, the stage initiated by the separation of sister chromatids, during which the daughter chromosomes move to opposite poles of the cell; in meiosis I, marked by separation of replicated homologous chromosomes.

angiosperms The flowering plants, one of five phyla of seed plants. In angiosperms, the ovules at the time of pollination are completely enclosed by tissues.

anterior (L. *ante,* before) Located before or toward the front. In animals, the head end of an organism.

anther (Gr. *anthos,* flower) The part of the stamen of a flower that bears the pollen.

antibody (Gr. *anti,* against) A protein substance produced in the blood by a B cell lymphocyte in response to a foreign substance (antigen) and released into the bloodstream. Binding to the antigen, antibodies mark them for destruction by other elements of the immune system.

anticodon The three-nucleotide sequence at the end of a tRNA molecule that is complementary to, and base pairs with, an amino acid-specifying codon in mRNA.

antigen (Gr. *anti,* against + *genos,* origin) A foreign substance, usually a protein, that stimulates lymphocytes to proliferate and secrete specific antibodies that bind to the foreign substance, labeling it as foreign and destined for destruction.

apical meristem (L. *apex,* top + Gr. *meristos,* divided) In vascular plants, the growing point at the tip of the root or stem.

aposematic coloration An ecological strategy of some organisms that "advertise" their poisonous nature by the use of bright colors.

appendicular skeleton (L. *appendicula,* a small appendage) The skeleton of the limbs of the human body containing 126 bones.

archaea A group of prokaryotes that are among the most primitive still in existence, characterized by the absence of peptidoglycan in their cell walls, a feature that distinguishes them from bacteria.

asexual Reproducing without forming gametes. Asexual reproduction does not involve sex. Its outstanding characteristic is that an individual offspring is genetically identical to its parent.

association neuron A nerve cell found only in the CNS that acts as a functional link between sensory neurons and motor neurons. Also called interneuron.

atom (Gr. *atomos,* indivisible) A core (nucleus) of protons and neutrons surrounded by an orbiting cloud of electrons. The chemical behavior of an atom is largely determined by the distribution of its electrons, particularly the number of electrons in its outermost level.

atomic mass The atomic mass of an atom consists of the combined mass of all of its protons and neutrons.

atomic number The number of protons in the nucleus of an atom. In an atom that does not bear an electric charge (that is, one that is not an ion), the atomic number is also equal to the number of electrons.

autonomic nervous system (Gr. *autos,* self + *nomos,* law) The motor pathways that carry commands from the central nervous system to regulate the glands and nonskeletal muscles of the body. Also called the involuntary nervous system.

autosome (Gr. *autos,* self + *soma,* body) Any of the 22 pairs of human chromosomes that are similar in size and morphology in both males and females.

autotroph (Gr. *autos*, self + *trophos*, feeder) Self-feeder. An organism that can harvest light energy from the sun or from the oxidation of inorganic compounds to make organic molecules.

axial skeleton The skeleton of the head and trunk of the human body containing 80 bones.

axon (Gr., axle) A process extending out from a neuron that conducts impulses away from the cell body.

B

bacterium, *pl.* bacteria (Gr. *bakterion*, dim. of *baktron*, a staff) The simplest cellular organism. Its cells are smaller and prokaryotic in structure, and they lack internal organization.

basal body In cells that contain flagella or cilia, a form of centriole that anchors each flagellum.

base Any substance that combines with H$^+$ ions thereby reducing the H$^+$ ion concentration of a solution. Having a pH value above 7.

Batesian mimicry After Henry W. Bates, English naturalist. A situation in which a palatable or nontoxic organism resembles another kind of organism that is distasteful or toxic. Both species exhibit warning coloration.

B cell A lymphocyte that recognizes invading pathogens much as T cells do, but instead of attacking the pathogens directly, it marks them for destruction by the nonspecific body defenses.

bilateral symmetry (L. *bi*, two + *lateris*, side; Gr. *symmetria*, symmetry) A body form in which the right and left halves of an organism are approximate mirror images of each other.

binary fission (L. *binarius*, consisting of two things or parts + *fissus*, split) Asexual reproduction of a cell by division into two equal, or nearly equal, parts. Bacteria divide by binary fission.

binomial system (L. *bi*, twice, two + Gr. *nomos*, usage, law) A system of nomenclature that uses two words. The first names the genus, and the second designates the species.

biomass (Gr. *bios*, life + *maza*, lump or mass) The total weight of all of the organisms living in an ecosystem.

biome (Gr. *bios*, life + *-oma*, mass, group) A major terrestrial assemblage of plants, animals, and microorganisms that occur over wide geographical areas and have distinct characteristics. The largest ecological unit.

C

calorie (L. *calor*, heat) The amount of energy in the form of heat required to raise the temperature of 1 gram of water 1 degree Celsius.

calyx (Gr. *kalyx*, a husk, cup) The sepals collectively. The outermost flower whorl.

cancer Unrestrained invasive cell growth. A tumor or cell mass resulting from uncontrollable cell division.

capillary (L. *capillaris*, hairlike) A blood vessel with a very small diameter. Blood exchanges gases and metabolites within capillaries. Capillaries join the end of an arteriole to the beginning of a venule.

carbohydrate (L. *carbo*, charcoal + *hydro*, water) An organic compound consisting of a chain or ring of carbon atoms to which hydrogen and oxygen atoms are attached in a ratio of approximately 1:2:1. A compound of carbon, hydrogen, and oxygen having the generalized formula $(CH_2O)_n$ where n is the number of carbon atoms.

carcinogen (Gr. *karkinos*, cancer + -gen) Any cancer-causing agent.

cardiovascular system (Gr. *kardia*, heart + L. *vasculum*, vessel) The blood circulatory system and the heart that pumps it. Collectively, the blood, heart, and blood vessels.

carpel (Gr. *karpos*, fruit) A leaflike organ in angiosperms that encloses one or more ovules.

carrying capacity The maximum population size that a habitat can support.

catabolism (Gr. *katabole*, throwing down) A process in which complex molecules are broken down into simpler ones.

catalysis (Gr. *katalysis*, dissolution + *lyein*, to loosen) The enzyme-mediated process in which the subunits of polymers are positioned so that their bonds are stressed.

catalyst (Gr. *kata*, down + *lysis*, a loosening) A general term for a substance that speeds up a specific chemical reaction by lowering the energy required to activate or start the reaction. An enzyme is a biological catalyst.

cell (L. *cella*, a chamber or small room) The smallest unit of life. The basic organizational unit of all organisms. Composed of a nuclear region containing the hereditary apparatus within a larger volume called the cytoplasm bounded by a lipid membrane.

cell cycle The repeating sequence of growth and division through which cells pass each generation.

cellular respiration The process in which the energy stored in a glucose molecule is released by oxidation. Hydrogen atoms are lost by glucose and gained by oxygen.

central nervous system The brain and spinal cord, the site of information processing and control within the nervous system.

centromere (Gr. *kentron*, center + *meros*, a part) A constricted region of the chromosome joining two sister chromatids, to which the kinetochore is attached.

chemical bond The force holding two atoms together. The force can result from the attraction of opposite charges (ionic bond) or from the sharing of one or more pairs of electrons (a covalent bond).

chemiosmosis The cellular process responsible for almost all of the adenosine

triphosphate (ATP) harvested from food and for all the ATP produced by photosynthesis.

chemoautotroph An autotrophic bacterium that uses chemical energy released by specific inorganic reactions to power its life processes, including the synthesis of organic molecules.

chloroplast (Gr. *chloros*, green + *plastos*, molded) A cell-like organelle present in algae and plants that contains chlorophyll (and usually other pigments) and is the site of photosynthesis.

chiasma, *pl.* chiasmata (Gr. a cross) In meiosis, the points of crossingover where portions of chromosomes have been exchanged during synapsis. A chiasma appears as an X-shaped structure under a light microscope.

chromatid (Gr. *chroma*, color + L. -*id*, daughters of) One of two daughter strands of a duplicated chromosome that is joined by a single centromere.

chromatin (Gr. *chroma*, color) The complex of DNA and proteins of which eukaryotic chromosomes are composed.

chromosome (Gr. *chroma*, color + *soma*, body) The vehicle by which hereditary information is physically transmitted from one generation to the next. In a eukaryotic cell, long threads of DNA that are associated with protein and that contain hereditary information.

cilium, *pl.* cilia (L. eyelash) Refers to flagella, which are numerous and organized in dense rows. Cilia propel cells through water. In human tissue, they move water over the tissue surface.

cladistics A taxonomic technique used for creating hierarchies of organisms that represent true phylogenetic relationship and descent.

class A taxonomic category ranking below a phylum (division) and above an order.

clone (Gr. *klon*, twig) A line of cells, all of which have arisen from the same single cell by mitotic division. One of a population of individuals derived by asexual reproduction from a single ancestor. One of a population of genetically identical individuals.

codominance In genetics, a situation in which the effects of both alleles at a particular locus are apparent in the phenotype of the heterozygote.

codon (L. code) The basic unit of the genetic code. A sequence of three adjacent nucleotides in DNA or mRNA that codes for one amino acid or for polypeptide termination.

coelom (Gr. *koilos*, a hollow) A body cavity formed between layers of mesoderm and in which the digestive tract and other internal organs are suspended.

coenzyme A cofactor that is a nonprotein organic molecule.

coevolution (L. *co-*, together + *e-*, out + *volvere*, to fill) A term that describes the long-term evolutionary adjustment of one group of organisms to another.

commensalism (L. *cum*, together with + *mensa*, table) A symbiotic relationship in which one species benefits while the other neither benefits nor is harmed.

community (L. *communitas*, community, fellowship) The populations of different species that live together and interact in a particular place.

competition Interaction between individuals of two or more species for the same scarce resources. Intraspecific competition is competition between individuals of a single species. Interspecific competition is competition between individuals of different species.

competitive exclusion The hypothesis that if two species are competing with one another for the same limited resource in the same place, one will be able to use that resource more efficiently than the other and eventually will drive that second species to extinction locally.

complement system The chemical defense of a vertebrate body that consists of a battery of proteins that insert in bacterial and fungal cells, causing holes that destroy the cells.

concentration gradient The concentration difference of a substance as a function of distance. In a cell, a greater concentration of its molecules in one region than in another.

condensation The coiling of the chromosomes into more and more tightly compacted bodies begun during the G_2 phase of the cell cycle.

conjugation (L. *conjugare*, to yoke together) An unusual mode of reproduction in unicellular organisms in which genetic material is exchanged between individuals through tubes connecting them during conjugation.

consumer In ecology, a heterotroph that derives its energy from living or freshly killed organisms or parts thereof. Primary consumers are herbivores; secondary consumers are carnivores or parasites.

cortex (L. bark) In vascular plants, the primary ground tissue of a stem or root, bounded externally by the epidermis and internally by the central cylinder of vascular tissue. In animals, the outer, as opposed to the inner, part of an organ, as in the adrenal, kidney, and cerebral cortexes.

cotyledon (Gr. *kotyledon*, a cup-shaped hollow) Seed leaf. Monocot embryos have one cotyledon, and dicots have two.

countercurrent flow In organisms, the passage of heat or of molecules (such as oxygen, water, or sodium ions) from one circulation path to another moving in the opposite direction. Because the flow of the two paths is in opposite directions, a concentration difference always exists between the two channels, facilitating transfer.

covalent bond (L. *co-*, together + *valare*, to be strong) A chemical bond formed by the sharing of one or more pairs of electrons.

crossing over An essential element of meiosis occurring during prophase when nonsister chromatids exchange portions of DNA strands.

cuticle (L. *cutis*, skin) A very thin film covering the outer skin of many plants.

cytokinesis (Gr. *kytos*, hollow vessel + *kinesis*, movement) The C phase of cell division in which the cell itself divides, creating two daughter cells.

cytoplasm (Gr. *kytos*, hollow vessel + *plasma*, anything molded) A semifluid matrix that occupies the volume between the nuclear region and the cell membrane. It contains the sugars, amino acids, proteins, and organelles (in eukaryotes) with which the cell carries out its everyday activities of growth and reproduction.

cytoskeleton (Gr. *kytos*, hollow vessel + *skeleton*, a dried body) In the cytoplasm of all eukaryotic cells, a network of protein fibers that supports the shape of the cell and anchors organelles, such as the nucleus, to fixed locations.

D

deciduous (L. *decidere*, to fall off) In vascular plants, shedding all the leaves at a certain season.

dehydration reaction Water-losing. The process in which a hydroxyl (OH) group is removed from one subunit of a polymer and a hydrogen (H) group is removed from the other subunit, forming a water molecule as a by product.

demography (Gr. *demos*, people + *graphein*, to draw) The statistical study of population. The measurement of people or, by extension, of the characteristics of people.

density The number of individuals in a population in a given area.

deoxyribonucleic acid (DNA) The basic storage vehicle or central plan of heredity information. It is stored as a sequence of nucleotides in a linear nucleotide polymer. Two of the polymers wind around each other like the outside and inside rails of a circular staircase.

depolarization The movement of ions across a cell membrane that wipes out locally an electrical potential difference.

deuterostome (Gr. *deuteros*, second + *stoma*, mouth) An animal in whose embryonic development the anus forms from or near the blastopore, and the mouth forms later on another part of the blastula. Also characterized by radial cleavage.

dicot Short for dicotyledon; a class of flowering plants generally characterized by having two cotyledons, netlike veins, and flower parts in fours or fives.

diffusion (L. *diffundere*, to pour out) The net movement of molecules to regions of lower concentration as a result of random, spontaneous molecular motions. The process tends to distribute molecules uniformly.

dihybrid (Gr. *dis*, twice + L. *hibrida*, mixed offspring) An individual heterozygous for two genes.

dioecious (Gr. *di*, two + *eikos*, house) Having male and female flowers on separate plants of the same species.

diploid (Gr. *diploos*, double + *eidos*, form) A cell, tissue, or individual with a double set of chromosomes.

directional selection A form of selection in which selection acts to eliminate one extreme from an array of phenotypes. Thus, the genes promoting this extreme become less frequent in the population.

disaccharide (Gr. *dis*, twice + *sakcharon*, sugar) A sugar formed by linking two monosaccharide molecules together. Sucrose (table sugar) is a disaccharide formed by linking a molecule of glucose to a molecule of fructose.

disruptive selection A form of selection in which selection acts to eliminate rather than favor the intermediate type.

diurnal (L. *diurnalis*, day) Active during the day.

division Traditionally, a major taxonomic group of the plant kingdom comparable to a phylum of the animal kingdom. Today divisions are called phyla.

dominant allele An allele that dictates the appearance of heterozygotes. One allele is said to be dominant over another if an individual heterozygous for that allele has the same appearance as an individual homozygous for it.

dorsal (L. *dorsum*, the back) Toward the back, or upper surface. Opposite of ventral.

double fertilization A process unique to the angiosperms, in which one sperm nucleus fertilizes the egg and the second one fuses with the polar nuclei. These two events result in the formation of the zygote and the primary endosperm nucleus, respectively.

E

ecdysis (Gr. *ekdysis*, stripping off) The shedding of the outer covering or skin of certain animals. Especially the shedding of the exoskeleton by arthropods.

ecology (Gr. *oikos*, house + *logos*, word) The study of the relationships of organisms with one another and with their environment.

ecosystem (Gr. *oikos*, house + *systema*, that which is put together) A community, together with the nonliving factors with which it interacts.

ectoderm (Gr. *ecto*, outside + *derma*, skin) One of three embryonic germ layers that forms in the gastrula; giving rise to the outer epithelium and to nerve tissue.

ectothermic Referring to animals whose body temperature is regulated by their behavior or their surroundings.

electron A subatomic particle with a negative electrical charge. The negative charge of one electron exactly balances the positive charge of one proton. Electrons orbit the atom's positively charged nucleus and determine its chemical properties.

electron transport chain A collective term describing the series of membrane-associated electron carriers embedded in the inner mitochondrial membrane. It puts the electrons harvested from the oxidation of glucose to work driving proton-pumping channels.

element A substance that cannot be separated into different substances by ordinary chemical methods.

endergonic (Gr. *endon*, within + *ergon*, work) Describing reactions in which the products contain more energy than the reactants and require an input of usable energy from an outside source before they can proceed. These reactions are not spontaneous.

endocrine gland (Gr. *endon*, within + *krinein*, to separate) A ductless gland producing hormonal secretions that pass directly into the bloodstream or lymph.

endocrine system The dozen or so major endocrine glands of a vertebrate.

endocytosis (Gr. *endon*, within + *kytos*, cell) The process by which the edges of plasma membranes fuse together and form an enclosed chamber called a vesicle. It involves the incorporation of a portion of an exterior medium into the cytoplasm of the cell by capturing it within the vesicle.

endoderm (Gr. *endon*, outside + *derma*, skin) One of three embryonic germ layers that forms in the gastrula; giving rise to the epithelium that lines internal organs and most of the digestive and respiratory tracts.

endoskeleton (Gr. *endon*, within + *skeletos*, hard) In vertebrates, an internal scaffold of bone to which muscles are attached.

endosperm (Gr. *endon*, within + *sperma*, seed) A nutritive tissue characteristic of the seeds of angiosperms that develops from the union of a male nucleus and the polar nuclei of the embryo sac. The endosperm is either digested by the growing embryo or retained in the mature seed to nourish the germinating seedling.

endosymbiotic (Gr. *endon*, within + *bios*, life) theory Proposes that eukaryotic cells arose from large prokaryotic cells that engulfed smaller ones of a different species, which were not consumed but continued to live and function within the larger host cell. Organelles that are believed to have entered larger cells in this way are mitochondria and chloroplasts.

endothermic Referring to the ability of animals to maintain a constant body temperature.

energy The capacity to bring about change, to do work.

enhancer A site of regulatory protein binding on the DNA molecule distant from the promoter and start site for a gene's transcription.

entropy (Gr. *en*, in + *tropos*, change in manner) A measure of the disorder of a system. A measure of energy that has become so randomized and uniform in a system that the energy is no longer available to do work.

enzyme (Gr. *enzymos*, leavened; from *en*, in + *zyme*, leaven) A protein capable of speeding up specific chemical reactions by lowering the energy required to activate or start the reaction but that remains unaltered in the process.

epidermis (Gr. *epi*, on or over + *derma*, skin) The outermost layer of cells. In vertebrates, the nonvascular external layer of skin of ectodermal origin; in invertebrates, a single layer of ectodermal epithelium; in plants, the flattened, skinlike outer layer of cells.

epistasis (Gr. *epistasis*, a standing still) An interaction between the products of two genes in which one modifies the phenotypic expression produced by the other.

epithelium (Gr. *epi*, on + *thele*, nipple) A thin layer of cells forming a tissue that covers the internal and external surfaces of the body. Simple epithelium consists of the membranes that line the lungs and major body cavities and that are a single cell layer thick. Stratified epithelium (the skin or epidermis) is composed of more complex epithelial cells that are several cell layers thick.

erythrocyte (Gr. *erythros*, red + *kytos*, hollow vessel) A red blood cell, the carrier of hemoglobin. Erythrocytes act as the transporters of oxygen in the vertebrate body. During the process of their maturation in mammals, they lose their nuclei and mitochondria, and their endoplasmic reticulum is reabsorbed.

estrus (L. *oestrus*, frenzy) The period of maximum female sexual receptivity. Associated with ovulation of the egg. Being "in heat."

estuary (L. *aestus*, tide) A partly enclosed body of water, such as those that often form at river mouths and in coastal bays, where the salinity is intermediate between that of saltwater and freshwater.

ethology (Gr. *ethos*, habit or custom + *logos*, discourse) The study of patterns of animal behavior in nature.

euchromatin (Gr. *eu*, good + *chroma*, color) Chromatin that is extended except during cell division, from which RNA is transcribed.

eukaryote (Gr. *eu*, good + *karyon*, kernel) A cell that possesses membrane-bounded organelles, most notably a cell nucleus, and chromosomes whose DNA is associated with proteins; an organism composed of such cells. The appearance of eukaryotes marks a major event in the evolution of life, as all organisms on earth other than bacteria and archaea are eukaryotes.

eumetazoan (Gr. *eu*, good + *meta*, with + *zoion*, animal) A "true animal." An animal with a definite shape and symmetry and nearly always distinct tissues.

eutrophic (Gr. *eutrophos*, thriving) Refers to a lake in which an abundant supply of minerals and organic matter exists.

evaporation The escape of water molecules from the liquid to the gas phase at the surface of a body of water.

evolution (L. *evolvere*, to unfold) Genetic change in a population of organisms over time (generations). Darwin proposed that natural selection was the mechanism of evolution.

exergonic (L. *ex*, out + Gr. *ergon*, work) Describes any reaction that produces products that contain less free energy than that possessed by the original reactants and that tends to proceed spontaneously.

exocytosis (Gr. *ex*, out of + *kytos*, cell) The extrusion of material from a cell by discharging it from vesicles at the cell surface. The reverse of endocytosis.

exoskeleton (Gr. *exo*, outside + *skeletos*, hard) An external hard shell that encases a body. In arthropods, comprised mainly of chitin.

experiment The test of a hypothesis. A successful experiment is one in which one or more alternative hypotheses are demonstrated to be inconsistent with experimental observation and are thus rejected.

F

facilitated diffusion The transport of molecules across a membrane by a carrier protein in the direction of lowest concentration.

family A taxonomic group ranking below an order and above a genus.

feedback inhibition A regulatory mechanism in which a biochemical pathway is regulated by the amount of the product that the pathway produces.

fermentation (L. *fermentum*, ferment) A catabolic process in which the final electron acceptor is an organic molecule.

fertilization (L. *ferre*, to bear) The union of male and female gametes to form a zygote.

fitness The genetic contribution of an individual to succeeding generations, relative to the contributions of other individuals in the population.

flagellum, *pl.* flagella (L. *flagellum*, whip) A fine, long, threadlike organelle protruding from the surface of a cell. In bacteria, a single protein fiber capable of rotary motion that propels the cell through the water. In eukaryotes, an array of microtubules with a characteristic internal 9 + 2 microtubule structure that is capable of vibratory but not rotary motion. Used in locomotion and feeding. Common in protists and motile gametes. A cilium is a small flagellum.

food web The food relationships within a community. A diagram of who eats whom.

founder principle The effect by which rare alleles and combinations of alleles may be enhanced in new populations.

frequency In statistics, defined as the proportion of individuals in a certain category, relative to the total number of individuals being considered.

fruit In angiosperms, a mature, ripened ovary (or group of ovaries) containing the seeds.

G

gamete (Gr. wife) A haploid reproductive cell. Upon fertilization, its nucleus fuses with that of another gamete of the opposite sex. The resulting diploid cell (zygote) may develop into a new diploid individual, or in some protists and fungi, may undergo meiosis to form haploid somatic cells.

gametophyte (Gr. *gamete*, wife + *phyton*, plant) In plants, the haploid (n), gamete-producing generation, which alternates with the diploid ($2n$) sporophyte.

ganglion, *pl.* **ganglia (Gr. a swelling)** A group of nerve cells forming a nerve center in the peripheral nervous system.

gastrulation The inward movement of certain cell groups from the surface of the blastula.

gene (Gr. *genos*, birth, race) The basic unit of heredity. A sequence of DNA nucleotides on a chromosome that encodes a polypeptide or RNA molecule and so determines the nature of an individual's inherited traits.

gene expression The process in which an RNA copy of each active gene is made, and the RNA copy directs the sequential assembly of a chain of amino acids at a ribosome.

gene frequency The frequency with which individuals in a population possess a particular gene. Often confused with allele frequency.

genetic code The "language" of the genes. The mRNA codons specific for the 20 common amino acids constitute the genetic code.

genetic drift Random fluctuations in allele frequencies in a small population over time.

genetic map A diagram showing the relative positions of genes.

genetics (Gr. *genos*, birth, race) The study of the way in which an individual's traits are transmitted from one generation to the next.

genome (Gr. *genos*, offspring + L. *oma*, abstract group) The genetic information of an organism.

genomics The study of genomes as opposed to individual genes.

genotype (Gr. *genos*, offspring + *typos*, form) The total set of genes present in the cells of an organism. Also used to refer to the set of alleles at a single gene locus.

genus, *pl.* **genera (L. race)** A taxonomic group that ranks below a family and above a species.

germination (L. *germinare,* to sprout) The resumption of growth and development by a spore or seed.

gland (L. *glandis,* acorn) Any of several organs in the body, such as exocrine or endocrine, that secrete substances for use in the body. Glands are composed of epithelial tissue.

glomerulus (L. a little ball) A network of capillaries in a vertebrate kidney, whose walls act as a filtration device.

glycolysis (Gr. *glykys,* sweet + *lyein,* to loosen) The anaerobic breakdown of glucose; this enzyme-catalyzed process yields two molecules of pyruvate with a net of two molecules of ATP.

gravitropism (L. *gravis,* heavy + *tropes,* turning) The response of a plant to gravity, which generally causes shoots to grow up and roots to grow down.

greenhouse effect The process in which carbon dioxide and certain other gases, such as methane, that occur in the earth's atmosphere transmit radiant energy from the sun but trap the longer wavelengths of infrared light, or heat, and prevent them from radiating into space.

guard cells Pairs of specialized epidermal cells that surround a stoma. When the guard cells are turgid, the stoma is open; when they are flaccid, it is closed.

gymnosperm (Gr. *gymnos,* naked + *sperma,* seed) A seed plant with seeds not enclosed in an ovary. The conifers are the most familiar group.

H

habitat (L. *habitare,* to inhabit) The place where individuals of a species live.

half-life The length of time it takes for half of a radioactive substance to decay.

haploid (Gr. *haploos,* single + *eidos,* form) The gametes of a cell or an individual with only one set of chromosomes.

Hardy-Weinberg equilibrium After G. H. Hardy, English mathematician, and G. Weinberg, German physician. A mathematical description of the fact that the relative frequencies of two or more alleles in a population do not change because of Mendelian segregation. Allele and genotype frequencies remain constant in a random-mating population in the absence of inbreeding, selection, or other evolutionary forces. Usually stated as: If the frequency of allele A is p and the frequency of allele a is q, then the genotype frequencies after one generation of random mating will always be $(p + q)^2 = p^2 + 2pq + q^2$.

Haversian canal After Clopton Havers, English anatomist. Narrow channels that run parallel to the length of a bone and contain blood vessels and nerve cells.

helper T cell A class of white blood cells that initiates both the cell-mediated immune response and the humoral immune response; helper T cells are the targets of the AIDS virus (HIV).

hemoglobin (Gr. *haima,* blood + L. *globus,* a ball) A globular protein in vertebrate red blood cells and in the plasma of many invertebrates that carries oxygen and carbon dioxide.

herbivore (L. *herba,* grass + *vorare,* to devour) Any organism that eats plants.

heredity (L. *heredis,* heir) The transmission of characteristics from parent to offspring.

heterochromatin (Gr. *heteros,* different + *chroma,* color) That portion of a eukaryotic chromosome that remains permanently condensed and therefore is not transcribed into RNA. Most centromere regions are heterochromatic.

heterokaryon (Gr. *heteros,* other + *karyon,* kernel) A fungal hypha that has two or more genetically distinct types of nuclei.

heterotroph (Gr. *heteros,* other + *trophos,* feeder) An organism that does not have the ability to produce its own food. *See also* autotroph.

heterozygote (Gr. *heteros,* other + *zygotos,* a pair) A diploid individual carrying two different alleles of a gene on its two homologous chromosomes.

hierarchical (Gr. *hieros,* sacred + *archos,* leader) Refers to a system of classification in which successively smaller units of classification are included within one another.

histone (Gr. *histos,* tissue) A complex of small, very basic polypeptides rich in the amino acids arginine and lysine. A basic part of chromosomes, histones form the core around which DNA is wrapped.

homeostasis (Gr. *homeos,* similar + *stasis,* standing) The maintaining of a relatively stable internal physiological environment in an organism or steady-state equilibrium in a population or ecosystem.

homeotherm (Gr. *homeo,* similar + *therme,* heat) An organism, such as a bird or mammal, capable of maintaining a stable body temperature independent of the environmental temperature. "Warm-blooded."

hominid (L. *homo,* man) Human beings and their direct ancestors. A member of the family Hominidae. *Homo sapiens* is the only living member.

homologous chromosome (Gr. *homologia,* agreement) One of the two nearly identical versions of each chromosome. Chromosomes that associate in pairs in the first stage of meiosis. In diploid cells, one chromosome of a pair that carries equivalent genes.

homology (Gr. *homologia,* agreement) A condition in which the similarity between two structures or functions is indicative of a common evolutionary origin.

homozygote (Gr. *homos,* same or similar + *zygotos,* a pair) A diploid individual whose two copies of a gene are the same. An individual carrying identical alleles on both homologous chromosomes is said to be homozygous for that gene.

hormone (Gr. *hormaein,* to excite) A chemical messenger, often a steroid or peptide, produced in a small quantity in one part of an organism and then transported to another part of the organism, where it brings about a physiological response.

hybrid (L. *hybrida,* the offspring of a tame sow and a wild boar) A plant or animal that results from the crossing of dissimilar parents.

hybridization The mating of unlike parents of different taxa.

hydrogen bond A molecular force formed by the attraction of the partial positive charge of one hydrogen atom of a water molecule with the partial negative charge of the oxygen atom of another.

hydrolysis reaction (Gr. *hydro,* water + *lyse,* break) The process of tearing down a polymer by adding a molecule of water. A hydrogen is attached to one subunit and a hydroxyl to the other, which breaks the covalent bond. Essentially the reverse of a dehydration reaction.

hydrophobic (Gr. *hydro*, water + *phobos*, hating) Refers to nonpolar molecules, which do not form hydrogen bonds with water and therefore are not soluble in water.

hydroskeleton (Gr. *hydro*, water + *skeletos*, hard) The skeleton of most soft-bodied invertebrates that have neither an internal nor an external skeleton. They use the relative incompressibility of the water within their bodies as a kind of skeleton.

hypertonic (Gr. *hyper*, above + *tonos*, tension) Refers to a cell that contains a higher concentration of solutes than its surrounding solution.

hypha, *pl.* hyphae (Gr. *hyphe*, web) A filament of a fungus. A mass of hyphae comprises a mycelium.

hypothalamus (Gr. *hypo*, under + *thalamos*, inner room) The region of the brain under the thalamus that controls temperature, hunger, and thirst and that produces hormones that influence the pituitary gland.

hypothesis (Gr. *hypo*, under + *tithenai*, to put) A proposal that might be true. No hypothesis is ever proven correct. All hypotheses are provisional—proposals that are retained for the time being as useful but that may be rejected in the future if found to be inconsistent with new information. A hypothesis that stands the test of time—often tested and never rejected—is called a theory.

hypotonic (Gr. *hypo*, under + *tonos*, tension) Refers to the solution surrounding a cell that has a lower concentration of solutes than does the cell.

I

inbreeding The breeding of genetically related plants or animals. In plants, inbreeding results from self-pollination. In animals, inbreeding results from matings between relatives. Inbreeding tends to increase homozygosity.

incomplete dominance The ability of two alleles to produce a heterozygous phenotype that is different from either homozygous phenotype.

independent assortment Mendel's second law: The principle that segregation of alternative alleles at one locus into gametes is independent of the segregation of alleles at other loci. Only true for gene loci located on different chromosomes or those so far apart on one chromosome that crossing over is very frequent between the loci.

industrial melanism (Gr. *melas*, black) Phrase used to describe the evolutionary process in which initially light-colored organisms become dark as a result of natural selection.

inflammatory response (L. *inflammare*, to flame) A generalized nonspecific response to infection that acts to clear an infected area

of infecting microbes and dead tissue cells so that tissue repair can begin.

integument (L. *integumentum*, covering) The natural outer covering layers of an animal. Develops from the ectoderm.

interneuron A nerve cell found only in the CNS that acts as a functional link between sensory neurons and motor neurons. Also called association neuron.

internode The region of a plant stem between nodes where stems and leaves attach.

interoception (L. *interus*, inner + Eng. [re]ceptive) The sensing of information that relates to the body itself, its internal condition, and its position.

interphase That portion of the cell cycle preceding mitosis. It includes the G$_1$ phase, when cells grow, the S phase, when a replica of the genome is synthesized, and a G$_2$ phase, when preparations are made for genomic separation.

intron (L. *intra*, within) A segment of DNA transcribed into mRNA but removed before translation. These untranslated regions make up the bulk of most eukaryotic genes.

ion An atom in which the number of electrons does not equal the number of protons. An ion carries an electrical charge.

ionic bond A chemical bond formed between ions as a result of the attraction of opposite electrical charges.

ionizing radiation High-energy radiation, such as X rays and gamma rays.

isolating mechanisms Mechanisms that prevent genetic exchange between individuals of different populations or species.

isotonic (Gr. *isos*, equal + *tonos*, tension) Refers to a cell with the same concentration of solutes as its environment.

isotope (Gr. *isos*, equal + *topos*, place) An atom that has the same number of protons but different numbers of neutrons.

J

joint The part of a vertebrate where one bone meets and moves on another.

K

karyotype (Gr. *karyon*, kernel + *typos*, stamp or print) The particular array of chromosomes that an individual possesses.

kinetic energy The energy of motion.

kinetochore (Gr. *kinetikos*, putting in motion + *choros*, chorus) A disk of protein bound to the centromere to which microtubules attach during cell division, linking chromatids to the spindle.

kingdom The chief taxonomic category. This book recognizes six kingdoms: Archaea, Bacteria, Protista, Fungi, Animalia, and Plantae.

L

lamella, *pl.* lamellae (L. a little plate) A thin, platelike structure. In chloroplasts, a

layer of chlorophyll-containing membranes. In bivalve mollusks, one of the two plates forming a gill. In vertebrates, one of the thin layers of bone laid concentrically around the Haversian canals.

ligament (L. *ligare*, to bind) A band or sheet of connective tissue that links bone to bone.

linkage The patterns of assortment of genes that are located on the same chromosome. Important because if the genes are located relatively far apart, crossing over is more likely to occur between them than if they are close together.

lipid (Gr. *lipos*, fat) A loosely defined group of molecules that are insoluble in water but soluble in oil. Oils such as olive, corn, and coconut are lipids, as well as waxes, such as beeswax and earwax.

lipid bilayer The basic foundation of all biological membranes. In such a layer, the nonpolar tails of phospholipid molecules point inward, forming a nonpolar zone in the interior of the bilayers. Lipid bilayers are selectively permeable and do not permit the diffusion of water-soluble molecules into the cell.

littoral (L. *litus*, shore) Referring to the shoreline zone of a lake or pond or the ocean that is exposed to the air whenever water recedes.

locus, *pl.* loci (L. place) The position on a chromosome where a gene is located.

loop of Henle After F. G. J. Henle, German anatomist. A hairpin loop formed by a urine-conveying tubule when it enters the inner layer of the kidney and then turns around to pass up again into the outer layer of the kidney.

lymph (L. *lympha*, clear water) In animals, a colorless fluid derived from blood by filtration through capillary walls in the tissues.

lymphatic system An open circulatory system composed of a network of vessels that function to collect the water within blood plasma forced out during passage through the capillaries and to return it to the bloodstream. The lymphatic system also returns proteins to the circulation, transports fats absorbed from the intestine, and carries bacteria and dead blood cells to the lymph nodes and spleen for destruction.

lymphocyte (Gr. *lympha*, water + Gr. *kytos*, hollow vessel) A white blood cell. A cell of the immune system that either synthesizes antibodies (B cells) or attacks virus-infected cells (T cells).

lyse (Gr. *lysis*, loosening) To disintegrate a cell by rupturing its cell membrane.

M

macromolecule (Gr. *makros*, large + L. *moliculus*, a little mass) An extremely large molecule. Refers specifically to carbohydrates, lipids, proteins, and nucleic acids.

macrophage (Gr. *makros*, large + -*phage*, eat) A phagocytic cell of the immune system able to engulf and digest invading bacteria, fungi, and other microorganisms, as well as cellular debris.

marrow The soft tissue that fills the cavities of most bones and is the source of red blood cells.

mass In chemistry, the total number of protons and neutrons in the nucleus of an atom. Approximately equal to the atomic weight.

mass flow The overall process by which materials move in the phloem of plants.

meiosis (Gr. *meioun*, to make smaller) A special form of nuclear division that precedes gamete formation in sexually reproducing eukaryotes.

Mendelian ratio After Gregor Mendel, Austrian monk. Refers to the characteristic 3:1 segregation ratio that Mendel observed, in which pairs of alternative traits are expressed in the F_2 generation in the ratio of three-fourths dominant to one-fourth recessive.

menstruation (L. *mens*, month) Periodic sloughing off of the blood-enriched lining of the uterus when pregnancy does not occur.

meristem (Gr. *merizein*, to divide) In plants, a zone of unspecialized cells whose only function is to divide.

mesoderm (Gr. *mesos*, middle + *derma*, skin) One of the three embryonic germ layers that form in the gastrula. Gives rise to muscle, bone, and other connective tissue; the peritoneum; the circulatory system; and most of the excretory and reproductive systems.

mesophyll (Gr. *mesos*, middle + *phyllon*, leaf) The photosynthetic parenchyma of a leaf, located within the epidermis. The vascular strands (veins) run through the mesophyll.

metabolism (Gr. *metabole*, change) The process by which all living things assimilate energy and use it to grow.

metamorphosis (Gr. *meta*, after + *morphe*, form + *osis*, state of) Process in which form changes markedly during postembryonic development—for example, tadpole to frog or larval insect to adult.

metaphase (Gr. *meta*, middle + *phasis*, form) The stage of mitosis characterized by the alignment of the chromosomes on a plane in the center of the cell.

metastasis, *pl.* metastases (Gr. to place in another way) The spread of cancerous cells to other parts of the body, forming new tumors at distant sites.

microevolution (Gr. *mikros*, small + L. *evolvere*, to unfold) Refers to the evolutionary process itself. Evolution within a species. Also called adaptation.

microfilament (Gr. *mikros*, small + L. *filum*, a thread) In cells, a protein thread composed of parallel fibers of actin cross-connected by myosin. Their movement results from an ATP-driven shape change in myosin. The contraction of vertebrate muscles and many other kinds of cell movement in eukaryotes result from the movements of microfilaments within cells.

microtubule (Gr. *mikros*, small + L. *tubulus*, little pipe) In eukaryotic cells, a long, hollow cylinder about 25 nanometers in diameter and composed of the protein tubulin. Microtubules influence cell shape, move the chromosomes in cell division, and provide the functional internal structure of cilia and flagella.

mimicry (Gr. *mimos*, mime) The resemblance in form, color, or behavior of certain organisms (mimics) to other more powerful or more protected ones (models), which results in the mimics being protected in some way.

mitochondrion, *pl.* mitochondria (Gr. *mitos*, thread + *chondrion*, small grain) A tubular or sausage-shaped organelle 1 to 3 micrometers long. Bounded by two membranes, mitochondria closely resemble the aerobic bacteria from which they were originally derived. As chemical furnaces of the cell, they carry out its oxidative metabolism.

mitosis (Gr. *mitos*, thread) The M phase of cell division in which the microtubular apparatus is assembled, binds to the chromosomes, and moves them apart. This phase is the essential step in the separation of the two daughter cell genomes.

mole (L. *moles*, mass) The atomic weight of a substance, expressed in grams. One mole is defined as the mass of 6.0222×10^{23} atoms.

molecule (L. *moliculus*, a small mass) The smallest unit of a compound that displays the properties of that compound.

monocot Short for monocotyledon; flowering plant in which the embryos have only one cotyledon, the flower parts are often in threes, and the leaves typically are parallel-veined.

monosaccharide (Gr. *monos*, one + *sakcharon*, sugar) A simple sugar.

morphogenesis (Gr. *morphe*, form + *genesis*, origin) The formation of shape. The growth and differentiation of cells and tissues during development.

motor endplate The point where a neuron attaches to a muscle. A neuromuscular synapse.

multicellularity A condition in which the activities of the individual cells are coordinated and the cells themselves are in contact. A property of eukaryotes alone and one of their major characteristics.

muscle (L. *musculus*, mouse) The tissue in the body of humans and animals that can be contracted and relaxed to make the body move.

muscle cell A long, cylindrical, multinucleated cell that contains numerous myofibrils and is capable of contraction when stimulated.

muscle spindle A sensory organ that is attached to a muscle and sensitive to stretching.

mutagen (L. *mutare*, to change) A chemical capable of damaging DNA.

mutation (L. *mutare*, to change) A change in a cell's genetic message.

mutualism (L. *mutuus*, lent, borrowed) A symbiotic relationship in which both participating species benefit.

mycelium, *pl.* mycelia (Gr. *mykes*, fungus) In fungi, a mass of hyphae.

mycology (Gr. *mykes*, fungus) The study of fungi. A person who studies fungi is called a mycologist.

mycorrhiza, *pl.* mycorrhizae (Gr. *mykes*, fungus + *rhiza*, root) A symbiotic association between fungi and plant roots.

myofibril (Gr. *myos*, muscle + L. *fibrilla*, little fiber) A contractile microfilament, composed of myosin and actin, within muscle.

myosin (Gr. *myos*, muscle + *in*, belonging to) One of two myosin components of myofilaments. (The other is actin.)

N

natural selection The differential reproduction of genotypes caused by factors in the environment. Leads to evolutionary change.

nematocyst (Gr. *nema*, thread + *kystos*, bladder) A coiled, threadlike stinging structure of cnidarians that is discharged to capture prey and for defense.

nephron (Gr. *nephros*, kidney) The functional unit of the vertebrate kidney. A human kidney has more than 1 million nephrons that filter waste matter from the blood. Each nephron consists of a Bowman's capsule, glomerulus, and tubule.

nerve A bundle of axons with accompanying supportive cells, held together by connective tissue.

nerve impulse A rapid, transient, self-propagating reversal in electrical potential that travels along the membrane of a neuron.

neuromodulator A chemical transmitter that mediates effects that are slow and longer lasting and that typically involve second messengers within the cell.

neuromuscular junction The structure formed when the tips of axons contact (innervate) a muscle fiber.

neuron (Gr. nerve) A nerve cell specialized for signal transmission.

neurotransmitter (Gr. *neuron*, nerve + L. *trans*, across + *mitere*, to send) A chemical released at an axon tip that travels across the synapse and binds a specific receptor protein in the membrane on the far side.

neurulation (Gr. *neuron*, nerve) The elaboration of a notochord and a dorsal nerve cord that marks the evolution of the chordates.

neutron (L. *neuter*, neither) A subatomic particle located within the nucleus of an atom. Similar to a proton in mass, but as its name implies, a neutron is neutral and possesses no charge.

neutrophil An abundant type of white blood cell capable of engulfing microorganisms and other foreign particles.

niche (L. *nidus*, nest) The role an organism plays in the environment; realized niche is the niche that an organism occupies under natural circumstances; fundamental niche is the niche an organism would occupy if competitors were not present.

nitrogen fixation The incorporation of atmospheric nitrogen into nitrogen compounds, a process that can be carried out only by certain microorganisms.

nocturnal (L. *nocturnus,* night) Active primarily at night.

node (L. *nodus,* knot) The place on the stem where a leaf is formed.

node of Ranvier After L. A. Ranvier, French histologist. A gap formed at the point where two Schwann cells meet and where the axon is in direct contact with the surrounding intercellular fluid.

nonrandom mating A phenomenon in which individuals with certain genotypes sometimes mate with one another more commonly than would be expected on a random basis.

notochord (Gr. *noto,* back + L. *chorda,* cord) In chordates, a dorsal rod of cartilage that forms between the nerve cord and the developing gut in the early embryo.

nucleic acid A nucleotide polymer. A long chain of nucleotides. Chief types are deoxyribonucleic acid (DNA), which is double-stranded, and ribonucleic acid (RNA), which is typically single-stranded.

nucleosome (L. *nucleus,* kernel + *soma,* body) The basic packaging unit of eukaryotic chromosomes, in which the DNA molecule is wound around a ball of histone proteins. Chromatin is composed of long strings of nucleosomes, like beads on a string.

nucleotide A single unit of nucleic acid, composed of a phosphate, a five-carbon sugar (either ribose or deoxyribose), and a purine or a pyrimidine.

nucleus (L. *a kernel,* dim. Fr. *nux,* nut) A spherical organelle (structure) characteristic of eukaryotic cells. The repository of the genetic information that directs all activities of a living cell. In atoms, the central core, containing positively charged protons and (in all but hydrogen) electrically neutral neutrons.

O

oocyte (Gr. *oion,* egg + *kytos,* vessel) A cell in the outer layer of the ovary that gives rise to an ovum. A primary oocyte is any of the 2 million oocytes a female is born with, all of which have begun the first meiotic division.

operon (L. *operis,* work) A cluster of functionally related genes transcribed onto a single mRNA molecule. A common mode of gene regulation in prokaryotes; it is rare in eukaryotes other than fungi.

order A taxonomic category ranking below a class and above a family.

organ (L. *organon,* tool) A complex body structure composed of several different kinds of tissue grouped together in a structural and functional unit.

organelle (Gr. *organella,* little tool) A specialized compartment of a cell. Mitochondria are organelles.

organism Any individual living creature, either unicellular or multicellular.

organ system A group of organs that function together to carry out the principal activities of the body.

osmoconformer An animal that maintains the osmotic concentration of its body fluids at about the same level as that of the medium in which it is living.

osmoregulation The maintenance of a constant internal solute concentration by an organism, regardless of the environment in which it lives.

osmosis (Gr. *osmos,* act of pushing, thrust) The diffusion of water across a membrane that permits the free passage of water but not that of one or more solutes. Water moves from an area of low solute concentration to an area with higher solute concentration.

osmotic pressure The increase of hydrostatic water pressure within a cell as a result of water molecules that continue to diffuse inward toward the area of lower water concentration (the water concentration is lower inside than outside the cell because of the dissolved solutes in the cell).

osteoblast (Gr. *osteon,* bone + *blastos,* bud) A bone-forming cell.

osteocyte (Gr. *osteon,* bone + *kytos,* hollow vessel) A mature osteoblast.

outcross A term used to describe species that interbreed with individuals other than those like themselves.

oviparous (L. *ovum,* egg + *parere,* to bring forth) Refers to reproduction in which the eggs are developed after leaving the body of the mother, as in reptiles.

ovulation The successful development and release of an egg by the ovary.

ovule (L. *ovulum,* a little egg) A structure in a seed plant that becomes a seed when mature.

ovum, *pl.* **ova** (L. egg) A mature egg cell. A female gamete.

oxidation (Fr. *oxider,* to oxidize) The loss of an electron during a chemical reaction from one atom to another. Occurs simultaneously with reduction. Is the second stage of the 10 reactions of glycolysis.

oxidative metabolism A collective term for metabolic reactions requiring oxygen.

oxidative respiration Respiration in which the final electron acceptor is molecular oxygen.

P

parasitism (Gr. *para,* beside + *sitos,* food) A symbiotic relationship in which one organism benefits and the other is harmed.

parthenogenesis (Gr. *parthenos,* virgin + Eng. *genesis,* beginning) The development of an adult from an unfertilized egg. A common form of reproduction in insects.

partial pressures (P) The components of each individual gas—such as nitrogen, oxygen, and carbon dioxide—that together constitute the total air pressure.

pathogen (Gr. *pathos,* suffering + Eng. *genesis,* beginning) A disease-causing organism.

pedigree (L. *pes,* foot + *grus,* crane) A family tree. The patterns of inheritance observed in family histories. Used to determine the mode of inheritance of a particular trait.

peptide (Gr. *peptein,* to soften, digest) Two or more amino acids linked by peptide bonds.

peptide bond A covalent bond linking two amino acids. Formed when the positive (amino, or NH_2) group at one end and a negative (carboxyl, or COOH) group at the other end undergo a chemical reaction and lose a molecule of water.

peristalsis (Gr. *peri,* around + *stellein,* to wrap) The rhythmic sequences of waves of muscular contraction in the walls of a tube.

pH Refers to the concentration of H^+ ions in a solution. The numerical value of the pH is the negative of the exponent of the molar concentration. Low pH values indicate high concentrations of H^+ ions (acids), and high pH values indicate low concentrations (bases).

phagocyte (Gr. *phagein,* to eat + *kytos,* hollow vessel) A cell that kills invading cells by engulfing them. Includes neutrophils and macrophages.

phagocytosis (Gr. *phagein,* to eat + *kytos,* hollow vessel) A form of endocytosis in which cells engulf organisms or fragments of organisms.

phenotype (Gr. *phainein,* to show + *typos,* stamp or print) The realized expression of the genotype. The observable expression of a trait (affecting an individual's structure, physiology, or behavior) that results from the biological activity of proteins or RNA molecules transcribed from the DNA.

pheromone (Gr. *pherein,* to carry + [hor]mone) A chemical signal emitted by certain animals that signals their reproductive readiness.

phloem (Gr. *phloos,* bark) In vascular plants, a food-conducting tissue basically composed of sieve elements, various kinds of parenchyma cells, fibers, and sclereids.

phosphodiester bond The bond that results from the formation of a nucleic acid chain in which individual sugars are linked together in a line by the phosphate groups. The phosphate group of one sugar binds to the hydroxyl group of another, forming an—O—P—O bond.

photon (Gr. *photos,* light) The unit of light energy.

photoperiodism (Gr. *photos,* light + *periodos,* a period) A mechanism that organisms use to measure seasonal changes in relative day and night length.

photorespiration A process in which carbon dioxide is released without the production of ATP or NADPH. Because it produces neither ATP nor NADPH, photorespiration acts to undo the work of photosynthesis.

photosynthesis (Gr. *photos,* light + *-syn,* together + *tithenai,* to place) The process by which plants, algae, and some bacteria use the energy of sunlight to create from carbon dioxide (CO_2) and water (H_2O) the more complicated molecules that make up living organisms.

phototropism (Gr. photos, light + trope, turning to light) A plant's growth response to a unidirectional light source.

phylogeny (Gr. phylon, race, tribe) The evolutionary relationships among any group of organisms.

phylum, pl. phyla (Gr. phylon, race, tribe) A major taxonomic category, ranking above a class.

physiology (Gr. physis, nature + logos, a discourse) The study of the function of cells, tissues, and organs.

pigment (L. pigmentum, paint) A molecule that absorbs light.

pinocytosis (Gr. pinein, to drink + kytos, cell) A form of endocytosis in which the material brought into the cell is a liquid containing dissolved molecules.

pistil (L. pistillum, pestle) Central organ of flowers, typically consisting of ovary, style, and stigma; a pistil may consist of one or more fused carpels and is more technically and better known as the gynoecium.

plankton (Gr. planktos, wandering) The small organisms that float or drift in water, especially at or near the surface.

plasma (Gr. form) The fluid of vertebrate blood. Contains dissolved salts, metabolic wastes, hormones, and a variety of proteins, including antibodies and albumin. Blood minus the blood cells.

plasma membrane A lipid bilayer with embedded proteins that control the cell's permeability to water and dissolved substances.

plasmid (Gr. plasma, a form or something molded) A small fragment of DNA that replicates independently of the bacterial chromosome.

platelet (Gr. dim of plattus, flat) In mammals, a fragment of a white blood cell that circulates in the blood and functions in the formation of blood clots at sites of injury.

pleiotropy (Gr. pleros, more + trope, a turning) Describing a gene that produces more than one phenotypic effect.

polarization The charge difference of a neuron so that the interior of the cell is negative with respect to the exterior.

polar molecule A molecule with positively and negatively charged ends. One portion of a polar molecule attracts electrons more strongly than another portion, with the result that the molecule has electron-rich (−) and electron-poor (+) regions, giving it magnetlike positive and negative poles. Water is one of the most polar molecules known.

pollen (L. fine dust) A fine, yellowish powder consisting of grains or microspores, each of which contains a mature or immature male gametophyte. In flowering plants, pollen is released from the anthers of flowers and fertilizes the pistils.

pollen tube A tube that grows from a pollen grain. Male reproductive cells move through the pollen tube into the ovule.

pollination The transfer of pollen from the anthers to the stigmas of flowers for fertilization, as by insects or the wind.

polygyny (Gr. poly, many + gyne, woman, wife) A mating choice in which a male mates with more than one female.

polymer (Gr. polus, many + meris, part) A large molecule formed of long chains of similar molecules.

polymerase chain reaction (PCR) A process by which DNA polymerase is used to copy a sequence of interest repeatedly, making millions of copies of the same DNA.

polymorphism (Gr. polys, many + morphe, form) The presence in a population of more than one allele of a gene at a frequency greater than that of newly arising mutations.

polynomial system (Gr. polys, many + [bi]nomial) Before Linnaeus, naming a genus by use of a cumbersome string of Latin words and phrases.

polyp A cylindrical, pipe-shaped cnidarian usually attached to a rock with the mouth facing away from the rock on which it is growing. Coral is made up of polyps.

polypeptide (Gr. polys, many + peptein, to digest) A general term for a long chain of amino acids linked end to end by peptide bonds. A protein is a long, complex polypeptide.

polysaccharide (Gr. polys, many + sakcharon, sugar) A sugar polymer. A carbohydrate composed of many monosaccharide sugar subunits linked together in a long chain.

population (L. populus, the people) Any group of individuals, usually of a single species, occupying a given area at the same time.

posterior (L. post, after) Situated behind or farther back.

potential difference A difference in electrical charge on two sides of a membrane caused by an unequal distribution of ions.

potential energy Energy with the potential to do work. Stored energy.

predation (L. praeda, prey) The eating of other organisms. The one doing the eating is called a predator, and the one being consumed is called the prey.

primary growth In vascular plants, growth originating in the apical meristems of shoots and roots, as contrasted with secondary growth; results in an increase in length.

primary nondisjunction The failure of homologous chromosomes to separate in meiosis I. The cause of Down syndrome.

primary plant body The part of a plant that arises from the apical meristems.

primary producers Photosynthetic organisms, including plants, algae, and photosynthetic bacteria.

primary structure of a protein The sequence of amino acids that makes up a particular polypeptide chain.

primordium, pl. primordia (L. primus, first + ordiri, begin) The first cells in the earliest stages of the development of an organ or structure.

productivity The total amount of energy of an ecosystem fixed by photosynthesis per unit of time. Net productivity is productivity minus that which is expended by the metabolic activity of the organisms in the community.

prokaryote (Gr. pro, before + karyon, kernel) A simple bacterial organism that is small and single-celled, lacks external appendages, and has little evidence of internal structure.

promoter An RNA polymerase binding site. The nucleotide sequence at the end of a gene to which RNA polymerase attaches to initiate transcription of mRNA.

prophase (Gr. pro, before + phasis, form) The first stage of mitosis during which the chromosomes become more condensed, the nuclear envelope is reabsorbed, and a network of microtubules (called the spindle) forms between opposite poles of the cell.

protein (Gr. proteios, primary) A long chain of amino acids linked end to end by peptide bonds. Because the 20 amino acids that occur in proteins have side groups with very different chemical properties, the function and shape of a protein is critically affected by its particular sequence of amino acids.

protist (Gr. protos, first) A member of the kingdom Protista, which includes unicellular eukaryotic organisms and some multicellular lines derived from them.

proton A subatomic particle in the nucleus of an atom that carries a positive charge. The number of protons determines the chemical character of the atom because it dictates the number of electrons orbiting the nucleus and available for chemical activity.

protostome (Gr. protos, first + stoma, mouth) An animal in whose embryonic development the mouth forms at or near the blastopore. Also characterized by spiral cleavage.

protozoa (Gr. protos, first + zoion, animal) The traditional name given to heterotrophic protists.

pseudocoel (Gr. pseudos, false + koiloma, cavity) A body cavity similar to the coelom except that it forms between the mesoderm and endoderm.

punctuated equilibrium A hypothesis of the mechanism of evolutionary change that proposes that long periods of little or no change are punctuated by periods of rapid evolution.

Q

quaternary structure of a protein A term to describe the way multiple protein subunits are assembled into a whole.

R

radial symmetry (L. radius, a spoke of a wheel + Gr. summetros, symmetry) The regular arrangement of parts around a central axis so that any plane passing through the central axis divides the organism into halves that are approximate mirror images.

radioactivity The emission of nuclear particles and rays by unstable atoms as they decay into more stable forms. Measured in curies, with 1 curie equal to 37 billion disintegrations a second.

radula (L. scraper) A rasping, tonguelike organ characteristic of most mollusks.

recessive allele An allele whose phenotype effects are masked in heterozygotes by the presence of a dominant allele.

recombination The formation of new gene combinations. In bacteria, it is accomplished by the transfer of genes into cells, often in association with viruses. In eukaryotes, it is accomplished by reassortment of chromosomes during meiosis and by crossing over.

reducing power The use of light energy to extract hydrogen atoms from water.

reduction (L. *reductio*, a bringing back; originally, "bringing back" a metal from its oxide) The gain of an electron during a chemical reaction from one atom to another. Occurs simultaneously with oxidation.

reflex (L. *reflectere*, to bend back) An automatic consequence of a nerve stimulation. The motion that results from a nerve impulse passing through the system of neurons, eventually reaching the body muscles and causing them to contract.

refractory period The recovery period after membrane depolarization during which the membrane is unable to respond to additional stimulation.

renal (L. *renes*, kidneys) Pertaining to the kidney.

repression (L. *reprimere*, to press back, keep back) The process of blocking transcription by the placement of the regulatory protein between the polymerase and the gene, thus blocking movement of the polymerase to the gene.

repressor (L. *reprimere*, to press back, keep back) A protein that regulates transcription of mRNA from DNA by binding to the operator and so preventing RNA polymerase from attaching to the promoter.

resolving power The ability of a microscope to distinguish two lines as separate.

respiration (L. *respirare*, to breathe) The utilization of oxygen. In terrestrial vertebrates, the inhalation of oxygen and the exhalation of carbon dioxide.

resting membrane potential The charge difference that exists across a neuron's membrane at rest (about 70 millivolts).

restriction endonuclease A special kind of enzyme that can recognize and cleave DNA molecules into fragments. One of the basic tools of genetic engineering.

restriction fragment-length polymorphism (RFLP) An associated genetic mutation marker detected because the mutation alters the length of DNA segments.

retrovirus (L. *retro*, turning back) A virus whose genetic material is RNA rather than DNA. When a retrovirus infects a cell, it makes a DNA copy of itself, which it can then insert into the cellular DNA as if it were a cellular gene.

ribose A five-carbon sugar.

ribosome A cell structure composed of protein and RNA that translates RNA copies of genes into protein.

RNA polymerase The enzyme that transcribes RNA from DNA.

S

saltatory conduction A very fast form of nerve impulse conduction in which the impulses leap from node to node over insulated portions.

sarcoma (Gr. *sarx*, flesh) A cancerous tumor that involves connective or hard tissue, such as muscle.

sarcomere (Gr. *sarx*, flesh + *meris*, part of) The fundamental unit of contraction in skeletal muscle. The repeating bands of actin and myosin that appear between two Z lines.

sarcoplasmic reticulum (Gr. *sarx*, flesh + *plassein*, to form, mold; L. *reticulum*, network) The endoplasmic reticulum of a muscle cell. A sleeve of membrane that wraps around each myofilament.

scientific creationism A view that the biblical account of the origin of the earth is literally true, that the earth is much younger than most scientists believe, and that all species of organisms were individually created just as they are today.

secondary growth In vascular plants, growth that results from the division of a cylinder of cells around the plant's periphery. Secondary growth causes a plant to grow in diameter.

second messenger An intermediary compound that couples extracellular signals to intracellular processes and also amplifies a hormonal signal.

seed A structure that develops from the mature ovule of a seed plant. Contains an embryo surrounded by a protective coat.

selection The process by which some organisms leave more offspring than competing ones and their genetic traits tend to appear in greater proportions among members of succeeding generations than the traits of those individuals that leave fewer offspring.

self-fertilization The transfer of pollen from an anther to a stigma in the same flower or to another flower of the same plant, leading to self-fertilization.

sepal (L. *sepalum*, a covering) A member of the outermost whorl of a flowering plant. Collectively, the sepals constitute the calyx.

septum, *pl.* septa (L. *saeptum*, a fence) A partition or cross-wall, such as those that divide fungal hyphae into cells.

sex chromosomes The X and Y chromosomes, which are different in the two sexes and are involved in sex determination.

sex-linked characteristic A genetic characteristic that is determined by genes located on the sex chromosomes.

sexual reproduction Reproduction that involves the regular alternation between syngamy and meiosis. Its outstanding characteristic is that an individual offspring inherits genes from two parent individuals.

shoot In vascular plants, the aboveground parts, such as the stem and leaves.

sieve cell In the phloem (food-conducting tissue) of vascular plants, a long, slender sieve element with relatively unspecialized sieve areas and with tapering end walls that lack sieve plates. Found in all vascular plants except angiosperms, which have sieve-tube members.

soluble Refers to polar molecules that dissolve in water and are surrounded by a hydration shell.

solute The molecules dissolved in a solution. *See also* solution, solvent.

solution A mixture of molecules, such as sugars, amino acids, and ions, dissolved in water.

solvent The most common of the molecules dissolved in a solution. Usually a liquid, commonly water.

somatic cells (Gr. *soma*, body) All the diploid body cells of an animal that are not involved in gamete formation.

somite A segmented block of tissue on either side of a developing notochord.

species, *pl.* species (L. kind, sort) A level of taxonomic hierarchy; a species ranks next below a genus.

sperm (Gr. *sperma*, sperm, seed) A sperm cell. The male gamete.

spindle The mitotic assembly that carries out the separation of chromosomes during cell division. Composed of microtubules and assembled during prophase at the equator of the dividing cell.

spore (Gr. *spora*, seed) A haploid reproductive cell, usually unicellular, that is capable of developing into an adult without fusion with another cell. Spores result from meiosis, as do gametes, but gametes fuse immediately to produce a new diploid cell.

sporophyte (Gr. *spora*, seed + *phyton*, plant) The spore-producing, diploid ($2n$) phase in the life cycle of a plant having alternation of generations.

stabilizing selection A form of selection in which selection acts to eliminate both extremes from a range of phenotypes.

stamen (L. thread) The part of the flower that contains the pollen. Consists of a slender filament that supports the anther. A flower that produces only pollen is called staminate and is functionally male.

steroid (Gr. *stereos*, solid + L. *ol*, from oleum, oil) A kind of lipid. Many of the molecules that function as messengers and pass across cell membranes are steroids, such as the male and female sex hormones and cholesterol.

steroid hormone A hormone derived from cholesterol. Those that promote the development of the secondary sexual characteristics are steroids.

stigma (Gr. mark) A specialized area of the carpel of a flowering plant that receives the pollen.

stoma, *pl.* **stomata (Gr. mouth)** A specialized opening in the leaves of some plants that allows carbon dioxide to pass into the plant body and allows water and oxygen to pass out of them.

stratum corneum The outer layer of the epidermis of the skin of the vertebrate body.

substrate (L. substratus, strewn under) A molecule on which an enzyme acts.

substrate-level phosphorylation The generation of ATP by coupling its synthesis to a strongly exergonic (energy-yielding) reaction.

succession In ecology, the slow, orderly progression of changes in community composition that takes place through time. Primary succession occurs in nature on bare substrates, over long periods of time. Secondary succession occurs when a climax community has been disturbed.

sugar Any monosaccharide or disaccharide.

surface tension A tautness of the surface of a liquid, caused by the cohesion of the liquid molecules. Water has an extremely high surface tension.

surface-to-volume ratio Describes cell size increases. Cell volume grows much more rapidly than surface area.

symbiosis (Gr. syn, together with + bios, life) The condition in which two or more dissimilar organisms live together in close association; includes parasitism, commensalism, and mutualism.

synapse (Gr. synapsis, a union) A junction between a neuron and another neuron or muscle cell. The two cells do not touch. Instead, neurotransmitters cross the narrow space between them.

synapsis (Gr. synapsis, contact, union) The close pairing of homologous chromosomes that occurs early in prophase I of meiosis. With the genes of the chromosomes thus aligned, a DNA strand of one homologue can pair with the complementary DNA strand of the other.

syngamy (Gr. syn, together with + gamos, marriage) Fertilization. The union of male and female gametes.

T

taxonomy (Gr. taxis, arrangement + nomos, law) The science of the classification of organisms.

T cell A type of lymphocyte involved in cell-mediated immune responses and interactions with B cells. Also called a T lymphocyte.

tendon (Gr. tenon, stretch) A strap of cartilage that attaches muscle to bone.

tertiary structure of a protein The three-dimensional shape of a protein. Primarily the result of hydrophobic interactions of amino acid side groups and, to a lesser extent, of hydrogen bonds between them. Forms spontaneously.

test cross A cross between a heterozygote and a recessive homozygote. A procedure Mendel used to further test his hypotheses.

theory (Gr. theorein, to look at) A well-tested hypothesis supported by a great deal of evidence.

thigmotropism (Gr. thigma, touch + trope, a turning) The growth response of a plant to touch.

thorax (Gr. a breastplate) The part of the body between the neck and the abdomen.

thylakoid (Gr. thylakos, sac + -oides, like) A flattened, saclike membrane in the chloroplast of a eukaryote. Thylakoids are stacked on top of one another in arrangements called grana and are the sites of photosystem reactions.

tissue (L. texere, to weave) A group of similar cells organized into a structural and functional unit.

trachea, *pl.* **tracheae (L. windpipe)** In vertebrates, the windpipe.

tracheid (Gr. tracheia, rough) An elongated cell with thick, perforated walls that carries water and dissolved minerals through a plant and provides support. Tracheids form an essential element of the xylem of vascular plants.

transcription (L. trans, across + scrībere, to write) The first stage of gene expression in which the RNA polymerase enzyme synthesizes an mRNA molecule whose sequence is complementary to the DNA.

translation (L. trans, across + latus, that which is carried) The second stage of gene expression in which a ribosome assembles a polypeptide, using the mRNA to specify the amino acids.

translocation (L. trans, across + locare, to put or place) In plants, the process in which most of the carbohydrates manufactured in the leaves and other green parts of the plant are moved through the phloem to other parts of the plant.

transpiration (L. trans, across + spirare, to breathe) The loss of water vapor by plant parts, primarily through the stomata.

transposon (L. transponere, to change the position of) A DNA sequence carrying one or more genes and flanked by insertion sequences that confer the ability to move from one DNA molecule to another. An element capable of transposition (the changing of chromosomal location).

trophic level (Gr. trophos, feeder) A step in the movement of energy through an ecosystem.

tropism (Gr. trop, turning) A plant's response to external stimuli. A positive tropism is one in which the movement or reaction is in the direction of the source of the stimulus. A negative tropism is one in which the movement or growth is in the opposite direction.

turgor pressure (L. turgor, a swelling) The pressure within a cell that results from the movement of water into the cell. A cell with high turgor pressure is said to be turgid.

U

unicellular Composed of a single cell.

urea (Gr. ouron, urine) An organic molecule formed in the vertebrate liver. The principal form of disposal of nitrogenous wastes by mammals.

urine (Gr. ouron, urine) The liquid waste filtered from the blood by the kidneys.

V

vaccination The injection of a harmless microbe into a person or animal to confer resistance to a dangerous microbe.

vacuole (L. vacuus, empty) A cavity in the cytoplasm of a cell that is bound by a single membrane and contains water and waste products of cell metabolism. Typically found in plant cells.

variable Any factor that influences a process. In evaluating alternative hypotheses about one variable, all other variables are held constant so that the investigator is not misled or confused by other influences.

vascular bundle In vascular plants, a strand of tissue containing primary xylem and primary phloem. These bundles of elongated cells conduct water with dissolved minerals and carbohydrates throughout the plant body.

vascular cambium In vascular plants, the meristematic layer of cells that gives rise to secondary phloem and secondary xylem. The activity of the vascular cambium increases stem or root diameter.

ventral (L. venter, belly) Refers to the bottom portion of an animal.

vertebrate An animal having a backbone made of bony segments called vertebrae.

vesicle (L. vesicula, a little (ladder) Membrane-enclosed sacs within eukaryotic organisms created by weaving sheets of endoplasmic reticulum through the cell's interior.

vessel element In vascular plants, a typically elongated cell, dead at maturity, that conducts water and solutes in the xylem.

villus, *pl.* **villi (L. a tuft of hair)** In vertebrates, fine, microscopic, fingerlike projections on epithelial cells lining the small intestine that serve to increase the absorptive surface area of the intestine.

vitamin (L. vita, life + amine, of chemical origin) An organic substance required in minute quantities by an organism for growth and activity but that the organism cannot synthesize.

viviparous (L. vivus, alive + parere, to bring forth) Refers to reproduction in which eggs develop within the mother's body and young are born free-living.

voltage-gated channel A transmembrane pathway for an ion that is opened or closed by a change in the voltage, or charge difference, across the cell membrane.

W

water vascular system The system of water-filled canals connecting the tube feet of echinoderms.

whorl A circle of leaves or of flower parts present at a single level along an axis.

wood Accumulated secondary xylem. Heartwood is the central, nonliving wood in the trunk of a tree. Hardwood is the wood of dicots, regardless of how hard or soft it actually is. Softwood is the wood of conifers.

X

xylem (Gr. *xylon,* wood) In vascular plants, a specialized tissue, composed primarily of elongate, thick-walled conducting cells, that transports water and solutes through the plant body.

Y

yolk (O.E. *geolu,* yellow) The stored substance in egg cells that provides the embryo's primary food supply.

Z

zygote (Gr. *zygotos,* paired together) The diploid ($2n$) cell resulting from the fusion of male and female gametes (fertilization).

Credits

Photo Credits

Chapter One
Opener: © Frans Lanting; 1.1 (bottom left): © PhotoDisc/BS Vol. 15; (top left): © R. Robinson/Visuals Unlimited; (top center): © Alfred Pasieka/Science Photo Library/Photo Researchers; (top right): © Corbis/Vol. 64; (bottom center): © Corbis/Vol. 46; (top right): © Michael Fogden/DRK Photo; (bottom right): © PhotoDisc/Vol. 44; 1.2: © T.E. Adams/Visuals Unlimited; 1.3: Corbis/Vol. 6; 1.4a (top left): From C.P. Morgan & R.A. Jersid, Anatomical Record, 166: 575–586, 1970 © John Wiley & Sons; 1.4b (bottom left): Photo Lennart Nilsson/Albert Bonniers Forlag AB, Behold Man, Little, Brown & Co.; 1.4c (top center): © Ed Reschke; 1.4d (bottom center): © PhotoDisc/Vol. 4; 1.4f (right 2nd from top): © Corbis; 1.4g (right 2nd from top): © PhotoDisc/Vol. 4; 1.4e (top right): © PhotoDisc/Vol. 44; 1.4h (right 3rd from top): © KirtleyPerkins/Visuals Unlimited; 1.4i (bottom right): © Robert & Jean Pollock; page 7 (top left): © Joe McDonald/Animals Animals/Earth Scenes; (top center): © Kenneth Fink/Photo Researchers; (top right): © Tom McHugh/Photo Researchers; (middle): © Royalty-Free/Corbis; bottom left): © Royalty-free/Corbis; (bottom right): © Runk/Schoenberger/Grant Heilman Photography; 1.5: Courtesy of Bill Ober; 1.8: Ozone Processing Team, Goddard Space Flight Center, NASA; 1.9: © Laurel Hungerford; 1.10: © Dennis Kunkel/Phototake; 1.13: © Leonard Lessin/Peter Arnold, Inc.; 1.14: © Photo Disc/Vol. 6; 1.15c (bottom right): © Dennis Kunkel Microscopy, Inc.; 1.15 (bottom left): © Dr. T.J. Beveridge/Visuals Unlimited; 1.15 (bottom center): © John D. Cunningham/Visuals Unlimited; 1.15 (left 3rd from top): © Royalty-Free/Corbis; 1.15(right 4th from top):© PhotoDisc/Website; 1.15 (left 2nd from top): © Royalty-Free/Corbis; 1.15 (middle right): © Royalty-Free/Corbis; 1.15 (middle left): © Royalty-Free/Corbis; 1.15 (top left): Nature Picture Library/Alamy Images; 1.15 (far left): © Royalty-Free/Corbis; 1.15 (top center): James Gregory/Alamy Images; 1.15(top right): © Royalty-Free/Corbis; 1.15 (right 2nd from top):© Corbis/Vol. 6; 1.15 (right 3rd from top):© Corbis/Vol. 6.

Chapter Two
Opener (top left, bottom right): Dr. Robert H. Rothman, Department of Biological Sciences, Rochester Institute of Technology; (top right): © Tui De Roy/Minden Pictures: (bottom left): © D. Parera & E. Parera-Cook/Auscape/Minden Pictures; 2.1: Huntington Library/Superstock; 2.2: From Darwin, The Life of a Tormented Evolutionist, by Adrian Desmond; 2.8: © Mary Evans Picture Library/Photo Researchers; 2.9: Smithsonian Institution Libraries © 2000 Smithsonian Institution; Page 29: © Chas. McRae/Visuals Unlimited; 2.13: © Cleveland Hickman; 2.15 (all): J.B. Losos; 2.19: Centers for Disease Control; 2.20: © Frans Lanting/Minden Pictures.

Chapter Three
Opener: © Michael Viard/Peter Arnold, Inc.; 3.6: © Mary Evans Picture Library/Photo Researchers; Page 47: © Kevin R Morris/Corbis; Page 49: Fort Worth Zoo; 3.11:© PhotoDisc/Vol. 6; 3.12a: © Royalty-Free/Corbis; 3.12b: © Hermann Eisenbeiss/National Audubon Society Collection/Photo Researchers; 3.23a: © Manfred Kage/Peter Arnold, Inc.; 3.23b: © PhotoDisc/Vol. 6; 3.23c: © Scott Blackman/Tom Stack & Associates; 3.23d: © PhotoDisc/Vol. 6; 3.31: © J.D. Litvay/Visuals Unlimited; Page 64 (top, second from top): © Royalty-Free/Corbis; (3rd from top): © PhotoDisc; (4th from top): © Royalty-Free/Corbis; (bottom): © Scott Johnson/Animals Animals/Earth Scenes.

Chapter Four
Opener: © Manfred Kage/Peter Arnold, Inc.; Page 73 (top left): © David M. Phillips/Visuals Unlimited; (left 2nd from top): © M. Abbey/Visuals Unlimited; (left 3rd from top): © David M. Phillips/Visuals Unlimited; (left 4th from top): © Mike Abbey/Visuals Unlimited; (top right): © K.G. Murti/Visuals Unlimited; (right 2nd from top): © David Becker/Science Photo Library/Photo Researchers; (right 3rd from top): © Microworks/Phototake; (right fourth from top): © Stanley Flegler/Visuals Unlimited; Page 76: © SIU/Visuals Unlimited; 4.9a: © Andrew Syred/Science Photo Library/Photo Researchers; 4.9b: © Microfield Scientific, Ltd./Science Photo Library/Photo Researchers; 4.9c: © Alfred Paseika/Science Photo Library/Photo Researchers; 4.13: © R. Bolender & D. Fawcett/Visuals Unlimited; 4.14b: Courtesy of Dr. Charles Flickinger, Medical Cellular Biology, W.B. Saunders, 1979; 4.16b: © Don W. Fawcett/Visuals Unlimited; 4.17: Courtesy of Dr. Kenneth Miller, Brown University; 4.21b: © Stanley Flegler/Visuals Unlimited; 4.23, 4.24: © Biophoto Associates/Photo Researchers, Inc.; 4.28a, b: © David M. Phillips/Visuals Unlimited; 4.30b: Courtesy of Dr. Birgit H. Satir; 4.31: Courtesy of M.M. Perry & A.B. Gilbert, Cell Science, 39–257, 1979.

Chapter Five
Opener: © Jane Burton/Bruce Coleman, Inc.; 5.3: © Spencer Grant/Photo Edit.

Chapter Six
Opener: © Skip Moody/Dembinsky Photo Associates; 6.1a: © Manfred Kage/Peter Arnold, Inc.; 6.1b: Courtesy of Dr. Kenneth Miller, Brown University; 6.5 (both): © Eric Soder/Tom Stack & Associates; 6.15: © Robert A.Caputo/Aurora & Quanta Productions, Inc.; Page 133 (top): © Corbis/Vol. 145; (2nd from top): © PhotoDisc/Vol. 44; (3rd from top): © PhotoDisc/Vol. 44;(4th & 5th from top): © Edward S. Ross.

Chapter Seven
Opener: © L. Maziarski/Visuals Unlimited; 7.1a: © Lee D. Simon/Photo Researchers; 7.4: © Science Photo Library/Photo Researchers; 7.6: © Biophoto Associates/Photo Researchers, Inc.; 7.07(all): © Dr. Andrew S. Bajer; 7.8a: © David M. Phillips/Visuals Unlimited; 7.12: © Moredun Animal Health LTD/Science Photo Library/Photo Researchers, Inc.; 7.17: © David Cavagnaro/Peter Arnold; 7.21: © Dr. Andrew S. Bajer; 7.23 (all): © C.A. Hasenkampf/Biological Photo Service; Page 163: Sir John Tenniel.

Chapter Eight
Opener: © Richard Gross/Biological Photography; 8.1: Courtesy of American Museum of Natural History; 8.4: Courtesy R.W. Van Norman; 8.12a: From Albert & Blakeslee "Corn and Man," Journal

of Heredity V. 5, pg. 511, 1914, Oxford University Press; 8.15 (both): © Fred Bruemmer; 8.17 (top left): © Richard Hutchings/Photo Researchers; (bottom left): © Cheryl A. Ertelt/Visuals Unlimited; (bottom right): © William H. Mullins/Photo Researchers; (top right): Gerard Lacz/Peter Arnold, Inc.; 8.18: © H. Kellams; 8.20 (both): © CBS/Phototake; 8.24a: Courtesy of Loris McGavaran, Denver Children's Hospital; 8.24b: © Joseph Sohm; ChromoSohm, Inc./Corbis; 8.28: © Corbis; 8.30 (both): © Stanley Flegler/Visuals Unlimited; 8.35: © Yoav Levy/Phototake.

Chapter Nine

Opener: N. Ban, P. Nissen, J. Hansen, P.B. Moore & T.A. Steitz, "The Complete Atomic Structure of the Large Ribosomal Subunit at 2.4 Å Resolution," reprinted with permission from Science v. 289 #5481, p. 917, © American Association for the Advancement of Science; 9.4a&c: From J.D. Watson, The Double Helix Atheneum, New York, 1968. Cold Spring Harbor Lab; 9.4b: © A.C. Barrington Brown/Photo Researchers, Inc.; 9.22: © David Scharf.

Chapter Ten

Opener: © The Roslin Institute; 10.1: Nancy Federspiel; 10.3 (top left): © Dr. Gopal Murti/Science Photo Library/Photo Researchers; (bottom right): Courtesy of Monsanto Corporation; (bottom left): Robert H. Devlin/Fisheries & Oceans Canada; (top right): Courtesy of Dr. Ken Culver, Photo by John Crawford, National Institutes of Health; page 229: Courtesy of Lifecodes Corp., Stamford, CT; 10.6: R.L. Brinster, U. of Pennsylvania Sch. of Vet. Med.; 10.8: Courtesy of Dr. John Sanford, Cornell University; 10.9: Courtesy of Monsanto Corporation; 10.10: Courtesy of Ingo Potrykus and Peter Beyer; Page 237: AP/Wide World Photos; 10.13: © University of Wisconsin-Madison News & Public Affairs; 10.14: © SUW-Madison News & Public Affairs, Photo by Jeff Miller.

Chapter Eleven

Opener: © Breck P. Kent/Animals Animals/Earth Scenes; 11.7: Courtesy of Michael Richardson and Ronan O'Rahilly; 11.11: Courtesy of Dr. Victor A. McKusick, Johns Hopkins University; 11.16: Courtesy of the University of Chicago Library/Dept. of Special Collections and Todd L. Savitt; 11.20: Courtesy of H. Rodd; 11.21a: © Royalty-Free/Corbis; 11.21b: © Corbis/Vol. 6; 11.21c: © Porterfield/Chickering/Photo Researchers, Inc.; 11.22 (top left): © Rob & Ann Simpson/Visuals Unlimited; (top right): © Suzanne L. Collins & Joseph T. Collins/National Audubon Society Collection/ Photo Researchers; (bottom left): © Phil A. Dotson/National Audubon Society Collection/Photo Researchers; (bottom right): © John Shaw/Tom Stack & Associates.

Chapter Twelve

Opener: © Dave Watts/Tom Stack & Associates; 12.2 (top left): © Dwight R. Kuhn; (top center): © Heather Angel; (bottom center): © S. Maslowski/Visuals Unlimited; (bottom left) © Henry Ausloos/ Animals Animals/Earth Scenes; (top right): © John Cancalosi/Peter Arnold, Inc.; (bottom right): © Manfred Danegger/Peter Arnold, Inc.; 12.4 (top left): © Gerard Lacz/Peter Arnold, Inc.; (top right): © Ralph Reinhold/Animals Animals/Earth Scenes; (below): © Grant Heilman/Grant Heilman Photography; 12.7 (top): © PhotoDisc; (2nd from top): © Royalty-Free/Corbis; (3rd from top): © PhotoDisc/ Website; (left 3rd from top): © PhotoDisc; (center far right): © Corbis/Vol. 6; (inset): © PhotoDisc/Vol. BS24; (4th from top): © PhotoDisc; (right 4th from top): © Royalty-Free/Corbis; (5th from top):© PhotoDisc/Vol. 6; (bottom far left): © Royalty-Free/Corbis; (bottom left): © Royalty-Free/Corbis; (bottom right): © Royalty-Free/Corbis; (inset): © Adam Jones/Visuals Unlimited; (bottom far right): © PhotoDisc; (inset): © PhotoDisc; Page 281: © PhotoDisc/ Vol. 44; 12.11b: © OSF/Animals Animals/Earth Scenes.

Chapter Thirteen

Opener: © Dwight R. Kuhn; page 292 (top): NASA; (bottom) NASA/JPL/Cornell; 13.5: Courtesy of Dr. Charles Brinton; page 295 (above): © Abraham & Beachey/BPS/Tom Stack & Associates; (below): © F. Widell/Visuals Unlimited; page 297 (top left): Royalty-Free/Index Stock; (bottom left): Royalty-Free/Corbis; (top right): © Gene Ott; (middle right): © China Photo/Reuters/ Corbis; (bottom right): © Tim Zurowski/ Corbis; (middle left): © Photodisc/ EP073; 13.8: © John D. Cunningham/Visuals Unlimited; 13.9: © John D. Cunningham /Visuals Unlimited; 13.11 (both): © Corbis/ R-F Website; 13.12: © Bill Keogh/Visuals Unlimited; 13.13: © L. West/Photo Researchers; 13.14: © Corbis/R-F Website.

Chapter Fourteen

Opener: © Royalty-Free/Corbis; 14.1: © Royalty- Free/Corbis; 14.2 © Terry Ashley/Tom Stack & Associates; 14.4a: © Edward S. Ross; 14.4b: © Richard Gross/Biological Photography; 14.6b: © Edward S. Ross; 14.7: Courtesy of Hans Steur, The Netherlands; 14.8: © E.J. Cable/Tom Stack & Associates; 14.9a: © Edward S. Ross; 14.9b: © Royalty-Free/Corbis; 14.9c: © Kingsley R. Stern; 14.9d: © Edward S. Ross; 14.9e: © Rod Planck/Tom Stack & Associates; 14.11: © Kingsley R. Stern; 14.12a: © Runk/Schoenberger/Grant Heilman Photography; 14.12b: © Kevin & Betty Collins/Visuals Unlimited; 14.13a: T. Walker; 14.13 (inset): © R.J. Delorit, Agronomy Publications; 14.14: © Carolina Biological/Visuals Unlimited; 14.15a: © Walter H. Hodge/Peter Arnold, Inc.; 14.15b: © Kjell Sandved/Butterfly Alphabet; 14.15c: © Runk/Schoenberger/ Grant Heilman Photography; 14.17b: © Ed Pembleton; 14.18a,b: © Thomas Eisner; 14.18c: © OSF/Animals Animals/Earth Scenes; 14.19: © Michael & Patricia Fogden; 14.22a: © Patrick Johns/ Corbis; 14.22c: © Kingsley R. Stern; 14.22b: © James L. Castner.

Chapter Fifteen

Opener: © James H. Robinson/Animals Animals/Earth Scenes; 15.1: Photodisc Red/Getty Images; 15.3a: © Alamy Images; 15.3b: © RoyaltyFree/Corbis; 15.4: © Gwen Fidler/Tom Stack & Associates; 15.8a: © T.E. Adams/Visuals Unlimited; 15.8b: © Stan Elems/Visuals Unlimited; 15.10a: © Larry Jensen/Visuals Unlimited; 15.10b: © T.E. Adams/Visuals Unlimited; 15.11a: © Image Ideas, Inc./PictureQuest; 15.11b: © Kjell Sandved/Butterfly Alphabet; 15.11c: © Fred Bavendam/Peter Arnold Inc.; 15.12a: © David M. Dennis/Tom Stack & Associates; 15.12b: © Kjell Sandved/Butterfly Alphabet; 15.15a: © Kjell Sandved/Butterfly Alphabet; 15.15b: © Edward S. Ross; 15.15c: © Don Valenti/Tom Stack & Associates; 15.15d: © John Gerlach/Visuals Unlimited; 15.15e: © Edward S. Ross; 15.15f: © J.A. Alcock/Visuals Unlimited; 15.15g: © Cleveland P. Hickman; 15.17a: © Randy Morse/Tom Stack & Associates; 15.17b: © Carl Roessler/Tom Stack & Associates; 15.17c: © William C. Ober; 15.17d: © Jeff Rotman; 15.17e: © Daniel W. Gotshall; 15.18a: © Jim & Cathy Church; 15.18b: © Heather Angel; 15.18c: © Stephen J. Krasemann/DRK Photo; 15.20: © Stephen Frink Collection/Alamy Images; 15.22: © John Shaw/Tom Stack & Associates; 15.28a: © Erwin & Peggy Bauer/Tom Stack & Associates; 15.28b: © Charles Philip/ Corbis; 15.28c: © Corbis/Vol. 5.

Chapter Sixteen

Opener: © Bill Ross/Corbis; 16.3a: © Dave G. Houser/Corbis; 16.3b: © Royalty-Free/Corbis; 16.3c: © Corbis/Vol. 86; 16.3d,e: © Edward S. Ross; 16.8: © Doug Sokell/Tom Stack & Associates; 16.19a: © Digital Vision/PictureQuest; 16.19b: © W. Gregory Brown/Animals Animals/Earth Scenes; 16.20a: © Jim Church; 16.20b: Courtesy of J. Frederick Grassel, Woods Hole Oceanographic Institution; 16.20c: © Kenneth L. Smith; 16.21: © Edward S. Ross; 16.22a: © Fred Rhode/Visuals Unlimited; 16.22b: © Dwight Kuhn; 16.26: © Michael Graybill & Jan Hodder/Biological Photo

Service; 16.27: © E.R. Degginger/Photo Researchers; 16.28: © S.J. Krasemann/Peter Arnold, Inc.; 16.29: © J. Weber/Visuals Unlimited; 16.30: © IFA/Peter Arnold, Inc.; 16.31: © Charlie Ott/The National Audubon Society Collection/Photo Researchers; 16.32: © John Shaw/Tom Stack & Associates; 16.33: © Tom McHugh/Photo Researchers; 16.34: © Dave Watts/Tom Stack & Associates; 16.35: © E.R. Degginger/Animals Animals/Earth Scenes; 16.36: © Gunter Ziesler/Peter Arnold, Inc.

Chapter Seventeen
Opener: © ABPL/Gavin Thomson/Animals Animals; 17.1b: © Manfred Danegger/Peter Arnold, Inc.; 17.6: Courtesy of National Museum of Natural History, Smithsonian Institution; 17.8a: © Vanessa Vick/Photo Researchers; 17.8b: © Ken Lucas/Visuals Unlimited; 17.8c,d: © Edward S. Ross; 17.13: © Merlin D. Tuttle, Bat Conservation International; 17.14: © N&C Photography/Peter Arnold, Inc.; 17.15: © Michael Fogden/DRK Photo; 17.16b: Courtesy of Steve J. Upton, Division of Biology, Kansas State University; 17.16a: © Edward S. Ross; 17.16c: © Roger Wilmshurst/ The National Audubon Society Collection/Photo Researchers; 17.17: © Jim Harvey/Visuals Unlimited; 17.18: © Gunter Ziesler/ Peter Arnold; 17.19: Courtesy of Rolf O. Peterson; 17.21a: © Tom J. Ulrich/Visuals Unlimited; 17.22: © Gary Milburn/Tom Stack & Associates; 17.23: © Edward S. Ross; 17.24: © Lincoln P. Brower, Sweet Briar College; 17.25: © James L. Castner; 17.27a: © Edward S. Ross; 17.27b: © Paul A. Opler; 17.28 (all): © Edward S. Ross; 17.30 (all): © Tom Bean; 17.31: © Edward S. Ross.

Chapter Eighteen
Opener: © Steve McCurry/National Geographic Society; 18.2: © Grant Heilman/Grant Heilman Photography; 18.3: Courtesy of Richard Klein; 18.4: NASA; page 417: © 1999 Ed Ely/Biological Photo Service; 18.9: © Byron Augustine/Tom Stack & Associates; 18.10: Sovfoto/Eastfoto; 18.11: © Gary Griffen/Animals Animals/ Earth Scenes; 18.12a: © James Blair/National Geographic Society; 18.12b: NASA; 18.12c: © Frans Lanting/Minden Pictures; 18.14: © Stephanie Maze/Woodfin Camp & Associates; 18.16: University of Wisconsin-Madison Arboretum; 18.17: © Kennan Ward/Corbis; 18.18: Merlin D. Tuttle/Bat Conservation International; 18.19a,b: Courtesy of Nashua River Watershed Association; 18.19c: © Seth Resnick; 18.20: © T. Henneghan/ImageState.

Chapter Nineteen
Opener: © Anthony Bannister/Animals Animals/Earth Scenes; 19.4: © Jim Merli/Visuals Unlimited; page 509 (top): © Ed Reschke/ Peter Arnold, Inc.; (2nd, 3rd from top): © Ed Reschke; (4th from top): © Fred Hossler/Visuals Unlimited; (bottom): © Ed Reschke/ Photo Researchers; (2nd from top): © Biophoto Associates/Photo Researchers, Inc.; (3rd from top): © Chuck Brown/Photo Researchers; (4th from top): © Ken Edward/Science Source/Photo Researchers; (bottom): © Biophoto Associates/Photo Researchers, Inc.; 19.6: © Dr. Michael Klein/Peter Arnold, Inc.; Page 513 (all): © Ed Reschke; 19.10: © Cleveland P. Hickman, Jr.; 19.11: © David M. Dennis/Tom Stack & Associates; Page 519: © Royalty-Free/Corbis.

Chapter Twenty
Opener: © David M. Phillips/Visuals Unlimited; 20.5, 20.6: © Ed Reschke; 20.11: © Manfred Kage/Peter Arnold, Inc.; 20.18 (all): Courtesy of Frank P. Sloop, Jr.

Chapter Twenty-One
Opener: © John Gerlach/Animals Animals/Earth Scenes; 21.14b: © David M. Phillips/Visuals Unlimited; 21.14c: © Biophoto Associates/Science Source/Photo Researchers, Inc.; 21.29: © Larry Brock/Tom Stack & Associates; 21.32: © Prof. P. Motta/Dept. of

Anatomoy/University "La Sapienza," Rome/Science Photo Library/ Photo Researchers.

Chapter Twenty-Two
Opener: © CDC/Science Source/Photo Researchers; 22.1: © Joe McDonald/Visuals Unlimited; 22.4: © Manfred Kage/Peter Arnold, Inc.; 22.6: Courtesy of Dr. Gilla Kaplan; 22.7: From Alan S. Rosenthal, New England Journal of Medicine 303: 1153, 1980; 22.19: © Visuals Unlimited; 22.20: National Library of Medicine; 22.22: © Stuart Fox; 22.23: © Oliver Meckes/Photo Researchers; page 603: © J. Cavallini/Custom Medical Stock Photo.

Chapter Twenty-Three
Opener: © Dennis Kunkel/Visuals Unlimited; 23.6: © Dennis Kunkel/Phototake; 23.7b: © John Heuser, Washington University School of Medicine, St. Louis, MO; 23.8b: © E.R. Lewis, YY Zeevi, T.E. Everhart, University of California/Biological Photo Service; 23.11: © Ernest Wilkinson/Animals Animals/Earth Scenes; 23.18: Photo Lennart Nilsson/Albert Bonnier Forlag AB, Behold Man, Little Brown & Co.; 23.23: © Wendy Shatil/Bob Rozinski/Tom Stack & Associates; 23.31: © Stephen Dalton/Animals Animals/ Earth Scenes; 23.17: Dr. Marcus E. Rachle, Washington University, McDonnell Center for High Brain Function; 23.33 (left): © David M. Dennis/Tom Stack & Associates; 23.33 (center,right): Kjell Sandved/Butterfly Alphabet; 23.33 (right): Kjell Sandved/Butterfly Alphabet; 23.35e: Courtesy of Beckman Vision Center at UCSF School of Medicine; D. Copenhagen, S. Mittman, M. Maglio; 23.38: Reproduced from Ishihara's Tests for Colour Deficiency published by Kanehara Trading, Inc., Tokyo, Japan; 23.41 © Leonard L. Rue, III.

Chapter Twenty-Four
Opener: © E.R. Degginger/Animals Animals; 24.11b: © John Paul Kay/Peter Arnold, Inc.

Chapter Twenty-Five
Opener: © Photo Researchers; 25.2a: © Chuck Wise/Animals Animals/Earth Scenes; 25.2b: © Fred McConnaughey/The National Audubon Society Collection/Photo Researchers; 25.4: © David Doubilet; 25.6: © Hans Pfletschinger/Peter Arnold, Inc.; 25.8: © Cleveland P. Hickman; 25.9b: © Frans Lanting/Minden Pictures; 25.10a: © Jean Phillippe Varin/Jacana/Photo Researchers; 25.10b: © Tom McHugh/The National Audubon Society Collection/Photo Researchers; 25.10c: © Corbis/Vol. 86; 25.13b: © David M. Phillips/Photo Researchers; 25.19b: © Ed Reschke; 25.20a-c: Photo by Lennart Nilsson/Albert Bonniers Forlag AB, A Child is Born, Dell Publishing Co.; 25.20d: Photo Lennart Nilsson/Albert Bonniers Forlag AB,Behold Man, Little Brown & Co.; 25.23: © McGraw-Hill Higher Education Group, Inc./Bob Coyle, photographer.

Chapter Twenty-Six
Opener: © Scott T. Smith/Corbis; 26.2: © Biophoto Associates/Photo Researchers, Inc.; 26.3: © George Wilder/Visuals Unlimited; 26.4: © Lawrence Mellinchamp/Visuals Unlimited; 26.5a: © Dr. Jeremy Burgess/Science Photo Library/Photo Researchers, Inc.; 26.5b: © Andrew Syred/Science Photo Library/Photo Reseachers, Inc.; 26.6d: Courtesy of Wilfred Cote, SUNY College of Environmental Forestry; 26.7a: © Randy Moore/Visuals Unlimited; 26.8b: © Terry Ashley/Tom Stack & Associates; 26.9a: © Kingsley R. Stern; 26.9b: Photomicrograph by G.S. Ellmore; 26.10: © E.J. Cable/Tom Stack & Associates; 26.12a, b: © Ed Reschke; 26.13b: © John D. Cunningham/ Visuals Unlimited; 26.14: © CBS/Phototake; 26.15a: © Michael P. Gadomski/Dembinsky Photo Associates; 26.15c: © Steven J. Baskauf; 26.15d: © Jack Wilburn/Animals Animals; 26.15b, e, f : © Royalty-Free/Corbis; 26.16a: © Kjell Sandved/Butterfly Alphabet; 26.16b: © Pat Anderson/Visuals Unlimited; 26.18: © Ed Reschke; 26.22:

© Kingsley R. Stern; 26.25a: © J.A.L. Cooke/Animals Animals/ Earth Scenes; 26.25b: © Robert Mitchell/Tom Stack & Associates; 26.25c: © Kjell Sandved/Butterfly Alphabet.

Chapter Twenty-Seven

Opener: © Adam Hart-Davis/SPL/Photo Researchers; 27.1: © Jerome Wexler/Photo Researchers; 27.3a: © Corbis; 27.3b: Courtesy of B.A. Wilson, Las Pilitas Nursery; 27.5a: © Alamy Images; 27.5b: USDA photo by Jack Dykinga; 27.5c: © Alamy Images; 27.5d: © Bill Bonner; 27.5e: © Royalty-Free/Corbis; 27.5f: © Ed Reschke/Peter Arnold, Inc.; 27.6a: © Holt Studios International, Ltd/Alamy Images; 27.6b: © Helmut Gritscher/Peter Arnold, Inc.; 27.12: © Runk/Schoenberger/Grant Heilman Photography; 27.13a, b: © Malcolm Wilkins; 27.14: © Kingsley R. Stern; 27.18a,b: © Runk/Schoenberger/Grant Heilman Photography; 27.18c: John D. Cunningham/Visuals Unlimited.

Line Art and Text

Chapter 1

Figure 1.1: Copyright © 1988 From *Essential Cell Biology: An Introduction to the Molecular Biology of the Cell, 1st Edition* by Bruce Alberts, et al. Reproduced by Permission of Routledge, Inc., part of The Taylor & Francis Group.

Chapter 2

Figure 2.16: Data from E.J. Heske, et al., *Ecology,* 1994. Figure 2.24: Data from *National Geographic,* July 2001.

Chapter 12

Figure 12.3: From Niles Eldredge, "Life in the Balance," in *Natural History,* June 1998. Reprinted by permission of Patricia J. Wynne.

Chapter 17

Figure 17.10: Data from Begon et al., *Ecology,* 1996. After W.B. Clapham, *Natural Ecosystems,* Clover, Macmillan. Figure 17.12: Data from Schluter & McPhail, 1993. After Begon, Harper, Townsend, *Ecology: Individuals, Populations, and Communities, 3rd edition,* Blackwell Publishing.

Chapter 18

Figure 18.5: Data from Geophysical Monograph, American Geophysical Union, National Academy of Sciences, and National Center for Atmospheric Research.

Chapter 22

Figure 22.17: From "Immunity and the Invertebrates," *Scientific American,* November 1996. Copyright Roberto Osti Illustrations. Reprinted with permission.

Index

Communication, by cells, 96–98
Communities, 4, 5 *fig.*, 31, 358, 390–409
    animal defenses and, 400–401, 401 *fig.*
    climax, 404
    coevolution and, 394, 394 *fig.*, 401
    commensalism and, 397, 397 *fig.*
    mimicry and, 402 *fig.*, 402–403, 403 *fig.*
    niche and competition and, 391–393
    parasitism and, 396
    pioneering, 404
    plant defenses and, 400
    predator-prey interactions and, 398 *fig.*, 398–399, 399 *fig.*
    stability of, 404–408
    symbiosis and, 394
Community ecologists, 31
Compact bone, 440
Companion cells, 610, 611 *fig.*
Competition, 391
    reduction by predation, 399, 399 *fig.*
Competitive exclusion, 33, 391–392
Competitive inhibitors, 107, 107 *fig.*
Complementarity, 202
Complement system, 517 *fig.*, 517–518
    in invertebrates, 526
Complex carbohydrates, 63, 63 *fig.*, 64 *table*
Complex multicellular organisms, 300
Compound leaves, 617, 617 *fig.*
Compound microscopes, 72
Concentration gradient, 90
Condensation, 146
Condoms, 603, 603 *fig.*
Cones
    of conifers, 320
    of eye, 561–562, 562 *fig.*
Confocal microscopes, 73 *table*
Conidiophores, 304
Coniferophyta, 313 *table*
Conifers, 320, 320 *fig.*
Conjugation, bacterial, 294, 294 *fig.*
Connective tissues, 440–442, 441 *table*, 442 *fig.*
    of vertebrates, 434, 434 *fig.*
Connell, J. H., 391
Conservative replication, 202, 202 *fig.*
Consumers, 358
Consumption in developed world, 37–38, 38 *fig.*
Continuous variation, 178, 178 *fig.*
Contraception, 603 *fig.*, 603–604
Contractile vacuoles, 88
Control experiments, 11
Conus arteriosus of fishes, 464, 464 *fig.*
Convergent evolution, 252, 253 *fig.*
Cooke, Howard, 602
Cooksonia, 310, 315
Cooperation, 6, 7 *table*
Coprophagy, 496
Cork cambium, 608, 614–615, 615 *fig.*
Cork cells, 615
Corn
    color of, 180
    genetically modified, 233
Corn borer, 235
Cornea, 561, 561 *fig.*
Coronary arteries, blocked, 469
Corpus luteum, 595, 595 *fig.*
Correns, Karl, 185
Cortex of stems, 614
Corticotropin-releasing hormone (CRH), 574, 578 *table*
Cortisol, 577, 578 *table*
Cotton, genetically modified, 233
Cotyledons, 318, 318 *fig.*, 629, 629 *fig.*
Countercurrent flow, 471, 471 *fig.*

Countercurrent heat exchange, 457, 457 *fig.*
Coupled channels, 95, 96 *fig.*
Coupled transport, 97 *table*
Covalent bonds, 46, 48, 48 *fig.*
C phase of cell cycle, 143, 143 *fig.*
C$_3$ photosynthesis, 122, 123 *fig.*
C$_4$ photosynthesis, 123 *fig.*, 123–124, 124 *fig.*
Crassulacean acid metabolism (CAM), 124
Crenicichla alta, 262
CRH (corticotropin-releasing hormone), 574, 578 *table*
Crick, Francis, 14, 200, 201 *fig.*
Cristae, 84, 84 *fig.*
Crocodiles, 465
Crops. *See* Agriculture
Crosscurrent flow, 473, 473 *fig.*
Crossing over, 156, 156 *fig.*, 159, 162
Cryptic coloration, 401, 401 *fig.*
Ctenophora, 335
Cuboidal epithelium, 439 *table*
Cuenot, Lucien, 178
Cuscuta, 396 *fig.*
Cutaneous respiration, 465
    of amphibians, 350
Cuticle of plants, 310, 610
Cyanide in Kentucky grass, 47
Cycadophyta, 313 *table*
Cycling, 363
Cyst(s), protist, 299
Cystic fibrosis (CF), 76, 189 *table,* 243
Cystic fibrosis transmembrane conductance regulator (CFTR), 76
Cytokinesis, 143 *fig.*, 148, 148 *fig.*
Cytokinins, 633 *table,* 636, 636 *fig.*
Cytoplasm, 71
Cytoskeleton, 79, 81 *table,* 86 *fig.*, 86–88
Cytotoxic T cells, 519 *table,* 521, 521 *fig.*

Daltons, 41
Danaus plexippus, 402 *fig.*
Darevsky, Ilya, 554
Dark-field microscopes, 73 *table*
Darwin, Charles Robert, 16, 21, 22 *fig.*, 22–23, 248
    Beagle voyage of, 22–24, 23 *fig.*
    Galápagos Islands research of, 21, 23, 27 *fig.*, 27–28, 28 *fig.*, 33, 33 *fig.*
    natural selection theory of, 22, 25–28 *fig.*, 25–30, 30 *fig.*
    plant studies of, 634, 634 *fig.*
Darwin, Francis, plant studies of, 634, 634 *fig.*
Daughter cells, 142
DDT (dichlorodiphenyltrichloroethane), 425
Deamination, 498
Death(s). *See also* Mortality
    cancer-related, 150
    of cells, 148, 148 *fig.*
Death rate, 36, 36 *fig.*
Deciduous forests, 378, 378 *fig.*
Decomposers, 305, 360, 360 *fig.*
Deductive reasoning, 8, 8 *fig.*
Deep-sea waters, 372–373, 373 *fig.*
Defenses, 513–535
    of animals, 400–401, 401 *fig.*
    cellular counterattack as, 514, 516–518, 516–518 *fig.*
    immune system and. *See* Immune response; Immune system
    of plants, 400
    skin as, 514 *fig.*, 514–515, 515 *fig.*
Defensive coloration, 401, 401 *fig.*
Dehydration synthesis, 54, 54 *fig.*

Deletion, 217, 217 *fig.*
Demography, 388, 388 *fig.*
Denaturation of protein, 59, 59 *fig.*
Dendrites, 445, 445 *fig.*, 540
Density-dependent effects, 386, 386 *fig.*
Density-independent effects, 386
Deoxyribonucleic acid. *See* DNA (deoxyribonucleic acid)
Depression, 544
Derived characters, 278
Dermal tissue of plants, 610, 610 *fig.*
Dermis, 514
Descent of Man, The (Darwin), 26
Deserts, 377, 377 *fig.*
Detrivores, 360, 360 *fig.*
Deuterostomes, 344, 344 *fig.*, 432–433, 433 *table*
Development, 629
    cellular, 300
Devonian period, 347
Diabetes, 230, 531, 575
    embryonic stem cell therapy for, 238
    obesity and, 487, 575
Diaphragm (muscle), 353, 474
Diaphragms (contraceptive), 603, 603 *fig.*
Diastolic pressure, 468
Diatom(s), 302 *table*
Diazepam, 543
Dichlorodiphenyltrichloroethane (DDT), 425
Dicots, 325, 325 *fig.*
    development of, 631, 631 *fig.*
Dictyostelium discoideum, genome of, 223 *table*
Diencephalon, 547
Dieting, 131
Differential-interference-contrast microscopes, 73 *table*
Diffusion, 90, 90 *fig.*, 97 *table*
    facilitated, 94, 94 *fig.*, 97 *table*
    selective, 94
Digestion, 488
    extracellular, 335
Digestive systems, 488–499, 499 *fig.*
    accessory digestive organs and, 498, 498 *fig.*, 499 *fig.*
    defense of, 515
    esophagus and stomach of, 492 *fig.*, 492–493
    intestines of, 494, 495 *fig.*
    mouth and teeth of, 490 *fig.*, 490–491
    types of, 488, 488 *fig.*
    variations in, 496, 496 *fig.*, 497 *fig.*
    of vertebrates, 435, 437 *fig.*, 489 *fig.*, 489–498
Dihybrid individuals, 175, 175 *fig.*
Dileptus, 69
Dilobderus abderus, 343 *fig.*
Dinoflagellates, 302 *table*
Dioecious plants, 627
Dionaea muscipula, 622, 622 *fig.*
Diploid cells, 154, 154 *fig.*
Dipodomys panamintensis, 507 *fig.*
Directional selection, 258 *fig.*, 259, 259 *fig.*
Disaccharides, 63, 63 *fig.*
Diseases and disorders
    allergic, 531, 531 *fig.*
    autoimmune, 517, 531
    bacterial, 294
    cardiovascular, 469, 469 *fig.*
    fungal, 305
    genetic, 76, 189 *table,* 190–192, 190–192 *fig.*, 193–194, 243
    immunodeficiency, 297, 513, 528–529, 532 *fig.*, 532–534
    malignant. *See* Cancer
    Mendelian inheritance and, 184

Applications Index

Bringing Biology to Life!

George Johnson knows how hard it can be to make science relevant and interesting to nonscientists. As a professor with his own issues-based nonmajors biology course, and past author of a weekly science column for the *St. Louis Post-Dispatch,* he has spent years looking for ways to make science matter. He brings this experience to *Essentials of The Living World* in three critical ways.

Helping Students Visualize It

NEW! Biology Digitized Videos

McGraw-Hill is pleased to offer adopting instructors a new presentation tool—digitized biology video clips on DVD! Licensed from some of the highest-quality science video producers in the world, these brief segments range from 15 seconds to two minutes in length and cover all areas of general biology from cells to ecosystems. Engaging and informative, McGraw-Hill's digitized biology videos will help capture students' interest while illustrating key biological concepts and processes such as how cilia and flagella work and how some plants have evolved into carnivores.

Alex the African Grey Parrot correctly answers complex questions in a video segment that vividly introduces the subject of animal cognition and behavior.

Vorticella, pulled back on its coiled stalk.

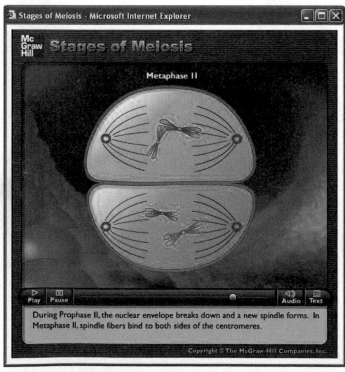

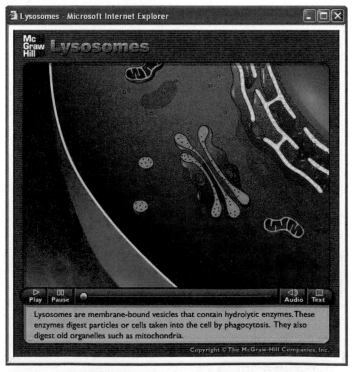

NEW! General Biology Animations and General Biology Animations-Spanish Version

Full-color presentations of key biological processes have been brought to life via animation. These animations offer flexibility for instructors and were designed to be used in lecture. Instructors can pause, rewind, fast forward, and turn the audio off or on to create dynamic lecture presentations. The animations are now also available with Spanish narration and text.